COASTAL DYNAMICS '94

Proceedings of
An International Conference on the Role of the Large Scale Experiments in Coastal Research

Universitat Politècnica de Catalunya
Barcelona, Spain
February 21-25, 1994

Sponsored by the
Ministerio de Educación y Ciencia (Spain)
Office of Naval Research, Ocean Engineering Division (USA)
Commission of the European Communities, DGXII
Generalitat de Catalunya
Universitat Politècnica de Catalunya
American Society of Civil Engineers
Japan Society of Civil Engineers
E.T.S. d'Enginyers de Camins, Canals i Ports de Barcelona, UPC
Laboratori d'Enginyeria Marítima, UPC

Generalitat de Catalunya
Departament
de Medi Ambient

Approved for publication by the Waterway, Port, Coastal and Ocean Division of the American Society of Civil Engineers

Edited by A.S.-Arcilla, M.J.F. Stive, and N.C. Kraus

Published by the
American Society of Civil Engineers
345 East 47th Street
New York, New York 10017-2398

ABSTRACT

During the past decade, numerical models of coastal dynamics have provided far more information than could be validated with field and laboratory measurements. Lack of data has restricted advances in predictive capability of many such models. This situation has begun to change because of the growing number of large-scale measurement campaigns ranging from laboratory and field experiments carried out in the United States to the laboratory tests performed in many European laboratories in the context of the European Community (EC) Large Installations Plan (LIP). Because of the wealth of experimental information that has become available a conference on coastal dynamics was organized to bring together the producers and consumers of these data sets. This proceedings, Coastal Dynamics '94: Proceedings of an International Conference on the Role of the Large Scale Experiments in Coastal Research, contains the papers given at this conference organized under 14 different categories: 1) Bar dynamics; 2) field data and modeling; 3) sea-estuary interaction; 4) SUPERTANK; 5) surf zone hydrodynamics; 6) dynamics of shore nourishment and natural beaches; 7) aeolian sediment transport; 8) Delta flume; 9) waves near structures; 10) surf zone sediment and bed dynamics; 11) wave nonlinearities; 12) new facilities; 13) sea-bed dynamics; and 14) morphodynamics near structures. Engineers interested in the modeling of coastal dynamics will find this proceedings a useful collection of information on field and laboratory tests from the standpoint of the underlying dynamics.

Library of Congress Cataloging-in-Publication Data

Coastal dynamics '94: proceedings of an international conference on the role of the large-scale experiments in coastal research, Universitat Politecnica de Catalunya, Barcelona, Spain, February 21-25, 1994 / sponsored by the Ministerio de Educacion y Ciencia (Spain) ... [et al.]; edited by A.S.-Arcilla, M.J.F. Stive, and N.C. Kraus.
p. cm.
Includes indexes.
ISBN 0-7844-0043-1
1. Coast changes—Congresses. I. Arcilla, A.S.- II. Stive, M. J. F. (Marcel J. F.) III. Kraus, Nicholas C. IV. Spain. Ministerio de Educacion de Catalunya.
GB450.2.C595 1994 94-5424
551.4'57—dc20 CIP

Library of Congress Catalog Card No: 94-5424
ISBN 0-7844-0043-1
Manufactured in the United States of America.

PREFACE

The coastal zone is experiencing ever increasing and conflicting pressures for both development and preservation. Because of this pressure and the unexpected behavior of the coast as perceived by those who tend to think of it as static, resolution of coastal problems continues to require better prediction tools. These tools range from numerical simulation models to small- and large-scale laboratory facilities for specialized investigation, having, as a reference, field measurements.

During the past decade, numerical models of coastal dynamics have provided far more information than could be validated with field and laboratory measurements. Lack of data has restricted advances in predictive capability of many such models. Recently, this situation has begun to change because of the growing number of large-scale measurement campaigns, ranging from the field and laboratory experiments carried out in the United States to the laboratory tests performed in many European laboratories in the context of the European Community (EC) Large Installations Plan (LIP).

Because of the wealth of experimental information that has become available, it was considered timely to organize a "Coastal Dynamics" conference to bring together the producers and consumers of these data sets. Considering the schedule of already existing American-based coastal technical specialty conferences, the on-going EC research on Marine Sciences and Technology (MAST-Program), and the opening up of Eastern Europe, it seemed natural to organize Coastal Dynamics as a European-based conference addressing the whole international scientific community. It is envisioned that the Coastal Dynamics conferences will provide complimentarity to the Coastal Sediments conference series and International Coastal Engineering Conferences.

The 1994 inaugural conference in this new Coastal Dynamics series emphasized field and laboratory tests from the stand-point of the underlying dynamics. This focused approach resulted in a lively and fruitful exchange which justifies, in our opinion, continuation of this conference series. The next Coastal Dynamics conference will be held in Poland in the late summer of 1995, and it will provide a forum to discuss the evolution of the future field and laboratory campaigns to be carried out in the U.S. and Europe among other countries.

We would like to acknowledge all the sponsors who have made possible the 1994 session of Coastal Dynamics: Ministerio de Educacion y Ciencia

(Spain), Office of Naval Research (USA), the EC, Generalitat de Catalunya (Dept. de Medi Ambient, Commissionat per a Universitats i Recerca and Dept. de Politica Territorial i Obres Publiques), Autoritat Portuaria de Barcelona, Fundacio "La Caixa," American Society of Civil Engineers, Colegio de Ingenieros de Caminos, Canales y Puertos de Catalunya and Universitat Politecnica de Catalunya (ETSECCPB and LIM/UPC), and the Conrad Blucher Institute for Surveying and Science at Texas A&M University-Corpus Christi. We also thank all who have contributed to the conference organization, in particular, the Paper Review Committee, the International Scientific Committee, Ms. M. Ruiz, Mrs. G. Comas, and Mr. J. M. Vilalta of the LIM/UPC staff, and Ms. Deidre Williams at Texas A&M University-Corpus Christi.

Agustín S.-Arcilla
Univ. Polytecnica
de Catalunya

Nicholas C. Kraus
Texas A&M University-
Corpus Christi

Marcel J. F. Stive
Delft Hydraulics

April 15, 1994

CONTENTS

Volume I

BAR DYNAMICS

FIELD DATA AND MODELING

SEA–ESTUARY INTERACTION

SUPERTANK

SEDIMENT TRANSPORT TRACING

SURF ZONE HYDRODYNAMICS

DYNAMICS OF SHORE NOURISHMENT AND NATURAL BEACHES

AEOLIAN SEDIMENT TRANSPORT

DELTA FLUME

WAVES NEAR STRUCTURES

SURF ZONE SEDIMENT AND BED DYNAMICS

WAVE NONLINEARITIES

NEW FACILITIES

SEA-BED DYNAMICS

MORPHODYNAMICS NEAR STRUCTURES

THEORY AND OBSERVATION OF CURRENTS AND SETUP OVER A SHALLOW REEF

Graham Symonds[1]

ABSTRACT: Over a shallow reef the absence of a shoreline allows both setup and mean cross shore flow to result from gradients in the radiation stress through the surf zone on the outer part of the reef. The relative magnitude of setup and currents depends on the magnitude of the forcing and the geometry of the reef. In this paper observations of currents and setup over natural reefs and in the laboratory are presented revealing interesting features of these flows. A one dimensional analytic model is described which explains the observed features and is able to simulate the observed currents on a natural reef by adjusting friction and the reef width.

INTRODUCTION

Coral reefs are common throughout the Pacific and Indian Oceans and often protect low lying atols by dissipating incident wave energy through the surf zone over the reef. These low lying atols are particularly threatened by long term sea level rise which may also lead to reduced wave breaking over the reef and an increase in wave energy transmitted across the reef with an associated change in wave driven currents and setup.

The Great Barrier Reef off north eastern Australia contains over 2,000 reefs over an area of 250,000 square kilometres along a 1600 kilometre stretch of the continental shelf. The offshore facing edge of the reefs are often steep (slopes of 0.1 or greater) resulting in narrow surf zones with a wide reef flat (width of 200 to 300 metres) leading into a lagoon which is typically 5 to 10 metres deep. The reefs are often exposed at low tide when they resemble a steep beach, while at high tide the maximum water depth may be a few metres and significant wave energy may

---

[1]Department of Geography and Oceanography, University College, Australian Defence Force Academy, Northcott Drive, Canberra, Australia, 2600.

maximum water depth may be a few metres and significant wave energy may propagate across the reef flat with little or no breaking.

The relative importance of wave driven flow in exchange of water and nutients from offshore and the associated flushing of the lagoon is not well understood. Hearn and Parker (1988) reported a significant correlation between longshore currents in a lagoon behind Ningaloo Reef in Western Australia with offshore wave height (see figure 1). A volumetric flushing time of the lagoon was estimated to be 5 to 20 hours suggesting wave-induced currents across the reef flat may provide an important mechanism of exchange between the lagoon and offshore.

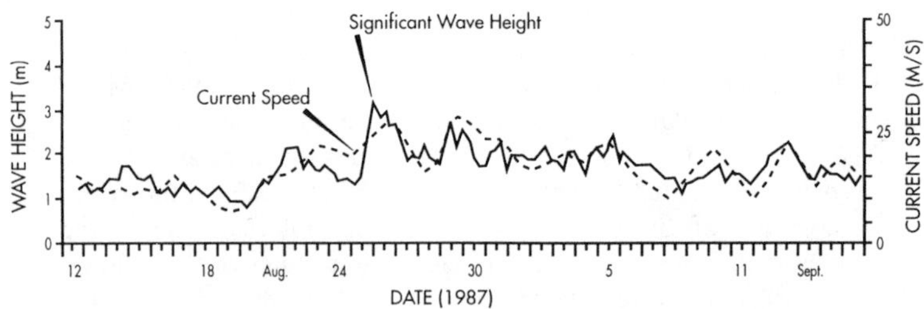

Figure 1. Lowpass filtered longshore current in the lagoon behind Ningaloo Reef, Western Australia and the significant wave height offshore (after Hearn and Parker (1988) with permission from P. E. Davies (ed)).

In addition to wave driven currents, wave induced setup over reefs has also been reported by Tait (1972) and Jensen (1991). This setup has been explained in a manner analagous to setup on a plane beach described by Longuet-Higgins and Stewart (1964). This explanation assumes there is no cross reef flow and the sea level sets up uniformly across the reef flat and lagoon with additional small setup associated with wave breaking at the shoreline. However many reefs are broken by channels which allow flow from the lagoon to the open ocean while other reefs do not surround an atol and are completely submerged at high tide so there is no shoreline.

Gourlay (1993) measured cross reef flow and setup over an idealized reef in a laboratory wave basin. His experimental configuration consisted of a reef and reef flat without a shoreline and the water level behind the reef was maintained at the same level as in front by means of channels either side of the reef. Some results from his experiment are reproduced in figure 2. In figure 2a is shown the sea surface elevation over the reef as a function of wave height for different still water depths over the reef. The results have been scaled but we are only interested in the relative magnitudes here. In figure 2b is shown the corresponding results for transport (depth × velocity) across the reef. As the wave height increases both the setup and cross reef transport increase. Assuming the water level behind the reef is at the still water level, the presence of setup over the reef implies a pressure gradient must exist over the reef flat. As the still water depth over the reef decreases the setup increases (see fig 2a). However, as the still water depth decreases, the transport decreases (see fig 2b) while the gradient in sea surface elevation across the reef is increasing. The pressure gradient across the reef has been attributed to bottom friction associated with the cross reef flow but this does not explain the opposite pressure gradient which must exist through the surf zone.

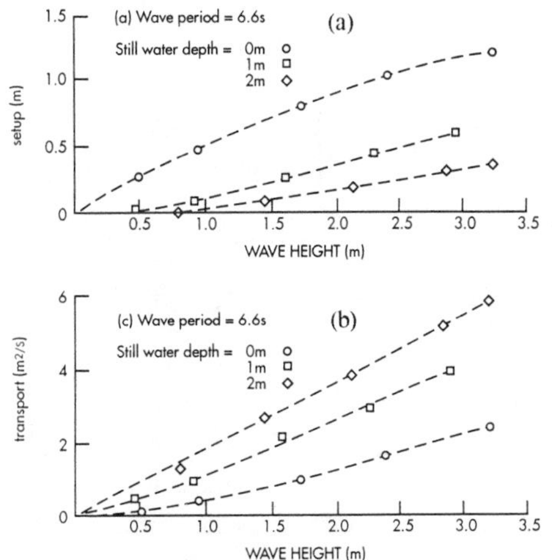

Figure 2. Laboratory measurements of setup (a) and cross reef transport (b) as functions of incident wave height and still water depth over the reef (after Gourlay (1993) with permission from The Institute of Engineers, Australia).

A linear, one dimensional model which includes wave forcing over an idealized reef and including both pressure driven flow and bottom friction has been described by Symonds et al (1994), hereafter referred to as **I**. The purpose of this paper is to present observations of currents and sea level over a reef, some of which were presented in **I**, and to use the theoretical model to simulate the observed currents.

JOHN BREWER REEF

John Brewer Reef is located 70 kilometres north east of Townsville, Australia (see figure 3), and is part of the Great Barrier Reef. The reef is approximately 5 kilometres long and 3 kilometres across and the dominant wave direction is from the south east. The reef flat is 200 to 300 metres wide and is exposed at low spring tides while at high spring tides the depth is approximately 3 metres over the reef flat. The outer part of the south east facing reef has a slope of 0.1 to a depth of 10 metres and then drops almost vertically to a depth of 50 metres.

In 1988 a month long experiment was conducted on John Brewer Reef to investigate wave attenuation and transformation across the reef. Details of the experiment and results have been reported by Hardy et al (1990) and Hardy (1993). In this paper attention is focussed on the variation in currents and sea level at much longer time scales. In particular, the dependence of cross reef currents on tidal elevation and offshore wave conditions associated with passing weather systems is examined.

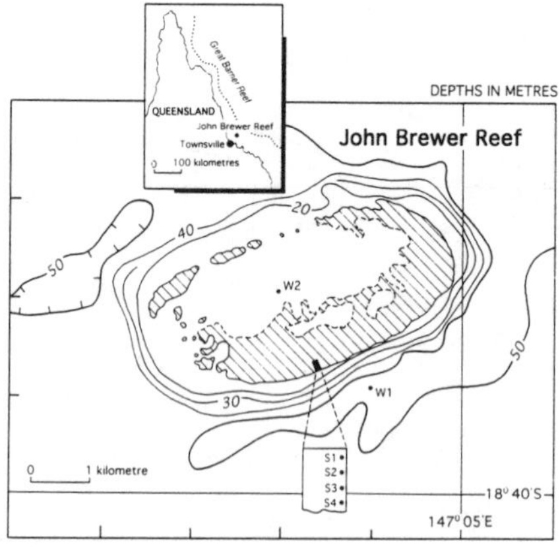

Figure 3. Location of John Brewer Reef measurement sites.

Cross reef and along reef currents were measured at site S4 (see figure 3) using an s4 electromagnetic current meter which recorded 1 minute averages every 10 minutes for the duration of the experiment. Offshore wave conditions were recorded by a wave rider buoy (W1 in figure 3) which sampled 2048 points with a sampling rate of 2.56Hz every hour. Using these data a time series of hourly rms wave height was constructed.

The cross reef currents and rms wave height time series were lowpassed filtered with a half power point at .04 cycles per hour to remove the tidal component from the currents. Shown in figure 4 is a plot of the filtered cross reef current versus the filtered rms wave height both decimated to hourly values. The correlation coefficient is 0.92 and the solid line is the best fit straight line through the data points. The cross reef currents are almost always negative which, in this case, is directed into the lagoon. Positive currents do occur at low wave heights and probably are being dominated by some other forcing not directly associated with the incident waves and resulting in a mean offshore flow across the reef.

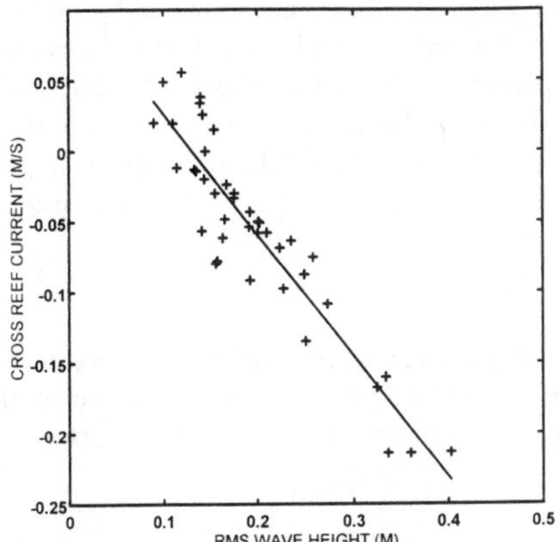

Figure 4. Observed cross reef currents versus offshore rms wave height. The straight line is given by $u=0.11-0.86H_{rms}$.

In addition to waves and currents, sea surface elevation was measured using surface piercing wave staffs (Zwartz, 1974). The wave staffs had a sampling rate of 4Hz and initially recorded for twenty minutes every hour. Due to data logging

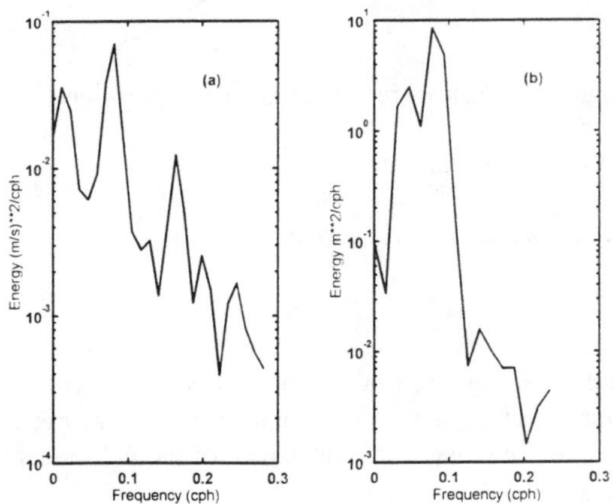

Figure 5. Spectra of cross reef currents (a) at site S1 and sea surface elevation (b) at site S4.

limitations the interval between data runs was increased to 2 hours several days into the experiment. Mean sea level was determined from each record and combined to produce a time series of sea level at two hour intervals for the duration of the experiment. Energy spectra of the 10 minute cross reef currents at site S4 and the 2 hourly sea surface elevations at site S1 are shown in figure 5. Both the current and elevation spectra show significant peaks at the semi-diurnal tidal frequency. A diurnal peak is also apparent in the elevation but has been smoothed out in the velocity spectrum. However, the spectrum of cross reef currents also has a significant peak at twice the semi-diurnal tidal frequency. A similar peak at this frequency is not observed in the spectrum of sea surface elevation.

THEORY

A detailed description of the theoretical model is described in **I** and only a brief summary is given here. An idealized one dimensional reef is shown in figure 6 where the breakpoint is at $x=L$ and the inner edge of the surf zone is at $x=X_r$, the edge of the reef flat.

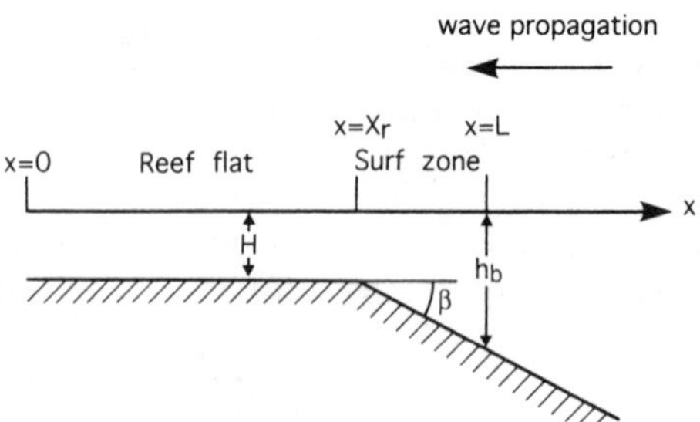

Figure 6. Idealized reef defining the model parameters.

The equations governing the flow are

$$g\frac{\partial \zeta}{\partial x} = -\frac{1}{\rho h}\frac{\partial S_{xx}}{\partial x} - \frac{ru}{h} \tag{1}$$

and

$$\frac{\partial (hu)}{\partial x} = 0 \tag{2}$$

where ζ and u are the sea surface elevation and cross reef current respectively, h is the depth, ρ is the density, g is the gravitational acceleration, r is a linear friction coefficient and S_{xx} is the cross shore component of the radiation stress which, in shallow water is given by

$$S_{xx} = \frac{3}{2}\rho g a^2 \tag{3}$$

where a is the incident wave amplitude (Longuet-Higgins and Stewart; 1962, 1964). Through the surf zone the incident wave amplitude is assumed to be given by

$$a = \gamma h \tag{4}$$

where γ is a constant. The effects of wave shoaling are ignored and seawards of the surf zone and over the reef flat the wave height is assumed constant and the radiation stress forcing term in (1) is zero.

On a plane beach the cross shore flow is zero, the friction term in (1) does not appear, and the gradient in the radiation stress is entirely balanced by a pressure gradient. On the idealized reef the radiation stress gradient is balanced by the combination of a pressure gradient and friction. In this case the relative magnitude of these two terms in the momentum balance depends on the magnitude of the forcing and the geometry of the reef.

In **I** the offshore distance is scaled by the surf zone width, $L - X_r$, the depth is scaled by the breakpoint depth, h_b, the sea surface elevation is scaled by the equivalent plane beach setup at the shoreline with the same surf zone width and the velocity scale is found by assuming all of the radiation stress gradient is balanced by friction in (1). After scaling it can be shown (see **I**) the solutions depend on the two parameters

$$R_1 = \frac{H}{h_b} \tag{5}$$

and

$$R_2 = \frac{X_r}{L}. \tag{6}$$

These two parameters vary between zero and one and the limiting form of the solutions are discussed in **I**. It suffices to note that R_1=0 corresponds to the plane beach case (H=0), while R_1=1 corresponds to no forcing ($H = h_b$). The case R_2=0 corresponds to a truncated beach where the still water depth is non-zero at the shoreline, while R_2=1 corresponds to a reef width, X_r which is infinitely wide compared to the surf zone width.

The solutions for the non-dimensional sea surface elevation and cross reef transport at $x = X_r$ are shown in figure 7 as functions of R_1 and R_2. A non-dimensional sea surface elevation of one is equivalent to the shoreline setup on a plane beach with the same surf zone width. When $R_1 \to 0$ the depth over the reef is small compared to the depth at the breakpoint and the setup approaches one while the cross reef transport approaches zero, equivalent to the plane beach solution. When $R_1 \to 1$ the depth over the reef approaches the depth at the breakpoint when the waves would cease to break. In this case there is no forcing and both the setup and cross reef transport approach zero. Between these two limits the cross reef transport goes through a maximum (in the negative x-direction) producing a modulation in the cross reef transport with depth over the reef. For a given incident wave height, varying R_1 is equivalent to varying the depth H over the reef flat. If, between low tide and high tide, the depth over the reef varies such that the cross reef transport goes through a maximum then the wave driven flow will be modulated at twice the tidal frequency.

The setup through the surf zone occurs because a pressure gradient must be established across the reef flat, where there is no gradient in radiation stress, to drive a transport which matches the transport through the surf zone. Part of the radiation stress gradient through the surf zone drives a cross reef current while the remainder is balanced by a pressure gradient such that the resulting setup and pressure gradient across the reef flat forces a transport which satisfies (2) across the model domain. Across the reef flat the transport is given by

$$Hu = -\frac{gH^2}{r}\frac{\partial \zeta}{\partial x} \tag{7}$$

The model results show that as the depth H decreases the setup at $x = X_r$, and hence $\partial\zeta/\partial x$ across the reef flat, increases. However, if the depth is small enough the transport in (7) will decrease with H while the sea surface slope increases as observed by Gourlay (1993). For larger depths, increasing H decreases the sea surface slope with a corresponding decrease in transport.

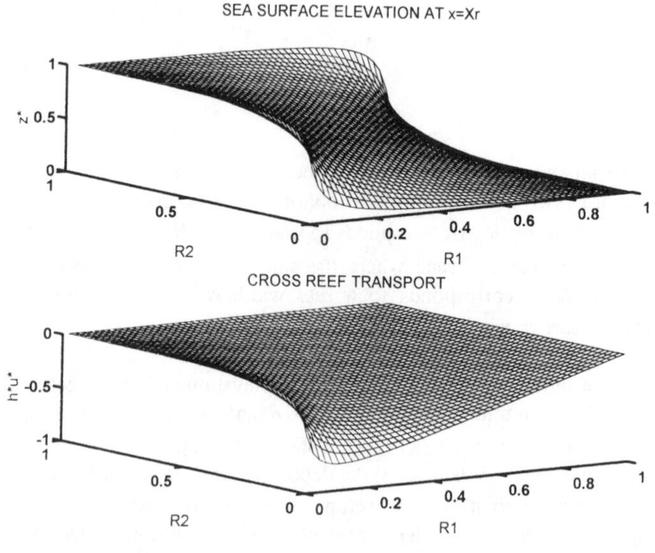

SEA SURFACE ELEVATION AT x=Xr

CROSS REEF TRANSPORT

Figure 7. Non-dimensional model solutions for sea surface elevation and cross reef transport at the seaward edge of the reef flat.

The parameter R_2 can be written as

$$R_2 = \frac{X_r \tan\beta}{h_b(1 - R_1) + X_r \tan\beta} \tag{8}$$

In this case, varying R_1 is equivalent to varying H for given values of h_b, X_r and $\tan\beta$ with R_2 then given by (8). The dimensional setup at $x = X_r$, is shown in figure 8 as a function of H for a range of breakpoint depths. As the depth H approaches the breakpoint depth the setup (fig 8a) approaches zero. As $H \to 0$ the setup approaches the plane beach setup which increases with increasing breakpoint depth. The corresponding cross reef transport is shown in figure 8b which shows the transport goes to zero as the depth H approaches the breakpoint depth and as H approaches zero. The cross reef transport (in the negative x-direction) increases as the breakpoint depth increases and the depth at which maximum transport occurs depends on the breakpoint depth.

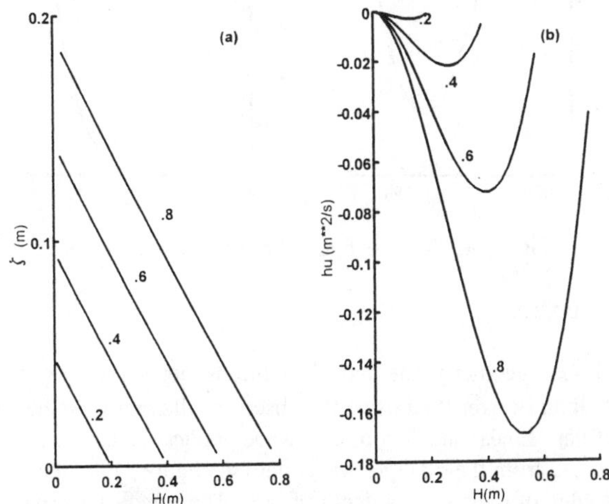

Figure 8. Dimensional solutions for setup (a) and cross reef transport (b) at the seaward edge of the reef flat. The different curves correspond to breakpoint depths as indicated in metres.

The corresponding solutions at $x=L$ are shown in figure 9 which shows the cross reef transport (fig 9b) is identical to the transport at $x = X_r$ as required by continuity. However, the sea surface elevation (fig 9a) now mirrors the transport curves with set-down which increases with increasing breakpoint depth and approaches zero as H approaches zero and one. Recal wave shoaling was not included in the formulation of the model so this set down is not due to a gradient in the radiation stress seawards of the surf zone. The set down develops to produce a pressure gradient seawards of the surf zone which drives a transport to match the transport through the surf zone. Thus, while the transport may be modulated at twice the tidal frequency across the entire model domain ($0<x<\infty$), a similar modulation in sea surface elevation will only be observed seawards of the surf zone.

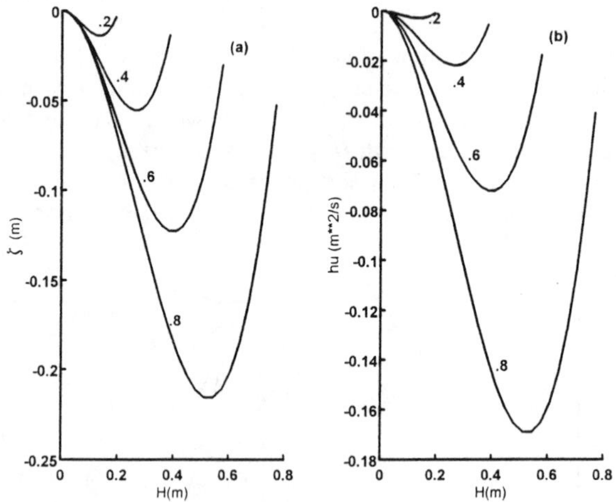

Figure 9. Same as figure 8 but at the breakpoint.

MODEL VS DATA

For a given reef geometry the model solutions depend on R_1 which can be estimated on John Brewer Reef using the observed offshore wave height and depth over the reef flat. Linear shoaling of a 5s wave predicts an increase in wave height by a factor of 2.2 from the offshore waverider (see figure 3) in 50m water depth to the offshore edge of the reef at a depth of 2m. The significant wave height at the offshore edge of the reef can then be estimated from the observations at the waverider and the breakpoint depth estimated using (4). Using the observed wave heights over the reef Hardy (1993) estimates $\gamma \approx 0.25$. This value is somewhat smaller than the more comonly used value of 0.4 and may be associated with wave transformation over the steep reef face. In this paper the value reported by Hardy (1993) is used to construct a time series of breakpoint depth. The depth over the reef flat was measured at site S1 (see figure 3) which, combined with the breakpoint depth, defines the forcing parameter R_1 to be used in the model with R_2 given by (8). In the results presented here $\tan\beta$=0.1, consistent with the measured slope at the top of the reef face. The magnitude of the cross reef current depends on the reef width X_r and the friction coefficient r, both of which can be adjusted to obtain the best fit between the model and the observed currents.

A comparison of the lowpass filtered observed and model derived currents is shown in figure 10. In this case X_r=300m and r=0.25. The gaps in the model results correspond to low wave conditions when, according to (4), no breaking was occurring and the model currents are zero. At these times the observed currents are directed offshore and are not associated with wave breaking over the reef. The fit shown in figure 10 is not unique in that a similar fit can be obtained by suitably changing X_r and r. For example, increasing X_r to 500m with a corresponding decrease in r to 0.1 produces a similar fit to the one shown in figure 10.

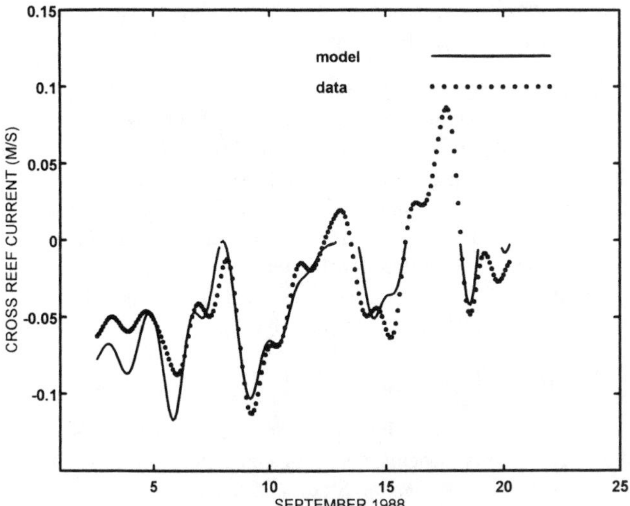

Figure 10. Lowpass filtered model currents (solid) and observed currents (dotted) with X_r=300m and r=0.25.

High friction factors over reefs have been reported previously (Black and Hatton, 1990) consistent with the high values used here. The relatively large X_r used above suggests the sea surface slopes across the reef flat would be of order 2×10^{-4} which would be too small to resolve in the John Brewer data.

CONCLUSIONS

Observations from natural reefs and laboratory studies show significant currents and setup associated with wave breaking on the outer slope of the reefs. A one dimensional analytic model has been used to show how the radiation stress forcing, due to wave breaking, is partitioned between balancing a pressure gradient through the surf zone and driving a cross reef current. This partitioning is determined by matching the transport through the surf zone with the transport across the reef flat which is forced by a pressure gradient which, in turn, depends on the magnitude of the setup through the surf zone. The solutions converge to the plane beach setup with no cross reef transport as the depth over the reef approaches zero and as the reef becomes infinitely wide. In the latter case the pressure gradient across the reef flat approaches zero and the transport must also vanish. As the depth over the reef approaches the depth at the breakpoint the waves stop breaking and the setup and transport both approach zero. Between these two limits the cross reef transport is a maximum leading to the possibility that the cross reef transport may be modulated at twice the dominant tidal frequency if the depth during half a tidal cycle causes the transport to pass through its maximum.

A good comparison between the model and observed currents is achieved by adjusting the linear friction coefficient and reef width parameter. This fitting exercise leads to high friction factors (0.1 to 0.3) and small sea surface slopes across the reef flat. A consequence of using linear friction is the friction coefficient r only appears as a scaling parameter for velocity (see Symonds et al, 1994). The magnitude of the setup, and hence the sea surface slope across the reef flat, is independent of friction; doubling the friction halves the velocity while the setup remains constant. If quadratic friction is used an additional setup would be necessary to balance friction across the reef flat.

ACKNOWLEDGEMENTS

This work has been supported by a grant from the Australian Research Council under the Small Grants Scheme and was completed while the author was on sabbatical at the Institute of Marine Studies, University of Plymouth. The author would like to thank C. Hearn and M. Gourlay for allowing reproduction of their figures and T. Hardy for providing the John Brewer Reef data.

REFERENCES

Black, K.P., and Hatton, D., 1990. Dispersal of larvae, pollutants and nutrients on the Great Barrier Reef, at scales of individual reefs and reef groups, in 2 and 3 dimensions, *Great Barrier Reef Park Authority Report*, 77pp.

Gourlay, M.R., 1993. "Wave Setup and Wave Generated Currents on Coral Reefs", *11th Australasian Conf. on Coastal and Ocean Engineering*, Institution of Engineers, Aust., Conf. Publ. No 93/4, 479-484.

Hardy, T.A., 1993. "The Attenuation and Spectral Transformation of Wind Waves on a Coral Reef", PhD thesis, James Cook University, Australia, 336pp.

Hardy, T.A., Young, I.R., Nelson, R.C., and Gourlay, M.R., 1990. "Wave Attenuation on an Offshore Coral Reef", *Proc. 22nd Int. Coastal Engineering Conf.*, Delft, The Netherlands, **1**, 330-344.

Hearn, C.J., and Parker, I.N., 1988. "Hydrodynamic Processes on the Ningaloo Reef, Western Australia", *Proc. 6th Int. Coral Reef Symposium*, Australia, **2**, 497-502.

Jensen, O.J., 1991. "Waves on Coral Reefs", *Coastal Zone '91*, 2668-2680.

Longuet-Higgins, M.S., and Stewart, R.W., 1962. "Radiation Stress and Mass Transport in Gravity Waves", *J. Fluid Mech.*, **13**, 481-504.

Longuet-Higgins, M.S., and Stewart, R.W., 1964. "Radiation Stresses in Water Waves; a Physical Discussion with Applications", *Deep Sea Research*, **11**, 529-562.

Symonds, G., Black, K.P., and Young. I.R., 1994. "Wave Driven Flow Over Shallow Reefs", submitted *J. Geophys. Res.*

Tait, R.J., 1972. "Wave Setup on Coral Reefs", *J. Geophys. Res.*, **77**, 2207-2211.

Zwartz, C.M.G., 1974. "Transmission Line Wave Height Transducer", *Proc. Int. Symposium on Ocean Wave Measurement and Analysis*, New Orleans, ASCE, **1**, 605-620.

SEDIMENT TRANSPORT BY WIND WAVES, LONG WAVES AND MEAN CURRENTS: AN EXPERIMENT ON NEARSHORE MORPHODYNAMICS, LAKE HURON, CANADA.

Troels Aagaard[1] and Brian Greenwood[2]

ABSTRACT: A large scale field experiment was undertaken to monitor suspended sediment transport and nearshore bar dynamics in southern Lake Huron, Canada, as part of the Canadian Coastal Sediment Transport Programme (C-COAST). Two storms of approximately equal magnitude induced significantly different morphological responses. During the first event, sediment transport rates were large and resulted in an offshore bar migration ≈ 25 m. This indicates that the pre-storm bar location was not in equilibrium with storm wave conditions. The single most important sediment transport mechanism was the offshore directed mean current, with significant spatial transport gradients set up by long infragravity waves. Sediment transport rates and morphological change were much smaller during the second event mainly due to smaller undertow velocities, lower sediment concentrations and a broad-banded infragravity wave field.

INTRODUCTION

Recently, detailed field measurements have been made of suspended sediment transport in wave-dominated nearshore environments (e.g. Hanes and Huntley, 1986; Huntley and Hanes, 1987; Doering and Bowen, 1988; Green and Vincent, 1990; Osborne and Greenwood, 1992a&b) and a general consensus is beginning to appear with regard to sediment transport processes and directions. Under shoaling waves outside the breakpoint, a predominantly onshore directed transport is induced by high

[1] Institute of Geography, University of Copenhagen, Oster Voldgade 10, DK-1350 Copenhagen K., Denmark.

[2] Scarborough College Coastal Research Group, University of Toronto, 1265 Military Trail, Scarborough, Ontario M1C 1A4, Canada.

frequency wind waves which is balanced to some extent by offshore transports due to mean flows and group-bound long waves. In contrast, within the surf zone sediment transport magnitudes and directions appear exceedingly complex. Generally, transport directions and magnitudes are determined by interactions between mean flows and long infragravity waves, with incident wind waves assuming secondary importance (Beach and Sternberg, 1988, 1991; Davidson et al., 1993; Foote et al., 1993; Russell, 1993; Aagaard and Greenwood, 1994); the resultant net transport, however, depends critically upon cross-shore location and general hydrodynamic conditions (i.e. incident wave characteristics and the amount of energy dissipation). A further complexity involves the role of bedforms in the resuspension process (Osborne and Greenwood, 1992b, 1993; Davidson et al., 1993) and the variable nature of the bedform regime (Sherman and Greenwood, 1984).

In this paper, we report on a large scale field experiment undertaken to monitor suspended sediment transport and nearshore bar dynamics over storm cycles. The experiment encompassed two storms of approximately similar magnitudes with respect to incident wave energy and storm duration. Net sediment transport rates and directions, and the ensuing morphological responses were significantly different during the two events. These differences are examined in terms of the relative importance and interactions of the various sediment transport mechanisms.

EXPERIMENTAL DESIGN

Field site

The experiment was part of the Canadian Coastal Sediment Transport Programme (C-COAST; Greenwood et al., 1990). The field site was located at Burley Beach within Pinery Provincial Park, southeastern Lake Huron, Canada. The shoreline is oriented southwest-northeast and is exposed to a relatively long fetch (up to 300 km) towards the northwest. The nearshore profile is very gently sloping, on the order of 0.009, and includes 3 nearshore bars. Sand with mean grain sizes of 180-200µ is abundant; no major lithological variations occur across the profile. The wave climate is dominated by storm-generated wind waves and tidal effects are negligible.

Instrumentation

The main feature of this experiment was the deployment of a High Resolution Remote Tracking Sonar (HRRTSIII; Greenwood et al., 1993; Richards and Greenwood, 1993) on the lakeward slope of the second bar, at a distance of approximately 115 m from the baseline (x = 111.5-116.5m), see Figure 1. The sonar was used to measure not only the two-dimensional bedform geometry and the influence of this geometry on sediment resuspension, but also to provide continuous records of average bed elevation change. A total station (Geodimeter, System 400) was used for surveys of topographic change in 5 survey lines (ranging 140 m south and north of the instrument transect).

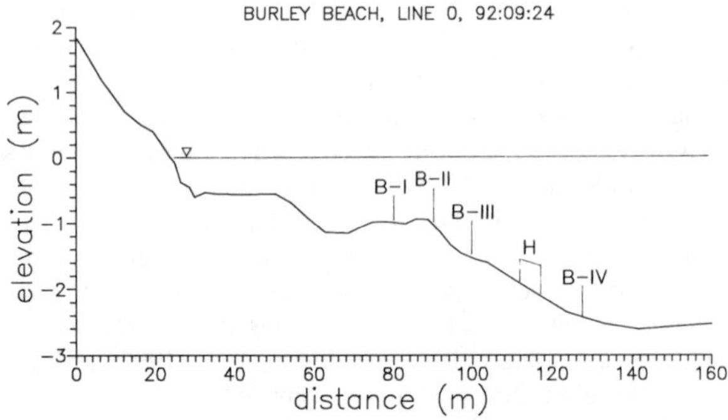

Figure 1. The nearshore profile at Burley Beach. H marks the location of the HRRTS-frame, while B-I through B-IV are the locations of the instrument stations at x = 80, 90, 100 and 127 m.

Six electromagnetic current meters (MarshMcBirney OEM512) and fifteen optical backscatter sensors (OBS-1P) were deployed at the four corner posts of the support frame for HRRTSIII to provide measurements of local sediment transport gradients. To record larger scale cross-shore sediment transport gradients and the associated morphological response (e.g. direction and rate of bar migration), four additional instrument stations were deployed across the second bar at x = 80, 90, 100 and 127 m (see Figure 1). These stations were each equipped with a single current meter and one or two optical backscatter sensors. The current meters were oriented with positive axes onshore (u) and to the northeast (v), and the backscatter sensors were deployed at nominal elevations of z = 0.05, 0.10 and 0.15 m above the bed. In the case of one (or two) OBS-sensors, they were mounted at z = 0.05 (and 0.10) m. All sensors were precalibrated, the current meters in a large tow tank and the optical sensors in a sediment recirculating facility using sand from the field deployment locations. All sensors were hardwired to an underwater data acquisition and transmission system and sampled at 4 Hz for periods of 55 minutes separated by 5 minute breaks.

Data analysis

Instrument records were truncated to 8192 data points, corresponding to approximately 34 minutes. Sensor outputs were screened to check data quality, and records which contained obvious errors or noise were discarded. Local time-averaged oscillatory ($<uc>_{osc}$), mean ($<uc>_{mean}$), and net ($<uc>_n$) suspended sediment transport were determined from the velocity and sediment concentration vectors:

$$<uc>_n = <uc>_{mean} + <uc>_{osc}$$

$$<uc>_{mean} = 1/n \, \Sigma u * 1/n \, \Sigma c$$

$$<uc>_{osc} = (\Delta f \,/\, F) \, \Sigma C_{uc} \, (f)$$

where u is instantaneous cross-shore velocity (ms^{-1}), c is instantaneous suspended sediment concentration (kgm^{-3}), n is number of data points in the records, C_{uc} is cospectral density of velocity and concentration ($kgm^{-2}s^{-1}$), f is frequency, Δf is cospectral resolution and F is cospectral frequency range. The oscillatory transport rate was further separated into incident wave and infragravity wave transports at a frequency of 0.067 Hz (15 s).

THE STORM EVENTS

Data were obtained from two intense storms which occurred on October 16-17 (Event 1), and October 24-25 (Event 2) 1992, respectively. During Event 1, a deep low pressure system crossed the research site generating winds of 17-20ms^{-1} out of the southwest quadrant (fetch 50-120 km) and waves approached the beach at large angles. Wind waves grew rapidly with periods increasing from 4.8 s to 7 s as the storm winds decreased. Maximum orbital velocities reached 1.75 ms^{-1} on the lakeward slope of the bar and strong longshore currents (up to 1.20 ms^{-1}) were generated in a wide surf zone, which contained up to 9-10 dissipative surf bores. Mean cross-shore currents attained velocities of -0.34 ms^{-1} and were consistently directed offshore across the second bar, resembling an undertow. The longshore and cross-shore currents peaked at 1930 h on October 16.

Event 2 was characterized by somewhat smaller wind speeds (up to 12 ms^{-1}), but winds were northwesterly and the fetch was much longer (300 km). Wave heights were similar to those of Event 1, although approach angles were now essentially shore-normal and wave periods were consistently around 7 s. Maximum orbital velocities on the upper lakeward slope of the bar were also similar ($\approx$ 1.55 ms^{-1}) as was the surf zone width and storm duration ($\approx$ 24 h). However, the longshore and cross-shore currents were significantly weaker, peaking at 0.46 ms^{-1} and -0.19 ms^{-1}, respectively, the latter declining to -0.10-0.15 ms^{-1} after the storm peak.

MORPHOLOGICAL RESPONSE

Event 1: As a result of this storm, the entire second bar was displaced lakeward; in the instrument transect (Line 0), the crest migrated approximately 25 m offshore (Figure 2, 92:10:15-20). The profiles responded in a similar manner in all five survey transects with the bar appearing linear after the storm. Erosion in the pre-storm bar crest region (when averaged over all five survey transects) was 13.9 m^3m^{-1}; accretion on the lakeward slope amounted to 12.7 m^3m^{-1}. Considering the errors associated with such surveys, the sediment budget balanced without any significant losses or gains to/from longshore and/or cross-shore sources. Maximum depth of erosion in the pre-storm bar crest region was $\approx$ 0.92 m, while the maximum accretion on the lakeward slope was 0.63 m (see Figure 2).

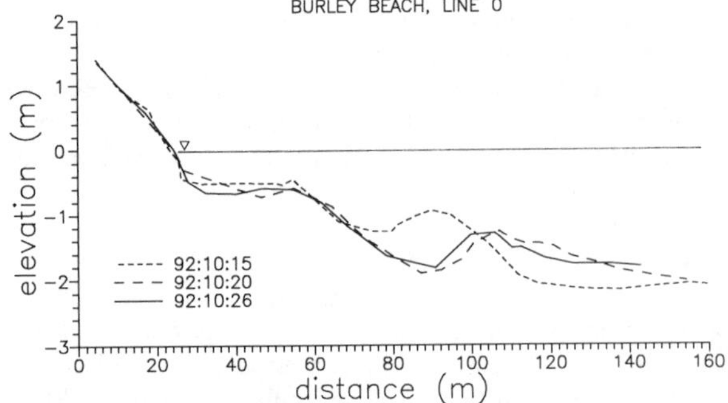

Figure 2. The instrument transect at Burley Beach, surveyed on October 15, 20 and 26, 1992.

The increase in storm wave energy was very rapid. Prior to 1200 h on October 12, the winds were light and offshore; at 1200 h, the winds veered southwest and increased. By 1400 h, the significant orbital velocity on the lakeward slope of the bar reached 1.55 ms^{-1}. Orbital velocities peaked at 1.75 ms^{-1} at 0100 h on October 17.

The morphological response was equally rapid. Figure 3 illustrates the mean bed elevation on the lakeward slope as a function of time; the accretion rate was large and virtually constant (≈ 5 cmh^{-1}), from the beginning of the storm, until approximately 2200 h, October 16. This rapid accretion rate was undoubtedly a reflection of the offshore migration of the bar. At 2200 h, the accretion rate was reduced sharply to ≈ 1 cmh^{-1} and a small amount of erosion was recorded in the morning of October 17; the latter probably reflected a small onshore migration of the bar as the storm decayed. It is of interest that the accretion rate was reduced several hours before the peak of the incident wave energy which occurred at 0100 h, October 17. If we can assume that the bar position at the end of the storm was similar to that at 2200 h, October 16 (Figure 3), then the offshore bar migration rate would have been ≈ 2.5 mh^{-1}, and the erosion rate in the trough would have been ≈ 1.5 $m^3m^{-1}h^{-1}$. These rates are similar to the maximum rates measured in an oceanic environment at Duck, NC (Birkemeier, 1985; Sallenger et al., 1985).

Event 2: Profile changes associated with Event 2 were much less dramatic and more subtle. While the second bar remained stable in position in the southern part of the surveyed area, sediment was transported onshore exclusively in the northern part. Thus, during this storm the bar developed a more sinuous form. In the central instrument transect, a sediment transport divergence appeared to exist over the upper lakeward slope of the bar. The result of this divergence was sediment accretion on the lower lakeward slope, and a shoreward displacement of the bar crest ≈ 5-10 m (Figure 2, 92:10:20-26). Net bed elevation change at the HRRTS-III station (upper lakeward slope of the bar) was -0.19 m, see Figure 3. The sonar record illustrates an almost linear decrease in mean bed elevation at the upper lakeward slope and thus

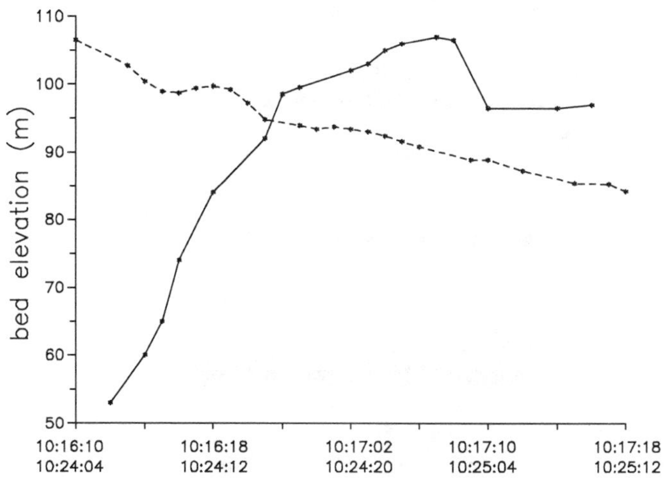

Figure 3. Mean bed elevation change on the lakeward slope of the second bar. Solid line = Event 1 (10:16:10-10:17:18); dashed line = Event 2 (10:24:04-10:25:12).

an almost constant erosion rate (≈ 0.65 cm/h) throughout this storm. Volumetric changes were much smaller than for Event 1, with an average erosion of 2.3 m^3m^{-1} and an average accretion of 2.0 m^3m^{-1} in the region of the second bar-trough region. The local sediment budget was again balanced.

SUSPENDED SEDIMENT TRANSPORT

Event 1: During this event, sediment resuspension and transport was spatially homogeneous, both horizontally and vertically. Figure 4 illustrates time series of cross-shore velocity, low-passed cross-shore velocity (using a filter cut-off of 0.02 Hz), and sediment concentration at elevations of z=0.05, 0.10 and 0.15 m recorded on the lakeward slope of the bar (x=111.5 m) on October 16, 1402 h. At this time (only two hours after the onset of the event), the lakeward slope was accreting extremely rapidly. Immediately apparent in the time series of sediment concentration is a strong low-frequency modulation. Concentration fluctuations coincident with the wind waves are present, but the low-frequency modulation is dominant. Moreover, concentrations at all elevations follow nearly identical trends, the only difference being a decrease in magnitude with elevation above the bed.

The low-passed cross-shore velocity exhibits a modulation similar to that of the concentrations, with a periodicity of approximately 1.79 min (0.009 Hz, Figure 4). Concentration maxima are generally associated with the offshore phases of these infragravity oscillations; this is confirmed by examining expanded time series

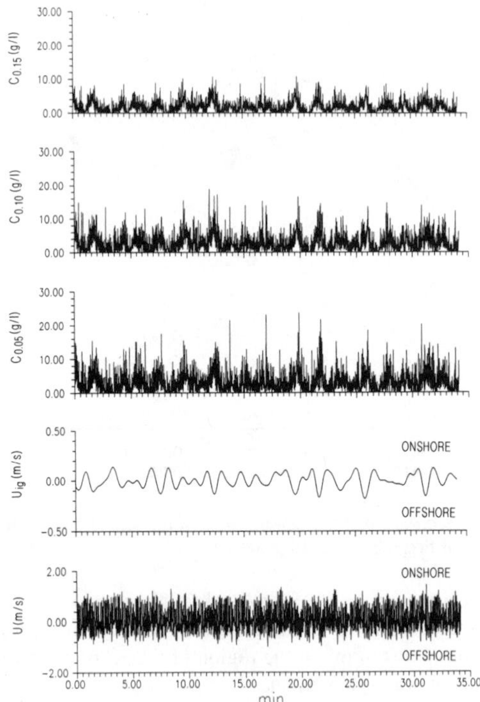

Figure 4. Time series of (from bottom upwards) cross-shore velocity, low-passed velocity, and sediment concentrations at z=0.05, 0.10 and 0.15 m recorded on the lakeward slope of the bar, October 16, 1402 h.

recorded on October 16, 1402 h at this location (x = 111.5 m; Figure 5). Time series recorded simultaneously on the bar crest (x = 90 m; Figure 5) also illustrate low-frequency modulations of the sediment concentration. However, in this case concentration maxima are associated with the onshore phases of the infragravity wave oscillations and resuspension events at incident wave frequencies are even less important. The latter undoubtedly reflect the increased dissipation of incident wave energy at this station closer to the shoreline, and/or the higher sensor elevation. The true elevation of the OBS at this station was estimated as z' = 0.14 m (Aagaard and Greenwood, in prep.).

Spectra of cross-shore velocity and sediment concentration together with cospectra from time series recorded on the lakeward slope (x = 111.5 m) on October 16, 1402 h are illustrated in Figure 6a&b. The cross-shore velocities reveal a broad-banded incident wave field with a peak period of ≈ 5.5 s (0.18 Hz), and a rather narrow sharp peak at 0.009 Hz (111 s) reflecting the peak frequency of the long infragravity waves. In contrast, the sediment concentrations display a variance increase of an

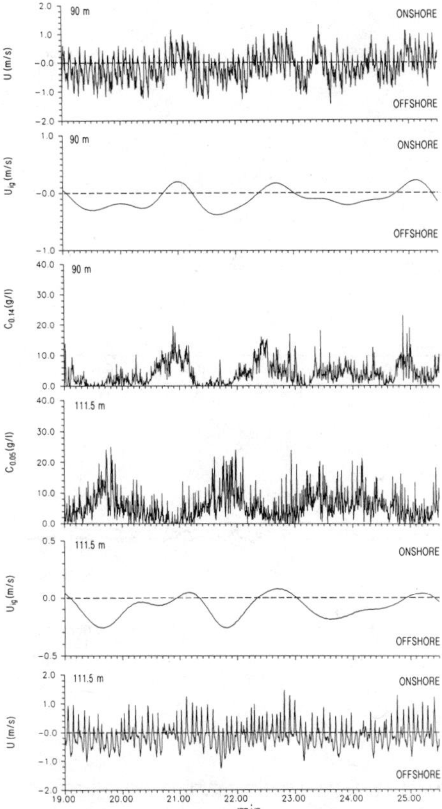

Figure 5. Expanded time series recorded on October 16, 1402 h. From bottom upwards: cross-shore velocity, low-passed velocity and sediment concentration (z=0.05m), all from the lakeward slope of the bar (x=111.5 m), and sediment concentration (z=0.14 m), low-passed velocity and raw cross-shore velocity from the bar crest (x=90 m).

order of magnitude towards low frequencies and a single significant peak coincident with that of the infragravity waves (0.009 Hz). The cospectra indicate relatively small onshore transport rates at incident wave frequencies (and only in the lower parts of the water column; z = 0.05 m), while the maximum oscillatory sediment transport rates occurred at 0.009 Hz and were directed offshore.

Figure 7 illustrates the sediment transport rates (z = 0.05 m) induced by oscillatory infragravity and incident wave motions, mean currents ($\bar{u}\bar{c}$) and finally net transport rates computed for the bar crest and lakeward slope stations at October 16, 1402 h. The mean transport rates have been adjusted for the effects of a logarithmic current boundary layer (Aagaard and Greenwood, in prep.). A large net transport gradient

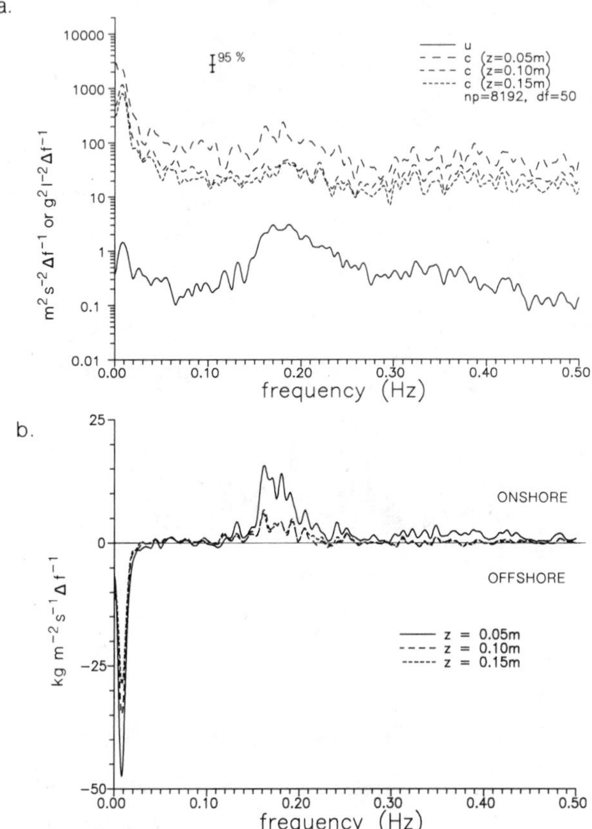

Figure 6. Variance density spectra (a) and cospectra (b) of cross-shore velocity and sediment concentration recorded on the lakeward slope of the bar on October 16, 1402 h. The spectra have 50 degrees of freedom (df).

appeared to exist between these two station at this time; while the net transport at the bar crest was essentially zero (+0.004 kgm^{-2}s^{-1}), it was large and directed offshore on the lakeward slope (-0.490 kgm^{-2}s^{-1}). This spatial gradient was mainly caused by the infragravity wave transports which were directed onshore at the bar crest. At this location, oscillatory transports balanced the offshore directed mean flux due to the undertow. On the lakeward slope, infragravity transports were directed offshore, augmenting the mean flux. Thus, while the undertow was the single most important transport mechanism, the net transport divergence which resulted in the rapid erosion/accretion and offshore migration of the bar can be attributed to the presence of infragravity waves. The latter were standing in the cross-shore direction and the

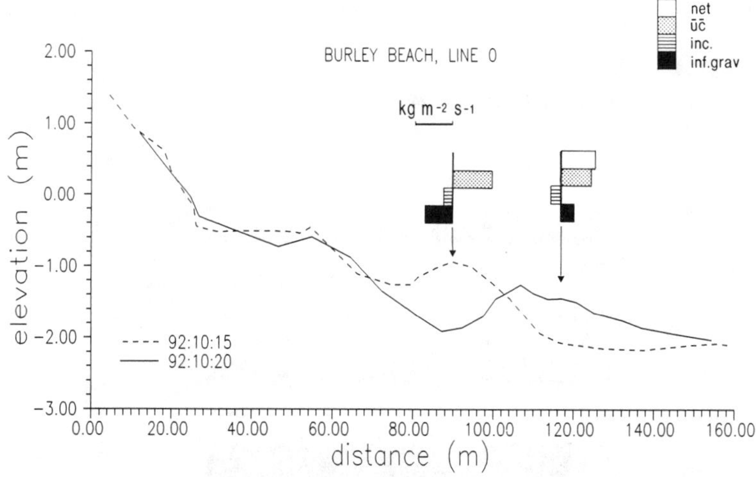

Figure 7. Suspended sediment transport rates due to infragravity waves, incident waves and mean currents, as well as the net transport rates, measured at the crest and lakeward slope of the second bar during Event 1 (October 16, 1402 h).

transport reversals at infragravity frequencies were probably a result of a node in surface elevation between the two stations (Aagaard and Greenwood, in prep.).

Event 2: Unfortunately, only sensors around the HRRTS were deployed during Event 2 and thus horizontal transport gradients cannot be documented. Nevertheless, the characteristic transport patterns from the lakeward slope reveal significant differences from the earlier event. In general, sediment transport rates were significantly smaller than those in Event 1, resulting in the much smaller morphological response noted earlier. Time series of cross-shore velocity and suspended sediment concentration ($z = 0.05$, 0.10 and 0.15 m), recorded on the lakeward slope of the second bar ($x = 111.5$ m) at the peak of this storm (October 24, 1500 h) are illustrated in Figure 8. Sediment resuspension was dominated by incident waves and low-frequency modulations were much smaller than in Event 1. Spectra of sediment concentration (Figure 9a) lack the dramatic increase in variance at low frequencies, appearing more "white" than "red", although small spectral peaks at incident wave frequencies confirm the greater relative importance of the latter. Cross-shore velocity spectra again indicate the presence of considerable kinetic energy at infragravity frequencies; at this station, the significant orbital velocity at infragravity frequencies ($f<0.067$ Hz) was 0.34 ms^{-1} for Event 2, as opposed to 0.25 ms^{-1} for Event 1 (Figure 6). However, the infragravity band appeared much broader during Event 2; the spectral valley at $\approx$ 0.02 Hz was probably an artifact of sensor location relative to a standing wave structure.

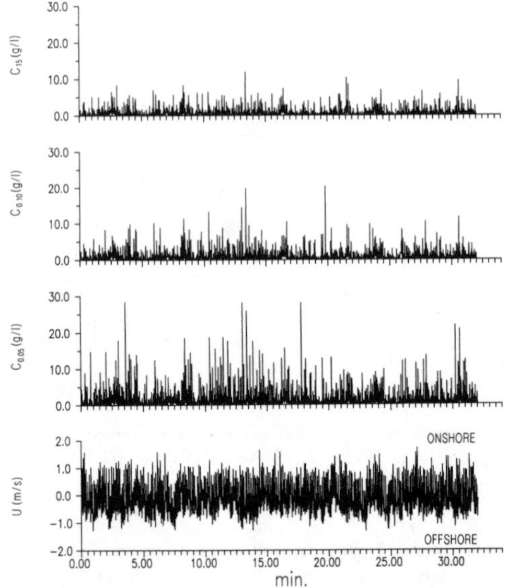

Figure 8. Time series recorded on the lakeward slope of the bar during the peak of Event 2 (October 24, 1500 h). From the bottom upwards the panels illustrate cross-shore velocity and sediment concentrations (at z=0.05, 0.10 and 0.15 m).

Cospectra from these time series (Figure 9b) are less clearly structured than those from Event 1 and reveal other significant differences: (**a**) the largest oscillatory transport rates occurred at wind wave frequencies and significant transport reversals were recorded within the incident wave band (0.067-0.5 Hz); (**b**) sediment transport at infragravity wave frequencies was significantly smaller and distributed over a larger bandwidth; (**c**) in some instances, transport reversals occurred within the infragravity band, indicating the possible presence of several standing wave modes (Aagaard and Greenwood, 1994); (**d**) finally, sediment transport at harmonic frequencies (0.2-0.4 Hz) was important and appeared to be predominantly directed offshore (Figure 9b).

The individual transport components for October 24, 1500 h (x=111.5 m; z=0.05 m) were: Mean transport: -0.103 $kgm^{-2}s^{-1}$; oscillatory infragravity transport: +0.030 $kgm^{-2}s^{-1}$; oscillatory transport at wind wave frequencies: +0.037 $kgm^{-2}s^{-1}$; oscillatory transport at harmonic frequencies: -0.075 $kgm^{-2}s^{-1}$. The net suspended sediment transport rate at the storm peak (at z = 0.05 m) was -0.111 $kgm^{-2}s^{-1}$ and again directed offshore. In general, transport rates were significantly smaller than those recorded during Event 1 (see Figure 7). Mean currents and infragravity waves in particular were much reduced in importance; in contrast oscillatory transport at harmonic frequencies increased considerably in relative importance.

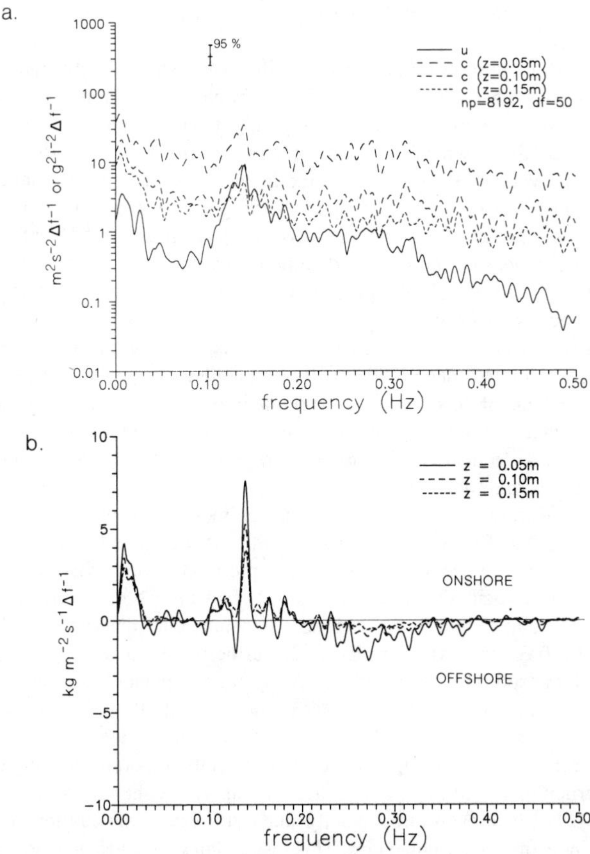

Figure 9. Variance density spectra (a) and cospectra (b) of cross-shore velocity and sediment concentration recorded on the lakeward slope of the bar; October 24, 1500 h. The spectra have 50 degrees of freedom (df).

DISCUSSION AND CONCLUSIONS

During Event 1, the morphological response of the nearshore bar-trough system was dramatic (offshore migration ≈ 25 m) with a very short relaxation time (≈ 10 h). This illustrates the morphodynamic behaviour when a shoreface is significantly out of equilibrium with a new set of wave and current conditions. Morphological adjustment occurred virtually instantaneously and as the new equilibrium was approached, the rate of change became significantly smaller. This interpretation is supported by the morphological response to Event 2 which occurred one week later over a topography which was little changed from that at the end of Event 1; as a result, sediment transport rates and morphological change were significantly reduced.

At least four reasons may account for the smaller sediment transport rates during Event 2:

(**a**) The mean offshore current (undertow) velocities were significantly smaller, by ≈ 50%. This was somewhat surprising, as offshore wave heights appeared broadly similar during the two events and wave set-up might have been expected to be similar. However, as Svendsen (1984) and Greenwood and Osborne (1990) have suggested, undertow velocities depend upon the local setup gradient. The lakeward slope of the second bar was much more gentle during Event 2; this would reduce local radiation stress and setup gradients relative to the beginning of Event 1. The rapidly decreasing gradient on the lakeward slope of the bar as a result of accretion during Event 1 could also explain the peaking of mean currents several hours prior to the maximum incident wave energy.

(**b**) Suspended sediment concentrations were significantly lower during Event 2, reaching mean values approximately 50% of those recorded during Event 1. A major difference was the lack of low-frequency modulations during Event 2. It would appear therefore that infragravity waves can contribute significantly to the entrainment as well as transport of sediment by enhancing near-bed velocities, providing these waves are coherent.

(**c**) The lack of a well-organized forcing of the sediment resuspension at specific low frequencies during Event 2 was due to the broad-banded nature of the infragravity wave field. The occurrence of a single, preferred infragravity peak in the spectra of cross-shore velocity during Event 1 gave rise to a much more effective flux coupling (between u and c) and hence sediment transport than that associated with the broad-banded infragravity waves in Event 2. Furthermore, some cospectra from Event 2 revealed transport reversals within the infragravity band, suggesting the presence of several interacting standing wave modes with the net transport at infragravity frequencies being small (Aagaard and Greenwood, 1990).

(**d**) Finally, the role of bedforms might have been critically important. During Event 1, the vertical structure of sediment concentration was very coherent with a strong phase coupling between cross-shore velocity and sediment concentration at both infragravity and incident wave frequencies. This resulted in very consistent cospectral shapes, incident wind wave transport being in the direction of incident wave skewness (i.e. onshore; see also Foote et al., 1993). It is possible that bedforms had either a very low amplitude, or were largely absent during this Event. During Event 2, sediment transport direction within the incident wave band was strongly frequency-dependent (Figure 9b); this suggests the existence of phase lags with distance above the bed induced by convection of separation vortices ejected from the lee of ripple crests (e.g Vincent and Green, 1990; Osborne and Greenwood, 1993). The comparatively large sediment transport rates at harmonic frequencies (Figure 9b) suggest the presence of ripple-induced vortices, ejected twice every incident wave cycle. Furthermore, the direction of the net oscillatory transport exhibited a temporal dependency, indicative of changing bedform controls. It is not possible in this paper to examine the bedform dependencies in detail.

During Event 1, the bar would appear to have been significantly out of equilibrium with storm-wave conditions (too close to shore). Large offshore sediment transport rates were associated with quasi-steady mean currents (undertow); however, the spatial gradients in net transport responsible for the morphological adjustment (a uniform offshore displacement to produce a linear bar form) were induced largely by standing infragravity waves. As a result of Event 1, the bar was closer to an equilibrium position for Event 2, and net sediment transport rates were smaller. Morphological change during this second event was spatially inhomogeneous. There was virtually no change in the southern part of the experimental area while the bar migrated 5-10 m shoreward in the northern part. The instrument transect at the center of the area experienced a temporal net transport divergence with an offshore transport occurring during the storm peak, which further decreased the gradient of the lakeward slope of the bar. The net result of this transport divergence was an onshore migration of the bar crest, on the order of 5 m and the lakeward slope developing a more concave form.

ACKNOWLEDGEMENTS

This is a contribution from the Canadian Coastal Sediment Transport Programme (C-COAST) supported by Strategic and Operating Grants awarded to B.Greenwood from the Natural Sciences and Engineering Research Council of Canada. T.Aagaard was supported by an International Postdoctoral Fellowship from the same body. We would like to thank all of those who assisted with the data collection, particularly R.W.Brander, D.Rockwell, K.Jagger, R.Atkins and R.Richards (University of Toronto) and Dr.A.J.Bowen, D.G.Hazen and S.McLean (Dalhousie University). Thanks are also due to the staff at Pinery Provincial Park for their assistance.

REFERENCES

Aagaard, T. and Greenwood, B., 1994. "Suspended sediment transport and the role of infragravity waves in a barred surf zone", *Marine Geology*, 118 (in press).

Aagaard, T. and Greenwood, B., in prep. "Suspended sediment transport and morphological response on a dissipative beach".

Beach, R.A. and Sternberg, R.W., 1988. "Suspended sediment transport in the surf zone: response to cross-shore infragravity motion", *Marine Geology*, 80, pp.61-79.

Beach, R.A. and Sternberg, R.W., 1991. "Infragravity driven suspended sediment transport in the swash, inner and outer-surf zone", *Proceedings Coastal Sediments '91*, ASCE, NY, pp.114-128.

Birkemeier, W.A., 1985. "Time scales of nearshore profile change", *Proceedings 19th Coastal Engineering Conference*, ASCE, NY, pp.1507-1521.

Davidson, M.A., Russell, P.E., Huntley, D.A. and Hardisty, J., 1993. "Tidal asymmetry on a macrotidal intermediate beach", *Marine Geology*, 110, pp.333-353.

Doering, J.C. and Bowen, A.J., 1988. "Wave-induced flow and nearshore suspended sediment", *Proceedings 21st Coastal Engineering Conference*, ASCE, NY, pp.1452-1463.

Foote, Y., Huntley, D.A., Davidson, M., Russell, P., Hardisty, J., and Cramp, A., 1993. "Incident wave groups and long waves in the nearshore zone", *Proceedings 23rd Coastal Engineering Conference*, ASCE, NY, pp.974-989.

Green, M.O. and Vincent, C.E., 1990. "Wave entrainment of sand from a rippled bed", *Proceedings 22nd Coastal Engineering Conference*, ASCE, NY, pp.2200-2212.

Greenwood, B. and Osborne, P.D., 1990. "Vertical and horizontal structure in cross-shore flows: an example of undertow and wave set-up on a barred beach", *Coastal Engineering*, 14, pp.543-580.

Greenwood, B., Osborne, P.D., Bowen, A.J., Hazen, D.G. and Hay, A.E., 1990. "Nearshore sediment flux and bottom boundary dynamics: The Canadian Coastal Sediment Transport Programme C-Coast", *Proceedings 22nd Coastal Engineering Conference*, ASCE, NY, pp.2227-2240.

Greenwood, B., Richards, R.G. and Brander, R.W., 1993. "Acoustic imaging of sea-bed geometry: a High Resolution Remote Tracking Sonar (HRRTS II)", *Marine Geology*, 112, pp.207-218.

Hanes, D.M. and Huntley, D.A., 1986. "Continuous measurements of suspended sand concentration in a wave dominated environment", *Continental Shelf Research*, 6, pp.585-596.

Huntley, D.A. and Hanes, D.M., 1987. "Direct measurement of suspended sediment transport", *Proceedings Coastal Sediments '87,* ASCE, NY, pp.723-737.

Osborne, P.D. and Greenwood, B., 1992a. "Frequency dependent cross-shore suspended sediment transport. 1. A non-barred shoreface", *Marine Geology*, 106, pp. 1-24.

Osborne, P.D. and Greenwood, B., 1992b. "Frequency dependent cross-shore suspended sediment transport. 2. A barred shoreface", *Marine Geology*, 106, pp.25-51.

Osborne, P.D. and Greenwood, B., 1993. "Sediment suspension under waves and currents: time scales and vertical structure", *Sedimentology*, 39, pp.599-622.

Richards, R.G. and Greenwood, B., 1993. "An acoustic sensor for measuring bedform geometry and dynamics", *Proceedings Canadian Coastal Conference 1993*, NRC, Ottawa, pp.557-569.

Russell, P.E., 1993. "Mechanisms for beach erosion during storms", *Continental Shelf Research*, 13, pp.1243-1265.

Sallenger, A.H., Holman, R.A. and Birkemeier, W.A., 1985. "Storm-induced response of a nearshore-bar system", *Marine Geology*, 64, pp.237-257.

Sherman, D.J. and Greenwood, B., 1984. "Boundary roughness and bedforms in the surf zone", *Marine Geology*, 60, pp.199-218.

Svendsen, I.A., 1984. "Mass flux and undertow in a surf zone", *Coastal Engineering*, 8, pp.347-365.

Vincent, C.E. and Green, M.O., 1990. "Field measurements of suspended sand concentration profiles and fluxes and of the resuspension coefficient γ_0 over a rippled bed", *Journal of Geophysical Research*, 95, pp.11591-11601.

LABORATORY EXPERIMENT OF LONGSHORE BARS PRODUCED BY BREAKER-INDUCED VORTEX ACTION

Da Ping Zhang[1*], Tsuguo Sunamura[1],
Shigenobu Tanaka[2] and Koji Yamamoto[2]

ABSTRACT: Mechanisms and processes of longshore bar formation were investigated in the light of vortices induced by breaking waves in the laboratory using a two-video-camera system. Vortices associated with wave breaking can be largely classified into two: oblique and horizontal vortices through a fixed-bed wave-flume experiment. Conditions for seven types of vortices reaching bottom in the surf zone can be described by a combination of Galvin's breaker type index and Reynolds number of breaking waves. A movable-bed experiment using six kinds of non-uniform beach profiles as the initial boundary condition, indicated that single bar, double, triple, and quadruple bars were formed. It is found that these bars were formed only by the vortices reaching bottom; the number of bars formed coincides with that of such vortices. The vortices reaching bottom, acted the bed material and lifted the sediment up into suspension; the suspended sediment is transported offshore by the mean offshore flow field, causing net offshore sediment movement to form a bar. The breaker-induced vortices reaching bottom is crucial for the bar formation. Occurrence conditions for single and multiple bars using data of small-scale and prototype-scale experiments were described quantitatively.

INTRODUCTION

Major hypotheses regarding longshore bar formation are: the breaking wave

1) Institute of Geoscience, University of Tsukuba, Ibaraki 305, Japan.
2) Coastal Engineering Division, Public Works Research Institute, Ministry of Construction, Ibaraki 305, Japan.
* Present Address: Coastal Engineering Division, Public Works Research Institute, Ministry of Construction, Ibaraki 305, Japan.

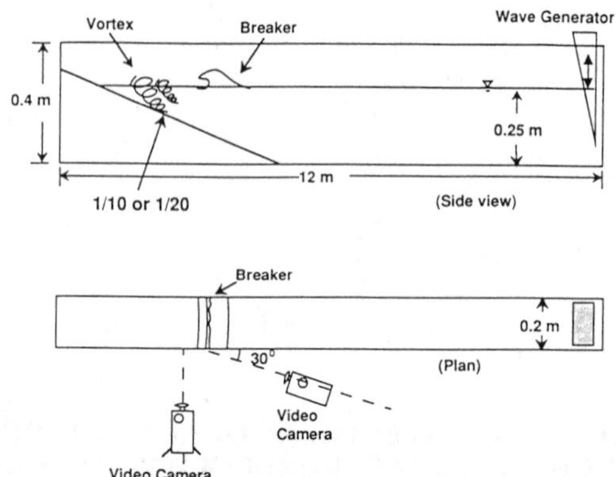

Fig. 1 Experimental equipment

hypothesis (e.g., Miller 1976, Dolan and Dean 1985, Dally 1987, Sunamura and Maruyama 1987) and the long-wave hypothesis (e.g., Aagaard 1990, 1991, and references therein). Difficulties in field measurements and morphological surveys during the severe stormy weather have hindered the examination of the validity of these hypotheses. This study, conducted in the laboratory from the standpoint of the breaking-wave hypothesis, is to investigate the effect of vortices induced by breaking waves on the bar formation.

With the recent progress of measuring techniques of flow velocity, studies from a micro-scopic point of view have been conducted with the purpose of elucidating the nearshore hydrodynamics. Nadaoka et al. (1987, 1989) found the presence of "oblique vortex", which is a vortex with an oblique rotating axis, occurred under spilling breaker conditions, and reconfirmed that this oblique vortex induced sediment suspension in the surf zone. Although they have not noted the interaction of bar formation and this oblique vortex, it is suggested that such sediment suspension could induce the net sediment movement in the surf zone and possibly produce the bar formation, since the bar formation depends on the net offshore sediment movement.

This study examines the characteristics of vortices in the surf zone through a wave-flume experiment, elucidates the relationship between the vortex action and bar formation, and discusses occurrence conditions for single and multiple bars using experiment data.

EXPERIMENT ON BREAKER-INDUCED VORTICES REACHING BOTTOM

In order to examine the characteristics of vortices produced by breaking waves, a fixed-bed experiment was conducted at the Institute of Geoscience, University of Tsukuba, using a 1/10 or 1/ 20 uniform beach in a small-scale wave flume (12 m long,

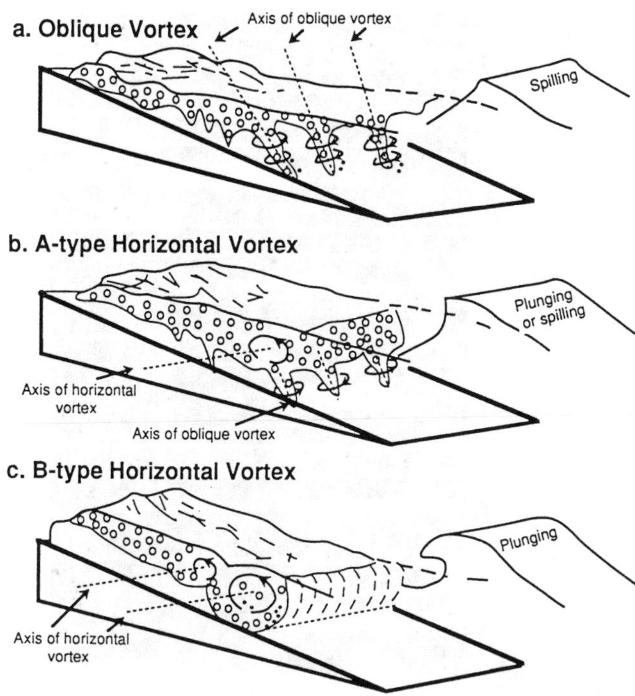

Fig. 2 Three vortex type

0.2 m wide, and 0.4 m deep, Fig. 1). The period of laboratory waves ranged from 0.6 to 2.4 sec and the height of breaking waves from 5.0 to 14 cm. To examine the characteristic vortices distributed in the whole surf zone, two video cameras were used (Fig. 1). One was set up normal to the side of a glass window of the flume, and the other was installed with an angle of 30 degrees behind the breaking point. The three-dimensional characteristics of vortices were examined on reproduced video pictures using both breaker-induced air bubbles and neutrally buoyant particles (1.2mm in diameter) as tracers. Total experiment runs were 110.

Types of Breaker-induced Vortices

Vortices formed just after wave breaking are classified into three types according to the direction of vortex axis: oblique vortex and two types of horizontal vortex, i.e., A-type and B-type horizontal vortices. The oblique vortex is like a tornado that has an obliquely stretched axis of rotation (Fig. 2a). The B-type horizontal vortex, which is shown by the lower diagram (Fig. 2c), looks a cultivator that has a horizontal axis of rotation. The A-type horizontal vortex, which is shown in the middle, is a hybrid

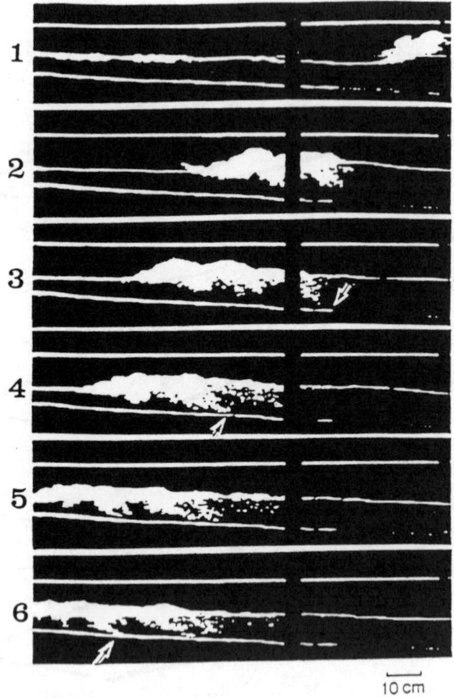

Fig. 3 Development of triple vortices reaching bottom

between horizontal and oblique vortices (Fig. 2b). Namely, the horizontal vortex forms first in the upper part of water column and then it changes to the oblique vortex. Considering that the vortices reaching bottom are very important when we study sediment motion, we examined this kind of vortices.

Types of Breaker-induced Vortices Reaching Bottom

The vortices reaching bottom are not only formed at the break-point, but also in the whole surf zone depending on experimental conditions. Figure 3 is an example of a sequence of video-pictures that show developmental processes of the triple vortices reaching bottom. The time between one stage to the next is 0.3 sec. The first oblique vortex develops at Stage 2 just after wave breaking and it touches the flume bottom as indicated by the arrow at Stage 3. As the bore propagates, the second oblique vortex develops inshore and reaches bottom at Stage 4, and finally the third oblique vortex reaching bottom forms further inshore at Stage 6 as indicated by the arrow.

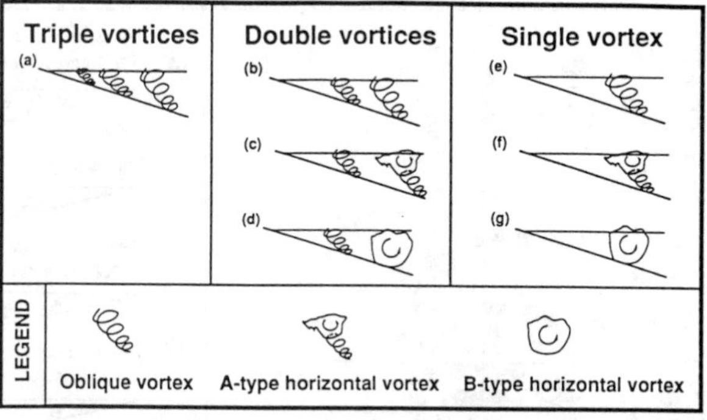

Fig. 4 Schematic diagram showing vortices reaching bottom

Results obtained through the present experiment indicate that vortices reaching bottom are largely classified into seven: (1) the triple vortices (Fig. 4a), (2) three types of the double vortices (Fig. 4b), and (3) three types of the single vortex (Fig. 4c). The first vortex, i.e., the vortex formed just after wave breaking, is an oblique vortex or an A- type or a B- type horizontal vortex depending on experimental conditions, but the inshore vortex is always oblique vortices. In this experiment it was also observed that there exist the vortices not reaching bottom; this phenomenon was found whenever the breaker height was extremely small.

Occurrence Conditions for the Vortices Reaching Bottom

In order to investigate systematically the occurrence condition for the seven types of characteristic vortex, two dimensionless parameters were introduced. One is the breaker-type index given by Galvin (1968), which is denoted as B_t in this study:

$$B_t = H_b / gT^2 \tan\beta \tag{1}$$

where H_b is the breaker height, T is the wave period, $\tan\beta$ is the bottom slope, and g is the acceleration due to gravity. The reason for the selection of this parameter is that the type of breaking waves is closely related to vortex types. The other parameter selected here is Reynolds number for breaking waves, which will be used as an approximate index to evaluate the viscous effect on fluid motion in the surf zone. The breaker height and the horizontal water-particle velocity at the crest of breaking waves were taken into account as a characteristic length and velocity, respectively. This particle velocity, U_b, can be assumed to be equal to wave velocity at the breaking point, i.e., $U_b \sim L_b/T$, where L_b is the length of breaking waves. Therefore, Reynolds number for breaking waves, Re, can be expressed as (Zhang and Sunamura 1990):

$$Re = H_b L_b / \nu T \tag{2}$$

where ν is the kinematic viscosity of fluid.

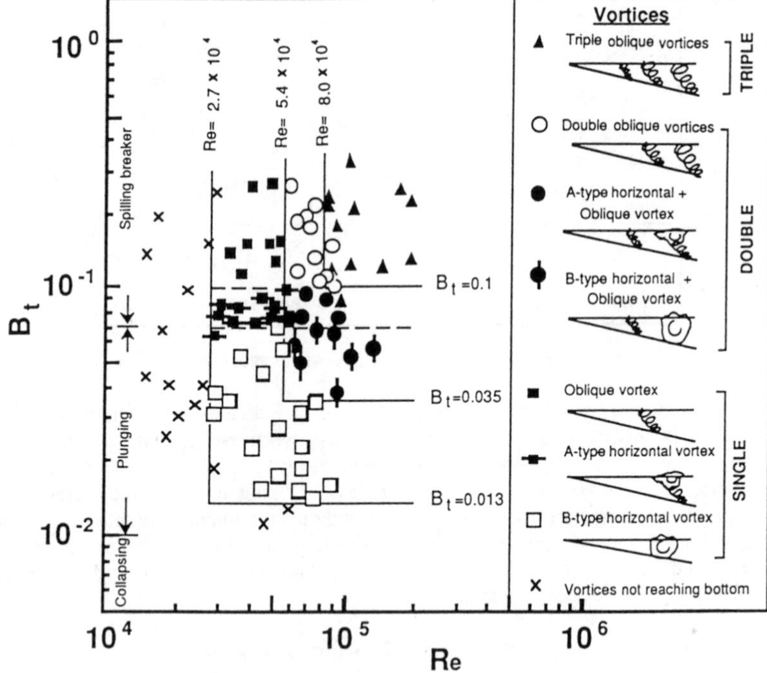

Fig. 5 Conditions for the occurrence of vortices reaching bottom

Conditions for the vortex occurrence are plotted using B_t and Re in Fig. 5. The three L-shaped solid lines are boundaries for the area of triple vortices, double vortices, and single vortex, respectively.

Based on Galvin's (1968) criteria, breaker types were classified on reproduced video pictures in this study; the result is plotted along the y-axis in Fig. 5. It is indicated that the breaker type does not always coincide with the type of vortices, but spilling breakers provide a necessary condition for the occurrence of the triple oblique vortices. It is also indicated that the number of multiple vortices increases with increasing values of B_t and Re.

EXPERIMENT ON BAR FORMATION DUE TO BREAKER-INDUCED VORTICES REACHING BOTTOM

To investigate the relation of vortex action to bar formation, a movable-bed

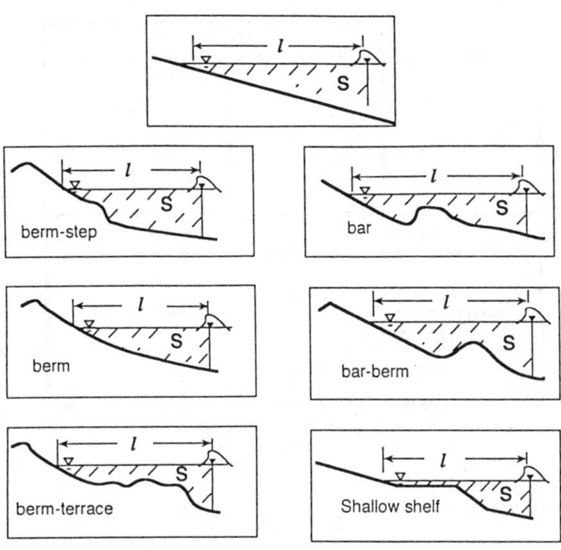

Fig. 6 1/10 uniform beach and 6 non-uniform beach profiles used as the initial boundary condition

experiment was conducted using 1/10 uniform beach profile and six kinds of non-uniform beach profiles as the initial boundary condition (Fig. 6) set up in the same wave flume shown in Fig.1. To examine the vortex characteristics and vortex-topography interaction, the two-video camera system was again used. Changing the combination of these initial beach slopes, the grain size of beach material (0.22, 0.69, 1.3, and 2.4 mm), wave period (0.7, 0.8, 0.9, and 1.0 sec) , and breaker height (5.0～10cm), 61 experiment runs were conducted. Waves continued to act until bar morphology attained the equilibrium; the time required for it ranged from thirty minutes to two hours depending on experiment runs. The beach topography was measured at the center of the flume by an automatic profiler every five minutes as a general rule.

Figure 7 shows the resultant morphology, which is largely classified into four types according to the number of bars formed: single bar, double, triple, and quadruple bars. It was found that the vortices reaching bottom can produce these bar morphologies. The double bars can be subclassified into two types depending on the relative depth of bar crests: I-type and II-type.

Taking a break-point bar as an example, the mechanism for bar initiation is shown in Fig. 8. Figure 8a shows that the oblique vortex reaches bottom, whirls the bed material and lifts the sediment up into suspension acting as a tornado. The suspended sediment is

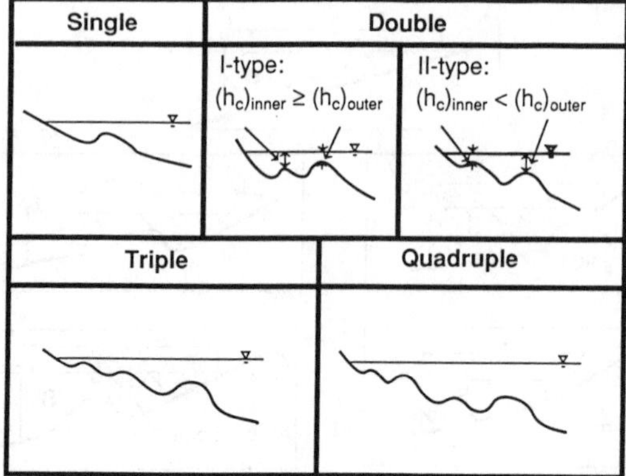

Fig. 7 Bar types

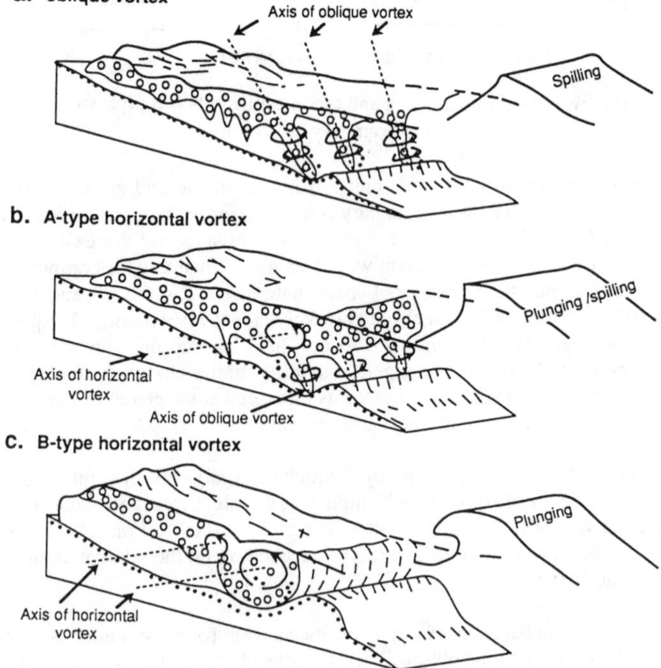

Fig. 8 Schematic diagrams showing bar formation by vortices reaching bottom

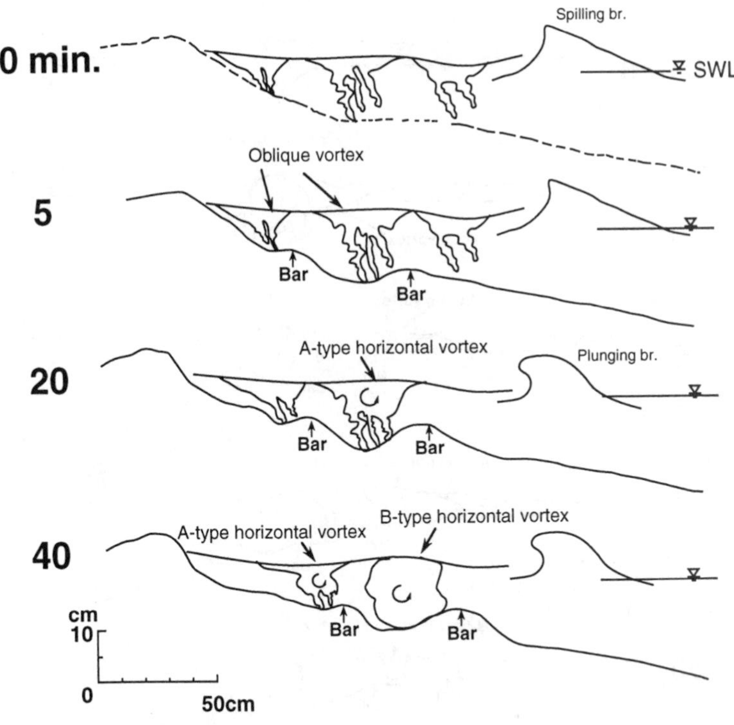

Fig. 9 An example of development of double bars

transported offshore by the mean offshore flow field, causing a net offshore sediment movement to form a small bar. Figure 8b shows that the action of A-type horizontal vortex. Because this action is somewhat stronger than that of oblique vortex, a middle-sized bar forms. Figure 8c shows that the B-type horizontal vortex digs the bed material like a cultivator to form a large bar.

Some examples will be shown to illustrate multiple bar formation. Figure 9 shows the case of double bars. The initial morphology was a berm profile. During the first 5-minute wave action, two bars started to form at the same time by two oblique vortices reaching bottom, respectively. The break-point bar, located offshore, grew to a large bar as the vortex type changed from the oblique vortex to the A-type and finally to the B-type horizontal vortex after 40-minute wave action. At the same time, the vortex type near the inner bar changed from the oblique to the A-type horizontal vortex. This type of bar growth was observed in the cases in which the initial morphology was a berm or step.

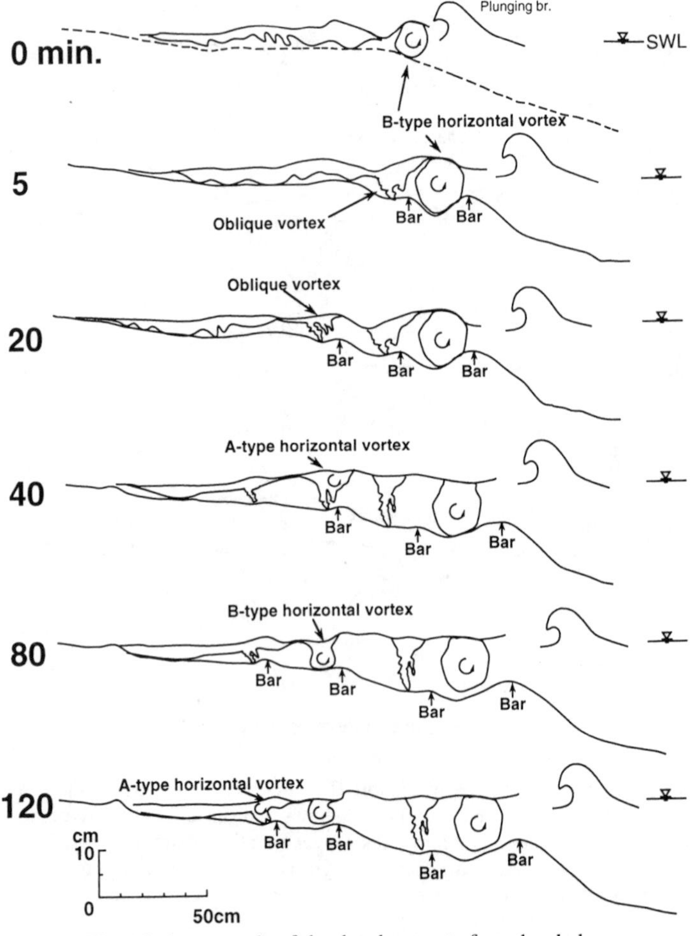

Fig. 10 An example of the development of quadruple bars

Figure 10 shows the development of four bars. The initial morphology had a profile of a shallow shelf. During the first 5 minutes, the break-point bar was formed by the B-type horizontal vortex, and the second bar was formed inshore by the action of oblique vortex. At 20 minutes, the third bar appeared by the action of oblique vortex formed further inshore. At 40 minutes this oblique vortex changed to the A-type horizontal vortex, which finally changed to the B-type horizontal vortex at 80 minutes. At the same time the second breakers were observed above the third bar. At this stage, the fourth bar was formed most landwards by the oblique vortex. Finally, the equilibrium state of four bars was achieved when most vortices could not touch bottom. This type of bar growth was observed in the cases in which the initial morphology had a shallow shelf.

The size of inshore bars is smaller than the break-point bar, because the strength of vortex decreases towards the shore. The present experiment showed that (1) bars are formed only by the vortex reaching bottom, and (2) the number of bars formed coincides with that of such vortices. Therefore, it is concluded that the vortex reaching bottom provides a necessary condition for bar formation.

OCCURRENCE CONDITIONS FOR SINGLE AND MULTIPLE BARS

Quantitative evaluation of morphological effects on bar formation first requires a parameter to describe quantitatively various shapes of nearshore bottom profiles shown in Fig. 6, because they must affect whether the vortex can reach bottom or not. Considering that the wave breaking is an important dynamic factor, we define that the average beach slope shoreward of the wave break-point, i , as:

$$i = 2h/l \quad (3)$$

where h is the average depth of the surf zone and l is the width of the surf zone. The average depth h is given by:

$$h = 1/l \int_0^l h(x)dx = S/l \quad (4)$$

where S is the cross-sectional area of the surf zone as shown in Fig. 6. Substituting Eqs. (4) into (3), we obtain the following equation for the average slope of the surf zone:

$$i = 2S/l^2 \quad (5)$$

In order to examine the occurrence condition for the bar types shown in Fig. 7, the two dimensionless parameters, H_b/gT^2 and iD/H_b (D=grain size of beach material), were used for plotting bar data of small-scale experiments (the present experiment and Yokotsuka 1985) and the existing prototype experiments (Sunamura and Maruyama 1987, Kraus and Larson 1988) in Fig. 11. The boundaries for single or multiple bar formation are indicated by the solid lines in this figure. A family of the lines can be expressed by:

$$H_b/gT^2 = K(iD/H_b)^2 \quad (6a)$$

or

$$K = H_b/gT^2 \, (i)^{-2} \, (D/H_b)^{-2} \quad (6b)$$

where K is a dimensionless parameter which takes different values depending on the number of bars formed.

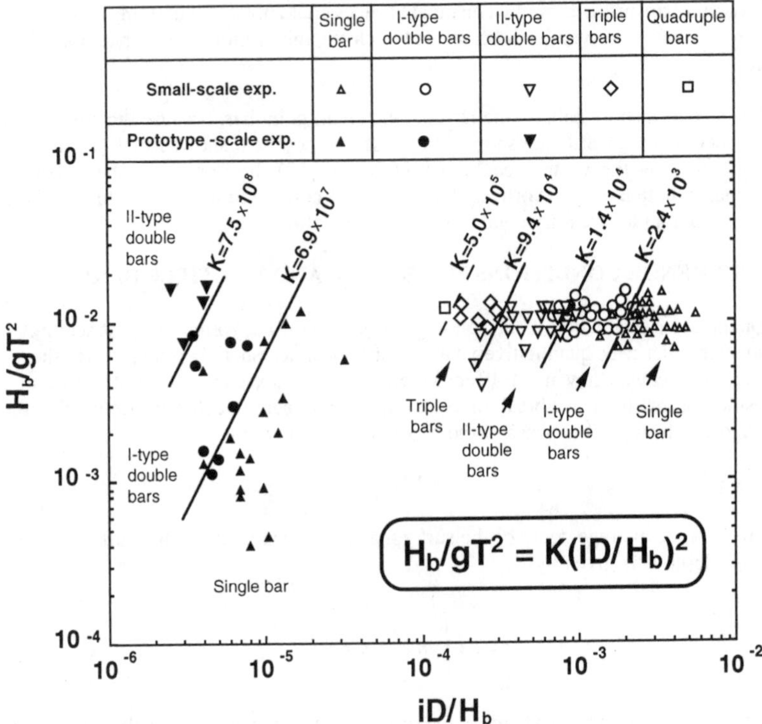

Fig. 11 Conditions for the occurrence of single and multiple bars in the small-scale and prototype-scale laboratory environments

For the boundary of the single bar and the I-type double bars in the small-scale experiment,

$$K = 2.4\text{x}10^3 \tag{7}$$

For the boundary of the I-type double bars and II-type double bars,

$$K = 1.4\text{x}10^4 \tag{8}$$

For the boundary of the II-type double bars and the triple bars,

$$K = 9.4\text{x}10^4 \tag{9}$$

For the boundary of the triple bars and the quadruple bars,

$$K = 5.0\text{x}10^5 \tag{10}$$

As to the demarcation of bar morphologies in the prototype experiment, it is found that the same equation as Eq. (6) can be applied. The K-value for delimiting the single and the I-type double bars is given by:

$$K = 6.9\text{x}10^7 \tag{11}$$

Demarcation of the I-type and II-type double bars is:

$$K = 7.5\text{x}10^8 \tag{12}$$

Comparisons between Eqs. (7) and (11), and between Eqs. (8) and (12), indicate that the K-value for the prototype-scale experiment is four orders of magnitude greater than that for the small-scale laboratory tests. This discrepancy is due to the scale effect, being greatly different in the value of D/H_b between for the small and prototype experiments.

In either small-scale or prototype-scale experiments, the number of bars increases with increasing value of K. Equation (6b) shows that K-value increases with increasing breaker height (H_b), or with decreasing wave period (T), surf zone gradient (i) or sediment grain size (D). Because i is strongly dependent on D (i generally decreases with decreasing D), this laboratory finding suggests that the number of bars increases when higher breaking waves with shorter period attack on a gentler beach (i.e., with finer material). This is well concordant with the field findings that multiple bears are favorably found along gentle beaches of closed waters (i.e., under no swell conditions) with episodic assault of storm waves; such beaches are located along Canadian lakes and bays (Greenwood and Davidson-Arnott 1975, Davidson-Arnott 1988), Chesapeake Bay (Aagaad 1990), and Sea of Japan (Hom-ma and Sonu 1962, Mogi 1963, Katoh 1984).

CONCLUSIONS

(1) Vortices associated with wave breaking can be largely classified into three: oblique vortex, and A-type and B-type horizontal vortices. Seven kinds of vortices reaching bottom in the surf zone of the present experiment are demarcated on the B_t-Re plane (Fig. 5).

(2) The oblique vortex reaches bottom, whirls the bed material and lifts the sediment up into suspension acting as a tornado. The suspended sediment is transported offshore by the mean offshore flow field, which causes net offshore sediment movement to form a small bar. The A-type horizontal vortex, which is somewhat stronger than the oblique vortex, form a middle-sized bar. The B-type horizontal vortex digs the bed material like a cultivator to form a large bar.

(3) Bars are only formed by the vortices reaching bottom; the number of bars formed

coincides with the number of vortices reaching bottom. The occurrence of vortices reaching bottom is crucial for the bar formation.

(4) Occurrence conditions for single and multiple bars (Fig. 11) were described by K-value in Eq. (6).

(5) The number of bars increases as breaker height increases, or as wave period or surf zone slope decreases.

ACKNOWLEDGMENTS

Part of this study was made under the Grant-in-Aid for Scientific Research from Ministry of Education, Science and Culture to TS (B-04452329).

REFERENCES

Aagaard, T. 1990. "Infragravity waves and nearshore bars in protected, storm-dominated coastal environments," *Marine Geology*, 94, PP. 181-203.

Aagaard, T. 1991. "Multiple-bar morphodynamics and its relations to low-frequency edge wave," *J. Coastal Res.*, 7, pp. 801-813.

Dally, W. R. 1987. "Longshore bar formation - surfbeat or undertow? *Proc. Coastal Sediments' 87*, ASCE, pp. 71-86.

Davidson-Arnott, R. G. D. 1988. "Controls on formation and from of barred nearshore profiles," *Geogr. Review*, 78, pp. 185-193.

Dolan, T. J. and Dean, R. G.1985. "Multiple longshore sand bars in the upper Chesapeake bay," *Estuarine Coastal Shelf Science*, 21, pp. 727-743.

Galvin, C. J., Jr. 1968. "Breaker type classification on three laboratory beaches," *J. Geophys. Res.*, 73, pp. 3651-3659.

Greenwood, B. and Davidson-Arnott, R. G. D. 1975. "Marine bars and nearshore sedimentary processes, Kouchibouguac Bay, New Brunswick," in: Hails, J. and Carr, A., (eds.), *Nearshore Sediment Dynamics and Sedimentation*, Wiley-Science, London, pp. 123-150.

Hom-ma, M. and Sonu, C.J. 1962. "Rhythmic patterns of longshore bars related to sediment characteristics," *Procs. 8th Conf. Coastal Eng.*, Council on Wave Research, pp. 248-278.

Katoh, K. 1984. "Multiple longshore bars formed by long period standing waves," Japan: Report Port and Harbour Res. Inst., Ministry of Transport, 23, pp. 3-46.

Kraus N. C. and Larson, M. 1988. "Beach profile change measured in the tank for large waves 1956-1957 and 1962," Coastal Engineering Research Center, Technical Report 88-6, 39 p.

Miller, R. 1976. "Role of vortices in surf zone prediction: sedimentation and wave forces," in: Davis, R.A. Jr. and Ethington, R.L., (eds.), *Beach and Nearshore Sediment,* Society Economic Paleontologists Mineralogists, Special Pub., 24, pp. 92-114.

Mogi, A. 1963. "On the shore types of the coasts of Japanese islands," *Geogr. Review Japan*, 36, pp. 245-266 (in Japanese with English abstract).

Nadaoka, K., Ueno, S., and Igarashi, T. 1987. "Characteristics of bottom velocity and sediment suspension in the surf zone," *Proc. 34th Japan. Conf. Coastal Eng.*, pp.

256-260 (in Japanese).

Nadaoka, K., Hino, M., and Koyano, Y. 1989. "Structure of the turbulent flow field under breaking waves in the surf zone," *J. Fluid Mech.*, 204, 359-389.

Sunamura, T. and Maruyama, K. 1987. "Wave induced geomorphic response of eroding beaches with special reference to seaward migrating bars," *Proc. Coastal Sediments '87*, ASCE, pp. 788-801.

Yokotsuka, Y. 1985. "A wave-tank study on topographical change of sandy beaches," Unpublished BSc Thesis, Institute of Geoscience, University of Tsukuba, 51 p. (in Japanese with English abstract).

Zhang D. P. and Sunamura T. 1990. "Conditions for the occurrence of vortices induced by breaking waves," *Coastal Eng. Japan*, 33, pp. 145-155.

THE RESPONSE OF A BARRED COAST TO A SEQUENCE OF STORMS

Felix C.J. Wolf [1]

ABSTRACT: The storm induced response and the following recovery of a barred coast was daily monitored during a period of six weeks. Five storms were encountered during this period. The shoreline varied only little and no clear relation with the individual storms was present. The resulting net shoreline displacement was mainly caused by the welding of an inter-tidal bar. The storms were reflected in the behaviour of the inner bar. The net result of the entire period was an onshore migration of the inner bar over 70 meter. The outer bar showed little, though, significant changes. The migration direction of the outer bar varied alongshore, stressing the three dimensionality of the surf zone morphology.

INTRODUCTION

There has been a growing understanding of the storm-induced morphologic behaviour of longshore bars in the last decades. However, apart form the measurements in the United States (e.g. the Duck studies), there were few studies that intensively investigated the *morphodynamics* of nearshore bars. Rather, large time intervals (several days to weeks) are present between two succeeding surveys. Earlier studies (e.g. Sallenger et al., 1985) have shown that this monitoring should be very intensive because bars react very quickly on changing hydraulic conditions. The morphodynamics of the nearshore bars and the role of bars in the longer term coastal development (several storms) is still unclear. In order to understand these morphodynamics, a synoptic monitoring of the sediment-transport, the hydraulics and the morphologic development is necessary.

Nearshore bars are present along the central Dutch coast. The behaviour of these bars has been monitored since 1964. Cross-shore profiles with a length of 800-1200 meter and a longshore spacing of 250 meter have been surveyed yearly. The bars show a net offshore movement when the behaviour of the bars over years is considered. However, these yearly surveys do not reveal the morphodynamics of an individual bar. Aliasing effects and the

---

[1] Coastal Researcher, Utrecht University, Institute for Marine and Atmospheric Research (IMAU), Department Physical Geography, P.O. Box 80.115, 3508 TC Utrecht, The Netherlands (E-mail: f.wolf@frw.ruu.nl).

lack of detailed hydraulic and sediment transport measurements are due to this. Therefore in 1987 a field programme was set-up with the aim to improve the understanding of the beach and bar morphodynamics. During 1988-1990 four field campaigns were executed near Egmond aan Zee (Kroon, 1990, 1994). These campaigns revealed that characteristic time- and spatial scales differed for the swash bar and two nearshore bars present in the Egmond field site. The outer bar was fairly stable and moved offshore only during very large storms ($H_{s,\ offshore} > 4$ m). Nevertheless, the outer bar's behaviour seemed to be linked to larger time-scales than that of single storms. The inner bar was only measured during periods of low waves, i.e. with intervals of several days to weeks. These inner bar measurements showed a large variation of the bar location. The swash bar was highly moveable and linked to short time scales of hours to days. The swash bar eroded during storms and built up during calm weather. This storm related behaviour of the swash bar was likewise to the accretion/erosions cycle described by e.g. Orme and Orme (1988) and Owens and Frobel (1977).

The 1988-1990 studies made clear that, to further reveal the morphodynamics of the surf zone at Egmond, a more intensive and detailed monitor programme had to be executed. Especially the morphodynamics of the inner bar and its relation to the beach and outer bar morphodynamics should be studied. Hence, the new field studies should focus on short time scales (one storm to several storms). Because of the complexity of the processes involved it was also decided to concentrate on cross-shore morphodynamics.

New field campaigns were conducted in 1991 and 1992. Some of the results of the 1991 field campaign are presented in Wolf (1993). This paper deals with the October-November 1992 field study which has investigated the hydraulics, the sediment-transport and the morphologic response of a barred profile during a period of 6 weeks and on a daily basis. The purpose of this paper is to investigate how the beach and bars change during different wave conditions. Furthermore, the influence of individual storms on the net morphological development over weeks will be discussed. Hence, this paper focuses on the morphologic response with the hydraulics only briefly outlined here.

STUDY SITE AND MEASUREMENTS

The data was collected from 1 October 1992 to 11 November 1992 at the Egmond field site. This site is an 1 kilometer stretch located at the central part of the sandy Dutch coast near Egmond aan Zee (Figure 1). Man-made structures are absent within the field area. The central Dutch coast is a storm dominated coast. The annual significant offshore wave height is 1.3 m. The tide is asymmetrical with a flood period of about 4 hours and an ebb period of about 8 hours. The tidal range varies between 1.2 m at neap tide and 2.1 m at spring tide.

The cross-shore profile commonly shows three bars: one swash bar and two nearshore bars. The swash bar is located near the shoreline, the inner nearshore bar at a distance of 200 m and the outer nearshore bar at a distance of 550 m (Figure 2.). The inner and outer bar are about SSE-NNW oriented and make an angle of about 8 degrees with the coast. Both have a shore attachment point north of the study area. The outer bar is mainly straight, while the inner bar sometimes shows rhythmic characteristics. The two nearshore bars remain present during all seasons but show a net offshore migration of about 15-20 m per year (de Vroeg et al., 1989). The mean grainsize is about 270 μm.

Hydraulic measurements

A Wavec-buoy, located at a water depth of about 15 m, measured the offshore wave height, -period and -direction. The sampling frequency was 1.28 Hz, de burst length 20 minutes and the burst interval 30 minutes. Three poles with a pressure gage and capacitance wire were located in the nearshore zone (Figure 2). The two most seaward poles also contained a Delft Hydraulics type of current meter which measured the longshore and cross-shore currents in a horizontal plane. All instruments connected to the poles were sampled with a sampling frequency of 4 Hz. The burst duration of the instruments on the most seaward pole was 40 minutes. The instruments on the two inner poles were sampled continuously.

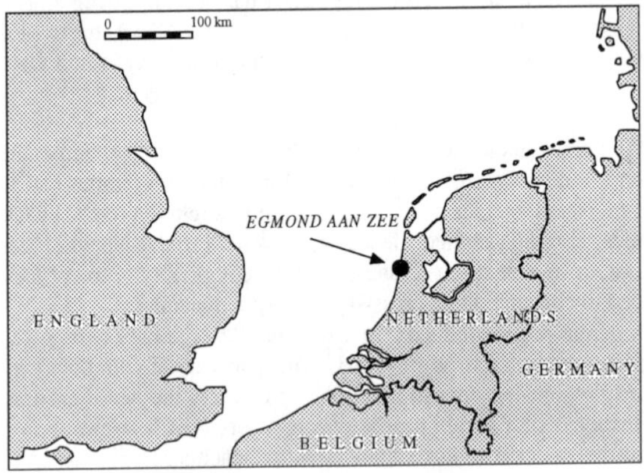

Figure 1. Geographical location field site

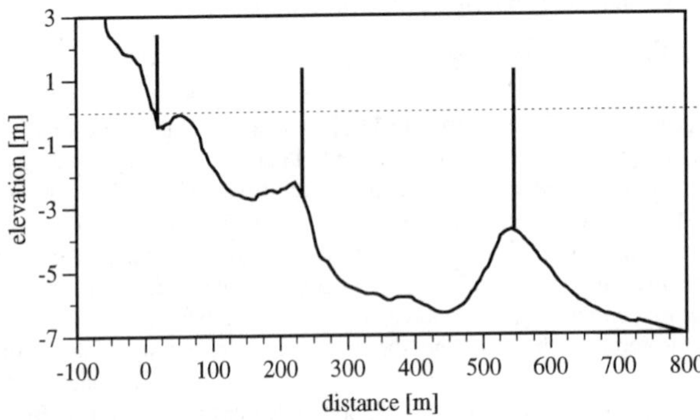

Figure 2. Cross-shore profile, at the centre of the study area, and locations of instrumented poles; elevation relative to Dutch ordnance datum (NAP≈MSL), distance relative to beach pole reference line (RSP).

Morphologic Measurements

Beach profiles were measured perpendicular to the beach pole reference line (RSP). The RSP is a longshore line which makes a slight angle with the shoreline, resulting in a more landward location in the northern part of the field area compared with the southern part. The lines are numbered in coherence with the RSP number. Profile 39.000 marks the northern border of the field area and profile 40.000 is the southern border of the field area.

One beach profile was monitored every day. Eight other beach profiles spaced over a longshore distance of 1 kilometer, were bi-daily surveyed. All profiles were measured with either an Elta-20 total station or with an Polartrack total station.

One cross-shore profile, extending 250 m in the nearshore, was daily measured with the so-called Sub-Aquatic-Profiler (SAP). This cross-shore profile was located 50 meter north of the line with the instrumented poles. The profile incorporated the swash bar, the foreshore and the inner nearshore bar.
The SAP is a 3 x 2.5 x 1.2 meter (l x w x h) 'sea-sled' with a 7 meter long mast. A measuring wheel was attached under the SAP to control the logging of the SAP's instruments. One watertight PVC cylinder, containing two inclinometers and a compass, are placed on the SAP. A second watertight cylinder with a logger and batteries is crosswise attached. The position of the SAP is obtained by pointing either an Elta-20 total station or with an Polartrack total station to a reflector in the top of the mast.. This position is then adjusted for the longitudinal and transverse tilt of the SAP with the inclinometer values. The SAP is pulled, back and forth, through the inner nearshore zone with a capstan and a cable in a closed loop between the capstan and 2 pulleys: one pulley at a beach pole, the other at a pole in the surfzone (Figure 3). The capstan is attached to a tractor.

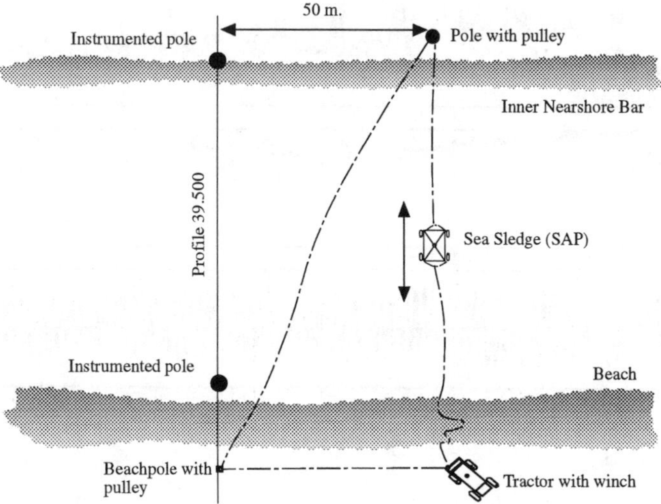

Figure 3: Field Layout Sub-Aquatic-Profiler (SAP). The 39.500 profile marks the centre of the study area

In order to monitor, qualitatively, the longshore variation of the beach and of the nearshore bars, a photo camera was placed on a flat building located on top of the dunes. This camera took, during daylight low tide, 10 minute averaged photo's of foam associated with wave breaking. This technique is based on the video system of Lippmann and Holman (1989).

Nearshore surveys were executed with a boat with echosounder. Twelve cross-shore profile lines, covering the entire surfzone and the 1 km longshore stretch, were to be surveyed on a weekly basis. Two main morphologic developments could be monitored in this way. First, the longshore variation of the inner bar which makes it possible to estimated the significance of the SAP profile. Second, the storm related changes of the outer bar. These changes are important because a change in the outer bar characteristics is reflected in the behaviour of the inner nearshore (Lippmann et al, 1993; Wijnberg and Wolf, this volume)

Sediment-transport measurements were conducted in the inner nearshore on 14 days but are not considered here.

HYDRAULIC CHARACTERISTICS DURING STUDY PERIOD

The Wavec buoy recorded, with some interruptions, the offshore wave characteristics in October and the start of November. The offshore wave height after 2 November was

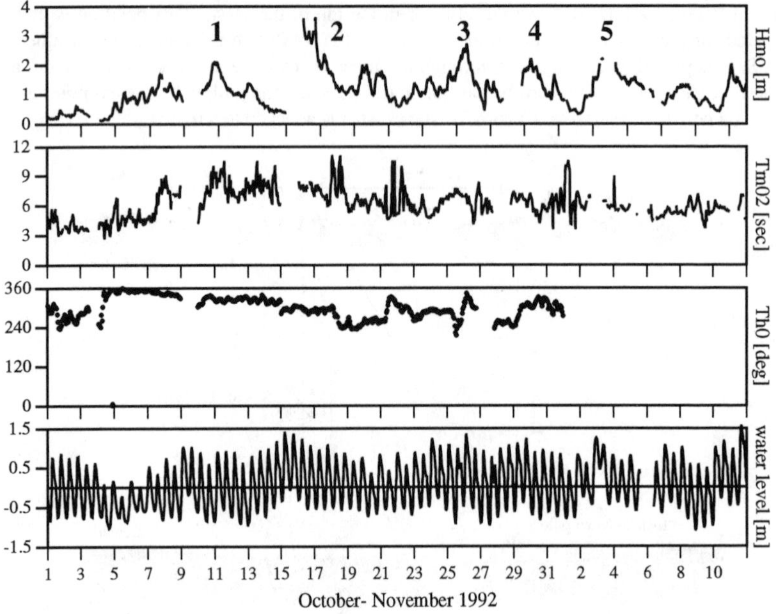

Figure 4. Wave height (Hmo) wave period (Tm02), wave direction (Th0) and water level during field period.

obtained from the capacitance wire on the pole located near the outer bar. Figure 4 shows the offshore wave characteristics.

The entire field period shows moderate to high wave conditions. During six occasions the offshore wave heights exceeded values above two meters. These occasions are designated as a storm. The final storm passed during the last day of the field measurements. As hydraulic and morphologic measurements continued during all storms, it is possible to evaluate both the impact of five storms (numbered 1-5 in Fig. 4) and the following morphologic recovery. The observed wave periods and the offshore wave spectra (not shown) reveal the dominance of wind waves over swell. This was also affirmed by comparing the wind- and wave directions. During the period 5-7 October the wind direction was north-east to north with daily averaged wind velocities up to 12 m/s. The wave direction during this period was also north-east to north. This resulted in a suppression of the flood current and in a reduced high tide water levels. These water level recordings are the mean of two tide gauges located 17 kilometer south and 16 km north of the field site.

The recordings from the nearshore located poles (not shown) exhibit small interruptions and some instruments were damaged resulting in larger data gaps. Nevertheless, the overall data return from these pole was about 60%. In the surf zone, the highest significant wave heights were measured during the second storm (14-15 October). During this storm burst averaged wave heights of around 2 meters were measured on the two inner poles while at the outer pole significant wave heights of 2.5 meters were measured.

MORPHOLOGIC RESPONSE

cross-shore response of the beach

The morphologic analysis of the beach and bars first concentrates on the cross-shore developments near the instrumented transect in the centre of the study area. Thereafter, the longshore variability is considered.

In total, more than 200 beach profiles were surveyed along the one kilometer stretch. The variability of the beach and foreshore is examined by studying the time-series of the position of the shoreline and the development of a swash bar. The shoreline is defined as the intersection of the profile with the NAP datum. The 39.500 profile, in the centre of the field area, was the only beach profile monitored on a daily basis and is therefore first analysed. A complicating factor was that the former swash bar (Kroon, 1990; Wolf, 1993) had moved offshore, as a result of the yearly net offshore movement (Wijnberg and Wolf, this volume). This former bar now showed more resemblance with an inter-tidal bar or low tide terrace (LTT).

The time series of the shoreline position (figure 5), indicates that the first storm (9 Oct.) resulted in an *seaward* shifting of the shoreline position of 10 meter. This is caused by the erosion of the berm in the upper part of the profile (Figure 6a/b). The storm also resulted in the erosion of the inter-tidal bar. Between the first and second storm, the shoreline position hardly changes. As the shoreline already moved seaward as a result of the first storm, a recovery of the beach, i.e. a (further) seaward movement was also not expected.

The second storm results in a retreat of the shoreline of 5 meter. The Figures 6c and 6d show that the berm is now completely eroded. In contrast to the first storm, however, a new inter-tidal bar emerged. Possibly, the sediment from the berm is accumulated in this inter-tidal bar. In the period 16-20 October, the inter-tidal bar moves initially onshore (16-18 October) under moderate conditions, but is moved seaward (19-20 October) under higher wave conditions (1.5 meter) and finally transforms in a low tide terrace (LTT) prior to the third storm.

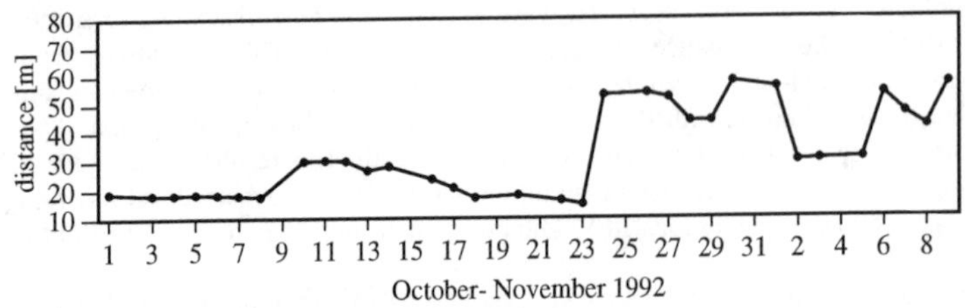

Figure 5: position of shoreline at profile 39.500. distance relative to RSP.

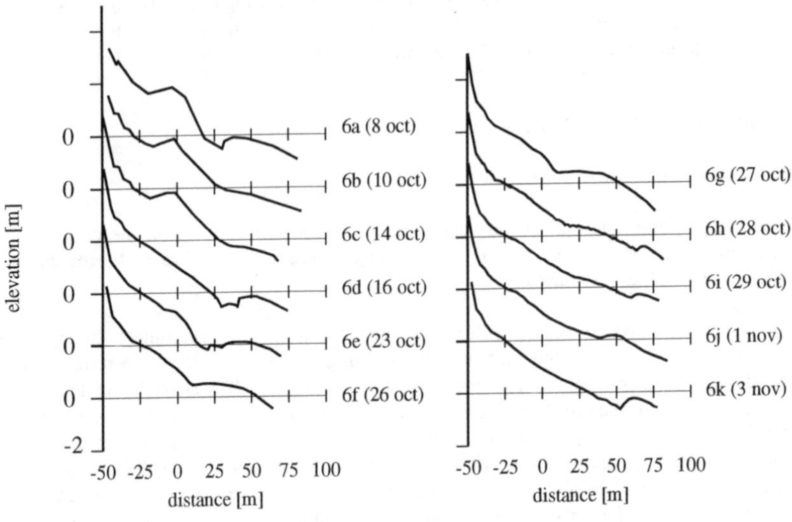

Figure 6. morphologic development of the 39.500 profile.

The third storm, like the first storm, results in a seaward displacement of the shoreline (Fig. 6e/f). The level of the LTT before the storm was just under NAP datum, while after the storm, the elevation of LTT was about 0.5 meter NAP. Part of the large displacement is, however, related to the definition of the shoreline position. However, it remains unusual that the net result of the third storm is a rise of the LTT. The LTT remains largely unchanged during the interval between the end of the third and the start of the fourth storm.

The fourth storm flattens the LTT, resulting is a small landward displacement of the shoreline (Fig. 6g-i). The profile of 28 October shows the looming of a inter-tidal bar although offshore significant wave heights still exceed 1.5 meter. The net result of the period 28 October to 1 November is that the inter-tidal bar has welded onshore and has increased its size. As a result, the shoreline has moved over 10 meter seaward.
The final storm erodes the small inter-tidal bar and moves the shoreline 25 meters in a landward direction. Like the previous storm, a new inter-tidal bar is formed just after the initial erosion of the previous one. In this case the new bar emerges even before the peak of the storm (Fig. 6j/k). The following welding of this bar to the shore results in a considerable (25 m) seaward displacement of the shoreline.

cross-shore responses of the inner bar

The SAP-profile was surveyed on every day except on 19 October, and 2 November due to mechanical difficulties. The profile of 2 October showed very strange values which could not be directed to a certain cause. Apart from these three days, profiles are available for the remainder of the field period. The position of the SAP was determined with the Elta 20 from 1 - 14 October. As the Elta 20 Total station has no self tracking option, the SAP was stopped every 10 meter to obtain a position. The profiles of 1-14 October, therefore, are based on discrete points. The remaining profiles (14 October - 11 November) were surveyed with the Polartrack automatic tracking total station. The position of the SAP was obtained every second. The resulting profiles were smoothed using a moving average filter.

In the analysis of the SAP-profiles only the inner bar characteristics are considered. Figure 7 shows the course of the elevation and distance of the inner bar's crest. The crest is defined as the location where the slope of the bar changes from a seaward slope to a landward slope. The elevation of the crest is given relative to Dutch ordnance datum (NAP≈MSL). It assumed here, that the behaviour of these variables represent the behaviour of the entire bar (Birkemeier, 1985).

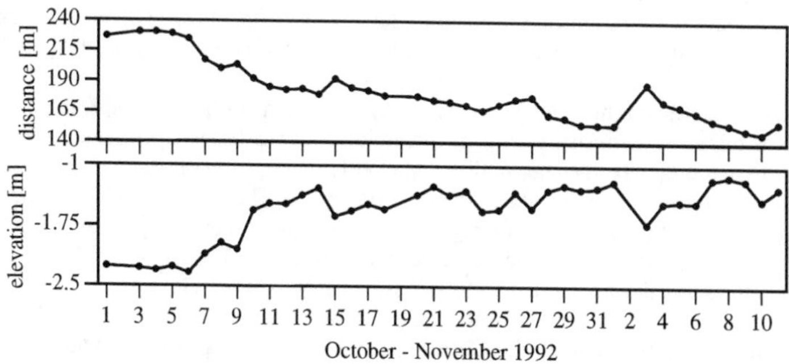

Figure 7. Distance and elevation of inner bar crest. Distances relative to RSP line, elevation relative to NAP (Dutch ordnance datum).

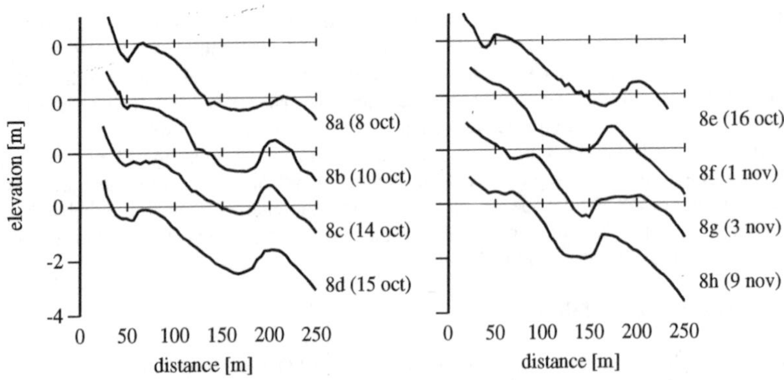

Figure 8. Morphologic development of inner nearshore

Prior to the first storm, the inner bar became better developed with the crest of the inner bar moving onshore. The first storm further amplified the inner bar. The height between the crest and trough increased from 0.3 meter to 1.1 meter while the steepness of the bar also increased (fig 8a/b). In the period between the first and second storm, the inner bar continued to grow in height and moved onshore.

The morphologic response of the inner bar on the second storm was opposite to the first storm. The bar migrated offshore in the first part of the storm and decreased in height. (14-15 October, (Fig. 8c/d). The averaged offshore migration rate was about 0.5 m/hr. During the second part of the storm (15 -16 October, Fig 8d/e), the inner bar properties showed almost no change. After the second storm the distance of the bar crest moved gradually further onshore. The elevation of the crest showed more variation is this period. However, no clear relation exists between the elevation of the crest and the offshore wave conditions.

The third storm initiated a period of three days with offshore movement of the crest. The average offshore migration rate was about 3.5 m/day (0.15 m/hr). The gradual change increase in height of the crest was interrupted by the short increase in wave height on the 26th.

The period of offshore movement is stopped by the fourth storm. Moreover, the offshore trend was turned in an onshore trend. In the period after the fourth storm, the elevation of the crest remains largely unchanged, dispite the onshore migration of the crest.

The fifth storm had a large impact on the crest properties. The crest of the bar migrated 34 meter offshore in two days. The average offshore migrating rate being about 0.7 m/hour. The elevation of the crest decreased more than 0.5 meter. The shape of the inner bar remained the same during the period 16 October to 1 November. However, the fifth storm did not only cause large changes of the crest, but also changed the form of the bar (Fig. 8f/g). In the recovery period after the final storm, the bar again regains his asymmetric form (Fig. 8g/h)

cross-shore response of the outer nearshore bar

Due to the rather rough weather conditions, only three echo-sounding surveys were executed during the field period. These surveys were executed on 2 October, 13 October and 31 October. The outer bar crest related results are, for the whole field area, shown in the table below. However, only the results of profile 39.500 are considered in this paragraph.

Table 1. Outer bar crest changes alongshore. Distances relative to RSP, elevation relative to NAP.

Survey date	distance crest [m]					elevation crest [m]				
	39000	39250	39500	39750	40000	39000	39250	39500	39750	40000
2 October	492	528	549	561	566	-3.70	-3.75	-3.95	-4.05	-4.20
13 October	487	518	544	556	561	-3.49	-3.65	-3.90	-4.10	-4.10
31 October	-	508	539	571	576	-	-3.35	-3.75	-4.35	-4.40

- = not surveyed

The distance and elevation changes between 2 and 13 October are approximately equal to the accuracy of the positioning system, and the echosounder, respectively. Hence, the outer bar crest near profile 39.500 did not significantly change between these dates. The changes between 13 and 31 October are significant and show an onshore moving outer bar crest, while the elevation of the crest increases.

longshore variability of beach and nearshore

In order to evaluate to which extend the 39.500 profile and the SAP-profile represent the morphologic response of entire field area, eight other beach profiles, the time averaged photographs and the echo-soundings were (further) analysed. The eight beach profiles were approximately surveyed every other day. The photographs cover the three final storms. As mentioned before, an (3D-) echo-sounding was executed on three days during the field period.

The longshore variability of the beach is examined with the most comprehensive data set, namely the shoreline variations. The crest of the inter-tidal bar is less useful because in the case of a flat profile or a LTT, a distinct crest does not exists. Figure 9 shows the shoreline variations at the profiles. The profiles may be divided in two groups. The 39.000 to 39.500 profiles show a large displacement of the shoreline which is related to a storm. This storm event, however, has not the same impact on all profiles. The large displacement of the shoreline is for all five profiles caused by the welding of the inter-tidal bar to the shore. In contrast to the five profiles mentioned above, the profiles 39.583 to 40.000 show no clear impact of an individual storm on the shoreline position.

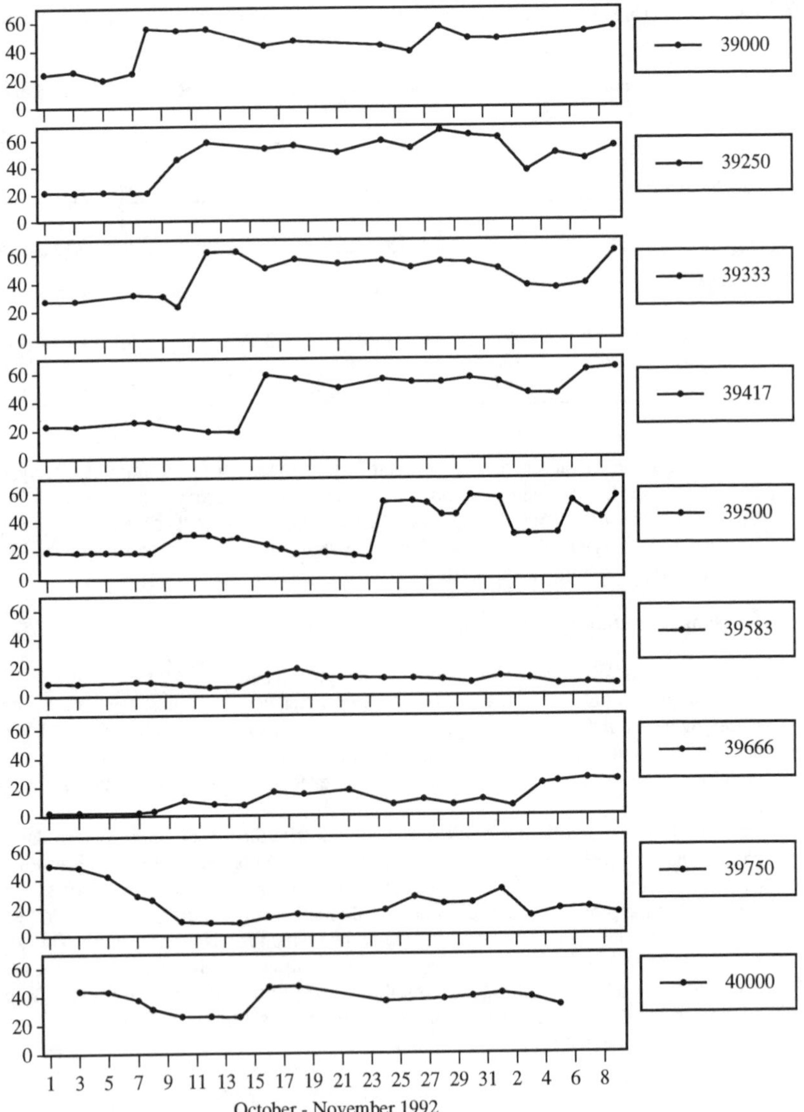

Figure 9. Shoreline response at different locations along the field site. Y-axis: distance relative to RSP line.

The time- averaged photos of the breaking waves on the inner bar show that this bar had a rhythmic character alongshore. The longshore wavelength was about 250 meter. The

longshore variability is further caused by the obliqueness of the bar. The third storm did not change the initial rhythmic character, while the final two storms straightened the bar. Thus the three storms had a different impact of the longshore variability.
Although the crest of the outer bar did not change along profile 39.500, other profiles do show significant changes (Table 1). Both the elevation and the distance of the outer bars crest show no uniform behaviour along the coast. The profiles 39.750 and 40.000, 250 and 500 m south of profile 39500, respectively, show an offshore movement while the 39.250 profile shows an onshore movement. The elevation changes concur with this movement. The photographs of the outer bar did not show any signs of an a crescentic configuration.

DISCUSSION AND CONCLUSIONS

Spatial changes

The observed morphologic variations of the beach and bars reflect different spatial scales. The response of the morphology is characterised by a relative stable shoreline, a highly movable inner bar, and a relatively stable outer bar.

Except for the welding of the inter-tidal bar to the beach, the storms seem to have little influence on the shoreline position. The standard deviation of the shoreline, along the entire study area, is about 10 meter. The welding of the inter-tidal bar, partly because this bar also increased its volume, coincides with a storm. Generally, a (inter-tidal) bar tends to weld to the shore under low wave conditions (e.g. Short and Aagaard, 1993). The apparent contradiction can be explained by the erosion of the berm. Through this, sand was transported seaward. This sand was accumulated in the (enlarged) inter-tidal bar, resulting in seaward displacement of the shoreline. A similar process is described by Lippmann et al. (1993). The described process is sustained by the sediment transport measurements. These measurements show that during comparable conditions, the net cross-shore suspended sediment transport near the foreshore is offshore directed.
The importance of a berm in determining the direction of the shoreline movement during storms, is further supported by the longshore variations. A small berm was present at profile 39750 and no berm was present at profile 40.000. This resulted in a retreat of the shoreline during the first week of October. A seaward moving of the shoreline was observed in the northern profiles during the same period. However, the longshore variation may also have been influenced by the obliqueness of the inner bar. If this is the case, then the standard deviations of the shoreline movement should have shown an longshore trend. A trend in the longshore variation of the shoreline is present, showing larger variation in the northern part than in the southern part of the field area. However, this is solely due to the large displacement of the coast in the northern part during one storm.

The inner bar showed larger variations than the shoreline. This is reflected in a larger standard deviation (23 m). The large variation is not the result of a single event but of a gradual onshore movement of the bar crest. The inner bar variability, especially as the result of (large) storms, was less than observed at Duck (Holman and Sallenger, 1993) This may be due to the persistent and pronounced presence of the outer bar which limits the wave height near the inner bar.

The variability of the outer bar was less compared to the inner bar. This is a common observation along multi barred coasts (e.g. Short and Aagaard, 1993; Kroon, 1994). For

instance, between 2 and 13 October the outer shows hardly any changes. During the same time interval, the crest of the inner bar migrates over 43 m while the crest of the inner bar increases 0.8 meter.
The time averaged photos showed no rhythmic behaviour of the outer bar. However, the 13 and 31 October surveys reveal longshore differences in the migration direction of the outer bar crest. These longshore differences are probably caused by the obliqueness of the bar. This is obliqueness is, in turn, related to large time scales (Wijnberg and Wolf, this volume).

The observed cross-shore variability of the inner and outer bar may be the result of an longshore migration of the morphology (Kroon, 1990). However, this migration occurs gradual, involves larger time scales (several months to years), and probably occurs during quiet weather . It is unlikely, therefore, that the rapid cross-shore migration are the result of the longshore migration. However, the longshore migration of the bar has some influence on the net cross-shore changes over the entire study period

Temporal changes

The individual storms are not clearly related to morphologic changes of the beach. None of the five storm events caused a uniform reaction for all beach profiles. Neither an longshore trend can be distinguished in response of the shoreline for each of the individual storms. It seems, therefore, that the offshore wave conditions do not correlate linear with the hydraulics near the beach. The large distance between the Wavec and the beach and the existence of two (breaker) bars are probably due to this.
The changing offshore conditions are better correlated with the inter-tidal morphology. An temporal trend, observed along the entire shoreline, was the transformation of the initial inter-tidal bar into a more terrace like feature. A similar observation was made by Lippmann et al. (1993) during extreme (Hs > 3m) storms. The response of the inter-tidal features on storms, but also the recovery thereafter, was very quick. The latter is reflected in the emerge of a new inter-tidal bar even before the peak of the storm (Fig 6k).

The offshore wave conditions are best reflected in the distance of the inner bar crest. The observation, that an nearshore bar moves offshore during a storm (e.g. Short, 1979; Sallenger et al, 1985) is largely confirmed by this study. The storm periods are less correlated with the crest elevation (Birkemeier, 1985). After the first the storm, the crest elevation varies only 0.3 meter and has no clear correlation with the offshore wave conditions.
The inner bar responded very quickly on the changing offshore wave conditions. The adjustments of the inner bar to the changing wave conditions generally occurred within one day. The offshore displacement of the inner bar during a particular storm, is more determined by cross-shore location of the bar then by the magnitude of the storm. This is envisaged by comparing the response of the inner bar to the first, second and fifth storm.
The storm changes were very rapid, but so was the initial recovery of the inner bar. During the first day of onshore movement after a storm, the onshore movement was clearly larger than during the remainder of the recovery period. The recovery period seems to increase with every next storm. During the first storm, the inner bar regained his pre-storm location within one day, while during the final storm the recovery period lasted 5-6 days.
The fast initial recovery of the inner bar is reflected in the averaged onshore migration rates. The average values of these rates hardly differ from the offshore migration rates, being 5.0

m/day (onshore) and 6.2 m/day (offshore). The initial onshore migration rates observed during the storms are somewhat lower than the observations of Orme (1985) and Sallenger et al. (1985). The average onshore migration rates for the whole period are in line with Japanese measurements (e.g. Sunamura and Takeda, 1984). The offshore migration rates during the initial storm phase are less than observed at Duck (e.g. Sallenger et al., 1985).

The inner bar observations already indicate that the response of the bar and beach morphology on an individual storm is not the same for each individual storm. The impact of the fourth storm on the beach, for instance, is smaller than the impact of the first storm. Duration and maximum offshore wave height, though, are more or less the same for both storms. It is clear that the morphologic response depends also on the previous wave conditions i.e. on the position of the individual storm in a sequence of storms. This stresses the importance of the pre-storm morphology in determining the nature of the response, since the pre-storm morphology is the result of the previous wave conditions.
Besides the temporal variation of the morphologic response to storms there is also a spatial variability in response. Moreover, the spatial and temporal variability may be encountered at the same time. For instance, during the first storm, the inter-tidal bar was eroded, while the inner bar became better developed. During the second storm the inter-tidal bar increased its size, while the outer bar was lowered (Fig 8a/8b and 8c/d) .

The net result of the succession of five storms is a seaward displacement of the shoreline of 38 meter for the beach. The beach is flattened and transformed from a profile with a berm and inter-tidal bar into a profile with less pronounced characteristics. The seaward movement during high energy period was also noted for the Duck field site (Lippmann et al, 1993). The individual storms and the rather rough conditions between the storms do not lead to a significant trend in the position of the shoreline.
The net result for the inner bar is an shoreward displacement of 70 meter. The individual storms reduced or changed the direction of the longer term inner bar development. A wave height above 2 meter was noticed on 12 days,. However, only on 8 of the 37 days being reflected in an offshore migration. This, and the fast recovery of the inner bar after the storms, result in the net shoreward migration.

Limitations

The data has some limitations which may influence the analysis. These limitation are mainly related to the limited spatial and temporal sampling of the morphologic features. Some of the morphologic changes, for instance the shoreline changes, occur well within one day. Therefore, aliasing effects can not be totally ruled out. The same accounts for the sampling distance of the SAP during the first half of October and the limited information about the longshore variation of the inner bar. The latter makes it hard to asses the significance of the of the SAP-profile for the entire field area. Moreover, the SAP profile did not include the entire bar and the crest is only *one* indicator for the bar development.

ACKNOWLEDGEMENTS

This research is part of the Coastal Genesis Programme (contract DG-476) which is funded by the National Institute for Coastal and Marine Management (RIKZ) of *Rijkswaterstaat*. The instrumental set-up and maintenance was executed by the Directorate North Holland of *Rijkswaterstaat*. Technical support was given by the Section Instruments and Automation of the Physical Geography Laboratory of the Utrecht University and Delft Hydraulics. All support, in and out of the field, is gratefully acknowledged by the author.

REFERENCES

Birkemeier, W. A., 1985. Field data on the seaward limit of profile change. *Journal of Waterway, Port, Coastal and Ocean engineering,* Vol. 111, pp. 598-602.

de Vroeg, J. H., Smit, E. S. P. and Bakker, W. T., 1989. Coastal Genesis. *Proceedings 21th Coastal Engineering Conference*, ASCE, pp. 2825-2839.

Holman, R. A. and Sallenger, A. H., 1993. Sand Bar Generation: A discussion of the Duck Experiment Series. *Journal of Coastal Research,* Special Issue no. 15, pp. 76-92.

Kroon, A., 1990. Three dimensional morphologic changes of a nearshore bar system along the Dutch coast near Egmond aan Zee. *Proceedings International symposium on Coastal Geomorphology*, Skagen, Denmark, pp. 430-451.

Kroon, A., 1994. Sediment-Transport and Morphodynamics of the Beach and Nearshore Zone near Egmond, The Netherlands, Ph. D.-Thesis Utrecht University, Department of Physical Geography.

Lippmann, T. C. and Holman, R. A., 1990. The spatial and temporal variability of sand bar morphology. *Journal of Geophysical Research,* Vol. 95, pp. 11575-11590.

Lippmann, T. C. and Holman, R. A., 1993. Episodic, Nonstationary behaviour of a Double bar System at Duck, North Carolina, U.S.A., 1986-1991. *Journal of Coastal Research,* Special Issue no. 15, pp. 49-75.

Orme, A. R., 1985. The behaviour and migration of longshore bars. *Physical Geography,* Vol. 5, pp. 142-164.

Orme, A. R. and Orme, A. J., 1988. Ridge and runnel enigma. *The Geographical Review,* Vol. pp. 169-184.

Owens, E. H., 1977. Temporal variations in beach and nearshore dynamics. *Journal of Sedimentary Petrology,* Vol. 47, pp. 168-190.

Owens, E. H. and Frobel, D. H., 1977. Ridge and runnel systems in the Magdalen Islands, Quebec. *Journal of Sedimentary Petrology,* Vol. 47, pp. 191-198.

Sallenger, A. H., Holman, R. A. and Birkemeier, W. A., 1985. Storm-induced response of a nearshore-bar system. *Marine Geology,* Vol. 64, pp. 237-257.

Short, A., 1979. Three dimensional beach state model. *Journal of Geology,* Vol. 87, pp. 553-571.

Short, A. D. and Aagaard, T., 1993. Single and Multi-bar Beach Change Models. *Journal of Coastal Research,* Special Issue no. 15, pp. 141-157.

Sunamura, T. and Takeda, I., 1984. Landward migration of inner bars. *Marine Geology,* Vol. 60, pp. 63-78.

Wijnberg, K. M. and Wolf, F. C. J., 1994. Three Dimensional behaviour of a nearshore bar system.

Wolf, F. C. J., 1993. The use of profile response parameters in describing the impact of a single storm on the nearshore zone near Egmond aan Zee, the Netherlands. *Proceedings Hilton Head Island International Symposium*, Hilton Head, SC, USA, pp. 485-502.

THREE-DIMENSIONAL BEHAVIOUR OF A MULTIPLE BAR SYSTEM

Kathelijne M. Wijnberg[1] and Felix C.J. Wolf[2]

ABSTRACT: The analysis of a large bathymetric data set of a part of the central Dutch coast reveals that a multiple bar system as a whole may exhibit three-dimensional behaviour. This behaviour is cyclic and consists of a net offshore migration of the bars, with the outer bar fading away and with a new bar being generated near the shoreline. The bar behaviour becomes three-dimensional because parts of the 25 km long bar system are not in the same phase of the 15 year cycle. Parts of the bar system are in the same phase over alongshore distances of typically 6 km. Transition areas between the 'in phase' parts have a typical length of 2 km. It is hypothesised that the cyclicity in the behaviour of the bar system is governed by the behaviour of the outer bar. Further, it is proposed to adapt the length of the coastal stretch that is monitored for morphological changes during field experiments to the length scales of the three-dimensional bar system behaviour.

INTRODUCTION

Nearshore breaker bars are common features along sandy coastlines. These bars may exhibit three-dimensional (3D) characteristics such as crescentic configurations (e.g. Komar, 1976; Wright and Short, 1984; Sallenger et al., 1985). These types of 3D characteristics apply to individual bars. Wijnberg and Terwindt (in prep.) analysed an extensive bathymetric data set to quantify decadal morphological developments of the central Dutch coast. This analysis revealed that a multiple bar system as a whole may exhibit highly organised 3D behaviour as well.

1, 2) Coastal researcher, Institute for Marine and Atmospheric Research Utrecht, Dept. of Physical Geography, Utrecht University, PO box 80115, 3508 TC Utrecht, The Netherlands.

In this paper we will describe the 3D behaviour of a multiple bar system, including the temporal and alongshore scales inherent to that behaviour. Some considerations are given on the cyclic nature of the bar behaviour. In addition, the consequences of the 3D behaviour for the monitoring of morphological changes in field experiments will be discussed.

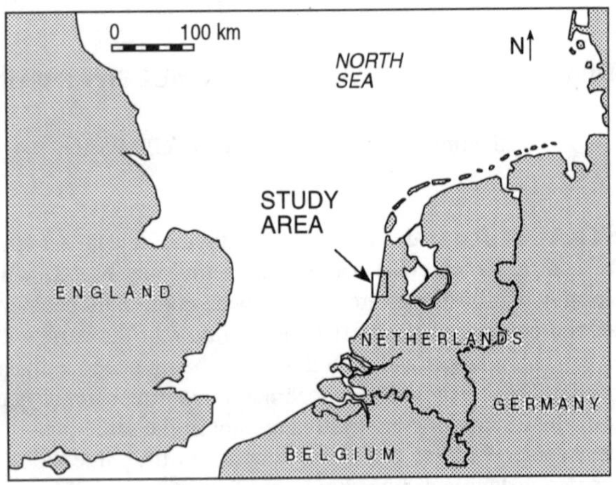

Figure 1: Location of the study area

The multiple bar system that is described in this paper is located along the central part of the Dutch coast (figure 1). The alongshore length of this bar system is about 25 km. To the south the bar system is bounded by long harbour moles and to the north the multiple bar system changes into a single bar system in front of a seawall. In the studied area man made structures are absent. The bar system generally consists of 2 to 3 breaker bars and the mean slope in the surf zone is about 1:110. The bar system is located along a storm dominated type of coast. The mean annual significant wave height in this area is about 1.3 meter and the period is about 5 seconds. The largest waves approach the coast from the NW and the most frequently occurring waves approach from the SW. Mean tidal range is about 1.7 meter. The median grain size of the sediment is in the fine to medium sand range (i.e. 125-500 μm).

Near Egmond aan Zee several field experiments have been conducted (Kroon, 1990; Wolf, 1993; Kroon, 1994). The Egmond site is located in the middle of the multiple bar area indicated above. The data set of this location is used to illustrate how 3D bar system behaviour becomes apparent in the observations of morphological developments in these types of field experiments.

DATA SET AND METHOD OF ANALYSIS

The bathymetric data set

The description of the bar system behaviour is based on a data set containing 27 year of annual bathymetric surveys (from the *Rijkswaterstaat* data base called 'JARKUS'). An annual survey consists of soundings of the coastal profile every 250 meter alongshore. The bar system that is discussed in this paper is located between km 28.00 and km 53.00. These figures refer to a shore-parallel datum line that is used to mark the alongshore position of cross-shore survey lines.

The cross-shore profiles are analysed over a distance of 750 m, starting at the +1m NAP contour. (NAP is the Dutch vertical ordnance datum which approximates mean sea level.) The mean depth at 750 meter offshore is about -6 m NAP. Depth contour plots (figures 6 to 9) extend to depths up to -9 m NAP. The cross-shore interval between depth measurements ranges from 10 m near the shoreline to 20 m in the seaward part. This cross-shore interval is made equidistant at 15 m by a cubic spline interpolation.

Empirical eigenfunction analysis combined with a moving window approach

The data set of coastal profiles is analysed with the objective to characterise the basic morphological changes. This has been achieved by combining empirical eigenfunction analysis with a moving window approach (Wijnberg and Terwindt, 1993). The analysis technique is described in detail by Wijnberg and Terwindt (in prep.) and is only briefly explained in this paper.

Each profile is described by 49 depth values at a 15 m equidistant interval, from 30 m offshore to 750 m offshore. Within a 1 km stretch of coast, 5 survey lines are present that have been surveyed 27 times. This set of 135 profiles is analysed with the empirical eigenfunctions technique. In simplified terms, this statistical technique separates the variation in a rectangular data matrix into two sets of orthogonal functions. In case of a data matrix containing coastal profiles, one set of functions describes the modes of cross-shore correlation between depth values (the eigenvectors or eigenfunctions). The other set of functions describes to what degree each of these modes of cross-shore correlations is present in each of the observed profiles (the eigenvector weightings). The first eigenfunction, the eigenfunction with the largest eigenvalue, explains most of the variation in depth values. Each successively higher eigenfunction explains most of the depth variation left unexplained by the preceding eigenfunctions. So, usually only a few eigenfunctions are required to represent the basic characteristics of the original data.

The application of the empirical eigenfunction technique on coastal profiles was introduced by Winant et al. (1975) and has been adopted by several authors. Some of them changed the type of correlation matrix that is used in the analysis technique to derive the eigenfunctions (e.g. Zarillo and Liu, 1988; Pruszak, 1993). In this

study the two sets of orthogonal functions are derived from an uncorrected sums of products matrix [U] which is defined by: $[U] = 1/n \times [D] \times [D]^T$.
In which n = number of profiles = 135; [D] = (49 x 135) matrix of depth values; $[D]^T$ = transpose of [D].

The eigenfunction analysis is repeated for every kilometre of the 25 km long stretch of coast. Hence, 25 sets of eigenfunctions have been calculated.

RESULTS

The general interpretation of a set of empirical eigenfunctions

The first empirical eigenfunctions usually explain over 97 % of the total sums of squares (table 1). These eigenfunctions represent the characteristic shape of the local mean profile. The second and third eigenfunctions explain nearly 70 % of the remaining depth variation, i.e. 70 % of the part of the total sums of squares that is not explained by the first eigenfunction. The second and third function describe general characteristics of the bar topography, such as the position and the relative size of the bars. Figure 2 shows, as an example, the first three eigenfunctions at km section 31-32 and the weightings for the profiles from the survey line at km 31.50.

Table 1: Percentage of sums of squares explained by the first 4 eigenfunctions (averaged over the 25 sets of eigenfunctions ± standard deviation).

	% of total sums of squares	% of remaining sums of squares
eigenfunction 1	97.3 ± 0.4	
eigenfunction 2	1.06 ± 0.19	39.4 ± 6.9
eigenfunction 3	0.81 ± 0.16	30.1 ± 6.0
eigenfunction 4	0.21 ± 0.03	7.7 ± 1.1

A positive (negative) weighting of the first eigenfunction on a profile means that this profile is flatter (steeper) than the local mean profile. A positive weighting of the second (third) eigenfunction on a profile means that the bar topography in that profile compares well to that represented by the second (third) eigenfunction. A negative weighting of the second (third) eigenfunction on a profile means that the bar topography in that profile compares well to the mirror image of the topography represented by the second (third) eigenfunction. This means that bars are at trough locations and troughs at bar locations (figure 3a).

The weightings of the second and third eigenfunction both exhibit the same periodicity in time (T ≈15 year) but with a phase shift of 90 degrees. Maxima in weightings of the second eigenfunction then coincide with near-zero values of the

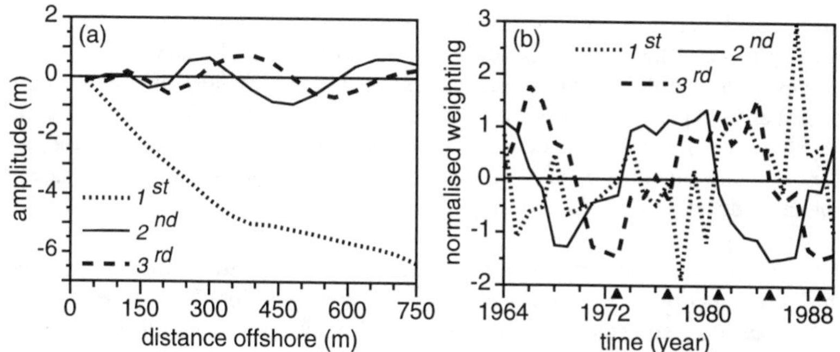

Figure 2: Results of the empirical eigenfunction analysis on profiles located between km 31 and km 32. (a) First 3 eigenfunctions. (b) Weightings of first 3 eigenfunctions on profiles in the survey line at km 31.50. (Black triangles mark the profiles that are plotted in figure 3b.)

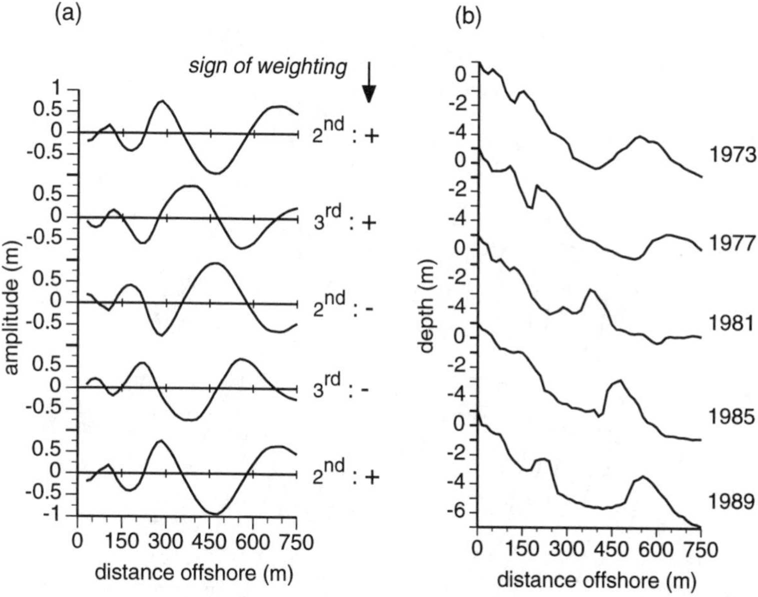

Figure 3: Cyclic bar behaviour in a multiple bar system, demonstrated by (a) the second and third eigenfunction (km 31-32), (b) 5 selected profiles with high positive or negative weightings of either the second or the third eigenfunction (km 31.50).

third eigenfunction and vice versa. The weightings of the second eigenfunction are leading those of the third eigenfunction. The morphological meaning of these two curves is that on the long term the bars move in a net offshore direction with the outer bar fading away at the seaward end, and with a new bar being generated near the shoreline (figure 3). As a consequence, the former inner bar becomes the outer bar. This new outer bar ends up at approximately the same location as its predecessor and reduces in height at its turn, starting a new cycle.

The periodicity in the curves of the weightings, 15 year in this example, represents the 'return period' of a certain multiple bar configuration. This type of cyclic behaviour has also been observed in multiple bar systems along other parts of the Dutch coast although the lengths of the 'return periods' are different (Wijnberg and Terwindt, 1993; Ruessink and Kroon, in prep.). The recurrence of profile configurations after 3.5 year near Duck, USA (Birkemeier, 1984; Lippmann et al., 1993), may also be regarded as cyclic bar system behaviour.

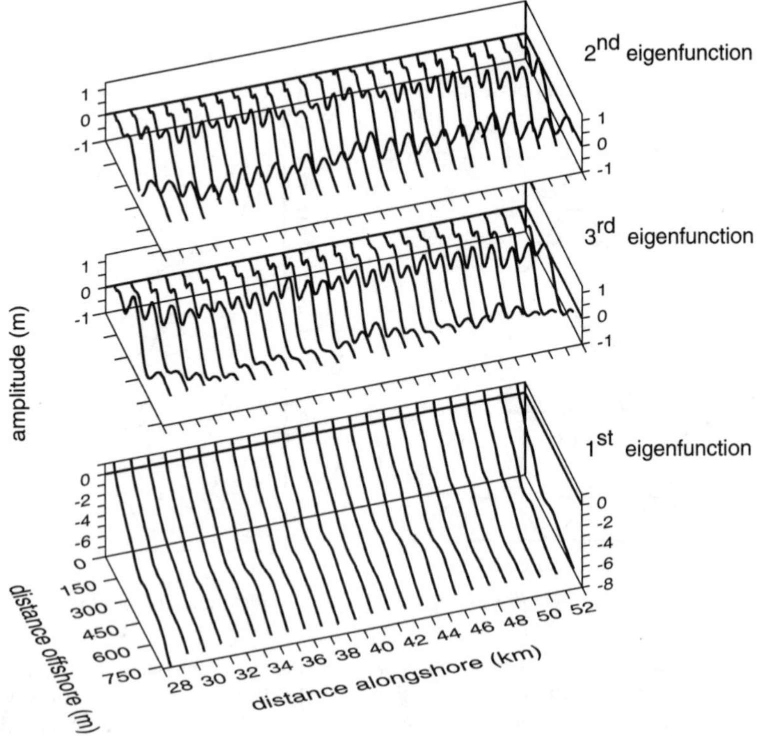

Figure 4: Alongshore variation in the shapes of the first 3 eigenfunctions

In summary, the second and third eigenfunction describe two modes of cross-shore bar positioning. The position of the bars at a given time can be derived from the weightings of these eigenfunctions. The weightings thus tell in which phase of the above described cycle the bar system is at a particular time.

The 3D behaviour of the multiple bar system

The alongshore variation in the shapes of the first three empirical eigenfunctions is small (figure 4). In the following, only the second and third eigenfunction are discussed because these functions give insight in the behaviour of the bars.

The weightings of the second and third eigenfunction on the observed profiles are presented in contour plots (figure 5). The horizontal axis (x) represents the alongshore location of the profile and the vertical axis (t) represents the year of

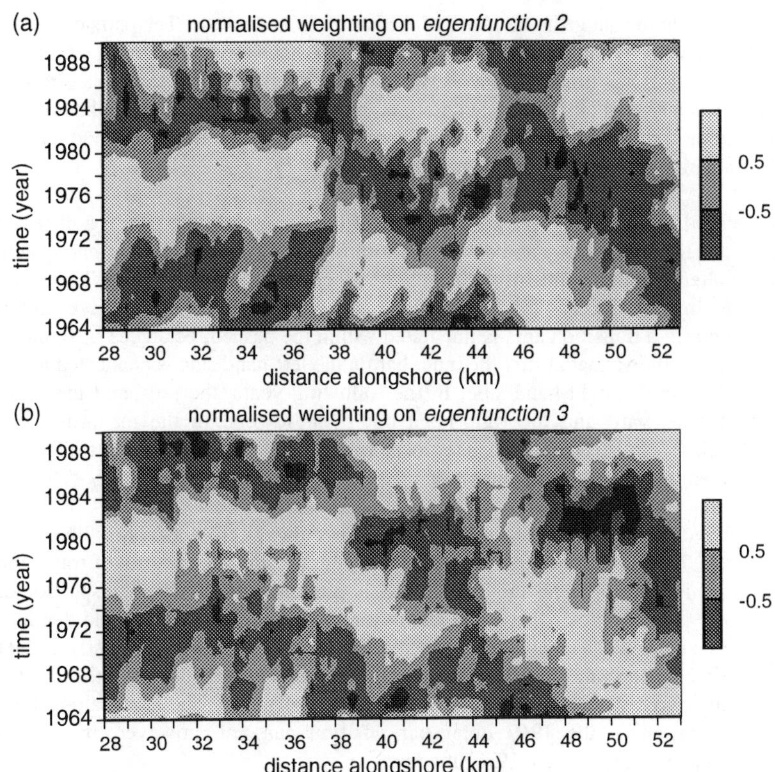

Figure 5: Normalised weightings of (a) the second eigenfunctions, (b) the third eigenfunctions

measurement. The contours connect points with equal (normalised) weightings. For reasons of clarity only two contours are plotted (viz. +0.5 and -0.5). The patterns emerging from these simplified plots are representative for the patterns emerging from more detailed contour plots.

When the two contour plots are overlaid, the maxima and minima of figure 5a appear to coincide with the near-zero values in figure 5b and vice versa. This implies that for each survey line (constant x) the curves of the weightings of the second and third eigenfunction are 90 degrees out of phase. This pattern is similar to the pattern shown in figure 2b for the survey line at km 31.50. The morphological meaning of figure 5 is that the multiple bar system exhibits cyclic behaviour everywhere along this part of the coast. At a given time (constant t), however, parts of the bar system are in different phases of the multiple bar cycle. As a consequence, the behaviour of the bar system as a whole is three-dimensional.

The alongshore change in the phase of the bar cycle is stepwise. The parts of the bar system that are in the same phase of the bar cycle have a typical alongshore length of about 6 km. The transition area between two parts of the bar system in a different phase has a typical alongshore length of about 2 km. An example of 3D bar system behaviour is shown in figure 6. Note that the parts of the bar system left and right of the transition area are in a different phase of the bar cycle.

Adjacent to a transition area are two parts of the bar system that are in a different phase of the bar cycle. The bars from those out of phase parts of the bar system make alternating attachments with each other across the transition area. Therefore, the bars in a transition area do not exhibit a net offshore movement. This type of bar behaviour in a transition area is illustrated within the dashed rectangles of figure 6. This figure shows that at first the inner bar on the left-hand side is connected to the outer bar on the right-hand side. In the following years, the left-hand inner bar migrates seaward and the right-hand outer bar lowers. In the meantime, the connection between these two bars loosens and the left-hand inner bar becomes connected to the right-hand inner bar.

The above described type of transition area is not always present. This is illustrated in figure 7, which shows that in 1967 the characteristic bar patterns of transition areas are lacking, whereas they are obviously present in 1973. Closer inspection of the 1967 picture reveals that small alongshore differences in position and depth of the outer bar exist. For example, in 1967 the outer bar is located slightly farther offshore between km 37 and km 41 than in the neighbouring areas. This seems a sign that parts of the bar system are getting out of phase. The small alongshore differentiation in the 1967 outer bar position can not, however, be simply extrapolated to arrive at the 1973 situation.

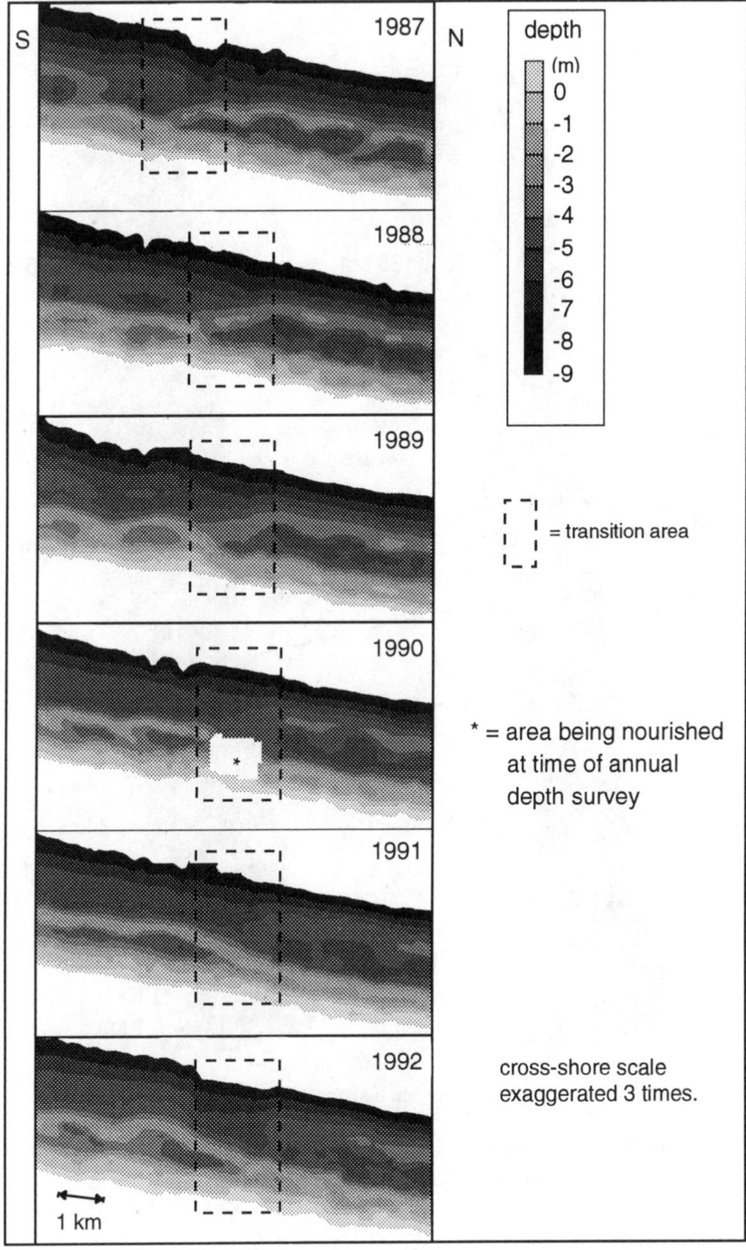

Figure 6: Bar behaviour between km 32.50 and km 42.75.

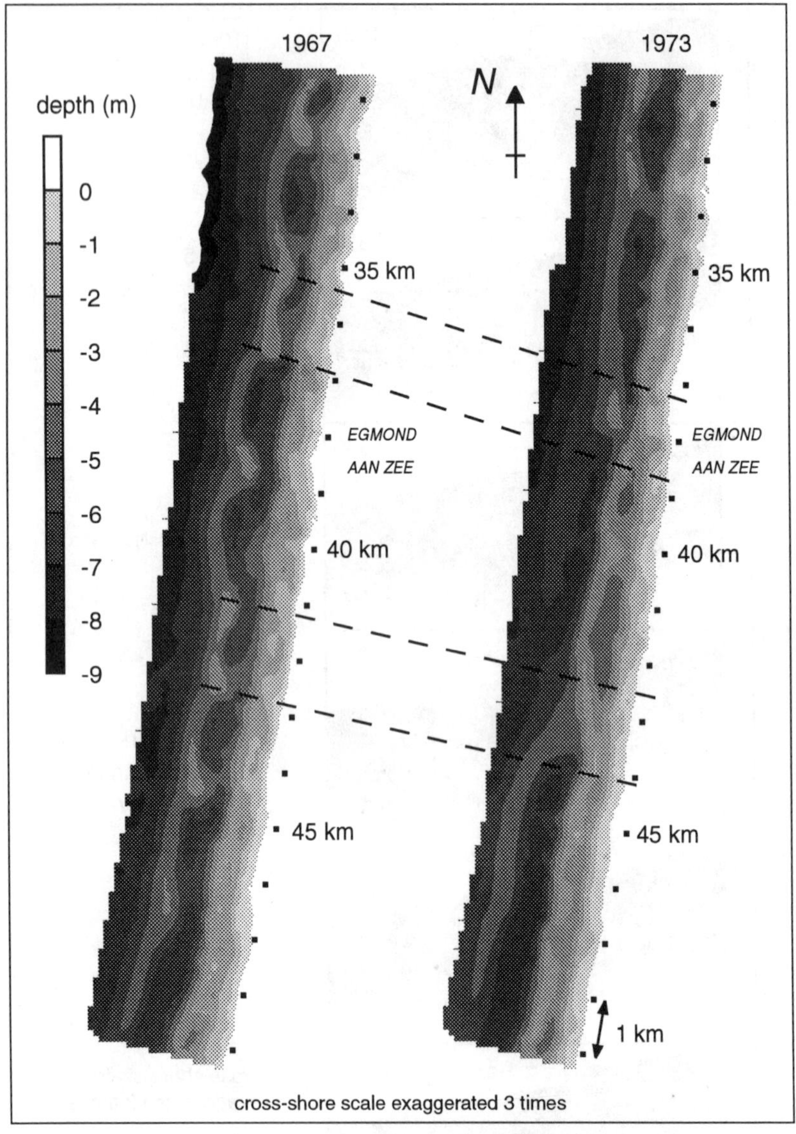

Figure 7: Depth contours in 1967 and 1973. The dashed lines indicate the development of two transition areas.

The dashed lines in figure 7 and the tilting to the right of the contours in the eigenfunction contour plots (figure 5) show that the transition areas may migrate alongshore in time. The dashed lines in figure 7 also suggest that the alongshore length of a transition area is related to the length scale of the rhythmic configuration of the outer bar.

DISCUSSION

Some considerations on the cyclicity in the behaviour of the bar system

The longer term cyclic behaviour of bars in a multiple bar system was first recognised by De Vroeg (1987). The present study extended the knowledge about the cyclic behaviour by revealing the alongshore coherence in the cyclicity. The reason for the cyclic behaviour of the bar system is still unknown. In this paragraph some considerations are given on this intriguing subject.

The behaviour of the multiple bar system has been schematised by two empirical eigenfunctions (figure 4 and 5). These eigenfunctions reveal, as well as bathymetric maps of the annual surveys, that a lower elevation of the crest of the outer bar is accompanied by a seaward shift of the inner bar. The lowering may be caused by degeneration of the outer bar as well as by a seaward migration of the outer bar to deeper water. A lower elevation of the crest of the outer bar allows higher waves to pass the outer bar unbroken. Consequently, the level of mean and maximum wave energy experienced by the inner bar increases. On the short time-scale of hours to days, bars tend to move offshore under conditions of increased wave energy, (e.g. Sallenger et al., 1985). By analogy, it is suggested that the longer term (> 1 year) increase in mean or maximum level of wave energy on the inner bar results in a more seaward position of this bar. When the outer bar has disappeared, the former inner bar will migrate seaward towards a position that is determined by the energy level of the offshore wave climate.

Another hypothesis was put forward by Kroon and Hoekstra (1993). They suggested that the lowering of the outer bar reduced the influence of this bar on the selection or forcing of long wave structures. Hence, with a lowering of the outer bar, the inner bar became free to move offshore.

Both hypotheses state that the behaviour of the outer bar influences the hydrodynamics near the inner bar and consequently governs the behaviour of the inner bar. The key role played by the outer bar in explaining the behaviour of the inner bar was also recognized by Lippmann et al. (1993). Neither of the two hypotheses explains, however, why the outer bar lowers and finally disappears. To complete the hypotheses on the occurrence of cyclic bar system behaviour, the mechanism that drives the long-term behaviour of the outer bar should be revealed.

Implications of 3D bar system behaviour for field experiments

In field experiments on bar morphodynamics, the morphological changes are generally monitored over a period of only a few weeks. On a superficial view, there is no relation with the 3D behaviour of the bar system that emerges from a data set of annual surveys. A relation does exist, however, because the net offshore bar movement in a multiple bar system is a discontinuous process. The outer bar probably changes its cross-shore position and height during just a few storm events per year (Houwing, 1991; Kroon and Hoekstra, 1993). A storm event that alters the outer bar may occur during a field experiment. The bar behaviour observed on that occasion will be related to the overall 3D behaviour of the bar system.

The maximum alongshore distance that is monitored for morphological changes in field experiments, is about 1 km (e.g. Howd and Birkemeier, 1987; Kroon, 1990; Lippmann et al., 1993). This length scale is usually adequate for recognising the development or alongshore migration of rhythmic patterns in individual bars. It is inadequate, however, for recognising the development or alongshore migration of transition areas.

The bar behaviour in a transition area differs from that in the rest of the bar system. Therefore, conclusions on bar dynamics reached from field experiments in a transition area may only be valid for a limited part of the bar system. In addition, the position of the field site relative to the location of transition areas may strongly influence the bar behaviour observed within a 1 km stretch. This is illustrated by figure 8. Figure 8a shows the morphologic developments at the Egmond field site (km 39-40). Figure 8b shows the developments over the same 5 year period in the adjacent kilometre (km 38-39). In km section 39-40, the bar system exhibits the 'normal' net offshore movement of the bars. In km section 38-39, the behaviour of the bars is more irregular because this particular stretch is located at the boundary of an 'in phase' part of the bar system. Consequently, the bar behaviour includes several characteristics of the behaviour in a transition area.

To put morphological developments observed near the measurement site into the right framework, the length of the monitored stretch should be adapted to the typical alongshore scales of the 3D bar system behaviour. For future field experiments on bar morphodynamics in multiple barred surf zones, it is suggested to increase the alongshore length of the stretch that is monitored for morphological changes. For the multiple bar system near Egmond aan Zee the length should be about 8 km, because stretches with uniform offshore bar movement have a typical length of about 6 km and a transition area has a typical length of about 2 km.

The optimal time interval for morphological surveys of the large coastal stretch is more difficult to determine than the length of this stretch. The determination of the time interval requires more knowledge of the type of conditions that are likely to change the outer bar or the bars in a transition area.

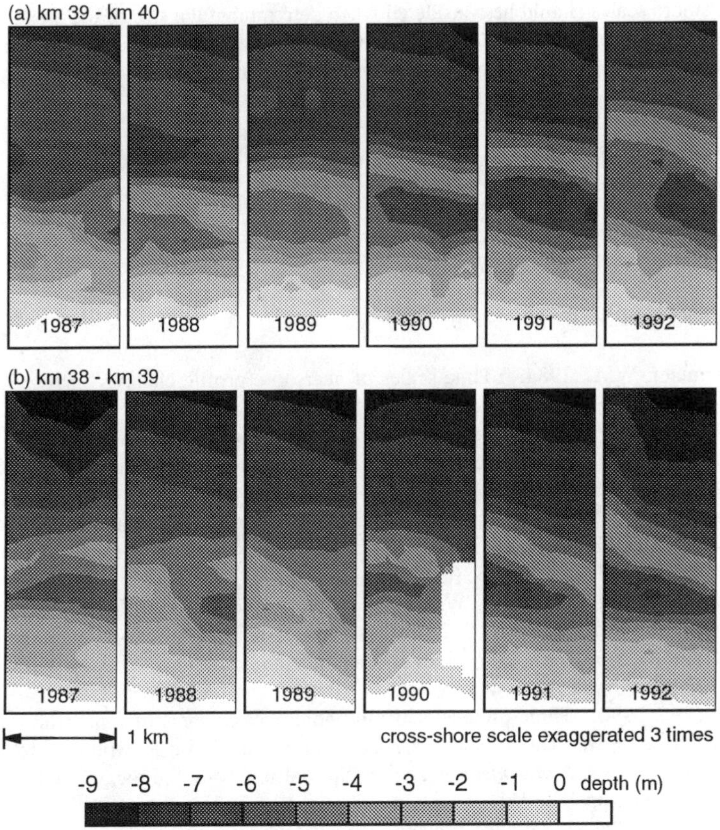

Figure 8: Bar behaviour (a) between km 39 and 40, (b) between km 38 and 39.

CONCLUSIONS

Multiple bar systems may exhibit cyclic behaviour consisting of net offshore movement of all bars with the outer bar lowering and finally fading away, and with a new bar being generated near the shoreline. This behaviour is not necessarily in phase alongshore, causing three-dimensionality in the behaviour of the bar system. Probably, the behaviour of the outer bar is governing the cyclic behaviour of the bar system.

For the multiple bar system near Egmond aan Zee, a typical scale for stretches with uniform offshore bar movement is 6 km. A typical alongshore scale for the transition area between two parts of the bar system with a different phase is 2 km. In field experiments on morphodynamics in a multiple bar environment, these alongshore scales should be considered when determining the size of the area to be monitored for morphological changes. This will prevent erroneous conclusions on bar behaviour and at the same time may help to increase our knowledge on this large-scale behaviour of multiple bar systems.

ACKNOWLEDGEMENTS

The Ministry of Transport, Public Works and Watermanagement (*Rijkswaterstaat*) division RIKZ, is kindly thanked for providing the bathymetric data from the JARKUS data base on which this study is based.

REFERENCES

Birkemeier, W.A., 1984. " Time scales of nearshore profile changes", *Proc. 19th Coastal Engineering Conf.*, Houston, USA, ASCE, pp. 1507-1521.

De Vroeg, J.H., 1987. "A schematic representation of surf zone bars based on annual coastal surveys", Report Dept. of Civil Engineering, Technical University Delft, 37 pp. (in Dutch)

Houwing, E.J., 1991. "Analysis of TAW profiles, Egmond aan Zee and Katwijk aan Zee", GEOPRO-Report 1991.019, Dept. of Physical Geography, Utrecht University, 20 pp. + figures. (in Dutch)

Howd, P.A. and Birkemeier, W.A., 1987. "Storm-induced morphology changes during DUCK85", *Coastal Sediments '87*, ASCE, pp. 834-847.

Komar, P.D., 1976. Beach processes and sedimentation. Prentice-Hall, New Jersey, 429 pp.

Kroon, A., 1990. "Three-dimensional morphological changes of a nearshore bar system along the Dutch coast near Egmond aan Zee", in: P. Bruun and N.K. Jacobsen (eds.), *J. of Coastal Research,* Special Issue No. 9, Proc. of the Skagen Symposium, pp. 430-451.

Kroon, A., 1994. "Sediment transport and morphodynamics of the beach and nearshore zone near Egmond, The Netherlands", PhD-thesis, Utrecht University, The Netherlands. 282 pp.

Kroon, A. and Hoekstra, P., 1993. "Nearshore bars and large scale coastal behaviour", in: J.H. List (ed.), *Large Scale Coastal Behaviour '93*, U.S. Geological Survey Open-File Report 93-381, pp. 92-95.

Lippmann, T.C., Holman R.A. and Hathaway, K.K., 1993. "Episodic, nonstationary behavior of a double bar system at Duck, North Carolina, U.S.A., 1986-1991", *J. of Coastal Research*, Vol. 15, pp. 49-75.

Pruszak, Z., 1993, "The analysis of beach profile changes using Dean's method and empirical orthogonal functions", *Coastal Engineering*, Vol. 19, pp. 245-261.

Ruessink, B.G. and Kroon, A., in prep.. "The autonomous behaviour of a multiple bar system in the nearshore zone of Terschelling, The Netherlands", submitted to *Marine Geology.*

Sallenger, A. H., Holman, R. A. and Birkemeier, W. A., 1985. "Storm-induced response of a nearshore-bar system", *Marine Geology*, Vol. 64, pp. 237-257.

Wijnberg, K.M. and Terwindt, J.H.J., 1993. "The analysis of coastal profiles for large-scale coastal behaviour", in: J.H. List (ed.), *Large Scale Coastal Behaviour '93*, U.S. Geological Survey Open-File Report 93-381, pp. 224-227.

Wijnberg, K.M. and Terwindt, J.H.J., in prep.. "Quantification of decadal morphological behaviour of the central Dutch coast", submitted to *Marine Geology.*

Winant, C.D., Inman, D.L. and Nordstrom, C.E., 1975. "Description of seasonal beach changes using empirical eigenfunctions", *J. of Geophys. Research* Vol. 80, No.15, pp. 1979-1986.

Wolf, F.C.J., 1993. "The use of profile response parameters in describing the impact of a single storm on the nearshore zone near Egmond aan Zee, the Netherlands", *Proc. Hilton Head Island International Symp.*, Hilton Head, USA, pp. 485-502.

Wright, L.D. and Short A.D., 1984. "Morphodynamic variability of surf zones and beaches: a synthesis", *Marine Geology*, Vol. 56, pp. 93-118.

Zarillo, G.A. and Liu, J.T., 1988. " Resolving bathymetric components of the upper shoreface on a wave-dominated coast. *Marine Geology*, Vol. 82, pp. 169-186.

THE ROLE OF LONG WAVES IN SAND BAR FORMATION - A MODEL EXPLORATION

Tim O'Hare*

ABSTRACT: Model tests are presented which examine the form of the sediment transport contribution due to coupling between groupy short waves (which mobilise sediment) and associated long wave motions (which transport sediment) for regular *and* random period wave groups. The roles of the bound and breakpoint-forced long wave components are also examined. It appears that there is a continuum of behaviour in which multiple bars may form when the wave group periods are narrow banded and the bound long wave is relatively small, while bars may tend to migrate offshore and reduce in size when the group periods are more broad banded and the bound long wave dominant.

INTRODUCTION

Longshore bars are one of the most common features found offshore of sandy beaches around the world, yet there is still considerable uncertainty surrounding the mechanisms which contribute towards their formation. The importance of bars is readily acknowledged, for as potential reflectors of long wave energy and dissipaters of higher-frequency waves they may play a crucial role in protecting the beachface from the ravages of storms.

Interest in sand bar systems first intensified during the 2nd World War in connection with Allied landing operations on European beaches (e.g. Williams, 1947). In the following two decades bar dynamics were investigated by workers at

---

* Lecturer in Ocean Science, Institute of Marine Studies, University of Plymouth, Drake Circus, Plymouth, PL4 8AA, United Kingdom.

the U.S. Army Corps of Engineers, both in the field (e.g. Shepard, 1950) and in controlled laboratory experiments (e.g. Keuleghan, 1948). The majority of these early studies attempted to find simple relationships between the basic wave parameters and the bar topography and did not look in detail at the *processes* involved in shaping the bed. Early attempts to *explain* the formation of bars linked their presence to the most obvious dynamic feature of the coastal zone - breaking waves. Initially, bar formation was attributed to the action of plunging breakers, the falling crests of which scour out a hollow (trough) in the bed sediment leaving a bar just offshore of the breakpoint (Evans, 1940). A more subtle breaking-related mechanism involves the movement of sediment by steady currents induced by the waves (King & Williams, 1949; Dyhr Nielsen & Sorensen, 1970). Offshore of the breakpoint the residual flow close to the bed is shorewards due to wave asymmetry and onshore of the breakpoint the near-bed flow is seawards (the undertow). Thus there is a region of flow convergence, and hence sediment accumulation, close to the breakpoint. It has been argued (Dyhr Nielsen & Sorensen, 1970; Exon, 1975; Dolan & Dean, 1985) that if broken waves re-form and continue to propagate towards the beach, there may be a succession of breakpoints and a series of bars. However, in many cases bars are observed in fully-saturated surf-zones (i.e. where waves break over a wide region close to the shoreline) in which there are no easily identifiable breakpoints (e.g. Sallenger & Howd, 1989). In such cases, it is difficult to see how a breakpoint-bar hypothesis can be valid. In addition, bars have been observed in water depths too great for wave breaking to be an important factor (Zenkovich, 1967). Consequently, attention has turned to a variety of other mechanisms which may be responsible for bar formation, including standing waves and edge waves (see later), and the decomposition of wave-trains in shoaling water (Boczar-Karakiewicz & Davidson-Arnott, 1987).

When a standing wave component is present in the wavefield (e.g. caused by reflection of waves at the shoreline), mean currents are set up which may transport sediment to fixed positions beneath the standing wave envelope (Hunt & Johns, 1963; Carter *et al.*, 1972; Lau & Travis, 1973; Short, 1975; Nielsen, 1979; Bowen, 1980). It has been suggested that such bars may themselves cause wave reflection, strengthening the residual flows and enhancing bar growth, and may lead to the formation of new bars in the up-wave direction (Davies, 1982; Davies & Heathershaw, 1984; Mei, 1985; Brooke Benjamin *et al.*, 1987). The formation of bars by partially-standing waves has been demonstrated in the laboratory (O'Hare & Davies, 1990) and modelled successfully (O'Hare & Davies, 1993).

In the field, bar spacings are of the order of a hundred metres and so the standing wave component must have a similar spatial scale (e.g. Bowen & Huntley, 1984). Longuet-Higgins & Stewart (1962, 1964) explained how such long waves or "surf beat" may result from the radiation stress variations beneath incoming wave groups, and Symonds *et al.* (1982) described long wave generation by time-variation of the breakpoint, again caused by the group structure of the incident waves. Suhayda (1974) showed the presence of "leaky-mode" long standing waves on a barred beach and Short (1975) linked the locations of multiple bars with the positions of the nodes (or antinodes) of the standing wave envelope, depending on whether sediment was transported as bed load or suspended load. Bar formation has also been attributed to a further class of long wave motions known as edge waves (Huntley, 1976) which are essentially long waves trapped close to the shoreline by refraction. Bowen (1980) developed a simple model which related the known structure of long waves on *plane* beaches to the spacing of multiple bars, and Holman & Bowen (1982) extended this model to indicate how combinations of long wave modes could lead to a range of bar topographies. Recent field measurements by Aagaard (1990) and, in particular, Bauer & Greenwood (1990) provide considerable support for the idea that long waves play a role in determining beach morphology.

Despite the close correspondence between bar spacings and the spatial scales associated with long waves, long wave mechanisms have come in for considerable criticism (e.g. Dally, 1987; Dean *et al.*, 1992). Specific objections to long wave mechanisms of bar formation are that :

1. there is no field evidence of partially-standing long wave fields and the apparent matching between theory and field observations of bars just reflects the fact that it is easy to get things to agree if you make the right (or wrong!) assumptions.

2. there is a convenient ambiguity over whether bars form at nodes or antinodes such that if one doesn't work, the other surely will. Dally (1987) indicates that suspended load might be expected to dominate in the surf zone whereas bed load may be more important offshore - thus the location of bars with respect to the wave envelope should change.

3. studies of several bar systems have failed to find an appropriate correspondence between the long wave wavelength and bar separation.

4. they depend upon there being a well-defined, stable standing wave envelope. This can only occur if the long wave spectrum is narrow banded in frequency. However, most field measurements reveal a broad-banded spectrum of long waves for which nodal and antinodal positions would be continually changing.

Whilst recent field studies have gone some way to answering the first of these criticisms, there can be no doubt that there are important questions concerning bar formation which remain unanswered. These doubts are fuelled by Dally's (1987) description of laboratory experiments designed to *favour* bar formation by the standing long wave mechanism which according to the author "provide very little support for the surf beat mechanism". This does indeed appear to be the case, but Dally also comments that "Perhaps unexpectedly, for the breakpoint varied widely across the surf zone, bars did appear to form in response to the breakpoint/undertow scenario, with sand being scoured from the bottom by turbulence induced by wave breaking, carried offshore by undertow, and deposited outside the breakpoint of the largest wave".

Thus there is no clear consensus regarding the formation of longshore bars, with two schools of thought dominating work on the subject - breakpoint bar hypotheses and long wave residual hypotheses. In a recent paper, O'Hare & Huntley (1994) indicated an additional potential mechanism for bar formation which should be considered, namely *coupling* between sediment entrained by individual waves within incident wave groups and the first-order long wave motions resulting from the "bound long wave" (Longuet-Higgins & Stewart, 1962; 1964), the "breakpoint-forced long wave" (Symonds *et al.*, 1982), and the free long waves resulting from reflection of these forced motions at the shoreline. Using a model of wave group propagation and long wave generation developed by List (1992), the possibility that such a mechanism could result in convergence and divergence of sediment and the formation of bars was demonstrated for the case when the wave group period was *fixed*. The model predicted that the inner bar would form close to the outer surf-zone and thus be complimentary to any breakpoint bar generation. This new mechanism was expected to be a more potent cause of bar formation than the standing long wave mechanism but was still restricted to a narrow-banded (in fact, monochromatic) spectrum of long waves.

In this paper the previous modelling work of O'Hare & Huntley (1994) is extended by allowing the wave group period and the magnitude of the incoming bound long wave to vary. Attention is focused on exploring the possible role of coupling between long and short wave motions and, to this end, other factors such as undertow and wave asymmetry are ignored here. In the following section the model is summarised and its mode of operation outlined. The results of a series model tests are then discussed commencing with the simple case of constant period wave groups (as per O'Hare & Huntley, 1994). Results are presented which indicate that variation of the wave group period does *not* inhibit the growth of bars. These results are compared with the predictions of a more complex model of beach evolution produced by Roelvink (1993). This comparison involves consideration of the magnitude of the incoming bound long wave which is found, in this study, to be an important factor determining the predicted pattern of erosion and deposition, and hence bar formation, occurring in the model. Finally, the conclusions of this study are presented.

THE MODEL

General

O'Hare & Huntley (1994) used the numerical model of List (1992) to examine the propagation and transformation of normally-incident gravity waves and their associated long wave motions over the nearshore. This model assumes that the beach is uniform in the alongshore direction so that the problem reduces to the consideration of a two-dimensional slice perpendicular to the shoreline. The two components of the wavefield, short-period waves which comprise the incident wave groups and longer-period motions induced by the groups, are modelled separately but coupled together via the description of the radiation stress in the model. The short waves are parameterised in terms of a time-varying envelope, which propagates through the nearshore undergoing linear shoaling and breaking as appropriate. The long wave motions are computed using the depth-integrated, linearized, shallow water equations of continuity and momentum which allow for the forcing of long waves by variations in radiation stress and the reflection of wave energy at the shoreline. Details of the model are given in the original paper of List (1992) and a brief description is incorporated in O'Hare & Huntley (1994). Here, only the key points are outlined.

Short waves

For short-wave motions, the model employs a floating grid scheme designed to minimise numerical dispersion. The representative wave height (H) at each grid location is propagated shorewards in water of depth h, with the local (shallow water) phase speed for the timestep Δt. This height is then adjusted for linear shoaling/de-shoaling, for which

$$H \propto \left(\frac{1}{h}\right)^{1/4} \qquad [1]$$

If the waves enter water shallow enough for breaking to occur (determined by a depth-dependent breaking criterion) the wave height is reduced appropriately; in this case using

$$H = 0.7h \qquad [2]$$

The resulting short-wave field is then re-gridded back onto a regular grid for use in the long wave calculations.

Long Waves

Long wave motions are modelled using a finite difference representation of the depth-integrated, linearized, shallow water equations of momentum

$$\frac{\partial u}{\partial t}+g\frac{\partial \eta}{\partial x}=-\frac{1}{\rho h}\frac{\partial S_{xx}}{\partial x} \qquad [3]$$

and continuity

$$\frac{\partial \eta}{\partial t}+\frac{\partial (hu)}{\partial x}=0 \qquad [4]$$

where $\eta(t)$ is the sea surface elevation averaged over the short-wave period, $u(t)$ is the depth-integrated, cross-shore velocity averaged over the short-wave period, $S_{xx}(t)$ is the radiation stress perpendicular to the wave crests, ρ is the fluid density, t is time and x is the horizontal co-ordinate (+ve onshore).

Solutions are obtained using a semi-implicit finite difference scheme in which elevations at the new timestep are calculated using velocities at the previous one using the continuity equation [4] and the new velocities are then obtained from the momentum equation [3] using the new elevations. The scheme employs a staggered grid of alternating η and u points, and solutions are subject to the Courant condition which ensures stability when $\Delta x/\Delta t > (gh)^{1/2}$.

Interaction between short and long waves

The short wave and long wave motions may be coupled in two ways :-

1. spatial and temporal variations in the short-wave height, which arise as the incoming groups propagate shorewards, produce radiation stress variations which, in turn, drive long wave motions

2. surface elevation changes due to long waves alter the local water depths through which the short wave groups propagate.

Away from the shoreline, the model results are insensitive to the inclusion of the latter coupling mechanism, and since the very-nearshore region is not particularly well modelled by the linearized equations used, only the first coupling mechanism is considered here.

Boundary conditions

Two boundary conditions are required to allow solutions to the long wave equations to be obtained.

At the shoreline, it is assumed that long wave motions are reflected. In the model, this requirement is met by setting the product $hu = 0$ at the landward-most velocity grid point - effectively, this forces a long wave antinode at a point close to the shoreline where the water is just a few centimetres deep.

At the offshore boundary, there are two factors which must be incorporated in the boundary condition used in the model. There must be an incoming bound long wave with a form dependent on the group structure of the short waves, and any offshore-propagating long waves reflected at the shoreline or generated within the model domain must pass freely through the boundary. To accomplish these requirements, the long wave field close to the boundary is split into onshore- and offshore-propagating components using the procedure suggested by Guza *et al.* (1985) and the long wave at the boundary is then obtained as a superposition of a suitable incoming bound long wave and an appropriate offshore-propagating component calculated from the decomposed long wave field at the previous timestep. The instantaneous surface elevation at the offshore boundary due to the bound long wave, η_{BLW}, is specified in terms of the instantaneous height of the incoming waves $H(t)$, their mean height $\langle H \rangle$, and a constant B by the expression

$$\eta_{BLW} = -B\left(H(t)^2 - \langle H \rangle^2\right) \qquad [5]$$

Thus for a given set of input parameters, B determines the magnitude of the bound long wave.

Sediment transport

Sediment transport due to coupling between short waves (which mobilise sediment) and long waves (which move it) is inferred from the pattern of the correlation between the short wave height $H(t)$ and the long wave velocity $u(t)$ across the nearshore C_{SED}.

Solution procedure

The model is run first in "set-up" mode, whereby the incoming wave height is set to a constant value equal to the mean of the required offshore wave height time series and with the bound long wave is omitted. The resulting variations in the radiation stress produced by the shoaling and breaking of the incident waves are balanced by changes in the mean water level (set-up and set-down). Once an equilibrium is reached, the offshore wave height is allowed to vary about its mean value to simulate the propagation of short wave groups into the model and the appropriate bound long wave forcing is switched on. The model then generates time-varying short-wave heights, long wave elevations and depth-averaged currents across the nearshore profile which are used to compute the form of the transport correlation C_{SED}.

DISCUSSION OF MODEL TEST RESULTS

Regular wave groups with low B value

Initial tests (O'Hare & Huntley, 1994) investigated the action of *regular* wave groups incident on a plane beach with an incoming bound long wave computed using B = 0.02m/m² (as suggested by List, 1992). Figure 1 shows a typical pattern of C_{SED} under these circumstances, for which Δx = 10m, Δt = 1s, beach slope β = 0.025 and the period of the incident wave groups T_g = 60s. The wave height at the offshore boundary is assumed to vary sinusoidally :-

$$H(t) = H_0 + \Delta H \sin\left(\frac{2\pi t}{T_g}\right) \qquad [6]$$

with H_0 = 0.8m and ΔH = 0.2m.

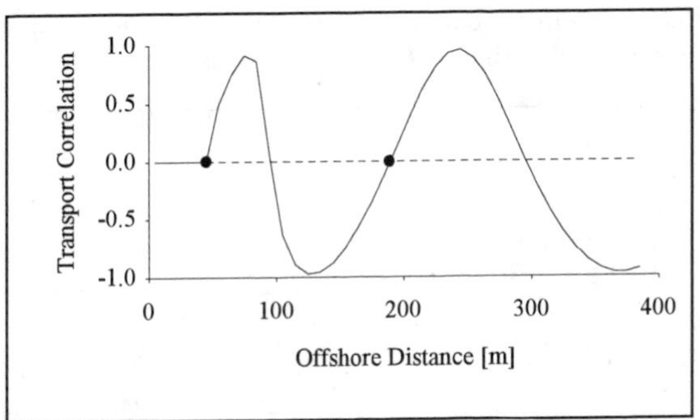

Figure 1. Sediment transport correlation pattern for regular wave groups with a bound long wave constant B = 0.02m/m²

There is a region of zero correlation close to the shoreline corresponding to the inner surf zone (where all waves are breaking). Here, the short wave groupiness has been removed by the wave breaking and by the neglect of water depth changes due to the presence of long waves in the short wave height calculations. Offshore of the inner surf zone, the transport correlation ranges from +1 to -1 in a systematic pattern, suggesting alternating regions of onshore- and offshore-directed sediment motion. Regions of negative correlation represent possible offshore transport of sediment whereas positive correlation implies onshore transport. Note that this sign convention is the reverse of the results shown in O'Hare & Huntley (1994) for consistency with other studies. Also shown (•) are the approximate locations of bars consistent with this sediment transport pattern. These *potential* bars arise as a result of the presence of a standing long wave through which the "stirring" pattern of wave groups propagates.

Random wave groups with low B value

Although the results of tests with regular wave groups (e.g. Figure 1; O'Hare & Huntley, 1994) are interesting, in that they suggest a possible mechanism of bar formation which has been largely overlooked, they have limited application. Field observations (e.g. Guza & Thornton, 1985; Davidson, 1991) indicate that wave groups do not tend to have a regular period but, instead, follow each other in an apparently random sequence (i.e. the spectrum of long wave frequencies is broad-banded). This randomness of the long wave motion present on beaches might be expected to wipe out the transport correlation pattern found with regular groups (Figure 1) since different frequency long wave components would produce convergences and divergences of sediment at different locations.

To address this important factor, further model runs were conducted in which the period of successive groups in the model was varied randomly within a specified range. Figure 2 shows the offshore variation of C_{SED} for the same basic configuration as Figure 1 but with a randomly-varying group period in the (integer) range $T_g \pm \Delta T_g$ with T_g = 60s and ΔT_g = 40s.

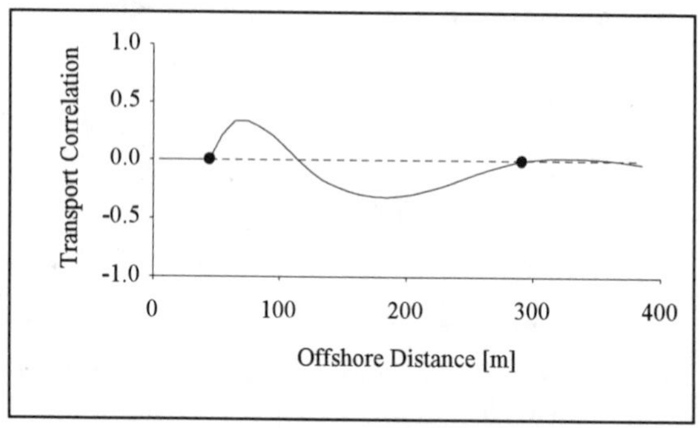

Figure 2. Sediment transport correlation pattern for random period wave groups with a bound long wave constant B = 0.02m/m^2

Surprisingly, the transport correlation for such a sequence of groups converges rapidly (within tens of groups), giving a similar pattern of inferred sediment transport to the regular group case. Despite the wide range of group periods present, and the inherent randomness of the long wave motion, there is still considerable variation of the inferred sediment transport, and the possibility of bar formation remains. C_{SED} magnitudes remain relatively high in a region offshore of the outer surf-zone. This region widens as the range of wave group periods increases (i.e. as ΔT_g increases), and is the region in which the period of outgoing long waves generated in the outer surf zone may be equal to that of the incoming long wave motions (i.e. the distance travelled by the longest period breakpoint-forced long wave away from its zone of generation in half its period). In this region, at least part of the long wave motion is narrow banded at any instant and thereby produces water motions similar to those of the regular group case.

Wave groups with high *B* value (enhanced bound long wave)

On the basis of Figure 2, and other model tests not presented here, it appears that the broad-banded nature of the long wave motion present on natural beaches does not necessarily preclude the formation of multiple longshore bars. Indeed in this respect, and in contrast to other bar-forming mechanisms involving long waves, the mechanism proposed by O'Hare & Huntley (1994) seems surprisingly robust. However, the results presented thus far are at odds with the findings of Roelvink (1993) using an alternative, more complex model for cross-shore transport due to random wave groups, which suggest that the transport correlation should decrease steadily offshore to a constant negative value (and hence there will no significant bar-generating fluctuations in the sediment transport).

In Roelvink's model, the size of the bound long wave at the offshore boundary is determined using the expression

$$\eta_{BLW} = R_{rep}\left(E - \langle E \rangle\right) / \rho g \qquad [7]$$

where E is the energy of the short waves, g is the gravitational acceleration and R_{rep} is a constant, found be linear regression to be of the order of -4m/m^2. Substitution for the energy E, using the linear wave relation $E = \rho g H^2/8$ and comparison with the expression [5] making the assumption that $<H>^2 = <H^2>$, indicates that List's constant B, and Roelvink's constant R_{rep}, are related by the expression

$$B = -\frac{R_{rep}}{8} \qquad [8]$$

(Note that the exact relationship between $<H>^2$ and $<H^2>$ depends upon the spectral distribution of short wave heights, but that the assumption of equality which is made here for convenience is not likely to produce a qualitative or significant quantitative error). Thus it appears that Roelvink's study uses a parameter setting equivalent to a value of B some 25 times greater than that suggested by List (1992), used by O'Hare & Huntley (1994), and in the model runs used to produce Figures 1 and 2 in the present work.

Further model tests to investigate the influence of the magnitude of the constant B on the predictions of List's model have been carried out. It should be borne in mind here that the value of B dictates the size of the bound long wave as it enters the model domain. For a given set of wave parameters, increasing B (artificially) enhances the bound long wave while the internally-generated breakpoint-forced long wave remains essentially unchanged. Thus, altering B

allows the relative roles of the bound long wave and breakpoint-forced long wave to be examined. Figure 3 shows the variation of C_{SED} for the same basic parameters as Figure 1, but now with B = 0.3 (still less than the equivalent of Roelvink's value). Note that the wave group period is fixed (60s). Increasing B, and thereby enhancing the bound long wave and its effects, results in a rapid decrease in the size of the positive peaks of the transport correlation, such that for $B \geq 0.25$, the transport correlation is always negative (i.e. offshore transport dominated by the presence of the bound long wave).

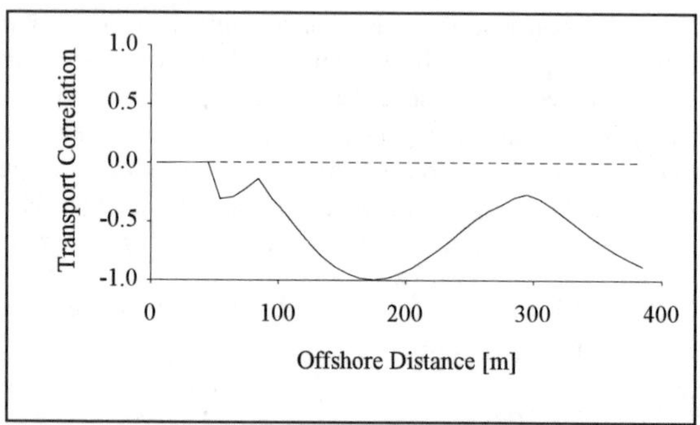

Figure 3. Sediment transport correlation pattern for regular wave groups with an enhanced bound long wave for which B = 0.3m/m^2

Encouragingly, if the wave group period is allowed to vary when B is large, the form of the transport correlation strongly resembles that predicted by Roelvink (1993). For example, Figure 4 shows the results of a model run similar to that of Figure 2, but with T_g = 60s, ΔT_g = 40s and B = 0.3.

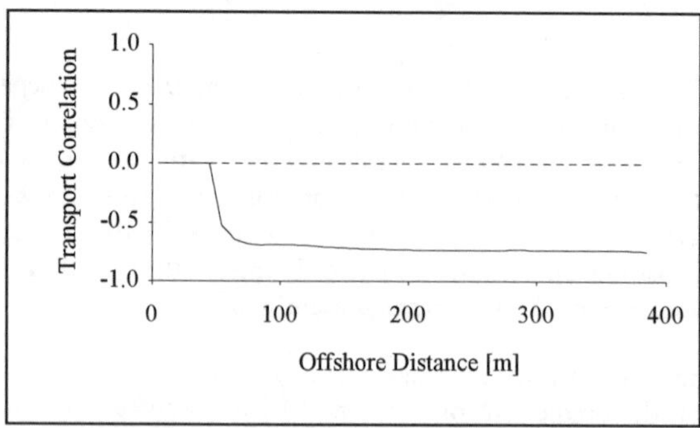

Figure 4. Sediment transport correlation pattern for random wave groups with an enhanced bound long wave for which B = 0.3m/m^2

CONCLUSIONS

On the basis of the results presented here, it appears that there are *two* important considerations for the formation of bars due to wave groups and associated long waves, namely the bandwidth of the spectrum of wave group periods and the relative strengths of the bound and breakpoint-forced long waves. When the wave group spectrum is broad-banded and the bound long wave dominates, short-wave/long-wave coupling tends to move sediment offshore (i.e. bars may migrate offshore and reduce in size - Figure 4, the "Roelvink" case). However, if the breakpoint-forced long wave is sufficiently large relative to the bound long wave, multiple bars may be formed by the mechanism proposed by O'Hare & Huntley (1994), particularly if the group spectrum is more narrow-banded (e.g. Figure 1). This situation is summarised in Figure 5.

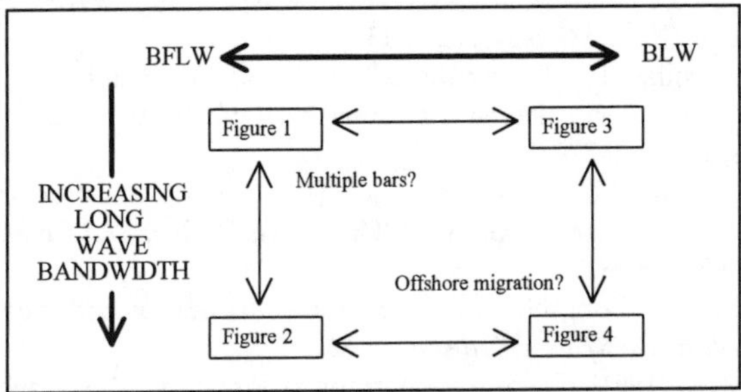

Figure 5. Potential beach response to wave groups and associated long waves

It is hoped that further work with List's model will address how *evolving* bars influence the form of the nearshore water motion and sediment transport. Once formed, bars could play an important role in modifying the form of the surface wavefield both at short wave frequencies (through wave breaking) and for longer waves (through reflection and trapping, e.g. Kirby *et al.*, 1981; Symonds & Bowen, 1984). However, future work must also examine the extent to which the size of the bound long wave can vary in natural conditions. Since the bound long wave is carried into coastal regions from deeper water, it is possible that wave transformations and bottom topography occurring beyond the usual region of study may have a profound influence on beach morphodynamics.

ACKNOWLEDGEMENTS

The author wishes to acknowledge the support of the Natural Environment Research Council in funding the initial stages of this study through their Research Fellowship scheme, and the Royal Society for providing assistance with the cost of attending the Coastal Dynamics '94 conference. He is also indebted to Professor David Huntley for encouragement, help, and stimulating discussion.

REFERENCES

Aagaard, T. (1988). A study of nearshore bar dynamics in a low-energy environment: Northern Zealand, Denmark, *Journal of Coastal Research*, **4**, 115-128.

Bauer, B.O., & Greenwood, B. (1990). Modification of a linear bar trough system by a standing edge wave, *Marine Geology*, 92, 177-204.

Boczar-Karakiewicz, B., & Davidson-Arnott, R.G.D. (1987). Nearshore bar formation by nonlinear wave processes - a comparison of model results and field data, *Marine Geology*, 77, 287-304.

Bowen, A.J. (1980). Simple models of nearshore sedimentation; beach profiles and longshore bars, in S.B. McCahn (editor), "The Coastline of Canada", *Geological Survey*, **Paper 80-10**, 1-11.

Bowen, A.J., & Huntley, D.A. (1984). Waves, long waves and nearshore morphology, *Marine Geology*, 60, 1-13.

Brooke Benjamin, T., Boczar-Karakiewicz, B., & Pritchard, W.G. (1987). Reflection of water waves in a channel with corrugated bed, *Journal of Fluid Mechanics*, **185**, 249-274.

Carter, T.G., Liu, P.L.-F., & Mei, C.C. (1972). Mass transport by waves and offshore sand bedforms, *Journal of Waterways, Harbors and Coastal Engineering Division*, 99, 165-184.

Dally, W.R. (1987). Longshore bar formation - surf beat or undertow?. *Proceedings Coastal Sediments '87, New Orleans*, 71-86.

Davidson, M.A. (1991). Field investigations of infragravity oscillations on a high energy dissipative beach (Llangenith, S.Wales, UK), *Ph.D. thesis, , University of Wales (Cardiff).*

Davies, A.G. (1982). The reflection of wave energy by undulations on the seabed, *Dynamics of Atmospheres and Oceans*, 6, 207-232.

Davies, A.G., & Heathershaw, A.D. (1984). Surface-wave propagation over sinusoidally varying topography, *Journal of Fluid Mechanics*, **144**, 419-443.

Dean, R.G., Srinivas, R., & Parchure, T.M. (1992). Longshore bar generation mechanisms, *Proceedings of the 23rd International Conference on Coastal Engineering, Venice*, 2001-2014.

Dolan, T.J., & Dean, R.G. (1985). Multiple longshore sand bars in the Upper Chesapeake Bay, *Estuarine, Coastal and Shelf Science*, 21, 727-743.

Dyhr Nielsen, M., & Sorensen, T. (1970). Some sand transport phenomena on coasts with bars, *Proceedings of the 12th International Conference on Coastal Engineering, Washington D.C.*, 855-866.

Evans, O.F. (1940). The low and ball of the eastern shore of Lake Michigan, *Journal of Geology*, **48**, 476-511.

Exon, N.F. (1975). An extensive offshore sand bar field in the western Baltic Sea, *Marine Geology*, **18**, 197-212.

Guza, R.T., & Thornton, E.B. (1985). Observations of surf beat, *Journal of Geophysical Research*, 90, 3161-3172.

Guza, R.T., Thornton, E.B., & Holman, R.A. (1985). Swash on steep and shallow beaches, *Proceedings of the 19th International Conference on Coastal Engineering, New York*, 708-723.

Holman, R.A., & Bowen, A.J. (1982). Bars, bumps and holes: Models for the generation of complex beach topography, *Journal of Geophysical Research*, **87**, 457-468.

Hunt, J.N., & Johns, B. (1963). Currents induced by tides and gravity waves, *Tellus*, **15**, 343-351.

Huntley, D.A. (1976). Long period waves on a natural beach, *Journal of Geophysical Research*, **81**, 6441-6449.

Keuleghan, G.H. (1948). An experimental study of submarine sandbars. *US Army Corps of Engineers, Beach Erosion Board,* **Technical Report 3,** 40pp. Reprinted in M.L.Schwarz (editor), "Spits and Bars", *Dowden, Hutchinson & Ross* (1972).

King, C.A.M., & Williams, W.W. (1949). The formation and movement of sand bars by wave action, *Geographical Journal*, **112**, 70-85.

Kirby, J.T., Dalrymple, R.A., & Liu, P.L.-F. (1981). Modification of edge waves by a barred-beach topography, *Coastal Engineering*, **5**, 35-49.

Lau, J., & Travis, B. (1973). Slowly varying Stokes waves and submarine longshore bars, *Journal of Geophysical Research*, **78**, 4489-4497.

List, J.H. (1992). A model for the generation of two-dimensional surf beat, *Journal of Geophysical Research*, 97, 5623-5635.

Longuet-Higgins, M.S., & Stewart, R.W. (1962). Radiation stress and mass transport in gravity waves, with an application to surf beats, Journal of Fluid Mechanics, 13, 481-504.

Longuet-Higgins, M.S., & Stewart, R.W. (1964). Radiation stress in water waves: A physical discussion with applications, *Deep-Sea Research*, 11, 529-562.

Mei, C.C. (1985). Resonant reflection of surface water waves by periodic sandbars, *Journal of Fluid Mechanics*, **152**, 315-335.

Nielsen, P. (1979). Some basic concepts of wave sediment transport. *Technical University of Denmark, Institute of Hydrodynamic and Hydraulic Engineering*, **Series Paper** 20, 160pp.

O'Hare, T.J., & Davies, A.G. (1990). A laboratory study of sand bar evolution, *Journal of Coastal Research*, **6**, 531-544.

O'Hare, T.J., & Davies, A.G. (1993). Sand bar evolution beneath partially-standing waves: laboratory experiments and model simulations, *Continental Shelf Research*, **13**, 1149-1181.

O'Hare, T.J., & Huntley, D.A. (1994). Bar formation due to wave groups and associated long waves, *Marine Geology*, **116**, 313-325.

Roelvink, J.A. (1993). Surf beat and its effect on cross-shore profiles, *Ph.D. thesis, Delft Technical University*.

Sallenger, A.H., & Howd, P.A. (1989). Nearshore bars and the breakpoint hypothesis, *Coastal Engineering*, **12**, 301-313.

Shepard, F.P. (1950). Longshore-bars and longshore-troughs. *US Army Corps of Engineers, Beach Erosion Board,* **Technical Memorandum 48**, 37pp.

Short, A.D. (1975). Multiple offshore bars and standing waves, *Journal of Geophysical Research*, **80**, 3838-3840.

Suhayda, J.N. (1974). Standing waves on beaches, *Journal of Geophysical Research*, **79**, 3065-3071.

Symonds, G., & Bowen, A.J. (1984). Interactions of nearshore bars with incoming wave groups, *Journal of Geophysical Research*, **89**, 1953-1959.

Symonds, G., Huntley, D.A., & Bowen, A.J. (1982). Two-dimensional surf beat. Long wave generation by a time varying breakpoint, *Journal of Geophysical Research*, **87**, 492-498.

Williams, W.W. (1947). The determination of the gradient of enemy held beaches, *Geographical Journal*, 109, 76-93.

Zenkovich, V.P. (1967). "Processes of Coastal Development", *Oliver and Boyd, London*, 738pp.

FIELD STUDIES OF VARIABILITY OF UNDERWATER COASTAL RELIEF AND BED SEDIMENT COMPOSITION DURING STORM

Ruben D.Kos'yan*, Nicolay V.Pykhov **

*Director of the Southern Branch of the P.P.Shirshov Institute of Oceanology. Russian Academy of Sciences. Gelendzhik-7, Krasnodar Region, 353470 Russia. Tel.(86141) 232 61

**Senior scientist of the P.P.Shirshov Institute of Oceanology. Russian Academy of Sciences. 23, Krasikova str., Moscow. 117218 Russia. Tel.(095) 124 6394

ABSTRACT

The composition of bed surface sediment can change in different points of submerged slope both as a result of redeposited sediment shift or as a result of erosion of lower sedimentary layers. The purpose of field works described in this paper was the definition of changeability of surface sediment composition subject to the change of surface wave parameters and the deformation of bottom relief during the storm.

INTRODUCTION

When the underwater slope is acted upon by the frontal storm waves approaching the shore with uniform conditions along it, the only sediment transport is normal to isobaths. As a result of this transport, the underwater slope profile is changed in the zone of active wave effect.

In contrast to the longshore sediment transport characterized by unidirectional movement at all levels above sea bottom, cross-shore movement of sand occurs both shoreward and seaward, the sign of movement direction reversing with depth at various points of the underwater slope.

Unfortunately, the correlation between grain-size variations, and short-term profile deformations is poorly studied. Thus, the accurate quantitative prediction of the underwater slope deformation under the known surface wave parameters is still impossible correctly.

When modeling the deformation of submerged profile during the storm it is usually supposed that the sediment composition is invariable. But we assume that the composition of surface sediments can change in different points of submerged slope both as a result of redeposited sediment shift or as a result of erosion of lower sedimentary layers. Therefore, one of the purpose of these field works was the definition of changeability of surface sediment composition subject to the change of surface wave parameters and the deformation of bottom relief during the storm.

SHORT-TERM VARIATIONS

During the international experiments "Shkorpilovtzy 85 - -88", the great attention was paid to the studies of short-term deformations of underwater slope profile accompanied by variations of bed sediment composition relative to wave regime. The field observations enabled us to widen the existing ideas of the processes concerned and to obtain the data on sediment transport along the bottom profile during a storm (Kos'yan,1987; Nikolov, Kos'yan,1991).

The water depths were measured during storms repeatedly along the trestle bridge of the "Shkorpilovtzy" testing site from water edge to a 5-*m* depth, measuring points being spaced 5 *m* apart. Long metallic rod graduated to 1 *cm* was used. Metallic plate welded to its lower end prevented the rod from biting deeper into the bottom sediment. The accuracy of measurements varied within 2 *cm* providing for reliable measurement of relatively small deformations of bottom relief. The bed sediment was sampled concurrently with depth measurements. For this purpose the surface sand layer 3 to 5 *cm* thick was scraped off the bottom with a cylindrical sampler 100 *cm* in diameter.

During one month of "Shkorpilovtzy-85" experiment, deformations of the underwater profile along the trestle bridge were measured 16 times and 17 series of bottom sediment were sampled. Profile variations ($\Delta \mathbf{h}$) were registered in cm relative to the some bottom level, taken as a reference mark. Dried samples of solid particles were subjected to 18-fraction grain-size analysis the results of which were used for calculation of distribution parameters of sediment composition. A mean size of solid particles ($\mathbf{d}$), assumed as mathematical expectation; root-mean square deviation of particle size (δ); coefficients of asymmetry (α); excess (τ) and variations of sediment composition ($\beta = \mathbf{d}/\delta$) have been calculated for every sample. The data obtained allow one to follow the temporal variations of these parameters in the same points of the underwater profile relative to wave regime (Figs. 1-3).

Unfortunately, constant measurements of wave field parameters were not made, so that the time intervals could be characterized only by separate measurements of wave elements. In some intervals (at times amounting to several days) wave regime was estimated visually (details are in Nikolov,Kos'yan,1991).

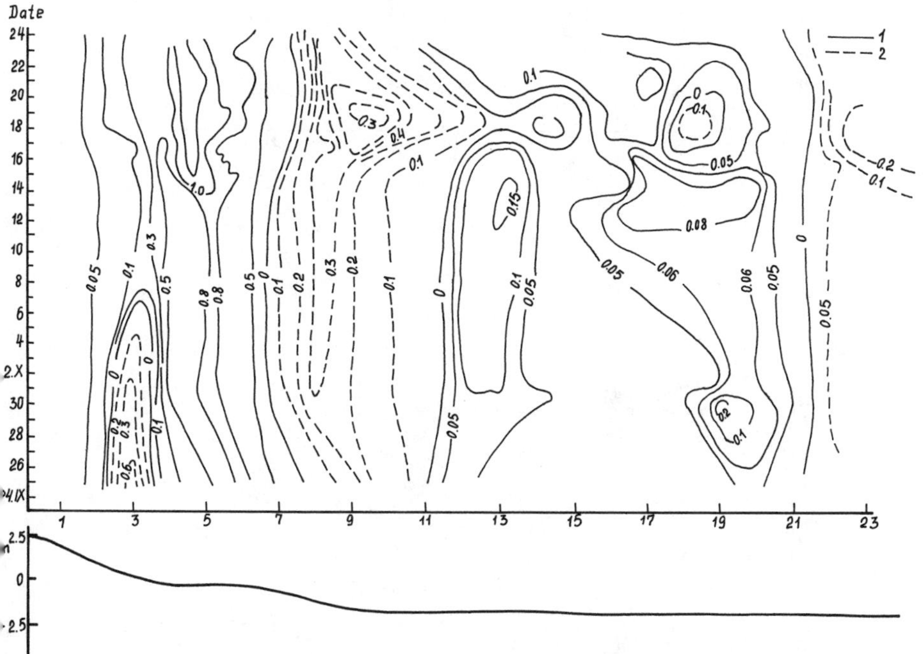

Fig. 1. Short - term alterations of bottom relief (Δh , m) during storm. 1- accumulation; 2- erosion.
Below profile of the underwater shore slope.

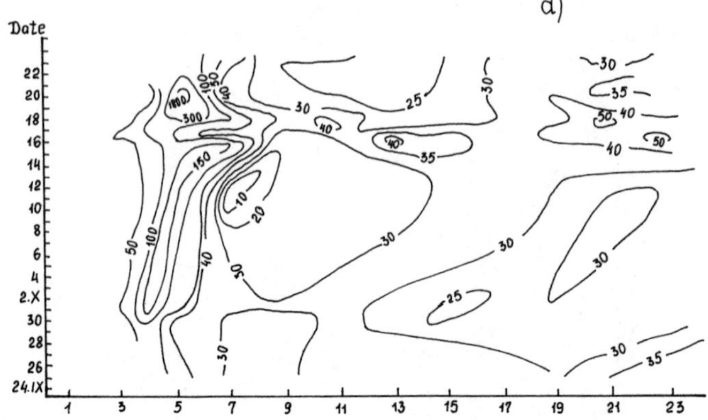

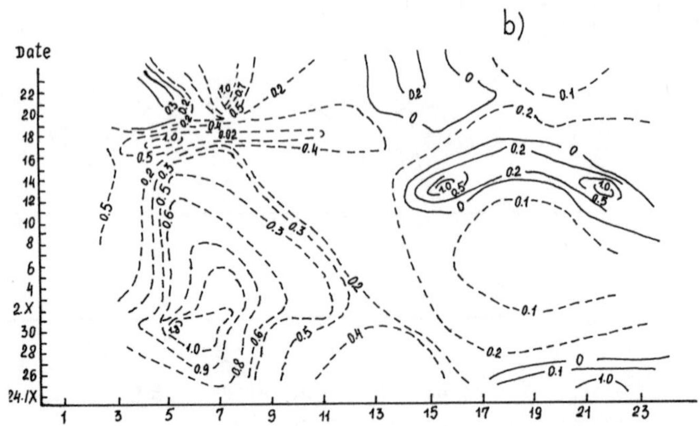

Fig. 2. Temporal variations: a) of mean grain size (d , mm) of bottom sediment; b) of coefficients of composition variation (β , %).

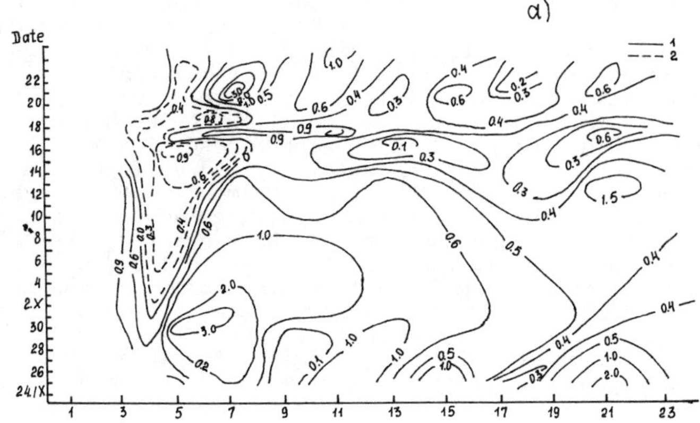

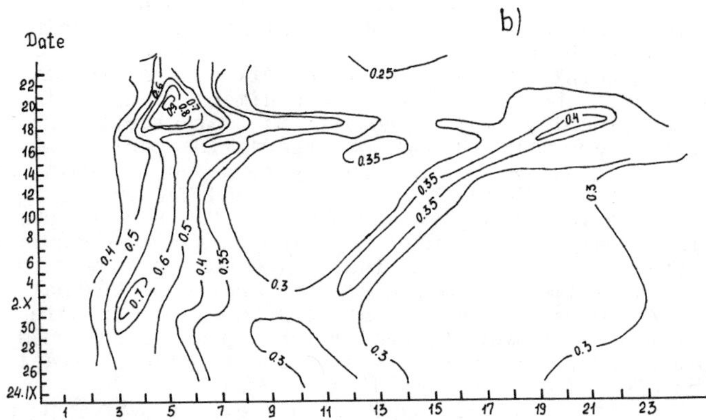

Fig. 3. Temporal variations: a) of coefficients of asymmetry (α); b) of coefficient of excess (τ). 1- positive changes; 2- negative changes.

The following conclusion can be made from the data analysis. Under slight sea the bottom aggradated gradually at point 5. The waves were plunging close to water edge where the bar was formed by input of coarser material. The sediment grading was worsening simultaneously: from good to moderately good by R.Folk's (1974) classification. Under higher sea the plunge point migrated shoreward resulting in further aggradation of the bar accompanied with particle fining and grading enhancement. When sea calmed, the plunge point migrated to point 5 where the bottom aggradation recommenced, the sediments became coarser and their grading worsened.

Around point 7, the bottom was eroded slowly under slight sea with insignificant fining of bed sediment and better grading. The coarser material was evidently transported shoreward closer to plunge point. Under higher sea, the migration of plunge point to this area induced accumulation of coarser sediment which became moderately well sorted.

The bottom between points 9 and 14 was eroded insignificantly under slight sea, the sediment composition remaining stable. Under stronger storm and migration of plunge zone to points 7 and 8, bottom aggradation began with preferential wash out of finer fractions and relative growth of coarser particles concentration. When sea calmed the initial bottom level and sediment composition restored. During observations, the sediment grading remained good at points 9 and 11.

At point 13, accumulation of bed sediment continued under calm and smooth sea, the sediment composition varying insignificantly. The composition characteristics generally remained constant throughout observation period, while under higher surface waves the grading worsened insignificantly.

At point 15 the bottom level and sediment composition varied insignificantly. At points 17-21 the tendency towards slight sediment accumulation with constant composition was observed. At higher sea, the bottom was eroded in this area, the bed sediment becoming coarser and worse graded.

At point 23 located at the seaward trestle end the bottom erosion was observed which intensified under higher sea, with preferential wash out of finer fractions and worsening of sediment grading.

Coefficients of asymmetry and excess were generally less than unit at all points indicating the small difference of grain-size distribution curve from lognormal. It should be noted that asymmetry coefficients were usually negative, while the excess coefficients - positive. This means that the maximum of particle distribution density corresponds to fractions coarser than the mean size. If compared with normal law, the curve of sediment composition distribution has higher and sharper peak.

Under slight sea the coefficients of asymmetry and excess decreased with distance from plunge zone (seaward of point 7) and in plunge zone itself, where the excess coefficient even took on negative values. Coefficients of

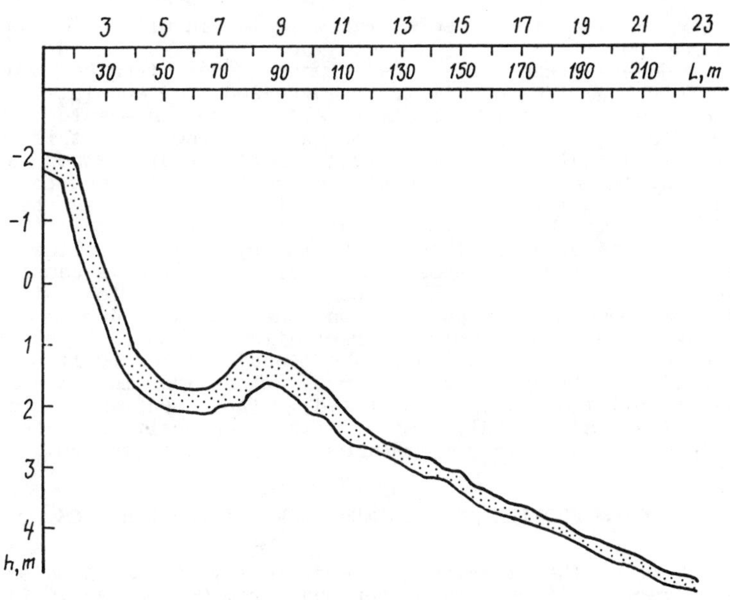

Fig.4. Maximal variations of the underwater profile during observation period (Shkorpilovtzy-88).

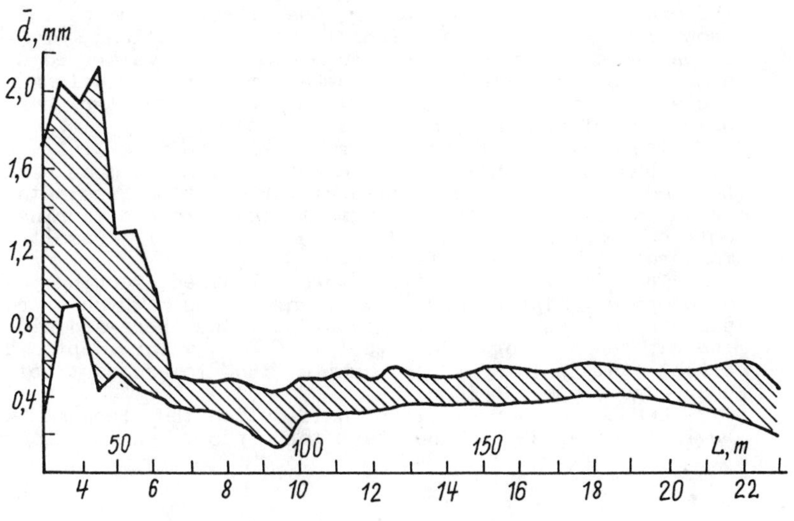

Fig.5. Maximal and minimal d values at various points of the underwater profile during observation period (Shkorpilovtzy-88).

asymmetry and excess were maximal in magnitude close to plunge point.

Under higher sea the plunge zone migrated seaward (points 7-9). The coefficients decreased in absolute value and remained nearly constant along the underwater slope profile seaward of plunge zone, while shoreward of this zone (point 5) their absolute values grew rapidly. When storm abated, the waves were breaking close to point 5 and in this case the coefficients of asymmetry and excess behaved similarly to those typical of slight sea.

The results obtained can be interpreted as follows. In the nearshore zone the sediment differentiation is caused by turbulence generated at both borders of the flow: in the bottom boundary layer and on the surface during wave breaking. When sediment differentiation is caused by a single mechanism, the distribution of sediment composition is described well by lognormal curve. When the mechanisms of differentiation are integrated, the sediment grading deteriorates rapidly and its composition becomes heterogeneous, with polymodal asymmetric distribution.

BOTTOM PROFILE AND SEDIMENT COMPOSITION DYNAMICS

During the experiment "Shkorpilovtzy-88" the studies of 1985 were continued and elaborated. From October 11 to 24, a total of 23 measurements of underwater profile were made and 19 series of bed sediment samples were collected (Kos'yan,Pykhov,1991). While turning to interpretation of the underwater profiles obtained during this experiment, it should be noted that the measurements were generally made during the periods of maximal storm development. Maximal and minimal depths and **d** values registered at various points of the underwater profile during the observation period are shown in Figs.4-5. Most significant fluctuations of these values were observed on the slope close to water edge and around the underwater bar. Somewhat smaller variations were registered in the trough between the underwater bar and near-edge slope, and seaward of this bar.

The examples of underwater slope profiles in various storm phases are shown in Fig.6. Variations of mean size of bed sediment along the profile at various time instants are shown in Fig.7, and that in some typical points during the entire observation period - in Fig.8. All results of measurements are shown in Figs.9-10.

The first 8 profiles were obtained during storm development. There are some variations which proceed successively as the storm intensifies. The shoreward side of the bar tends to aggradate (points 6-8), water depth grows slightly at points 10-14, while the lower part of the near-edge slope subsides.

Due to further swell intensification the largest waves begin to break above the bar top (p.8), which gradually

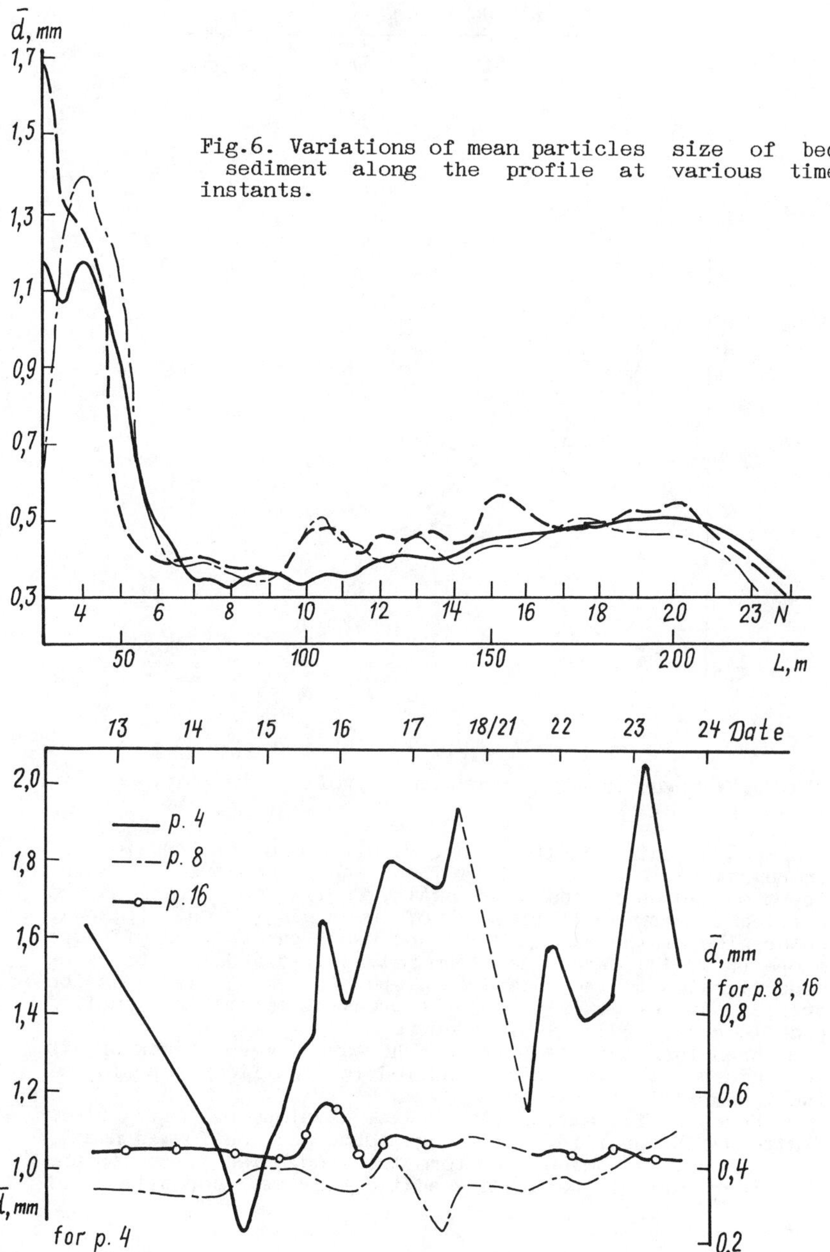

Fig.6. Variations of mean particles size of bed sediment along the profile at various time instants.

Fig.7. Variations of mean particles size of bed sediment in some typical points during the entire observation period.

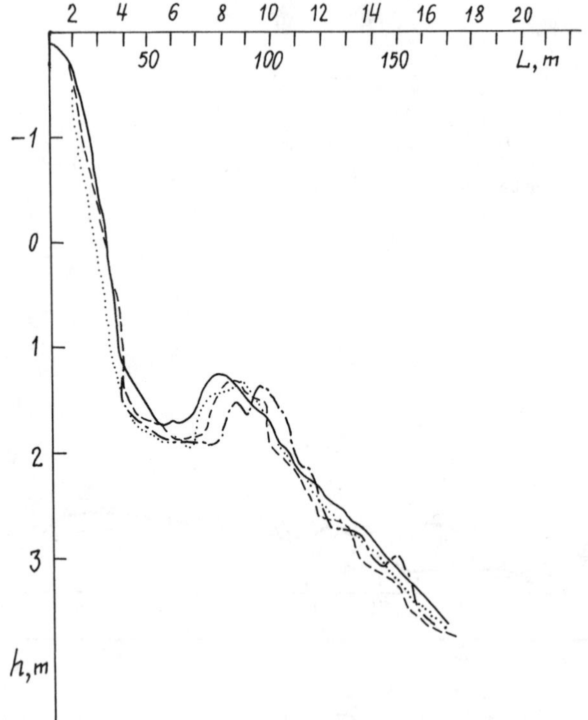

Fig.8. Examples of the underwater profile in various storm phases.

becomes the main breaking zone, resulting in destruction of shoreward bar side (p.6-8). Then the reconstructed waves approach the shore and break again. Profile modifications are related to growing thickness of near-edge slope (p.3-4), outer side of the bar (p.8-10) and upper part of the offshore slope (p.10-13). When the storm intensifies and the breaking zone migrates to the seaward slope (p.8-15), wave action creates shallow dents behind the breaking point. This part of profile resembles a saw-like curve.

When the storm stabilized, the waves were breaking in the same point (p.9) which resulted in significant erosion of the inner part of the bar.

More rapid attenuation that development of storm flattened the profile. The trough submerged and formed nearly regular arc. By the end of storm the trough was filled in and the inner part of the bar aggradated due to deposition of suspended particles and sediment displacement from the bar top. The longer the phase of storm attenuation, the greater the trough filling and the closer to shore the bar is.

Comparison of **d** distribution curves along the profiles of various series shows that the material coarseness changes abruptly close to p.5-6. Shoreward of this line the sediment size **d** did not exceed 1,0 mm, while seaward of p.5-6 **d** value was generally below 0,5 mm. This boundary was dynamic during storm and tended to migrate shoreward during storm development and to return to its seaward position during its attenuation.

At the initial phase of storm development only the finest particles were displaced. When rising up the slope, they reduced the sediment coarseness close to water edge (p.3-5) and on the bar (p.7-10).

When the compensating outflow develops, finer fractions from the water edge are carried offshore and the mean grain size of sediment increases. Further development of swell induces suspension of finer sands and involves coarser fractions in long-slope motion between points 8 and 15. At greater wave height the same tendency is observed on the deeper section of the profile resulting in growth of sediment coarseness at points 20-23 during maximal storm intensification. When the storm abates, mean size of the sediment decreases on the seaward bar slope and in the trough due to settlement of finer fractions from water column. The coarser sediment is concentrated at the base of the slope close to water edge. Sediment thickness at points 5 to 9 grows as a result of solid particle precipitation.

SOME REGULARITIES OF TRANSVERSAL SEDIMENT TRANSPORT

In order to consider the bottom relief variations during storm, the underwater slope profile was divided into three zones, differing in intensity of bottom deformation: trough zone of the slope close to water edge ; bar zone and seaward bar slope zone.

At relatively slight sea, surface waves are breaking only on the slope close to water edge. Surf and compensating outflow thus formed, induce circulation which transports finer fractions (if present in this zone) toward the trough, in which they precipitate or remain suspended. The slope being composed of coarse sand, water depth does not change significantly at this stage.

As storm develops and wave height increases, their deformation and breaking on the seaward bar slope generates more complex motion of water masses. In these conditions finer particles can be transported seaward, beyond the bar top. The equilibrium profile is developed near the water edge. For a short period sand particles are tracted shoreward by surf. Compensating outflow acts for a longer period, during which finer particles cover a longer distance seaward and are accumulated there. The sediment is differentiated in a particular pattern along the slope. Sea calming causes

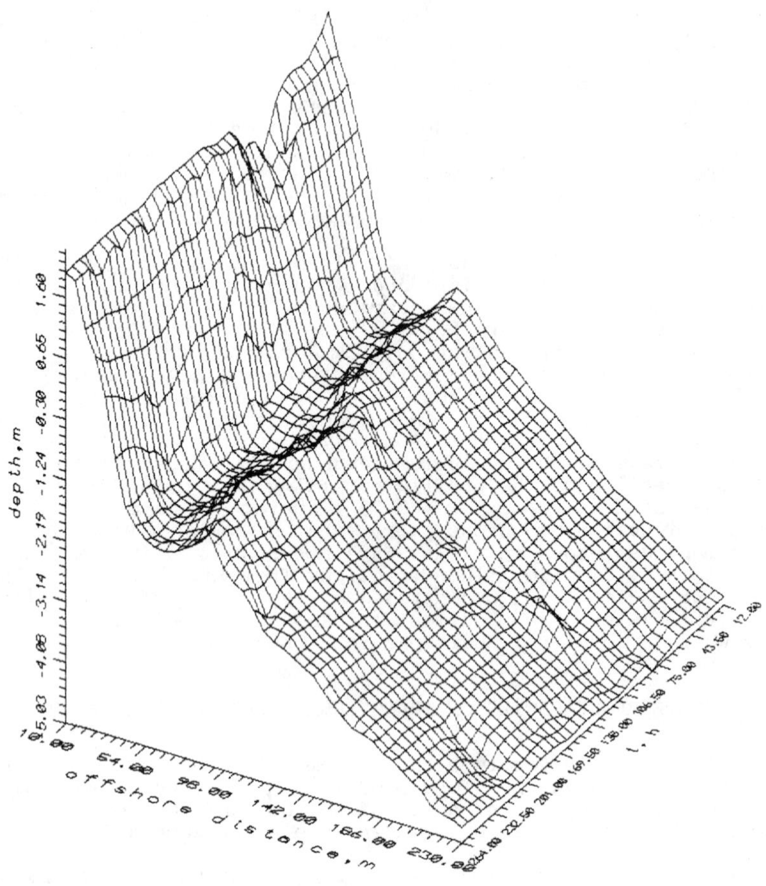

Fig. 9. Variability of underwater coastal relief during storm

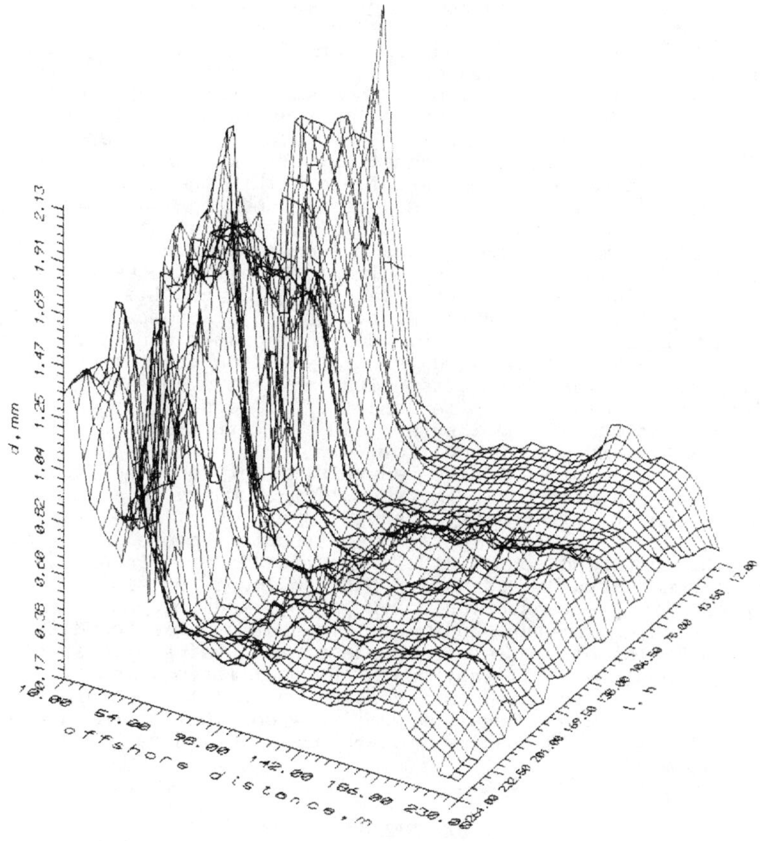

Fig. 10. Variability of mean grain size (d, mm) of bottom sediment during storm

precipitation of suspended particles on near-edge slope, reducing the mean diameter of bed sediment.

The underwater bar zone is characterized by high grading and uniformity of bed sediment. Insignificant growth of **d** value was caused by erosion of shoreward bar slope due to transport of finer fractions on its seaward side. Alteration of bar height and its migration along the profile depend on preferential action of compensating outflow upon the sediment of shoreward bar side, on turbulence within the breaking wave suspending large amounts of sediment, and on asymmetry of wave velocities. Predominance of the latter at the initial stage of storm development causes shoreward migration of the bar due to transport of bed sediment from its seaward side onto a shoreward one. Storm intensification results in wave breaking at the seaward bar slope, which is accompanied with seaward migration of its top, so that sand particles are transported to its seaward side by stronger compensating outflow. When the storm attenuates and wave breaking zone migrates shoreward, this tendency is reversed. Absence of finer fractions around the bar can be explained by their removal by compensating outflow.

The sediment particles behave similarly on the seaward bar slope and within the bar zone. The boundary between these zones can be traced by abrupt drop of finer fraction percentage. Being the function of bottom slope and wave length, the third zone width is dynamic. Under higher waves this zone expands seaward. Bed sediment particles are preferentially transported up the slope and become coarser when reaching the bar. Under certain hydrodynamic conditions the particles of similar coarseness are being transported along the slope until they reach the equilibrium state.

It should be noted as a conclusion that the seaward part of the bar retains the slope profile when the storm abates abruptly, while the sediment coarseness everywhere tends to pre-storm values.

REFERENCES

Folk R.L. Petrology of sedimentary rocks Austin. 1974. 250p.

Kos'yan R.D. About ripple formation and existence under wave action in the coastal zone. Water resources. 1987. N 1. P. 52-60 (in Russian).

Kos'yan R.D.,Pykhov N.V. Hydrogenous sediment shift in the coastal zone. Moscow. Science. 1991. 280p. (in Russian).

Nikolov H.I., Kos'yan R.D. Short term changes of relief and composition of bottom sediments along the profile of submerged coastal slope. Oceanology. Sofia. 1991. N 21. P. 43 - 47 (in Russian).

3-DIMENSIONAL PROCESSES ALONG SANDY BEACHES - DIMENSIONS OF A COASTAL FACILITY

Ida Brøker[1], Hakeem Johnson[1], Erik Asp Hansen[1], Ann Skou[1], Peter Justesen[1], Rolf Deigaard[1]

ABSTRACT: Both in the case of a natural, undisturbed beach and in situations where man-made structures interact with the environment, the idea of large scale facilities for the study of coastal zone processes has been the topic for discussions over the last decades. Up till now a number of large scale wave flumes have been constructed, and useful experiments for the study of pure cross-shore processes have been performed. In the present paper the possibilities of construction of a 3D coastal facility are discussed. By available numerical tools, the required dimensions for achieving transport regimes and hydrodynamic phenomena, in the facility similar to those encountered in nature, are discussed. The conclusions are that the required dimensions are such that the advantages of controllable conditions become too small compared to the efforts to establish and run such a facility.

INTRODUCTION

The recent developments in numerical modelling of coastal morphodynamics have been very rapid. The need for verification or calibration of theories and models is significant. Field experiments play an important role, but the well controlled conditions of laboratory experiments are often preferable. Cross-shore processes and the morphological development of coastal profiles for direct incoming waves have been studied in large wave flumes since the 1950's and many useful results obtained.

With the increasing capabilities of the numerical models there is a need for experimental results from three-dimensional situations. Examples of areas where such experiments would be valuable are:

---

1) Danish Hydraulic Institute, Agern Alle 5, DK 2970 Hørsholm, Denmark

- Experimental verification of coastal morphological models, i.e. the dynamic coupling of wave models, wave, tidal and wind driven current models and sediment transport models used as an engineering tool for the modelling of the morphological evolution for instance around coastal structures.

- Local scour around coastal structures caused by 3-dimensional flows and wave reflection.

- Combined longshore and cross-shore processes, hydrodynamics as well as sediment transport. On open, sandy coasts the cross-shore processes determine the shape of the coastal profile. This shape influences the longshore transport which in many cases is responsible for the overall evolution of the coastal area.

- The formation of rip channels, effects of wave groups and long waves

- Shear waves

In the following the dimensions of a coastal facility are discussed. The requirements of the facility are that the above phenomena must be similar in the facility and in nature and that for instance the natural beach stages described by Lippman and Holman (1990) should be reproducible in the facility. The facility is a large wave basin equipped with wave-makers capable of producing directional, irregular waves.

The aspects discussed cover open versus closed boundaries of the facility, wave generated bed forms, ie. sediment transport regime, cross-shore width and longshore length of the facility. The discussions are quantified by the use of numerical models.

THE NUMERICAL MODELLING TOOLS

The phenomena in question concern waves and wave driven currents. Two models from DHI's modelling complex MIKE 21 are used for the investigations.

Further, the time scales for the morphological evolution in the facility compared to characteristic locations are illustrated by some calculations performed by the aid of DHI's model complex LITPACK.

The applied models are briefly described below.

Wave Model (MIKE21 PMS)

MIKE21 PMS is based on a simplification of the mild slope equations assuming wave propagation in one predominant direction, neglecting diffraction in this direction and back scattering in the opposite direction. Using the approach of Kirby (1986), this model is extended to allow large-angle (up to 60 deg.) propagation to the predominant direction. Furthermore, MIKE21 PMS includes the effect of a frequency spectrum and directional spreading using the method of linear superposition. Breaking is included in the model according to Battjes and Janssen (1978). The results from the model are (stationary) fields of wave parameters, Hrms, Tp, mean wave period of waves and radiation stresses.

Flow Model (MIKE21 HD)

MIKE21 HD solves the vertically integrated equations of conservation of mass and momentum in two horizontal dimensions. MIKE21 HD is a general model for calculation of dynamic flow fields driven by waves, winds and tide taking into account the effect of the Coriolis force. The equations are cast in terms of fluxes and solved using an implicit finite difference technique with variables on a space-staggered rectangular grid. A 'fractioned-step' technique combined with an Alternating Direction Implicit (ADI) algorithm is used to avoid the necessity for iteration. Details about the model can be found in Abbott et al. (1973) and Abbott et al. (1981).

The wave-generated current and wave set-up along the open boundaries of the model area are calculated by assuming steady, uniform conditions in the longshore direction and negligible flow in the cross-shore direction.

In the general situation, the contribution of the waves to flux or set-up is added to the tidal boundary conditions in order to get a combined boundary condition for the simulation.

The increased resistance due to the higher levels of turbulence that exist close to the bed because of the presence of the wave boundary layer in the case of combined wave-current flow is accounted for according to the theory of Fredsøe (1984).

LITPACK

LITPACK is a modelling complex for the calculation of the longshore littoral drift along quasi-uniform sandy beaches and the corresponding coastline evolution. For description of the calculation of the littoral drift, reference is given to Deigaard et al (1986). It is noted that LITPACK and MIKE21 give identical waves, currents and sediment transport in case of uniform conditions along the shore.

OPEN VERSUS CLOSED BASIN

The limited area of a laboratory basin implies that there might be some areas near the boundaries where the waves and flow are influenced by the type of boundary conditions. The first test is made to give indications of the extent of the areas affected by boundary conditions in a wave basin where obliquely incoming waves generate a longshore current on an open coast.

The model set-up is shown in Figure 1 together with the calculated wave field. Figure 2 shows the calculated stationary flow fields in the two extreme cases, 'A', where the up- and downdrift boundaries are closed and 'B', where optimal flow recirculation is applied as boundary conditions. (These calculations are made with grid spacings of 0.5m and 1.0m in the wave and flow models and it is noted that stationary conditions in the flow model are achieved after three hours of wave action). Figure 2A clearly shows that large parts of the flow fields are disturbed by the closed boundaries. The test implies that efforts should be put into construction of recirculation facilities and that both inflow and outflow discharge and their distribution along the boundary should be adjustable.

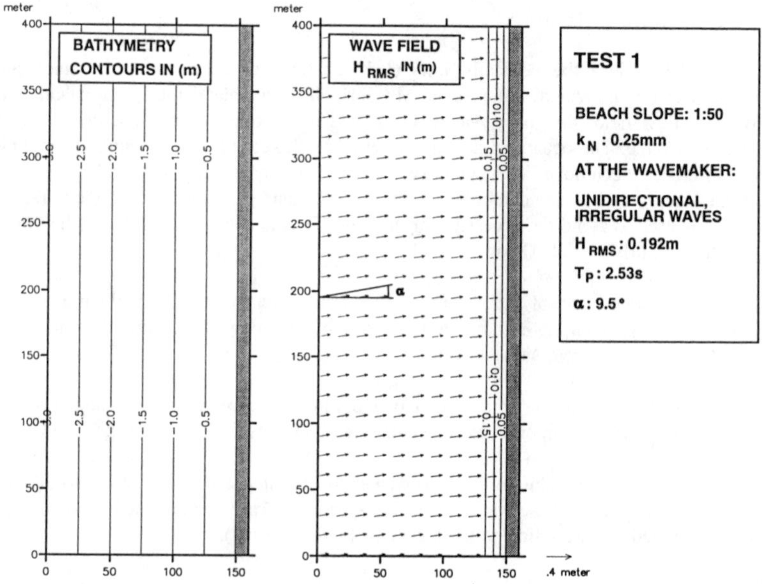

Figure 1. Bathymetry and wave field, test 1 closed and open boundaries.

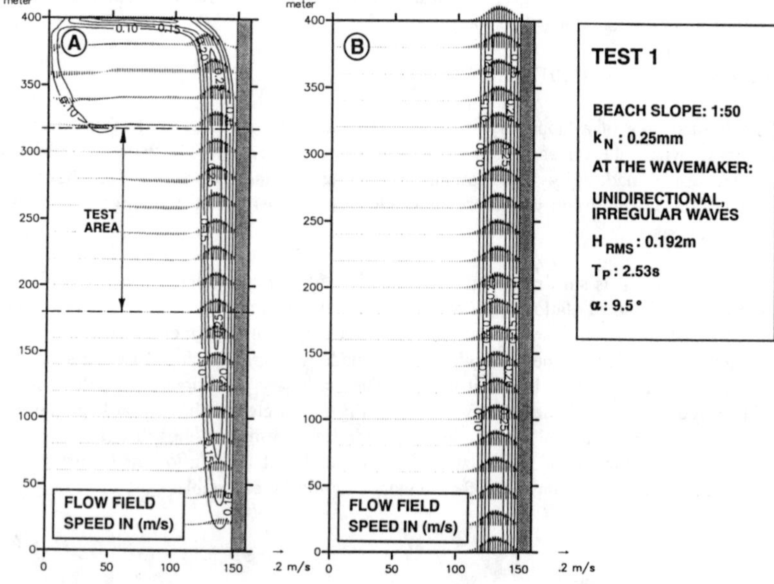

Figure 2. Wave-generated flow patterns, test 1, A: Closed basin B: Open basin

Figure 3 A1 and B1 show the spatial variability of the hydraulic roughness as calculated by the boundary layer model for the combined wave current motion. It appears that the apparent roughness reaches nearly 10cm just inside the breaker line where the current speed is maximum. This roughness, due to the presence of the turbulent wave boundary layer, should be compared to the geometrical roughness which is set to 0.25 mm. The bed is assumed to be plane with an average grain size of the sediment of 0.1 mm in these calculations.

Under prototype conditions, the bed is often plane (ie. without wave ripples) during the storms which are the dominant morphological events. One of the strongest requirements for the size of an experimental facility is often based on the requirement that the bed must also be plane in the experiments. The criterion for the disappearance of wave ripples is normally that the maximum Shields' parameter (dimensionless bed shear stress) shall be larger than about 1.

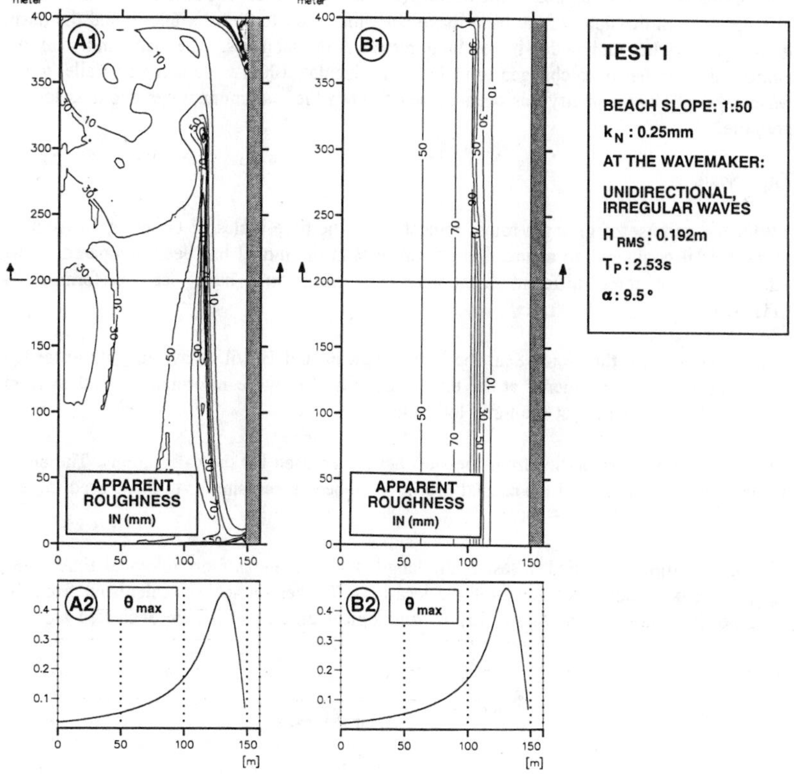

Figure 3. Apparent hydraulic roughness and max. dim. bed shear stress theta, test 1
A1 & A2: Closed basin, B1 & B2: Open basin

The spatial variation of the local maximum bed shear stress over the wave period is calculated in the boundary layer model. Assuming a relative density of the sand of 2.65, the spatial variation of the maximum Shields' parameter is found from the maximum shear stress distribution. The variations along the central sections of the basin are shown in Figure 3 A2 and B2. It appears that the assumption of plane bed is in contradiction with the results. The parts of the wave basin, where sediment transport can take place, will be covered by wave ripples.

SEDIMENT TRANSPORT REGIME/BED FORMS

If the requirement of plane bed conditions in the nearshore zone should be met, and if it is a fixed assumption that the sediment is fine, non-cohesive sand, the wave heights must be increased compared to the above test.

The below Figure 4 shows results from a test where the maximum Shields' parameter is well above one in a wide area in the nearshore zone. This result has required a wave height of 0.4m. The dimensions of the basin have been increased to 250m x 600m, and the basin is assumed to be equipped with optimal recirculation facilities. The bathymetry of the numerical basin has been changed to include a horizontal, 100m wide section parallel to the wavemakers. This geometry has been chosen to keep the maximum water depth as low as possible.

Time Scale

With the purpose of giving a rough indication of the time scales of the coastal evolution in the model compared to nature, the littoral drift in the model has been compared to the littoral drift at three different locations. The calculations have been performed by LITPACK.

The locations are the North Sea, the Mediterranean and Brazil. Wave heights exceeded 2% of the time are considered at the three locations. The wave information is taken from 'Global Wave Statistics' at the three locations.

Typical sediment properties from the sites have been used for the calculations. The angle of the incoming waves and the coastal profile have been taken similar to those used in Test 2, see Figure 4.

For the conditions applied in test 2, the length scale, λ, and **a** morphological time scale, λ_{morph}, have been calculated for the three locations. The length scale is calculated from the wave heights using Froude's law. The morphological time scale is here evaluated as:

$$\lambda_{morph.} = \frac{\lambda^3}{Q_{nature}/Q_{model}} \tag{1}$$

where Q_{nature} and Q_{model} are the total littoral drift in the model and in nature.

The results are given in Table 1.

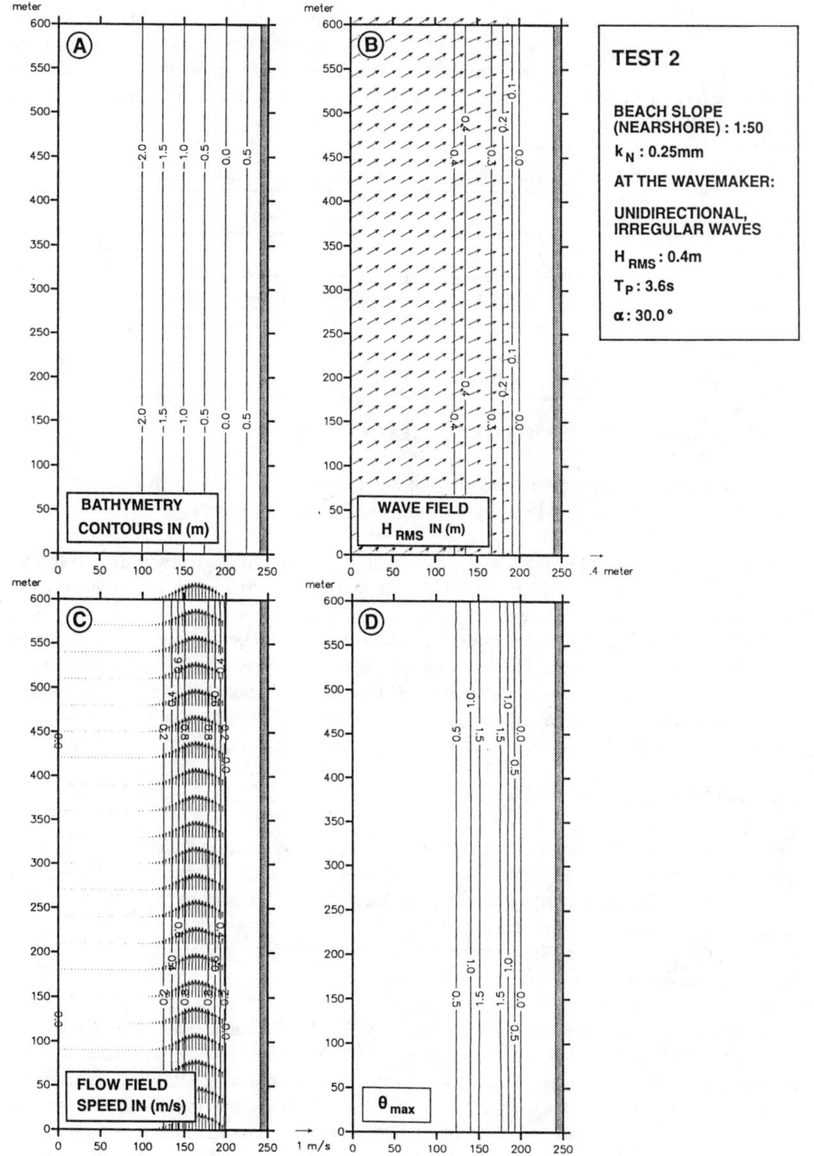

Figure 4. A: Bathymetry, B: Wave field, C: Flow field, D: θ_{max}, test 2

It is emphasised that these simple considerations are made assuming that the coastal profiles are geometrically identical in model and in nature. The results indicate that the morphological time scales vary strongly with the natural grain sizes. In a case with fine sand in nature, a trench across the surf zone will even silt up faster in nature than in the model basin.

Table 1. Estimates of morphological time scales considering littoral drift

	H_{rms}	$d_{50\%}$	λ	Q	λ_{morph}
	m	mm	-	m^3/s	-
Model	0.4	0.1	1	0.353	1
North Sea	4.24	0.1	10.6	1541	0.3
Mediterranean	3.54	0.15	8.8	173	1.4
Brazil	3.61	0.40	9.0	2	125

CROSS-SHORE WIDTH OF THE FACILITY

The large coastal facility must provide the possibility for the study of rip currents, both naturally developed and provoked by structures in the nearshore zone.

If the cross-shore width of the basin is too small, the outer boundary might disturb the rip currents. The formation of rip currents has been studied by introduction of two groynes in the numerical test basin. The hydrodynamic boundary conditions are identical to the ones used for the above test 2, but the groynes block large parts of the surf zone resulting in rip currents on the updrift sides. The bathymetry and the calculated stationary flow field are shown in Figure 5. The results show an undisturbed flow field near the outer boundary indicating that rip currents can be reproduced without boundary effects in the sketched set-up where the width of the entire basin is approximately twice the width of the surf zone.

LENGTH OF THE FACILITY

From the above tests it appears that with perfect control of the inflow and outflow boundaries, steady conditions can be achieved in the numerical basin. In nature, however, it is observed that even with constant forcing, i.e. wave action, on uniform beaches the longshore current does not become steady but develops shear waves as described by Bowen and Holman(1989) and Oltman-Shay, Howd and Birkemeier(1989). The coastal facility should be capable of reproducing such instability phenomena.

The phenomena of shear waves have been studied by the same numerical models as applied to the above tests by Deigaard et al (1994). Figure 6 shows a calculated current field driven by constant wave forcing and with constant inflow and outflow conditions. After some time shear waves appear in the numerical basin but the distance from the updrift, constant boundary is considerable. These numerical tests have been performed for a coastal profile with one bar as indicated in Figure 6 and with wave heights and periods somewhat larger than in the previous Test 3. Assuming the length for evolution of the shear waves can be scaled following Froude's model law, the length for the shear waves to be established is 250m - 500m under the wave conditions used in Test 3.

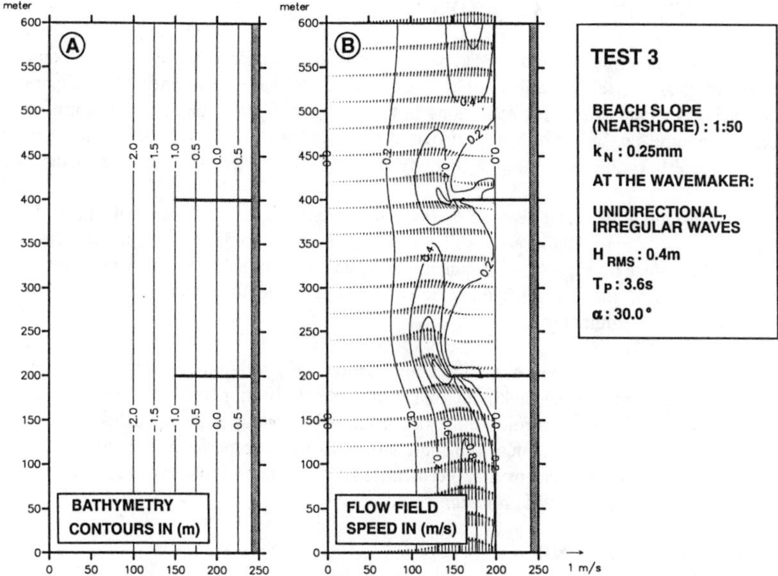

Figure 5. A: Bathymetry, B: Flow field, Test 3

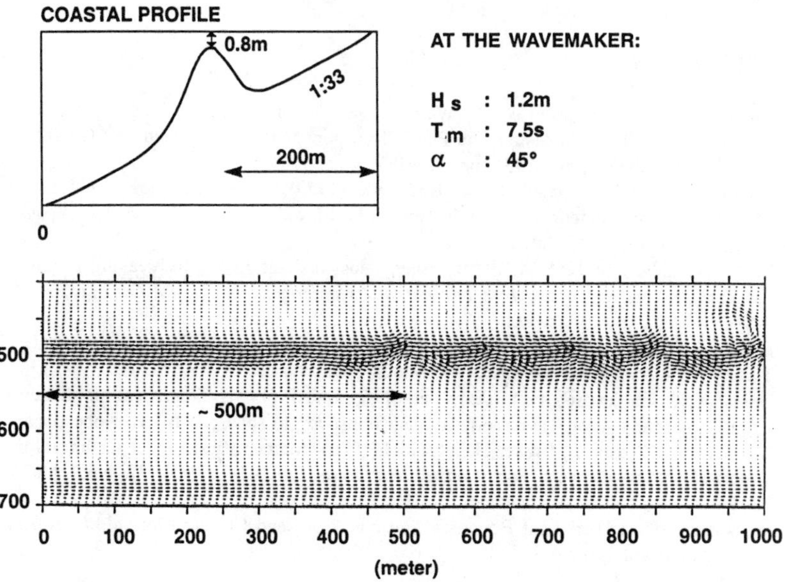

Figure 6. Illustration of modelled shear waves, from Deigaard et al (1994)

COMMENTS AND CONCLUSIONS

The necessary dimensions of a 3D coastal facility with movable bed have been examined by existing numerical modelling tools. The requirements have been partly to be capable of producing sediment transport conditions in the facility similar to those under storm conditions in nature, i.e. sediment transport over a plane bed without wave ripples, and partly to achieve a test section large enough to study rip currents and natural longshore instabilities.

The considerations have lead to horizontal dimensions of the basin of the order of magnitude 250m x 1000m with still water depth of the order of magnitude 2m. The wavemaker must be capable of producing irregular, directional sea with wave heights up to approx. 0.7m, and the basin must be equipped with flexible recirculation systems for both water and sediment.

The costs of such a wavemaker, the complications of operation of such a huge facility, the problems of measuring techniques all lead to the realisation that for now high quality measurements of hydrodynamics must be performed in small scale fixed bed basins, and experimental studies of sediment transport processes must be performed in special laboratory facilities. Data for the overall verifications of coastal morphodynamic model complexes must still be collected in nature.

ACKNOWLEDGEMENT

This work was carried out as part of the G8 coastal Morphodynamics Research Programme. It was funded jointly by the Danish Technical Research Council (STVF) and the Commission of the European Communities, Directorate General for Science, Research and Development under contract no. Mas2-0027.

REFERENCES

Abbott, M.B., A. McCowan & I.R. Warren (1981) 'Numerical Modelling of Free-surface flows that are Two-dimensional in Plan', Transport Models for Inland and Coastal Waters, *Academic Press*, 1981, pp. 222-283.

Abbott, M.B., A. Damsgaard & G.S. Rodenhuis (1973) 'System 21, Jupiter, A Design System for Two-dimensional Nearly Horizontal Flows', *Journal of Hydraulic Research,* Vol. 11, 1973, pp. 1-28.

Battjes J.A., Janssen J.P.F.M.(1978) ' Energy loss and set-up due to breaking in random waves' *Proceedings of the 16th International Conference on Coastal Eng.* pp 569-587, ASCE

Bowen A.J.,Holman R.A. (1989) 'Shear instabilities of the mean longshore current, 1, Theory', *J. Geophysical Research, 94*, 18.023-30

Deigaard R., Christensen E.D., Damgaard S. Fredsøe J. (1994) 'Numerical simulation of finite amplitude shear waves and sediment transport' to appear at ICCE, Kobe, Japan

Deigaard R., Fredsøe J., Brøker Hedegaard I.(1986) 'Mathematical model for littoral drift' *Journal of the Waterway, Port, Coastal and Ocean Eng.* ASCE, Vol 112, No.3, pp 351 - 369.

Fredsøe J. (1984) 'Turbulent Boundary Layer in Wave and Current Motion', *J. Hydraulic Eng.* ASCE vol 110 No 8

Hogben N., Dacanha N.M.C. & Oliver G.F., *Publ. for British Maritime Technology* by Unwin Brothers Ltd., Surrey, 1986.

Kirby, J.T. (1986) 'Rational approximations in the parabolic equation method for water waves', *Coastal Engrg.*, Vol. 10, pp. 355-378

Kirby (1986) Lippmann T.C., Holman R.A. (1990) 'The spacial and Temporal Variability of Sand Bar Morphology', *Journal of Geophysical Research*, vol. 95, No. C7, pp 11,575-11,590, July 15

Oltman-Shay J., Howd, P.A., Birkemeier, W.A. (1989) 'Shear instabilities of the mean longshore current, 2, Field data', *J. Geophysical Research*, 94,18.031-43

VALIDATION OF COASTAL MORPHODYNAMIC MODELS WITH FIELD DATA

Helen M. Wallace and Tim J. Chesher[1]

ABSTRACT: Requirements for validation of coastal morphodynamic models are discussed, with reference to a review of available field data. A validation exercise is then performed, comparing results from two commonly used types of model (coastal profile and coastal area) with berm monitoring data from Silver Strand beach, California. A quantitative comparison is made between model results and data, concentrating on the most important aspects of the prediction. The models reproduce the measurements reasonably well, given the particular assumptions on which each type of model is based. This kind of validation exercise should be repeated after any model improvement, to assess the effects of such improvements on the final prediction and its engineering objectives. It is argued that more such data sets are needed, covering different sites, conditions and predictive objectives.

INTRODUCTION

It is widely acknowledged that a need exists for more extensive validation of coastal numerical models against field data, so that they can be applied with more confidence to practical engineering problems. A recent review of field data on coastal sand transport and morphodynamics (Wallace, 1993) lists over 150 field experiments, 24 of which are reviewed in detail. In spite of the large amount of data available, very few experiments were found to be suitable for validation of the coastal profile and area models currently used for sediment transport and morphodynamic predictions.

Some important model requirements can be identified, that are often lacking in the data:

---

[1]HR Wallingford Ltd, Howbery Park, Wallingford, Oxfordshire, OX10 8BA, UK

1. Continuous measurement of the model boundary conditions, in particular offshore directional wave spectra. However detailed the process measurements within the model boundaries, predictions are based on the assumption that the hydrodynamic driving forces are known throughout the experiment.

2. Accurate, frequent bathymetric surveys. These must be frequent, accurate, and extensive enough to resolve the morphodynamic changes occurring over the time and length scales of interest.

3. Speedy analysis of data and its availability and presentation in a form easily used by modellers.

Results from a comparison of HR Wallingford's COSMOS-2D profile model (Southgate and Nairn, 1993; Nairn and Southgate, 1993) with field data from Silver Strand Beach, USA (Andrassy, 1991) are presented. The comparison is used to illustrate how data satisfying the above requirements may be used to provide valuable test cases for such models, even where detailed process measurements within the model area are not available.

The usefulness of the Silver Strand data for validation of coastal area models is also assessed. Results from HR Wallingford's PISCES model (Chesher et al, 1993) are presented.

THE MODELS

Predictive numerical models of coastal morphodynamics include line models of coastline evolution, cross-shore profile models and coastal area models. Three dimensional and quasi-3D models are at a much earlier stage of development, are expensive to run and as yet are not commonly used for morphodynamic predictions in the coastal zone. The same applies to intra-wave models of sediment transport, which are commonly used at shorter time scales. This paper therefore concentrates on profile and area models which might be applied to the Silver Strand site as part of an engineering study. These models are wave-period averaged and suited for making medium scale predictions (time scales of weeks to months, length scales of a few kilometres).

COSMOS-2D is HR Wallingford's coastal profile model. It is a 2DV model (two dimensional in the vertical and cross-shore horizontal directions) which assumes that the coastline is locally straight, with depth contours parallel to the coast. The physical processes included in the model are:

(i) Wave transformation by refraction (by depth variations and currents), shoaling, Doppler shifting, bottom friction and wave breaking.

(ii) Wave set-up and driving forces for wave-induced currents, determined from values of wave radiation-stress gradients.

(iii) Longshore currents from pressure-driven tidal forces and wave-radiation stress forces, and the interaction between the two types of current.

(iv) Cross-shore undertow velocities using a three layer model of the vertical distribution of cross-shore currents.

(v) Cross-shore and longshore sediment transport rates using an "energetics" approach.

(vi) Seabed level changes due to cross-shore sediment transport.

PISCES is HR Wallingford's coastal area model. It is a 2DH model, which can deal with general depth contours but doesn't calculate the vertical flow structure. It includes:

(i) Wave transformation by refraction, diffraction, shoaling, bottom friction and wave breaking.

(ii) Wave set-up and driving forces for wave-induced currents, determined from values of wave radiation-stress gradients.

(iii) Depth-integrated currents from pressure-driven tidal forces and wave-radiation stress forces, by solution of the depth-averaged shallow water equations.

(iv) Sediment transport rates by an advection-diffusion method, including a wave-stirring enhancement.

(v) Seabed level changes.

Descriptions of COSMOS-2D and PISCES in comparison with other coastal models of the same type are given in Brøker Hedegaard et al (1992) and De Vriend et al (1993), along with intercomparisons of some results from these models.

MODEL VALIDATION REQUIREMENTS

Proper validation of any morphodynamic model of the coastal zone requires that the boundary conditions (required as model inputs) are known, along with the final bed configuration that we are trying to predict. This leads to the following set of *minimum* measurements.

1. The wave height, period and direction offshore (beyond the breaker point) for the full duration of the experiment.

2. Storm surge levels, tidal elevations and velocities at the model boundaries.

3. Sediment grain size and bed state (for friction).

4. Initial and final bed levels and mean water level.

These minimum requirements allow a quantitative evaluation of model performance, even in the absence of more detailed process measurements within the area of interest. However, they are obviously not sufficient to identify the parts of the model that require improvement.

In addition, morphodynamic models currently in use are often restricted to certain sites, because of the simplifications and assumptions they contain. Table 1 summarises the site requirements for COSMOS-2D and PISCES. For simplicity we also restrict ourselves to sandy sites.

Table 1 Site Requirements for Model Validation

Profile Model	Area Model
Parallel depth contours	General depth contours
Large cross-shore transport rates and changes	Dominated by longshore transport
No structures that destroy longshore uniformity	Structures are possible
Measurements on one or more cross-shore profiles	Spatial measurements on over model area
Fine enough grid to resolve longshore bars (a few metres spacing)	Medium grid
Measurable bed evolution over timescale of experiment	Measurable bed evolution over timescale of experiment

SILVER STRAND BEACH DATA

In December 1988, material dredged from the entrance channel to San Diego harbour was placed off Silver Strand beach in the form of a rectangular berm, approximately 360m alongshore and 180m across shore, with an average relief of about 2m (see Andrassy, 1991, Larson and Kraus, 1992). Nine bathymetric surveys were carried out along seven profile lines at intervals between one and three months. Directional wave data was recorded at three hour intervals between January and May 1989, covering four survey dates. Figure 1 shows the profile changes near the centre of the berm during the period of wave measurement. The data shows onshore movement and flattening of the berm under the action of fairly low, long period waves (H_s < 1m 93% of the time, with approximately half the peak periods between 13 and 15 seconds). The median grain size of the surveyed area was 0.25mm, with the berm sand being slightly finer at 0.20mm.

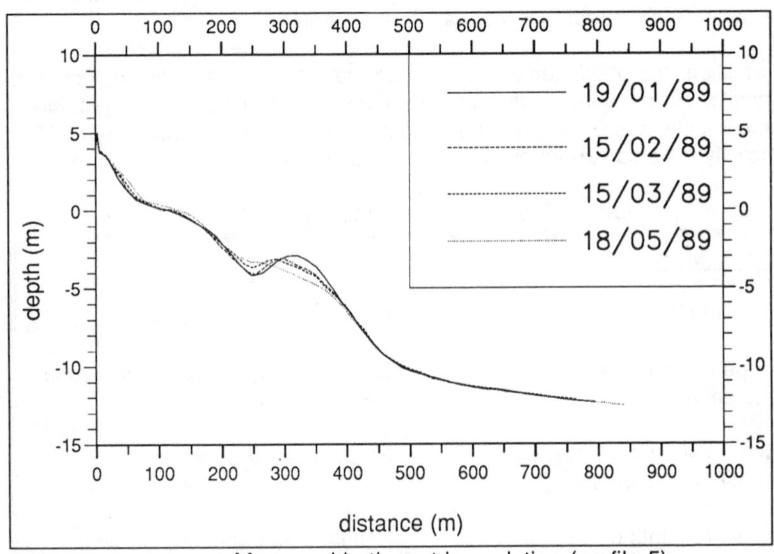

Figure 1 Measured bathymetric evolution (profile 5)

This data satisfies most of the COSMOS-2D validation requirements, although water level variations were not measured. For PISCES validation, it is important to include tidal elevations and currents (which give longshore transport rates) so these boundary conditions must be predicted from tide tables. A depth integrated model such as PISCES cannot reproduce cross-shore changes such as the observed onshore movement of the berm. At Silver

Strand, the assumption of longshore uniformity required by COSMOS-2D seems to be better satisfied than the PISCES requirement that the bed changes be dominated by longshore transport. However, it is still useful to compare initial deposition and erosion rates from PISCES with observed longshore transport rates (calculated from the loss or gain of sand along each profile).

COSMOS-2D VALIDATION

COSMOS-2D is commonly used to predict erosion due to a single storm, hence validation exercises have concentrated on this aspect. The Silver Strand data is thus very valuable as a much longer term data set, showing onshore movement.

Figure 2 shows the predicted evolution of the profile near the centre of the berm, assuming a tidal range of 1.3m. Comparison with Figure 1 shows that the model has quite successfully predicted the berm behaviour. The details of profile development on the upper part of the beach are not so consistent with the data. In the model a bar forms where the waves break, a feature absent from the measurements.

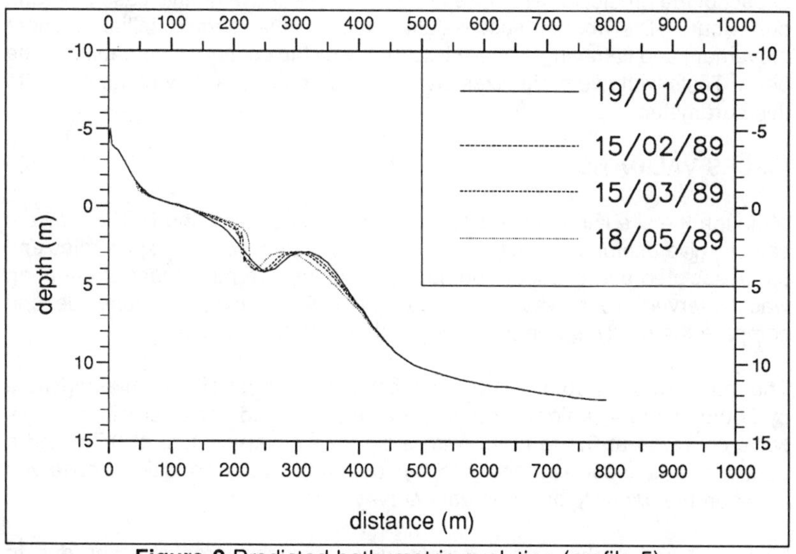

Figure 2 Predicted bathymetric evolution (profile 5)

Table 2 shows comparisons of the measured and predicted berm position and its height and volume relative to the equilibrium profile suggested by Larson and Kraus (1992).

Table 2 Measured and Predicted Berm Parameters

	Measured	Predicted (D_{50}=0.20 mm)	Predicted (D_{50}=0.25 mm)
Date	Hght Rel (m) Area Posn	Hght Rel (m) Area Posn	Hght Rel (m) Area Posn
19/01/89	1.00 1.00 326	1.00 1.00 326	1.00 1.00 326
15/02/89	0.82 0.91 324	0.94 1.01 320	0.95 1.01 321
15/03/89	0.77 0.95 319	0.89 0.84 314	0.91 0.85 328
18/05/89	0.61 0.85 311	0.76 1.03 295	0.82 0.79 320

The heights and volumes are given relative to those on the January survey date and the position is that of the centre of mass. This kind of calculation allows a quantitative comparison of model and data, concentrating on the aspect of the prediction which we are most interested in (in this case the berm behaviour). The model successfully predicts the berm's initial onshore movement and flattening, but predicts less well the continued spreading of the berm. Tests with two grain sizes give a measure of sensitivity of the model to this parameter.

PISCES VALIDATION

Although the site is dominated by cross-shore processes, the PISCES model can still give useful additional information about longshore transport rates and the associated bed changes. Some slight alongshore movement of the berm was observed, so it was decided to run PISCES over a single tide and compare the predicted berm erosion with the observations.

The model was set up on a 10m grid, covering an area of 1240m alongshore by 800m offshore. A tide of range 1.3m was imposed, together with a single wave condition at 37° to north, with a significant wave height of 0.7m and a period of 14s - corresponding to the dominant wave conditions observed between the January and February survey dates.

Figures 3 and 4 show the resultant wave field, the large stresses due to breaking occurring shorewards of the berm.

Figures 5 and 6 show the velocity field at peak flood and peak ebb.

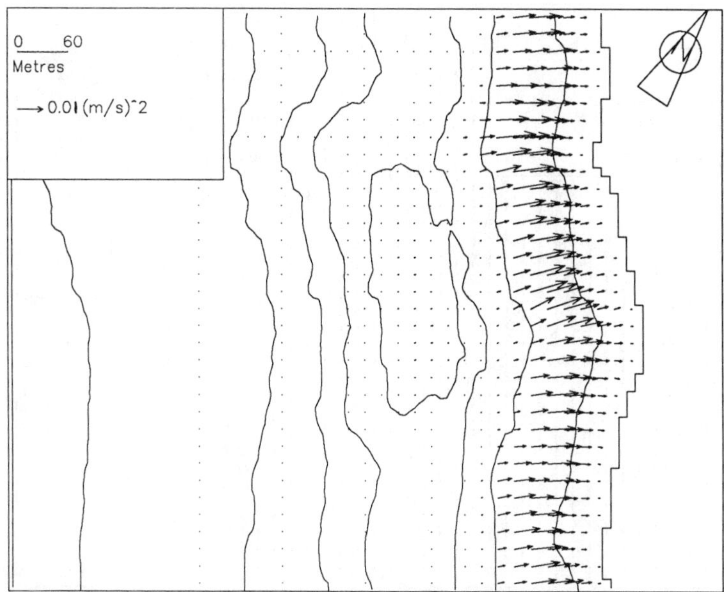

Figure 3 Predicted wave breaking stresses (waves at 37°N)

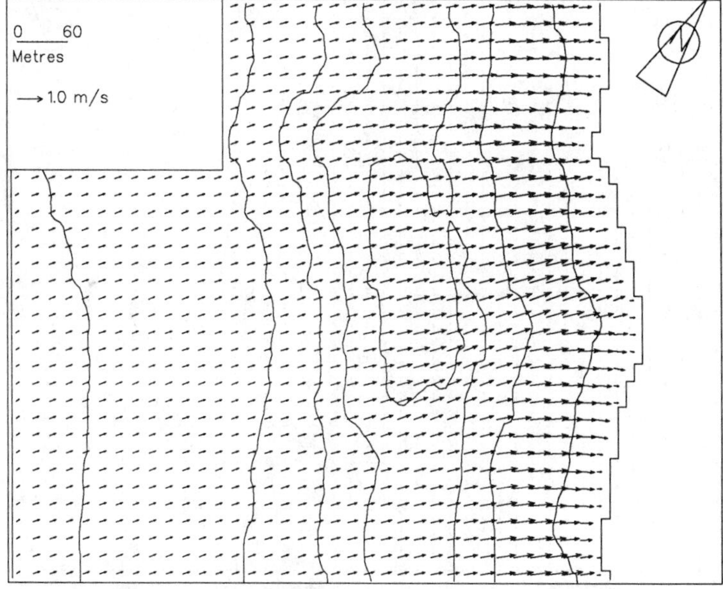

Figure 4 Predicted wave orbital velocities (waves at 37°N)

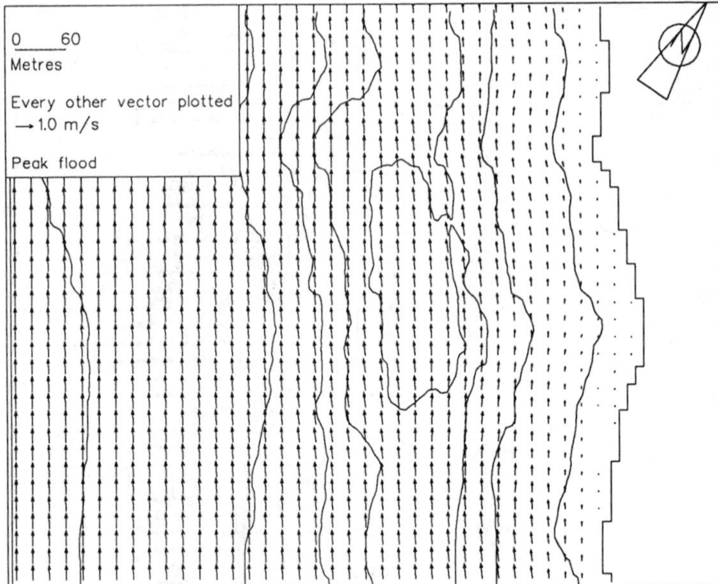

Figure 5 Predicted velocities at peak flood (waves at 37°N)

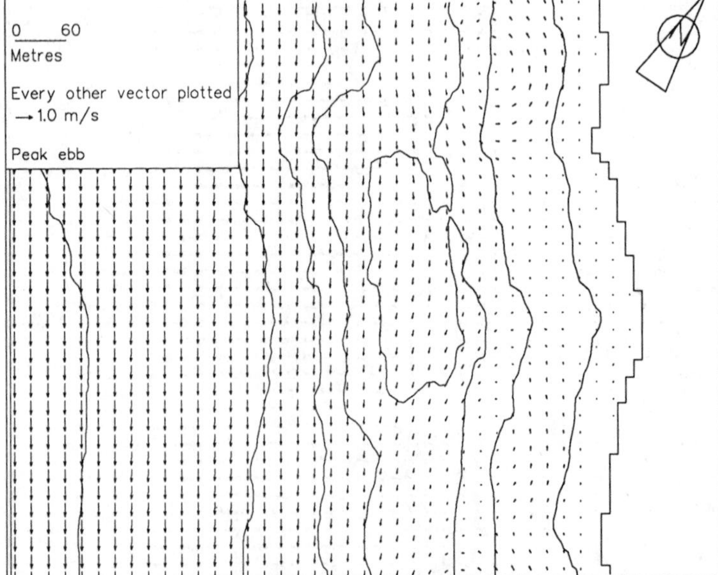

Figure 6 Predicted velocities at peak ebb (waves at 37°N)

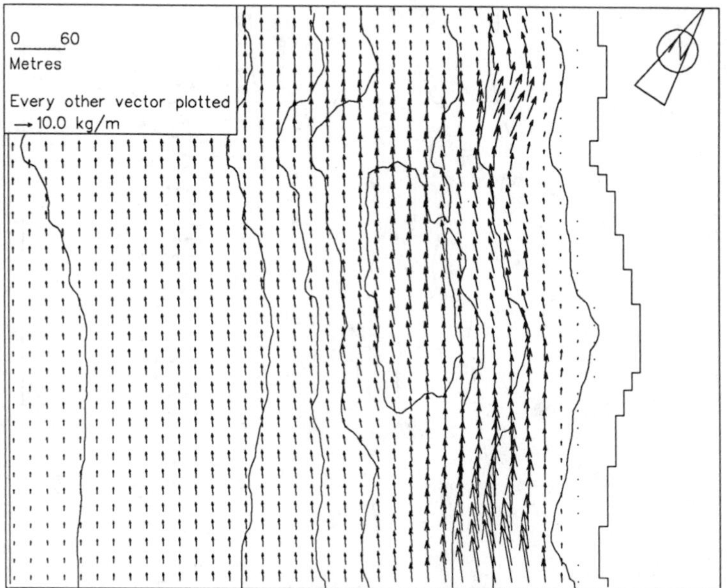

Figure 7 Predicted residual sand transport field over a tide (waves at 37°N)

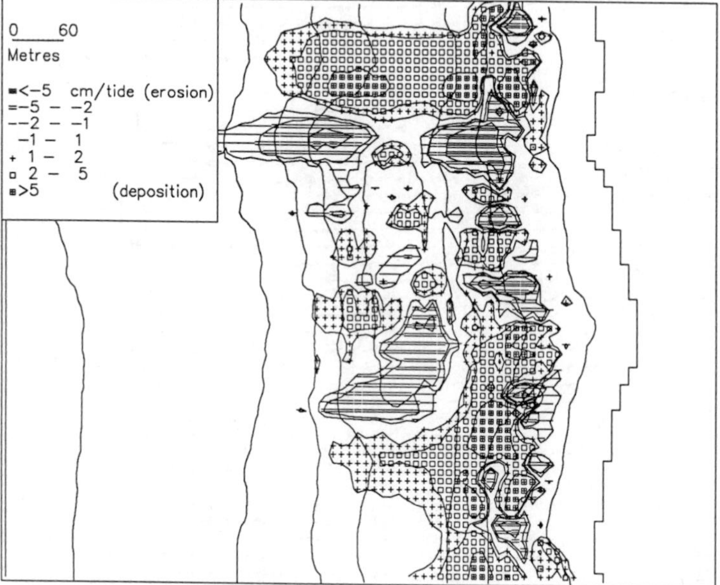

Figure 8 Predicted bed level changes over a tide (waves at 37°N)

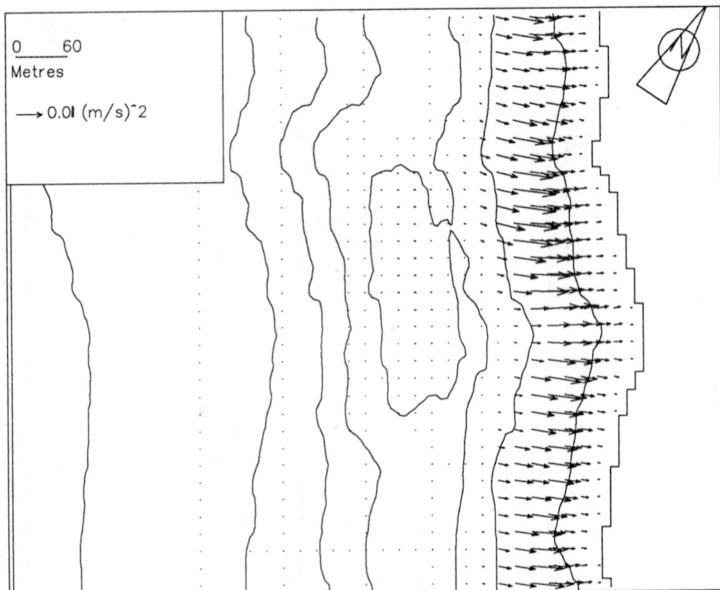

Figure 9 Predicted wave breaking stresses (waves at 77°N)

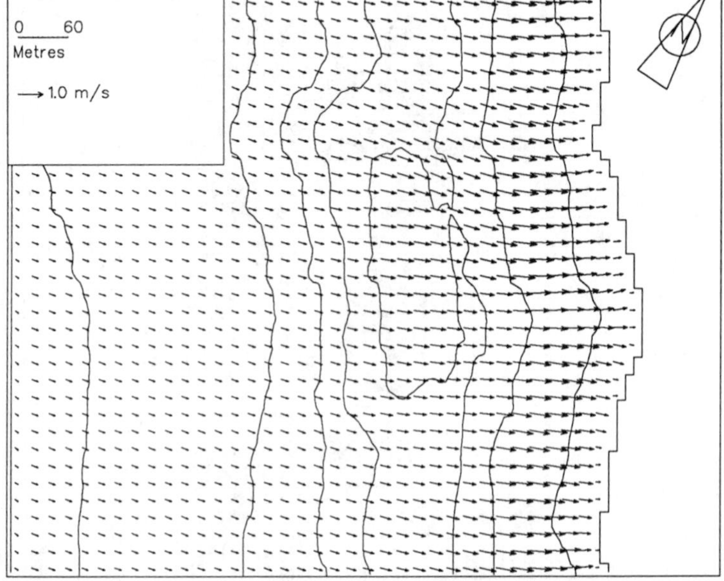

Figure 10 Predicted wave orbital velocities (waves at 77°N)

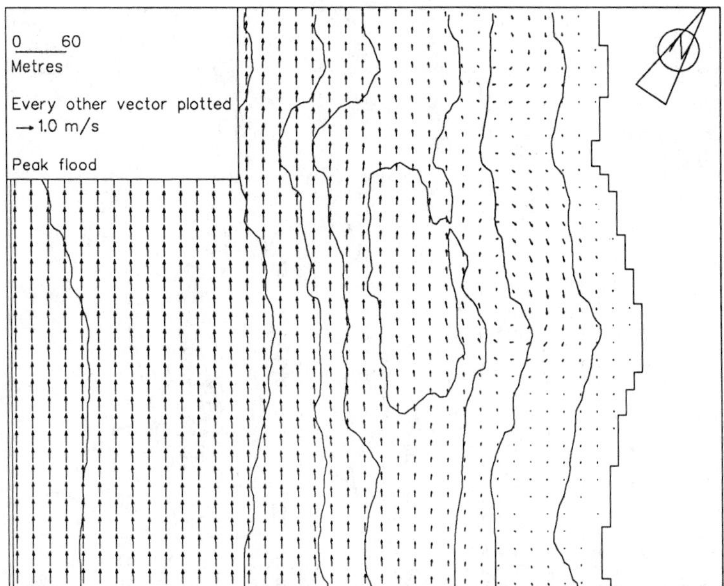

Figure 11 Predicted velocities at peak flood (waves at 77°N)

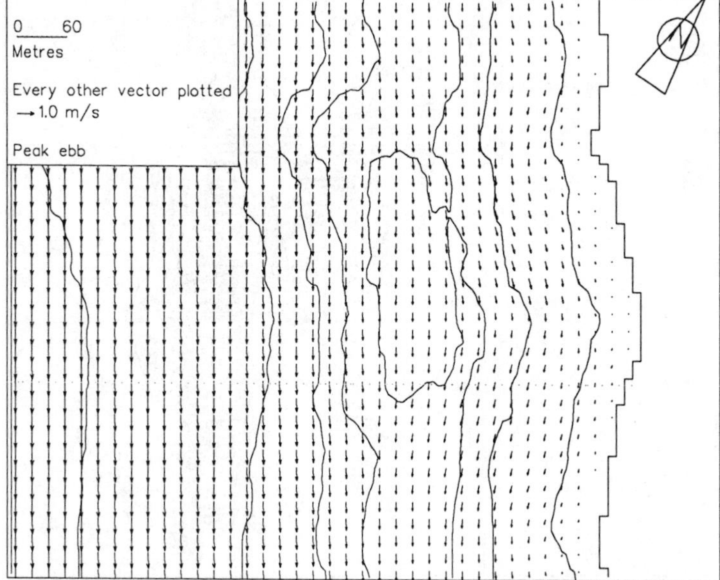

Figure 12 Predicted velocities at peak ebb (waves at 77°N)

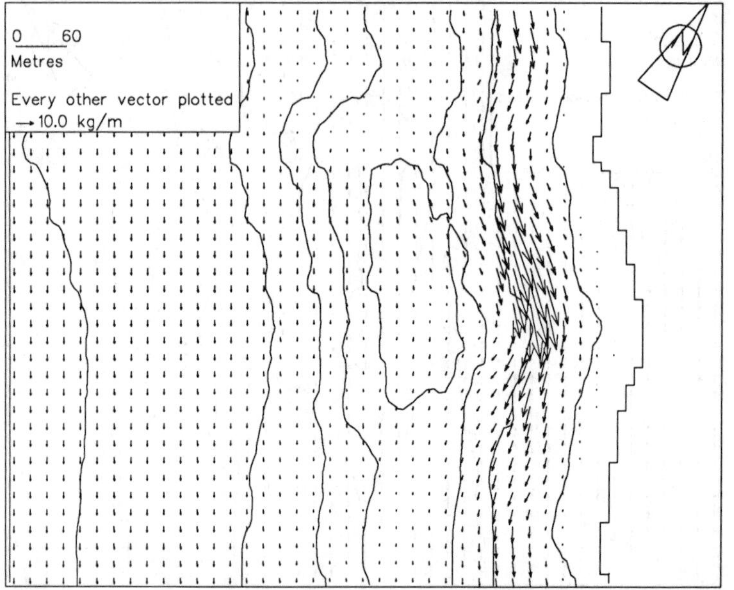

Figure 13 Predicted residual sand transport over a tide (waves at 77°N)

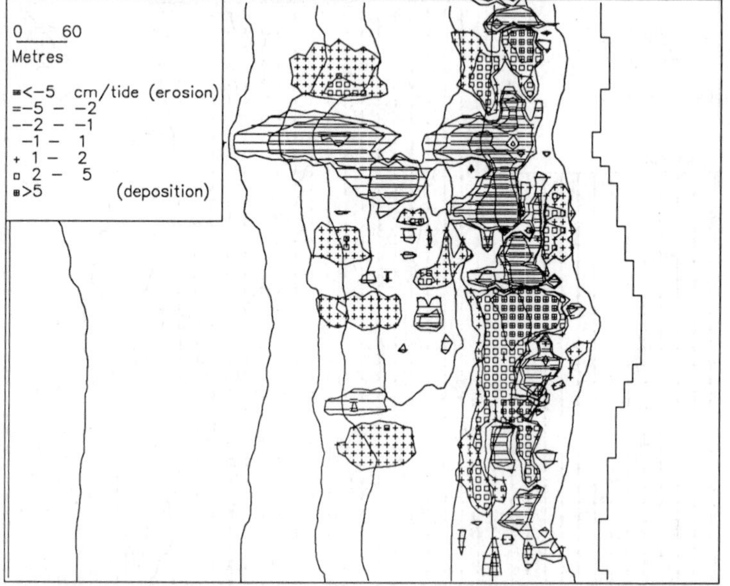

Figure 14 Predicted bed level changes over a tide (waves at 77°N)

Figures 7 and 8 show the residual sand transport field and resulting bed changes over the tide. Gradients in transport rate over the berm are small, but result in some erosion at the southern end of the berm, equivalent to a sand loss of roughly 300m^3 of sand per tide under these wave conditions. This corresponds well with the measured sand losses over the berm between the January and February surveys - a volume of 270m^3 per tide.

Figures 9 to 14 show the equivalent results for a wave at 77° N, for comparison.

CONCLUSIONS

The Silver Strand data set satisfies the minimum requirements for profile model validation and is a very useful data set for long-term accretive conditions. It is distinguished from many similar, less useful data sets by the three requirements outlined in the introduction, and the relative simplicity of the site.

The profile model validation exercise has given confidence that we can predict onshore movement and flattening of the berm over a 5 month timescale, but has shown some quantitative differences between the model predictions and the real world. More detailed process measurements from different sources are required to improve the model. Following these improvements the model may be compared again with the Silver Strand data and the success of the improvements evaluated.

The data is less useful for depth integrated area model validation, since such models cannot reproduce the cross-shore processes dominant at this site. However, longshore processes are well represented and the associated initial bed changes agree well with a crude comparison with the data. The description of the berm evolution is therefore enhanced by the combined use of both the PISCES and COSMOS-2D models.

There is a need for more data to be collected with model validation requirements in mind, on relatively simple sites where the basic model assumptions are satisfied as far as possible. There is also obviously a need to extend the models to real, complex sites, particularly by including 3D effects, and to improve our understanding of the important physical processes. Field and prototype-scale laboratory data have important roles to play, the former particularly in investigating 3D effects, and the latter in detailed process measurements.

ACKNOWLEDGEMENTS

The authors acknowledge the help of Darren Price in producing the PISCES model results and the advice of Howard Southgate.

Separate parts of this work were funded by the UK Ministry of Agriculture, Fisheries and Food, and by the Commission of the European Communities Directorate General for Science, Research and Development, under contract no MAS2-CT92-0027 as part of the G8 Coastal Morphodynamics Research Programme.

REFERENCES

Andrassy, CJ, (1991), 'Monitoring of a nearshore disposal mound at Silver Strand State Park', Proc. Coastal Sediments '91, Seattle, Washington, June 1991, Vol. 1, 1970-1984, ASCE.

Brøker Hedegaard, I, Roelvink, JA, Southgate, HN, Pechon, P, Nicholson, J, Hamm, L, (1992) 'Intercomparison of coastal profile models', 23rd International Conference on Coastal Engineering, Venice, Italy, October 1992.

Chesher, TJ, Wallace, HM, Meadowcroft, IC, Southgate, HN, (1993), 'PISCES: A morphodynamic coastal area model. First annual report', HR Wallingford Report SR 337.

De Vriend, HJ, Zyserman, J, Nicholson, J, Roelvink, JA, Pechon, P, Southgate, HN, (1993), 'Medium-term 2DH coastal area modelling', Coastal Engineering, **21**, 193-224.

Larson, M, Kraus, NC, (1992), 'Analysis of cross-shore movement of natural longshore bars and material placed to create longshore bars', US Army Corps of Engineers, Technical Report DRP-92-5, September 1992.

Southgate, HN, Nairn, RB, (1993), 'Deterministic profile modelling of nearshore processes. Part 1. Waves and currents', Coastal Engineering, **19**, 27-56.

Nairn, RB, Southgate, HN, (1993), 'Deterministic profile modelling of nearshore processes. Part 2. Sediment transport and beach profile development.', Coastal Engineering, **19**, 57-96.

Wallace, HM, (1993),'Coastal sand transport and morphodynamics: A review of field data.', HR Wallingford Report SR 355 (Issue A), April 1993.

FIELD AND PROTOTYPE MEASUREMENTS FOR NUMERICAL COASTAL MODELLING

Hans-H. Dette[1], M.ASCE, Jürgen Newe[2] and Yongjun Wu[2]

ABSTRACT: Various data sets on dune erosion and beach profile changes have been obtained from prototype experiments in the Large Wave Flume in Hannover and from field measurements carried out at the high energy coast of the Island of Sylt/North Sea. Based upon these data aspects like reliability and comprehensiveness of data and their value for validation of numerical models is discussed.

INTRODUCTION

A large number of numerical models exists for the simulation of various coastal morphological processes. Field or prototype data are frequently used for calibration of the numerical model. Progress in the development of such numerical models, however, depends on the availability of reliable data sets. In order to model coastal morphology and to verify the results, not only the data of bed evolution but also simultaneously recorded data about waves, currents and sediment transport are required.

The Prototype experiments have been carried out in the Large Wave Flume (LWF) in Hannover, which is a joint Central Research Facility for coastal engineering studies by both the University of Hannover and by the Technical University of Braunschweig in Germany. The Large Wave Flume has a length of about 300 m, a depth of 7 m and a width of 5 m, in which prototype experiments in coastal engineering can be carried out with wave heights up to 2.0 m.

---

1) Academic Director 2) Research Assistant, Department of Hydrodynamics and Coastal Engineering, Leichtweiss-Institute for Hydraulics, Technical University Braunschweig, Beetovenstr. 51a, 38106 Braunschweig, Germany.

Abundant data sets have been produced from the Large Wave Flume, which can be used for developing numerical beach evolution models. Four data sets (1986/87, 1990, 1991 and 1993) will be briefly introduced in the present considerations. Besides a comparison of results between above prototype measurements and some numerical evolution models was done.

Field measurements have been carried out at the west coast of the Island of Sylt/North Sea between 1986 and 1990. More than 7000 measurements of waves at intervals of 4 hours have been recorded, so that data for a great variety of boundary conditions from calm sea up to extreme storm surge with significant wave heights over 5 m became available for further studies. On this behalf a numerical simulation of dune erosion is compared with field data from Sylt as further example.

PROTOTYPE EXPERIMENTS

Coastal processes ideally should be investigated in the field itself. Since the severe storm conditions only do occur occasionally in the nature, so it would take long time to collect representative data sets. All these disadvantages are avoided when a prototype facility can be used for testing all imaginable constellations of interactions between the sea parameters and the coastal morphology.

Four test data sets are presented here covering the period of 1986 to 1993. The sand size and the initial beach shapes are given in Table 1. The wave parameters and water depth are tabulated in Table 2. For the monochromatic wave tests, the wave steepness calculated by application of linear wave theory is also provided in this table for reference. The test duration and the types of the random waves in each experiment are summarized in Table 3. The initial profiles for the four test sets are shown in Figure 1.

In 1986/87, the purpose of the experiments was concentrated on the study of dune erosion under regular and irregular waves. The same initial profile was exposed at first to regular waves and afterwards to irregular waves with wave height (H_o or H_s) of 1.5 m.

In 1990 and 1991, the main concern of the experiments was the investigation of energy dissipation and the mode of the sediment transport in case of an equilibrium profile.

In 1993, experiments were carried out in the Large Wave Flume to simulate erosional profile evolution under short waves and accretional profile evolution under long waves. Tests were carried at constant levels and with varying tide.

Table 1. Grain size and initial beach slope used in prototype experiments

Tests	Median grain size D_{50} (mm)	Initial beach slope	Slope of dune area
1986/87	0.33	1:20	1:4
1990	0.22	in equilibrium form, $h=A\ x^{2/3}$, $A\approx0.07$	-
1991	0.33	in equilibrium form, $h=A\ x^{2/3}$, $A\approx0.10$	-
1993	0.22	1:30	1:6

Table 2. Initial wave parameters in 1986/87, 1990/91 and 1993 tests*

Tests	Waves	T_o or T_p (sec.)	H_1 or H_s (m)	h_1 (m)	H_o/L_o
1986/87	M+S	6	1.5	5.0	0.028
1990	M+S	4-6	0.7-0.8	2.0	0.0065-0.029
1991	M+S	4-8	0.9-1.0	2.5	0.0085-0.0386
1993	M	5-10	1.2	4.5	0.0068-0.0325
	S	5-10	1.2	4.5	
	S+t	5-10	1.1	4.0±0.5	

* M=monochromatic waves, S=wave spectrum, T_o=wave period, T_p=peak period, H_s=significant wave height, t=tidal effect, H_1 and h_1 are wave height and water depth in front of the wave generator, respectively.

Table 3. Test duration of experiments in 1986/87, 1990, 1991 and 1993

Tests	With regular waves (hour)	With random waves (hour)
1986/87	4.3	9.8
1990	9.0	9.0
1991	10.6	9.6
1993	40.08	23.5 (no tide) 84.0 (with tide)

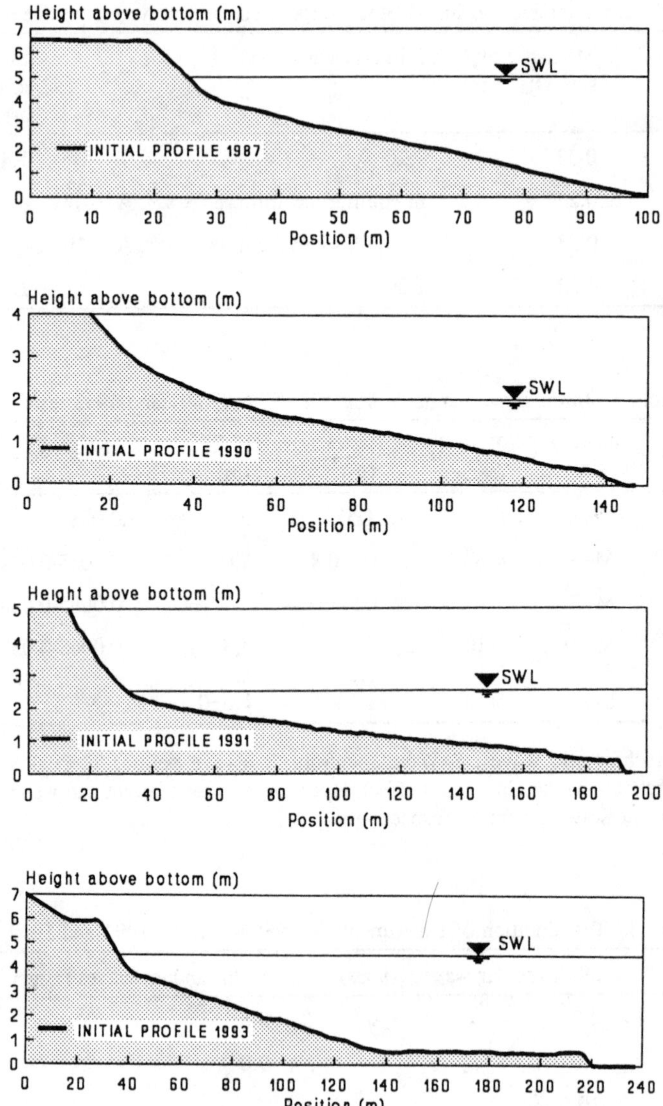

Figure 1. Initial profile in test-1986/87, test-1990, test-1991 and test-1993.

The onshore and offshore movement of the bar position during the test-1993 is shown in the Figure 2 to 4.

Figure 2 shows the beach profile change during the first test-phase in 1993, the initiation of the bar under regular waves within 17 hours of test duration, the onshore movement of the bar under longer waves within about 16 hours, and the offshore movement of the bar under short waves within about 7 hours. The range of the bar movement is about 10 m.

In the second test phase, random waves were run on the preceding beach profile, the offshore movement of the bar under short waves within 13 hours and the onshore movement of the bar under longer waves within 10.5 hours are shown in Figure 3. Under action of longer waves, the form of the breaking bar was significantly changed.

Finally, in the third test phase, the beach profile from last test-phase was exposed to random wave with tidal simulation (12 hours, 1 m tidal range). The offshore movement of the bar with short waves and the onshore movement of the bar with longer waves are shown in Figure 4. Clearly the bar profile was smoothed and the final profile approached quasi-equilibrium shape.

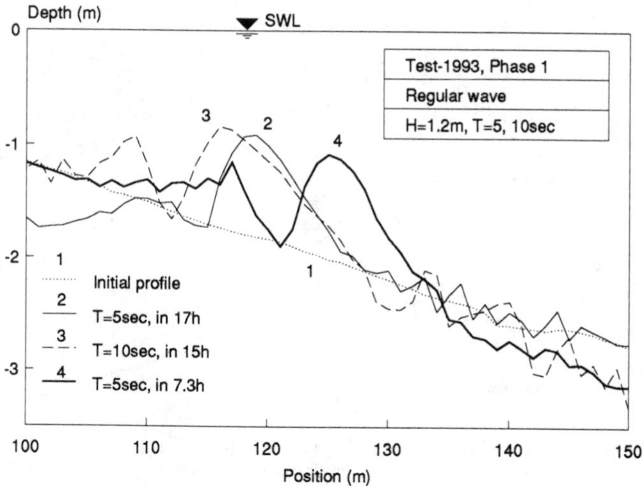

Figure 2. Details of offshore and onshore movement of the bar under short and long regular waves in the 1993 tests (with shifted x coordinate).

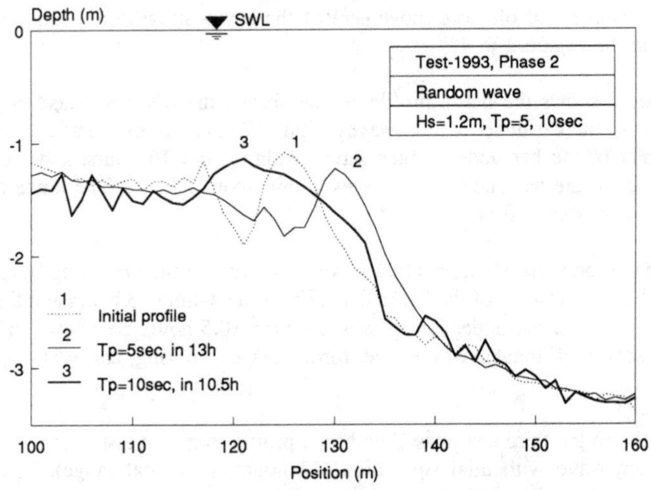

Figure 3. Details of profile development under random waves in the 1993 tests (with shifted x coordinate)

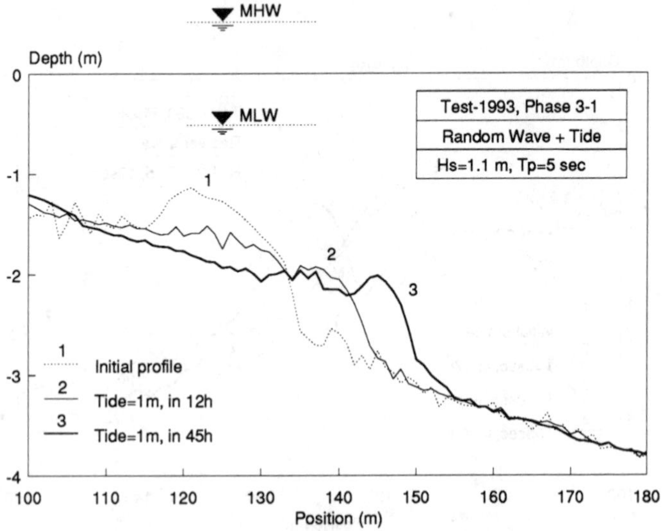

Figure 4. Details of profile development in the 1993 tests (test duration = 45 hours, tide = 1 m, with shifted x coordinate)

COMPARISON OF PROFILE MODELLING WITH PROTOTYPE DATA

The experimental results from LWF-1986 test series (regular waves) were compared with numerical results from six numerical models, which is shown in Figure 5. Generally it can be concluded that all models are able to predict the breaker position and bar-trough profile well, whereas dune erosion and erosion in the bar-trough area are less characterized by most of the models.

To simulate the beach profile evolution under random waves is still one of the major problems in morphological models. An own numerical model taking irregular wave into account was developed based upon various existing numerical models for regular waves. In order to couple the beach profile model with irregular waves three approaches are theoretically possible:

1. Time series transformation. The time series can be achieved from: a. from real sea; b. simulated from a given spectrum; c. simulated from a given distribution. Use of real time series is more realistic but very time consuming. In other way, the beach profile response is not significant with individual wave.

2. Spectral transformation. This approach will have problem in surf zone.

3. "Parametric" transformation of wave height distribution. In this case the time averaged wave deformation is taken into account for the calculation of beach profile change.

Since profile response is expected to be sensitive to the wave sequence as well as to the magnitude of the time step, approach c was at first chosen. On this behalf, a quasi-time series will be produced through a random-procedure and the wave heights are Rayleigh-distributed. In each time step the beach profile change will be calculated with regular wave.

A comparison of measured (LWF-1987) and simulated beach profile evolution is shown in Figure 6. Both the form of the profile and the volume loss along the profile is properly simulated.

The comparison of numerical results with the data from Large Wave Flume 1993 gives different issues. In the erosional cases, the profile evolution was quite well reproduced. But in the accretional cases, the onshore sediment transport was not taken placed in the numerical simulation with the 10 sec waves. How to simulate the beach profile evolution under accretional cases is still a problem to be solved.

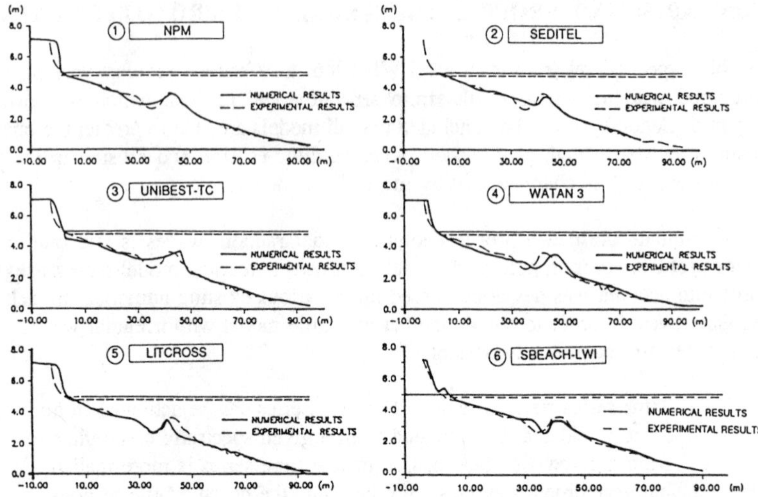

Figure 5. Comparison measured and calculated beach profiles evolution (model no. 1 to 5, HEDEGAARD et al., 1992; no. 6 is a modified SBEACH model elaborated by the authors)

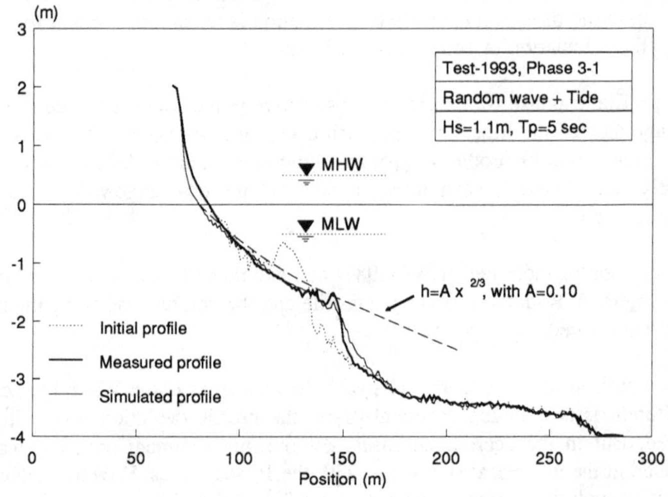

Figure 6. Comparison of measured (LWF-1993) and numerical simulated beach profile evolution.

COMPARISON OF DUNE EROSION MODELLING WITH FIELD DATA

The high energy coast of the Island Sylt/North Sea which is about 40 km long is characterized by sandy beaches in front of dunes and cliffs (up to 25 m high). After 1950 with increased storm surge frequencies the recession of dunes and cliffs has nearly doubled from an average yearly rate of 0.7 m to 1.5 m/year. Beach nourishments are carried out regularly since the early 80ies in order to stop or at least to retard the recession. Profile data from the dunes and cliffs at 500 m spacings along the coast are available since 1970. Together with the nourishments regular profile measurements are carried out nearly once in the year and additionally after major storm surges. In order to support the planning and management of dune protection a numerical dune erosion model for practical engineering application is desirable. The KRIEBEL dune erosion model EDUNE (KRIEBEL, 1990) proved the best agreement between measured and predicted erosion rates above MHW level. For a selected profile (figure 7) recession was investigated for two cases, recession of the nourished

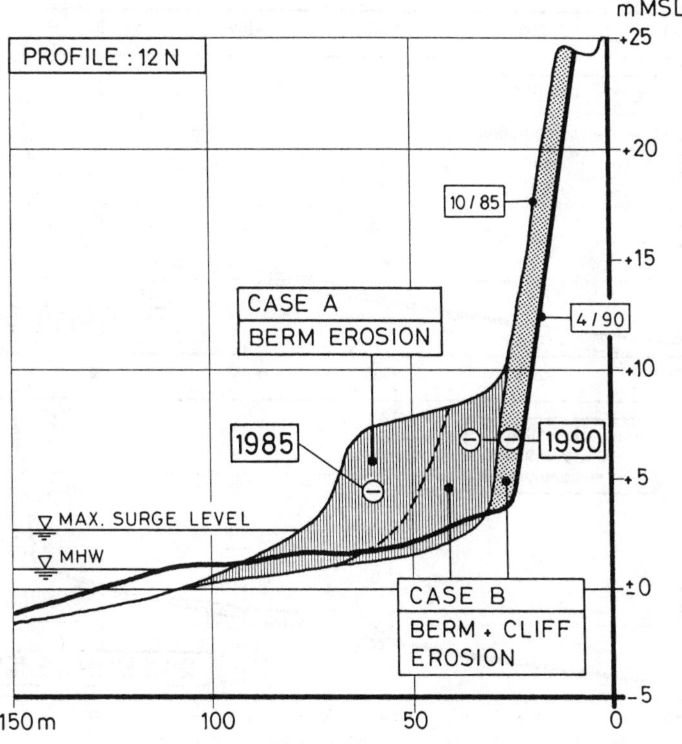

Figure 7. Dune recession during three storm surges in the northern part of Sylt

berm, shortly after a storm in 1985 (case A) and combined recession of berm and natural cliff (case B) after a sequence of storm surges in January and February 1990. Figure 8 shows a comparison between observed and predicted erosion profiles for these cases. The time series of water level and of the H_{rms} wave heights during the three storm surges are illustrated in figure 9.

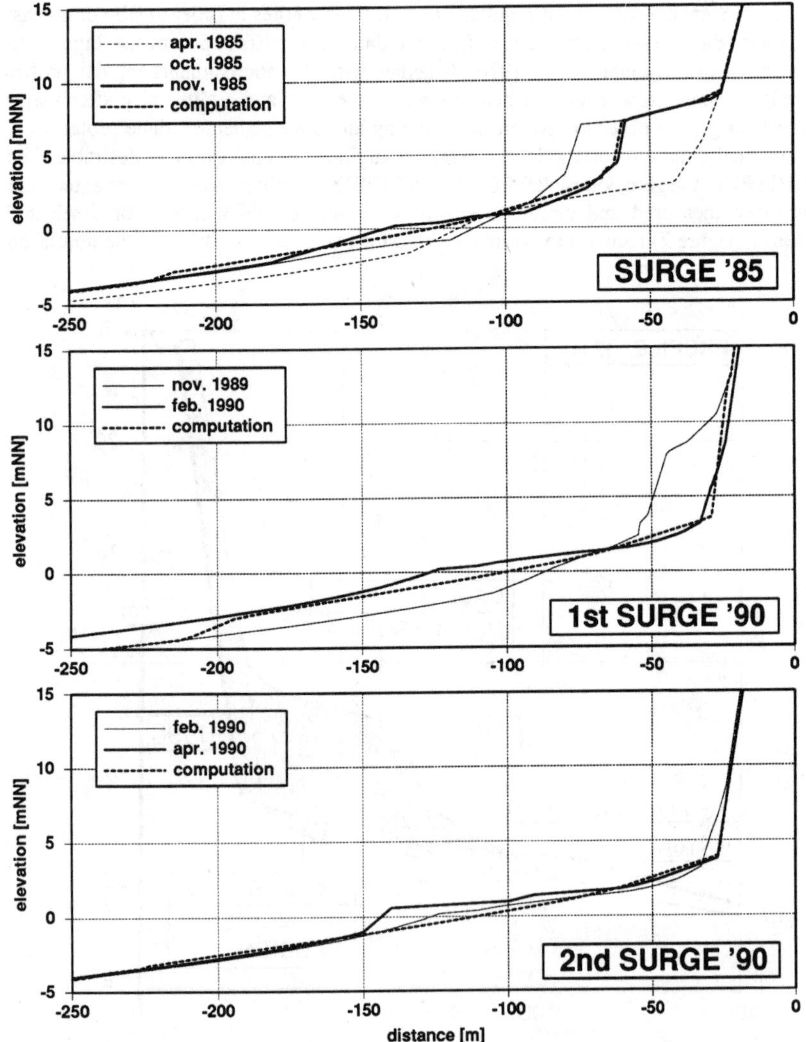

Figure 8. Comparison of observed and predicted erosion profiles

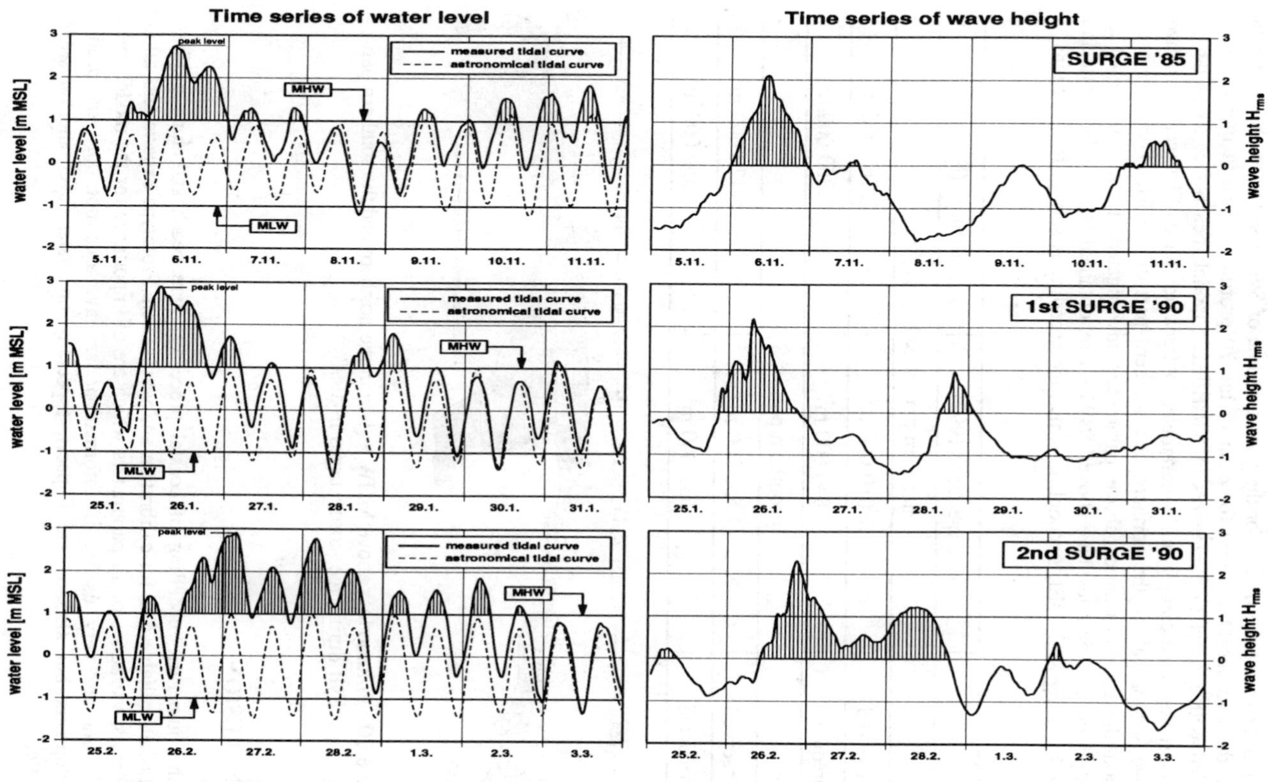

Figure 9. Time series of the three storm surges (water level and wave heights) in 1985 and 1990

In order to use the model for the total coastline of Sylt based upon the field data from the two surges in 1990 for each of the 70 profiles a 'best-fit method' was applied with respect to parameters in KRIEBEL's model (beach face slope tan β, transport rate coefficient k and A-parameter). Afterwards all profiles were classified in two sections along the northern and southern part of the island and for each section the combination of above parameters investigated. Figure 10 shows as example the erosion rates, here above MHW + 1 m, obtained from the field data and those by using KRIEBEL's model, in which the coastline is split in four sections.

	1st surge (Jan. 1990)	2nd surge (Feb. 1990)
	northern part of Sylt	
observed	1 010 000 m³	360 000 m³
predicted	840 000 m³	350 000 m³
	southern part of Sylt	
observed	550 000 m³	170 000 m³
predicted	530 000 m³	350 000 m³

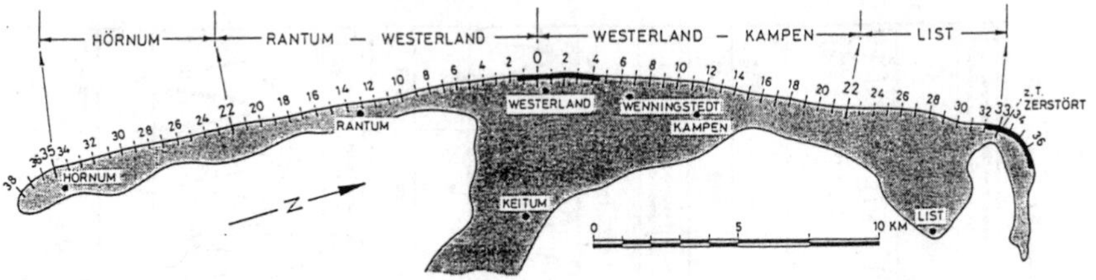

Figure 10. Volume loss above MHW + 1 m in the northern and the southern part of Sylt during the storm surges in 1990

CONCLUSION

The numerical modelling of coastal processes, here beach and dune profiles, necessitate reliable data for calibration and verification. Prototype experiments allow systematic variations of all parameters which are of importance for the model, even field data are very useful, if informations are available not only on profile development but also on the impacts, especially on water level and on wave climate.

ACKNOWLEDGEMENT

This work was carried out as part of G8 Coastal Morphodynamics research program. It was funded by the Commission of the European Communities, Directorate General for Science Research and Development, under contract no. MAS 2 - CT 92-0027. The prototype experiments have been carried out in 'Sonderforschungsbereich 205', inaugurated by Deutsche Forschungsgemeinschaft.

REFERENCES

Dette, H.H. and Uliczka, K. (1987) "Prototype Investigation on Time-Dependent Dune Recession and Beach Erosion," Proceeding of Coastal Sediments'87, ASCE, pp 1430-1444.

Hedegaard, I.B.; Roelvink, J.A.; Southgate, H.; Pechon, P.; Nicholson, J.; Hamm, L. (1992) "Intercomparison of Coastal Profile Models", Report of MAST-Project.

Kriebel, D.L; Dean, R.G. (1985) "Numerical Simulation of Time-Dependent Beach and Dune Erosion", Coastal Engineering Vol. 9, 1985.

Kriebel, D.L. (1990) "Advances in Numerical Modelling of Dune Erosion", Proceedings of the 22nd Coastal Engineering Conference.

Larson, M. and Kraus, N. (1989) "SBEACH: Numerical Model For Simulating Storm-Induced Beach Change", Technical Report CERC-89-7, US Army Corps of Engineers.

Technical Reports (prepared for EC-MAST Project "Coastal Morphodynamics" (EG-M))

Dette, H.H. and Oelerich, J. (1991) "Measurements in the Big Wave Flume in Hannover - Individual Breaking Waves, Breaker Distributions", Report No. 1.

Dette, H.H. and Rahlf, H. (1992a) "Time-Dependent Dune and Beach Transformations - Prototype Experiments with Monochromatic Waves", Leichtweiß-Institut, Report No. 734.

Dette, H.H.; Oelerich, J.; Peters, K. (1992b) "Time-Dependent Dune and Beach Transformations -Prototype Experiments with JONSWAP-Spectrum", Leichtweiß-Institut, Report No. 735.

Dette, H.H.; Rahlf, H.; Peters, K. (1992c) "Suspension Measurements Outside the Surf Zone -Prototype Experiments in the Large Wave Flume", Leichtweiß-Institut, Report No. 739.

Dette, H.H.; Rahlf, H.; Wu, Y.; Peters, K. (1992d) "Wave Measurements across the Surf Zone at Equilibrium Beach Profile - Prototype Experiments (1990/1991) with Monochromatic Waves and Wave Spectra", Leichtweiß-Institut, Report No. 762.

Dette, H.H.; Rahlf, H.; Peters, K. (1992e) "Suspension Measurements across the Surf Zone -Prototype Experiments (1990/1991) with Monochromatic Waves and Wave Spectra", Leichtweiß-Institut, Report No. 763.

FIELD MEASUREMENTS IN A TIDAL RIVER ESTUARY ON THE CANTABRIAN COAST, SPAIN.

D. A. Huntley[1], M.A. Losada[2], R. Medina[2], C. Vidal[2] and K. Stapleton[1]

ABSTRACT: The estuary mouth at Suances, northern Spain, has a section with training walls on either side in order to maintain a navigable channel. This channel section has developed a sand bank on one side filling about half the channel width, and a study is underway to investigate the hydrodynamics and sediment dynamics associated with this topography. Preliminary measurements of waves and currents over the sand bank were made in November 1991 using electromagnetic current meters and pressure sensors. The data reveal the presence of a low frequency (approx. 75 seconds period) cross-channel seiche over the sand bank which becomes more pronounced as the tidal water level over the bank decreases. Cross-correlations between the seiche motion and the envelope of the incident waves shows that the seiche is driven by the incident wave groupiness.

INTRODUCTION.

In many parts of the world, estuary mouths are regions of considerable economic importance, particularly by providing navigation links to harbours and inland industries and transport routes. On sandy coastlines navigation channels in estuary mouths are liable to movement as offshore sediment banks shift position, and there is therefore a need to predict sediment transport, erosion and deposition in these environments.

---

1)Institute of Marine Studies, University of Plymouth, Drake Circus, Plymouth, Devon, PL4 8AA, UK.
2) Departamento de Ciencias y Tecnicas del Agua y del Medio Ambiente, Universidad de Cantabria, Avda. de los Castros, 39005 Santander, Espana.

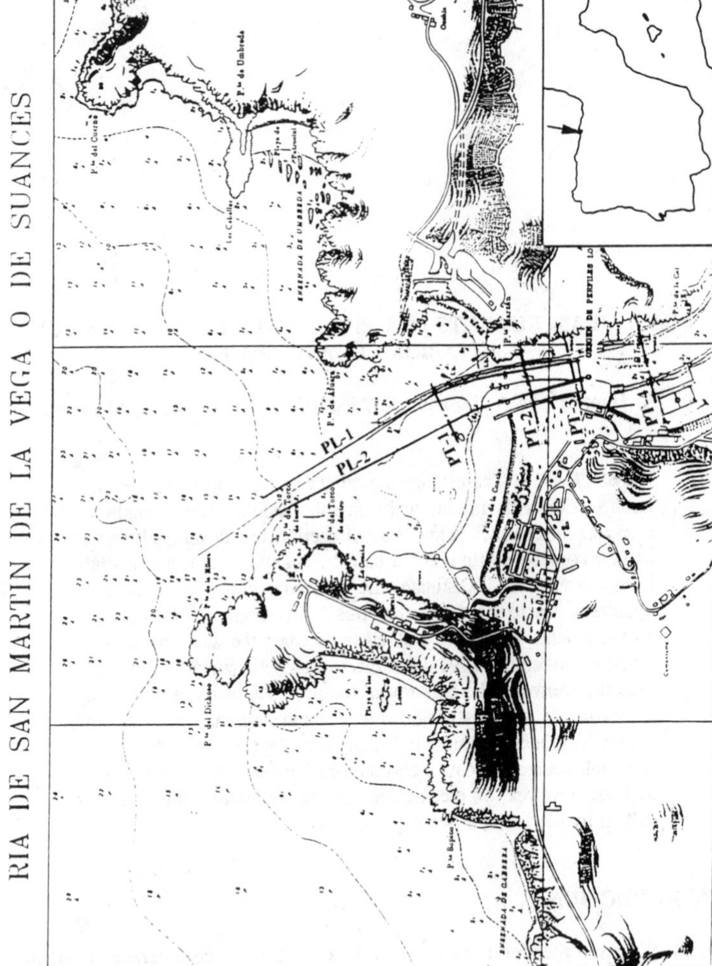

Figure 1. Location map

Estuary mouths are also regions of considerable scientific interest, as meeting places of steady-flow-dominated fluvial conditions and wave-dominated coastal conditions, of fresh water and sea water, and of nearshore (& beach) and estuarine sediment processes.

This paper describes measurements made in the mouth of the Ria de San Martin de la Vega at Suances on the northern coast of Spain. This estuary mouth has extensive sand banks offshore and an adjacent beach. The estuary itself acts as a navigable channel to industries up-river. The present study of the hydrodynamics in the mouth forms part of a larger study aimed at mapping and modelling the processes responsible for sediment movement.

SITE DESCRIPTION.

Figure 1 shows the location of Suances and a detailed map of the estuary. Suances is approximately 20 km to the west of Santander, Cantabria, with a north-facing coastline exposed to Atlantic swell through the Bay of Biscay. The estuary mouth itself has training walls on both sides at the inner end which are designed to maintain the navigable channel into the river. However, an extensive sand bank has formed on the western side of this channelised section, resulting in a narrow navigation channel on the eastern side.

To show this topography more clearly, figure 2 shows depth profiles measured along two longitudinal and four transverse sections (PL-1 and 2, and PT-1 to 4 respectively) which are marked on figure 1. Section PL-1 shows an essentially constant depth from offshore through the eastern side of the channel, with the suggestion of sand waves in the inner part of the section. In contrast, section PL-2 shows a steadily decreasing depth from offshore to the training wall section, where the depth remains almost constant and above mean water level. The transverse sections confirm this trend, though the innermost section, PT-4, shows the deep channel to have shifted to the western side as the river curves, upstream from the training walls. One of the objectives of the overall study at Suances is to determine the causes of this apparently stable narrowing of the constrained channel by the western sand bank.

FIELD MEASUREMENTS.

Field measurements of waves, currents and water depths were made within the walled channel in November 1991. Unfortunately attempts to deploy a tripod of sensors within the deep channel proved unsuccessful due to the high mobility of the bed (perhaps evidenced by the suggestion of sand waves in Section PL-2; Figure 2). We therefore concentrate on measurements made on the sand bank on the western side of the channel.

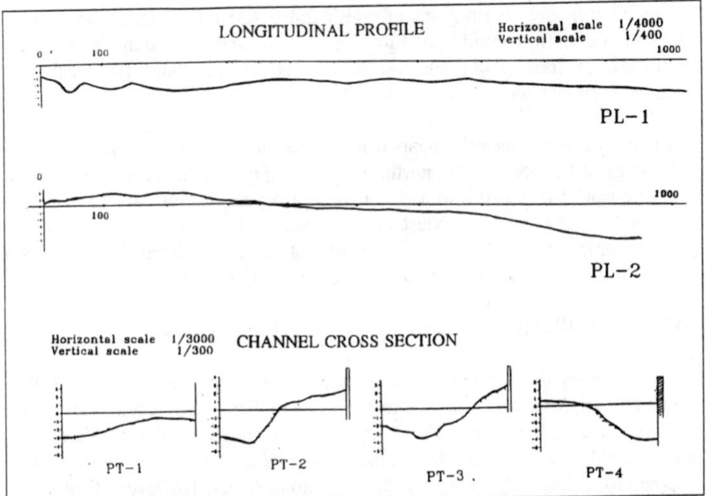

Figure 2. Depth profiles alongthe longitudinal lines Pl-1 & 2 and transverse lines PT-1 to 4 as shown in figure 1

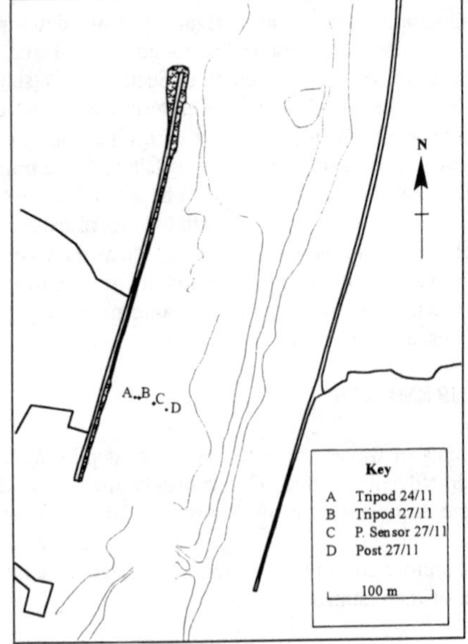

Figure 3. Plan view of the field site

Figure 3 shows the locations of the measurements. Three types of sensor station were used. A tripod placed on the bed held two electromagnetic current meters (EMCM) to measure the three-dimensional components of the flow (with two measurements of the vertical component) at a single height above the bed, 0.47m on 24th November and 0.25m on 27th November, a third EMCM at 0.675m (27th November only) to measure the two hrizontal components of flow, and a horizontal pressure sensor to measure surface elevations. This tripod also held optical sensors to measure suspended sediment load in the water, but these measurements are not discussed here. A second station consisted of a single pole driven into the sand bank on which were mounted a pair of EMCM's, also oriented to measure the three flow components at a height of 0.5m (27th November only), along with a pressure sensor. A third station consisted only of a pressure sensor, attached to a low post driven into the sand.

The present paper considers measurements made on two days. On the 24th November, only the tripod was deployed, at location A in figure 3, but on 27th November all three stations were deployed, at locations B,C and D in figure 3.

On both days, deployment took place near low water, when the bank was exposed, and measurements were then made over flood and ebb phases of the tide, from three hours before high water to three hours after high water on 24th November, and from 2 hours before high water to 4.5 hours after high water on 27th November. Maximum water depths over the bank were approximately 2.4m on 24th November and 2.1m on 27th November.

Offshore wave conditions were measured at Gijon, aproximately 100 km to the west of Suances, but only overall significant wave heights and zero-crossing periods are available, with no separate recording of swell conditions. On November 24th, wave conditions were slight, with the significant wave height only 0.6m and wave period 6 seconds. On the 27th November significant wave heights of about 2m and a zero-crossing period of 13 seconds were recorded at the beginning of the day, but these diminished through the morning on the waning stages of a minor storm.

RESULTS.

Figure 4 shows spectra for along-channel (fig. 4a) and cross-channel (fig. 4b) flows on 27th November when the mean depth over the bank was 0.95m. Both spectra are dominated by a low frequency peak at approximately 0.015 Hz. A second peak at around 0.08 Hz shows evidence of the incident swell, partially refracted onto the bank to be present in both the along-channel and cross-channel spectra.

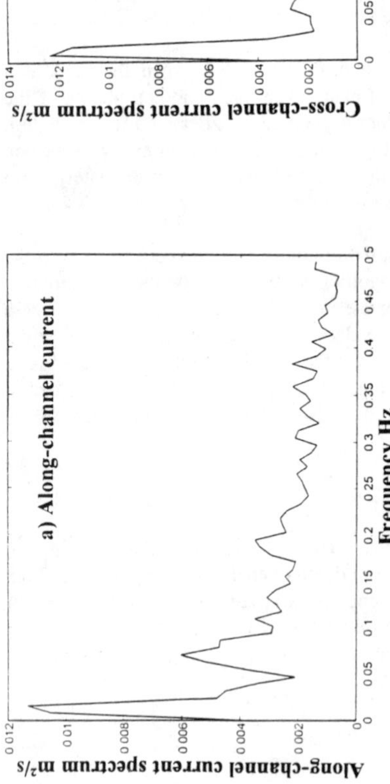

b) Cross-channel current

Cross-channel current spectrum m²/s

Frequency Hz

Figure 4. Spectra of currents on 27th November, ebb tide, with mean water depth over the bank of 9.5m

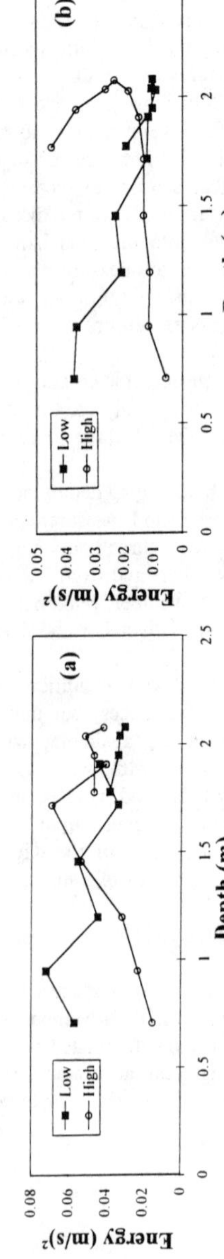

Figure 5. Dependence of low frequency (0 - 0.06 Hz) and high frequency (0.06 - 0.11 Hz) energy on mean water depth, for along-channel (a) and cross-channel (b) currents on 27th November

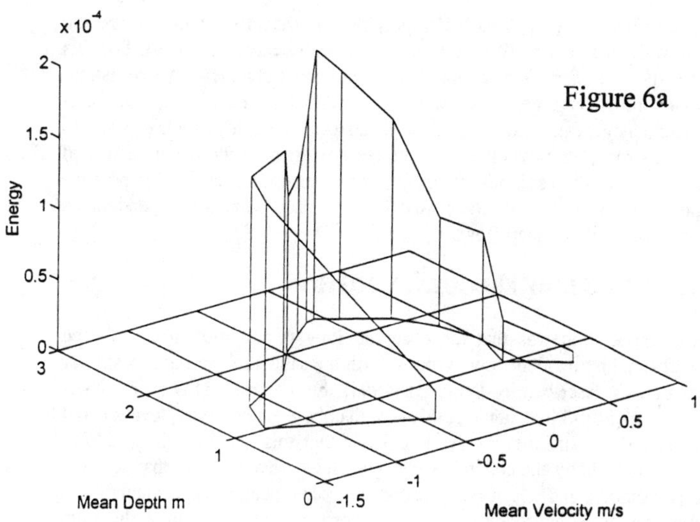

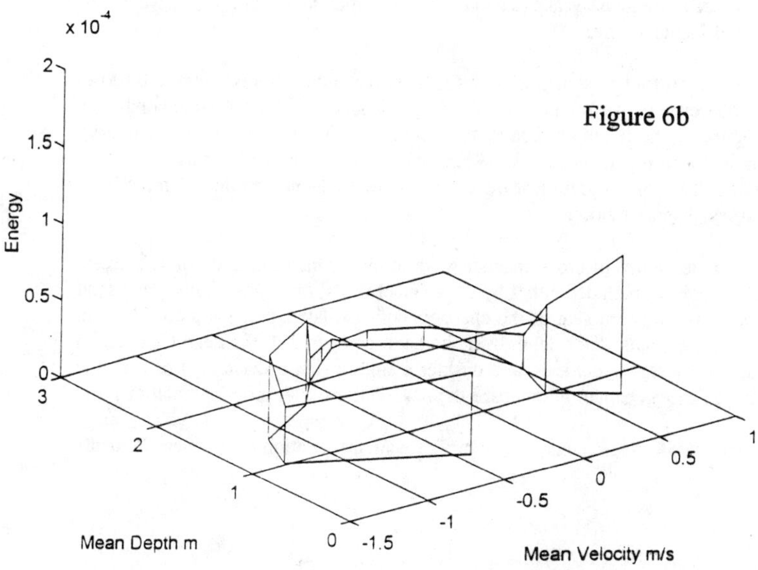

Figure 6. Dependence of low and high frequency energies on mean depth and mean velocity on 24th November
a) High frequency b) Low frequency

Figure 5 confirms that the relative importance of the low frequency motion compared to the incident swell motion increases with decreasing depth. For this figure, the spectral energies of along-channel and cross-channel flows measured on 27th November have been integrated over 0.055 Hz bandwidths, centred at the low frequency peak ('Low') and the swell peak ('High'). Figure 6 shows similar results for 24th November, but plotted against both the mean depth and the mean velocity; the essential symmetry of the spectral energies for positive and negative flows shows that their variations are predominantly dependent on mean depth rather than mean flows.

NATURE OF THE LOW FREQUENCY MOTION.

Our initial expectation was that the observed long period motion was forced along the channel by incident wave groups, with the sand bank acting in a similar way to a reef, as described by Symonds (1994; this volume). However, along-channel motion of this kind should result in high coherence between elevation and along-channel flow, and the observations show that this was not the case. For example, figure 7 shows the coherence between elevation and along-channel flow for 24th November, and it can be seen that the coherence at low frequencies is well below the 95% confidence bound for zero coherence, despite the high coherence at frequencies above about 0.05 Hz which indicates progressive along-channel incident waves.

The situation for cross-channel flows is however quite different. Figure 8 shows the coherence between elevation and cross-channel flow for the same conditions as figure 7, and it can be seen that the coherence is high for the low frequency peak and also for the peak at 0.08 Hz. but is low for most of the incident wave band. It therefore appears that the coherent low frequency motion is in the form of cross-channel motion.

The simplest form of cross-channel wave motion which might occur is a cross-channel seiche, bounded either by the channel walls, or by the width of the sand bank on the western side of the channel, with the latter perhaps more likely in view of the small water depth over the bank compared to the deep channel. In order to test this hypothesis we consider a simple first-order model of a seiche in the Suances channel to compare the observations with model predictions.

The observed cross-channel depth profile near the location of the measurements

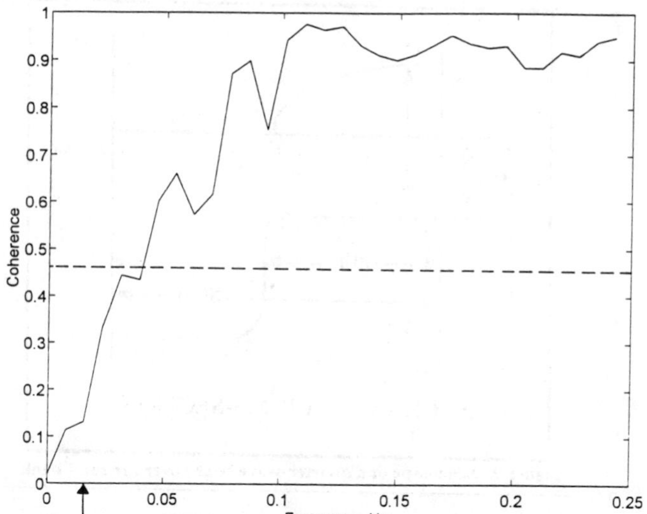

Figure 7. Typical coherence between elevation and along-channel current (24th November, mean depth 1.8m). The horizontal dashed line is the 95% confidence level for zero coherence; the vertical arrow indicates the frequency of the low frequency spectral peak.

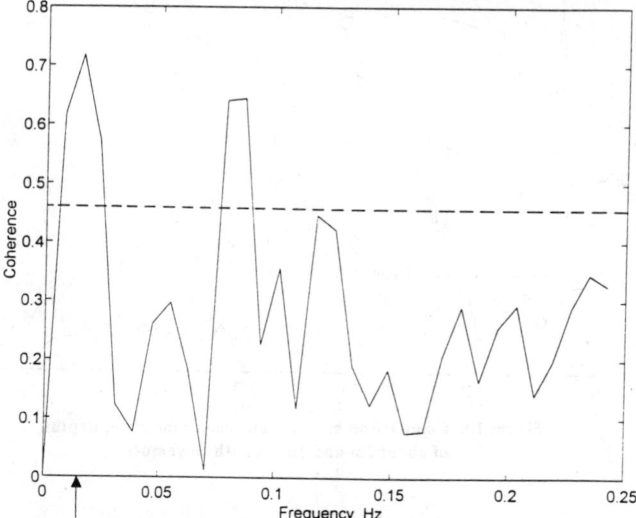

Figure 8. Coherence between elevation and cross-channel current for the same conditions as figure 7

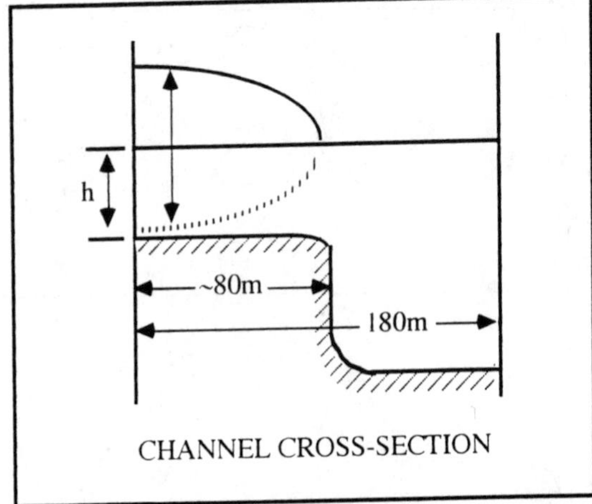

Figure 9. Schematic of a quarter-wave seiche over the sand bank

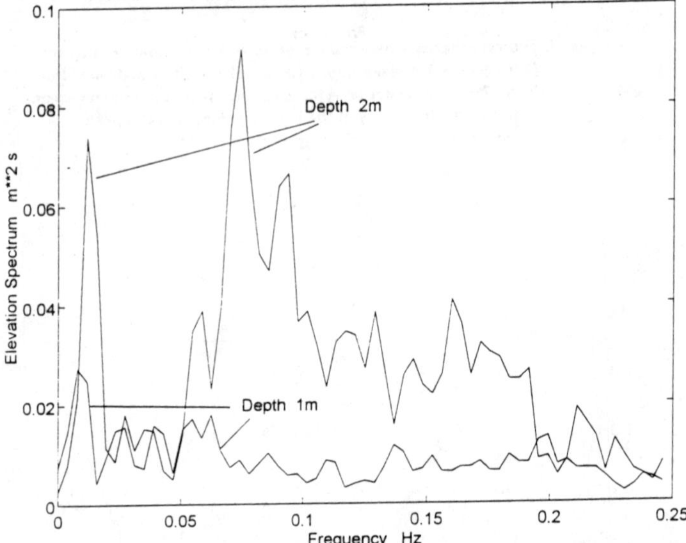

Figure 10. Comparison of elevation spectra for mean depths of about 2m and 1m on 27th November

(PT-2; figure 2) suggests that to first order we can consider the sand bank to be of constant depth across the channel to about 80 m away from the wall, followed by a rapid drop to the deep channel. We therefore consider a situation shown schematically in figure 9. We assume that the deep channel is very much deeper than the water depth over the bank (it is typically 7-8 m deep compared to 1-2 m over the bank) so that a node of elevation and antinode of cross-channel current occurs at the edge of the sand bank. The lowest order seiche will then have a quarter wavelength over the bank.

With this model, a number of characteristics are expected for the low frequency motion:

1. The full wavelength of the seiche, L, should be:

$$L = 4 \times 80 = 320 \text{ m}$$

and the frequency of this motion, f_1, will be given by:

$$f_1 = \sqrt{gh} / L \qquad (1)$$

where h is the mean water depth. For a water depth of 2m (relevant, for example, to figures 7 and 8), this gives a seiche frequency of 0.014 Hz., in very good agreement with the observed low frequency peak and the peak in the coherence (figure 8).

2. Equation 1 also suggests that the frequency of the low frequency peak should vary as the square root of the mean depth, thus decreasing as the tide level over the bank decreases.

Unfortunately the relatively short measured record lengths (17 minutes) give only coarse frequency resolution at the low frequency end of the spectrum when standard FFT techniques are used. Nevertheless the low frequency peak is still clearly defined even for a very low number of spectral degrees of freedom and hence high frequency resolution, so some estimate of the change in frequency of the peak from maximum to minimum measurement depths over the sand bar is possible. Figure 10 shows spectra of pressure measurements (converted to elevation by a simple linear calibration) taken on 27th November, near the time of high water when the depth over the sand bank was approximately 2m, and near the minimum depth for sensor cover at approximately 1m. The low frequency peak has clearly moved to a lower frequency with the smaller water depth. The actual peak frequencies for these runs were 0.0127 ± 0.0010 Hz for a mean depth of 2.03m, and 0.0108 ± 0.0011 Hz for a mean depth of 0.95m.

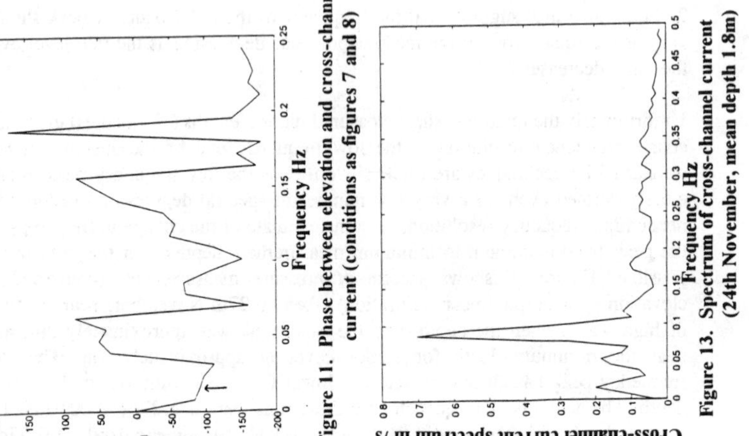

Figure 11. Phase between elevation and cross-channel current (conditions as figures 7 and 8)

Figure 13. Spectrum of cross-channel current (24th November, mean depth 1.8m)

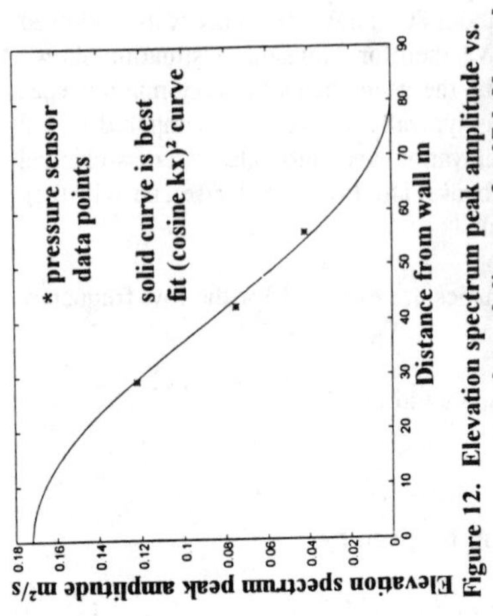

Figure 12. Elevation spectrum peak amplitude vs. cross-channel distance for 27th November

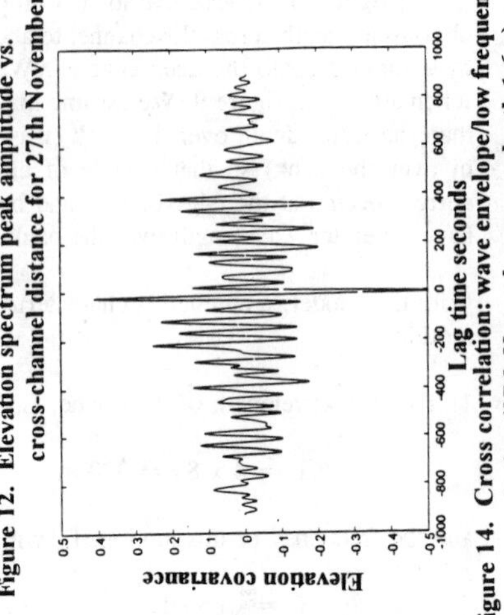

Figure 14. Cross correlation: wave envelope/low frequenc motion (27th November, mean depth 1.6m)

Thus the frequency ratio:

$$f_{2.03}/f_{0.95} = 1.2 \pm 0.2$$

compared with a predicted ratio:

$$\sqrt{(2.03/0.95)} = 1.46$$

Thus the observed reduction in frequency with decreasing mean water depth is in reasonable agreement with the predicted change.

3. For a cross-channel seiche we would expect phase quadrature between elevation and cross-channel current.

Figure 11 shows the phase spectrum associated with the coherence spectrum of figure 8. The phase at the low frequency peak (denoted by an arrow figure 11) is 91 degrees (with 95% confidence bounds of approximately ±30°), in very good agreement with this prediction.

4. The spectral amplitude of the seiche motion should vary across the sand bank approximately as $\cos^2(kx)$, where $k = 2\pi/320\ m^{-1}$ and x is the cross-channel distance from the wall.

The three pressure sensors deployed on 27th November allow the cross-channel amplitude variation to be measured, and figure 12 shows an example. In this figure the solid line shows the predicted shape and has been scaled to provide the best regression fit to the three observations. The observed points do not match the predictions exactly, suggesting a rather slower decay away from the wall than predicted. However it is clear that the amplitude of the motion does decay in a similar way to that expected.

The conclusion to be drawn from these observations is that the low frequency motion takes the form of a cross-channel seiche of one quarter wavelength, controlled by the topography of the sand bar. For 24th November there is even a hint of a 3/4 wavelength seiche over the sand bar, in the form of a small (not significant at the 95% level) but persistent peak at a frequency three times the frequency of the main low frequency peak. Figure 13 shows an example for the cross-channel flow. This rather speculative suggestion receives some support from the correlation with the pressure signal, which also shows quadrature in phase, though of + 90° rather than the -90° observed for the lower frequency peak, suggesting that the observations were made on the channel side of a node in elevation, as predicted for this case.

In order to investigate possible driving mechanisms for the low frequency motion, the relationship between the low frequency motion and the incident wave groups has been investigated by comparing the time series of low frequency motion with the time series of the incident wave envelope. The measured time series was digitally filtered using a 9th order Butterworth filter with a cut-off frequency of 0.05 Hz to separate high and low frequency time series. The wave envelope time series was then calculated by taking absolute values of the high frequency series and low-pass filtering again, with the same cut-off frequency of 0.05 Hz.

An example of the cross-correlation between the wave envelope and low frequency motion, using the pressure sensor closest to the wall and data from 27th November, is shown in figure 14. For this run the mean water depth over the sand bank was 1.8m, with unbroken incident waves. The peak negative correlation at zero time lag suggests that the low frequency motion was locally forced by the incident wave groups, and implies a relatively low Q-factor for the cross-channel seiche. This observation clearly requires further investigation.

DISCUSSION.

Lateral seiching in estuaries has been previously observed, notably in the context of internal wave motion in stratified water (Dyer, 1982; New and Dyer, 1987). However, to the authors' knowledge, these observations are the first to show lateral surface seiching in an estuary mouth linked to incident wave groups. The relevance of this observation to the development and maintenance of the sand bank is at present unknown, though the seiche will enhance motion at the scale of the bank, and may be especially strong when a quarter-wave over the bank matches a resonant frequency of the whole channel width; this possibility is being investigated.

The present observations are clearly preliminary. Further work is in progress on a number of fronts:

i) Higher frequency resolution spectral techniques (maximum likelihod and maximum entropy) are being used to determine seiche frequencies more accurately, and to compare the spectra of seiche response to the spectra of incident wave envelope forcing in order to determine a Q for the seiche resonance.

ii) More accurate modelling of the estuarine topography is being undertaken, to improve the description of the seiche motion across the whole estuary.

CONCLUSIONS.

1. Low frequency motion in the Suances estuary is found to be in the form of a cross-channel seiche, with a quarter wavelength over the sand bar.

2. This seiche is driven by the envelope of the incident swell, which propagates obliquely up onto the bar.

3. There is the hint of a 3/4 wavelength seiche across the bar for low wave conditions.

4. Further work is needed to characterise the seiche motion more fully, to determine its persistence and to assess its significance for sand migration in the estuary.

ACKNOWLEDGEMENTS.

Gareth Lloyd, as instrumentation technician, was invaluable during the fieldwork and we thank him for his versatility. We also especially thank the Universidad de Cantabria "beach team" for their unstinted and good-natured help, often under atrocious weather conditions! The project was funded partly through a British Council Accion Integrada, partly through the Spanish Ministry of Education and Science under Contract PB89.0381 and partly by EC MAST project MAS2-CT92-0030.

REFERENCES.

Dyer, K.R., 1981. Mixing caused by lateral internal seiching within a partially mixed estuary. Estuarine, Coastal and Shelf Science, **15**, 443-457.

New, A.L. and Dyer, K.R., 1987. On the generation of lateral internal waves by a surface seiche in a partially mixed estuary. Estuarine, Coastal and Shelf Science, **24**, 557-566.

Symonds, G., 1994. Theory and observation of currents and setup over a shallow reef.

SMALL SCALE DYNAMIC STRUCTURES AND THEIR SEDIMENTATION EFFECTS IN ESTUARINE ENVIRONMENT

Chaoyu Wu[1] and Jiaxue Wu[2]

ABSTRACT On the basis of field data, theoretical consideration and mathematical simulations and modeling, the present paper gives a general description of several small scale dynamic structures and their interaction common in estuaries in South China coast with attention fixed to long-term simplest processes of importance. Mathematical models were applied to simulate the dynamic structures and reveal the characteristics of the associated sedimentation. Qualitative analyses of the topographic response to the dynamic and sedimentation processes are also presented.

INTRODUCTION

Small scale dynamic structures (SSDS) in estuarine environment in this study refer to dynamic structures with Rossby Number from 1 to 10 including density driven circulation, jet system with entrainment currents, river mouth horizontal circulation caused by Kelvin wave or vorticity variation and large scale eddies, etc. (Wu,1992, Christiansen and Kirby,1991). Some of these flow patterns or circulation, for example, the density-driven gravitation circulation and their sedimentation effects have been under intensive study (Pritchard,1955; Postma, 1967; Festa and Hansen, 1978; Allen et al,1977). Some others, e.g., horizontal circulation, received less attention. These active dynamic structures are driven by different forces whose relative role varies in space and time. The SSDS often dominate different parts of an estuary alternately and interact with each other. It is very unlikely that a single dynamic structure dominates the entire reach of a large scale estuary. The combination and interaction of these dynamic structures construct the overall dynamics of an actual estuary and have profound effects on sediment transport and sedimentation. Because of the irregular boundaries, bottom topography alternating with channels and shoals, and numerous islands in the estuaries in the study area, the dynamics in estuaries is extremely complicated compared to open ocean and shelf waters.

Estuarine morphology is to a large extent determined by the residual sediment transport patterns which, in turn are closely related to the dynamic processes. The evolution of an estuary depends essentially on two processes:

-the long-term averaged sediment supply from inland or coastal origin, and the directions

1) Deputy director, Institute of Coastal and Estuarine Studies, Zhongshan University, Guangzhou, 510275, China.

2) Institute of Coastal and Estuarine Studies, Zhongshan University, Guangzhou, 510275, China.

and magnitude of the long-term averaged sediment transport,

-abrupt changes in the estuarine morphology caused by storm surges, river floods or by engineering works.

The present study is concerned with the first process. Sediment fluxes are governed by sedimentation and resuspension processes, which are subject to strong spatial and temporal fluctuations. From isolated time series and incidental surveys it is not possible to reconstruct a reliable picture valid for different time and space scale. Sedimentation records the long-term and cumulative effects of the dynamics and sedimentation processes, with the help of some elaborately calibrated statistic models, some of these effects can be recovered from sediment records. Theoretical concepts and numerical simulation are necessary in order to arrive a better understanding of estuarine processes.

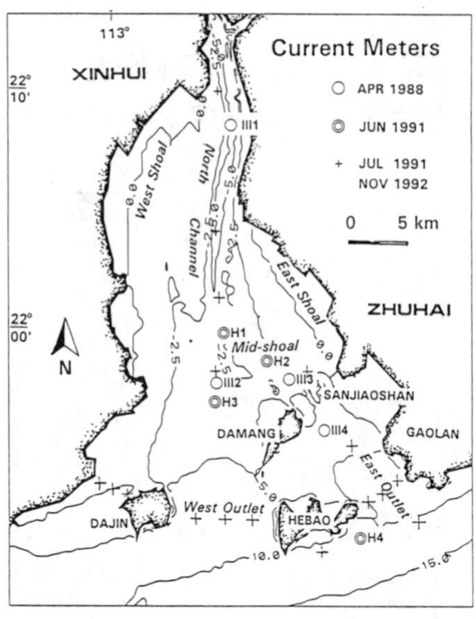

Figure 1. Huangmaohai estuary and current meter mooring.

DATA SETS

Field data used in this study include: (1) Five hydrographic surveys with totally 56 current meter moorings in 1987,1988,1990 and July and December 1992 and water stage of the same periods in Huangmaohai estuary (Figure 1) and totally 90 moorings in 1978 and 1979 in Lingdingyang estuary (Figure 2) ; (2) Repeat echo-sounding in approximately two month intervals along the experimental navigation channel in 1991;(3) Totally 245 bottom sediment samples collected using grab samplers in 1987 and 1992;(4) Navigation charts of Huangmaohai and Lingdingyang estuaries from 1861 to the present and large scale (1:5,000) subaqueous topography maps of Huangmaohai and Nandujaing estuaries collected for the study of modern evolution of estuary topography; (5) Seven cores for ^{210}Pb analysis to determine the natural sedimentation rate.

SMALL SCALE DYNAMIC STRUCTURES

In large scale estuaries like Huangmaohai and Lingdingyang estuaries in the Pearl River estuarine complex, the dominant circulation alternate in different portions of the estuary because of the transition of the driving forces along the estuary. Based on field observations and theoretical consideration, the characteristics of SSDS in estuarine environment can be summarized as following:(1) They are relatively stable with spatially and temporally variations

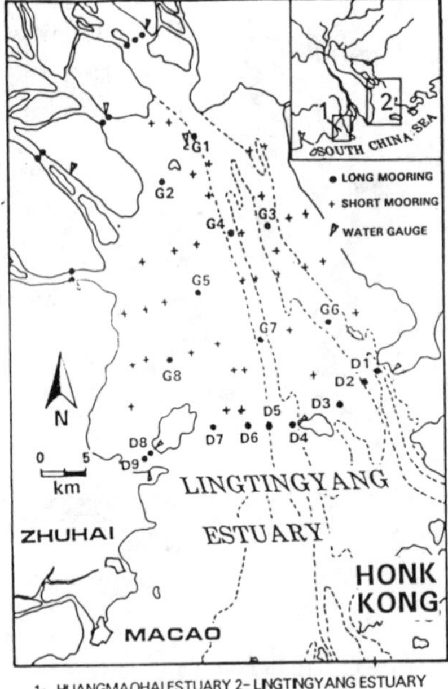

Figure 2. Lindingyang estuary and current meter mooring.

and exist along the estuary alternately. (2) With Rossby number ranges from 1 to 10 and Ekman number from 10^3 to 10^4(with exception), they show the nature of small scale structure and complexity, meaning the nonlinear effect and Coriolis force cannot be neglected. (3) Because of the limit water depth in estuaries, their horizontal scale of the structure, ranging from several tens of meters to several tens of kilometers, is much greater than their vertical scale which however, should not be ignored. They are rather sensitive to the boundary conditions. As a matter of fact, a bottom Ekman layer may develop in the lower reach of estuaries. (4) Residual circulation often forms within these structures. Very little is known about the energy and mass transition between structures. Discontinue surface or front may be formed between systems. (5) Finally, these dynamic structures are induced or driven by different forcing including density, momentum and buoyancy, boundary effects, Coriolis force and variation of vorticity etc.

Low Velocity Zone in the Mid-estuary

The low velocity zone is not considered as one of the small scale dynamic structures, rather, it is actually an important overall dynamic feature of the estuary, a combined results of the SSDS. In Huangmaohai estuary, the spatial distribution of current velocity based on the four cruises of hydrographic survey and the verified 2-D and 3-D (Wu, 1994) models indicate the existence of a low energy (velocity) zone in the mid-estuary, which is coincide with the river mouth bar, or mid shoal. The maximum ebb current exceeds 110 cm/s in the North Channel and 90 cm/s in the lower estuary while it is less than 60 cm/s in the mid-estuary. During most of the time in a tidal cycle, current velocity over the mid-bar is substantially lower than that in both upper and lower estuary. Figure (3) shows the velocity distribution from a 2D numerical model. This low energy zone acts as a sediment trap for both suspended and bed load from inland and marine sources.

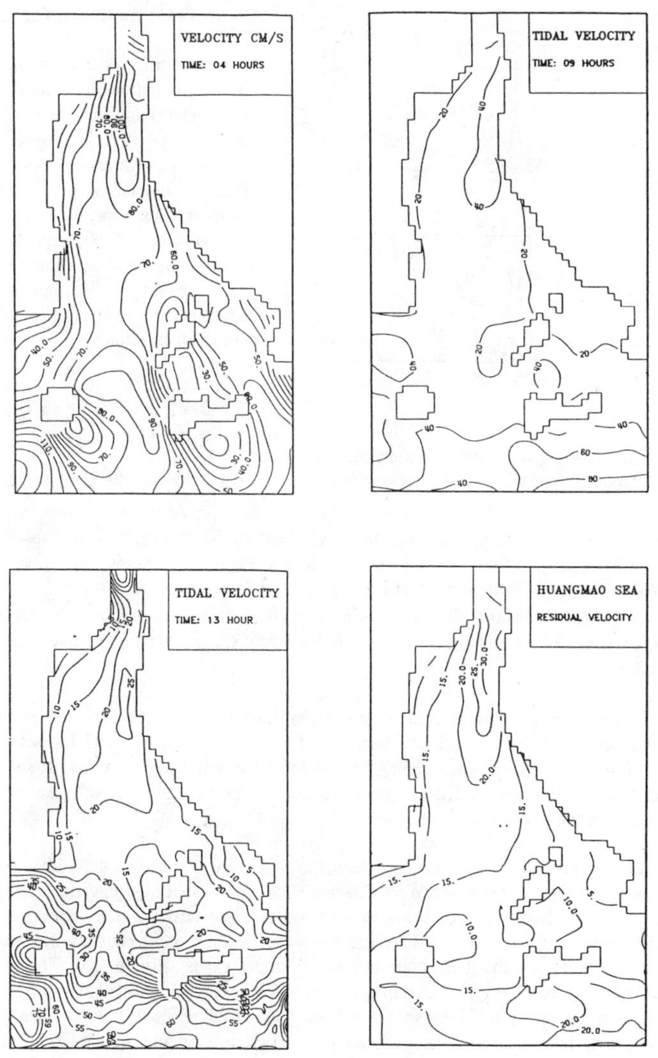

Figure 3. Tidal current velocity distribution.

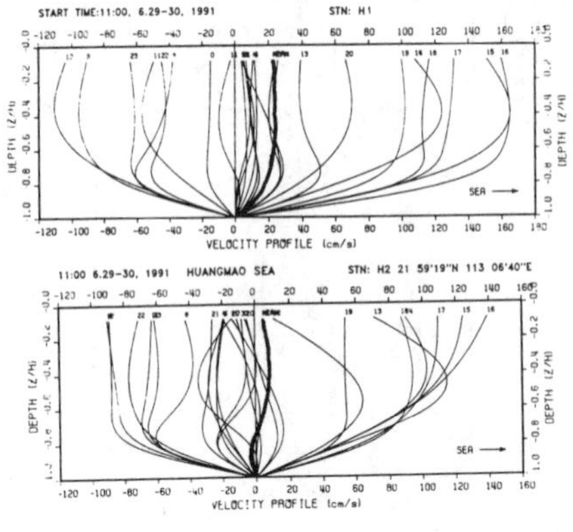

Figure 4. Velocity profile along water column at station H_1 and H_2 in Huangmaohai estuary

Vertical Density Driven Circulation

Vertical density driven circulation is a typical circulation in estuaries that has been under intensive study. In all the case study in this paper, density circulation plays an important role in estuary dynamics and sediment transport. The circulation is driven by the variation of horizontal pressure alone depth between the baroclinic and barotropic components. The horizontal pressure gradient at any depth z in an estuary is

$$\frac{\partial p}{\partial x} = g\int_{-\zeta}^{z} \frac{\partial \rho}{\partial x}dz - g\,\rho_s \frac{\partial \zeta}{\partial x} \qquad (1)$$

where ρ_s is the density at the sea surface. At the toe of the density circulation near the bottom, the upstream flow meets the downstream flow, which is often called the 'null point' where serious siltation occurs. Figures (4) show the velocity profiles along water column at stations H_1 and H_2. The gross line represent the profiles averaged over a tidal cycle. The density circulation is mainly affected by estuarine bathymetry, river discharge, coastal water body and winds. The develop of the circulation is most sensitive to fresh water run-off.

The 'null point' can be considered as the upper limit of the gravitation circulation in Huangmaohai estuary along the East Channel, but where is the lower limit? It does not go all the way down to the continental shelf. According to surveys taken in different seasons, the gravitation circulation is replaced by the efflux system near the Sanjiaoshan Gorge which will be discussed in the next section.

The gravitation circulation also has the lateral limits. Figure (5a) shows the multi-layered low passed velocity time series at the West Channel (Station D4) of Lingdingyang estuary, it is a typical pattern of density circulation with fresh water flows seaward and saline water flows landward. In station D5 next to D4, where the water depth is approximately 3.5 meters less than that in the trough, the density circulation no longer exists Figure (5b). The gravitation circulation seems exist only in the deep channels in Lingdingyang and Huangmaohai estuaries as a dominant flow pattern. In the shoals on both sides of the channel, the turbulent generated from the strong bottom shear creates more tense vertical mixing and reduces the first term in the RHS of Equation (1) and the stratification. Since the current is mainly bi-directional, there is a ***zero velocity zone*** in the lower water column between station D4 and D5. This zero velocity zone serves as an indication to define the dynamic structures of gravitation circulation.

Tidal Jets and Lateral Entrainment

Turbulent plane jets that issue from river outlets and discharge into quiescent bays have been studied from both field observation (Wright and Colman,1974) and theoretical principles (Eillison and Turner,1959;Wang,1984;Ozsey,1986). Because of the different geological and morphological setting, several types of tidal jet systems develop in South China coast. In Huangmaohai and Lingdingyang Estuaries, as a dominant dynamic feature, a strong tidal jet system with distinctive entrainment currents develop in the head of the bell-shaped estuary (Figure 1,2). Another type of tidal jet systems is commonly found on both sides of a gorge between rocky islands in the middle of the estuary or between island and the mainland. Because of the bi-direction flows the jet system forms on each side of the gorge alternately during ebb and flood tides.

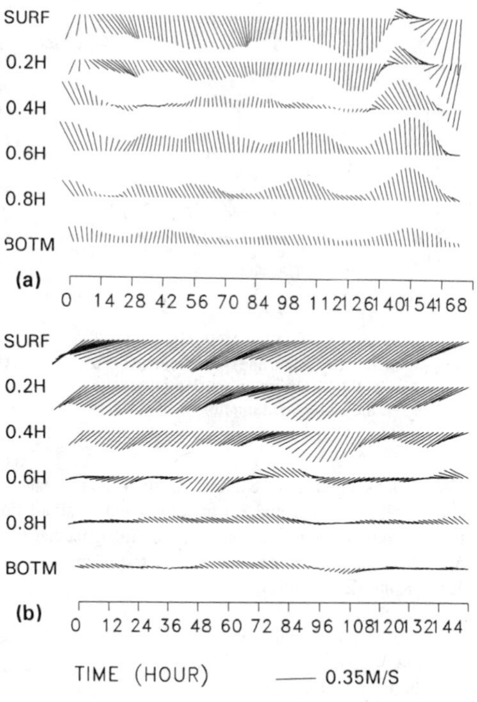

Figure 5. Low passed velocity at (a) station D4 and (b) station D5 in Lingdingyang estuary

Take the jet system in Huangmaohai as an example, the vertical circulation in the East Channel of Huangmaohai estuary is strongly disturbed by the jet current system near the narrow gorge between Sanjiaoshan and Damang Islands. The gorge is approximately 1500 m wide. The water depth exceeds 5m in the trough and decreases to 2.5-3 m to both sides of the gorge which separates the mid shoal in the East Channel into two parts. Based on observations taken near the gorge, the Eulerian residual current is directed upstream in all depths meaning that the efflux system has restrained and actually replaced the density driven circulation near the gorge as a dominant dynamic process. Tidal current velocity exceeds 1.0 m/s in the upper layer and 0.8 m/s in the lower layer in station III 3 near the gorge, and decreases rapidly in both directions. Shown in Figure (6) are the gradual changes of tidal current velocities calculated from the 2-D model. The central current velocity decreases with the increase of the distance from the gorge. The fully developed lateral entrainment current in several locations adjacent to central jet is another distinctive dynamic feature. The velocity v_e caused by the entrainment of the surrounding fluid into the jet is proportional to the jet centerline velocity which is expressed (Eillison and Turner, 1959) by

$$v_e = eu_c(x) \tag{2}$$

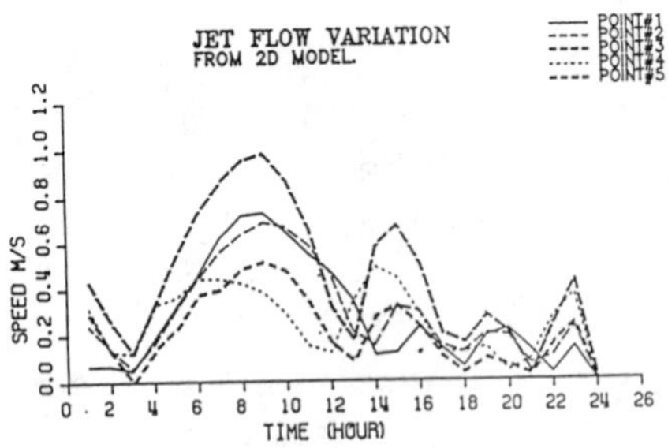

Figure 6. The centerline curent velocity in a two-way tidal jet based on a 2D model, with P#3 at the mid-gorge,sampling intervel 800 m

where e is the entrainment coefficient, and $u_c(x)$ the centerline velocity of the jet. The instantaneous jet velocity easily exceeds 1.5-1.8 m/s and a strong entrainment current is expected. The lateral entrainment into the jet and the flood-flow combine to yield a net residual transport toward the inlet entrance which is opposite to the general seaward flow. Shown in Figure (7) is the fully developed entrainment current in Lingdingyang estuary at station G1.

Horizontal Circulation

Horizontal circulation with spatial scale of several tens of kilometers in the lower reach on the right hand side (facing landward) near continental shelf in several estuaries has been reported (Wu,1993; Zhao,1991). The 2D and 3D numerical model simulation conducted in this study also reveal this horizontal residual circulation. In Huangmaohai estuary, the mean residual current ranges 5-20 cm/s. A similar anticlockwise circulation also exists in the Lingdingyang estuary near Hong Kong (Zhao,1991).

Take the Huangmaohai estuary as example. Residual current flows into the estuary from East Outlet between Gaolan and Herbao Islands. One branch flows north passing through the Damang Gorge, the main stream turns west and joints the West Channel. The mean velocity of the residual current ranges 5-15 cm/s and reaches 20 cm/s outside the East Outlet. The mechanism of this circulation is not yet completely clear. Kelvin wave, potential vorticity conservation and topography may all have their contribution to the development of the circulation. Kelvin wave is the balance between Coriolis force and the horizontal gradient when tidal waves enter an estuary. Water

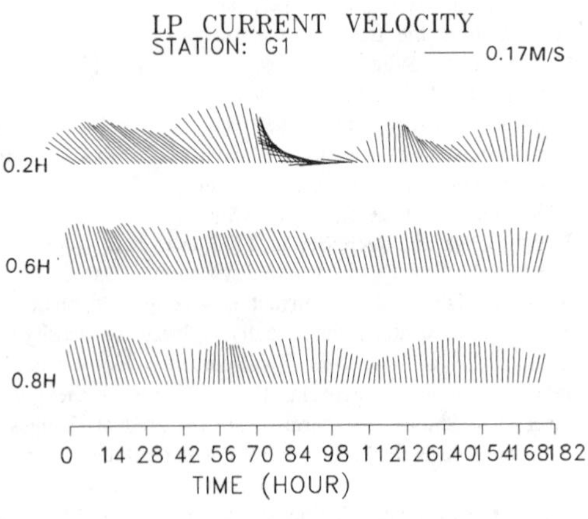

Figure 7. A fully developed entrainment current at station G1 in Lingdingyang estuary

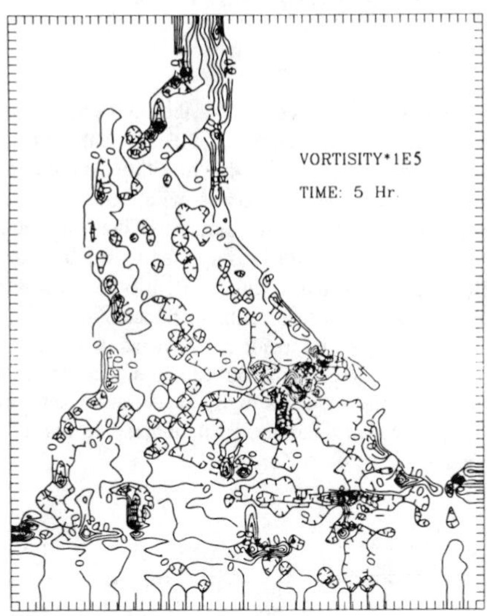

Figure 8. Distribution of relative vorticity ζ in Huangmaohai estuary from a 2D model

elevation and velocity under the Kelvin wave are given by

$$\zeta = R_0 e^{f/c} \cos(\sigma t - \frac{a}{c}x)$$

$$U = \frac{g}{c} R_0 e^{f/c} \cos(\sigma t - \frac{a}{c}x) \qquad (3)$$

where c is the phase celerity of the tidal wave, σ is angular frequency of the tidal waves, f the Coriolis parameter and R_0 the amplitude of the tidal wave. Under the influence of Kelvin wave, both tidal range and current velocity in the eastern part are greater than that in the western part. An anticlockwise horizontal circulation is then prone to develop. West Guangdong nearshore drift flows from ENE to WSW along the coast. It can be considered as a geostrophic flow which is the result of the dynamic balance between the Coriolis force and the water elevation gradient with the high water along the coast.When the drift reaches the estuary, the gradient no longer exists and the drift flows in the estuary through the East Outlet. Another consideration is vorticity conservation,

$$\frac{d}{dt}\left(\frac{\zeta + f}{D}\right) = 0, \qquad (4)$$

where ζ is relative vorticity, f is planetary vorticity and D is the thickness of water layer. Both Huangmaohai and Lingdingyang Estuaries have the north-south orientation, when water moves meridionally toward south, f decreases by the order of 10^{-5}- 10^{-6} s^{-1} based on the dimension of the estuaries. Relative vorticity in Huangmaohai estuary ranges from -10^{-5} ~ 10^{-5} according to a 2D numerical model (Figure 8). Compared to the vorticity in open continental shelf, vorticity variations in estuary are more intensive and complicated due to velocity shear. Since D also increases in lower estuary, the water column tends to more positive (anticlockwise) rotation.

The residual circulation has certain effects on the water and sediment transport in the lower estuary. The upstream residual flow restrains the sediment entering the sea through East Outlet. Since the shelf water with low contain of suspended sediment flows in, it is favored for the East Outlet to maintain a deep channel free from serious siltation.

SEDIMENT TRANSPORT AND SEDIMENT DISTRIBUTION PATTERNS

Turbidity Maximum

Turbidity maximum was found based on several field surveys in the mid-estuary in Huangmaohai which is spatially coincident with the low energy zone. Vertical density circulation and flocculation may contribute to the formation of the turbidity maximum, however, it is believed the turbidity maximum in Huangmaohai estuary is mainly the result of the velocity distribution with low velocity in the mid-estuary. A two-dimension shallow water numerical model incorporated with suspended sediment dispersion was applied to simulate the suspended sediment distribution patterns in Huangmaohai estuary (Xin,1993). It is found that a distinctive turbidity maximum is formed in the mid-estuary with or without flocculation effects (Figure 9). Flocculation and gravitation circulation are the two major mechanisms proposed to explain the formation of turbidity maximum (Schubel,1971). It is an example that velocity distribution mainly due to morphological boundary could induce a high concentration zone in an estuary without the appearance of flocculation and gravitation circulation, since the TM aroused when both were excluded from the model.

Bottom Sediment Distribution Patterns

Although a single sediment sample is not ordinarily a good indicator of the deposition environment, it may be useful if it represents a group of sediments which share the same or similar characteristics. To divide sediments into such 'groups', a fuzzy cluster analysis (FCA) was applied to the 235 bottom sediment samples. Of the 235 samples 102 were collected in April,1988 the rest were collected in flood season (July and August) and dry season (November) in 1992 at the same locations in Huangmaohai estuary. Grain-size distribution ranges from -2ϕ to 10ϕ in 1ϕ interval.

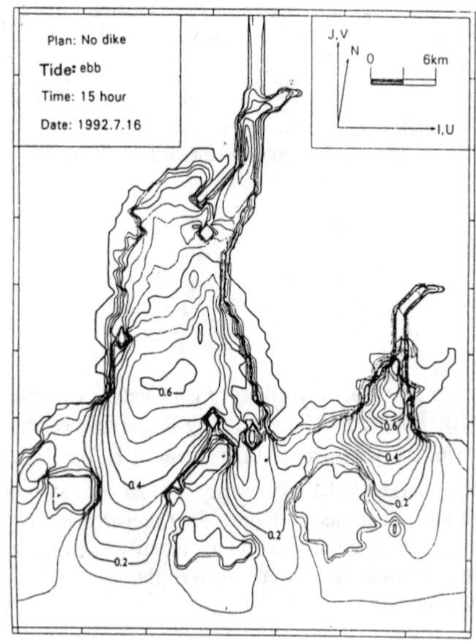

Figure 9. Suspended sediment distribution shows a high concentration zone in Huangmaohai estuary from 2D model

Fuzzy cluster analysis provides a statistical method for handling large data sets and extracting sedimentologically meaningful and environmentally significant results (Wu, 1985,1987). The cluster analysis technique finds 'natural classification' or grouping of the sediment characteristics without prior knowledge or assignment of any arbitrary limits. However, interpretation of the results requires integrating knowledge of

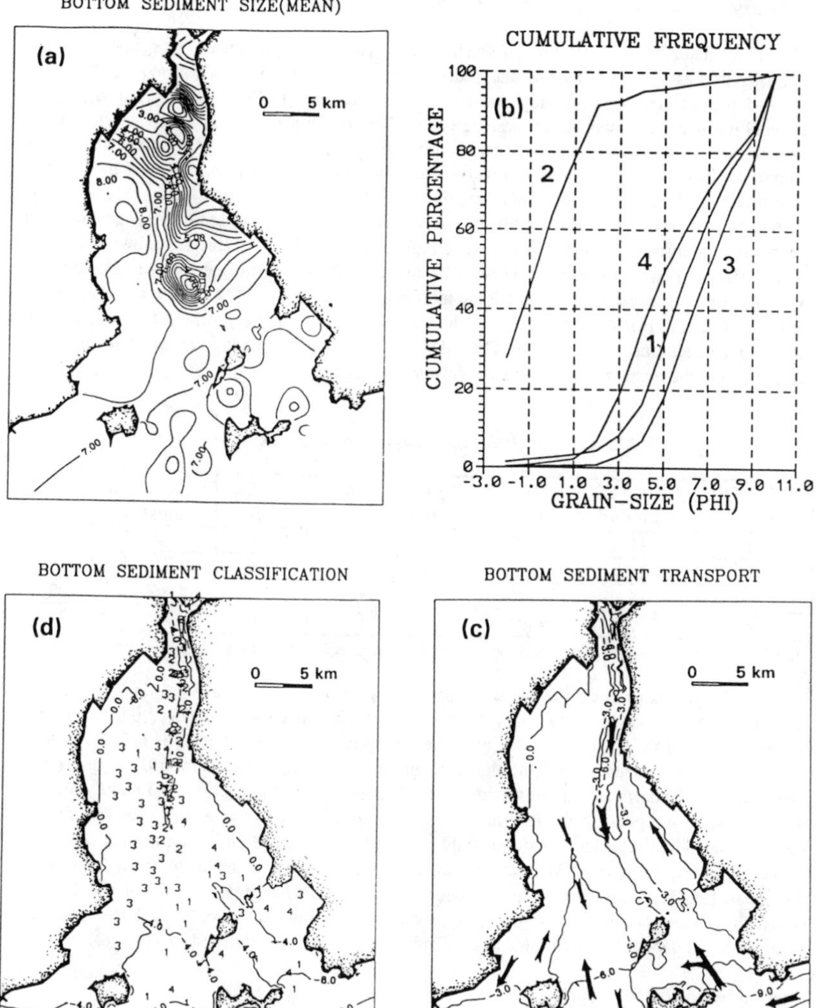

Figure 10. (a) Mean grain-size distribution, (b) cumulative frequency for each cluster, (c) distribution of clusters, (d) the most likely bottom sediment transport

the study area, its physical processes and geomorphic features.

The four major clusters or groups, namely scoured channel sediment, marginal tidal flat deposit, mid-estuary bar deposit and sediment from marine source identified by the FCA method proved sedimentologically meaningful and environmentally significant. It should be pointed out that never was the geographic position of the samples used to assist in determining the clusters, but that after the clusters were determined, sample location was used as an important factor in relating the cluster to environments of deposition. Also of interest for interpretation purpose is the grain-size distribution of each cluster. Figure (10a) is the distribution of sediment mean grain-size. Cumulative frequency of sediment grain-size and sample location for each cluster is shown in Figure (10b and 10c). Table (1) shows the grain-size statistics of each cluster for data set collected in 1988. The settling velocity and competent velocity, or critical velocity to initiate the motion of sediment, for each cluster are also listed. It is worth noting that the fines are more difficult (Uc is much higher) to be set to motion than the coarse sands once they deposit. The calculation was based on semi-empirical formulas of this area and the mean grain-size of each cluster was used.

Table 1. Grainsize Statistics of Clusters

CLUSTER	SAMPLES NUMBER	MD (Φ)	QD (Φ)	SK (Φ)	MQ (Φ)	ω (cm/s)	Uc (cm/s)
1	31	7.08	1.7	0.0	7.0	2.84	71~77
2	21	0.21	1.5	0.09	-1.0	10.81	33~44
3	36	8.00	1.2	0.10	7.0	2.56	95~102
4	19	6.01	2.1	0.49	5.0	3.54	48~52

MD-Mean, QD-Standard deviation, SK-Skewness, MQ-Median, ω-Settling velocity, Uc-Competent velocity.

Measured by mean value, the order of grain-size ranging from coarse to fine is 2-4-1-3. Cluster 2, coarse sand and the only group that falls into the sand category, poorly sorted, distributes along the scoured North Channel exclusively. The poorly sorting is because part of the sediments is carried down from the river and the others are scoured from the channel bed which is a very poorly sorted weathering layer. Cluster 3 is the finest group and mainly located in the western marginal shoal. It is moderately sorted, the best sorted group, probably due to its riverine source. Cluster 1 and 4 mainly occupied the lower reach of the estuary and may come from marine source. Material in the mid-shoal consists of sediment from Cluster 1, poorly sorted. Most sediment samples collected in the eastern marginal shoal belong to Cluster 4 which is very poorly sorted. Clusters produced from the 102 samples and the dynamic structures and deposition environments they are mostly related to are listed in Table (2).

Bottom Sediment Transport

The environmental interpretation of grain-size distribution found in sedimentary deposits has been, and still is, a fundamental goal of sedimentology. There have been many attempt to relate the grain-size distribution directly to the environment of deposition for the purpose of differentiating transport pathways, transport mechanisms, and types of stratigraphic sequences (Friedman,1967; Wu,1987). In this section changes in statistics (mean, sorting and skewness) describing grain-size distributions are used to speculate on the direction of sediment transport. McLaren (1985) proposes a simple model whereby the distributions of sediment in transport

Table 2. Clusters and Related Dynamic Environments

CLUSTER	GRAINSIZE (Φ)	LOCATIONS	DEPOSITION ENVIRONMENTS AND DYNAMIC STRUCTURES
1	7.08	North Channel	strong turbulent jet.
2	0.21	western marginal shoal	friction (wind and bottom) dominant flow
3	8.00	lower estuary, mid-shoal	density circulation, gorge tidal jet systems in low velocity zone
4	6.01	lower estuary, eastern shoal	horizontal circulation, friction dominant flow

are related to their source by a *sediment transfer function* which defines the relative probability that a grain within each particular class interval will be eroded and transported. According to Mclaren (1985), when any two samples are compared with respect to their mean size, sorting, and skewness, eight possible trends exist. Of these trends, only two are indicative of transport, namely finer(F),better sorted(B),more negative(-) (case B) and coarser(C), B and more positive(+) (case C) for which there is one-eighth probability of either occurring at random(0.125). To determine if the number of occurrences of a particular case exceeds the random probability of 0.125, the model tests the following two hypotheses:

$\mathbf{H_0}$: $p \leqslant 0.125$, and there is no preferred direction; and

$\mathbf{H_1}$: $p > 0.125$, and transport is occurring in preferred direction;

Using the Z-score in a one-tailed test, $\mathbf{H_1}$ is accepted if

$$Z = \frac{x-Np}{\sqrt{Npq}} \geqslant 1.645 \text{ (0.05 level of significance)}, \tag{5}$$

where x = observed number of pairs representing a particular case in one of the two opposing directions; and N= total number of possible unidirectional pairs. $N=(n^2-n)/2$ where n=number of samples in the sequence; p=0.125 and q=1.0-p=0.875. The readers are referred to McLaren (1987) for the principles and details of the model.

Mclaren model was applied on six sets of bottom sediment samples to examine the most possible directions of sediment transport on the channels, mid-shoal, marginal shoals and secondary tidal channels on shoals. The results are shown in Table 3 and Figure (10d).

The examination of grain-size trends of the six sample sets taken along the dominant flow axis suggests the following:

(1) In the North Channel, of 78 possible pairs, case C in the south direction is the only significant trend (Table 1). The sediment trends predict accurately the transport direction. They also suggest the transport processes produce high-energy transfer function and the energy is generally decreasing in the transport direction both of which agree with the general dynamic

Table 3. Summary of The Numbers of Pairs of The Six Sets of Sediment of the Huangmaohai estuary Producing Transport Trends, N,x and Z are defined in text

		North Trend	South Trend
No. 1*	Case B	N=78 x=13 Z=1.11	N=78 x=10 Z=0.09
	Case C	N=78 x=7 Z=-0.94	N=78 x=20 Z=3.51
		North Trend	South Trend
No. 2	Case B	N=45 x=12 Z=2.87	N=45 x=9 Z=1.52
	Case C	N=45 x=10 Z=1.97	N=45 x=3 Z=-1.18
		North Trend	South Trend
No. 3	Case B	N=36 x=3 Z=-0.76	N=36 x=8 Z=1.76
	Case C	N=36 x=9 Z=2.27	N=36 x=1 Z=-1.76
		North Trend	South Trend
No. 4	Case B	N=36 x=13 Z=1.76	N=36 x=7 Z=1.26
	Case C	N=36 x=7 Z=0.76	N=36 x=0 Z=-2.27
		North Trend	South Trend
No. 5	Case B	N=45 x=1 Z=-2.08	N=45 x=19 Z=6.03
	Case C	N=45 x=7 Z=0.62	N=45 x=1 Z=-2.08
		North Trend	South Trend
No. 6	Case B	N=36 x=8 Z=1.76	N=36 x=6 Z=0.76
	Case C	N=36 x=6 Z=0.76	N=36 x=3 Z=-0.76

*No.1-North Channel, No.1 - E Channel, No.3 - W Channel, No.4-West Dajin Channel, No.5 - Tidal Channel in W Shoal, No.6 - E Shoal.

setting. The North Channel is dominated by the strong fresh water jet current toward south.

(2) For the sediment samples in the lower East channel, case B and north is the preferred trend and indicate a low-energy regime. Considering the horizontal residual circulation in the lower estuary, the result is reasonable.

(3) For the sediment samples from the Lower West Channel both case B and case C in the preferred north trend are significant and in high energy regime. Test also indicates a south preferred direction in low-energy regime. Lower West Channel is the major outlet of the estuary discharging the fresh water, in the meantime it is the main pass of salt water intrusion. It is not a surprise that the results display a complex transport regime.

(4) Tests indicate a northward sediment transport on the eastern marginal shoal and a southward trend in the secondary tidal channel on the western marginal shoal.

In general, the preferred directions of sediment transport calculated from McLaren Model are consistent with the overall dynamic setting and the dominant dynamic structures in the estuary.

Sediment Inventory, Deposition Rate and Experimental Channel

The sediment in Huangmaohai estuary is mainly from both inland and marine sources carried by river discharge during flood season and residual current from the inlets. Based on calculation on the navigation charts published from 1940 to 1988 the annual mean deposition rate is approximately 1.38 cm/yr, or 7.4 million m^3/yr within which 77% is from inland and

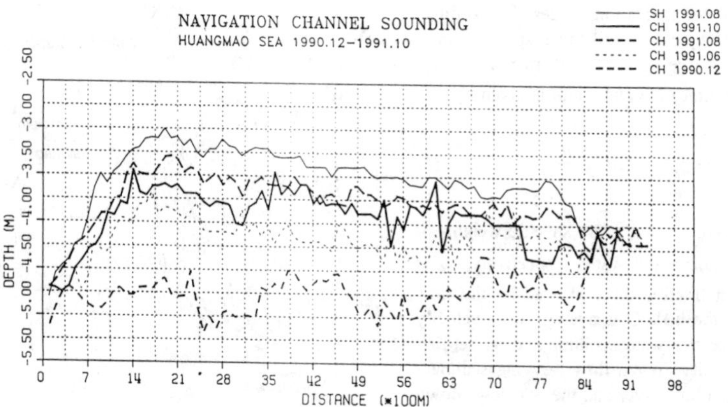

Figure 11. Repeat echo sounding of the experimental channel profile

23% from continental shelf and alongshore sediment transport. The ratio of suspended load and bed load is estimated 10:1. The sediment sources and sinks have been inventories by Yang and Wu (1988).

Based on the measurement of navigation charts from 1937 to 1977, the average deposition rate of the bay is 1.28 cm/yr. There are three zones with the highest deposition rate, the mid-shoal, the western marginal shoal and eastern marginal shoal. Since large-scale reclamation projects have been carried out in both the eastern and western marginal shoals, the maximum natural deposition rate occurred in the mid-shoal, which reaches 4.0 cm/yr. Negative deposition (scouring) rate occurred in both the North Channel and the Lower estuary. The high deposition zone in the mid-estuary is considered the result of the low energy zone and to less extent, the gravitation circulation.

In October 1990 a experimental navigation channel was opened in the mid-shoal connected the North Channel and the East Channel. The total length of the artificial channel is 7.5 Km, the bottom width is 60 m, the bottom elevation is 4.0 m below the local datum and average dredged depth is 1.14 m. Echo sounding surveys were taken in a regular base to monitor the redeposition. Figure (11) shows the echo sounding longitudinal profile of the channel. The survey indicates that severe deposition occurred during river floods and storm surges. Slight scouring took place in low water season in winter. The redepositing rate is more than 60 cm/yr. It is not surprise that the artificial channel was completely refilled in less than one and half hydrographic years when no maintenance dredging was taken. The experimental channel is located mainly in the low energy zone and the reach of upstream end of the gravitation circulation.

TOPOGRAPHY RESPONSE TO DYNAMIC STRUCTURES AND SEDIMENTATION

Mid-estuarine Bar

The decrease of current velocity significantly affects the transport rate of suspended sediment and sedimentation process. According to Bagnold (1966) the suspended load discharge q_s expressed as dry weight per unit time and width is given by:

$$\frac{\gamma_s-\gamma}{\gamma}q_s = 0.01\tau_0\frac{U^2}{\omega} \qquad (6)$$

where γ_s and γ are specific weights of sediment and water respectively, τ_0 is unit tractive force exerted by the flow on the bed, U is average flow velocity over the water column, ω is settling velocity. When flow carrying sediment from both river and the sea slows down in the mid estuary, the sediment transport capacity decreases rapidly which in turn may cause suspended sediment to accumulate in the water column over the mid shoal. Satellite picture, hydrographic surveys and numerical models reveal a high sediment concentration zone in the mid estuary during flood season. The mid-estuary bar is mainly the topographic response to the low energy zone.

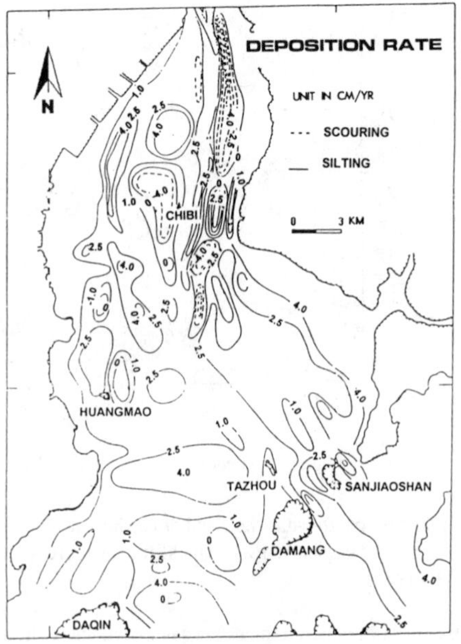

Figure 12. Natural deposition rate in Huangmaohai estuary (modified from Yang, 1988)

Scoured Channel

The average water depth in the upper Huangmaohai estuary is about 5 meters. The most outstanding feature is the deep and narrow scoured channel of 20 kilometers in length and with depth exceeding 11 meters. The strong jet current issuing from the rocky gorges in the north may reach 1.5~1.8 m/s. Consider that the competent velocity for sediments in the channel is about 33-44 cm/s, it is not a surprise that the channel has been under continuous scouring. The bell-shape of the estuary also increases the tidal velocity.

Deposition Features Response to Gorge Tidal Jets

The gorge efflux system is sedimentologically significant. The two deposition shoals developed to the north and south of the gorge between Damang Island and Sanjiaoshan Islands is a typical saddle-shaped deposition feature caused by the two-way tidal jets. The seaward and landward changes in the water depth is mainly considered to be the morphological response to the two efflux systems developed in each side of the gorge. The shoal to the north of the gorge is the combined result of density circulation in the low energy zone that acts as a sediment trap and the deceleration of the jet flows. The shoal to the south of the gorge is mainly the result of the efflux system. Since the jet systems are physically stable the shoals in both sides of the gorge have not shown significant changes since at least the 1860's based on the navigation charts published from 1861 to 1988. As a comparison, the shoal in the West

Channel has experienced significant seaward migration in the same period. The inner slope of the shoal has been scoured and the iceboat of 5 m migrated 40 m per year seaward in the last 40 years (Yang and Wu,1988), at the mean time the outer slope advanced 36 m/yr to the sea (Figure 12).

Topography in River Mouth

Both Huangmaohai and Lingdingyang estuaries have the north-south orientation. Fresh water flows in from the north and sea water flows up from the south. Because of the Coriolis force and the resultant horizontal anti-clockwise circulation, the east side of the estuary is more or less dominated by shelf water with low sediment contents and high salinity, the water in the west side of the estuary caring by far more suspended sediment (0.05~0.2 chi/m^3) and is less saline. As a result, the deposition rate in the west side is much higher than that in the east side in the lower estuary. Water depth in the mouth of the eastern Huangmaohai estuary exceeds 18-20 meters compared to 5-8 meters in the west. This topographic asymmetry in the lower estuary is one of the striking features displaying how the modern subaqueous topography can be affected by dynamic and sedimentation processes.

DISCUSSION AND CONCLUSIONS

This study is in the initial stage, many critical questions remain to be answered and the others have been raised from this study: (1) What and how is the mass and energy transport between dynamic structures?(2) How the SSDS affect the Lagranjian motion of water parcels and sediments?(3) What is the relation of estuary fronts and SSDS interface?(4) What criteria can be applied to define the SSDS?

However, there is no doubt about that SSDS play an important role in estuary dynamics and sedimentation processes in view of both theoretical consideration and practical application. The following will conclude this paper:

(1) The overall dynamic feature of an large scale estuary can be studied by dividing it into small scale dynamic structures. The transition of the role of driving forces provides a physical base for defining the SSDS. The SSDS have profound effects on sedimentation and produce typical morphological response.

(2) Sediment records are the results of long-term and accumulative effects of the dynamics and sedimentation processes, with the help of some elaborately calibrated statistic models, some of these effects can be recovered from sediment records. Sediment clusters and pathways of sediment transport produced from fuzzy cluster analysis and Mclaren model proved to be sedimentologically meaningful and environmentally significant. These results are consistent with dynamic analysis and can be corrobarated with each other. The 'dynamics-sedimentation-morphology' approach is an effective way to arrive a general understanding of physical processes of estuaries.

REFERENCES

(1)Allen,G.P.,G. Sauzay,P.Casting and J.M.Jouanneau,1977,Transport and deposition of suspended sediment in the Gironde estuary, France. In: M.Wiley. Estuarine processes Vol.2,

Academic Press N.Y.:63-81.
(2)Bagnold, R.A., 1966 An approach to the sediment problem from general physics, U.S. Geol. Survey, Prof. Paper 422-I, p37.
(3)Christiansen,H. and R.Kirby,1991, Fluid mud intrusion and evaluation of a passive device to reduce mud deposition,Proceedings of the CEDA-PIANC CONFERENCE, November 13-14, Amsterdam.
(4)Coleman,J.M.,1976, Delta, Processes of Deposition and Models for Exploration. Bergess Minneapolis Minn. 1976.
(5)Eillison,T.H. and S.Turner, 1959, Turbulent entrainment in stratified flows, J. Fluid Mech. 6(3),423-488.
(6)Festa,J.P. and D.V. Hansen, 1978, Turbidity maxima in partially mixed estuaries: A two-dimensional numerical model, Estuar. coast. mar. Sci. 7:347-359.
(7)Friedman,G.M.,1981,Distinction between dune, beach and river sands from textural characteristics, J. Seed. Petrology,Vol 27,p514-529.
(8)Geyer, W. R., 1988, The advance of a salt wedge front: observations and dynamical model; In J. Dronkers and van Leuseen. Physical Processes in Estuaries. Springer-Verlag,p181-195.
(9)McLaren, P., 1985, The effects of sediment transport on grain-size distribution, J. Sedimentary Petrology, Vol. 55, No. 4, p0457-0470.
(10)Ozsoy, E. 1986, Ebb-tidal jets: A model of suspended sediment and mass transport at tidal inlets, Estuarine, Coastal and Shelf Science 22, 45-62.
(11)Partch, E. and J. D. Smith, 1978, Time-dependent mixing in a salt wedge estuary. Estuar. Coast. Mar. Res., No. 6, p3-19.
(12)Postma H., 1967, Sediment transport and sedimentation in the marine environment, In: G.H.Lauff, Estuaries, Am. Ass. Adv. Sci. Washington, p158-179.
(13)Pritchard, D.W.1955, Estuarine circulation patterns, Proceedings, American Society Civil Engineers, 81,717/1-717-11.
(14)Schubel, J.R.,1971, The Estuarine Environment, American Geological Institute, Washington D.C.
(15)Wang, F.C., 1984, The dynamics of a river-bay-delta system, J. Geophy. Res. Vol.89, No C5 p8054-8060.
(16)Wright,L.D. and J.M.Coleman, 1974, Mississippi river mouth processes: Effluent dynamics and morphologic development, J Geol. 82, 751-778.
(17)Wu.C.Y.,1986, A dynamics and sedimentology study of eastern Atchafalaya Bay, Louisiana State Univ. PH.D. dissertation
(18)Wu,C.Y.,1987, Sediment distribution patterns and the related deposition environments (A cluster analysis), Geotechnology 1987 Vol.VII(2).
(19)Wu, C.Y., 1994, Dynamic structures and their sedimentation effects in Huangmaohai Estuary, China, J. Coastal Research (accepted, in press).
(20)Wu,C.Y.,1994, Dynamic and morphologic processes of Huangmaohai estuary, National '85' Key Project Technique Report /No:85-404-01-02B, Inst. /Coast. & Estuar. Studies, Zhongshan Univ.
(21)Xin,W.J., 1993, A two-dimensional numerical model of the suspended sediment in Huangmaohai estuary, National '85' Key Project Technique Report /No:85-404-01-02A (in Chinese), Nanjing Institute of Hydraulics Research.
(22)Yang, G.R and C.Y.Wu,,1988, A feasibility study of the proposed navigation channel in Huangmaohai estuary, Technique Report /No:8808 (in Chinese), Institute of Coastal & Estuarine Studies, Zhongshan Univ.
(23)Zhao, H.T. 1991, Evolution of the Pearl River Estuaries, Science press, Beijing. pp. 356.

EXPERIMENTAL FIELD INVESTIGATIONS OF SMALL DELTA FORMATION

Maria Mikhailova[1]

ABSTRACT: The main objects of investigations were small new protruding deltas at the nontidal mouths of the Terek and Sulak Rivers, being formed at the deep offshore of the Caspian Sea with the changing level. The complex of special experimental observations at the Terek and Sulak mouths helped us to study the history of delta evolution, to establish the relations between the delta formation factors and the changes in delta morphology, to examine the variations in delta channel and sea bottom profiles and to set up the sediment balance equations. These studies allowed us to assess the role of the combined influence of water flow and sediment yield, sea wave action, anthropogenic factors and sea level changes in the delta formation processes. As a result of the field experiments we have obtained the method of calculation, which permits to evaluate and to predict the delta formation. This method is based on the sediment balance equation with the use of the criterion of the river and sea interaction.

INTRODUCTION

Usually deltas are formed at the mouths of rivers with a large sediment yield. These deltas actively

[1]Doctor, Water Problems Institute, Russian Academy of Sciences, Novobasmannaya 10, P.O.Box 524, Moscow, 107078, Russia

protrude into the open sea and are formed under the combined influence of riverine and marine factors.

The study of protruding delta formation is of a large scientific and practical interest because of its importance for rational use and protection of natural resources at the river mouths.

The main objects of investigations were small new deltas at the nontidal mouths of the Terek and Sulak Rivers, being formed at the deep offshore of the Caspian Sea with the changing level. The study of these deltas gives a rare chance to examine delta formation processes in almost experimental conditions. New deltas of the Terek and Sulak Rivers started to be formed quite recently in the open sea after artificial diversion of the river flow in new directions into the Caspian Sea. Field investigations of these new deltas allowed us to obtain the experimental material, giving possibility to study some features of delta formation processes qualitatively and quantitatively.

The complex of special observations has been carried out at the Terek and Sulak mouths. It includes 27 topographic and sounding surveys at the Terek mouth (from 1973 to 1990) and 9 topographic and sounding surveys at the Sulak mouth (from 1958 to 1987). Also the measurements of the delta formation factors such as the river water flow, the sediment yield, the sea wave energy, the river and sea levels have been carried out. On the basis of topographic and sounding surveys, the evolution of these deltas have been studied and morphometrical characteristics such as length, width and area of a delta, fan volume and delta coast length have been calculated by the author. The relations between the above mentioned delta formation factors and delta morphology changes have been established (Mikhailova 1993 a,b).

THE TEREK DELTA FORMATION

In 1973 a new delta at the Terek mouth began to advance into the open deep middle part of the Caspian Sea as a result of diversion of the river flow through artificial cutoff across the Agrakhan peninsula (Figure 1). Formerly the river flowed into the shallow semienclosed Agrakhan Bay .

The evolution of this new delta includes three periods. The first one lasted from January to October 1973, when the river flowed into the sea and the pioneer mouth bar and delta were formed (Figure 2a). The second period began in November 1973, when the cutoff was artificially blocked and the delta was eroded by sea

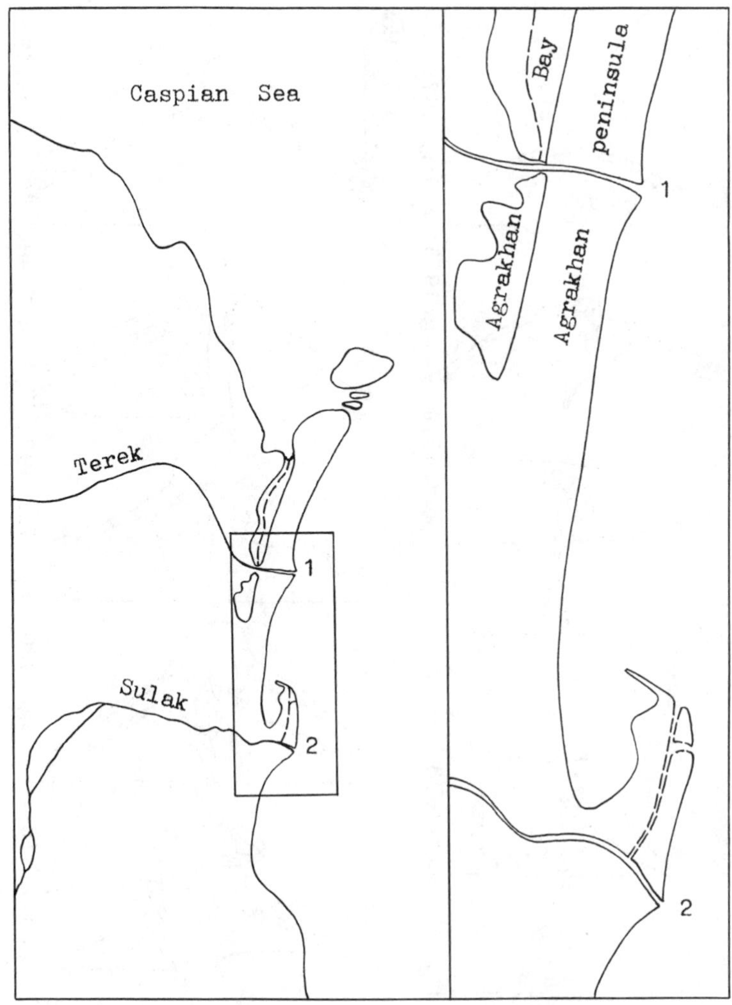

Figure 1. The New Terek (1) and Sulak (2) Deltas

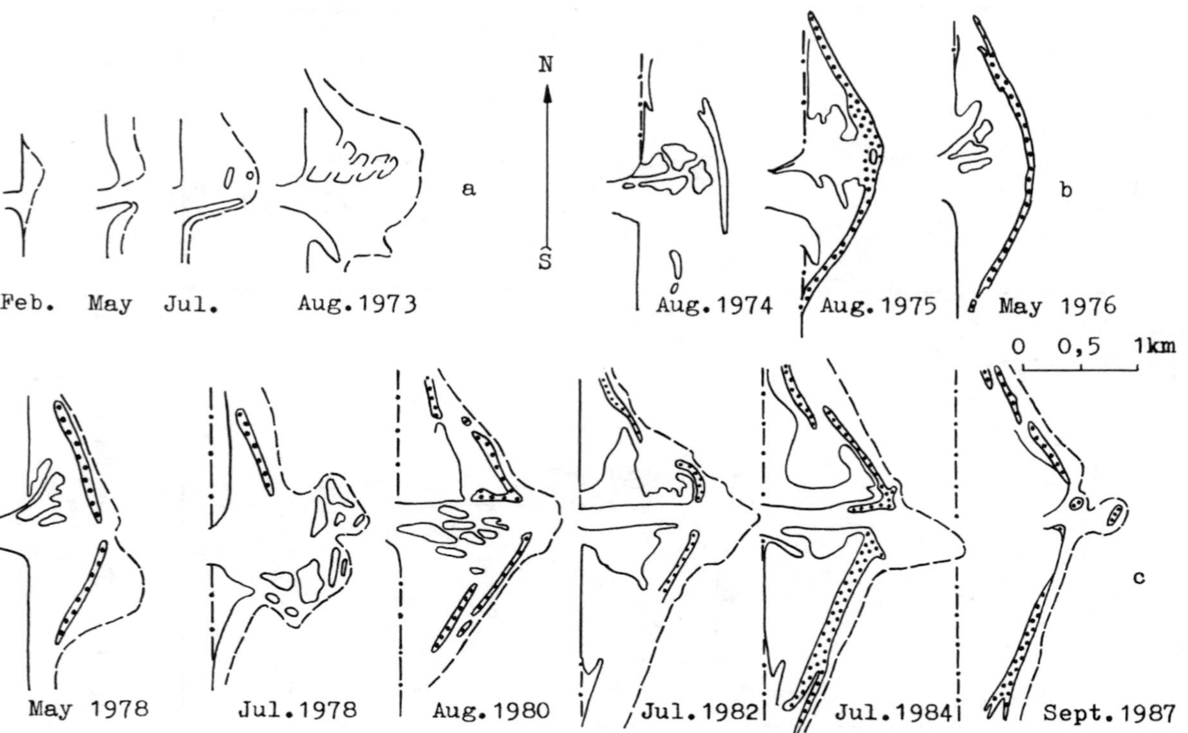

Figure 2. Scheme of the Terek Delta Formation

waves (Figure 2b). The third period began in August 1977, when cutoff was opened again and the river restored the flow into the open sea. Now the delta continues to advance into the deep offshore (Figure 2c).

The average water and sediment discharges in the new Terek delta are presented in Table 1.

Table 1. Conditions of the New Terek and Sulak Deltas Formation

Delta	Period	Mean Water Discharge m^3/s	Mean Sediment Discharge kg/s	Mean Sediment Concentration kg/m^3
New Terek Delta	1973, 1977-1986	136	253	1,86
New Sulak	1958-1974	145	415	2,86
Delta	1975-1985	127	49,3	0,39

During the first period of the Terek delta formation (February- October 1973) the values of the transit river sediments and products of the channel and the cutoff banks erosion were equal to 8,37 and 1,79 million m^3 correspondingly. The part of the sediment yield consisted of particles coarser 0,05 mm was 3,129 million m^3. Increase of the part of the delta fan composed by the coarse sediments comprised 3,140 million m^3. Hence longshore sediment supply was equal only to 0,011 million m^3. By the end of the period the delta length, the delta area and the fan volume reached 0,67 km, 1,032 km^2 and 4,822 million m^3 accordingly.

During the second period (November 1973-August 1977) the sediment yield was absent. Because of the wave action erosion the delta length and the fan volume decreased by 0,03 km and 1,322 million m^3, the delta began to spread along the coast, a beach barrier was formed and lagoons occurred between the barrier and the old shore. During this period the Caspian Sea level fall was favorable for the Terek delta fan preservation.

During the third period of the Terek delta

formation (August 1977-August 1987) the value of sediments derived to the offshore was 36,399 million m^3. The coarse sediment yield was 6,970 million m^3. Increase of the delta fan part composed by the coarse sediments comprised 5,052 million m^3. So the volume of the sediments carried away by the waves was 1,918 million m^3. By September 1987 the delta length, the delta area and the fan volume reached 1,40 km, 3,162 km^2 and 15,583 million m^3 accordingly. During the years with a large river sediment yield the Terek delta actively protruded into the sea. On the other hand, during the years with a low sediment yield Terek delta protruding slowed down and began to be eroded by thewave action. The sea level rise from 1978 to 1993 led to the fan erosion and the flooding of lagoons.

The analysis of the water surface and bottom longitudinal profiles allowed us to study the influence of the Caspian Sea level changes on channel processes at the Terek mouth and their intensity and direction. For example, as we can see in Figure 3, the sea level rise since 1978 led to the according water level and bottom elevation increase at the Terek mouth near the sea. The delta surface and the bar bottom (from 1978 to 1987) rose approximately parallel themself by 1 m. In spite of the sea level rise, the coastline cross-section (A, B, C points) was displaced seaward nearly by 1 km.

THE SULAK DELTA FORMATION

The formation of a new delta at the Sulak mouth also began after diversion of the river flow through artificial cutoff in a new direction into the open sea (Figure 1). From 1957 to 1989 the Sulak delta evolution included the pioneer mouth bar formation at the cutoff mouth, the formation of subaerial delta and its channel network (Figure 4).

The development of the Sulak delta was also determined by the changes in the delta formation factors. As a result of the Chirckeysk reservoir construction in 1974 the sediment yield decreased by about 10 times (Table 1). The decrease of the sediment yield at the Sulak mouth and the Caspian Sea level rise since 1978 led to the following. During the period from 1982 to 1987 the delta length, the delta area and the fan volume decreased by 0,08 km, 0,73 km^2 and 2 million m^3 accordingly. The beach barrier began to be

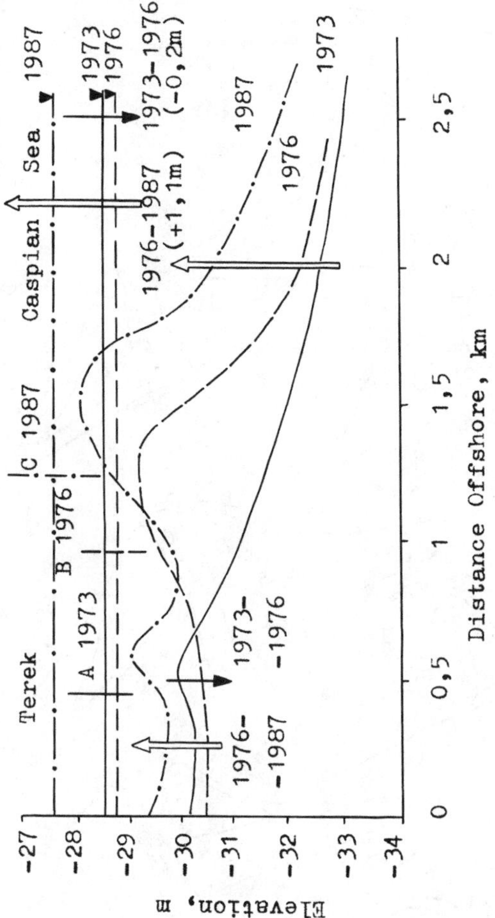

Figure 3. Longitudinal Profiles of Water and Bottom Surface

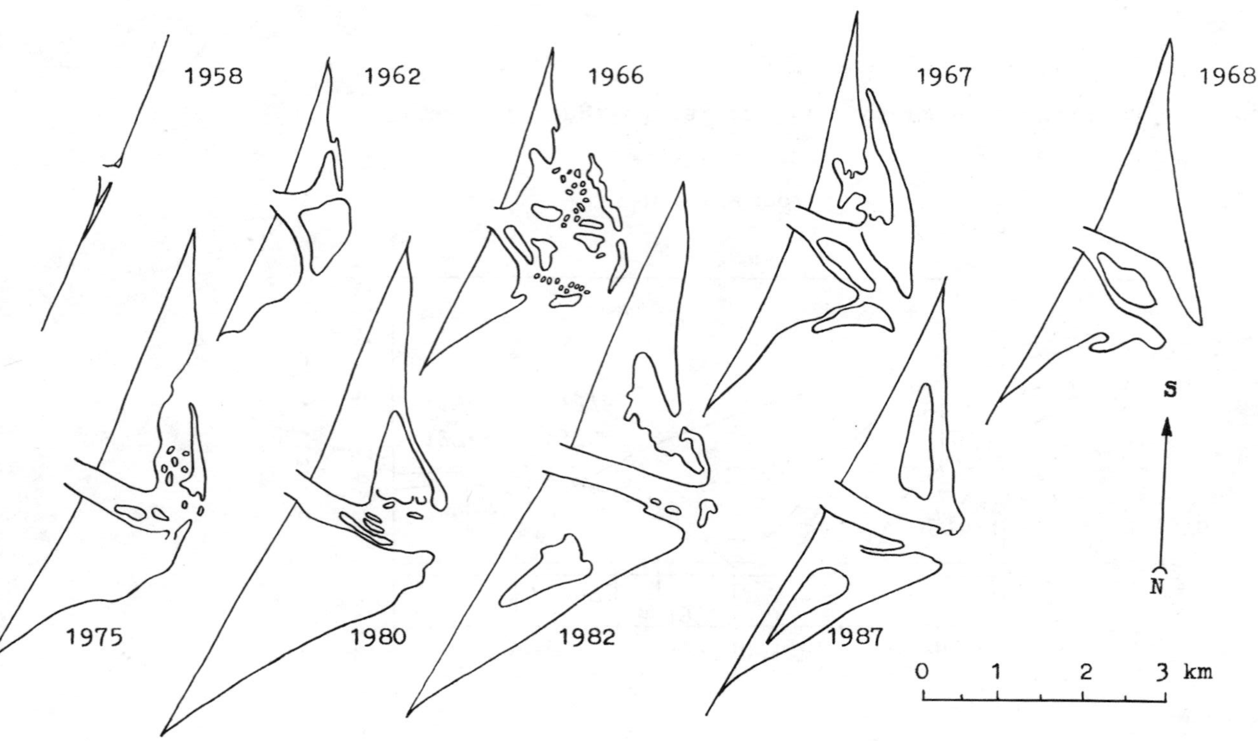

Figure 4. Scheme of the Sulak Delta Formation

displaced landward and the mouth lagoons started to be deepened and extended. The delta length, the delta area and the fan volume at the Sulak mouth by 1987 were equal to 1,8km, 4,9 km^2 and 0,05 km^3 correspondingly.

SEDIMENT BALANCE EQUATION

The sediment balance at the river mouth provides a physical basis for the delta changes.

The sediment balance equation for the river mouth during the time interval Δt is as follows:

$$\pm\Delta W_f = W_r \pm W_w - W_m , \qquad (1)$$

where $\pm\Delta W_f$ = change in the fan (delta) volume; W_r = yield of suspended and bed river sediments; W_w = volume of coarse sediments brought (+) or carried away into the sea (-) by the longshore drift; W_m = volume of fine sediments carried away to the large marine depths outside the fan.

The equation of the balance of only coarse sediments (> 0,05 mm), forming the part of the delta and the river mouth bar, can be simplified as:

$$\pm\Delta W_f' = W_r' \pm W_w , \qquad (2)$$

where $\pm\Delta W_f'$ = change in the part of the fan (delta) volume, formed by coarse sediments; W_r' = yield of coarse river sediments.

The observation data for the investigated river mouths allow us to find the components of the sediment balance equation (1). The results of the calculation are given in Table 2.

The analysis of the sediment balance equations for different periods of the delta formation shows the following: 1) the main role in the Terek and Sulak delta formation belongs to the river sediments; 2) the greater the delta size, the greater part of the river sediments remains in the fan (delta); 3) the large part of the rest of the sediments is carried away into the sea outside the fan; as a rule these sediments are fine ones; 4) the part of the sediments carried by the longshore drift is not large in the delta formation; 5) usually the delta does not intercept the longshore drift of marine sediments but it is destroyed by the waves; it also feeds the longshore sediment drift.

Table 2. Components of the Sediments Balance Equation (%)

Delta	Period	W_r	ΔW_f	W_w	W_m
New Terek Delta	3 Jan.-31 Oct. 1973	100	47,5	0,1	-52,6
	11 Aug.1977- -27 Jul.1978	100	47,5	-8,9	-43,6
	22Aug.1979- -23 Aug.1981	100	67,8	12,8	-45,0
	11 Aug.1977- -1 Sept.1987	100	33,2	-5,3	-61,5
New Sulak Delta	1958-1987	100	11,4	-88,6	

METHOD OF DELTA FORMATION EVALUATION

On the basis of the sediment balance equations we have developed a model, which connects changes in the fan volume and a criterion of the river and sea interaction.

The assumption that the morphometry and the morphology of the protruding deltas and the river mouth bars are depended on the relative influence of the riverine and marine factors have been proposed long ago. The water and sediment flow were considered as the main riverine factors and the wave energy was considered as a main marine factor. This idea belongs to L.D. Wright and J.M. Coleman (1973). These authors suggested the quantitative criterion called "river discharge index". This criterion is equal to the ratio between the specific water discharge (per unit of the delta coast length) and the specific power of the sea waves.

It was shown by the other authors (Mikhailov et al. 1981, Mikhailov et al. 1986, Polonskiy 1981) that it is more correct to use the specific sediment discharge instead of the specific water discharge, because the sediment load is the major factor forming the protruding deltas. But the investigations mentioned above have some drawbacks. They are carried out without taking into account the sediment balance. Besides, the specific sediment discharge is estimated as the whole suspended and bed sediment load. The waves are considered as the factor of the transport of only coarse sediments.

These limitations and contradictions can be eliminated by using of the balance equation (2) for coarse sediments during the time interval Δt .

The balance equation for coarse sediments can be presented as:

$$\Delta W'_f = (R' - R_W)\Delta t \ , \tag{3}$$

where R' and R_W = volume sediment discharges of coarse sediments (m^3/s) brought by the river to the delta coastline and carried away from the coastline by the waves. The sediment discharges can be expressed as the specific sediment discharges (per unit of the delta coastline length L_{dcl}):

$$R' = r' L_{dcl} \ , \tag{4}$$

$$R_W = r_W L_{dcl}, \tag{5}$$

where r' and r_W = specific discharges of coarse sediments brought by the river and carried away by the waves.

The ratio between r' and r_W is defined as the criterion of the river and sea interaction $\lambda_O = r'/r_W$. The value r' is not complicated to determine. For this purpose we should know the river discharge of coarse sediments averaged during the time interval Δt and the delta coastline length. The value r_W is complicated to evaluate by observations. Thus it is assumed that the value r_W is in proportion to the specific power of wave energy E , averaged during the time interval Δt , $r_W = aE$. Here E is equal to *1/16* $\rho g h_W^2 c$, where ρ is water density (kg/m^3), g is gravity acceleration (m/s^2), h_W is wave height (m), c is wave propagation velocity (m/s). The wave direction and the angle of wave approach to the beach is not taken into account.

Thus the sediment balance equation (3) can be written as:

$$\Delta W'_f = (r' L_{dcl} - aEL_{dcl})\Delta t \quad (6)$$

or

$$\Delta W'_f/(EL_{dcl}\Delta t) = r'/E - a \quad (7)$$

After dividing to *a* the equation (7) becomes

$$\Delta W'_f/(aEL_{dcl}\Delta t) = r'/aE - 1 = \lambda_0 - 1 , \quad (8)$$

where λ_0 = nondimensional criterion of the river and sea interaction.

As we can see from the equation (8), at $\lambda_0 = 1$ the value $\Delta W'_f$ is equal to 0 and part of the fan (delta),composed by coarse sediments, is invariable. The total volume of the fan ΔW_f is correlated with $\Delta W'_f$, so in this case the delta as a whole is stable. At $\lambda_0 > 1$ the value ΔW_f and $\Delta W'_f$ are greater than 0 and the delta protrudes into the open sea. At $\lambda_0 < 1$ the value $\Delta W'_f$ and ΔW_f are less than 0 and the delta is eroded by the waves and degrades.

The coefficient *a* in the equation (7) can be obtained by the regression analysis of the observation data.

By analogy with the equation (7) we can receive the similar equation for the total volume of the fan and the total sediment load:

$$\Delta W_f/(EL_{dcl}\Delta t) = b\ (r/E) - c. \quad (9)$$

The coefficients *b* and *c* also can be obtained by the regression analysis of the observation data.

The method can be applied for the description and evaluation of delta evolution (protruding or degradation).

This method allows us to , evaluate the changes in the fan volume (ΔW_f and ΔW_f) and these values themselves.

The considered method has been used for evaluation and prediction of the protruding delta evolution at the Terek mouth.

The measurements of coarse sediment yield, volume of the fan part, formed by coarse sediments and data for the wave action intensity at the Terek mouth during several years allow us to establish an empirical relationship for coarse sediments:

$$\Delta W_f' / E L_{dcl} \Delta t = 0{,}992\, r' / E - 0{,}222\, 10^{-6}, \qquad (10)$$

(the correlation coefficient equals 0,985).

The similar equation has been obtained for all sediments and full fan volume:

$$\Delta W_f / E L_{dcl} \Delta t = 0{,}478\, r / E - 0{,}482\, 10^{-6}, \qquad (11)$$

(the correlation coefficient equals 0,976).

The relationships (10) and (11) help us with known r, r' and E to determine the tendency of delta evolution and the value of delta advancing or eroding ($\Delta W_f'$ and ΔW_f).

Using this method, the possible changes in the fan volume at the Terek mouth and some delta morphometrical characteristics in the future under different combined impact of the river sediment yield and the wave energy have been calculated.

For calculation of the changes in the delta length L_d and the delta area F_d, the empirical relationships $W_f = 0{,}0073 L_d^{1{,}540}$ and $W_f = 0{,}0036\, F_d^{1{,}106}$ can be applied.

The specific sediment discharge (r and r') and the power of wave energy (E) in the future were given equal to average modern values ($r = 5{,}6\ 10^{-6}\ m^2/s$, $r' = 0{,}59\ 10^{-6}\ m^2/s$, $E = 4{,}23\ J/(m.s)$) and greater and smaller values than these ones. The sea level has been taken as unchangeable and equal to a modern one.

The result of the calculation showed that the direction and intensity of the Terek delta formation would depend on the relationship between the sediment yield and the wave energy.

CONCLUSIONS

Thus, the extensive field experiment has been carried out in new deltas at the Terek and Sulak mouths at the offshore of the nontidal Caspian Sea. It made a contribution to the study of the complicated delta formation processes. Among other things this experiment allowed us to assess the role of the combined influence of the water flow and sediment yield, the sea wave action, anthropogenic factors and the sea level changes in the delta formation processes. Simultaneously the sediment balance at the investigated mouths has been studied and a method of calculation of the delta changes under the combined influence of riverine and marine factors has been developed.

REFERENCES

Mikhailov,V.N., Ivanov, A.N., Lyutikov, A.V., and Polonskiy, V.F. 1981. "Delta Coastline as a Result of River and Sea Interaction", *Coastal Zone of the Sea,* Nauka Press, pp.26-32. (in Russian)

Mikhailov, V.N., Rogov, M.M., and Chistyakov, A.A. 1986. "River Deltas. Hydrological-Morphological Processes", Hydrometeorological Press, 280 pp. (in Russian)

Mikhailova, M.V., 1993 a. "The New Terek Delta Formation", *Hydrology of the Terek and Sulak Rivers Mouths*, Nauka Press, pp.70-88. (in Russian)

Mikhailova, M.V., 1993 b. "The Sulak Delta Formation", *Hydrology of the Terek and Sulak Rivers Mouths*, Nauka Press, pp.116-132. (in Russian)

Polonskiy, V.F. 1981. "New Prediction Method of River Mouth Bar Formation and Fan Protruding", *Coastal Zone of the Sea*, Nauka Press, pp.148-156. (in Russian)

Wright, L.D., and Coleman, J.M., 1973. "Variations in Morphology of Major River Deltas as Functions of Ocean Waves and River Discharge Regimes", *Bull. Amer. Assoc. Petrol. Geologists.*, Vol.57, N 2, pp. 370- 398.

ASSESSMENT AND RECOMMENDATIONS FOR THE ENHANCEMENT OF THE BARDAWIL LAGOON OUTLETS

Alfy M. Fanos[1], Ahmed A. Khafayg[1], Mohamed N. Anwar[2], and Mary G.Naffaa[1]

ABSTRACT: El Bardawil lagoon 600 square kilometers of surface area, lies along the northern coast of Sinai. The only source of water feeding this lagoon is the Mediterranean sea through two artificial boughazes and a natural outlet.. Serious problems related to shoaling and/or closure of these boughazes which transfer the lagoon to a salt pan and destroy fish productivity are addressed.. This paper reports on the. field program launched by the Coastal Research Institute (CRI) and the main findings, conclusions and recommendations of the control measures of such problem.

INTRODUCTION

Bardawil lagoon figure 1, is the largest lagoon in Egypt, situated in the north of the Sinai peninsula.. The 60000 hectare of lake Bardawil yield some 1500-2500 ton of fish yearly. This production could be raised by fisheries management and some in- lake aquaculture development to a sustainable 3000 ton. The significance of Bardawil lagoon fisheries is not only in supplying of fish for regional domestic consumption but especially in employment and export earnings.

Bardawil lagoon is a man-made hypersaline lagoon, separated from the sea by a long narrow curving sand barrier about 500 m wide (Levy 1974, 1977 a, 1977 b & 1980). Its geographical boundaries are 32° 41' & 33° 30' East and 31° 02' &

[1] Coastal Research Institute, 15 El-Pharaana St., El-Shallalat 21514 Alexandria,. Egypt.

[2] Faculty of Engineering, Alexandria University, Alexandria, Egypt.

31° 14' North. This lagoon differs from the Nile delta lagoons (Idku, Burullus and Manzala) in that it is of tectonic origin (Neev 1967), and not a deltaic lagoon, that receives its water from the drainage of the agriculture land and from the sea. The only source of water feeding El-Bardawil lake is the Mediterranean through three outlets (boughazes):. two artificial ones and the third is a natural one, figure 1. These outlets are always subjected to casual siltation; due to sediments transported by combined action of waves and currents; which tends to transfer the lagoon to a closed salt pan (Neev 1967).

When this situation takes place the lake fisheries are totally destroyed. That is why these outlets should be kept opened to ensure good circulation and renewal of water inside the lake by the continuous inflow and outflow during flood and ebb tide respectively. Thus the water salinity in the lake would be reduced and consequently providing a good environment for the fish growth and developing the lake for a higher productivity.

Several dredging operations have been carried out since 1927 and the most recent one took place in 1992(Commission of. European Community 1993). In order to investigate this siltation problem of the outlets and erosion on the eastern side of both boughazes, the Coastal Research Institute in Alexandria (CRI) conducted a comprehensive field monitoring program during the period from 1985 to 1987. which was re-evaluated and updated in 1990, 1991 and 1992 (Khafagy et al 1988, 1990 and Khafagy & Fanos 1990 & 1992).

The main objectives of this paper are: i- to summarize the collected field data, their analysis and the main findings, ii-. to present the results obtained from a developed numerical model and iii- to highlight the recommended protective works which based on the above information.

FIELD WORK MONITORING PROGRAM

In order to get a better understanding for the hydraulic & dynamic factors and the predominant conditions in the study area, a comprehensive program has been started in 1985 and continued through 1986 and 1987. These data are also renewed in 1990. 1991 and 1992. The program included collecting data on: profile survey, currents, waves, water level variation, discharge measurements through each boughaz, water properties such as salinity and temperature and surface bottom sediment samples. The following sections give brief summary of each items and the important results drawn from these data. Full details are shown and given in the technical reports published by the Coastal Research Institute (Khafagy et al 1988, 1990).

PROFILE SURVEYS AND CONTOUR MAPS

The hydrographic profiles covering the boughazes no.1 and no.2 and the adjacent areas are connected to each other by a baseline which extends more or less parallel to the shoreline (zero level line). The survey data are corrected to the same zero datum. for comparison. Contour maps have been drawn up from these data after correction. Figure. 2 shows the comparison of the shoreline for the four surveys. It is noticed that there is difference of about 900 & 1000 meters at boughazes no.1 and no.2. respectively, between the western and eastern barriers. This variation is mainly attributed to the erection of two jetties on both sides of each boughaz, hence causing accretion to the west of the western jetty and erosion on the eastern barrier. The erosion of the eastern barrier destroyed. the eastern jetty. Also deposition has been taking place near the upper part of the eastern side of the western jetty causing the shift of contour lines in this area towards the east. At the same time, further to the south of this deposited materials it was observed that erosion took place and depths reached about 5.0 meters below the mean sea level due to the ebb currents flowing to the sea. On the eastern side, the land barrier has been eroded and shifted towards the south by distances of 100 and 400 meters in the period from December 1986 to August 1992 for boughazes no.1 and no.2 respectively. The erosion and southward. shift of the barrier has created a very wide outlet at boughaz no. 2. The narrowest cross-section became about 800 m width in 1992 as opposed to 300 m in 1986. This. can be attributed to the effect of littoral current which has a predominant direction towards the east(Fanos 1986). At boughaz no. 1, the cross-section area remained almost the same because the eroded material was transported to the west and deposited in the eastern side of the boughaz. Part of the eroded material had been transported in the offshore direction thus causing accretion of the depth contours beyond 3.00 m.

LONGSHORE CURRENTS

Longshore currents inside the breaking zone were measured at two points on each side of each boughaz with special floats. two times per day. The collected data are subjected to statistical analysis and it was found that:

- The predominant longshore current direction is towards the east causing siltation on the western side of boughaz no. 1 & no. 2.

- Maximum velocity recorded at boughaz no. 1 is 56 cm/sec towards the east and 67 cm/sec towards the west. The average velocity is about 34 cm/sec in both directions.

- Maximum velocity recorded at boughaz no. 2 is 59 cm/sec towards the east and 65 cm/sec towards the west. While the average velocity is 28 cm/sec and 23 cm/sec towards east and west respectively.

DISCHARGE THROUGH THE OUTLETS

The current distribution was measured across the narrowest cross-section of the two boughazes and the discharges through each one are. computed. This is carried out by measuring the average velocity V_{av} and its direction for each element of width Dx. The discharge through each element Dq was computed as $Dq=Dx.d.V_{av}$, where d is water depth of each element. Then the total discharge through each boughaz was calculated from the following formula:

$$Q=\sum Dq=\sum Dx\ d\ V_{av}$$

The results from the field measurements and the computations using the formula could be summarized as follows:

At Boughaz no.1:

- The maximum inflow is about 660 m^3/sec with average velocity of 0.91 m/sec and maximum one is 1.15 m/sec.
- The maximum outflow is about 390 m^3/sec with average velocity of 0.54 m/sec and maximum one is 0.80 m/sec.

At Boughaz No. 2:

- The maximum inflow is about 1115 m^3/sec with average velocity of 0.82 m/sec and maximum one is 1.25 m/sec.
- The maximum outflow is 1225 m^3/sec with average velocity of 0.96 m/sec and maximum one is 1.20 m/sec.

It is clear from the above figures that the discharges through boughaz no.2. is much bigger than the corresponding ones of boughaz no. 1. This is because the cross-sectional area of boughaz no. 1 is less than that of boughaz no. 2.

SURFACE BOTTOM SEDIMENT SAMPLES

The collected surface bed sediment samples were washed, dried and mechanically analyzed to get their grain size distributions. It was found that coarser sediments of D_{50}, ranging between 0.3 and 0.5 mm are found along the beach, while the adjacent deeper parts are characterized by finer sediments between 0.05 and 0.3 mm.

CIRCULATION PATTERN

The circulation pattern in the nearshore zone of El-Bardawil outlets, limited by 6 meter contour, is deduced from currents, temperature. and salinity measurements. This pattern shows that the flow is in the longshore direction towards the east in the western side. of the outlet, while it is in the offshore / onshore directions on the eastern-side. The longshore flow causes the siltation of the western side of the outlet itself(beside the eastern side of the western jetty), while the offshore / onshore flows cause the erosion of the eastern barrier of the boughaz and accretion in both offshore contours and inside the outlet itself. The current velocity is ranging between 2&45 cm/sec with most frequent value of 25 cm/sec.

SALINITY AND WATER TEMPERATURE

The water salinity is found to be as follows:

- In the sea it varies from 38.6 to 39.78 p.p.t., with high values on the east side of the outlet due to the ebb flow from the lake and the lower ones are on the west side.
- In the middle of the lagoon, it varies from 45 to 55 p.p.t.
- In the fringes of the lagoon, it varies from 60 to 65 p.p.t.

Figure (3) shows the variations of the mean (over area and depth) salinities in six sections, of the lagoon during the period from 1969 to 1973. It should be noticed that the recommended salinity inside the lagoon is ranging from 45 to 55 p.p.t. (Commission of European Community).

The temperature in the sea is ranging between 15.4 to 22.7 degrees depending on the season of the measurements.

VARIATIONS OF WATER LEVEL

The water level was measured in the sea and at the Southern border of the lagoon at El-Talool, figure 1. The measurements in the sea showed that the maximum water level is 60 cm, while the average one ranges between 21 and 26 cm. At El-Talool the maximum water level is 26 cm and the average is about 8 cm. This means that the water entering the lagoon during flood period does not reach its southern end because of the huge area of the lagoon and the long distance between the sea and El-Talool.

WATER WAVES

The waves in the Easter Mediterranean comes mainly from directions between West and North; in accordance with the winds and fetch (Khafagy et al 1988). It was found that the predominant wave direction is NW while the swells are from NE. The maximum wave height observed is 7.5 meters in the deep water and occurs for 0.2% of the times. The following table summarizes the parameters that are to be taken into consideration for the design purposes.

Wave Height in Deep Water (H_o) in meters	0.75	1.5	2.5	3.5
Wave Period in seconds	5	6	7	8

Also there are few swells in summer with wave period of about 13 seconds.

LITTORAL PROCESSES

The prevailing waves from directions between West and North cause a resultant eastward sand drift along the entire barrier. It is composed of a component to the East (E) and a component to the West (W). Estimates of these transports of sand were carried out by different methods (Commission of European Community 1993), as follows:

i- The accumulation rate of sand near the western jetties of the outlets was observed by Inman 1990 .

ii- The monthly rate of transport was calculated by Delft Hydraulics 1990, using the available wave data and the CERC formula.

iii- The monthly rate of transport was calculated by Suez Canal Authority 1983, using the recent wave data and the CERC formula.

The results are given in the following table:

Method	Direction	Boughaz No.1	Boughaz No.2
Inman 1990	Towards E (Net)	300000	500000
Delft Hydraulics 1970.	Towards E Towards W Net E Total	340000 140000 200000 480000	600000 100000 500000 700000
Suez Canal Authority	Towards E Towards W Net E Total	576000 149500 426500 725500	360500 54500 306000 415000

These estimates clearly show the uncertainty about the movements of sand along the coast but the differences are acceptable.

In addition to those longshore transports, there are probably losses to the off-shore and on-shore by waves and wind. Moreover, erosion of the barrier is a supply to the stream of sand along the coast. Coastal Research Institute 1988 found that accretion occurs on the western side of the boughazes and erosion takes place on the eastern barrier of each boughaz which causes its retreat southward.

NUMERICAL MODEL FOR STABILIZED CROSS-SECTION

A numerical model was developed and tested for estimating velocities, discharge and lagoon water level as a function of the geometry of the system and water level fluctuation in the sea. The continuity and motion equations, governing the system are numerically solved by a variable marching time step Runge -Kutta Gill procedure (Fanos et al 1989). The model was calibrated and verified against field data. Implementation of this model for the stabilized cross-section leads to:

- The cross-sectional area of boughaz no. 1 is about 725 m^2. leading to bed width of 110 m, water depth of 5 m and side slopes of 7:1.
- The cross-sectional area of boughaz no. 2 is about 1000 m^2, leading to bed width of 165 m, water depth of 5 m and side slopes of 7:1.

PROPOSED PROTECTIVE MEASURES

The main purpose of the proposed protective measures is to create two narrow, deep and stable fixed boughazes capable to circulate and renew the water of El-Bardawil lagoon. This could be achieved by the construction of a jetty on each side of each boughaz. These jetties are expected to prevent the sand from passing the northern end of the existing western jetty (which causes the siltation of the boughaz) and to strengthen the eastern barrier. There are two alternatives for boughaz no. 1 and three ones for boughaz no.2. A brief description of those alternatives is given in the following sections (Khafagy and Fanos 1990), see figures (4) and (5).

Alternative no. 1: This alternative consists of the following items for the two boughazes:

i- Construction of two jetties , one on each side of the boughaz, the western one reaches contour (-5.00), while the eastern one reaches contour (-3.00). The cross-sectional area of the outlet gorge confined between the two jetties is about 725 and 1000 m^2., for boughaz no. 1 and 2 respectively with maximum water depth of about 5.00 meters.

ii- At the Southern end of each jetty a sandy core embankment protected by basalt stones is provided.

iii- The eastern coastal barrier is strengthened by a protective embankment with sand filling behind it up to a level +(1.5).

Alternative No. 2 : This alternative is the same as alternative no. 1 except the protective measures on the western side would be as follows for the two boughazes :

i- Renewing the old western jetty.

ii- Extending it up to water depth of 5 meters.

Alternative No. 3: The purpose of this alternative is to block the existing boughaz no.2 and dredge a new one in the eastern narrow part of the barrier at a distance of about 3.5 km to the east of the existing one.This new boughaz will be protected by two jetties, figure (6).

Dredging of the channel of each boughaz should take place in the above proposed three alternatives.The choice of the suitable solution will depend on the total cost of each alternative.

SUMMARY AND CONCLUSION

The two artificial boughazes of El-Bardawil lagoon; 600 square km ; are subjected to siltation problems which tend to transfer the lagoon to a hypersaline one that causes serious hazards to the fisheries industry. A comprehensive field monitoring program, as well as continuous updating of collected data, have been carried out during the period from 1985 to 1992. Also, a numerical model was developed to get the stabilized cross-section.It was found that:

i- The predominant longshore current is towards the east with maximum velocity of 0.56 and 0.40 cm/sec at boughazes no.1 and no.2 respectively.

ii- The predominant wave direction is NW with maximum wave height of 7.5 m, while the swells are from NE.

iii- The flow through boughaz no.2 is larger than those through boughaz no.1 because of the smaller cross-sectional area of the latter.

iv- The sediments near the shoreline are coarser (D_{50} = 0.3 - 0.5 mm) than those in deeper water (D_{50} = 0.05 - 0.3 mm).

v- The resulted littoral drift towards the east is the main reason for the siltation of the two boughazes and the erosion of the eastern barrier.

vi- The stabilized cross-sectional area was found to be 725 and 1000 square meters respectively.

Based on the above results, three alternatives have been proposed to improve the existing hydraulic conditions of the lagoon and renew its water. The choice will depend on the cost of each.

REFERENCES

Commission of European Community, 1993. Bardawil Lagoon. Development Project,Management of Environmental parameters. Project submitted from Euroconsult in association with Delft Hydraulics, EC:476.7105/DH:H1460.

Fanos,A.M.,1986. Statistical analysis of longshore current along the Nile Delta coast. Water Science Journal,Cairo,1:45-55.

Khafagy, A.A.,and Fanos, A.M.,1990. Protective works for El-Bardawil outlets:. Alignment and Conceptual Design. Technical Report no.2, Coastal Research Institute Alexandria.

Fanos, A.M. and Khafagy,A.A., 1991. Detailed protection works for El-Bardawil outlets. Technical Report no. 23,(3 volumes in Arabic) Coastal

Research Institute, Alexandria.

Khafagy,A.A., and Fanos, A.M.,1992. Alignment of El Bardawil outlets and dredging works. Technical Report no.4, Coastal Research Institute; Alexandria.

Fanos,A.M., Khafagy, A.A. and Elwani, M.H.,1989. Numerical Model for El -Bardawil tidal inlets. Estuarine and Coastal Modelling and Pollutant Transport, Modelling Conference, ASCE, 15-17 November 1989, Sheraton Islander Newport, R.I.1989.

Khafagy,A.A., Fanos,A.M., Frihy,O.E., El-Fishawi,N.M., and Naffaa,M.G.,1988 Project of development El-Bardawil Fish Resources. Final Technical Report, 3 volumes, Coastal Research Institute, Alexandria.

Khafagy,A.A., Fanos,A.M., Frihy,O.E., El-Fishawi,N.M., and Naffaa,M.G., 1990.
Up-dating the coastal processes data of El-Bardawil outlets. Technical Report No. 1, Coastal Research Institute, Alexandria.

Levy, Y., 1974. Sedimentary reflection of depositional environments in the Bardawil lagoon, Northern Sinai. J. Sed. Pet.,vol.44, No.1,pp.219-227.

Levy,Y.,1977a. The origin and evolution of different types of brine in coastal sabkas, northern Sinai. J.Sed. Pet., vol. 47,pp. 451-462.

Levy, Y., 1977b. Description and mode of formation of the supratidal evaporite facies in Northern Sinai coastal plain J.Sed.,vol.47,pp 463-474.

Levy, Y. , 1980. Evaporatic environments in Northern Sinai, in:A.Nissenbaum, hypersaline brines and evaporitic environment. Developments in Sedimentology Series, No.28,pp. 131-143.

Neev, D, 1967.Geol. Survey Israel Marine Geol Div., report No.1767, pp. 15.

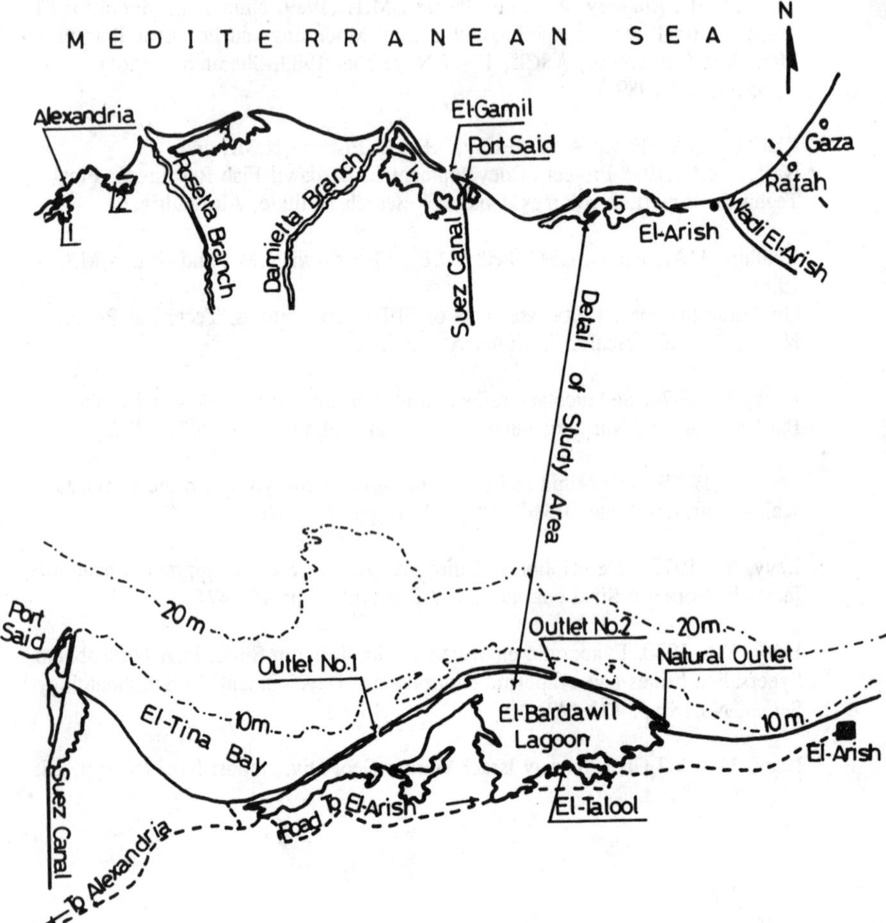

Fig 1 : General Location Map for Study Area.

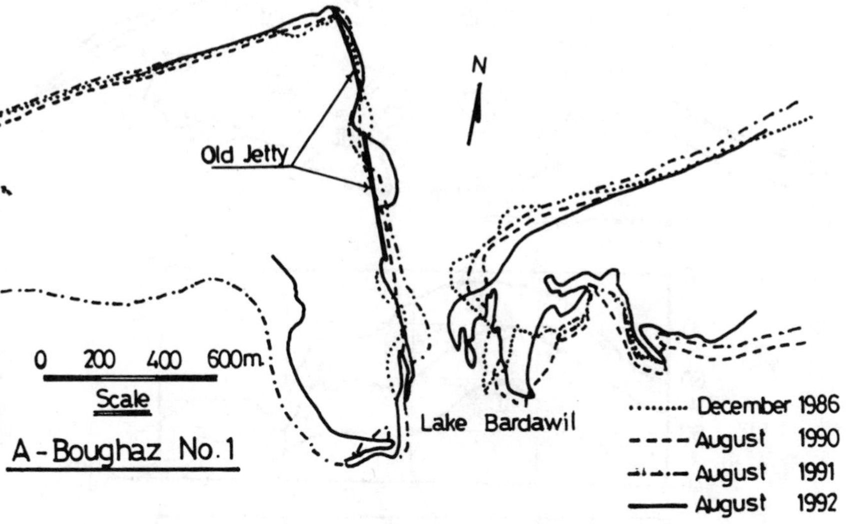

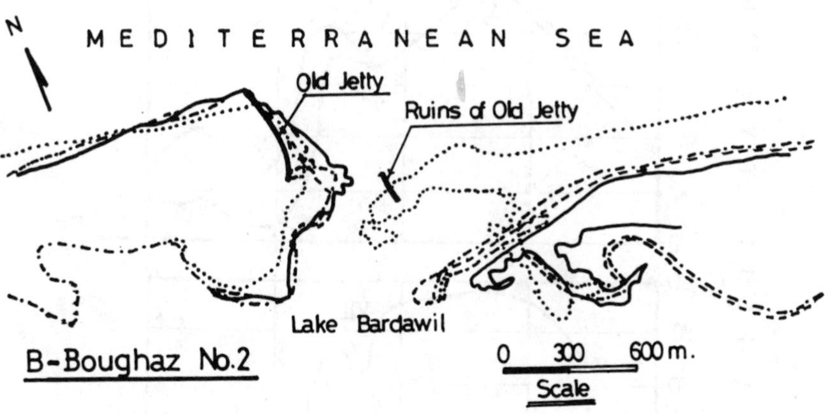

Fig 2 : Shoreline Changes for Boughazes No.1 & No.2

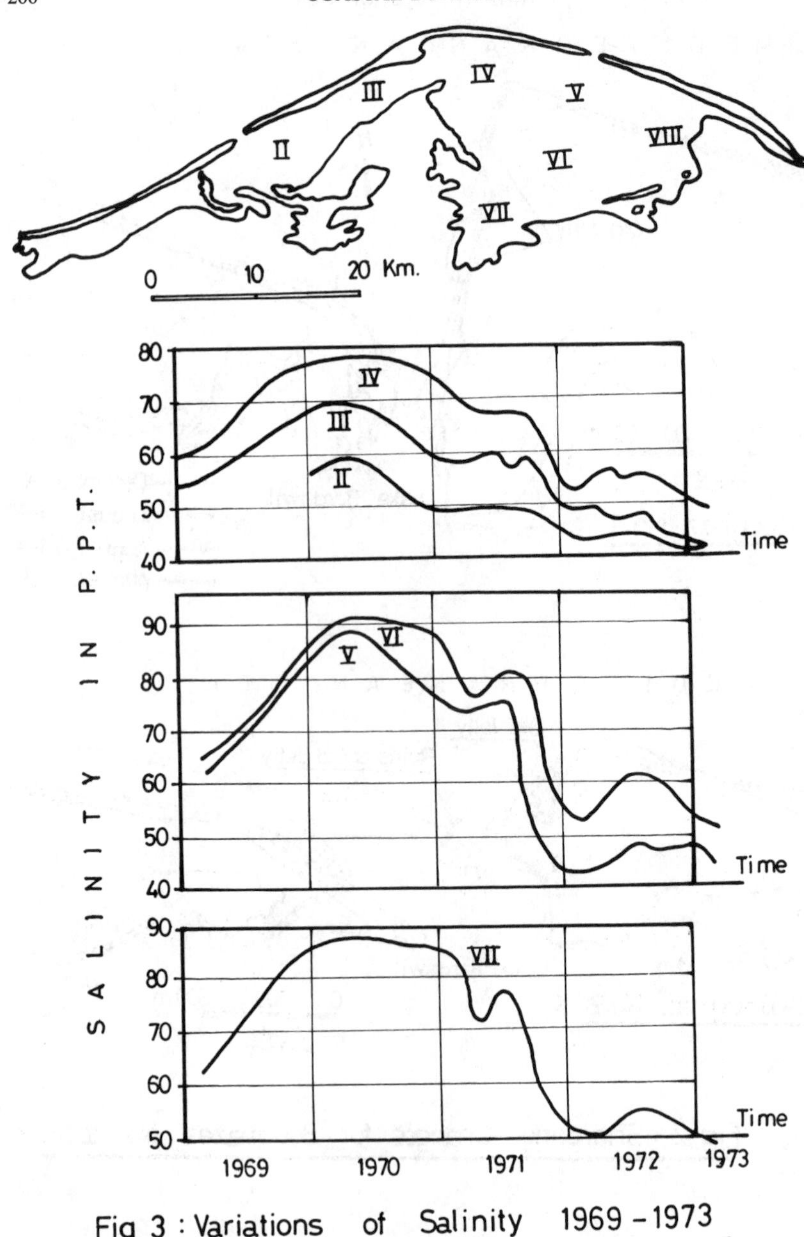

Fig 3 : Variations of Salinity 1969 - 1973

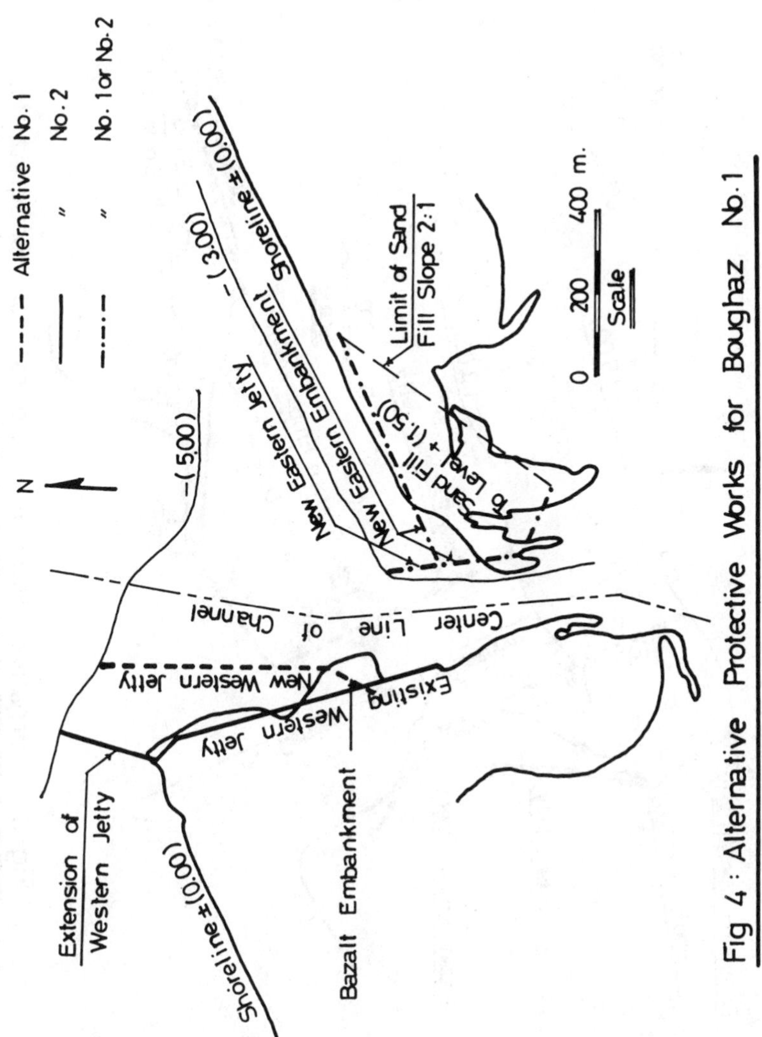

Fig 4 : Alternative Protective Works for Boughaz No. 1

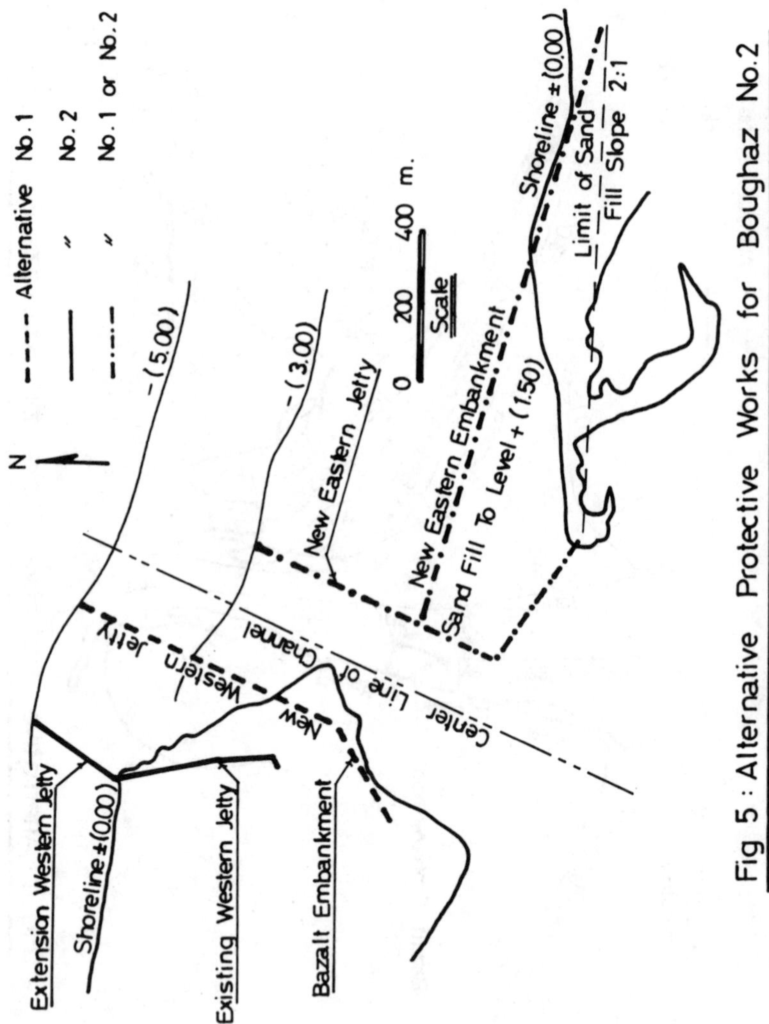

Fig 5 : Alternative Protective Works for Boughaz No.2

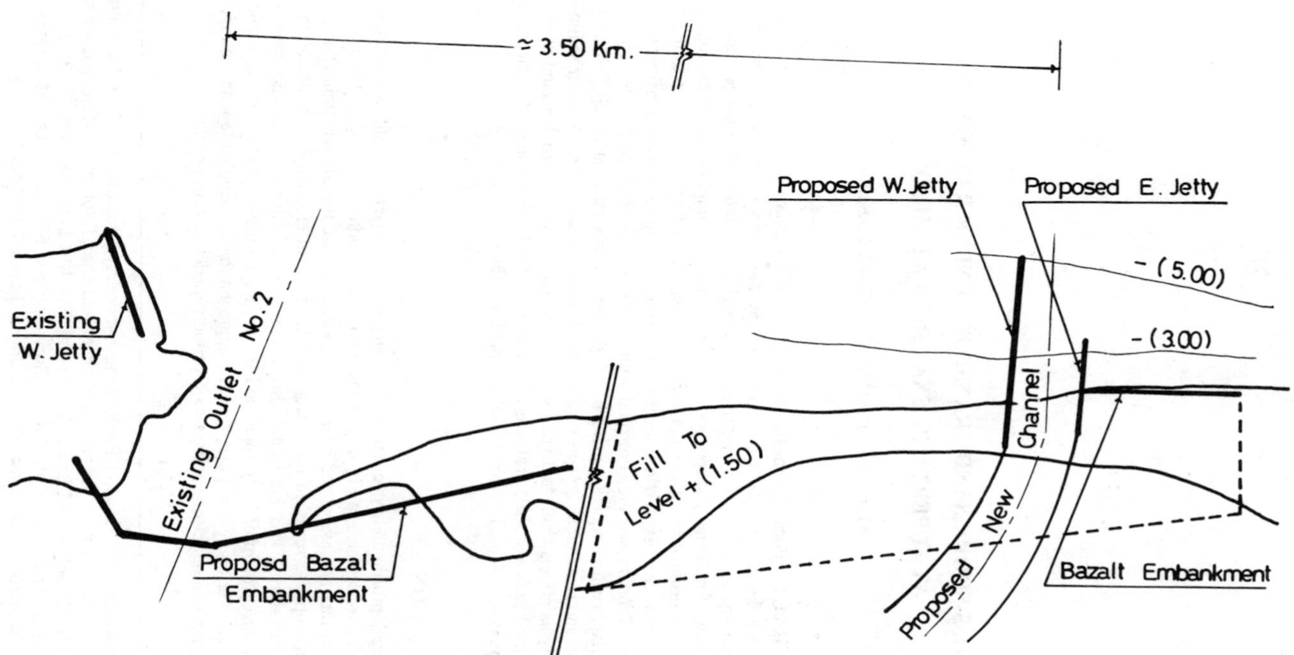

Fig 6 : Alternative No. 3 for Boughaz No. 2

CROSS-SHORE SAND TRANSPORT UNDER RANDOM WAVES AT SUPERTANK EXAMINED AT MESOSCALE

Magnus Larson[1] and Nicholas C. Kraus[2]

ABSTRACT: Formulas for the net cross-shore transport rate by random breaking waves are derived based on previous work by the authors addressing transport by monochromatic waves. Starting from a Rayleigh wave height distribution, analytic expressions for the net transport rate are obtained under simplifying assumptions that include representation of the random wave field by a large number of individual waves. Predictive expressions for the net direction of transport and its magnitude are derived. Calculations are compared with transport rates obtained by integrating the mass conservation equation using beach profile surveys made at the SUPERTANK Project, and reasonable agreement is found. It was necessary to account for sediment porosity as affects total sand volume, and it is concluded that variable sand porosity along the beach and shelf profile is a process that should be considered in theoretical developments of sand transport and morphology change.

INTRODUCTION

The beach profile undergoes cross-shore sand transport over many different temporal and spatial scales, as depicted in Fig. 1. Depending on the phenomenon or application of interest, the transport rate and associated beach profile evolution are usually determined as averages over the appropriate scale. In numerical models of storm erosion, beach profile change is predicted for storms and hurricanes that occur on the order of days and span several kilometers across the profile. In such models the sand transport rate is preferably calculated as an average over many wave periods in an approach that is compatible with the input data, reduces computational effort, and promotes robust model performance.

---

1) Guest Researcher, Coastal Engineering Laboratory, Department of Civil Engineering, University of Tokyo, Tokyo, Japan. Permanent address: Dept. of Resources Engineering, Lund Institute of Technology, University of Lund, Box 188, Lund, Sweden S-221 00.
2) Director, Conrad Blucher Institute for Surveying and Science, Texas A&M University – Corpus Christi, 6300 Ocean Drive, Corpus Christi, Texas 78412-5599.

Several studies have inferred cross-shore transport rates from measured beach profiles in the laboratory, where longshore transport and offshore sand losses are absent (cf., Larson and Kraus 1989). These *mesoscale* transport rate distributions represent averages over hundreds of individual waves and describe morphological features that evolve at this time scale (e.g., dune erosion; bars and berms). The SUPERTANK Laboratory Data Collection Project (Kraus et al. 1992; Kraus and Smith 1994) generated high-accuracy data on profile change with negligible scale effects. SUPERTANK also provides information on the mesoscale transport rate for both monochromatic and random waves, allowing direct comparison between these wave types in examining profile evolution and cross-shore transport rate.

The objectives of this paper are to derive a mesoscale model of the net cross-shore transport rate under random waves and to validate the model using the SUPERTANK data. Calculated rates are compared with transport rate distributions determined from sediment conservation between consecutively surveyed beach profiles at SUPERTANK.

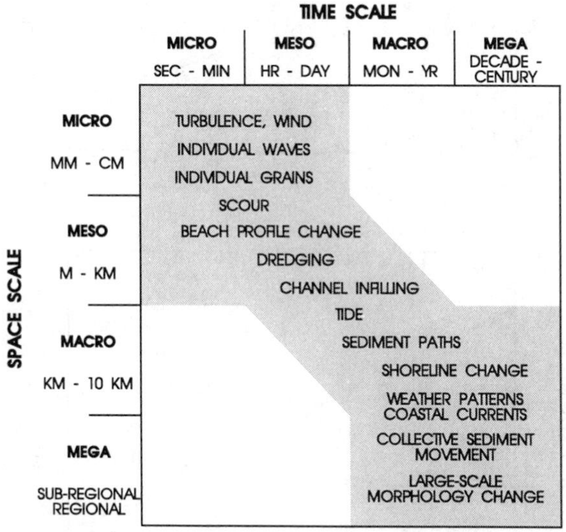

Fig. 1. Compatible temporal and spatial scales for calculating sediment transport and beach morphology change (from Larson and Kraus 1994).

CROSS-SHORE TRANSPORT UNDER MONOCHROMATIC AND RANDOM WAVES

Larson and Kraus (1989) studied beach profile change measured in large wave tanks (LWT) (Saville 1957, Kajima et al. 1982, Kraus and Larson 1988) with monochromatic waves. They inferred four different transport zones based on the wave hydrodynamics and analysis of net cross-shore transport rate distributions obtained from profile surveys, which are the prebreaking, breaker transition, broken wave, and swash zones. Semi-empirical transport relationships were derived for each zone, for which the magnitude of the transport is primarily a function of the wave energy dissipation produced by breaking waves.

Such a simplified description of the cross-shore transport is a first step toward characterizing field conditions and random waves. Random waves typically break across a much wider portion of the profile than monochromatic waves, producing a more evenly distributed energy dissipation across shore. Thus, the net cross-shore transport rate distribution is expected to be more uniform under random waves and profile evolution smoother. Boundaries between transport zones should be less distinguishable for random waves with no easily identifiable average break or plunge point. However, in an individual-wave approach, it is still possible to identify the point and plunge point of the individual waves in a random wave field.

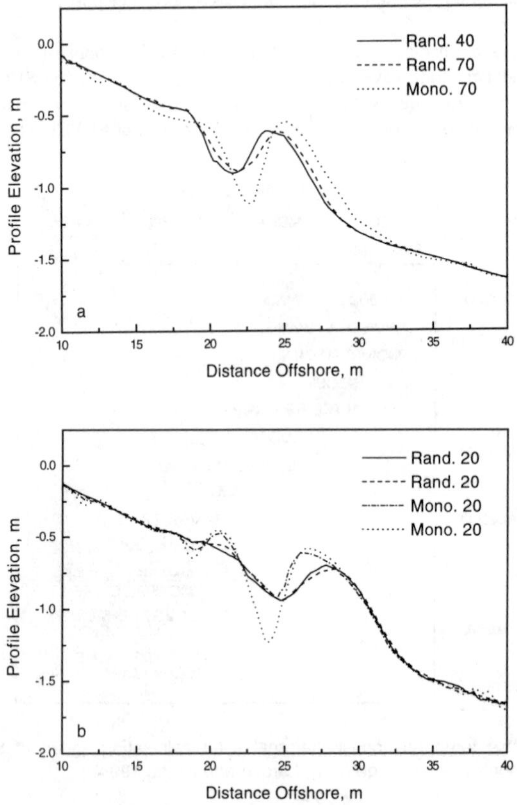

Fig. 2. Bar shape under random and mono. waves measured during SUPERTANK Case ST_10 for (a) Runs A0611A-A0613A, and (b) Runs A0815A-A0817A.

Figs. 2a and 2b show comparisons between the action of random and monochromatic waves on profile development using data from SUPERTANK Case ST_10 (Kraus et al. 1992, Kraus and Smith 1994). For the runs shown in both figures the beach profile was first exposed to random waves until near equilibrium, after which monochromatic waves were employed. A monochromatic wave height of H=0.8 m, corresponding to the energy-based significant wave height H_{mo} of the random wave field, was used in the runs shown, and the monochromatic wave period was 3.0 and 4.5 sec for the respective runs in Figs 2a and 2b, the same as the peak spectral period T_p. The figures only display that portion of the profile with the bar, because the inshore region did not change notably during the discussed runs.

It is clearly seen in both figures how the shape of the bar changed after it was exposed to monochromatic waves; bar size increased simultaneously as a pronounced trough developed. The seaward slope of the bar did not change, whereas the shoreward slope steepened markedly as the trough formed. Fig. 2b shows that for the longer period waves a small secondary feature appeared directly shoreward of the trough, probably created by the splash-jet motion of plunging waves pushing material onshore.

NET CROSS-SHORE TRANSPORT RATE

A standard method of determining the net cross-shore transport rate is to integrate the sediment conservation equation between profile surveys. Assuming the density of the sediment particles is constant, the conservation equation applied across shore is,

$$\frac{\partial q_s}{\partial x} = \frac{\partial}{\partial t}[h(1 - n)] \tag{1}$$

where q_s is the volumetric transport of sediment particles, h the elevation of the bottom topography, and n the sediment porosity. In numerical models of beach profile change, the volumetric transport of water and sediment particles $q=q_s/(1 - n)$ is typically employed, because this transport directly gives the corresponding change in the bottom elevation. The porosity is assumed constant irrespective of location across the profile; without this simplifying assumption, spatial and temporal changes in porosity must be modeled.

In determining the net cross-shore transport rate from measured profile change, n has also typically been assumed constant, implying conservation of sediment volume so that q is obtained directly by integrating elevation changes between two profile surveys. However, n is in general not constant, and small changes in n can lead to large deviations in sediment volume. If n varies appreciably, Eq. 1 should be employed with n as a function of x and t. To derive representative cross-shore transport rates from profile measurements, assumptions (or data) are needed on how the porosity changed between surveys. It is of interest to determine such transport rates and examine their properties, because numerical models of beach profile change typically calculate a response based on an average porosity.

There may be several causes for an apparent or actual variation in measured sediment volume, such as inadequate survey accuracy and three-dimensionality in profile response. Larson and Kraus (1989) used LWT data to derive net cross-shore distributions for q, and they assumed survey accuracy to be the major factor causing non-conservation of volume. They modified calculated q-values for those cases where sediment volume was not conserved so that q=0 at the seaward boundary, obtaining transport rates that, in effect, correspond to a characteristic porosity of the beach profile. Larson and Kraus (1989) displaced one of two compared profiles with respect to the other so that volume conservation was obtained. The magnitude of this displacement was typically smaller than the estimated survey accuracy (order of 1 cm or less).

Calculated net transport rates from profile surveys using the SUPERTANK data also showed variation in sediment volume between profiles. In the SUPERTANK project, however, survey accuracy was higher than the earlier LWT experiments studied by Larson and Kraus (1989), and it is expected that survey errors are minor. During SUPERTANK some measurements of sediment density were performed from which porosities were derived, giving values for n in the range 0.28 to 0.47 (Sollit et al. 1994). This considerable variation in n may explain the lack of sediment volume conservation in spite of high survey accuracy.

Based on the above considerations, several different methods for obtaining representative transport rates that ensure volume conservation were investigated in the present study. If porosity is a major cause of variation in sediment volume, some assumption must be made about how the porosity changes between surveys. By introducing a conceptual model for porosity change, Eq. 1 provides constraints on the variation in the porosity values for the specific case under study. For sufficiently simple models, this constraint uniquely determines the porosity variation, and q_s is known everywhere. By dividing q_s with an average porosity, a representative transport rate is obtained that conserves sediment volume. Most wave tank experiments have focussed on profile surveying, and few measurements exist of the variation in sediment properties such as grain size or porosity, giving little guidance on how to select an appropriate model of porosity change.

In addition to the method employed by Larson and Kraus (1989) discussed above, more physically based methods related to assumptions about the porosity changes were employed to obtain representative transport rates: (1) porosity is uniform along the tank but changes between surveys, (2) porosity is constant, but different, for eroding and accreting portions of the beach, and (3) porosity varies from its average across the profile according to a specified function, which is assumed proportional to the elevation change. Change in porosity could result from compaction of the entire sediment volume or because sediment is deposited with a different porosity than what it had in the region from where it originated. Method 1 is a simple description of general compaction, whereas Methods 2 and 3 are based on porosity changes related to the material actually transported along the profile.

Figs. 3a and 3b show computed net transport rates from SUPERTANK profile surveys based on elevation change and application of the different methods to obtain transport rates that conserve volume. The case in Fig. 3a has a large difference in calculated sand volume between consecutive surveys, and Fig. 3b is a case with volume almost conserved. In the latter case, the selection of method is seen to be of minor importance for determining a representative net transport rate distribution, because sand volume is almost conserved. In Fig. 3a, the particular method influences the distribution of q more markedly, although the difference between methods is not great in the region of major profile change, especially considering the fairly large discrepancy in sediment volume for which an adjustment must be made. Because it is the gradient in q that produces depth change, the effective difference between methods is smaller than what the different q-distributions may indicate. For example, the various q-distributions in Fig. 3 a have similar local gradients in q across the profile, even though the absolute values may differ considerably at certain locations.

Thus, the representative q is an average volumetric rate that is based on a porosity characteristic of the transition from one profile to the other. Derivation of a transport rate based on average porosity is necessary if modeling the spatial and temporal changes in the porosity is to be avoided. However, derivation of a q for an average n implies that the profile change

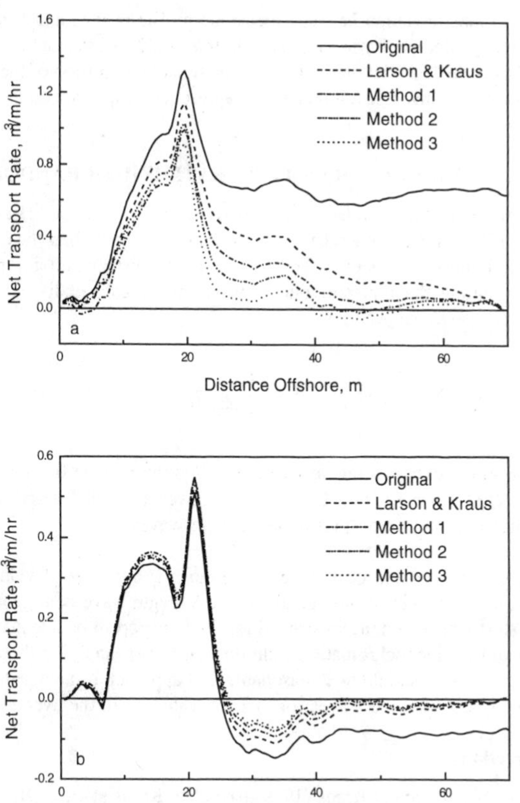

Fig. 3. Net cross-shore transport rates calculated from profile change: SUPERTANK Case ST_10: (a) A0510A, and (b) A0512A.

between the two surveys used in the transport rate computations can no longer be exactly reproduced. We have found the deviations to be small, but they do depend on the method employed. If it is assumed that changes in sediment volume are mainly due to compaction of the total volume (Method 1), measured changes in profile elevation are not greatly modified because the modifications of q are distributed over the entire profile. A method applied only to portions of the profile where elevation changes occur (Method 2 or 3) could produce larger deviations with respect to the measured elevation change.

In summary, for studying properties of the cross-shore transport rate and implications for the resultant profile evolution, the different methods investigated gave similar results, in particular with regard to the crude assumptions made in deriving representative transport rates.

However, to minimize deviation between measured profile change and change produced by the representative q, Method 1 was employed in this study to calculate transport rate distributions for comparison to model predictions. In the remaining portion of the paper, where the inferred transport rate is discussed, it refers to a representative q based on the average porosity for the profile.

MODEL OF CROSS-SHORE TRANSPORT RATE UNDER RANDOM WAVES

The cross-shore transport relationships derived by Larson and Kraus (1989) from monochromatic LWT data are generalized to random waves by treating the random wave field as a collection of individual waves. The transport rate distribution for random waves is obtained by computing the transport rate distribution for each individual wave and then averaging over all waves according to

$$\bar{q} = \frac{1}{N}\sum_{i=1}^{N} q_i \tag{2}$$

where $\bar{q}$ is the average transport rate at location x, N is the number of individual waves, and q_i the transport rate for wave i at x. Larson (1994) gives a detailed description of the model used to calculate cross-shore transport under random waves.

The transport relationships of Larson and Kraus (1989) were developed to represent averages over a large number of individual waves. A single wave is in general not expected to develop the steady-state conditions needed for the transport to be adequately described by the equations valid for monochromatic conditions. However, applying these transport equations for individual waves should be a reasonable first approach if the equations give the correct proportionality between the waves for their contribution to the average transport rate.

Transport Direction

The criterion of Larson and Kraus (1989) (see also, Kraus et al. 1991) to distinguish accretionary and erosional waves (transport onshore or offshore, respectively) is used to determine the smallest monochromatic wave height in deep water that will yield erosion,

$$H_{oc} = \sqrt{\frac{1}{M}\frac{(wT)^3}{L_o}} \tag{3}$$

where H_{oc} is critical deep-water wave height that separates erosion and accretion, w is sediment fall speed, T is wave period, L_o is deepwater wavelength, and M is an empirical coefficient equal to 0.00070. If $H_o > H_{oc}$, erosion is predicted, whereas $H_o < H_{oc}$ indicates accretion. A Rayleigh distribution for the wave height in deep water contains a certain number of accretionary and erosional waves, and the net change in profile shape will be the sum of all these waves. Eq. 3 was developed to predict profile response primarily to breaking waves, which should be taken into account in judgement of the overall computed change along the profile.

Generalization of Eq. 3 to random waves is first considered for a profile where the depth increases monotonically with distance offshore. For such a profile the ratio of breaking waves

α may be determined at all points across shore by truncating the local Rayleigh probability density function (pdf). At location x, a certain portion δ_e of the breaking waves will be erosional, which may be determined by

$$\delta_e = \frac{e^{-\left(\frac{H_{oc}}{H_{rmso}}\right)^2}}{e^{-\left(\frac{H_{bo}}{H_{rmso}}\right)^2}} \qquad \delta_e \le 1 \tag{4}$$

where H_{rmso} is the deep-water rms wave height, and H_{bo} is the wave height at incipient breaking transformed to deep water. Eq. 4 is obtained by comparing, in deep water, the ratio of waves for which $H > H_{oc}$ with the ratio of breaking waves. If the expression for H_{oc} from Eq. 3 is substituted into Eq. 4, the following is obtained:

$$\delta_e = \frac{e^{-\frac{1}{M}\frac{H_{rmso}}{L_o}\left(\frac{wT}{H_{rmso}}\right)^3}}{e^{-\left(\frac{H_{bo}}{H_{rmso}}\right)^2}} \qquad \delta_e \le 1 \tag{5}$$

Thus, the portion of the breaking waves that are erosional may be predicted using a similar set of parameters as for distinguishing profile response under monochromatic waves; however, the deep-water wave steepness and dimensionless fall speed should be expressed in terms of H_{rmso} as the characteristic wave height. Note that in both Eq. 4 and Eq. 5, δ_e cannot exceed 1.0, which states all breaking waves are erosional, that is, $H_{bo} > H_{co}$.

Eq. 5 predicts only the portion of the breaking waves that are erosional at points across the profile, and the direction of transport is not obtained. The simplest approach for arriving at a predictive equation for the transport direction from Eq. 5 is to assume that at a certain depth each individual wave transports about an equal magnitude of material. If a portion δ_e of α consists of erosional waves, then the remaining waves, $\delta_a=1-\delta_e$, of α are accretionary. Assigning equal weight to each individual wave in summing to determine the net direction yields $\xi=\delta_e-\delta_a=2\delta_e-1$, where ξ is a *transport function* that gives the net direction and a weight that includes the variability in wave height defined by the Rayleigh pdf. Thus, using δ_e from Eq. 5 gives an expression for ξ:

$$\xi = 2\frac{e^{-\frac{1}{M}\frac{H_{rmso}}{L_o}\left(\frac{wT}{H_{rmso}}\right)^3}}{e^{-\left(\frac{H_{bo}}{H_{rmso}}\right)^2}} - 1 \qquad -1 \le \xi \le 1 \tag{6}$$

Figs. 4a and 4b illustrate the dependence of the transport function ξ on normalized water depth and deep-water wave steepness for selected values of the dimensionless fall speed. The breaker ratio γ_b was set to 0.78. Fig. 4a shows an example where most of the waves are accretionary close to shore, and it is only at larger depths that the transport function ξ becomes positive, which indicates erosion. In contrast, Fig. 4b displays the dependence of ξ for wave conditions where erosion is also predominant close to the shoreline.

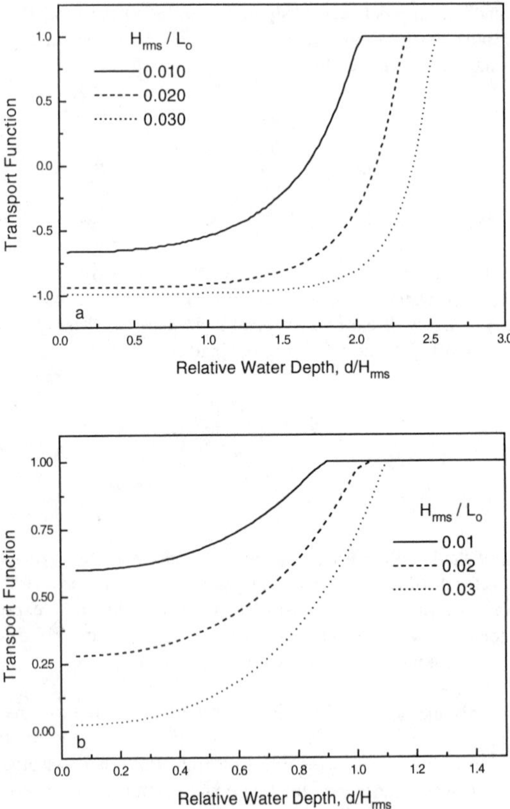

Fig. 4. Dependence of transport function ξ on normalized water depth and deep-water wave steepness for dimensionless fall speeds of (a) 2.0, and (b) 4.0.

One important condition for deriving the transport function ξ and the criterion for transport direction under random waves was that the profile be monotonic. A monotonic profile implies a unique relationship between the ratio of breaking waves α and depth. For a non-monotonic profile, such as a barred profile, α is not a simple function of depth because wave reforming may occur. A random wave model (for example, see Larson 1993) may provide predictions of α for an arbitrary profile; however, to estimate ξ across a barred profile the portion of the reforming waves that is either erosional or accretionary must be specified. An assumption compatible with the approach presented here is that there is equal probability that erosional and accretionary waves reform shoreward of the bar, suggesting that ξ is constant along negatively sloping sections of the beach as α decreases. This means that ξ always increases with distance offshore.

Transport Under Breaking Waves

The average transport rate $\overline{q_b}$ under breaking waves may be written

$$\overline{q_b} = \frac{1}{N}\sum_{i=1}^{N} K\left[D_i - (D_{eq} - \frac{\epsilon}{K}\frac{dh}{dx})\right] \tag{7}$$

where N is a large number of waves, D is the wave energy dissipation per unit water volume and D_{eq} its equilibrium value (Dean 1977), and K and ε are empirical transport coefficients. Because only breaking waves contribute to the transport at a certain location, N in Eq. 7 may be replaced by the number of breaking waves m, which is a function of x:

$$\overline{q_b} = \frac{1}{N}\sum_{i=1}^{m} K\left[D_i - (D_{eq} - \frac{\epsilon}{K}\frac{dh}{dx})\right] \tag{8}$$

A certain number of the breaking waves, m_e, will be erosional and the remainder, m_a, will be accretionary at a location x, where $m(x)=m_e(x)+m_a(x)$. Separating erosional and accretionary waves in Eq. 8 yields, after some rearranging:

$$\overline{q_b} = \frac{1}{N}\sum_{i=1}^{m_e} KD_i - \frac{1}{N}\sum_{i=1}^{m_a} KD_i - K\left(D_{eq} - \frac{\epsilon}{K}\frac{dh}{dx}\right)\frac{m_e - m_a}{N} \tag{9}$$

The last factor of the right-hand term in Eq. 9 is identified as $\xi\alpha$.

An assumption must be made regarding partitioning of average energy dissipation between erosional and accretionary waves. A random wave model will predict the average energy dissipation, but no information is obtained on the amount of dissipation that contributes to onshore and offshore transport. The simplest approach is to assume that each breaking wave at a specific water depth transports approximately the same amount of material. The wave energy dissipation terms in Eq. 9 can be written,

$$\sum_{i=1}^{m_e} D_i - \sum_{i=1}^{m_a} D_i = \frac{m_e - m_a}{m}\sum_{i=1}^{m} D_i \tag{10}$$

where $(m_e-m_a)/m$ is identified as ξ.

Substituting Eq. 10 into Eq. 9, the following expression is obtained,

$$\overline{q_b} = K\xi\left[\overline{D} - \alpha\left(D_{eq} - \frac{\epsilon}{K}\frac{dh}{dx}\right)\right] \tag{11}$$

where $\overline{D}$ is the average energy dissipation per unit water volume. Similar to the situation for monochromatic waves, the expression within the larger brackets has to be positive; otherwise the transport is set to zero. The function ξ automatically provides the direction of the transport, as well as weighting the influence of erosional and accretionary waves.

Transport Under Non-Breaking Waves

The net transport rate seaward of the break point of an individual wave is assumed to decay exponentially with distance offshore (Larson and Kraus 1989). For a random wave field, the contribution to the transport rate from non-breaking waves is estimated to be

$$\overline{q_u} = \frac{1}{N}\sum_{i=1}^{n} q_{bi} e^{-\lambda_i (x - x_{bi})} \tag{12}$$

where n is the number of non-breaking waves at x, x_{bi} the location of the break point for wave i, q_{bi} the transport rate at incipient breaking for wave i, and λ_i the exponential decay coefficient, the latter two variables evaluated at x_{bi}. Eq. 12 sums up the contributions to the transport rate from all waves that break inshore of x. The coefficient λ_i depends on the median grain size and incipient breaking wave height as for monochromatic waves.

Eq. 12 is most easily solved by dividing the profile shoreward of the studied point x in a number of grid cells n_s and adding together the contribution from each cell to the transport rate at x. Such a method of approximating Eq. 12 yields

$$\overline{q_u} = \sum_{j=1}^{n_s} q_{bj} e^{-\lambda_j (x - x_{bj})} \Delta \alpha_j \tag{13}$$

where $\Delta\alpha_j$ represents the increase in the ratio of breaking waves in cell j (index j denotes the grid cell number as opposed to i that denotes the number of a single wave). In Eq. 13, q_{bj} and λ_j must be estimated at all shoreward locations before the transport rate can be calculated. The transport rate q_{bj} may be calculated from the local wave energy dissipation per unit water volume D_{bj} at the depth d_{bj}. If the wave model by Dally et al. (1985) is used to estimate D_{bj}, the following expression results,

$$D_{bj} = \frac{1}{8}\rho g \kappa C_{gbj} (\gamma_b^2 - \Gamma_b^2) \tag{14}$$

where C_{gbj} is the wave group speed at x_{bj}. The transport rate q_{bj} is then estimated as,

$$q_{bj} = K\left[D_{bj} - (D_{eq} - \frac{\epsilon}{K}\frac{dh}{dx})\right] \qquad q_{bj} \geq 0 \tag{15}$$

In calculating q_{bj}, terms should only be included in the sum if $\Delta\alpha_j$ is positive; in the case of $\Delta\alpha_j < 0$, the contribution to the transport rate is set to zero.

COMPARISON BETWEEN TRANSPORT MODEL AND DATA

Predicted net cross-shore transport rate distributions were compared with distributions inferred from measured profile change during SUPERTANK. Selected results are presented for SUPERTANK Cases ST_10 and ST_K0 that are characteristic of the model performance. Case ST_10 involved foreshore erosion and offshore bar formation, and Case ST_K0 concerned response of a relatively large, artificially placed offshore mound to breaking waves.

In order to better assess model performance, comparisons were primarily made for consecutive runs where significant elevation change occurred. If a beach profile is close to equilibrium shape, as was the case for many of the SUPERTANK runs, elevation changes induced by wave breaking are minor. Even though a beach profile develops toward an equilibrium shape under stationary wave conditions, there is always a small-scale random component superimposed on the larger-scale deterministic profile evolution. For a profile close to equilibrium the effect of this random variation is relatively larger than for a profile far from equilibrium; thus, for runs with larger elevation change the deterministic component that is predicted by the model dominates, which is necessary for model evaluation.

The average energy dissipation was estimated using the random wave model of Larson (1993), which has been validated with data on random waves both from laboratory (SUPERTANK) and field (DELILAH; Smith et. al 1993). A smoothed bottom topography was employed in calculations of wave properties in accordance with Larson and Kraus (1989), who filtered the topography using a moving average to represent a more realistic interaction between the waves and bottom. The moving average was applied over a length $3H_{rms}/\gamma_b$ in analogy to Larson and Kraus (1989) for monochromatic waves.

The empirical coefficient K in Eq. 11 was used as a calibration parameter in comparing predicted and inferred transport rate distributions. Larson and Kraus (1989) determined an average value of $K=1.6\ 10^{-6}$ m^4/N based on LWT data for monochromatic waves, but for individual cases K varied between 1.0 - $2.5\ 10^{-6}$ m^4/N. The slope-dependent transport coefficient ε was set to zero because it is expected that this term will be less important for random as compared to monochromatic waves. Calculations of the transport rate were cut off just below the still-water level, and swash zone transport was not modeled.

Fig. 5 displays comparison between the calculated net transport rate with the model and the inferred transport rate for Run A0510A (same run as presented in Fig. 3a). This run occurred at the start of a case which involved offshore sediment transport and bar formation under random waves.

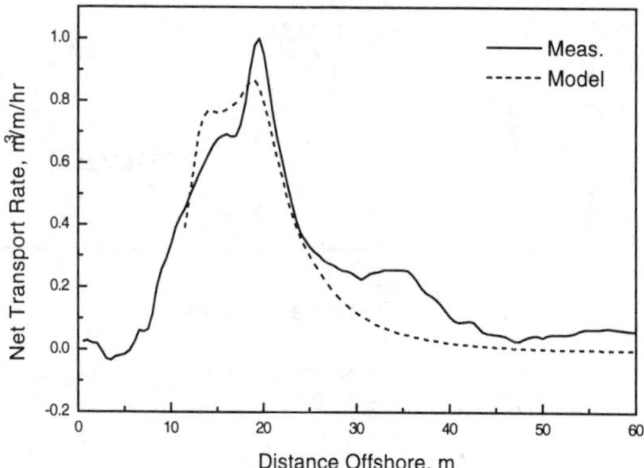

Fig. 5. Calculated net cross-shore transport rate distribution and distribution from profile surveys, Run A0510A, Case ST_10.

A coefficient of $K=2.2\ 10^{-6}$ m^4/N predicted the transport magnitude well; this value is somewhat higher than the average determined by Larson and Kraus (1989), but within the range of K-values found for monochromatic waves. The approach taken in deriving Eqs. 11 and 13 is supported by the fact that a K-value similar to monochromatic conditions may be employed directly in the transport model without modifications.

The overall shape of the transport rate distribution is also captured, although the model predicts a flatter distribution in the surf zone and a more steeply decaying transport rate in the offshore than the data indicate. However, Figs. 3a and 3b show that it is in the most offshore portion of the profile that the inferred transport rate deviates markedly, and it is in this region that the calculations become most uncertain. The potential for validating the model with inferred transport rate distributions in this region is thus more limited than in the surf zone. Because the model calculations were stopped just below the still-water level, no comparisons were made for the swash zone, but the inferred transport rate distribution in Fig. 5 shows that a simple linear decay provides a first approximation.

The next comparison is for Run A0512A, performed immediately following Run A0510A presented above (Case ST_10). A distinct bar developed, and the predicted transport rate distribution had a minimum in the surf zone because of wave reforming (Fig. 6). The outer peak in the transport rate distribution is well reproduced, whereas the inner peak predicted by the model is considerably more pronounced than what the data indicate. As in Fig. 5, distributions from model and data differ in the most seaward portion of the profile. A value of $K=1.4\ 10^{-6}$ m^4/N was used to calculate the transport rate distribution in Fig. 6.

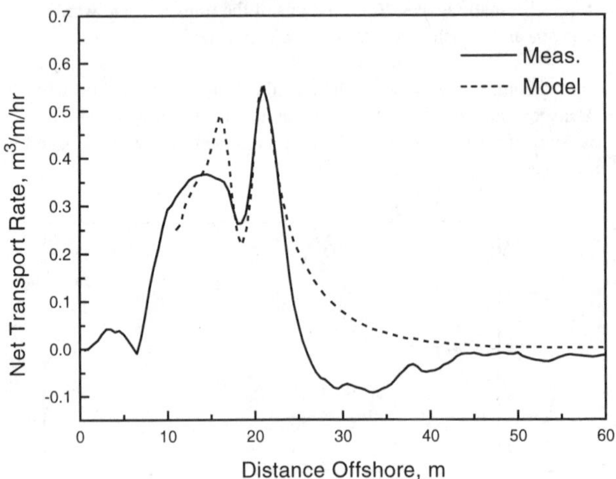

Fig. 6. Calculated net cross-shore transport rate distribution and distribution from profile surveys, Run A0512A, Case ST_10.

The final comparison between model and data concerns Case ST_K0, where an offshore mound was constructed so that a large portion of the waves broke on it. For the runs used in the model validation the input wave conditions were $H_{mo}=0.7$ m and $T_p=3.0$ sec. The net transport rate resulting from both Runs S1208B and S1209A was derived from the profile measurements, encompassing in total *40* min of wave action, to allow for significant changes in profile elevation along the mound. The sediment conservation equation was integrated from a point just shoreward of the mound to the end of the profile to obtain the most reliable estimate of the net transport rate. There were no indications in the data of transport between the mound and the inner part of the profile.

Fig. 7 illustrates predicted and inferred transport rate distributions for the mound. The model gives the correct overall magnitude and shape, but the predicted distribution is somewhat displaced towards the inshore. A small amount of material was pushed onshore by the waves on the lee side of the mound, indicated by the negative transport rates on the shoreward side of the inferred distribution, that the model could not predict. The transport rate coefficient was set to $K=1.4\ 10^{-6}$ m^4/N.

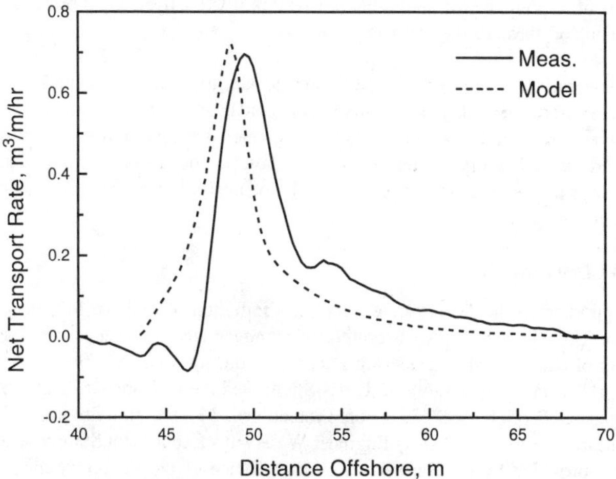

Fig. 7. Calculated net cross-shore transport rate distribution and distribution from profile surveys, Runs S1208B & S1209A, Case ST_KO.

CONCLUDING DISCUSSION

Extension to random waves of formulas for the net cross-shore sand transport rate under monochromatic breaking waves was successfully made by representing the random wave field as a number of individual waves Rayleigh distributed in height in deep water. Formulas were developed for predicting both the net direction and the magnitude of transport under random waves, and calculations agreed reasonably well quantitatively and qualitatively with measurements made at SUPERTANK.

With minor modifications of *K*-values within the range of values expected for monochromatic waves, transport rate distributions determined from profile surveys were reproduced with the transport model developed for random waves. The approach of generalizing the transport model from monochromatic to random waves by superimposing the action of individual waves seems justified. In determining optimal *K*-values, there was a tendency for large *K*-

values to occur in the beginning of a run when the profile was further from equilibrium. Model improvement could include such a dependence; another possibility would be to choose a nonlinear relationship, such as $(D - D_{eq})^p$, where the value of the power p is greater than unity, between transport and energy dissipation, although this would not allow a simple superposition technique in going from monochromatic to random waves.

Feedback to the transport rate from changes in the beach profile during a run was not incorporated. Inclusion of such feedback and the averaging of number of calculated transport rate distributions during a run would produce a smoother shape, and possibly reduce some of the discrepancies in comparisons between model and data as the inner peak shown in Fig. 6. Another improvement to the model would include a description transport under non-breaking waves outside the surf zone. For example, because only the effect of breaking waves is modeled, the slow onshore transport of material that may occur under non-breaking waves, and which is of importance for predicting seasonal profile change, can at present only be conceptually modeled through the exponential decay.

By consideration of measurements of sand density and volume balance between beach profile surveys performed at SUPERTANK, it is concluded that sand porosity can vary substantially across shore under breaking waves. It appears that sand porosity must be examined in future field and laboratory studies, and temporal and spatial variability in bed porosity may have to be taken into account in theoretical developments for bottom boundary layer quantities and sediment transport.

ACKNOWLEDGEMENTS

The support from the Japan Society for the Promotion of Science for the research visit of ML to the University of Tokyo is gratefully acknowledged, as well as the assistance from all members of the Coastal Engineering Laboratory during his stay. The work of ML was conducted under the Calculation of Cross-Shore Sediment Transport and Beach Profile Change Processes Work Unit 32530, Shore Protection and Restoration Program, Coastal Engineering Research Center, U.S. Army Engineer Waterways Experiment Station. Contract coordination was provided by the European Research Office of the US Army in London under contract DAJA45-93-C0013. We appreciate a critical review of the manuscript by Ms. Cheryl Brown, Conrad Blucher Institute.

REFERENCES

Dally, W.R., Dean, R.G., and Dalrymple, R.A. 1985. "Wave Height Variation Across Beaches of Arbitrary Profile," *J. Geophys. Res.*, 90(C6): 11917-11927.

Dean, R.G. 1977. "Equilibrium beach Profiles: U.S. Atlantic and the Gulf Coasts," Dept. of Civil Eng., Ocean Eng. Rep. No. 12, U. of Del., Newark, Del.

Kajima, R., Shimizu, T., Maruyama, K., and Saito, S. 1982. "Experiments of Beach Profile Change with a Large Wave Flume," *Proc. 18th Coastal Eng. Conf.*, ASCE, 1385-1404.

Kraus, N.C. and Larson, M. 1988. "Beach profile change measured in the tank for large waves, 1956-1957 and 1962," Tech. Rep. CERC-88-6, U.S. Army Engineer Waterways Expt. Station, Coastal Eng. Res. Center, Vicksburg, Miss.

Kraus, N.C., Larson, M., and Kriebel, D.L. 1991. "Evaluation of Beach Erosion and Accretion Predictors," *Proc. Coastal Sediments '91*, ASCE, pp 572-587.

Kraus, N. C., and Smith, J. M. (editors) 1994. "SUPERTANK Laboratory Data Collection Project, Volume 1: Main Text," Tech. Rep. CERC-94-3, U.S. Army Engineer Waterways Expt. Station, Coastal Eng. Res. Center, Vicksburg, Miss.

Kraus, N.C., Smith, J.M., and Sollitt, C.K. 1992. "SUPERTANK Laboratory Data Collection Project," *Proc. 23rd Coastal Eng. Conf.*, ASCE, pp 2191-2204.

Larson, M. 1993. "Model for Decay of Irregular Waves in the Surf Zone," Report 3167, Dept. of Water Resources Eng., Lund Inst. of Technology, U. of Lund, Lund, Sweden.

Larson, M. 1994. "Prediction of Cross-Shore Transport Rate and Beach Profile Change at Mesoscale Under Random Waves," Report 3172, Dept. of Water Resources Eng., Lund Inst. of Technology, U. of Lund, Lund, Sweden.

Larson, M. and Kraus, N.C. 1989. "SBEACH: Numerical model for simulating storm-induced beach change, Report 1: Empirical foundation and model development," Tech. Rep. CERC-89-9, U.S. Army Engineer Waterways Expt. Station, Coastal Eng. Res. Center, Vicksburg, Miss.

Larson, M., and Kraus, N.C. 1994. "Prediction of Cross-Shore Sediment Transport at Different Spatial and Temporal Scales," submitted to special issue of *Mar. Geol* on Large-Scale Coastal Behavior.

Saville, T. 1957. "Scale Effects in Two Dimensional Beach Studies," *Trans. 7th Gen. Meeting IAHR*, 1: A3.1-A3.10.

Sollitt, C.K., McDougal, W.G., Standley, D.R., Dibble, T.L., and Hollings, W.H. 1994. "Pore Pressure and Sediment Density Measurements at SUPERTANK," in Kraus, N.C. and Smith, J.M., (eds.) SUPERTANK Laboratory Data Collection Project, Vol. 1: Main Text, Tech. Rep. CERC-94-3, U.S. Army Engineer Waterways Expt. Station, Coastal Eng. Res. Center, Vicksburg, Miss.

Smith, J.M., Larson, M., and Kraus, N.C. 1993. "Longshore Current on a Barred Beach: Field Measurements and Calculation," *J. Geophys. Res.*, 98(C12): 22,717-22,731.

UNDERTOW AT SUPERTANK

Jane McKee Smith[1], M.ASCE

ABSTRACT: Cross-shore currents were measured as part of the large-scale SUPERTANK project. These measurements will be used to evaluate a numerical model of time-varying undertow and support sediment transport modeling efforts. This paper describes collection and analysis of cross-shore flow at SUPERTANK with emphasis on wave reflection, mean cross-shore flow, and time and depth variation of undertow. Wave reflection coefficients in the wave channel were typically 0.3 for the total spectrum and 0.7 for low frequencies. Mean cross-shore current magnitudes were up to 0.5 m/sec, and the spatial variation was similar to field measurements. For a narrow incident wave spectrum, there was a strong negative correlation between the low-frequency water surface and cross-shore velocity outside the surf zone and a positive correlation inside the surf zone.

INTRODUCTION

Undertow is driven by the balance among radiation stresses, the pressure gradient from the sloping mean water surface, and turbulent shear stresses. The prediction of this cross-shore flow is a critical step in advancing the modeling of storm-induced beach erosion and post-storm recovery. Numerical models of undertow have focused on regular waves and idealized beach profiles (Svendsen 1984; Dally and Dean 1984; Stive and Wind 1986; Svendsen et al. 1987; Svendsen and Hansen 1988; and Okayasu et al. 1988). Data for evaluating these models have included laboratory measurements of undertow, which have generally been at small scales with plane, smooth beach profiles and regular waves (Stive and Wind 1982; Nadaoka and Kondoh 1982; Hansen and Svendsen 1984; Okayasu et al. 1986; and

1) Research Hydraulic Engineer, USAE Waterways Experiment Station, Coastal Engineering Research Center, 3909 Halls Ferry Rd., Vicksburg, MS 39180-6199, USA, e-mail: smith%cr1.decnet@coafs1.wes.army.mil.

Okayasu et al. 1988). Field measurements of undertow are scarce (Wright et al. 1982; Guza and Thornton 1985; Greenwood and Osborne 1990; and Smith et al. 1992). Field data include the complicating factors of unknown three-dimensional effects, non-synoptic measurements, and lack of control over incident hydrodynamic conditions.

High-quality measurements of cross-shore flow were made at the SUPERTANK Laboratory Data Collection Project (Kraus et al. 1992; Kraus and Smith 1994). The SUPERTANK project was conducted by the U.S. Army Engineer Waterways Experiment Station, Coastal Engineering Research Center (CERC), in the large wave channel at Oregon State University (OSU). This project provided the facility to measure undertow on realistic beach profile shapes (0.22-mm grain size) at prototype scale with a broad range of irregular and regular waves. The purpose of this paper is to describe collection of data and analysis of cross-shore flow at SUPERTANK with special emphasis on wave reflection calculated from surface elevation and cross-shore flow measurements, mean cross-shore flows, and time and depth variations of the low-frequency component of the flow.

SUPERTANK MEASUREMENTS

SUPERTANK was conducted in the large wave tank (LWT) at OSU during August and September 1991. The channel of the LWT is 104 m long, 3.7 m wide, and 4.6 m deep. A 76-m-long sand beach was constructed in the channel for the project. Figure 1 is a view of the LWT as configured during SUPERTANK.

Figure 1. View of LWT with SUPERTANK instrumentation

The cross-shore and vertical flows were measured with 18 two-component, Marsh McBirney (model 512) electromagnetic current meters. The current meters were deployed in the nearshore from just outside the incipient breaker zone to the mid surf zone in vertical stacks of one to four current meters with vertical spacing of approximately 0.3 m and horizontal spacing of 3.7 m. The timing pulse was shared by the current meters in each vertical array to reduce cross-talk between gauges.

Figure 2. Vertical array of electromagnetic current meters

Fourteen of the current meters were mounted to the channel wall (left tank wall in Figure 1) with aluminum plates and rods. The plates were bolted to the wall, and spacers were welded to the back of the plates to set the plates approximately 5 cm out from the wall. A vertical array of slots was machined into the plates at 15.2-cm spacing. Aluminum rods were positioned in the slots and were fastened with nuts on the backside of the plate (the space between the wall and plate allowed access behind the plate). The opposite end of the rod was machined flat on one side, and the mounting bar of the current meter was attached to the flat portion of the rod with two small hose clamps. Current meter cables were secured to the plate with cable ties attached through small holes punched into the plate. The current meters extended approximately 0.6 m out from the channel wall. The elevation of the current meters could be adjusted between runs by repositioning the aluminum rods in the vertical array of slots. Figure 2 shows the current meter mounting configuration. Four meters were mounted on a mobile carriage that was easily re-positioned between runs. The current meters deployed off the carriage were hose clamped to angles bolted to an aluminum wing-shaped section hanging from the carriage.

Wave characteristics were measured throughout the channel at a spacing of 3.7 m with resistance and capacitance wave gauges (right tank wall in Figure 1). Wave transformation measurements at SUPERTANK are discussed by Kriebel and Smith (1994). Cross-shore flow and wave measurements were made for over 200 SUPERTANK runs with wave heights in the range 0.2 to 1.0 m and periods in the range 3 to 10 sec. Wave and current data were collected at a frequency of 16 Hz. Wave characteristics included monochromatic waves and narrow- and broad-band TMA spectra. Special runs consisted of time-varied wave amplitude and dual-peaked spectra. The beach profile was surveyed before and after each wave run. Beach profile evolution over the SUPERTANK runs provided another variable.

DATA ANALYSIS

Analysis of the current meter data included calibration, preprocessing, filtering, spectral analysis, and time series analysis. Details of the current meter data analysis is given in Chapter 3 of Kraus and Smith (1994).

Current meters were calibrated immediately after the project in September 1991 at the U.S. Geological Survey Indoor Hydraulic Laboratory Facility in Bay St. Louis, Mississippi, USA. The calibration gains were used in the analysis procedure. Gauge offsets were calculated in the analysis software as the average meter reading during the quiescent period at the beginning of each run. For gauges that were not submerged at the beginning of the run, the post-project calibration offsets were applied.

The preprocessing consisted of analog (10-Hz) and digital (2-Hz) filters to avoid aliasing and eliminate high-frequency noise. High-frequency noise was present in some data records for deployments with the current meters spaced less than 30 cm apart. Also in the preprocessing, the elevation of each current meter was compared with the water surface elevation (estimated from the co-located wave gauges) to determine if the current meter was submerged. Current meter readings were set to zero when the gauge was not submerged.

To separate incident-band wave motions from low-frequency motions due to time-varying undertow and tank seiche, a non-recursive, low-pass filter was applied. The period cutoff of the filter was set to twice the peak period of the incident waves (the peak of the long-period motion was generally 20 sec or longer). Spectral and time series analyses were performed on the total, low-pass, and high-pass data to quantify the mean flow, variance, peak period, skewness, and kurtosis of the velocity. Figure 3 shows an example of a spectrum and time series for the cross-shore current (offshore-directed flows are positive and onshore-directed flows are negative). The time-series plot shows both the total current signal and the low-pass signal. For the run shown in Figure 3, the incident wave height was 0.7 m, the peak period was 3 sec, and the current measurement was made at approximately 0.5 m below the mean water elevation in a water depth of 1.1 m. Wave groups are

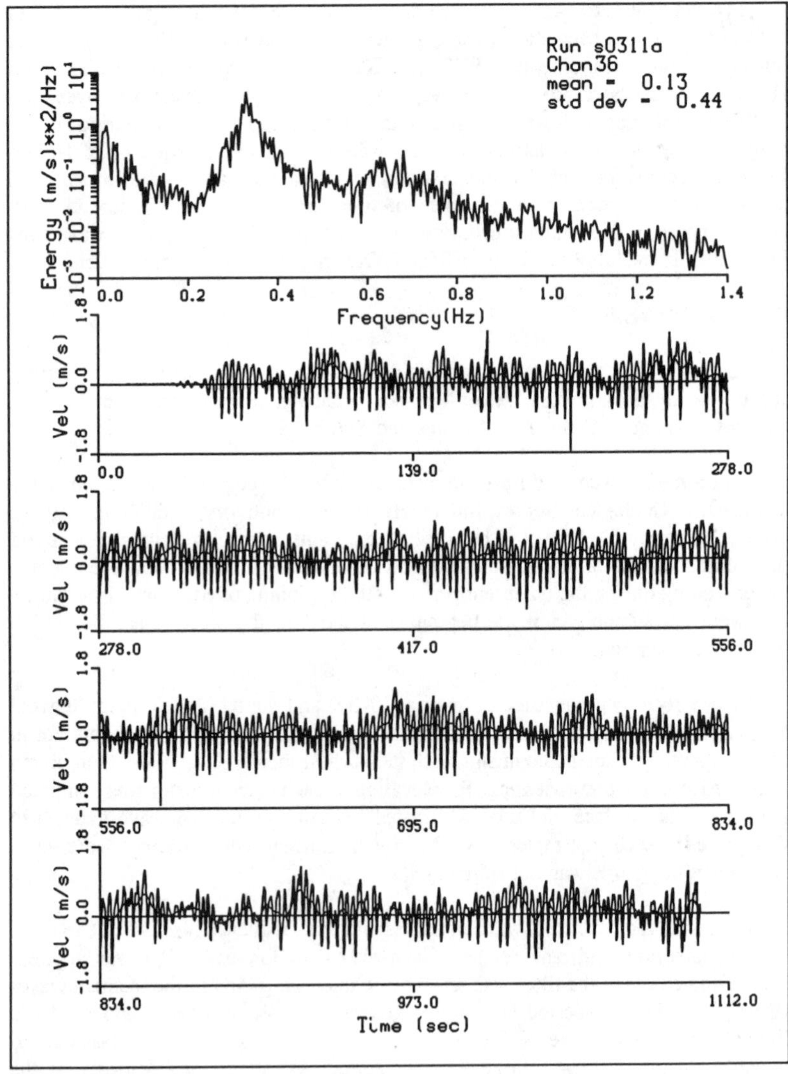

Figure 3. Example of current meter raw spectrum and time series of total signal and low-pass signal

evident in the time series, as would be expected with the relatively narrow incident spectrum ($\gamma = 20$). The mean velocity at this gauge is 0.13 m/sec, signifying a net offshore flow at this elevation. The primary peak of the energy spectrum is at 0.33 Hz with harmonics at 0.7 Hz and 1.0 Hz. Approximately 14 percent of the energy is contained in the low-pass signal, and the low-frequency peak is at 0.02 Hz (50 sec). Similar spectral and time series analyses was performed on the wave gauge data.

WAVE REFLECTION

Wave reflection was calculated for selected wave runs at several locations in the channel. Reflection estimates were calculated with a frequency-domain linear theory analysis technique using synoptic time series of cross-shore velocity and water surface elevation (Hughes 1993). This method assumes normally-incident waves. An example of the reflection analysis output is shown in Figure 4. The figure shows time series of the total, incident, and reflected low-pass water surface elevation for a run with a narrow-banded (γ = 100) incident wave spectrum (wave height of 0.7 m and peak period of 5 sec).

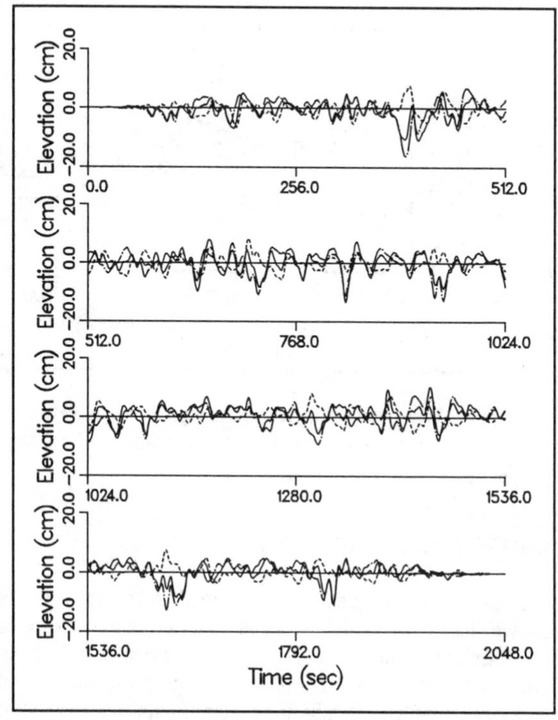

Figure 4. Low-pass water surface elevation (total -- solid line, incident -- chain-dot line, reflected -- dashed line)

Reflection coefficients were generally 0.20 to 0.45 (all frequencies) for the SUPERTANK runs. Coefficients for the low-pass frequencies were in the range 0.4 to 1.0. The reflection coefficient was strongly related to the incident peak period, with greater reflection for longer peak periods. The wave generator used at the LWT was equipped to absorb reflected waves at the peak frequency, so reflected wave energy was suppressed. Figure 5 shows wave height calculated in 5-min segments for the low-pass signal over a 70-min run. The solid lines are surf zone wave gauges (depth less than 1 m), the chain-dot lines are gauges in the deepest portion of the channel (depth approximately 3 m), and the dotted and dashed lines are at intermediate depths. There is a variation in long-wave height with time, but there is not a consistent increase in height with time from re-reflection of waves from the wavemaker, as might be expected.

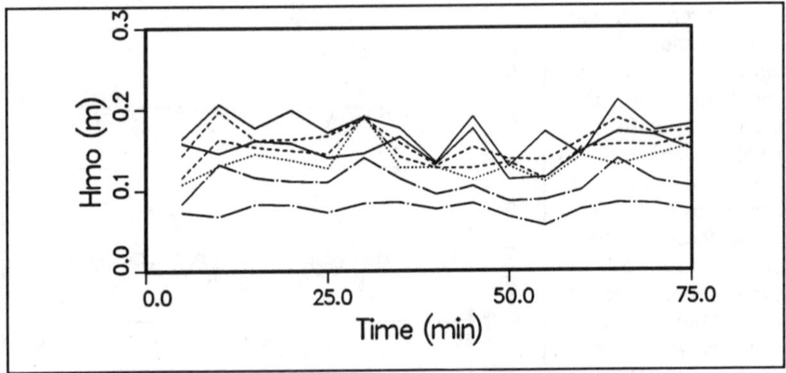

Figure 5. Wave height of low-pass signal as a function of time (surf zone -- solid lines, deep -- chain-dot lines, intermediate depths -- dashed and dotted lines)

CROSS-SHORE CURRENTS

A sample of the mean cross-shore flow is shown in Figure 6. The figure also shows the beach profile and the transformation of the significant and maximum wave height. The undertow velocities are relatively strong shoreward of the bar crest and weak seaward of the bar, where the water depth increases. The mean undertow profile is parabolic in shape, with the greatest velocities near the bottom and lower velocities near trough level. The maximum mean cross-shore velocities at SUPERTANK were 0.5 m/sec, shoreward of the bar. Velocities in the offshore region of the channel were typically less than 0.05 m/sec, in the range 0.2 to 0.5 $(H/h)^2 (gh)^{0.5}$ (an estimate of the flow required to balance the mass transport). The velocity profiles in the offshore region are linear in shape (constant over depth or slightly increasing offshore-directed velocity from the bottom to the surface).

The SUPERTANK cross-shore flow measurements show trends similar to field measurements of undertow at the DELILAH experiment (Smith et al. 1992), conducted at the CERC Field Research Facility in North Carolina, USA. The velocity profile shape offshore of breaking and across the bar crest are similar in

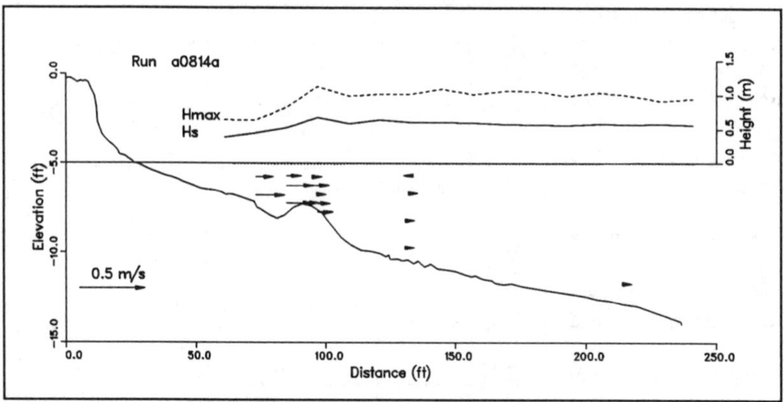

Figure 6. Mean flow, wave height, and beach profile at SUPERTANK

the laboratory and field measurements. However, measured undertow velocities in the trough at DELILAH were generally smaller than those measured at SUPERTANK. This may be because the trough was deeper and wider and the foreshore steeper at DELILAH than at SUPERTANK.

WAVE GROUPS

The incident wave spectra at SUPERTANK included narrow-banded spectral shapes (TMA spectra with γ = 20 and 100). The narrow spectra produced wave groups which induced time variations in the low-pass surface elevations and velocities. The time series of high-pass wave energy (*E*) and low-pass surface elevation (*Eta*) are given for three locations in the tank in Figure 7. The high-pass wave energy was calculated by squaring the high-pass surface elevation and then low-pass filtering the squared elevation to get the envelop of the energy. The top panel of Figure 7 is a location near the wave generator, the middle panel is a location just seaward of the incipient breaking zone, and the bottom panel is in the mid surf zone. The middle panel shows strong correlation between the high-pass wave energy and the low-pass surface elevation. The bottom panel shows that in the surf zone, the high-pass wave energy is dissipated due to wave breaking, but the low-pass surface elevation maintains a group structure.

As shown in the middle panel of Figure 7, the strongest correlation between the high-pass wave energy and the low-pass water surface elevation occurs just seaward of the breaker zone. The correlation of these two signals is shown in Figure 8 as a function of lag time. A strong negative correlation occurs at a time lag of approximately zero (high-pass wave energy leads slightly), as would be expected from radiation stress forcing (Longuet-Higgins and Stewart 1964). The negative correlation means that high wave groups drive a setdown of the low-pass water surface elevation, and low wave groups drive a setup.

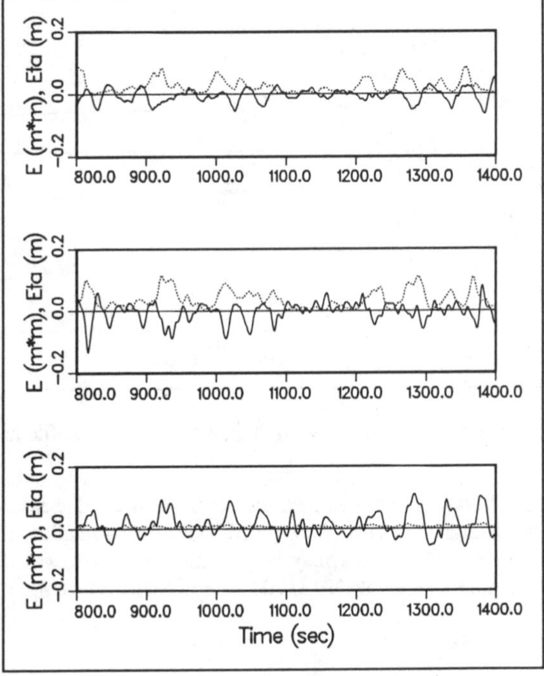

Figure 7. Incident wave energy (*E* -- dotted line) and low-pass surface elevation (*Eta*-- solid line) offshore (top), seaward of incipient breaking (middle), and surf zone (bottom)

Figure 9 shows a repeat of the time series of high-pass wave energy (*E*) and low-pass water surface elevation (*Eta*) from the center panel of Figure 7 with the low-pass velocity measured with a co-located current meter (just outside the incipient breaker zone). Offshore flow is correlated with the high wave groups and depressions in the low-pass surface elevation. The time series show there is a mean setdown in the water surface in this location and a mean offshore flow at this mid-depth current measurement location. Figure 10 shows the correlation between the low-pass water surface elevation and the low-pass cross-shore velocity corresponding to Figure 9. The negative correlation is fairly strong, with the water surface elevation lagging the velocity by a few seconds. The correlation of the low-pass water surface elevation and velocity in the mid surf zone is shown in Figure 11. In the surf zone, the correlation is positive, and the water surface leads the cross-shore velocity by approximately one fourth of the low-pass wave period.

SUMMARY

High-quality measurements of the cross-shore flow and water surface elevation were made at the SUPERTANK project. These measurements will be used to evaluate a model of the time-varying undertow, to gain understanding of nearshore hydrodynamic processes, and support studies of sediment transport and beach profile change.

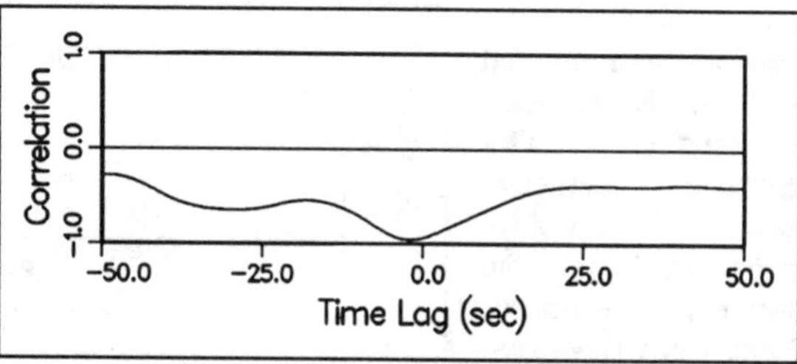

Figure 8. Correlation of the incident wave energy and the low-pass surface elevation at a location just seaward of the incipient breaker zone

SUPERTANK was conducted in a LWT with a 0.22-m sand beach. The cross-shore flow was measured with 18 electromagnetic current meters, and the water surface elevation was measured with resistance and capacitance wave gauges. Cross-shore flow and wave measurements were made for over 200 SUPERTANK runs with wave heights of 0.2 to 1.0 m and periods of 3 to 10 sec. Wave characteristics included monochromatic waves and narrow- and broad-band TMA spectra.

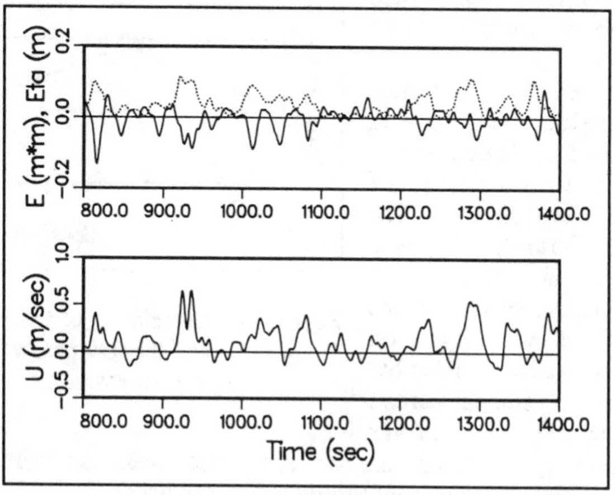

Figure 9. Time series of incident wave energy (*E* -- dotted line), low-pass surface elevation (*Eta* -- solid line), and low-pass velocity (*U*)

Analysis of the current meter data included calibration, preprocessing, filtering, spectra analysis, and time series analysis. The meters were calibrated immediately after the project. Preprocessing of the data consisted of a 10-Hz analog and 2-Hz digital filter to avoid aliasing and eliminate high-frequency noise. The data were filtered with a low-pass filter with a cutoff of twice the peak period. Finally, spectral and time series analyses were performed on the total, high-pass, and low-pass signals for each run. Similar analyses were performed on the wave gauge data.

Wave reflection was analyzed from co-located current and wave gauges using the method given by Hughes (1993). Typically the reflection coefficients were approximately 0.3 for the total spectrum and 0.7 for the low-pass spectrum. The reflection coefficient was strongly dependent on the incident wave period. Long-wave (low-pass) energy varied over the wave runs but did not systematically increase through the runs (runs up to 70 min were performed).

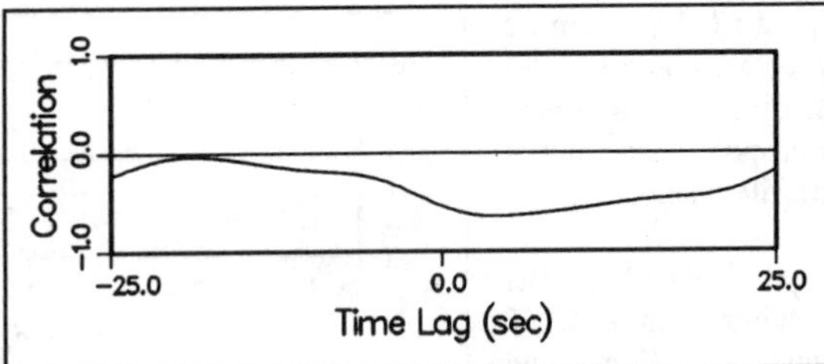

Figure 10. Correlation of the low-pass surface elevation and the low-pass velocity at a location just seaward of the incipient breaker zone

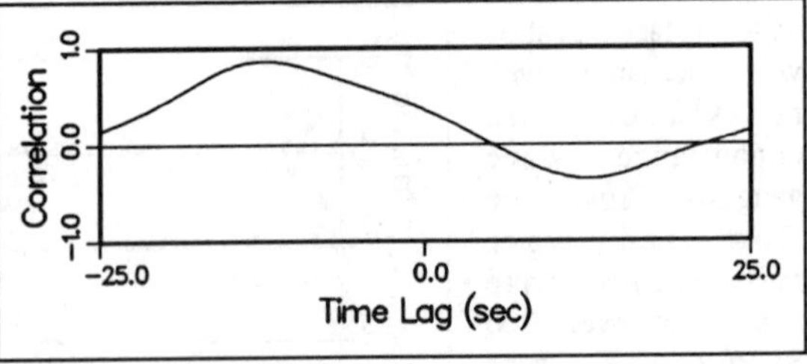

Figure 11. Correlation of the low-pass surface elevation and the low-pass velocity at a location in the mid surf zone

Mean cross-shore velocities up to 0.5 m/sec were measured at SUPERTANK. The undertow profiles were similar to measurements in the field, with the greatest velocities over the bar and a parabolic current profile shape in the surf zone. Outside the surf zone, mean velocities were typically less than 0.05 m/sec and the current profile shape was linear.

Narrow spectra generated at SUPERTANK produced wave groups which induced variation in the low-pass surface elevations and velocities. The strongest correlation between the high-pass wave energy and the low-pass surface elevation is in the region just seaward of the surf zone (correlation of approximately -1.0 with lag time of zero). The low-pass surface elevation and velocity are negatively correlated seaward of the surf zone (surface elevation lags velocity slightly), and the surface elevation and velocity are positively correlated in the mid surf zone (surface elevation leads velocity by about one-fourth the wave period).

Additional information on the SUPERTANK measurements is given by Kraus and Smith (1994).

ACKNOWLEDGMENTS

The success of the SUPERTANK current and wave measurements was a result of the cooperation and assistance of the entire SUPERTANK team. Special acknowledgement is given to Terry Dibble (OSU) and Bill Grogg (CERC) for designing the electronics and overcoming electronic problems, Bill Hollings (OSU) for designing mounts, Dave Standley (OSU) for collecting the data, Ed Thornton (Naval Postgraduate School) and Reggie Beach (OSU) for supplying current meters, and Gray Smith and Tom Wendell (CERC) for assisting in the data analysis. Mary Cialone, Randy Wise, and Bruce Ebersole (CERC) are acknowledged for their helpful review comments on the draft paper. Permission was granted by the Office, Chief of Engineers, U.S. Army Corps of Engineers, to publish this information.

REFERENCES

Dally, W. R., and Dean, R. G. 1984. "Suspended Sediment Transport and Beach Profile Evolution," *J. Waterway, Port, Coastal, and Ocean Engrg.*, 110(1), 15-33.

Greenwood, B., and Osborn, P. D. 1990. "Vertical and Horizontal Structure in Cross-shore Flows: An Example of Undertow and Wave Set-up on a Barred Beach," *Coastal Engrg.*, 14, 543-580.

Guza, R. T., and Thornton, E. B. 1985. "Local and Shoaled Comparisons of Sea Surface Elevations, Pressures, and Velocities," *J. Geophysical Res.*, 85(C3), 1524-1530.

Hansen, J. B., and Svendsen, I. A. 1984. "A Theoretical and Experimental Study of Undertow," *Proc. 19th Coastal Engrg. Conf.*, ASCE, 2246-2262.

Hughes, S. A. 1993. "Laboratory Wave Reflection Analysis Using Co-located Gages," *Coastal Engrg.*, 20, 223-247.

Kraus, N. C., and Smith, J. M. 1994. "SUPERTANK Laboratory Data Collection Project, Volume I: Main Text," Technical Report CERC-94-3, U.S. Army Engineer Waterways Experiment Station, Vicksburg, MS.

Kraus, N. C., Smith, J. M. and Sollitt, C. K.. 1992. "SUPERTANK Laboratory Data Collection Project," *Proc. 23rd Coastal Engrg. Conf.*, ASCE, 2191-2204.

Kriebel, D. L., and Smith, J. M. 1994. "Wave Transformation at SUPERTANK," *Proc. Coastal Dynamics'94*, this volume.

Longuet-Higgins, M. S., and Stewart, R. W. 1964. "Radiation stresses in Water Waves; a Physical Discussion, with Applications," *Deep-Sea Res.*, 11, 529-562.

Nadaoka, K., and Kondoh, T. 1982. "Laboratory Measurements of Velocity Field Structure in the Surf Zone by LDV," *Coastal Engrg. in Japan*, 25, 125-145.

Okayasu, A., Shibayama, T., and Horikawa, K. 1988. "Vertical Variation of Undertow in the Surf-zone," *Proc. 21st Coastal Engrg. Conf.*, ASCE, 478-491.

Okayasu, A., Shibayama, T., and Mimura, N. 1986. "Velocity Field Under Plunging Waves," *Proc. 20th Coastal Engrg. Conf.*, ASCE, 660-674.

Smith, J. M., Svendsen, I. A., and Putrevu, U. 1992. "Vertical Structure of the Nearshore Current at DELILAH: Measured and Modeled," *Proc. 23rd Coastal*

Engrg. Conf., ASCE, 2825-2838.

Stive, M. J. F., and Wind, H. F. 1982. "A Study of Radiation Stress and Set-up in the Nearshore Region," *Coastal Engrg.*, 6(1), 1-26.

Stive, M. J. F., and Wind, H. F. 1986. "Cross-shore Mean Flow in the Surf Zone," *Coastal Engrg.*, 10(4), 325-340.

Svendsen, I. A. 1984. "Wave Heights and Setup in a Surf Zone," *Coastal Engrg.*, 8(4), 303-329.

Svendsen, I. A., and Hansen, J. B. 1988. "Cross-shore Currents in Surf-zone Modelling," *Coastal Engrg.*, 12, 23-42.

Svendsen, I. A., Schäffer, H. A., and Hansen, J. B. 1987. "The Interaction Between the Undertow and the Boundary Layer Flow on a Beach," *J. Geophysical Res.*, 92(C11), 11845-11856.

Wright, L. D., Guza, R. T., and Short, A. D. 1982. "Waves and Longshore Currents: Comparison of a Numerical Model with Field Data," *J. Geophysical Res.*, 90, 4951-4958.

WAVE TRANSFORMATION MEASUREMENTS AT SUPERTANK

David L. Kriebel[1] and Jane M. Smith[2]

ABSTRACT: Twenty-six wave gages were used to document wave transformation across a large-scale laboratory beach profile in the SUPERTANK Laboratory Data Collection Project. This paper gives an overview of data available from this project for both statistical and spectral (energy-based) wave heights. Additional emphasis is placed on separating the wave records into low-frequency (long wave) and high-frequency (short wave) components. A unique aspect of this data set is that wave measurements were made in all regions of the beach profile between the wavemaker and the runup limit, with the highest resolution in the inner surf and swash zones. As a result, wave transformation is defined on the active beach-face where little data has traditionally been available.

INTRODUCTION

As part of the SUPERTANK Data Collection Project, twenty-six wave gages were used to document wave transformation across a sandy beach profile in a large wave tank. Water surface elevations were measured at these 26 locations for over 129 hours of wave action and more than 66 unique sets of wave conditions throughout the seven-week long project. These wave gages complement other instruments used during the SUPERTANK project, including 18 electromagnetic current meters and 34 optical backscatter sensors. A review of the entire SUPERTANK project is given by Kraus et al. (1992) and a thorough documentation of all experimental procedures is given by Kraus and Smith (1994). The purpose of this paper to summarize some basic results of the SUPERTANK wave measurements, including the transformation of spectral and statistical wave parameters across the surf zone and swash zone.

---

[1] Ocean Engineering Program, U.S. Naval Academy, Annapolis, MD 21402

[2] Coastal Engineering Research Center, U.S. Army Corps of Engineers, Vicksburg, MS 39180

While numerous descriptions of wave transformation may be found in the literature, several aspects of the SUPERTANK measurements are noteworthy. These include: (1) the large number of wave gages used to define the spatial variations in wave conditions, (2) the number and range of wave conditions tested, and (3) the emphasis placed on obtaining wave measurements at all locations across the beach profile, beginning outside the surf zone and extending through the surf zone to the swash zone and runup limit. Special emphasis was placed on documenting wave transformation in the inner surf zone and swash zone where the gradients in wave height and beach profile change are typically greatest. An additional unique aspect of the data analysis is the separation of wave records into low-frequency and high-frequency components in order to define wave transformation characteristics of long-waves and short-waves respectively.

DESCRIPTION OF EXPERIMENTS AND ANALYSIS METHODS

The SUPERTANK Project was conducted in the large wave tank at Oregon State University during the summer of 1991. The wave tank is 104 meters long, 3.7 meters wide, and 4.6 meters deep. A sand beach with median grain size of 0.22 mm was initially placed in an "equilibrium" form (concave-upward) with a length of about 76 meters and a maximum water depth of about 3.0 meters. The wave tank was equipped with a digitally-controlled hydraulic wavemaker with reflection-compensation to absorb reflected waves at the peak spectral frequency.

Numerous wave conditions were tested during the seven-week experiment, including both regular and irregular waves producing both erosional and accretional conditions. Overall, 66 different sets of wave conditions were tested during 129 hours of wave action; and, of these, 70 percent involved irregular waves. Each set of wave conditions was generally tested for between 20 and 350 minutes. For the longer tests, multiple runs of 20 to 70 minutes each were used with some down time between runs for beach profiling. In general, wave heights ranged from 0.2 to 1.0 meters and wave periods ranged from 3 to 8 seconds, with selected tests exceeding these ranges. Due to space limitations, this paper presents results from one set of irregular wave tests from the first two days of the SUPERTANK Project.

As noted, 26 wave gages were used to document both mean water levels and wave heights. These gages included 16 resistance-wire wave gages, spaced at 3.7 meter intervals and placed from the wavemaker shoreward into the mid-surf zone, along with 10 capacitance-based wave gages, spaced at 0.9 to 1.8 meter intervals in the inner surf zone and swash zone. The capacitance wave gages consisted of a single sensing wire, the lower portion of which was partially buried in the sand beach. These gages recorded the time-varying water surface elevation as each wave (bore) passed each gage during the runup-rundown sequence. In between successive swash events, these gages recorded the saturated sand bed elevation. Similar measurements in the swash zone have been reported by Waddell (1973) and Sonu et al. (1974).

Wave properties discussed in this paper include wave heights defined according to two methods. First, standard zero-crossing analysis was performed to define statistical wave height parameters such as H_{RMS}, $H_{1/3}$, $H_{1/10}$, and H_{MAX}. Second, standard spectral analysis was performed to define the zero-moment wave height, *Hmo*, and other spectral parameters. In each case, the wave setup and setdown (mean water levels) were also determined relative to the initial mean water level datum.

For those wave gages located in the swash zone, some additional interpretation was required since the wave records show a flat or clipped wave trough when the sand bed was exposed. As the sand bed eroded or accreted, an upward or downward trend was introduced in the measured wave record. To remove these trends, an algorithm was developed to identify the sand bed during the exposures, to interpolate sand beds in between exposures, and then to remove the sand bed record from the original signal. Once this accomplished, wave setup and wave heights were defined relative to the initial sand bed elevation.

The water surface records were also filtered to separate the signal into low- and high-frequency wave components. The filter cut-off frequency was selected as one-half of the peak wave frequency based on the target wave spectrum input to the wavemaker. Following this filtering, three time-series were analyzed: (1) the original or total (but de-trended) wave record, (2) the low-frequency wave record containing long-wave and tank seiching frequencies, and (3) the high-frequency wave record containing frequencies higher than one-half the peak frequency, including the primary wavemaker frequencies and their higher-harmonic components.

WAVE TRANSFORMATION

Results of the SUPERTANK wave measurements are first considered in dimensional form for two sample data sets. Figures 1 and 2 show examples of the wave height transformation based on the first set of wave conditions tested during the SUPERTANK Project. Target wave conditions consisted of a narrow TMA spectrum (spectral width parameter equal to 20) with a peak period of 3 seconds and a zero-moment wave height of 0.8 meters near the wavemaker. These conditions produce a random sea with a spectral steepness, Hmo_O/L_O, equal to 0.056, where Hmo_O denotes the zero-moment wave height measured at the wave gage closest to the wavemaker.

In Figures 1 and 2, the beach profiles at the beginning and end of the data run are shown in the bottom panel by the solid and dashed curves respectively. Statistical wave heights, based on zero-crossing analysis, are shown in the top panel. Wave setdown/setup and energy-based zero-moment wave heights are shown in the middle panel. In general, the most significant overall feature of Figures 1 and 2 is that wave transformation is documented across the entire active beach profile to the runup limit, with the finest resolution in the inner surf zone and swash zone where the gradients in wave height are greatest.

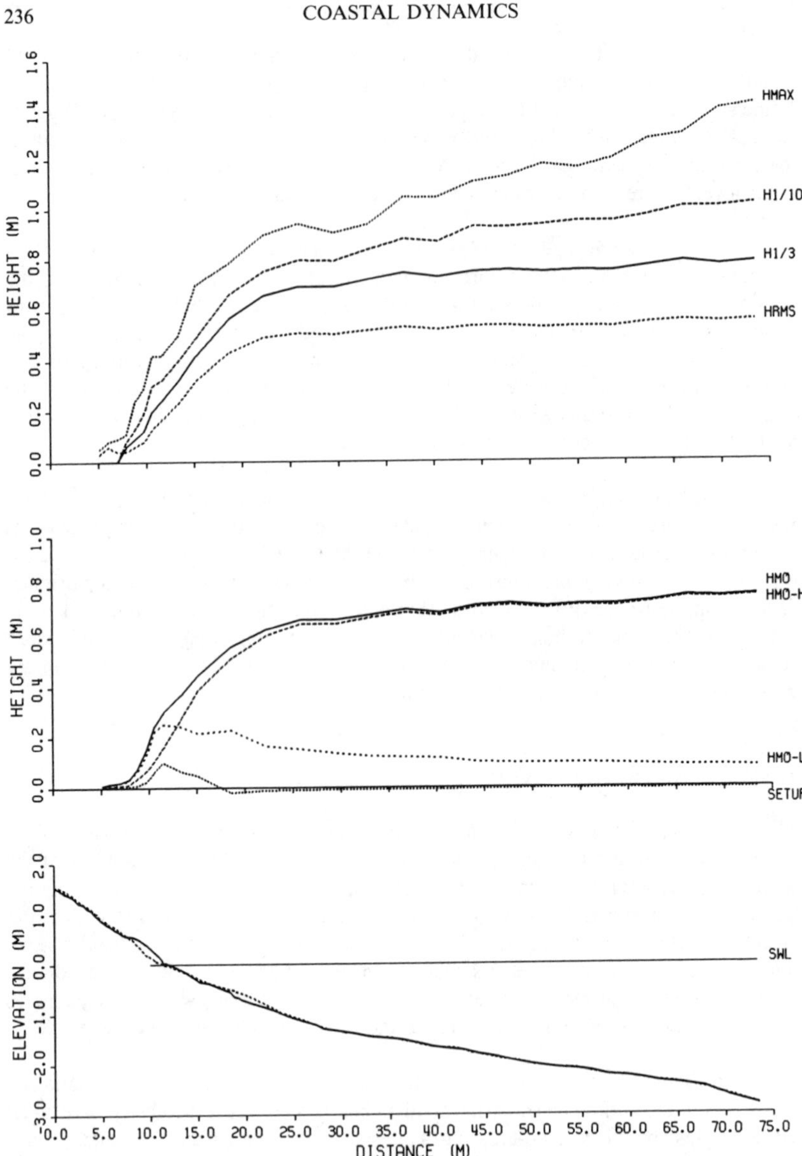

Figure 1. Wave transformation for selected wave height parameters, initial run with $Hmo_O/L_O = 0.056$ for duration of 20 minutes.

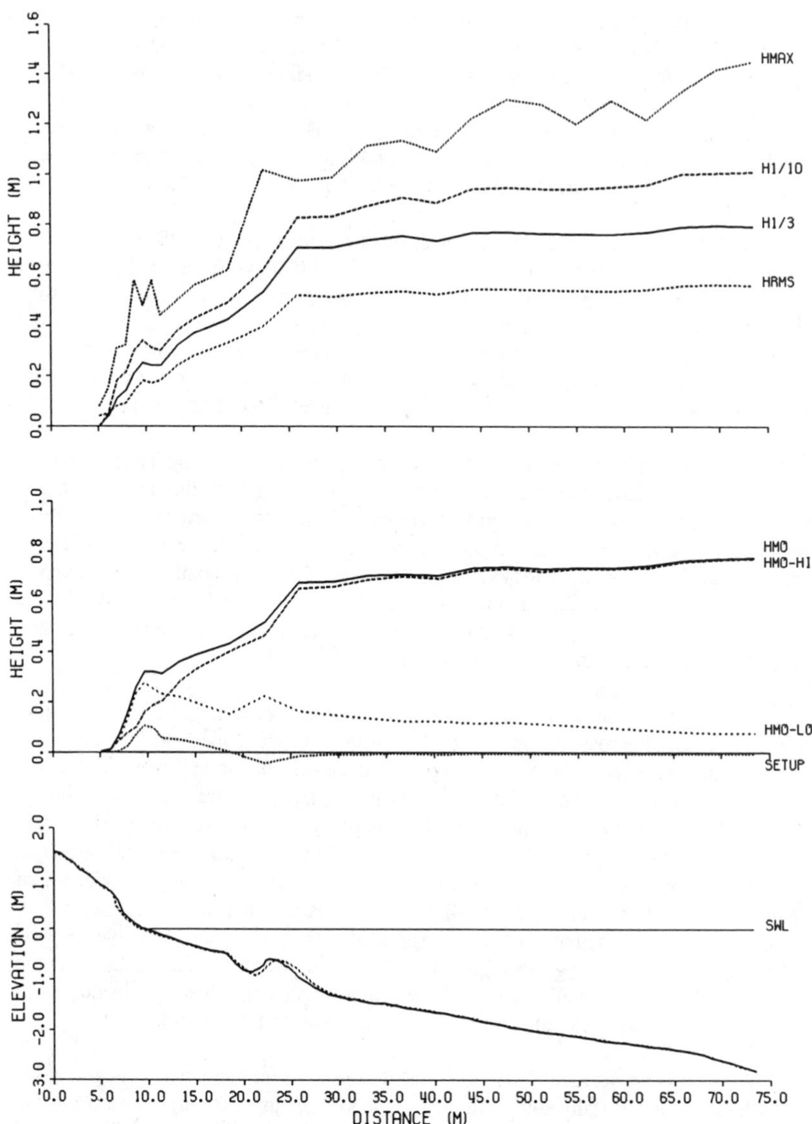

Figure 2. Wave transformation for selected wave height parameters, final run with $Hmo_O/L_O = 0.056$ for duration of 70 minutes. Final profile is the result of 270 minutes of cumulative wave action.

Results in Figure 1 are based on the first 20 minutes of wave action of the SUPERTANK Project where the beach profile was initially molded to a concave "equilibrium" form with a planar foreshore slope. Sediment transport was offshore, resulting in erosion of the foreshore and deposition offshore producing an emergent offshore bar. Results in Figure 2 are based on the last 70 minutes of wave action of the same irregular wave conditions. The cumulative duration of wave action between the initial profile in Figure 1 and the final profile in Figure 2 is 270 minutes.

During the evolution of the beach profile, the mean water levels and wave heights across the profile also evolve significantly. As shown in the middle panel of Figures 1 and 2, the width of the surf zone increases during the tests and, as a result, the distance between the maximum setdown and maximum setup increases as the foreshore recedes and as the bar moves offshore. The wave setup reaches a maximum at the still-water shoreline and has nearly the same maximum value at the beginning and end of the wave action regardless of the presence of the offshore bar.

The statistical wave height parameters, depicted in the top panels of each figure, show dramatically different transformation behavior from the smooth concave profile (Figure 1) to the barred profile (Figure 2). In both cases, the relationships between H_{RMS}, $H_{1/3}$, $H_{1/10}$, and H_{MAX} are in general accordance with the Rayleigh distribution at the seawardmost wave gage. Wave heights decay continuously as waves approach the bar crest due to the selective breaking of the largest waves in the random sea. As a result, $H_{1/10}$ and H_{MAX} decrease substantially while both H_{RMS} and $H_{1/3}$ show relatively little decay seaward of the bar location.

In Figure 1, where the bar is poorly developed, wave height decay is smooth and well-behaved across the emerging bar, through the surf zone, and into the swash zone. Gradients in wave height are largest just seaward of the still-water shoreline. One curious feature is that, at the runup limit, H_{RMS} and H_{MAX} are non-zero while the $H_{1/3}$ and $H_{1/10}$ are both zero. This occurs because only two large runup events actually reached this wave gage. In Figure 2, by contrast, waves shoal somewhat and H_{RMS} actually increases in height over the outer face of the bar. There is also evidence of increased wave reflection in the offshore region, particularly in H_{MAX}. Waves then break abruptly over the bar producing large gradients in wave height. Wave decay is much more gradual through the surf zone until waves reach the shoreline. Near the shoreline, there is a second maximum in wave heights (partly due to reflection from the steeper beach face slope) followed by rapid decay to the runup limit.

As may be seen in Figures 1 and 2, the zero-moment wave height Hmo_O (middle panel) is generally similar in magnitude to $H_{1/3}$ (upper panel) over most of the profile. In the swash zone, however, differences between these definitions of significant wave height become most pronounced. In general, the energy-based wave heights are more well-behaved near the runup limit since these are based on the variance of the water surface time-series rather than on the chance occurrence of individual waves.

In the middle panel of Figures 1 and 2, the zero-moment wave height is further decomposed into low- and high-frequency components. Results are typical of highly energetic wave conditions in that high-frequency components dominate in the offshore region and throughout most of the surf zone, while low-frequency components dominate in the inner surf zone and swash zone. In both figures, irrespective of the presence of a well-developed bar, the low-frequency wave heights exhibit two maxima: one at the location of the breakpoint bar and a second larger maximum at the still-water shoreline. The high-frequency wave heights decay continuously throughout the surf and swash zones producing no local maxima.

DIMENSIONLESS WAVE TRANSFORMATION

Additional details of the wave height transformation from the first set of wave conditions tested during the SUPERTANK Project are shown in Figures 3 through 7 in dimensionless form. In these figures, mean water level and wave height parameters are shown for four wave runs associated with the same random wave conditions described above. Results depicted in Figures 1 and 2 form the first and last of these runs. As a result, the wave transformation shown in Figures 3 through 7 occur over a beach profile that was initially barless (results from Figure 1 depicted by dashed line) but which evolved into a well-developed stable barred profile (results from Figure 2 depicted by dotted line).

The abscissa in Figures 3 through 7 is the local sand-bed elevation relative to the still water level datum, normalized by the zero-moment wave height from the seawardmost wave gage, here denoted as Hmo_0. With this convention, the initial still-water shoreline is located at a relative depth of zero, while sand-bed elevations located landward of (and above) the shoreline are denoted by negative depths. The ordinate in each figure is either the mean water level or the localized energy-based wave height, again normalized by the zero-moment wave height from the seawardmost gage. The only exception is Figure 7 in which the ordinate represents the local ratio of the low- and high-frequency wave heights from each gage.

Figure 3 shows the normalized wave setdown-setup distribution across the profile. In the cases shown, the wave setdown outside the breakpoint is essentially the same for each of the four data runs since the profile did not change in this region. Near the bar crest (relative depth of about 0.9), the maximum setdown is highly variable and difficult to resolve accurately. The general form of the wave setup through the surf zone is again quite consistent until the location of maximum setup just outside the still-water shoreline. In this region, it appears that the spacing of the capacitance gages was not close enough to fully resolve the highly localized setup maximum. On the other hand, the setup across the swash zone seems to be well-defined, showing a concave distribution between the point of maximum setup near the shoreline and the runup limit (near a relative depth of -1.1 in this case).

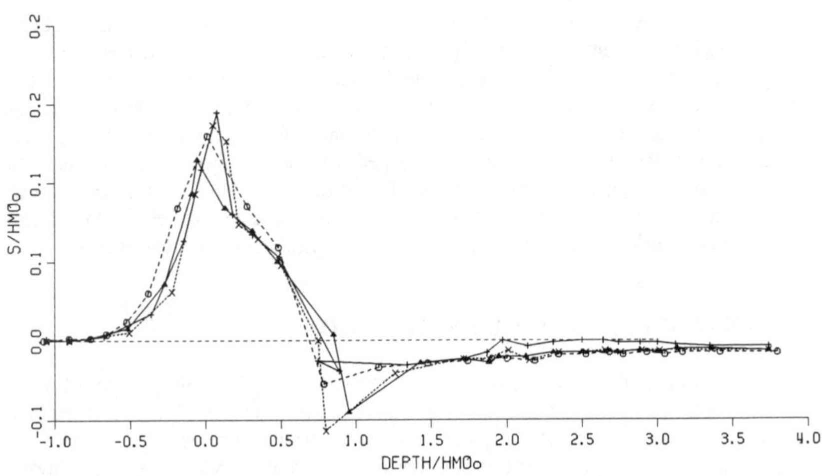

Figure 3. Dimensionless mean water levels, $Hmo_o/L_o = 0.056$.

In Figure 4, the total zero-moment wave height is shown for each of the four data runs. Outside the breakpoint, wave heights decay slightly and show some evidence of wave reflection. In the bar-trough region, wave height decay is initially well-behaved but becomes more abrupt as the developing bar causes localized shoaling and forces more waves to break. Because the resistance wave gages were located at fixed positions, with a spacing of 3.6 meters, some of the variation evident in Figure 4 is due to the gage locations not coinciding exactly with the bar or trough. Inside the surf zone, however, wave heights on all four data runs converge to almost exactly the same value, equal to about 40 percent of their offshore height, by the time waves reach the still-water shoreline. From the shoreline to the runup limit, wave height transformation is very similar in all four runs but with somewhat larger gradients for the first run where the foreshore eroded most rapidly. Overall, it appears as though the offshore bar has only a minor effect on wave heights in the swash zone in this case.

In Figures 5 and 6, the zero-moment wave heights depicted in Figure 4 are decomposed into high- and low-frequency components based on filtering the wave record at one-half the peak spectral frequency. Because high-frequency wave energy is so dominant from the wavemaker across the bar crest and into the mid-surf zone, Figure 5 is nearly identical to Figure 4 in this region. Near the shoreline, however, several interesting features are found in the high-frequency wave heights in Figure 7 that differ from those found in the total wave heights in Figure 4.

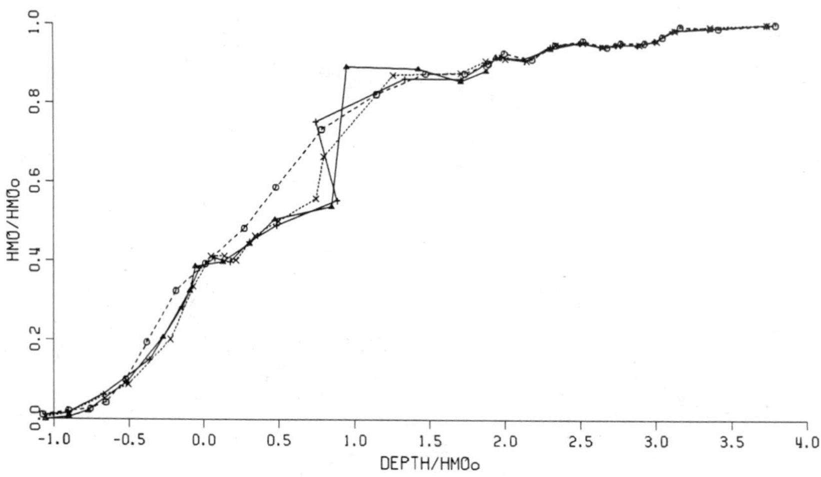

Figure 4. Dimensionless zero-moment wave height, $Hmo_O/L_O = 0.056$.

First, high-frequency wave heights at the shoreline are only about 20 percent of the deep-water zero-moment wave height (total wave height was about 0.40 Hmo_O). As a result, the distribution of the high-frequency wave heights has a concave shape throughout the surf zone in comparison to the convex shape displayed in the distribution of total wave heights in Figure 4. In addition, high-frequency wave heights in the swash zone show somewhat more sensitivity to profile development. As the beach-face slope erodes and flattens, it is evident in Figure 5 that the wave height distributions between the shoreline and the runup limit are significantly affected. Gradients in wave height decay are initially very large in the swash zone but diminish as the beach-face flattens and as the beach-face slope equilibrates.

Low-frequency wave heights, shown in Figure 6, exhibit several consistent features regardless of whether there is a well-developed bar offshore. First, it is apparent that outside the bar-trough region, the low-frequency wave heights are initially equal to about 0.1 Hmo_O. In this region, these low-frequency motions consist of forced incident long waves generated by the incident wave groups, free incident long waves generated by the wavemaker, and free long waves that result from reflection of long waves from the beach and re-reflection of these waves from the wavemaker (including the seiching modes of the tank), as discussed by Mansard and Barthel (1984) and others. Despite the initial absence and subsequent formation of the bar, the magnitude of the long waves increases by about a factor of three in the vicinity of the breaking region to a magnitude of about 0.3 Hmo_O.

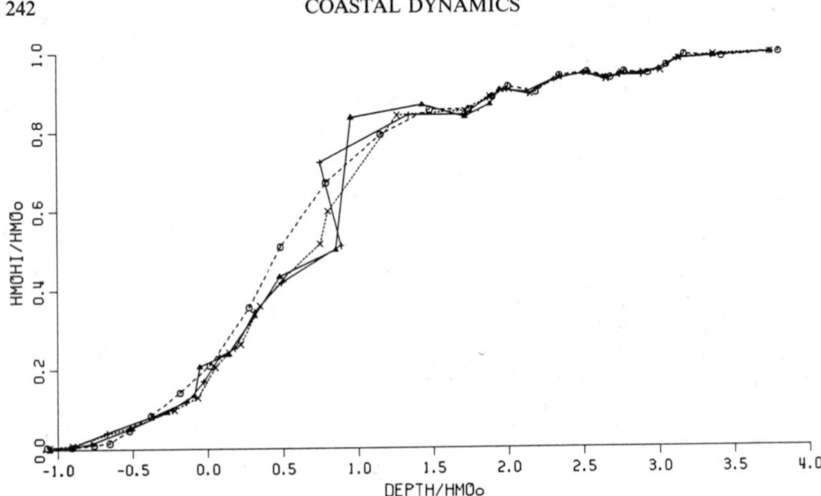

Figure 5. High-frequency zero-moment wave height, Hmo_O/L_O = 0.056.

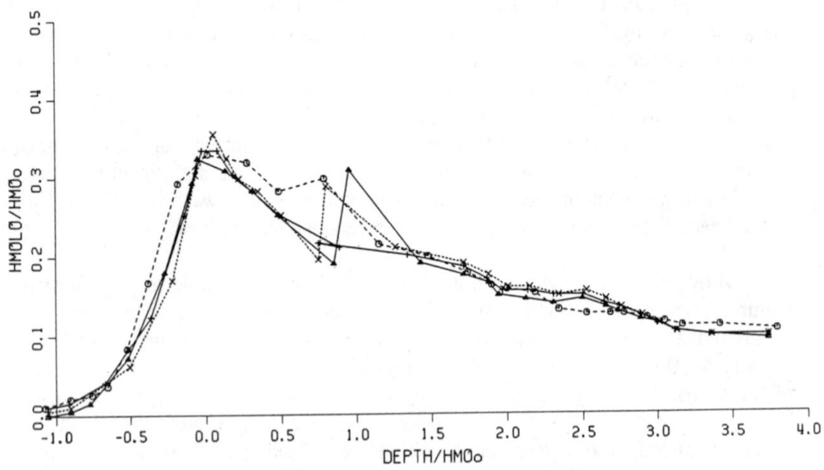

Figure 6. Low-frequency zero-moment wave height, Hmo_O/L_O = 0.056.

Once the bar is well-developed, low-frequency wave heights in the trough region are diminished to about 20 percent of Hmo_O. As a result, there is a local antinode in the long waves at the bar crest, as suggested by the hypothesis that long waves are forced by the moving breakpoint proposed by Symonds et al. (1982). Inside the surf zone, the height of these long waves then continues to increase to a maximum value of about 0.30 to 0.35 Hmo_O. at the shoreline. In general, the heights of these low-frequency waves are of similar magnitude to those measured in the field by Guza and Thornton (1985).

In the swash zone, the low-frequency wave heights diminish and exhibit decreasing gradients toward the runup limit. Based on visual observations, and based on spectral analysis (not shown), these low-frequency motions are coherent standing waves in the swash zone. Unlike the high-frequency waves in Figure 5, which show large gradients in the swash zone due to breaking and energy dissipation, the low-frequency waves in Figure 6 exhibit these gradients due to the presence of a reflected wave antinode at the sloping shoreline.

Figure 7 shows the ratio between the low- and high-frequency zero-moment wave heights at each wave gage. This ratio seems to be among the most consistent and well-behaved of the dimensionless wave height relationships considered. In the offshore region, the ratio of low-to-high frequency wave heights is approximately 0.1 near the wavemaker and increases to about 0.3 to 0.4 near the bar crest. Despite the fact that the low-frequency wave heights diminish in the trough region, the ratio of low-to-high wave heights continues to increase through the trough region and through the entire surf zone due to the strong dissipation of the high-frequency waves. At the shoreline, for this case of very energetic random waves, the ratio of low-to-high frequency wave heights is then about 2.0. In the swash zone, this ratio seems to decrease but is difficult to resolve accurately since both the low- and high-frequency wave heights are small in this region.

The magnitude of the low-frequency wave heights, and the ratio of low-to-high frequency heights, are of particular interest because the low-frequency waves are strongly reflected from the beach and re-reflected from the wavemaker in the wave tank. Despite these effects, it appears from a preliminary analysis that the magnitudes of the long waves in the tank are similar to those observed in the field. In Figures 8 and 9, results of another SUPERTANK data set having a spectral wave steepness of 0.01 are compared to field data from Thompson and Briggs (1993) from the DELILAH experiment conducted at the Corps of Engineers' Field Research Facility at Duck, North Carolina. During one day of the DELILAH experiment, when waves were approximately shore-normal, the spectral steepness was also approximately 0.01, in agreement with the laboratory conditions. In the laboratory tests, $Hmo_O = 0.64$ meters and $T_p = 6$ seconds, while the field conditions included values of Hmo_O in the range of 2.0 to 2.5 meters and T_p equal to about 12 to 13 seconds.

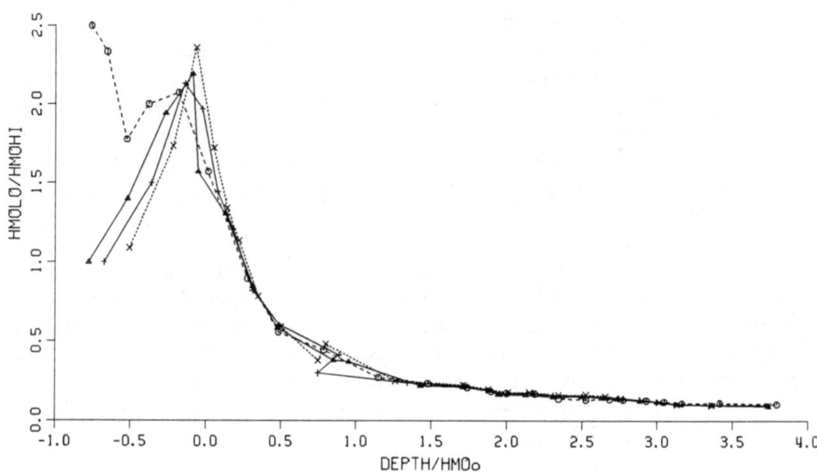

Figure 7. Ratio of low- and high-frequency zero-moment wave heights, $Hmo_O/L_O = 0.056$.

As may be seen in Figures 8 and 9, the field data from Thompson and Briggs, and the laboratory data from SUPERTANK, follow essentially the same trends when expressed in dimensionless form. In Figure 8, the dimensionless magnitudes of the low-frequency waves are nearly identical in the range of relative water depths were field data is available. In both the laboratory tests and in the field data, this range of depth encompasses an area extending from seaward of the bar crest to the inner surf zone. In Figure 9, the ratio of low-to-high frequency wave heights is also very similar in both the laboratory and field data. Despite the fact that the laboratory experiment contained a reflective boundary (wavemaker) in the offshore region, it appears that long wave magnitudes, and the relationship between long and short waves, may not differ significantly from those found under field conditions.

Based on their field data, Thompson and Briggs (1993) conclude that the long-wave heights may reach a maximum of about 1.2 times the short-wave heights at the shoreline. This seems to be confirmed by the laboratory data, however, comparison of Figures 7 and 9 shows that the maximum ratio of low-to-high frequency wave heights depends strongly on the incident wave steepness. For example, this ratio is about 1.2 for a steepness of 0.01 but is about 2.0 for a steepness of 0.056. Other SUPERTANK results (not shown) indicate that for a much lower spectral steepness of 0.005, high-frequency waves are actually larger than low-frequency waves near the shoreline such that the ratio of low-to-high frequency wave heights is only 0.5.

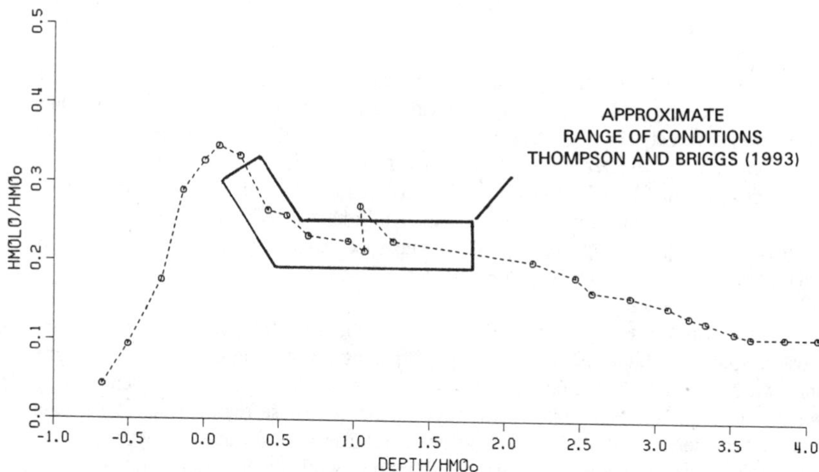

Figure 8. Comparison of low-frequency zero-moment wave heights from SUPERTANK tests and from field data of Thompson and Briggs (1993).

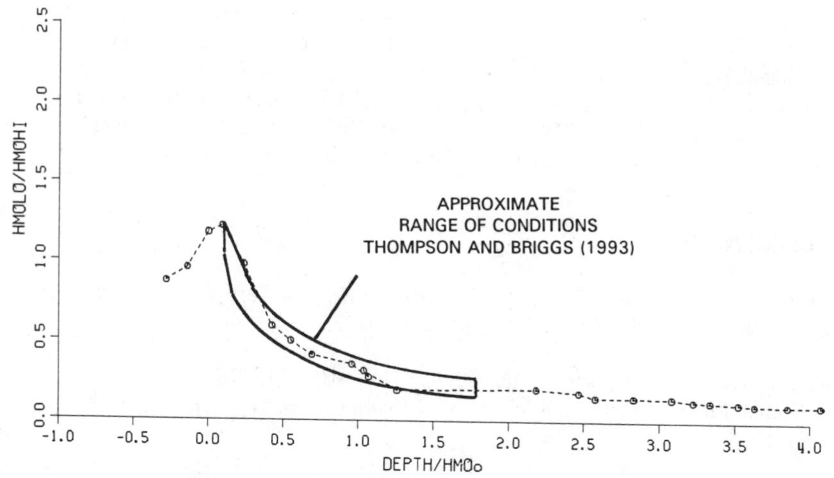

Figure 9. Comparison of ratio of low-to-high low-frequency wave heights from SUPERTANK tests and from field data of Thompson and Briggs (1993).

CONCLUSIONS

This paper has summarized measurements of mean water levels and wave heights made using 26 wave gages during the SUPERTANK Laboratory Data Collection Project. Due to the extensive data set (66 different sets of wave conditions), only selected results, primarily from the first set of wave conditions tested during the SUPERTANK Project, have been presented. Results have been shown in dimensional form for both statistically-based and spectrally-based wave heights across an eroding beach profile. The spectral (energy-based) wave heights have then been further considered in dimensionless form.

Emphasis in each case has been placed on those aspects of the data set that are most unique, specifically: (1) the close gage spacing used through the inner surf zone and swash zone and (2) the de-composition of wave heights into low-frequency (long wave) and high-frequency (short wave) components. These two aspects of the data sets are considered of fundamental importance because several wave height parameters have their maximum values very near the shoreline, while other parameters have their largest gradients near the shoreline or in the swash zone. In either case, these values have not been documented by most previous experimental data sets on wave transformation.

ACKNOWLEDGEMENTS

This work was supported by the Coastal Engineering Research Center (CERC) of the U.S. Army Corps of Engineers. The first author was also supported by a Presidential Young Investigator Award from the National Science Foundation. Permission to publish this paper was granted by the Chief of Engineers. The authors would like to thank Dr. Nick Kraus, formerly of CERC, for his encouragement and support for these measurements and of the entire SUPERTANK Project.

REFERENCES

Guza, R., and Thornton, E., 1985, "Observations of Surf Beat," Journal of Geophysical Research, Vol., 90, C2, pp. 3161-3172.

Kraus, N., Smith, J., and Sollitt, C., 1992, "SUPERTANK Laboratory Data Collection Project," Proc. 23th Intl. Conference on Coastal Engineering, ASCE, pp. 2191-2204.

Kraus, N. and Smith, J., 1994, "SUPERTANK Laboratory Data Collection Project, Volume 1," Waterways Experiment Station, U.S. Army Corps of Engineers, Tech. Rpt. CERC-94-3, 274 pp.

Mansard, E. and Barthel, V., 1984, "Shoaling Properties of Bounded Long Waves," Proc. 19th Intl. Conf. on Coastal Engineering, ASCE, pp. 798-814.

Sonu, C., Pettigrew, N., and Fredericks, R., 1974, "Measurement of Swash Profile and Orbital Motion on the Beach," Proc. Ocean Wave Measurement and Analysis Conf., ASCE, pp. 621-638.

Symonds, G., Huntly, D., and Bowen, A., 1982, "Two-Dimensional Surf Beat: Long Wave Generation by a Time-Varying Breakpoint," Journal of Geophysical Research, Vol. 87, C1, pp. 492-498.

Thompson, E. and Briggs, M., 1993, "Surf Beat in Coastal Waters," Waterways Experiment Station, U.S. Army Corps of Engineers, Tech. Rpt. CERC-93-12, 66 pp.

Waddell, E., 1973, "Dynamics of Swash and Implication to Beach Response," Coastal Studies Institute, Louisiana State University, Tech. Rpt. No. 139, 49 pp.

Numerical Model Verification using SUPERTANK Data in Surf and Swash Zones

Daniel T. Cox[1], Nobuhisa Kobayashi[2], and David L. Kriebel[3]

ABSTRACT: A previously developed one-dimensional time-domain model is modified at the seaward boundary to allow specification of the measured total free surface elevation, eliminating uncertainties with separating incident and reflected waves especially inside the surf zone. Computed free surface oscillations are compared with surf and swash zone measurements from the SUPERTANK Laboratory Data Collection Project. Comparisons of measured and computed results for regular and irregular waves show that the time-dependent model can predict free surface oscillations in the surf and swash zones, whereas conventional models based on the time-averaged equations predict only wave heights and setup and do not predict swash dynamics.

INTRODUCTION

For typical beaches, the flow in the swash zone occurs in a thin sheet and may be dominated by friction. Swash dynamics may play a significant role in shoreline erosion since sediment transport in this region can be large. Therefore, a quantitative understanding of the flow in the swash zone is essential for predictive sediment transport models.

Kobayashi *et al.* (1989) developed a numerical model based on the nonlinear shallow water wave equations including the effects of bottom friction. This is probably the simplest one-dimensional, time-dependent model for predicting the nonlinear wave characteristics in the surf and swash zones in a unified manner. Kobayashi and Wurjanto (1992; K+W hereafter) showed that their extended model could predict available field data on shoreline oscillations fairly well, provided that the incident wave train including the low frequency motion is known. The model's seaward boundary is modified herein to allow input of the measured free surface oscillation, eliminating the uncertainty associated with the separation of incident and reflected waves using linear theory.

To assess the predictive capability of the modified numerical model, RBREAK2, comparisons are made with the swash data from the SUPERTANK Laboratory Data Collection Project (Kraus *et al.*, 1992). Two runs, one of regular waves and one of irregular waves, were selected considering the small change in beach profile over the duration of each run since the numerical model assumes a fixed bed. The measured free surface oscillations at the first gage, located inside the surf zone, were specified to the numerical model; and the computed free surface oscillations are compared with measurements in the surf and swash zones. For

[1] Center for Applied Coastal Research, University of Delaware, Newark, DE 19716 USA Tel: 1 302 831 8477; Fax: 1 302 831 1228; Email: dtc@rad.coastal.udel.edu

[2] Center for Applied Coastal Research, University of Delaware, Newark, DE 19716 USA

[3] Ocean Engineering Program, 590 Hollaway Road, Stop 11D, United States Naval Academy, Annapolis, MD 21402-5000 USA

the regular wave run, a phase averaging analysis is used examine the fluctuations associated with regular wave breaking and imperfect wave generation. The irregular wave run is compared using time series to show that the numerical model can predict the cross-shore variations of the broken wave shapes in the surf and swash zones. Spectral comparisons show that the model can predict the evolution of the spectral peak and low frequency components throughout the model domain. A further analysis shows that the model can predict the cross-shore variations of time-averaged quantities such as setup and root-mean-square wave height up to the point of maximum uprush, unlike conventional time-averaged models. Additionally, the model predicts the cross-shore variations of time-averaged energy dissipation rates due to wave breaking and bottom friction.

MODIFIATION OF SEAWARD BOUNDARY CONDITION

The modification to the seaward boundary condition of K+W is presented concisely. The two-dimensional coordinate system is defined as follows: x' = horizontal coordinate taken to be positive landward with $x'=0$ at the seaward boundary of the computational domain; z' = vertical coordinate taken to be positive upward with $z'=0$ at the still water level (SWL). Primes indicate physical variables. The instantaneous free surface is located at $z'=\eta'$, and the depth is denoted by h'. The seabed is located at $z'=\eta'-h'$, and the local angle of the bed is given by θ'. Assuming the vertical pressure distribution to be approximately hydrostatic, the governing equations for mass and x' momentum integrated from the fixed seabed to the free surface may be expressed as

$$\frac{\partial h'}{\partial t'} + \frac{\partial}{\partial x'}(h'u') = 0 \tag{1}$$

$$\frac{\partial}{\partial t'}(h'u') + \frac{\partial}{\partial x'}(h'u'^2) = -gh'\frac{\partial \eta'}{\partial x'} - \frac{\tau_b'}{\rho} \tag{2}$$

where t' = time; u' = depth-averaged horizontal velocity; τ_b' = bottom shear stress; ρ = fluid density that is assumed constant; and g = gravitational acceleration. The bottom shear stress may be expressed

$$\tau_b' = \frac{1}{2}\rho f'|u'|u' \tag{3}$$

where f' is the constant friction factor. The following dimensionless variables are introduced:

$$t = \frac{t'}{T'} \; ; \; x = \frac{x'}{T'\sqrt{gH'}} \; ; \; u = \frac{u'}{\sqrt{gH'}} \; ; \; z = \frac{z'}{H'} \; ; \; h = \frac{h'}{H'} \; ; \; \eta = \frac{\eta'}{H'} \; ; \; d_t = \frac{d_t'}{H'} \tag{4a}$$

$$\theta = (2\pi)^{1/2}\xi \; ; \; \xi = \frac{\sigma \tan\theta'}{(2\pi)^{1/2}} \; ; \; f = \frac{1}{2}\sigma f' \; ; \; \sigma = T'\sqrt{g/H'} \tag{4b}$$

where T' and H' are the reference wave period and height, respectively, used for normalization; θ = normalized gradient of the slope; ξ = local surf similarity parameter; f = normalized friction factor; and σ = ratio of the horizontal and vertical length scales where $\sigma^2 >> 1$ is assumed in the model.

Substitution of (4a) and (4b) into (1) and (2) yields

$$\frac{\partial h}{\partial t} + \frac{\partial}{\partial x}(hu) = 0 \tag{5}$$

$$\frac{\partial}{\partial t}(hu) + \frac{\partial}{\partial x}(hu^2 + \frac{1}{2}h^2) = -\theta h - f|u|u \tag{6}$$

which are solved numerically in the time domain to obtain the variations of h and u with respect to t and x for given θ, f, and initial and boundary conditions.

To derive the appropriate seaward boundary condition, (5) and (6) are expressed in characteristic form

$$\frac{\partial \alpha}{\partial t} + (u + c)\frac{\partial \alpha}{\partial x} = -\theta - \frac{f|u|u}{h} \quad \text{along} \quad \frac{dx}{dt} = u + c \tag{7a}$$

$$\frac{\partial \beta}{\partial t} + (u - c)\frac{\partial \beta}{\partial x} = \theta + \frac{f|u|u}{h} \quad \text{along} \quad \frac{dx}{dt} = u - c \tag{7b}$$

with

$$c = \sqrt{h} \quad ; \quad \alpha = u + 2c \quad ; \quad \beta = -u + 2c \tag{8}$$

where α and β are the characteristic variables. It is assumed that $u < c$ in the vicinity of the seaward boundary. This assumption is satisfied for the computed results presented in the following.

For the modified seaward boundary condition, the measured free surface elevation is specified directly, $h(t) = d_t + \eta(t)$, at $x = 0$ where d_t is the normalized depth below the SWL, eliminating additional assumptions: the linear long reflected wave and uncertainties in separating incident and reflected waves. Fig. 1 is a schematic of the model boundary with the seaward characteristic of (7b). Node 1 (indicated by subscript 1) is the boundary node at $x = 0$, node 2 (subscript 2) is the first interior point, and the values at the next time level are indicated by an asterisk. K+W used a simple first-order finite difference based on (7b) with $f = 0$ where Δx =constant node spacing, and Δt =variable time step. However, this approximation has been improved as follows. Fig. 1 shows the seaward advancing characteristic β through node 1 at the next time level. The values of u_1, h_1 u_2, and h_2 at the present time level are known. The value of u_1^* at the next time level is unknown, but the value h_1^* is known from the seaward boundary condition. The linear approximation of $dx/dt = (u - c)$ is expressed as

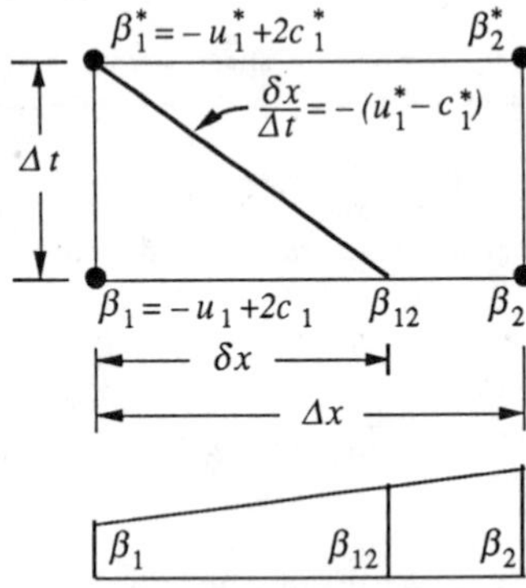

Figure 1: β characteristic at seaward boundary and linear interpolation

$$\frac{\delta x}{\Delta t} = -(u_1^* - c_1^*) > 0 \quad \text{with} \quad \delta x > 0 \tag{9}$$

where the positive spatial increment $\delta x < \Delta x$ to satisfy the numerical stability criterion. Making use of (9), the value of β_{12} between nodes 1 and 2 is found by linear interpolation,

$$\beta_{12} = \beta_1 - \frac{\Delta t}{\Delta x}(u_1^* - c_1^*)(\beta_2 - \beta_1) \tag{10}$$

With $f=0$, Eq. (7b) is simplified as

$$\frac{d\beta}{dt} = \theta \quad \text{along} \quad \frac{dx}{dt} = u - c \tag{11}$$

The finite difference approximation of (11) along $dx/dt = u_1^* - c_1^*$ using (10) yields

$$\beta_1^* = \beta_1 - \frac{\Delta t}{\Delta x}(u_1^* - c_1^*)(\beta_2 - \beta_1) + \theta_1 \Delta t \tag{12}$$

Eq. (12) differs from that of K+W with $(u_1^* - c_1^*)$ replacing $(u_1 - c_1)$. Essentially, this gives an implicit scheme dependent on u_1^* and h_1^* rather than u_1 and h_1. Solving (12) for u_1^* gives

$$u_1^* = \frac{2\,c_1^* - \beta_1 - \frac{\Delta t}{\Delta x}(\beta_2 - \beta_1)\,c_1^* - \theta_1\,\Delta t}{1 - \frac{\Delta t}{\Delta x}(\beta_2 - \beta_1)} \tag{13}$$

where $c_1^* = (h_1^*)^{1/2}$ and h_1^* is known from the measured free surface oscillation η_1^* at x=0. The numerical model with the seaward boundary condition modified by (13) is called RBREAK2 and is used for the following computations.

EXPERIMENTAL CONDITIONS

To assess the predictive capability of the modified numerical model, comparisons are made with the swash data from the SUPERTANK Laboratory Data Collection Project (Kraus *et al.*, 1992). This project provided a comprehensive set of cross-shore hydrodynamic and sediment transport data at prototype scale. Ten capacitance wave gages, partially buried in the sand, provided time histories in the surf and swash zones (Kriebel, 1993). Gages 1–5 measured the free surface oscillations of broken waves, and Gages 6–10 measured the wave uprush and downrush along with the time history of the wetted sand surface during the backrush phase when the beach was exposed. Two runs were selected: one of regular waves, A1615A, and one of irregular waves, A1610A, as listed in Table 1. These runs were selected considering the small change in beach profile over the duration of each run since the numerical model assumes a fixed bed. The bottom profile is specified using the measurements taken at each gage location at the end of the run.

Although the duration of both runs was 40 min, only the last part of the measurements, approximately 15 min, are specified to the model as follows. The computation duration is specified for $-10 < t < T_{max}$, where t is the normalized time, and $t = -10$ is the start of the computation and corresponds to approximately 25 min after the start of the wavemaker. To eliminate transitional effects starting with no wave action at $t = -10$, the initial part of the computed time series is truncated for $-10 < t < 0$. All computed results herein are for the truncated time series $0 < t < T_{max}$. The reference wave height, H', for normalization is based on the spectral estimate of the significant wave height of the measured free surface oscillation at $x = 0$. The reference wave period, T', is based on the period specified to the wavemaker for the regular wave run and the spectral peak period of the target spectrum for the irregular wave run. The surf similarity parameter, ξ, given in (4b) is computed using the typical value $\tan\theta' = 0.0298$, based on the outer surf zone slope for both runs. The bottom friction factor f'=0.05 was used previously by K+W. Figs. 7a and 7b (upper panels) show the beach profiles and gage locations for the two runs selected with the gage letters in parenthesis as used by Kriebel (1993). The beach profile for the regular wave run has a well defined bar at $x = -0.5$ (not shown) and approximately a 1:33 slope over most of the surf zone and approximately a 1:5 slope for the swash zone. The beach profile for the irregular wave run was similar but with a remnant bar at $x = -1.0$.

Table 1: Wave characteristics for Runs A1615A and A1610A.

Run	Wave Type	T_{max}	H' (m)	T' (s)	ξ	σ	f'
A1615A	Regular	130	.2446	7.00	.527	44.33	0.05
A1610A	Irregular	120	.3185	7.00	.462	38.85	0.05

Fig. 2 shows the normalized spectra and time series of the input wave train for the two runs. The spectra are plotted with normalized frequency, f^*, where $f^* = 1$ roughly corresponds to the spectral peak of the input wave spectrum. The spectra presented herein are smoothed using ensemble averaging to give 16 degrees of freedom and corresponding bandwidths of $\Delta f^*_{sm} = 0.0615$ for the regular wave run and $\Delta f^*_{sm} = 0.0667$ for the irregular wave run. Fig. 2 also shows the last ten normalized wave periods for each run, indicating that the measured free surface oscillations at Gage 1 specified to the numerical model are broken waves.

MEASURED AND COMPUTED COMPARISONS

The phase averaged results for the regular wave run are shown in Figs. 3a–3c for the measured and computed free surface oscillations at Gages 1–8. In this figure, η^* is the normalized mean-adjusted free surface displacement and t^* is the normalized time over one period. The middle curve is the ensemble average of 130 waves and the envelope is this average $\pm$ one standard deviation. The measured profiles are well predicted by the model in the surf zone, including small features such as the reflected wave in the region $0.6 < t^* < 0.8$ at Gage 4. The model slightly underpredicts the profiles in the swash zone at Gages 6–8 in Figs. 3b and 3c. The measured results show that fluctuations associated with breaking regular waves and imperfect wave generation are more or less constant for $0 < t^* < 1$ with no clear indication of turbulence at the steep front of a bore. It is difficult to estimate the effect of the aerated bores on the capacitance gages, but this would also contribute to the discrepancy between the measured and computed results. Larger fluctuations were expected for the measured results because the numerical model does not include turbulence. The computed results show larger fluctuations near the bore fronts and on the reflected waves in Figs. 3a and 3b which may be attributed to small errors in the computed phase speed. The deviation at Gages 7 and 8 of the measured waves in Fig. 3c for $0 < t^* < 0.2$ and $0.7 < t^* < 1$ are due to the change in beach profile of the moveable bed. The computed results show no deviation in this range since the bed is assumed fixed.

Fig. 4 shows normalized spectral densities, S, for the measured and computed free surface elevation, η, plotted for the range $\Delta f^*_{sm} < f^* < 3$ at Gages 2–7 for the regular wave run. The spectral energy near the peak and higher harmonics are well predicted throughout the surf and swash zones although the model underpredicts the peaks at some gage locations. For the lower frequencies, $\Delta f^*_{sm} < f^* < 0.5$, the model predicts some spurious low frequency energy, but this low frequency energy is two orders of magnitude less than the peak for the regular wave run.

For the irregular wave run, Fig. 5 shows measured and computed η plotted for the range $110 < t < 120$ at Gages 2–7. The time series at Gage 7 shown in Fig. 5 is the instantaneous water depth h above the measured movable bed. The cross-shore variations of the broken wave shape are well predicted by the model, that is, the saw-tooth shape in the inner surf zone and the more parabolic shape in the swash zone. The arrival of each bore is also predicted fairly well.

Fig. 6 shows normalized spectral densities, S, for the measured and computed η plotted for the range $\Delta f^*_{sm} < f^* < 3$ at Gages 2–7 for the irregular wave run. The model predicts the evolution of the spectral peak and low frequency components in the surf and swash zones. It is noted that the low frequency components for the irregular wave run are not negligible unlike the regular wave run.

Figs. 7a and 7b show the cross-shore variations of several quantities for the two runs. The upper panel shows the bottom profile, measured setup, setup computed by RBREAK2, and setup computed by the model of Battjes and Stive (1985; B+S hereafter). The model of B+S solves the time-averaged energy and momentum equations for random waves, so the comparisons in Fig. 7a for regular waves are not strictly applicable but are included for comparison with Fig. 7b. The second panel is a detail of the upper panel showing that the setup, $\bar{\eta}$, is well predicted by both models in the surf zone. However, the model of B+S is not applicable in the swash zone where the wave breaking criterion based on the local water depth below the mean water level appears to be too crude, and the relationship between the local wave height and fluid velocity may not be approximated by linear theory.

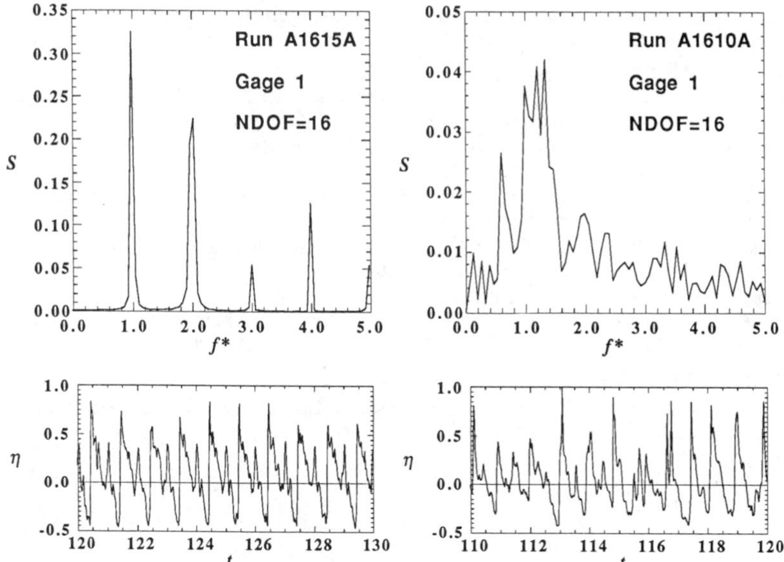

Figure 2: Normalized spectral densities, S, of input wave train, η, at Gage 1 for regular wave run (upper left panel) and irregular wave run (upper right panel). Corresponding time series (lower panels) for last ten normalized wave periods, $(T_{max}-10) < t < T_{max}$.

The third panel shows that the spectral estimate of the root-mean-square wave height, H_{rms}, is well predicted by RBREAK2 even in the swash zone. The bottom panel shows the computed cross-shore variations of the time-averaged rates of energy dissipation per unit area due to wave breaking and bottom friction, $\overline{D_B}$ and $\overline{D_f}$, respectively. Due to the segmented bottom profile at each gage location, the curve for $\overline{D_B}$ contained some noise and was smoothed using a 5 point running average. Since the input waves at the seaward boundary were essentially broken, this curve rises sharply for $0 < x < 0.2$, indicating the shortcoming of RBREAK2 which estimates $\overline{D_B}$ using the energy equation without modeling it explicitly and physically (K+W). The interesting result is from the still water shoreline, $x = 0.8$, and landward which shows the increasing importance of the bottom friction energy dissipation in the swash zone. Fig. 7b is qualitatively similar to Fig. 7a. In Fig. 7b it is possible to compare directly the RBREAK2 and the model of B+S in predicting setup and H_{rms}.

CONCLUSION

A quantitative understanding of the variations of hydrodynamic quantities over a wave period is essential in the swash zone for improving the landward boundary condition for cross-shore sediment transport models. Time-averaged models such as B+S can predict root-mean-square wave height and setup for irregular waves in the shoaling and surf zones, but B+S underpredicts these quantities near the still water shoreline for the present data set and gives no results in the swash zone. Time-dependent models will be necessary in predicting the instantaneous swash dynamics and resulting sediment transport rates from which the net sediment transport rates can be computed.

REFERENCES

Battjes, J.A. and Stive, M.J.F. (1985) "Calibration and Verification of a Dissipation Model for Random Breaking Waves" *J. Geophys. Res.,* 90(C5), 9159–9167.

Kobayashi, N. and Wurjanto, A. (1992) "Irregular Wave Setup and Run-up on Beaches" *J.W.P.C. and O. Engrg.,* ASCE, 118(4), 368–386.

Kobayashi, N., DeSilva, G.S., and Watson, K.D. (1989) "Wave Transformation and Swash Oscillation on Gentle and Steep Slopes" *J. Geophys. Res.,* 94(C1), 951–966.

Kraus, N.C., Smith, J.M., and Sollitt, C.K. (1992) "SUPERTANK Laboratory Data Collection Project" *Proc. 23rd ICCE,* ASCE, 2191–2204.

Kriebel, D.L. (1993) ''SUPERTANK Swash Measurements,'' *SUPERTANK Laboratory Data Collection Project , Volume 1, Chapter 4,* N.C. Kraus and J.M. Smith (eds.), U.S. Army Corps. of Engineers, CERC–WES, Vicksburg, MS (in preparation).

ACKNOWLEDGMENT

This work was sponsored by the U.S. Army Research Office, University Research Initiative, under contract No. DAAL03–92–G–0016.

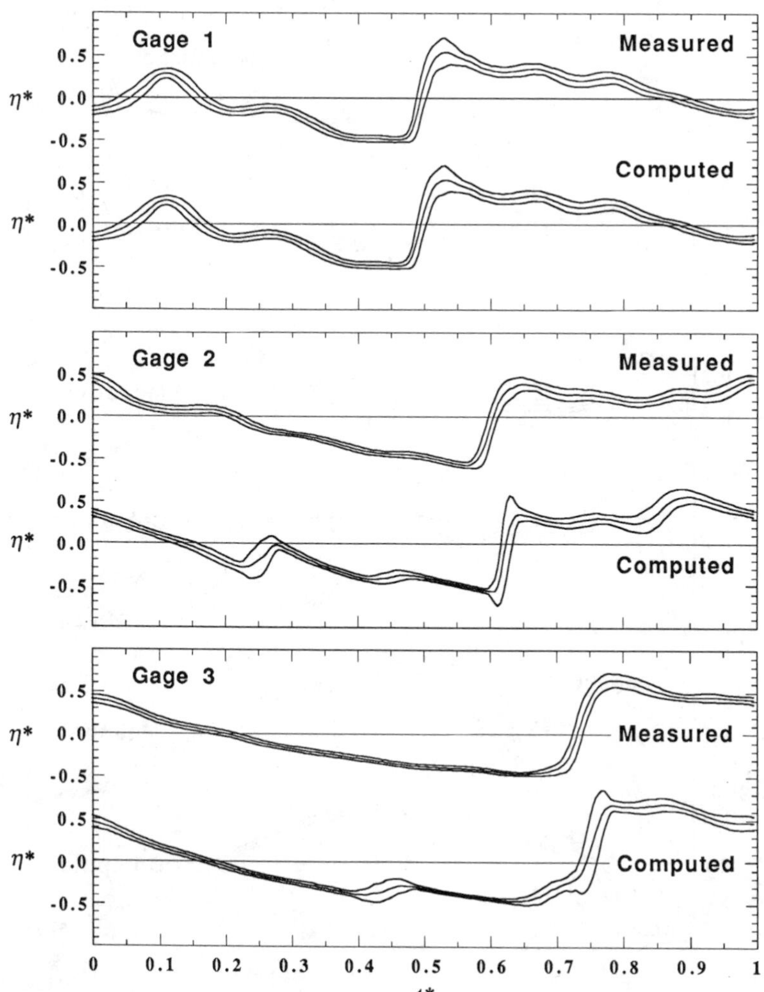

Figure 3a: Measured and computed normalized free surface oscillations, η^*, mean-adjusted and phase-averaged with ± one standard deviation for normalized one wave period $0 < t^* < 1$ at Gages 1–3 in the surf zone for regular wave run.

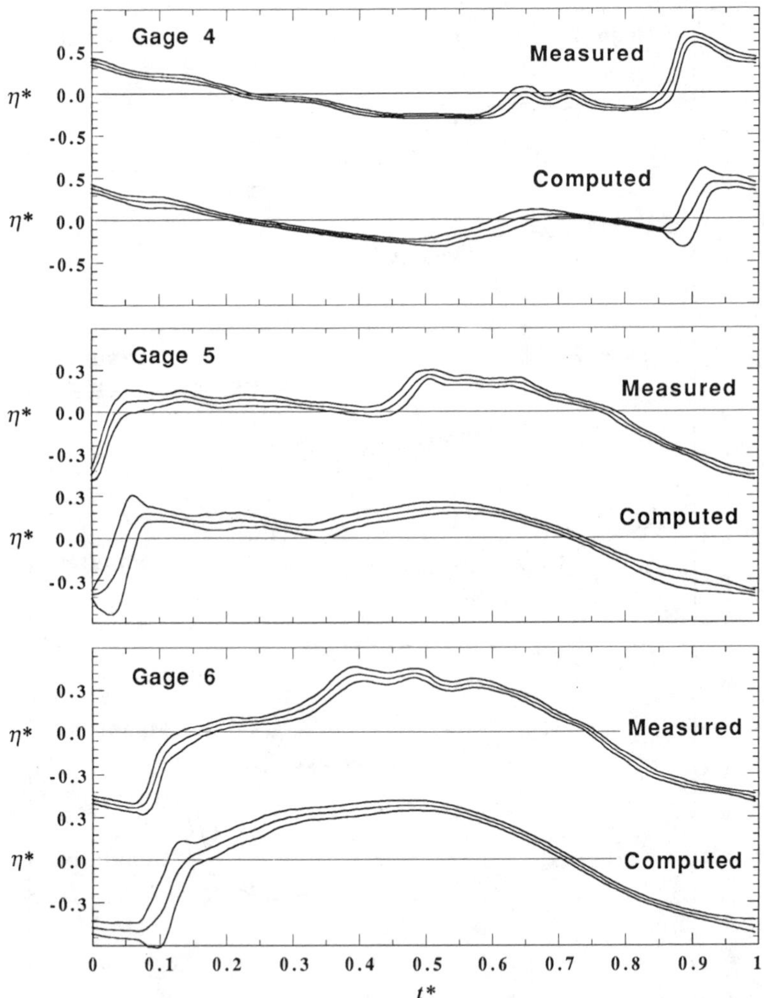

Figure 3b: Measured and computed normalized free surface oscillations, η^*, mean-adjusted and phase-averaged with ± one standard deviation for normalized one wave period $0 < t^* < 1$ at Gages 4–6 in the surf and swash zones for regular wave run.

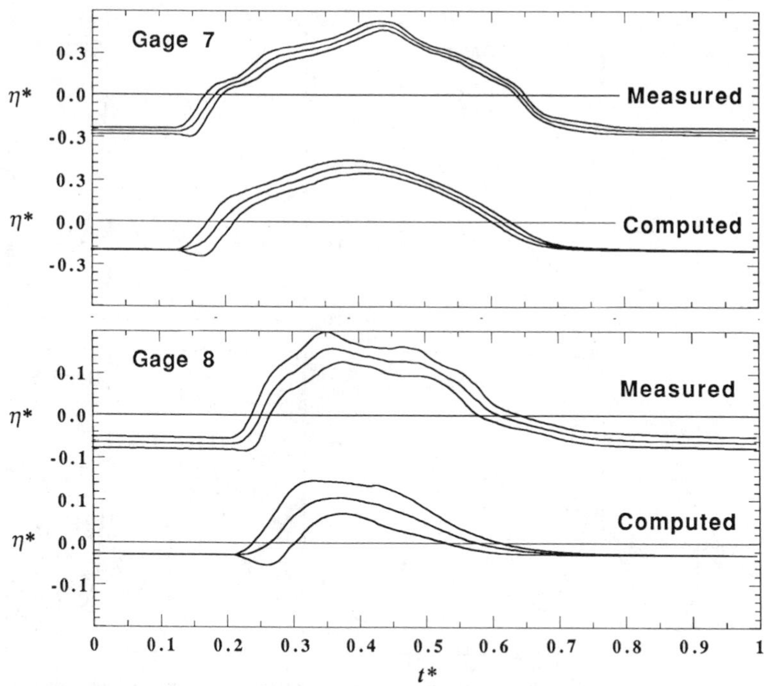

Figure 3c: Measured and computed normalized free surface oscillations, η^*, mean-adjusted and phase-averaged with ± one standard deviation for normalized one wave period $0 < t^* < 1$ at Gages 7 and 8 in the swash zone for regular wave run.

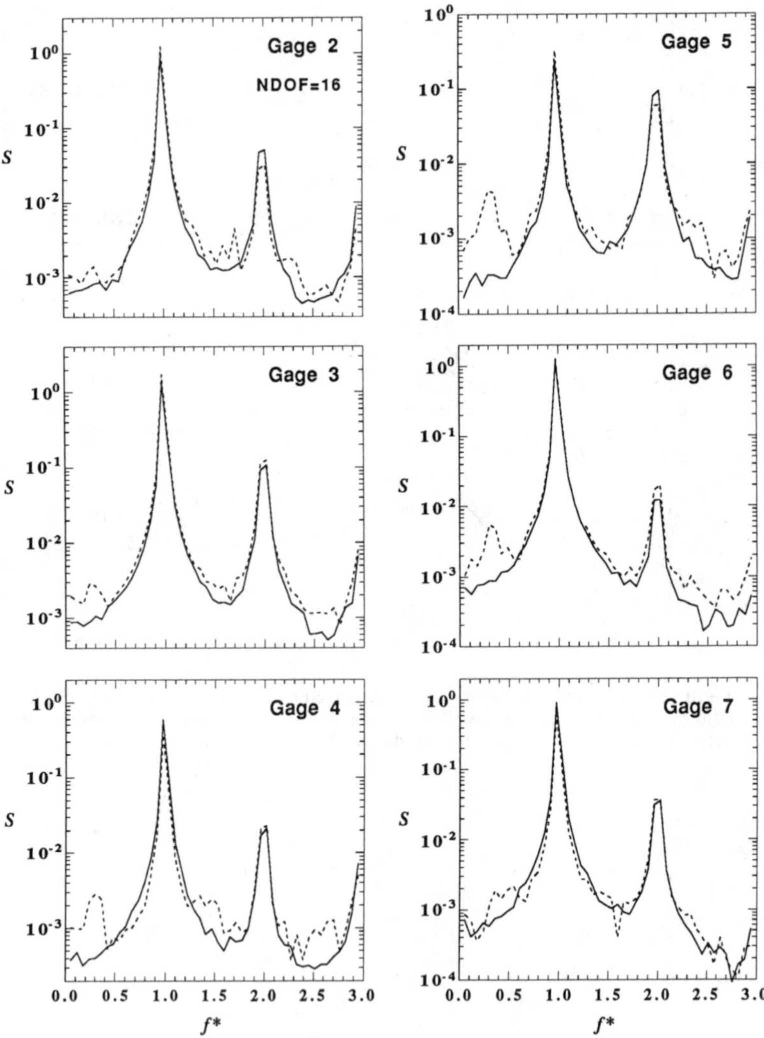

Figure 4: Measured (solid line) and computed (dashed line) smoothed spectral densities, S, of normalized free surface elevation, η, as a function of normalized frequency, f^*, at Gages 2–7 for regular wave run.

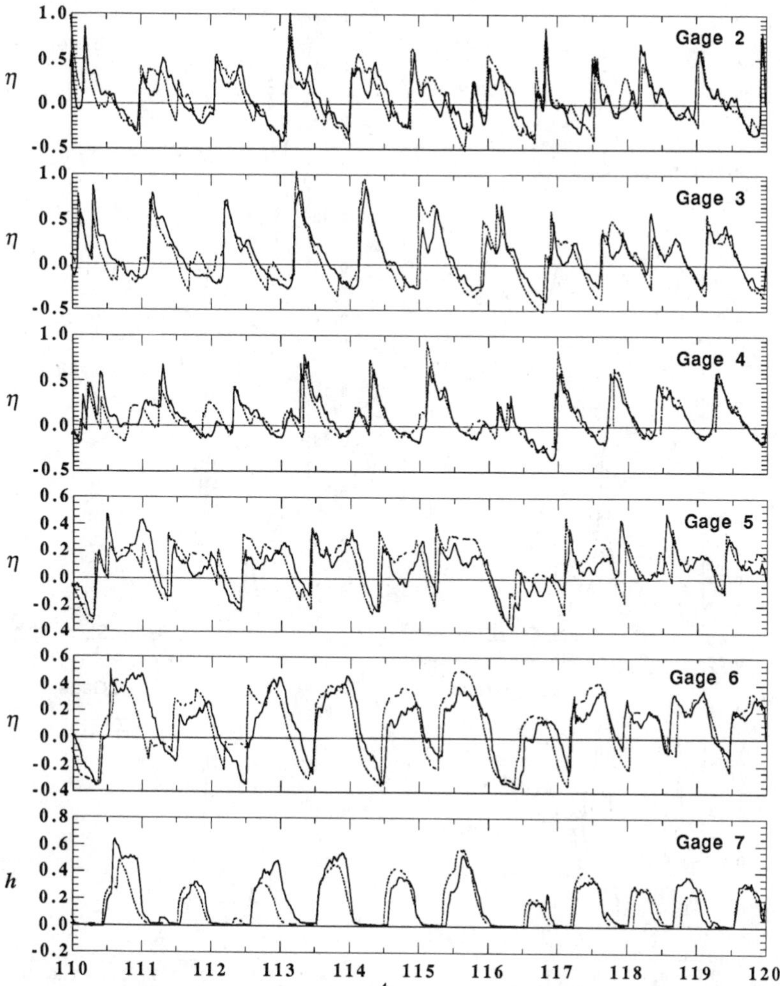

Figure 5: Measured (solid line) and computed (dashed line) normalized free surface elevation, η, or water depth, h, as a function of normalized time, t, at Gages 2–7 for $(T_{max}-10) < t < T_{max}$ for irregular wave run.

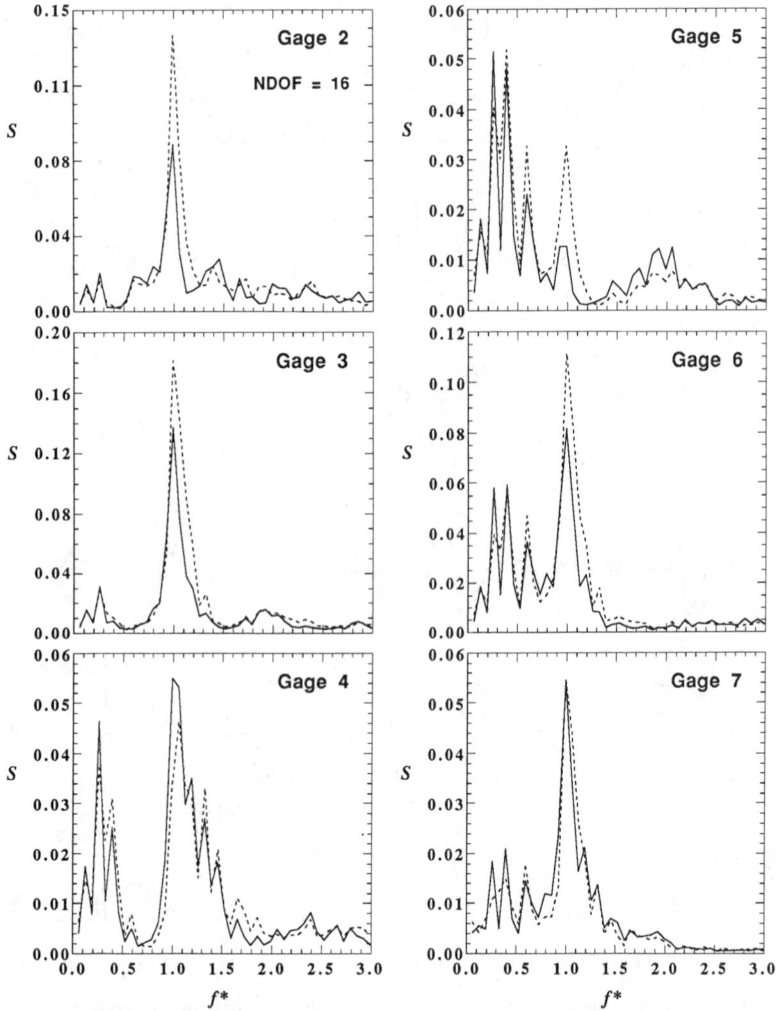

Figure 6: Measured (solid line) and computed (dashed line) smoothed spectral densities, S, of normalized free surface elevation, η, as a function of normalized frequency, f^*, at Gages 2–7 for irregular wave run.

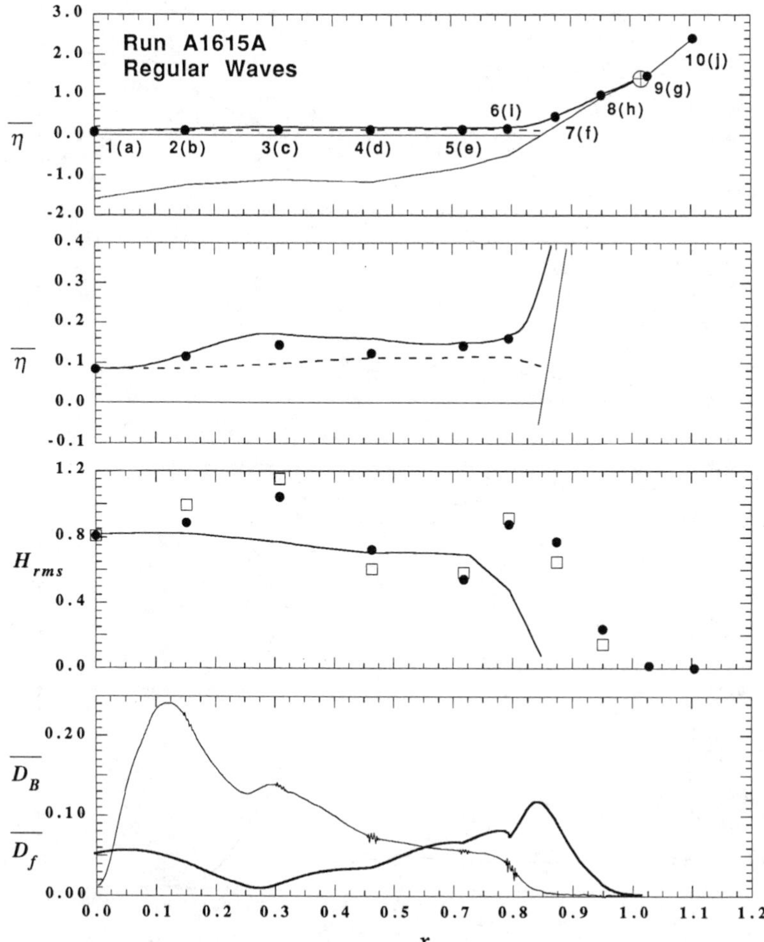

Figure 7a: Measured and computed cross-shore quantities for regular wave run. Upper panel shows model domain including gage locations, bottom profile (light solid line), and setup, $\overline{\eta}$, with measured (solid circle), computed RBREAK2 (heavy solid line) with computed maximum runup (circle with cross), and computed B+S (dashed line). Second panel is detail of setup. Third panel shows spectral estimate of root-mean-square wave height, H_{rms}, with measured (solid circle), computed RBREAK2 (open square), and computed B+S (solid line). Bottom panel shows computed time-averaged energy dissipation rates due to breaking, $\overline{D_B}$, (light solid line) and bottom friction, $\overline{D_f}$, (heavy solid line) for RBREAK2.

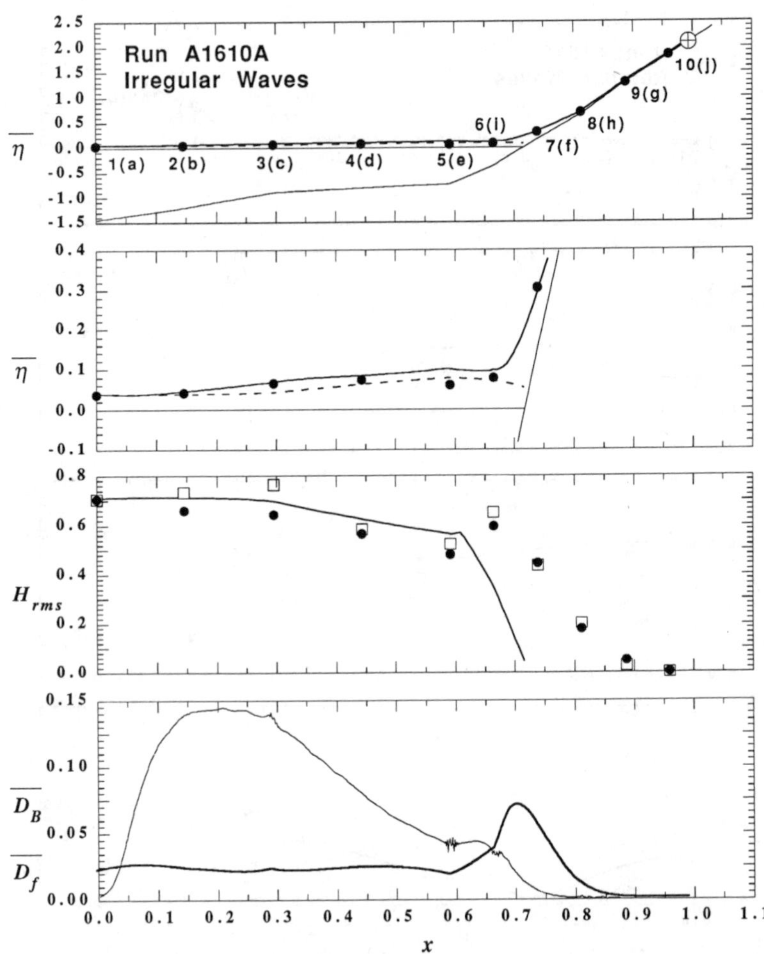

Figure 7b: Measured and computed cross-shore quantities for irregular wave run. Upper panel shows model domain including gage locations, bottom profile (light solid line), and setup, $\overline{\eta}$, with measured (solid circle), computed RBREAK2 (heavy solid line) with computed maximum runup (circle with cross), and computed B+S (dashed line). Second panel is detail of setup. Third panel shows spectral estimate of root-mean-square wave height, H_{rms}, with measured (solid circle), computed RBREAK2 (open square), and computed B+S (solid line). Bottom panel shows computed time-averaged energy dissipation rates due to breaking, $\overline{D_B}$, (light solid line) and bottom friction, $\overline{D_f}$, (heavy solid line) for RBREAK2.

HIGH-RESOLUTION MEASUREMENTS OF SAND SUSPENSION BY PLUNGING BREAKERS IN A LARGE WAVE CHANNEL

William R. Dally[1], M.ASCE and Stephen F. Barkaszi Jr.[2]

ABSTRACT: A tightly spaced array of Optical Backscatterance Sensors was deployed in the outer surf zone of a large wave channel, to measure sediment suspension under regular, plunging breakers at high temporal and spatial resolution. Results show that in moving from the zone of shoaling, to the plunge point, and into the surf zone (a distance of only 7 m), the time series, spectrum, mean vertical profile, and depth-averaged suspended load all undergo remarkable change in structure and intensity. The data support a conceptual model in which 1) sand is convectively entrained at the plunge point to create suspension clouds, 2) the clouds are then advected back-and-forth by the oscillatory wave motion and offshore by the undertow, 3) the clouds are remixed by turbulence that is either locally generated or advected by the undertow, and 4) shoreward of the plunge point where breaking is fully developed, entrainment becomes a diffusive process.

INTRODUCTION

A fundamental feature of the surf zone is the suspension of sediment by breaking waves, and accurate depiction of the vertical and cross-shore distributions of the suspended sediment load is an important component of processes-based beach evolution modeling. Field observations indicate that distinct regions of high mean concentration occur in the zone of initial wave breaking, especially if the breakers are plunging (see e.g. Kana, 1979). Establishing the intensity and distribution of these high-concentration zones, which also appear at

---

[1]Associate Professor, Ocean Engineering Program, Florida Institute of Technology, !50 W. University Boulevard, Melbourne, FL 32901, U.S.A

[2] Graduate Assistant (currently Research Associate, Florida Solar Energy Center, 300 State Road 401, Cape Canaveral, FL 32920)

the toe of the swash, is essential to the successful modeling of 1) depositional and erosional features created by cross-shore transport, such as the longshore bar/trough and the swash step, and 2) longshore transport, particularly its distribution across the surf zone.

Also crucial to the development of generic sand transport models is establishing the dependence of sand suspension across a wide range of wave and sediment characteristics. However, the data required to address these issues is extremely difficult, time-consuming, and expensive to collect in the field due to the number of instruments required, and the due to the lack of control over the water level, wave climate, sand composition, etc.

OBJECTIVES

The major objective of this study was, by exploiting the tightly controlled environment afforded by a large wave channel facility, to measure time series of suspended sand concentration and relevant fluid flow parameters in the region of incipient and initial wave breaking. Observations were to be at high vertical, cross-shore, and temporal resolution, and encompass a wide range of (regular) wave conditions.

APPARATUS AND PROCEDURE

During the SUPERTANK Data Collection Project (Kraus and Smith, 1994), two days were devoted to an experiment that focussed on sand suspension in the outer surf zone. For this effort, an instrument array consisting of thirty-three Optical Backscatterance Sensors (OBS), four electromagnetic current meters (EMCM), a capacitance wave staff, and a video camera was attached to a mobile carriage that spanned the walls of the facility. The OBS were organized in four vertical arrays of 7-9 sensors, with the arrays spaced at 0.75 m intervals. An EMCM accompanied each of the vertical arrays. The instrument array is shown schematically in Figure 1.

The beach installed in the tank was composed of well-sorted quartz sand with mean fall speed of 3.3 cm/s (roughly 0.25 mm diameter). A histogram of fall speed, measured with an automated settling tube, is presented in Figure 2. The OBS were calibrated on site using the same calibration tank described in Downing and Beach (1989). Sand used in the calibration was taken directly from that stockpiled for constructing and modifying the beach during the SUPERTANK experiment.

Twenty-two experiments were conducted during a two-day test, with thirteen test conditions that encompassed incident wave heights ranging from 0.3 to 0.8 m, and periods from 3 to 10 s. The water depth at the array was nominally 1.3 m. All experiments were conducted with regular waves.

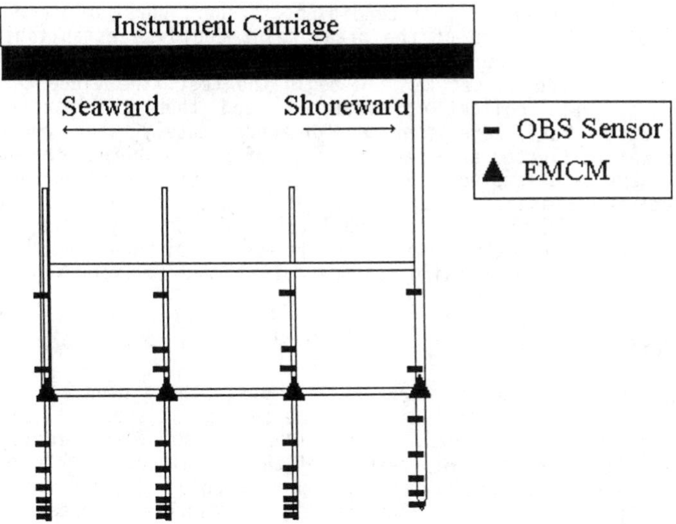

Figure 1 - Schematic of instrument array. Vertical stacks of 7-8 OBS are spaced at 0.75 m intervals (from Barkaszi, 1993).

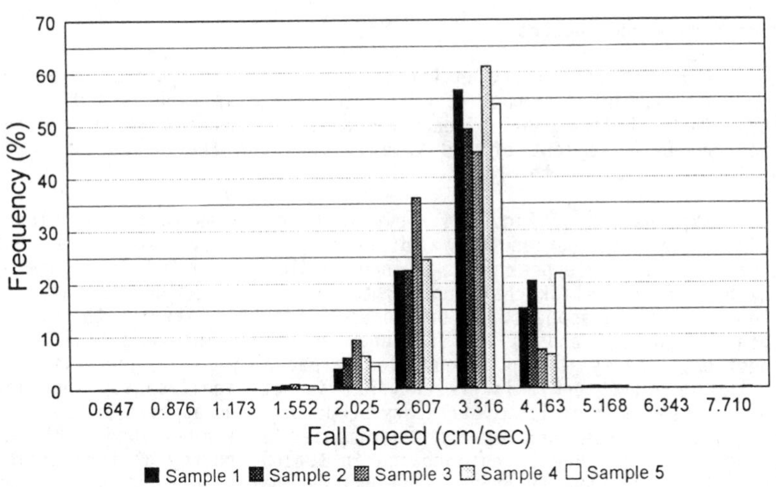

Figure 2 - Results of settling tube analysis of SUPERTANK sediment (from Kraus and Smith, 1994).

For each set of test conditions, a short burst of waves was run so that the location of the break point could be established. The instrumented carriage was then positioned accordingly, and the wave channel allowed to settle. To begin the test, the video camera was started, data collection initiated, and then the wavemaker was restarted. Waves were generated for approximately twenty cycles. OBS data were collected at a sampling rate of 10 Hz, whereas the wave gage and EMCMs were sampled at 16 Hz. Data collection continued well after wavemaking ceased, so that the settling of the channel could be documented. Exploiting the mobility of the carriage, the array was deployed at locations spanning the zones of shoaling, incipient breaking, initial breaking (plunge point), and fully broken waves.

RESULTS AND DISCUSSION

The OBS data collected in this effort have been examined in a variety of formats, including 1) time series, 2) spectral, 3) running average, 4) time-averaged, 5) depth-averaged, and 6) horizontally-and-vertically averaged. All results of these analyses exhibit a notable cross-shore structure in passing from the region of incipient breaking and into the surf zone. The results and discussion presented below focus on a single set of test conditions, for which the wave period was 10 s and the measured breaker height was 0.6 m. Successive runs were made with the carriage at three different positions, and the measurements from six OBS stacks are discussed below. These six stacks are located in Figure 3, along with the bottom profile and still water level.

Time Series and Spectra

For each of the six positions identified in Figure 3, a set of time series from the stack of OBS is presented in Figures 4-9. Also shown in the bottom panel of each figure is the concentration spectrum produced by a segment of the time series from the sensor that was nominally 7 cm from the bed.

Examination of Figure 4 shows that well seaward of the plunge point, the time series has a distinct oscillatory structure that appears as a single peak in the concentration spectrum. In Figure 5, which is from a stack that was 1.8 m closer to the plunge point, low frequency signal appears. As one moves even further towards the plunge point (Figures 6 and 7) low frequency energy begins to dominate the spectrum. Energy at high frequencies, especially at integer multiples of the wave frequency, is present, but is not significant. The time series loses most of its oscillatory structure, and appears as a string of variable suspension events superimposed on a mean load. At the plunge point (Figure 8; note change in scale), energy at frequencies lower than the wave frequency (0.1 Hz) dominates fully. Landward of the plunge point (Figure 9), although low frequency energy still exists, a peak again appears at the wave frequency.

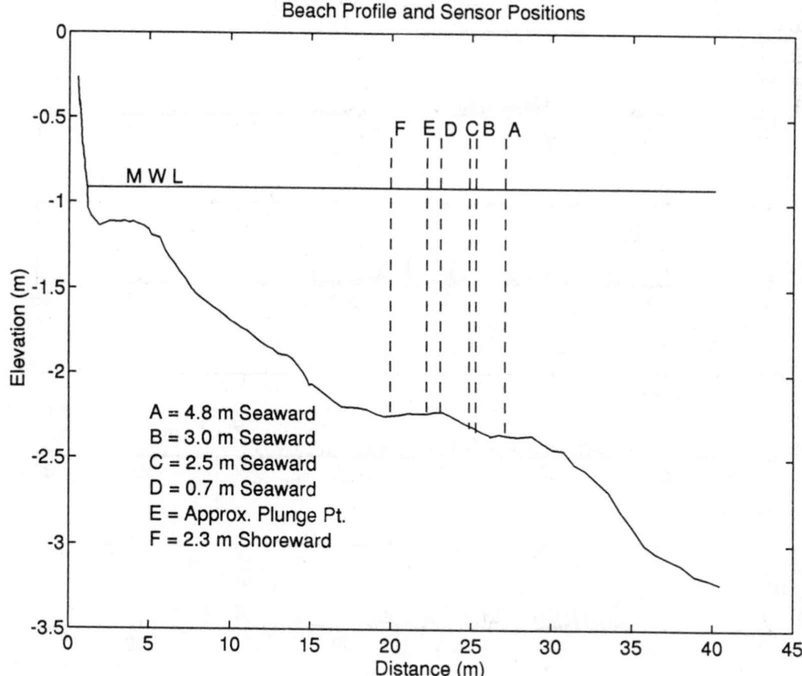

Figure 3 - Bottom profile, still water level, and locations of six OBS stacks during the test.

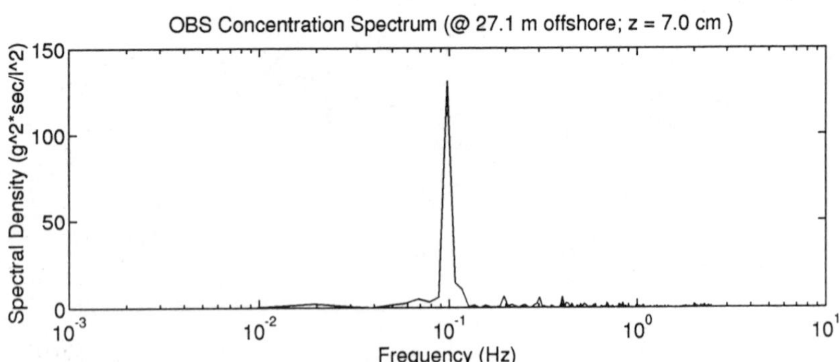

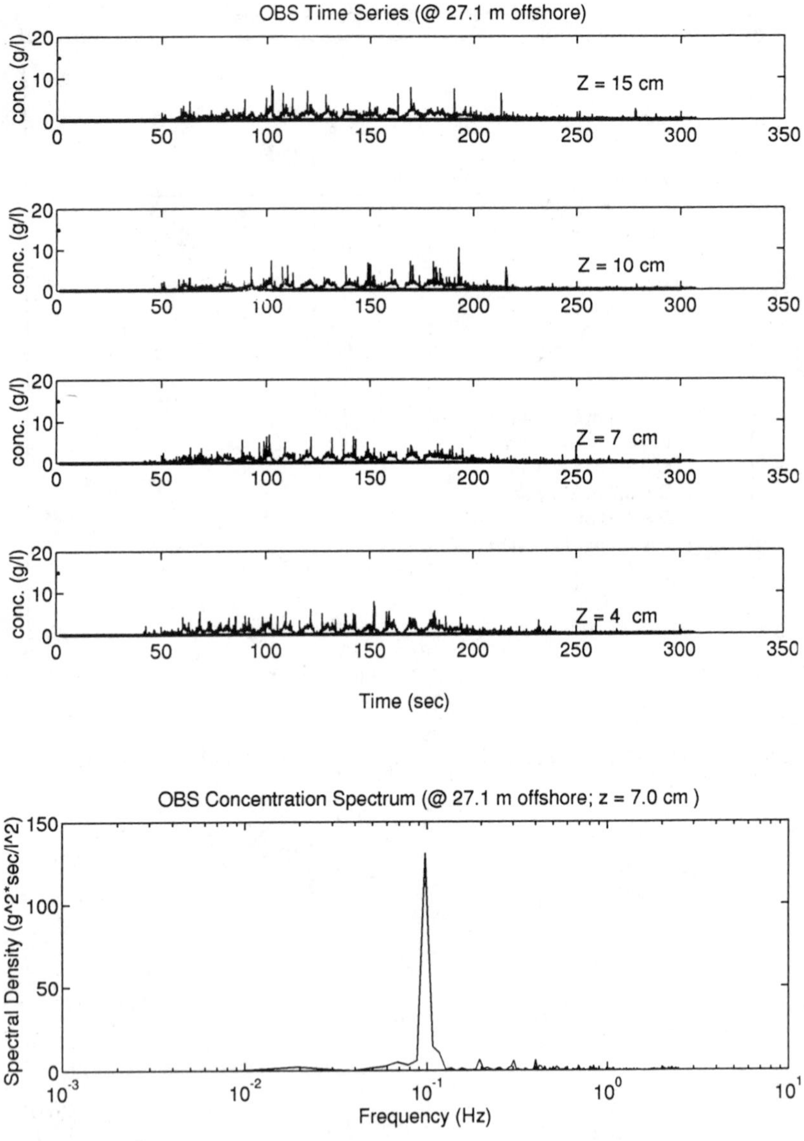

Figure 4 - Time series and sample concentration spectrum from OBS stack located 4.8 m seaward of plunge point.

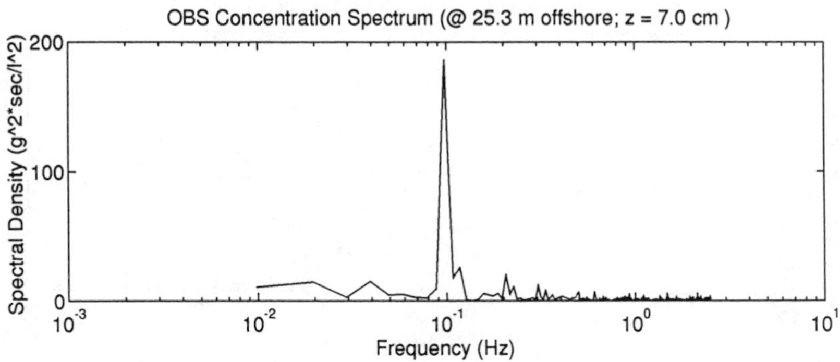

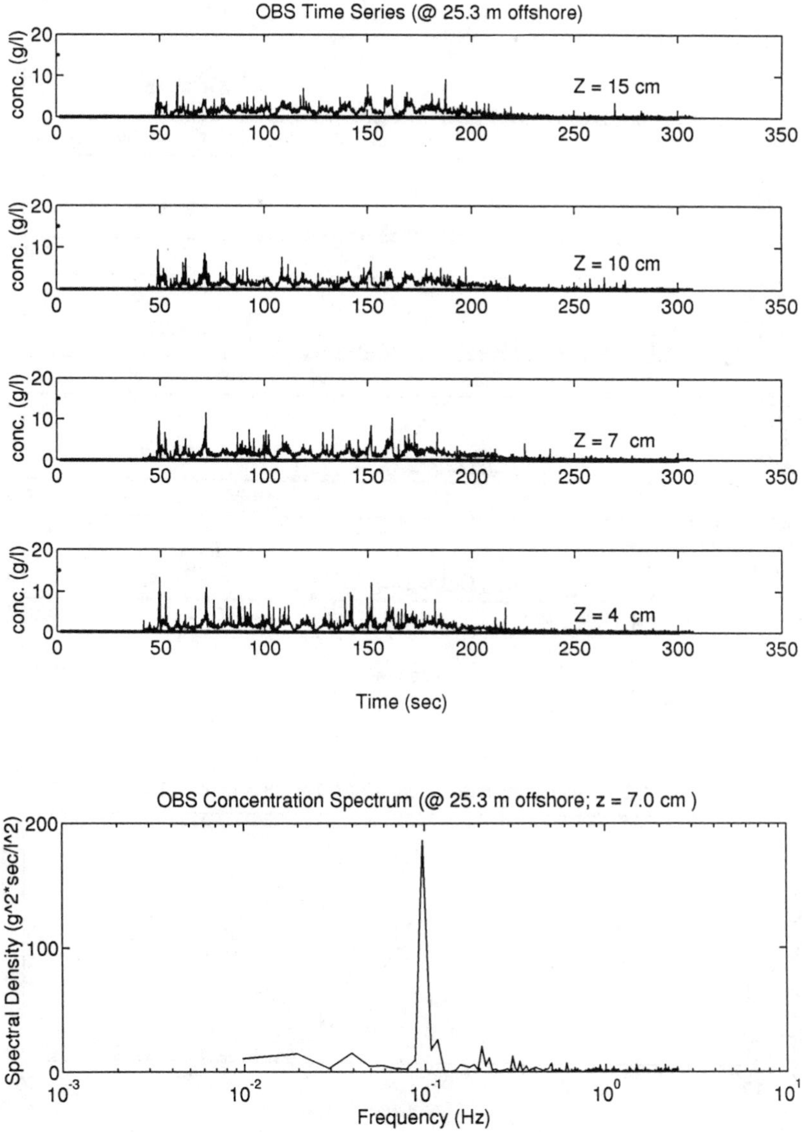

Figure 5 - Time series and sample concentration spectrum from OBS stack located 3 m seaward of plunge point.

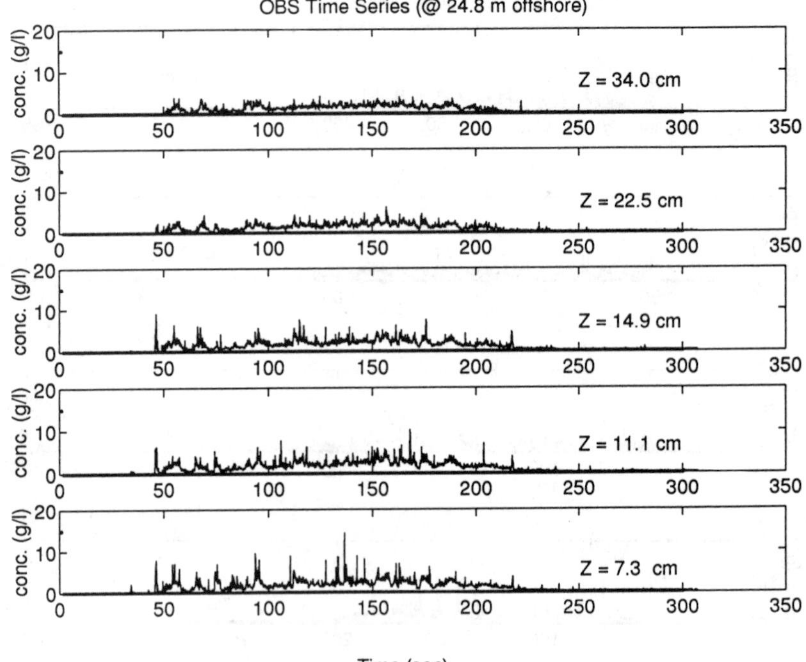

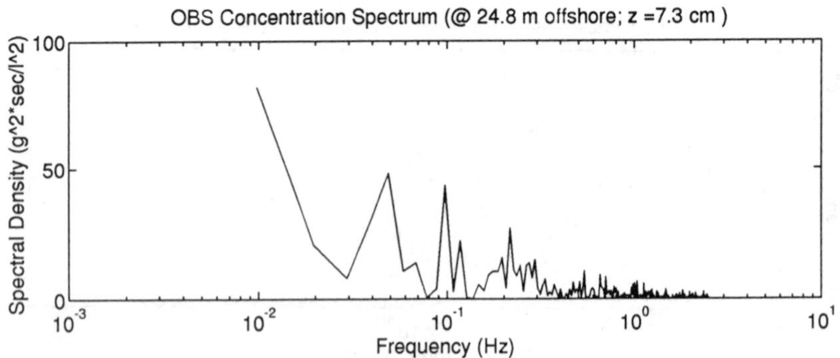

Figure 6 - Time series and sample concentration spectrum from OBS stack located 2.5 m seaward of plunge point.

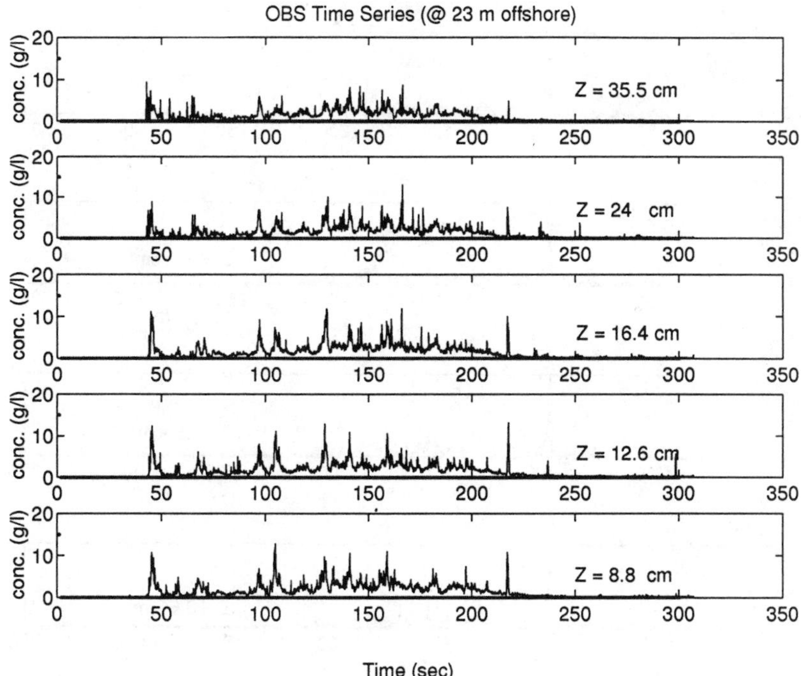

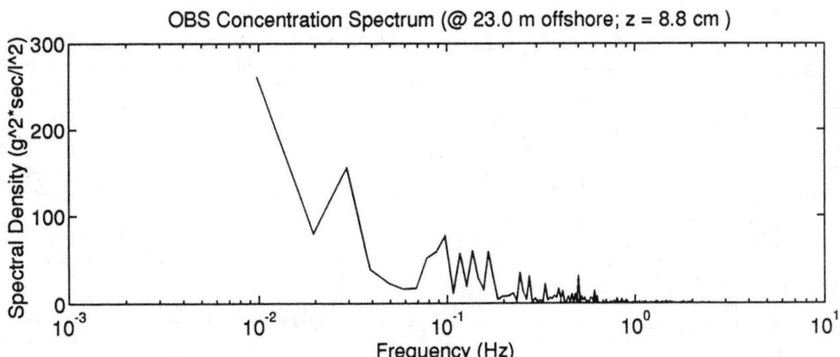

Figure 7 - Time series and sample concentration spectrum from OBS stack located 0.7 m seaward of plunge point.

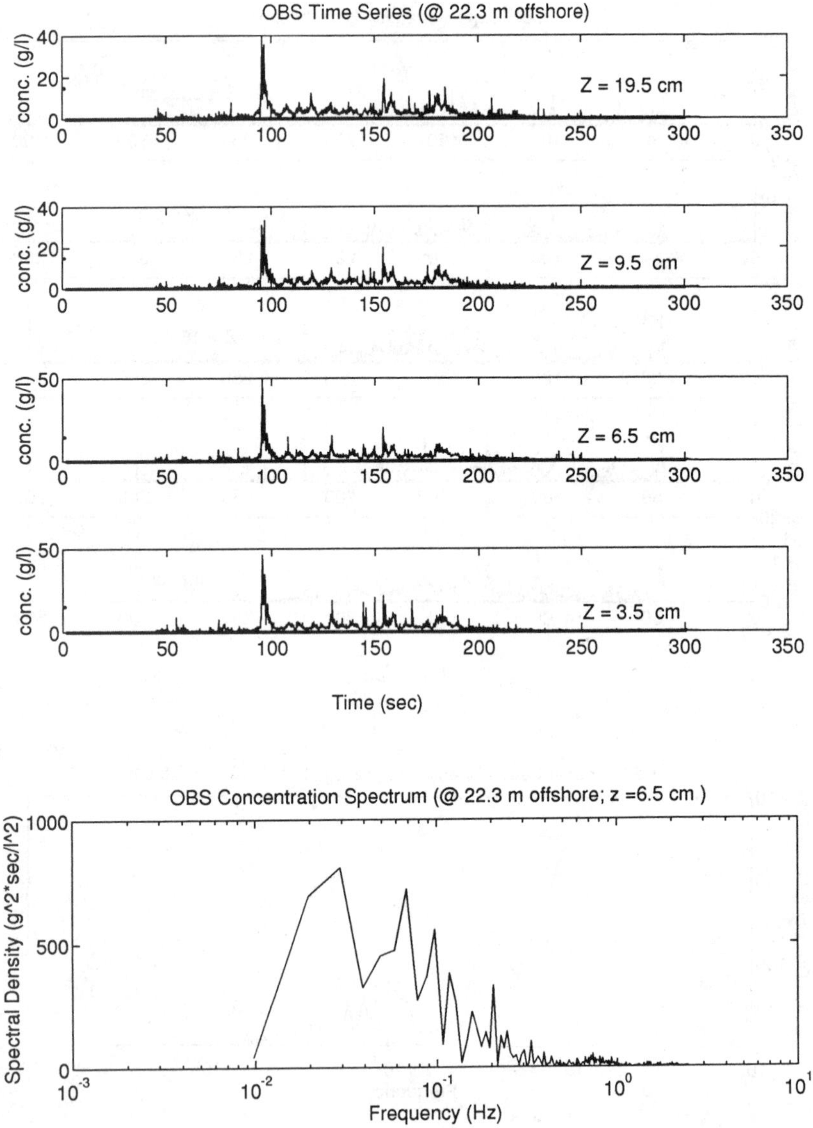

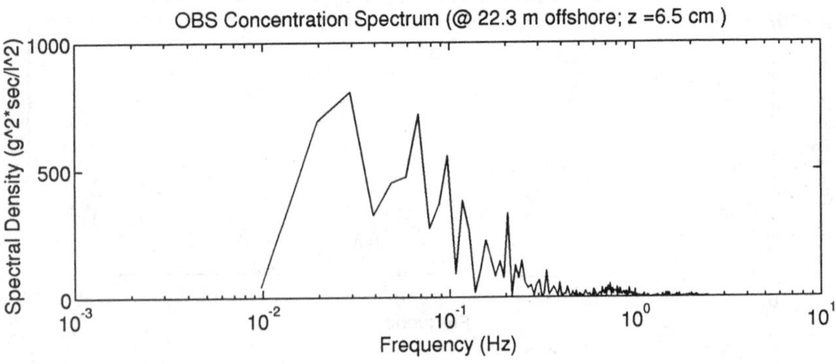

Figure 8 - Time series and sample concentration spectrum from OBS stack located near plunge point.

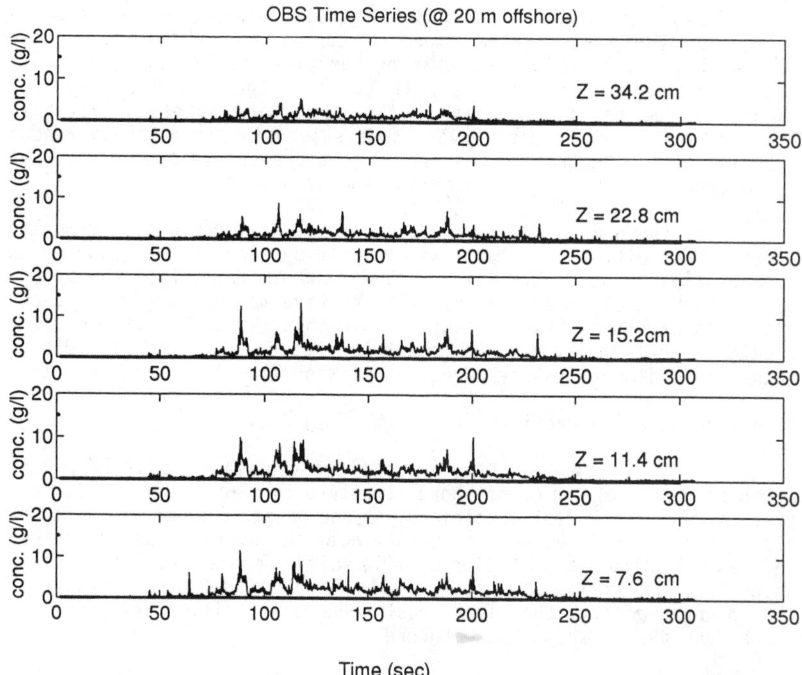

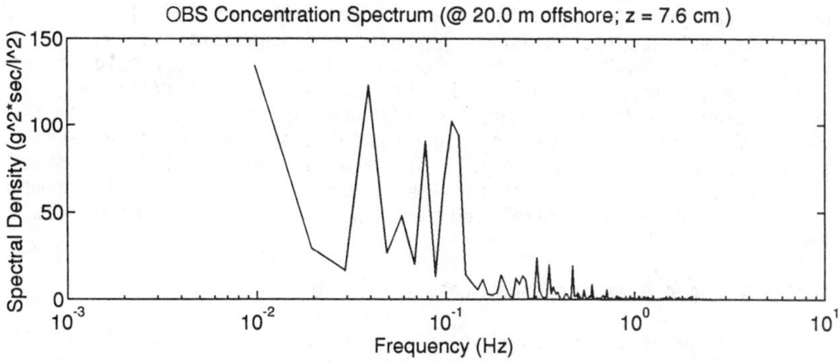

Figure 9 - Time series and sample concentration spectrum from OBS stack located 2.3 m shoreward of plunge point.

Because the waves in the channel were regular, the low frequency energy in the concentration spectra cannot be attributed to low frequency fluid motion. In examining the time series, and especially the video records of the tests, it is clear that the low frequency signal is due to irregularity in the suspension events themselves. That is, even with regular waves, sand suspension is a highly random process, as the "source" cloud of sediment created by each breaker at the plunge point varies in intensity and structure.

Near the plunge point, the source cloud is continually reworked by turbulent diffusion as well as directly by convection. However as time progresses, undertow spreads the cloud offshore into the region of low turbulence, where the oscillatory wave motion carries it back-and-forth past the offshore sensors. In this manner the strong, very regular signal apparent in Figure 4 is created, i.e. not by locally entrained sediment, but by advection of sand from the plunge point.

Mean Concentration Profiles

Figure 10 presents time-averaged concentration profiles, from which the vertical and cross-shore structure in mean suspended load can be seen. For these test conditions, as one passes from offshore to the plunge point (profiles A-E), not only does the overall suspended load increase, but the concentration profile shifts from a linear/concave-up shape to a convex-up shape. At profile F, which is 2.3 m shoreward of the plunge point, the load has reduced significantly and the linear/concave-up shape has returned.

According to Nielsen (1991), the change from convex-up to concave -up profile shape indicates a shift from a convective entrainment process to one of diffusion. This assessment makes sense in the present situation, as suspension at the plunge point is clearly a convective process associated with the presence of large-scale vortices, whereas smaller-scale turbulence dominates after the roller is developed. Small-scale turbulence, created by the dissolution of the plunger vortex, is also advected seaward by the undertow. In conjunction with the rapid settling of the larger size classes from the cloud, the profile outside the plunge point adopts a linear/concave-up shape. It is interesting to note that whereas Nielsen (1991) examined the response of different grain size classes to the same fluid regime, here we examined the "same" sediment in response to different fluid regimes.

Depth-Averaged Mean Suspended Load

To examine the cross-shore structure of the suspended load, the profiles from the twelve stacks available for these test conditions are depth-averaged below 25 cm from the bed, and the results plotted in Figure 11. The increase from a mean load of 0.75 g/l to 2.0 g/l in a distance of 5 m is noteworthy. Although not perfectly repeatable, the results of the three tests are consistent with one another.

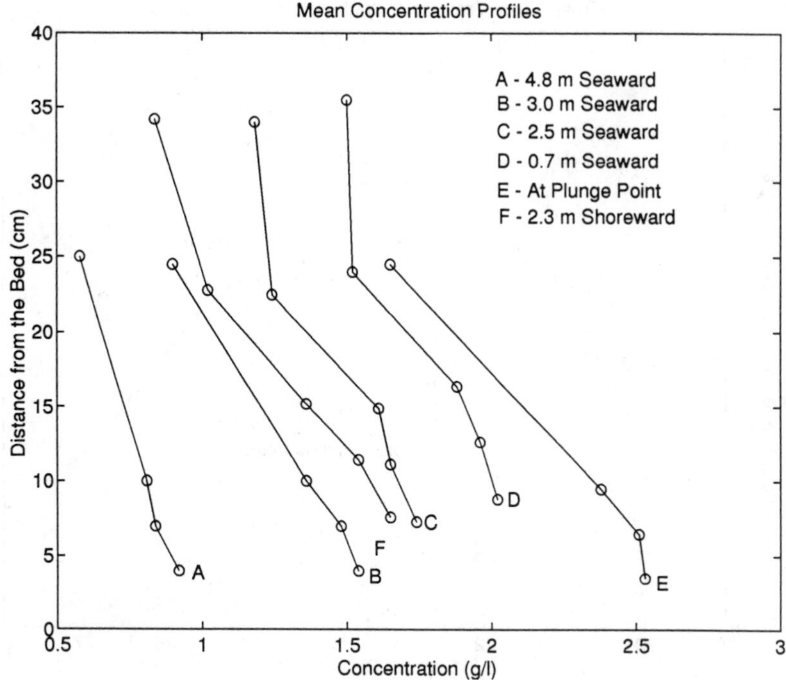

Figure 10 - Time-averaged sediment concentration profiles.

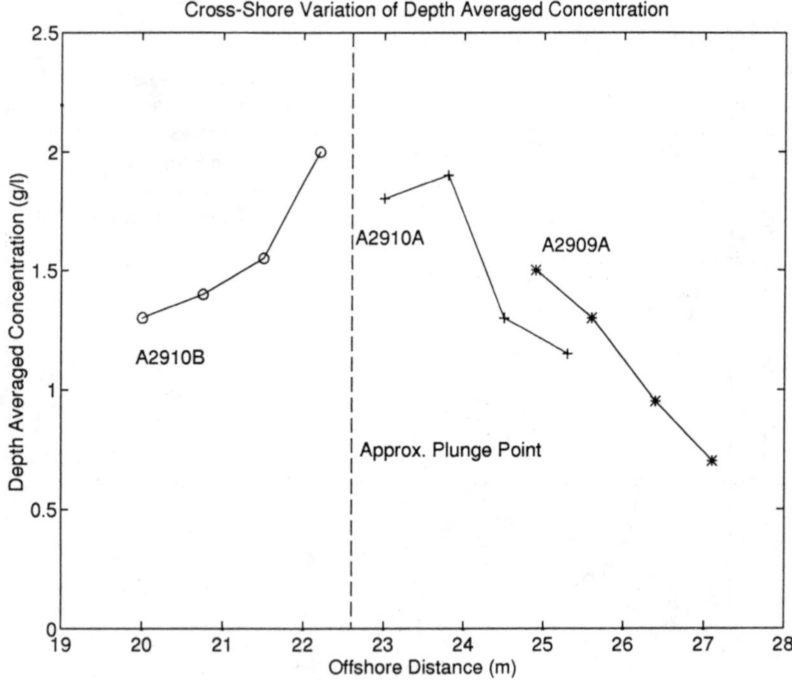

Figure 11 - Mean concentration averaged over region 25 cm from the bed.

SUMMARY AND CONCLUSIONS

Exploiting the tightly controlled conditions provided by a large wave channel facility, detailed measurements of sediment suspension due to plunging breakers were made in the region of transition from a nonbreaking to a fully broken state. Findings can be summarized as follows:

1) As one passes from shoaling into breaking, the time series of suspended concentration changes from a regular, oscillatory signature, to one of irregular events superimposed on a background load. Because the events are irregular in intensity, low frequency energy appears in the concentration spectrum.

2) Also in passing from shoaling to breaking, the mean concentration profile changes from a linear/concave-up shape to a convex-up shape, indicating a shift from diffusive to convective entrainment processes. In passing into the region where the breaking wave roller is fully developed, the profile returns to a linear/concave-up shape.

3) For the plunging wave conditions examined herein, a strong cross-shore gradient in mean suspended load (~0.25 g/l/m) exists in the zone of initial wave breaking.

4) Even for regular wave conditions, sediment suspension is a random process, and probably should be modeled as such. Conceptually, clouds of sediment are created by each wave, with these clouds being reworked by small-scale turbulence. The suspended sediment boundary at the outer surf zone is gradually advected offshore by the undertow, as well as carried back-and-forth in the oscillatory wave motion.

REFERENCES

Barkaszi, S.F. Jr., 1993, "Measurement of Sand Suspension by Breaking Waves in a Large Wave Channel," M.Sc. Thesis, Ocean Engineering Program, Florida Institute of Technology, 112 pp.

Downing, J.P. and Beach, R.A., 1989, "Laboratory Apparatus for Calibrating Optical Suspended Solids Sensors," Marine Geology, V86, pp. 243-249.

Kana, T.W., 1979, "Suspended Sediment in Breaking Waves," Tech. Rep. No. 18-CRD, Dept. of Geology, U. of South Carolina.

Kraus, N.C. and Smith, J.M., 1994, "SUPERTANK Data Collection Project - Volume 1: Main Text," Tech. Rep. CERC-94-3, U.S. Army Corps of Engineers, Waterways Experiment Station, 274 pp.

Nielsen, P.N., 1991, "Combined Convection and Diffusion: a New Framework for Suspended sediment Modeling," Proceedings of Coastal Sediments '91, ASCE, V1, pp.418-431.

SIMULATION OF WAVE AND BEACH PROFILE CHANGE IN SUPERTANK SEAWALL TESTS

William G. McDougal[1], M.ASCE, Nicholas C. Kraus[2], M.ASCE
and Harman Ajiwibowo[3]

ABSTRACT: A numerical profile change model is developed for a beach fronting a seawall. The model closely follows the SBEACH model developed for open coastlines (Larson and Kraus 1989). The model consists of three main components; wave transformation, cross-shore sediment transport, and profile change. The most significant enhancement of SBEACH is in the wave transformation, in that the present model includes the reflected wave from the seawall. Reflection produces a partial standing wave which modifies the radiation stresses, wave setup, and breaking. Model results are compared with measurements from SUPERTANK. The breaking wave heights predicted by the model are in a good agreement with the measurements. Scour depth at the seawall for the measured and calculated results are in fair agreement. Profile responses calculated with the present model, which include the reflected wave, are not significantly different from the SBEACH model estimates.

INTRODUCTION

The interaction among coastal structures, waves and beach profiles are of major concern in shore protection and harbor planning. Coastal defense structures such as

---

1) Professor, Ocean Engineering Program, Department of Civil Engineering, Oregon State University, Corvallis, OR 97331, USA.
2) Director, Conrad Blucher Institute for Surveying and Science, Texas A&M University-Corpus Christi, 6300 Ocean Drive, Corpus Christi, TX 78412-5599, USA.
3)Graduate Research Assistant, Ocean Engineering Program, Department of Civil Engineering, Oregon State University, Corvallis, OR 97331, USA.

seawalls, revetments, breakwaters, and coastal dikes are commonly constructed for preventing landward retreat of the shoreline and upland inundation. These structures modify the interactions between the beach and waves. Several numerical models have been developed to simulate beach profile changes. One of the most widely used is SBEACH developed by the US Army Corps of Engineers (Larson and Kraus 1989). SBEACH was developed for beaches backed by berms or dunes. SBEACH does not include reflected waves which exist if a seawall is present. SBEACH applies the D^3 model (Dally et al. 1984) to evaluate the wave transformations across the surf zone. Since the presence of a coastal structure facing the beach influences the wave field and wave-induced transport, it is necessary to develop a model that includes reflected waves. The D^3 model is modified to include the partial or complete wave reflection induced by a seawall. The radiation stress is, of course, influenced by the reflected waves. Therefore, an expression for the radiation stress which includes reflected waves is derived. The new expressions are used in the cross-shore momentum equation to determine the wave setup/setdown. These results are used in evaluating the wave transformations across-shore. The sediment transport model applied in this model is the same as in SBEACH. This sediment transport, when combined with conservation of sediment mass, leads to the determination of profile change. To verify the numerical model, tests were conducted as one component of the large-scale experimental study, SUPERTANK (Kraus, et al. 1992). Three seawall tests were conducted in which wave heights, water levels, and profile changes for a sand beach fronting a seawall were measured.

WAVE TRANSFORMATION MODEL

As waves propagate toward the shore, they refract, shoal, break and may reflect from a seawall. An empirical equation is applied across the profile which relates the change in energy flux to the difference between the actual energy flux and a stable energy flux. The steady two-dimensional conservation of energy equation, incorporating energy dissipation associated with wave breaking, is written as follows,

$$\nabla \bullet \vec{F} = -\frac{\kappa}{d} \left(|\vec{F}_i| + |\vec{F}_r| - |\vec{F}_s| \right) \tag{1}$$

in which $\vec{F}$ is total wave energy flux, $\vec{F}_i$ is the incident wave energy flux, $\vec{F}_r$ is the reflected wave energy flux, $\vec{F}_s$ is the stable wave energy flux, κ is an empirical wave decay coefficient and d is the total water depth. The total water depth is the sum of the still-water level plus the setup/setdown, $d = h + \bar{\eta}$. A seawall located in the surf zone contributes additional energy because of reflection in front of it. As a result, more energy dissipation occurs in the surf zone. It is assumed that this increased dissipation occurs in the incident wave. This equation is a modification of the D^3 model (Dally, et al. 1984) to include reflection. Figure 1 is a plan view which shows the incident and reflected waves relative to the shoreline. In Figure 1, θ_i is the incident wave angle, θ_r is the reflected wave angle and the x coordinate is positive offshore. The reflected wave angle is related to the incident wave angle as,

$\theta_r = -\theta_i$, so the angles will simply be denoted as θ with the appropriate sign. Dally, et al. (1984) experimentally observed that the stable wave height, H_s, is a function of depth,

$$H_s = \Gamma\, d \tag{2}$$

where Γ is the stable wave index, and d is the total depth including setup. It is assumed that there are no longshore gradients, so

$$\frac{\partial}{\partial y} = 0 \tag{3}$$

The reflected wave energy flux is assumed to be constant and is taken to be equal to the value at the location of the seawall. So let,

$$A = H_r^2\, C_{gr} \tag{4}$$

Since at each location, the magnitudes of the group velocities C_{gr}, C_{gs} and C_{gi} are equal, they can just be referred to C_g. The incident wave height, H_i, will now be denoted by H. Substituting back into the energy equation gives,

$$\frac{d}{dx}\left(-H^2 C_g \cos\theta + A \cos\theta\right) = -\frac{\kappa}{h+\overline{\eta}}\left[H^2\, C_g + A - \Gamma^2\, (h+\overline{\eta})^2\, C_g\right] \tag{5}$$

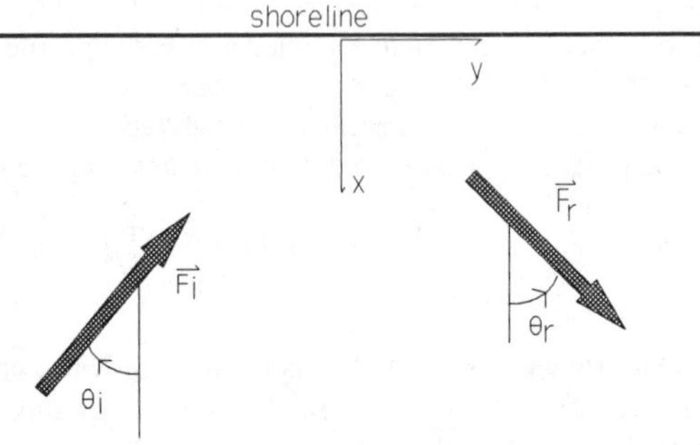

Figure 1. Incident and reflected wave coordinates.

Equation (5) is valid across the entire profile. For the region outside the breaker line, no dissipation occurs and κ is set to zero. In this region, the wave height and direction gradually change due to shoaling and refraction. Inside the breaker line, κ is set to a positive non-zero value and the wave height will decay due to breaking. Larson and Kraus (1989) empirically determined the wave decay coefficient, κ =

0.15 based on laboratory data (Visser 1982) and field data (Wu, Thornton, and Guza 1985). Wave breaking occurs when the ratio between the wave height and the water depth exceeds a certain value which is a function of the surf similarity parameter expressed in terms of the deep water steepness and profile slope prior to breaking (Larson and Kraus 1989)

$$\gamma = \alpha \, \zeta^{0.21} \tag{6}$$

where $\alpha = 1.14$, γ is the breaker ratio, H_b/d_b. Based on the analysis of the SUPERTANK data, α is found here to be 0.75. The surf similarity parameter ζ given by,

$$\zeta = \tan\beta \left[\frac{H_o}{L_o} \right]^{-0.5} \tag{7}$$

in which $\tan\beta$ is the local beach slope seaward of the breakpoint, H_o and L_o are the deep-water incident wave height and wave length, respectively.

To use (5) it is necessary to determine $\bar{\eta}$, C_g, and θ. The setup/setdown is determined from the cross-shore momentum equation,

$$-\rho g(h+\bar{\eta}) \frac{d}{dx}\bar{\eta} = \frac{d}{dx} S_{xx} + \tau_{sx} \tag{8}$$

in which S_{xx} is the onshore-onshore component of the radiation stress, ρ is mass density of seawater, g is the acceleration due to gravity, and τ_{sx} is the wind stress. Assuming that a partial standing wave develops in front of the seawall, an expression for S_{xx} is derived which includes the reflected wave. The result is written as

$$S_{xx} = \left(E_{ii} + E_{rr}\right) \left[n\,(\cos^2\theta + 1) - \frac{1}{2} \right] + E_{ir} \cos\left(2kx \cos\theta + \epsilon_r\right) \left[2\,n \sin^2\theta - 1 \right] \tag{9}$$

where E_{ii} and E_{rr} are the incident and reflected wave energies, E_{ir} is

$$E_{ir} = \frac{1}{8} \rho g \, H_i \, H_r \tag{10}$$

and n is the ratio between group velocity and celerity. To simplify (9), the spatial oscillation term, $\cos(2kx \cos\theta + \epsilon_r)$ is separated. This term is denoted as f, so that,

$$f = \cos\left(2kx\cos\theta + \epsilon_r\right) \tag{11}$$

where ϵ_r is the phase between reflected and incident wave. Temporarily, this phase term will be considered constant. The actual value cross-shore phase is not a constant. It will be integrated separately using an iteration technique. Replacing H_i with H and using (4), the expression for S_{xx} is,

$$S_{xx} = \frac{1}{8}\rho g\left[H^2 + \frac{A}{C_g}\right]\left[\frac{C_g k}{\omega}\left(\cos^2\theta + 1\right) - \frac{1}{2}\right] + \frac{1}{8}\rho g H\left[\frac{A}{C_g}\right]^{\frac{1}{2}} f\left[\frac{2\,C_g\omega}{k}\sin^2\theta_o - 1\right] \tag{12}$$

The cosθ term in (5) is determined based on the Snell's law as follows,

$$\frac{\sin\theta}{C} = \frac{\sin\theta_o}{C_o} \Rightarrow \cos^2\theta = 1 - \frac{\omega^2}{k^2\,C_o^2}\sin^2\theta_o \tag{13}$$

where C_o is the deep-water wave celerity, and θ_o is the deep-water wave angle. The group velocity, C_g , is

$$C_g = n\,C = n\,\frac{\omega}{k} \tag{14}$$

in which n is modified to use the total depth rather than just the still-water depth,

$$n = \frac{1}{2}\left[1 + \frac{2k(h+\bar{\eta})}{\sinh 2k(h+\bar{\eta})}\right] \tag{15}$$

The linear wave theory dispersion equation is also modified to include setup,

$$\omega^2 = gk\,\tanh\left[k(h+\bar{\eta})\right] \tag{16}$$

Seven equations and seven unknowns for the wave transformation model have been developed. The seven unknowns are,

$$H\ ,\ \bar{\eta}\ ,\ S_{xx}\ ,\ C_g\ ,\ k\ ,\ f\ ,\ \cos\theta \tag{17}$$

As a matter of convenience, $\cos\theta$ rather than θ is defined as an unknown. One additional unknown, ϵ_r, will be determined by iteration. The phase is determined by the amount of time it takes the wave to propagate from an offshore location x to the seawall and back to x. This is an integral term given by

$$\epsilon_r = \frac{4\pi}{T} \int_x^{x_s} \frac{k^2\ C_o\ dx}{\omega \left(k^2 C_o^2 - \omega^2 \sin^2\theta_o \right)^{1/2}} \tag{18}$$

in which x_s is the seawall location.

The equations are modified to develop a set of seven coupled first-order ordinary non-linear differential equations. These are integrated using a 4th order Runge-Kutta routine. At the seaward boundary, where the numerical integration is started, the initial values for each of the unknowns are analytically estimated. Since the reflected wave and phase are not known, these initial estimations are iterated until the solution converges. The algorithm to calculate the cross-shore wave transformations may be summarized as:

1. Estimate a value of H_{rs}, and C_{gs} for the first iteration. Then A can be determined from (4).
2. Estimate the cross-shore phase for the first iteration.
3. Do the cross-shore Runge-Kutta 4th order integration using the boundary values for each unknown.
4. With the results from the integration, calculate new values for cross-shore phase using (18). Check the difference between the old phase and the new phase.
5. The new group velocity at the seawall (new C_{gs}) and the incident wave height at the seawall (H_{is}) are also obtained from the integration in Step 3. H_{rs} is then obtained by multiplying H_{is} with K_r. Check the difference of the new values of H_{rs} and C_{gs} with the old values.
6. If the difference between old values and new values in Step 4 and 5 exceeds the convergence criteria, then proceed to the next iteration starting from Step 3 with new values of H_{rs}, C_{gs} and ϵ_r. These are taken to be the average of the previous two iterations.

CROSS-SHORE TRANSPORT AND PROFILE CHANGE MODEL

The cross-shore sediment transport model is based on energy dissipation. Therefore, an expression for energy dissipation due to wave breaking is needed. In this macroscale approach for beach profile modelling, it is satisfactory to derive the energy dissipation from wave height decay directly (Larson and Kraus 1989). A formulation relating the cross-shore transport rate in the zone of fully broken waves with the wave energy dissipation per unit volume of water has previously been successfully used by Kriebel and Dean (1985). Analyses of large-scale wave tank data also substantiate this type of formulation in the zone of broken waves (Larson and Kraus 1989). The transport relationships in the other zones of the surf zone are empirical and primarily based on large-scale wave tank experiments (Larson and Kraus 1989). The wave energy dissipation D per unit water volume is,

$$D = \frac{1}{(h+\bar{\eta})}(\vec{\nabla}\cdot\vec{F}) \qquad (19)$$

This is easily computed from (1).

Changes of profile are calculated at each time step from the conservation of sediment mass using the net cross-shore transport rate q. Assuming no changes in porosity, sediment sources or sinks, the one dimensional mass conservation equation takes the form,

$$\frac{dh}{dt} = \frac{dq}{dx} \qquad (20)$$

The standard boundary conditions in the model are zero sand transport rates at the seawall and the seaward limit of the computational domain.

Occasionally, local oversteepening at a computational grid occurs. To avoid this, an avalanching routine is included. Local profile slopes exceeding the angle of repose are adjusted to attain a stable slope. There are two limiting profile slopes, the angle of initial yield (angle of repose) and the residual angle after failure. For the numerical model, the value for initial yield is 28° and the residual angle after failure is 18°, as cited from SBEACH (Larson and Kraus 1989).

RESULTS AND EVALUATION

The information required to run the model consists of wave period, deep-water wave height and angle, sediment median grain size and fall velocity, seawall reflection coefficient, and the initial bottom profile.

Plane Beach Case

Table 1 shows the conditions selected to evaluate bottom profile changes for an initially plane beach. Monochromatic waves were used in all conditions. The model can be used to evaluate the profile change with input data time history of wave height, wave period, incident wave angle, and water depth at the structure. Figure 2 shows the profile at 6 hour intervals. A bar forms at the breaker line and the beach face tends to flatten. This is the general response observed for all plane beach runs. Figure 3 shows the incident and reflected wave heights, the reflection coefficient and the setup/setdown after 24 hours. The incident wave refracts, shoals and breaks over the bar. Inside the breaker line it decreases to the value at the seawall. The reflected wave height is equal to the incident wave height at the seawall because, for this example, the seawall reflection coefficient is 1.0. The reflected wave tends to decrease with distance offshore due to "reverse" shoaling and refraction. The local increase at the bar is due to the depth change. The reflection coefficient at the wall is 1.0 as specified, but decreases to an offshore value of less than 0.4. The reflection coefficient tended to increase with time as more of the seawall was exposed due to scour. The reflected wave results in a partial standing wave and a modulation of the setup. The length of the modulation increases with distance offshore because the wave length increases as the depth increases. As the profile approaches equilibrium the net cross-shore transports goes to zero. In this case the transport at 24 hours was down to approximately 2.5% of the initial transport.

Table 1. Sensitivity analysis reference data

H_o (m)	h_s (m)	T (sec)	K_r	θ (deg)	D_{50} (mm)	m
1.5	0.3	5	1	0	0.17	0.05

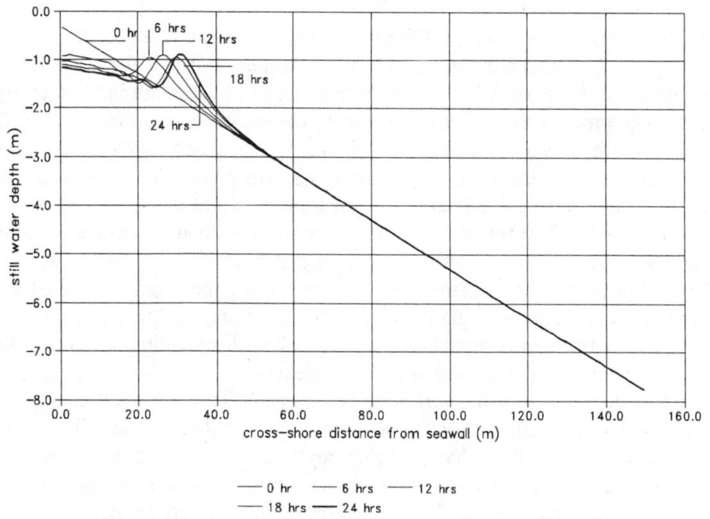

Figure 2. Beach profile change for plane beach case.

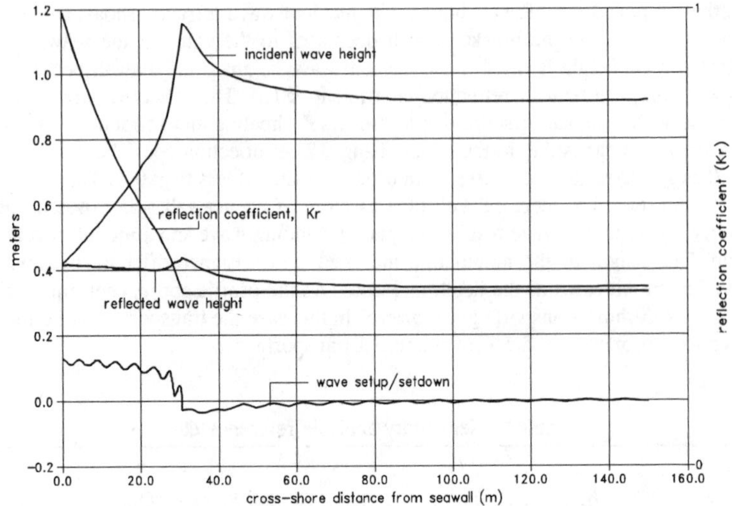

Figure 3. Wave heights, reflection coefficient, and setup/setdown after 24 hours

Sensitivity Analysis

Figure 4 shows the beach profile response for different values of H_o calculated after 24 hours. The larger the wave height, the deeper the scour depth. This seems reasonable since the larger waves have more energy to initiate motion and transport sediment. The profiles show that the larger H_o causes the waves to break and form a bar at a more seaward location. In Figure 4, the scour depth at the seawall increases as the offshore wave height increases. However, the increase is rather minor. This is because the waves are depth limited. Since the larger waves cause more setup, the depth at the seawall, and therefore the depth limited wave heights are slightly larger. The incident wave height and setup are shown for $K_r=0$ and $K_r=1.0$ in Figure 5. The influence of reflection is as anticipated. Reflected waves cause the incident waves to begin breaking farther offshore and to be smaller across the surf zone. The setup is modulated for the case with reflection. Results indicate that the scour depth at the seawall is almost independent of K_r. The breaking model assumes that the energy in the surf zone is saturated. This saturated energy is the sum of incident and reflected and is a function of the local depth. The sediment transport model is based on changes in energy and is not sensitive to whether the energy is incident or reflected. Figure 6 shows the profiles as a function of median grain size. As expected, the coarse grained beach tends to erode less than the fine grained. On the fine beach, the upper profile has a milder slope and the bar is created farther offshore. Figure 7 shows variation in the profile as a function of the slope. For a steeper beach, the wave energy dissipates in a narrower region, therefore a larger gradient in wave energy and hence sediment transport causes more

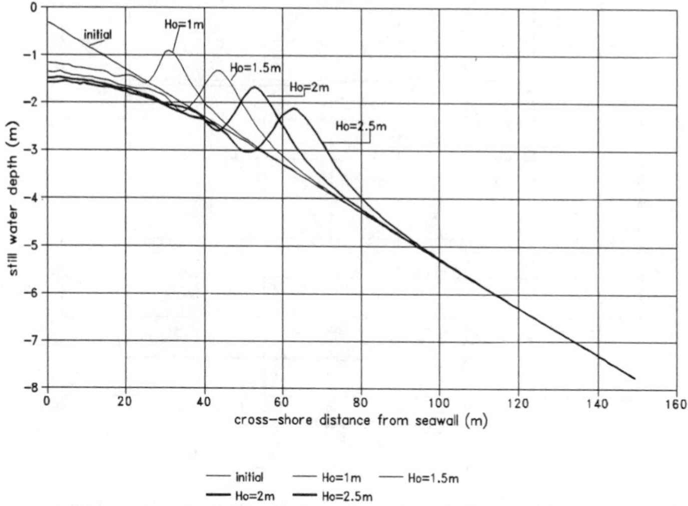

Figure 4. Beach profile change after 24 hours for cases with different deep-water incident wave heights.

scour. It can be seen from Figure 7 that the steeper bed slope gives a deeper scour depth and the bar is created closer to the seawall. The wave period has a small influence on the scour depth. Approximately a 15% increase in scour depth at the seawall occurred by increasing wave period from 4 to 10 seconds. There is not a significant variation in the scour depth as a function of the still water depth at the seawall. The greater water depth simply results in a more shoreward breaking point. The model can include the wave angle influence on the cross-shore processes. However, the longshore current is not included, so the profile calculations for waves which are not normally incident would be misleading.

A simple dimensional analysis was conducted based on the sensitivity analysis results. An approximate equation for scour depth S_w at the seawall is given by,

$$\frac{S_w}{H_b} = 6775 \left[\frac{\nu^6 \, m^7}{g^3 \, D_{50}^4 \, H_b^5} \right]^{1/10} \tag{21}$$

in which ν is the kinematic viscosity. This result is compared with computed results for the sensitivity analysis cases in Figure 8. The solid line corresponds to perfect agreement. It is seen that this approximation is quite good. An equilibrium scour

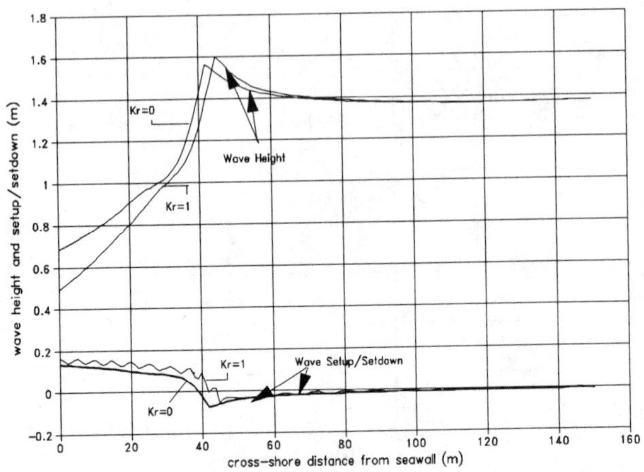

Figure 5. Cross-shore wave transformation for cases with K_r=0 and K_r=1.

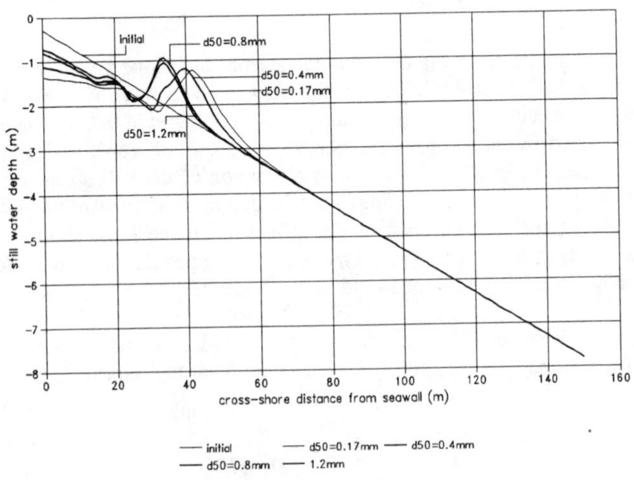

Figure 6. Beach profile change after 24 hours of waves with different median grain sizes.

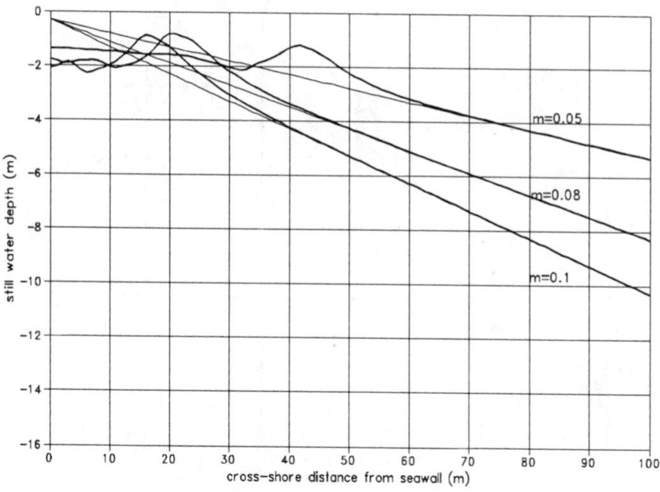

Figure 7. Beach profile change after 24 hours of waves with different bed slopes.

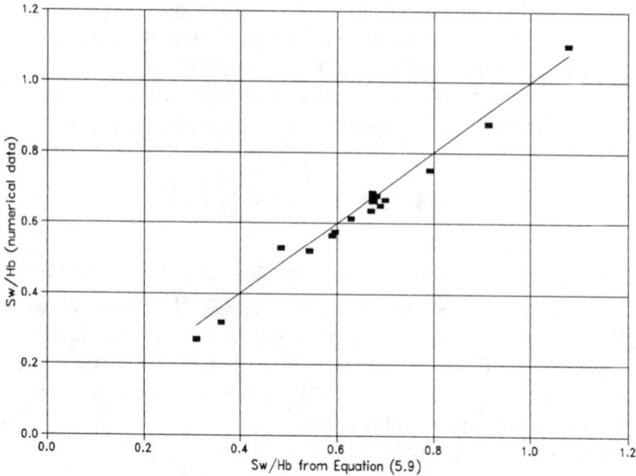

Figure 8. Comparison of the best-fit non-dimensional equilibrium scour depth with the equilibrium scour depth obtained from the numerical model.

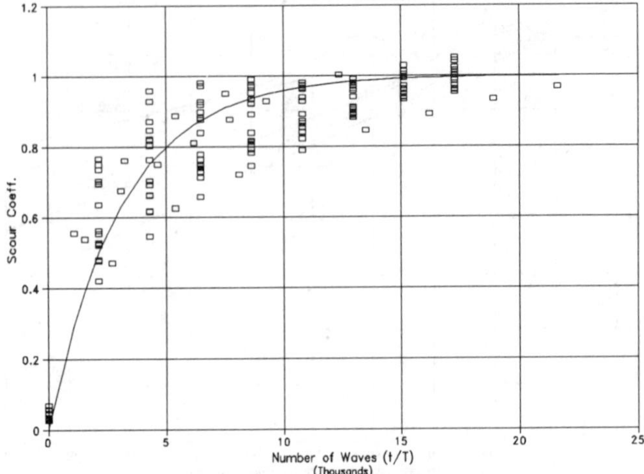

Figure 9. Scour coefficient with respect to number of waves for all plane beach computations.

coefficient can be defined from (21) by dividing through by S_w/H_b. The scour coefficient value of 1.0 corresponds to the equilibrium scour depth. The equilibrium scour coefficient is calculated for all of the computer runs used in the parameter sensitivity analysis. Figure 9 shows the scour coefficient as the profile approaches equilibrium. Although the data are scattered, most cases reach equilibrium by 20,000 waves. The time dependency is approximately given by,

$$S(t) = S_{eq}\left(1 - e^{-\alpha \frac{t}{T}}\right) \tag{22}$$

where S_{eq} is the scour coefficient corresponding to the equilibrium scour depth, which is 1.0. A best fit to the data gives $\alpha = 0.000321$. This is shown as a line in Figure 9. The scour depth increases rapidly during the first few hours and then the erosive processes slow down and reach an equilibrium depth. Based on this simple result, initial toe scour occurs rather rapidly. Approximately 99% of the equilibrium scour occurs by 14,000 waves. This corresponds to 39 hours of 10 second waves. However, 50% of this scour would occur in the first 6 hours.

EXPERIMENTAL RESULTS

Seawall-beach interaction was addressed as one component of the SUPERTANK Laboratory Data Collection Project (Kraus, Smith, and Sollitt 1992). The beach was composed of approximately 600 cubic meters of uniform quartz sand with a 0.22-

mm median diameter. Three large-scale seawall tests were conducted corresponding to Test Numbers ST-70, ST-80 and ST-CO. A rigid, vertical impermeable seawall was installed just seaward of the initial shoreline. The profile was then exposed to

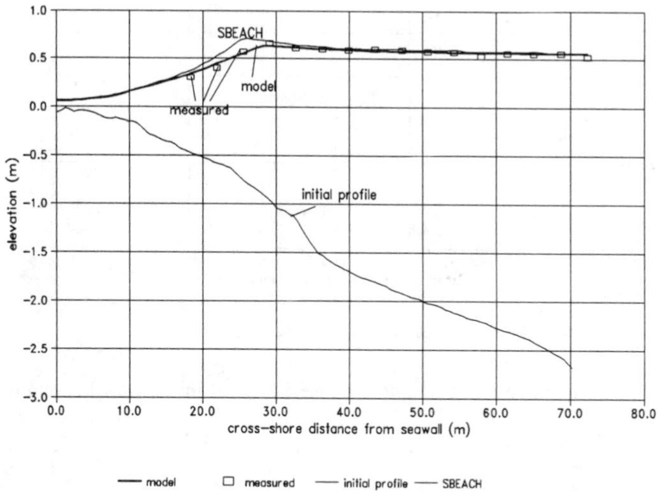

Figure 10. The incident wave height on case ST-70 at initial run.

waves for intervals of 10 to 40 minutes and then surveyed. Shorter intervals were used at the beginning of the tests and longer intervals toward the end. In total, the profiles were exposed to 3 to 4 hours of waves. In Test Numbers ST-70 and ST-CO, the wave heights were changed over the course of the tests to simulate constructive and destructive conditions. The numerical model was compared with SUPERTANK data Test Numbers ST-70, ST-80 and S-TC0.

Figures 10 through 12 show a comparison of the measured and computed wave height and setup. The results using the present model were computed using $K_r = 1.0$. The results in all three cases are in reasonable agreement with the SUPERTANK measurements. Also shown are results using SBEACH. Although not as accurate as the present model, the results are surprisingly good. Figures 13 through 15 show the calculated and measured results of the beach profile evolution for SUPERTANK Test Numbers ST-70, ST-80 and ST-C0, respectively. The present model and SBEACH are both in general agreement with the SUPERTANK measurements. However, the present model results are not significantly different from the SBEACH model.

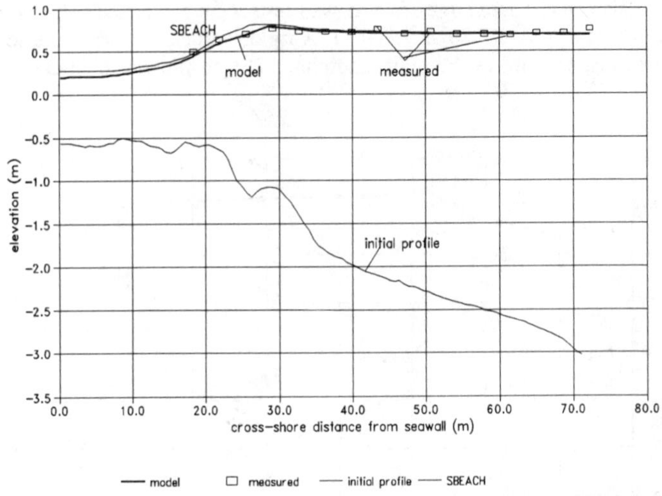

Figure 11. The incident wave height for case ST-80 at initial profile conditions.

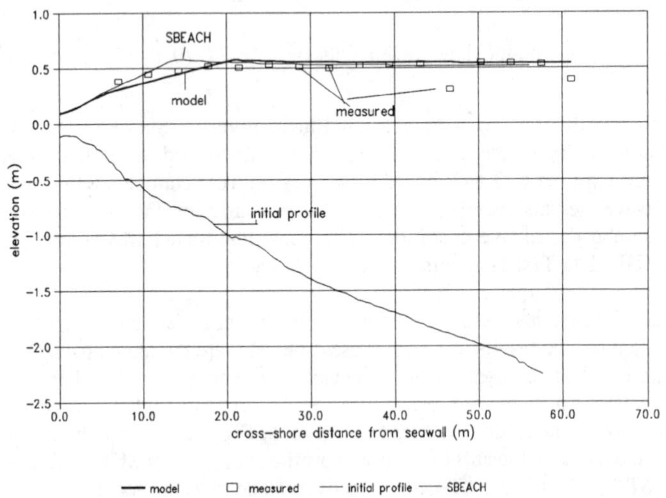

Figure 12. The incident wave height for case ST-CO at initial profile conditions

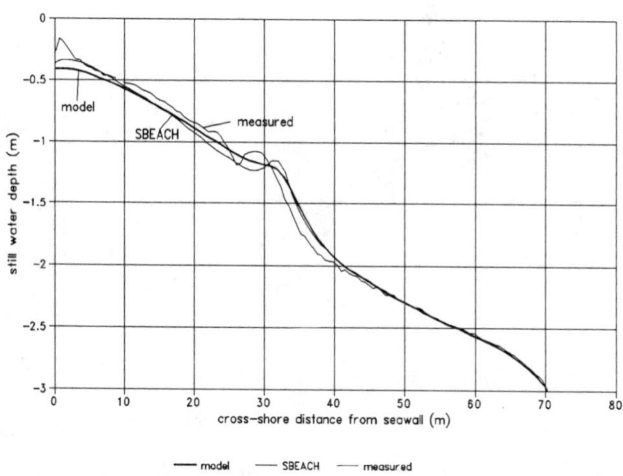

Figure 13. The profile bars on ST-70 after 3.5 hours of waves calculated by the model and the SBEACH.

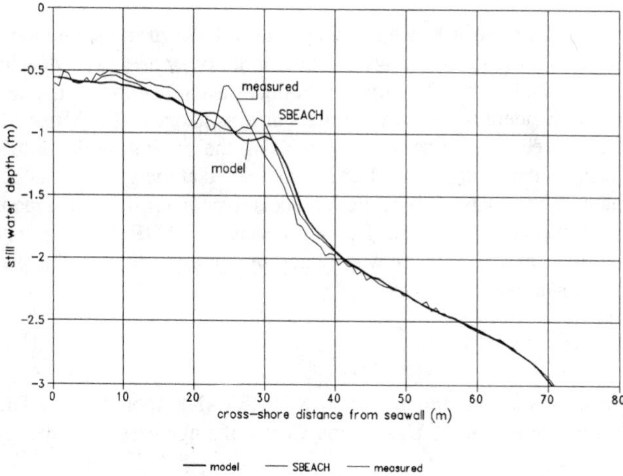

Figure 14. The bar profiles on the case ST-80 after 2.7 hours of waves calculated by the model and the SBEACH.

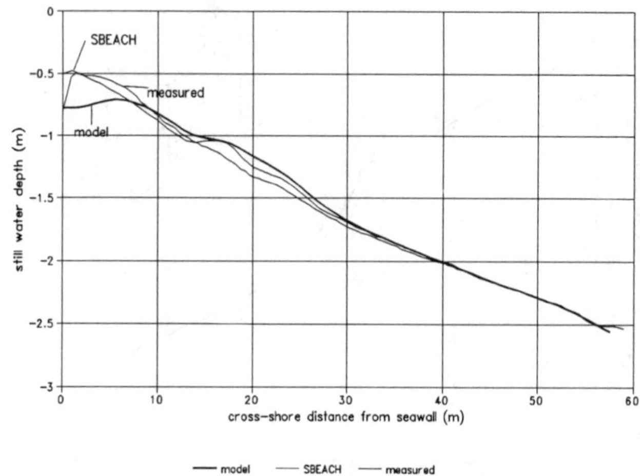

Figure 15. The bar profiles on case ST-CO after 4.2 hours of waves calculated by the model and the SBEACH.

CONCLUSIONS

The profile response of a beach fronting a seawall was examined numerically and experimentally in this study. A wave transformation model was developed that includes the reflected wave. Reflection introduces a modulation of the setup and a reduction of the incident wave. Wave reflection was added to the SBEACH model. Comparisons between the numerical results using the present model and SBEACH with the measured data from SUPERTANK indicate that the present model provides better estimates for the wave height. Both models were in reasonable agreement with profile measurements for seawalls fronting beaches in SUPERTANK. The scour observed in front of the seawall was rather minor even though the profile was exposed to rather large waves.

ACKNOWLEDGMENTS

This project was one component of the SUPERTANK sponsored by The Coastal Research Engineering Center, U.S. Army Corps of Engineers, Vicksburg, MS.

REFERENCES

Dally, W.R., Dean, R.G. and Dalrymple, R.A. (1984), *Modelling Wave Transformation in The Surf Zone*, Miscellaneous Paper CERC-84-8, U.S Army Corps of Engineers, Waterways Experiment Station, Vicksburg, Mississippi.

Kraus, N.C and Larson, M (1990), *SBEACH: Numerical Model for Simulating Storm-Induced Beach Change*, Report II: Numerical Formulation and Model Tests, CERC Technical Report, U.S. Army Corps of Engineers, Waterways Experiment Station, Vicksburg, Mississippi.

Kraus, N.C., J.M. Smith and C.K. Sollitt (1992), *SUPERTANK Laboratory Data Collection Project,* Proceedings of the 23rd International Conference on Coastal Engineering, American Society of Civil Engineers, New York, New York.

Kriebel, D.L. and Dean, R.G. (1985), *Numerical Simulation of Time-Dependent Beach and Dune Erosion*, Coastal Engineering, Vol.9, pp 221-245.

Larson, M. and Kraus, N.C (1989), *SBEACH: Numerical Model for Simulating Storm-Induced Beach Change*, Report I: Theory and Model Foundation, CERC Technical Report CERC-89-9, U.S. Army Corps of Engineers, Waterways Experiment Station, Vicksburg, Mississippi.

Visser, P.J. (1982), *The Proper Longshore Current in a Wave Basin*, Report No. 82-1, Department of Civil Engineering, Delft University of Technology, Delft, The Netherlands.

Wu, C.S., Thornton, E.B., and Guza, R.T. (1985), *Waves and Longshore Currents: Comparison of a Numerical Model with Field Data*, Journal of Geophysical Research, Vol 90, No. C3, pp 4951-4958.

PRINCIPAL COMPONENTS TIME SPECTRA OF SUSPENDED SEDIMENT IN RANDOM WAVES

Onyx W.H. Wai[1], Keith W. Bedford, M. ASCE[2], and Sean T. O'Neil[3]

Abstract: Sediment turbulent length and time scales are essential elements in modeling the fate and transport of contaminated mud. However, the sediment particle turbulent behavior in the water column is one of the least understood phenomena in sediment transport. Based on dimensional arguments, this paper discusses the spectrum shapes of the sediment concentration measured in the SUPERTANK experiment during broad banded and narrow banded random wave forcings. The sediment spectra used in this study are the fast fourier transforms of the temporal eigenfunctions computed by the Empirical Othogonal Function Analysis Method (Wai *et al.*, 1993) in which the eigenfunctions are the time correlation between profiles. With low sediment concentration and small settling velocity, the sediment dynamics in the fluid has little effect on the flow dynamics, and the turbulent cascade of the sediment particles in the water column follows an Eulerian frequency, f, spectrum with a $f^{-5/3}$ slope (Tennekes, 1975). If the sediment characteristic length and velocity scales are comparable to the Komogorov scales in the flow field, time spectrum variations with f^{-2} and f^{-1} are also possible.

INTRODUCTION

Interactions between sediments and water are of great interest in many scientific fields. Civil engineers are interested in knowing the suspended sediment transport load, which is the integral of the product of the momentum velocity

[1]University Lecturer, Department of Civil and Structural Engineering, Hong Kong Polytechnic, Hung Hom, Kowloon, Hong Kong.

[2]Professor, Department of Civil Engineering, The Ohio State University, Columbus, Ohio 43210, USA.

[3]Graduate Research Associate, Department of Civil Engineering, The Ohio State University, Columbus, Ohio 43210, USA.

and suspended sediment concentration, for coastal planning and monitoring. It is also increasingly aware that sediment particles in the water column can serve as an agent for storage and transport of heavy metals and chemical wastes which can be environmentally hazardous.

The two major sediment transport processes that are greatly influenced by the way that the sediments and water are being interacted are the entrainment ($\mathcal{E}$) and deposition ($\mathcal{D}$) processes. The excessive shear approach is most commonly used for the computation of $\mathcal{E}$ and $\mathcal{D}$ and the equations are given below.

$$\mathcal{E} = \alpha\left(\frac{\tau - \tau_{cr}}{\tau_{cr}}\right)^{\beta} \qquad \tau > \tau_{cr} \;, \tag{1}$$

$$\mathcal{D} = \gamma w_s c \qquad \tau \leq \tau_{cr} \;, \tag{2}$$

where α, β, and γ are constants determined empirically. Also, w_s and c are the sediment particle settling velocity and sediment concentration, respectively. This approach is based on the assumption that within a constant stress layer sediment entrainment takes place when the momentum generated shear stress (τ) is larger than the critical shear stress (τ_{cr}) of the sediment particle, and sediment deposition occurs when τ is less than or equal to τ_{cr}. Although the formulation of this approach is simple to use, it is highly empirical and lacks of considerations of the spatial and temporal constituents of the momentum and sediment turbulent fields. This approach should be applied only to steady and homogenous momentum-sediment flows which are seldom seen in the complex geophysical flow field. Bedford *et al.* (1987) developed a control volume approach for insituly estimating the terms $\mathcal{E}$ and $\mathcal{D}$. This conceptual approach has taken into account the unsteadiness and vertical variation of sediment concentration within the control volume. Still, the spatial scale of the size of the control volume and the temporal scale of the averaging period for the conservation of mass equation have to be resolved before employing this approach. With limited knowledge on the temporal and spatial turbulent scales of the momentum-sediment field, the application of this approach is ofter hindered.

To improve the accuracy and extend the applicability of Eqs. (1) and (2) or the control volume approach, it is necessary to have more investigation on the sediment turbulent temporal and spatial scales. This study aims at spectrally analysing high frequency sediment concentration at various elevations subjected to broad banded (BBR) and narrow banded (NBR) random waves to gain better understanding of the sediment fluctuation behavior in different temporal (frequency) scales. Having this knowledge, the spectral frequency regions (or the turbulent temporal scales) of sediment entrainment inferred by energy containning eddies, equilibrium between entrainment and deposition, and deposition dominated by dissipation can possibly be determined.

SUPERTANK MEASUREMENTS

Two sets of suspended sediment concentration data are used in this study. One data set was obtained during NBR waves were in command while the other was obtained during BBR waves were in command. The data were chosen from the data file that were collected by the Ohio State University's (OSU) coastal engineering team in August 1991 in the Supertank Laboratory Data Collection Project (SUPERTANK). The coastal team participated in SUPERTANK as part of the offshore group responsible for measuring and analysing offshore momentum and sediment data outside the breaker zone. The OSU Acoustics Resuspension Measurement System (ARMS), originally designed for long-period high-frequency in-situ sediment concentration and velocity measurements, was for the first time employed in such a large scale laboratory experiment. There are three major components of the system which are (i) a 3 MHz acoustic sediment concentration profilometer which can measure up to 110 equi-spaced intervals of sediment concentration every second in the water column, (ii) the benthic acoustic stress sensor system which can obtain 4 Hz three dimensional velocity information in four levels in the water column, and (iii) a 2 Hz pressure transducer. Figure 1 shows the arrangement of ARMS used in the Supertank experiment. Table 1 summarizes the specifications of the instruments. A detailed description of ARMS implementation in SUPERTANK is presented by O'Neil (1993).

SUSPENDED SEDIMENT TIME SPECTRA

In fluid dynamics, spectral analysis is a powerful analytical tool for studying temporal and spatial turbulent energy cascades. A flow energy spectrum can be obtained by fourier transforming the energy autocorrelation function in time or space domain to frequency or wavenumber domain. The spectrum is then a compilation of different-sized fourier components decomposed from the turbulent fluid motion. The components can be associated with energy bearing waves, e.g. turbulent vortices or eddies, of different wave periods or wavelengths. Thus, the spectral shapes reveal the manners that the energy of the eddying waves transfers at different turbulent scales and sizes.

Similarly, the above rationales of spectral analysis are also valid in sediment dynamics. Let c' denotes the sediment concentration fluctuations, the variance of c and its frequency spectrum $\Psi_c(f)$ are related as follows,

$$\begin{aligned} c'(z,t)c'(z,t+\check{\tau}) &= R_c(\check{\tau}) && (3) \\ &= \int_{-\infty}^{\infty} \exp(if\check{\tau})\Psi_c(f)df, && (4) \end{aligned}$$

$$\Psi_c(f) = \frac{1}{2\pi}\int_{-\infty}^{\infty} \exp(if\check{\tau})R_c(\check{\tau})d\check{\tau}, \qquad (5)$$

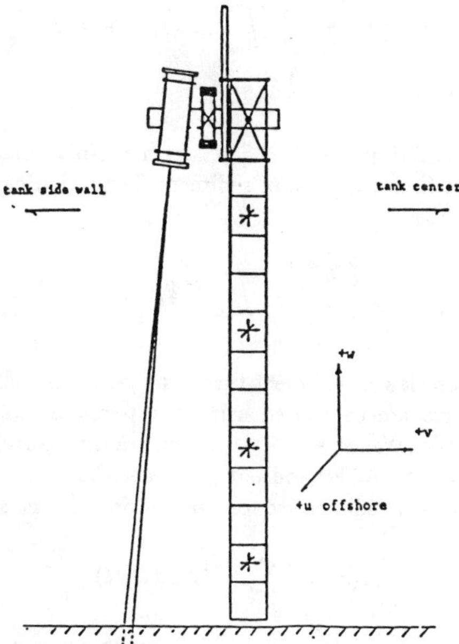

Figure 1: SUPERTANK Version of ARMS

Instrument	Operational Description	Sampling Frequency
Profilometer or Acoustic Concentration Profiler (ACP)	3 MHz Acoustic Backscatter Device Profile signal range gated to 0.5 inch "bins"	Variable 32 Hz averaged to 1 Hz for Supertank
Benthic Acoustic Stress Sensor (BASS)	1.75 MHz Acoustic Time-of-Travel Device Four vertically aligned 3-D velocity sensors	Variable 4 Hz for Supertank
Pressure Transducer	Amplified Piezoresistive Device	Variable 2 Hz for Supertank
ARMS Controller	Data Logger and Controlling Circuitry	Controls sampling frequency and data transmission of all instruments

Table 1: Instrumentation Summary

$$\Psi_c(0) = \frac{1}{2\pi}\int_{-\infty}^{\infty} \exp(if\check{\tau})R_c(\check{\tau})d\check{\tau} \tag{6}$$

$$= \frac{1}{2}\frac{\overline{c'^2}}{\pi}T, \tag{7}$$

where T is the integral time scale and $\check{\tau}$ is the time displacement. Also, the mean square value of the time variation of sediment fluctuation can be estimated by the following equation.

$$\overline{\left(\frac{\partial c'}{\partial t}\right)^2} = \int_{-\infty}^{\infty} f^2\Psi_c(f)df. \tag{8}$$

To have a better view of sediment turbulent variations in the water column, the sediment spectra are calculated using the temporal empirical orthogonal eigenfunctions, EOFs (Wai *et al.*, 1993), which are time correlation functions of sediment concentration profiles and not point correlation functions. Eq. (9) is the EOF expansion of a sediment concentration profile time series.

$$c(z,t) = \sum_{j=1}^{k} \lambda_j^{1/2}\mathcal{Z}_j(z)\mathcal{T}_j(t), \tag{9}$$

where $z = 1, 2, \ldots, n$, $t = 1, 2, \ldots, m$, and k = number of nonzero eigenvalues. λ, $\mathcal{Z}$, and $\mathcal{T}$ represent the eigenvalue, spatial EOF, and temporal EOF, respectively. Besides, n is the total number of bins in each concentration profile, and m is the total number of time steps in the series. Despite the temporal EOFs are independent to the space domain, each temporal EOF associates with its same order spatial EOF. Thus, each $\mathcal{T}$ is the temporal variations of the particular (corresponding) $\mathcal{Z}$ according Eq. (9). Using temporal EOFs for spectral analysis allows the time spectra to carry spatial information.

Figures 2 and 3 show the 1st and 2nd normalized temporal EOF concentration spectra with BBR and NBR waves as driving forces. The 1st and 2nd EOFs together with their corresponding λ account for the first two largest uncorrected variances in the time series. The 1st EOF of the BBR waves accounts for 57.51 % of the total variance, and the 2nd accounts for 25.74 %. For the NBR waves, the 1st and 2nd EOFs account for 45.22 % and 30.86 % of the total variance, respectively.

The wave conditions for the BBR waves are 3 s wave period and 0.8 m wave height, and the NBR waves are 4.5 s wave period and also 0.8 m wave height. The BBR and NBR spectrally significant wave periods are clearly revealed on the corresponding sediment spectral plot at the maxima of the humps. The spectral slopes are consistent in all plots which are approximately '-2' spanning from a frequency of 0.02 (50 second wave) to the wave periods of the driving forces.

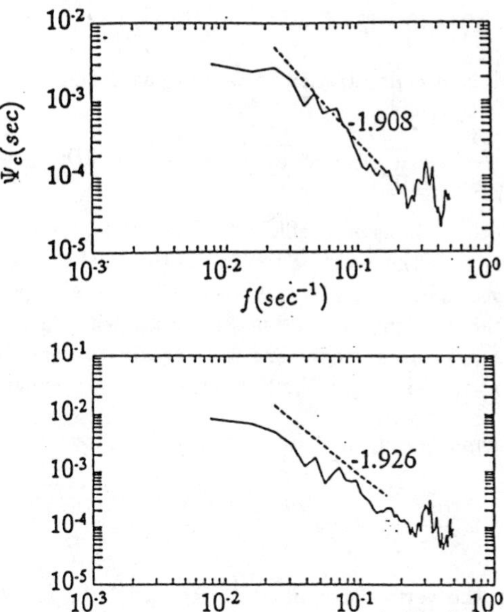

Figure 2: 1st & 2nd Normalized Temporal EOF Spectra In BBR Waves

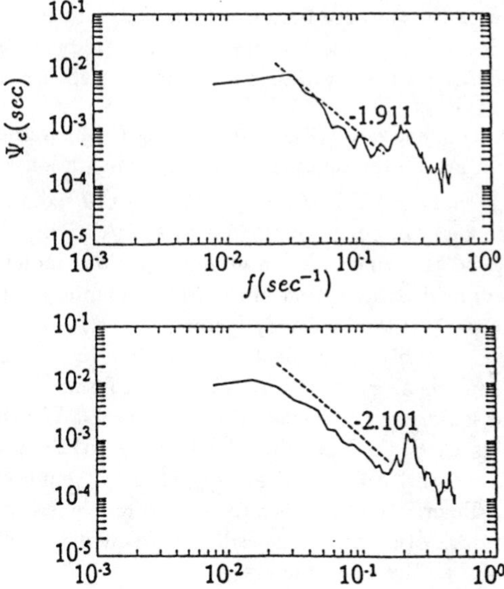

Figure 3: 1st & 2nd Normalized Temporal EOF Spectra In NBR Waves

DIMENSIONAL ANALYSIS

Consider the sediment concentration variance dynamic equation,

$$\frac{\partial \overline{c'^2}}{\partial t} + \overline{U_j}\frac{\partial \overline{c'^2}}{\partial x_j} - \overline{w_s}\frac{\partial \overline{c'^2}}{\partial x_3} = -2\overline{c'u'_j}\frac{\partial \overline{c}}{\partial x_j} - \frac{\partial \overline{u'_j c'^2}}{\partial x_j} - 2D_s\overline{\left(\frac{\partial c'}{\partial x_j}\right)^2}. \tag{10}$$

Here D_s is the molecular diffusion coefficient. The overbars indicate Reynolds averages. The first term in the equation represents the local storage of sediment concentration variance. The second term is the convection of sediment variance. The third term is the gradient production associated with turbulent motions within a mean concentration gradient. The forth term represents the turbulent transport of sediment variance. And the last term is the molecular dissipation.

In a horizontal homogeneous flow, Eq. (10) is reduced to

$$\frac{\partial \overline{c'^2}}{\partial t} - \overline{w_s}\frac{\partial \overline{c'^2}}{\partial z} = -2\overline{c'w'}\frac{\partial \overline{c}}{\partial z} - \frac{\partial \overline{w'c'^2}}{\partial z} - 2D_s\overline{\left(\frac{\partial c'}{\partial z}\right)^2}. \tag{11}$$

Where z stands for the vertical coordinate. Similarly, the left hand side of the equation is the Eulerian rate of change of sediment variances $\overline{c'^2}$, and the right hand side accounts for the production, turbulent transport, and dissipation of $\overline{c'^2}$. The gradient production, including the negative sign, usually has a positive contribution to the variance because the mean sediment concentration typically increases downward to the bottom (negative gradient) and utilizing mixing length considerations the covariance of c' and w' is then positive.

According to the sediment variance equation, the critical parameters governing the sediment field are (i) the sediment molecular dissipation rate (the last term in the variance equation, $\Upsilon = D_s\overline{(\partial c'/\partial z)^2}$), (ii) the sediment eddy diffusivity (ε_s) in which $\overline{c'w'}$ can be approximated by $\varepsilon_s \partial c/\partial z$, (iii) the molecular diffusion coefficient, (iv) the concentration gradient ($\partial c/\partial z$) indicating the degree of suspended sediment stratification in the water column, (v) the diameter size (d) of the sediment, (vi) the buoyancy factor ($\gamma_s = g(\rho_s - \rho)/\rho$), (vii) the settling velocity (w_s), and (viii) the ambient concentration (c_o) which is included in the dimensional analysis as suggested by Yalin (1971) and also for Eulerian measurements where sediment is convected through sensors. For the flow field, the critical parameters are (i) the kinematic viscosity (ν), (ii) the viscous energy dissipation rate (ϵ), and (iii) $q = \sqrt{u_i u_i}$ as suggested by Tennekes (1975) for scaling the spectra of Eulerian observations to account for the eddy advection is scaled by the shear velocity (u_*) here. The sediment frequency spectrum can be written as a function of all these parameters,

$$\Psi_c = F(f, \nu, \epsilon, u_*, \Upsilon, \varepsilon_s, D_s, \frac{\partial c}{\partial z}, \gamma_s, d, w_s, c_o). \tag{12}$$

In general, molecular diffusion of non-cohesive sediment in water is small compared to viscous diffusion. The Kolmogorov scales of velocity $(\nu\epsilon)^{1/4}$, time $(\nu/\epsilon)^{1/2}$, and length $(\nu^3/\epsilon)^{1/4}$ for typical coastal flow environments are 10^{-1} cm/sec, 10^{0} sec, and 10^{-1} cm, respectively with typical values of $\epsilon \sim 10^{-2}$ cm^2/sec^3 and $\nu \sim 10^{-2}$ cm^2/sec. The velocity and length scales are the same order of magnitude as the settling velocity. For a sediment size of sand particles of 100 micron in diameter for instance, w_s and d can be scaled by the Kolmogorov values. Eq. (12) can now be reduced to,

$$\Psi_c = F(f, \nu, \epsilon, \Upsilon, -\frac{\varepsilon_s^2}{u_*^4}\frac{\gamma_s}{c_o}\frac{\partial c}{\partial z}). \tag{13}$$

The last term in the equation is the stability parameter for sediment induced stratification (Grant and Glenn, 1987) which can inhibit turbulent transport of mass and momentum and increase the rate of energy dissipation. The negative sign is to account for the negative concentration gradient.

DISCUSSION

In a steady and horizontal uniform flow field with negligible w_s, Eq. (11) will be reduced to the equilibrium equation where the rate of sediment variance dissipation is balanced by the gradient production. The corresponding scalar spectrum in an inertial-convective subrange is only dependent on the parameters ϵ and Υ if the field of turbulence is homogeneous and isotropic as suggested by Corrsin (1951), the frequency spectrum will have the following form,

$$\Psi_c(f) \sim \Upsilon f^{-2}. \tag{14}$$

In the high frequency region where the fluid viscosity becomes predominant, the critical parameters are either the strain rate $(\epsilon/\nu)^{1/2}$ in a viscous-convective subrange or the molecular diffusion time scale $(\nu/\epsilon)^{1/2}$ in a viscous-diffusive subrange (Tennekes and Lumley, 1972). The sediment frequency spectrum may now carry the form,

$$\Psi_c(f) \sim \Upsilon(\epsilon/\nu)^{1/2} f^{-2} \tag{15}$$

or

$$\Psi_c(f) \sim \Upsilon(\nu/\epsilon)^{1/2} f^{-1}. \tag{16}$$

Eqs. (15) and (16) indicate the spectra are directly proportional to the strain rate and the diffusion time scale, respectively. It should note that the two equations are crude estimations of spectral cascade and the forms are not universal.

Furthermore, if the scales of the dominant sediment parameters, w_s and d, are only comparable to the Kolmogorov velocity and length scale and D_s is appreciably smaller than ν, in an inertial-convective subrange the sediment fluctuations are strongly dominated by the large-scale eddies and not likely to be separated from the bulk of the liquid. Thus, it is reasonable to believe that the sediment spectral cascade in the inertial-convective subrange will just follow the energy cascade in the flow field. According to the Eulerian frequency energy spectrum derived by Tennekes (1975) in this subrange, it is likely that the sediment spectrum also cascades with a $-\frac{5}{3}$ slope in the following form,

$$\Psi_c(f) \sim \left(\frac{\partial c}{\partial z}\right)^2 (\epsilon u_*)^{2/3} f^{-5/3}. \tag{17}$$

As Tennekes (1975) pointed out for observations at a fixed point there will be a spectral 'smearing' effect caused by random advection of dissipative eddies. This effect also extends to the high frequency end of the Eulerian spectrum. It is speculated that this spectral broadening may be intensified in spectra of near bottom sediment concentration where the settling velocity is higher than the average and one or two orders larger than the Kolmogorov velocity scale. This may be one of the reasons that -1 slopes are also observed by O'Neil (1993). For the sampling resolution of ARMS, it is expected that the frequency sediment spectra cascade with slopes range between '-2' to '-1'.

Acknowledgements

The involvement of the OSU coastal team in SUPERTANK was supported by a US. Army Corps Dredging Research Program Grant, contract no. DACW-39-88-K-0040. Dr. Nicholas Kraus was the Army Corps project monitor. The support is gratefully appreciated. The authors would like to thank Mr. Rob Van Evra III and Mr. Jongkok Lee for their great effort to make the Supertank deployment so successful.

REFERENCES

Bedford, K. W., O. Wai, C. Libicki, and R. Van Evra III (1987). "Sediment Entrainment And Deposition Measurements In Long Island Sound," *J. Hydraulic Engineering,* Vol. 113, No. 10, pg. 1325-1342.

Corrsin, S. (1951). "On The Spectrum Of Isotropic Temperature Fluctuations In Isotropic Turbulence," *J. of Applied Physics*, **22**, pp. 469-473.

Glenn, S. and W. Grant (1987). "A Suspended Sediment Stratification Correction For Combined Wave And Current Flows," *J. Geophys. Res.*, **92**, pp. 8244-8264.

O'Neil, S. T. (1993). "Comparison Of Sediment Transport Due To Monochromatic and Spectrally Equivalent Random Waves," M.S. thesis, The Ohio State University, Columbus, Ohio, 155 pp.

Tennekes, H. and J. L. Lumley (1972). *A First Course In Turbulence.* M.I.T. Press. 300 pp.

Tennekes, H. (1975). "Eulerian and Lagrangian Time Microscales In Isotropic Turbulence," *J. Fluid Mech.*, **67(3)**, pp. 561-567.

Wai, O., K. Bedford, and M. Abdelrhman (1993). "Long Term Structural Variations in Suspended Sediment Profiles," in: *Nearshore and Estuarine Cohesive Sediment Transport*, ed. Ashish J. Mehta, American Geophysical Union, pg. 92-107.

Yalin, M. S. (1971). *Theory of Hydraulic Models*, MacMillan Press, 294pp.

RADIOGENIC HEAVY MINERALS AS TOOLS IN SELECTIVE COASTAL TRANSPORT

R.J. de Meijer, I.C. Tánczos and C. Stapel

ABSTRACT: Natural radioactivity in certain heavy minerals provides the possibility to locate and to follow in time and space these components in coastal sands. In this paper, the sensitivity of radiometric techniques is demonstrated in measurements carried out in the laboratory and in the field. In the laboratory, on the beach and on the seafloor indications were found that transport processes acting on light and heavy minerals often result in net transport modes in opposite directions. A simplified transport model will be discussed that describes this selective transport semi-quantitatively. The model calculates trajectories of individual grains using physical expressions for forces acting on them. The model incorporates turbulence and vortex motion in an effective viscosity coefficient.

1. INTRODUCTION

Sparked by the discovery of Bonka (Bonka, 1982) on the beach of the German Frisian Island of Nordeney of patches of sand with enhanced natural radioactivity, we started a search of such areas along the Dutch coast. One of the areas where we found enhancements in natural radioactivity was the Dutch Frisian Island of Ameland. Similar to Bonka's results we found that these enhancements were due to concentrations in radiogenic heavy minerals. Various investigations followed on the dependance of the concentrations of radionuclides, ^{40}K and the decay series of ^{238}U and ^{232}Th, on grain size and magnetic susceptibility (Schuiling et al, 1985; de Meijer et al, 1985; de Meijer et al, 1987; de Meijer et al, 1988).

Kernfysisch Versneller Instituut, Rijksuniversiteit Groningen, Zernikelaan 25, 9747 AA Groningen, The Netherlands

These investigations showed that the concentrations of radionuclides in the decay series of U and Th increased with decreasing grain size and that these activities were present in small grain size, high density minerals. Calculations indicated that these grains have a similar settling velocity in water as the larger quartz grains.

Radionuclides as ^{40}K, and ^{235}U, ^{238}U and ^{232}Th have been present in the crust of our planet since its origin and since their half-life is similar or longer than the age of our planet their total concentrations have only slightly diminished. Of importance for this paper are mainly ^{40}K and gamma-ray emitting nuclei in the decay series of ^{238}U and ^{232}Th. For both the half-life of the decay products is considerably shorter than of the mother nuclei. In isolated systems, meaning that no nuclei disappear otherwise than by nuclear decay, the activity concentrations of the nuclei in the chain reach so-called secular equilibrium (Evans, 1969). In secular equilibrium the activity concentrations of all nuclei is the same. For a closed system, this means that measuring the activity concentration of one member of the decay chain provides information on the presence of all members. For the ^{238}U decay series measurements of the activity concentrations of ^{214}Bi and ^{214}Pb, both gamma-ray emitting decay products yield information on ^{238}U (which emits no gamma-rays) provided no elements dissolve or escape like the gaseous member ^{222}Rn.

The relation between the number of atoms N of a certain species and its activity A is defined as

$$N = T_{1/2}\,A/\ln 2 \qquad (1)$$

where $T_{1/2}$ is the half-life of the radionucleus. Since activities are expressed in Becquerel (1 Bq corresponds to one decaying atom per second), $T_{1/2}$ has to be expressed in seconds. With equation (1) and Avogadro's number, one can calculate that 1 ppm U and Th correspond to 12.3 and 4.0 $Bq.kg^{-1}$, respectively. Similarly, one may calculate that 1% K_2O corresponds to 257 $Bq.kg^{-1}$ ^{40}K. In this paper concentrations will be given in $Bq.kg^{-1}$.

Using the above relation we have checked to what extent mineral grains may be considered as closed systems. For that purpose activity concentrations of ^{214}Pb and ^{214}Bi were compared with U concentrations based on X-ray Fluorescence measurements (XRF). Within the uncertainties in the two techniques no evidence was found that the systems were not closed. Nevertheless we will report our activity concentrations in the decay chain of ^{238}U as ^{214}Bi.

Coastal sands are the resistant products of mechanical weathering of rock. Sand mineralogy therefore reflects the mineralogy of the parent rock. Although dependent on the amount of sediment mixing during transportation from the source area to the site of deposition, sand mineralogy generally provides a reliable reflection of the mineral composition of the source rocks.

Sand may be divided into light- and heavy-mineral fractions. This division is based on the traditional density-separation method in which sand is settled in organic liquids of high specific density, such as bromoform (sp. dens. 2.82 g cm^{-3}). Light minerals consist mainly of quartz and feldspar. Quartz is nearly pure SiO_2 and contains little U, Th or K. Feldspar is a potassium-rich mineral. Heavy minerals present in Dutch beach sands include garnet, zircon, epidote, ilmenite and magnetite (de Meijer et al, 1990). On average the heavy-mineral fraction of Dutch beach sands has a 100 to 200 times higher specific activity concentration than light minerals (de Meijer et al, 1990). This large difference allows the detection of heavy-mineral concentrations in sediments and in principle, provides a method to determine radiometrically a coarse composition of the sediment by deriving the relative mass of a number of groups of minerals.

In de Meijer et al, (1990) the activity concentrations of groups of minerals with similar values of the magnetic susceptibility are presented; one notices the enhancement by two to four orders of magnitude in Bi and Th for various groups with respect to the light minerals. In their article de Meijer and Donoghue (1994) show that for the beaches of the Dutch, German and Danish coasts the total heavy-mineral activity concentrations may vary depending on the origin of the sands. These concentrations may therefore be used to identify regions with the same source area for the heavy minerals in coastal sands (radiometric fingerprint). They thereby reflect the effects of sediment transport in geological time scales.

Traditionally heavy minerals are used as indicators of transport processes. Stapor (1973) notices a difference in the grain size at sheltered and open beaches. Finer grains are removed from open beaches resulting in a coarser grain size. This is in agreement with the findings of Komar and Wang (1984). The reason for this difference is not yet clear. Stapor (1973) points out that there are indications that on these beaches heavy minerals are deposited that were concentrated at deeper waters off shore.

This conclusion of Stapor is contrary to the more general opinion that heavy-mineral deposits result from eroding beaches and dunes leading to deposits. Sorting processes were studied by Slingerland (1984). He distinguishes four sorting processes:

- entrainment sorting due to selective removal from the sediment bed due to size, density and shape;
- suspension sorting due to differences in settling velocities;
- shear sorting due to grain to grain interactions in a moving sediment layer and
- transport sorting due to differences in transport velocities of light and heavy minerals. In transport sorting also entrainment and suspension sorting play a role.

Although the above investigations indicate the potential importance of heavy minerals to decipher sediment transport processes, in practice, however, such studies require extensive sampling in the field followed by a quite elaborous analysing procedure. Using radiometric techniques could change this situation. A feasibility study was carried out by Greenfield et al (1989) on the beach of the Dutch Island Texel. At a number of test sites the changes in beach elevation and γ-radiation intensity were monitored in time for sites located in the different erosion sensitive areas. For sites dominated by aeolian transport the changes in elevations showed a good inverse correlation with changes in count rate, meaning that removal and deposition of sand mainly occured via light minerals. At some sites where erosion was caused by wave action sometimes an opposite tendency was observed.

During this investigation some major storm events took place whereafter heavy-mineral concentrations were observed on the beach with volumes that seemed to exceed the heavy-mineral content of the removed dune and beach sands. These observations, which support the ideas of Stapor (1973), triggered a research programme to investigate the processes that selectively remove and deposit light and heavy minerals, respectively, in the same event. For that purpose we started to develop a towed seabed detector to map radiometrically the seafloor, set up laboratory experiments in a wave flume and extended our beach investigations at two larger test sites on the island of Ameland. This paper describes the methodology of the experiments, the first series of results and modelling of these results in terms of a simplified selective transport model.

2. EXPERIMENTS: METHODOLOGY AND RESULTS

Field experiments have been carried out near the Dutch Frisian Island of Ameland. Up to a couple of years ago the island was eroding in the middle part and stable at the two ends of the island. In recent years erosion is occurring also on the west end of the island. Beach and dune profiles in the middle part of the island were restored temporarily by beach renourishment; the latest one occurred in the summer of 1992.

2.a Test sites on the beach

Following the experience with small scale test sites at the beach of Texel on the changes in elevation and γ-ray count rates, it was decided to set up two larger test sites on the island of Ameland. First a 800 x 300 m^2 site was chosen at a rectangular grid with a grid size of 25 m. This site is located near the east end of the island, named Oerd. Later, when it was noted that rapid erosion was occurring at the west end of the island, named Bornrif, a 200 x 150 m^2 site was selected. After initial measurements at a grid size of 25 m it was noticed that essential information was missed and a 12.5 m grid size was taken.

Measurement positions were marked by stalks. Days prior to measurements, missing stalks were replaced by personnel of Rijkswaterstaat. At each grid point the elevation and the total-counts rate were measured and recorded. The elevation measurements were carried out by Rijkswaterstaat by standard levelling techniques. The inaccuracy of these measurements depends on the softness of the sand and is believed to be at most 2 cm. A hand-held γ-monitor (Scintrex) was used to measure

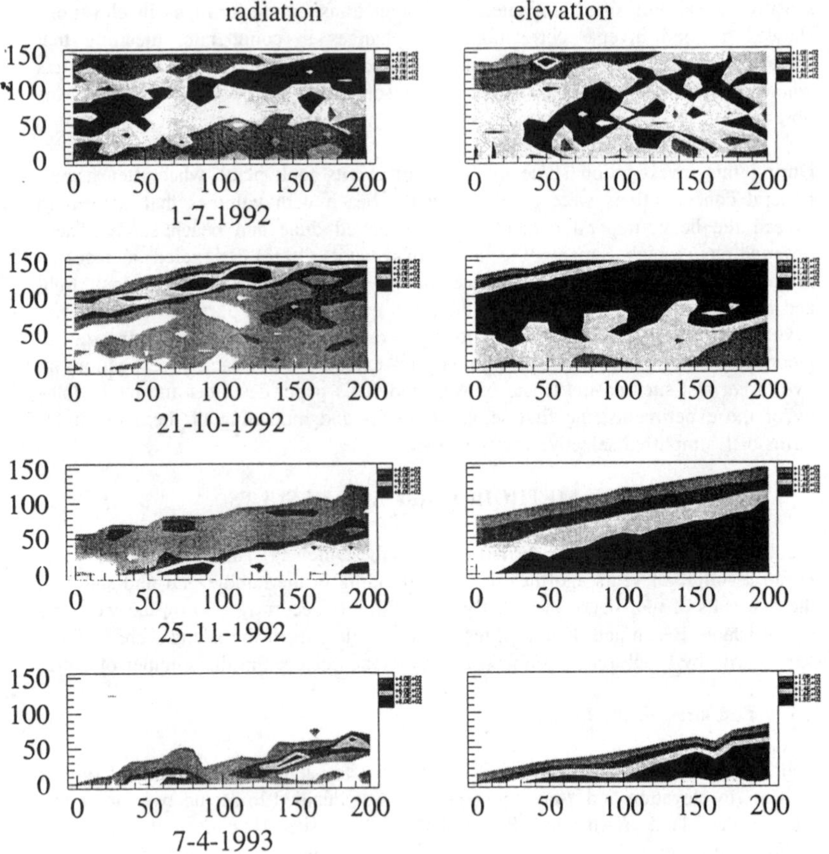

Figure 1. Smoothed data on radiation intensity and elevation for the test site at the location Bornrif on the island of Ameland, recorded at four dates. The shades of gray range from < 400 counts per 10 seconds to > 800 counts per 10 seconds for radiation intensity and for elevation from < 100 to > 180 cm + NAP.

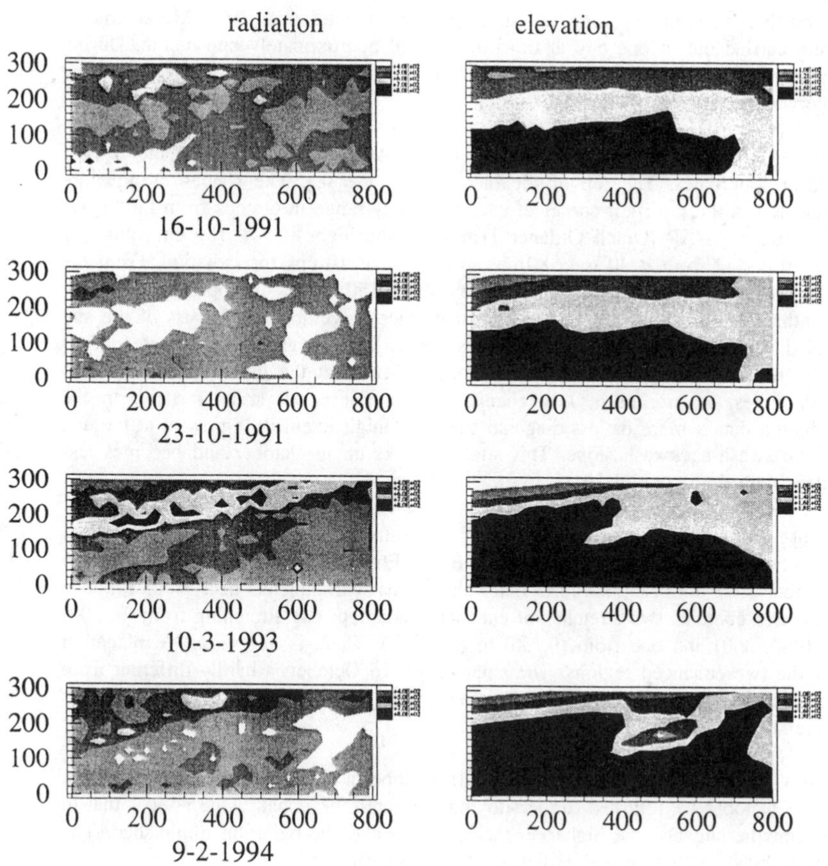

Figure 2. Smoothed data for the location Oerd presented in a similar way as in figure 1.

at hip height, point-by-point, time-integrated count rates over 10 s. Measurements were carried out in one day at time-intervals of approximately one month. During holiday seasons and times that Rijkswaterstaat personnel was occupied the time intervals became longer.

Figure 1 presents the smoothed results of radiation and elevation for the Bornrif site at four dates. The top of the blocks faces the sea. The shades of gray are indicated in the top right corner of each box; they range in 6 steps from the lowest < + 100 cm NAP (Dutch Ordance Datum) to the highest: > + 180 cm value for elevation and from < 40 to > 80 cps in 6 steps of 10 cps for radiation. From the figure one sees a retreat of the beach by about 150 metres in approximately nine months. This erosion rate was so high that since November 1992 parts of the site had disappeared. For the April data this means that the grid actually extended to only 100 m. From the elevation data one notices that the quite irregular pattern with dunes, as present in July changes into a flat beach at later dates. In the radiation data a more or less diagonal stretch of high intensity is present at the end of the swash-backwash slope. This stretch breaks up in October and becomes less intens with the retreating beach front.

Figure 2 presents a selection of smoothed results for the Oerd site at four dates. Also here the top of the blocks faces the sea. From the figure one sees that on 16 October 1991 the beach is rather flat with an area with dunelets near (400, 50). The total area contains two stretches of enhanced radiation: one stretching from (0, 200) to (400, 300) and one from (0, 50) to (300, 50). There is no "visual" explanation for the two enhanced regions. The situation on 16 October is hardly different from the situations measured in September and August 1991, when the measurements were started.

A few days after our measurements on 16 October a high energy event took place; on 23 October we returned to the site and measured the impact. One notices that in general the site became higher in the elevation and the radiation diminished. The two stretches with enhanced radiation had almost vanished.

To investigate these observations further we chose a 1 m^2 area just west of the site near (0, 225). Contact measurements with the radiation monitor were made every time a thin layer of sand was removed. From these measurements it became clear that a layer of sand with enhanced radioactivity was present at about 20 cm depth. This result indicates that the vanishing two stretches were buried by low-activity sand during the storm event; the reduction in radiation intensity is than due to absorption of radiation by the top layer. (A layer of about 7 cm reduces the radiation intensity by a factor of two (Greenfield et al, 1989)).

In the next period until February 1993, the situation in the site hardly changes. A new high energy event with northwesterly winds changes both the elevation and radiation patterns.

In the data of March 1993 these changes show up as a new ridge of accumulated sand from (0, 160) to (400, 240), whereas the beach near (100, 300) was lowered. In the radiation pattern a higher intensity ridge stretching from (0, 160) to (750, 300) shows up. This ridge was visually observed as a thin layer of reddish coloured heavy minerals.

In this case the data are consistent with deposition of sand and heavy minerals from the off-shore area. This ridge occurs more southwesterly than the ridge in the top box and since the elevation increased with respect to 23 October 1991 the contribution from the deeper layer, observed earlier, is further diminished by absorption. The presently observed enhancement in the radiation ridge is due to freshly deposited heavy minerals. This observation is consistent with the ideas of Stapor (1973).

After March 1993 the situation at the seaside part of the site changes gradually due to accretion with new sand that covers the thin layer of heavy minerals. Thereby the radiation ridge diminishes. In the winter of 1993-1994 many high energy events take place mainly with SW-W winds; in January-February 1994 a more N-NW event takes place that again modifies the site considerably. The west side of the site increases in height especially the middle and southern part, whilst a small inlet is formed near x = 500. The changes in elevation are accompagnied by the building of a weak radiation ridge stretching from (0, 120) to (500, 300).

From the data presented above, it may be concluded that changes in radiation patterns are correlated to changes in beach morphology. Without information on the actual location of the heavy-mineral layers (top or deeper) the correlation is not always easy to understand. From the data it becomes evident that heavy-mineral concentrations are formed on the top of the swash-backwash zone. This is the region where the up and down running water experiences rapid deceleration and acceleration, providing thereby optimal conditions for selective grain transport.

2.b **Seafloor mapping**

In collaboration with the British Geological Survey (BGS) a project was started to design and build a towed seabed detector based on a prototype detector of BGS developed more than a decade ago. In comparison the KVI detector is equipped with a γ-ray detector with an order of magnitude larger efficiency for high-energy γ-rays. Such a detector thereby becomes more sensitive to locate heavy-mineral concentrations near the surface.

In May-June 1993 a trial tow was executed in which both detectors were anticipated to be compared on the seafloor north of the island of Ameland in a 20 x 10 km^2 area. A line pattern was set out with distances of 1 km and angles of about 45° with respect to the coast. The detectors were towed over the seafloor behind a vessel of Rijkswaterstaat ("Blauwe Slenk").

For that purpose the detectors were placed in water-tight casings that were connected to the ship by an armored coaxial cable. The cable serves for as well the physical as the electrical connection with the ship.

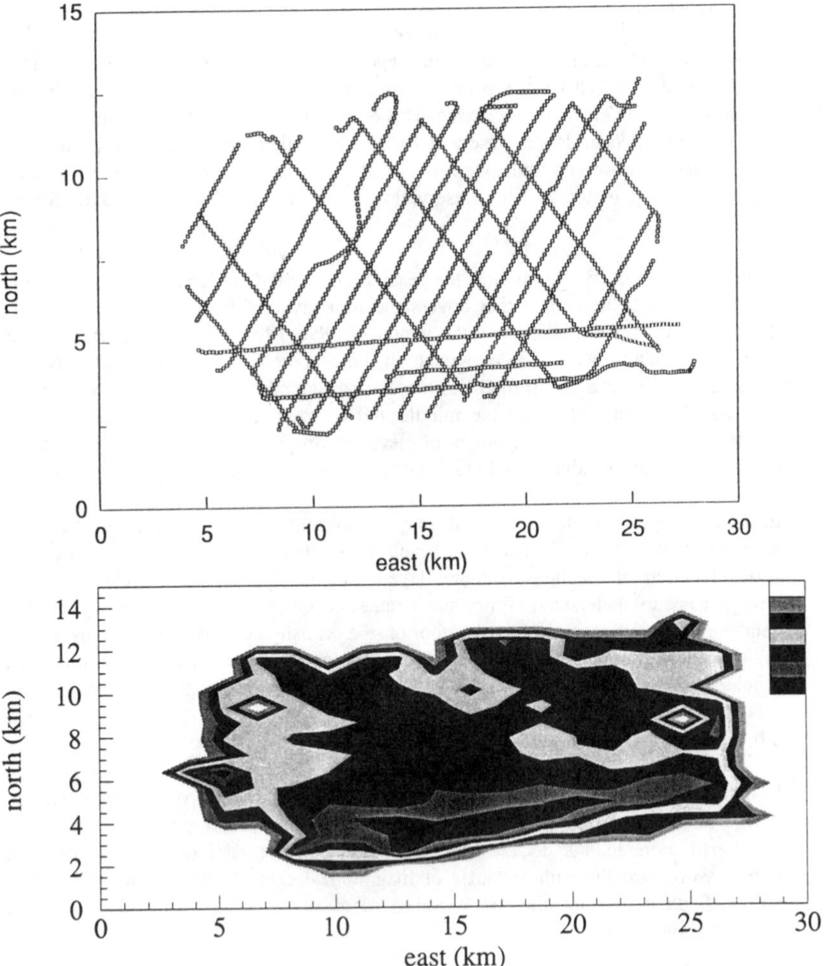

Figure 3. Top: Lines sailed during our seabed experiment. The coordinates are given with respect to the origin of our research area. Bottom: Map of smoothed total count rates. The rapid variation at the edges are artifacts of the smoothing procedure. The two spots in the figure reflect the lack of data. The shades of gray are indicated in the top right corner; they from < 26 to > 56 cps in steps of 6 cps.

To prevent the detector from being snatched by underwater obstacles like wrecks, the detector was placed at the end of a 30 m long PVC hose. The front site of this "eel" is kept upwards by giving just the correct length of cable to keep the detector in contact with the seafloor. The cable is mounted on a winch placed at the stern of the ship. Signals from the detector are transmitted via the cable to an onboard computer for data handling and storage. The vessel is standard equipped with a position and depth logging system (Syledis). For our measurements side-scan sonar was installed. Results of the sonar were recorded on paper, for possible off-line analysis of the data with bottom topography.

Data from the logging system were stored every 0.1 s, total counts of the γ-ray detector every 15 seconds and γ-ray spectra every 5 minutes. The towing speed was kept rather constant at about 10 km h^{-1}. Due to weather conditions only the BGS prototype could be used on the North Sea. Test comparisons of both systems were made during bad weather conditions in the sheltered Wadden Sea. During the tests the improvements in the KVI detector, measured already on test slabs at BGS and the beach of Ameland were confirmed.

Figure 3 shows the lines sailed with the BGS-detector in the top part and the smoothed radiation data in the bottom part. The shades of gray indicated in the top left corner range from < 26 cps to > 56 cps in steps of 6 cps. The data are "raw" data, not corrected for (varying) background due to cosmic radiation and K in seawater. The scales along the x- and y-axis denote the distance to an arbitrarily choosen origin; the y-axis points due north.

In addition to the measurements grab samples were taken of the top 20 cm with a Van-Veen sampler. Locations were chosen to cover the full range of radiation-intensity variations. Radiometric analysis of the samples yielded total heavy-mineral concentrations from < 1 to about 20%. Based on this result and assuming a layer thickness of 20 cm the area of the two highest intensities in the bottom part of figure 3 comprises an estimated volume of 0.2×10^6 m^3 of pure heavy minerals.

The bottom part of figure 3 shows that heavy minerals are concentrated in an area elongated parallel to the coast. The shape of the concentrated area and the fact that the highest concentrations occur in the middle of the enhanced area are consistent with a cross-shore transport process. Its location support the ideas of Stapor (1973) that heavy minerals are deposited on the beach from concentrations off-shore.

Figure 4 shows details of part of the most northern east-west lines. The detail corresponds to the west part of the line; here the bathymetry shows banks that are oriented northwest-southeast. The figure shows depth information together with the total-count rate in 15 seconds. The figure shows that the radiation is not evenly distributed but seems to be concentrated on the seaward slopes of some sand bars. This phenomenon is the same as seen on the beach where heavy minerals concentrate on the swash-backwash slope of the shore face.

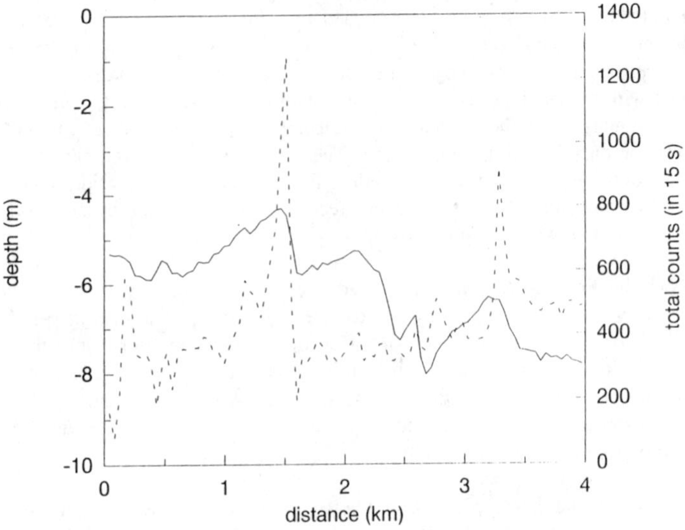

Figure 4. Bathymetry (——) and total γ-ray count rate (- - - - -) measured along part of one of the lines. The distance along the horizontal axis refers to the distance to the point on the line nearest to the shore.

2.c Laboratory experiments

A study was carried out in the 14 x 0.8 x 0.5 m^3 wave flume of the laboratory of Fluid Mechanics at the University of Technology Delft in The Netherlands. The wave generator can be adjusted in tilt angle and frequency to allow for a wave height, amplitude, frequency and asymmetry within a certain range of values. At the end of the flume a damping slope is present to avoid reflection. In addition to waves, a current may be induced opposite to the wave direction. In the experiments a layer of about 30 x 50 cm^2 and a thickness of about 5 cm of beach sand, naturally enhanced in heavy minerals (~ 30%), was placed at about the middle of the flume. For a number of wave conditions the transport of sand was followed for periods between 1 and 17 hours. Prior and after the experiments sand samples were taken for density, grain-size and radioactivity-concentration measurements. Water velocities (0.2 - 0.5 m s^{-1}) were measured at about 3 cm above the sediment bed and were recorded on paper.

In this study it was observed that after some time "ripple" areas were formed at both sides of the initial position. Usually the extent of the ripples in the direction of the largest peak velocity was considerable larger than in the opposite direction. Moreover the colour of the ripples became darker with increasing distance in that direction.

From the analysis of the sand samples it followed that, under all conditions heavy minerals were transported in the direction of the largest velocity near the bottom and that light minerals either stayed at the same place or were transported in opposite direction. It turned out that the characterisation of the minerals by radionuclide concentrations was a hundred times more sensitive tool than either grain size or density (see figure 5). This observation may be a basis for a non-destructive method for sediment transport studies.

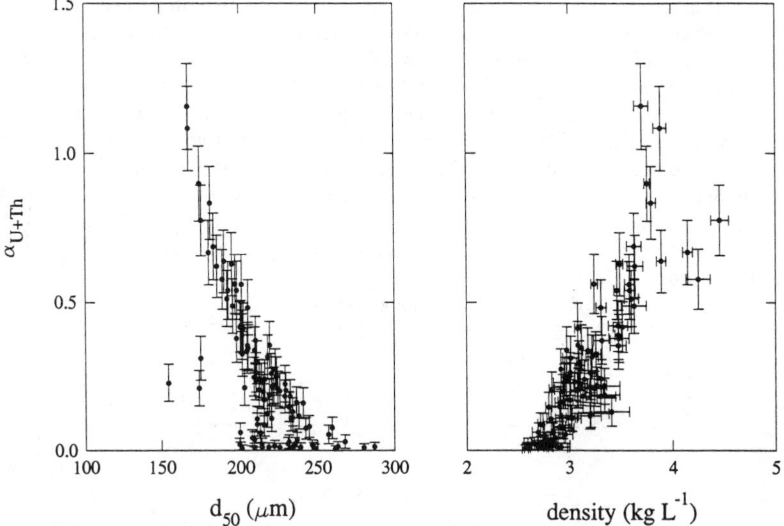

Figure 5. Radioactivity concentration ratio α as function of median grain size, d_{50} , and density. The ratio α is defined as the sum of the Bi and Th activity concentrations (Bq kg^{-1}) divided by 118076 $\pm$ 120 Bq kg^{-1}, the value for the total heavy-mineral fraction of the reference sample of De Hors, Texel.

DISCUSSION, CONCLUSIONS AND OUTLOOK

3.a Selective transport

In this paper we demonstrated the sensitivity of radiometric techniques in measuring heavy minerals in sands. The main reason for this sensitivity is the two orders of magnitude higher concentration of radionuclides of the decay series of ^{238}U and ^{232}Th in the group of heavy minerals relative to the group of light minerals.

Two restricting remarks should be made: one is that absorption of γ-radiation by sand restricts the depth of sight into the sediment bed to about 50 cm and the other is that this application is still in its exploratory phase such that we are not yet convinced that the methodology that works so nice in our research area will work equally well in environments with much less or much more (wave)energy.

With these restrictions in mind, we like to point out that originally we thought that concentrations of heavy minerals were a rather rare phenomena, occurring occasionally at certain beaches under or after certain weather conditions. Our experiments in the wave flume therefore were originally set up to see if we could reproduce conditions under which the segregation of heavy and light minerals could be involved. To our surprise, it turned out that is was hardly possible to avoid segregation and that the phenomenon was rather a rule than an exception. This lead us to the idea that, despite all complications that occur in sediment movement, apparently the selective transport is an "eigenmode". It resembled us, as nuclear physicist, to the motion of protons and neutrons in an atomic nucleus, where certain properties of the nucleus are being much simpler described by collective behaviour of groups of protons and neutrons than by the motion of individual nucleons. In nuclear physics one distinguishes microscopic models, starting from individual nucleons, and macroscopic models in which the collective behaviour is described. An example is a giant resonance mode in nuclei where protons and neutrons as groups are thought to oscillate collectively against each other despite the fact that the motion of each nucleon is well described by moving in a potential well generated by the other nucleons.

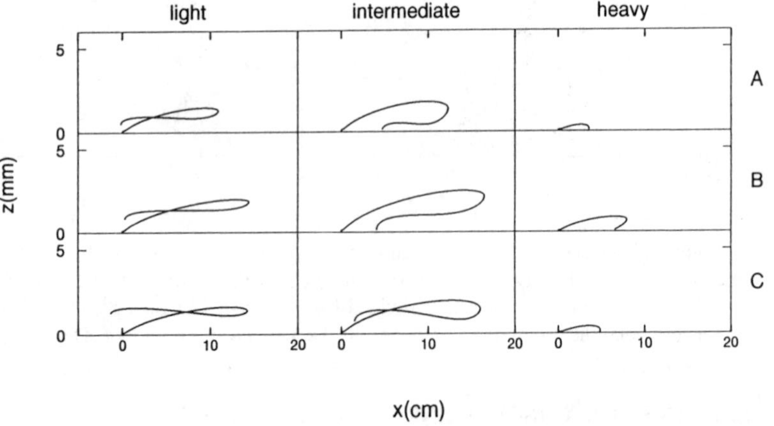

Figure 6. Trajectories of a light, intermediate and heavy grain under wave conditions with different peak velocities under crest and trough, respectively. The values for U_c and U_t are A: 0.5 and 0.3 m s^{-1}, B: 0.6 and 0.3 m s^{-1} and C: 0.5 and 0.4 m s^{-1}.

This experience has triggered us to see if such a dualism could also be found in sediment movement. On the microscopic scale the situation requires an extremely complicated approach. All particles interact with each other and due to these interactions the boundary conditions (i.e. the sediment bed) change continuously. Turbulence or vortex motions generate complications that can only be handled in detail by the use of enormous computer capacity. Nevertheless, nature confronts us with rather simple features.

So the challenge becomes to find macroscopic physical approaches in which the complexity may be reduced to a couple of quantities. In this section we like to present a first step in that direction. This step is strongly influenced by the observations through the glass wall of the wave flume, where we visually followed the motion of grains due to undulation.

Our model describes selective transport in a semi-quantitative way in terms of gravity, a lift and a friction force. The description uses almost only physical relations and needs one overall effective quantity to cover chaotic motions like turbulence and vortices and an emperical description of the velocity in the boundary layer. With this model a successful description of the difference in the motion of heavy and light minerals in sand can be given.

The model calculates grain trajectories under oscillatory flow within a boundary layer. It is assumed that the fluid motion is only horizontal (in the x-direction) and is a logarithmic function of the distance, z, from the sediment bed. The boundary conditions are given by:

$$U(z) = 0 \text{ at } z = z_o \text{ and} \qquad (4)$$
$$U(z) = U(t)_{max} \text{ at } z = z_o + d$$

where z_o is zero velocity level, d is the boundary layer thickness and $U(t)_{max}$ is the velocity at the edge of the boundary layer; the velocity is a function of time following the wave motion.

The trajectories follow from the forces acting on the grain. A friction force with an effective viscosity coefficient is used to describe in first-order turbulence and vortices. These chaotic motions are in first-order circles that have no net horizontal displacement and lead effectively to a slower settling of the grains. This effect on the vertical motion is incorporated in the model by introducing an "effective viscosity coefficient" in the friction force according to Stokes law. Since this coefficient is larger than unity the friction force is enhanced. By including all viscosity effects into the friction force, the two other forces are acting in non-viscous fluids, according to the separation of variables. Therefore for the lift force a Bernoulli description can be used.

Figure 6 presents a number of trajectories for a light (quartz), an intermediate and a heavy grain with density and grain size of 2.65 kg L^{-1}, 0.20 mm; 3.0 kg L^{-1}, 0.25 mm and 4.0 kg L^{-1} , 0.15 mm, respectively. The results are given for fixed values of the viscosity coefficient (20) and the thickness of the boundary layer and various combinations of crest and through velocities, U_c and U_t. As can be seen the light grains are moved forward and back by the wave motion leading to an almost net zero displacement, whilst the heavy move only forward. For details on the experiments and the model we temporarily refer to an internal report available from the authors (Tánczos and De Meijer, 1993).

This result can qualitatively be understood by realising that the threshold velocity for moving light grains is lower than for heavy grains and that under asymmetric waves the bottom velocities for the largest peak exceed the threshold velocity for both light and heavy grains and for the lowest peak-velocity period only for the light grains. Based on this model it is to be expected that segregation of light and heavy minerals on the seafloor will predominantly take place in the domain where waves and swells, in combination with current, will lead to asymmetric velocity patterns in the boundary layer.

Presently we are preparing experiments in an oscillating wave flume to further test and develop the model. In those experiments it is anticipated to monitor the displacement of the minerals radiometrically rather than sampling the sediment itself. This allows us to obtain intermediate results without too much interference with the experiment.

3.b Conclusions

The results of our studies and the success of our simplified model have convinced us of the role that radiometric methods may contribute to understanding certain aspects of coastal dynamics. Resticting ourselves to selective transport features, we have become convinced that differences in grain size and density play a much more important role than is usually acknowledged in coastal studies.

Measuring techniques as the towed seabed γ-ray detector system will enable one to measure the result of transport processes in time and place. These measurements may be carried out in the laboratory under controlled conditions, at the beach and on the seafloor. Such measurements in the field may provide supplementary information to experiments carried out with e.g. rigs. The advantage of the proposed technique is that information may be obtained over large areas and shortly after storm events, provided shiptime is available.

In our opinion the advantage of large scale mapping over spot measurements is that one obtains macroscopic information that may help to understand and indentify eigenmodes in coastal dynamics. One clear example is that results from spot measurements often indicate the importance of longshore transport whereas our large scale mapping seem to point into the importance of cross-shore transport. Again a difference in microscopic and macroscopic behaviour?

3.c Outlook

Stimulated by the present results we intend to extend our activities in this field. One of our nearby goals will be to validate the radiometric fingerprinting model, mentioned in the introduction, with detailed geological and sedimentological analyses. One of the possibilities is to further exploit the data obtained from Bi, Th and K concentrations.

As pointed out by de Meijer and Donoghue (1994), and de Meijer et al (1994), these three concentrations together with the trivial relation that relative masses sum to unity allow a decomposition of the mineral content of a sediment into four mineral groups. Such a decomposition needs quite some calibration work in the laboratory.

Combining the method of radiometric mineral decomposition with our Bi, Th and K measurements from laboratory analysis or from measurements in the field with our towable seabed detector system may help us to interpret the data in changes of mineral composition. Here we hope to unravel the segregation of quartz and feldspar and of e.g. garnet, zircons and monazite. Such data may shed light on intricate processes occurring near the seabed.

One of the questions we hope to resolve on a longer time scale is the question if the concentration of heavy minerals is due to a local winnowing process or if there is a source from which the minerals are transported.
If the latter is the case, is the source an ancient deposit (pocket) which is presently eroding or is the source more diffuse and is concentration the result of selective transport over quite some distance? A way to find out would be to remove the present concentration and follow the behaviour of the area in time. Such a large scale experiment requires quite some infrastructure and funds. Removal of the minerals could be used to either generate funds by selling the minerals as a mining operation or use the enhanced body of sand to carry out a renourishment at another location. Preferentially such a renourishment should be carried out under water, such that the effect of the renourishment may be followed by tracing the migrating minerals by radiometric techniques.

Summarising it seems fair to state that radiogenic heavy minerals provide a new tool to coastal dynamics research. The tool is worth to be exploited on a large scale and to be further developed.

ACKNOWLEDGEMENTS

This work is part of the research programme "Environmental Radioactivity Research" of the KVI, University of Groningen. This work is also financially supported by the by the programme "Kustgenese" (Coastal Genesis) of Rijkswaterstaat and by the Ministry of Economic Affairs under the programme "ISP-IV", administrated by the Noordelijke Ontwikkelings Maatschappij.

The collaboration with BGS (dr. D.G. Jones and P.R. Roberts) and the support of the personnel of Rijkswaterstaat on Ameland and on "Blauwe Slenk", is gratefully acknowledged. The authors also would like to thank dr. ir. J.S. Ribberink, Prof. dr. ir. H.J. de Vriend, prof. dr. ir. G.J.F. van Heijst, dr. J. Wiersma and dr. ir. J. van de Graaff for their critical support in the development of the transport model.

REFERENCES

Bonka, H., 1982, "Enhanced natural radiation exposure due to enriched heavy minerals at the coast of Northern Germany", *Natural Radiation Environment*, Ed. K.G. Vohra et al, Wiley Eastern Limited, New Delhi, 58-66 and references therein.

Evans, R.D., 1969, "The atomic nucleus", *McGraw-Hill*, New York.

Greenfield, M.B., de Meijer, R.J., Put, L.W., Wiersma, J. and Donoghue, J.F., 1989, "Monitoring beach sand transport by use of radiogenic heavy minerals", *Nuclear Geophysics*, Vol. 3, pp 231-244.

Komar, P.D. and Wang, C., 1984, "Processes of selective grain transport and the formation of placers on beaches", *Journal of Geology*, Vol. 92, pp 637-655.

de Meijer, R.J., Put, L.W., Bergman, R., Landeweer, G., Riezebos, H.J., Schuiling, R.D., Scholten, M.J. and Veldhuizen, A., 1985, "Local variations of outdoor radon in The Netherlands and physical properties of sand with enhanced natural radioactivity", *The Science of the Total Environment*, Vol. 45, pp 101-109.

de Meijer, R.J., Put, L.W., Schuiling, R.D., de Reus, J.H. and Wiersma, J., 1988, "Provenance of coastal sediments using natural radioactivity of heavy-mineral sands", *Radiation Protection Dosimetry*, Vol. 24, pp 55-58.

de Meijer, R.J., Put, L.W., Schuiling, R.D., de Reus, J.H. and Wiersma, J., 1989, "Natural radioactive heavy minerals in sediments along the Dutch coast", *Proceedings KNGMG Symposium "Coastal Lowlands, Geology and Geotechnology, 1987*, Kluwer Academic Publishers, Dordrecht, pp 355-361.

de Meijer, R.J., Lesscher, H.M.E., Schuiling, R.D., Elburg, M.E., 1990, "Estimate of the heavy-mineral content in sand and its provenance by radiometric means", *Nuclear Geophysics*, Vol. 4, pp 455-460.

de Meijer, R.J., and Donoghue, J.F., 1994, "Radiometric fingerprinting of sediments on the Dutch, German and Danisch coasts", submitted to *Quaternary Geology*.

de Meijer, R.J., Tánczos, I.C. and Stapel, C., 1994, "Radiometric techniques in heavy-mineral exploration and exploitation", submitted to *Exploration and Mining Geology*.

Schuiling, R.D., de Meijer, R.J., Riezebos, H.J. and Scholten, M.J., 1985, "Grain size distribution of different minerals in a sediment as function of their specific density", *Geologie en Mijnbouw*, Vol. 64, pp 199-203.

Slingerland, R., 1984, "Role of hydraulic sorting in the origin of fluvial placers", *Journal of Sedimentary Petrology*, Vol. 54, pp 137-150.

Stapor, F.W., 1973, "Heavy mineral concentrating processes and density/shape/size equilibria in the marine and coastal dune sands of the Apalachicola, Florida, Region", *Journal of Sedimentary Petrology*, Vol. 43, pp 396-407.

Tánczos, I.C. and de Meijer, R.J., 1993, "Selective transport of heavy minerals by oscillatory experiments", *KVI Internal Report Z-25* (available from the authors).

MAGNETIC TRACING OF BEACH SAND: PRELIMINARY RESULTS

Kuno D. van der Post[1], Frank Oldfield[2] and George Voulgaris[3]

ABSTRACT: The use of magnetically enhanced sands can provide rapid field data on the direction of movement of beach materials. This is shown from the results of a preliminary experiment performed on the Belgian Coast. Further laboratory analysis of grid samples has shown that it is possible to discriminate low concentrations of tracer material from both background heavy mineral assemblages and anthropogenic material.

INTRODUCTION

The aim of this study was to find out using an active magnetic tracing technique, how rapidly and reliably initial directional data could be obtained on sand movement in a coastal environment, and whether, at low concentrations, the tracer material could be distinguished from naturally occurring heavy mineral assemblages.

Magnetic tracing techniques have previously been used in fluvial, lacustrine and estuarine environments under two general classifications:

'Passive' Tracing

Oldfield *et al* (1979) and Walling *et al* (1979) traced the suspended sediment sources in streams using the natural magnetic enhancement known to occur in soils on a wide variety of substrates (Le Borgne 1955; Mullins 1977). Additionally Rummery *et al* (1979a) were able to show the downstream movement of magnetic minerals formed in the soils, through the action of a forest fire. In estuarine (Yu and Oldfield, 1989) and reservoir (Yu and Oldfield, 1993) systems, the magnetic properties of sediments have

[1]Research Assistant, Department of Geography, University of Liverpool, P.O. Box 149, Liverpool, L69 3BX, UK.

[2]John Rankin Professor of Geography, Department of Geography, University of Liverpool, P.O. Box 149, Liverpool, L69 3BX, UK.

[3]Research Fellow, Department of Oceanography, University of Southampton, Southampton, SO17 1BJ, UK.

been used to provide quantitative estimates of contributions from different catchment sources.

'Active' Tracing

Work by Arkell (1984), Arkell *et al* (1982) and Rummery *et al* (1979b) showed that artificially enhanced (through heating, Oldfield *et al* 1981) iron rich stream bedload could be suitably characterised so as to provide material for sediment tracing in fluvial systems.

Three types of magnetic measurement have been used in this present study and each is described briefly below. Further, more detailed explanations can be found in O'Reilly (1984) and Thompson and Oldfield (1986).

1. Magnetic Susceptibility (K) and (χ): This is used as an indicator of how easily a sample can be magnetised. It can be defined by K=M/H where H is the applied field and M is the volume magnetisation induced in a material of susceptibility K. Susceptibility values can be used for general indicators of the relative concentrations of ferrimagnetic minerals such as magnetite and maghaemite. It is the property use here for recording the movement of the tracer sands and can be measured *in situ* in the field using a 20 cm diameter surface loop sensor attached to a portable meter. Magnetic susceptibility measurements can also be made in the laboratory or motel/apartment room using sensors designed to scan cores of sediment or make precise measurements on 10 ml subsamples. These latter can be made mass specific by drying the sample and dividing the K reading by the mass to give units in m^3kg^{-1}, denoted by the symbol χ.

2. Isothermal Remanent Magnetisation (IRM): This is the remanence which can be acquired by a sample at room temperature from a given applied DC field. The field at which saturation (SIRM) is reached is dependant on the composition of the mineral and its grain size. In the present study the common convention of describing the remanence acquired in a field of 1 Tesla (1000 mT) as the Saturation IRM (SIRM) is used. The relative ease with which IRM's are acquired or demagnetised/reversed in a succession of increasing or decreasing magnetic fields can be used to discriminate between materials on the basis of their magnetic grain size (not the same as particle size) and mineralogy.

3. Anhysteretic Remanent Magnetisation (ARM): If a sample is kept in a constant, small DC field (for example the earth's magnetic field), and is then exposed to a smoothly increasing then decreasing AC field, the resulting acquisition of remanence is different from that described above and provides a basis for discriminating between materials.

Susceptibility measurements have been used primarily in the tracing experiments as the main indicator of the direction of tracer movement and of spatial and temporal changes in tracer concentrations. The surface loop measurements in the field cannot be fully quantitative where any burial has occurred, since the strength of the signal is distance dependant following an inverse square law. However, detection of buried tracer is possible provided large enough concentrations are present close enough to the surface. In this respect, the method has some advantages over the use of fluorescent tracers depending on optical detection. Confirmation of the presence of buried tracer can readily be made by scanning and measuring subsamples from small cores taken in plastic tubes.

Used in combination with each other or with the susceptibility values, the various simple remanence measurements made, provide a series of quotients that can be used to characterise more fully the magnetic 'signatures' of both the tracer and the 'host' sands. They therefore provide a basis for discriminating between elevated susceptibility values arising from tracer sands, heavy mineral assemblages or anthropogenic material.

TRACER MANUFACTURE

This whole technique depends on converting iron coatings on silica sand grains to magnetite/maghaemite. Suitable sands are readily available from quarries in red glacial outwash sands derived from Triassic rocks in North Wales, Cheshire and Shropshire, and from Cretaceous Greensand quarries in Bedfordshire. The magnetic enhancement process involves heating the sand at high temperatures (~ 700°C) for two hours in a reducing atmosphere (achieved by mixing flour into the sand), followed by rapid cooling in air. This procedure was chosen after a series of experiments in which the variables used were materials, period of heating, peak temperature and atmosphere (Figures 1 and 2).

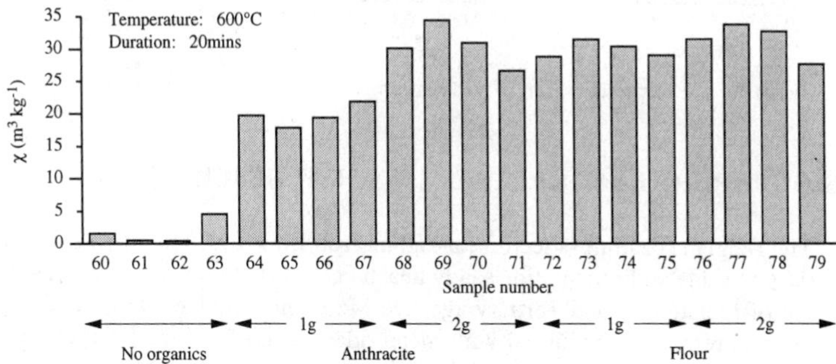

Figure 1. Comparison of the effect different reducing agents on tracer enhancement.

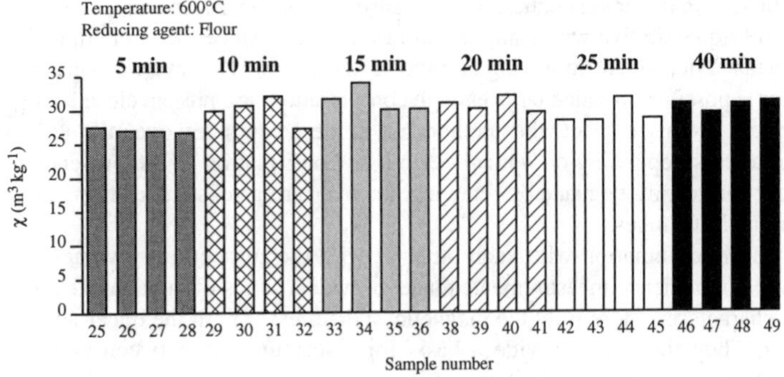

Figure 2. Effect of duration on tracer enhancement using one reducing agent

The material used for this experiment was sieved, and K values for each particle size were obtained. These values were multiplied by the percentage weight of each phi size to establish particle size contributions to the bulk sample (Figure 3).

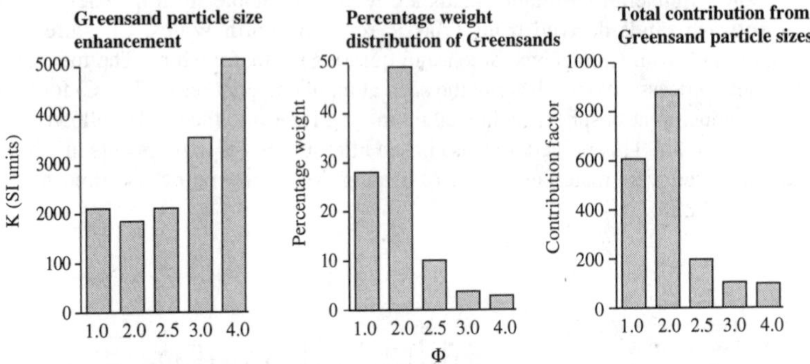

Figure 3. Particle size results of tracer material.

PHYSIOGRAPHIC BACKGROUND AND SITE DESCRIPTION

The Belgian coastline is rectilinear and macrotidal. Where groynes are absent, it comprises of ridge and runnel, fine sandy beach backed first by a dune belt then by rising land of Pleistocene and Tertiary age (De Moor and Bloome 1988). The tidal currents in this region are bidiurnal with mean tidal range of 4.5 m. Tidal currents continually change their orientation and velocity with the majority of sand transport being in an Easterly direction through residual flood displacement and both elongate

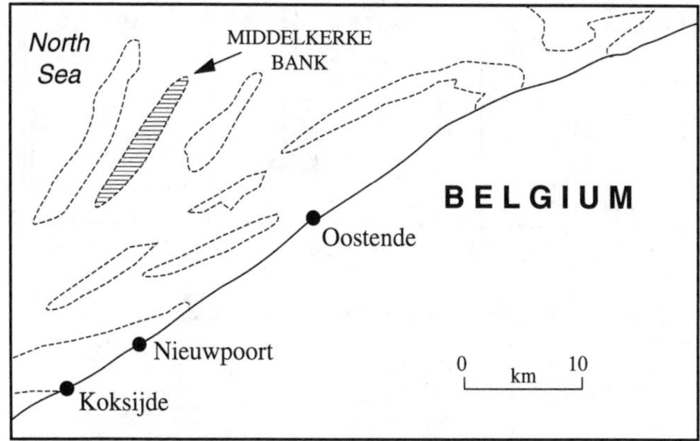

Figure 4. Location of tracing experiment and sampling sites for background susceptibility studies

and asymmetrical current ellipses (De Moor 1991). The location for this experiment was Niewpoort an Zee, Belgium (Figure 4). This site was chosen as part of the MAST2/CSTAB (Circulation and Sediment Transport Around Banks) Project (MAS2-CT92-0024-C). An important consideration in the choice of the site was the absence of groynes. A transect of the beach profile can be seen in Figure 5a and the particle size distribution from selected points along the transect can be seen in Figure 5b.

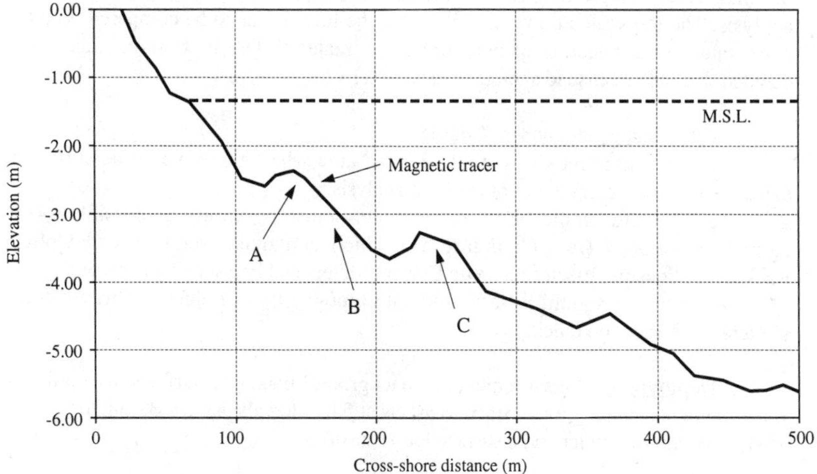

Figure 5a. Transect of beach profile

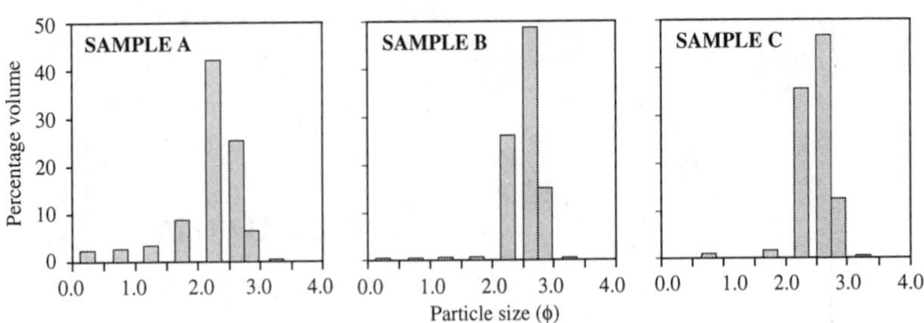

Figure 5b. Particle size distribution of background samples

METHODOLOGY

Pre-Deployment Assessment: Before setting up a magnetic tracing experiment, account must be taken of background magnetic mineral assemblages. The most likely sources for these are industrially derived particles and heavy minerals derived from reworked seafloor sediment deposited in the Upper Pleistocene by fluvioperiglacial activity (De Moor 1991). Therefore samples from Middelkerke Sandbank (courtesy of Dr Lankneus, Gent University) and various sites along the coast between Oostende and Koksijde were taken and measured in the laboratory, in order to establish the variability of the natural 'background' magnetic concentrations. The simplest indication of this is the magnetic susceptibility (K) of the material. Figure 6 shows the results of this analysis. The log scale allows the values for the local sands to be compared with the peak values of a selection of potential tracer material. On the basis of this it was decided that the Greensand should be used.

Establishing Deployment Criteria:

1. The tracer pit site should be in an area where there would be sufficient exposure to ensure completion of the field analysis.

2. The site should be in as reasonably close proximity as possible to the equipment stations (A - C; Figure 5a) which contained Acoustic and Optical Backscatter Sensors, Electromagnetic Current Meters and Pressure Transducers.

3. The site should have a moderate topography to reduce the tracer being scattered or buried too quickly.

Deployment: Measurements of background magnetic surface susceptibility values were made along three beach transects at 5 m intervals using a Bartington MS1 susceptibility meter and a surface loop sensor.

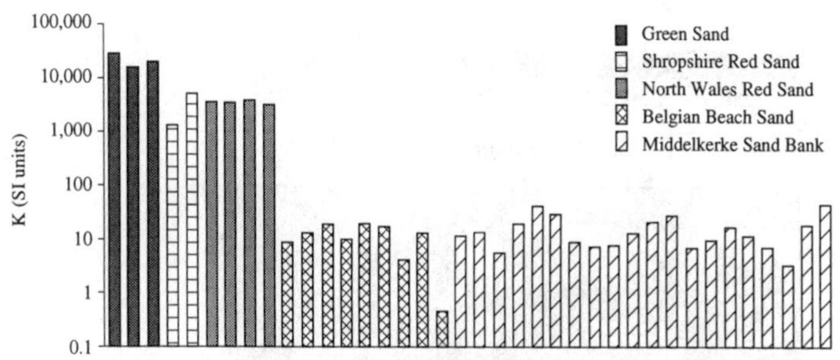

Figure 6. Comparison of enhanced tracer sands and background samples

Approximately 100 kg of tracer material was placed in a pit 0.5 x 1.5 x 0.06 m. The sand was wetted with a solution of local sea water and detergent. The tracer was then compacted to replicate as well as possible the surrounding conditions. A 12 x 12m grid was constructed around the pit (this was later expanded) with theodolite coordinates being noted for the centre of the pit and the four corners of the grid. At each subsequent low tide these coordinates were relocated using a theodolite and staff and the grid was reconstructed. This method was adopted to prevent any alterations to the hydrodynamics of the tracer grid area that may be caused from the use of fixed markers. K measurements were taken from the centre of each grid square using a Bartington MS1 Susceptibility Meter and Surface Loop Sensor. The tracer was emplaced at low tide on Saturday morning May 1993 and monitored every low tide until Monday afternoon 24th May. On Sunday morning 23rd May, surface scrape samples were taken from each grid square in the grid using a trowel, then placed in sealed polythene bags to await analysis back in the laboratory. Susceptibility values were measured using a Bartington MS1 Meter and a Dual Frequency Sensor. IRM values were measured using a Molspin Pulse Magnetiser and a Molspin Fluxgate Magnetometer. ARM values were measured using Molspin Demagnetiser and a Molspin Fluxgate Magnetometer.

RESULTS

Figure 7 shows the surface K values, once they have been converted to logarithms and used to construct contour maps. Figure 7a shows two areas with high K values with the first lying shoreward Southerly from the tracer pit and the other area being shoreward South Westerly. Whilst there is a small zone of high K values lying on the North Easterly side of the tracer pit in both contour maps, the largest area of high K values in Figure 7b is shoreward South Westerly.

Figure 8 shows the results of laboratory measurements on 10 ml scrape subsamples. From there can be seen two areas of high K values at Grid Squares: a (x=3-4; y=7-8 m) and b (x=5-6; y=8-9 m), both lying North Westerly of the tracer pit.

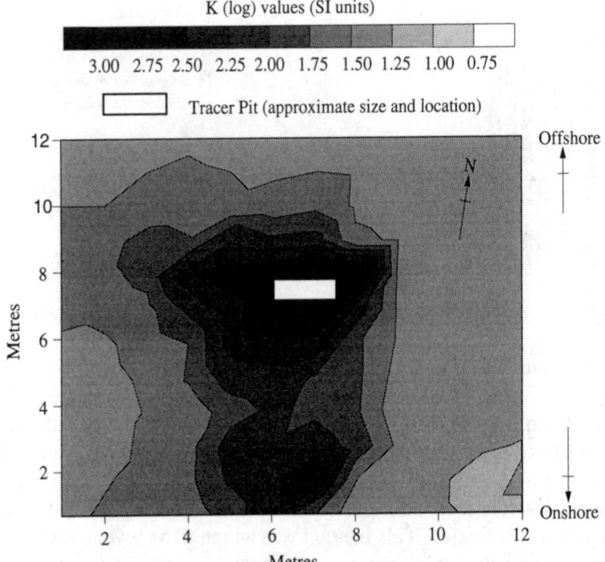

Figure 7a. Magnetic tracing of the Belgium coast Saturday 22nd May 1993 PM

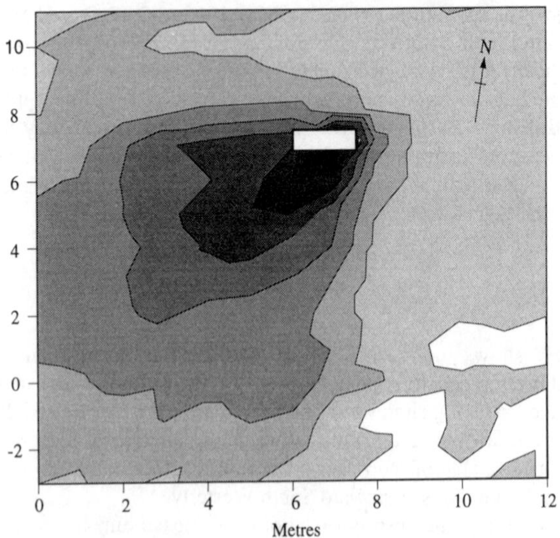

Figure 7b. Magnetic tracing of the Belgium coast Sunday 23rd May 1993 PM

After three low tides a core was taken from the centre of the tracer pit which showed that there was approximately 3.2 cm of sand deposition and additionally at the end of the tracing experiment the pit was re-excavated showing that after three further low tides, there was still approximately 3.2 cm of sand deposition.

Figure 9 summarises in simplified form, the results of additional measurements on the laboratory subsamples. Clearly there is a contrast between tracer derived material and background dominated material.

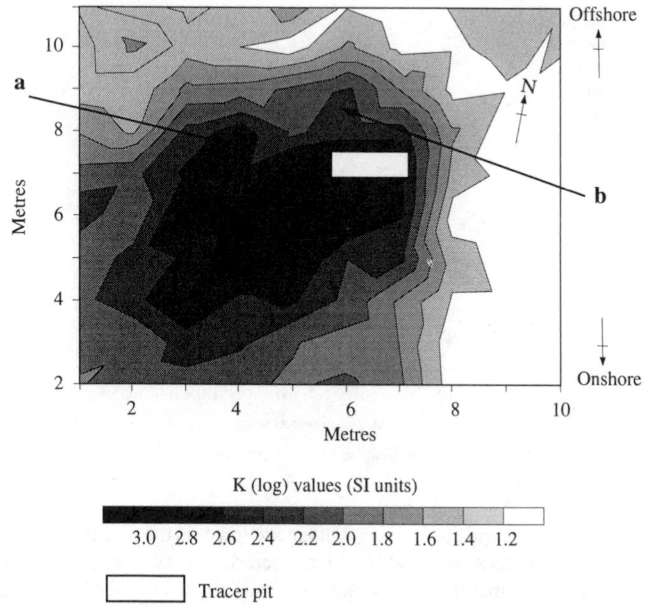

Figure 8. K (log) values from scrape samples confirming the directional trends supplied by field measurements

DISCUSSION

The contour maps in Figure 7 show the dominant direction for sand movement. Whilst there has been minor seaward movement in a North Easterly direction, the majority of the mobilised tracer material has been transported and deposited in a South Westerly direction. Figure 7a does also show that there has been movement in a North North West direction reflecting the effects of the ebbing tide. The shift in direction from a Southerly to a South Westerly can be explained by the change in wave direction with wave angle becoming more oblique by the third low tide (CSTAB unpublished data).

The results of the field measurements have been confirmed by laboratory measurements of magnetic susceptibility on 10ml subsamples taken from the scrape samples (Figure 8). The two North Westerly zones of high concentration in a and b may suggest greater transportation by ebb currents than has been shown in Figure 7b.

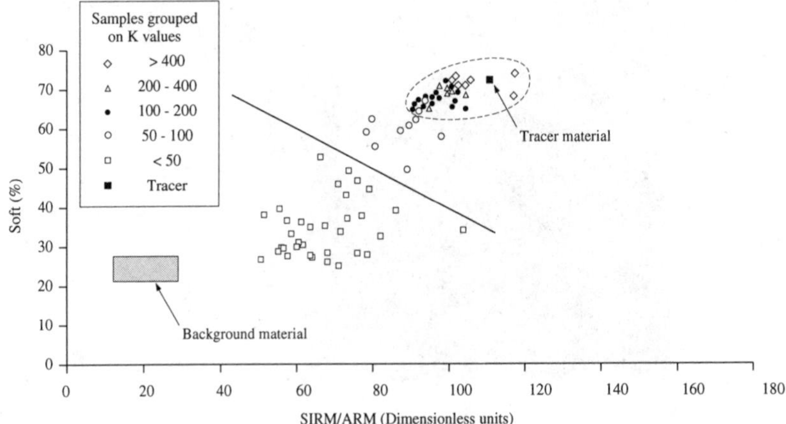

Figure 9. Discrimination and verification of tracer material

An important prerequisite in using a magnetic tracer is the ability to confirm or disregard anomalous values detected as tracer material or extraneous background material. Figure 9 shows that this can be done successfully using the aforementioned parameters. The two properties used in the bivariate scatter plot are quotients derived from IRM and ARM values. These properties are therefore normalised and independent of simple concentration changes alone. The values plotted reflect concentration changes only in so far as they represent a graded series of mixtures between tracer material with its distinctive combination of normalised properties and the 'host' sand. The heat enhanced tracer material has distinctive magnetic properties reflecting the fine magnetic grain sizes involved. By contrast the natural magnetic mineral assemblages have 'signatures' reflecting relatively coarser grains (NB- magnetic grain size does not reflect particle size). The symbols used in Figure 9 are differentiated to reflect the range of magnetic susceptibility values. Samples which exceed 100 (10^{-5} SI Units) form a relatively narrow envelope dominated by 'tracer'- derived properties, whereas values below 100 (10^{-5} SI Units) show a gradation between the 'tracer' and 'background' properties. All this accords with expectations and confirms that the samples recovered reflect the movement of the tracer material. A sole exception in the form of an outlier on the graph suggests that contamination by magnetic non-tracer material can be readily detected using this approach. This is an important consideration where, as in a second tracing experiment in a higher energy environment in Southern Portugal (van der Post and Oldfield unpublished data), there was no continuity of high K values between the site of emplacement and the peak values revealed by subsequent surface scan surveys.

The effect of deposition on the tracer pit can be seen by the diminished values in Figure 7b and shows that a major attribute in using a magnetic tracer is that it can still be detected at depth. It is assumed from the core and excavation data that only approximately 50kg of tracer material was mobilised during this experiment. This does suggest that there has been effective sealing of the tracer for the greater part of this study, and may partially explain why the directional information does not accord with that of De Moor (1991) based on larger scale and longer term analyses. It therefore becomes doubtful whether the short term movement recorded at the site chosen for this first field experiment is representative of the broader scale of movements along the shoreline as a whole.

CONCLUSIONS

Magnetic tracers can provide rapid information on the direction of sand movement in beach environments. Measurements can be made on site using a surface loop sensor linked to a portable susceptibility meter. In the present experiment, subsequent laboratory measurements confirmed (i) the directional data obtained from the surface scans, (ii) the field detection of tracer buried beneath 3 - 4 cm of sand and (iii) the distinctive properties of the tracer sands compared with the host material. Further development of the strategies used in this, the first experiment of its kind, can be made by adapting the field methods for detecting buried material in the field and adding automatic logging to the survey techniques.

ACKNOWLEDGMENTS

Thanks to J. Bloemendal, J. Dix, E.H. Hull, D. Huntley, R. Jude, J. Lankneus, S.E. Mather, A.J. Plater, D. Simmonds, R. Wake, M. Wilkins and S. Yee. This project is being funded by the EC under the MAST2/CSTAB project (MAS2-CT92-0024C).

REFERENCES

Arkell, B 1984. Magnetic tracing of river bedload. Unpublished PhD thesis. University of Liverpool.

Arkell, B., Leeks, G., Newson, M., and Oldfield, F. 1982. Trapping and tracing: some recent observations of supply and transport of coarse sediment from upland Wales. Spec. Publ. Int. Assoc. Sediment. 6, pp 117-129.

De Moor, G., 1991. The beach nourishment of Bredene-De Haan and its impact on the beach morphology and the coastal evolution of the Belgian Coast East of Ostend. Proceedings IGU Symposium "Coastal Protection". Nantes, October 1991.

De Moor, G and Bloome, E. 1988. Belgium. In "Artificial Structures and Shorelines", pp 115-126. Kluwer Academic Publ.

Le Borgne, E. 1955 Susceptibilite magnetique anormale du sol superficiel. Annales de Geophysique. 11, pp 399-419.

Mullins, C.E. 1977 Magnetic susceptibility of the soil and its significance in Soil Science: a review. Journal of Soil Science 28, pp 223-246.

Oldfield, F., Rummery, T.A., Thompson, R. and Walling, D.E. 1979. Identification of suspended sediment sources by means of magnetic measurements: some preliminary results. Water Resources Research. 15, pp 211-218.

Oldfield, F., Thompson, R., and Dickson, D.P.E. 1981. Artificial enhancement of stream bedload: a hydrological application of superparamagnetism. Physics of the Earth and Planetary Interiors. 26, pp 107-124.

O'Reilly, W. 1984. Rock and mineral magnetism. Glasgow: Blackie.

Rummery, T.A., Bloemendal, J., Dearing, J., Oldfield, F., and Thompson, R. 1979a. The persistence of fire induced magnetic oxides in soils and lake sediments. Annales de Geophysique. 35, pp 103-107.

Rummery, T.A., Oldfield, F., Thompson, R. and Newson, M. 1979b. Magnetic tracing of stream bedload. Geophysical Journal of the Royal Astrological Society. 57, pp 278-279.

Thompson, R. and Oldfield, F. 1986. Environmental Magnetism. London: Unwin and Allen.

Walling, D.E., Peart, M.R., Oldfield, F., and Thompson, R. 1979. Suspended sediment sources identified by magnetic measurements. Nature 281, pp 110-113.

Yu, Lizong and Oldfield, F. 1989. A multivariate mixing model for identifying sediment source from magnetic measurements. Quaternary Research 32, pp 168-181.

Yu, Lizong and Oldfield, F. 1993. Quantitative sediment source ascription using magnetic measurements in a reservoir-catchment system near Nijar, S.E. Spain. Earth Surface Processes and Landforms, 18, pp 441-454.

THE CONTINENTAL SHELF: A SOURCE FOR NATURALLY-DELIVERED BEACH SAND

Robert H. Osborne, Kelly A. Ahlschwede, Sean D. Broadhead, Kyung Cho, Joshua R. Feffer, Arthur C. Lee, Jinyou Liu, Craig Magnusen, Janette M. Murrillo de Nava, Rory A. Robinson, Chia-Chen Yeh and Yi Lu[1]

ABSTRACT: Fourier shape analysis was performed on detrital quartz grains from the medium sand fraction (0.25 to 0.50 mm) of 608 upper foreshore samples collected from November 1956 to April 1992 along the southern California coast from Point Conception to the United States-Mexico border, as well as an additional 52 samples collected from October 1990 to July 1991 along the Creciente Island barrier complex, Baja California Sur, Mexico. The grain-shape composition in the Ventura Harbor area is overwhelmingly fluvially dominated to a depth of at least -17 m MLLW. No significant change was observed from June 1969 through March 1993. The Dockweiler-El Segundo area is starved with respect to terrestrial sand input. A nine percent decrease (63 to 54 percent) in rougher and more elongate shelf-derived sand from November 1956 to January 8, 1988 is attributed to the placement of mostly fluvially-derived dredge spoil from Marina del Rey. A 22 percent increase (54 to 76 percent) in the amount of shelf sand occurred from January 8 to April 21, 1988, which is attributed to the great storm of January 17-18, 1988. The percentage of shelf sand increased to 88 percent through October 1988, then decreased to its pre-storm percentage (54 percent) by April 1992. In the Oceanside-Carlsbad area, a 14 percent increase in the amount of shelf-derived sand occurred between November 1956 and August 1983, and remained almost constant (37 to 41 percent) through April 1992. The observed increase in August 1983 and consistency thereafter is attributed to continued beach nourishment using spoil mostly derived from Oceanside Harbor and longshore transport. A 16 percent increase observed in October 1991 is attributed to operation of the Oceanside Experimental Sand Bypass System, which also uses spoil from Oceanside Harbor.

---

[1]Department of Earth Sciences, University of Southern California, Los Angeles, CA 90089-0740

Analysis of regional sample sets collected in November 1956 and October 1991 from Point Conception to the United States-Mexico Border indicates that the amount of shelf-derived sand remains much the same (42 and 43 percent, respectively). Therefore, on a regional basis, the contribution of inner shelf-derived medium quartz sand essentially is equal to the amount of river- and cliff-derived sand delivered to the shore zone during this 35 year period.

INTRODUCTION

When considering the budget of littoral sediment, one of the most difficult components to quantify and therefore one of the most likely sources of error concerns the character of the cross-shore transport vector. In most analyses, net onshore or offshore sand transport at the seaward boundary of the littoral zone is either inferred when estimates for all other elements in the budget have been made, or net transport at this boundary is assumed negligible. Neither of these approaches generates great confidence in the accuracy of the resultant sediment budget. This problem exists because the exact character of the forcing agents, sediment transport pathways and transport rates associated with the cross-shore vector are poorly understood.

Shepard (1973) recognized the likely importance of the inner continental shelf as a source for littoral sand when he stated "The direct source of almost all beach sand is the shallow sea floor, although previously the bulk of the shelf sand was carried into the ocean by runoff from the land ... Beaches may also receive their supply from sands that were deposited on the continental shelf during Pleistocene stages of low sea level when the shelves were largely dry land. Hence, here the terrestrial sources may be remote in time."

Along the low-gradient Atlantic and Gulf coasts, McMaster (1954, New Jersey), van Andel and Poole (1960, Texas Gulf Coast), Saville (1960, Long Island), Curray (1960, northwest Gulf of Mexico), Giles and Pilkey (1965, southeastern United States) and Pierce (1968, Cape Hatteras) all concluded that much of the beach sand was derived from the erosion of older strata associated with the adjacent continental shelves. Along the higher-gradient Pacific coast, Trask (1952, Santa Barbara), Bowen and Inman (1966, Sebastian Vizcaino Bay), Inman and others (1966, Point Arguello), Clemens and Komar (1988, Cape Blanco), and Osborne and Yeh (1991, Oceanside) all have either suggested or documented net onshore sand transport from the shelf to the adjacent littoral zone. For the most part, these studies have been qualitative; however, Pierce (1968) estimated that about 42 percent of the annual sand input along a 100-km coastal reach from Hatteras Inlet to Cape Lookout was derived from the adjacent shelf. Osborne and Yeh (1991) report that an average of 36 percent of the foreshore sand in the Oceanside area is shelf derived. The Clemens and Komar (1988) model may be considered one spectral end member where no new shelf sand is currently transported to the beach, whereas Pierce's model suggests more continuous additions of shelf sand in response to long-period waves and/or storm-generated currents.

The purpose of this paper is to present preliminary Fourier grain-shape results concerning the character of sand exchange between the upper foreshore and the inner continental shelf along the southern California bight (Fig. 1) and Creciente Island, Baja California Sur, Mexico.

PROCEDURES AND RESULTS

Fourier Grain-shape Analysis

Although mineralogic data is useful to identify the types of ultimate crystalline sand sources (e.g. granite, granodiorite, gneiss, etc.) and to trace associated sediment transport pathways, differences in grain size, specific gravity and grain shape among minerals derived from even one type of source rock make it difficult to quantify such information to provide accurate estimates concerning the rate of supply from local sand sources (rivers, sea cliffs, dunes, etc.):

Fourier grain-shape analysis (Schwarz and Shane, 1969; Ehrlich and Weinberg, 1970) has proven to be most useful in identifying the amount of sand derived from a given local source. Furthermore, the analysis of sample sets collected at different times permits estimates of at least the average relative rate of supply from each local source. Ehrlich and Weinberg (1970) described a closed-form Fourier method to examine the observed variation of two-dimensional, maximum projection, grain-shape area. Such shape analysis usually is performed with detrital quartz grains of a restricted size range, therefore variations in grain size and specific gravity are largely eliminated, leaving grain shape as the primary variable of interest. Sample preparation and the Fourier procedure used are discussed at length in Osborne and Yeh (1991) and Osborne and others (1993). For the interested reader, these two papers also contain a large number of references concerning sedimentologic applications of Fourier grain-shape data.

Sampling

Fourier grain-shape analysis has been completed on the medium sand fraction (0.25 to 0.50 mm) of at least 200 grains for each of 608 upper foreshore samples collected along the coast and inner shelf from Point Conception to the U.S.-Mexico border (Fig. 1), and an additional 52 samples in the Creciente Island area of Baja California Sur, Mexico (latitudes 24°15' to 24°25' North; longitudes 111°30' to 111°45' West) (Murrillo de Nava, 1993). Shape analysis was performed on three regional sample sets collected at 60 stations from Point Conception to the United States-Mexico border during November 1956, October 1991 and April 1992. The November 1956 sample set was collected by K. O. Emery, and the stations shown in Figure 1 correspond to those in Figures 162 and 163 of Emery (1960). Additional, more temporally and geographically closely-spaced sample sets were collected in the Ventura Harbor area (Fig. 1, E14.5 to E18; nine sets from June 1969 through March 1992), Dockweiler-El Segundo beach area (Fig. 1, E27 to E28; eight sets from January 1988 through April 1992) and the Oceanside-Carlsbad area (Fig. 1, E45 to E46; 12 sets from August 1983 through April 1992).

Results

The percentages of shelf-derived sand as determined by grain-shape analysis for the Dockweiler-El Segundo and Oceanside-Carlsbad areas are presented in Figures 2 and 3, respectively. The results for the regional sample sets from Santa

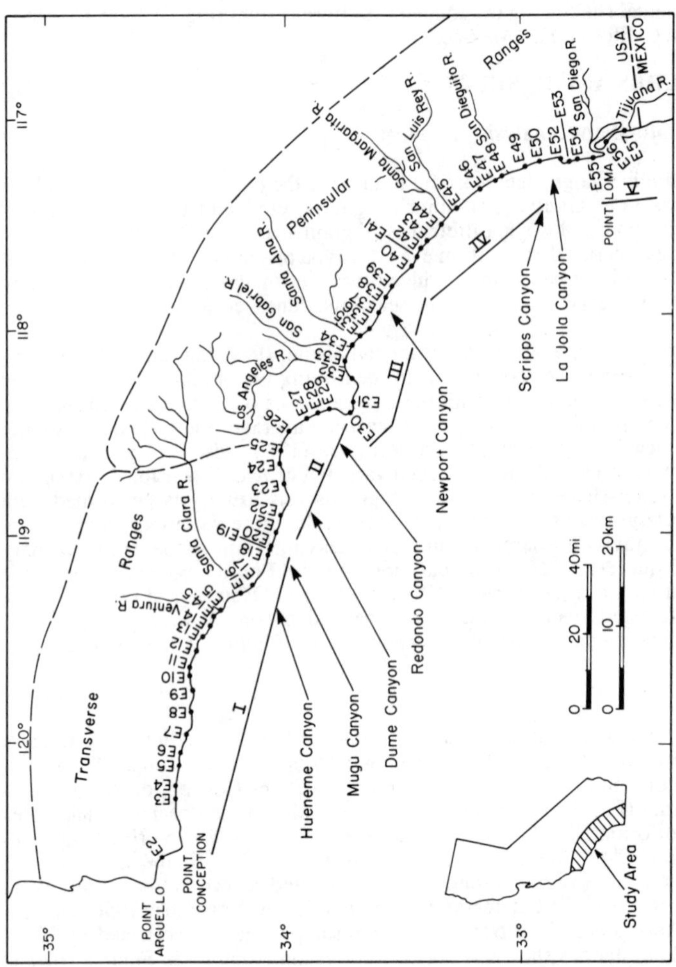

Figure 1. Map showing the locations of foreshore (0 m MLLW) sample stations E2 through E57 along the southern California bight. Roman numerals I through V represent the Santa Barbara, Santa Monica, San Pedro, Oceanside and Silver Strand Littoral Cells, respectively.

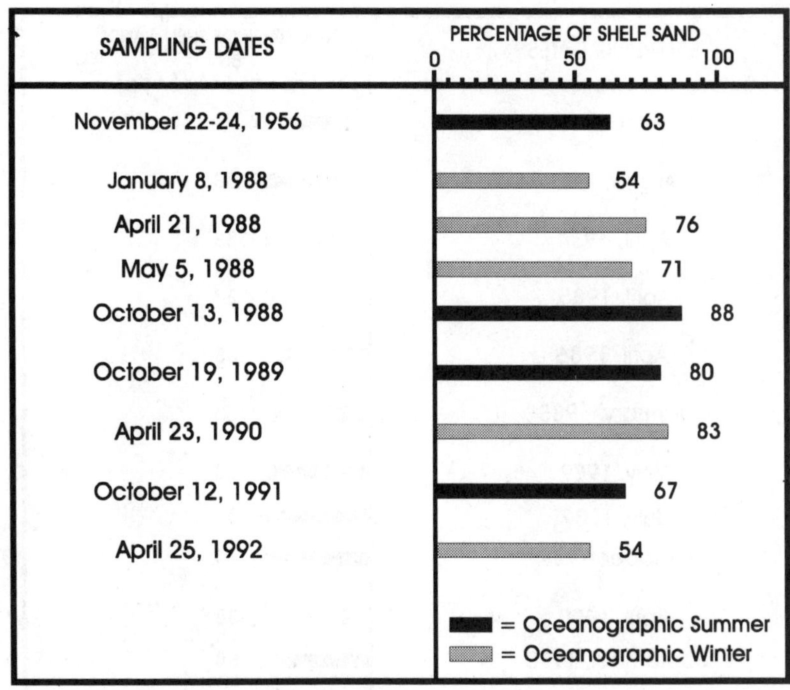

Figure 2. Percentages of inner shelf-derived sand at Dockweiler-El Segundo beaches, Santa Monica Bay, California from November 1956 to April 1992.

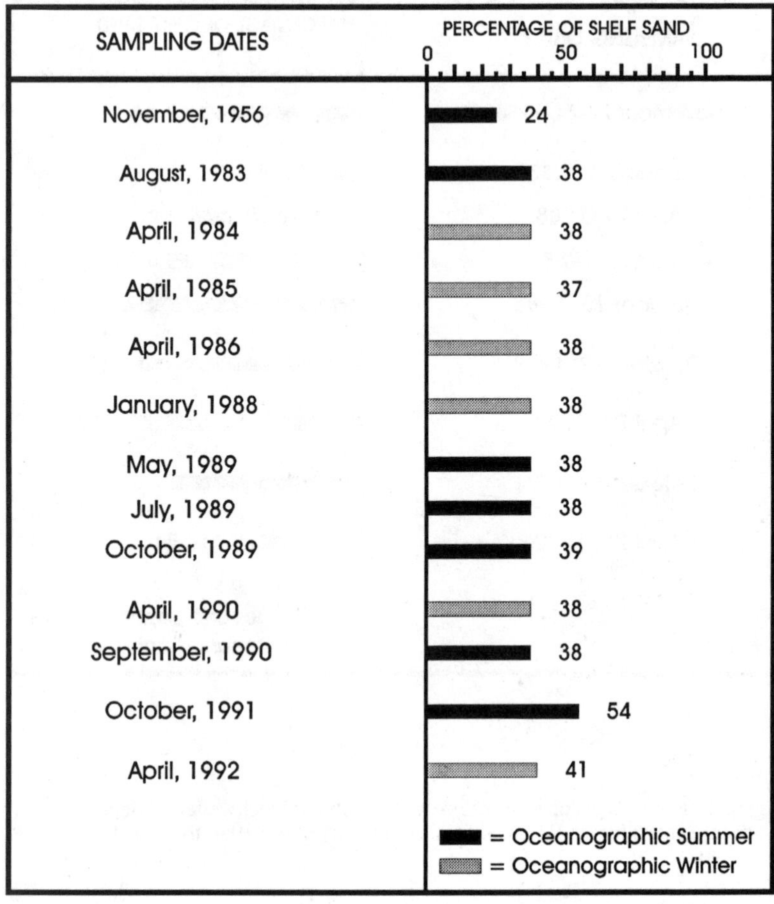

Figure 3. Percentages of inner shelf derived sand at Oceanside-Carlsbad beaches, San Diego County from November 1956 to April 1992.

Monica Bay to the United States-Mexico border (Fig. 1, E23 through E57) and Creciente Island are shown in Figure 4.

DISCUSSION OF RESULTS

Ventura Area

The most recent and comprehensive sediment budget analysis of the Santa Barbara and Ventura County coastline was performed by Noble Consultants, Incorporated (1989). The Santa Clara and Ventura Rivers occur along the northern part of the Oxnard Plain, and are the only rivers in this area known to contribute appreciable volumes of sand to local beaches (Kolpack, 1986; Noble Consultants, Incorporated, 1989). The high sediment yield principally is due to the facts that these rivers mostly drain relatively unresistant sedimentary strata exposed in the Santa Ynez Mountains, and they are among the least controlled rivers in southern California. The Santa Clara River drainage area is 4,219 km^2 of which only 36.5 percent is controlled, and the Ventura River drainage area is 585 km^2 of which 41.5 percent is controlled (Brownlie and Taylor, 1981).

Robinson (1993) discusses the results of Fourier grain-shape analysis in the eastern and central Santa Barbara Littoral Cell, with special reference to the greater Ventura Harbor area (Fig. 1, E14.5 to E18). Sand samples from the Santa Clara and Ventura Rivers are statistically homogeneous with respect to quartz grain shape composition. Episodic input from these two rivers overwhelms the adjacent beaches and inner shelf to a depth of at least -17 m MLLW, which is the limiting depth of sampling. No significant change in grain-shape composition of upper shoreface (0 m MLLW) samples occurs within nine sample sets collected during the 23-year period from June 1969 through March 1992. For the period studied, the greater Ventura area represents one sedimentologic end member among the coastal and inner shelf settings of southern California, namely a fluvially-dominated system.

Dockweiler-El Segundo Area

The beaches along central and southern Santa Monica Bay (Fig. 1, E27 to E29) were supplied principally by the Los Angeles River during late Pleistocene and early Holocene time. The absence of significant channeling in the Ballona Gap area (Fig. 1, E27) during middle and late Holocene time led Poland and others (1959) to conclude that the Los Angeles River drained southward into San Pedro Bay during much of the Holocene. Josselyn and Chamberlain (1993) report that when Portola's expedition arrived in 1769, the Los Angeles River meandered through an extensive network of swamps, lakes and marshes before entering Santa Monica Bay. Thereafter the Los Angeles River must have reentered San Pedro Bay, because in 1815, flooding caused the river to change its course and flow westward where it joined Ballona Creek and again emptied into Santa Monica Bay. It maintained this course until 1825, when flooding forced the river to establish a well-defined channel draining south into San Pedro Bay. Intense flooding in 1862 and 1864 may have caused additional discharge through Ballona Creek, but this has not yet been verified in historical records.

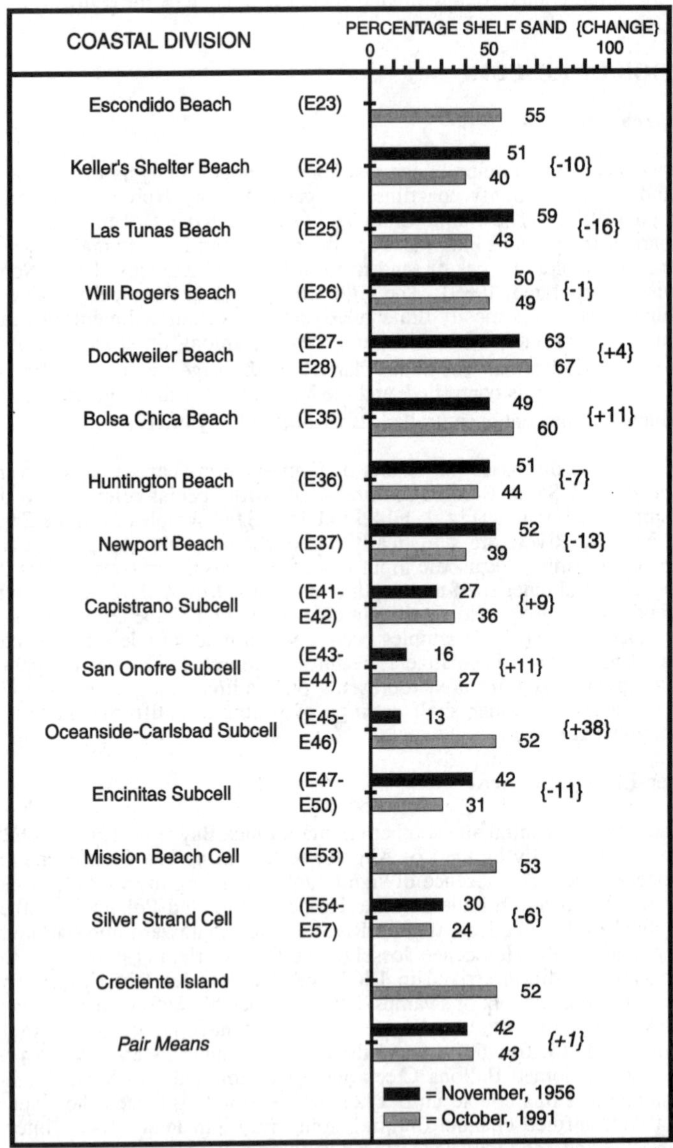

Figure 4. Percentages of inner shelf-derived sand at coastal divisions from Santa Monica Bay to the United States-Mexico border and Creciente Island, Baja California Sur. The locations of sample stations E23 through E57 are shown on Figure 1.

With the loss of sediment from the Los Angeles River, streams draining the Malibu Creek (282 km^2) and Ballona Creek (239 km^2) watersheds in the Santa Monica Mountains became the principal sediment sources for Santa Monica Bay (Lu, 1992). Flood control structures were built in the Malibu Creek watershed from 1881 to 1925, whereas Ballona Creek flowed uninterrupted until 1938. Even with flood control structures in place, these streams, ordinarily of negligible size, produced coarse-grained deltaic deposits following major runoff events (Shepard and MacDonald, 1938). These deposits were incorporated into downcoast beach and perhaps adjacent coastal dune deposits following marine reworking of the deltaic lobes.

Prior to 1871, there were no engineered structures extending into Santa Monica Bay. Comparison of the U.S. Coast & Geodetic Surveys of 1876 and 1933 indicates an average shoreline retreat rate of 0.27 m/year (Page, 1950). From 1925 to 1941, Dockweiler and El Segundo beaches receded to the base of the Hyperion coastal dune field. Such shoreline retreat reflected, in part, the interruption of downcoast longshore transport by a 235 m rubble mound jetty constructed at the mouth of Ballona Creek in 1938 as well as the occurrences of intense regional ocean storms in 1939 and 1941. However, the most important reason for such shoreline retreat was the loss of a copious supply of coarse-grained terrigeneous sediment dating back to middle-to-late Holocene time, when the Los Angeles River first diverted to San Pedro Bay. Erosional rates must have increased even more following the reduction of sediment from the Malibu Creek and Ballona Creek watersheds as well as the introduction of upcoast engineered structures. From a geologic perspective, Santa Monica Bay is in the initial stage of a sediment-starved basin, and, as such, may be considered a sedimentologic end member exactly opposite of the fluvially-dominated condition present in the greater Ventura area.

As pointed out by Woodell and Hollar (1991), Coastal Frontiers (1992), Flick (1993) and Wiegel (1994), the widths of the present beaches in the Dockweiler-El Segundo area are largely due to artificial nourishment. Such projects have placed nearly 20.2 x 10^6 m^3 of sediment in this area since 1938. Although approximately 8.2 x 10^6 m^3 of sand derived from the construction of Marina del Rey was placed on northern Dockweiler Beach from 1960-62 to 1987, the shoreline retreat rate for this 25-year period has averaged approximately 0.59 m/year, which is about 2.2 times that observed for the period from 1876 to 1933.

Due to the paucity of coarse-grained terrigeneous influx, the lack of substantive natural dune erosion, and the availability of sound information concerning the sediment sources, placement locations, placement times and sediment volumes involved with numerous beach nourishment projects, the Dockweiler-El Segundo area is an excellent site to monitor changes in the percentage of inner shelf sand present in upper foreshore sand samples. Figure 2 shows that the percentage of rougher and more elongate inner shelf sand decreased from 63 to 54 from November 1956 to January 8, 1988 (Lee, 1993). This decrease most likely reflects the placement of approximately 7.7 x 10^6 m^3 of dominantly fluvial sand dredged from Marina del Rey from 1960 to 1963. Following the 100- to 200-year storm of January 17-18, 1988 (Egense, 1989; Seymour, 1989a), there was a 22 percent increase in shelf-derived sand by April 21, 1988, and a 34 percent increase by October 13, 1988. Such enrichment continued through April 23,

1990. Then associated foreshore strata were buried and/or diluted by longshore drift or were eroded to expose older backshore sediment by April 25, 1992, when the percentage of inner shelf sand in upper foreshore samples again equaled that of January 8, 1988.

Refraction analysis using deep-water wave records indicates that the storm of January 17-18, 1988 produced significant wave heights of 7.6 to 8.8 m in 9.1 m of water and breaking waves farther offshore (Egense, 1989). The observed increase in the abundance of inner shelf grains continued for a period of about two years (at least from April 21, 1988 through April 23, 1990) following this extraordinary storm, and its effect on the grain-shape composition of upper foreshore samples persisted at least 3.5 years (until at least October 12, 1991). Dayton and others (1989) report that during the storm of January 17-18, 1988, two concrete-filled scuba tanks were transported from a water depth of -12 m to a point inshore of the -8 m -- a distance of at least 0.5 km from their original location. Likewise reef-derived boulders were transported from a depth of -15 m as much as 10 m shoreward, and a rock reef projecting about 2.5 m above the substrate at a depth of -18 m was reduced to rubble and much of this debris was projected shoreward. There is ample evidence of landward transport of even boulder-size sediment at inner shelf depths during intense ocean storms.

Oceanside-Carlsbad Area

The 86-km long Oceanside Littoral Cell extends from Dana Point southeast to Point La Jolla (Fig. 1). Flick (1993) reports estimates for the average total actual sand input from rivers as ranging from 112,000 to 203,000 m^3/year for the entire Oceanside Cell. Depending upon which of these fluvial estimates is more accurate, the ratio of annual coarse-grained river input/annual nourishment input ranges from 0.59 to 1.07. Therefore the Oceanside Cell as a whole may be considered intermediate between the Ventura and Dockweiler-El Segundo end members with respect to the relative importance of fluvial sand.

The Oceanside coastal reach has experienced persistent shoreline erosion and suffered millions of dollars in damage and clean-up costs following severe storms that occurred during the winter of 1982-83 and January 17-18, 1988 (Armstrong and Flick, 1989; Dayton and others, 1989; Seymour, 1989b). In addition to such natural disasters, the harbor facility at Oceanside altered the longshore transport system since its inception in 1942 (Hales, 1978). Harbor construction resulted in severe and prolonged erosion of the downcoast beaches, which required massive and somewhat continuous nourishment. Graves (O'Hare and Graves, 1991) provides an informative discussion concerning the scientific, engineering, legal and political issues associated with the development of Oceanside Harbor as well as maintenance of the downcoast beaches. As part of the Coast of California Storm and Tidal Waves Study, the Los Angeles District of the U.S. Army Corps of Engineers published the State of the Coast Report for the San Diego Region in 1991. The report summarizes much pertinent information dealing with the theOceanside Littoral Cell, as well as the Mission Beach and Silver Strand Cells farther south. Osborne and Yeh (1991) report the results of a Fourier grain-shape analysis based on April and October 1986 sample sets, which were collected throughout the Oceanside Littoral Cell.

The percentages of inner shelf-derived sand in the Oceanside-Carlsbad beach area (Fig. 1, E45 and E46) from November 1956 through April 1992 are shown in Figure 3. Inasmuch as these beaches are located downcoast of Oceanside Harbor, they largely owe their existence to numerous beach nourishment projects mostly using dredge spoil from the entrance channel at Oceanside Harbor (Flick, 1993). As in Ventura Harbor (Robinson, 1993), the sand grains in the Oceanside Harbor entrance channel are rougher and equally or more elongate than the beach sand in this area, and such grains are believed to be of inner-shelf derivation. Figure 3 shows that the percentage of shelf sand increased from 24 to 38 percent from November 1956 to August 1983; and, except for October 1991, remained almost constant (37 to 39 percent) through April 1992. The 14 percent increase from November 1956 to August 1983 reflects the placement of approximately 5.3 x 10^6 m^3 of dredge spoil on Oceanside Beach starting in 1962. A total of at least another 0.75 x 10^6 m^3 was placed during 1983, 1986 and from January to March 1992. Therefore the 14 percent increase from November 1956 to August 1983 as well as the observed consistency thereafter may be attributed to continued beach nourishment, where uniquely-sourced harbor sediment, mostly silt containing shelf-derived, more rough and elongate sand grains, frequently was placed on adjacent downcoast beaches coupled with a net downcoast longshore transport rate of approximately 165,154 m^3/year (Los Angeles District, U.S. Army Corps of Engineers, 1991). The occurrence of proportionately more rough and elongate sand grains in the entrance channel most likely represents selective onshore transport from inner shelf deposits. The 16 percent increase in inner shelf sand from September 1990 to October 1991 is somewhat enigmatic, but most likely represents operation of the Oceanside Experimental Sand Bypass System, which also derives its sand from the entrance channel at Oceanside Harbor.

Regional Sample Sets (Point Conception to United States-Mexico Border)

Inner shelf samples for the coastal reach north of Point Dume (Fig. 1, E22.5) are not available at this time, therefore sourcing in this area has not been completed. Results for the November 1956 and October 1991 sets (end of oceanographic summer) for the shorezone from Santa Monica Bay south are summarized in Figure 4. Coastal areas with a negative value for percentage change in Figure 4, e.g. -10 at Keller's Shelter Beach, represent localities where the terrigenous sediment yield (river discharge and/or sea cliff erosion) exceeded the inner shelf yield during this period. Positive values for percentage change indicate that the inner shelf yield exceeded terrigenous yield for this period. Space does not permit a detailed discussion of the coastal setting and sedimentologic history of each area shown in Figure 4. In general, however, the negative values occur in areas associated with significant river and/or active cliff sources, whereas positive values occur in areas lacking significant terrigenous influx or in artificially nourished areas.

For the November 1956 sample set, the amount of inner shelf-derived sand in upper foreshore samples ranged from 13 to 63 percent, and averaged 42 percent. For the October 1991 set, the amount of shelf sand ranged from 24 to 67 percent, and averaged 43 percent. There was a net increase of one percent in the relative amount of inner shelf sand. This value indicates that the inner shelf contribution to the upper foreshore sand along the southern California coast equaled that of terrigenous sources during this 35-year period. The overall similarity of pair means summed over this coastal reach in part reflects the reduction of terrigenous input due to the presence of one or more flood control structures on each of the major rivers. As the supply of channel sand below such structures continues to decrease, the relative percentage of inner shelf sand supplied to the littoral zone should increase. Such increases also may have occurred during the climatically benign periods in southern California from 1892 to 1934 and from 1945 to 1977 (Emery and Kuhn, 1982).

The selective onshore transport of elongate and rough sand grains requires the existence of associated inner shelf lag deposits enriched in more equant and smoother quartz grains. Three such deposits have been located seaward of the -12, -17 and -20 m MLLW isobaths offshore of the Surfside-Newport Beach area, the Oceanside-Carlsbad area, and Creciente Island, respectively. One can properly question whether such inner shelf deposits are enriched in more equant and smoother grains because of the derivation and seaward transport of such grains from the upper foreshore, which may reflect the greater pivotability of more spherical grains; or whether more elongate and rougher grains are selectively transported from inner shelf sources and deposited in the upper foreshore as suggested in this paper. We favor the second alternative for the following three reasons. More elongate and rough grains are more abundant in upper foreshore samples at the end of oceanographic summer (October) when ridge and runnel systems have accreted onto the shoreface. This relationship has been observed at the Dockweiler-El Segundo, Oceanside-Carlsbad and Surfside-Newport Beach areas. At Newport Beach, upper foreshore samples have been enriched in more elongate and rougher grains following the placement of dredge spoil from the mouth of the Santa Ana River at a depth of -3 m MLLW. Furthermore, the Holocene dune ridge adjacent to Dockweiler-El Segundo Beach is enriched in more elongate and slightly smoother grains than the known potential sources (Pre-Flandrian dunes, Los Angeles River, and smaller rivers draining the Malibu and Ballona Creek watersheds). There is little question concerning the net transport direction in these cases, and it appears that grain form (sphericity) is the dominant shape factor controlling both selective entrainment and selective deposition of medium-grained quartz sand.

CONCLUSIONS

The results presented in this paper document the role of a large ocean storm and of beach nourishment in supplying inner continental shelf sand to the shoreface as well as the regional addition of shelf sand equal in amount to that of terrigenous sources during the 35-year period from 1956 to 1991. Although episodic, regional ocean storms which occurred in southern California from 1978 to 1983 as well as 1988 probably contributed to the observed increase in the amount of inner shelf sand during this period, a more continuous onshore transport mechanism such as that described by Dean (1987) also may have been involved, but cannot be demonstrated with the available data. Although considerable research is required to document the dynamics of sediment exchange between the shoreface and inner shelf, the obtained results indicate that the inner shelf should not be ignored in sediment budget analysis, especially in areas with relatively low terrestrial sand yields.

ACKNOWLEDGMENTS

The authors wish to thank Abraham Golik and Malcolm Webster for reviewing a draft of this paper. Sue Turnbow typed the manuscript, and Janet Dodds drafted Figure 1.

REFERENCES

Armstrong, G.A., and Flick, R.E., 1989. Storm damage assessment for the January 1988 storm along the southern California shoreline, *Shore & Beach,* v. 57, No. 4, p. 18-23.

Bowen, A.J., and Inman, D.L., 1966. Budget of littoral sands in the vicinity of Point Arguello, Coastal Engineering Research Center, U.S. Army Corps of Engineers, Technical Memorandum, No. 19, 41 p.

Brownlie, W.R., and Taylor, B.D., 1981. Sediment management for southern California mountains, coastal plains and shoreline, California Institute of Technology, Pasadena, California EQL Report No. 17-C, p. C26-C110.

Clemens, K.E., and Komar, P.D., 1988. Oregon beach-sand compositions produced by mixing of sediments under a transgressing sea, *Journal of Sedimentary Petrology,* Vol. 59, p. 519-529.

Coastal Frontiers, 1992. Historical changes in the beaches of Los Angeles County, Malaga Cove to Topanga Canyon, 1935-1990, County of Los Angeles, Department of Beaches and Harbors, 109 p.

Curray, J.R., 1960. Sediments and history of Holocene Transgression, continental shelf, northwest Gulf of Mexico, *in* Shepard, F.P., Phleger, F.B., and van Andel, T.H., eds., Recent sediments, northwest Gulf of Mexico, *American Association of Petroleum Geologists*, Tulsa, Oklahoma, p. 221-266.

Dayton, P.K., Seymour, R.J., Parnell, P.E., and Tegner, M.J., 1989. Unusual marine erosion in San Diego County from a single storm, *Estuarine, Coastal and Shelf Science,* Vol. 29, p. 151-160.

Dean, R.G., 1987. Additional sediment input to the nearshore region, *Shore & Beach*, Vol. 55, Nos. 3-4, p. 76-81.

Egense, A.K., 1989, Southern California beach changes in response to extraordinary storm, *Shore & Beach*, Vol. 57, No. 4, p. 14-17.

Ehrlich, R., and Weinberg, B., 1970. An exact method for characterization of grain shape, *Journal of Sedimentary Petrology,* Vol. 40, p. 205-212.

Emery, K.O., 1960. The sea off southern California: A modern habitat of petroleum, New York, New York, John Wiley and Sons, 366 p.

Emery, K.O. and Kuhn, G.G., 1982. Sea cliffs: their processes, profiles and classification, *Geological Society of America Bulletin,* Vol. 93, p. 644-654.

Flick, R.W., 1993. The myth and reality of southern California beaches, *Shore & Beach*, Vol. 61, No. 3, p. 3-13.

Giles, R.T., and Pilkey, O.H., 1965. Atlantic beach and dune sediments of the southern United States, *Journal of Sedimentary Petrology,* Vol. 35, p. 900-910.

Hales, L.Z., 1978. Coastal processes study of the Oceanside, California littoral cell, U.S. Army Corps of Engineers, Miscellaneous Paper H-78-8, 60 p. plus 13 appendices.

Inman, D.L., Ewing, G.C., and Corliss, J.B., 1966. Coastal sand dunes of Guerro Negro, Baja California, *Geological Society of America Bulletin*, Vol. 77, p. 787-802.

Josselyn, M., and Chamberlain, S., 1993. The way it was: *Coast & Ocean*, Vol. 9, No. 3, p. 20-23.

Kolpack, R.L., 1986. Sedimentology of the mainland nearshore region of Santa Barbara Channel, California: in Knight, R.J., and McLean, J.R., eds., Shelf sands and sandstones, *Canadian Society of Petroleum Geologists*, Alberta, p. 57-72.

Lee, A.C., 1993. Sources and relative fluxes for foreshore sand from Point Arguello to the United States-Mexico border, Fourier grain-shape analysis, Unpublished Ph.D. dissertation, University of Southern California, Los Angeles, California, 193 p.

Los Angeles District, U.S. Army Corps of Engineers, 1991. Coast of California storm and tidal waves study, state of the coast report, San Diego region - Main Report, 877 p.

Lu, Y., 1992. Fourier grain-shape analysis of beach sand samples and associated sedimentary processes, Dockweiler and El Segundo beaches, Santa Monica Bay, southern California: Unpublished Ph.D. dissertation, University of Southern California, Los Angeles, California, 200 p.

McMaster, R.L., 1954. Petrography and genesis of the New Jersey beach sands, New Jersey State Department of Conservation Series, Bulletin 63, 239 p.

Murillo de Nava, J.M., 1993. Characteristics and sources of the Creciente Barrier Island sediments within Margarita lagoonal complex, Baja California Sur, Mexico, Unpublished M.S. thesis, University of Southern California, Los Angeles, California, 370 p.

Noble Consultants, Incorporated, 1989. Coastal sand management plan, Santa Barbara/Ventura County coastline - Main Report: Noble Consultants, Incorporated, Irvine, California, 186 p.

O'Hara, S.P.,. and Graves, G., 1991. Saving California's coast: Spokane, Washington, Arthur Clark Company, 277 p.

Osborne, R.H., and Yeh, C.-C., 1991. Fourier grain-shape analysis of coastal and inner continental-shelf sand samples, Oceanside Littoral Cell, southern Orange and San Diego Counties, southern California, *in* Osborne, R.H., ed., From shoreline to abyss, Contributions in marine geology in honor of Francis Parker Shepard, SEPM (Society for Sedimentary Geology) Special Publication No. 46, Tulsa, Oklahoma, p. 51-66.

Osborne, R.H., Bomer, E.J., III, Wang, Y.-C., and Lu, Y., 1993. Application of a tumbler experiment using granodioritic grus to examine the character of quartz-grain fracture in high-gradient streams, *in* Johnsson, M.J., and Basu, A., eds., Processes controlling the composition of clastic sediments, Boulder, Colorado, *Geological Society of America*, Special Paper 284, p. 211-234.

Page, G.B., 1950. Beach erosion and composition of sand dunes Playa del Rey-El Segundo area California, Unpublished M.A. thesis, University of California at Los Angeles, Los Angeles, California, 53 p.

Pierce, J.W., 1968. Sediment budget along a barrier island chain, *Sedimentary Geology*, Vol. 3, p. 5-16.

Poland, J.F., Garrette, A.A., and Sinnott, A., 1959. Geology, hydrology and chemical character of the ground water in the Torrance-Santa Monica area, California, U. S. Geological Survey, Water Supply Paper 1461, 425 p.

Robinson, R.A., 1993. Fourier grain-shape analysis of quartz sand from the eastern and central Santa Barbara littoral cell, southern California, Unpublished M.S. thesis, University of Southern California, Los Angeles, California, 104 p.

Saville, T., 1960. Sand transfer, beach control, and inlet improvements, Fire Island Inlet to Jones Beach, New York, 7th Conference on Coastal Engineering, p. 785-807.

Schwarcz, H.P., and Shane, K.C., 1969. Measurement of particle shape by Fourier analysis, *Sedimentology*, Vol. 13, p. 213-231.

Seymour, R.J., 1989a. Wave observations in the storm of 17-18 January, 1988, *Shore & Beach*, Vol. 57, no. 4, p. 14-17.

Seymour, R.J., 1989b. Unusual damage from a California storm, *Shore & Beach*, Vol. 57, No. 3, p. 31.

Shepard, F.P., and MacDonald, G.A., 1938. Sediments of Santa Monica Bay, California, *Bulletin of the American Association of Petroleum Geologists*, Vol. 22, p. 201-216.

Shepard, F.P., 1973. Submarine Geology, New York Harper & Row, Publishers, 3rd ed., 517 p.

Trask, P.D., 1952. Source of beach sand at Santa Barbara, California as indicated by mineral grain studies, Beach Erosion Board, U.S. Army Corps of Engineers, Technical Memorandum No. 28, 24 p.

van Andel, T.H., and Poole, D.M., 1960. Sources of recent sediments in the northern Gulf of Mexico, *Journal of Sedimentary Petrology,* Vol. 30, p. 91-122.

Wiegel, R.L., 1994. Ocean beach nourishment on the USA Pacific coast, *Shore & Beach*, Vol. 62, No. 1, p. 11-36.

Woodell, G., and Hollar, R., 1991. Historical changes in the beaches of Los Angeles County: American Society of Civil Engineers, Coastal Zone '91, Vol. 2, p. 1342-1354.

IMAGE ANALYSIS OF SURF ZONE HYDRODYNAMICS

J.M.Redondo[b], A.Rodriguez[a], E. Bahia[a], A.Falqués[b]
V. Gracia[a], A.Sánchez-Arcilla[a] and M.J.F. Stive[a]

ABSTRACT: A serie of relatively novel techniques is presented to study surf-zone hydrodynamics by means of digital processing of video image recordings of the sea surface near the coast. Image analysis may be used to estimate spatial and temporal characteristics of wave fields, surface circulation, mixing and morphodynamics in the surf zone. Preliminary field measurements were conducted in May 1993 along the Ebro Delta (Spanish Mediterranean Coast), to test the methodology. Spectral analysis on the images was applied used to estimate energetic wave frequency bands as well as dispersion relations of shear instabilities. These results are compared with numerical model predictions. Results from a recent campaign at the same site in December 1993 were used to validate the analysis. Results from three days with quite different sea conditions are compared and longshore current and dispersion measurements from the tracking of dye blobs are presented.

INTRODUCTION

The study of near-shore processes, and especially the interaction between wave fields, longshore currents, turbulence characteristics and beach morphology, needs detailed measurements of the simplest possible events to understand the complexity of coastal dynamics. A step by step approach starting from a well behaved (linear) coastal area was selected.

The generation of longshore currents by wave fields has been measured and discussed in detail by e.g. Thornton (1970). Since then, many authors have shown the complexity of the wave-current-turbulence interactions which has helped to gain insight into the time and space variability of longshore currents due to e.g. edge waves and shear instabilities.

The study of coastal hydrodynamics and morphodynamics by means of aerial images has also been treted by several researchers. Sonu (1969) utilized a 35mm motor-driven photocamera to study nearshore current patterns and mixing. Sasaki and Horikawa (1972/76) improved the resolution and accuracy

Universidad Politécnica de Cataluña, U.P.C. Campus Nord, 08034 Barcelona, Spain, *a*: Lab. Ing. Marítima, and *b*: Dept. Física Aplicada.

for mapping the quantitative flow field in the surf zone applying the idea of Stereo-Bacs (with two baloons and 70mm motor-driven cameras). Maresca and Seibel (1976) have used land based phogrametry to measure breaking waves and longshore currents; buoys and fluorecein dyes were the tracers used. Recent work by Kuriyama and Ozaki (1993) has compared numerical model results with extensive field measurements using a 400 m long offshore pier perpendicular to a sandy beach in Hazaki, Japan. In this work the longshore current was measured by tracking tracers on the sea surface.

Since the 80's, video and digital procesing techniques have allowed flow visualization in accurate and efficient ways, for a complete description of these techniques see Hesselink (1988). The Army corps of Engs. during Duck, N.C. experiments, replaced photographic material by video, and began to explore the potential of this technique (Holman and Sallenger, 1986). The expected capabities of video recording and digital processing for the study of coastal processes include horizontal velocity fields, dye mixing, angle and period of incident waves, wave runup, surf zone width, sand bar morphology, shoreline response, etc.

Holman and Lippmann (1987/89/93) have employed extensively video techniques in the quantification of sand bar morphology. Breaking waves characteristics such as overwash bore velocities (Hollad et al., 1991), or phase speed and angle (Lippmann and Holman, 1991) have also been measured using video techniques. These authors have also quantified the long period time scales associated with breaking waves during DELILAH experiment on North Carolina, (Lippmann and Holman, 1992). Recently, the wave runup has also been measured using video records of a beach on Maryland by Walton (1993). This list shows the extensive use of these techniques, which appear to be an useful complementary tool for remote sensing (field and laboratory measurements) of coastal dynamics.

The paper starts with a description of the site, the set up of the DELTA 93/94 field experiments, and the conditions on the different days. In section three results from the image analysis are presented and compared with other measurements. The numerical model used is then discussed and its results compared with the experimental results.

FIELD WORK AND METHODOLOGY

The site chosen for this study was the Trabucador bar in the Ebro delta which has a long history of data gathering and simple linear geometry. In figure 1 a map of the area is shown together with other zones considered for the study. Two sets of measurements will be discussed in the paper, corresponding to May and December 1993. The bathymetry was measured both before and after the field campaign. A bottom elevation plot at the start of measurements on the 26th of May is shown in figure 2. On this day, the profile showed both a primary and a secondary bar. The profiles measured in December showed only a single bar, with no trace of the inner bar. This difference is due to the varying weather and sea conditions described below.

The DELTA 93/94 measurements include simultaneous time series of waves, velocities and sediment concentrations from the surf zone of this

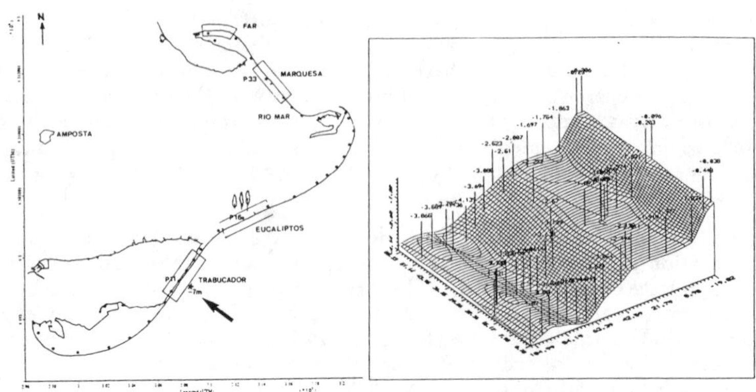

Figures 1 and 2: Map of the Ebro Delta showing the Trabucador studied area (P11), and the P11 bathymetry for 26/May/93, respectively.

microtidal barred beach, during calm and storm conditions. The main characteristics of the incident wave field were measured by means of a video camera (VTR), a directional wave rider (DWR) placed just in front of the Trabucador site, (1500 m offshore at 7 m water depth) and a X-band radar.

In the December measurements, in addition to the DWR, a conductivity type wave gauge (WG) was placed on a movable "sledge" and used to measure surf zone wave transformation. The vertical structure of the velocity field was measured by means of six electromagnetic current meters ($EMCM$) deployed in the vertical pole of the sledge. Suspended sediment concentrations were measured with three optical back-scatter sensors (OBS) and the time averaged suspended sediment transport was measured by means of two portable sediment traps (PST). These sediment transport measurements will not be discussed in here.

During the experiments video images were recorded in a S-VHS system. The BW video camera was placed in a waterproof housing at 20 m above the sea surface by means of a crane. Several fiducial points were marked with white flags both at the coastline and 60 m offshore. An elevation view, with the shore details, the equipment and the camera position is shown in figure 3. The height and orientation of the camera could be changed by means of the crane controls. Two types of video images were collected, viz. frontal and lateral views of the shoreline. The frontal imaged area ranges from the beach to the horizon in cross-shore directon (with an effectively resolvable area from the shoreline to the outer bar -200m-). In the longshore direction the images cover 150 m along the shoreline and 300 m along the inner bar (50 m from the shoreline). The most detailed measurements were restricted to the points inside the fiducial points, where the camera distortion was lower.

Quantitative information from the video images is acomplished using the *DigImage* video processing system, and an arithmetic frame grabber (DT2861)

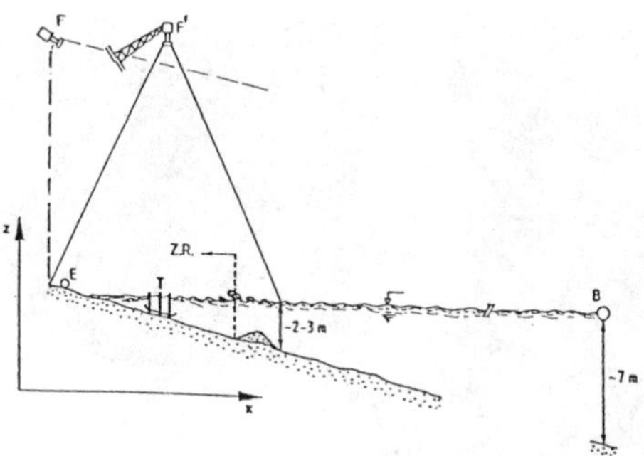

Figure 3.: A sketch of the area and equipment: E: measurement station, T: sledge with EMCM, OBS, PST and WG, B: DWR, F: video camera.

on an IBM compatible computer, which allows a resolution of 512 by 512 pixels and 256 grey levels. Images can be digitally enhanced to stretch the contrast and filter "noise" before analysis. The video may be controlled by the computer, allowing, remote control of the processing.

The images were corrected for optical deformations due to camera angle and lens distortion. World coordinates are asigned to known points marked in the video image and pixel values may be easily converted to real calibrated coordinates. The resolution in frontal view image pixels is less than 0.4 m in the cross-shore direction, and 0.1-0.4 m in the longshore direction up to the outer bar. The video recordings show clearly areas of breaking-induced turbulence and foam, which correspond to the swash zone and to first and second breaklines. See figure 4 for a false colour sequence of time averaged (equivalent to a long exposure) and instantaneous images of the area. The first bar and the swash are seen as dark bands, and contour lines are also shown.

The analized video series correspond to the three days indicated below, displaying fairly different sea conditions:

i) a post storm situation with double peaked waves coming from NE during May 26 1993. The spectrum corresponding to the time of video measurements is shown in figure 5a. The observed peaks of the spectrum were $T_{p1} = 13$ s and $T_{p2} = 3$ s. The evolution of wave heights during the day is shown in figure 6a. The storm which developed during the previous day reached a maximum in wave height in the early hours of the morning $H_{1/3} = 1.10$m. At the time of the video measurements $H_{1/3}$ was 0.5 m.

ii) a calm situation with waves coming from SE during December 15 1993. The spectra of low energy incident waves shows two close peaks corresponding to $T_{p1} = 8.3$ s and $T_{p2} = 6.6$ s ($H_{1/3} = 0.25$m according to DWR data). See figure 5b for the wave spectrum and figure 6b for the evolution of the $H_{1/3}$. An arrow indicates the time of detailed

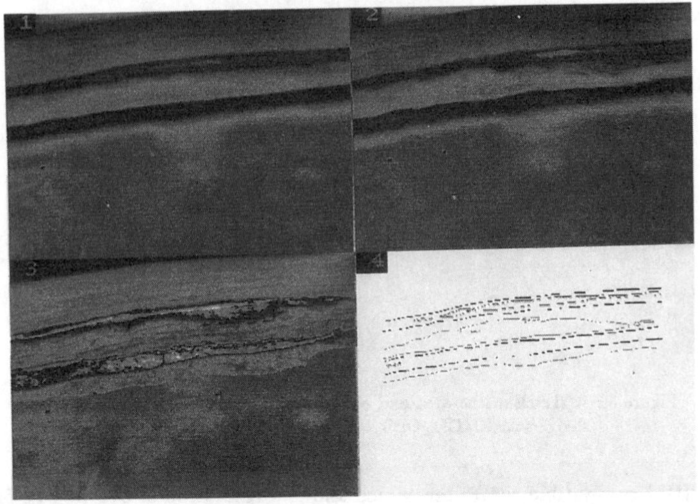

Figure 4: (1-2) Averaged images of the surf-zone, (3) single frame and (4) isolines of intensity.

measurements.

iii) a moderate situation storm with waves coming from SE during December 16 1993. The spectra shows a dominant peak at 7.1 s with lower energy signatures corresponding to a low frequency oscillation (T=18.3 s) and to a higher wave frequency with period of T=5.1 s, see figure 5c. The wave height evolution during the day is shown in figure 6c where the passage of the storm produce an increase of $H_{1/3}$ between 0.50 - 0.80 m. The video and WG data where obtained around 16 h GMT (shown with an arrow). The $H_{1/3}$ during the experiment was 0.6 m.

Spectral analysis of VTR/WG/DWR time series

Three types of time series were used in the spectral analysis: intensities from video images (VTR), water level oscilations from the WG and the DWR. The sampling intervals were respectively 0.20-0.60-1.64 s, 0.25 and 0.78 s. The length of the time series corresponding to video images were 100-300-840, and 1200 s for both WG and DWR.

The time series were detrended using a simple linear regression tapered with a Welch window prior to applying FFT. The raw spectra have been calculated over series with a minimum of 512 points grouped in subseries of 128 records (50 % overlapped). After that a high pass filter with a folding frequency of 0.000001 Hz has been used to eliminate the undesired frequencies. The associated spectral variance was 14 %.

The intermediate frequency wind-wave components, and the infragravity

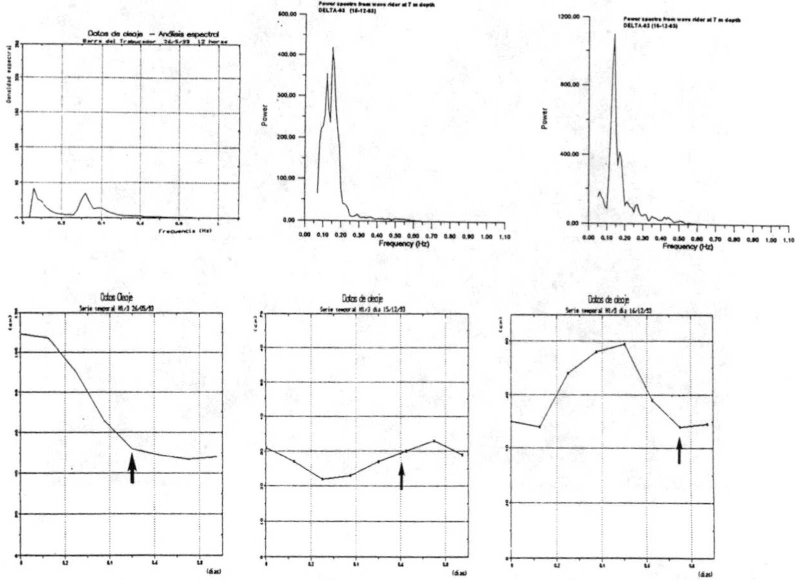

Figures 5 and 6:Measured wave spectra and $H_{1/3}$ evolution respectively for the three selected days: 26 May, 15 and 16 December 1993.

and shear instabilities were investigated. Due to the short time series used in the video data (up to 15 min), the low frequency ranges had to be handled with extreme caution.

FIELD RESULTS

The use of time series of horizontal and vertical pixel intensities allowed to measure wave frequencies, as well as the longshore component of the velocity thanks to the traces of advected surf (foam). In figure 7(1-4) an example of the time evolution (vertical) of the color coded reflected light intensities of a line is shown at four locations offshore.

Fast Fourier transform (FFT) methods applied to the digitized video images and to their correlations were used to detect the dominant wavenumbers and wave directions. Time averaged spatial images and time series of a transect (see figure 8) are seen to be convenient techniques for the characterization of wave fields. The relationship between light intensity, which has been the main parameter studied, and the surface elevation needs further study, but it is clear that for the non breaking waves, the angle between the wave surface, (bisecting the sun direction), and the video camera, will produce a maximum reflected intensity. This is also true for cloud covered skies. In addition, wave breaking,

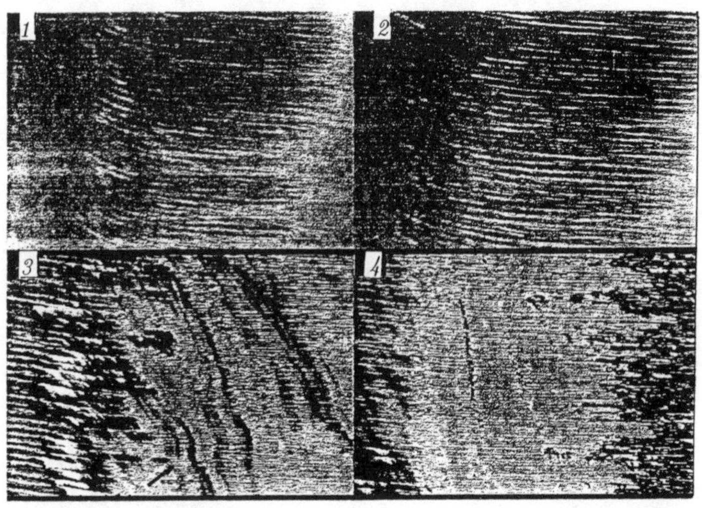

Figure 7: Time series of colour coded reflected light intensity showing the drift of surf. Ligth areas indicates wave breaking, (1) area behind the bar, (2) inside near the bar, (3) between the bar and the swash, (4) swash area. See in figure (3) the oscillations of the longshore current.

can be clearly seen and whitecaps may be easily traced.

Power spectra across-shore

Analysis of breaking and non breaking time series of 100, 300 and 840 s length and their intensity power spectra are presented as a function of cross-shore distance. These image spectra are then compared with WG and DWR spectra. In figure 9 the intensity spectra of four points near the WG structure are presented, together with the spectrum obtained from the WG data. The agreement, as seen in this figure, is very good.

i) 26/May/93: Four long time series of 300 s are presented in figure 10(a), and their respective spectra in figure 10(b). The intensity series are presented with an offset of 20 in the range 0-256 given by the digitizer resolution. The spectra from the longer series (including several 840 s series) are compared with low frequency shear waves obtained from the numerical model described below. The sites chosen were the outer bar, the inner bar, the trough and the swash zone. The dominant frequencies were: a low frequency oscillation with period of 77 s and a wave induced peak of period 6 s, which appeared in all positions. The DWR spectra for the incident wave field showed dominant waves of periods 13 s and 2 s. The incident wave groups appeared to merge, giving a dominant period of 6 s. At the swash zone the intensity variations are much larger, as seen in figure 11, and other significant periods are found at 19 s, 11 s and 7.7 s.

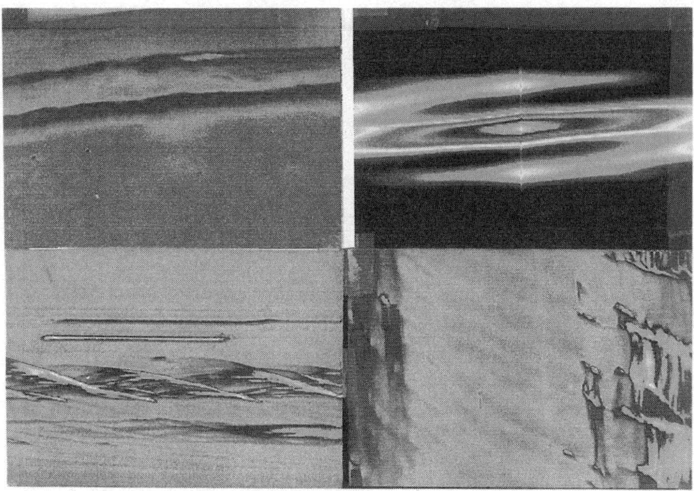

Figure 8: **(Top/left) Colour coded intensity of a time averaged image showing the average surf zone width, (Top/right) Auto correlation function of the previous image showing the prefered directions, (Bottom/left) Time series of the central vertical line of the frave showed above showing period and direction of the incident waves, (Bottom/rigth) Time series of the central horizontal line. showing surface velocities across the surf zone.**

ii) 15/December/93: The evolution of wave frequencies with cross-shore distance was investigated using 100s video intensity time series in order to resolve wave generated frequencies. The seven selected cross-shore positions are described in figure 11(a), and the agreement with the DWR was also good, measuring a period of 6.4 s from the video images, and a period of 6.6 from the DWR showed in figure 11(b). Other periods detected at the DWR were 8.3 s and 10.5 s. A transition from the 6.6 s period, (DWR dominant incident waves) to the recorded periods at the WG of 7.9 s and 2.8 s could be followed with the aid of the video images.

As mentioned above, the dominant waves at a position offshore, in the vicinity of the DWR have a period of 6.4 s. At the outer bar, where there is very little breaking, during calm conditions, the 6.4 s waves are mantained, but some period doubling occurs and a 12.8 s peak appears. In addition, waves of period 4.2 s are also generated near the outer bar. During these experiments, the position of the WG was just outside the surf zone and the video images near it show dominant periods of 6.6 s and 2.9 s just before to the surf zone. The observed periods are also 6.6 s and 2.9 s, with a new developing period of 4 s. There is some breaking at the surf zone which generates some new spectral peaks. Inside the surf zone, the dominant peak corresponds to 8.5 s. and there are smaller peaks corresponding to wave periods of 5.1 s and 3.6 s.

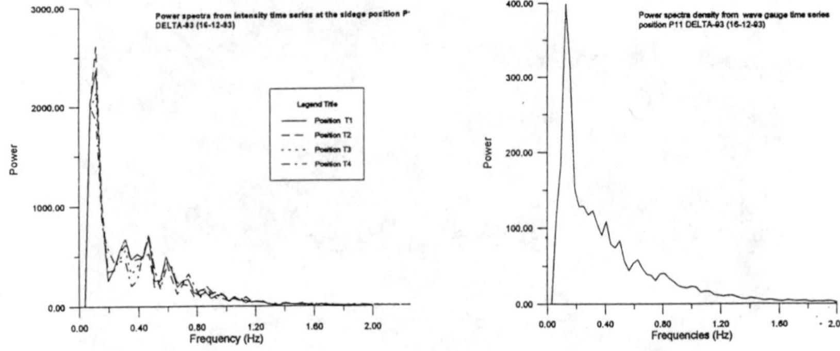

Figure 9: Comparison of spectra from WG data and video intesities at the same position.

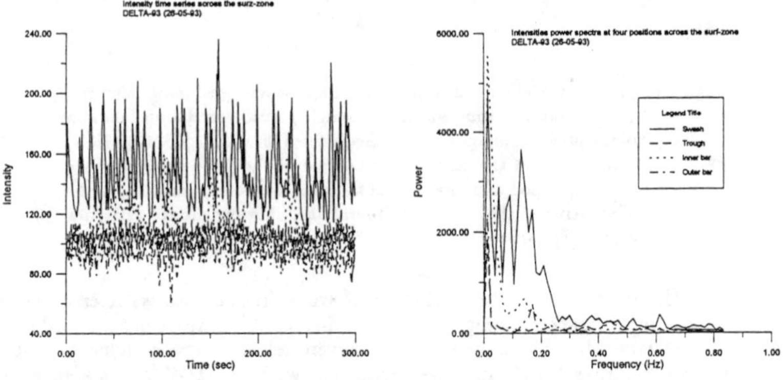

Figure 10: (a) Intensity time series,(b) Intensity spectra of four cross-shore positions.

It is interesting to note waves of period 8.5 s, which are close to some of the incident wave periods at the DWR. At the swash zone the dominant period is of 12.8 s, with a secondary one at 2.8 s, and at the shore line, the periods are 5.1 s and again 12.8 s.

The interpretation of this spectral cross-shore evolution is as follows: due to the low energy of the incident wave field, the dominant wave field interacts with the outer bar producing a reinforcement of every other wave. The intermediate 7.9 - 8.5 s waves appear only outside the surf zone and inside the trough, where the breaking distorts the periodicity, as may be seen in a video

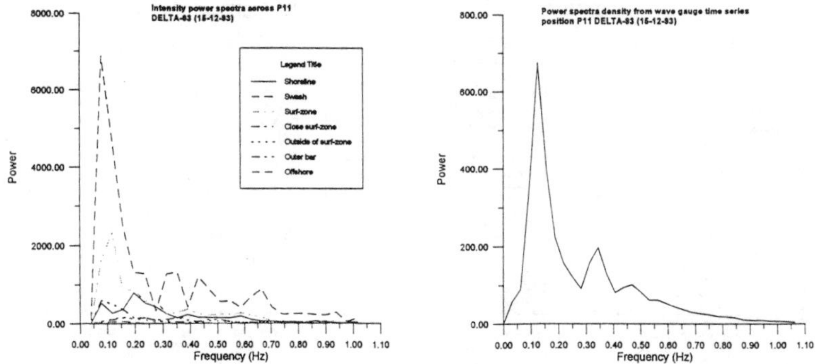

Figure 11: (a) Intensity spectra of seven cross-shore positions, (b) spectra from the WG.

image of a time evolution of a line inside the surf zone (figure 12). Finally the long (slow) waves ($T = 12.8$ s) break against the shore with some superimposed faster wave components.

Figure 12: Intesity time serie at a horizontal line showing wave distortion across the surf zone.

***iii*) 16/December/93: The evolution of wave frequencies with cross-shore distance was investigated using 100s video intensity time series. The sites chosen are described in figure 13, where the spectra are also shown.**

The agreement with the *DWR* was also good, and a transition from the

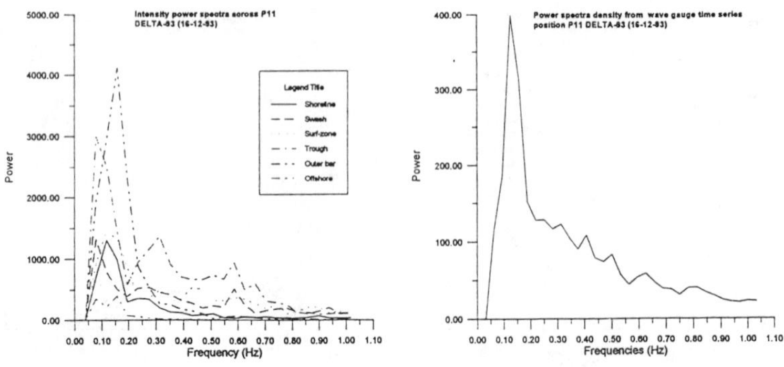

Figure 13: (a) Intensity spectra of six cross-shore positions,(b) spectra from the WG.

6.4 s period waves to the 8.0 s waves detected at the *WG* could be followed as waves approached the coastline. The offshore video spectra, (closer than the *DWR*) shows dominant periods of 1.8 s 6.4 s and 12.8 s , as in the previous day. The *DWR* during the time of video measurements showed periods of 5.8 s and 7.1 s, which are close enough to the 6.4 s. This period is seen at several locations, but not at the trough, where the *WG* was placed. The *WG* measurements show wave periods of 3.2 s and 8 s, just offshore the position where the *WG* video measurements showed periods of 1.7 s , 3.2 s and 12.8 s. Closer to the surf zone, periods of 2.3 s, 3.2 s and 8.5 s are detected. At the swash zone, the periods are 3.6 and 12.8, again a multiple of the offshore one, and at the shoreline 1.9 s , 4.2 s and 8.5 s periods are detected.

The main aim of the spectral analysis presented here for the December campaign is to show that much more spatial information may be quantitatively obtained at wave spatial and temporal ranges. The use of light intensity data in order to describe wave frequencies and their structure was used by Lippmann and Holman mostly for breaking waves, but due to the good resolution of the images, changes in the sea surface angle could also be detected with the present data. In order to justify the use of the video time series, a detailed comparison was made with the *WG* data and the video intensity series for the december campaign.

The measurements with longer video time series have not been as successful in revealing the infragravity wave range and the longer periodicities, mostly due to the lack of sufficiently long and steady measurements.

Dye dispersion measurements

Measurements of dye dispersion at different distances from the shore (inside the surf zone) have been used to obtain estimates for longshore current and turbulence characteristics, such as the mean Lagrangian trajectories, mean and r.m.s. velocity integral lengthscales, shape and anisotropy of blobs, etc.

The tracers were selected after field intercomparisons of different substances (fluoresceine, rhodamine, MnO_4K_2, SO_4Ca and milk). The best tracer varies depending on weather conditions because of the difference of constrast between dye and marine water, but milk and fluoresceine were the best ones because of their good contrast and persistance. Figure 14 shows a time averaged image during 30 seconds of dye trajectory (20 gr of fluoresceine).

Figure 14: Average picture obtained integrating point video intensities during 30s as one of the dye blobs is released. The trajectory may be identified near the center of the image.

From dye dispersion measurements and from the evolution of intensity isolines, fractal measures of selfsimilarity ranges and of dominant scales are also obtained. The spatial and time evolving characteristics of the surface velocity fields, are used to complement the vertical velocity profiles obtained by means of electromagnetic current meters.

NUMERICAL MODEL RESULTS

Numerical longshore currents models have been used to predict the most relevant modes taking into account the measured boundary conditions. The dispersion relations and pattern of edge waves for the actual bathymetry and longshore current have been computed using the numerical method described in Falques and Iranzo (1992). The steady longshore current profile has been computed with an incoming wave height of 0.46 m, and an incidence angle of 23.4^o. Under the assumption of alongshore uniformity, the occurrence of shear instabilities of the longshore current has also been investigated by means of a numerical simulation. For this purpose, the basic current profile (Figure 15 c) and the lateral momentum diffusion distribution (Figure 15 b) have been obtained from the incident wave field applying a propagation and circulation model (Rivero and S.Arcilla, 1991) and the measured bathymetry (Figure 15 a).

The longshore current profile has two peaks: the strongest one near the shoreline with $V_m \simeq 0.46m/s$ and maximum backshear $f_s \simeq 0.08s^{-1}$ and the weaker one over the inner longshore bar with $V_m \simeq 0.13m/s$ and maximum backshear $f_s \simeq 0.0007s^{-1}$. Similarly, the eddy viscosity coefficient distribution has also two peaks of about $0.04m^2s^{-1}$. A constant drag coefficient of $c_d = 0.0015$ has been taken for bottom friction.

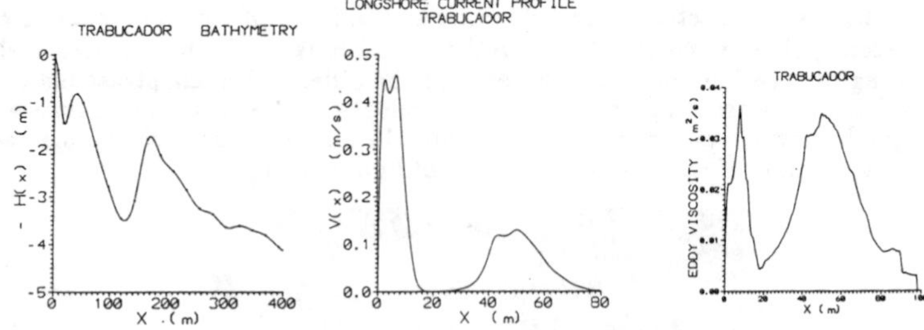

Figure 15: (a) Bathymetry, (b) eddy viscosity and (c) computed longshore velocity profiles at P11 considered in the shear instabilities simulation.

The shear instability analysis started from the shallow water equations

$$\frac{\partial v_i}{\partial t} + v_j v_{i,j} + g\eta_{,i} = -\frac{c_d}{h}|\vec{v}|v_i + \frac{1}{h}S_{ij,j} + \frac{1}{h}[\epsilon h(v_{i,j} + v_{j,i})]_{,j} \qquad i = 1,2 \quad (1)$$

$$\frac{\partial \eta}{\partial t} + [hv_j]_{,j} = 0 \qquad (2)$$

followinf the approach of Falqués and Iranzo, 1994 (hereinafter FI94). A linear stability analysis has been performed by assuming perturbations with the form

$$e^{(\omega_r + i\omega_i)t} e^{iky} v'(\vec{x}),$$

where $k = 2\pi/\lambda$. The same spectral method of FI94 based on rational Chebyshev expansions has been used here. However, the present calculation uses bathymetry, current and eddy viscosity interpolated from the experimental data. The considered eddy viscosity coefficient, $\epsilon(x)$, has therefore a cross-shore variation.

Two linear unstable shear modes have been found. As it can be seen in Figures 16(a-b) ,the dominant one has a fastest growing wavelength about $\lambda = 31m$, a period $T = 104s$ and a growth rate $\omega_i = 0.006s^{-1}$. The other one has $\lambda = 70m, T = 1000s$ and $\omega_i = 0.0018s^{-1}$. Figure 16(c) shows the cross-shore structure of both modes.

Since the dominant mode is located near the shoreline it appears to be associated with the strongest current peak. On the other hand, the other mode is distributed in a wider region centered at the inner bar. Therefore, it is likely associated to the lowest current peak. Assuming this correspondence, both solutions agree quite well with the rough predictions for plane beach and inviscid flow in FI94. For instance, the phase speed to peak longshore current ratio, is $c_1/V_{m1} = 0.65$ for the first mode and $c_2/V_{m2} = 0.54$ for the second one. Further,

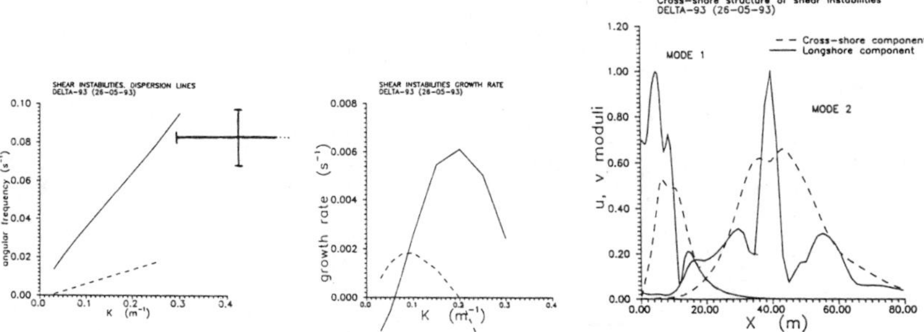

Figure 16: Linear unstable shear modes.

the wavelength to current width ratio is about 1.7 for the first mode and 2.3 for the second one. According to the estimates for inviscid flow in FI94, growth rates should be around $\omega_i \sim 0.1 f_s$. For the second mode this would give 0.008 s^{-1} whereas the present calculation gives 0.006 s^{-1}, which shows a small decrease due to dissipation. However, for the second mode the inviscid estimate gives 0.0007 s^{-1} and the present calculation (which takes dissipation into account) gives 0.0018 s^{-1}. Dissipation seems to de-stabilize the flow in this case. Which is in line with some recent calculations for a more general situation where it was found that some eddy viscosity distributions may in some cases destabilize the basic flow.

In order to estimate dispersion relations from the field measurements, two methods were used: the first one was to consider the low frequency amplitude spectrum peaks as indications of slow sea surface undulations. The estimates range between 70 s and 84 s. A clearer indication of the shear instability may be the recorded oscillations in the longshore current, in which dye (milk and fluorescein) blobs sometimes showed large mean velocity oscillations as noted in figure 17 (a). In this figure the evolution of dye spots may be seen for two days with opposing longshore currents.

Quantitative measurements for the period of oscilation of the longshore current can be seen in figure 7(3). The oscillations of the longshore current on the 26th of May were of the order of 100 s. The lengthscales corresponding to the largest waves obtained by means of spatial fourier transforms in the trough lie between 10 and 20 m, which give an estimate of T and λ. These data are also plotted for comparison in figure 16 (c), together with the dispersion relation for the 1st and 2nd shear instability modes.

CONCLUSIONS

Digitized video images of field events seems a promising technique for

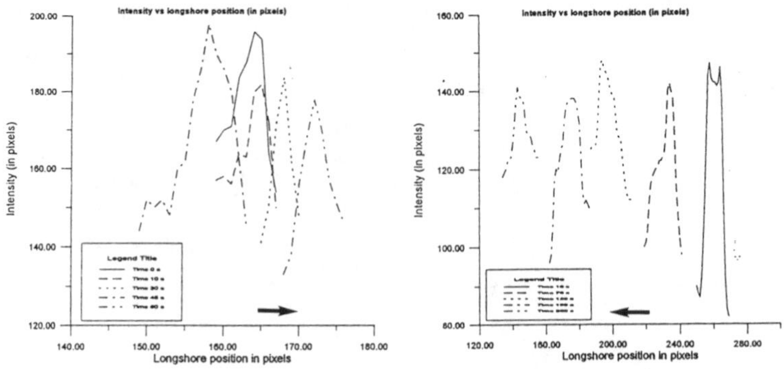

Figure 17: Blob evolution from video images, (a) 26/May/93 and (b) 15/December/93.

extended quantitative measurements of sea surface instabilities. The agreement between Video spectra and DWR and WG is qualitatively good and allows to describe in more detail the global characteristics of the surf zone.

The evolution of wave characteristics as they interact with the surf zone may be followed in detail. The outer and the inner bars act as a filter to some frequencies that reapear near the shoreline or at the trough, the variation of frequencies being clearer for the more energetic sea conditions.

The non-linear effect of the surf zone is apparent in enhanced images, such as figure 14. There is a process of wave dislocations, between the surf and swash zones that needs further study. This is also reflected in the changes in dominant wave period as the waves approach the coastline, from measurement with image analysis. The dynamic response of the surf zone to different incoming wave fields may be, thus, investigated in detail.

The numerical model shows some agreement in time scales for the dispersion relationship, but the spatial scales are somewhat lower than predicted. Considering that the model is linear and several approximations (e.g. constant bottom drag) have been used, there seems to be ground for improvement. In future models other important effects, like variable friction should be taken into account.

The measured longshore current at the position where the dye blobs were released is of $0.5\ ms^{-1}$, which agrees well with the predicted maximum longshore currents, see figure 17. The detailed longshore velocity distribution needs further comparison with field data. In the December campaign, several series of simultaneous blobs were released at different cross-shore locations and further work will include comparisons between these measurements and the numerical models.

AKNOWLEDGEMENTS

This work was undertaken as part of the Surf Zone Research project of LIM-UPC. It was funded jointly by the Programa de Clima Marítimo PCM-MOPTMA and the Ministerio de E. y C. (DGICYT) of Spain, with some support from the MAST-II Euromarge contract of the EU. We want to thank the research staff of LIM-UPC, particularly to J. Gomez, F. Rivero, J. Sospedra and all those who endured the field work. Thanks are also due to G. Voulgaris and M.A. Tenorio from Southampton University for their OBS and radar, and research staff of DFA-UPC, particularly to V. Iranzo and D. Crespo.

REFERENCES

Falqués,A. and Iranzo,V., 1992. "Edge waves on a longshore shear flow", *Physics Fluids A 4 (10),* 2169-2190.

Falqués,A. and Iranzo,V., 1994. "Numerical Simulation of Vorticity Waves in the nearshore, *J.Geophys Res.* 99, (C1), 825-841 (1994).

Hesselink,L., 1988. "Digital Image processing in flow visualization" *Ann. Rev. Fluid Mech.*, 20, 421-485.

Holland,K.; Holman,R. and Sallenger,A., 1991. "Estimation of overwash bore velocities using video techniques" *Coastal sediments* ASCE, 489-497.

Holman,R. and Sallenger,A., 1986. "High-energy nearshore processes" *EOS* Trans. Americ. Geoph. Union, December 9, 1369-1371.

Holman,R. and Lippmann,T., 1987. "Remote sensing of nearshore bar systems - making morphology visible" *Coastal sediments* ASCE, 929-944.

Kuriyama,Y. and Ozaki,Y., 1993. "Longshore current distribution on a bar-trough beach" Report of Port and Harbour Research Institute. Japan. Vol. **32. 3**, 3 - 37.

Lippmann,T. and Holman,R., 1989. "Quantification of sand bar morphology: a video technique based on a wave dissipation", *J.Geophys.Res.* Vol 94, No C1, 995-1011.

Lippmann,T. and Holman,R., 1991. "Phase speed and angle of breaking waves measured with video techniques" *Coastal sediments* ASCE, 542-556.

Lippmann,T. and Holman,R., 1992. "Wave group modulations in cross-shore breaking patterns", Proc. I.C.C.E., ASCE, 918-931.

Lippmann,T., Holman,R. and Hathaway, K., 1993. "Episodic, nonstationary behavior of a double bar system at Duck, N.C., USA, 1986-1991" Jour. of Coastal Reseach, **SI,15**, 49-75.

Maresca,J. and Seibel,E., 1976. "Terrestrial photogrametric measurements of breaking waves and longshore currents in the nearshore zone", Proc. I.C.C.E., ASCE, 681-700.

Sonu,C., 1969. "Tethered balloon for study of coastal dynamics", Amer. Soc. Photogram., Tech. Rep., 66, 91-103.

Thornton,E.B., 1970.- "Variation of longshore current across the surf zone" Proc. I.C.C.E., ASCE, . **12.** , 291-308.

Sasaki,T., Horikawa,K. and Hotta,S., 1976. "Nearshore current on a gently sloping beach" Proc. I.C.C.E., ASCE, . **36.**, 626-644.

Rivero,F. and Sánchez Arcilla,A., 1991. "Quasi-3D nearshore current modelling", *Computer Modelling in Ocean Engineering 91*, Balkema, 171-178.

Walton,T., 1993. "Ocean City, Maryland, wave runup study" Jour. of Coastal Reseach, **9,1**, 1-10.

ADVANCES IN 3D MODELLING OF WAVE-DRIVEN CURRENTS

Philippe Péchon [1]

ABSTRACT : The numerical modelling of time-averaged three-dimensional currents due to breaking waves is presented. The basic equations and the closures are described. The driving terms in the momentum equations are the radiation stresses derived from "organized" velocity of waves and the roller contribution. The flow computed in the case of a reactilinear beach with oblique incidence wave displays a helicoidal circulation. In the situation of a detached breakwater with a normal wave, depth-integrated currents show two large symetrical cells in the lee of the structure whereas streamlines near the bottom converge spirally toward the centers of the eddies.

INTRODUCTION

The final purpose of this work is to build a compound system which is able to predict accuretaly waves, currents and sediment transport in coastal areas. The first module computes the wave filed in the nearshore region by solving time-varying equations of Serre and the third module calculates sand transport with Bailard's formula and bed-evolution. This paper describes the second module for time-averaged currents induced by breaking waves. Since the code is 3D, no assumption is required for the coupling of horizontal and vertical velocity components. So the model will be

[1] Research Engineer, Electricité de France, Laboratoire National d'Hydraulique, 6, quai Watier 78400 Chatou FRANCE.

applicable in complex areas, for instance with irregular bathymetries or in the vicinity of coastal works.

THE EQUATIONS

The velocity is separated into three contributions (Svendsen and Lorentz 1989) : A time-averaged current induced by breaking waves (U, V, W), a purely periodic current (u_w, v_w, w_w) corresponding to wave motion, and turbulent fluctuations (u', v', w'). These components are three-dimensional. It is noted that W is not neglected here.

The time-averaged equations reads in a 2 DV flume for sake of clarity (Péchon 1994) :

$$\frac{\partial U^2}{\partial x} + \frac{\partial UW}{\partial z} + \frac{\partial\,(\overline{u_w^2} - \overline{w_w^2})}{\partial x} = -\,g\,\frac{\partial \bar{\xi}}{\partial x} - \frac{\partial \overline{u'w'}}{\partial z} - \frac{\partial \overline{u_w w_w}}{\partial z} + t$$

$$\frac{\partial U}{\partial x} + \frac{\partial W}{\partial z} = 0$$

where t is the contribution of the roller of the breaking waves, $t = -\partial \overline{u'^2} / \partial x$, and ξ is the free surface level.

This system differs from the generaly used 1 DV equation in that the momentum equation includes the advection term and the mass conservation equation is considered.

The equations are solved with the code TELEMAC-3D (Janin et al. 1992) using the closures outlined hereafter. It is pointed out that the model predicts only the currents due to breaking effects whereas it does not give the contribution due to wave propagation. This one is supposed to be given by solution of time-varying equations such as Boussinesq or Serre's.

THE CLOSURES

The terms which depends on the periodic current of breaking waves are expressed in function of the input wave characteristics and the energy dissipation (Longuet-Higgins 1970, Deigaard and Fredsoe 1989) :

$$\frac{\partial\,(\overline{u_w^2}-\overline{w_w^2})}{\partial x}=\frac{1}{\rho h}D$$

$$\frac{\partial \overline{u_w w_w}}{\partial z}=\frac{1}{2\rho h}D$$

with D : energy dissipation
C : wave celerity
h : mean water depth

It is noted that the contribution of non-breaking waves in the expression of the radiation stress is not included here.

Apart from the roller, Reynolds stresses are given using the eddy viscosity concept :

$$\overline{u'w'}=-\nu_t\frac{\partial U}{\partial z}$$

A uniform turbulence viscosity distribution is adopted, following Svendsen et al (1987) :

$$\nu_t=Mh\,(\frac{D}{\rho})^{1/3}$$ The constant M is taken to be 0.03

The contribution of the roller in the term $-\partial\overline{u'^2}/\partial x$ can be expressed by approximating the horizontal velocity profile as suggested by Svendsen (1984) :

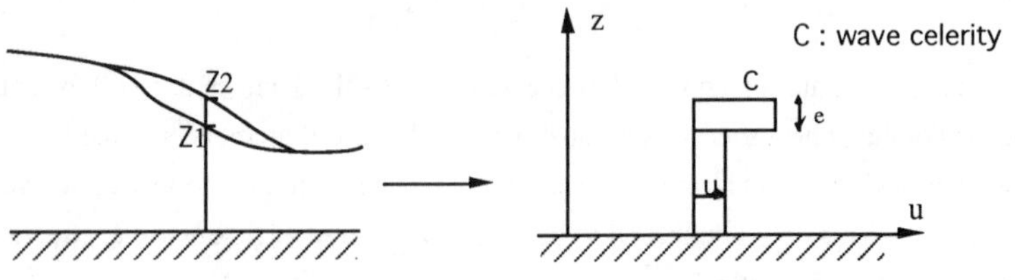

Svendsen proposed : $\overline{e}=0.9\,H^2/L$
with H : wave height
L : wave length

So we obtain $$\overline{\int_{Z_1}^{Z_2} u'^2\, dz} = C^2 \overline{e} = 0.9 \frac{C^2}{L} H^2$$

The time-averaged entrainment force t due to the roller is specified uniform from the wave trough to the mean water level, on a layer thickness of $H/2$. It follows :

$$t = -\frac{2}{H}\frac{\partial}{\partial x}\left(0.9\,\frac{C^2 H^2}{L}\right) = -\frac{1.8}{H\,T}\frac{\partial\, C H^2}{\partial x}$$

Introducing the flux of energy E_f, in shallow water :

$$E_f = \frac{1}{8}\rho\, g\, H^2 C_g \approx \frac{1}{8}\rho\, g\, H^2 C$$

$$t = \frac{14}{\rho\, g}\frac{D}{HT} \quad \text{where } D = -\frac{\partial E_f}{\partial x} \text{ energy dissipation}$$

T : wave period

Remarks : - The contribution of the organized wave velocity is not included in the term t because it was already taken into account.

- In quasi-3D models (Wind and Stive 1987, Sanchez-Arcilla et al. 1992) the equations are generaly solved from the bottom to the trough level and a shear stress is specified at the boundaries. Since the present model is three-dimensional, the time-averaged equations have to be solved form the bottom to the mean water depth and the force t is specified on a thickness $H/2$.

APPLICATIONS

Some tests were conducted in order to validate the model (Péchon 1994) or to simulate qualitative circulation on schematical cases. Two of them are described hereunder.

Three-dimensional circulation along a rectilinear beach.

The studied domain had a constant bottom slope equal to 2/100 , the off-shore water depth was 9.0 meters. The wave height at the deeper limit was 2.0 m, the period 8.0 s and the incidence 25° (fig. 1). All boundaries were closed in this test, with a slipping condition.

Wave propagation was computed with a simple refraction model and the decay in the surf-zone was deduced from the energy equation where the energy dissipation in the bore is assumed to be similar to an hydraulic jump one.

In this situation an helicoidal wave-driven flow occurs in the domain (fig. 2) and the velocity near the bottom is orientated offshore. This qualitative result is confirmed by field observations (Ingle 1966) but measurements are not available for validation.

Currents near a the detached breakwater

The bottom slope is constant and equal to 3/100. The still water depth is 5.0 meters at the off-shore frontier and it is 0.5 meter at the shoreline. The off-shore boundary is open while the three others are closed and combined with slipping conditions. The detached breakwater is 100 meters long and it is located at 100 meters from the shoreline. The still water depth at the toe of the structure is 3.5 meters.

The wave at the open boundary is 2.5 meters high, its period is 8.0 seconds and its incidence is normal to the coast. For this preliminary simulation the wave field is schematized, the wave-driven terms are nil in the lee of the structure and the wave height decay elsewhere in the surf-zone is given by a formula. Of course this wave field is very crude but the aim of this test is to display vertical heterogeneity of the velocity pattern. A more accurate computation of the wave field is planned in the future.

The depth-averaged flow pattern shows two large and symetrical eddies behind the breakwater (fig.3). The maximum velocity reaches 1.5 m/s. In order to see three-dimensional effects, particles tracks computed with the bidimensional velocity fields near the bed, near the surface and with depth-averaged one are also visualized on figure 3. The streamlines near the bottom are orientated toward the center of the eddy while the others describe large cells. This difference between the depth-averaged and the near-bed flow patterns is crucial for sediment transport modelling.

CONCLUSION

The computation of time-averaged wave-driven currents in the surf-zone for schematical 3D cases gives satisfying qualitative results. The model have already been compared succesfully with measurements collected in experimental flumes but additional validation tests are needed in three-dimensional situations.

The wave pattern in practical studies can be complex and its prediction requires a sophisticated wave model, especially if hydrodynamic results are used to determine sediment transport. It can be provided by bidimensional infragravity wave models (Boussinesq, Serre) which also simulate wave height decay in the surf-zone (Hamm et al. 1993, Karambas et al. 1992, Schäffer et al. 1992). The total velocity in the surf zone is the combination of the three-dimensional currents induced by breaking waves and the instantaneous wave velocity. However some care have to be taken in the summation of the two velocity fields because the current resulting from the wave model also partly includes the breaking effect. So the componant which comes from breaking in the wave model must be substract before adding the two velocity fields.

ACKNOWLEDGEMENT

This work was carried out as part of the G8 Coastal Morphodynamics research programme. It was funded partly by the Service Central Technique du Secrétariat d'Etat à la Mer and the Commission of the European Communities, Directorate General for Science, Research and Development, under contract n° MAS2-CT92-0027.

REFERENCES

De Vriend, H.J. and Stive M.J.F, 1987. Quasi-3D modelling of nearshore currents. *Coastal Engineering,* Vol. 11, 5 & 6, pp. 565-601.

Ingle, 1966. "The movement of beach sand", *Development in sedimentology* , Edited by Elsevier.

Janin, J.M., Lepeintre, F. and Péchon, P. , 1992. "TELEMAC-3D : a finite element code to solve 3D free surface flow problems". *Computer Modelling of Seas and Coastal Regions*, Southampton.

Hamm, L., Madsen, P. A. and Peregrine, D. H., 1993. "Wave transformation in the nearshore zone : A review ", *Coastal Engineering*, Vol. 21, 1-3, pp 5-39.

Karambas, Th. and Koutitas, C., 1992. "A breaking wave propagation model based on the Boussinesq equations". *Coastal Engineering*.Vol. 18,1-19.

Longuet-Higgins, M.S., 1970. "Longshore currents generated by obliquely incident sea waves," *Journal of Geophysical Research*, Vol. 75, pp. 6778-6801.

Péchon, P., 1994, "Numerical modelling of wave-driven currents in the surf-zone," *International Coastal Engineering Conference.* (to be published).

Sanchez-Arcilla,A., Collado, F. and Rodriguez, A., 1992. "Vertically varying velocity field in Q-3D nearshore circulation," *International Coastal Engineering Conference.* pp. 2811-2824.

Schäffer, H.A., Madsen P.A. and Deigaard, R., 1992. A two-dimensional surf-zone model based on the Boussinesq equations, *International Coastal Engineering Conference.*, pp. 576-589.

Svendsen, L.A., 1984. "Waves heights and set-up in a surf zone". *International Coastal Engineering Conference.* pp. 303-329.

Svendsen, I.A., Schäffer, H.A. and Buhr Hansen, J., 1987. "The interaction between the undertow and the boundary layer flow on a beach," *Journal of Geophysical Research*, vol 92, n° C11.

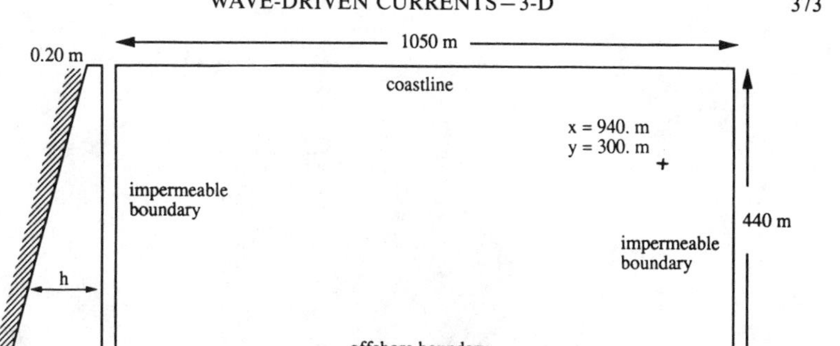

The computational domain

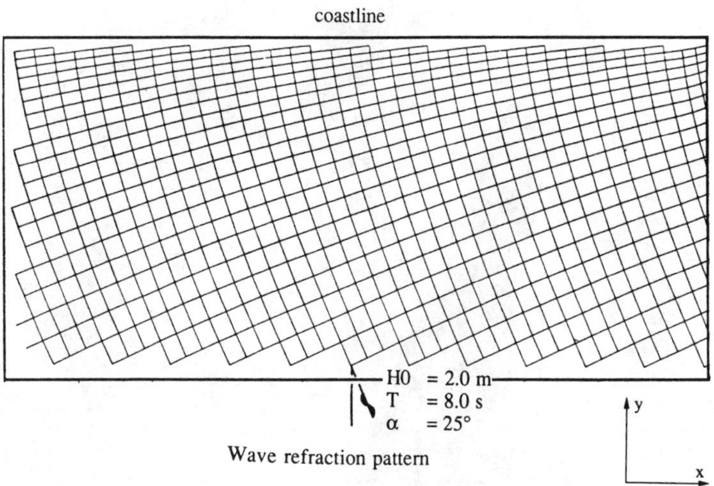

Wave refraction pattern

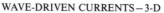

Figure 1. Wave propagation along a rectilinear coast

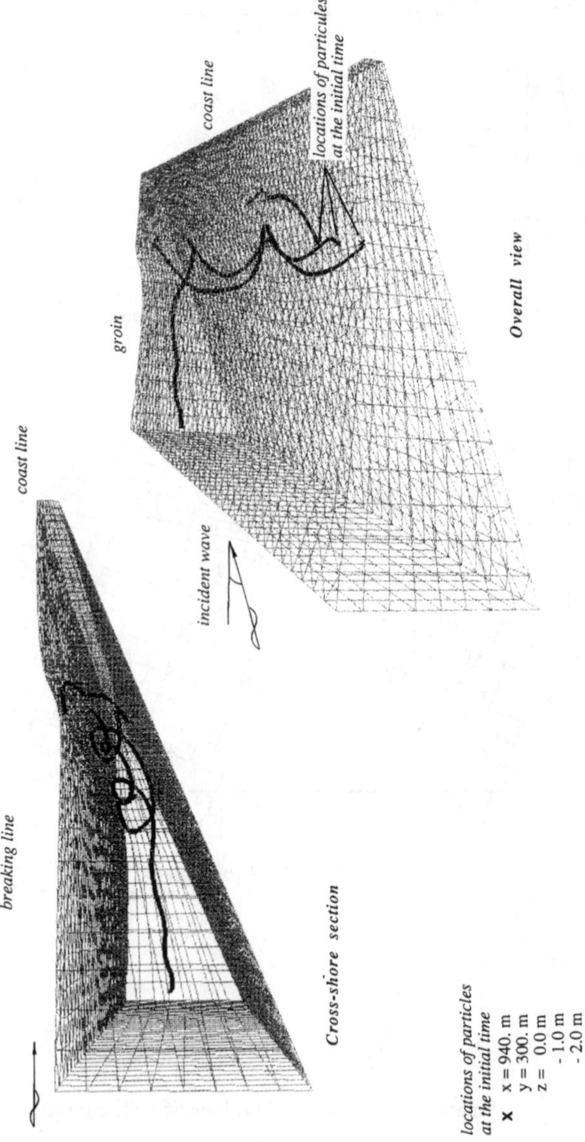

Figure 2. Currents induced by breaking waves - Particles tracks

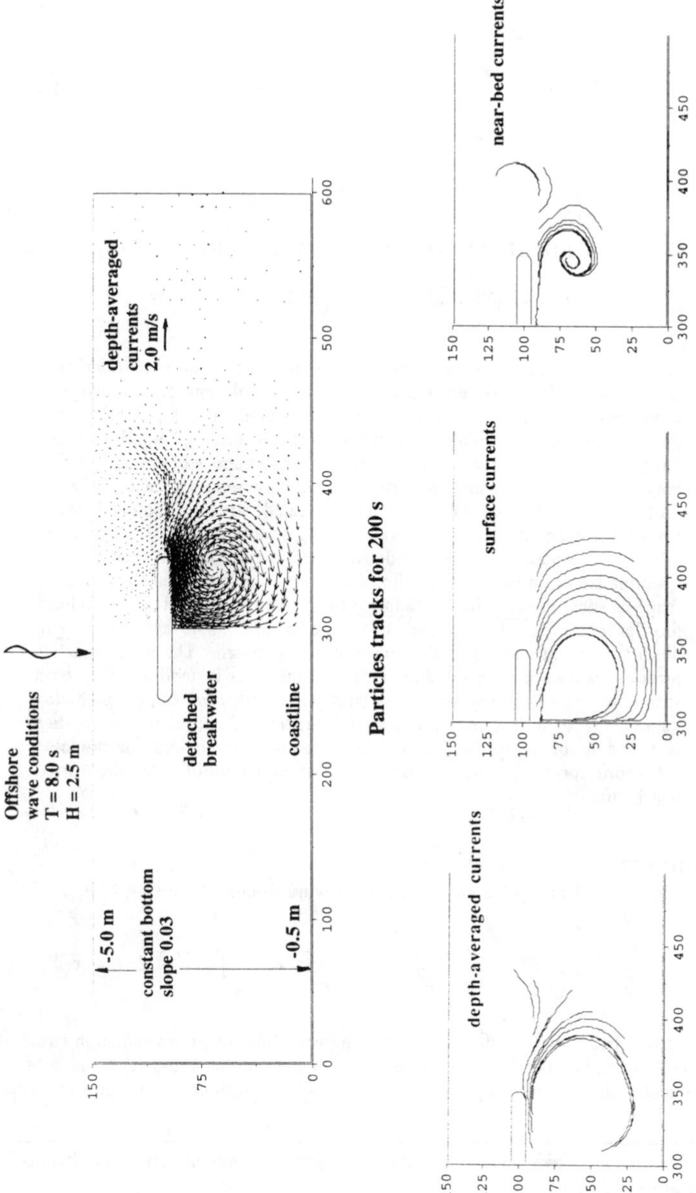

Figure 3. Circulations near a detached breakwater

MIXING BY SHEAR INSTABILITIES OF THE LONGSHORE CURRENT

J.C. Church[1], E.B. Thornton, and J. Oltman-Shay

ABSTRACT: Shear instabilities of the longshore current (nominally frequencies < 0.01 Hz) are examined as a possible source of horizontal turbulent momentum mixing (i.e. cross-shore gradients of $\rho\overline{u'v'}$ with x positive offshore) within the surfzone. Such mixing is of potential importance in the surfzone of a barred beach where existing models of longshore current produce poor agreement with observations. $\overline{u'v'}(x)$ profiles are calculated using model generated stream functions whose amplitudes are calibrated via observed energy density spectra. Data from the DELILAH experiment, conducted over the barred beach at Duck, North Carolina are examined. The predicted and observed wavenumber/frequency range of shear instabilities are found to be in good agreement. The modeled profiles of $\frac{1}{2}(\overline{u'^2+v'^2})$ and $\overline{u'v'}$ are in reasonable agreement with observations, although slightly compressed shoreward. The modeled $\overline{u'v'}$ profiles, calibrated with the data, indicate that shear instabilities may be a strong source of mixing within the surf zone with a maximum predicted value for $\overline{u'v'}$ of 0.07 m^2s^{-2}. The mixing predicted due to shear instabilities is found to be in qualitative agreement with that required for modeled longshore current profiles to agree with observed profiles for the barred beach studied.

INTRODUCTION

Modeling of longshore currents without wind forcing, following Mei (1983), may be written as a:

$$\frac{\partial}{\partial x}(\tilde{S}_{yx} + S'_{yx}) = \frac{\partial}{\partial x}\left(\rho \int_{depth} \tilde{u}\tilde{v}\,dz + \rho \int_{depth} \overline{u'v'}\,dz\right) = \overline{\tau_y}^{B} \qquad \textbf{(1)}$$

where the first two terms are the gradients of the vertically integrated radiation stress of the waves (~) and turbulence('), and the last term is the bottom stress (u being cross-shore velocity and v alongshore and ρ is density). Predicted longshore current profiles without

---

1) Military Deputy, Naval Research Laboratory, Marine Geoscience Division, Stennis Space Center, MS 39529-5000.

2) Professor, Department of Oceanography, Naval Postgraduate School, Monterey, CA 93943

3) Northwest Research Associates, 300 120th Ave NE, Bldg 7, Suite 220, Belleview WA 98005.

S'_{yx} (mixing, which is typically parameterized) show poor agreement with observations, as the longshore current velocity maximum is typically observed during DELILAH over the trough, a region modeled with near-zero forcing (see f.e. Church and Thornton (1993)). The wave induced radiation stress and bottom friction terms are thought to be reasonably modeled and therefore mixing, or more correctly the gradient of the radiation stress associated with turbulence due to shear instabilities, is examined as a possible explanation.

Energy density distributions represented by gray shading in wavenumber-frequency space for 2213, 9 October 1990, during the DELILAH experiment, are shown in Fig. 1. The theoretical dispersion curves for trapped edge waves, modes 0, 1, and 2 are shown for the appropriate beach slope. Significant energy is seen outside of these edge wave curves; this energy is linear in f-K space (where f is frequency in Hz and K, cyclic alongshore wavenumber, is equal to k(wavenumber)/2π), indicating that these oscillations, considered to be alongshore progressive waves, are non-dispersive. This energy was observed by Oltman-Shay *et al.* (1989) with Bowan and Holman (1989) providing a theoretical framework for using a conservation of potential vorticity equation in which the vorticity of the longshore current shear functioned as the restoring force. They pointed out that the phase shift in the stream functions produced by the instabilities produces non-zero $\overline{u'v'}$ values which were suggested as possible sources of mixing in the nearshore. In a numerical study of shear instabilities over various topography, Putrevu and Svendsen (1992) carried out an order of magnitude analysis and concluded that even a weak shear in the longshore current might be capable of producing significant mixing.

There are three principle approaches to the estimation of the stream function amplitudes of shear instabilities. Dodd and Thornton (1992) apply weakly non-linear theory to the simplified case of an analytical longshore current profile over a planar beach. Such mathematically rigorous non-linear models contain the stream function amplitudes explicitly. Linear models (which are better suited to observed current profiles over measured bathymetry) utilize stream functions which are of arbitrary amplitude; thus the magnitudes of the predicted velocities, which are based on the gradients of the stream functions, are likewise arbitrary. A second method was employed in Dodd *et al.* (1992), assuming that the growth rates predicted by the model may be taken as an indication of the ultimate distribution of energy across the wavenumber spectrum. Linear theory is then used to relate energy to amplitude squared. This relative method allows for the inter-comparison of different wavenumbers, but lacks an absolute reference. A third method is to measure the shear instability kinetic energy density, S_{ske}, (the sum of the u and v energy densities associated with shear instabilities where the vertical velocity is assumed zero through the rigid lid approximation) over the frequency range of interest and then scale the model-produced stream function amplitudes such that the predicted and observed energy densities match. This approach produces an absolute reference and was initially explored by Church *et al.* (1992), but is treated here in a more detailed method. S_{ske} is used for calibration, instead of the cross-spectra, S_{uv}, as S_{ske} is invariant with current meter alignment errors. Once the stream function amplitudes are calibrated, one may obtain a model predicted alongshore averaged profile of $\overline{u'v'}(f,x)$ for each wavenumber/frequency for which growth is predicted. The cross-shore gradient of $\overline{u'v'}(x)$ (integrated over frequency) produces a profile which represents the net mixing associated with the shear instabilities. Data obtained during the 1990 DELILAH experiment are used to evaluate the magnitude and structure of this term across a barred beach. The effects of this term on the longshore current profile are discussed.

SHEAR INSTABILITY THEORY

Linear wave theory is utilized, with the x-axis positive seaward. Mean and perturbation current velocities are vertically integrated and the mean current is assumed steady state. The longshore current and bathymetry are assumed uniform in the alongshore direction.

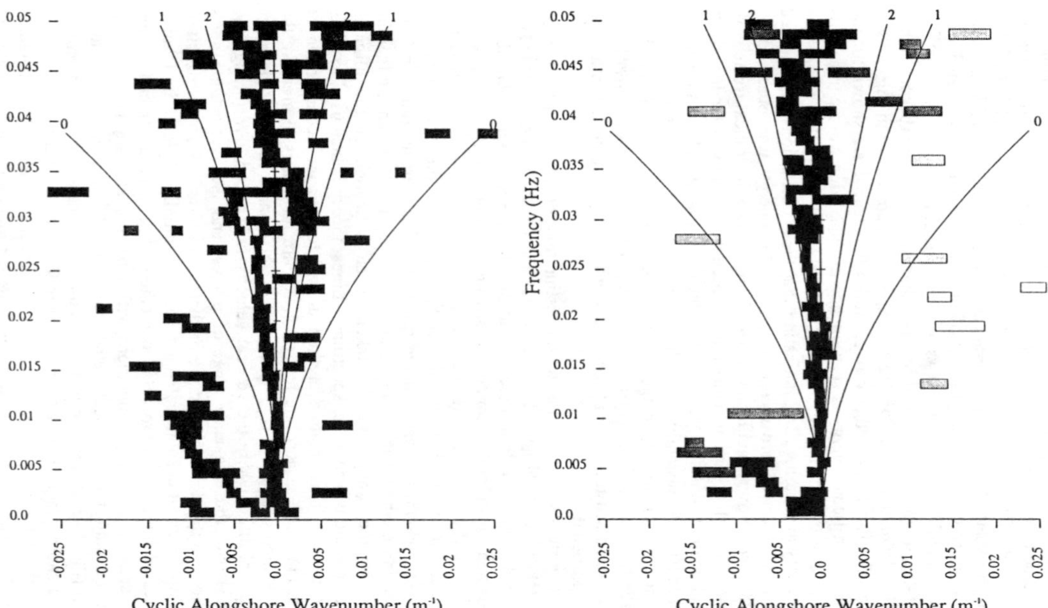

Fig 1. IMLE estimated frequency-alongshore cyclic wavenumber spectra from a) inner alongshore array, located coincident with current meter 30, and b) outer alongshore array, coincident with meter 70. The asterisk identify dispersion curves of infragravity edge wave modes, determined from the numerical solutions for depth and current profiles on this day. Δf=0.00195 Hz, ΔK=0.0005m^{-1}.

Bowen and Holman (1989) developed a theoretical basis for shear instabilities. Using conservation of potential vorticity as the restoring force, they were able to relate the mean longshore current shear to observed oscillations. The momentum and mass continuity equations, with the velocity consisting of perturbations (u', v') and a mean longshore current (V) are:

$$\frac{\partial u'}{\partial t} + V\frac{\partial u'}{\partial y} = -g\frac{\partial \eta}{\partial x} \tag{2}$$

$$\frac{\partial v'}{\partial t} + u'\frac{\partial V}{\partial x} + V\frac{\partial v'}{\partial y} = -g\frac{\partial \eta}{\partial y} \tag{3}$$

$$\frac{\partial \eta}{\partial t} + \frac{\partial (hu')}{\partial x} + \frac{\partial (hv')}{\partial y} = 0 \tag{4}$$

where η is surface elevation. These equations are linearized and the non-divergent (rigid lid) approximation is applied allowing the use of stream functions to represent the transport, such that:

$$u' = -\frac{1}{h}\frac{\partial \Psi}{\partial y} \qquad v' = \frac{1}{h}\frac{\partial \Psi}{\partial x} \tag{5}$$

Cross differentiating to combine equations and eliminate η, gives:

$$\overset{1}{\left(\frac{\partial}{\partial t}\right.} + \overset{2}{\left.V\frac{\partial}{\partial y}\right)}\overset{3}{\left(\frac{\Psi_{yy}}{h} + \left(\frac{\Psi_x}{h}\right)_x\right)} = \overset{4}{\Psi_y\left(\frac{V_x}{h}\right)_x} \tag{6}$$

where the subscripts denote differentiation. Term 1 represents the local rate of change. Term 2 is the advection by the mean longshore current. Term 3 is the relative potential vorticity of the perturbations. Term 4 represents the advection of the background vorticity of the mean longshore current (V_x/h), by the perturbations. This potential vorticity equation is comparable to the barotropic Rossby equation used for planetary scale flow with the exception that the background vorticity of the current shear is used in place of the Coriolis parameter.

A solution is then assumed of the form:

$$\Psi = Re\{\phi(x)\, e^{i(ky-\omega t)}\} \tag{7}$$

where ϕ is a cross-shore structure function. The alongshore wavenumber, k, is taken to be real, but ω, the angular frequency ($\omega = f/2\pi$), and ϕ may be complex. The form of the solution which allows growth with time is then:

$$\Psi = \exp(\omega_{im} t)\, Re\{\phi(x) \exp[i(ky - \omega_{re} t)]\} \tag{8}$$

Inserting this solution in (5) yields:

$$(V-c)\left(\phi_{xx}-k^2\phi-\frac{\phi_x h_x}{h}\right)-h\phi\left(\frac{V_x}{h}\right)_x=0 \tag{9}$$

where Re{c} is the phase speed of the shear wave, equal to ω/k.

Dodd *et al.* (1992) included the dissipative effects of bottom friction through a parameterization, $\mu=2c_f U_0/\pi$, where c_f is a friction coefficient and U_o is the magnitude of the incident swell orbital velocity. The resulting modification of the basic equation produces:

$$\left(V-\frac{i\mu}{kh}-c\right)\left(\phi_{xx}-k^2\phi-\frac{\phi_x h_x}{h}\right)-h\phi\left(\frac{V_x}{h}\right)_x+\frac{i\mu}{kh}\left(\frac{\phi_x h_x}{h}\right)=0 \tag{10}$$

The principle result of the inclusion of dissipation is a dampening effect on instabilities as indicated through the model by the reduced range over which growth is predicted. Model sensitivity to the value chosen for c_f will be examined.

After inserting known topography and an *a priori* longshore current profile, this equation takes the form of a quadratic equation in ω. This may be written in matrix form as $[A]\{\phi\}=c[B]\{\phi\}$ which produces the eigenvalues, c, for each wavenumber. Using c=ω/k the real and imaginary parts (should ω be complex) may be found. It is the cases when ω_{im} is positive that growth is predicted for an instability of that particular wavenumber.

EXPERIMENT

The 1990 DELILAH experiment was conducted at the U.S. Army Corps of Engineers Field Research Facility at Duck, North Carolina, (the same site as SUPERDUCK), with one of the specific goals being to measure shear instabilities. Two alongshore arrays composed of 5 and 6 current meters were used to identify shear instabilities. One array was located in the trough, in approximately 1.5 meters of water, and the other was located on the seaward face of the bar in approximately 3 meters of water. Phase lagged analysis was conducted on three-hour data blocks centered on the period of interest. f-K spectra were produced for the arrays using an Iterative Maximum Likelihood Estimator (Pawka 1982). Frequency resolution for the f-K processing was 0.001 and cyclic alongshore wavenumber resolution 0.00098. Daily bathymetric measurements were surveyed using an autonomous Coastal Research Amphibious Buggy (CRAB), described in detail by Birkemeier and Mason (1984). The longshore current was measured using a cross-shore array of 9 current meters and wave sensors extending across the surf zone. These three principle arrays, shown in Fig. 2, were used to acquire near-continuous data for three weeks at a sampling rate of 8 Hz. A wide variety of wave conditions occurred during the experiment, including a northeaster which drove broad-banded waves, and two distant hurricanes which generated narrow banded swell incident at large angles to the beach. These events resulted in strong longshore currents and concomitant shear instabilities.

MODEL CALIBRATION

Model verification is addressed in three sections. First, the wavenumber/frequency ranges predicted by the shear instability model are compared with the estimated f-K spectra obtained using the two alongshore arrays. Second, a method is described to remove infragravity contamination from the energy density spectra measured at the individual current meters using coherence between horizontal velocities and surface elevation; the results are compared with the energy partitioning provided by f-K spectra estimated for each of the two alongshore arrays (i.e. where infragravity energy is identified via the dispersion curves). Finally, the stream functions are calibrated, and the $\overline{u'v'}$ profile is compared with values measured at each of the nine current meters in the cross-shore array.

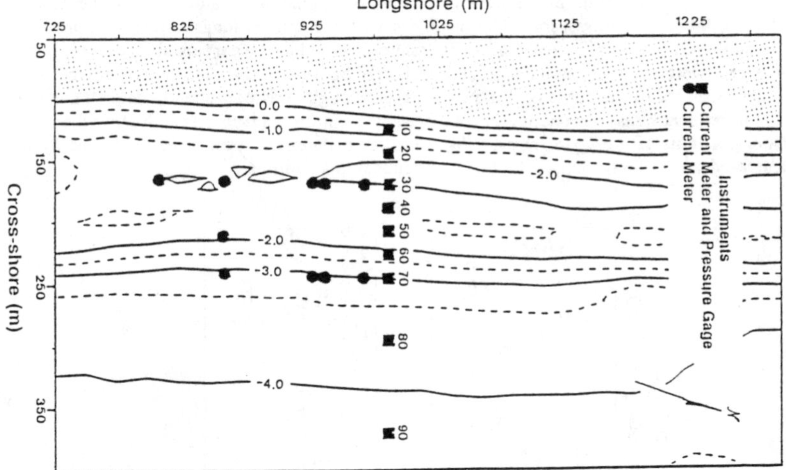

Fig.2 Current meter and pressure gage locations overlaying bathymetry for 12 OCT 90.

The cross-shore profile of the longshore current is required as input to the shear instability model and in the present work is obtained through application of a cubic spline to the two-hour mean observations at the nine locations. Dodd *et al.* 1992 noted the sensitivity of their shear instability model to the "smoothness" of the longshore current profile and opted to use a model predicted current profile as input in order to minimize any discontinuous derivatives. In the present study, using a spline with subjectively chosen weighting assigned to individual points produces a smooth profile in sufficient agreement with the observations. The splined profile is shown in Fig. 3 together with the measured bathymetry and the calculated background vorticity. The background vorticity exhibits a relative minima, corresponding to an elimination of the restoring force, just shoreward of the bar crest.

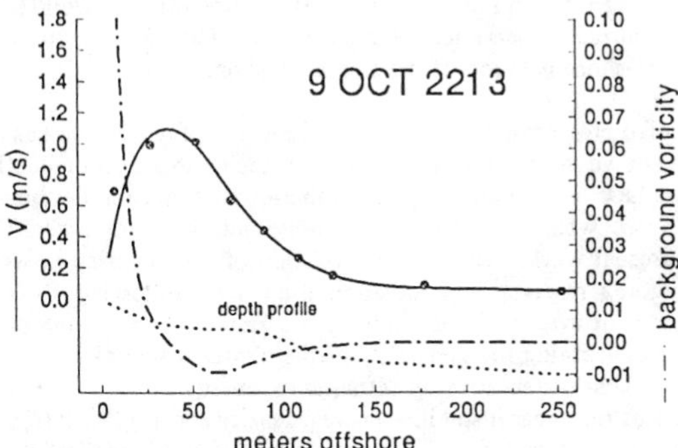

Fig. 3. Mean current observations, splined profile with calculated background potential vorticity, and measured bathymetry.

Two examples of model predicted stream functions (cyclic alongshore wavenumbers 0.00125, 0.005) are shown for 9 Oct, in Fig. 4 with the splined longshore current profile is overlaid. Differences in the frequency range over which shear instability energy was observed were routinely noted during DELILAH between the trough array and the array located seaward of the bar (Fig. 1 a & b) with the trough array consistently detecting energy over a frequency range extending beyond that of the more seaward array. Noting the cross-shore positions of these two arrays (Fig. 4), these differences seem to be explained by the decreased cross-shore span of the higher frequency instabilities.

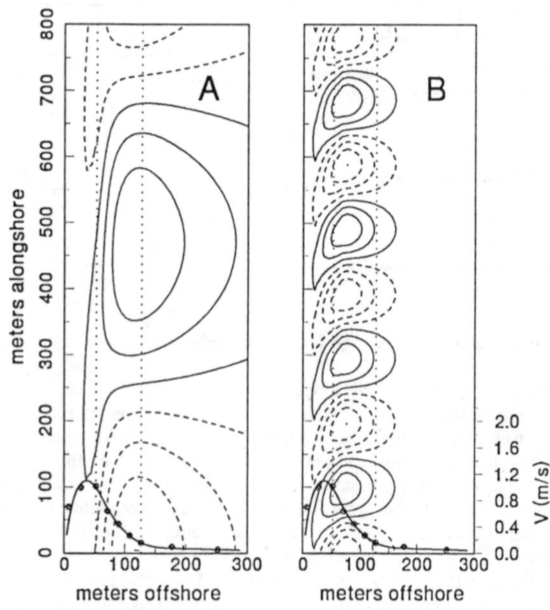

Fig. 4. Model generated stream functions, a) f=.001Hz/K=.00125, b)f=.004Hz/K=.005 (aspect ratios 1:1). Mean current observations and spline superimposed at bottom. Vertical dotted lines indicate cross-shore position of the two alongshore arrays.

Model predicted growth rate, ω_{im}, and frequency, $f=\omega_{re}/2\pi$, versus cyclic alongshore wavenumber are shown in Fig. 5 for two choices of the bottom friction coefficient, (c_f =0.002 and 0.004). The lack of predicted growth in the lowest wavenumber bins for the c_f=0.004 case does not agree well with the observations and therefore c_f=0.002 has been used throughout the present work. The parameterization of bottom friction within the surfzone is a topic of ongoing research and the form used here, following Dodd *et al.* (1992) is considered an order of magnitude estimate. As a result it is not appropriate to attempt to draw conclusions concerning the generality of the chosen value of c_f.

Contamination of shear instabilities by infragravity energy

Estimates of the shear instability energy density spectra (S_{ske}), from each of the nine cross-shore current meters are required to calibrate the amplitudes of the model-produced stream functions. The two-hour record length was broken up into 8 sub-records, with 50%

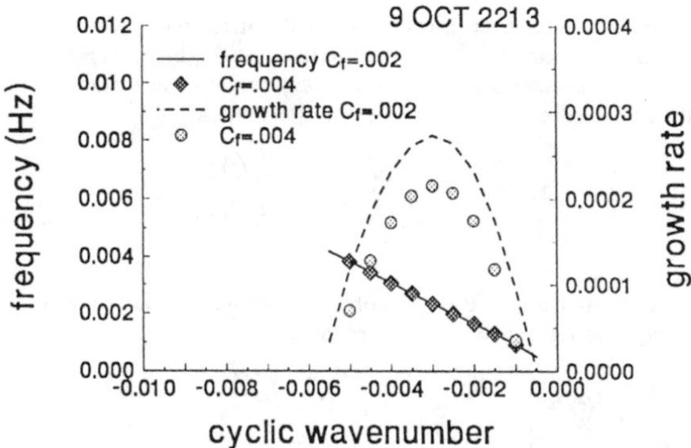

Fig. 5. Cyclic alongshore wavenumber versus both frequency and growth rate predicted for bottom friction coefficient values of 0.002 and 0.004.

overlap, based on the required record length necessary to produce the desired frequency resolution of 0.0005. The resulting degrees of freedom (16) produce large confidence intervals. Conversely, a record length long enough to produce significantly more degrees of freedom strains the steady-state assumption. While the shape of the estimated spectra is used within the calibration process, the final comparisons between model and data use variances obtained by integrating over frequency.

It is assumed that the measured spectra, S_{ke}, calculated for each of the 9 current meters in the cross-shore array, includes both shear instability energy, S_{ske}, as well as infragravity wave contamination, S_{ike}. Separation of these two categories may be done at the two locations shared with the alongshore arrays by examining the f-K analysis. For the other 7 current meters in the cross-shore array, this may not be done so readily, as a method to partition shear instability energy from infragravity energy using u,v, and pressure is not known. In the present work the coherence between the velocity components and the surface elevation signature is used. Bowen and Holman (1989) demonstrated that the rigid lid assumption was reasonable in the case of shear instabilities and so the cross-spectral coherence between η and the horizontal velocity components is assumed to be zero. Conversely, the coherence for infragravity waves is assumed to be unity. This later assumption is somewhat tenuous owing to possible mixed-mode effects, which might produce significantly reduced coherences, (particularly in the case of alongshore velocity component). Although this method requires rigorous testing, an alternative method is presently lacking. Comparison of the energy partitioning so produced with that done through the f-K analysis will be shown.

Following Thornton (1979) in which the coherence between surface elevation and velocity, i.e.

$$\gamma^2_{\eta u}(f) = \frac{|S_{\eta u}(f)|^2}{S_\eta(f)\, S_u(f)} \tag{11}$$

was suggested as a means to separate the wave energy from alongshore component of the horizontal turbulence, the velocity spectrum in the frequency range of interest is now considered composed of an infragravity component, ui, and a shear instability component, us, where it is assumed the shear instabilities and infragravity waves are statistically independent. The cross-spectra of u and η is written as $S_{\eta u}(f)=S_{\eta ui}(f)+S_{\eta us}(f)$. Again it is expected that the cross-spectra between η and u is zero for shear instabilities, and the last term is neglected. Substituting, the coherence may be written:

$$\gamma_{\eta u}^{2}(f) = \frac{|S_{\eta ui}(f)|^2}{S_{\eta}(f)S_{ui}(f)}\left[1+\frac{S_{us}(f)}{S_{ui}(f)}\right]^{-1} \tag{12}$$

The first term on the right hand side is the coherence between η and ui which is assumed equal to one. Finally, the ratio of S_{ui} to S_u is obtained

$$\gamma_{\eta u}^{2}(f) = \left[1+\frac{S_{us}(f)}{S_{ui}(f)}\right]^{-1} = \left[\frac{S_{ui}(f)+S_{us}(f)}{S_{ui}(f)}\right]^{-1} = \frac{S_{ui}(f)}{S_{u}(f)} \tag{13}$$

The same process is carried out for v, ultimately yielding

$$S_{ike}(f) = \gamma_{\eta u}^{2}(f)S_{u}(f) + \gamma_{\eta v}^{2}(f)S_{v}(f) \tag{14}$$

It should be noted that the assumptions made lump any turbulent energy at the same frequencies into the category of shear instability energy.

The shear instability spectrum, S_{ske}, is calculated by subtracting the magnitude of the infragravity contamination, S_{ike}, from the measured S_{ke}. The method of calculating the shear instability spectrum using a single meter location is checked at the two alongshore arrays where the estimated f-K spectra may be used to partition energy by both frequency and alongshore wavenumber, where energy with K within the zero mode dispersion curves is assumed associated with infragravity waves. S_{ike} spectra calculated using these two methods are compared in Fig. 6.

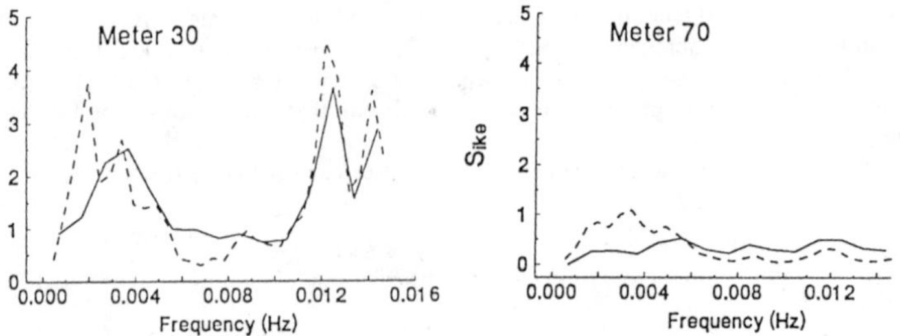

Fig 6. Infragravity energy density calculated using surface elevation-horizontal velocity coherence method (dashed) and f-K spectral method using alongshore arrays (solid).

Stream function calibration

Calibration of the model predicted stream functions is performed by assuming that the growth term in Eq.(8), (e.g. $\exp(\omega_{im}t)$) can be represented in a steady-state form by an amplitude variable, A(f). A(f) is then solved for by fitting the modeled kinetic energy density surface, $\frac{1}{2}(\overline{u'^2+v'^2})(f,x)$, to the measured data in f-x space. The modeled $\overline{u'^2}(f,x)$ and $\overline{v'^2}(f,x)$, averaged over one wavelength in the alongshore direction are re-written:

$$\overline{u'(f,x)^2} = \frac{A(f)^2 k(f)^2}{2h(x)^2}\left(\phi_r(f,x)^2 + \phi_i(f,x)^2\right) \quad (15)$$

$$\overline{v'(f,x)^2} = \frac{A(f)^2}{2h(x)^2}\left[\left(\frac{\partial\phi_r(f,x)}{\partial x}\right)^2 + \left(\frac{\partial\phi_i(f,x)}{\partial x}\right)^2\right] \quad (16)$$

The model predicted $\frac{1}{2}(\overline{u'^2+v'^2})(f,x)$ is the sum of Eq. (15) & (16) and is represented spectrally as S_{mske}. Using 9 OCT as an example, the 9 current meters produce 9 lines of $S_{ske}(f)$, one at each of the cross-shore positions of the meters. This surface is splined over the x direction such that the grid density is increased to match that of the model output (Fig. 7). For the model output, there are 11 frequencies which have growth predicted, and so there are 11 profiles of $S_{mske}(x)$ distributed over frequency. Each of these 11 profiles of $S_{mske}(x)$ is scaled independently to the corresponding profile from S_{ske} such that one A(f) value is obtained for each frequency for which the model predicts growth. Above the maximum frequency for which growth is predicted A(f) is set to zero. Observed energy above this cut-off is treated as noise and no fitting between S_{mske} and S_{ske} is attempted. Scaling is done at each frequency for which growth is predicted in a best fit manner based upon equal areas under the curves over the cross-shore region between the second and seventh meters. These profiles taken collectively produce the calibrated S_{mske} surface.

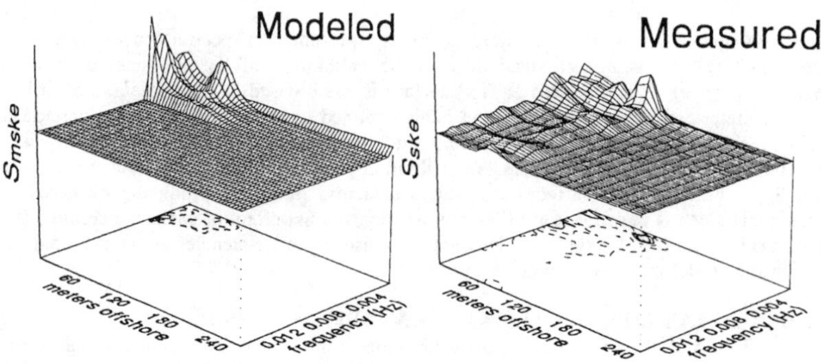

Fig. 7. Kinetic energy density f-x surfaces, modeled (S_{mske}) and measured (S_{ske}), splined, after removal of infragravity contamination.

Comparison of the net effect of shear instabilities is done by integrating S_{ske} and S_{mske} across frequency producing observed and predicted $\frac{1}{2}(\overline{u'^2+v'^2})(x)$ variance profiles (Fig. 8). Both surfaces are integrated over the entire frequency range considered (0.0005 - 0.015) and therefore the observed profile contains energy which has been regarded as noise because of being above the frequency range predicted by the model to experience growth.

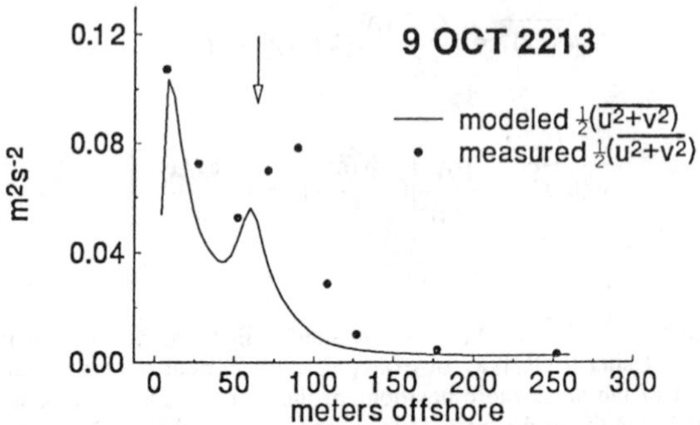

Fig. 8. Frequency integrated $\frac{1}{2}(\overline{u'^2+v'^2})$ profile (modeled) and observations, arrow indicates cross-shore position of the bar.

The model predicted $\overline{u'v'}_m$(f,x) profile is obtained using the previously calibrated stream functions:

$$\overline{u'v'}_m(f,x) = -\frac{A(f)^2k}{2h(x)^2}\left(\phi_r(f,x)\frac{\partial\phi_i(f,x)}{\partial x} - \phi_i(f,x)\frac{\partial\phi_r(f,x)}{\partial x}\right) \quad (17)$$

where m again denotes model predicted. This is represented in spectral density as S_{muv}(f,x). Inserting the A(f) values obtained through the calibration of S_{ske} produces a calibrated S_{muv}(f,x) (Fig. 9). The observed S_{uv}(f,x) surface is again based on spectra obtained at the 9 current meter cross-shore positions and is also splined over the x direction to increase the grid density to match that of the model output. It is emphasized that splining of S_{uv} is for comparison of the surfaces only as the calibration process is only carried out for S_{mske} and not S_{muv}. Finally, these surfaces are integrated across frequency producing observed and predicted $\overline{u'v'}$(x) covariance profiles representing the net effect of shear instabilities (Fig. 10). Once again the observed profile contains "noise" from frequencies above the band over which the model predicts growth.

SHEAR INSTABILITY MODEL/DATA COMPARISON

Two methods of evaluating the calibration of the stream function magnitude are the comparison of the observed and modeled S_{ske} and S_{uv} spectral shapes (in f-x space) and the integrated $\frac{1}{2}(\overline{u'^2+v'^2})$ and $\overline{u'v'}$ profiles. Comparison of the observed S_{ske} and modeled S_{mske} surfaces (Fig. 7) indicates at least qualitative agreement with some apparent "compression" of the model's cross-shore structure. The calibration term, A(f), does not vary in the cross-shore direction and so this spatial compression cannot be remedied through calibration. This same shoreward shift was observed by Church *et al.* (1992). At this time

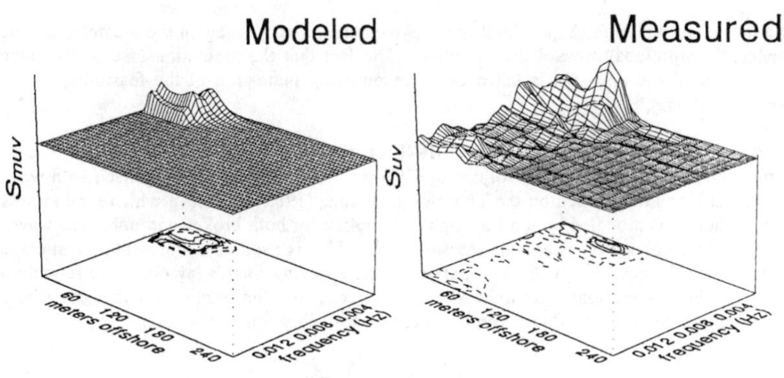

Fig. 9. Covariance energy density (S_{uv}) f-x surfaces, modeled and measured.

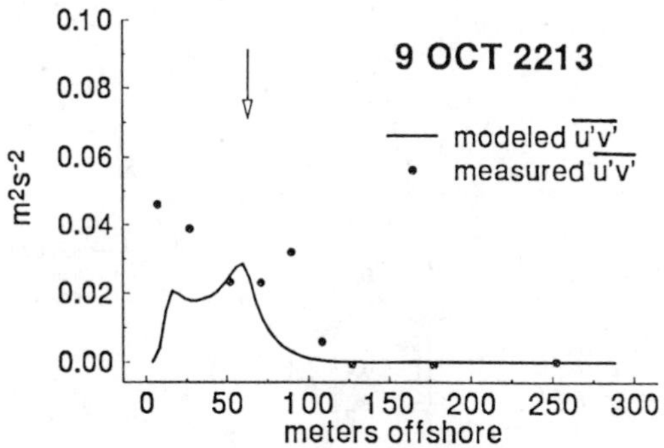

Fig. 10. Frequency integrated $\overline{u'v'}$ profile (modeled) and observations, arrow indicates cross-shore position of the bar.

it cannot be determined if this is an indication of some deficiency within the physics of the model. This compression is perhaps more easily seen in the integrated $\frac{1}{2}(\overline{u'^2+v'^2})$ profile (Fig. 8), which has been annotated with arrows marking the cross-shore location of the bar.

Examining the observed/predicted S_{uv} surface (Fig. 9) and $\overline{u'v'}$ profile (Fig. 10) reveals similar behavior. It is emphasized that the $\frac{1}{2}(\overline{u'^2+v'^2})$ profile was used to calibrate the stream functions and therefore good agreement should be expected. Comparison of the S_{uv} surface and $\overline{u'v'}$ profile test the modeled relationship between $\frac{1}{2}(\overline{u'^2+v'^2})$ and $\overline{u'v'}$

via the stream function shape. Qualitative agreement is seen, but again the predicted profile appears "compressed" toward the shoreline. The fact that the predictions are of the same general magnitude and sign is taken as an encouraging indication of the feasibility of the overall method.

APPLICATIONS TO LONGSHORE CURRENT MODELING

Longshore currents are modeled following a Thornton and Guza (1986) approach with wave height transformation based on the Thornton and Guza (1983) model, in which randomness in wave height is modeled by the Rayleigh distribution for both broken and unbroken waves inside and unbroken waves outside the surf zone. H_{rms} is used as a representative statistic of the ensemble wave height transformation. After applying Snell's law for wave refraction based on the assumption of straight and parallel contours, the gradient of the alongshore wave-induced momentum flux given by linear wave theory may be written:

$$\rho\frac{\partial}{\partial x}\int_{-h}^{0}\overline{\tilde{u}\tilde{v}}dz = \frac{\sin\alpha_0}{c_0}\frac{\partial}{\partial x}(EC_g\cos\alpha) \qquad \textbf{(18)}$$

in which the ensemble averaged wave energy, $E=\frac{1}{8}\rho g H_{rms}^2$, α is the wave incidence angle from shore normal, c is the wave phase speed (the subscript 0 denotes offshore conditions), and C_g is the wave group velocity. The H_{rms}-wave height transformation, bathymetry and H_{rms} observations are plotted in Fig. 11.

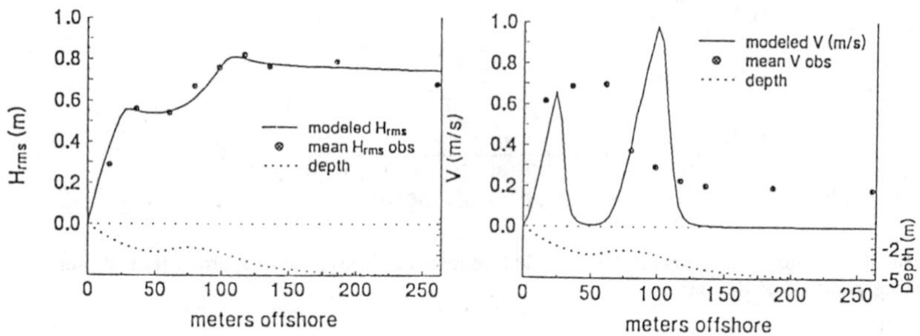

Fig. 11. H_{rms} transformation model.

Fig. 12. Longshore current model without mixing.

The linearized bottom stress is written

$$\overline{\tau^b_y} = c_f\rho u_m V \qquad \textbf{(19)}$$

with u_m as the maximum near-bottom wave induced orbital velocity.

With the turbulent radiation stress term omitted from Eq. (1) (i.e. no mixing) the predicted longshore current profile obtained shows poor agreement with observations (Fig. 12). A natural question is what $\overline{u'v'}$ profile would be required to balance the difference between the wave generated radiation stress and the bottom shear stress. The turbulent radiation stress profile required to balance Eq. (1) is calculated by assuming that the wave-induced radiation stress term is reasonably modeled and inserting the observed (splined) longshore current profile into the bottom stress term. The "required" profile (Fig. 13) is compared with the profile predicted previously by the shear instability model. Qualitative agreement can be seen with respect to both magnitude and structure, with the same "compression" toward the shoreline previously discussed.

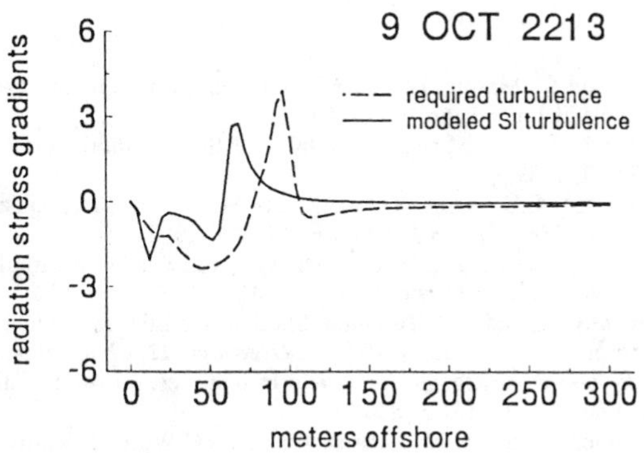

Fig. 13. Turbulent radiation stress gradient required to justify wave forcing and observed longshore current profile (dashed) and turbulent radiation stress gradient associated with integrated shear instability $\overline{u'v'}$ profile (solid).

SUMMARY AND CONCLUSIONS

The magnitude and cross-shore structure of the $\overline{u'v'}$ associated with shear instabilities of the longshore current have been examined. A "smoothing" spline has been used to obtain the longshore current profile required as model input. Shear instability frequencies and growth rates are predicted. Using measured energy density spectra to calibrate the model-generated stream function amplitudes, an absolute reference is used to obtain dimensional values of $\overline{u'^2}$, $\overline{v'^2}$ and $\overline{u'v'}$. It is again acknowledged that the coherence method applied to the partitioning of shear instability and infragravity energy is susceptible to mixed-mode effects. Although the limited data examined shows reasonable comparison between the coherence method and the f-K partitioning, an improved method is required.

Agreement is found between both modeled and observed $\frac{1}{2}(\overline{u'^2+v'^2})$ and $\overline{u'v'}$, although the model's consistent shoreward compression of the stream function structure is not presently understood. $\overline{u'v'}(x)$ associated with shear instabilities appears to be a significant source of mixing in the nearshore. The modeled mixing produced by the shear instabilities appears to be in qualitative agreement with that required to reconcile the disparity between model predicted longshore current profiles and observations.

ACKNOWLEDGEMENTS

The authors wish to express their appreciation to all those who participated in the DELILAH experiment and in particular the staff of the Coastal Engineering Research Center's Field Research Facility, under the direction of William Birkemeier. Bob Guza, Scripps Institution of Oceanography, is gratefully acknowledged for the use of the data acquired by his two alongshore arrays. Nick Dodd of the Institute for Marine and Atmospheric Research, Utrecht University, is thanked for his initial development of the shear instability model code while formerly associated with the Naval Postgraduate School. In addition, special appreciation is expressed to Rob Wyland and Tim Stanton, Naval Postgraduate School, and Katie Scott, University of California, Santa Cruz, for their roles in acquisition of wave and current data, and to Mary Bristow, Naval Postgraduate School, for initial processing of the data. EBT was funded by Office of Naval Research Coastal Sciences Grant N00014-92-AF-0002, JOS was funded by ONR Coastal Sciences Grant N00014-92-C-0014, and JCC is an ONR fellow.

REFERENCES

Birkemeier, W.A., and C. Mason, The CRAB: a unique nearshore surveying vehicle. *J. Surv. Eng.* ASCE, vol. 110, no. 1, pp. 1-7, 1984.

Bowen, A.J., and R.A. Holman, Shear instabilities of the longshore current. *J. Geophys. Res.* **94**, 18,023-18,030, 1989.

Church, J.C., Thornton, E.B., and J. Oltman-Shay, Mixing by shear instabilities of the longshore current. *Proc. 23rd Coastal Engr. Conf., ASCE*, 1992.

Church, J.C., and E.B. Thornton, Effects of breaking wave induced turbulence within a longshore current model. *Coastal Engng.* (20), 1-28, 1993.

Dodd, N., J. Oltman-Shay, and E.B. Thornton, Shear instabilities in the longshore current: A comparison of observation and theory. *J. Phys. Oceanogr.* **22**, 62-82, 1992.

Dodd. N., and E.B. Thornton, Longshore current instabilities: growth to finite amplitude., *Proc. 23rd Coastal Engr. Conf., ASCE*, 1992.

Mei, C.C, The applied dynamics of ocean surface waves, World Scientific, New Jersey, pp740, 1989.

Oltman-Shay,J., P.A. Howd, and W. A. Birkemeier, Shear instabilities of the mean longshore current, 2, Field Observations. *J. Geophys. Res.* **94 (C12)**, 18,031-18,042, 1989.

Pawka, S.S., Wave directional characteristics on a partially sheltered coast. Ph.D. dissertation, Scripps Inst. of Oceanography, UCSD, 1982.

Putrevu U., and I.A. Svendsen, Shear instabilities of longshore currents: a numerical study. *J. Geophys. Res.* **97 (C5)**, 7283-7303, 1992.

Thornton,E.B., Energetics of breaking waves within the surfzone, *J. Geophys. Res.* **84 (C8)**, 4931-4938, 1979.

Thornton,E.B., and R.T. Guza, Transformation of wave height distribution, *J. Geophys. Res. 88*, 5925-5938, 1983.

Thornton, E.B., and R.T. Guza, Surf zone longshore currents and random waves: Field data and models. *J. Phys. Oceanogr.*, **16,** 1165-1178, 1986.

APPLICATION OF A PARAMETRIC LONG TERM MODEL CONCEPT TO THE DELRAY BEACH NOURISHMENT PROGRAM

Michele Capobianco[1], Huib J. De Vriend[2], Robert J. Nicholls[3], Marcel J.F. Stive[4]

ABSTRACT: At scales larger than those where processes are well understood (≥ years), it is difficult to set up traditional process-based models. However, there is a growing demand for predictions of shoreline behaviour at these longer timescales, particularly for beach nourishment. Therefore, we are developing a framework based on parametric (*behaviour-oriented*) modelling which may be applied to areas where limited observations are available. With the present paper we examine the application of our concepts to a real world case, viz. the beach response at Delray Beach, Florida, to nourishment over a 20 year period. The problems of dealing with profile data are considered. The simple diffusion-based model we utilize cannot reproduce non-uniform behaviour. However if beach volume above a reference contour is considered, the model works well and predicts the erosive losses between renourishments quite effectively. Further improvements and application of the model is considered.

INTRODUCTION

Shore nourishment is increasingly being applied as a method of erosion control as it maintains a wide beach, providing both coastal protection and recreation. Of

---

1) System Engineer, R&D Div., Tecnomare S.p.A., San Marco 3584, 30124, Venezia, Italy.

2) Professor, University of Twente, Section of Civil Engineering and Management, PO Box 217, 7500 AE Enschede, The Netherlands.

3) Assistant Research Scientist, Lab. for Coastal Res., Univ. of Maryland, Lefrak Hall, College Park, Maryland 20742, USA.

4) Professor, Netherland Centre for Coastal Research, c/o Delft Hydraulics, PO Box 152, 8300 Ad Emmeloord, The Netherlands.

course, sand placed in this manner along a uniform sandy beach can be viewed as a perturbation which tends to be smoothed out by long-shore transport and modified by cross-shore transport. Models of this longshore smoothing phenomenon on the medium and long term range (scales of years to decades) are usually *diffusion-type* models which are able to take into account the long-shore evolution. For coastal management applications and for realistic evaluation of nourishment performance, the cross-shore evolution is also of prime interest. As an example, the study of (Lippman & Holman, 1990) underlines the importance of cross-shore processes for the coastal response to hydrodynamic forces. Practical questions related to nourishment are:

- How much sand has to be placed and where?
- What will the directions of transport be?
- How will be the beach affected?
- How will the Nourishment affect the behaviour of the beach in time?
- When will it be necessary to renourish?

In order to quantify project benefits and the cost of restoration it is also important to define the speed of response of profile nourishments. Erosion of the nourished volume can be described by the sum of a linear and an exponential term (Verhagen, 1992). Present design methods for nourishments are simple and reliable enough for many applications and no complex models are in principle required for this purpose. The disadvantage is that time history of beach profile data need to be available. Unfortunately sufficient experimental data are not always available and, in order to assess something about the distribution of nourishment across the profile and to possibly have some quantitative evaluation of beach erosion when cross-shore processes are important, it is necessary to use models for the evaluation of cross-shore evolution. Examples of optimum beach fill design cross-section through the use of a numerical model to simulate storm induced beach changes are given by (Hansen & Byrnes, 1991).

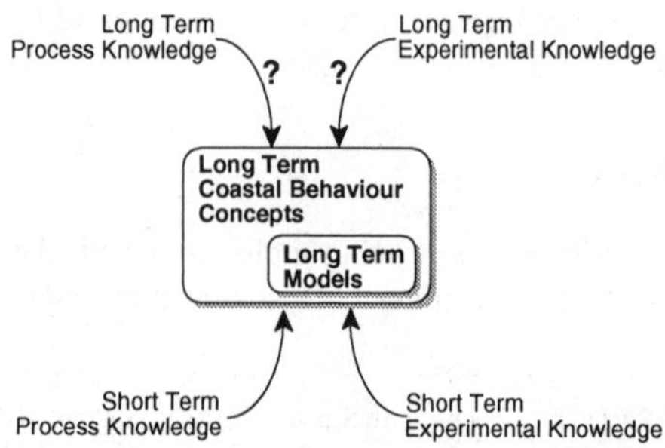

Figure 1. Approach for the development of long term models.

(Larson & Kraus, 1993) examine calculation procedures for the cross-shore transport rate at different scales. However, at scales larger than existing detailed process understanding (≥ years) it is difficult to set up traditional process-based models. Therefore, we are working to obtain a framework, based on some parametric (*behaviour-oriented*) modelling tools, that may be applied in a context where little experimental information (especially historical data) is available, with the aid of validated short term process-based models (Capobianco et al., 1993). The idea is to link the available quantitative knowledge about short term processes with the available qualitative knowledge about long term processes (Figure 1). In the application to the evaluation of nourishment performance, they should be able to reproduce both static conditions and give an assessment of the transitions between different static conditions.

DELRAY BEACH

First we tested the validity of our concept for a hypothetical test case against a detailed process-based model used with real-life input (Stive et al., 1992; De Vriend et al., 1993). The idea was to generate extensive sets of reference profile evolution data to be reproduced by the parametric model. With the present paper we examine the application of our concepts to a real world case, thus making a further step toward the development of a practical and usable tool.

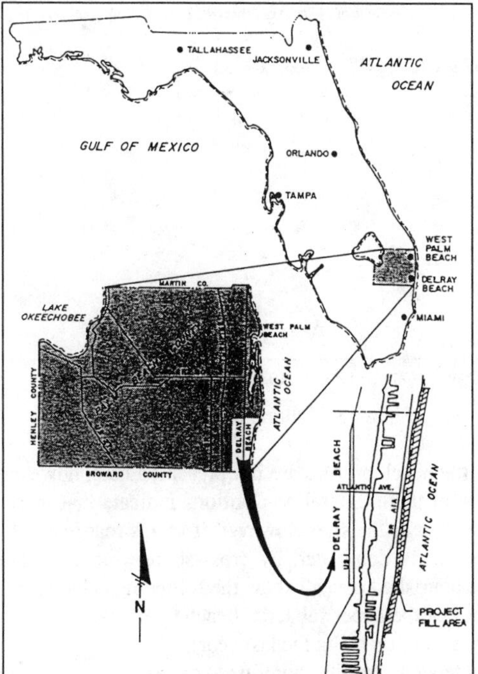

Figure 2. Delray Beach, location map (from Beachler, 1993)

We consider the beach response to the 20-year renourishment program performed at the City of Delray Beach, Florida (Figure 2). A description of the nourishment program can be found in (Beachler, 1993); we briefly review some basic information. The first beach nourishment was put into place in 1973. A total of about 1.250.000 cubic meters of sand were placed along 4.350 meters of shoreline. In 1978 the beach was renourished by placing 550.000 cubic meters of sand along 2.700 meters fronting public beach areas. In 1984 a second renourishment was carried out and placed approximately 1 million cubic meters of sand along the original 4.350 meters of shoreline.

As part of the monitoring program, beach profile surveys have been conducted from a point landward of the dune seaward to beyond the 18-feet (5.5-meters) depth contour. Volumetric information concerning the evolution of the nourishment has been derived from these data (Beachler, 1993; see Figure 3). More detailed data are available concerning the second periodic nourishment (post-1984). This consists of the evolution of 16 profiles from plus 10 to 15 feet to minus 20 to 30 feet. The 16 profiles are labelled from R175, corresponding to the north limit, and R190, corresponding to the south limit.

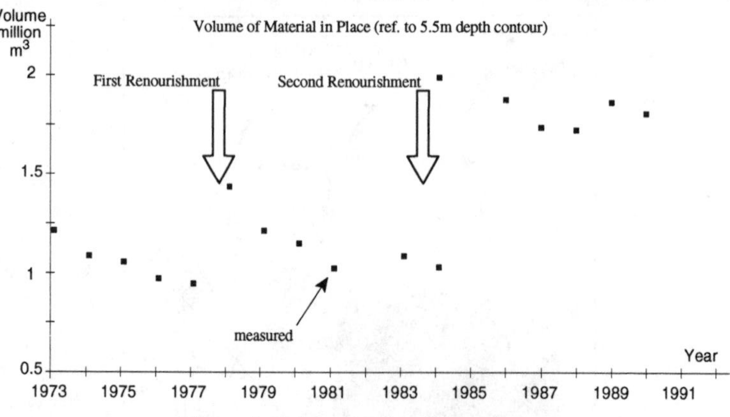

Figure 3. Volume of material in place.

The details of the beach profiles are not presented here, however we review some qualitative aspects of them. Shoreline positions indicate that there are some effects of longshore spreading at the site. However, it seems reasonable to assume that the central beach section is dominated by cross-shore processes. The profiles in the central part of the nourished area all show the following behaviour:

- a tendency of the upper subaerial beach to accrete
  (the only exception being the last year),
- a general tendency of the shoreline to erode,
- a general tendency of the upper subacqueous profile to erode,

- a systematic accretion of the middle part of the profile,
- a generally erosive tendency interrupted by occasional accretive periods such as in 1989 (also visible in Figure 3).

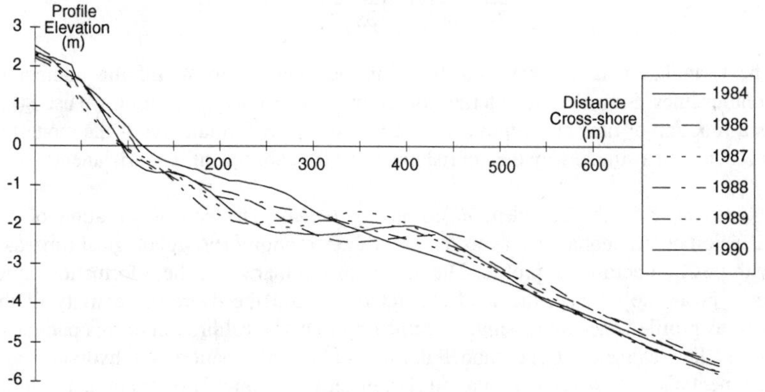

Figure 4. Mean Profiles (R179-R186)

Figure 4 shows the evolution in time of the averaged central part of the nourished area (which we intended to model). It is still possible to note the general erosive tendency and the tendency to form a bar in the upper part of the profile. Unfortunately the possibility to compare with the pre-nourishment profile is missing.

APPLICATION OF A DIFFUSION-TYPE MODEL

Formulation

Using this set of profile data we explored the application of our diffusion-type formulation for a behaviour oriented model of coastal profile evolution. The reference scheme is defined in Figure 5.

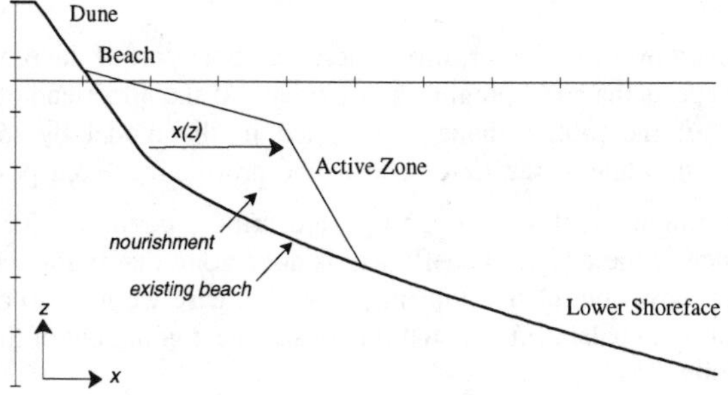

Figure 5 - Schematics of the Beach Nourishment

With appropriate initial and boundary conditions the *displacement of the cross-shore position with respect to the initial profile (x)* can be described as a function of *profile depth (z)*:

$$\frac{\partial x}{\partial t} = \frac{\partial}{\partial z}(D(z)\frac{\partial x}{\partial z}) + S(t,x,z)$$

The formulation is an extension of the n-line model with an infinite number of contour lines. S(t,x,z) is an external source function. In our application we use S(t,z) as a function of time and depth (thus we use the linear formulation of the model) to reproduce in a simple way the nourishment and the subsequent renourishments.

D(z) is a depth dependent *diffusion coefficient*. The spatial variation of the diffusion coefficient allows us to represent the variation of morphological timescale with position across the profile. The idea is to summarise all the information about the typical site climate, the sand characteristics and the degree of activity of the various profile zones into a single parameter: D(z). The calibration of this parameter is the key element of the model definition: all information, on hydraulic and sediment characteristics as well as on shorter-term dynamics is stored in it. The long-term objective of this work is to be able to directly express the parameters that give shape and value to D(z) as functions of mean environmental parameters (wave input and water level variations) and geometric characteristics of the profile.

A practical limitation to the approach is represented by the choice of a stationary diffusion term (and resulting transport) in the formulation. What we may expect in terms of output is a sort of "mean evolution" for the modelled period; in other words the model as such can only reproduce constantly accreting or constantly eroding situations. The stationarity also has strong implications on the behaviour of the solution and, in the ultimate analysis, on the character of the profile evolution that has to be reproduced. The evolution in the diffusion model is in fact influenced by the way we compare the profiles (modelled and measured ones), viz. raw data or "filtered/weighted" data. The implicit assumption is that the prevailing wave energy or the wave climate is approximately constant over time, so that a constant erosion rate is a meaningful representation of the actual profile behaviour.

When averaged over the shorter time scales, the upper part of the profile tends to maintain its shape as the coast progrades or retreats. At the upper end of the profile, the invariance of the profile shape is reflected in the model by the boundary condition $\partial x/\partial z = 0$, while at the lower end of the profile, the fixed position of the shoreface root implies that x = 0. This is the critical condition for the overall sediment balance. If the diffusion coefficient is not zero at that point, then we allow for sediment to move (out of the domain). In such a case we need to quantify this movement in order to balance it with the flux of sediment going out of the domain in the real world situation.

Practical Problems

The first problem using the profile data was the lack of pre-nourishment profiles or of a description of the spatial distribution of the nourishment. This was solved empirically by assuming an equilibrium reference profile. Then we conducted an exploratory calibration of D(z) against the available profile data (1984-1990).

The available set of data is far from having the ideal characteristics that we require for the application of our model as it is currently formulated. In order to briefly show the reasons, in Figures 6 to 9 we present some Empirical Orthogonal Functions (EOF) analysis of the data set, a technique which we also normally use for comparison with the model results. Figures 6 and 7 show the comparison of the first two eigenfunctions (both in cross-shore and longshore directions) for 1984 and 1990; they register the situation of the nourished area in the two periods.

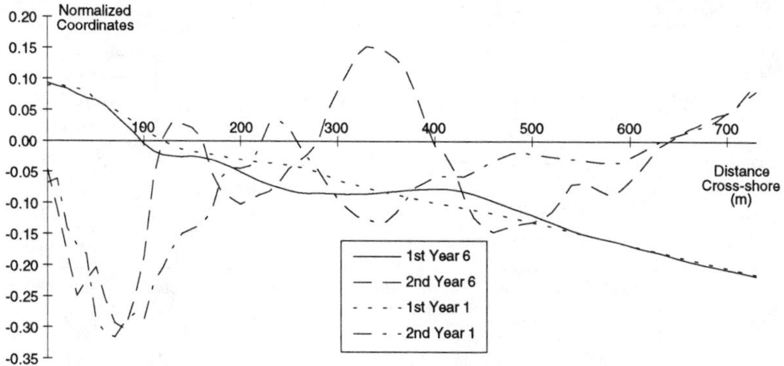

Figure 6. Cross-shore Eigenfunctions 1984 (Year 1) and 1990 (Year 6).

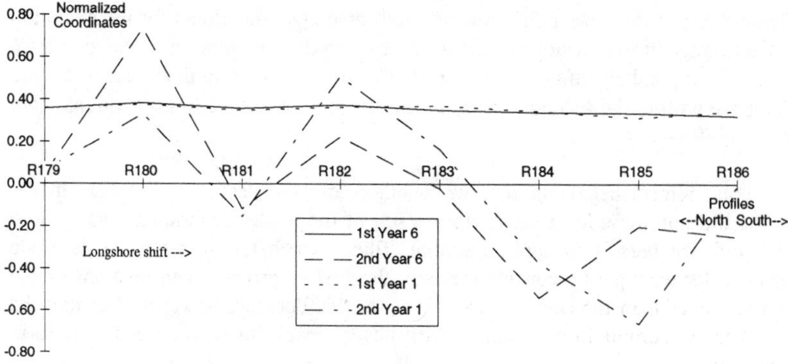

Figure 7. Longshore Eigenfunctions 1984 (Year 1) and 1990 (Year 6).

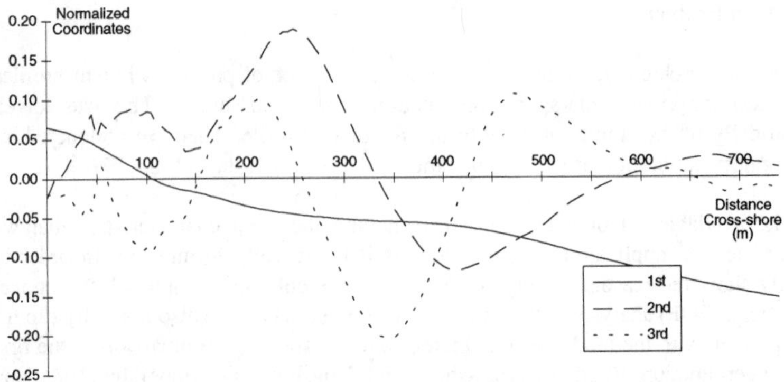

Figure 8. Cross-shore Eigenfunctions (Mean Profiles R179-R186).

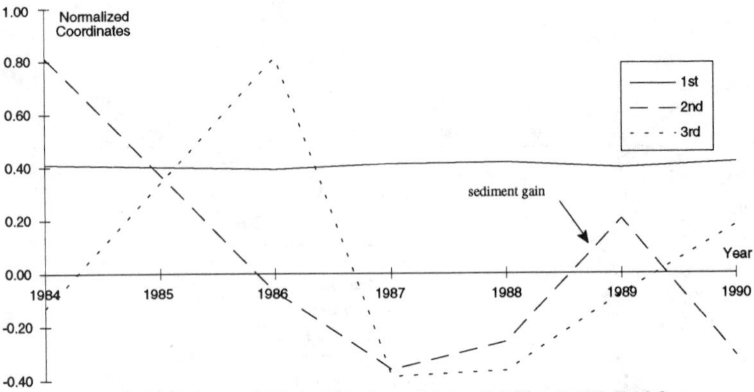

Figure 9. Temporal Eigenfunctions (Mean Profiles R179-R186).

Figure 8 and 9 show the EOF analysis (first three eigenfunctions) for the evolution of the average profile, which is what we try to reproduce with our model. Apart from the third temporal eigenfunction which shows a modification in the structure of bars along the profile, the second temporal eigenfunction clearly shows the volume gain of the 1989 survey.

The non uniform behaviour in time together with the occurrence of bars in the profile did not allow for a satisfactory result of the model calibration. One way to deal with the bars is to apply a spatial "filter" which removes the smaller-scale features, for example by considering only the first empirical eigenfunctions of the profile, rather than the actual shape. However this does not solve the fact that the model in its current formulation is intrinsically unable to reproduce non-uniform behaviour.

Application to Volume Data

We then decided to reduce our objectives to a more reasonable goal, thus considering as a way to calibrate the scale of the diffusion coefficient and to verify the model, the total sand volume between the dune crest and the reference 18 feet (5.5 m) depth contour. At Delray Beach, the total volume (and the sediment supply) is known as a function of time between 1973 and 1990. We might look at the volumetric data as an extreme synthesis of the profile data to be used to compare model results and field data. Figure 11 shows a comparison of the time evolution of the measured volume of sand and the model results with the maximum value of the diffusion coefficient calibrated at 1 m^2/day (Figure 10).

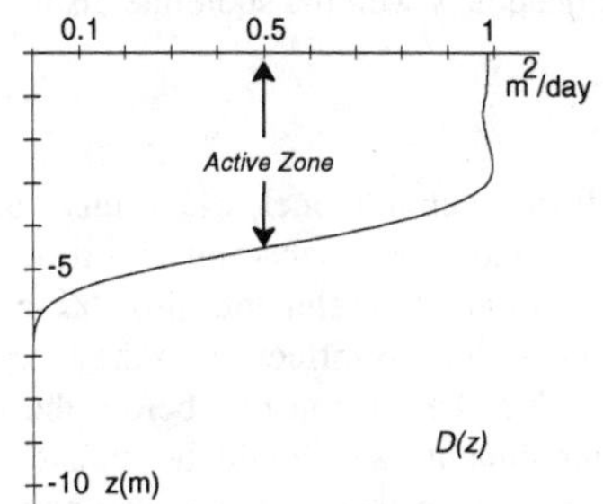

Figure 10. Applied shape of diffusion coefficient.

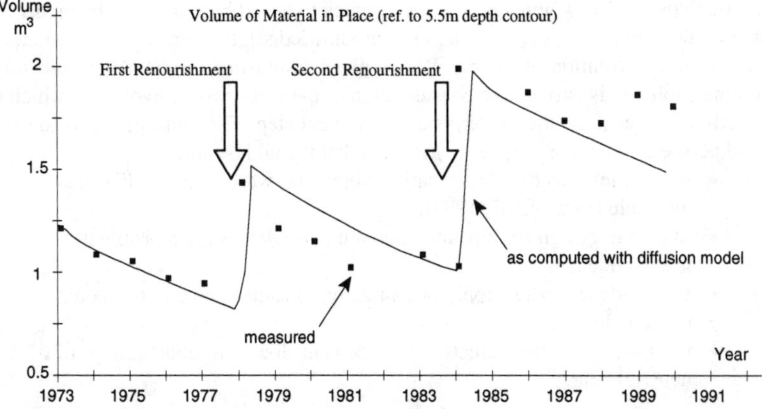

Figure 11. Evolution of the Volume of Material in Place (measured vs. computed)

The supply of sediment due to nourishment and renourishment is known. The influence of a sea-level rise is also considered in the model by using a time varying transformation of the z-domain. Together with possible longshore losses, wind losses and volumetric losses due to sink, these have been quantified in a *equivalent sea-level rise* of about 2 mm/year. This is the average value of sea-level rise for the Atlantic coast of Florida from 1950 to 1986 (Lyles et al., 1988).

Given the scatter of the basic data, the result is rather satisfactory, in that the rate of erosion is approximately correct whenever the coast is losing sand. In agreement with the observation of (Verhagen, 1992), such erosion is basically the sum of a linear and a non-linear (exponential) term, at least in the initial response to each renourishment (also see Kriebel & Dean, 1993). While the model used here fails to reproduce accretion, such events might be seen as perturbations around the long-term trend. Further, we model the evolution of the mean profile quite well and in future work we will investigate how well the shoreline displacements are reproduced.

CONCLUSION

The diffusion-type profile evolution model, when run as an initial value problem, is basically restricted to monotonous behaviour in time, i.e. either erosion or accretion. In the case of shore nourishment, this basic behaviour is erosion. Assuming that the nourishment has no influence on the long-term natural behaviour of the coast, which means that the erosion rate before the nourishment equals the erosion rate after the nourishment, we should be able to consider the diffusion coefficient to be only dependent on the reference (say equilibrium) profile and on extrinsic conditions (wave climate).

Although in every new application there is much to be calibrated about the model (in its present formulation and with the present knowledge), the verification of the total sand volume evolution at Delray Beach gives promising results. Once the total volume evolution is correct, the consideration of the mean profile evolution, which is directly reproduced by the model, is a logical next step. The concept seems to offer good perspectives in the long-term, assumed that it could be able:

- to take into account information about the wave climate, like that available from CERC (1993),
- to help in design and operate nourishments when limited profile data is available,
- to include the extra supply or extraction of sand to take into account possible losses,
- to investigate the effects of placement location and timing of nourishment.

In particular, with reference to the first point, we expect to be able to define correlations between maximum values and shape of the diffusion coefficient and the wave climate. Further availability of beach profile surveys and wave and water level information will provide the necessary basis to make the next step in the model development and to test beach fill design alternatives.

ACKNOWLEDGEMENTS

Delray Beach Data were kindly provided by Coastal Planning and Engineering, Inc., Boca Raton, Florida. This work has been carried out as part of the European Community MaST G8 Coastal Morphodynamics Projects (Directorate General for Science, Research and Development, MaST Contract no. MAS2-CT92-0027) and with the support of the Coastal Genesis Research Programme of The Netherlands (Rijkswaterstaat, Tidal Waters Division) and of the Researchers of Excellence Programme for Catalan Universities of the Generalitat de Catalunya.

REFERENCES

Beachler K.E., 1993. "Delray Beach Nourishment Program", *International Symposium on Coastal Geomorphology, Beach Erosion, Preservation of Beaches by New Technology*, Hilton Head Island, USA, June 6-9, pp. 151-161.

Capobianco M., De Vriend H.J., Nicholls R.J., Stive M.J.F., 1993. "Behaviour-Oriented Models Applied to Long-Term Profile Evolution", in List J.H., ed., *Large Scale Coastal Behaviour '93*, U.S. Geological Survey Open-File Report 93-381, pp. 21-24.

CERC, 1993. "Hindcast Wave Information for the US Atlantic Coast", Coastal Engineering Research Center, WIS Report 30.

De Vriend H.J., Stive M.J.F., Nicholls R.J., Capobianco M., 1993. "Cross-Shore Spreading of Beach Nourishment", *International Symp. on Coastal Geomorphology, Beach Erosion, Preservation of Beaches by New Technology*, Hilton Head Island, USA, June 6-9.

Hansen M.E., Byrnes M.R., 1991. "Development of Optimum Beach Fill Design Cross-Section". *Proc. Coastal Sediments '91*, American Society of Civil Engineers. New York, NY, pp. 2067-2080.

Kriebel D.L., Dean R., 1993. Convolution Method for Time-Dependent Beach-Profile Response. *Journal of Waterway, Port, Coastal and Ocean Engineering*, v. 119, n. 2, pp. 204-226.

Larson M., Kraus N.C., 1993. "Prediction of Cross-Shore Sediment Transport at Different Spatial and Temporal Scales", in List J.H., ed., *Large Scale Coastal Behaviour '93*, U.S. Geological Survey Open-File Report 93-381, pp. 96-99.

Lippman T.C., Holman R.A., 1990. "The Spatial and temporal Variability of Sand Bar Morphology". *Journal Geophys. Res.*, v. 95(C7), pp. 11575-11590.

Lyles S.D., Hickman L.E., Debaugh H.A., 1988. "Sea Level Variations for the United States: 1855-1986", *U.S. Department of Commerce* (NOS-NOAA), Rockville MD, 182 pp.

Stive M.J.F., De Vriend H.J., Nicholls R.J., Capobianco M., 1992. "Shore Nourishment and the Active Zone: A Time Scale Dependent View". *Proc. 23rd Int. Conf. on Coastal Engineering*, Venice, Italy, pp. 2464-2473.

Verhagen H.J., 1992. "Methods for Artificial Beach Nourishment". *Proc. 23rd Int. Conf. on Coastal Engineering*, Venice, Italy, pp. 2475-2485.

THE NOURTEC EXPERIMENT OF TERSCHELLING: PROCESS-ORIENTED MONITORING OF A SHOREFACE NOURISHMENT (1993-1996)

P. Hoekstra[1], K.T. Houwman[1], A.Kroon[1], P.van Vessem[2] and B.G. Ruessink[1]

ABSTRACT: A nearshore profile nourishment has been implemented in the surfzone of the barrier island of Terschelling. Sand is supplied in the trough between 2 nearshore bars. A research programme (NOURTEC) has started to monitor and explain the behaviour of the nourishment in the coming years. Process-oriented field experiments are carried out to explain the role of the hydrodynamic processes and the morphological conditions.

This paper gives a description of the project and a first impression of the results of a morphological and hydrodynamical analysis.

INTRODUCTION

Beach nourishment programmes have become common practice in order to combat coastal erosion, to maintain the existing coastline, to protect expensive coastal defence works and for widening beaches for recreational purposes. At present there is substantial knowledge about the engineering and environmental aspects involved in artificial beach nourishment (Van de Graaff et al. 1991 and Stauble and Kraus 1993). Recently shoreface nourishment has been gaining acceptance as a possible, cost-effective alternative for beach nourishment, e.g. by combining dredging activities with nearshore dumping. Shoreface nourishment, either shaped like a blanket of sediment or as an artificial submerged shore-parallel feeder berm is expected to be beneficial for the littoral system in two ways. Firstly, the onshore movement of sediment provides material to the beach- and surfzone system. Especially the wave-induced streaming mechanism, as proposed by Longuet-Higgins, in combination with wave asymmetry are processes which are held

[1,2] Netherlands Centre for Coastal Research (NCK)

[1] Institute for Marine and Atmospheric Research Utrecht, Utrecht University, P.O. Box 80.115, 3508 TC UTRECHT, the Netherlands

[2] Rijkswaterstaat, National Institute for Coastal and Marine Management (RIKZ), P.O. Box 20907, 2500 EX 's-GRAVENHAGE, the Netherlands

responsible for this onshore movement of sediment (compare e.g. Roelvink and Stive 1990).

Secondly, the change in nearshore morphology results in a decrease of the mean water depths. This increases the bottom friction and wave breaking. The associated increase in wave energy dissipation reduces the erosive wave action.

SHOREFACE NOURISHMENT

Experiments with shoreface nourishments have been carried out as early as the mid thirties (Atlantic City, 1935; reported by Mclellan and Imsand 1989). In literature a great number of studies are reported dealing with the behaviour of nearshore berms or offshore mounds composed of dredged material (e.g. Healy et al. 1991). These submerged features, however, are usually not created to nourish the beach or protect the coast. Recent shoreface nourishment projects have been carried out in the USA (Mobile Harbor; Burke et al. 1991) and in Australia (Kirra Beach Gold Coast; Jackson and Tomlinson 1990). Most of these projects have in common that the monitoring and evaluation of the nourishments or the submerged offshore structures remains limited to an analysis of the morphological and sedimentary developments of the offshore features in both time and space. Generally no explanations are offered - and no data are available - on the hydrodynamic processes determining the morphological changes. As a result no general conclusions or conceptual models on the behaviour of shoreface nourishments as a function of environmental conditions, are available; this seriously limits the predictability of the efficiency of the shoreface nourishment. Design criteria e.g. related to the location of the nourishment, the shape, and the dimensions are hard to define. One of the first attempts to derive a general guidance for design, based on engineering practice and elementary wave and grain size statistics, is presented by Mclellan and Kraus (1991). In addition, modelling can be an efficient tool to arrive at a series of more generalized conclusions with respect to the design and behaviour of nearshore profile nourishments (Van Alphen et al. 1990).

NOURTEC PROJECT

In 1992 the initiative was taken to set up the NOURTEC project. The name NOURTEC is an acronym for "Innovative Nourishment Techniques Evaluation". The project started in 1993 and is jointly funded by the EC MAST-2 programme and local authorities in Denmark, Germany and the Netherlands. The general objectives of NOURTEC are to study and explain the feasibility, effectiveness and optimum design characteristics of shoreface nourishment techniques for different environmental conditions. As a first step the behaviour of shoreface nourishments has to be determined. An important aspect of the programme is to explain this behaviour in terms of relevant hydrodynamic processes and conditions. In addition the feedback of the nourishment on these processes and the morphological features in the natural nearshore environment has to be understood.

In the final phase, NOURTEC aims at generalized conclusions with respect to the applicability and behaviour of nearshore profile nourishments.

In the framework of NOURTEC comparative and process-oriented, full-scale field experiments are carried out in three different nations for three different types of coastal environments around the North Sea basin. In West Denmark, along the closed-barrier coast of the Jutland peninsula, a combined beach- and shoreface nourishment has been carried out by the Danish Coastal Authority (DCA). The two nourishments - the two sites are located at a longshore distance of 2 km - are subject to wave-dominated conditions in a microtidal environment. In Germany the Coastal Research Station on the East Frisian Wadden island of Norderney is responsible for the research and evaluation of a shoreface nourishment on the west coast of this island, near a tidal inlet. The beaches are locally protected by groins. Both waves and tides (mesotidal conditions) are expected to play a major role in the behaviour of the nourishment.

The largest experiment and the most extensive field measuring programme is executed on the barrier island of Terschelling, the Netherlands. This experiment implies a joint research effort of the Ministry of Transport, Public Works and Water Management, Directorate General Rijkswaterstaat (RWS; also the European coordinator of the NOURTEC project) and the Institute for Marine and Atmospheric Research (IMAU), Utrecht University.

In this paper, the NOURTEC experiment of Terschelling is discussed. An outline is given of the local hydrodynamical, morphological and sedimentological conditions. The design characteristics of the nourishment are briefly summarized. Also included is an overview of the scientific programme that accompanies the experiment. The second part of the paper presents the first scientific results on an analysis of the autonomous morphological developments in the coastal zone of Terschelling prior to the nourishment. In combination with results from the initial hydrodynamic measurements, the paper ends with preliminary conclusions on longshore and cross-shore dynamics.

ENVIRONMENTAL CONDITIONS

The Wadden island of Terschelling (Fig. 1) is part of a chain of barrier islands along the Dutch, German and Danish coast. The north coast of the island consists of sandy beaches and dunes. In the study area (Fig. 1; km section 10-22) the orientation of the shoreline is mainly WSW-ENE. On the seaward side the morphology of the nearshore zone is characterized by the presence of 2 or 3 breaker bars (Fig. 2) and the total surfzone width varies from approximately 1200 to 1500 m. The central part of the island is chronically suffering from erosion and the average annual coastal retreat is ca. 2-3 m/yr. This corresponds with an estimated annual loss of sediment of about 110.000 m^3 over a stretch of about 5 km.

Tides are semidiurnal with a neap tidal range of about 1.2 m and a spring tidal range of ca. 2.8 m (mesotidal). Eastward flowing flood tidal currents are somewhat stronger than westward flowing ebb currents. Average annual significant wave height (offshore at the -15 m isobath) is in the order of 1.5 m, the mean offshore significant wave period is about 8 s. However, the offshore wave height varies greatly throughout the year. During severe storms in the autumn or winter wave heights and associated wave periods increase to 5 or 6 m and 10-15 s respectively. Highest waves are commonly incident from the NW.

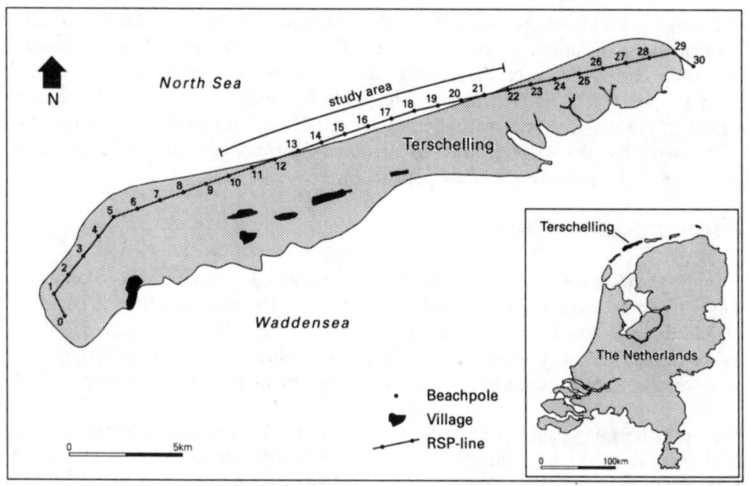

Figure 1. The barrier island of Terschelling along the northern part of the Dutch Coast. Cross-sections defined in this paper are related to the individual beach poles of the RSP-line (beach pole line).

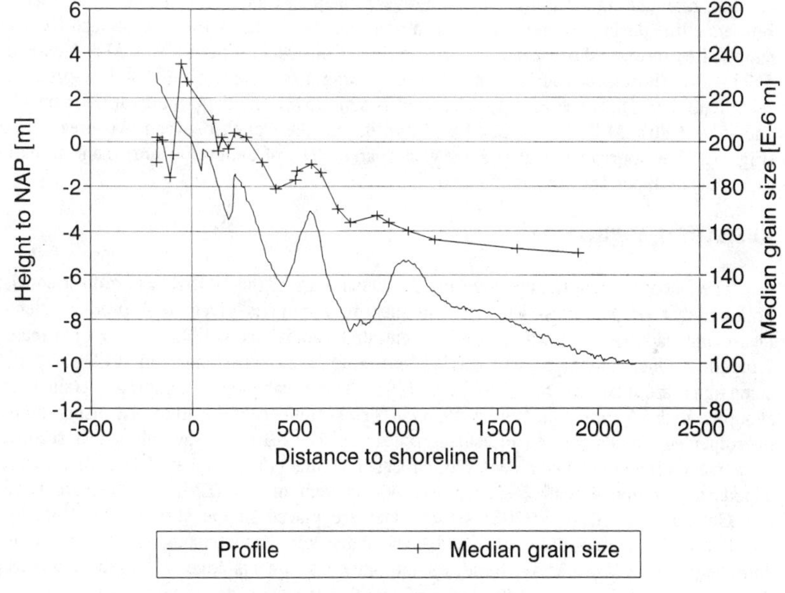

Figure 2. Nearshore bar morphology and median grain size distribution in cross section 18.

Median grain sizes (D_{50}) vary from 220 to 260 μm on the intertidal beach and show a gradual fining towards the lower shoreface (D_{50} 150 to 160 μm). Simultaneously there is a strong correlation with the morphological features. A representative profile (km section 18, see Fig. 1) is presented in Fig. 2. In the inner nearshore zone sediments on the crest of the bars are systematically coarser (10-40 μm) than sediments in the troughs. For the outer bar the coarsest samples commonly correspond with the landward facing slip face of the bar, showing a kind of spatial lag.

DESIGN CONSIDERATIONS

In 1990 the government of the Netherlands instituted a new national coastal defence policy: "dynamic preservation of the 1990 coastline". The present policy is based on the principle that the actual coastline may not retreat behind a fixed reference coastline, the basal coastline. The continuous erosion along the northern coast of Terschelling required the use of nourishment techniques to counteract the erosion and to maintain the coastline.

In a forthcoming paper (Van Vessem et al. in prep.) many aspects of the final design and dimensions of the nourishment will be treated in more detail. Some facts and figures are briefly summarized here. Results from modelling (Van Alphen et al. 1990) suggest that nearshore nourishments carried out landward of -7 m isobath have positive effects on the coastline. A shoreface nourishment, however, requires twice as much sediment as a conventional beach nourishment programme to achieve the same result. The model computations were essentially based on conditions observed along the central part of the Dutch coast, the Coast of Holland (Van Alphen et al. 1990). Considering the existing erosion rate and the requirement to prevent the coastline from retreating behind the reference line during the next 10 years, a total volume of 2.5 million m^3 of sand has been supplied to the nearshore zone of Terschelling. Construction period was May-November 1993. The shoreface nourishment at Terschelling has a length of 4.4 km (from km section 13.7 to 18.1; see Fig. 1). The sand is supplied in the depth interval between -7 m and -5 m below MSL (Mean Sea Level), filling up the trough between two breaker bars (Fig. 3). The nourished sand is somewhat coarser (D_{50} of about 200 μm) than the native sand (D_{50} of about 180 μm) in the area.

RESEARCH APPROACH

The scientific programme covering the Dutch part of the NOURTEC project consists of a number of activities. Emphasis in the programme is given to 4 process-oriented measuring campaigns. Each of the concentrated campaigns will last for approximately 2 months. One campaign was carried out after the nourishment in 1993 and new campaigns are planned for 1994 and 1995. These campaigns comprise detailed and synoptic hydrodynamic measurements at a great number of locations by using buoys, instrumented tripods (6) and measuring poles (2). The offshore wave climate is recorded by a wave-directional buoy (WAVEC) located at the -15 m isobath. The instrumented tripods are equipped with 2 electro-magnetic current meters (EMF), a pressure sensor and Optical Back Scatter (OBS) sensors and are placed in the surfzone. Instrumented poles, carrying a pressure sensor and capacitance wire, and tripods are organized in a dominantly cross-shore array; besides some other instruments cover a longshore transect. A Field Research Facility on the Beach of Terschelling supports the measurements.

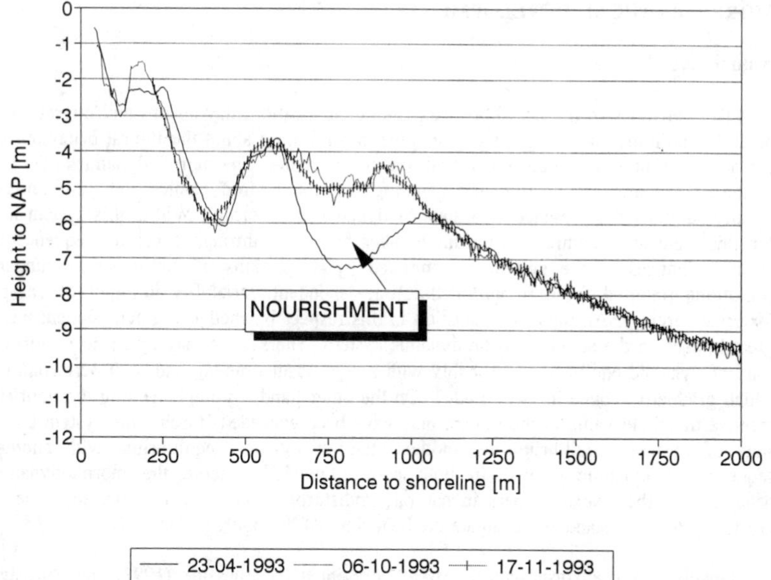

Figure 3. Nearshore profiles measured in cross-section 17 before and after the nourishment. The sand is supplied in the depth interval between -7 m and -5 m below Dutch ordnance datum (NAP), filling up the trough between 2 bars.

Measurements of waves, currents and suspended sediment concentrations are used to unravel the role of different hydrodynamic processes and their effect on resulting sediment transport patterns. The net result of these processes will finally determine the response of the nourishment.

In the NOURTEC monitoring scheme detailed soundings of the morphology are scheduled at approximately 2 month-intervals. Morphological information of the nearshore zone is also obtained in an experimental way by using radar images of the European Remote Sensing (ERS-1) satellite, in combination with a data-assimilation model (see Calkoen et al. 1993). Sediment samples are taken regularly. Meanwhile, meteorological and hydrodynamical boundary conditions are being monitored continuously during the project (1993-1996).

Prior to the nourishment a study has been carried out to analyse the autonomous, long-term morphological developments in the nearshore zone. Next to this, a T_0 campaign was launched in February-March 1993. An inventory survey of the bathymetry, hydrodynamic processes and sediments was executed to document and understand the natural environmental conditions in the nearshore zone just before the nourishment.

MORPHOLOGICAL DEVELOPMENTS

Autonomous bar development

The dynamics of the nearshore bars, like their location, dimensions and migration in natural conditions may, to some extent, give an indication about the futural behaviour of the nourishment and its potential impact on the shoreline. The morphodynamics of a bar system, the mutual interaction and coupling between hydrodynamics and morphology, usually suggests the presence of a state of dynamic equilibrium. Within this system and for this state of dynamic equilibrium breaker bars pass through a certain sequence of events. Changes in the system are initiated by adaptations of the internal boundary conditions (internal forcing) and/or the time-dependent variability in input of energy (external forcing). An important question in this respect is whether the nourishment has a great impact on the stability of the dynamic system. The system may return to its former state of dynamic equilibrium, probably with a significant time-lag and with perturbations which gradually reduce in "amplitude". On the other hand, by implementing the nourishment, a threshold value in the system may have been exceeded, forcing the system into a state of unstable equilibrium or another state of dynamic equilibrium with another sequence of dynamic events. In both cases, knowledge about the morphodynamic behaviour of the nearshore bars in natural, undisturbed conditions is expected to be of great help to understand and evaluate the behaviour of the nourishment.

Details of the analysis and results are presented by Ruessink (1992) and Ruessink and Kroon (in prep.). The analysis is based on series of echo-sounding data obtained in several bathymetrical monitoring programmes, in particular the JARKUS and TAW programmes. In the JARKUS programme annual soundings are carried out since 1965 with an alongshore spacing of 200 m. Survey lines are oriented perpendicular to the coast extending to a depth of 6 m below NAP (Dutch Ordnance Datum) until 1984. After 1984 the profiles reach the -8 m isobath. Individual profiles may even extend to the -12 m isobath. Soundings carried out in the TAW programme usually cover the nearshore zone until the -8 m depth contour.

Summarizing the results of the cross-shore profile study it appears that the nearshore bars have a highly dynamic character (Fig. 4). The bars are generated in the inner nearshore zone and show a net seaward migration in time. In fact each bar commonly passes through a "life-cycle", following a morphological sequence with 3 stages of development (Ruessink 1992):

Stage 1: Generation of a bar.
The bar is generated close to the shore. At this stage the bar remains close to the coast within a range of 300 m, often for several years, and does not show a net seaward migration. The planform is frequently characterized by rhythmic patterns.

Stage 2: Net seaward migration of a bar.
At this stage the bar migrates in a net seaward direction from about 300 m towards 1300 m from the local shoreline at rates varying from 20-40 m/yr to 120-140 m/yr. The height of the bar remains about the same whereas the seaward located trough experiences erosion. The bar becomes straight crested.

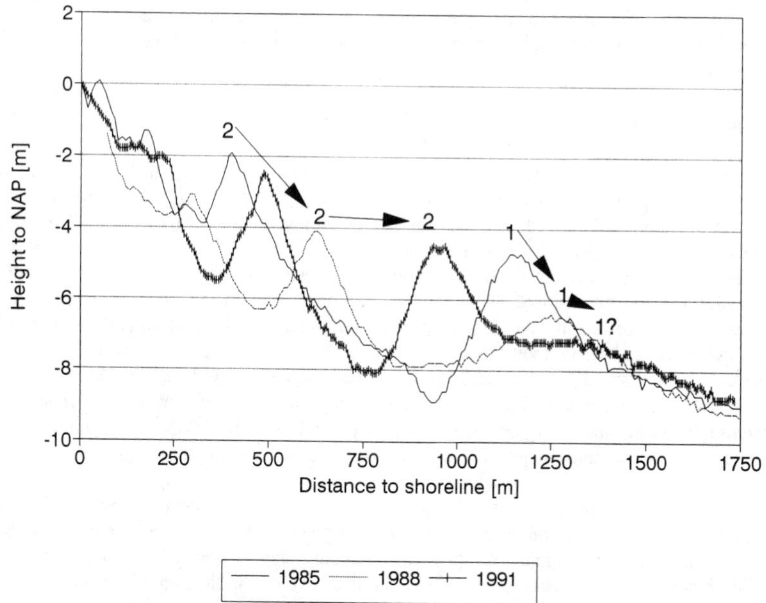

Figure 4. Nearshore profiles measured in cross-section 17 for a number of years. The bars show a net seaward migration.

Stage 3: Degeneration of a bar.
At the seaward limit of the surfzone the bar has lost its mobile character, decreases in size, with a crest located at greater depth, and gradually disappears.

Based on the present data set (1965-1993), the estimated length of the "life-cycle" happens to be in the order of 12-15 years. Ruessink and Kroon (in prep.) have observed a strong coupling between the behaviour of the individual bars. Crucial factors in this respect are the presence and dimensions of the outer bar, i.e. the state of the bar in stage 3. As long as the outer bar is clearly present as a seaward perturbation, it prevents the transition from an inner bar from stage 1 to the next. A similar development can be noticed if a new outer bar appears in a cross-shore profile due to alongshore migration. This alongshore migration has a strong episodic character and varies from almost zero to approximately 1400 m/yr (Ruessink 1992). If the outer bar starts degenerating, the first landward located bar shows a transition from stage 1 to 2. Simultaneously, a new bar will be generated close to the shore. At this stage the presence of the outer bar still affects the dynamics of the inner surfzone. As long as the outer bar remains an obstacle for incoming waves, the average, annual net-seaward migration rate of the inner bar (stage 2) is strongly reduced. Though, as soon as the outer bar has disappeared, seaward migration rates for bars in stage 2 increase by a factor 3 or 4.

In conclusion, prior to the nourishment the bar patterns are characterized by highly dynamic developments in which both longshore and cross-shore mechanisms play an important role.

Impact of the nourisment

By filling up the trough between the middle and the outer bar (Fig. 3), the bar system will display a greater degree of inertia. By "enhancing" the outer bar, the migration rates will be reduced and the bar system may be fixed in place. This effect alone must already have a positive influence on the shoreline. Energy dissipation due to wave breaking and bottom friction is expected to reach a maximum in conditions with a well-defined and pronounced 3 bar system (e.g. the situation in 1985, see Fig. 4). In a transitional phase, during the degeneration of the outer bar (Fig. 4; conditions in 1988), the outer bar is largely reduced in size and the height of its crest is located at greater depth. The landward located bar has also already moved to greater depths and a well-pronounced inner bar is still lacking. In these conditions, the waves will experience a less intense breaking and energy dissipation (Carter and Balsillie 1983). Consequently, the wave attack on the intertidal beach and beach-dune area may temporarily increase.

According to the previous discussion, the futural development of the nourishment and its effect on the shoreline will very much depend on the way the former bar-trough pattern may recover again. At present, sounding data taken just shortly after the nourishment (Fig. 3; date: 17-11-1993) do not give an indication for erosion and a renewed deepening of the former trough.

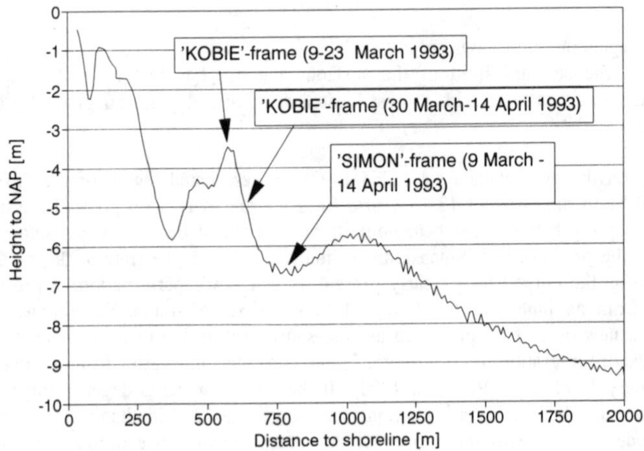

Figure 5. Positions of the instrumented tripods during the T_0-campaign.

TIDE- AND WIND-DRIVEN FLOW

Measurements

A first attempt to evaluate the role of longshore and cross-shore current processes is based on the results of the T_0 - campaign. In March and April of 1993 two instrumented tripods were deployed in section 16 (Fig. 1). Measurements were carried out on the crest of the middle bar, on the seaward flank of the same bar and in the trough, respectively (Fig. 5). Current measurements with the tripod KOBIE have been executed according to a burst-mode sampling scheme with a burst length of 34 minutes, a burst interval of 1 hour and a sampling frequency of 2 Hz. From these bursts, burst-averaged current velocities were computed and decomposed into longshore and cross-shore components. Velocities were measured at about 0.25 and 0.50 cm above the bed. Frame SIMON (type S4; in the trough) measured with a sampling frequency of 2 Hz and every 5 minutes an average velocity was computed. Flow was measured at 0.5 m above the bed.

Offshore wave conditions during the measurements are characterized by both fair weather and storm events. Low energy events with swell waves are dominant in the first 160 hours (=bursts). Significant wave heights (H_s) are in the order of 0.9 m. A small storm event is observed after this period with a maximum significant wave height varying from 1.15 (around hour 200) to 1.80 m (hour 238 to 245). The maximum measured wave height is about 3.5 m. During the storm a set-up in water levels is observed of 0.5 - 1.0 m. After hour 250 the value of H_s rapidly drops again to about 0.4 m. The second period of observations is in general characterized by low energy conditions (H_s < 1.0 m). A temporary increase in wave heights (to 1.6 m) can be noticed between hour 560 and 580.

From the data set a number of events are selected to illustrate the various effects of wind, waves and tides. A selection is made of measurements carried out in the trough since this area has been nourished. Occasionally, a comparison is made with measurements performed on the middle bar. A summary of the selected events and associated conditions is given in table 1.

Table 1. Hydro-meteo scenarios T_0-campaign

Event No.	Tidal Regime Δh (m)	Set Up (m)	Wave Height H_s (m)	Wind Direction	Magnitude (Bft)	Burst Number - Hour
1	ST 2.5	-	0.2-0.4	E	2	30-42
2	NT 1.7	-	0.2	E-S	2-3	446-458
3	~NT 1.3	+1.0	±1.0	W	7-5	181-204
4	~ST 2.4	+0.2-0.3	1.8-3.0*	N-W	5	-
5	- 2.0	-0.5	0.3	S	6-9	533-556

ST = spring tide; NT = neap tide; Δh = tidal range; H_s = significant wave height
*) estimated wave height (WAVEC)

Tidal flow

A first impression of the tidal effects can be obtained by comparing the velocities measured during both spring tide and neap tide (case 1 and 2; table 1). For neap tide conditions the tidal ellipse is very flat (rectilinear) and simply shore-parallel. Flow in the trough is only weak and peak flood velocities are somewhat larger than peak ebb velocities (0.28 m/s and 0.23 m/s, respectively). As expected flow velocities during spring tide are higher and have almost doubled in comparison to neap tide. Again the maximum flood currents (0.55 m/s) are slightly stronger than the maximum ebb currents (0.50 m/s). For spring tidal conditions there is a difference of about 20% between currents measured on the crest of the bar and those measured in the trough. The flow in the trough is consistently larger. No such difference is present during neap tide.

Wind-driven flow

Characteristic flow conditions for winds with a dominantly longshore direction from the west are illustrated in Fig. 6 (case 3, table 1). Due to the fact that flood tidal velocities are stronger and ebb tidal velocities are smaller the tidal ellipse has shifted in a positive longshore direction. In addition the flood tidal period is considerably prolonged

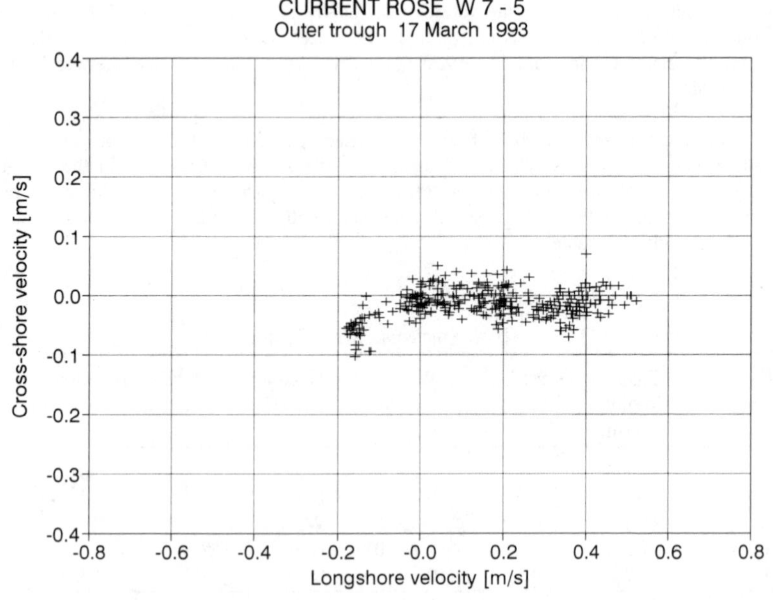

Figure 6. Longshore and cross-shore velocities during two tidal cycles with a shore-parallel wind; positive longshore corresponds with flow to the east, positive cross-shore is a landward flow.

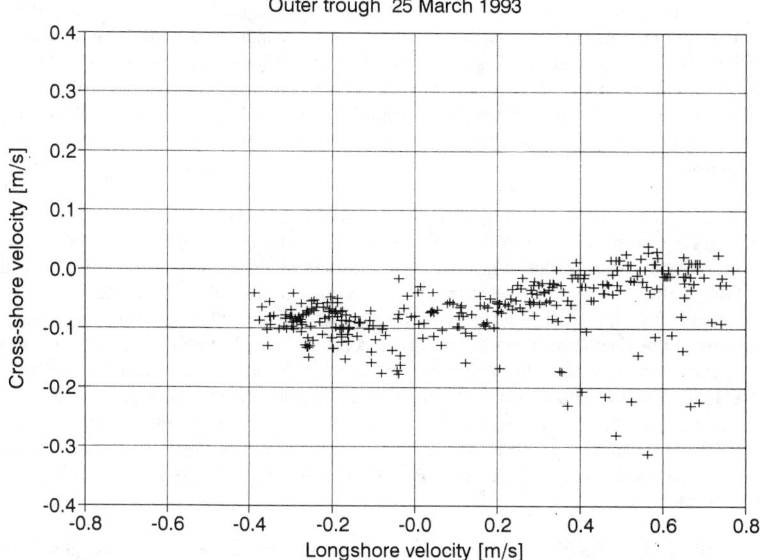

Figure 7. Longshore and cross-shore velocities during two tidal cycles with onshore winds (same definitions as in Fig. 6).

and, consequently, the ebb stage is reduced in both length and magnitude. There are several reasons to assume that wind-driven flow is the dominant factor in this case. The western wind (=shore-parallel) has an initial strength of 7 Bft which gradually reduces to 5 Bft at the end of the day. Wave-induced longshore currents are expected to be relatively small. According to the radiation stress principle, driving the longshore momentum flux, an oblique wave approach is required. On this specific day the waves were incident almost perpendicular, making only a small angle with the shore-normal of 10-20 degrees. A second argument in favour of wind-driven flow is that the maximum wave-driven longshore current ought to be present in the area with the strongest wave height transformation, being the area between the seaward flank and the crest of a bar. In this specific case velocities are measured in the trough.

Besides, breaking waves should be present to drive the longshore current. The relative wave height on the middle bar remains continuously below 0.35 which suggests that only during low tide, the highest waves are likely to break on the middle bar, landward of the trough. Indirect evidence for wave breaking at low tide may be the existence of an offshore-directed near bottom flow especially during one of the ebb stages (Fig. 6; V=0.10 m/s). This current is thought to be the result of the breaking-induced undertow mechanism. Even stronger offshore flows are recorded during both low tides above the crest of the landward located middle bar. Mean offshore flows are observed in the order of 0.2 to 0.3 m/s, shortly before low tide. This supports the idea of the wave-driven cross-shore flow.

Based on existing records of tidal flow and assuming a simple linear superimposition of wind- and tide-driven currents, the magnitude of the wind-driven longshore flow in the previous case is estimated to be ca. 0.25 m/s. For somewhat stronger winds (6-7 Bft) this flow increases up to 0.35 to 0.40 m/s. In futural campaigns a low-pass filtering of continuous time series is planned. This method is expected to be a more reliable and elegant method to deduce the wind-driven flow. The present time-series are too short in this respect.

The results presented in Fig. 7 (case 4, table 1) are typical for onshore winds (5 Bft), varying from North to West. Most of the time negative or offshore directed near bottom currents are recorded (Fig. 7). The magnitude of these currents again varies with the tide. Towards low tide, velocities exceed values of 0.2 m/s. This time- and tide-dependent behaviour of the cross-shore flow again suggests that the breaking-induced undertow is the predominant process. This is confirmed by the limited amount of set-up in the coastal zone: 0.20-0.30 m with respect to the predicted astronomical tides. Case 5 (table 1) is related to conditions with strong offshore winds (from the South, 6-9 Bft) and correspondingly low wave heights (H_s is ca. 0.3 m). Now, onshore velocities are dominant. Simultaneously, mean water levels seaward of the middle bar show a set-down: on average about 0.5 m. In this case a wind driven cross-shore circulation is clearly present.

In conclusion, the observations do show the important influence of the wind already for moderate wind conditions (< 8 Bft). Wind-driven flow velocities frequently have the same order of magnitude as the tidal currents. For stronger winds, wind-driven longshore and cross-shore flows in combination with wave-driven flow processes will dominate the nearshore flow patterns.

It is clear that for the Terschelling site wind-driven flow has to be considered as an important factor for the transport and dispersion of sediment and the behaviour of the nourishment.

Although few examples are known in recent literature, some authors have pointed out the great influence of the wind force on especially the longshore current (Hubertz 1986 and Whitford and Thornton 1993).

Hubertz (1986), based on field experiments at the DUCK-site (North Carolina), analysed empirical data of nearshore currents, wind velocity and direction and atmospheric pressure. From these observations it is clear that for similar wave conditions (wave height and angle of wave incidence), the longshore current could vary by a factor 3, depending on the strength of the wind. A time-series analysis and computations of power spectra from nearshore currents and wind speed showed significant spectral peaks at corresponding periods, supporting the evidence that longshore currents are strongly coupled to local weather patterns. Whitford and Thornton (1993) determined the role of the surface wind stress contribution to the longshore current by using a theoretical model based on the wind and wave forcing terms in the alongshore momentum equation, integrated across the surfzone. A coupling is, of course, expected between the wind and wave effects. The wave forcing can be a function of the local wind and, consequently, the wind directly forces longshore currents via surface stress and indirectly forces the current via the radiation stress. By comparing the total wave and wind force for increasing wind speeds at constant angle of wave and wind direction, they noticed that the wind force increases more rapidly than the wave force for increasing wind speed (Whitford and Thornton 1993). They also expected that the wind would be relatively more important for barred beaches than for planar beaches at lower wind speeds.

DISCUSSION AND CONCLUSIONS

The shoreface nourishment along the central coast of the island of Terschelling is implemented in a highly dynamic coastal environment. Prior to the nourishment the bar patterns did show a regular sequence of events with a certain life cycle of 12-15 years, in which individual bars passed from one stage to the next. Both longshore and cross-shore mechanisms are playing a major role in this behaviour. In a cross-shore direction bar dynamics is essentially determined by certain threshold values with respect to the state of the outer bar. The presence of the nourishment, stabilizing and enhancing the existing bar(s) in the outer surfzone most probably leads to a cross-shore fixation of the present bar topography. Maintaining the present bar patterns, with 3 bars causing wave breaking and dissipation, may already have a positive effect on the coastline. Longshore currents in the former trough, however, are higher than those measured on the middle bar, especially around spring tide. The trough has been very efficient in transporting water, probably due to a reduction in flow resistance. Since the outer trough is filled with sand, now the longshore flow pattern and the flow strength may have changed considerably. Either flow above the former trough has increased (based on continuity) or has decreased (due to an increase in friction). In both cases, the cross-shore distribution of the longshore current changes in the area of the nourishment. Therefore, it is not clear whether the former trough will erode again or that the nourishment will act as a stable platform between the middle and outer bar. From the present data set it is obvious that longshore currents in general can be very efficient agents in transporting water and sediment. Western winds are dominant in the area and the angle of wave incidence commonly varies between North and West. Consequently, wind- and wave- driven transport is expected to be mainly directed towards the East.

To what extend cross-shore processes will restore the former bar-trough morphology again, by causing a divergence and convergence of sediment fluxes e.g. due to the observed breaking-induced undertow and wave asymmetry, is at present unknown. The role of cross-shore processes is an important aspect of the research programme and the evaluation of the nourishment. If the nourishment has to act as a feeder berm, these processes, whether or not in close cooperation with longshore processes, have to be responsible for a dominantly onshore transport.

ACKNOWLEDGMENTS

This work was carried out as part of the project NOURTEC: Innovative Nourishment Techniques Evaluation. It was funded jointly by the Ministry of Transport, Public Works and Water Management in the Netherlands and by the Commission of the European Communities, Directorate General for Science, Research and Development under the Marine Science and Technology programme contract no. MAS2-CT93-0049.

REFERENCES

Burke, C.E., McLellan, T.N., and Clausner, J.E., 1991. "Nearshore berms - update of the US experience," *Proceedings CEDA-PIANC Conference*, Amsterdam.

Calkoen, C.J., Wensink, G.J., and Hesselmans, G.H.F.M., 1993. ERS-1 SAR imagery to optimize the NOURTEC shipbased bathymetric survey: feasibility study, Delft Hydraulics, report H1875, 25 pp.

Carter, R.W.G., and Balsillie, J.H., 1983. "A note on the amount of wave energy transmitted over nearshore sand bars," *Earth Surface Processes and Landforms,* Vol. 8, pp. 213-222.

Healy, T., Harms, C., and de Lange, W., 1991. "Dredge spoil and inner shelf investigations off Tauranga Harbour, Bay of Plenty, New Zealand," *Proceedings Coastal Sediments '91, Seattle*, ASCE, New York, pp. 2037-2051.

Hubertz, J.M., 1986. "Observations of local wind effects on longshore currents," *Coastal Engineering*, vol. 10, pp. 275-288.

Jackson, L.A., and Tomlinson, R.B., 1990. "Nearshore nourishment implementation, monitoring and model studies of 1.5 Mm^3 at Kirra Beach," *Proceedings 22nd Coastal Engineering Conference*, ASCE, New York, vol. 3, pp. 2241-2254.

McLellan, T.N., and Imsand, F.D., 1989. "Berm Construction Utilizing dredged Materials," *Proceedings of WODCON XII*, Western Dredging Ass., pp. 811-820.

McLellan, T.N., and Kraus, N.C., 1991. "Design Guidance for nearshore berm construction", *Proceedings Coastal Sediments '91*, Seattle, ASCE, New York, pp. 2000-2011.

Roelvink, J.A., and Stive, M.J.F., 1990. "Sand transport on the shoreface of the Holland Coast," *Proceedings 22nd Coastal Engineering Conference*, ASCE, New York, vol. 2, pp. 1909-1921.

Ruessink, B.G., 1992. "The Nearshore morphology of Terschelling (1965-1991)," Institute for Marine and Atmospheric Research Utrecht, IMAU report R92-11, 30 pp.

Ruessink, B.G., 1993. "Morphology, sediment distribution, local hydrodynamics and sediment transport in the Terschelling study area in a natural situation," Institute for Marine and Atmospheric Research Utrecht, IMAU report R93-09, 38 pp.

Ruessink, B.G., and Kroon, A., (in prep.). "The autonomous behaviour of a multiple bar system in the nearshore zone of Terschelling, the Netherlands: 1965-1993," submitted to *Marine Geology*.

Stauble, D.K., and Kraus, N.C., 1993. "Beach nourishment engineering and management considerations," Coastlines of the World, *Proceedings Coastal Zone '93*, New Orleans, ASCE, New York, 245 pp.

Van Alphen, J.S.L.J., Hallie, F.P., Ribberink, J.S., Roelvink, J.A., and Louisse, C.J., 1990. "Offshore sand extraction and nearshore profile nourishment," *Proceedings 22nd Coastal Engineering Conference*, ASCE, New York, pp. 1998-2009.

Van de Graaff, J., Niemeyer, H.D., and Van Overeem, J., 1991. "Special Issue on Artificial Beach Nourishment," *Coastal Engineering*, vol. 16.

Van Vessem, P., Van de Kreeke, J., Mulder, J.P.M., Laustrup, C., and Niemeyer, H.D., 1994. "Design of a shoreface nourishment NOURTEC experiment at Terschelling, The Netherlands," *Abstract 24th International Conference on Coastal Engineering*, Kobe Japan.

Whitford, D.J., and Thornton, E.B., 1993. "Comparison of wind and wave forcing of longshore currents," *Continental Shelf Research*, Vol. 13, 11, pp. 1205-1218.

THE INFLUENCE OF BOUNDARY CONDITIONS ON BEACH ZONATION

Ignacio Alonso[1] and Federico Vilas[2]

ABSTRACT: A new data set of beach profiles on the foreshore is used to propose a morphodynamical model based on the alongshore variations on the arriving wave energy. Such changes are due to the boundary conditions, and determine the existance of different sectors on the beach. Each one of this sectors follow a certain pattern on the long-term volume change, the foreshore slope variability, the magnitude of the subaerial sand bars, and the presence/absence of beach cusps. Additional information on wave conditions and direction of sediment transport is obtained to characterize each sector.

INTRODUCTION

It is well known that the presence of lateral and offshore estructures, headlands, river mouths and dunes, as well as the bottom topography, determine the amount of the incomming wave energy at a certain beach. As the arriving wave energy is not constant on time, several models have been proposed to describe the morphodynamical evolution followed by the beach. Such models consist on a certain sequence of beach stages, where the change from one stage to another depends on a certain parameter closely related to wave energy. In other words, such models are based on temporal fluctuations of wave energy.

---

1) Assistant professor, Dept. of Física, Univ. of Las Palmas de Gran Canaria, P.O. Box 550, Las Palmas, Spain.
2) Director, Dept. of Recursos Naturales y Medio Ambiente, Univ. of Vigo, P.O. Box 874, Vigo (Pontevedra), Spain.

Probably the first three-dimensional sequential model of beach change was proposed by Sonu (1973), which was subsequently expanded by Short (1978, 1979) and Sunamura (1985). Short's model consists on six consecutive beach stages where the extremes are the dissipative and reflective beach stages. Wright and Short (1984) found that the threshold between the dissipative, intermediate and reflective types could be defined using Dean's (1973) dimensionless fall velocity

$$\Omega = H_b/wT$$

where H_b is breaker height, w is sediment fall velocity and T is wave period. Sunamura's model is composed of two extreme stages, erosional and accretionary, connected by six transitory stages. A dimensionles parameter K, originally derived from wave-tank experiments (Sunamura, 1984), explains stage movement through the model. The parameter is expresed by

$$K = H_b^2/gT^2D$$

where H_b and T are daily average values of breaker height and period; D is the representative grain size; and g the acceleration due to gravity. Apart from Ω and K, the distinction between different beach stages has been usually made by means of the surf scaling parameter (e.g., Wright and Short 1983, Lorang *et al.* 1993), defined by Guza and Inman (1975) as

$$\epsilon = a_b w^2/g tan^2\beta$$

where a_b is wave amplitude at the breaking point, w is incident wave radian frecuency (2π/T; T = period), g is acceleration of gravity, and β is beach/surf zone gradient. Masselink and Short (1993) have proposed a conceptual beach model which takes into account the combined effect of wave height and tide range on beach morphodynamycs.

There is no doubt that all previously mentioned models are very useful in case of open and pocket beaches with different energy conditions, as well as on microtidal and macrotidal environments. But, are these models correct in case of beaches were a big alongshore variation of the incoming wave energy takes place?. Short (1979) refers to that question, but the boundary conditions at Narrabeen Beach are not so strong as they are at Las Canteras Beach.

Many authors (e.g., Bascom 1951, Oertel *et al.* 1989, Martínez *et al.* 1990, Nafaa and Omran 1993) have pointed out that such alongshore variations on a certain area provokes spatial changes on the foreshore slope, grain size, and volume of transported sediments along the beach, as well as different characteristics on morphological features like bars, ridge-runnel systems, scarps and cusps. Present research focuses on that spatial variability, and proposes a morphodynamical zonation at a certain beach.

STUDY SITE

The study site is Las Canteras Beach, a nearly 3 km long sandy beach located at the north coast of Gran Canaria (Canary Islands, Spain). The beach is delimited by a rocky headland on the north end, and by an small dam at the south end. The north sector of the beach is very well sheltered from the prevailing northern waves by the shoreline configuration, and by a natural offshore rocky bar whose height is very close to MSL. This bar is partially fragmented and extends parallel to the shoreline 200 m off (fig. 1). On the contrary, the south end of the beach is completely exposed to waves.

The tidal range exceeds 2.5 m at spring tides, and it is around 1 m at neap tides. The average significant wave height is 1.4 $\pm$ 0.6 m, with an spectral peak period of 10.2 s (Alonso, 1993). Sediment mean size (D_{50}) ranges from 0.54 to 2.56 phi (from coarse to fine sands according to the scale proposed by Krumbein, 1934), but most grains are medium and fine sands ($1.6 < D_{50} < 2.3$ phi). The sorting (σ_I) of the sediment samples ranges from very well sorted to poorly sorted ($0.3 < \sigma_I < 1.14$) following the classification proposed by Folk and Ward (1957).

This beach is an urban beach located into the city of Las Palmas de Gran Canaria, which holds nearly 400.000 inhabitants. The building of the town has affected the natural dynamic of the sediments, which arrives from the bottom of the Confital Bay pulled by waves and currents, and after drying on the beach, grains were blown to the south by trade winds. Since 1960 the beach front was rebuilt, and a new seawall and higher buildings were constructed. The result was that wind is not able to blow the sediments over such fence, and therefore grains acumulate on the beach face (Araña and Carracedo 1975, Martín Galán 1984, Alonso 1993).

DATA COLLECTION

Field data consists on 14 profile lines surveyed with an standard levelling method (see fig. 1 for profiles position). Surveys were conducted approximately at monthly intervals from June 1987 until June 1992. Furthermore, several surveys were carried out just after selected storms in order to know the foreshore behaviour under different wave conditions. In overall, the data set includes 67 surveys conducted during a 5 years period. The monthly rate is very good to show seasonal changes, while superimposed surveys permit to obtain any beach variability related to particular events. Profile 1 was not surveyed during first 20 surveys, which represents a certain gap on the whole data set.

The profiles were backed by a seawall and surveyed down to about 1 m below MSL. This is not, of course, a closure depth, but allows for the inclusion of the foreshore, where short term sediment transport between the beach and the inshore zone is most active.

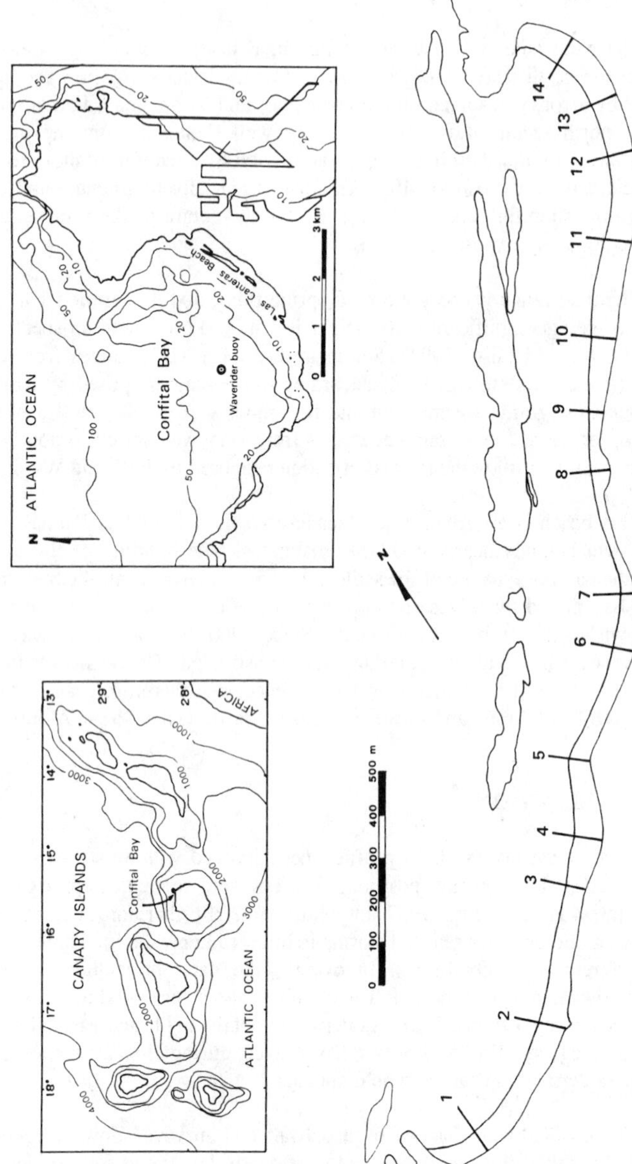

Figure 1.- Location map of the study site showing position of the profile lines.

Sediment samples were collected on three occasions, in order to identify any possible variation on grain size and grain composition, which migth be related to wave climate. For that reason sampling campaigns took place at different seasons all over the year (October 3rd, 1991; February 21st, 1992 and June 16th, 1992).

Wave data were recorded from a waverider buoy installed at a water depth of 40 m off the beach. The buoy records data 8 times per day every 3 hours, during 17 minutes each record, except when wave height exceeds 2 m. In that cases the recording interval is 1 hour.

TIME DEPENDENCE OF THE MORPHODYNAMIC ZONATION

It is perfectly known that beach changes are mainly due to two phenomena related with wave energy. Seasonal wave climate changes and stormy events are responsible of most of the sediment transport happening at a certain beach. For that reason, any variation on morphologycal features like bars, ridge and runnel systems, cusps and foreshore slope, as well as changes on grain properties, are due to the above mentioned phenomena.

Both phenomena operate at different time scales, since wave climate changes with an strong seasonal dependence, while stormy situations are occasional events which may happen all over the year. For that reason any morphodynamic zonation should be established based on observations covering a period long enough to account for the different beach conditions. In that way, both the seasonal variations and the extreme situations will contribute to stablish the morpodynamic zonation.

The enlapsed time between consecutive surveys is of crucial importance, since the beach morphology corresponds to a certain amount of erosional and/or depositional conditions. Data based on larger enlapsed time have the effect of averaging-out many happened events, while data based on shorter enlapsed times, more closely reflect event-related changes (Oertel *et al.*, 1989). On the other hand, the time of beach response to any change on wave power is relatively large, since whereas wave power can change markedly in the order of 1-10 hours, morphology has a lag on the order of 10-100 hours (Short 1979, Wright *et al.* 1984).

MORPHODYNAMIC CRITERIA

Keeping in mind that the main purpose of this work is to group profiles according with their response to the arriving wave energy, we have focussed on certain beach conditions related to the sedimentary dynamic. These conditions are morphological features formed on the beach face (the volume of the subaerial sand bars, the presence/absence of beach cusps and the scarps magnitude) as well as other dynamical criteria such as the volume change per unit width and the foreshore slope variability. All them have been related with the amount of sediment available wave conditions and dominant beach type (e.g. Short, 1979).

Volume of Transported Sediments

The sediment dynamics of the studied area was calculated starting from the volume per unit width for each profile. The Beach Profile Analysis System method (Fleming and DeWall, 1982) was used down to a seaward bound according to the shorter profile. Erosions and accretions were computed for each profile from the change per unit width relative to the volume in the first survey. In that way, possitive values are indicative of accretions relative to the situation on the first survey (June 26, 1987), while negative values show erosions.

Figure 2 shows the evolution of the volume changes per unit width for all the profiles during the surveying period. It can be noted that the area between profiles 2 and 5 is characterized by a very important erosion. Such area of negative values become wider with time, which means that it presents a certain erosive trend. On the contrary, the other side of the beach (profiles 11 to 14) presents mostly positive values, denoting the existance of a net accretionary trend on this sector. Finally, the central area of the beach (profiles 6 to 10) presents a null trend on its volume changes, since most of the values ranges between -15 to 15 m^3/m.

In order to verify such apparent similarity between profiles in the three sectors, a cross-correlation study was performed (table 1). Only the group formed by profiles 11, 12, 13 and 14 presents relatively high correlation coefficients between them ($0.69 < r < 0.76$), which means that most of the volumetric changes along this sector take place in a simultaneous way on the different profiles.

	P 1	P 2	P 3	P 4	P 5	P 6	P 7	P 8	P 9	P 10	P 11	P 12	P 13	P 14
P 1	1.00													
P 2	.28	1.00												
P 3	-.07	.38	1.00											
P 4	.02	.28	.71	1.00										
P 5	-.29	-.22	.21	.51	1.00									
P 6	.30	.55	.31	.40	-.06	1.00								
P 7	.17	.49	.22	.17	-.26	.43	1.00							
P 8	.29	.38	.23	.24	-.03	.32	.07	1.00						
P 9	.29	.02	.11	.05	.03	-.06	-.28	.58	1.00					
P 10	.27	.38	.10	.09	-.18	.27	.33	.51	.50	1.00				
P 11	.25	.12	-.17	-.27	-.45	.14	.20	-.28	-.45	.01	1.00			
P 12	.31	.13	-.09	-.12	-.36	.30	.36	-.36	-.59	-.11	.74	1.00		
P 13	.39	.32	.03	-.01	-.41	.44	.50	-.08	-.35	.17	.73	.76	1.00	
P 14	.48	.12	-.24	-.22	-.47	.22	.15	-.07	-.20	.08	.70	.69	.71	1.00

Table 1.- Correlation matrix of the volume changes relative to the first survey. (Profile 1 → 47 surveys; Profiles 2-14 → 67 surveys.

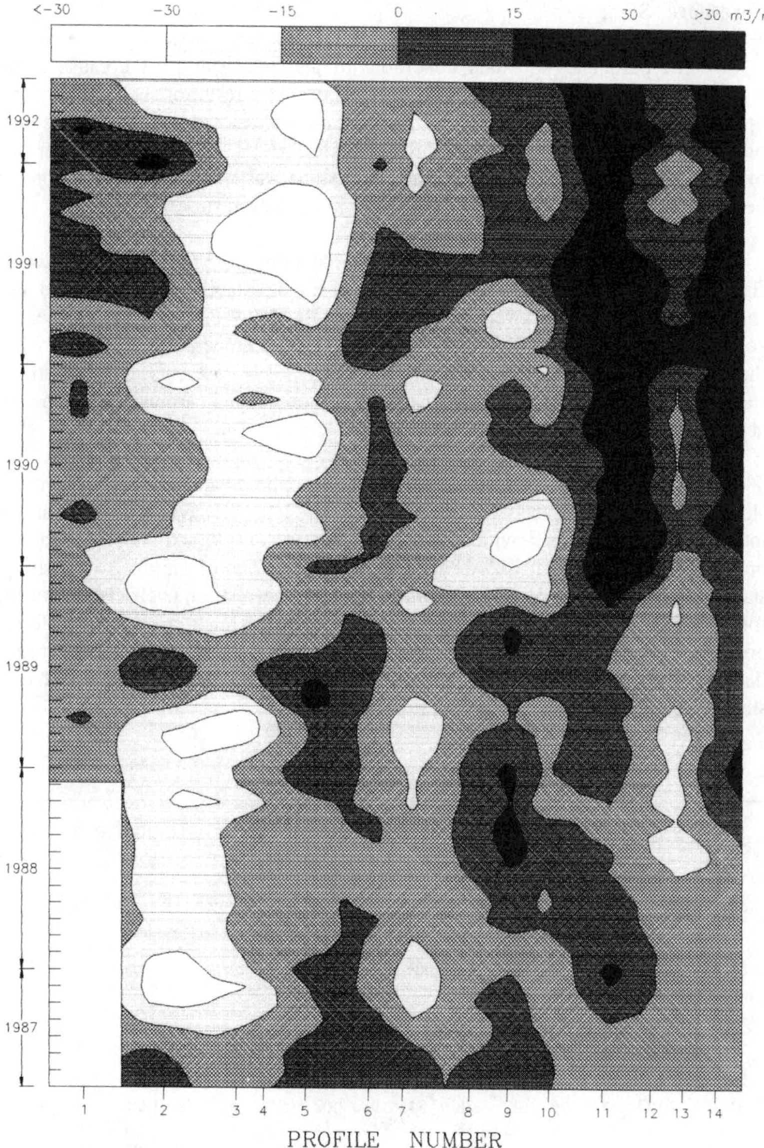

Figure 2.- Evolution of the volume changes on each profile, which indicate the existance of erosive and accretionary sectors on the beach.

Foreshore Slope

The particular boundary conditions of the study site are the cause of the wave energy gradient on the alongshore direction. As a result of such gradient, the uprush limit changes along the beach, so that on the exposed zone waves impinge on the whole profile, while on the protected area waves only affect the outer part of the profile. For that reason, the foreshore slope was calculated for each profile between the low water level and the uprush limit at each survey time.

The time evolution of the foreshore slope alongshore the beach is shown on fig. 3, and a very similar zonation of the beach can be established: the exposed area (profiles 1 to 5) presents a very gentle slope ranging between 3 and 5%, the central zone (profiles 6-10) has an almost constant foreshore slope between 6-7%, while the most protected area covered by profiles 11-14 presents a very strong stacionality, so that on summer periods the foreshore slope in around 10%, whereas at winter time drops at 5% or even less.

A simple statistical analysis shows that the exposed area presents an average slope of 4.5% with an standard deviation of 1.1; the average slope at the intermediate zone is 6.4% with an standard deviation of 0.9; and the protected area presents a mean slope of 7.2%, but an standard deviation of 1.5 due to the strong stacionality. Focussing on that variability, the cross-correlation coefficientes for the foreshore slope data point out that there is not any correlation between different profiles, except for that of the protected area, where $0.73 < r < 0.86$. It means that the strong variability along this sector takes place at the same time and with similar magnitude on the different profiles.

	P 1	P 2	P 3	P 4	P 5	P 6	P 7	P 8	P 9	P 10	P 11	P 12	P 13	P 14
P 1	1.00													
P 2	.11	1.00												
P 3	.57	.21	1.00											
P 4	.38	.04	.61	1.00										
P 5	.37	-.16	.46	.59	1.00									
P 6	-.06	.05	.26	.60	.34	1.00								
P 7	.32	.27	.26	.41	.22	.30	1.00							
P 8	.13	.31	.18	.17	.02	.38	.33	1.00						
P 9	-.33	.19	-.06	.06	.04	.41	.17	.55	1.00					
P 10	-.12	-.21	.02	.11	.24	.41	.10	.30	.25	1.00				
P 11	-.10	-.45	-.05	.29	.28	.49	.05	.05	.13	.61	1.00			
P 12	-.10	-.33	.06	.42	.38	.54	.08	-.02	.10	.55	.85	1.00		
P 13	-.07	-.15	.16	.55	.38	.66	.14	.14	.21	.50	.73	.80	1.00	
P 14	-.22	-.32	-.07	.26	.22	.59	.08	.20	.27	.60	.85	.82	.77	1.00

Table 2.- Correlation matrix of the foreshore slope data. (Profile 1 → 47 surveys, Profiles 2-14 → 67 surveys).

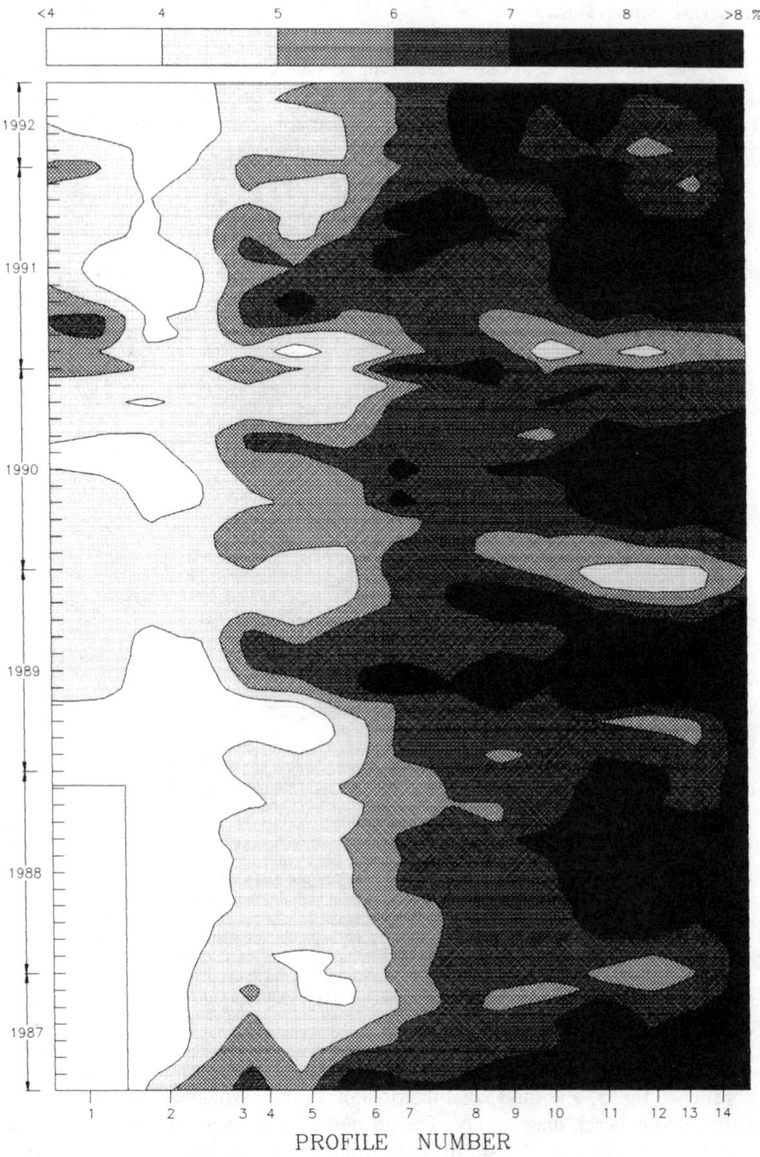

Figure 3.- Evolution of the foreshore slope alongshore the beach.

Subaerial Sand Bars

The subaerial sand bars observed at Las Canteras Beach are seasonal structures originated during calm periods that migrate up the beach face as a result of the onshore sediment transport. Figure 4 presents the evolution of one of this structures, where the onshore movement can be observed.

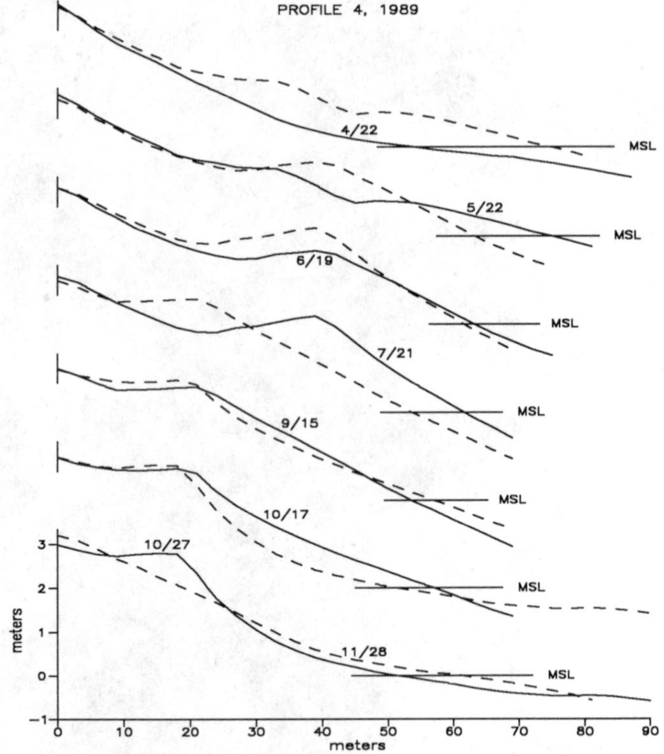

Figure 4.- Subaerial sand bar migration between May and October, 1989 at profile 4.

The volume per unit width of each one of this structures was computed according to fig. 5, and the spatial distribution of these volumes has led to a new zonation of the beach under study. The greatest systems were present on the zone covered by profiles 2, 3 and 4, with volumes up to 30, 19 and 21 m^3/m respectively, which points out an important sediment transport on the cross-shore direction (Short, 1979). Smaller structures were formed on the central area (profiles 6-10), where the average volume of these bars is around 4 m^3/m, which indicates a very weak cross-shore transport. Finally, no subaerial sand bar were observed on the sheltered zone (profiles 11-14), as a result of the predominant longshore sediment transport along this sector.

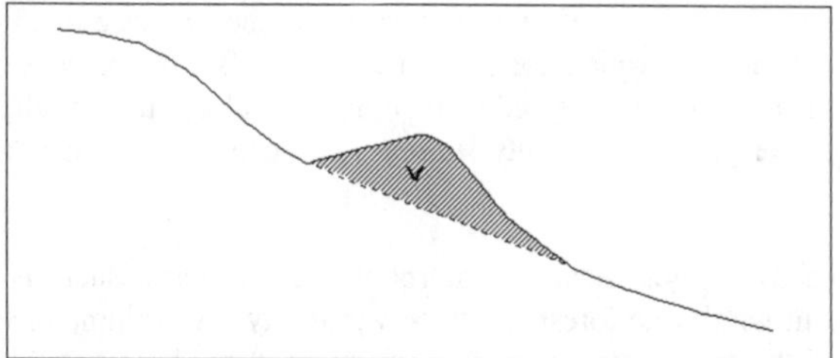

Figure 5.- Representation of the method followed to compute the volume of the subaerial sand bars (V).

Presence of Beach Cusps

On 33 surveys the amplitude of the cusps placed in front of the profiles was measured. The average cusp spacing is 25.8 m with an standard deviation of 6.4 m. This data set allows to divide the beach on three sectors analogous with that obtained previously.

Only on three occasions cusps were observed at the exposed area, but always at the seaward slope of a subaerial sand bar. It confirms that along this sector conditions are too dissipative for edge waves formation, except when the foreshore slope increases due to the presence of a subaerial sand bar (Komar 1976, Short 1979). The same result is obtained by Werner and Fink (1993), who states that on gentle beaches the local depressions formed in a single swash cycle are too small to deflect water particles.

Cusps were present along the central area in 8 of the 33 surveys, but used to be poorly developed and quite irregularly spaced. It is due to the small amount of sediment available, as well as to a foreshore slope not steep enough for edge waves resonance.

In contrast, the sheltered sector of the beach presented quite regularly spaced cusps on 28 of the 33 occasions. The reason of this almost continuos presence of beach cusps is found on three aspects: i) the large amount of sediments available on this sector, ii) the foreshore slope relatively steep, and iii) the headland that limits this sector, which helps the developing of trapped waves (Sunamura, 1989).

DISCUSSION AND CONCLUSIONS

There are certain beaches throughout the world where the particular boundary conditions determine a very strong longshore variability on the arriving wave energy, and in consecuence, the simultaneous presence of reflective and dissipative conditions along different sectors of the beach. One of this beaches is Las Canteras Beach, in which the presence of an offshore rocky bar determines pronounced differences on the sediment dynamics along the beach. A data set consisting of five years of monthly surveys has been used to characterized such differences.

By means of dynamical and morphologycal criteria, such us the volume change per unit width, the foreshore slope variability, the volume of the subaerial sand bars, and the presence/absence of beach cusps, it has been possible to separate the beach under study into three homogeneous sectors. The exposed one is under the influence of incident waves that break ~100 m from the shoreline due to the gentle slope of the surf/swash zone. The central sector is partially protected by the two main fragments of the offshore rocky bar, but the opening between them is large and deep enough for waves to come in without breaking, but dissipating part of their enery flux by diffraction and refraction. The north end of the beach is very protected not only by the offshore bar, where waves break on the seaward edge, but also by the shoreline configuration (see fig. 1).

In order to assign each profile to one of the three sectors, specially for those profiles that are in between two sectors, the representation of the average volume change rate followed by each profile during the surveying period, versus the average foreshore slope, permits to distinguish each profile according with their morphodynamic behaviour (fig. 6).

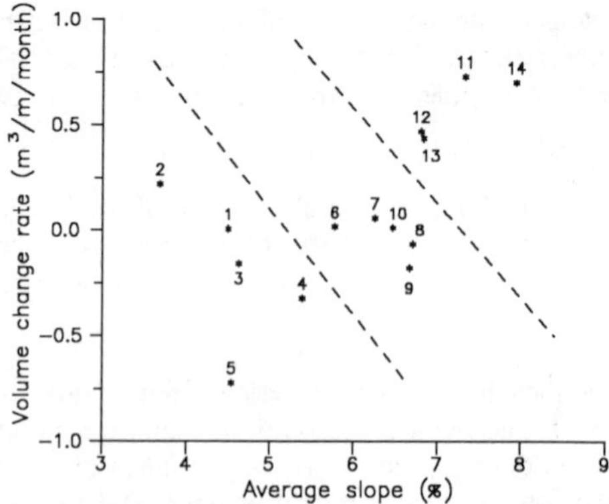

Figure 6.- Average volume change rate vs. mean foreshore slope for each profile. Numbers correspond to profiles.

From fig. 6 it is possible to observe that profiles 1, 2, 3, 4 and 5, characteristics of the exposed sector, have a gentle slope and a certain tendency to erode. On the other side, profiles 11, 12, 13 and 14 present steeper slopes and a possitive volume change rate at an over-annual time scale. It is indicative of an accumulative tendency, and agrees perfectly with Sunamura (1989), since he states that beach cusps are purely accretionary features and need steep slopes to develop. In between these two groups lay profiles 6, 7, 8, 9, and 10, representative of the intermediate zone.

Each of this sectors behaves in a different way, with strong differences on dominant wave conditions, direction of sediment transport and beach type according with the well known morphodynamic classification of Short (1978) and Wright and Short (1983). Table 3 summarizes the main characteristics of each sector.

SECTOR (profiles)	Volume Change	Foreshore slope (%)	subaerial sand bars	Cusps Spacing *Occurrence*	Transport direction	Beach type	Dominant waves
Exposed (1-5)	tendency to erode	< 4	big	*occasional*	cross-shore	dissipative	incident waves
Intermediate (6-10)	no change	5 - 8	small	irregular *25%*	mixed	intermediate	incident and edge waves
Protected (11-14)	accretion	winter: 4-6 summer: > 9	no bars	regular *almost continuous*	longshore	reflective	edge waves

Table 3.- Summary of the main characteristics of the three sectors determined at Las Canteras Beach.

Finally this work illustrates the big differences that can be found on a certain beach as a result of the effect of the boundary conditions. Furthermore, it has to be taken into account that data used in this work are from a narrow strip as the foreshore. It allow us to conclude that even if it is desirable to handle data from the whole profile, there is a lot of information on wave conditions and sediment transport just from the beach face.

ACKNOWLEDGEMENTS

This study was partially funded by the company UNELCO through the Fundación Universitaria de Las Palmas. Special thanks are due to R. Alvarez and many students for surveying assistance.

REFERENCES

Alonso, I. 1993. "Procesos sedimentarios en la playa de Las Canteras (Gran Canaria)", *Ph.D. Disertation (umpubl.)*, Dept. of Física, Univ. of Las Palmas de Gran Canaria, 333 pp. (in Spanish)

Araña, V. and Carracedo, J.C. 1975. "Los volcanes de las Islas Canarias. III. Gran Canaria", Ed. Rueda, 175 pp. (in Spanish)

Bascom, W.N. 1951. "The relationship between sand size and beach-face slope", *Trans. Am. Geophys. Union*, Vol. 32, pp. 866-74.

Dean, R.G. 1973. "Heuristic models of sand transport in the surf zone", *Proc. Conf. on Eng. Dynamics in the Surf Zone,* Sydney, pp. 208-214.

Fleming, M.V. and DeWall, A.E. 1982. "Beach Profile Analysis System (BPAS)", *Tech. Rep. Nº 82-1 (VIII)*. U.S. Army Corps of Engineers, CERC.

Folk, R.L. and Ward, W.C. 1957. "Brazos river bar: A study in the significance of grain parameters", *J. Sed. Petrol.*, Vol. 27, pp. 3-26.

Guza, R.T. and Inman, D.L. 1975. "Edge waves and beach cusps", *J. Geophys. Res.*, Vol. 80, No 21, pp. 2997-3012.

Komar, P.D. 1976. "Beach processes and sedimentation", Prentice Hall, Inc., 429 pp.

Krumbein, W.C. 1934. "Size frecuency distribution of sediments", *J. Sed. Petrol.*, Vol. 4, pp. 65-77.

Lorang, M.S.; Stanford, J.A.; Hauer, F.R. and Jourdonnais, J.H. 1993. "Dissipative and reflective beaches in a large lake and the physical effects of lake level regulation", *Ocean & Coastal Management*, Vol. 19, pp. 263-287.

Martín Galán, F. 1984. "La formación de Las Palmas: ciudad y puerto. Cinco siglos de evolución", Ed. Junta del Puerto de La Luz y Las Palmas, Gobierno de Canarias, Cabildo Insular de Gran Canaria and Ayuntamiento de Las Palmas de Gran Canaria, 324 pp. (in Spanish)

Martínez, J.; Alvarez, R.; Alonso, I. and Del Rosario, M.D. 1990. "Analysis of sedimentary processes on the Las Canteras Beach (Las Palmas, Spain) for its planning and management", *Eng. Geol.*, Vol. 29, pp. 377-386.

Masselink, G. and Short, A.D. 1993. "The effect of tide range on beach morphodynamics and morphology: A conceptual beach model", *J. Coastal Res.*, Vol. 9, No 3, pp. 785-800.

Nafaa, M.G. and Omran, E.F. 1993. "Beach and nearshore features along the dissipative coastline of the Nile Delta, Egypt", *J. Coastal Res.*, Vol. 9, No 2, pp. 423-433.

Oertel, G.F.; Ludwick, J.C. and Oertel, D.L.S. 1989. "Sand accounting methodology for barrier islands sediment budget analysis", *Proc. Coastal Zone '89*, Charleston, ASCE, pp. 43-61.

Short, A.D. 1978. "Wave power and beach-stages: a global model", *Proc. 16th Coastal Eng. Conf.,* ASCE, pp. 1045-1062.

Short, A.D. 1979. "Three dimensional beach-stage model". *J. Geol.*, Vol. 87. pp. 553-571.

Sonu, C.J. 1973. "Three-dimensional beach changes", *J. Geol.*, Vol. 81, pp. 42-64.

Sunamura, T. 1984. "Onshore-offshore sediment transport rate in the swash zone of laboratory beaches", *Coastal Eng. in Japan*, Vol. 27, pp. 205-212.

Sunamura, T. 1985. "Morphological change of beaches", *Lecture Notes 21st Summer Seminar on Hydraulics*, JSCE, B7, pp. 1-17.

Sunamura, T. 1989. "Sandy beach geomorphology elucidated by laboratory modeling", In: *Applications in coastal modeling*, V.C. Lakhan and A.S. Trenhaile (Eds), Elsevier, pp. 159-213.

Werner, B.T. and Fink, T.M. 1993. "Beach cusps as self-organized patterns", *Science*, Vol. 260, pp. 968-971.

Wright, L.D. and Short, A.D. 1983. "Morphodynamics of beaches and surf zones in Australia". In: *CRC Handbook of Coastal Processes an Erosion.* P.D. Komar (Ed), pp. 35-64.

Wright, L.D. and Short, A.D. 1984. "Morphodynamic variability of surf zones and beaches: a synthesis", *Mar. Geol.*, Vol. 56, pp. 93-118.

Wright, L.D., May, S.K. and Short, A.D. 1984. "Beach and surf zone equilibria and time response", *Proc. 19th Int. Conf. on Coastal Eng.*, pp. 2150-2164.

EQUILIBRIUM BEACH PROFILES ON THE CATALAN COAST

J.P. Sierra, A. Lo Presti and A. Sánchez-Arcilla[1]

ABSTRACT: In the last decades, numerous studies directed to the obtention of analytical expressions for beach equilibrium profiles have been carried out. As a consequence of these studies, several empiric equations have been proposed. In this work, data of 82 beach profiles collected at different points of the Catalan Coast, have been used in order to get the best possible representation of beach profiles in the area. In most of the 82 profiles, it has been observed that two newly proposed expressions improve the fit to the measured data with respect to other state of the art expressions.

1. INTRODUCTION

Equilibrium beach profile definition is essential in order to solve a number of coastal engineering problems. For example, whenever beach nourishment operations are performed, a large volume of sand is placed on the shore-face to increase the emerged beach surface. This sand will be molded by wave and current action until a situation with a more or less stable profile is reached. Moreover, shoreline evolution models, which simulate shoreline changes need to know the beach profile shape.

Obviously, nature nearly never develops an equilibrium profile, in a strict sense. The term equilibrium profile refers to a curve of given shape, which adequately represents the average profile regardless of any storm or seasonal perturbation. The concept is highly dependent on the averaging process. The expressions proposed in this paper and the applied methodology are developed in this context.

The aim of this work is thus the obtention of an analytical expression which provides the best possible representation of the equilibrium beach profile in the Catalan Coast. This exercise will be performed on a purely statistical basis, without any in-depth analysis of the underlying physics.

1 Laboratori d'Enginyeria Marítima (LIM/UPC), Universitat Politècnica de Catalunya. C/ Gran Capitá s/n, mòdul D-1, 08034 Barcelona, SPAIN.

2. BACKGROUND

In the last decades, numerous studies directed to the obtention of analytical expressions for equilibrium beach profiles have been carried out. As a consequence of these studies, several empiric curves have been proposed. In this section, a review of the most representative ones is presented.

The first studies on this subject were carried out by Fenneman (1902), who presented qualitative considerations of several mechanisms related with equilibrium beach profiles. In this analysis, the author concluded that the underwater beach profile would be concave upward in the nearest to shore part and concave downward in the landward part.

Bruun (1954) studied beach profiles located on the North Sea (Danish Coast) and Pacific Ocean (Mission Bay, California) and proposed an empirical equation to describe their geometry. This equation gives a relationship between water depth (h) and distance from the shoreline (y):

$$h = A\, y^{2/3} \tag{1}$$

Bruun sugested two mechanisms which could lead to an equilibrium beach profile. The first mechanism is based on the adoption of an uniform onshore component of shear stress and a constant onshore component of the gradient of transported wave energy. The second mechanism consideres bottom friction as the only agent causing loss of wave energy and requires that the energy loss per unit area is constant. Both mechanisms lead to an approximate equation of the form of equation 1.

Dean (1976), using linear wave theory in his analysis, identified some possible mechanisms which lead to equilibrium beach profile forms. His physical considerations asume the existence of both constructive and destructive forces in the surf zone. Constructive forces would tend to transport sand onshore, while destructive ones would transport sand offshore. His study focuses on destructive forces without identifying the constructive ones. Dean points out that a given sand of particular size and specific gravity is stable under a given level of destructive forces. As a result of his study he arrives to an equilibrium beach profile of the form

$$h = A\, y^{m} \tag{2}$$

where A is a shape factor, which depends on the stability characteristics of the bed material. Depending on the type of destructive force considered, m can be equal to 2/3 or 2/5.

Following this research line, Dean (1977) analysed 502 beach profiles (described in Hayden et al., 1975) from the United States Atlantic Coast and Gulf of Mexico. He stablished correspondences between A and sediment diameter and suggested a value of $m = 2/3$.

An expression of the beach profile which is proportional to the offshore distance raised to 2/3 was also obtained by Bowen (1978). He performed an analysis of the balance between the down slope sediment transport due to gravity forces and the onshore sediment transport produced by wave action on suspended sediment.

McDougal and Hudspeth (1983) carried out a dimensionless analysis of the beach profile proposed by Bruun and Dean:

$$H = AY^{2/3} \tag{3}$$

where all the variables (written in capitals in this case) have been made dimensionless with the width of the surf zone y_b, so that: $Y = y/y_b$, $H = h/y_b$ and $A = \alpha/y_b$.

These authors pointed out that the main difficulty for the aplication of equation (3) is the singularity that it presents at $Y = 0$, since in natural beaches there is no point with an infinite slope. Nevertheless they affirm that this type of profile is a good model because it provides the least squares best fit for numerous natural beaches.

The equilibrium beach profile of Dean (1977) has been used for instance in the GENESIS model (Hanson and Kraus, 1989) applications, where shoreline responses to changes in the longshore sediment transport rate are examined. Coupled with Dean's profile, the equations of Moore (1982) have been employed. These equations relate the parameter A with the sediment diameter D_{50}:

$$\begin{array}{ll} A = 0.41(D_{50})^{0.94} & D_{50} < 0.4 \\ A = 0.23(D_{50})^{0.32} & 0.4 \leq D_{50} < 10.0 \\ A = 0.23(D_{50})^{0.28} & 10.0 \leq D_{50} < 40.0 \\ A = 0.46(D_{50})^{0.11} & 40.0 \leq D_{50} \end{array}$$

where D_{50} is expressed in mm and A in $m^{1/3}$.

McDougal and Hudspeth (1989) attempted to compatibilize the mathematical singularity that presents the Dean-type profile slope at the origin, with the uniform slope that is often shown by reflecting beach profiles (also called summer-type or swell-type profiles). As a result, a composite profile, which combines a planar beach with uniform slope in the beach face and a concave profile in the offshore portion, has been proposed.

Following a similar line of reasoning, Dean (1991) suggested a modification of his former equation of 1977, which yields infinite slopes at the shoreline. Taking into account that large slopes induce large gravity forces which are not represented in equation (1) and including these gravitational effects together with destabilizing forces due to turbulent fluctuations produced by wave energy disipation, he proposed:

$$y = \frac{h}{m} + \frac{1}{A^{3/2}}h^{3/2} \tag{4}$$

where h is the water depth, m is the beach face slope and A is a coefficient. In shallow water the term $h = my$ dominates, while in deeper water dominates the simplification $h = Ay^{2/3}$.

Similar expressions have been proposed by Larson (1988), Larson and Kraus (1989) and Kriebel, Kraus and Larson (1991), combining a planar profile with a parabolic one.

3. DATA EMPLOYED

The Catalan Coast has a total length of 580 Km, 270 of them being sandy beaches. The northern part shows a predominance of rocky cliffs and reefs. In the central and southern areas, sandy beaches are dominating.

In this work, a detailed analysis of 82 beach profiles from the Catalan Coast has been carried out. These profiles are representative of 105.2 Km of beach (37.6% of the total length). In figure 1, the profile locations are shown.

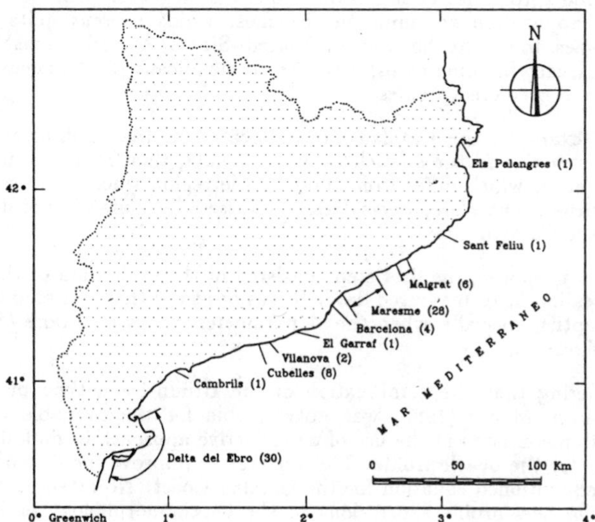

Figure 1. Location of the 82 beach profiles.

The profiles have been grouped taking into account the geographical location and morphological characteristics.

The profile data have been obtained from several field measurement campaigns. Some of them are described in Serra et al. (1989, 1990a, 1990b) and Jiménez et al. (1992).

4. APPLIED METHODOLOGY

As it was mentioned before, Bruun (1954) based on the analysis of beach profiles for the North Sea and California Coast, developed an empirical equation with a potential function which relates water depth h and shoreline distance y. Later on Bruun and other authors found theoretical justifications for this expression. Dean (1976, 1977), centering his study on linear wave theory and assuming that the ratio of breaking wave height to water depth is constant, justified from three different points of view this potential form for the equilibrium beach profile. He found a value of 2/5 or 2/3 for the exponent B and showed that the aforementioned equation with $B = 2/3$ is consistent with uniform wave energy dissipation per unit volume across the surf zone. Other authors, who also adopted this equation, suggested a potential relationship between the shape parameter A and the grain size (Moore 1982, Hanson and Kraus, 1989).

The wide spread use of this profile equation, together with its simplicity, led to its application to numerous beaches, which present quite different characteristics from the beaches of United States Atlantic Coast. These applications can, in some cases, give rise to significant discrepances between measured and predicted profiles.

As an example, the Catalan Coast, located in the Mediterranean Sea, presents distinctive features (such as wave climate, tidal range, beach slopes, grain size, etc.), which differ from those of the region where the formula was originally fitted. Therefore, it was decided to assess the suitability of this profile shape for the Catalan Coast.

For this purpouse, the first step consisted in the evaluation of the Bruun-Dean expression fit to measured data. It was observed that this equation over-predicts depths, specially when the A parameter values of Moore (1982) are used (see figure 2).

Considering that the combination of the Bruun-Dean type profile with A values from Moore (1982) was not suitable for the available data, the following step consisted in the use of an inductive approach to find alternative expressions for the beach profile. The aim was to improve the fit with respect to the aforementioned equation for the Catalan Coast. No attempt was made to relate the new profile expressions to the beach morphology or boundary conditions because of the limited number of surveyed profiles. The obtained expressions can be useful in the characterizacion of beach profiles for site-specific applications. It should be stressed that no attempt is here made to predict the actual beach shape parameters.

Obviously, the fit of all proposed expressions is poor for those profiles which have terraces or bars. In fact all these expressions overpredict the fore and back parts of the bar in compensation for the corresponding in-between smoothing.

The simpler expression to represent a beach profile is a straight line. This simplification could be admissible when the profile length is limited to a short distance from the shoreline. Otherwise, when the considered profile extension is longer, a curvilinear trend can be easily observed. In order to achieve the description of this nonlinear relationship and to find the best fit curve, a suitable

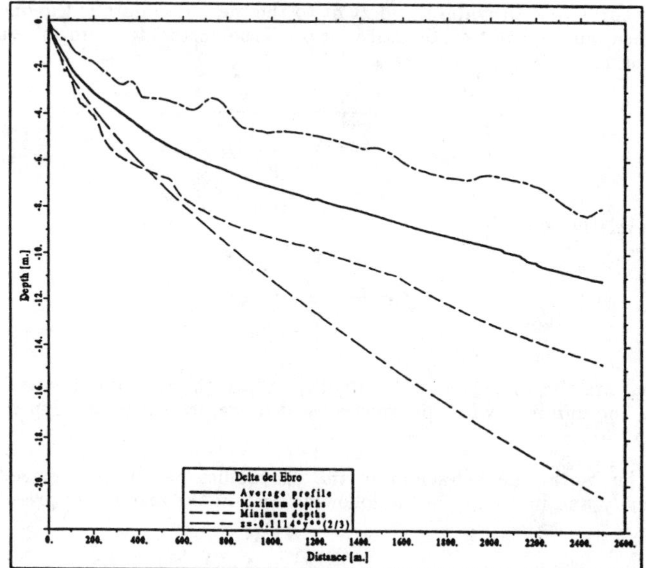

Figure 2. Bruun-Dean profile against measured data in the Delta del Ebro area.

transformation of variables has been carried out followed by a straight line regression analysis using the least squares method.

The selection of alternative expressions is based on the previous knowledge of general profile shapes. Two alternative expressions have been proposed for the equilibrium beach profile.

i) Exponential equation:

$$e^z = A(y + C)^B \tag{5}$$

ii) Rational equation:

$$z = \frac{y}{A + By} \tag{6}$$

Both equations will be compared with the power-law (potential) expression:

$$z = Ay^B \tag{7}$$

where z is the water depth, y the distance to the shoreline and A, B and C are parameters to be determined.

The curve that provides the best fit to the measured data is the one which shows the minimum vertical distance (or prediction error) to them. The average prediction error has been defined as

$$\epsilon = \sqrt{\frac{\sum_{i=1}^{n}(z_{m_i} - z_{c_i})^2}{n-p}} \tag{8}$$

and the relative error is

$$\delta = \frac{\epsilon}{\sqrt{\frac{1}{n}\sum_{i=1}^{n} z_{m_i}^2}} \tag{9}$$

where z_m are the measured water depths, z_c are the computed water depths and p is the number of free parameters which are present in the expression to be fitted.

In this work, the behaviour of the 82 profiles has been analyzed, both individually and in goups. In the following section the results are presented.

5. RESULTS AND DISCUSSION

Using the potential expression (given by equation 7), values for the two free parameters (A and B) have been obtained for each profile. A values range between 0.023 and 0.520, with an average value of 0.125. On the other hand, coefficient B ranges from 0.417 to 0.987 with an average value of 0.699, which is slightly greater than the value of 2/3 obtained by Bruun and Dean. The average prediction error is 0.75 m. Taking into account this figure, it is evident that the fit quality is limited and the potential expression may not always have the desired resolution for modelling beach profiles.

The analysis of the two newly proposed expressions has an empirical (statistical) character and is just an attempt to present alternative equations which improve the fit of beach profile shapes.

In the exponential expression (equation 5), and due to the special characteristics of this equation, the coefficient A takes very small values. For practical reasons it is better to work with $|\ln A|$. This parameter varies between 0 and 38.491 and the average value is 14.518. B ranges between 1.171 and 7.642, with an average value of 3.46. The range of C is very wide, between 5 and 155, and has an average value of 85.63. The average error is 0.36 m, which indicates that this equation fits better the measured profiles, although it does not offer the simplicity presented by the potential expression.

When the rational expression (equation 6) is used, parameter A varies between 3.350 and 87.296, with an average value of 35.76. Coefficient B ranges between 0.025 and 0.170, with an average value of 0.087. The prediction error is slightly greater than for the exponential fit (its average value is 0.40 m), but remarkablely lower than for the potential equation.

Taking into account the prediction errors of the three expressions for each of the 82 beach profiles, table 1 is easily obtained. In this table, the number of profiles which show the best fit (under column 1), the worst fit (under column 3) and the intermediate fit (under column 2) are indicated for each equation. The exponential expression is the one that provides a better fit in most cases (62.2%), while the potential one provides less good agreement in the majority of cases (84.1%). The rational equation works fairly well because for 25.6% of the profiles it presents the best fit and for 64.6% of cases the second best. Nevertheless, the behaviour of this last expression is quite satisfactory and comparable to the exponential one. In fact, the differences in fit quality between both expressions are minimal, as it can be verified in table 2. In this table, prediction error ranges and average errors (between parenthesis) for each studied sector are shown. The term C. Catalana refers to the whole 82 profiles. The other five names represent morphologically different sectors of the Catalan Coast.

Expression	1	2	3
Exponential	51 (62.2%)	26 (31.7%)	5 (6.1%)
Rational	21 (25.6%)	53 (64.6%)	8 (9.8%)
Potential	10 (12.2%)	3 (3.7%)	69 (84.1%)

Table 1. Profile fit quality.

Sector	ϵ potential	ϵ exponential	ϵ rational
Malgrat	0.91-1.59 (1.20)	0.69-1.11 (0.88)	0.67-1.22 (0.92)
Maresme	0.33-2.41 (1.14)	0.06-1.17 (0.45)	0.09-1.20 (0.49)
Barcelona	0.23-0.52 (0.41)	0.01-0.09 (0.04)	0.01-0.22 (0.10)
Cubelles	0.02-0.10 (0.06)	0.02-0.04 (0.03)	0.01-0.07 (0.03)
D. Ebro	0.29-1.64 (0.63)	0.12-0.83 (0.36)	0.12-0.91 (0.40)
C. Catalana	0.02-2.41 (0.75)	0.01-1.17 (0.36)	0.01-1.22 (0.40)

Table 2. Prediction errors for each coastal sector.

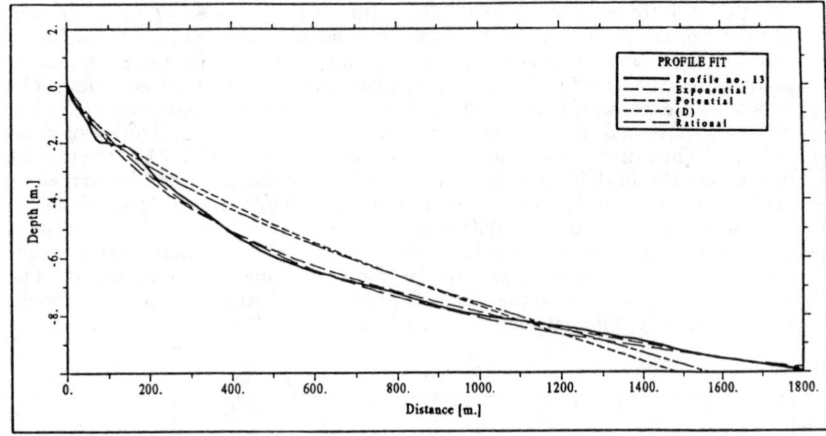

Figure 3. Fit of one profile in Delta del Ebro area.

Figures 3 and 4 show, as an example, the fit of the three expressions to two different measured beach profiles. Moreover, a potential expression with $B = 2/3$ has also been plotted and referred to with (D). The measured beach profiles are drawn with continuous line.

Likewise, average profiles for each studied sectór were obtained and fitted with the three aforementioned expressions. As it was observed in the analysis of individual profiles, exponential and rational-type curves reproduce better than the potential one the average profiles of each sector. Figures 5, 6 and 7 illustrate this behaviour by plotting the fit of the three expressions to the Maresme sector average profile.

Searching a possible explanation of why the two proposed equations improve the fit to measured data (with respect to the power-law expression), the local beach slope was obtained for each equation:

o Power-law equation:

$$z' = \frac{2}{3} A \; y^{-1/3} \tag{10}$$

o Exponential equation:

$$z' = \frac{B}{y + C} \tag{11}$$

o Rational equation:

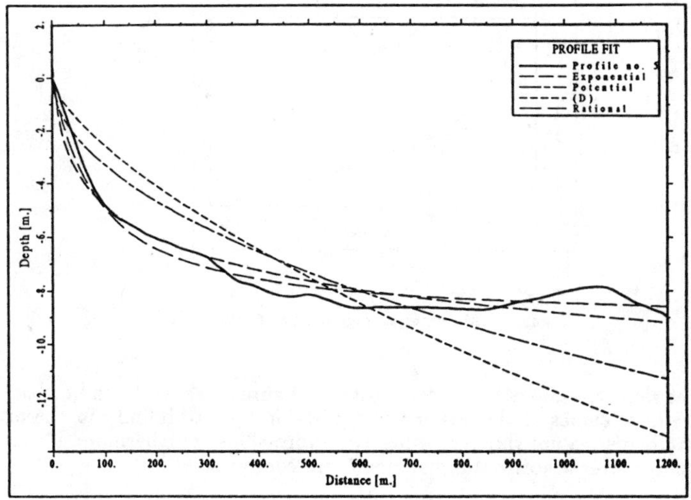

Figure 4. Fit of one profile in Maresme area.

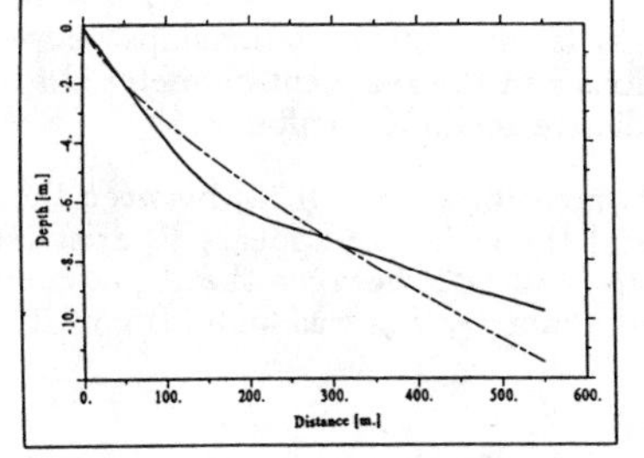

Figure 5. Maresme average profile. Potential fit.

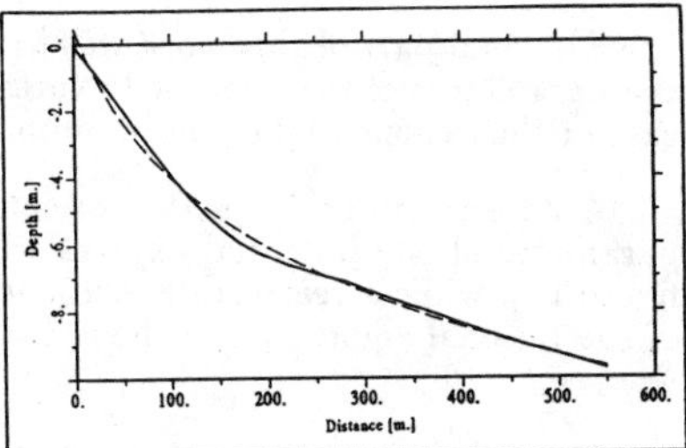

Figure 6. Maresme average profile. Exponential fit.

$$z' = \frac{A}{(A + By)^2} \tag{12}$$

Selecting now the mean values in the Catalan Coast for the free parameters,

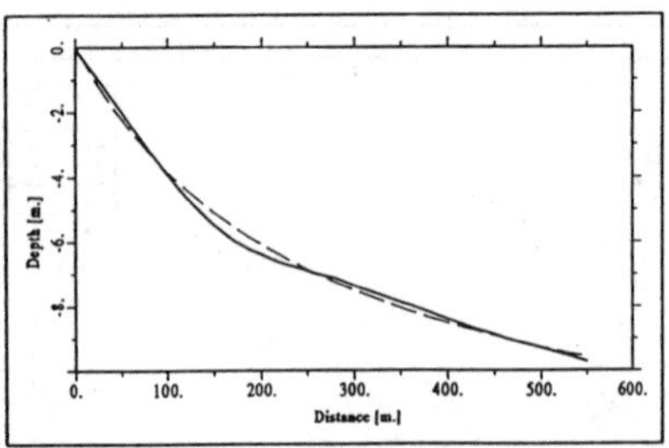

Figure 7. Maresme average profile. Rational fit.

the three slope equations have been plotted in figure 8, where it can be observed that the local slopes of the power-law expression tend to infinity in the origin and have bigger values than the other two expressions. Furthermore, if average beach slopes over a horizontal distance L are computed

$$\bar{z} = \frac{1}{L} \int_0^L z' dy \tag{13}$$

and plotted as a function of the offshore distance (see figure 9) the same trend is also observed. Therefore, the two newly proposed expressions (rational and exponential) have smaller (local and average) slopes than power law equations.

The final part of this work was to look for predictive relationships between the parameters of the proposed expressions and the sediment diameter. In the case of the exponential equation no predictive relation was found.

A poor relationship (with a correlation coefficient $r = 0.50$) between the A parameter of the potential expression and the gran size appears to exist (see figure 10), while a reasonable relationship ($r = 0.82$) between the A parameter of the rational equation and the sediment diameter D_{50} was found (figure 11).

6. SUMMARY AND CONCLUSIONS

82 beach profiles from the Catalan Coast have been employed to analyse equilibrium profile shapes in this zone. It has been shown that a potential expression like the ones proposed in the state of the art provides a limited quality fit to the aforementioned beach profiles. Moreover, it has been verified that when the curves of Moore (which give the parameter A of the potential

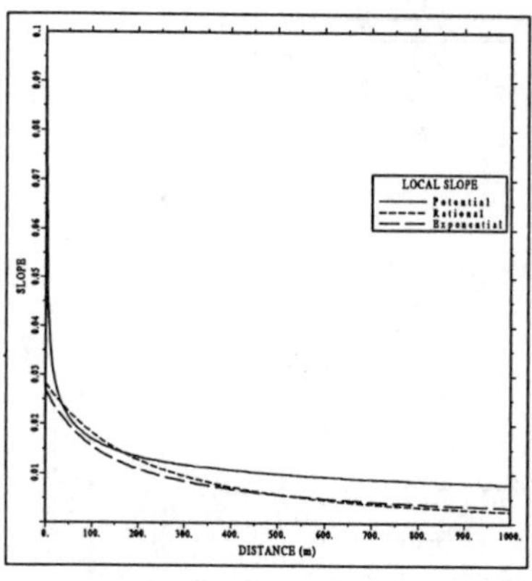

Figure 8. Comparison of local slopes.

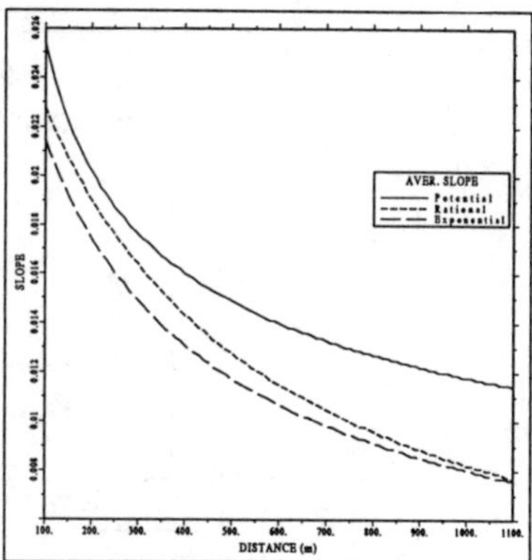

Figure 9. Comparison of average slopes.

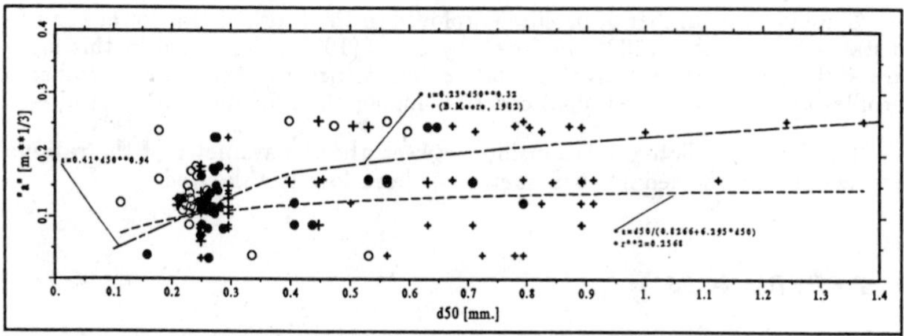

Figure 10. Relationship between A and D_{50}. Potential equation.

expression as a function of sediment grain size) are employed, they overpredict depths with regard to measured profiles. This overprediction effect is bigger when the average beach slope decreases.

As a consequence two alternative expressions, one exponential and another rational, have been proposed. These expressions improve the reproduction of

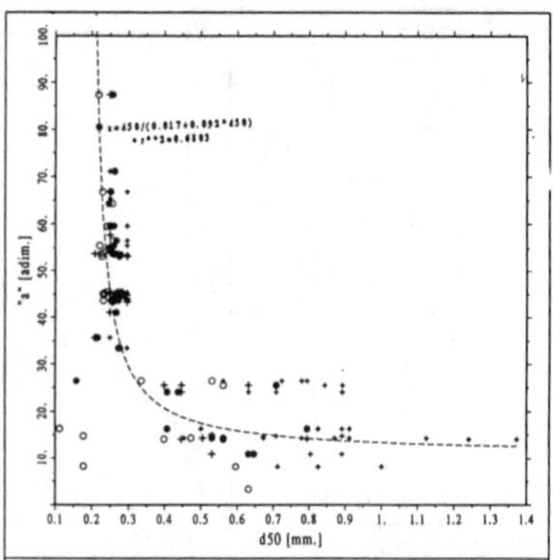

Figure 11. Relationship between A and D_{50}. Rational equation.

beach profile shapes in the Catalan Coast. In spite of the exponential equation yielding slightly better results, the rational one presents a similar predictive capability. Therefore both expressions are suitable for modelling beach profiles in the studied zone.

The two newly proposed expressions seem to fit better beach profiles with milder slopes, but further research is needed to confirm this hypothesis.

Though the number of profiles employed in this work (82) is lower than for instance the number (502) employed by Dean (1977), data used in this study are still quite representative in relative terms, because the ratio "number of profiles to length of the studied coast" is higher than for most other studies.

Finally, a predictive relationship between the A parameter of the rational expression and the sediment diameter has been locally validated.

7. REFERENCES

- Bowen, A.J. (1978). "Simple Models of Nearshore Sedimentation: Beach Profiles and Longshore Bars". *Proc. Conf. Coastlines of Canada.*

- Bruun, P. (1954). "Coast Erosion and the Development of Beach Profiles". *Beach Erosion Board, Tech. Memo. no. 44.*

- Dean, R.G. (1976). "Equilibrium Beach Profiles and Response to Storms". *Proc. 15th Conf. on Coastal Eng., Honolulu, Hawaii.*

- Dean, R.G. (1977). "Equilibrium Beach Profiles: U.S. Atlantic and Gulf Coasts". *Ocean Eng. Tech. Rep. no. 12, Univ. of Delaware, Newark.*

- Dean, R.G. (1991). "Equilibrium Beach Profiles: Characteristics and

Applications". *J. of Coastal Research, vol. 7.*

- Fenneman, N.M. (1902). "Development of the Profile of Equilibrium of the Subaqueous Shore Terrace". *J. of Geology, vol. X.*

- Hanson, H. and N.C. Kraus (1989). "GENESIS: Generalized Model for Simulating Shoreline Change". *CERC, Tech. Rep. 89-19.*

- Hayden, B., W. Felder, J. Fisher, D. Resio, L. Vincent and R. Dolan (1975). "Systematic variations in inshore bathymetry". *Department of Environmental Sciences, Tech. Rep. no. 10, University of Virginia, Charlotesville, Virginia.*

- Jiménez, J.A., V. Gracia, M.A. García and A. Sánchez-Arcilla (1992). "Análisis y propuesta de soluciones para estabilizar el Delta del Ebro: Balance sedimentario y esquemas de transporte". *Informe técnico LT-3/4, D.G. Ports i Costes, Generalitat de Catalunya.*

- Kriebel, D.L., N.C. Kraus and M. Larson (1991). "Engineering Methods for Predicting Beach Profile Response". *Coastal Sediments' 91, ASCE.*

- Larson, M. (1988). "Quantification of Beach Profile Change". *Rep. no. 1008, Dep. of Water Resources Eng., Univ. of Lund, Sweden.*

- Larson, M. and N.C. Kraus (1989). "Prediction of Beach Fill Response to Varying Waves and Water Level". *Proc. Coastal Zone '89, ASCE.*

- McDougal, W.G. and R.T. Hudspeth (1983). "Longshore Sediment Transport on Non-plane Beaches". *Coastal Engineering, vol. 7.*

- McDougal, W.G. and R.T. Hudspeth (1989). "Longshore Current Sediment Transport on Composite Beaches". *Coastal Engineering, vol. 12.*

- Moore, B. (1982). "Beach Profile Evolution in Response to Changes in Water Level and Wave Height". *M.S. Thesis, Dep. of Civil Eng., Univ. of Delaware, Newark.*

- Serra, J., A. Calafat, M. Canals, J.L. Casamor and J. Sorribas (1989). "Seguiment de l'evolucio de les platges del Maresme. Anys 1987 i 1988". *Informe técnico D.G. Ports i Costes, Generalitat de Catalunya.*

- Serra, J., J. Sorribas, A. Calafat and M.A. García (1990a). "Analisi del comportament de les platges regenerades al Maresme. Memoria anual (Set. 1989-Jul. 1990)". *Informe técnico D.G. Ports i Costes, Generalitat de Catalunya.*

- Serra, J., J. Sorribas, A. Calafat, M. Canals, J.L. Casamor, J. Argullós and F.J. Miguel (1990b). "Analisi del comportament de les platges regenerades al Maresme. 1a. Memoria". *Inf. Tec., Univ. Barcelona.*

MODELLING LARGE-SCALE COAST EVOLUTION

Zbigniew PRUSZAK[1] and Ryszard B. ZEIDLER[2]

ABSTRACT: Examples of millenial and centennial evolution of Poland's coast are shown to illustrate the effect of various types of large-scale factors. *IBW PAN* database and processing techniques for the field data accumulated over decades, and available for centuries (as maps) are presented as prerequisite to understanding and modelling of large-scale coastal behaviour. Two mathematical models are outlined, one in simplified yet closed form and another in general conceptual framework. The former, basing on the conservation law for sediment mass (volume), is derived and reformulated for a simple case of nontidal coast. Sediment transport and source terms appearing in that model are described in detail, followed by an example showing the application of the model for the Polish coast for both point and non-point sources. The other model consists of first-order nonlinear differential equations for overall erosion and accretion within a coastal cell. Potentials and limitations of large-scale modelling are discussed in closure.

1 INTRODUCTION

Evolution of the Polish coast in scales of millenia and centuries is exemplified in a recent paper by Zeidler et al. (1994), for various sites along 500 km of the coast. The transformation in the course of *millenia* is outlined in Figure 1. Line 1 dates back to some 8000 years ago and is believed to mark submerged cliffs (although various geologists provide different dating such as preLittorina, Ancyllus or even Yoldia).

In scales of *centuries*, the evolution of Baltic fjords, bays and estuaries is also well realized (Zeidler et al 1994). The case of Hel Peninsula is of concern as the spit might have been breached in recent centuries and split up into smaller

[1]Assoc.Prof. & [2]Professor, Polish Academy of Sciences' Institute of Hydro-Engineering *IBW PAN*, 80-953 Gdansk, Koscierska 7, Poland

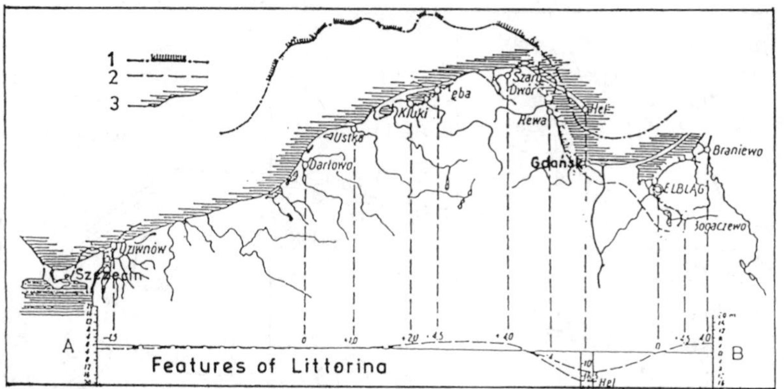

Figure 1. Polish coast at present (line 3) and in the past (line 2 ~ 4000 BP, line 1 ~ 8000 BP).

islands (still a hazard in view of the accelerated sea level rise due to the greenhouse effect, cf. Zeidler 1992). Several fairly well described examples of Polish shoreline migration over centuries are also known. Spectacular ruins of a church at Trzęsacz, now at the cliff edge, belong to the structure raised some 1800 m from shoreline in 1250. The major reasons of the *decelerating* cliff erosion are attributed to the convexity of the shore, the geologic structure and changes in groundwater flow regime (Zeidler et al. 1994).

Among the factors controlling the large-scale coast evolution, it is difficult to clearly identify *climatic* ones, at least in terms of directly measured data, which are simply unavailable for the remote past. The *geologic* factor (cf. geologic structure of shore at Trzęsacz, Jastrzębia Góra etc.) should never be overlooked — bed stratification with interlayers of softer and harder material often makes impossible the return to the original coastal form.

On the other hand, *anthropogenic* effects are often more than conspicuous. For example, the transformation of flow patterns in the Szczecin Bay (Stettiner Haff) after 1720, due to regulation of the Swine Strait, in the wake of political rivalry between Prussia and Sweden brought about clear erosion and retreat of some western banks of the bay. Aside from the Trzęsacz cliff and many more, two other cliffs of Orłowo and Redłowo (at Gdynia) display man-induced effects blamed responsible for their activation. The enhanced erosion of the former has been caused by extension of the Harbour of Gdynia after World War II, while the other was devastated once the drainage control has been neglected.

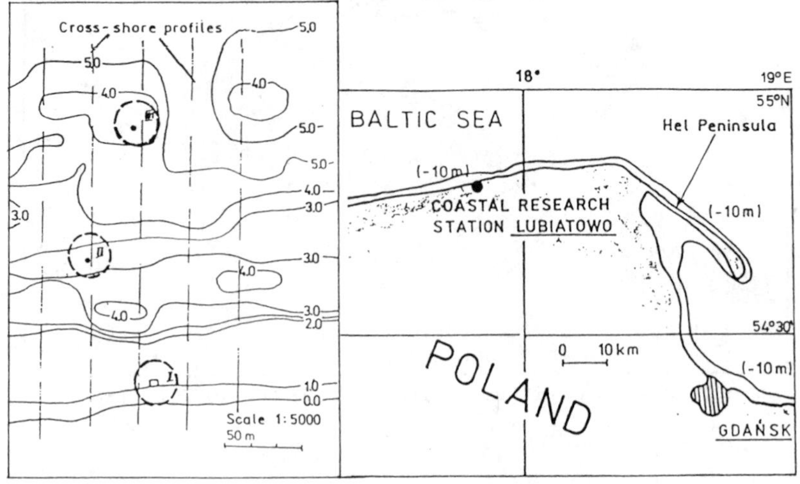

Figure 2. Site Topography at *IBW PAN* Coastal Research Station at Lubiatowo.

2 FIELD DATABASE; METHODOLOGY and FINDINGS

Under various programmes aiming at exploration of coast and its transformation, measurements of coastal processes have been continued at the Polish Academy od Sciences'(PAS) Institute of Hydro-Engineering (*IBW PAN*) Coastal Research Station at Lubiatowo (Polish coast of the Baltic Sea some 75 km from Gdańsk, cf. Figure 2). The shore belongs to multi-bar dissipative ones, with an average slope of 1–1.5% and sand quartz of grain size D_{50}=0.22 mm. High-energy wave input usually comes from the N–NW sector, although much rarer easterly events are more hazardous because of the long fetch. Usual high storm waves have H_s=3.5–4.0 m and T_s=7–8 s at the seaward edge of the surf zone ($h \approx 7$ m).

Along with parameters of wind, waves, currents and other hydrologic factors, sediment transport and bathytopographic features have been measured extensively at Lubiatowo since mid-sixties (embodying international expeditions as well). In 1983 a routine programme has been established for beach topography measurements on a 2.7-km beach and nearshore zone segment, coupled with less regular bathymetric surveys extending some 800 m from shoreline. The cross-shore profiles have been arranged every 100 m. The first systematic and mutually compatible records of beach and shore topography date back to 1964,

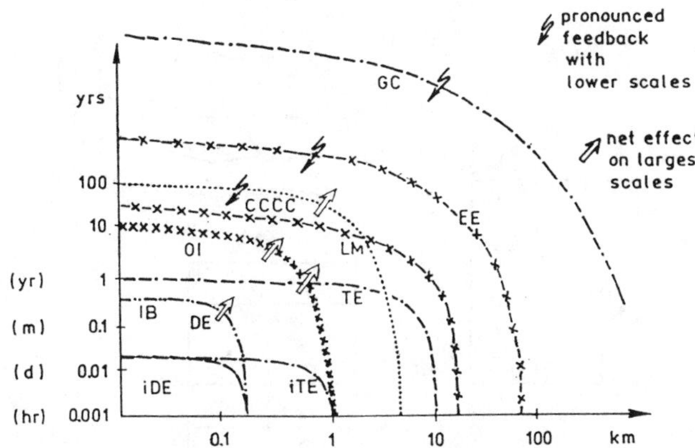

Figure 3. 'Cumulative diagram' of coast morphology change in various scales. (DE) dune erosion; (IB) inner bar–beach exchange; (TE) tidal induced coast evolution; (OI) outer bar–inner bar exchange; (LM) longwave morphologic features; (CCCC) climate change–induced coast change; (EE) extreme events; (GC) global change; (i) intensified.

and echosoundings plus tachimetry are continued until now.

Since several other Baltic, Black Sea, and West African coast datasets have also been acquired by the *IBW PAN* staff, we have insight into a considerable bulk of a reliable field database stretching over a reasonably long span of time. Various approaches to data processing have been attempted. *Statistical* and *spectral* tools have permitted us to identify the typical scales and their contribution to the overall changes. *Inter alia*, the celerities of alongshore propagation of macrocusps have been found to vary about 1 km per annum. Some long stretches of apparently fragile sandy beaches and shores possess the property of 'two-dimensional local stability' displayed as nodal points both along and across shore. Such nodes have been exposed at two different Baltic locations (Hel Peninsula and Mielno) as well as on the Atlantic coast of Senegal. *Empirical orthogonal functions (EOF)* have been harnessed to describe both cross-shore and longshore evolution of coast. The concept of *Dean profile* has been extended wherein the parameter A in the shore profile $y = Ax^{2/3}$ is exposed for Lubiatowo as a function of time — $A = A(t) \approx 0.075 + 0.022 \cos 0.23t(years)$ (Pruszak 1993).

3 GENERAL MODELLING CONCEPTS

Figure 3 outlines a certain idea of the evolution of coast and its controlling factors in different time and space scales. Dune erosion (DE) and inner bar–beach

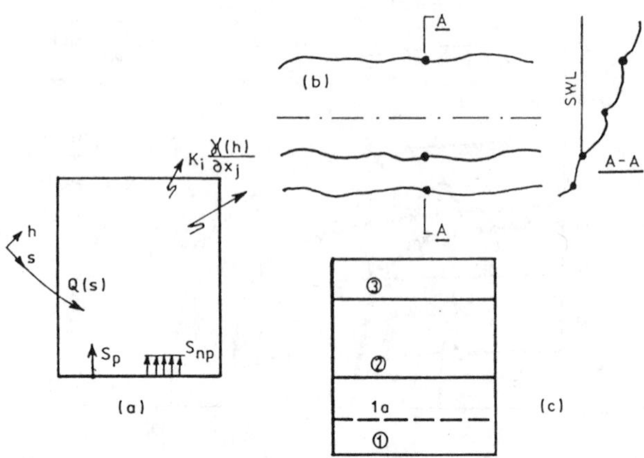

Figure 4. Control volume (coastal cell or control cell, a), schematization of coastal zone (b) and 3-line coastal cell (c).

exchange (IB) processes are seen as the smallest scales, in wave-dominated environments. Tide-induced evolution (TE) affects larger areas and time spans, and so forth. The largest scale in Figure 3 is delineated by global changes (GC) such as glaciation etc, which bring about revolutionary changes induced by factors not necessarily of hydrodynamic origin.

Since one faces different coastal conditions, it is not possible to derive a single model which would describe all environments. We are proposing two approaches which exhibit a good deal of generality (at the cost of internal fineness, though). A large-scale model basing on conservation laws for sediment volume, the concept of shore profile equilibrium, dispersion terms between coastal cells, and linkage between wave energy dissipation and shore transformation is put forward. The coastal cell (control volume) is delimited by different kinds of shore profiles along shore. The landward boundary can be assumed somewhere about the dune foot or shoreline, while the seaward boundary is far away from the location of perceptible sediment transport. The two other boundaries should be selected on the grounds of morphological distinction from the cell's neighbours. There occurs exchange of sediment through all boundaries.

The model in plan view is outlined in Figure 4; the testing of its sensitivity to various parameters is based on the shoreline migration and depth change. From the conservation law for sediment mass (or volume) within the control cell one has

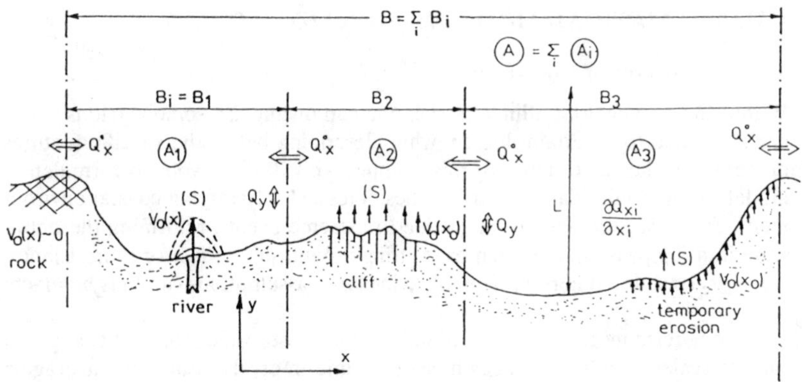

Figure 5. Sediment control volume and budget components adopted for modelling.

$$\int_B \left\{ \frac{\partial Q}{\partial s} ds - K_i \frac{\partial (Ch)}{\partial x_j} \right\} + \alpha S_p + \beta S_{np} dL_{np} = \gamma A \frac{\partial h}{\partial t} \quad (1)$$

in which B = boundaries of control cell [m], Q = sediment transport rate [m^3/s] per 1 m measured in direction normal to s (i.e. n), s = curvilinear coordinate (along sediment transport trajectory), S_p = strength of point source of sediment [m^3/s], S_{np} = strength of nonpoint source of sediment [m^2/s], L_{np} = active length of nonpoint source of sediment [m], K_i = sediment eddy diffusivity [m^2/s], C = sediment concentration [m^3/m^3], h = depth of water [m], A = surface area of control cell [m^2], i = 1, 2 = running index assigned to longshore (1) or cross-shore (2) direction.

The first term stems from net advection while the second term describes the spreading of sediment across the boundaries, in the direction i, due to existence of the concentration gradient in the direction j, which can principally occur even in the absence of advection. The two types of sources are continuous, i.e. are deemed to exist all over the lifetime of the control cell. The terms on the left-hand side are compensated by the change in the relative depth of water (that is, including all possible components of the vertical movement of bed and water surface). — Superimposed on the equilibrium profile can be sand bars, with geometrical properties (distance, spacing, height, volume, shape etc) specified on the basis of the available field data.

4 MATHEMATICAL MODEL for NONTIDAL COAST

4.1 Formulation of the Model

Within the framework outlined above, one can distinguish some versions which are simplified to a certain degree, while describing better the specific features modelled at the same time. In this chapter, we will endeavour to formulate a model fitting a nontidal, primarily cohesionless, slowly-varying coastal environment. *Inter alia*, dramatic changes due to extreme events, reforming the entire zone in a jumpwise mode, such as splitting a peninsula in two or more islands, are not considered herein. The following basic assumptions can thus be made:

1. Long-term morphological evolution of the coast results from net changes in smaller scales. The net changes in shorter time intervals result from averaging over the time span ΔT. For our purposes we can assume that ΔT is one year. Hence we neglect details of shore evolution processes occurring throughout one year, and concentrate attention on changes from year to year.

2. The characteristic alongshore dimension is taken as several, or even some dozens of kilometres. Hence we disregard the longshore migration of cuspate beach forms (as they become smoothed out), and generally — the longwave response to shortwave forcing altogether (sand waves being filtered out as well). The cross-shore dimension of the coastal cell is chosen in such a way that the exchange of sediment across the seaward boundary is permitted but generally minor bed changes are assumed to occur there.

Hence sediment transport can take place at all boundaries of the control volume (coastal cell) — the beach line, the seaward boundary, and the two longshore boundaries. The configuration of some possible sediment transport (denoted by Q) and source (labelled S) components is illustrated in Figure 5. The terms and symbols shown in the drawing are discussed in subsequent paragraphs.

One can distinguish three coastal subcells A_1, A_2 and A_3, which can be put together as one cell of total breadth B. The longshore and cross-shore rates of sediment transport, Q_x and Q_y, are gross quantities (in m^3/s) measured within the boundaries of the coastal cell, that is a strip of shore of breadth B and length L. The sources of sediment S can be either point- or nonpoint-type, with intensity V_0 (m^3/s) or ϑ_0 (m^2/s, per one metre of shoreline), respectively.

The conservation law for sediment mass (volume) can be formulated as follows

$$\sum_{k=1}^{n}\left[(\Delta Q_{x_{ik}} + \alpha_k \cdot \Delta S_k)\frac{T_k}{A} - \alpha_k \cdot \Delta h_k - \delta_{Z_k} - \delta_{D_k}\right] = 0 \qquad (2)$$

in which $\Delta Q_{x_i k}$ = net annual sediment transport rate along x_i in k-th year,

$\Delta S_{x_i k}$ = net annual transport rate from sediment source along x_i in k-th year, $k = 1...n$, = index of consecutive years of the long-term process change; i = 1, 2 = longshore or cross-shore direction, respectively: $(i = 1)$ for x or $(i = 2)$ for y; T_k = k-th 1-year time step; α_k = coefficient accounting for substrate type in a given time step (k), closely related to bed erosion mode and extent (1 for cohesionless bed, 0 for cohesive bed or rock, and between o and 1 for mixed bed).

It is seen that Equation 2 combines the inflow/outflow of sediment with changing depth of water occurring from year to year in the series of $k = n$ years, and resulting from the three major causes: bed changes Δh, long-term sea-level change δ_Z and eustatic/isostatic movements of the seabed δ_D.

4.2 Discussion of Sediment Transport Terms

The sediment transport rate gradient $\frac{\partial Q_{x_i}}{\partial x_i}$ is deemed to be controlled primarily by waves and currents. It is the resultant net annual value. It is traditionally split up into the longshore (x) and cross-shore (y) components, so that one has

$$\frac{\partial Q_{x_i}}{\partial x_i} dx_i = \frac{\partial Q_x}{\partial x} dx + \frac{\partial Q_y}{\partial y} dy \tag{3}$$

For the *longshore direction*, the gross sediment transport rate (across the entire shore profile measuring L and stretching from beach to the seaward boundary) in the k-th year will be different from the multiyearly equilibrium value Q_E:

$$\Delta Q_x = \Delta Q_{x_k} - Q_E \qquad [\tfrac{L^3}{\Delta T}] \tag{4}$$

Obviously, for $\Delta Q_x > 0$ the coastal cell becomes shallower in k-th year while for $\Delta Q_x < 0$ the depth of water increases (due to mere change in the longshore transport). The function $Q_x(y)$ is generally unknown; it depends on the annual wave climate (wave parameters, including incidence angle), bed topography etc. But it is known to be irregular and to decrease with distance from shoreline. As a rule of thumb, one can postulate the following linear decrease, from the maximum at shoreline to zero at the seaward boundary:

$$q_x(y) = \Delta Q_x \cdot \frac{2 \cdot (L - y)}{L^2} \qquad [\tfrac{L^2}{\Delta T}] \tag{5}$$

in which L = breadth of the active shore zone, usually of the order of $10^2 \div 10^3$ [m], $q_x(y)$ = net annual rate per cross-shore unit of length.

Within the coastal cell, one can assume a linear variation of the longshore gradient:

$$\frac{\partial q_x(y)}{\partial x} dx \approx \frac{\Delta Q_x}{B} \qquad [\mathrm{L}^2/\Delta T] \tag{6}$$

in which B = longshore dimension of the coastal cell, usually 10^3–10^4 [m].

For the cross-shore transport, one can adopt the relationship between the unit cross-shore sediment transport rate q_y and the wave energy dissipation D. For each and every year k one can write

$$q_{y_k} = \overline{M} \cdot (D_k - D_{E_k}) \qquad [\tfrac{L^2}{\Delta T}] \tag{7}$$

in which $\overline{M_0}$ = empirical coefficient of regional nature, having the dimension $[\frac{m^{\frac{3}{2}}}{\Delta T}]$, connected with the net annual cross-shore sediment transport rate, wherein $\overline{M_0} \neq \overline{M}$; D_k = mean annual net energy dissipation controlling the ultimate shore profile for the given time step k; D_{E_k} = mean annual energy dissipation corresponding to profile equilibrium (where one encounters no cross-shore sediment transport).

Making use of Equation 7 one obtains the following net annual cross-shore sediment transport rate

$$\Delta Q_{y_k} = \overline{M_0} B \cdot (A_k^{\frac{3}{2}} - \overline{A_E}^{\frac{3}{2}}) \tag{8}$$

The cross-shore gradient $\frac{\partial Q_y}{\partial y}$ in the area (B, L) for each time step k will then become

$$\frac{\partial q_y}{\partial y} dy \approx \frac{\Delta Q_{y_k}}{L} \qquad [\tfrac{L^2}{\Delta T}] \tag{9}$$

One may note that it is only q_x, not q_y, which varies with both x and y. Hence for each time step k one has

$$\frac{\partial q_x(y)}{\partial x} dx + \frac{\partial q_y}{\partial y} dy \approx \Delta \frac{Q_{xk}}{B} \frac{2(L-y)}{L^2} + \frac{\Delta Q_{yk}}{L} \tag{10}$$

4.3 Discussion of Source Terms

The source of sediment (having the strength S) can be either instantaneous or continuous, and either point (index p) or non-point type. In this paper we focus attention on continuous *non-point (linear)* sources, which are either confined to a certain alongshore segment (henceforth referred to as 'local') or stretch along the entire costline of interest (the entire coastal cell; and are labelled 'regional'). The sediment source strength S coupled with beach and dune erosion, riverine load or any other input from the land side, can be assumed in the following form for a point source (m³/s):

$$S_p = V + \overline{D_i} \cdot \frac{\partial \vartheta}{\partial x_i} \qquad [\tfrac{L^3}{\Delta T}] \tag{11}$$

The term V in Equation 11, which virtually stands for a continuous *point* source, can be replaced by $\vartheta(x_0)$, where x_0 denotes the length of a *non-point* source. Both V and $\vartheta(x_0)$ describe the advective movement of sediment in or out

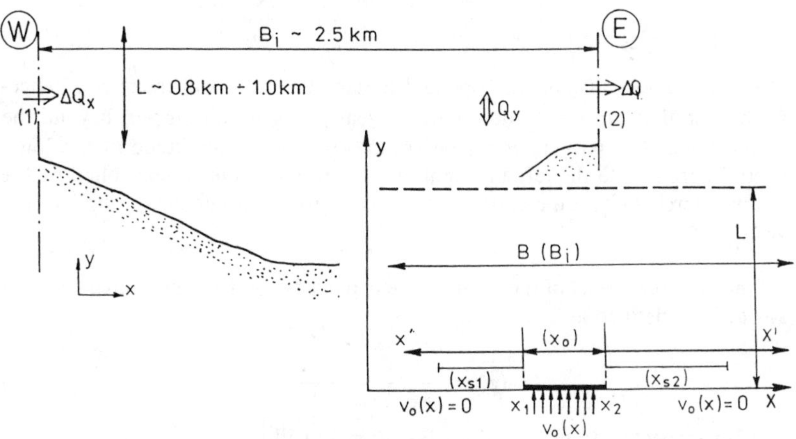

Figure 6. Computational scheme.

control cell, at speed u_0. The second RHS term schematizes the dispersion in the sea of the land-borne material, with the dispersion coefficient $\overline{D_i}$ (linear dimension [m] resulting from division of diffusivity-type quantity (m²/s) by u_0 (m/s)). D_i assumes values D_1 along shore and D_2 across shore.

One of the assumptions that can be made about the advection relies on the following exponential decrease in the seaward direction (although a power type relation might be better justified in view of solutions obtained from the theory of turbulent diffusion):

$$\vartheta\ (x, y) = \vartheta_0(x) \cdot exp(-a \cdot y) \qquad [\tfrac{L^3}{L \cdot \Delta T}] \tag{12}$$

The term $\overline{D_i} \cdot \frac{\partial V(x,y)}{\partial x_i}$ depends heavily on the diffusivity D_{x_i} (either D_x or D_y), where the latter can also be encompassed as

$$\overline{D_{x_i}}(\overline{D_x} \text{ or } \overline{D_y}) \approx \frac{D_x \text{ or } D_y}{(B) \text{ or } (L)} \cdot t_N \qquad [L] \tag{13}$$

in which $D_{x_i}(D_x$ or $D_y)$ = instantaneous diffusivity $[\frac{m^2}{s}]$ in the direction x_i (x or y); t_N = number of days per year with perceptible dispersion of sediment.

4.3.1 *Local Source*

We now confine ourselves to the local source along $x_1 \leq x_0 \leq x_2$, where the source strength is uniform and reads $\vartheta_0(x_0)$ (Fig. 6). Equation 11 will be modified to the form

$$S_{np} = \overline{\vartheta_0}(x) \cdot exp(-a \cdot y) \cdot (1 - \frac{|x'|}{x_{s_l}}) + \overline{D_{x_i}} \cdot \frac{\partial \vartheta\ (x, y)}{\partial x_i} \tag{14}$$

The index *np* will <u>not</u> be repeated hereinafter, so S will mean S_{np} henceforth. The function $1 - \frac{|x'|}{x_{sl}}$ describes the yearly longshore transport beyond the segment x_0. The parameter x_{s1} determines the yearly amplitude of the longshore migration, to the left and right of x_0. This parameter is controlled by the relative percentage of the opposite longshore currents occurring throughout the year.

Since the real effect of the local source extends beyond x_0, the source strength has to be redefined as

$$\overline{\vartheta_0}(x) = \frac{\int_{x_0} \vartheta_0(x_0) \cdot dx_0}{x_0 + \frac{1}{2} \cdot (x_{s_1} + x_{s_2})} \tag{15}$$

After respective transformations Equation 14 will read

$$\begin{aligned} S = \overline{\vartheta_0}(x) \cdot exp(-a \cdot y) \cdot \{[1 - \overline{D_y} \cdot a + \frac{\overline{D_x}}{\overline{\vartheta_0}(x)} \cdot \frac{\partial \overline{\vartheta_0}(x)}{\partial x}] - \frac{|x'|}{x_{s_l}} \cdot (1 - \overline{D_y} \cdot a) \\ + \frac{\overline{D_x}}{x_{s_l}} + \frac{|x'|}{x_{s_l}} \cdot \frac{\overline{D_x}}{\overline{\vartheta_0}(x)} \cdot \frac{\partial \overline{\vartheta_0}(x)}{\partial x}]\} = \overline{\vartheta_0}(x) \cdot exp(-a \cdot y) \cdot \{P[x_0] - P[x']\} \end{aligned} \tag{16}$$

in which $P[x_0], P[x']$ are functions connected with the areas (x_0) and (x'), where the latter (x') stretches from x_0 on one side to x_{s_1} and x_{s_2} on the left and right, respectively.

Note that Equation 16 describes the general case when the source function $\vartheta_0(x)$ is not uniform along the segment (x_0).

4.3.2 *Regional Source*

This case is simpler — the term $P[x']$ is neglected and the function S in Equation 14 becomes

$$S = \overline{\vartheta}_0(x) \cdot exp(-a \cdot y) \cdot \{P[x_0]\} \tag{17}$$

The gradient $\frac{\partial S}{\partial y}$ will also be reduced to

$$\frac{\partial S}{\partial y} = \overline{\vartheta}_0(x) \cdot exp(-a \cdot y) \cdot \{P'[x_0]\} \tag{18}$$

Further simplification follows if the source has constant strength $\overline{\vartheta}_0$ = const. along the entire shore of the coastal cell. All terms with the derivatives $\frac{\partial V_0(x)}{\partial x}$ and $\frac{\partial^2 V_0(x)}{\partial x^2}$ are then obviously zero.

5 SIMPLE EXAMPLE

5.1 Version I

The functioning of the simplified model outlined in Chapter 5 can be illustrated for the situation exercised at the *IBW PAN* coastal research station at Lubiatowo. In this Version I we assume that weak sediment supply persists over the entire alongshore beach segment due to erosion. The mean annual strength of that continuous non-point ('regional') source is assessed as $\vartheta_0(x) \approx 3\frac{m^3}{m \cdot 1yr} =$ const, corresponding to relatively slow beach (dune) erosion. The alongshore width of the coastal cell is of the order of L = 1 km (Fig. 6), thus yielding the constant $a \approx \frac{\alpha}{L} \approx 0.004$ [m^{-1}].

Hence the gradient $\frac{\partial S}{\partial y}$ reads

$$\frac{\partial S}{\partial y} = 3 \cdot exp(-0.004 \cdot y) \cdot \{P'[x_0]\} \tag{19}$$

in which the function $P'[x_0]$ becomes

$$P'[x_0] = [a \cdot (\overline{D_y} \cdot a - 1)] \tag{20}$$

The coefficient $(\overline{D_y})$ is assessed in terms of Equation 13, where the diffusivity D_y is taken as 0.05 m^2/s, while the quantity t is redefined as the number of days per annum with perceptible sediment transport; we assume t as 80–85%, with 11% of storms, as stemming from other Polish studies. Thus one comes to $t \sim 2 \cdot 10^7$ [s], the coefficient $\overline{D_y} \approx 10^3$ m and the function $P'[x_0] \approx 0.012$. Hence the gradient of S describing the annual variability of seabed along the axis y reads

$$\frac{\partial S}{\partial y} = 0.036 \cdot exp(-0.004 \cdot y) \qquad [\tfrac{m}{yr}] \tag{21}$$

If the erosion is halted one has $V_0(x)$=0 and $\frac{\partial S}{\partial x_i}$=0.

Estimation of the other term, i.e. ΔQ_{x_i} should be based on the available data on the annual sediment transport rates. Taken as our first approximation is what follows. The net resultant (multiyearly) longshore sediment transport is from west to east, without clear predominance of erosion or accretion, roughly $Q_E \approx 75 \quad thou \cdot \frac{m^3}{yr}$. The resultant yearly longshore sediment transport rate from west to east, with temporary predominance of erosion ($Q_{x_k} < Q_E$) taken as $Q_{x_k} \approx 50,000 \cdot \frac{m^3}{yr}$. The parameter $\overline{A_E} \approx 0.075$ (Pruszak 1993) is a mean multiyearly value, while for the single year one assumes $\overline{A_k} \approx 0.085$. The empirical factor $\overline{M_0} \approx 4.6 \cdot 10^3[\frac{m^{\frac{3}{2}}}{yr}]$ is regional.

Upon introduction of the above quantities in the governing equation one finds out that the annual bed changes Δh due to erosion (reflected in the source func-

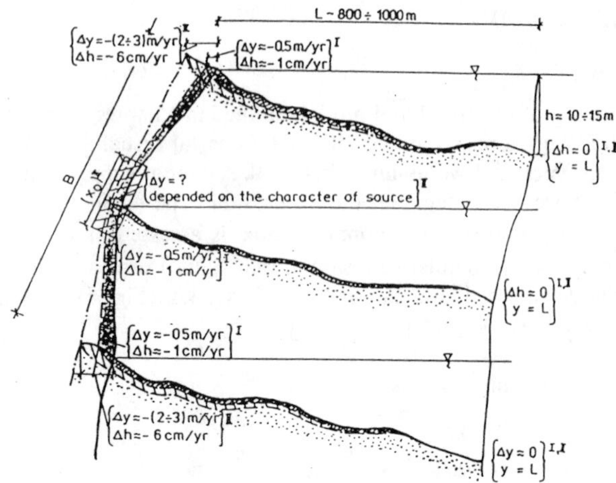

Figure 7. Overall results of simplified modelling, Versions I and II.

tion) range from -1.0 cm/yr at the shoreline to 0 at the depth h about 10 m. All values are small due to the weak erosion rate assumed. Putting 2% for the beach face slope yields the annual shore retreat of $\Delta y \approx 0.5$ m. Results are outlined in Figure 7.

5.2 Version II

In this version we assume local erosion of a shore segment stretching along shore on 500 m (x_0 = 500 m). The erosion rate on that segment is assumed as a uniform value of $V_0(x_0) = 50 \; \frac{m^3}{m \cdot yr}$ = const. (while 0 beyond that segment). Since the local source 'emanates' beyond the area x_0, Equation 15 should be taken into account.

By estimating the intervening quantities and parameters as in Version I one obtains

$$\frac{\partial S}{\partial x} = \overline{\vartheta_0}(x) \cdot exp(-a \cdot y) \cdot \{P[x_0] - P[x']\} = \underbrace{3.4 \cdot exp(-0.004 \cdot y) \cdot 0.012}_{area\ x\ \in\ (x_0)} - \underbrace{3.4 \cdot exp(0.004 \cdot y) \cdot P[x']}_{area\ x'\ \in\ (0, x_{s_l})} \tag{22}$$

Since $\frac{\partial \overline{\vartheta_0}(x)}{\partial x}$ is 0, the function $P[x']$ will be reduced to the form

$$P[x'] = [\frac{x'}{x_{s_l}} \cdot 0.012 + \frac{a \cdot (\overline{D_x} + \overline{D_y}) - 1}{x_{s_l}}] \tag{23}$$

so that the gradient $\frac{\partial S}{\partial x_l}$ will read

- in the area (x_0)

$$\frac{\partial S}{\partial x_l}\Big|_{x_0} = 0.041 \cdot exp(-0.004 \cdot y) \qquad [\frac{m}{yr}] \tag{24}$$

- in the area determined by the axis x', that is (0,x_{s_l}):

$$\frac{\partial S}{\partial x_l}\Big|_{x'} = \frac{\partial S}{\partial x_l}\Big|_{x_0} -3.4 \cdot \frac{exp(-0.004 \cdot y)}{x_{s_l}} \cdot \{0.012 \cdot x' + 4.6\} \qquad [\frac{m}{yr}] \tag{25}$$

Estimation of the terms $\frac{\partial Q_{x_i}}{\partial x_i}$, which describe the effect of the sediment transport rate on the nearshore bathymetry follows as in Version I but for slightly different external factors enforced by the coast morphodynamics (i.a. the sediment deficit is estimated at 50,000 m^3/yr and the term $\frac{\Delta Q_{y_k}}{L}$ is taken as ≈ -0.025 $\frac{m}{yr}$).

Hence the following results are obtained for Version II: the erosion of the seabed at the shoreline all along the segment x_0 is found to be -(2–3) cm/yr, with decrease towards greater depths; beyond the area x_0 the erosion Δh will grow to -6 cm/yr at the distance x_{s2} = 5.2 km (to the east) and x_{s1} = 2.2 km (to the west). This corresponds to the annual shoreline retreat in that area amounting to Δy = 2–3 m. Compare Figure 7 for gross results.

6 WHITE-BLACK-GREY MODEL

Another model can be formulated if one imagines a coastal cell as depicted in Figure 8a, with an area (volume) subject to accretion a and that of erosion measuring e, most conveniently in percentage of the total area (volume) of the cell. One is interested not in the configuration, or distribution, of a and e but just in their percentage. Since the quality of the coastal cell is either accretion ('white'), erosion ('black') or transient ('grey') one can label it a 'white, black & grey' (WB&G) box, and the resulting description of its inherent processes — a WB&G model. One can argue that the temporal change rates of both a and e depend on their size at time t, the capacity of the remaining control 'inactive' volume to transform into either accretion or erosion, and the exchange of mass with the adjacent coastal cells. This postulate can be written down as follows

$$\frac{da}{dt_*} = \alpha_1 a[(1 - a - e)\beta_1(E) - \gamma_1] \tag{26}$$

$$\frac{de}{dt_*} = \alpha_2 e[(1 - a - e)\beta_2(E) - \gamma_2] \tag{27}$$

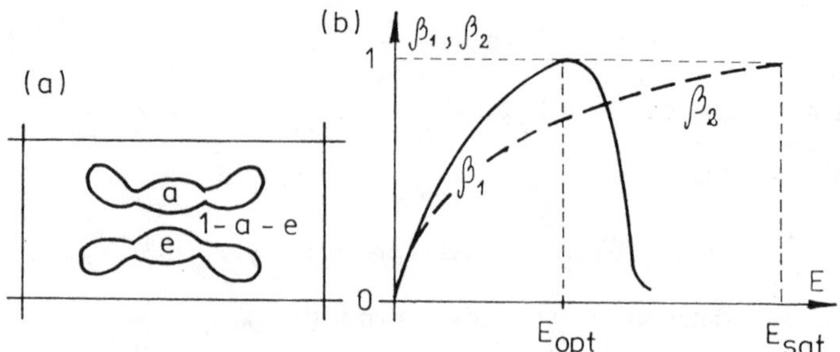

Figure 8a. Control volume (coastal cell) with arbitrary spatial layout of accretion (a) and erosion (e).

Figure 8b. Postulates regarding coefficients of WB&G model.

in which a = accretion volume (dimensionless), e = erosion volume (dimensionless), 1-a-e = transit volume (dimensionless), α_1, α_2 = proportionality factors, β_1, β_2 = accretion/erosion growth rate factors, γ_1, γ_2 = evacuation (dispersion) rate factors, $t_* = t/T$ = dimensionless time, T = characteristic large-scale time, E =energy input (dimensionless).

Equations 26 and 27 should be amended by the conservation law:

$$a = e + \alpha_1 \gamma_1 a - \alpha_2 \gamma_2 e \tag{28}$$

which is equivalent to

$$a(1 - \alpha_1 \gamma_1) = e(1 - \alpha_2 \gamma_2) \tag{29}$$

and some postulates concerning the coefficients β_1 and β_2 must also be added. In general, these coefficients should be linked to the energy input and can be suggested as depicted in Figure 8b.

One then arrives at three degrees of freedom as there are four unknown parameters $\alpha_1\beta_1$, $\alpha_1\gamma_1$, $\alpha_2\beta_2$ ans $\alpha_2\gamma_2$ and the conservation law as above (Eq. 28). The system of Equations 26 and 27 may be transformed to vector form typical of dynamic systems, and the respective analysis stemming from the stability theory for dynamical systems may be applied accordingly. The trajectory, phase space and the stationary point may then be tracked, with reference to traditional equilibrium concept for shore profile.

7 CLOSURE

The modelling proposals presented herein have the advantage of delineating some principal characteristics of large-scale behaviour. They disregard altogether the short-term changes within the control cell, although there are no obstacles to model those changes on the background of large-scale ones, as illustrated in Figure 4 for the 3-line model. Hence the large-scale model can be seen here as a kind of framework within which a variety of internal transformations would be permitted, and indeed looked forward to.

The authors are aware of the opposite option, wherein small-scale processes are reproduced in a mathematical model for a long time span, thus bringing about large-scale effects. Physically, there certainly is a long-term 'coast sculpturing' by minute yet 'patient' short-term processes. As mentioned earlier in this paper, some large-scale processes are controlled by slowly evolving day-to-day small-scale processes (such as CCCC). Emphasis herein has been placed on 'really' large-scale processes such as EE. Undoubtedly, a lot of interaction between various scales takes place, making the modelling even more complex.

Acknowledgements

This paper has been written in part under the research programme 6-P202-004-04 and in preparation for *Poland's Climate Change Country Study* programme (SE13). The authors gratefully acknowledge the sponsorship provided generously by the Polish *Committee for Scientific Research KBN*.

REFERENCES

Pruszak Z. 1993. The analysis of beach profile changes using Dean's method and empirical orthogonal functions. *Coastal Engineering* 19: 245–261

Pruszak Z. & R.B.Zeidler 1988. Estimates of cross-shore bedload and bed changes. *Proceed. 21st ICCE*, Malaga, ASCE.

Vriend, de H. J. 1992. Mathematical modelling and large-scale coastal behaviour. Part 1. Physical processes. Part 2. Predictive models. *J. Hydr. Research.*

Zeidler R.B. 1992. *Assessment of the Vulnerability of Poland's Coastal Areas to Sea Level Rise.* H*T*S Gdańsk. ISBN 83-85708-01-4.

Zeidler R., A. Mielczarski & Z. Pruszak 1994. Large-scale behavior of Poland's coast. *J. Mar. Geol.* (Special Issue LSCB '93, in press).

CARTOGRAPHIC CHARACTERIZATION OF THE LITTORAL CAMPS OF DUNES.

Dr. Jesús Martínez Martínez.[1]

ABSTRACT: This work is based on a study about the Dunes Camp of Maspalomas (figure 1), in Gran Canaria Island (Spain). Their sedimentary processes are detached and represented separately in order to delimit the different sedimentary and eolian sub-units to get an "spectral" analysis.

The developed analytic serie permits us to construct a sequence of thematic maps. Later, the cartographic puzzle is integrated, once the physical varieties of the sedimentary dynamics that intervene in the territory is well konwn and clearly understood.

The cartography of integration or the general vision about the processes of both transport and sedimentary deposits contain yet enough information, inside a physical perspective (based on the dune biotope), to decide in relation to the arrangement, planning and management of the territory.

INTRODUCTION: GEOGRAPHIC SCENERY AND METHODOLOGY.

This stady is about the Dunes Camp of Maspalomas, which is located in the southern part of Gran Canaria Island, Spain. (figure 1)

The eolian sedimentary formation covers a surface of 4 km^2. The highest length of the outcrop is 3 km. in NE-SW direction. It is 2

---

[1] Doctor in Geological Sciences and Professor of Littoral Management. Facultad de Ciencias del Mar. Universidad de Las Palmas de Gran Canaria. Campus Universitario de Tafira. Box 550. Postal Code 35080. Las Palmas de Gran Canaria. Spain.

km in width in the NS direction.

The Dunes Camp is delimitted by:

1. Two important sandy beaches with 5.3 km. of length:

   - Playa de El Inglés (the English Beach) in the eastern part, and

   - Playa de Maspalomas (Maspalomas Beach) in the southern one.

2. A sub-marine valley at the place where the above mentioned beaches confluence each other.(Bajeta Tip).

3. At the land side, by a slope with an almost vertical position in an elbow form, at height of 20 - 25 mt. It belongs to a blooded flatland.

4. And the Maspalomas Lagoon (Charca) and the course of the Fataga Gully (barranco), Westward. Both elements act as a final physical barricade of the eolian transportation of sands processes.

   The sands have an average D50 of 0.19 mm. They are formed by:

   - An important proportion of organic carbonate which represents a 48% in weight.

   - Phonolitic components without taking into consideration the trachytic contributions.

It has a witish and a yellowish index colour (blonde sands)

The sedimentary forms depict, basically, the transverse and barchan dunes, formed by winds that flow as from North-Eastern as from Southern. They represent the sedimentary sands process in answers to certain dominant trade winds (NE) and to the "South Weather".

The developed of an analytic serie of cartographies, about the sedimentary processes, permits us to construct the following sequence of thematic maps:

1.- Map about the intensities of the sedimentary eolian processes.

2.- Map about the distribution of the different dune forms.

3.- Modern map of individualization, localization and classification of the terminal zones with a sedimentary activity.

4.- Map which shows how were located, in the past, the terminal fronts connected with the sedimentary eolian processes.

5.- Map about a qualitative schematization of the eolian dynamics.

6.- Map about a semi-quantitative schematization of the eolian dynamics.

The above cartographies were taken from:

a.- Sistematic analysis and interpretations of the mosaic of the aerial photographies separated according to the time.

b.- Observations "in situ" (Martínez et al. 1986).

c.- And manipulations, representations and interpretations of:

- Certain physical varieties that condition the eolian processes (directional regimens of the winds and topographic and/or architectonic screens).

- Their morphologic effects in relation to the transport and deposit of sands (geometrical formation of both the dunes and the passege of eolian shadow).

RESULTS AND DISCUSSION.

The figure 2 is a descriptive map that just indicates observations. It is obtained from a mosaic of aerial photographies (January 1991). It delimits four sub-unites that depict the different kinds of intensities about the eolian sedimentary processes.

The figure 3 corresponds to another map, a descriptive one, but certainly with two extra connotations:

- The first one deals with the qualification procedure in which certain aspects are observable.

- The second has to do with the valuation procedure that takes into account the suitability of the territory for some appropriate usages.

The line and the spots of terminal activity, belonging to the eolian sedimentation processes, are isolated from the group of sedimentary sub - unites. Through that event, it is possible to visualize the fact that the spots are nothing but discontinious lines which also correspond with a terminal sedimentary activity. The whole of lines display a sub - parallel disposition in the NE - SW direction.

In a first scheme concerning to the relationship between the lines, it believes the following:

- They represent a recessive process, southerly, in the

development of the Dunes Camp, inside the Geological History in present-days.

- It identifies migrating passages.

The figure 4 shows another map about the terminal activity. It points up, particularly, the eolian transportation and deposit of sands. It was obtained from a mosaic of old aerial photographies made in March, 1962.

The above figure cartographies several sedimentary sub -unites, that correspond to the Western part of the Dunes Camp.

If we compared the figure 3 and 4, it would verify the southern recession of the terminal sedimentary activity. It is observed the bellow facts:

1. The zone that belongs to the denominated precocious terminal activity, in concern with the map 1991, is located southerly and inside the zone with an antropic intervention (map 1962).

2. The three terminal lines, observed in the modernest map, were developed in the zone of a proto - terminal activity that is identified in the oldest map.

3. It is also observed that as the eolian layer as the old zone of the terminal activity, identified in the oldest map, have been occupied by an urban building.

On the whole, in the Dunes Camp of Maspalomas, the evolution of the terminal and eolian sedimentary activity may determine the next story:

1. It develops the zone of a terminal activity that has the largest penetration (the oldest), where the wind lacks capability to build the dunes formation, although such a formation consists of a precarious feeding.

2. A line of the Dunes Camp has been used for agricultural purposes in the windward part of the before zone of terminal activity.

3. It has developed a second zone of terminal activity in the windward part of the occupied line, due to the previous intervention.

4. It occurs a retraction performed by the free dunes formation, southwards. It also produces sands invasion over the occupied line every time the cultivation is abandoned.

5. It decreases the sedimentary balance in the active zone. The persistence of that kind of circumstances imply certain consequences, such as the backward movements of the zones with

terminal activities. They progressively occupy the southernmost places.

6. It develops a particular type of vegetation that fixes the modernest zones of a terminal activity.

According to Paskoff (1985), the before established cronological sequence contradicts, at a first reading, the expected behavior in a camp of littoral dunes related to the borderline which presents a backward action as it happens in Maspalomas Beach. Due the dunes are in sympathy with the beaches to which they are associated, if the laters move back the formers will go backwards, too (towards land). That recession is necessary for the dunes fulfill their works as sedimentary reservation. On the contrary, it would break the physical equilibrium in this type of systems, or the littoral ecosystems.

The "anomalous" behavior, formulated and verified in the Dunes Camp of Maspalomas, would be explained in regard to the following premises:

1. The vegetable colonization advances southwards because of some determining factors, independient to the sedimentary processes.

2. There is a discreasement of the sedimentary contributions from Playa de El Inglés (The English Beach).

3. It produces a positive increasement in the discharge of the sedimentary reservation (contribution measure), in order to mitigate the recession towards land from the borderline in the Maspalomas Beach.

4. All the above causes happen in the Dunes Camp of Maspalomas, or at least some of them.

There is not a counteraction between:

- the prevailing tendency in the recessive process of the terminal and eolian sedimentary activity (to occupy places nearest to the borderline), in the Westwards part of the Maspalomas Dunes, and

- the dynamics of the winds that determine the sedimentary processes.

The fact is that in the sector where the above matter occurs, it is located the shadow zone or the progressive attenuation (Northwards), from the eolian capacity of transporting sands.

That shadowy zone is easily understood in the maps of the figures 5 and 6.

Besides, the figures 5 and 6 and the slope in the flooded flatland permit us to get a clear comprehension about the triangular geometry, that is acquired by the active zone of eolian and sedimentary processes.

Regarding the figures 3 and 4 and taking into consideration the sedimentary aspects of the biotope, it is possible to state that the terminal zones, previously described are able to support a soft antropic action for explotation. For instance, sunny spots for relaxation.

But, such actions must not destroy the identity of the zones of a terminal activity in a Dunes Camp. Supposing it happens, it would create other terminal zones depending on the surface of the active zones which have an amortized sedimentary activity. That process deals with the simile that contrasts the dynamics of a Dunes Camp with a dam whose function is to regulate the superficial channel of waters.

Nevertheless, the soft antropic actions for explotation would not be admitted in the active zone, since it would interfere, somehow, in the eolian transportation of sands.

CONCLUSIONS.

The synthetic, cualitative and valuation cartography of the figure 7 is useful to formulate the main conclusions related to the eolian sedimentology in the Dunes Camp of Maspalomas.

It is classified into four sectors:

1. The zone of incipient and sedimentary activity.

2. The active zone in the strict sense of the word.

3. The zone of the amortized sedimentary activity.

4. And the zone that has been swept by the terminal lines of eolian sedimentary processes, and that includes the spots that perform an amortized old activity.

The active zone contains the sedimentary reserves connected to its bordering beach environment. The mitigation of the shore backwards movement, in the Maspalomas Beach, depends on this deposit.

The zone of incipient sedimentary activity is basical in the Dunes Camp, due the sands reserves derive from them in the "active" zone. Here, the antropic actions must be carefully performed in order to avoid interfering with the eolian transportation of sands. If the negative physical impacts, in this aspect, will be extended into time:

- It would provoke the degradation of a place of several interests, such as landscape, scientific, didactic resources and psychological relaxation.

- And it also would accelerate indirectly the recession of Maspalomas Beach towards land.

The fourth cartographied zone is increasingly inactive from a sedimentology point of view. For that reason its southern limit is located in the modernest terminal line. That circumstance allows the fact that under a physical focus (about the biotope base, exclusively), it would be probable that the sub - unit can support the soft antropic interventions for explotation.

However the terminal zone must not lose its identity from the eolian sedimentology because of those types of interventions.

Supposing the destruction happens, it would be formed another terminal zone, or it would accelerate its appearence in territories that belong to the active zone. The before mentioned can be considered as a negative impact as far as the Geology field and the Environment are concerned, and whose effects would cause serious damages in the Dunes Camp of Maspalomas, and obviously it also would accelerate the recessive step of the borderline of Maspalomas Beach.

REFERENCES.

Cendrero, A. 1987. "Cartografia integrada de zonas litorales emergidas y sumergidas para la planificación". Seminario Internacional sobre Zonas Litorales. Consejo de Europa. Bilbao. 8 - 17 de Octubre. 50 pp. (in Spanish).

Martínez, J. et al. 1986. "Las Dunas de Maspalomas: Geología e Impacto del Entorno". Excmo. Cabildo Insular de Gran Carnaria. Universidad Politécnica de las Palmas. 151 pp. (in Spanish).

Martínez,J. 1900. "La Provincia morfodinámica de Morro Besudo - Faro de Maspalomas (Isla de Gran Canaria, España): Conocimiento y comprensión de sus procesos geomorfológicos y sedimentarios para la planificación y gestion de este litoral". I Reunion Nacional de Geomorfología (Teruel, 1990). pp 351 - 363. (in Spanish).

Martínez, J. and Casas, D. 1993. "La dinámica sedimentaria del litoral meridional de Gran Canaria (Islas Canarias - España)". pp 218 - 242, in: Losada, M. (Editor). "I Jornadas Españolas de Costa y Puertos".Universidad de Cantabria. Santander (Spain), 7 and 8 of May of 1992. 489 pp.

Paskoff, R. 1985. "Les littoraux, impact des aménagements sur leur évolution". Masson. Paris. (in French).

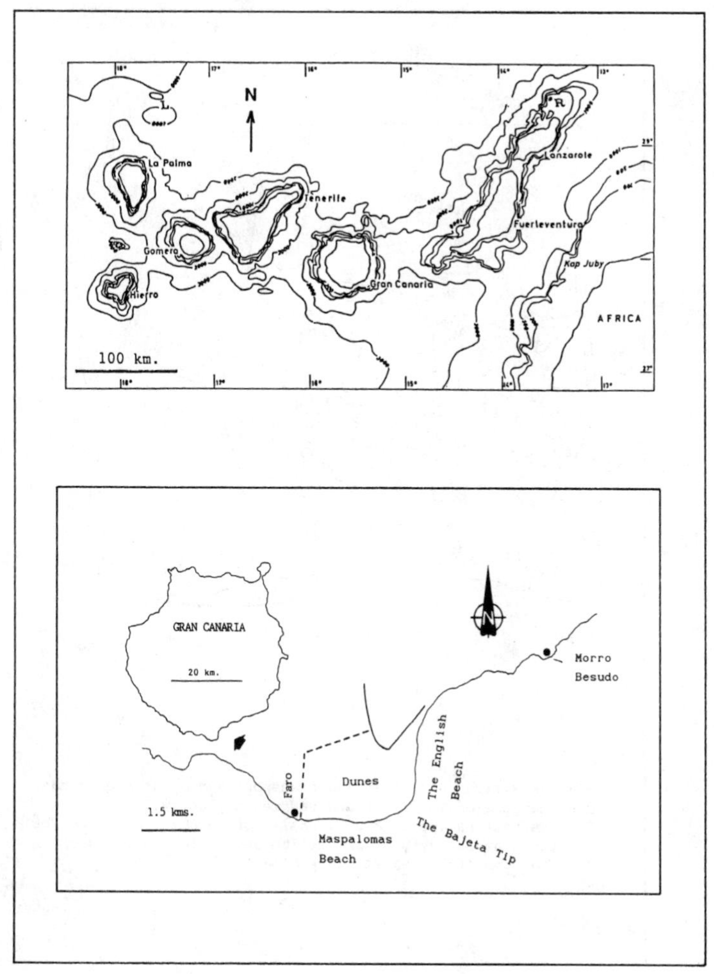

Figure 1

Geographic localization of de Dunes Camp of Maspalomas (Gran Canaria Island, Spain).

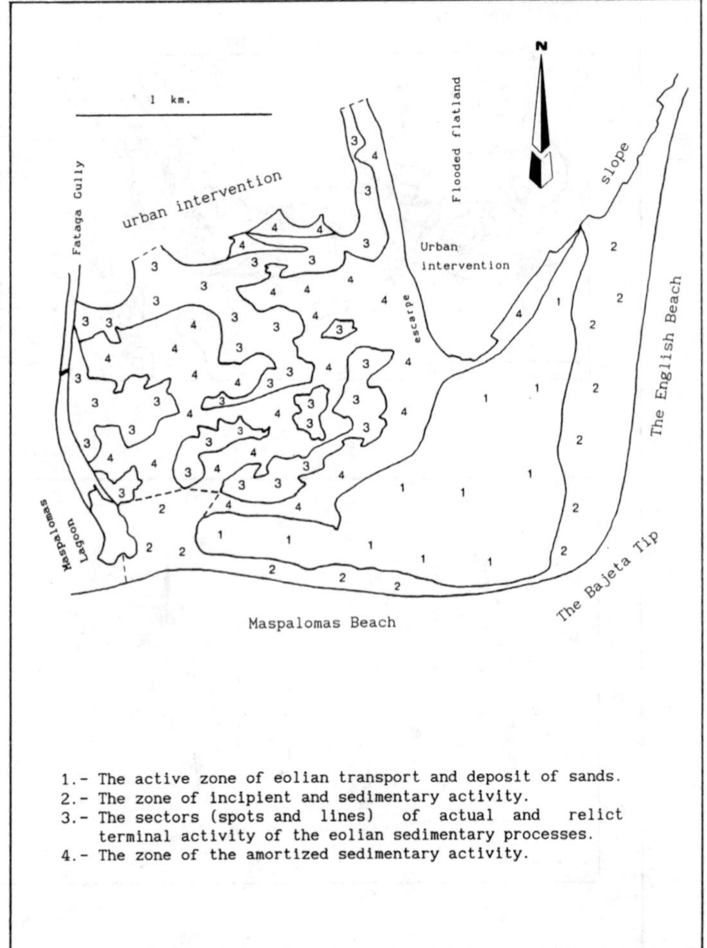
N
1 km.
Flooded flatland
Fataga Gully
urban intervention
slope
Urban
intervention
escarpe
The English Beach
Maspalomas
Lagoon
The Bajeta Tip
Maspalomas Beach
1.- The active zone of eolian transport and deposit of sands.
2.- The zone of incipient and sedimentary activity.
3.- The sectors (spots and lines) of actual and relict terminal activity of the eolian sedimentary processes.
4.- The zone of the amortized sedimentary activity.

Figure 2

Cartography concerning the Dunes Camp of Maspalomas, according to a mosaic of aerial photographies (January, 1991).

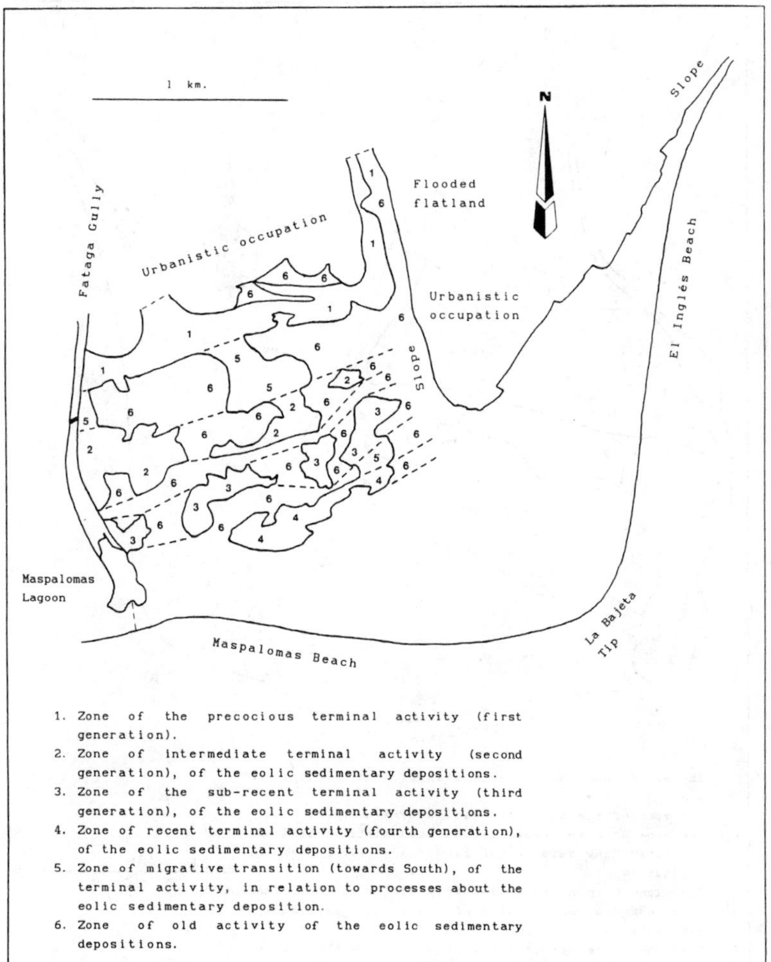
1 km.
N
Slope
Flooded
flatland
Fataga Gully
Urbanistic occupation
Urbanistic
occupation
Slope
El Inglés Beach
Maspalomas
Lagoon
Maspalomas Beach
La Bajeta
Tip
1. Zone of the precocious terminal activity (first generation).
2. Zone of intermediate terminal activity (second generation), of the eolic sedimentary depositions.
3. Zone of the sub-recent terminal activity (third generation), of the eolic sedimentary depositions.
4. Zone of recent terminal activity (fourth generation), of the eolic sedimentary depositions.
5. Zone of migrative transition (towards South), of the terminal activity, in relation to processes about the eolic sedimentary deposition.
6. Zone of old activity of the eolic sedimentary depositions.

Figure 3

"Spectral" Cartography of the Dunes Camp of Maspalomas, based on a mosaic of photographies, January 1991: Terminal lines of the eolic sedimentary depositions.

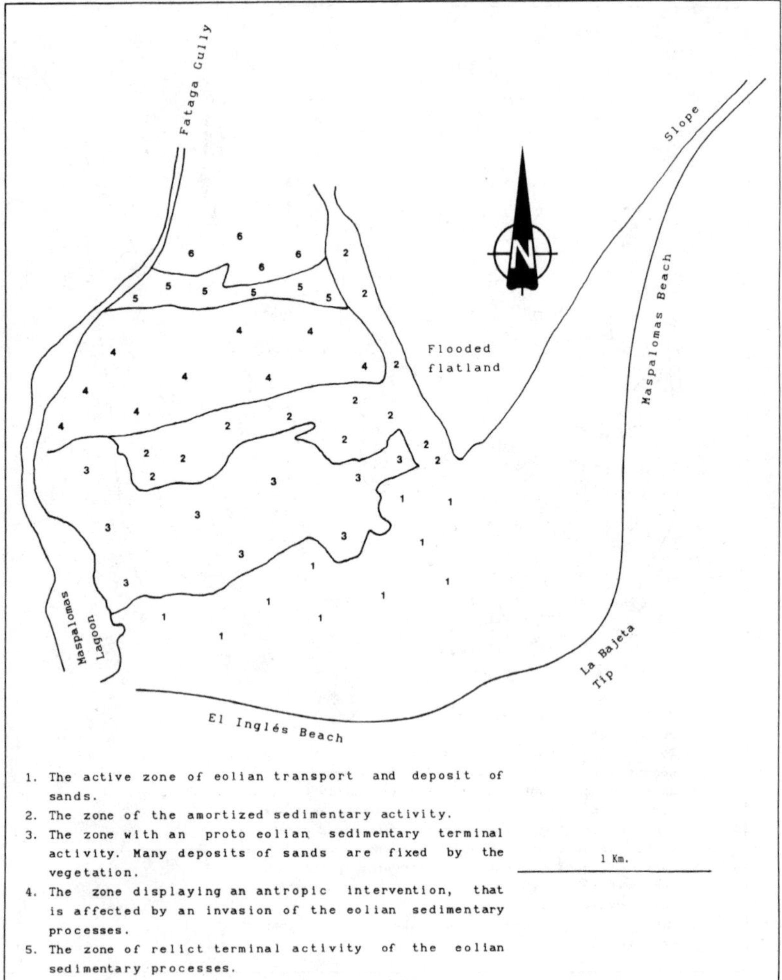
Fataga Gully
Slope
N
Flooded flatland
Maspalomas Beach
Maspalomas Lagoon
La Bajeta Tip
El Inglés Beach
1 Km.
1. The active zone of eolian transport and deposit of sands.
2. The zone of the amortized sedimentary activity.
3. The zone with an proto eolian sedimentary terminal activity. Many deposits of sands are fixed by the vegetation.
4. The zone displaying an antropic intervention, that is affected by an invasion of the eolian sedimentary processes.
5. The zone of relict terminal activity of the eolian sedimentary processes.
6. The zone of the eolian layer (flying sands).

Figure 4

Cartography of the terminal activity of the eolic sedimentary processes, in the Maspalomas Dunes Camp, based on a mosaic of aerial photographies (March, 1962).

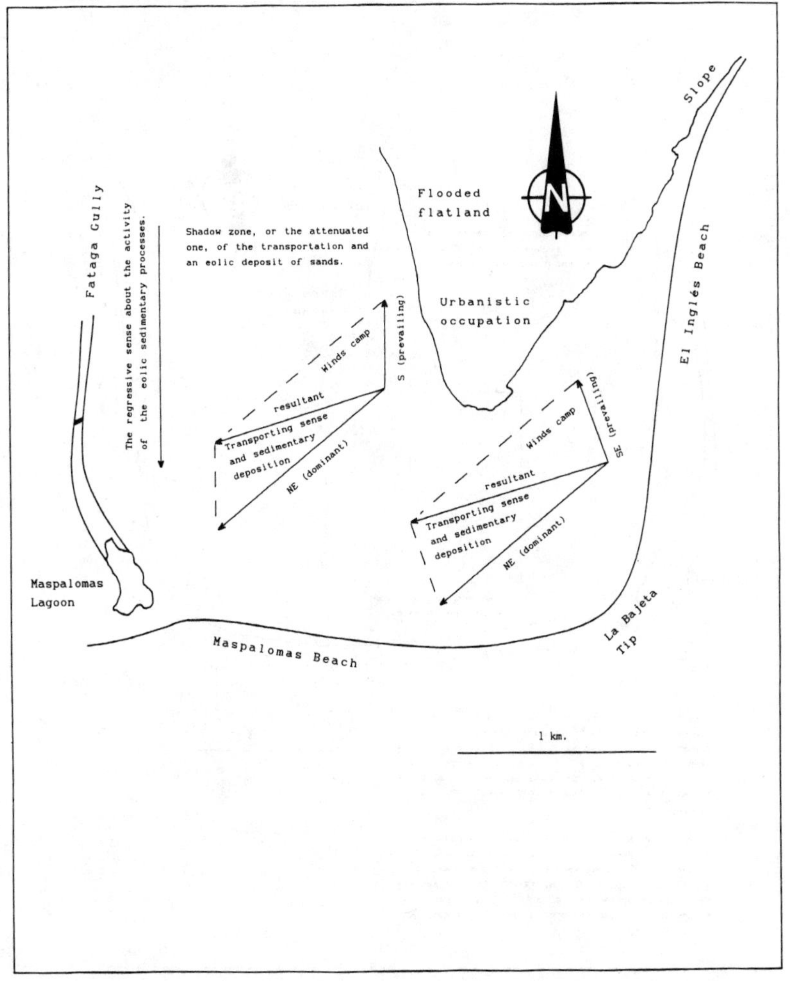

Figure 5

Qualitative schematization of the dynamic of the eolic sedimentary processes, in the Maspalomas Dunes Camp.

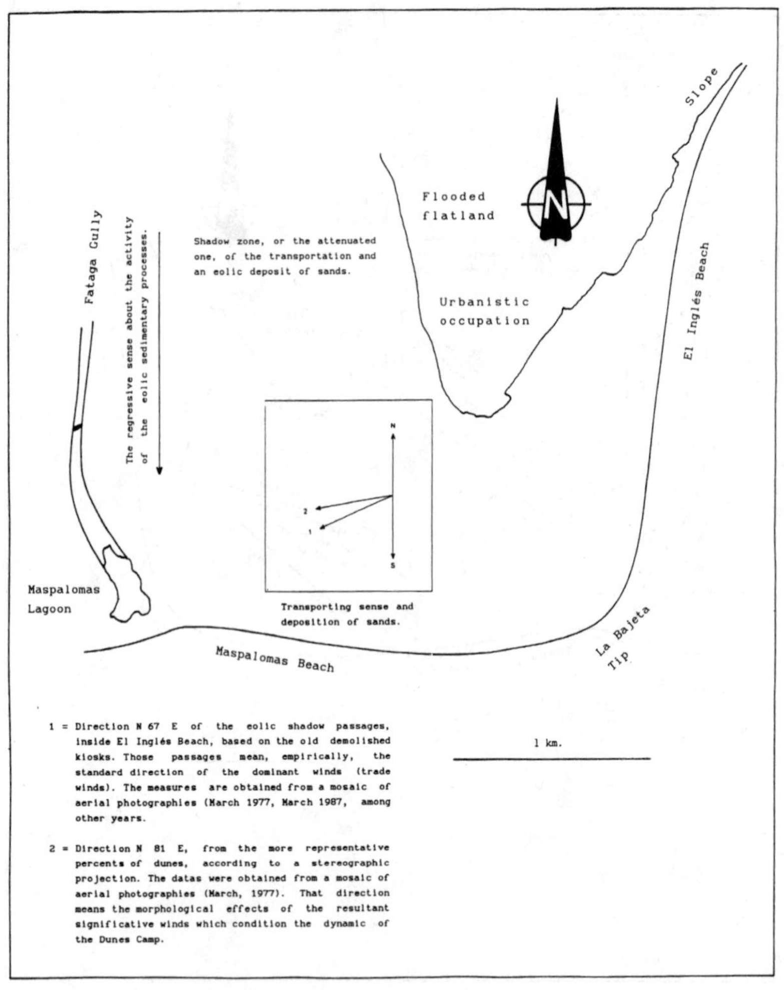

Slope
Flooded
flatland
N
Fataga Gully
The regressive sense about the activity of the eolic sedimentary processes.
Shadow zone, or the attenuated one, of the transportation and an eolic deposit of sands.
Urbanistic
occupation
El Inglés Beach
N
2
1
S
Maspalomas
Lagoon
Transporting sense and deposition of sands.
La Bajeta
Tip
Maspalomas Beach
1 = Direction N 67 E of the eolic shadow passages, inside El Inglés Beach, based on the old demolished kiosks. Those passages mean, empirically, the standard direction of the dominant winds (trade winds). The measures are obtained from a mosaic of aerial photographies (March 1977, March 1987, among other years.
2 = Direction N 81 E, from the more representative percents of dunes, according to a stereographic projection. The datas were obtained from a mosaic of aerial photographies (March, 1977). That direction means the morphological effects of the resultant significative winds which condition the dynamic of the Dunes Camp.
1 km.

Figure 6

Semi - quantitative schematization about the dynamic of the eolic sedimentary processes, in the Maspalomas Dunes Camp.

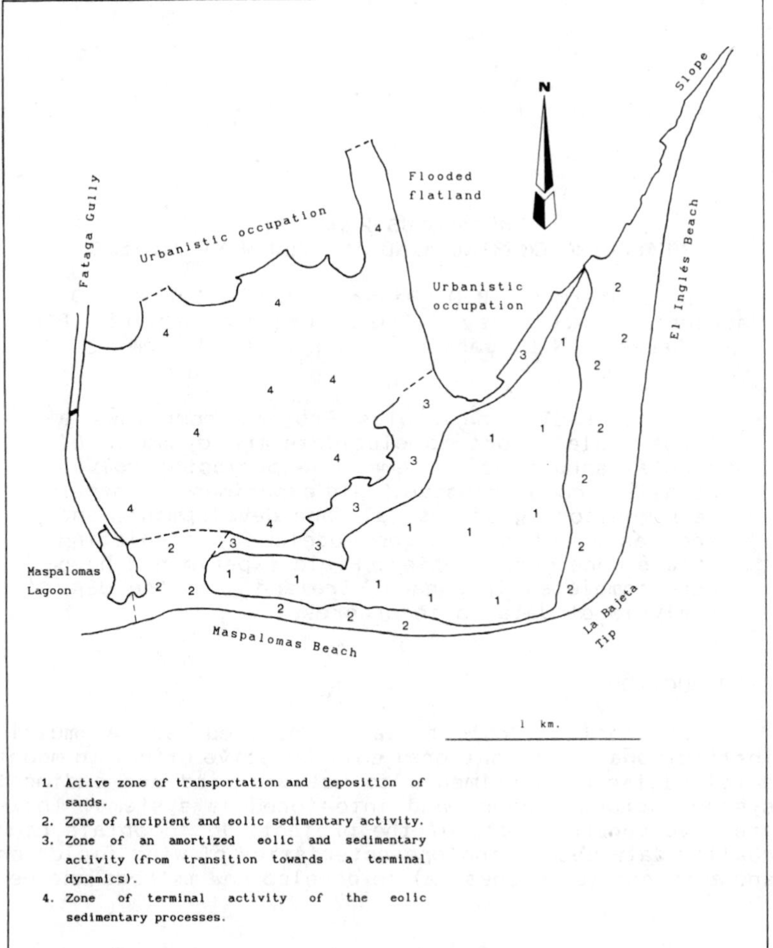
N
Slope
Flooded
flatland
Fataga Gully
Urbanistic occupation
Urbanistic
occupation
El Inglés Beach
Maspalomas
Lagoon
Maspalomas Beach
La Bajeta
Tip
1 km.
1. Active zone of transportation and deposition of sands.
2. Zone of incipient and eolic sedimentary activity.
3. Zone of an amortized eolic and sedimentary activity (from transition towards a terminal dynamics).
4. Zone of terminal activity of the eolic sedimentary processes.

Figure 7

Maspalomas Dunes: Generalized cartography of the eolic sedimentary processes, based on a mosaic of aerial photographies (January, 1991).

THE AEOLUS PROJECT: MEASURING COASTAL WIND AND SEDIMENT SYSTEMS

D.J. Sherman[1], B.O. Bauer[1], R.W.G. Carter[2], D. Jackson[2], J. McCloskey[2], R.G.D. Davidson-Arnott[3], P.A. Gares[4], N.L. Jackson[5] and K.F. Nordstrom[6],

ABSTRACT: The Aeolus Project comprises a large-scale effort to elucidate the dynamics of coastal sand dune systems. The project involves detailed, comprehensive field experiments, longer term monitoring of coastal dune development, and the design of simulation models for predicting future dune morphologies. Field experiments have been completed in Canada, Ireland, and the USA. Analysis of data is in progress.

INTRODUCTION

The Aeolus Project was conceived as a multi-institutional, international collaborative effort to mount detailed field experiments to measure wind and sediment systems across beaches and into foredune systems. There are five general goals of the project: 1) to obtain high quality data describing characteristics of wind fields on and adjacent to beaches; 2) to develop new methods and new

---

1) Department of Geography, University of Southern California, Los Angeles, CA, USA, 90089-0225
2) Department of Environmental Studies, University of Ulster, Coleraine, Northern Ireland
3) Department of Geography, University of Guelph, Guelph, Ontario, Canada
4) Department of Geography, East Carolina University, Greenville, North Carolina, USA
5) New Jersey Institute of Technology, Newark, New Jersey, USA
6) Institute of Marine and Coastal Studies, Rutgers University, New Brunswick, New Jersey, USA

equipment to measure sets of aeolian process and response parameters; 3) to develop improved field sampling designs, especially concerning the capture of spatial and temporal variability in the transport system; 4) to continue monitoring programs for long-term dune evolution at sites at Island Beach State Park, New Jersey, and at Castroville, California; and 5) to develop improved simulation models of the transport system and associated foredune development. The ultimate goal of the project is to improve understanding of the aeolian system on beaches to the degree that reasonable predictions can be made for aeolian sediment budgets and coastal dune development over year to decade time-scales.

The rationale for the initiation of the project originated from the increasing recognition that existing aeolian sand transport models do not replicate prototype conditions very well except by accident or through the manipulation of empirically derived constants that seem to exhibit site dependency. This problem arises from the assumtions and structure of these models. They are designed mainly from theory and laboratory work and are usually intended for application in arid environments. The models, for example those of Bagnold (1936), Kawamura (1951), Kadib (1967), Hsu (1974), Lettau and Lettau (1977), or White (1979), are based on a set of enabling assumptions that include steadiness of wind and a resulting equilibrium saltation field; a planar, horizontal, and unobstructed surface; clean, dry, and uniform sediments; and no vegetation. All of these assumptions are violated, some seriously, in coastal aeolian systems. Because of the complexity of the beach and dune aeolian systems, there have been few well-designed, comprehensive field experiments aimed at sand transport across a beach and into a fore dune. This results partly from the substantial personnel and equipment requirements associated with large-scale coastal experiments. It is also because there are aspects of the system that we still do not understand very well.

BACKGROUND

In many coastal environments, the development of dune systems represents an important factor in shoreline evolution. In particular, the volume of sand present in coastal dunes may comprise a significant element in local sediment budgets. The dunes are important from an environmental management perspective because of their value as ecosystems, natural resources, and the role they may play in affording protection against coastal flooding and beach erosion. The dynamics of aeolian transport and dune development are also important themes for scientific and engineering studies. Understanding the interactions between beach and dune systems represents a crucial step in producing workable models appropriate for meso-scale management schemes (e.g. Sherman and Bauer 1993).

In most aeolian models, the transport rate depends only upon shear velocity, u_*, sediment grain size, d, and sediment density, ρ_s. The latter parameters are critical for the estimation of the threshold shear velocity, u_{*t}, for sediment motion:

$$u_{*t} = A\left[gd\left(\frac{\rho_s - \rho}{\rho}\right)\right]^{0.5} \qquad (1)$$

where A is the square root of the Shields function (after Bagnold 1936), and ρ is air density. The threshold shear velocity is used explicitly in most transport rate equations. Kawamura's (1951) model is an example:

$$q = K\frac{\rho}{g}(u_* + u_{*t})^2(u_* - u_{*t}) \qquad (2)$$

where q is the sediment transport rate, and K is an empirical constant.

In many relatively simple aeolian environments, including some beach systems, the approach outlined above yields reasonable results when compared with empirical results (e.g. Horikawa et al. 1986). This is despite the difficulty in obtaining reliable measurements of sediment transport rates because of trapping problems, and because of multiple error sources associated with the derivation of shear velocity estimates from field measurements (Bauer et al. 1992). However, most beaches are subject to complicating factors that have the potential to reduce greatly the predictive vigor of standard models. Although there are many such complications, those of sediment moisture content (as per cent weight, w) and local surface slope are frequently of substantial influence. Belly's (1964) model of moisture content effects on the initiation of sand motion is a conservative approach to the problem relative to most alternatives:

$$u_{*tw} = u_{*t}(1.8 + 0.6\ \log_{10}w) \qquad (3)$$

where u_{*tw} is the moisture enhanced threshold for motion. The effects of even small quantities of moisture in the pore spaces between sediment grains is predicted to be a substantial element in increasing the threshold shear velocity. Using equation (3) with 0.2 mm sands, for example, shows that a 1% moisture content increases the threshold from 0.19 ms^{-1} to 0.32 ms^{-1}, an increase of about 70%. This increment, if correct, is similar to that expected as a result of tripling the grain diameter estimate used in equation (2). Unfortunately, there remains considerable disparity between the several moisture effect models, as demonstrated by Namikas and Sherman (in press). Their review showed that for a shear velocity of 0.80 ms^{-1}, again with 0.2 mm sands, the predicted transport rates with a 2% moisture content varied by about 300% between the highest and lowest estimates. This finding emphasizes the importance of the Aeolus Project focus on additional field data describing moisture content effects.

Similar, but less dramatic, changes in transport occur across sloping surfaces. The extreme model of Hardisty and Whitehouse (1988) predicts that a 1^0 upslope will reduce transport to about 65% of its value over a horizontal surface, although most slope models indicate a lesser impact. Again, the magnitude of potential effects requires closer empirical evaluation of the impacts of surface slope effects.

Capture of moisture content and surface slope effects in empirical data can only occur in experiments where the basic transport controls depicted in equations (1) and (2) can be accurately measured or derived, in conjunction with good measurements of the sediment transport rate. The series of field experiments have been designed to attack each of these problems either individually or as sets.

FIELD EXPERIMENTS

Four experiments have been completed (Figure 1): two at Castroville, California, USA (Jan. 1991 and Jan. 1993); at Carrick Finn Strand, Co. Donegal, Ireland (May, 1992), and at Long Point, Ontario, Canada (May, 1993). A future experiment is being planned for Ireland in 1994.

Castroville

The Castroville experiments were conducted over a three week period in January, 1991, and for four weeks in January, 1993. During the former experiment, wind speeds fast enough to move sediments did not occur. The second trial was more successful. The specific goals of the second Castroville experiment were (in addition to the general goals described above): 1) to measure sediment transport rates over a narrow, morphodynamically reflective beach (at high tide) with relatively coarse sands (ca. 0.4 mm diameter); 2) to measure the structure of the two dimensional wind field developing across the beach; and 3) to test a continuously-weighing sediment trap and an internally segregated, stocking-type vertical trap (the hose-type trap).

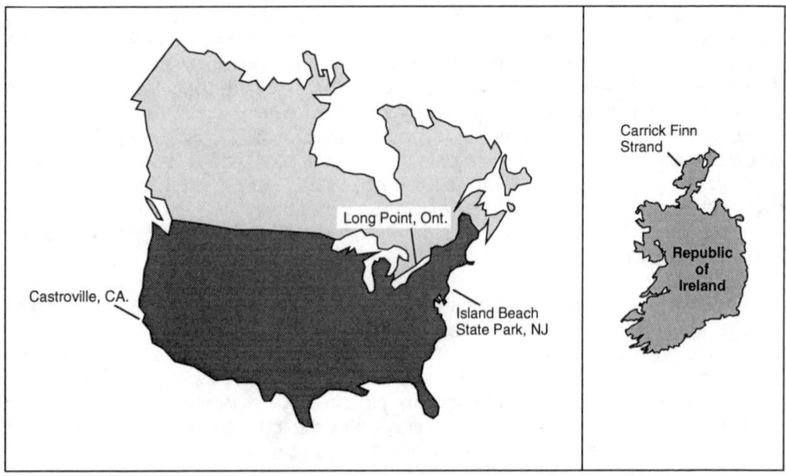

Figure 1: The Aeolus Project study locations.

Anemometers were installed in five vertical arrays across the foreshore. Figure 2 is a schematic showing the relative vertical and horizontal instrument spacings. Wind vanes were located at the tops of the seaward and shoreward arrays. Hose-type and cylindrical traps were installed adjacent to each anemometer array, and the continuous-weighing traps were adjacent the two landward locations.

Measurements were made for all of the pertinent parameters noted in equations (1), (2), and (3). Additionally, beach profiles were measured to obtain estimates of local surface slope elements. Sediment transport rates were obtained using trap data, although the continuous-weighing traps did not function correctly during data acquisition runs. Air temperature gradients were also measured.

The 1993 experiment was particularly successful, with strong winds during a rainstorm providing excellent information concerning moisture effects on transport rates. Further, wind speed data from this experiment is being used to investigate large-scale turbulence effects across the beach as illustrated through the mapping of wind measurements. An example of a set of such maps is

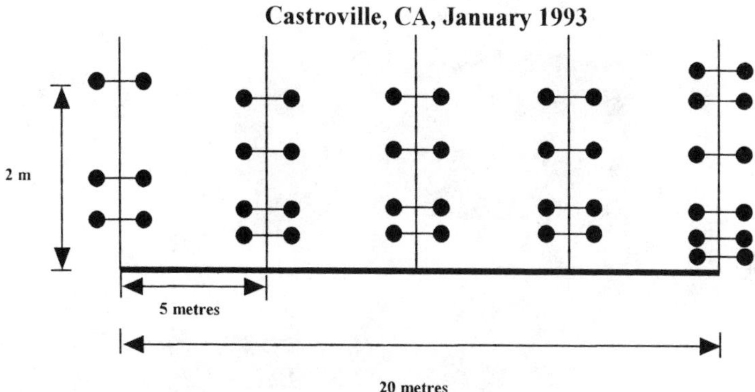

Figure 2: Schematic of the Castroville anemometer array, normalized with respect to beach slope. The ocean is to the right.

presented in Figure 3. These maps represent data obtained from the array shown in Figure 2. The three map sequence depicts the distribution of wind speeds at one second intervals. Points to note include the accelerations occurring near the beach surface as turbulent sweeps approach. Note that the accelerations indicated here may be conservative, as the frequency response of the cup anemometers will smooth high frequency fluctuations. One goal of this mapping approach is to attempt to link turbulent events with transport variations as recorded by the continuous-weighing traps. The system failed here, but represents a promising avenue for additional research.

Carrick Finn Strand

The Carrick Finn experiment was conducted over a two week period in May, 1992. There were four specific goals for the experiment: 1) to measure sediment transport rates across a wide, morphodynamically dissipative beach, with relatively fine sands (ca. 0.2 mm diameter), and backed by a high foredune system (ca. 10 - 15 m); 2) to test a portable electronic 'sediment moisture meter', based on a capacitance plate design; 3) to test and compare 'v-type',

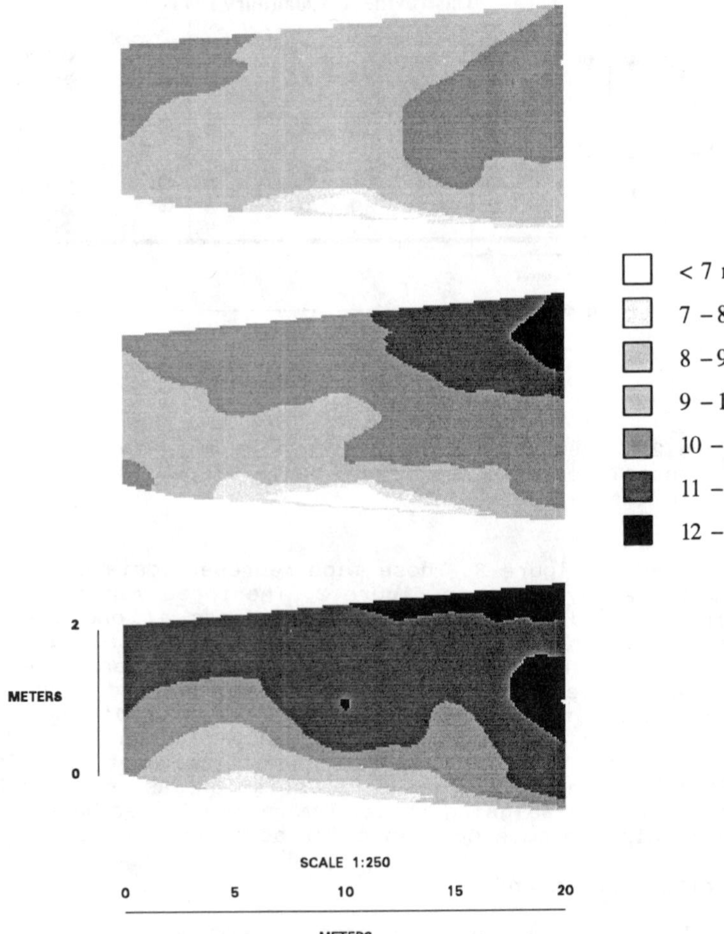

Figure 3: Maps of wind speed across the beach at Castroville. Maps are produced at one second intervals. The ocean is to the right.

'hoop-type', and cylindrical sand traps; and 4) to assess alongshore variability in transport rates.

Anemometers were installed at eight locations across the profile. Four high towers, each with arrays of six anemometers mounted at elevations of 0.25, 0.5, 1.00, 2.00, 3.00, and 5.00 m, were located at 0, 10, 20, and 35 m from a baseline. Another three 'short' towers, each with three anemometers mounted at 0.25, 0.50, and 1.00 m, were located 30, 40, and 45 m from the baseline. A final anemometer, with wind vane, was located mid-way up the foredune. Four other wind vanes were emplaced atop each of the four high towers. Pairs of v-type and cylindrical traps were set adjacent to each tower, and v-type traps were also installed in two shore parallel lines. The capacitance plate sediment moisture meter was used in an attempt to measure the _in situ_ moisture content of the surface sediments. Moisture content was to be indicated on a relative, calibrated scale, with 0 being completely dry sand, and 1.00 being saturated sand.

Measurements were made of wind speed, direction, and unsteadiness, and the wind profile data used to derive estimates of shear velocity. Surficial sediment measurements and samples were taken using a 10 x 10 m sampling grid. Sediment samples for size analysis, carbonate content, and direct moisture content measurement were taken from the grid, and the moisture meter was used in this same grid in order to calibrate the results against known moisture values. Elements of surface slopes were measured by profiling. Sediment transport rates were measured using trap data, and source areas for transport were indentified using erosion pins. Air temperature gradients were also measured.

Some preliminary findings of this study are reported in Gares et al. (1993). Other preliminary results show that the efficency of the cylindrical traps (first described by Leatherman 1978) decreases with increasing transport rates, falling perhaps as low as 50% when compared to trapping by v-type traps. Detailed data analysis of all the Carrick Finn data is still in process. Interpretation of results from this site are complicated by two developments. First, during the study period the only winds competent to move sand were blowing offshore. This means that wind speed measurements made close to the dune reflect strong lee effects, and it is not possible to use those velocity profiles to derive shear velocity estimates. The second problem is associated with the high carbonate content of the beach sediments, and the influence of the carbonates on apparent sediment moisture content. Carbonate content across this beach ranged from about 40% to about 90% by weight. At the same time, sediment moisture content values ranged from about 1% to about 50% by weight. For quartz sands with 40% porosity, the moisture content by weight should not exceed about 25% at saturation. Therefore the anomolously high moisture measurements from Carrick Finn Strand are enigmatic. It is likely that the high values result from the evacuation of moisture from pore spaces within shell fragments. This is also being investigated further.

Long Point

The Long Point experiment was conducted over a three week period in May, 1993. There were four specific goals for this experiment: 1) to measure sediment transport rates across a narrow, morphodynamically reflective beach with relatively fine sand (ca. 0.2 mm diameter), with substantial heavy mineral content and backed by dunes approximately 5 m high; 2) to assess the effects of transport 'fetch' lengths on measured transport rates; 3) to measure boundary-layer development off of the surf zone and across the beach and backshore; and 4) to test a second generation 'hose-type' trap against v-type traps.

A total of 38 anemometers were deployed in six towers across the beach, one additional tower in that same line but installed in the surf zone, and one three anemometer array offset alongshore from the main instrument line. Three wind vanes were mounted atop towers on the across shore line, and a fourth was deployed with the offset array.

Data analysis for this experiment is still in a form too preliminary form for detailed investigation. Finalized estimates of sediment transport, and subsequent modeling efforts, have been complicated at this site by a high beach water table (resulting from high lake levels), and frequent flooding of the beach by overwash during onshore winds. It is believed that the data set describing changes in the aeolian boundary level is particularly revealing, because the effort to obtain measurements over the surf zone was successful. Preliminary examination of the velocity profiles suggests, however, that problems similar to those associated with beach boundary layers occur over the water. Specifically, there appears to be growth of internal boundary layers even over the water, probably as a result of the wind encountering the surf zone itself. It was also demonstrated that the second generation hose-type sediment trap performed very well relative to the wind-tunnel tested v-type trap. This result is encouraging because of the low cost and easy deployment of the hose-type design, and because this trap can provide information concerning the vertical structure of the saltation field. Information concerning sediment source fetch effects is not yet available.

OTHER WORK

Other components of the Aeolus Project include long-term monitoring of dune morphology and the development of simulation models. The goal of the mapping project is to develop detailed measurements of dune development over decadal time frames, to be used with regional wind data to derive an aeolian "climatology". The information is also desired for improving geomorphological models of foredune systems and interaction of such dune systems with adjacent beaches and nearshore zones. A three-dimensional grid (5 m increments) within the foredune system at Castroville has been measured annually since 1987. A 1 m x 1m grid has also been measured four times in this period. Similarly, a series of transects has been monitored, at longer tme intervals, since 1980 at Island Beach State Park. However this dune system has also been the subject of several intensive monitoring efforts (e.g. Gares and Nordstrom 1988).

The simulation model is designed to model profile changes associated with aeolian sediment transport under conditions of varying grain size, sorting, moisture content, and local slope. One goal of the modeling is to predict meso-scale transport of sediments to the foredune. The first generation of this model has been tested for internal consistency, and has been used to appraise the physical validity of the Short and Hesp (1982) conceptual model of linkages between beach morphodynamics and dune development (Sherman and Lyons 1993).

CONCLUSIONS

The Aeolus Project is too young for the pronouncement of specific conclusions. However, several general findings (some preliminary) have been made: 1) over narrow (ca 50 m width) beaches, flow conditions are extremely complex, and difficult to characterize even with large instrument arrays. Internal boundary-layers are initiated over the surf zone, at the waterline, and at berm crests. More subtle boundary-layer changes are initiated across changes in sediment characteristics or over areas with ripples. The complexity of these flows makes the application of the law of the wall to measured velocity profiles of questionable value, even with the greatest attention to detail; 2) sand trapping technology is primitive relative to the wind measurement systems, although we have confidence in some new designs, especially the hose-type trap; 3) in these coastal environments, sediment moisture content plays a major role in controlling sediment transport rates, largely by increasing threshold shear velocities. It appears that moisture effects may overshadow sediment size effects, at least through the range of sand sizes; 4) divergence between ideal, predicted transport rates and measured rates is large, and significant challenges remain for fine-tuning these models; 5) total transport rates from the beach to foredune systems can be large, and represent a substantial component of local, coastal sediment budgets. Therefore the aeolian system demands additional attention if we wish to understand the coastal sediment system as a whole. This last finding suggests the rationale for a continuation of the Aeolus Project.

ACKNOWLEDGEMENTS

The authors gratefully acknowledge the support provided by the California Department of Boating and Waterways, The North Atlantic Treaty Organization, and the National Geographic Society. As well, our individual institutions provided necessary additional support.

REFERENCES

Bagnold, R.A. (1936). The movement of desert sand. *Proceedings, Royal Society of London*, A157: 594-620.

Bauer, B.O., D.J. Sherman, and J.W. Wolcott (1992). Sources of uncertainty in shear stress and roughness length estimates derived from velocity profiles. *Professional Geographer*, 44: 453-464.

Belly, P-Y. (1964). *Sand Movement by Wind.* US Army Corps of Engineers, Coastal Engineering Research Center, Tech. Memo. No. 1, Washington, D.C.

Gares, P.A., and K.F. Nordstrom (1988). Creation of dune depressions by foredune accumulation. *Geographical Review*, 78: 194-204.

Gares, P.A., K.F. Nordstrom, D.J. Sherman, R.G.D. Davidson-Arnott, R.W.G. Carter, D. Jackson, and N. Gomes (1993). Aeolian sediment transport under offshore wind conditions: implications for aeolian sediment budget calculations. In L.P. Hildebrand (ed.), *Coastlines of Canada*. New York, ASCE, pp. 59-72.

Hardisty, J., and R.J.S. Whitehouse (1988). Evidence for a new sand transport process from experiments on Saharan dunes. *Nature*, 322: 532-534.

Horikawa, K., S. Hotta, and N. Kraus (1986). Literature review of sand transport by wind on a dry sand surface. *Coastal Engineering*, 9: 503-526.

Hsu, S. (1974). Computing eolian transport from routine weather data. *Proceedings, 14th Conference on Coastal Engineering*. New York, ASCE, pp. 1619-1626.

Kadib, A.A. (1966). Mechanism of sand movement on coastal dunes. *Journal of the Waterways and Harbors Division, ASCE*, 92: 27-44.

Kawamura, R. (1951). *Study of Sand Movement by Wind*. University of California Hydraulics Engineering Lab Rep. HEL 2-8 (1964 translation).

Leatherman, S.P. (1978). A new eolian sand trap design. *Sedimentology*, 25: 303-306.

Lettau, K., and H. Lettau (1977). Experimental and micrometeorological field studies of dune migration. In K. Lettau and H. Lettau (eds.), *Exploring the World's Driest Climate*. University of Wisconsin-Madison, IES Rep. 101: 110-147.

Namikas, S.L., and D.J. Sherman (in press). Effects of surface moisture content on aeolian sand transport. In V. Tchakerian (ed.), *Desert Aeolian Processes*, Chapman Hall.

Sherman, D.J., and B.O. Bauer (1993). Dynamics of beach-dune systems. *Progress in Physical Geography*, 17: 413-447.

Sherman, D.J., and W. Lyons (1993). Beach state controls on aeolian sand delivery to coastal dunes. Third Int. Geomorphology Conference, Hamilton, Ontario, Canada (abstract).

Short, A.D., and P.A. Hesp (1982). Wave, beach and dune interactions in southeastern Australia. *Marine Geology*, 48: 259-284.

White, B.R. (1979). Soil transport by wind on Mars. *Journal of Geophysical Research*, 84: 4643-4651.

THE DELTA FLUME'93 EXPERIMENT

A. S.-Arcilla[1], J.A. Roelvink[2], B.A. O'Connor[3], A. Reniers[2] and J.A. Jiménez[1]

ABSTRACT: This paper presents the results and some preliminary analyses obtained from the Delta Flume'93 experiment. The main aim of this experiment, supported by the "Large Installations Plan" (LIP-code number 11D) of the European Union, was to generate high quality and resolution data on the hydro-/morpho-dynamics of a natural 2DV beach. This included tests for two different geometries (equilibrium parabolic Dean-type profile with and without dune) and three different "dynamic states", i.e. near-equilibrium, erosive and accretive conditions. The hydro-/morpho-dynamic data obtained from the resulting seven test conditions have already been pre-processed (standard statistical and spectral analyses) and distributed to the original group of european researchers responsible for the Delta Flume LIP-11D experiment plus the researchers from the MAST-G8M group.

On-going research based on this data-set includes wave transformation and decay (with emphasis on the lag associated to the transition zone and to the fraction of breaking waves) and the vertical distribution of mean current velocities (with emphasis on the near-bed solution, the eddy viscosity coefficient and the above trough-level conditions for the lower flow). From a morphodynamic stand point most of the analyses performed so far deal with transport rates and bar development. Predictions for accretive wave sequences rated comparatively worse, with the bar sometimes moving in the "wrong" direction. More information on these studies may be found in the other 6 papers within the "Delta Flume" session of this conference and in the proceedings of the MAST-G8M Overall Workshop held in Grenoble in September 1993.

1.- INTRODUCTION

The Delta Flume'93 experiment, carried out during the period April-June, 1993, at Delft Hydraulics is a large-scale wave-flume test supported by the "Large Installations Plan" (LIP) of the European Union (EU). The main aim of this experiment was to generate hydro- and morpho-dynamic data (including sediment transport) on a natural

---

[1] Maritime Engineering Laboratory (LIM/UPC), Catalunya University of Technology, Gran Capità s/n mòdul D1, 08034 Barcelona, Spain.

[2] Delft Hydraulics, P.O. Box 152, 8300 AD Emmeloord, The Netherlands.

[3] Civil Engineering Department, Liverpool University, P.O. Box 147, Liverpool, L69 3BX, U.K.

2DV beach under equilibrium, erosive and accretive conditions. The project was primarily conceived to provide high quality and resolution data to validate/calibrate the 2DV, Q-3D and 3D models for surf-zone processes that have been developed during the late'80s and early'90s. Apart from this, the wealth of experimental information obtained will contribute to enlarge the already existing base of large-scale flume data (U.S. Army Corps of Engineers tests; former Delta Flume tests; CRIEPI tests in Japan; GWK tests in Hannover; and Supertank tests in Oregon) (see e.g. Kraus et al, 1992 for a review).

The '93 Delta Flume tests (code number 11D and thus denoted LIP-11D tests) were designed and overseen by researchers from the Catalunya University of Technology (UPC), Liverpool University (UL), Padova University (UP) and Thrace University (UT). This team, together with researchers from Delft Hydraulics (DH) and other European institutions participating in the MAST G8-M project, are currently processing and further analysing the data (see LIP-11D papers in the Delta Flume Session within this conference for illustration).

The combination of low and high sediment transport conditions (corresponding to slightly erosive and fully erosive/accretive wave trains, respectively) provide an adequate set of tests to monitor hydrodynamics (in cases of low bottom mobility) and morphodynamics (in cases of high bottom mobility). This combination will, hopefully, be helpful in understanding surf-zone processes as well as providing another bench-mark test for 2DV surf-zone modelling efforts.

2. EXPERIMENTAL SET-UP

2.1. Geometry and Wave Conditions

The objective of the tests was, as mentioned before, the generation of high quality and resolution data on the hydro-/morpho-dynamics of a natural 2DV beach for two different geometries and three different "dynamic states", i.e. near-equilibrium, erosive and accretive conditions. Special attention was paid to long-wave effects and near bottom resolution.

The first geometry consisted of a Dean-type beach (the so-called equilibrium parabolic beach profile of Brunn-Dean-More) with a modified (more realistic) bottom slope near and above the water-line. The profile equation is given by:

$$h = A x^{2/3} \qquad \frac{\partial h}{\partial x} < m$$

$$h = h' - m(x' - x) \qquad \frac{\partial h}{\partial x} > m$$

(for symbols definition see figure 1 in which h', x' denote the coordinates of the transition point between a parabolic and a planar profile)

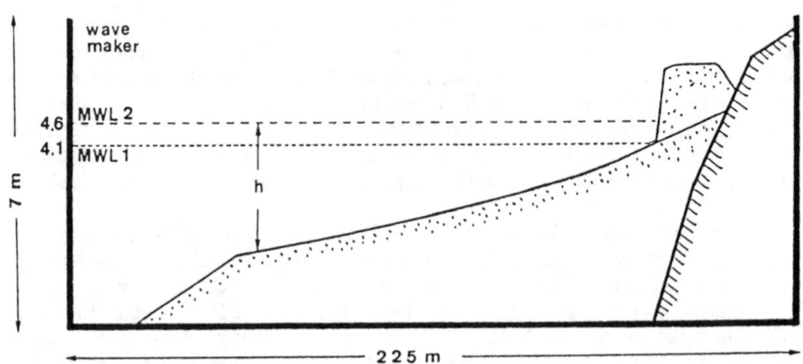

Figure 1.Flume geometry schematization with concrete bottom (\\\) and mobile profiles with and without dune (...). The two mean-water levels tested are also indicated.

The two free parameters, A and m, were mainly determined by the "flume boundary conditions" (i.e. flume dimensions which are 225 m length x 5 m width x 7 m depth and characteristics of available sediment which has a mean grain diameter of 200 μm).

From the grain diameter a value of $A = 0.10\ m^{1/3}$ was selected. Considering that the height of the walls in the DELTA-Flume is 7 m and taking into account the maximum wave height, run-up and dune height, the working range of water depths was between 4.0 and 5.0 m. For a waterdepth of 4.1 m and a "mobile bottom" length of 183 m (leaving the first 20m. from the wave maker without sand to avoid trouble with the wave generation equipment) the bottom slope was chosen to be:

$$m = \frac{1}{30} \quad \textit{for} \quad z\ \textit{bottom} > 3.7m \quad \textit{(above the flume bottom)}$$

$$m = \frac{1}{20} \quad \textit{for} \quad z\ \textit{bottom} < 1.6m \quad \textit{(above the flume bottom)}$$

The purpose of the second geometry with a dune profile was to investigate the effect of a low dune on the profile hydro- and morpho-dynamics (the dune acting as an upper boundary condition for the profile and bar development). The dune foot was chosen to be just above the still water level (see figure 1).

For each of these two geometries the following three wave conditions were tested: slightly erosive, highly erosive and strongly accretive. Narrow-banded random waves (generated by a random-phase, linear generator from a Jonswap spectrum) were chosen such that the combination of wave steepness (at peak frequency), bottom slope and water level would result in a stable, erosive and accretive beach, according to state-of-art criteria (e.g. Dalrymple, 1992). For the second geometry (dune profile) an extra wave series with an enhanced water level to promote further erosion was also included in the tests. This sequence of conditions could be interpreted as a schematization of the natural development of a storm, represented by three different sea-states (the slightly accretive case being disregarded due to the excessively long time required to produce appreciable bottom changes). During each condition the water level and wave parameters were kept constant and the corresponding duration (number of wave hours) was determined by the full test duration (given by the available budget) and the number of possible working hours per day. In all cases the selected duration was long enough to obtain accurate sediment transport estimates from profile measurements. A summary of all test conditions is presented in table 1, in which incident wave heights, periods, water levels and durations are described.

Test Code	Initial geometry	H_{m0} [m]	T_p [s]	Water level (m)	Duration (h)
1a	Dean-type	0.9	5	4.1	12
1b	result of 1a	1.4	5	4.1	18
1c	result of 1b	0.6	8	4.1	13
2a	Dean-type with dune	0.9	5	4.1	12
2b	result of 2a	1.4	5	4.1	12
2e	result of 2b	1.4	5	4.6	18
2c	result of 2e	0.6	8	4.1	21

Table 1. Summary of Tested Conditions.

Variable	Instrument	Accuracy	Sampling frequency
Water (dynamic) pressure	Pressure sensors	± 0.15%	10 Hz
Water velocity	EM sensors	± 2 cm/s	10 Hz
Free surface elevation	Surface following gauge	± 2.5 cm	10 Hz
	Resistance type gauge	± 1 cm	10 Hz

Table 2. Hydrodynamic Variables.

2.2 Instrumentation

Each sub-test (1a to 2e) was divided into a number of "wave hours". During each "wave hour" a time-series of exactly one hour of waves was generated. Measurements were taken by instruments attached to the flume wall (10 pressure sensors type PDCR 10/D/F-01, Druck Ltd, Leicester, and 5 electro-magnetic velocity-meters type EMF01, Delft Hydraulics), by instruments attached to a roving carriage (instrumented with an automatic sounding system -PROVO-, a sediment concentration sampler with 10 suction tubes, 5 electro-magnetic current-meters, 4 optical backscatter sensors, 1 bottom sediment transport meter -HARK- and 1 video camera) and by three movable wave gauges (2 surface-following gauges type WHM04, Delft Hydraulics, near the wave paddle and one resistance gauge type WHM10, Delft Hydraulics, near the carriage). The PROVO system (figure 2) was used to measure bottom profile elevations. In operation this system records the following signals: sounding echo (to measure submerged profile depths), angle of wheel arm, vertical rod displacement and electronic pulse (for indication of horizontal displacement). The angle of the wheel arm was used to guide the vertical rod displacement, with the echo sounder, rigidly attached to the rod right in front of the wheel, hovering over the bottom at a distance of about 10-20 cm. The combination of echo sounding and rod displacement yielded a very accurate and undisturbed profile measurement under water. In the vicinity of the water line and above it the combination of wheel angle and vertical rod displacement provided also an accurate profile measurement. The two measurements were combined by software in such a way that echodata were used where possible. Wheel data were used near and above the water line. The time-series so obtained were converted to spatial data on a 0.01 m grid.

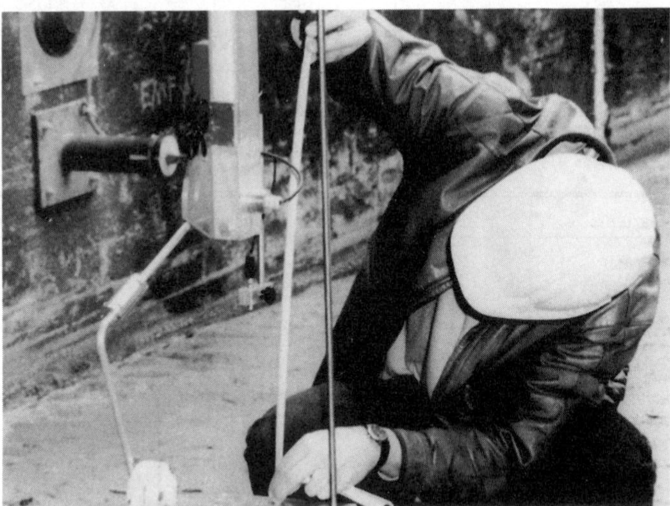

Figure 2. Snapshot showing the automatic sounding system PROVO installed in the roving carriage.

All movable instruments remained in a fixed position during each "wave hour". The position of the various instruments along the wave flume (13 longitudinal stations spaced at 7-40 m intervals and sampled at least once during each test series) was based on computational results obtained with UNIBEST-TC (e.g. vertical position of current meters close to the expected undertow maxima at about one third of the water depth, closer horizontal spacing of instruments in the shallower part of the flume where wave height gradients were larger, etc.). The resulting distribution for wall-fixed pressure sensors and electro-magnetic current meters is shown in figure 3. All instruments were located below the expected lowest trough level and above the bottom sheet-flow layer. Within these restrictions, the pressure sensors were positioned, nevertheless, as high as possible to avoid non-linear effects in the dynamic pressure measurements. Exceptionally, and because of inaccurate theoretical predictions (e.g. an unexpected bar migration burying a sensor), some of the fixed instruments had to be relocated during particular test series.

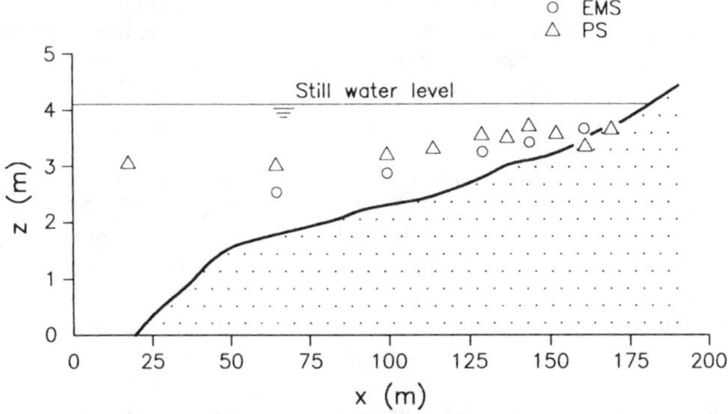

Figure 3. Schematization of wall-fixed instruments lay-out.

The layout of instruments on the roving carriage is shown in figure 4. The current meters were installed at the following distances from the bed: 0.10 m, 0.20 m, 0.40 m, 0.70 m and 1.10 m. This distribution was designed to capture the main features of the undertow distribution.

Figure 4. Snapshot showing a general flume perspective and the roving carriage.

3. OBTAINED RESULTS

For each of the seven test situations described in table 1, a range of hydro- and morpho-dynamic variables were recorded with the instrumentation described in the previous section. A summary of these variables appears in tables 2 and 3. This allowed a detailed monitoring of waves (short and long), currents, suspended and bed loads and bed levels at a number of wall-fixed and roving carriage stations (figure 3). The wall-fixed stations were sampled continuously during each test situation while the 13 roving carriage stations were sampled at least once during each test situation. A conventional video camera was also used to look at swash zone changes and visual features of breaking waves.

All instruments were sampled simultaneously at 10 Hz. In order to avoid aliasing each signal was low-pass filtered by an analog filter at 5 Hz before storage. Bed level surveys were done after each "wave hour" and covered three along-flume positions (one in the middle of the flume and two at a distance of 0.85 m from the flume walls). Two additional positions (+/- 0.80m from the center line) were added at the start and end of each test situation. The profile data were stored at 0.01 m intervals (horizontal bed resolution being dictated by the speed of the carriage and the sampling frequency).

In order to facilitate further analyses, the electromagnetic current-meters and optical backscatter sensors time-series were sub-divided into "high-frequency", "low-frequency" and "interaction" parts. The corresponding thresholds are schematized in table 4. The splitting was performed by the software package AUKE-PC (Delft Hydraulics).

Variable	Instrument	Accuracy	Sampling Frequency
Bed levels, bed forms. (3-5 long.lines)	Provo (Echo Sounder) (140mm dia wheel)	± 2 mm ± 2 mm	1cm spacing/each hr 1cm spacing/each hr
Bed load	HARK sediment meter		Every hour
Suspended load concentration	Pumping sampler 10 levels	± 20 %	Every hour
	OBS probes 4 levels,	± 0.05 g/l	10 Hz
Particle fall velocity	Visual accumulation tube (VAT)	± 0.1 mm/s	Each pumped sample
Grain size (suspended load) (bed material)	VAT		Each pumped/ bed sample

Table 3. Morphodynamic Variables.

Times-series component	Lower threshold	Upper threshold
High-frequency	1/2 f peak	f_{Nyq} = 5 Hz
Low-frequency	--	1/2 f peak

Table 4. Thresholds to split recorded time-series into high-frequency, low-frequency and interaction components.

Both the raw data and the computed parameters were stored in data files for each wave hour. The list of computed integral parameters, agreed beforehand, was:

i. Level-related parameters:

z_b (bottom profile), $\overline{\eta}$ (mean water level).

ii. Wave-related parameters:

mi, i= -1, 0, 1, 2, 4 (spectral moments), $H_{mo} = 4\sqrt{mo}$, $H_{rms,\,d}$ (down crossing), $H_{1/3}$, $H_{1/10}$ (10% exceeded), $H_{1/100}$, H_{max}, N (no. of waves in a record), T_p (peak period), T_{zd} (mean down crossing T), $T_{H_{1/3}}$ (mean T of highest 1/3 waves), skewness/kurtosis of η (free-surface elevation), groupiness factor, correlation between low-f waves and high-f wave envelope.

iii. Velocity-related parameters:

mean velocities in X and Z directions, rms orbital velocities in X and Z directions, total velocity moments $< | u(t) |^n >$, n=2, 3, 5, total (vector/modulus) moments $< u(t) | u(t) |^n >$ n=2, 3
third order velocity moments for

short waves $< u_s(t) | u_s(t) |^2 >$

long-short wave interaction $< 3\, u_l(t) | u_s(t) |^2 + u_l(t) | u_l(t) |^2 >$

mean current-wave interaction $< 3\, \bar{u} | u_s(t) + u_l(t) |^2 >$

mean current only $\bar{u} | \bar{u} |^2$

wave-induced equivalent stresses $< u_s w_s >$ and standard deviations of short-wave velocities at crest and trough in X and Z directions

iv. Sediment concentration related parameters

mean concentration, fall velocities (not exceeded by 10%, 50% and 90% of sampled volume) and grain diameters (not exceeded by 10, 50 and 90% of sampled volume) for bottom and suction samples.

All data were checked for spurious peaks and order of magnitude quality controls performed before storage. For each particular test situation the most important results were processed in the form of tables and graphs. Figures 5, 6 and 7 illustrate the wave height, set-up/down and bathymetry results obtained for each of the 12 "wave hours" of series 2a. The corresponding circulation pattern (showing also the H_{m0} decay and the estimated trough level) and vertical profiles of the mean current, $\overline{u(z)}$ and $\overline{w(z)}$, are shown in figures 8, 9 and 10. These results correspond to the full data set obtained for all "wave hours" within test series 2a. The presented figures illustrate the type of information available. Further analyses on these data can be found in companion papers within the "DELTA FLUME" Session in this conference.

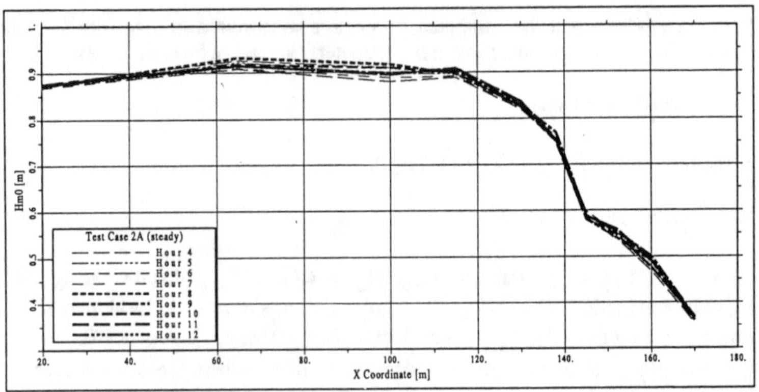

Figure 5. Hmo along-flume decay for each of the 12 wave hours in test series 2a.

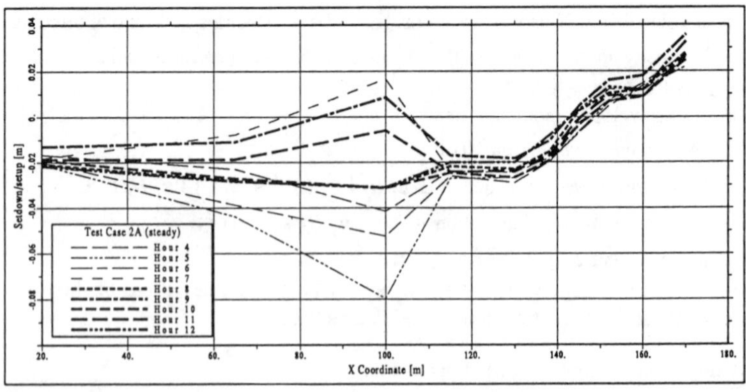

Figure 6. Mean water level variations along the flume for each of the 12 wave hours in test series 2a.

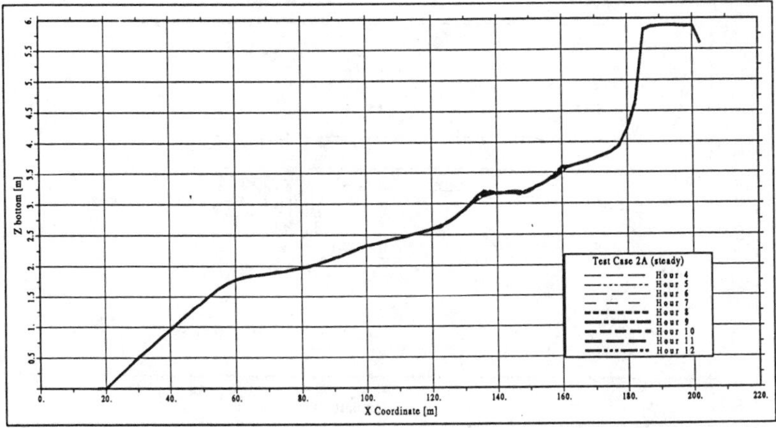

Figure 7. Bottom evolution along the flume for each of the 12 wave hours in test series 2a.

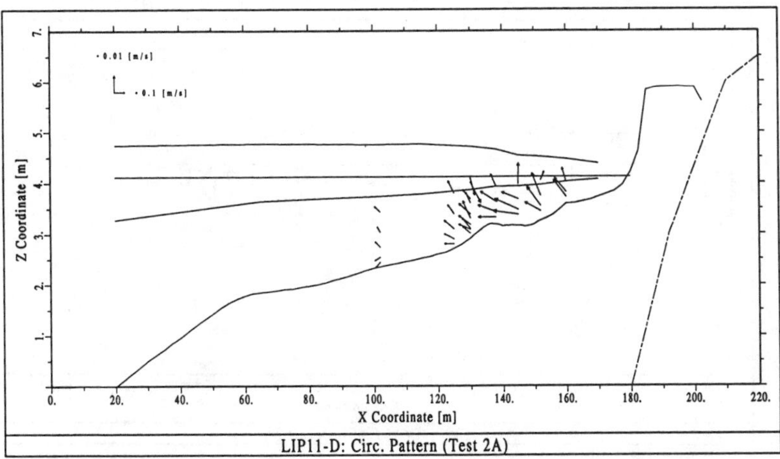

Figure 8. Circulation pattern measured along the flume for test series 2a. Also shown are the average Hmo decay law and an estimated trough level.

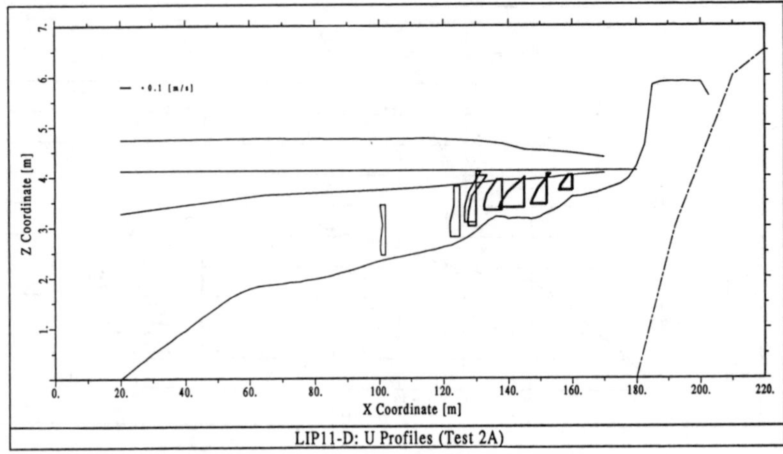

Figure 9. Along-flume distribution of vertical profiles for the horizontal mean-current velocity $\overline{u}$ (linear interpolation) in test series 2a. The thicker the line, the closer to the shore-line.

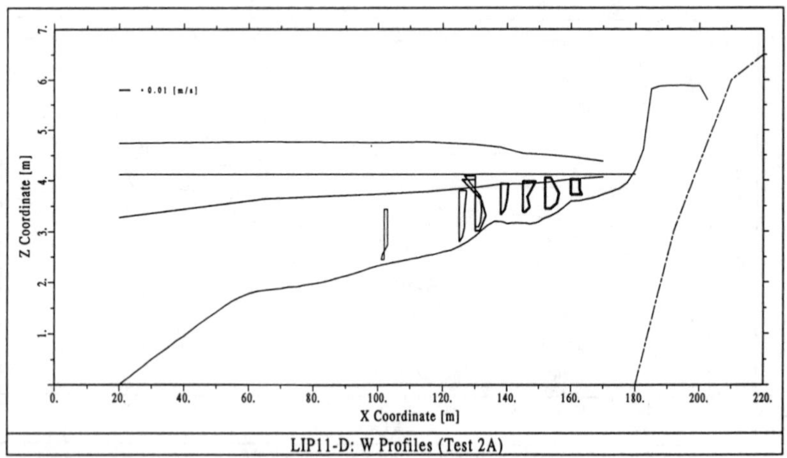

Figure 10. Along-flume distribution of vertical profiles for the vertical mean-current velocity $\overline{w}$ (linear interpolation) in test series 2a. The thicker the line, the closer to the shore-line.

4. PRELIMINARY PREDICTIONS

Apart from checks carried out prior to the test execution (to optimize the distribution of measuring equipment, etc.) some preliminary computations have already been executed, stimulated by various meetings amongst the involved researchers. Particular mention should be of the "modellers week" workshop held at the Delta Flume in June 1993 during the time the tests series 2b and 2e were actually being executed. The following researchers participated in this workshop: Luc Hamm (SOGREAH), John Nicholson and Brian O'Connor (UL), Irene Katopodi (UT), Helen Wallace and Howard Southgate (HR), Ad Reniers and Dano Roelvink (DH), José Jiménez and Marcel Stive (UPC), Marien Boers (Delft University of Technology), Claus Pedersen and Peder Clausen (Danish Hydraulic Institute) - all these from the MAST-G8M team - and Gary Mocke and Geoff Smith (CSIR, South Africa). Also available at the time were the measured time series of wave gauges, pressure sensors, em current meters and OBS sensors, as well as accurate data on profile development. This information, together with a simplified list of intergal parameters allowed most of the researchers to carry out comparisons between their models and the LIP-11D experiment.

Detailed comparisons between measured and predicted wave transformation and decay (paying special attention to the lag associated to the transition zone and to the fraction of breaking waves) are going to be presented by the DH, HR, SOGREAH and UPC researchers in this Coastal Dynamics'94 conference (see Rivero at al and Hamm and Roelvink papers). The vertical distribution of mean current velocities are also being currently analysed by CSIR, DH, HR and UPC researchers (see Collado et al., Stive and de Vriend and Reniers et al. papers). The emphasis, here, is on the cross-shore variation of vertical profiles and the accuracy and suitability of the required closure sub-models (bottom boundary layer solution, algebraic or differential closures for the eddy viscosity coefficients, mass fluxes in the crest-to-trough layer, and shear stresses at trough level).

Further analyses on the long-wave effects present in the flume are also foreseen. In the experiments, reflection compensation by "active wave absorption" was applied, which ensured a stationary random wave field during each wave hour. Long wave effects could be measured effectively because of this long sampling duration. During the tests, first-order wave generation was applied. The effect of including the incoming bound waves in the generated signal (second-order wave generation) can be studied by a surf beat model (see Hamm and Roelvink paper in this conference).

From a morphodynamic standpoint, most of the analyses performed so far deal with transport rates and bar development. Erosive wave sequences started developing a bar which was enhanced and transport inshore-wards during accretive wave sequences. In general, reasonable predictions have been obtained for the offshore bars during erosive tests (a and b) but less good results have been produced for the accretive sequence (c), with the bar often moving in the "wrong" - according to numerical results - direction. In some cases numerical transport rates were too large (measured transport rates being often in the order of 0.1 m^2/hr.), indicating the need for additional calibration of model constants. However, use of detailed K-ε models for mixing processes due to breaking waves and models for undertow velocities and suspended sediment concentration have produced very reasonable results for suspended load in test 1a. Most discrepancies appear to belong to the outer breaking zones, where experimental results also fluctuate more. An illustration of the sequence of profiles obtained in two erosive test sequences appears in figure 11, corresponding to test series 2b and 2e. The difference in behaviour, characterized by dune erosion and subsequent deposition (most pronounced at the dune toe itself although the profile and corresponding transport rates - mostly offshore directed - appear to be affected for a distance of almost 100 m from the dune face) shows the effect of increasing the mean water level by half a metre. The corresponding cross-shore transport rates, as derived from profile comparisons, are shown in figure 12. (See Nicholson and O'Connor paper in this conference for more details).

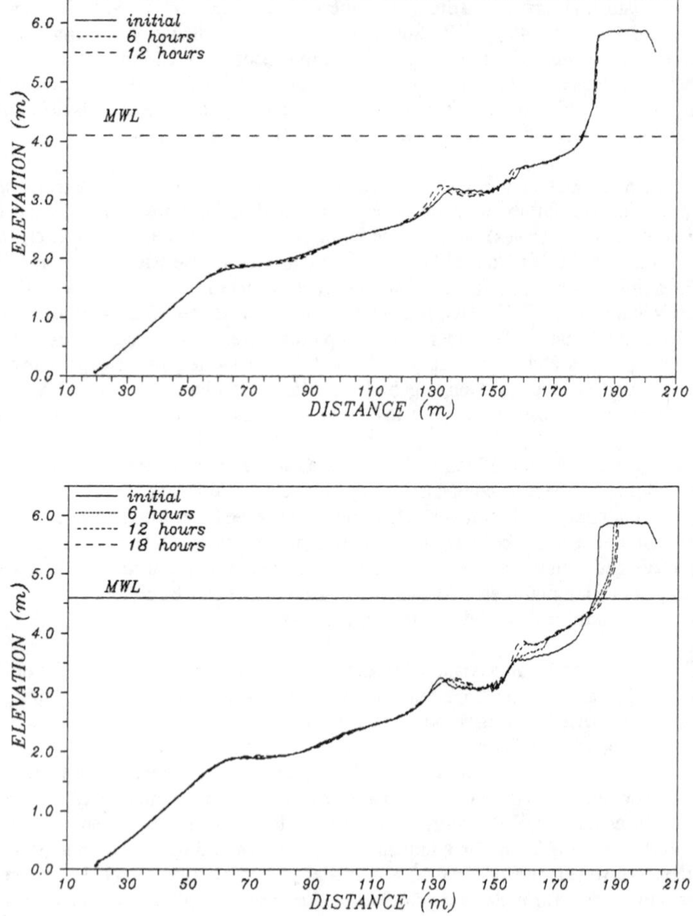

Figure 11. Sequence of profiles obtained in erosive test series 2b above and 2e below.

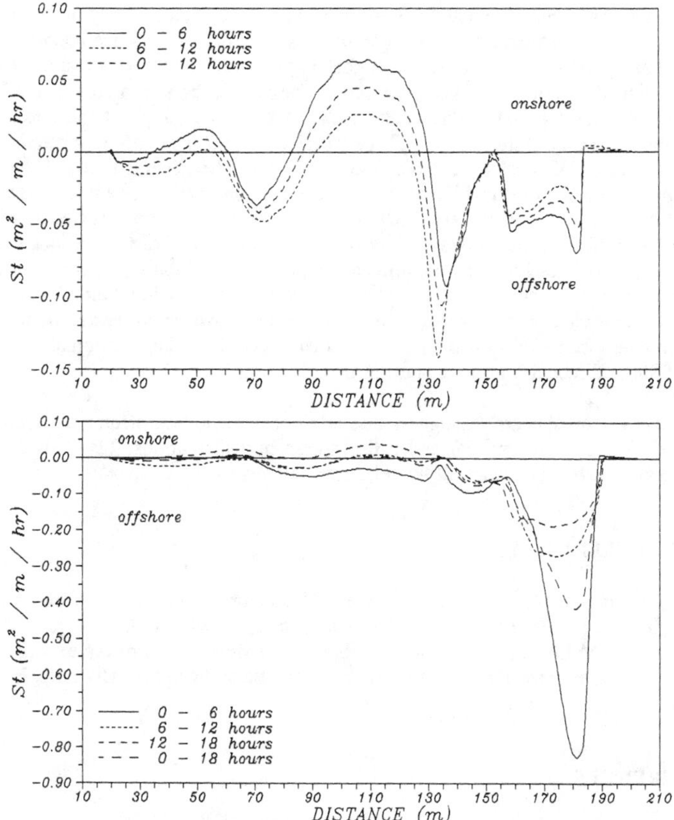

Figure 12. Along-flume (i.e. cross-shore) transport rates for test series 2b above and 2e below as derived from profile comparisons.

More in-depth analyses on these hydro- and morpho-dynamic topics can be, thus, found in the other 6 papers within the "Delta Flume" Session of this conference and in the proceedings of the MAST-G8M Overall Workshop held in Grenoble in September 1993.

5. CONCLUSIONS

Although conclusions at this stage must be considered preliminary, some "obvious" consequences may already be drawn. Hydrodynamic predictions outdid, in general, morphodynamic ones. The performed comparisons for accretive cases rated comparatively worse, with the bar often moving in the "wrong" direction. In some cases model transport rates were too large although undertow and suspended sediment concentrations have been predicted with reasonable accuracy, particularly in the "inner" breaker zone. The fluctuations occurring in the region closer to the wave paddle, added to the breaking of the highest waves in that region, hinders the establishment of a "suitable" boundary condition. The observed steepening of the swash zone (for all test conditions without a dune) is due to the fact that the initial slope of the beach was milder than the "equilibrium slope" (as derived from state-of-art criteria, e.g. (Sunamura'84) or (Kriebel et al.'91)). Apart from this, the general profile behaviour (according to the Sunamura and Horikawa'74 classification) was not as a priori expected: test conditions a and e were "erosive" (with cross-shore transport directed basically offshore and exhibiting the corresponding bar development) and test conditions c were "accretive" (with cross-shore transport directed on-shore and exhibiting the corresponding bar migration in this same direction). However, test conditions b, which should have been even more clearly "erosive" due to the increased wave height (wave period and MWL remaining constant) turned out to be of mixed type (featuring an alternation of on-shore/off-shore directed transports starting from the swash zone).

This behaviour appears to suggest that additional modelling efforts are required, particularly for turbulence/mixing and sediment transport simulations. It is likely that the newly obtained LIP-11D data will be helpful in this respect.

6. ACKNOWLEDGEMENTS

The authors wish to acknowledge the EU support from the LIP and MAST Programmes. The Spanish authors also want to acknowledge CIRIT and PCM-MOPTMA. Dr. M.J.F. Stive has also contributed at various stages of the experiments, providing valuable scientific support and personal encouragement. His help is duly appreciated.

7. REFERENCES

Collado, F.R.; Sánchez-Arcilla, A. and Coussirat, M.G. (1994) 2DV circulation-pattern: wave decay and undertow modelling with NEARCIR code. Proc. 1st Coastal Dynamics Conf. (this volume).

Dalrymple, J.A. (1992) Prediction of storm/normal beach profiles. Journal of Waterways, Port, Coastal and Ocean Engineering, 118, 2, 193-200.

Hamm, L. and Roelvink, J.A. (1994) Validation and Comparison of wave transformation models with large scale flume measurements. Proc. 1st Coastal Dynamics Conf. (this volume).

Hattori, M. and Kawamata, R. (1980) Onshore-offshore transport and beach profile change. Proc. 17th Coastal Eng. Conf., ASCE, 1175-1193.

Kraus, N.C.; McKee Smith, J. and Sollitt, C.K. (1992) Supertank Laboratory Data Collection Project, 23rd I.C.C.E., ASCE.

MAST G8M - Coastal Morphodynamics, Overall Workshop, Grenoble, September 1993.

Nicholson, J. and O'Connor, B.A. (1994) Calibration and validation of a cross-shore transport model. Proc. 1st Coastal Dynamics Conf. (this volume).

Reniers, A.; Mocke, G. and Smith, G. (1994) A surfzone parameter sensivity analysis using LIP11D suspended sediment and return flow measurements. Proc. 1st Coastal Dynamics Conf. (this volume).

Rivero, F.J.; Sánchez-Arcilla, A.; Southgate, H.N. and Wallace, H.M. (1994) Comparison of wave transformation models with LIP-11D data. Proc. 1st Coastal Dynamics Conf.(this volume).

Stive, M.J.F. and De Vriend, H.J. (1994) Verification of the Shearstress Boundary Conditions in the Undertow Problem. Proc. 1st Coastal Dynamics Conf. (this volume).

Sunamura, T. (1984) Quantitative predictions of beach-face slopes. Geological Society America Bulletin, 95, 242-245.

Sunamura, T. and Horikawa, K. (1974) Two dimensional beach transformation due to waves. Proc. 14th Coastal Eng. Conf., ASCE, 920-938.

SHORT AND LONG WAVE TRANSFORMATION MODELLING IN A LARGE SCALE FLUME

Luc HAMM[1] and J.A.(Dano) ROELVINK[2]

ABSTRACT: Comparison between three wave transformation models and measurements carried out in a prototype scale wave flume (LIP11D experiments) are presented. Parameters computed and discussed include statistical and energy-based wave parameters, the mean water level set-up, the fraction and distribution of breaking waves, and the long-wave motion.

INTRODUCTION

A number of experiments at prototype scale have been recently performed at the Delta-flume in de Voorst (The Netherlands) as part of the european LIP-project (Large Installation Plan). The objective of the experiments was the generation of high quality resolution data on hydrodynamics and sediment transport dynamics on a two-dimensional beach under equilibrium, erosive and accretive conditions. The experimental set-up and main results of these experiments are presented by Sánchez-Arcilla et al.(1994) in this conference. In the present paper, we shall focus on the numerical modelling of short and long wave transformation observed during these experiments.

Usual approaches included in coastal profile modelling to simulate wave transformation have been recently reviewed by Hamm et al.(1993). Three of those models will be used in the present study. The first one is the classical parametric spectral approach of Battjes and Janssen(1978) which is now widely accepted. Improvements of this model in relation with the LIP11D data are also reported by Rivero et al.(1994) in this conference. The second one is a major extension of this

---

[1]Senior Engineer. SOGREAH Ingénierie BP172, 38042 Grenoble CEDEX 09, France. Phone (33)76334188, Fax (33)7633 4296

[2]Senior Research Eng., DELFT HYDRAULICS, p.o. box 152, 8300 AD Emmeloord, The Netherlands. Phone (31)5274 2922, Fax (31)5274 3573

approach considering unsteady equations to describe the propagation of wave groups and generation of associated surf-beats (Roelvink, 1991, 1993b). An improvement of this model including the effect of the roller in the surf-zone will be presented and validated in the present paper. A probabilistic finite amplitude wave transformation model which predicts the evolution of an initial combined wave heights and periods distribution will also be used. This model is an extension of the work presented by Fornerino et al.(1992).

EXPERIMENTAL DATA

Beach profile and wave conditions

Two comparative tests have been performed from the same initial Dean's type underwater beach profile without (test 1) and with (test 2) a low dune (see figures 1 and 2). Each test has been subdivided into three (respectively four) sub-tests. A constant incident wave condition was applied during each sub-test over a duration of 12 to 18 hours (see Arcilla et al.,1994 for more details). Two sub-tests have been selected for this study: test 2A (H_s of 0.85 m, T_p of 5.0 s, still water level:0.0 m and duration of 12 hours) starting with the initial beach profile with the dune and test 1C (H_s of 0.60 m, T_p of 8.0 s, still water level:0.0 m and duration of 13 hours) starting with a barred profile. It should be noted that the dunefoot was chosen to be just above the maximum surfbeat level associated with a zero still water level sothat it stayed intact during test2A. A standard linear steering signal was used to drive the piston-type wave-maker but reflected long waves were absorbed at the wave-maker.

Instrumentation

A comprehensive description of the instrumentation is provided in Arcilla et al. (1994) and will not be repeated here. Hydrodynamic parameters were measured with ten fixed pressure sensors (DRO), five fixed electromagnetic current meters (EMS) attached to the flume and one measurement carriage equipped with one wave gauge (WHM), five electomagnetic currentmeters over a vertical and a video camera. This carriage was moved every hour in order to cover the entire surfzone. In addition, two wave gauges located at a distance of 20 and 40 m from the wave generator were used to determine the actual incident wave conditions. The following data have been used for the purpose of this study: wave heights along the beach profile have been estimated from the mobile wave gauge. The total energy level was checked against pressure sensors data. The set-up was derived from the pressure sensors. Velocity moments and short-long waves correlation coefficients were estimated from the lower EMS attached to the mobile carriage. The distance between this currentmeter and the bottom was about 0.20 m. Incident wave conditions were derived from the two wave gauges located near the wave-maker. Finally, fraction of breaking waves in the surfzone were derived from a visual analysis of video recordings (see Rivero et al.,1994).

Analysis of time-series

Time-series have been recorded during 60 mn each hour at a frequency rate of 10 Hz. A comprehensive analysis of the data was performed very quickly after each run and parameters stored in ASCII files. The low and high frequency parts of the signals have been separated at a cut-off frequency of 0.1 Hz. Velocity moments have been analysed following the procedure described in Roelvink and Stive(1989). The signal was decomposed into a mean current and a time-varying component including a low frequency and a high frequency part.

Measured data are presented in figures 1 (test 2A) and 2 (test 1C). Spectral and statistical estimates of the significant short-wave heights are very close for test 2A as opposed to test 1C where spectral estimates are lower. This is due to the the low steepness of the waves for which finite-amplitude effects are visible. The low frequency energy is kept at a reasonnable level thanks to the absorbing system at the wave-maker. Velocity moments related to the total third odd moment (gu2ux) which is considered to be of primary importance for sediment transport is also shown in figures 1 and 2. It includes contributions of short-waves (guss) and terms of interactions between short and long waves(guls) and short waves and the mean current(gusc). The other contributions (i.e. due to the mean current) were always negligible. The total moment is dominated by the assymetry of short waves (guss) offshore of the outer bar and by the return current term (gusc) further inshore. the short-long wave term (guls) plays mainly a role in test 2A where it contributes to counteract the short-wave term offshore leading to a total moment nearly nil.

THE SURFBEAT MODEL

Basic Formulation

The SURFBEAT model (Roelvink, 1993b) is a one-dimensional time-dependent model whoch describes the evolution of normally incident wave groups and associated long waves over an arbitrary beach profile, in a short-wave averaged sense. The basic equations are derived from Phillips (1977) by considering two time-scales: that of the individual waves and that of the wave groups and associated long waves. Both motions are separated by assuming that the time scale of short-wave fluctuations is much shorter than that of the wave groups and by time averaging the short wave motion. Then, the basic conservative equations are depth-integrated to lead to the non-linear shallow water equations which read:

$$\frac{\partial}{\partial t}h+\frac{\partial}{\partial x}Q_t=0 \qquad (1)$$

$$\frac{\partial}{\partial t}Q_t+\frac{\partial}{\partial x}\left[\frac{(Q_t^2-Q_w^2)}{h}+\frac{S_{xx}}{\rho}+\frac{1}{2}gh^2\right]=gh\frac{\partial}{\partial x}h_0-\frac{\overline{\tau_b}}{\rho} \tag{2}$$

where h is the total water depth, h_0 is the still water depth, Q_t is the total average flux, Q_w is the wave induced flux, S_{xx} is the radiation stress, ρ is the water density, τ_b is the bottom shear stress. The wave terms included in the above system (Q_w, S_{xx}, C_g)are derived from linear theory and the short-wave energy is computed from the conservation of the energy flux by assuming a narrow-banded spectrum and by neglecting all kinematic and dynamic effects of the long waves on the short waves. This equation reads:

$$\frac{\partial}{\partial t}E+\frac{\partial}{\partial x}EC_g=-D_w \tag{3}$$

The bottom friction is computed by:

$$\overline{\tau_b}=\frac{1}{2}\rho f_w\frac{|Q_t-Q_w|(Q_t-Q_w)}{h^2} \tag{4}$$

where f_w is a friction factor that has to be determined through calibration. Finally, the energy dissipation is computed following the classical approach of Battjes and Janssen(1978) with an improved empirical parametrization (Roelvink, 1993a) which reads as follows:

$$D_w=2f_pE\left(1-\exp\left(-\left(\frac{H}{0.55h}\right)^{10}\right)\right) \tag{5}$$

Equation (5) includes three empirical parameters which have been given constant values. It has been checked against a total of 39 datasets containing 548 data points (Roelvink, 1993a).

Equations (1) to (5) fully describe the motions under study and need boundary and initial conditions. At the seaward boundary, the (constant) frequency f_p is prescribed and the incoming short wave energy E is specified as a function of time. A so-called weakly reflective boundary condition is applied for the long-wave motion in order to generate the incoming bound wave and to let reflective free waves to propagate undisturbed out of the model area. Such a condition is detailed in Roelvink (1991, 1993b). The landward boundary is defined at the water line, a moving point where the water depth has a small fixed value. Here, the short-wave energy is set to zero.

This set of equations is solved numerically as following: First, the equations are transformed from the physical domain, which has a moving landward boundary, to a fixed computational domain including a fixed landward boundary and an equidistant grid spacing. A second-order Richtmeyer's predictor corrector numerical scheme is then applied (Roelvink, 1993b).

Inclusion of the roller effects

The observed lag between the maximum gradient of short wave energy and the maximum gradient of the set-up and return flow in breaking waves (the so-called transition zone) has been approached in different ways. Svendsen(1984a) pointed out the importance of surface rollers in breaking waves and derived a surf zone model able to predict the wave height decay and set-up in the inner surf zone when the roller is well established. Roelvink and Stive(1989) ascribed the lag in the return flow to the fact that the wave energy is first converted to large-scale turbulences before it is dissipated; using their model, they found no significant lag in the set-up. In Nairn et al.(1990), both approaches are compared and an alternative approach is suggested. Here, this approach is clarified and extended to the case of slowly-varying short wave energy.

Svendsen(1984a,b) derived the integral properties of a surface roller "riding" on a wave front which include a contribution to the energy (E_r), the energy flux(F_r), the radiation stress ($S_{xx,r}$) and the mass flux ($M_{x,r}$). These contributions could be expressed as a function of the roller energy as follows (Nairn et al.,1990):

$$E_r=\rho A\frac{C}{2T}; \qquad F_r=E_rC; \qquad S_{xx,r}=2E_r; \qquad M_{x,r}=2\frac{E_r}{C} \tag{6}$$

where C is the celerity of the wave approximated by its shallow water expression, A is the cross-sectionnal area of the roller, ρ is the density of water and T the wave period. Furthermore, the total energy dissipation in a breaking wave could also be expressed as a function of the roller energy (Nairn et al.,1990). This expression reads:

$$D_r=2\beta g\frac{E_r}{C} \tag{7}$$

where β is the mean slope under the roller defined by Deigaard(1989). Equation (3) is now reconsidered by including the roller contribution into the wave energy and by assuming that this total energy is totally dissipated in the roller. Under these assumptions, we can write:

$$\frac{\partial}{\partial t}(E+E_r)+\frac{\partial}{\partial x}(EC_g+E_rC)=-D_r \tag{8}$$

Then, we substract equation (3) from equation (8) to get an additional equation driving the formation and dissipation of the roller.

$$\frac{\partial}{\partial t}E_r+\frac{\partial}{\partial x}E_rC+2\beta g\frac{E_r}{C}=D_w \tag{9}$$

We now see that the roller energy is built up when the wave energy suddenly decreases; this means that the decay of the radiation stress just after breaking is delayed, so we can indeed expect a lag in the set-up.

Comparison against the LIP11D data

The improved SURFBEAT model has been used to simulate tests 2A and 1C presented in the preceding chapter. A grid of 101 points has been built. The friction factor and the mean slope under the roller have been set respectively to 0.02 and 0.10. The simulations have been run over one hour with a time step of 0.5 sec. Results are presented on figures 3 (test 2A) and 5 (test 1C) together with results obtained with the classical Battjes and Janssen model calibrated by Battjes and Stive(1985).

For both tests, the short wave energy decay is well reproduced by both models whereas the inclusion of the roller effect improves significantly the reliability of the set-up predictions by SURFBEAT. A discrepancy still remains in test 1C in the trough of the bar. Sensitivity tests have been performed by decreasing the β parameter appearing in equation (9) from its standard value of 0.1 down to 0.02 in order to get a good agreement with the measured data (figure 7). Further investigation is needed to understand that point. Long wave activity is fairly well reproduced with a slight overestimation of the increase of the energy in both tests. Three indexes (Roelvink,1993b) relating long and short wave motions are also compared on figures 3 and 5 namely, the long wave contribution to the third odd velocity moment (guls), the correlation coefficient between long wave elevation and short wave energy (C_r) and between long wave velocity and short wave velocity variance ($C_{r,u}$). A fair agreement is found except in very shallow water where computed indexes are too high in both tests.

THE PROBABILISTIC APPROACH

Introduction

Parametrized spectral models described above predict the evolution of the total potential energy (represented by H_{m0}) but give no information on the distribution of breaking and non-breaking wave heights. This distribution could be roughly estimated by using an analytical distribution function suitable for the nearshore like the Beta-Rayleigh distribution proposed by Hughes and Borgman(1987) or the Weibull formulation (Ahran and Ezraty,1975; Klopman and Stive, 1989). A better prediction could be generally obtained by using a wave-by-wave transformation model in which the incident distribution of zero-crossing wave heights and periods is discretized into a number of classes. Each class is propagated using a regular wave model assuming no wave-wave interaction. Statistical parameters like $H_{1/3}$ are rebuilt along the beach profile by applying to each class of waves its initial probability in the distribution.

This probabilistic approach has been introduced by Goda (1975) to estimate the transformation of wave statistics throughout the surf-zone on a plane beach. This first model includes most of the the basic physics i.e. finite-amplitude wave shoaling, a robust breaker index, a variable distribution of breakers along the beach, the inclusion of set-up to increase the mean water level and the effect of surf-beats which induce a fluctuation of this mean water level as pointed out in the preceding chapter. On the other hand, the weak point of this approach is the lack of proper wave energy balance which restricts its application to plane beaches. This point has since been improved (Mizuguchi,1982; Mase and Iwagaki,1982; Dally and Dean, 1986; Dally,1992). The other draw-back is the lack of proper modelling of the transition zone which is simply ignored by all the above cited authors.

Formulation of the present model

The present model used in this study (named REPLA in the figures) gathers most of the basic features described above. A finite wave amplitude transformation model based on ray kinematics, wave action balance and momentum conservation equations is used. The set of equations to be solved (Fornerino et al., 1992) includes a kinematic part which reads:

$$\frac{d(k\sin\theta)}{dx}=0 \tag{10}$$

$$\omega_a=\omega_r+k_x.U_x+k_y.U_y \quad with \quad k_x=k\cos\theta \quad k_y=k\sin\theta \tag{11}$$

$$F(\omega_r,k,H,h)=0 \tag{12}$$

where (k_x,k_y) is the wave number, θ is wave direction counted from the x-axis, ω_a is the absolute angular frequency, ω_r is the relative angular frequency, (U_x, U_y) is the wave-averaged, depth integrated current which is an input of this model, h is the water depth (including set-up), H is the wave height and F is the dispersion relationship function and a dynamic part:

$$\frac{dW_x}{dx}+\frac{E_d}{\omega_r}=0 \tag{13}$$

$$\frac{dS_{xx}}{dx}+\rho g(h_0+\bar{\eta})\frac{d\bar{\eta}}{dx}=0 \tag{14}$$

with

$$W_x=\frac{2K}{\omega_r}U_x+(3K-2V+\rho h\frac{\overline{u_b^2}}{2})\frac{k_x}{k^2} \tag{15}$$

$$S_{xx}=(3K-2V+\frac{1}{2}\rho h\overline{u_b^2})\cos^2\theta+(K-V+\frac{1}{2}\rho h\overline{u_b^2}) \tag{16}$$

where h_0 is the still water depth, K is the kinetic energy, V is the potential energy, ρ is the water mass density and u_b is the bottom orbital velocity (the square of u_b should be averaged over one period). The dispersion relationship and integral properties are computed with two different wave theories according to the value of the local Ursell number following Hardy and Kraus (1988). When the Ursell number is less than 25, Stokes third order theory derived by Cialone and Kraus (1987) is used. When it is more than 10, a cnoidal second order theory (Hardy and Kraus,1987) is used. These two theories have been derived with the second definition of Stokes celerity which is the most appropriate to simulate shoaling waves normally incident to a beach. It has been shown previously that the continuity of wave height and wave energy flux cannot be respected together at the matching point between these two theories (Hardy and Kraus, 1988). A smoothing procedure has thus been implemented in order to keep this continuity and avoid numerical noise in future morphodynamic computations. The wave breaking criteria derived by Weggel(1972) is used and the dissipation rate derived from a bore have been introduced in the following form:

$$E_d=\frac{1}{4}\rho g f_r\frac{H^3}{h} \tag{17}$$

where g is the acceleration of gravity and f_r is the relative wave frequency. Furthermore, the dissipation rate is put to zero when the wave height is less than half the local maximum wave height in order to simulate wave reformation after breaking on a bar.

Due to the dependence of the dispersion relationship with wave height, an iterative procedure is used to solve separately the kinematic and dynamic parts of the equations. The kinematic part is solved with a Raphson-Newton method by taking an initial value of the wave height. Then, the wave action conservation equation is solved with an explicit finite difference scheme. The kinematic part is solved again with the new wave height until convergence is reached (Fornerino et al., 1992). More recently, a special two-steps procedure has been implemented for the computation of the breaking and reformation positions in order to keep a good accuracy of the result. A second iterative loop was also implemented in order to include the contribution of the set-up to the mean water level. Modelling of the transition zone and of surf-beats effects are presently no included into this model. The method of resolution and validation against regular waves data could be found in Fornerino et al.(1992).

When random waves are to be simulated, an initial joint wave heights and periods distribution derived from data or from analytical distribution functions is discretized into a finite number of classes. Each class is represented by a quadruplet (H, T, θ,p) where H, T and θ are a representative wave height, period and direction respectively of the class considered and p is its weight in the distribution. Each quadruplet is propagated using the regular wave model described above. Then the wave height distribution, the set-up and the fraction of breaking waves are derived from the results at each grid point. Breaking and non-breaking wave height histograms could also be produced separately.

Validation against the LIP11D data

This probabilistic model has also been used to simulate tests 2A and 1C. A grid of 80 points has been built with a grid size of 2.5 m outside the surfzone reduced to 1.0 m further inshore. Around 50 wave classes have been used in the simulation and derived from the analysis of the time-series recorded by the wave gauges placed in front of the wave-maker. Results are presented on figures 4 (test 2A) and 6 (test 1C). For both tests, the significant and maximum wave heights are well reproduced except in the neighbourhood of the bar in test 1C where $H_{1/3}$ is too high whereas H_{max} is strongly underestimated. The latter point may be due to wave-wave interactions. Set-up results (without any roller) are surprisingly not too far from the data of test 2A (figure 4). This is no more the case in test 1C (figure 6) where plunging breakers are dominant. Finally, the fraction of breaking waves is compared to the results obtained from the Battjes and Janssen (BJ) model. The probabilistic model gives generally much higher and/or more convincing (top of the bar) results. Experimental data were available for test 1C and confirm the poor predictability of the BJ model. The probabilistic approach tends also to underestimate that parameter. It should be noted that no free empirical parameters are included into the probabilistic model so that no calibration is possible. On the other hand, sensitivity tests have been performed by ignoring the periods distribution and thus reducing the number of classes from 50 to 15 with only one representative period. Results are presented on figure 8 for test 2A and look reasonnable.

CONCLUSIONS

This comparison exercise has confirmed that probabilistic and parametrized spectral models presently used in coastal profile modelling give reliable estimates of wave heights transformation even with very low steepness waves (test 1C).

Reliable predictions of the set-up could only be obtained by inclusion of the modelling of the transition zone. The formulation described in this paper is based on the roller concept and has been found to be reliable and robust. It depends of one empirical parameter (β parameter) whose variation range requires more exploration.

The fraction of breaking waves is fairly well predicted by the probabilistic model. Predictions from the Battjes and Janssen model are rather poor as already noticed by Roelvink(1993a). This is due to the fact that the history of breakers is not kept into such a model. An improved approach aiming at including this history is described by Rivero et al. (1994) in the present conference.

Long wave activity is fairly well reproduced with the SURFBEAT model. Negative measured and predicted values of phase coupling between short and long wave motions found in both tests tend to indicate that the coupling between short waves and incident long waves is dominant leading to a seaward directed effect.

Apart from the validation of the different model concepts, an important result of this exercise is the generation of optimal cross-shore distributions of these parameters, which can be used as boundary conditions for computations of the vertical distribution of the mean current and of the sediment transport.

ACKNOWLEDGEMENTS

The flume data were collected during the LIP-11D research Project financed by the European Union within the Large Installation Programme(LIP). The Delta flume belongs to the research laboratory Delft Hydraulics in The Netherlands. Permission to make use of the data was kindly granted by the main investigator, Dr. A. Sánchez-Arcilla (from the Maritime Engineering Lab LIM/UPC in Spain) and Dr. J.A. Roelvink (from Delft Hydraulics in The Netherlands). The present research was undertaken as part of the MAST G8 Coastal Morphodynamics research program. It is supported by the Service Technique Central des Ports Maritimes et des Voies Navigables (French Sea State Secretary) and by the Commission of the European Communities, Directorate General for Science, Research and Development, under contract n° MAS2-CT-92-0027. Many thanks to D. Beyer (Delft Univ. of Techn. and LIM/UPC) who provided us with the measurements of the fraction of breaking waves and to C. Péronnard (SOGREAH) who carried out the programming of the present version of REPLA.

REFERENCES

Ahran, M. and R. Ezraty(1975). Saisie et analyse de données de courants particulaires et pressions au voisinage du fond dans la zone littorale proche du déferlement. La Houille Blanche, 7(8), 483-496

Battjes, J.A. and J.P.F.M. Janssen(1978). Energy loss and set-up due to breaking of random waves. Proc. Int. Conf. on Coastal Engineering, ASCE, 569-587.

Battjes J.A. and M.J.F. Stive (1985). Calibration and verification of a dissipation model for random breaking waves. J. of Geophys. Res.,vol90,n°C5, 9159-9167

Cialone, M.A. and N.C. Kraus (1987). A numerical model for the shoaling and refraction of third-order Stokes waves over an irregular bottom. Misc. Pap., CERC 87-10, US Army Eng. WES,CERC

Dally W.R.(1992). Random breaking waves: field verification of a wave-by-wave algorithm for engineering application. Coastal Engineering,16,369-397

Dally W.R. and R.G. Dean (1986). Transformation of random breaking waves on surf beat. Proc. Int. Conf. on Coastal Eng.,ASCE, 109-123

Deigaard, R. (1989). Mathematical modelling of waves in the surf zone. Prog. Rep. 69, Tech. Univ. Denmark, Inst. Hydrodyn. and Hydraulic Eng., 47-60

Fornerino M., P. Sauvaget and L. Hamm(1992). Numerical modelling of finite amplitude wave shoaling in presence of currents. In: Proc. 4th Int. Conf. on Hydraulic Eng. Software. Fluid flow modelling. Computational Science Publications, Southampton, 587-598

Goda, Y. (1975). Irregular wave deformation in the surf zone. Coastal Engineering in Japan, 18, 13-26

Hamm, L., P.A. Madsen and D.H. Peregrine (1993). Wave transformation in the nearshore zone: A review. Coastal Engineering, 21(1-3), 5-39.

Hardy T.A. and N.C. Kraus (1987). A numerical model for the shoaling and refraction of second-order cnoidal waves over an irregular bottom. Misc. Pap., CERC 87-9, US Army Eng. W.E.S.,CERC

Hardy T.A. and N.C. Kraus(1988). Coupling Stokes and cnoidal wave theories in a nonlinear refraction model. Proc. Int. Conf. on Coastal Eng.,ASCE, 588-601

Hughes, S.A. and L.F. Borgman(1987). Beta-Rayleigh distribution for shallow water wave heights. Proc. Conf. on Coastal Hydrodynamics, ASCE, 17-31

Klopman, G. and M.J.F. Stive(1989). Extremes waves and wave loading in shallow water. Proc. E&P Forum Workshop "Wave and current kinematics and loading", Paris, France, 25-26 Oct. 1989,

Mase H. and Y. Iwagaki(1982). Wave height distributions and wave grouping in surf zone. Proc. Int. Conf. on Coastal Eng.,ASCE,58-76

Mizuguchi M.(1982). Individual wave analysis of irregular wave deformation in the nearshore zone. Proc. Int. Conf. on Coastal Eng.,ASCE,485-504

Nairn R.B., J.A. Roelvink and H.N. Southgate(1990). Transition zone width and its implication for modelling surfzone hydrodynamics. Proc. Int. Conf. on Coastal Eng.,ASCE,68-81

Phillips, O.M.(1977). The dynamics of the upper ocean. Cambridge Univ. Press, Cambridge.

Rivero, F.J., A. Sánchez-Arcilla, H.N. Southgate and H.M. Wallace(1994). Comparison of wave transformation models with LIP11D data. Proc. Coastal Dynamics'94, ASCE, in print

Roelvink, J.A.(1991). Modelling of cross-shore flow and morphology. Proc. Int. Conf. Coastal Sediments'91, ASCE, New-York, 603-617.

Roelvink, J.A.(1993a). Dissipation in random wave groups incident on a beach. Coastal Engineering,19, 127-150

Roelvink, J.A.(1993b). Surf-beat and its effect on cross-shore profiles. Ph. D. Thesis, Delft Univ. of Technology, The Netherlands.

Roelvink, J.A. and M.J.F. Stive(1989). Bar generating cross-shore flow mechanisms on a beach. J. geophysical Res.,vol. 94, n°C4, 4785-4800.

Sánchez-Arcilla,A., B.A. O'Connor, D.A. Roelvink and J.A. Jiménez(1994). The Delta-flume'93 experiments. Proc. Coastal Dynamics'94, ASCE, in print

Svendsen I.A.(1984a). Wave heights and set-up in a surf zone. Coastal Engineering,8(4),303-329

Svendsen I.A.(1984b). Mass flux and undertow in a surf zone. Coastal Engineering,8(4),347-366

Weggel J.R. (1972). Maximum breaker height for design",Proc. Int. Conf. on Coastal Eng., ASCE, 419-432

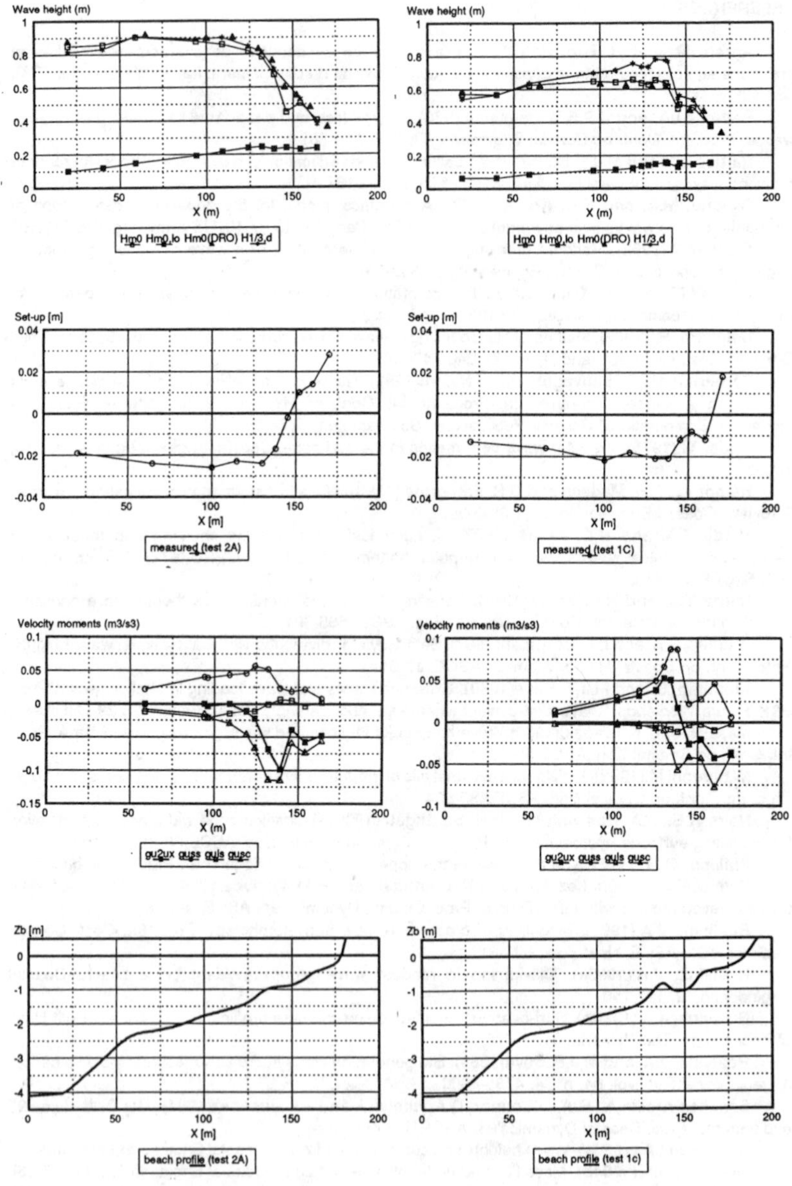

Figure 1 . LIP11D data
Test 2A

Figure 2 . LIP11D data
Test 1C

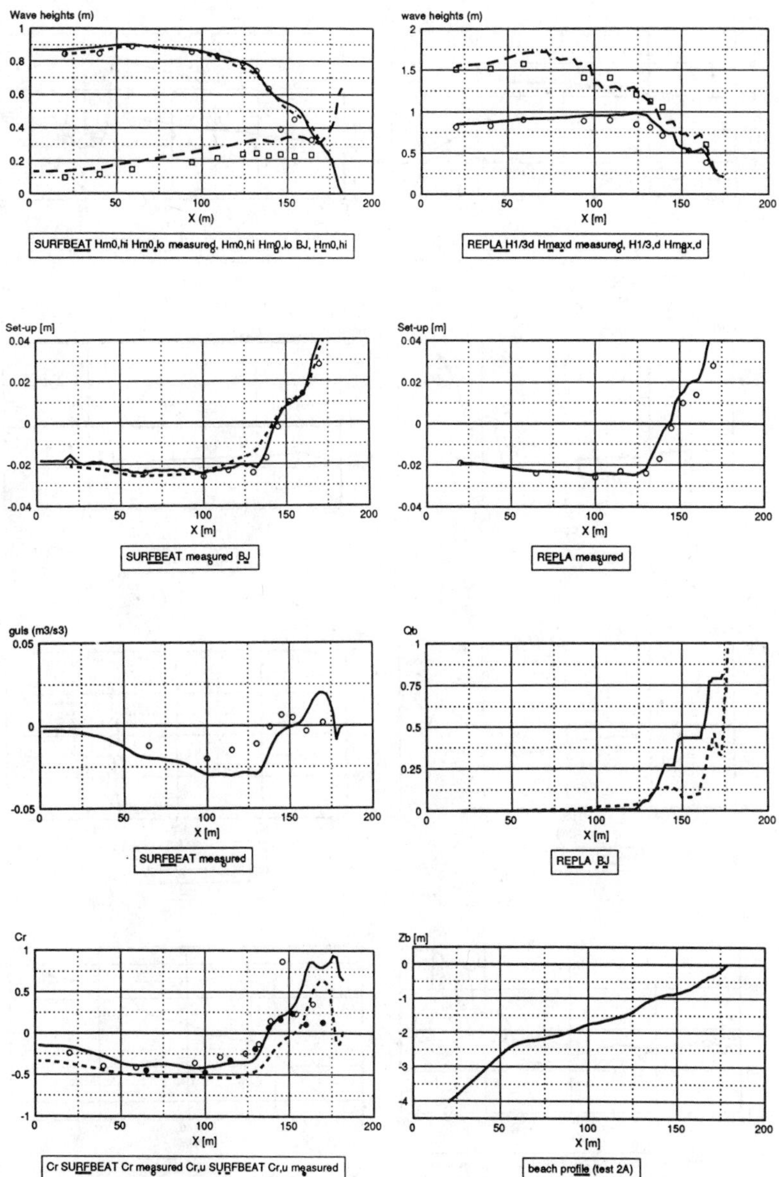

Figure 3 . Simulation of test 2A with parametrized spectral models

Figure 4 . Simulation of test 2A with a probabilistic model

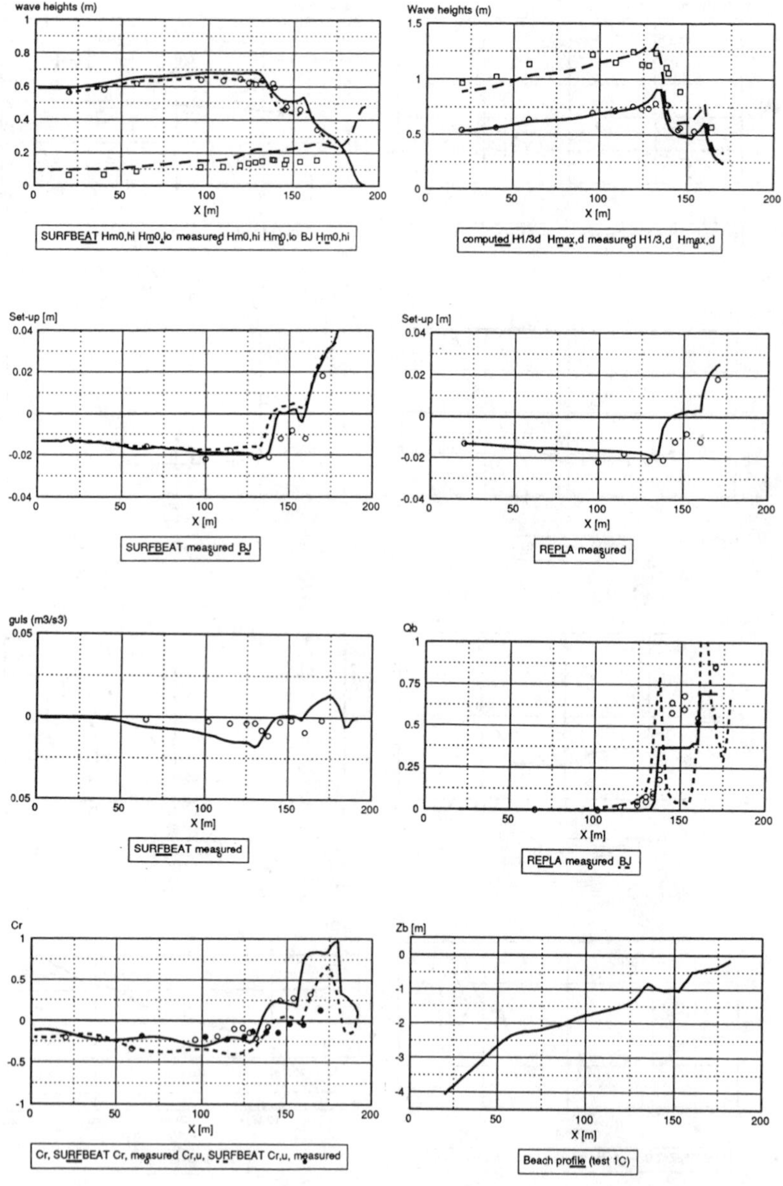

Figure 5 . Simulation of test 1C with parametrized spectral models

Figure 6 . Simulation of test 1C with a probabilistic model

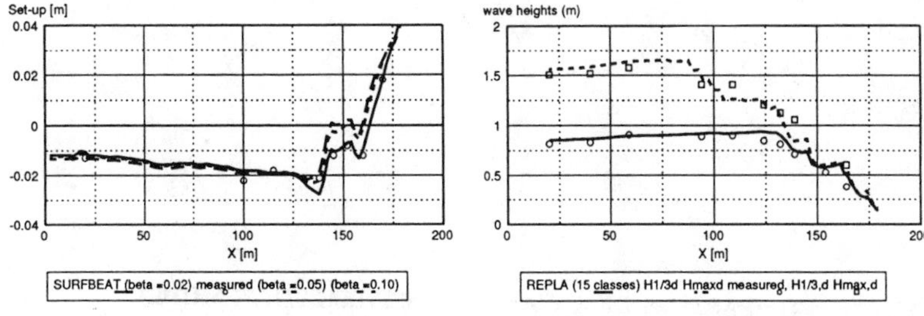

Figure 7 . Sensitivity tests with SURFBEAT (test 1C)

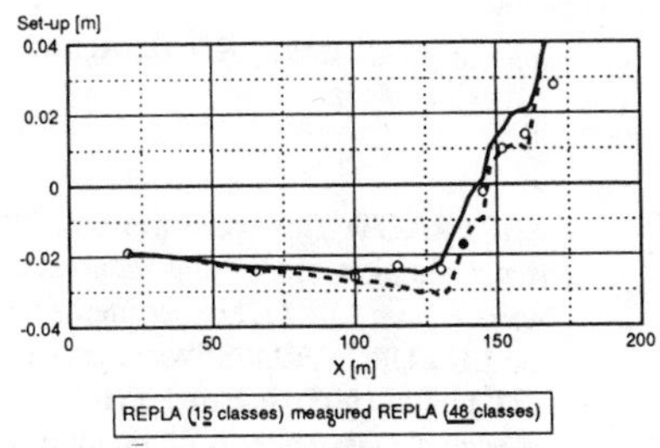

Figure 8 . Sensitivity tests with REPLA with 15 classes (test 2A)

COMPARISON OF A WAVE TRANSFORMATION MODEL WITH LIP-11D DATA

Francisco J. Rivero, Agustín S.-Arcilla[1] and *Dénes Beyer*[2]

ABSTRACT: This paper presents results of a comparison between a wave transformation model and measurements carried out in a prototype scale wave flume: LIP-11D experiments (Arcilla et al., 1994). The applied model, the LIM/UPC 'PROPS' wave propagation code (Rivero et al, 1993; Rivero and Arcilla,1993), is based on the formulation of (Battjes and Janssen, 1978) to evaluate wave energy dissipation. This model presents improvements with respect to the state-of-art in the calculation of the fraction of breaking waves, which are compared with data obtained from visual records of LIP-11D experiments. A slightly different way to include transition zone effects in the calculation of wave energy dissipation and radiation stress gradients is also proposed.

1. INTRODUCTION

The knowledge of the transformation of certain parameters of an incident random wave train across the surf zone is of paramount importance in the description of several hydrodynamic phenomena, such as wave-induced currents and sediment transport. Most of the existing formulations to estimate time-averaged properties for random waves separate the contribution of non-breaking waves to that of breaking waves. This makes it necessary to have knowledge of the fraction of breaking waves Q_b at a given location.

The dissipation model of Battjes and Jansen (1978) (hereafter referred to as B&J) for random breaking waves, which makes an implicit use of Q_b to calculate wave energy dissipation, has

---

[1] Associate professor and professor, Laboratori d'Enginyeria Maritima (LIM), Universitat Politecnica de Catalunya (UPC), Campus Nord (Mod. D-1), Gran Capita s/n,08034 Barcelona,Spain

[2] Graduate student, Faculty of Civil Engineering, Delft University of Technology, P.O.Box 5048, 2600 GA Delft, The Netherlands. On secondment for a Master thesis work to LIM/UPC as an Erasmus exchange student.

proved to be an excellent model to predict wave height decay across the surf zone. The separate prediction of Q_b, however, is not satisfactory.

This paper discusses the overall performance of the B&J parametric model to represent separately wave height decay and Q_b. A new approach to include transition zone effects in the wave transformation model is also presented.

2. THE BATTJES & JANSSEN (1978) DISSIPATION MODEL

Let us consider a record of N waves at a given location of mean water depth h, of which N_b are breaking waves. In a time-averaged sense, the wave energy dissipation rate is the average of the (N) energy dissipation rates of individual waves S_i:

$$S = \frac{\sum_{i=1}^{i=N} S_i}{N} \tag{1}$$

Assuming that dissipation takes place only for breaking waves (N_b), we have

$$S = \frac{\sum_{i=1}^{i=N_b} S_i}{N} = \left(\frac{\sum_{i=1}^{i=N_b} S_i}{N_b}\right)\left(\frac{N_b}{N}\right) = S_b Q_b \tag{2}$$

where S_b is the average energy dissipation rate of the N_b breaking waves only, and $Q_b = N_b/N$ is the fraction of breaking waves.

In the B&J approach, the following assumptions are made:

(a) S_b is equivalent to the energy dissipation rate of a periodic wave of the following characteristics:

- frequency f_p, which corresponds to the peak frequency of the incident wave spectrum.
- wave height H_m, which corresponds to the maximum stable wave height of periodic waves at the given mean water depth h. A Miche-type criterion was used in the original model to evaluate H_m:

$$H_m = \frac{0.88}{k_p}\tanh\left(\frac{\gamma}{0.88}k_p h\right) \tag{3}$$

Other formulations have also been considered in the literature (e.g. $H_m = \gamma h$ (Roelvink,1993)).

(b) The energy dissipation in a bore of height H can be written as:

$$S' = \frac{1}{4}\rho g \frac{H^3}{h}\sqrt{gh} \tag{4}$$

where ρ is the water density and g is the gravity acceleration.

(c) The energy dissipation of a periodic breaking wave in shallow water with frequency f_p and height H_m is modelled after an analogy with a bore:

$$S_b = \frac{S'}{L_p} = \frac{f_p S'}{\sqrt{gh}} = \frac{1}{4}\alpha' f_p \rho g \frac{H_m^3}{h} \tag{5}$$

α' is a calibration parameter of O(1) that accounts for the similarity between a fully breaking wave and a bore.

(d) The ratio H_m/h is of O(1) in the region where most dissipation occurs, reducing the above expression to:

$$S_b = \frac{1}{4}\alpha \rho g f_p H_m^2 \tag{6}$$

In this case, the parameter α may be interpreted as

$$\alpha = \alpha'(\frac{H_m}{h}) = (\frac{S_{breaking\ wave}}{S_{bore}})(\frac{H_m}{h}) \tag{7}$$

(e) To evaluate the fraction of breaking waves Q_b, the wave heights are assumed to have a Rayleigh distribution truncated at $H=H_m$, which means that all breaking (or broken) waves have the same height H_m:

$$\frac{1-Q_b}{-\log Q_b} = (\frac{H_{rms}}{H_m})^2 \tag{8}$$

in which $H_{rms}=(8*E/\rho g)^{1/2}$ is an energy-based measure of wave height.

The dissipation model of B&J gives, in general, a good representation of wave height decay due to breaking for a wide range of situations if appropriate parameters are used. Battjes and Stive (1985) calibrated the model using an extensive set of laboratory and field data for plane and barred beach profiles. In their calibration the parameter α was set to 1, and the parameter γ was determined to give an optimal fit to the experimental data. The obtained results indicated a dependence of γ on the deep-water wave steepness $s_o=(H_{rms}/L_p)_o$:

$$\gamma=0.5+0.4\tanh(33s_o) \tag{9}$$

However, it is important to remark that numerical simulations based on this model show that different combinations of α and γ (all compatible with the order of magnitude associated with the underlying physics of the problem) give similar, though not identical, wave height decay. As a matter of fact, a calibration procedure similar to the one used by Battjes and Stive with different α values could also be carried out (e.g. α=0.8, α=0.5, ...). This seems to indicate that, though the estimation of the total wave energy dissipation rate $S=S_bQ_b$ is correct in the model, it is not so for S_b and Q_b, separately. In such a case, what values of α and γ are the 'real' ones? This question will be adressed in section #4.

For practical purposes, the calibration coefficients as given by Battjes and Stive (α=1 and γ as in Eq.(9)) form an excellent basis to estimate wave energy dissipation, as long as the representation of wave height decay is the main objective. For other purposes a modified approach may be required.

3. INCLUSION OF TRANSITION ZONE EFFECTS

When waves break, it has been found that the wave energy dissipation does not occur immediately at the initial breaking point but at some distance beyond that breaker point. It is within this distance (known as the 'transition zone') that the decay of organized wave energy (identified as wave height decay) is converted into turbulent kinetic energy (the bulk of it being stored in the roller) which is not dissipated immediately (Roelvink and Stive, 1989). The inclusion of this lag -in time or in space- between production and dissipation of turbulent kinetic energy in wave propagation models has been shown to be important for the representation of several hydrodynamic phenomena in the surf zone (e.g. variation of mean water level, longshore current distribution, etc.).

As suggested by Svendsen (1984), the roller influence must be accounted for in the calculation of the total wave energy density E_{tot} and of the total wave momentum flux tensor $S_{ij,total}$:

$$E_{tot}=E_w+E_{roller}=\frac{1}{8}\rho g H_{rms}^2+E_{roller} \tag{10}$$

$$S_{ij}=S_{ij,w}+S_{ij,roller}=E_w[(1+G)\frac{k_ik_j}{k^2}+G\ \delta_{ij}]+\ 2E_{roller}\frac{k_ik_j}{k^2} \tag{11}$$

where G=2kh/sinh(2kh), ρ is the water density, g is the gravitational acceleration, H_{rms} is the *rms* wave height, **k** is the wavenumber vector, and δ_{ij} is the Kronecker delta.

The corresponding conservation equations for E_{tot}, E_w and E_{roller} are (Nairn et al, 1990):

$$\nabla\cdot[E_w\ C_g]+\nabla\cdot[E_{roller}C]=-D \tag{12}$$

$$\nabla\cdot[E_w\ C_g]=-S \tag{13}$$

$$\nabla\cdot[E_{roller}C]=S-D \tag{14}$$

where ∇ is the horizontal gradient operator, and C_g and C are the group and phase celerity, respectively, D is the total wave energy dissipation rate and S is the energy transfer rate from the organized wave motion (i.e. E_w) to the roller (i.e. E_{roller}). S is assumed to be a known function of the local wave parameters, as for instance in the B&J model.

The keypoint of the new approach presented here follows from the direct evaluation of E_{roller}, instead of expressing D in terms of E_{roller}, and solving the corresponding differential equation (12) afterwards [cf. (Roelvink and Stive, 1989; Nairn et al., 1990)]. An appropriate estimate for E_{roller} for random waves can be obtained by generalizing Svendsen's concept of a roller for regular waves, considering that the roller is generated only for the fraction of breaking waves (Q_b):

$$E_{roller}=\frac{1}{4\pi}\rho\frac{\omega^2}{k}AQ_b \tag{15}$$

where ω is the angular wave frequency and A is the area of the roller, which is assumed to be proportional to the energy density of organized motion (E_w):

$$A=\beta H_{rms}^2=\beta\frac{8E_w}{\rho g} \tag{16}$$

β is a calibration parameter of O(1), assumed to depend on the type of wave breaking (e.g. Iribarren's parameter).

Figure 1 shows a comparison of the resulting mean water level distribution (set-up/set-down) computed with the present model and that of (Nairn et al.,1990) with experimental data from (Stive, 1985). See also figure 4. Both models show improvements in the estimation of set-down, but fail in the evaluation of set-up near the waterline. This is probably due to an incorrect calculation of the radiation stress tensor $S_{ij,w}$ by linear wave theory.

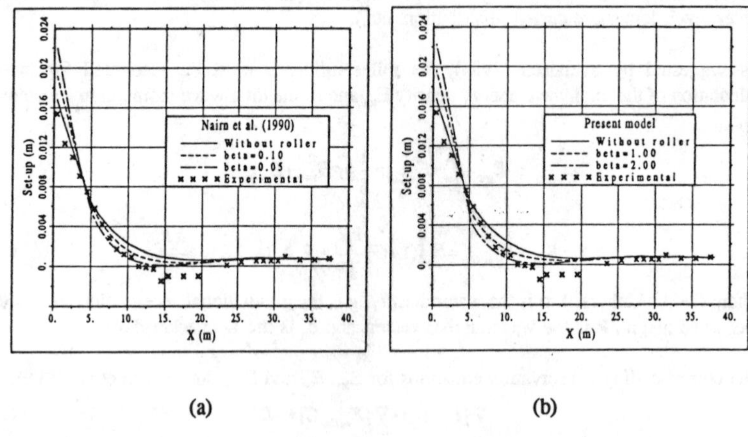

(a) (b)

Figure 1. Comparison between experiments (Stive,1985) and numerical results of set-up/set-down: (a) Nairn et al.,1990; (b) present model) . See also caption in figure 4.

The most relevant advantage of this model is its simplicity, since it does not need to solve any differential equation, as for instance in the aforementioned models. This makes it extremely simple to extend it to a 2DH formulation without increasing the complexity of the formulation (Eq.(11)). The main limitation comes from the need for a good representation of the fraction of breaking waves (Q_b).

For regular waves (H_{rms}=H), and to avoid a discontinuity of A (given by Eq.(13)) at the breaking point, the parameter β is allowed to vary continuously from 0 outside the breaking zone up to a given value β at the end of the transition zone. The width of the transition zone is estimated as in Southgate and Nairn (1993).

A comparison of the longshore current distribution computed with the 2DH mean flow model 'CIRCO' (Rivero, 1994) using this formulation with experimental results for Visser's (1984) test no.4 is given in figure 2. This case shows the importance of modelling transition zone effects to estimate the location of maximum longshore current velocity.

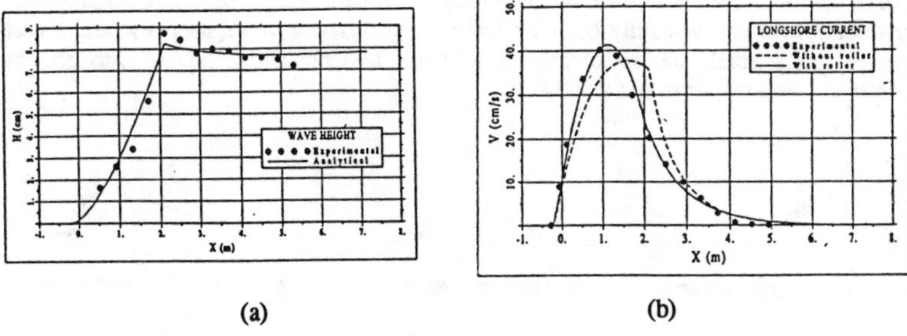

(a) (b)

Figure 2. Comparison between experimental and numerical results ('PROPS' and 'CIRCO' codes) for Visser's (1984) test no.4: (a) wave height, (b) longshore current velocity.

4. REPRESENTATION OF THE FRACTION OF BREAKING WAVES Q_b

As mentioned in section #2, different combinations of parameters α and γ in the B&J model give similar patterns of wave energy dissipation S (and, therefore, of wave height decay H_{rms} and set-up η), but a very different representation of the fraction of breaking waves Q_b. This latter variable appears to be strongly dependent on α. In the following a comparison between numerical and experimental results of H_{rms} and Q_b is presented for a limited number of cases (measurements of Q_b are very scarce). From this comparison some preliminary conclusions may be drawn with respect to the ability of B&J model to represent Q_b. Two different situations are considered:

(a) *Plane (or monotonic) beach profiles.* The B&J model with α=1 generally underpredicts Q_b values, especially for high steepness waves. Three test cases with measurements of H_{rms} and Q_b for random waves breaking on a 1:40 plane beach slope have been considered (Stive,1985; Stive,1986). Model simulations with α=1 and with an improved α value have been run; in each case the parameter γ was determined in each case so as to give the best fit to the experimental data. Table 1 gives a summary of the selected parameters.

Table 1. Selected values of α and γ for model comparisons with laboratory data of random waves breaking on a 1:40 plane beach slope (Stive,1985; Stive,1986).

Test	s_o	α	γ
MS10 (Stive, 1985)	0.010	0.55	0.55
Test 1 (Stive, 1986)	0.021	0.36	0.55
MS40 (Stive, 1985)	0.038	0.23	0.55

The obtained results appear to indicate that α is in general smaller than 1, and dependent on deep-water wave steepness $s_o=(H_{rms}/L_p)_o$ (i.e. increasing α for decreasing values of s_o). This is consistent with the physics and the definition of the parameter α (Eq.7), if one thinks on the resemblance between fully breaking waves with different wave steepness and a bore (figure 3). The small number of cases considered does not allow, however, to get any definite expressions for α and γ.

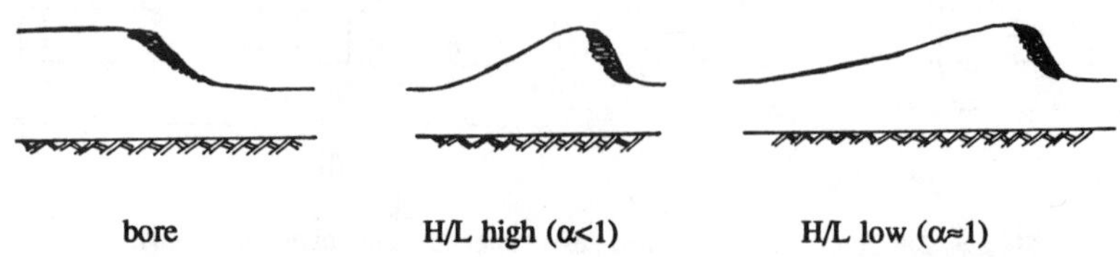

Figure 3. Sketch of a bore and breaking waves with high and low steepness.

The comparisons between model results and experiments show a good representation of Q_b if an appropriate value of α is used (see example in figure 4). In the shallower region (near the waterline), however, the model overpredicts Q_b.

(b) *Barred beach profiles*. The B&J model gives in general a wrong prediction of Q_b, even with improved values of α. Figure 5 shows a comparison between numerical results and experimental data for random waves breaking on a multiple bar beach profile (Stive,1986). Although the representation of wave height decay is quite satisfactory, the computed values of Q_b are not well predicted. Preliminary comparisons show two main discrepancies between predicted and measured Q_b, inherents to the B&J approach, that cannot be solved by any combination of the parameters α and γ:

(i) There exists, in general, a lag between the location of the maximum predicted Q_b -over the bar- and the maximum measured Q_b -onshorewards of the bar-, which means that some waves do not break over the bar, but after it. This lag appears to increase for decreasing wave steepness. A possible explanation for this could be the lag -in time- needed by waves to plunge and break once they 'feel the order of breaking' passing over the bar, which is larger for lower wave steepness.

(ii) The model is not able to reproduce the persistence of breaking that waves experience after the bar before reforming to non-breaking waves. Notice that the B&J method (Eq.8) determines Q_b entirely in terms of local parameters (essentially H_{rms}/h),and has no knowledge of wave behaviour at previous points along the profile. An attempt to overcome this limitation is given by Southgate and Wallace (1994).

5. COMPARISON WITH LIP-11D EXPERIMENTS

For testcases 1A, 1B and 1C of LIP-11D experiments the fraction of breaking waves Q_b has been determined from video records. During each hour of wave generation the video was positioned at different locations along the flume. At every location two estimates of Q_b were made: after ten and thirty minutes, respectively, from the start of wave generation. 100 waves were counted per estimate. The following criterion was adopted to recognize a

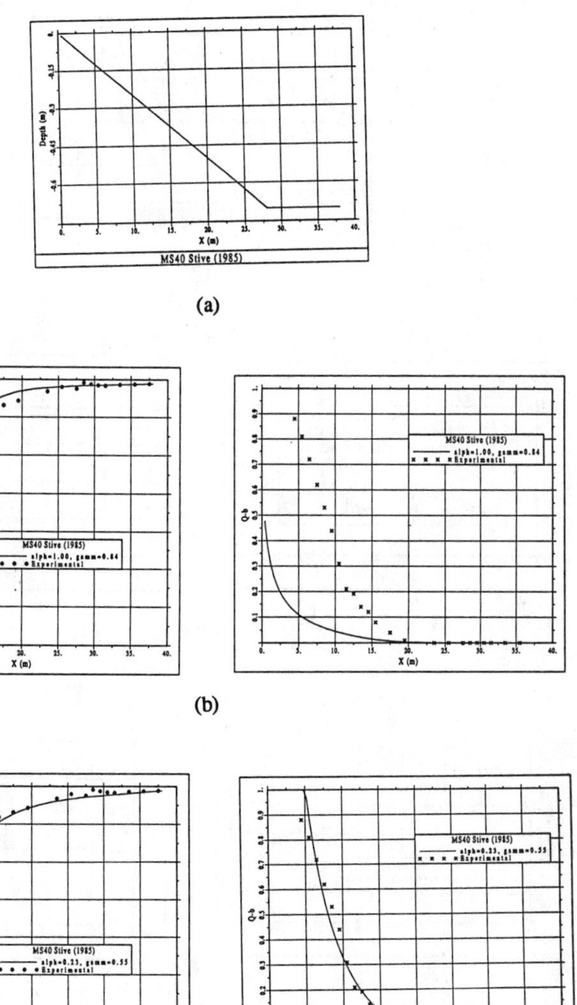

(a)

(b)

MS40 Stive (1985)

alph=0.23, gamm=0.55

Experimental

(c)

Figure 4. Comparison of numerical and experimental results for Stive (1985) test MS40: a) Bathymetry; b) H_{rms} and Q_b for α=1.0 and γ=0.84; c) H_{rms} and Q_b for α=0.23 and γ=0.55.

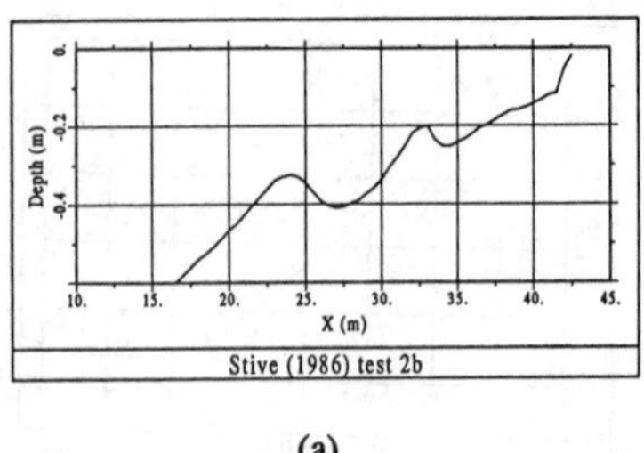

(a)

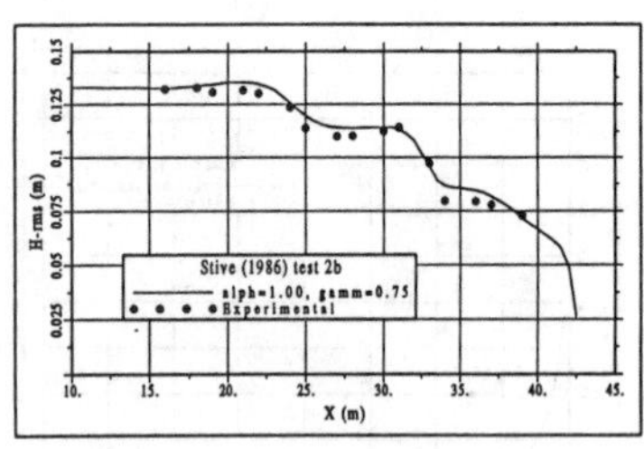

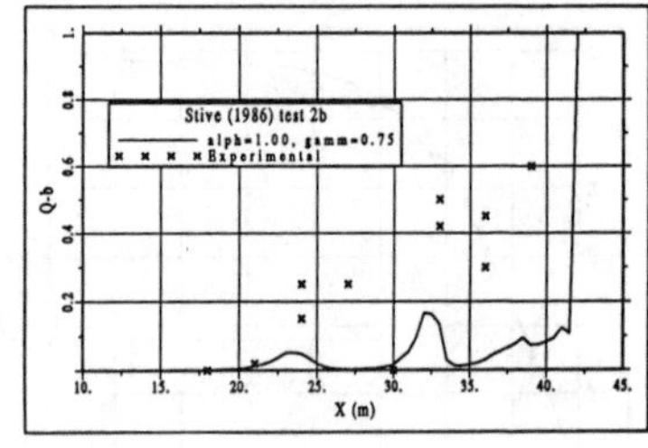

(b)

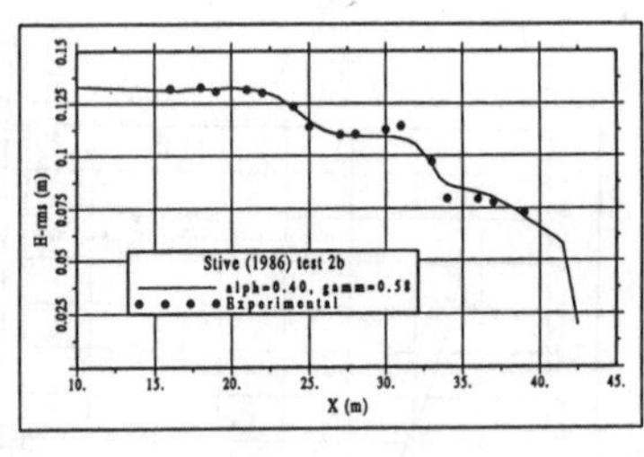

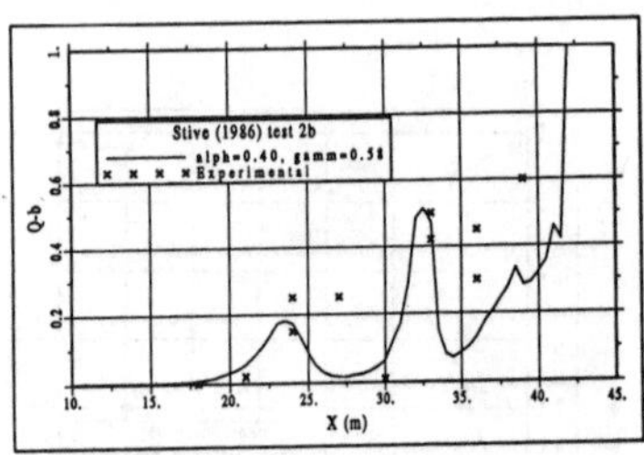

(c)

Figure 5. Comparison of numerical and experimental results for Stive (1986) test 2b: a) Bathymetry; b) H_{rms} and Q_b for α=1.0 and γ=0.75; c) H_{rms} and Q_b for α=0.40 and γ=0.58.

breaking (or broken) wave: if a wave crest passing a fixed point shows air-entrainment (i.e. there is foam on the wave crest), the wave is breaking (or broken). This is based on the finding that air-entrainment is a good indicator of potential energy dissipation.

It should be noted that some caution has to be taken when interpreting the data. Several factors influence the final result, of which subjectivity is the most evident. These include determining what kind of water movement can be called a wave, how to separate incident and reflected waves, and which waves break at the reference location.

For each testcase the comparison is performed for the first hour of the stationary period (hour 3 in test 1A, hour 7 in test 1B, hour 7 in test 1C). During these hours the wave generation was stationary and the bathymetry did not change significantly. In the first computation the parameterization was according to Battjes & Stive (1985). In the second computation the parameters were chosen in such a way that a good fit for both wave height decay and fraction of breaking waves could be achieved. For each testcase the results are discussed in the following.

Test 1A hour 3 (Figure 6)

α=1.00, γ=0.73:

The wave height distribution is predicted with a rms-error of 6.4%. The shoaling is slightly overpredicted, while the decay is clearly overpredicted. The Q_b is severely underpredicted.

α=0.35, γ=0.63:

Up to the bottom step (located at x=140 m) the wave height and Q_b are predicted well. After the step the Q_b decreases too rapidly, underpredicting the dissipation. The rms-error for the wave height is 8.3%.

Test 1B hour 7 (Figure 7)

α=1.00, γ=0.78:

The wave height is well predicted (rms-error 4.3%). Again the Q_b is severely underpredicted.

α=0.35, γ=0.58:

The prediction of the wave height remains good (rms-error 5.2%) but a good fit to the measured Q_b is difficult. It is not possible to predict both H and Q_b well up to the bar. After the bar the model shows a sudden decrease in Q_b while the measurements show an increase. The maximum and minimum Q_b at the trough are predicted offshorewards of the actual measurements(~10 m).

Test 1C hour 7 (Figure 8)

α=1.00, γ=0.55:

The prediction of the wave height is reasonable (rms-error of 8.4%). As in test 1A the wave shoaling is overpredicted and the wave height decay is too strong: it also starts too early. The Q_b values are predicted surprisingly well up to the bar, though the increase starts again too early.

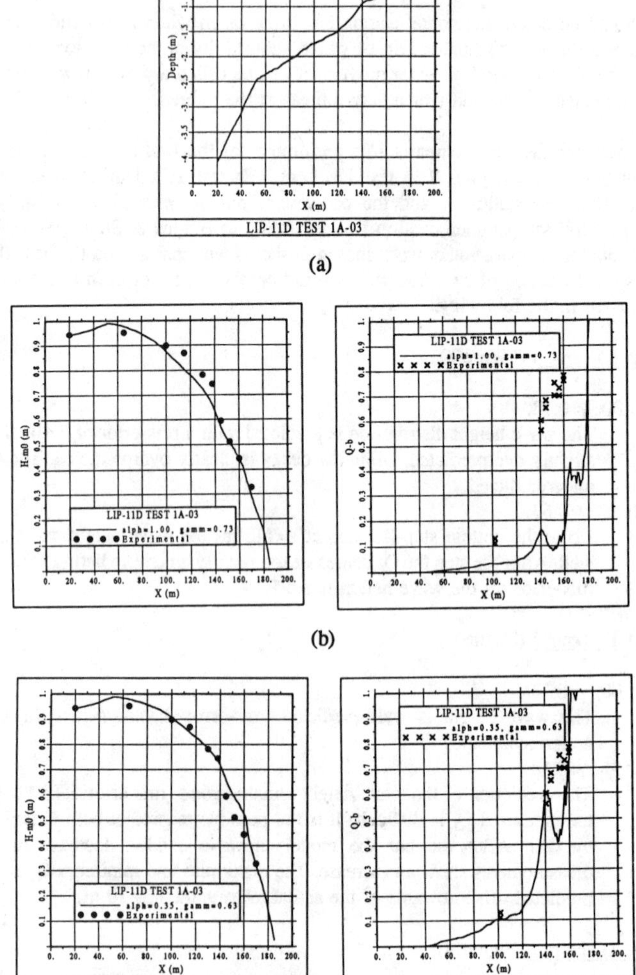

Figure 6. Comparison of numerical and experimental results for LIP-11D test 1A hour 3: a) Bathymetry; b) H_{m0} and Q_b for α=1.0 and γ=0.73; c) H_{m0} and Q_b for α=0.35 and γ=0.63.

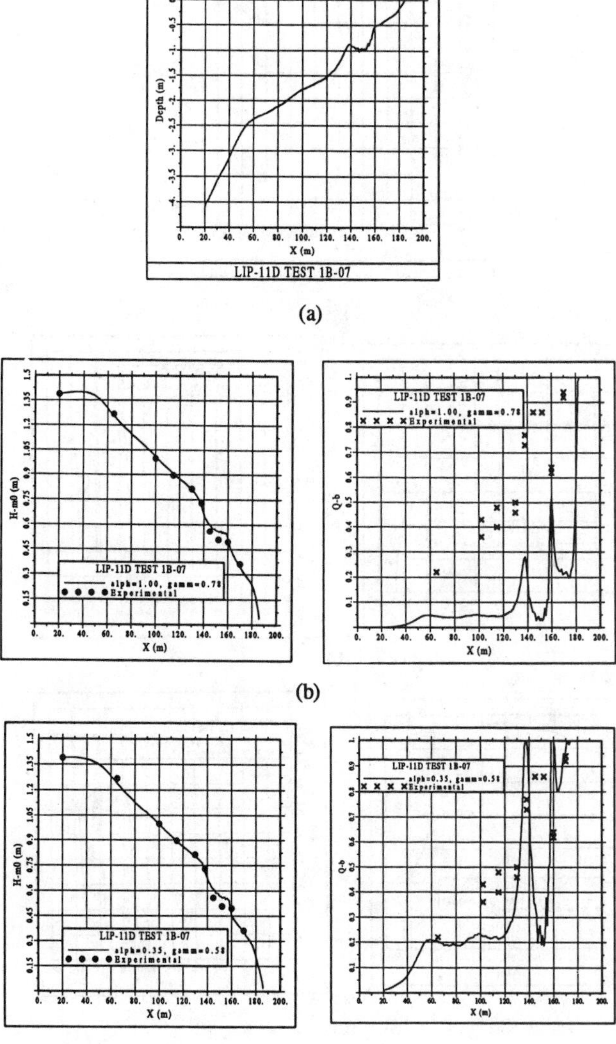

Figure 7. Comparison of numerical and experimental results for LIP-11D test 1B hour 7: a) Bathymetry; b) H_{m0} and Q_b for α=1.0 and γ=0.78; c) H_{m0} and Q_b for α=0.35 and γ=0.58.

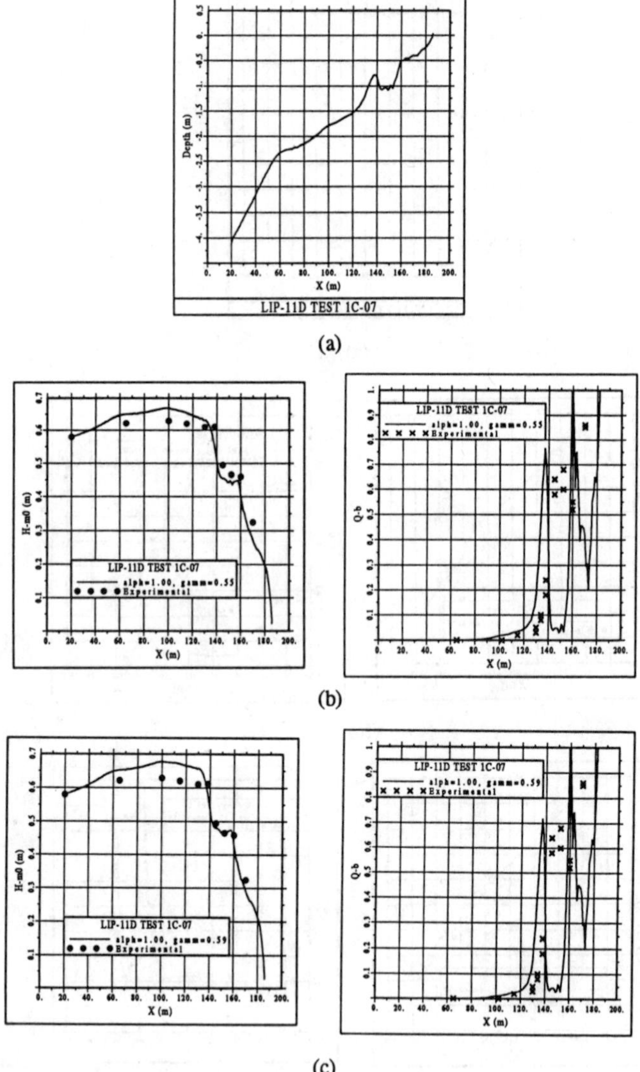

Figure 8. Comparison of numerical and experimental results for LIP-11D test 1C hour 7: a) Bathymetry; b) H_{m0} and Q_b for α=1.0 and γ=0.55; c) H_{m0} and Q_b for α=1.0 and γ=0.59.

α=1.00, γ=0.59:

The representation of Q_b is not improved for different values of the parameter α. The optimal fit of the wave height distribution gives a rms-error of 6.2%. The increase in Q_b is still predicted too far seawards (5-10 m). As in test 1B both the maximum (10-15 m) and the minimum (~ 10 m) values of Q_b are predicted too early.

6. CONCLUSIONS AND FUTURE RESEARCH

A new approach to include transition zone effects in a wave propagation model for random waves has been presented. It is based on a direct evaluation of the turbulent energy dissipation by generalizing Svendsen's concept of wave roller for the fraction of breaking waves Q_b. The proposed model is very simple to implement in a 2DH formulation since it does not need to solve additional differential equations.

A comparison between numerical simulations, based on the dissipation model of Battjes and Janssen (1978), and experimental data which include Q_b measurements has been performed. The obtained results seems to indicate that:

(1) For monotonic beach profiles a good representation of Q_b can be obtained if appropriate values of the parameters α and γ are used. Preliminary results indicate a dependence of α on deep-water wave steepness s_o (increasing α for decreasing s_o).

(2) For barred beach profiles model calculations show an incorrect prediction of Q_b over the bar, and the inability to reproduce the 'after bar' persistence of wave breaking observed in the measurements. These two features cannot be overcome by changing the model parameters α and γ.

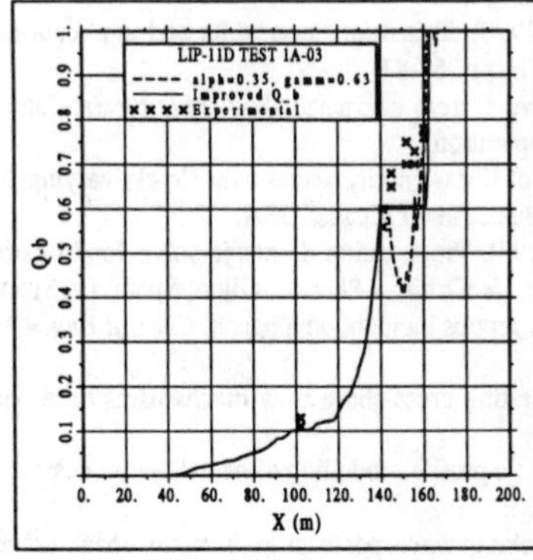

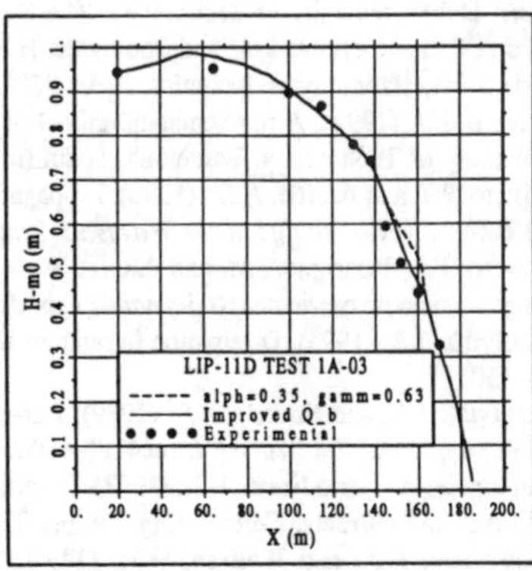

Figure 9. Comparison between numerical results and experimental data for LIP-11D test 1A with a heuristically improved Q_b representation.

Preliminary numerical results (Beyer,1994) suggest that the Battjes & Janssen model might be able to improve the representation of wave height decay if the calculation of Q_b could be improved, especially for barred beach profiles. Figure 9 shows a comparison between model calculations, with an 'ad hoc' improved representation of Q_b, and experimental data for the LIP-11D test 1A.

Another attempt to improve the overall model could be to consider the original formulation for the energy dissipation (Eq.5), in terms of H^3/h, instead of the utilized expression (Eq.6), in terms of H^2. For more details see Beyer (1994).

ACKNOWLEDGEMENTS

This research has been partly funded by the Commission of the European Communities Directorate General for Science, Research and Development, under contract no. MAS2-CT92-0027 as part of the G8 Coastal Morphodynamics Research Programme. The first two authors would also like to acknowledge financial support for research underpinning this work from the Programa de Clima Maritimo of the Spanish Ministry of Public Works. Mr. Dénes Beyer was visiting LIM/UPC with a grant of the ERASMUS exchange programme for foreign students of the EC.

REFERENCES

- Arcilla, A.S.-, Roelvink, J.A., O'Connor, B.A., Reniers, A. and Jiménez, J.A. (1994). The Delta Flume'93 Experiment. *Proc. of the Coastal Dynamics'94 Conference*, Barcelona, Spain.
- Battjes, J.A. and Janssen, J.P.F.M. (1978). Energy loss and set-up due to breaking of random waves, *Proc. 16th ICCE* pp. 569-587.
- Battjes, J.A. and Stive, M.J.F. (1985). Calibration and validation of a dissipation model for random breaking waves, *J. Geophys. Res., Vol.90, No.C5*, pp. 9159-9167.
- Beyer, D. (1994). Energy dissipation in random breaking waves: the probability of breaking. *Master thesis*, Delft University of Technology, The Netherlands.
- Nairn, R.B., Roelvink, J.A. and Southgate, H.N. (1990). Transition zone width and implications for modelling surfzone hydrodynamics, *Proc. 22nd ICCE*, pp. 68-81.
- Rivero, F.J. (1994). A mathematical model of wave-current interacted flow. *PhD thesis*, Catalonia University of Technology, Barcelona, Spain (in preparation).
- Rivero, F.J. and Arcilla, A.S.- (1993). Propagation of linear gravity waves over slowly varying depth and currents. *Proceedings of the Waves'93 Symposium*, New Orleans, USA.
- Rivero, F.J., Rodriguez, M. and Arcilla, A.-S. (1993). Propagación de oleaje sobre fondo variable y en presencia de corrientes, *II Jornadas Esp. de Ing. de Costas y Puertos*, Gijon, Spain. (in Spanish)
- Roelvink, J.A. (1993). Dissipation in random wave groups incident on a beach, *Coastal Eng. 19*, pp. 127-150.
- Roelvink J.A. and Stive, M.J.F. (1989). Bar-generating cross-shore flow mechanisms on a beach. *J. Geophys. Res., Vol.94,No.C4*, pp.4785-4800.
- Southgate, H.N. and Nairn, R.B. (1993). Deterministic profile modelling of nearshore processes. Part 1. Waves and currents, *Coastal Eng. 19*, pp. 27-56.
- Southgate, H.N. and Wallace, H.M. (1994). Breaking wave persistence in parametric surf zone models. *Proc. of the Coastal Dynamics'94 Conference*, Barcelona, Spain.
- Stive, M.J.F. (1985). A scale comparison of waves breaking on a beach, *Coast. Eng. 9*, pp. 151-158.
- Stive, M.J.F. (1986). A model for cross-shore sediment transport, *Proc. 20th ICCE*, pp. 1550-1564.
- Svendsen, I.A. (1984). Wave heights and set-up in a surf zone, *Coastal Eng. 8*, pp. 303-329.
- Visser,P.J. (1984). A mathematical model of uniform longshore currents and the comparison with laboratory data, *Comm. on Hydraulics,Dept. of Civil Eng.,Delft Univ. of Tech., Rep. 84-2*, 151 pp.

VERIFICATION OF THE SHEARSTRESS BOUNDARY CONDITIONS IN THE UNDERTOW PROBLEM

Marcel J.F. Stive[1)] and Huib J. De Vriend[2)]

ABSTRACT: As part of a revision of the quasi-3D approach for coastal currents, the two-dimensional undertow problem is being restudied. Since the first proposal for approaches in the 1980's progress has been made by several researchers (e.g. Deigaard and Fredsoe, 1989) on the potential importance of contributions neglected initially, such as the time-mean correlation between horizontal and vertical wave-induced velocities. The effects of wave-decay, sloping bottom and oscillatory bottom boundary layer on this term, initially neglected, have now been derived formally and included in the undertow description. Before checking the effects in comparisons with measurements of undertow profiles, it is considered essential to first concentrate on an improved, or at least verified, formulation of the shearstress boundary condition at wave trough level. Since it is momentarily not feasible to make direct measurements of the shear stress condition at wave trough level, only indirect verification is possible e.g. by expressing the shearstress condition in terms of the set-up gradient. Existing small scale laboratory and recently acquired large scale laboratory results provide the set-up gradient data for this approach. The verification leads to theoretical improvements, and provides insight into possible differences between large and small scale situations.

INTRODUCTION

The importance of the modelling of undertow in studies addressing the cross-shore hydrodynamics, sediment transport and morphology is now generally accepted. Since the first attempts of Dally (1980), Svendsen (1984), Stive and Wind (1986), De Vriend and Stive (1987), Stive and De Vriend (1987) in the 1980's, new findings have been introduced in the last years which may lead to improvements. Here it is attempted to highlight and describe some of these new findings, so as to enable undertow modellers to include the new ideas in the ongoing modelling efforts.

---

1) Laboratori d'Enginyeria Marítima, Universitat Politècnica de Catalunya, Barcelona, Spain; now at Netherlands Centre for Coastal Research, Delft University of Technology, c/o Delft Hydraulics, P.O. Box 152, 8300 AD Emmeloord, The Netherlands.

2) University of Twente, Section Civil Engineering and Management, P.O. Box 217, 7500 AE Enschede, The Netherlands.

PRINCIPLES OF SOLUTION FOR THE MEAN FLOW AND THE SET-UP

The present state-of-the-art local momentum equation most commonly used to solve the undertow reads:

$$\frac{\partial}{\partial z}\nu_t\frac{\partial U}{\partial z} = g\overline{\eta_x} + (\overline{u^2})_x - (\overline{w^2})_x + (\overline{uw})_z \tag{1}$$

where:
U is the undertow,
ν_t is the eddy viscosity,
η_x is the mean water level set-up, and
u,w are the horizontal and vertical wave-orbital velocities.

This equation is derived from the time-dependent horizontal and vertical momentum equations in a Eulerian framework, neglecting the advective acceleration terms for the time-mean flow. It is assumed to be valid in the full vertical domain, except for the region near the free surface. When assuming infinitely small wave amplitudes (e.g. assuming linear wave theory) it should be valid unto mean water level, but since the wave amplitudes are finite in the surfzone it may be more correct to limit its applicability to below the wave trough level.

Let us introduce the common assumption that the wave terms can be derived independently from the mean flow, i.e. we assume that there exists a sufficiently accurate wave theory to describe the wave terms which neglects the wave-current interaction. This of course a simplification, and we may state that a true break-through here would be achieved if we could tackle the problem with a Lagrangian approach in which waves and currents would be considered simultaneously, which at the same time would allow us to deal consistently with the near surface layer (NSL). However, until such an approach is developed we rely on the experience that the suggested approach is shown to yield sufficiently accurate results for the moment.

Let us further assume that the turbulence viscosity is constant over depth. Note that this is only for the clarity of the argumentation. The rationale which follows would also allow for a depth-varying viscosity, as long as this does not depend on the undertow solution, itself.

The above implies that we have one equation to solve for the undertow and the mean water level set-up. Integrating the equation twice yields:

$$\nu_t U = \frac{1}{2}g\overline{\eta_x}z'^2 + \int_0^{z'}\int_0^{z'}(\overline{u^2})_x\,dzdz - \int_0^{z'}\int_0^{z'}(\overline{w^2})_x\,dzdz + \int_0^{z'}(\overline{uw})\,dz' + C_1 z' + C_2 \tag{2}$$

This expression contains three unknowns, viz. the two constants of integration, C_1 and C_2 remaining in this expression, and the set-up gradient. We therefore need, in addition to this equation, three boundary conditions and/or constraints:

(1) no-slip condition at the bottom ($U = 0$), from which we can find C_2;
(2) shear stress condition at the transition from the middle layer (ML) to the NSL (τ_t given), from which we can find C_1;
(3) mass balance constraint (total mass flux in the lower layers balances that in the NSL), from which we can derive the set-up gradient.

The essential information which we need here is related to the NSL. In fact, condition (2) is based on the assumption that we know the shear stress at the lower end of the NSL from the momentum balance for this layer. Similarly, when formulating the constraint (3), we assume the total mass flux in the NSL to be known.

This route is suggested by Stive and De Vriend (1987) and also followed by Deigaard et al. (1991). The former derive a formal expression through a third depth integration, while the latter use an iteration procedure which sees to it that a set-up gradient is created such that the correct depth-average mass flux is created. The result should be the same.

Alternatively, one could use the depth-averaged horizontal mass and momentum equations, with the wave-induced radiation stresses and mass fluxes properly modelled, instead of giving τ_t and imposing constraint (3). In fact, the latter is even incorrect in 3-D situations, where the mass flux in the NSL is not necessarily compensated in the lower parts of the same water column (cf. De Vriend and Kitou, 1990). In section 3 we shall use the depth-averaged momentum equation, with a new expression for the radiation stresses, to derive an expression for τ_t. Note that in either case we face the problem of describing the NSL, via τ_t and the mass flux, or via the wave-related terms in the depth-averaged mass and momentum equations.

ROLE OF THE WAVE-RELATED TERMS IN THE MEAN MOMENTUM EQUATION

In De Vriend and Stive (1987) and in Stive and De Vriend (1987) the importance of the wave terms was not fully recognized. Deigaard and Fredsoe (1989) and De Vriend and Kitou (1990) have more fully pointed out the importance of the wave terms. A possible, zero-order approach to the magnitude of the wave-related terms is given below. This approach was essentially derived by Bijker et al. (1974).

Starting point is that -under the assumption that depth dependence of the orbital motion can be neglected- the horizontal orbital motion due to harmonic progressive waves (seaward positive) and the existence of an oscillatory turbulent boundary layer can be approximated by:

$$u = A[\cos(kx + \omega t) - e^{-\phi}\cos(kx + \omega t - \phi)] \tag{3}$$

where:

$A = a\,\omega\, /\, sinh\, (kh)$

$k = 2\,\pi\, /\, L$ is the wave number

$\omega = 2\,\pi\, /\, T$ is the wave frequency

$\phi = z'/\delta$

$\delta = \surd\,(2\,\nu_{tb}/\omega)$ the boundary layer thickness.

Note that the shallow water approximation is not essential to the rationale that follows. Essentially the same points can be proven without this assumption, it only takes more work (see De Vriend and Kitou, 1990).

Using the continuity equation and depth integration for the orbital motion by

$$w_z = -u_x \tag{4}$$

$$w = -\int_0^{z'} u_x dz \tag{5}$$

gives for the vertical orbital motion (cf. Bijker et al., 1974)

$$\begin{aligned} w = {} & -A_x\delta[\phi\cos(kx+\omega t)-\frac{e^{-\phi}}{2}\cos(kx+\omega t)(\sin\phi-\cos\phi) \\ & +\frac{e^{-\phi}}{2}\sin(kx+\omega t)(\sin\phi+\cos\phi)-\frac{1}{2}(\cos(kx+\omega t)+\sin(kx+\omega t))] \\ & +A(k+k_x x)\delta[\phi\sin(kx+\omega t)-\frac{e^{-\phi}}{2}\sin(kx+\omega t)(\sin\phi-\cos\phi) \\ & -\frac{e^{-\phi}}{2}\cos(kx+\omega t)(\sin\phi+\cos\phi)-\frac{1}{2}(\sin(kx+\omega t)-\cos(kx+\omega t))] \\ & +Ah_x[e^{-\phi}\cos(kx+\omega t)\cos\phi-e^{-\phi}\sin(kx+\omega t)\sin\phi-\cos(kx+\omega t)] \end{aligned} \tag{6}$$

where the first term is due to spatial variations in the orbital velocity amplitudes over a horizontal bottom, the second is due to the boundary layer streaming including a slope effect and the third term is due to spatial variations in the orbital velocity amplitudes over a sloping bottom. A term due to the spatial variation of the boundary layer thickness is neglected.

Based on these expressions for u and w one may -after time averaging- derive expressions for the wave terms in the momentum equation for the undertow. These results can be used to resolve the momentum equation for the undertow. We have done this according to the solution procedure described in section 1. Our analysis of these results is underway (Stive and De Vriend, 1994). Here, we concentrate on the derivation and verification of the resulting shear stress distributions.

SHEAR STRESS DISTRIBUTION OVER DEPTH AND AT THE TRANSITION TO THE NEAR SURFACE LAYER

The objective of this section is to compare the present approach with earlier results by Deigaard and Fredsoe (1989) and De Vriend and Kitou (1990). In the following we will consider two separate cases, i.e. (1) boundary layer dissipation and (2) breaking dissipation. If the problem is confined to boundary layer dissipation only, the shear stresses in the water column away from the bottom boundary layer should disappear, since the resulting wave motion can be described by potential flow (cf. Deigaard and Fredsoe, 1989 and 1992). In the case of wave dissipation due to breaking shear stresses can be generated where the energy dissipation is taking place. The latter is the most intense at the interface between the roller and the underlying body of water. We therefore can only introduce a shear stress condition at the transition between ML and NSL -as suggested in section 1- if this shear stress is related to the breaking-related energy dissipation.

CASE 1: Boundary layer dissipation, horizontal bottom, no breaking

If we consider the momentum equation outside the boundary layer, the wave-terms are to leading order:

$$\overline{u^2} = \frac{1}{2}A^2 \tag{7}$$

$$\overline{uw} = -\frac{1}{4}(A^2)_x\, z' \tag{8}$$

If we substitute this into the mean momentum equation (1) and integrate once, we find the shear stress distribution:

$$\tau(z') = \rho \nu_t \frac{\partial U}{\partial z} = \rho g \overline{\eta}_x\, z' + \frac{1}{4}\rho\, (A^2)_x z' + C_1 \tag{9}$$

Requiring the shear stress to be zero at the mean water level implies that we can solve for C_1 to yield:

$$\tau(z') = \rho g \overline{\eta}_x (z' - d_m) + \frac{1}{4}\rho (A^2)_x (z' - d_m) \tag{10}$$

This shows that no shear stresses will exist throughout the middle layer if the set-up gradient is given by:

$$\rho g \overline{\eta}_x d_m = -\frac{1}{4}\rho (A^2)_x d_m = -\frac{1}{2}E_x \tag{11}$$

(These equations are equivalent to equations (35) and (36) of Deigaard and Fredsoe, 1989). As pointed out by Deigaard and Fredsoe (1989), this is only one third of the equilibrium set-up commonly attributed to waves with boundary layer dissipation. By inserting the above result in the depth mean momentum balance equation:

$$\frac{dS_{xx}}{dx} + \rho g d_m \overline{\eta_x} + \tau_b = 0 \tag{12}$$

and using the shallow water approximation:

$$S_{xx} = S_{xx,p} + S_{xx,u} = \frac{1}{2}E + E \tag{13}$$

it follows that the bottom shear stress must balance the other two thirds of the radiation stress gradient. In the present case this can only be the shear stress due to the boundary layer streaming, which is acting in the same direction as the radiation stress gradient, indeed.

CASE 2: Wave breaking dissipation, horizontal bottom, no boundary layer dissipation

Again we use the local mean momentum balance equation (1) as the starting point, but now choose a shear stress τ_t at the transition between ML and NSL as a boundary condition. The idea is that in this case the shear stress can be maintained due to the NSL related dissipation due to wave breaking. Furthermore, we adopt the leading order approximations for the wave terms away from the bottom boundary layer:

$$\overline{u^2} = \frac{1}{2}A^2 \tag{14}$$

$$\overline{uw} = -\frac{1}{4}(A^2)_x z' \tag{15}$$

Assuming ν_t constant over the layers yields:

$$\tau(z') = \rho\nu_t \frac{\partial U}{\partial z} = \rho g \overline{\eta_x}(z' - d_m) + \tau_t + \frac{1}{4}\rho(A^2)_x(z' - d_m) \tag{16}$$

and

$$\tau_b = -\rho g d_m \overline{\eta_x} + \tau_t - \frac{1}{4}\rho(A^2)_x d_m \tag{17}$$

Since we consider wave breaking dissipation only, the mean bottom shear stress τ_b should be zero. Note that for the moment we assume that the middle layer is actually having an upper boundary at the mean water level d_m.

The shear stress τ_t may be resolved between the τ_b equation and the depth mean horizontal momentum balance equation. In order to do this we need to introduce an expression for the radiation stress which at least needs to be extended with the roller effect. Following Svendsen (1984) Deigaard and Fredsoe (1989) and De Vriend (1993) suggest the shallow water approximation

$$S_{xx} = S_{xx,p} + S_{xx,u} = \frac{1}{2}E + [E + \rho\frac{Rc}{T}] \tag{18}$$

where R is the roller area, empirically approximated by $R=0.9H^2$.

The above implies that the set of equations available to resolve the shear stress τ_t reads:

$$\frac{dS_{xx}}{dx} + \rho g d_m \overline{\eta_x} = 0 \tag{19}$$

$$\frac{dS_{xx}}{dx} = \frac{3}{2} E_x + \frac{\rho}{T} (Rc)_x \tag{20}$$

$$\tau_b = -\rho g \overline{\eta_x} d_m + \tau_t - \frac{1}{4} \rho (A^2)_x d_m = -\rho g \overline{\eta_x} d_m + \tau_t - \frac{1}{2} E_x = 0 \tag{21}$$

which yields:

$$\tau_t = -E_x - \frac{\rho}{T} (Rc)_x \tag{22}$$

and

$$\tau(z') = -E_x (1 + \frac{d_m - z'}{2d_m}) - \rho g \overline{\eta_x} (d_m - z') - \frac{\rho}{T} (Rc)_x \tag{23}$$

These equations are equivalent to equations (56) and (58) of Deigaard and Fredsoe (1989), when introducing the relation

$$E_x = \frac{1}{c} (E_f)_x = -\frac{D}{c} \tag{24}$$

where E_f is the energy flux and D is the dissipation due to wave breaking.

In the above the level of transition between ML and NSL is assumed to be at mean water level. In practice, the finite amplitudes in the surf zone make it more convenient for us to apply τ_t at the trough level, which will change the above derivations somewhat to yield:

$$\tau_t (1 + \frac{3(d_m - d_t)}{2d_t}) = -E_x - \frac{\rho}{T} (Rc)_x \tag{25}$$

EXPRESSING THE SHEAR STRESSES IN TERMS OF THE DISSIPATION

Since from the above we have shown that shear stresses above the wave boundary layer can only be generated in wave breaking related dissipative waves, it is useful and practical to express the shear stresses in terms of the dissipation rather than in terms of the energy density gradients. This will prevent driving mechanisms in the vorticity free situation, i.e. in the shoaling region possibly in combination with the boundary layer induced dissipation.

We start with noting that the now commonly used wave decay models, either for periodic or random waves, are calibrated and validated by using the observed decay of potential energy. In the stationary case these models solve the energy flux equation

$$\frac{\partial E c_g}{\partial x} + D_f = 0 \tag{26}$$

where D_f is a fictitious dissipation balancing the potential energy decay. So, effectively these models describe the potential energy decay due to wave breaking, in which we assume that this includes the potential energy due to the existence of the roller.

If we now adopt the common assumption that the kinetic energy flux due to the organized wave motion equals that of the potential energy flux we may express the energy balance equation including the roller related kinetic energy flux as

$$\frac{\partial 2E_p c_g}{\partial x} + \frac{\partial E_r c}{\partial x} + D_r = 0 \tag{27}$$

where D_r is now the real dissipation and the kinetic roller energy $E_r = {}^1/_2\, \rho\, Rc/T$.

By using relation (34) we may now express the real dissipation as follows

$$\frac{D_r}{c} = -E_x - \frac{\rho}{T}(Rc)_x = \frac{D_f}{c} - \frac{\rho}{T}(Rc)_x \tag{28}$$

Using the empirical result for the roller area, the roller decay related contribution can be expressed as:

$$\frac{\rho}{T}(Rc)_x \simeq 0.57k\,(Ed_m)_x \tag{29}$$

These results may be used to express the shear stresses in terms of the dissipation as follows

$$\tau_t = \frac{D_r}{c} = \frac{D_f}{c} - \frac{\rho}{T}(Rc)_x = \frac{D_f}{c} - 0.57k\,(E^* d_m)_x \tag{30}$$

and

$$\tau(z') = \frac{D_f}{c}\,(1 + \frac{d_m - z'}{2d_m}) - \rho g \overline{\eta}_x (d_m - z') - 0.57k\,(E^* d_m)_x \tag{31}$$

where E^* is the energy density of those waves which are breaking, i.e. which are effectively contributing to the breaking related dissipation. How to implement this in situations for random waves and for periodic waves will be addressed furtheron.

VERIFICATION OF THE SHEAR STRESS AT THE TRANSITION TO THE NEAR SURFACE LAYER

Since it is momentarily not feasible to make direct measurements of the shear stress at the transition between ML and NSL, only indirect verification is possible, e.g. by using relation (30) between τ_t and the mean wave set-up gradient. We emphasize that these two quantities are strongly related, which is not surprising since in both quantities the momentum properties of the NSL are embedded. Where τ_t translates the momentum decay of the NSL to the lower layers, the mean set-up includes the momentum decay of the NSL in the total decay. We may thus conclude that our ability to predict the mean water level set-up correctly indicates our ability for the prediction of the NSL momentum decay and therewith of τ_t. This is the topic addressed in this section both for random and periodic waves.

As a starting point for both cases we take the mean momentum balance and the expression derived for S_{xx}:

$$\frac{dS_{xx}}{dx} + \rho g d_m \overline{\eta_x} = 0 \tag{32}$$

$$\frac{dS_{xx}}{dx} = \frac{3}{2} E_x + \frac{\rho}{T} (Rc)_x = \frac{3}{2} E_x + 0.57k (E^* d_m)_x \tag{33}$$

Without the contribution of the roller to the radiation stress as expressed by the second term in equation (42) the problem reduces to the classical solution. For this it is well-known that the observed delay in the set-up gradients compared to the wave height gradients is not resolved. The roller term can be introduced to resolve this deficiency.

CASE 1: Periodic waves

In the case of periodic waves breaking on a slope there is a sharp transition in the fictitious energy flux as detected by the potential energy or wave height decay. If we would interpret this to be an instantaneous transition to a fully dissipative breaking mode, this implies that an instantaneous increase would be introduced in the radiation stress through the roller contribution to the radiation stress in expression (42). As described by Nairn et al. (1990) the physical mechanism to prevent this is the existence of a transition region, where the rollers are developed gradually as the potential energy decay sets in. It is only after the transition region that the fully developed breaking waves create rollers of which the properties, such as their kinetic energy, are proportional to the potential energy decay.

Here, we will not address possible modelling approaches to describe the transition zone phenomenon. Instead we will show that for random breaking waves the need to tackle the transition zone phenomenon is less important.

CASE 2: Random waves

In the case of random breaking waves there exists a gradually increasing fraction of breaking waves as the depth limitation starts to exert its influence on the wave field. Physically, this implies that while the total potential energy is still undergoing shoaling effects a fraction of the waves is dissipative due to wave breaking. If we assign a roller related kinetic energy to the breaking wave fraction, we will only gradually increase the radiation stress and therewith prevent an instantaneous increase of the radiation stress, an effect that would exist if all waves would be assumed to be in a full breaking mode. It so appears that to a first approximation there is no need to tackle the transition zone effect, because of the gradual increase of the dissipation.

So, in the case of random waves, we suggest to introduce $E^* = Q_b E_{br}$, so that:

$$\frac{dS_{xx}}{dx} = \frac{3}{2} E_x + k(Q_b d_m E)_x \tag{34}$$

where Q_b is the breaking wave fraction[1)], and E_{br} is the energy density of the breaking waves.

1) An initial parameterization of Q_b which we have used for a long time, and also here, reads:
$Q_b = 7\ Q_{b,B\&J}$ for $Q_{b,B\&J} \leq 0.1$
$Q_b = 1. - 548. (.3 - Q_{b,B\&J})^{4.67}$ for $0.1 < Q_{b,B\&J} < 0.3$
$Q_b = 1.$ for $Q_{b,B\&J} \geq 0.3$
where $Q_{b,B\&J}$ is the breaking wave fraction according to the Battjes and Janssen (1978) model, of which it is well-known that this significantly underestimates the breaking wave fraction.

Implementation of this in the mean horizontal momentum balance and applying this to existing small scale laboratory and recently acquired large scale laboratory results is being undertaken. An initial result is shown in Figure 1 (Battjes and Janssen, testcase 15). The upper part of Figure 1 indicates the increase of and shift in gradient change of the radiation stress due to the existence of the roller. As a consequence the set-up gradient undergoes a spatial delay leading to an improved prediction of the mean water level gradient, which indirectly implies a verification of the trough level shear stress (lower part of Figure 1).

Our result seems to contradict the findings of Nairn et al. (1990), who studied the so-named transition zone effects in detail. They found it necessary to introduce transition zone effects to arrive at a similar result. We suspect, however, that they haven't determined the relative effect of an approach without and with transition zone effects for the random wave case. Apparently, in the random wave case the present approach leads already to an important improvement. We note that the essential assumption in the present approach is that the roller energy can be expressed in locally determined variables, such as the fraction of breaking waves and the heights of the breaking wave fraction. Obviously, further improvements could be expected if nonlocal modelling approaches are introduced, such as the transition zone approximation of Nairn et al. (1990). Our present preference, however, would be to directly address the improvement of the breaking wave fraction with "transition" or memory effects such as addressed by Lippmann and Thornton (1993, these proceedings).

ACKNOWLEDGEMENTS

A large part of this paper is based on work in the "G8 Coastal Morphodynamics" research programme, which is funded partly by the Commission of the European Communities in the framework of the Marine Science and Technology Programme (MAST), under contract no. MAS2-CT92-0027. The work is co-sponsored by Delft Hydraulics, in the framework of the Netherlands Centre for Coastal Research.

REFERENCES

Battjes, J.A. and Janssen, J.P. (1978). Energy Loss and Set-up due to Breaking of Random Waves. Proc. 16th Conf. on Coastal Eng, ASCE, pp 569-588.

Bijker, E.W., Kalkwijk, J.P.Th. and Pieters, T. (1974). Mass transport in gravity waves on a sloping bottom. Proc. 14th Conf. on Coastal Eng, ASCE, pp 447-465.

Dally, W.R. (1980). A numerical model for beach profile evolution. M.Sc. Thesis, Dep. Civil Eng., Univ. Delaware.

Deigaard, R. and Fredsoe, J. (1989). Shear Stress Distribution in Dissipative Water Waves. Coastal Engineering, 13, 357-378.

Deigaard, R., Justesen, P. and Fredsoe, J. (1991). Modelling of undertow by a one-equation turbulence model. Coastal Engineering, 15, 431-458.

De Vriend, H.J. (1993). Quasi-3D nearshore currents revisited. Workshop Quasi-3D flow modelling, Barcelona.

De Vriend H.J. and Kitou, N. (1990). Incorporation of wave effects in a 3D hydrostatic mean current model. Proc. 22nd Coastal Eng. Conf., ASCE, 1005-1018.

De Vriend, H.J. and Stive, M.J.F. (1987). Quasi-3D Modelling of Nearshore Currents. Coastal Engineering, 11, 565-601.

Fredsoe, J. and Deigaard, R. (1992). Mechanics of coastal sediment transport. Advanced Series on Ocean Engineering, Vol. 3, World Scientific, pp 369.

Lippmann, T.C. and Thornton, E.B. (1994). The spatial distribution of wave rollers and turbulent kinetic energy on a barred beach. Proceedings Coastal Dynamics 1994.

Nairn, R.B., Roelvink, J.A. and Southgate, H.N. (1990). Transition zone width and implications for modelling surfzone hydrodynamics. Proc. 22nd Conf. on Coastal Eng., ASCE, pp 68-81.

Stive, M.J.F. and Wind, H.G. (1986). Cross-shore mean flow in the surf zone. Coastal Engineering, 10, 325–340.

Stive, M.J.F. and De Vriend, H.J. (1987). Quasi-3D nearshore current modelling: wave-induced secondary currents. Proc. Special Conf. on Coastal Hydrodynamics, ASCE, 356-370.

Svendsen, I.A. (1984). Wave Heights and Set-up in a Surf Zone. Coastal Engineering, 8, 303-330.

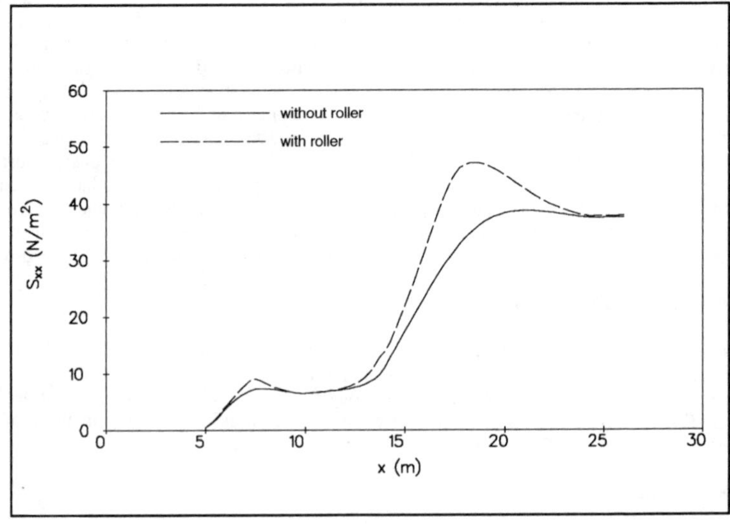

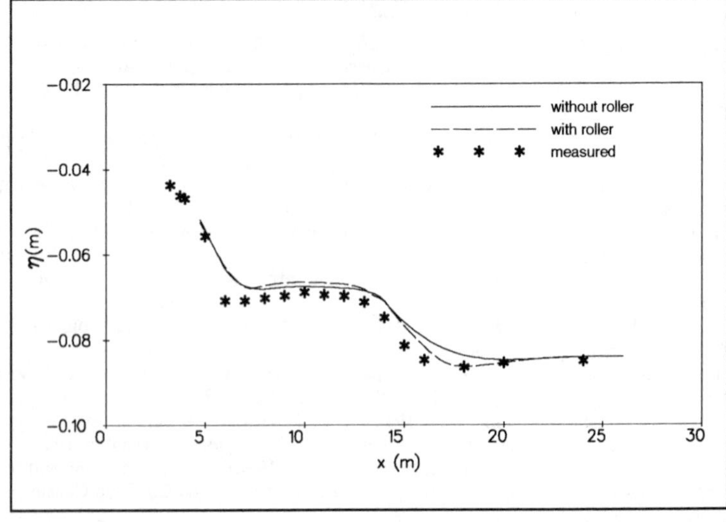

Figure 1 Radiation stress and set-up variation for test case 15 of Battjes and Janssen (1978)

BREAKING WAVE PERSISTENCE IN PARAMETRIC SURF ZONE MODELS

Howard N Southgate[1] and Helen M Wallace[1]

ABSTRACT: A weakness of parametric surf zone models of the Battjes and Janssen (1978) type for the dissipation of random breaking waves, is the prediction of the fraction of breaking waves over non-monotonic beach profiles. These formulations determine the fraction of breaking waves entirely in terms of local values of depth and wave height, and no account is taken of breaking behaviour that has occurred further seawards. The procedure described here attempts to overcome this limitation by introducing a 'persistence length' for breaking waves, beyond which the waves will reform to a non-breaking state. The method is simple and robust and can be easily incorporated into existing algorithms of the Battjes and Janssen type. Tests against measured values of the fraction of breaking waves in the LIP-11D experiment in the Delft Delta Flume are shown.

INTRODUCTION

The Battjes and Janssen (1978) (hereafter referred to as BJ) formulation of the dissipation of random breaking waves has proved to be a popular framework for parametric surf zone models of wave energy decay and morphodynamic development. The method assumes a Rayleigh distribution of wave heights with a truncation at the local breaker height. All broken waves are assumed to have a height equal to the breaker height, and therefore the broken waves appear as a delta function at the local breaker height in the wave height probability distribution, as shown in Figure 1. A tidal bore analogy is used for the dissipation of these broken waves. Computational models based on BJ involve the transformation of a single representative wave height (the root-mean-square value) across the nearshore zone.

---

[1] HR Wallingford Ltd, Wallingford, Oxfordshire, OX10 8BA, UK

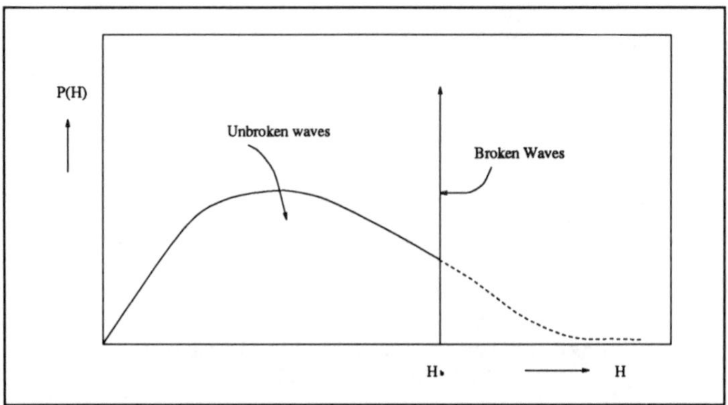

Figure 1 Rayleigh distribution of wave heights, with truncation at the breaker height

In the BJ method the fraction of breaking waves, Q_b, is calculated from known values of the rms wave height (H_{rms}) and the breaker height ($H_b = \gamma h$ where h is the water depth and γ the breaker index),

$$\frac{1 - Q_b}{-\ln Q_b} = \left(\frac{H_{rms}}{H_b}\right)^2 \qquad (1)$$

This expression works quite well for moderate or steep monotonic slopes. However, if the profile is non-monotonic or nearly flat there exists the possibility that waves, once broken (for example over a bar), will reform to non-breaking waves. To take account of the possibility of wave reformation in calculating Q_b, it is necessary to have knowledge of wave behaviour at previous points along the profile. This is not possible through Equation 1 since this equation determines Q_b entirely in terms of parameter values at the point of interest.

The following method attempts to improve Q_b predictions, taking account of previous breaking wave behaviour, in a simple algorithmic scheme that can be readily incorporated into existing models based on the BJ approach.

METHOD

One implication of determining Q_b entirely in terms of local variables (Equation 1) is that broken waves have zero persistence length. In other words, once a wave has started to break, it immediately stops breaking. This behaviour tends to be masked on monotonically decreasing bed slopes because the shallowing depths cause breaking to be continuously re-triggered, giving the appearance of a persisting breaking wave. However, the true nature of Equation 1 becomes apparent by considering a region of shallow water followed by a sharp increase in depth. Equation 1 will predict substantial breaking (high Q_b) in the shallow water region but there will be a sudden change to zero or very small Q_b in the deep water immediately following the shallow water region. This behaviour is contrary to observations that broken waves do persist for some distance even if depths are increasing.

A method that incorporates breaking wave persistence needs to make the distinction between newly breaking waves at a given point, and waves which have already broken further offshore and have persisted as far as the given point. The distinction is needed because the persistence length needs to be calculated for the latter types of wave to determine whether or not they are still broken at the given point. (In the BJ method, all breaking waves are newly breaking since the method implies zero persistence length).

Figure 2 shows the notation used. A nearshore profile (with a bar to illustrate the breaking persistence more clearly) is subdivided into a number of equally-spaced computational grid points of width Δx. The total fraction of breaking waves at a general point j is denoted by Q_j (in what follows the subscript b is dropped for clarity). The proportion of Q_j which represents newly breaking waves at point j is denoted by ΔQ_j, and the proportion which represents earlier breaking waves which have persisted to point j is therefore $Q_j - \Delta Q_j$.

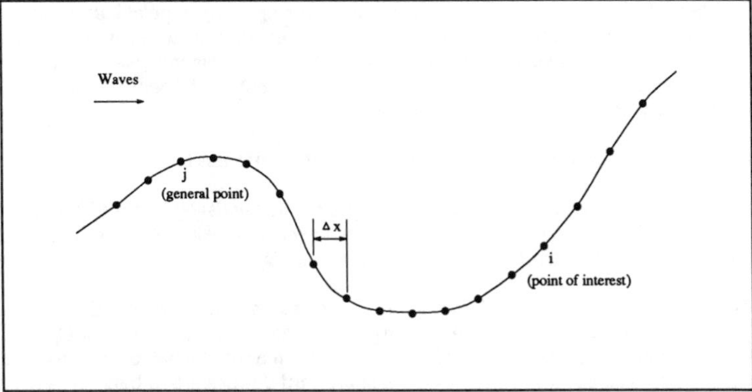

Figure 2 Notation of grid points along a profile

The method involves four steps which, taken together, constitute an algorithmic procedure that can be incorporated by simple adaptations to existing algorithms.

Physics of Breaking Wave Persistence

Knowledge of the physics of breaking wave persistence is needed to relate the persistence length to some properties of the waves when they are breaking initially, and their subsequent transformation. In the tests in this paper a simple plausible relationship is made, that the persistence length of a wave breaking at a point j is proportional to the breaker height at that point (H_{bj}). This is sufficient to illustrate the effect of inclusion of breaking persistence, and may well be sufficient for practical predictions as well. More detailed approaches, involving explicit consideration of breaker roller energy and momentum could be considered, (for example Rivero et al (1994), Lippmann and Thornton (1994) in this issue). However, all approaches would follow the same algorithmic procedure described in the following steps.

Determination of Active Grid Points for a Point of Interest, i

Relative to a point of interest i, all previous (ie seawards) grid points are designated either 'active' or 'inactive'. Active grid points are those in which

(a) new breaking takes place (ie $\Delta Q_j > 0$) <u>and</u>
(b) broken waves persist from that point to the point of interest, i.

A quantity $\Delta Q_{j,\,active}$ is defined as equal to ΔQ_j if point j is active, otherwise it is set to zero

$$\begin{aligned} \Delta Q_{j,\,active} &= \Delta Q_j \quad \text{if } (i-j)\Delta x < \beta H_{bj} \\ &= 0 \quad \text{if } (i-j)\Delta x \geq \beta H_{bj} \end{aligned} \qquad 2$$

The quantity βH_{bj} is the persistence length of waves breaking at point j, and β is the proportionality constant. The sum of $\Delta Q_{j,\,active}$ over all grid points from 1 to i-1 represents the fraction of breaking waves at point i that results from waves breaking earlier along the profile and persisting to point i. ΔQ_j is defined recursively by applying Equation 3 (see next section) to point j.

Determination of New Wave Breaking at Point of Interest i

In the previous step, breaking earlier along the profile persisting to point i has been calculated. To this needs to be added the new wave breaking at point i (denoted by ΔQ_i), to give the total fraction of breaking waves Q_i.

This is calculated as follows. Firstly the BJ value of Q_i is calculated from Equation 1. This is denoted $Q_{i,BJ}$ and represents a <u>minimum</u> value of Q_i. The value of Q_i could be greater than $Q_{i,BJ}$ if previous broken waves which have persisted to point i exceed $Q_{i,BJ}$. This could occur if point i is on the shoreward downside of a bar.

ΔQ_i is therefore determined as follows:

Define $a = Q_{i,BJ} - \sum_{j=1}^{j=i-1} \Delta Q_{j,active}$

$$\text{Then } \Delta Q_i = a \quad \text{if } a > 0; \qquad = 0 \quad \text{if } a \leq 0 \qquad 3$$

Determination of Total Fraction of Breaking Waves at Point of Interest, i

The final step is simply to add the new breaking at point i to the earlier breaking that has persisted as far as point i, to give the total fraction of breaking waves at i.

$$Q_i = \Delta Q_i + \sum_{j=1}^{j=i-1} \Delta Q_{j,active} \qquad 4$$

ASYMPTOTIC PROPERTIES OF THE METHOD

In this section, the behaviour of the method in various limiting cases is investigated.

Monotonic Profile

On a monotonic profile, which is sufficiently steep that breaking is continuously triggered, the method should simplify to standard BJ.

This is predicted by the method, since if breaking is continuously triggered, the quantity a in Equation 3 is positive at all grid points. Equation 4 then states that:

$$Q_i = Q_{i,BJ} \text{ for all } i \qquad 5$$

No Breaking Persistence (β=o)

For the case of zero breaking persistence length, the method should by definition default to standard BJ.

The condition of zero persistence length is represented by setting β to zero in Equation 2. This implies that no grid points are active, ie:

$$\Delta Q_{j,active} = 0 \text{ for all } i \text{ and } j \qquad 6$$

Equations 3 and 4 then show that:

$$Q_i = Q_{i,BJ} \text{ for all } i \qquad 7$$

The method then defaults to standard BJ, as shown in Figure 3.

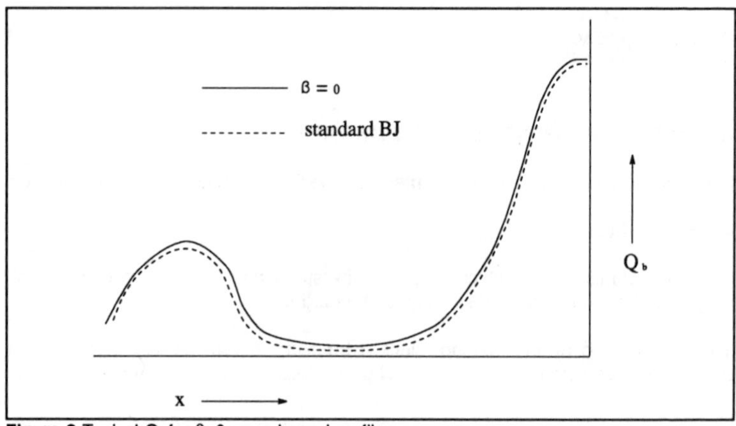

Figure 3 Typical Q_b for β=0, on a barred profile

No Breaking Reformation ($\beta = \infty$)

The case of no breaking reformation means that once a wave has started to break it continues indefinitely in a broken state and never reforms to an unbroken wave.

This implies an infinite persistence length and is represented by setting $\beta = \infty$ in Equation 2. In this case all grid points are active, ie:

$$\Delta Q_{j,active} = \Delta Q_j \text{ for all i and j} \tag{8}$$

This means that:

$$\sum_{j=1}^{j=i-1} \Delta Q_{j,active} \text{ never decreases as i increases} \tag{9}$$

and hence from Equation 3 and 4:

$$Q_i \text{ never decreases as i increases} \tag{10}$$

This behaviour is illustrated in Figure 4.

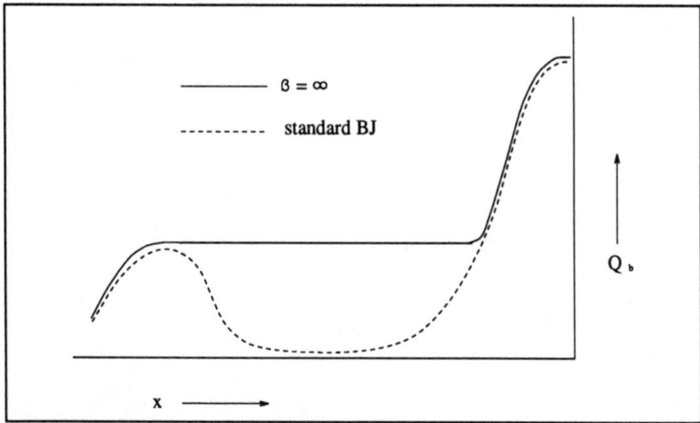

Figure 4 Typical Q_b for $\beta = \infty$, on a barred profile

General Case (β = general)

In the general case the behaviour of Q_i is intermediate between the extreme cases of no breaking persistence and no breaking reformation. The method shows that:

Q_i can decrease as i increases, but

$Q_i \geq Q_{i,BJ}$ for all i 11

Figure 5 illustrates this case.

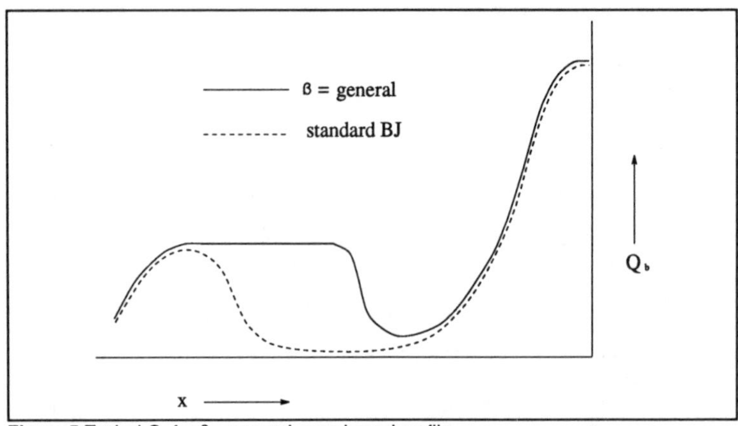

Figure 5 Typical Q_b for β = general, on a barred profile

COMPARISONS WITH LIP-11D DATA

The new method for calculating the fraction of broken waves was included in HR Wallingford's cross-shore profile model of the nearshore zone, COSMOS-2D, which formerly used the standard BJ method. Southgate and Nairn (1993) give a description of the hydrodynamic part of this model.

Tests were carried out to compare the model predictions with observations from the LIP-11D series of experiments, carried out in the prototype scale Delta wave flume at Delft Hydraulics (Sanchez-Arcilla et al, 1994). The first three tests are considered here, Test 1a (with, at the paddle, significant wave height H_s=0.9m, period T=5s), Test 1b (H_s=1.6m, T=5s) and Test 1c (H_s=0.55m, T=8s). Q_b was calculated on the profile measured 7 hours into each test, at which time the bed levels had settled down to a nearly constant profile. Videos from these experiments were analysed at the Universitat Polytècnica de Catalunya (UPC) by counting the number of broken and unbroken waves (see Rivero et al,1994). Average values of Q_b at each location are used here, the random error in Q_b is of the order of 0.05 (estimated based on the variability of results counted over different time intervals). Sources of error are discussed by Rivero at al (1994).

As shown in Rivero et al (1994), the use of standard values for the breaker index γ and the breaker coefficient α (which appears in the dissipation calculation) gives considerably lower values of Q_b than those observed in the first two LIP-11D tests. These coefficients have therefore been chosen as described in Rivero et al (1994) to give a reasonable fit to both the observed wave height and observed fraction of broken waves. The values used are slightly different to those used by Rivero et al (1994) due to some small differences in the models used. These coefficients were then fixed and a comparison made between the BJ method and the new method for calculating Q_b. Figures 6 to 11 show the results.

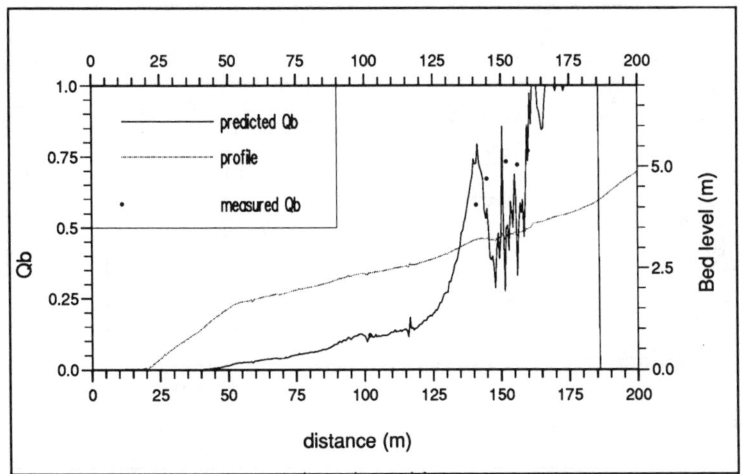

Figure 6 Test 1a, standard BJ method

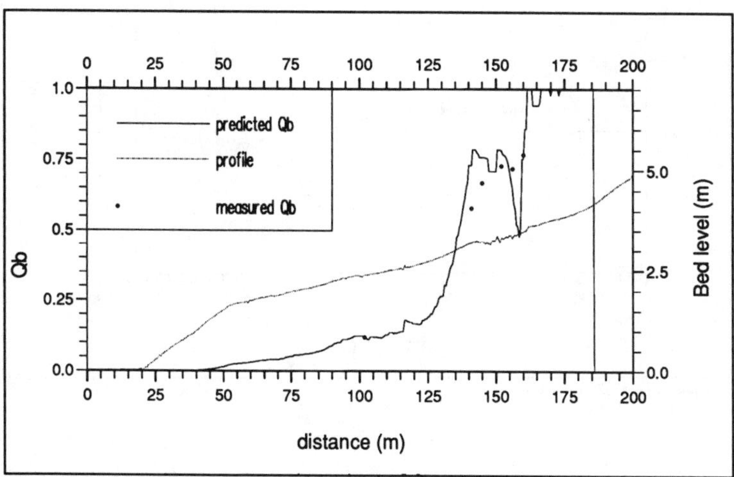

Figure 7 Test 1a, new method

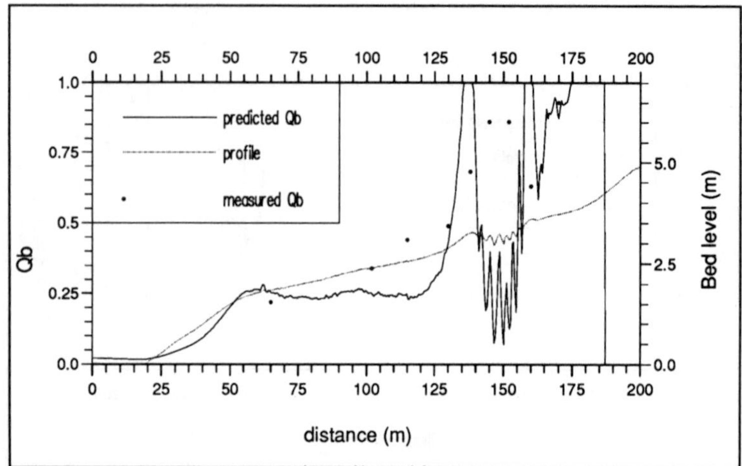

Figure 8 Test 1b standard BJ method

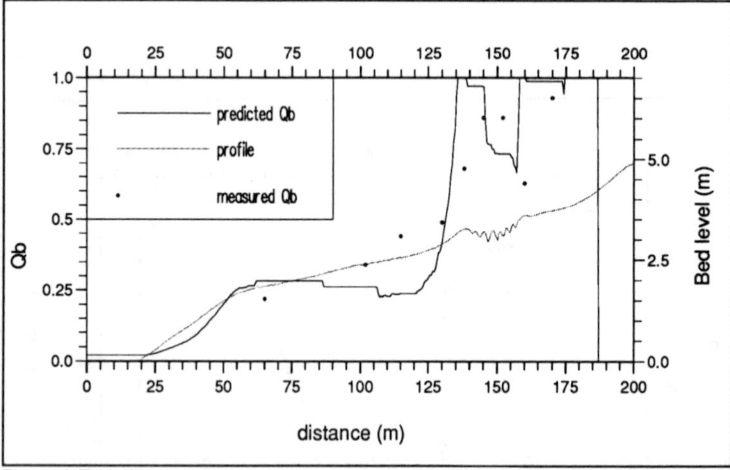

Figure 9 Test 1b, new method

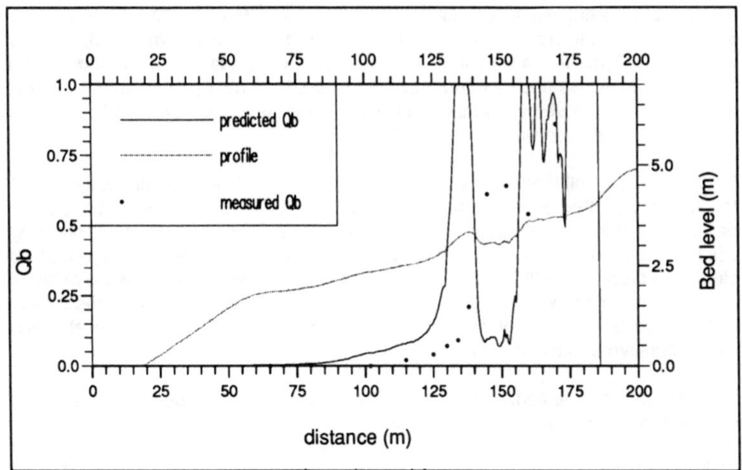

Figure 10 Test 1c, standard BJ method

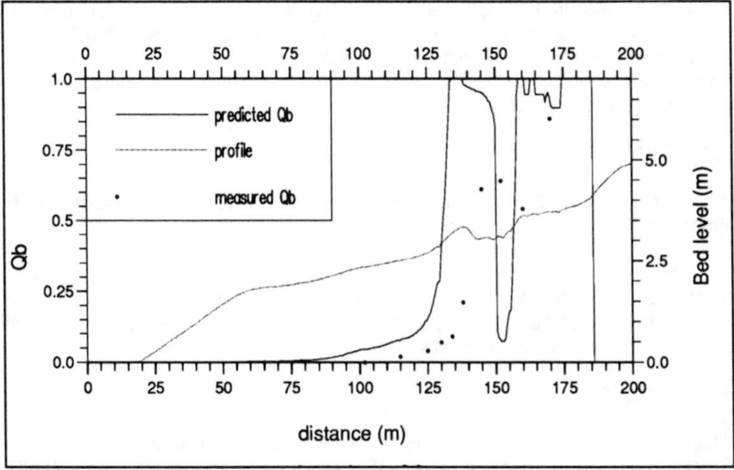

Figure 11 Test 1c, new method

The coefficient β used to calculate the persistence length has been set to 38 for all these tests (this is equivalent to saying that waves on a slope equal to or steeper than about 1 in 30 are continuously breaking). It can be seen from the results that the inclusion of this persistence length reproduces the observed type of behaviour, in terms of breaking waves persisting after a bar, for all the tests. A further property is that much of the high variability in Q_b over short distances is removed, and the variations in Q_b are much smoother. This is especially important for modelling derived surf zone processes such as undertow, longshore currents and sediment transport rates. Rapid cross-shore variations of these parameters are unrealistic and can cause numerical instabilities.

Some problems still remain, in particular breaking is predicted to occur earlier in the model than in the observations, particularly in Test 1c. This may be due to deficiencies in the modelling, or alternatively to the different criteria for wave breaking used in the model and in the video analysis. In the model a wave is regarded as starting to break when its height reaches a maximum, whereas in the video analysis the criterion used was when foam starts to appear at the wave crests. The appearance of white water generally occurs slightly later than the time at which waves achieve a maximum height.

The new method has addressed the problem of breaking wave persistence and is a clear improvement on the standard BJ method.

CONCLUSIONS

A new method for calculating the fraction of broken waves over non-monotonic beach profiles has been described. It takes into account the persistence of breaking over a certain distance from the point at which a wave first breaks. The method is simple and robust and can be easily incorporated into existing models of the BJ type. Comparisons with data show that the method does indeed improve predictions of the fraction of broken waves, with important consequences for the improvement of parametric surf zone models.

ACKNOWLEDGEMENTS

The authors are grateful to Dénes Beyer for the analysis of the videos and to Francisco Rivero for helpful discussions.

Separate parts of this work were funded by the UK Ministry of Agriculture, Fisheries and Food, and by the Commission of the European Communities Directorate General for Science Research and Development under contract no. MAS2-CT92-0027 as part of the G8 Coastal Morphodynamics research programme.

REFERENCES

Battjes, J. A., and Janssen, J. P. F. M., 1978. "Energy Loss and set-up due to Breaking of Random Waves", Proc. 16th Int. Conf. on Coastal Eng., ASCE, pp569-587.

Lippmann, T. C., and Thornton, E. B., 1994. "The Spatial Distributions of Wave Rollers and Turbulent Kinetic Energy on a Barred Beach", Proc. Coastal Dynamics '94, Barcelona, Spain.

Rivero, F. J., Sanchez-Arcilla, A., and Beyer, D., 1994. "Comparison of a Wave Transformation Model with LIP-11D Data", Proc. Coastal Dynamics '94, Barcelona, Spain.

Sanchez-Arcilla, A., Roelvink, J.A., O'Connor, B.A., Reniers, A., Jiménez, J.A., 1994. "The Delta Flume '93 Experiment", Proc. Coastal Dynamics '94, Barcelona, Spain.

Southgate, H.N., and Nairn, R.B., 1993. "Deterministic Profile Modelling of Nearshore Processes. Part I. Waves and Currents.", Coastal Engineering, Vol. 19, pp57-96.

2DV CIRCULATION-PATTERN: WAVE DECAY AND UNDERTOW MODELLING WITH NEARCIR CODE

F. Collado, A. Sánchez–Arcilla, M.G. Coussirat and J. Prieto [1]

ABSTRACT: *In this paper a hydrodynamic analysis of the 2DV circulation-pattern on beaches is presented. The study considers: i) comparative analyses of measured data under different wave conditions and bottom geometries, and ii) numerical model results for these same test cases. The main data set used comes from the LIP-11D experiment carried out in the Delta flume (Delft Hydraulics) during April, May and June of 1993. The experimental results configurate an excellent set of data, comparable to field measurements, that are used to obtain important conclusions of the hydrodynamics in the surf zone.*

INTRODUCTION

The main objective of this work is to get the circulation pattern for all LIP-11D test cases. The LIP-11D experiment consists of two series: without dune (series 1) and with dune (series 2). For each beach type there are equivalent test cases in each series (Sanchez-Arcilla et al., 1994).

The beach type is characterized by means of the bottom geometry and the wave steepness.

- A: Slighty erosive.
- B: Strongly erosive.
- C: Strongly accretive.
- E: Strongly accretive.

The data here analized are waves and currents from A, B, C and only waves from 2E. The test cases sequence was for series 1: A, B, C and for series 2: A, B, E and C. For both series, the initial bathymetry in cases A consisted of a Dean-type beach with constant slope near and above the water line. The initial bathymetry for the subsecuent test cases was the one resulting from the previous test.

For each test, a steady regime is searched for. Starting from this, the study is focused on wave height, setdown/setup and current vertical profiles measured along the flume.

[1] Laboratori d'Enginyeria Marítima, L.I.M., Univ. Politècnica de Catalunya, U.P.C. Av. Gran Capità s/n, 08034 Barcelona, Spain.

The measurements are compared with results obtained by means of a very simplified 2DV-version of the NEARCIR code (Sanchez-Arcilla et al., 1992). The main features of this model are: wave decay calculated solving the wave action equation, dissipation for irregular waves modelled following Battjes and Janssen (1978) approach, transition zone lag modelled according to Nairn et al. 1990, z-varying current component modelled using power series and a coupled parametrization for the middle and bottom boundary layers. Henceforth, the results are only illustated for test case 1A. A complete report for all cases can be seen in Coussirat et al., (1994).

ANALYSIS OF WAVE EXPERIMENTAL DATA

A steady regime from the standpoint of wave generation conditions (significant wave height H_{m0} and peak period T_p at x =20 m) is determined for each test case. In this steady time-interval no significant bathymetric nor SWL changes were detected. The criteria used to determine the stationarity was different depending on the variables. For H_{m0} and T_p the data for the different hours were plotted and then a visual inspection was carried out (see fig. 1 and 2 for test case 1A). For the bathymetry the main observed characteristics were bar migration and dune changes. Finally, the SWL was considered constant because the strongest variation of the SWL that has been found is only 0.04 m. The duration of the selected steady time-interval is long enough for all test cases, being the smallest one of about 6 hours (1C and 2E tests).

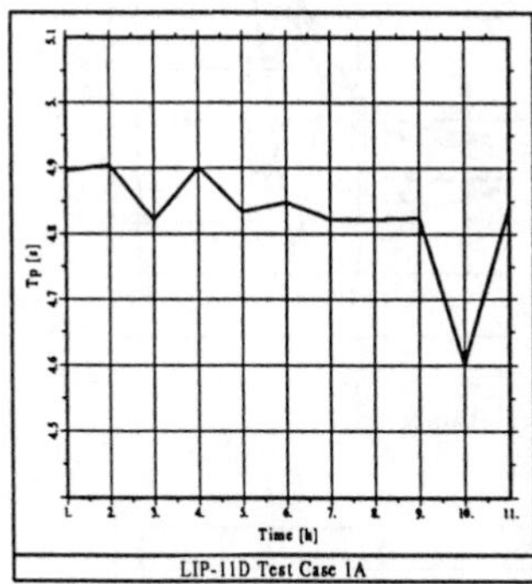

Fig. 1.– Time variation of spectral peak wave period ($x = 20$ m).

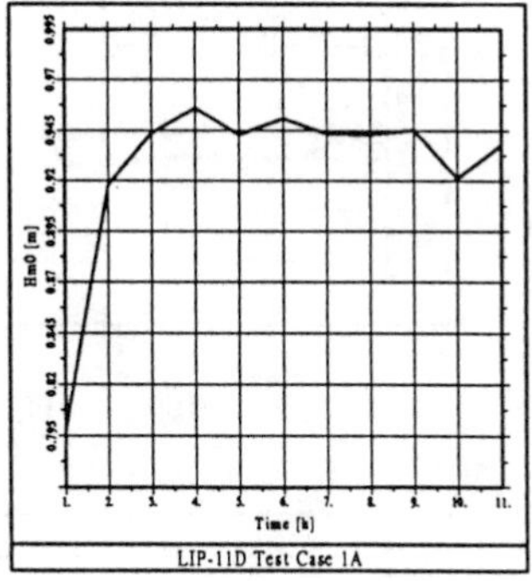

Fig. 2.– Time variation of wave height ($x = 20$ m).

A reasonable steady state interval for wave decay is found for all test cases. The situation for the setdown/setup data is very different, due to the important dispersion of data observed around $x = 100$ m (involving data from three pressure sensors). However, outside this region, experimental results become "reasonable" once again.

The influence of the beach dune has also been studied. The main characteristic found is its effect on the bar migration. For cases with dune, the bar remained farther offshore than without dune (see fig. 10).

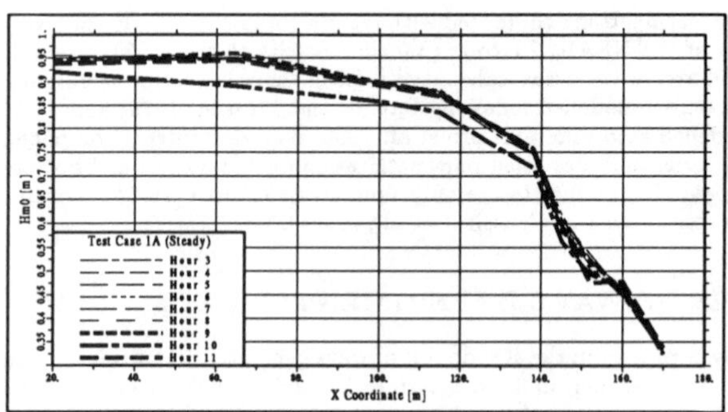

Fig. 3.– Steady significant wave height (case 1A).

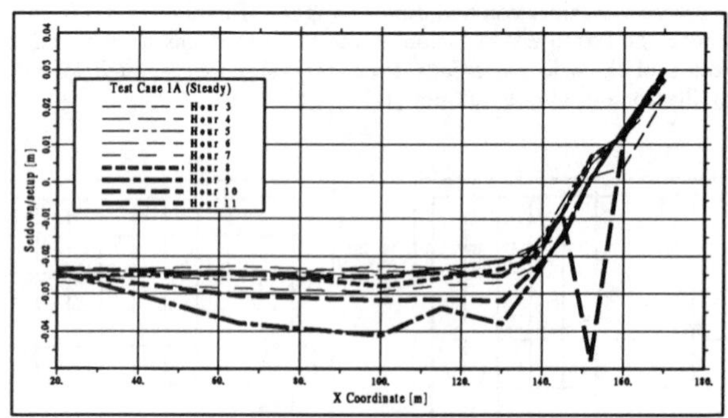

Fig. 4.– Steady setdown/setup (case 1A).

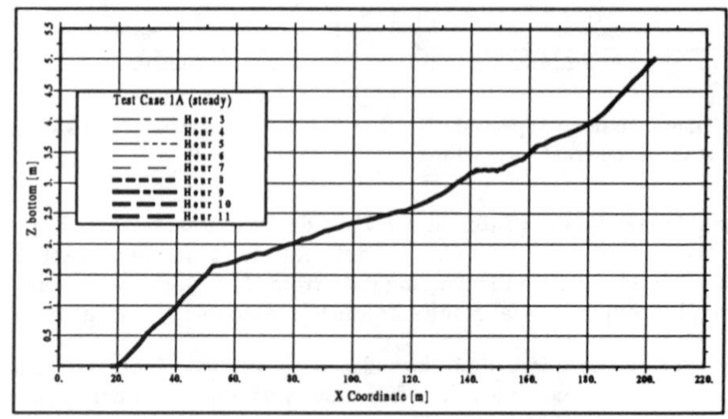

Fig. 5.– Steady bathymetry (case 1A).

To illustrate this situation, test case 1A is presented in what follow (see figs. 3, 4 and 5). Here, the steady interval goes from the third to the eleventh hour. The T_p variation range is 0.3 s approximately around 4.8 s and the H_{m0} variation range is 0.04 m around 0.94 m (see figs. 1 and 2). Their respectives standard desviations are 0.079 and 0.010. These values show the quality of the steady interval. The maximum wave amplitude oscillation is 5% approximately ($x = 70$ m, hour 10) and the measure decay is shown (see fig. 3). The setdown/setup present strong oscillations along the flume and becomes unrealistic around the bar trough zone (see fig. 4, $x = 150$ m). The origin of these anomalies is unknown yet. The bathymetry variation is smooth (see fig. 5) along the steady interval. Finally, the strongest variation of SWL is 0.04 m.

WAVE PROPAGATION MODEL

The depth-varying current module needs as input several wave-related parameters. The most important variables are the following:

- Wave setdown/setup ($<\eta>$).
- Trough level (z_{tr}).
- Breaking and bottom dissipation ($\mathcal{D}$).
- Orbital velocities ($\tilde{u}_{orb}$).
- Mass flux above wave trough level ($\vec{Q}_s$).

Hence, a suitable model concept such as NEARCIR is necessary. The following equations are solved in the 2DH wave propagation model included in the NEARCIR code:

Kinematic equations:

$$\frac{\partial \vec{K}}{\partial t} = -\vec{\nabla}_H \omega \qquad \textit{Kinematic Conservation Principle (KCP)} \qquad (1)$$

$$|\vec{K}|^2 = k^2 + \frac{1}{a}\Delta a \qquad \textit{Battjes' type relation} \qquad (2)$$

$$\omega = \sigma + \vec{K}\cdot\vec{U} \qquad \textit{Doppler relation} \qquad (3)$$

$$\sigma^2 = g\ k\ tanh(kH) \qquad \textit{Dispersion relation} \qquad (4)$$

where:

$$H = h + <\eta> \qquad (5)$$

The actual KCP solved is similar to the Yoo version, (1.986). For a longshore uniform beach (i.e. LIP-11D simulation) the problem is really 1D and expression (1) can be substituted by one equation imposing explicitly a constant value in the longshore component of $\vec{K}$.

Dynamic equations:

$$\frac{\partial}{\partial t}\left(\frac{E}{\sigma}\right) + \frac{\partial}{\partial x_i}\left(\frac{Cg_{x_i}E}{\sigma}\right) = \frac{\mathcal{D}}{\sigma} \qquad \textit{Wave action equation} \qquad (6)$$

where one component in the divergence term must be identically equal to zero due to longshore uniformity and in which:

- E : Wave energy density ($E = \frac{1}{2} \rho g a^2$).
- ω : Absolute frecuency.
- σ : Intrinsic frecuency.
- $\vec{C}g$: Group velocity vector.
- $\vec{K}$: Wave number vector.
- a : Wave amplitude.
- k : Separation factor.

The equations used to determine $< \eta >$ are the simplified momentum equations for a longshore uniform beach where the classical expression for the radiation stress has been used and the equations of Battjes and Janssen (1978) were selected to evaluate the dissipation rate. In the transition zone a roller contribution $\tilde{S}_{rxx}$ to $\tilde{S}_{xx}$ is considered:

$$\tilde{S}_{rxx} = c_{rs} E_r \tag{7}$$

The roller energy density E_r satisfies (Nairm et al, 1990):

$$D + \frac{\partial}{\partial x}\left(c_{rd} E_r C\right) = -2\beta g \frac{E_r}{C} \tag{8}$$

Where:

- c_{rd}: Dimensionless coefficient.
- c_{rs}: Dimensionless coefficient.
- C : Wave celerity.

WAVE MODELLING RESULTS

The wave submodel of NEARCIR described above has been used to model all test cases. Wave amplitude and setdown/setup have been computed and compared with the experimental data for wave amplitude and setdown/setup.

The wave decay results obtained show a reasonable fit to the measured data, except in the bar trough zone particulary for cases in which a significant bar existed. In spite of data dispersion, the mean setup values are suitably modelled. A general improvement is obtained when modelling the transition zone by introducing a lag in the energy dissipation, as suggested in Okayasu, 1989. The maximum differences in wave amplitude between data and model in case 1A ($x = 150\ m$, hour 11) is about 16%. This is due to an untimely prediccion of wave amplitude in the trough zone. For the comparison of setdown/setup data with model results, the zone around $x = 100\ m$ has not been considered because in there the data were not considered reliable (see fig. 8). On the other hand, a good fit is achieved in the bar zone ($x = 130 - 150\ m$), where the fit is further improved by means of incorporating a lag. The maximum relative error without lag which is 60% ($x = 130\ m$, hour 11), is reduced approximately to 25% when the transition zone lag is considered. Near the beach the approximation is not too good and the lag impairs the solution. In opposition to case 1A, test case 2A showed little discrepancy in wave amplitude results with and without lag, both being reasonably good. For this variable, the maximum value of relative error between data and model is approximately 4.5% ($x = 110\ m$, hour 12, figs. 6 and 7). On the other hand, in this case, the setdown/setup fit is not appreciably improved when the lag is introduced (see fig. 9).

A summary of parameters for the last hour of steady state condition is presented in table 1. The aim of this table is the comparison between initial and final values for some of the parameters used in data modelling. A study of the variation of the c_{rs} coefficient with the wave steepness for α and γ given, has been also carried out.

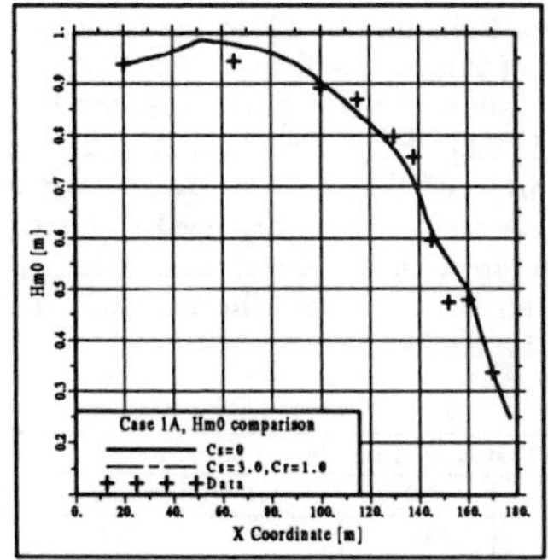

Fig. 6.– Wave height, case 1A
Hour 11

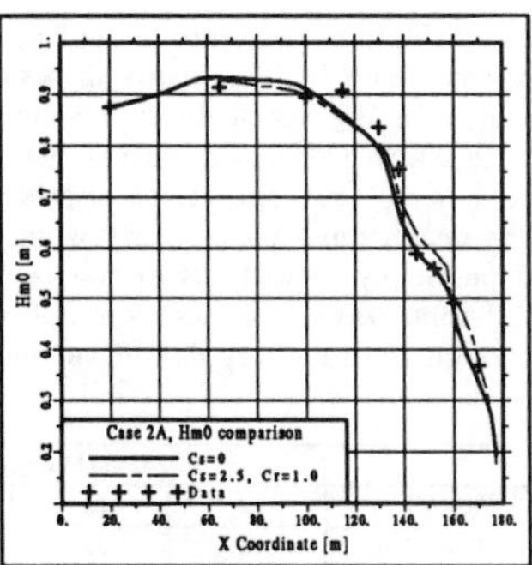

Fig. 7.– Wave height, case 2A
Hour 12

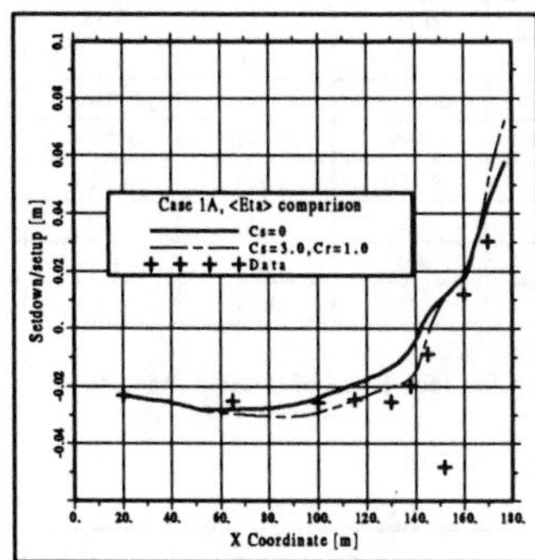

Fig. 8.– Setdown/setup, case 1A
Hour 11

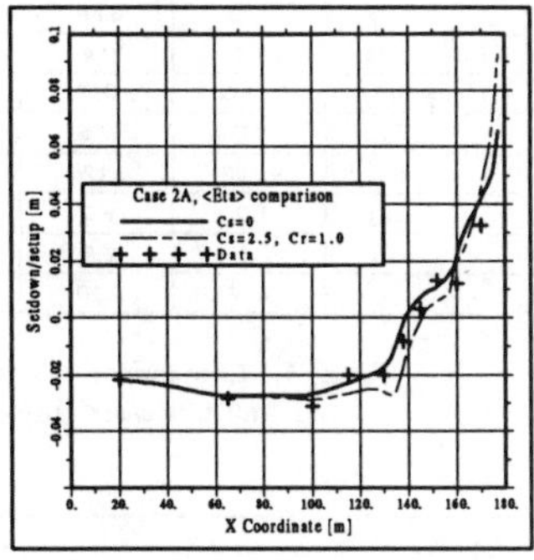

Fig. 9.– Setdown/setup, case 2A
Hour 12

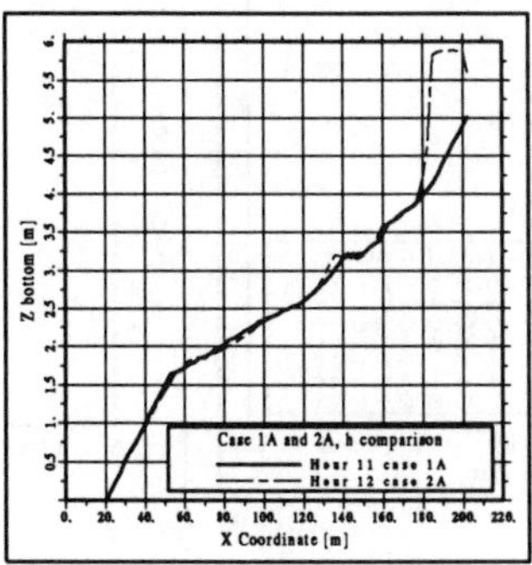

Fig. 10.– Bathymetry comparison, cases 1A and 2A

Coefficients α and γ are taken from Battjes and Janssen, 1978; and c_{rs} is the coefficient in equation (7). This study shows a trend of this c_{rs} coefficient with the wave steepness and with the beach dune presence. The values of the γ parameter are in accordance with Nairn (1990) and are shown (see fig. 11). For c_{rs} coefficient and identical wave steepness, the comparison shows that this coefficient is sensitive to the presence of the beach dune particularly when the wave steepness is increased. This confirms the dependency hypothesis of the setdown/setup with the beach dune (data points with different wave steepness are not comparable). However, the available data are not enough to draw very definitive conclusions, (see fig. 12).

LIP-11D Experiment (Last hour of steady state)							
Test Case	*1A*	*2A*	*1B*	*2B*	*1C*	*2C*	*2E*
Last Hour	*11*	*12*	*17*	*10*	*13*	*11*	*18*
H_{m0}	*0.9376*	*0.8744*	*1.3943*	*1.4136*	*0.5784*	*0.5854*	*1.4317*
T_p	*4.8428*	*4.9924*	*5.0220*	*5.0250*	*7.9706*	*7.9491*	*5.0239*
η_0	*-0.0231*	*-0.0216*	*-0.0492*	*-0.0446*	*-0.0112*	*-0.0097*	*-0.0322*
γ	*0.70*	*0.66*	*0.78*	*0.78*	*0.60*	*0.55*	*0.78*
α	*0.6*	*0.4*	*1.0*	*1.0*	*1.0*	*1.0*	*1.0*
Steepness	*0.0181*	*0.0159*	*0.0250*	*0.0253*	*0.0041*	*0.0042*	*0.0257*
a_{rms}	*0.3315*	*0.3091*	*0.4929*	*0.4998*	*0.2045*	*0.2070*	*0.5062*
swl	*4.120*	*4.120*	*4.100*	*4.110*	*4.100*	*4.130*	*4.610*
c_{rd}	*1.0*	*1.0*	*1.0*	*1.0*	*1.0*	*1.0*	*1.0*
c_{rs}	*3.0*	*2.5*	*2.0*	*4.0*	*4.0*	*4.0*	*4.0*
β	*0.10*	*0.10*	*0.10*	*0.10*	*0.10*	*0.10*	*0.10*

Table 1.– Comparison of parameters for last hour of steady state for all test cases (LIP-11D).

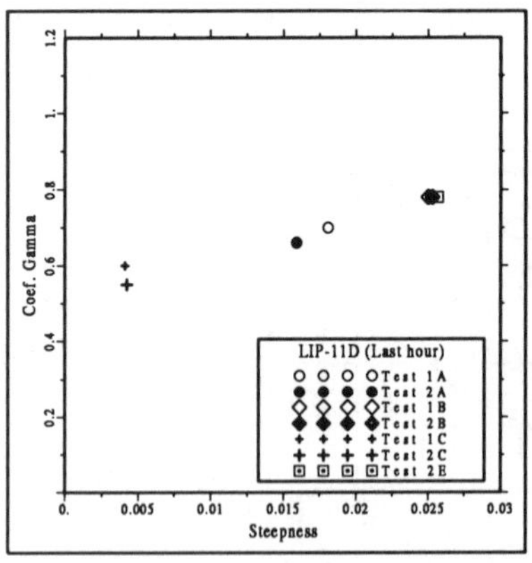

Fig. 11.– Variation of γ with steepness (steady).

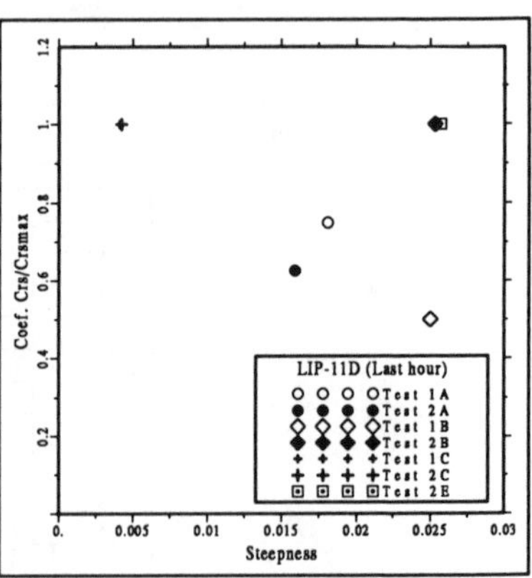

Fig. 12.– Variation of c_{rs} with steepness (steady, $c_{rsmax} = 4.0$).

ANALYSIS OF CURRENT EXPERIMENTAL DATA

Steady circulation patterns have been found in all cases except 2E. The results are shown (see figs. 13 – 18), where the vertical component of the current has been exaggerated about an order of magnitude with respect to the horizontal are. The bathymetry, the SWL and the z_{tr} level (eq. 15) are also shown. In these figures the H_{rms} (taken from the last hour) over the SWL has also been included to get an idea of the evolution of the decay along the flume. In general, the vertical resolution of the current velocity obtained in this experiment is good enough, with a maximum of five data in a spatial progressive. However, this number is often reduced since some current meters were above z_{tr} level or buried for some wave hours.

The general aspect of these patterns is in correspondence with what should be expected. A circulation cell can be appreciated in all cases. It can also be observed that if there is a positive gradient of the wave amplitude offshore of the bar, the vertical velocities are ascendent in this region. On the other hand, they are descendent when the gradient is negative which is in accordance with the prediction of the mass balance equation in a vertical volume control. The test cases 1A, 1B, 2B correspond to the second category and 2A, 1C, 2C to the former one. Although 1C does not show current data offshore of the bar, ascendent velocities can be observed in the data that does not fulfill the steady criterium. In test case 2A, the decay starts a slightly offshore of the bar ($x = 115\ m$) and it present ascendent velocities, whereas in test case 1A the decay starts well offshore of the bar (about $x = 80\ m$) and it shows descendent velocities; this is attributed to the presence of the dune (in case 2A).

All tests show a similar evolution of the undertow profile along the flume. Far away from the coast the profile gets its maximum value near the bottom. On the other hand it seems that the maximum is more relevant and located even closer to the bottom as we go onshore of the bar. In the bar trough zone, the undertow starts getting weaker and the profile adopts forms similar to what was observed offshore. The horizontal velocity against the dimensionless vertical coordinate for every spatial progressive in tests 1A and 2A is shown (see figs. 19 and 20); the thicker the lines, the closer is the undertow profile to the coast.

The mean horizontal velocity in test cases 1A and 2A (see figs. 21 and 22) get their minimum value after the incipient bar ($x = 154\ m$), where the wave decay shows a higher gradient. The mean vertical velocity reaches the maximum slightly onshorewards ($x = 154\ m$ and $x = 150\ m$ in test cases 1A and 2A respectively).

The steady current condition was also checked using the five wall-fixed current-meters placed along the flume. For each test, the dispersion of the module and angle of the current velocity vector during the steady period was studied. For example, the average error observed in all A test cases was about 0.03 – 0.04 m/s and this could represent approximately a 30 % of the average value of the module. The mean variation of the angle observed is of about 5 – 10 degrees (Coussirat et al, 1994).

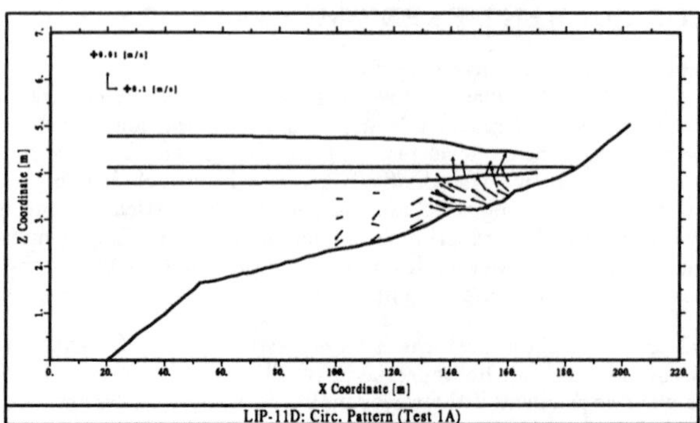

Fig. 13.– Circulation pattern for test case 1A.

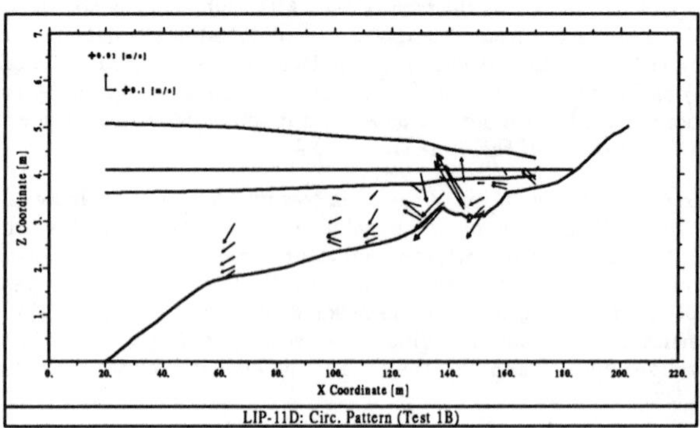

Fig. 14.– Circulation pattern for test case 1B.

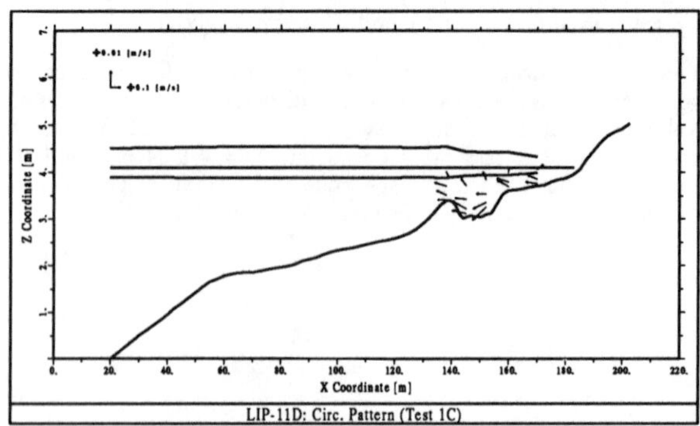

Fig. 15.– Circulation pattern for test case 1C.

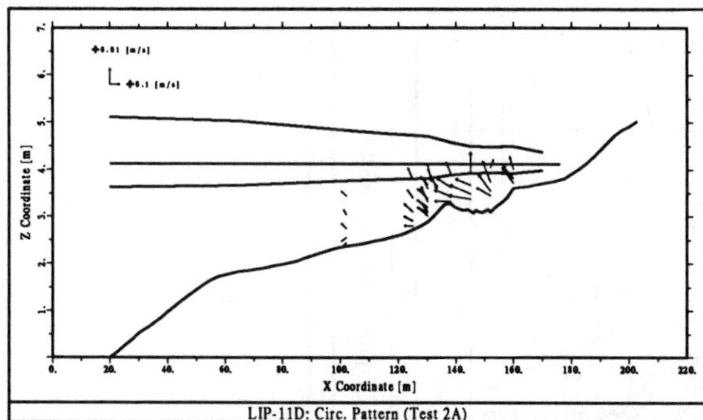

Fig. 16.– Circulation pattern for test case 2A.

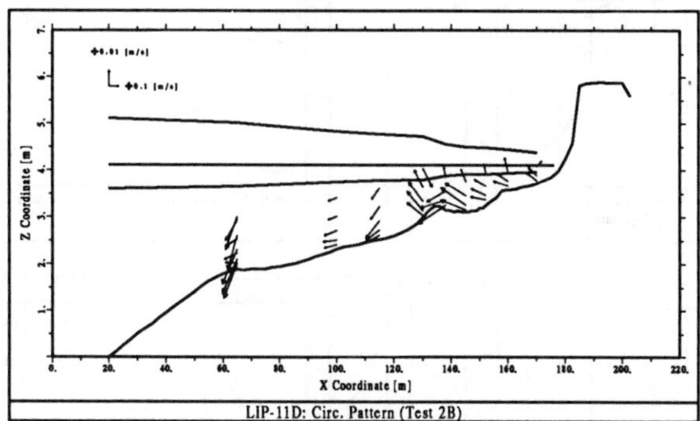

Fig. 17.– Circulation pattern for test case 2B.

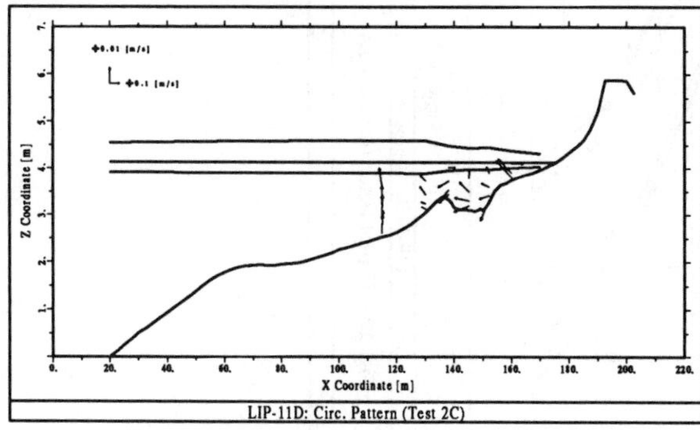

Fig. 18.– Circulation pattern for test case 2C.

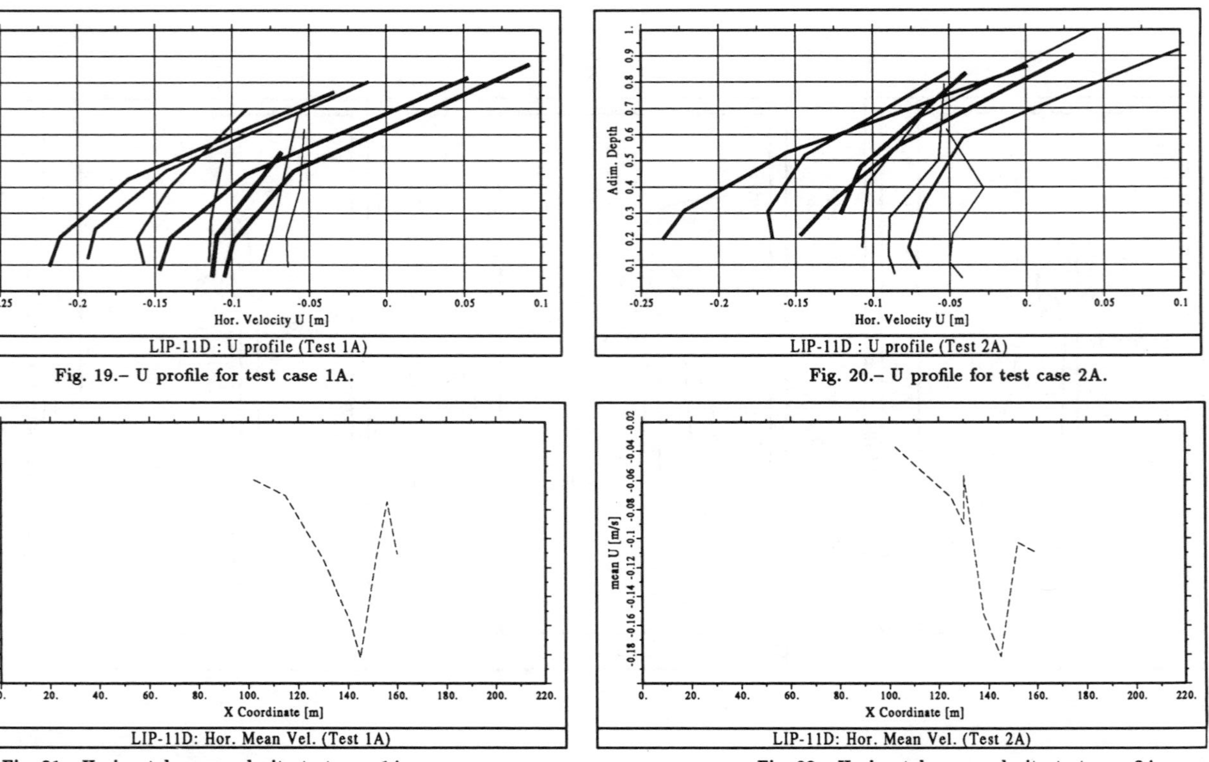

Fig. 19.– U profile for test case 1A.

Fig. 20.– U profile for test case 2A.

Fig. 21.– Horizontal mean velocity test case 1A.

Fig. 22.– Horizontal mean velocity test case 2A.

CURRENTS MODEL

The NEARCIR-1DV model here used solves in a coupled manner the Middle and Bottom-Boundary Layers. The Bottom-Boundary Layer (BBL) model, strongly inspired on Fredsøe (1984), has been parameterized to be compatible with the quasi-3D cost/efficiency philosophy.

The momentum equations in the middle layer are obtained by subtracting the depth-integrated equations from the general ones. A linearized version of these momentum equations may be written as follow:

$$\frac{\partial \vec{\tilde{u}}}{\partial t} + A\vec{\tilde{u}} - \frac{\partial}{\partial z}(\nu_v \frac{\partial \vec{\tilde{u}}}{\partial z}) = \vec{T}(z) \qquad z_l \leq z \leq z_{tr} \tag{9}$$

With

$$\vec{T}(z) = \frac{<\vec{\tau_0}>}{\rho e} - \frac{<\vec{\tau_{tr}}>}{\rho e} - 2G\nabla G\,(P_W^2 - \bar{P}_W^2) + \frac{\partial}{\partial z}<\vec{\tilde{u}}_H\tilde{w}> - <\vec{\tilde{u}}_H\tilde{w}>(z_{tr}) \tag{10}$$

$$A = \begin{pmatrix} \frac{\partial \bar{u}}{\partial x} & \frac{\partial \bar{u}}{\partial y} \\ \frac{\partial \bar{v}}{\partial x} & \frac{\partial \bar{v}}{\partial y} \end{pmatrix} \tag{11}$$

$$G = \nabla_H \vec{Q}_s \tag{12}$$

and $\vec{Q}_s$ being the wave plus current volume flux in the crest-to-trough layer and P_W the vertical distribution function of the vertical current velocity W. The thickness of the middle layer is given by e, $<\vec{\tau_{tr}}>$ and $<\vec{\tau_0}>$ are the shear stresses at trough and bottom levels and the A matrix are given by the depth-uniform current model.

The wave stresses effects induced by horizontal-vertical wave correlations present in vector $\vec{T}$ are being studied at present. They are produced by gradients in the wave amplitude field and current vertical velocity.

When the former effect is relatively more important, the following expresion obtained directy from the wave velocity field is used (Sánchez-Arcilla et al, 1992) :

$$<\vec{\tilde{u}}_H\tilde{w}> = \frac{1}{2}\,\frac{g^2}{\sigma^2}\,\frac{\partial Z}{\partial z}\,Z\,A\,\vec{\nabla}_h A \tag{13}$$

Z being the vertical shape function of the wave potential.

When the gradient in wave amplitude is negligible, an alternative differential equation is derived directly from the steady Reynold's equations:

$$\frac{\partial}{\partial z}(\nu_v \frac{\partial}{\partial z}<\vec{\tilde{u}}_H\tilde{w}>) = <\tilde{w}^2> \frac{\partial}{\partial z}\vec{\tilde{u}} \tag{14}$$

The more adequate expression for the general case is still an open point because the compatibility of both expressions is not easy to prove for the general case. Another option would be possible but expensive new approach consisting in solving coupledly $\vec{\tilde{u}}$ and $<\vec{\tilde{u}}_H\tilde{w}>$ using the system formed by (13) and (14) equations. The optimal strategy is not clear yet.

and $< \vec{\tilde{u}}_H \tilde{w} >$ using the system formed by (13) and (14) equations. The optimal strategy is not clear yet.

The model used for the BBL does not consider turbulence from the previous wave and assumes logarithmic velocity profiles for both waves and currents inside and outside the BBL. Two matching boundary conditions are applied at the top of the boundary layer, viz. continuity of current velocity and shear-stress (in the current direction). The momentum equation perpendicular to the current velocity is used to obtain an expression of the instantaneous friction velocity, u_*, as a function of the wave phase. The model needs additionally the orbital velocities at z_b. In the interval $[z_b, z_l]$ the outside BBL logarithm is used. For the time being the relation adopted between z_b and z_l is $z_l = 10 z_b$.

The middle layer equations are solved using a power series aproximation $(a_i z^i)$ to reproduce the vertical variation of the unknown $\vec{\bar{u}}$ and the right-hand-side of equation (9). Time is used as a marching variable since only stationary currents are here considered.

An important parameter in the model is z_{tr}. For the time being it is simply estimated as:

$$z_{tr} = max\{0.8\ h;\ h - 0.5\ H_{rms}\} \tag{15}$$

The global model needs, thus, three external closure submodels to determine the $\vec{\bar{u}}$ profile: The shear stress $< \vec{\tau_{tr}} >$ at trough level, the mass flux over this level $\vec{Q}_s$ and the eddy viscosity vertical distribution $\nu_v(z)$.

CURRENT MODELLING RESULTS

In general, the profile obtained from the NEARCIR model consist of a first logarithm in the BBL, a second one up to the z_l level and from there up to the trough level a power series $(a_i z^i)$ (of a certain optimal degree m). Considering that the data has an average vertical resolution of about four points is reasonable to select a low value for m. Because of that, the profile above z_l has been considered as parabolic. The additional advantage is that the boundary conditions are enough to determine the profile, and so the differential equation is not actually solved and there is no need to calibrate $\nu_v(z)$. The mean horizontal velocity obtained from the data has been used explicitely in the model because expression (9) for the mass flux does not fit well the experimental values (see fig. 23). Similarly, the boundary condition $\frac{\partial \bar{u}}{\partial z}\big|_{z=z_{tr}}$ has not been obtained by means of the closure submodels for $< \vec{\tau_{tr}} >$ and ν_v. It has been simply deduced from the data to circumvert the poor fit provided in the usual closure submodels.

The model results for test case 1A is presented (see fig. 24). The fit is good enough, considering the introduced simplifications and probably thanks to the use of measured data to by-pass the uncertainty of closure submodels. The obtained profile is able to reproduce the varying behaviour of the data; its maximum value gets relatively closer to the bottom when the undertow becomes greater. In order to get even higher values of the near surface current, it would be necessary to improve some aspects of the BBL model. A natural way to do this is to consider higher values for the vertical power series m. For this purpose, the $\nu_v(z)$ calibration is now being studied. The present results must, be, therefore considered as a very preliminary exercise to model this data set.

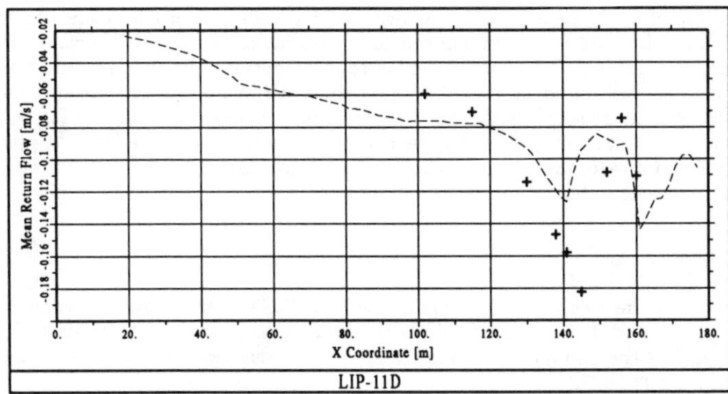

Fig. 23.– Experimental (dots)/theoretical (line) mean return flow comparison (case 1A).

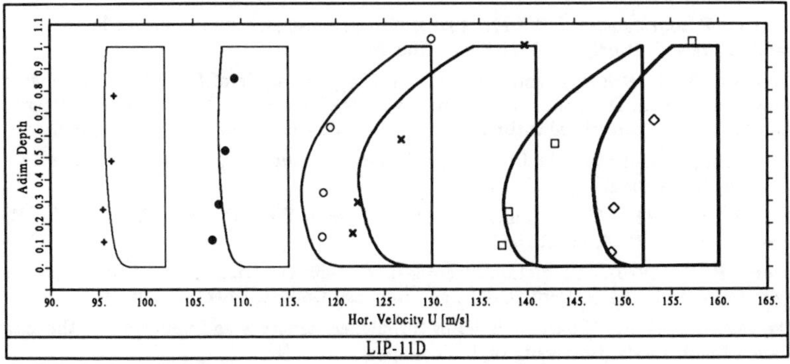

Fig. 24.– Illustration of the measured (dots) and modelled (line) results for horizontal velocity (case 1A).

CONCLUSIONS

The wave and current information from the LIP-11D experiment has been organized and stored in a data base. The preliminary analysis of experimental results carried out shows that they are an excelent set of data for wave and current models.

The NEARCIR wave module including a transitive zone lag has been able to reproduce wave decay and setdown/setup data reasonably well. The influence of the dune presence and wave steepness on the γ and c_{rs} parameters has also been studied with these data.

A methodology has been developed to obtain the steady circulation pattern. It appears to be applicable for all test cases, except 2E. The principal characteristic of the circulation patterns for different wave conditions and bottom geometry were also analyzed.

The evolution of the form and magnitude of the undertow profiles along the flume were, finally, also studied. The main conclusion is that undertow profiles can be fitted well using a parabolic function. Modelling results appear to be good, in spite of the simplifications introduced in the NEARCIR code and thanks to the use of measure values to circumvert some of the closure submodels. Closures submodels for $\vec{Q}_s$ and $< \vec{\tau_{tr}} >$ from the state of art do not appear to fit well to the measured data. At present, the model and measure results are being further analyzed seeking for better fits and to gain more understanding of the modelying physics.

ACKNOWLEDGEMENTS

This work was undertaken as part of the MAST-G8 Coastal Morphodynamics research programme. It was funded jointly by the Programa de Clima Marítimo (PCM-MOPT) and by the Commission of the European Communities D.G.XII under contract no. MAS2-CT92-0027.

REFERENCES

Battjes, F. and Janssen J.P.F.M. (1978), "Energy lost and set-up due to breaking of random waves", Proc. I.C.C.E., ASCE, Hamburg.

Coussirat M.G., Prieto J. and Collado F. (1994), "Analysis of LIP-11D Experiment: Wave and Current Data and its Modelation with the NEARCIR Code". Internal Report, Laboratori d'Enginyeria Marítima, L.I.M., Univ. Politècnica de Catalunya, U.P.C.

De Vriend, H. and Stive, M.J.F. (1987), "Quasi–3D modelling of nearshore currents". Coastal Engineering, 11, pp 565-601.

Fredsøe, J. (1984), "The turbulent boundary layer in Wave-Current Motion", J.H.E, ASCE, Vol 110, N° 8, pp 1103-1120.

Nairm R. B. (1990), "Prediction of cross-shore sediment transport and beach profile evolution", PhD thesis, Dept. Civil Eng., Imperial College, Univ. of London.

Okayasu, A. (1989), "Characteristics of turbulence structure and undertow in the surf zone", Ph.D. thesis, University of Tokio, Japan.

Sánchez-Arcilla, A.; Collado, F.; Lemos, C. and Rivero, F. (1990), "Another quasi-3D model for surf-zone Flows", ICCE, ASCE, Delft, pp. 316-329.

Sánchez-Arcilla, A., Collado, F. and Rodriguez A. (1992), "Vertically varying velocity field in Q3D nearshoe circulation", Proc. I.C.C.E., ASCE, pp 2811-2838.

Sánchez-Arcilla, A., Roelvink D., O'Connor B., Reniers A. and Jimenez J.A. (1994), "The Delta Flume'93 Experiment", Coastal Dynamics'94, Barcelona.

Yoo, D. (1986) "Mathematical modelling of waves-currents interacted flowin shallow waters". Ph.D. Thesis, University of Manchester, U.K.

A SURFZONE PARAMETER SENSITIVITY ANALYSIS ON LIP11D SUSPENDED SEDIMENT AND RETURN FLOW MEASUREMENTS

G Mocke[1], A Reniers[2], G Smith[1]

ABSTRACT: In the present analysis the principal parameters representing the constitutive processes associated with breaking are assessed in an effort to determine their influence on surfzone suspended sediment and undertow. Reference is made to continuous and time averaged velocity and suspended sediment profiles as well as video imagery recorded during the LIP11D large scale flume experiment. In a preliminary analysis covering the time-dependant data the primary importance of wave breaker induced turbulence is highlighted. Wave height and set-up measurements, as well as a vertical turbulence (k,ε) closure model, were exploited for quantifying this contribution. Computational modelling of mean flow and time averaged suspended sediment measurements is employed for the purpose of carrying out a parameter sensitivity analysis that incorporates inverse modelling. This exercise underlined the necessity for incorporating lag effects in determining the magnitude of breaker induced kinetic energy at any particular point across the surf-zone.

INTRODUCTION

Field and laboratory observations and modelling investigations (Dally and Dean,1984; Stive and Battjes, 1984) have shown that one of the more significant mechanisms for offshore sediment transport under active wave conditions is the time-mean 'seaward directed undertow current induced by wave breaking. As has been shown experimentally (Nadaoka *et al.*,1988) wave breaker induced turbulence also has a primary responsibility for the elevated sediment concentrations through the depth that are characteristic of the surf

[1] Research engineers, EMATEK, CSIR, P O Box 320, Stellenbosch 7599, South Africa

[2] Research engineer, Delft Hydraulics, P O Box 152, 8300 AD Emmeloord, The Netherlands

zone. In investigating bar generation mechanisms Roelvink and Stive (1989) found it necessary to introduce an empirical sediment "stirring" effect due to wave breaking within a morphological model so as to account for this important contribution to the undertow driven flux of sediment. An accurate determination of this sediment transport rate requires a detailed description of the hydro-sedimentary flow regime through the vertical, as has been illustrated by the vertically variant Reynolds stress and eddy viscosity magnitudes measured by Okayasu *et al* (1988).

The fundamental driving mechanism for sediment suspension and undertow generation is the wave height decay and dissipation of energy after breaking and the associated changes in mean water level. As has been shown by Battjes and Stive (1985), the model of Battjes and Janssen (1978) describing the average rate of energy dissipation in random waves satisfactorily predicts a set of measured H_{rms} values. Although Battjes and Stive (1985) find the maximum set-up of the mean water level (MWL) to be well described, they consistently found, along with Svendsen (1984), that the interval over which it has its steepest rise is predicted to be too far seawards. A like spatial shift between the maximum gradient in wave height and the undertow immediately following initation of breaking has been observed by Roelvink and Stive (1989). These effects have been ascribed to the presence of a transition zone following breaking over which organized wave energy is progressively converted into dissipative turbulent kinetic energy (TKE).

In the present study specific reference is made to a range of measurements recorded during the course of the Large Scale Installation (LIP11D) wave flume experiment (Sanchez-Arcilla et al., 1994). Continuous suspended sediment measurements as recorded by optical backscatter sensors (OBS) are examined in conjunction with simultaneous surface level and velocity measurements for determining the rôle of the breaker dynamic in the suspension process.

An explicit quantification of the breaker related contribution would require the measurement of the TKE field over depth. With the only limited availability of surfzone turbulence measurements, particularly in mixed hydro-sedimentary regimes such as at LIP11D, the present study evaluates model predictions of suspended sediment and undertow against measurements so as to make deductions regarding the actual rate of energy dissipation. In a complementary manner this parameter has also been quantified from the measured set-up gradient. Inverse modelling on undertow measurements provides further information on existing trough shear stress and mass flux formulations.

EXPERIMENTAL ARRANGEMENT

During the LIP11D experiment, as described in Sanchez-Arcilla *et al*

(1994), instrumentation was deployed along the length of the flume and on a mobile carriage. Instrumentation deployed from the carriage included five OBS sensors deployed in two vertical arrays, an array of sediment suction samplers and an array of electromagnetic current meters (EMF). Further use is made of measurements from EMF and pressure sensors attached to the wall of the flume as well as surface wave gauges.

The focus of the time series analysis is during the fifth hour of Test 2a, which incorporated a gently sloping beach with a low dune. The significant wave height during the test was 0.65 m, while the period of peak wave energy was 5 seconds; this led to mild erosion of the profile. During the one hour test, the mobile carriage was situated at 138 m from the wave generator, where approximately 63% of the waves were broken.

SEDIMENT SUSPENSION PROCESS

Suspension Relative to Incident Waves

Figure 1 illustrates a typical example of time series of the surf zone sediment concentrations at three elevations (filtered with a Kaiser filter - cutoff frequency 1 Hz) and the corresponding horizontal and vertical current velocity and water level traces (showing both the incident waves and the low-pass filtered water level). As can be seen, sediment suspension occurs in the form of intermittent events as has been found both in the surf zone and in the shoaling region (Hanes, 1988; Smith & Mocke, 1993). These events are apparently not strictly correlated with higher wave events and their associated stronger orbital velocities. For example, the high waves and correspondingly high shoreward orbital velocities occuring at 1070 seconds and at 1180 seconds cause no sediment suspension events at all. In contrast, the large event at 1020 seconds occurs during relatively low orbital activity.

In order to further explore the correlation of orbital currents with sediment suspension, the time series of orbital speeds and sediment concentration at 0,05 m elevation (duration 1 hour) were cross-correlated (at X = 138 m). The correlation was found to be extremely low, with the correlation coefficient of the peaks barely exceeding 0.05. This further supports the notion of random intermittent events in the surf zone.

Time Responses of Suspension Events

As can be seen from the time series (Figure 1) sediment suspension events are generally recorded at all three elevations *almost* simultaneously. Correlations were conducted to investigate the time lag between the occurrence of suspension events at the lowest versus the highest sensor, for both individual events and for the one hour time series. Figure 2 shows the typical correlation for a single event (dashed line) and that for the one hour

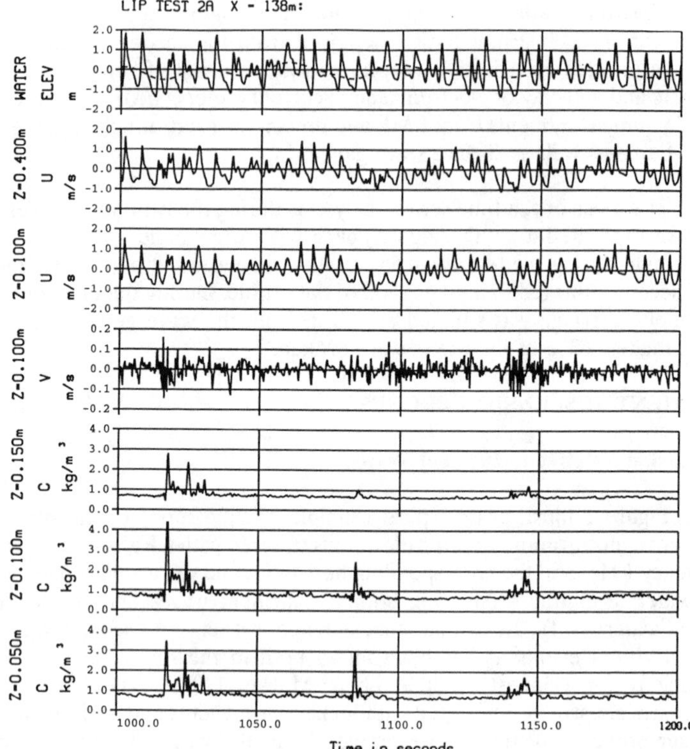

Figure 1: Time series of instantaneous water elevation (top - solid line and the low-pass filtered water level (dashed line), horizontal velocity (u) at two elevations (positive = shoreward) vertical velocity (v) and sediment concentrations (c) at three elevations. (Test 2A, X = 138 m; 63% broken waves)

time series (solid line). In the former the peak in the correlation occurs at 0,2 seconds while the peak in the latter is centered at about 0,05 seconds. Although the correlation at zero lag is large, indicating the simultaneous occurrence of sediment suspension at both the upper and lower sensors, the positive lag indicated by the correlation peak indicates that sediment concentrations propagating upward have more significance than those propagating horizontally. This was found to be the case by Osborne and Greenwood (1993); suggesting the applicability of a vertical advective-diffusion type of model (Hanes and Huntley, 1986) for predictive purposes.

Relationship to Bores and Low Mean Water-Levels

Previous analyses of measurements throughout the surf zone (Smith and Mocke, 1993) indicated a consistent correlation between sediment suspension and the passage of groups of higher waves. In the current analysis, however, this trend was not clear. Rather, video observations indicated that sediment suspension events are generally initiated during or directly after the passage of groups of turbulent bores co-incident with below average water levels. The occurrence of three suspension events co-incident (or soon after) low "mean" water levels is evident in Figure 1, where the water level, represented by a dotted line, has been filtered from the incident waves via a Kaiser filter (cutoff frequency 0.05 Hz).

In further exploring these relationships, the degree of correlation between the filtered level with the sediment concentration time series was assessed. The result is illustrated in Figure 3. The negative peak in the figure at a lag of about 7.5 seconds is due to the correlation of the water-level trough and sediment concentration at this lag. As mentioned, these periods of low mean water level coincide with the passage of bores. It is proposed that the reason for increased suspension at these times is due to the surface source of turbulence being in closer proximity to the bed. Furthermore, with the arrival of successive bores, a progressive stirring effect is likely.

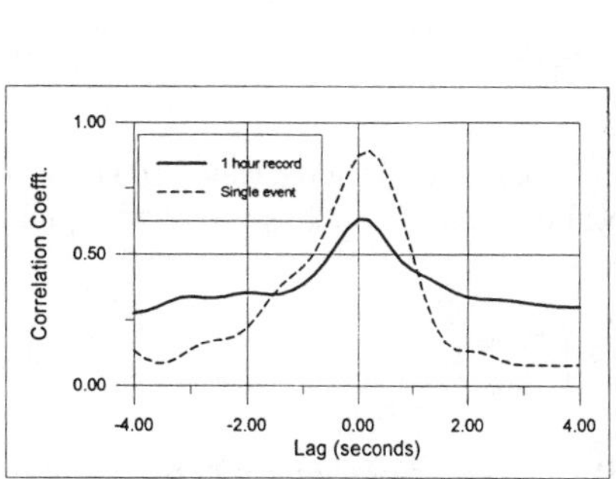

Figure 2: Cross-correlation of sediment concentration at 0,05 m and at 0,150 m (Test 2a,X = 138 m)

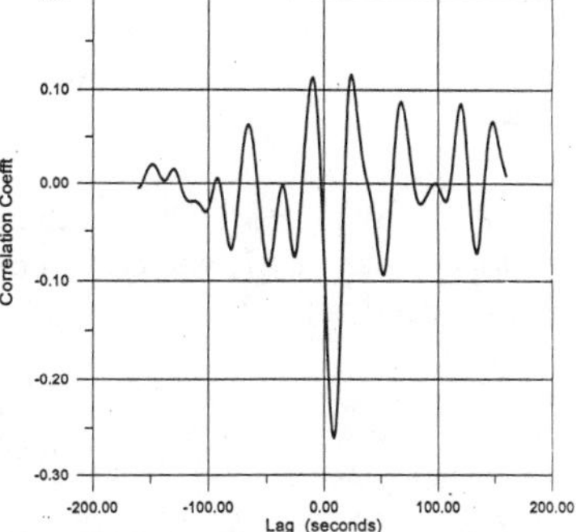

Figure 3: Cross-correlation of sediment concentration versus low-pass filtered water level

It is furthermore interesting to note the periodic trend in the correlation, with peaks repeating roughly every 30 seconds. This is similar to the peak period of infragravity wave energy (28 seconds), as determined from the wave energy spectrum. This correlation of sediment suspension with infragravity wave action has previously led to the conclusion that the infragravity fluid motions are directly responsible for the sediment suspension (Beach and Sternberg, 1988). Although it is accepted that this finding was made under different conditions, the results of the present study would appear to indicate that suspension is due to the condition of low water level created by the infragravity wave motion rather than the infragravity motion itself.

MODELLING

Wave Breaker Dissipation

The rate of wave energy dissipation is determined using two complementary approaches, drawing on measurements of wave height (*Hrms*) and set-up (η) respectively.

(a) Wave decay

The wave energy (E) balance equation:

$$\frac{\partial}{\partial x} Ec_g = D_b = \bar{\varepsilon} + \Gamma_x \tag{1}$$

is used to describe the change in wave energy flux across the surf zone with the time-averaged breaker dissipation approximated by the bore analogy of Battjes and Janssen (1978):

$$D_b = \frac{1}{4}\alpha \rho g f_p Q_b Hm^2 \tag{2}$$

where α = 0(1), *Hm* is the depth limited height of periodic waves in water of the local mean depth, and Q_b is the local fraction of breaking waves which may be expressed as:

$$\frac{1-Q_b}{-lnQ_b} = \left(\frac{Hrms}{Hm}\right)^2 \tag{3}$$

It may be noted that expression (1) includes a storage term Γ_x representing transition zone effects and the shorewards flux of TKE in the wave roller. In the present simulations this influence is neglected. In Figure 4 is presented for tests 1A-C comparisons between random wave height measurements and model predictions made over profiles measured midway through each of the tests - 1A-C. As may be noted the standard B&J formulation provides a good approximation of the measured *Hrms* values for most cases. Deviations of predictions from measurements appears to arise principally from abrupt changes in bottom configuration, such as following bar formation, which occurs from Test 1B. As will be explored later when assessing undertow predictions, this relates at least in part to inadequate Q_b and D_b representations.

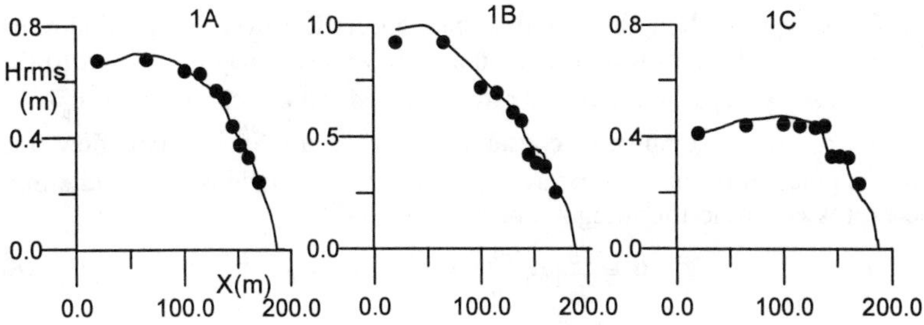

Figure 4: Intercomparison of measured and computed (B&J) wave heights for LIP11A tests 1A, 1B and 1C.

(b) Set-up

Assuming that we can use the total depth integrated momentum balance to compute the set-up gradient:

$$\frac{dS_{xx}}{dx} + \rho g h \frac{d\eta}{dx} = 0 \tag{4}$$

where S_{xx} is the radiation stress which is defined as:

$$S_{xx} = \left(2n - \frac{1}{2}\right)E \tag{5}$$

The radiation stress gradient can be written as:

$$\frac{dS_{xx}}{dx} = \left(2n - \frac{1}{2}\right)\frac{dE}{dx} \tag{6}$$

Combining eqs 1,4 and 6 gives:

$$D_b' = \frac{c_g}{\left(2n - \frac{1}{2}\right)}\left(\rho g h \frac{d\eta}{dx}\right) \tag{7}$$

The procedure is to fit a spline curve through the measured set-up values, from which is calculated the x-derivative to be substituted in equation 8.

Turbulence

Turbulence process modelling requires a description of the eddy viscosity coefficient ν_t describing the turbulent momentum flux over depth. Following the analysis of Prandtl-Kolmogorov the eddy viscosity may be related to the TKE density k ($\overline{u_i u_j} = 2k$) and its rate of dissipation ε such that $\nu_t = c_\mu k^2/\varepsilon$, where c_μ is an empirical constant. For a unidimensional flow the dimensionless form of the time-averaged transport equations for k and ε may be expressed in the following manner:

$$0 = \frac{\partial}{\partial z}\left(D_k \frac{\partial k}{\partial z}\right) + P_b + P_v - \varepsilon \tag{8}$$

$$0 = \frac{\partial}{\partial z}\left(D_\varepsilon \frac{\partial \varepsilon}{\partial z}\right) + \left[c_{1\varepsilon}\left(P_b + P_v\right) - c_{2\varepsilon}\,\varepsilon\right]\frac{\varepsilon}{k} \tag{9}$$

The forcing functions are the production P_b of TKE due to wave breaker induced turbulence and $P_v = \nu_t\,(\partial u/\,\partial z)^2$ due to mean shear. The choice of empirical constants, discussed in Mocke (1991), are close to classical values except for c_μ (= 0.3) which is adjusted to reflect a mainly diffusive flow regime.

The main forcing function for the flow is the production of turbulence in the surface roller. This production is assumed to be imposed at the upper boundary of the flow regime, which is taken at the mean water level. The surface boundary conditions for k,ε are controlled by the production P_b, whilst at the bottom magnitudes computed at the limit of the vicious sublayer are imposed. In reality P_b should be identical to $\bar{\varepsilon}$ representing the integrated TKE dissipation over depth. However, as the lateral KE flux term Γ_x remains unknown the model uses the dissipation term D_b as a first approximation.

Suspended Sediment : Time-Averaged

The computed eddy viscosity describes vertical diffusivity (i.e. $D_c = \nu_t$) in the time-averaged form of the classical advection-diffusion equation:

$$0 = \frac{\partial}{\partial z}\left[D_c \frac{\partial c}{\partial z} + w_s C\right] \tag{10}$$

where w_s is the sediment fall velocity. The suspended sediment bottom boundary condition corresponds to the reference concentration C_b, determined from a parametrization(Smith and Mocke, 1993) that incorporates breaker dissipation as well as a length scale related to the wave height.

Undertow

Local imbalance of mass and momentum flux over the vertical results in a seawards directed undertow current in compensation for the shoreward mass flux above wave trough level. The corresponding equation has the form:

$$\frac{\partial}{\partial z}\nu_t\frac{\partial U}{\partial z} = \bar{g}\eta_x + (\overline{u^2})_x - (\overline{w^2})_x + (\overline{uw})_z \tag{11}$$

Following the approach of De Vriend and Stive(1987) the water column is divided into three layers; i.e. a surface layer above the wave trough level (h_t), a middle layer, and a bottom layer corresponding to the wave boundary layer. The surface layer is taken into account via an effective shear stress τ_t at the trough level(DeVriend and Kitou,1990) which is directly proportional to the rate of energy dissipation:

$$\tau_t = \rho\nu_t K\frac{\tilde{u}_b^2}{c}\sinh 2Kh + \left(1+7\frac{h}{L}\right)\frac{D}{c} \tag{12}$$

where $\tilde{u}_b$ is the nearbed orbital velocity and a constant eddy viscosity (DVS) is assumed for the middle layer.

The shoreward mean mass flux has been estimated by DVS as:

$$m = \left(1 + Q_b\frac{7Kh}{2\pi}\right)\frac{E}{c} \tag{13}$$

Which is assumed equivalent to the seawards directed mass flux below trough level. The mean flow velocity for the middle and bottom layers is solved according to equation (12), with the orbital stress $\overline{uw}$ term excluded in the middle layer. Parametrized eddy viscosity magnitudes are assumed constant over the extent of each layer with the bottom layer value considerably lower than for the middle layer.

EMA model

An alternative approach exploits the depth-dependant eddy viscosity as computed by the k, ε turbulence model, together with the wave terms derived by Stive and De Vriend (1993), in a composite solution that does not differentiate between the lower two layers. This approach further employs an improvement to the shear stress term (Stive and De Vriend, 1993) in that it is formulated at the actual wave trough level:

$$\tau_t = \rho\nu_t K\frac{\tilde{u}_b^2}{c}\sinh 2Kh + \left[1+Kh+\frac{a^2}{h^2}(1+Kh)^2\right] \tag{14}$$

where a is the wave amplitude

Expression (19) is used as the surface boundary condition whilst a no slip condition (U = 0) is assumed at the bottom. The approach involves a double integration of equation 15 so as to arrive at a computationally efficient explicit expression in U(z).

SENSITIVITY ANALYSES

Suspended Sediment

In Figure 5 is shown an intercomparison of measured and computed suspended sediment distributions for a number of measuring stations in test 1A. As may be noted the model as well as the parametrized C_b provides a good approximation to the measured distributions. The one exception to this is in the vicinity of initation of breaking (X = 115 m) where the turbulence level determined from D_b results in a gross overprediction of the concentration distribution. Also plotted in the same figure are predictions where C_b or D_b have been factored to provide an improved fit.

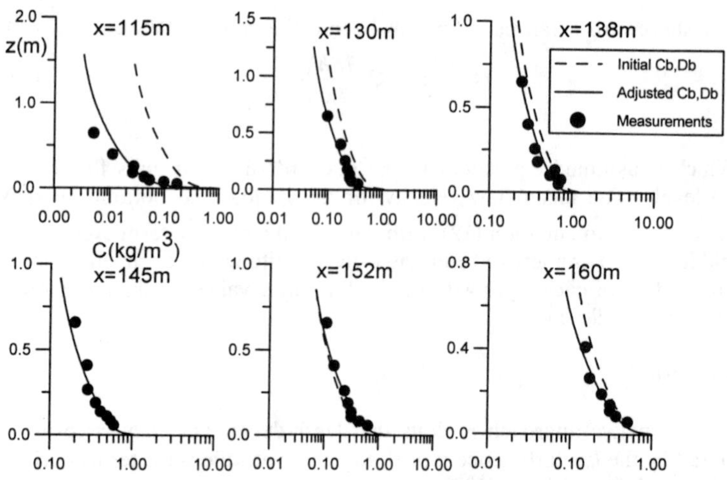

Figure 5: Intercomparison of measured concentrations with predictions using (---) parametrized and (—) adjusted C_b, D_b magnitudes (Test 1A)

In Figure 6 is shown for a number of LIP tests the values of D_b required for computing the best fit against $C(z)$ at various stations along the flume. Before evaluating these computed values against the theoretical (B&J) distribution, it should be recognized that any such comparisons should be made on a relative rather than absolute level. According to the deductive modelling approach employed the Battjes and Jansen formulation provides a generally good approximation of the required dissipation. There is a tendency, however, for the theory to overpredict the deduced values, particularly at the beginning of the surf-zone and where the theory predicts abrupt peaks. Also shown for Test 1A are dissipation values computed from the set-up gradient approach. Although generally not of the same magnitude as those deduced from the concentration profile predictions, these values display similar characteristics with respect to the theory. An example is the prominent second peak in D_b predicted for 1B and 2B by the B&J method, and which is not evident from the two deductive modelling approaches.

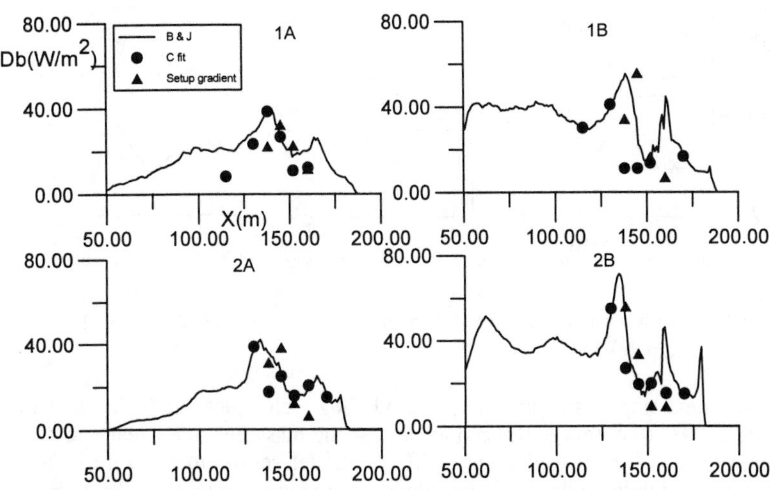

Figure 6: Intercomparisons of D_b as determined from (---) theory (B&J), (Δ) set-up gradient (D_b') and from (•) C(z) best fit analysis.

Undertow

DVS model

The DVS model was initially employed in a comprehensive analysis which quantitatively assessed the sensitivity of predictions to the following parameters: k_s (bottom roughness), h_t, Q_b, D_b, ν_t (middle layer). A random generator(normal distribution) is used to generate a number of input files of values about the mean for each of the above parameters. Computations were compared to measurements for the construction of a linear approximation of the DVS model, from which the sensitivity of return flow as a function of each parameter (du/df) was assessed. The analysis showed a particular sensitivity in the surf zone to D_b as well as to Q_b and $\overline{\nu_t}$ (Figure 7).

With dissipation (D_b') determined from the set-up gradient analysis, the

DVS model provides a good approximation to measurements (Figure 8). With the hypothesis being that dissipation is correctly computed, the observed discrepancies with measurements are assumed to arise from an incorrect formulation of mass flux and trough shear stress.

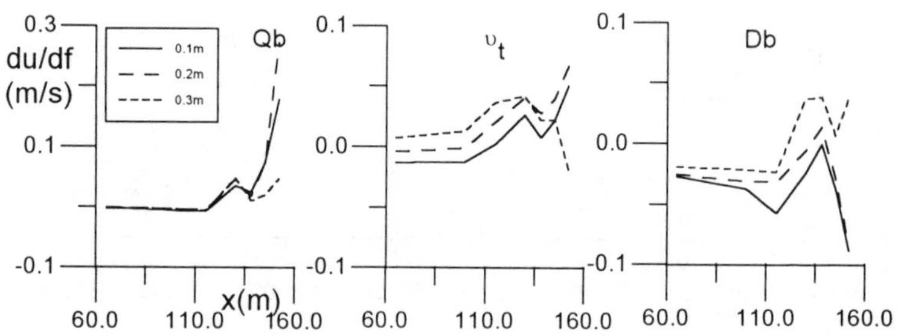

Figure 7: Sensitivity of undertow predictions at z = 0,1 m, 0,2 m and 0,3 m to (a) Q_b (b) $\overline{\nu}_t$ and (c) D_b

The next step was to use inverse modelling techniques (de Valk and van Hulzen,1993) to obtain optimized values for the mass flux and shear stress at trough level. The procedure is to fit the model results to the return flow measurement data by changing the values for the mass flux and the shear stress at trough level. Using a cost function to express the quality of the fit, the optimum values for both variables are obtained by minimising the cost function.

EMA model

In Figure 8 EMA undertow predictions made with the same input (τ_t,m) conditions as for the DVS case are also compared with test 1A measurements. Notwithstanding other differences between the DVS and EMA approaches, the influence of a depth-dependant ν_t is readily apparent in assessing the predictions using the composite model. Employing a similar approach to that used for the suspended sediment fitting exercise, τ_t and m are factored so as to arrive at an improved agreement with measurements. The results of such an exercise performed for a number of LIP tests are shown in Figure 9, where theoretical values are compared with "fitted" estimates along the flume axis. As τ_t is essentially driven by dissipation much the same observations may be made about this parameter as found in the suspended sediment exercise, i.e.

that where storage effects are not included, theory tends to overpredict. For the most part, however, theory was found to provide reasonable agreement. Much the same comments can be made regarding the mass flux term, with overly elevated Q_b values at initiation of breaking resulting in consistent overprediction over this area.

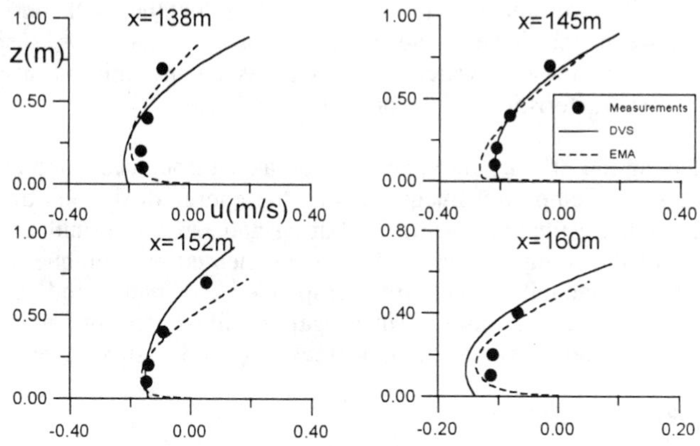

Figure 8: Undertow measurements versus predictions by (—) DVS and (---) EMA formulations using same τ_t and m input.

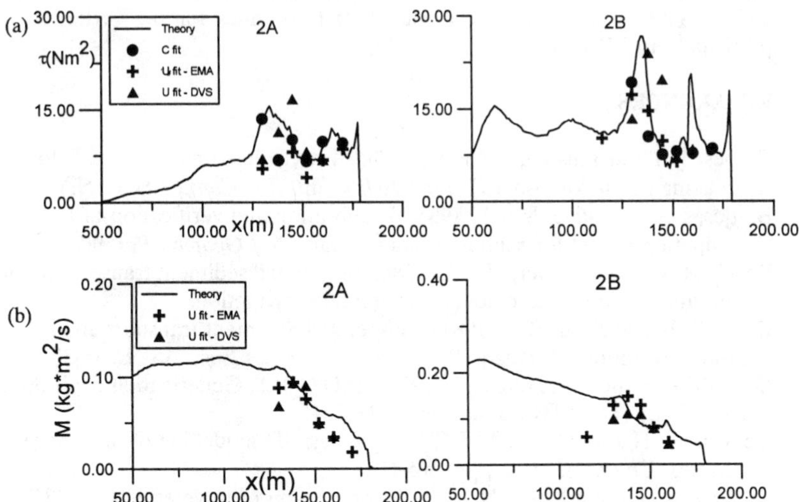

Figure 9: Theoretical values of (a) τ_t and (b) m as determined by D_b from B&J versus estimates derived from (•) $C(z)$ fit, (Δ) $U(z)$ fit (DVS) and (+) $U(z)$ fit (EMA)

CONCLUSIONS

The comprehensive LIP11D measurements effectively represent a valuable data set against which to assess the importance of variable surf-zone parameters for determining suspended sediment and return flow magnitudes. The time series measurements indicated minimal correspondence between sediment suspension and individual waves. This together with positive lag effects between the lowest and highest sensors as well as suspension correlations with lowered water levels suggests an intermittent suspension mechanism characteristic of wave breaker turbulence.

Reflecting the primary rôle played by breaker turbulence complementary approaches were employed for quantifying the energy dissipation across the surfzone. Initial estimations using the Battjes and Janssen formulation and a derivation of the set-up gradient were complemented by a unique approach that inversely modelled measured suspended sediment and undertow distributions. The ensemble of investigations illustrated the necessity of introducing lag effects when computing breaker related dissipation parameters.

ACKNOWLEDGEMENTS

The authors gratefully acknowledge the technical assistance of L Engelbrecht. The work done by Ad Reniers was carried out as part of the G8 Coastal Morphodynamics research program. It was funded by the commission of European Communities, Directorate General for Science, Research and Development, under contract no. MAST-CT-0027. G Mocke and G Smith thank MAST-G8 and Delft Hydraulics for the invitation to participate in LIP11D.

REFERENCES

Battjes, J A and Janssen, J P F M, 1978. "Energy loss and set-up due to breaking in random waves." *Proc 16th Coast. Eng Conf.*, ASCE, N.Y.

Battjes, J A and Stive, M J F, 1985. "Calibration and verification of a dissipation model for random breaking waves." *J Geophys Res*, 90.

Beach, R.A. and Sternberg, R.W. (1988). Suspended sediment transport in the surf zone: response to creoss-shore infragravity motion.

Dally, W R and Dean R G, 1984. "Suspended Sediment transport and beach profile evolution", *J Waterw Port Coast and Ocean Eng.*, ASCE, 15-33.

De Valk, C F and Van Hulzen, M 1993. "IMTOOLS, Generic tools for inverse modelling". Delft Hydraulics report H1224.

De Vriend, H J and Stive, M J F 1987. Quasi-3D modelling of nearshore currents. *Coastal Eng.*, 11, pp. 565-601.

De Vriend H J and Kitou, N 1990. "Incorporation of wave effects in a 3D hydrostatic mean current model." *Proc 22nd Coastal Eng Conference*, ASCE, NY, pp 1005-1018.

Hanes, D.M. and Huntley, D.A. (1986). Continuous measurements of suspended sand concentration in a wave dominated nearshore environment. *Cont. Shelf Res.*,6.,pp 585-596.

Hanes, D.M. (1988). Intermittent sediment suspension ad its implications to sand tracer dispersal in wave-dominated environments. Mar. Geol., 81.

Mocke, G P 1991. "Turbulence modelling of suspended sediment in the surf zone." *Proc Coastal Sediments* '91, ASCE.

Nadaoka, K, Ueno, S, Igarashi, T 1988. "Sediment suspension due to large scale eddies in the surf zone." *Proc 21st Intl. Conf. Coastal* Eng. ASCE.

Okayasu, A Shibayama, T and Horikawa, K 1988. "Vertical variation of undertow in the surf zone". Proc 21st Coast. Eng. Conf., ASCE, NY.

Osborne, P.D. and Greenwood, B. (1993). Sediment suspension under waves and currents: time scales and vertical structure. *Sedimentology*, 40., pp 599-622.

Roelvink, J A and Stive, M J F, 1989. Bar-Generating cross-shore flow mechanisms on a beach. *J. Geophysical Research*, 94, pp 4785-4800.

Sanchez-Arcilla, O'Connor, B A, Roelvink, D A and Jimenez, J A 1994. "The Delta Flume '93 Experiment." *Coastal Dynamics '94*, Barcelona, ASCE.

Smith, G G and Mocke, G P 1993. "Sediment suspension by turbulence in the surf-zone." *Proc. Euromech 310*, Le Havre.

Stive, M J F and Battjes, J A, 1984. "A model for offshore sediment transport", *Proc 19th Coast. Eng. Conf.*, ASCE, N.Y. pp 1420-1436.

Stive, M J F and De Vriend, H, 1993. "Quasi-3D modelling of nearshore currents revisited: the undertow formulation". *Workshop abstracts, G8 Coastal Morphodynamics*, MAST.

Svendsen, I A, 1984. "Wave heights and set-up in a surfzone." *Coastal Eng.*, 8, pp 303-329.

CALIBRATION AND VALIDATION OF A CROSS-SHORE TRANSPORT MODEL

J. Nicholson[1] and B.A. O'Connor[2]

ABSTRACT: This paper is concerned with a modified version of the Ohnaka and Watanabe (1990) one-dimensional cross-shore transport model. Following a brief description of the scheme, details are given of its calibration and validation against large-scale laboratory and field data. The results of the calibration demonstrate that the model is able to generate beach profiles which are both qualitatively and quantitatively correct. However, problems with the representation of irregular waves and the prediction of sediment transport rates are indicated by the results of the validation. Finally, possible causes of these problems are discussed and suggestions made for their solution.

INTRODUCTION

Morphological changes in the coastal zone are strongly influenced by cross-shore transport. In fact, this form of transport is the dominant process as far as the determination of coastal set-back lines is concerned. It also has a marked effect on "extreme event" accretion/erosion patterns associated with coastal structures and longer term changes in the coastline. There is, therefore, a pressing need for models which can predict cross-shore transport rates and their effect on coastal morphology.

Considerable advances have been made in the modelling of cross-shore transport in recent years (see Roelvink and Broker, 1993), the models employed having ranged from comparatively crude empirical approaches to comparatively sophisticated two-dimensional approaches. In between these two extremes lie one-dimensional models which are more accurate than the empirical schemes but more

[1] Research Associate, [2] Professor, Department of Civil Engineering, University of Liverpool, P.O. Box 147, Liverpool L69 3BX, UK.

economic than the two-dimensional schemes. In addition, the development and operation of one-dimensional cross-shore transport models provide useful pointers to the setting up of two-dimensional and three-dimensional area models. The work described in this paper concerns the calibration and validation of such a one-dimensional model, the scheme in question being a variant of that proposed by Ohnaka and Watanabe (1990).

MODEL DETAILS

Model Description

The model consists of wave, sediment transport and morphodynamic sub-models, full details of which are given in O'Connor et al (1992). The wave sub-model solves two equations, which together are equivalent to a time-dependent version of the mild slope equation and are as follows:

$$\partial q/\partial t + C^2 \partial \xi/\partial x + f_D q = 0 \quad \textbf{(1a)}$$

$$\partial \xi/\partial t + (1/n)\partial(nq)/\partial x = 0 \quad \textbf{(1b)}$$

where t = time (s); x = horizontal co-ordinate (m); C = phase velocity (m/s); n = ratio of the group velocity to the phase velocity; f_D = energy dissipation factor (s^{-1}); ξ = instantaneous water surface displacement from mean water level (m); q = instantaneous depth-integrated flow rate (m^2/s). The energy dissipation factor, in turn, is defined by the expression:

$$f_D = \alpha_D \beta_B \left[(g/h)\,(q_M - q_R)/(q_s - q_R)\right]^{1/2} \quad \textbf{(2)}$$

where α_D = coefficient; β_B = bed slope at the break point; g = gravitational acceleration (m/s^2); h = water depth (m); q_M = amplitude of q (m^2/s); q_R = amplitude of q for a recovered wave on a flat bed (m^2/s); q_s = amplitude of q for a breaking wave on a constant slope (m^2/s). In addition, the occurrence of breaking is determined by the criterion:

$$U_c/C > \gamma \quad \textbf{(3)}$$

where U_c = wave crest velocity (m/s); γ = constant.

An empirical sediment transport equation forms the basis of the sediment transport sub-model. This equation takes the form:

$$Q = F_D \left[A_W(\tau_B - \tau_C) + A_{WB}\,\tau_T\right]U_0/(\rho g) \qquad (4)$$

where Q = bulk sediment transport rate ($m^3/m/s$); F_D = dimensionless transport direction function; A_W and A_{WB} = coefficients; τ_B = maximum near-bed shear stress due to wave action (N/m^2); τ_C = threshold of movement shear stress (N/m^2); τ_T = additional near-bed shear stress generated by breaker turbulence (N/m^2); U_0 = maximum near-bed orbital velocity (m/s); ρ = fluid density (kg/m^3). The transport direction function is defined as (Watanabe and Dibajnia, 1988):

$$F_D = \tanh\left[K_D\,(II_c - II)/II_c\right] \qquad (5)$$

where K_D = coefficient; II = dimensionless transport direction indicator based on the ratio of the intensity of the orbital motion to the wave asymmetry; II_c = value of II at the null point for sediment transport. The value of the transport direction function ranges from -1 to +1, a negative sign indicating offshore transport (erosion) and a positive sign indicating onshore transport (accretion).

The final component of the scheme, the morphodynamic sub-model, solves the sediment mass continuity equation which contains a gravitational term, so that:

$$\partial z/\partial t + \partial(Q - \epsilon|Q|\beta)/\partial x = 0 \qquad (6)$$

where z = bed elevation above an arbitrary datum (m); ϵ = coefficient; β = local bed slope.

A refinement, which has been added to the original version of the model, is the inclusion of a breaker transition length. This feature allows for the fact that turbulence produced by the surface roller after breaking only penetrates to the bed, and hence enhances the transport rate, after the broken wave has advanced a certain distance, or breaker transition length, towards the shoreline. The effect of this refinement is to delay the contribution to the transport rate by the τ_T - term in Eq. 4.

Model Operating Technique

The wave sub-model is driven by sinusoidal waves at the seaward boundary and hence an operating technique is required to handle irregular waves. Based on the assumption that the wave heights conform to a Rayleigh distribution at all water depths, the first stage of the operating technique is to divide the wave height distribution at the seaward boundary into a number of representative values. A representative period is then assigned to each height using an empirical relationship between these two quantities, which was derived from joint height-period distributions measured in the field (Salih, 1989). This relationship takes the form:

$$T_R = 1.1\ \bar{T}\ (H_R/H_{RMS})^{0.64} \tag{7}$$

where T_R = representative wave period (s); $\bar{T}$ = mean wave period of the energy spectrum (s); H_R = representative wave height (m); H_{RMS} = root-mean-square wave height (m). Next, the wave and sediment transport sub-models are run for each representative wave condition to give the representative transport rates at each node point of the model grid. A composite transport rate is then computed for each grid point by averaging the representative transport rates, using the occurrence of the representative wave conditions as weighting factors. Finally, the morphodynamic sub-model is run once, with the composite transport rates as input, to yield the resulting bed level changes.

MODEL CALIBRATION

The model calibration was carried out in two stages. In the first stage, which was qualitative in nature, the aim was to check that the model was able to generate well formed accreting and eroding profiles under the appropriate wave and sediment conditions. The second stage, on the other hand, was quantitative in nature, the aim being to reproduce in detail a prototype-scale laboratory profile.

The qualitative stage of the calibration involved running the model for a variety of wave and sediment conditions and then comparing the resulting profile types with those given by the Kraus et al. (1991) accretion/erosion predictor. The latter, which is based on large-scale laboratory and field data, is shown in graphical form in Fig. 1. Expressed in terms of the profile parameter introduced by Dalrymple (1992), the boundary between accretion and erosion is given by the relationship:

$$\phi = gH_o^2 /(Tw^3) = 9000 \tag{8}$$

where ϕ = profile parameter; H_0 = deep water wave height (m); T = wave period (s); w = sediment fall velocity (m/s). This stage of the calibration indicated that the ability of the model to predict the correct type of beach profile was greatly improved if the II_c - parameter in Eq. 5 was made a function of ϕ, rather than being maintained at one fixed value. Consequently, the following relationship between these two quantities was adopted:

$$II_c = 0.2 \quad : \phi \geq 14000$$

$$II_c = 2.5 - 2.3\,(\phi - 5000)/9000 \quad : 5000 < \phi < 14000 \tag{9}$$

$$II_c = 2.5 \quad : \phi \leq 5000$$

Typical examples of eroding, 'neutral' and accreting profiles produced by the model are contained in Fig. 2.

The quantitative stage of the calibration was based on a beach profile generated in the University of Hannover prototype - scale wave flume (Dette and Uliczka, 1986). The flume is 324m long by 7m deep by 5m wide and the profile in question was formed of sand with a medium diameter of 330μm. The initial profile, which is shown in Fig. 3, consisted of a 1 in 20 offshore slope and a 1 in 4 onshore slope and was subjected to the action of regular waves with a period of 6s and an incident amplitude of 0.75m; the duration of the test was 4.6 hours (2730 waves). In order to avoid operating the model with the unnatural initial profile, that measured in the flume after 50 wave cycles was used as the starting condition. Following a sensitivity analysis, which involved the empirical coefficients contained in Eqs. 2 to 6, the best match between the computed and measured profiles is also shown in Fig. 3. As this figure shows, a satisfactory measure of agreement was obtained between the two profiles. The optimum values for the relevant coefficients, the majority of which closely resemble those used in other applications of the Ohnaka and Watanabe model, are listed in Table 1.

Table 1. Empirical Model Coefficients

Coefficient	α_D (Eq. 2)	γ (Eq. 3)	A_W (Eq. 4)	A_{WB} (Eq. 4)	K_D (Eq. 5)	ϵ (Eq. 6)
Value	7.0	0.45	0.20	0.02	0.5	7.0

MODEL VALIDATION

The model was validated against test data obtained from the Deltaflume (see Sanchez-Arcilla et al., 1994), the Delft Hydraulics prototype - scale wave flume. The latter is 250m long, 7m deep and 5m wide and contained 220μm sand which was subject to the action of irregular waves with a JONSWAP energy spectrum. Two runs were carried out during the test programme, the first run commencing with a Dean profile and a plane upper beach and the second run commencing with a similar profile except for the inclusion of a dune. Each run was divided into a series of tests, the aim of which was to simulate the growth and decay of a storm. Details of the tests are given in Table 2 and the expected transport directions, as given by the Kraus et al. accretion/erosion predictor, are plotted on Fig. 1.

Table 2. Deltaflume Test Details

Run 1 (without dune)				
Test	Duration (hr.)	Maximum Depth (m)	H_{RMS} (m)	T_p (s)
1A	12	4.10	0.66	4.83
1B	18	4.11	0.98	5.00
1C	13	4.12	0.41	7.97
Run 2 (with dune)				
2A	12	4.11	0.62	5.00
2B	12	4.12	1.00	5.03
2E	18	4.61	1.00	5.02
2C	21	4.12	0.41	7.94

A comparison of the computed and measured root-mean-square wave heights at the start of each run is shown in Fig. 4. Although the agreement between the two sets of values is reasonable, the computed wave heights were up to 15 per cent larger than their measured counterparts in the vicinity of the break point.

With regard to the transport rates, the computed values were also found to be larger than the measured values, a typical example being the situation existing at the start of test 2E and depicted in Fig. 5. In addition, the computed and measured transport sometimes took place in opposing directions, an example being the situation existing at the start of Test 1B which is contained in Fig. 6. In this case, the measured transport direction was essentially onshore but the computed direction was wholly offshore, the latter trend complying with the Kraus et al accretion/erosion predictor shown in Fig. 1. The computed transport rates also exhibited little or no tendency to decrease with time, in contrast to the measured values which decreased markedly as the beach profile approached equilibrium.

In spite of the discrepancies between the computed and measured wave heights and transport rates, there was broad overall agreement between the computed and measured profiles at the end of each run. These are shown in Fig. 7, the computed profile, as expected, exhibiting more erosion on the upper beach, and consequently more accretion on the lower beach, than their measured counterparts.

DISCUSSION

The divergence between the computed and measured root-mean-square wave heights in the vicinity of the break point (see Fig. 4) is probably due to the assumption that the Rayleigh wave height distribution is valid for all water depths. This is because a laboratory study carried out by Briand and Kamphuis (1993) has established that the wave height distribution in the breaker zone actually follows a distorted form of the Rayleigh distribution which results in a bias towards the larger wave heights. As larger waves break in deeper water, this bias, therefore, suggests that the use of a modified Rayleigh distribution would lead to a reduction in the computed wave heights in the vicinity of the break point.

The discrepancies between the computed and measured sediment transport rates, on the other hand, are almost certainly caused by the use of the Kraus et al. accretion/erosion predictor to determine the critical transport direction indicator. The latter is defined by Eq. 9 and not only affects the transport direction but its magnitude as well, through the transport direction function given by Eq. 5. No allowance is made by the Kraus et al. predictor for the shape of the profile under wave attack yet steep beaches are more likely to be eroded than flat beaches, all other factors being equal. This would account for the tendency of the model, which was calibrated against a steep laboratory profile (see Fig. 3), to bias the predicted transport rates in the offshore direction (see Figs. 5 and 6). A possible modification to the predictor would be the incorporation of a representative beach slope into the profile parameter currently defined by Eq. 8. Not only would this modification address the above sediment transport problems but it would also provide an answer to the problem caused by the inability of the scheme to predict decreasing transport rates as a profile approaches equilibrium. This improvement would be brought about by the tendency of eroding beaches to flatten with time and accreting beaches to steepen with time, because the modified profile parameter would then move towards its 'neutral' value, thereby reducing the transport rate.

CONCLUSIONS

The calibration of a modified version of the Ohnaka and Watanabe (1990) cross-shore transport model against prototype-scale laboratory and field data has shown that the model is able to generate beach profiles which are both qualitatively and quantitatively correct. However, the validation of the scheme against prototype-scale laboratory data has revealed deficiencies in the method of handling irregular waves and also in the prediction of the sediment transport rates. These deficiencies have been ascribed, respectively, to the lack of any allowance for changes in the Rayleigh wave height distribution in the breaker zone and to the omission of the effect of the beach profile on the transport direction function. It is suggested, therefore, that the accuracy of the model could be improved by using a modified form of the Rayleigh distribution in the breaker zone and by including a representative beach slope, for example, in the accretion/erosion predictor upon which the transport direction function is based.

ACKNOWLEDGEMENT

This work has been carried out as part of the G8 Coastal Morphodynamics research programme. It was funded by the Commission of the European Communities, Directorate-General for Science, Research and Development under MAST Contract No. MAS2-CT92-0027.

REFERENCES

Briand, M.H.G. and Kamphuis, J.W., 1993, "Waves and Currents on Natural Beaches: a Quasi 3-D Numerical Model", *Coastal Engineering*, Vol. 20, pp. 101-134.

Dalrymple, R.A., 1992, "Prediction of Storm/Normal Beach Profiles", *Journal Waterway, Port, Coastal and Ocean Engineering*, ASCE, NY, Vol. 118, pp. 193-200.

Dette, H.H. and Uliczka, K., 1986, "Velocity and Sediment Concentration Fields across Surf Zones", *Proceedings Twentieth Coastal Engineering Conference*, ASCE, NY, pp. 1062-1076.

Kraus, N.C., Larson, M. and Kriebel, D.L., 1991, "Evaluation of Beach Erosion and Accretion Predictors", in "*Coastal Sediments '91*", Kraus, N.C., Gingerich, K.J. and Kriebel, D.L. (eds), ASCE, NY, pp. 572-587.

O'Connor, B.A., Nicholson, J., MacDonald, N. and O'Shea, K., 1992, "Application of the Watanabe Cross-Shore Transport Model to Prototype - Scale Data", in "*Hydraulic and Environmental Modelling: Coastal Waters*", Falconer, R.A., Chandler-Wilde, S.N. and Liu, S.Q. (eds.), Ashgate Publishing Ltd., Aldershot, U.K., pp. 337-348.

Ohnaka, S. and Watanabe, A., 1990, "Modelling of Wave-Current Interaction and Beach Change", *Proceedings Twenty-Second Coastal Engineering Conference*, ASCE, NY, pp. 2443-2456.

Roelvink, J.A. and Broker, I., 1993, "Cross-Shore Profile Models", *Coastal Engineering*, Vol. 21, pp. 163-191.

Salih, B.A., 1989, "Probabilistic Properties of Wave Climates", *Ph.D. Thesis*, University of Liverpool, UK.

Sanchez-Arcilla, A., Roelvink, J.A., O'Connor, B.A., Reniers, A. and Jimenez, J.A., 1994, "The Deltaflume '93 Experiment", *this publication*.

Watanabe, A. and Dibajnia, M., 1988, "Numerical Modelling of Nearshore Waves, Cross-Shore Sediment Transport and Beach Profile Change", *Proceedings IAHR Symposium on Mathematical Modelling of Sediment Transport in the Coastal Zone*, pp. 166-174.

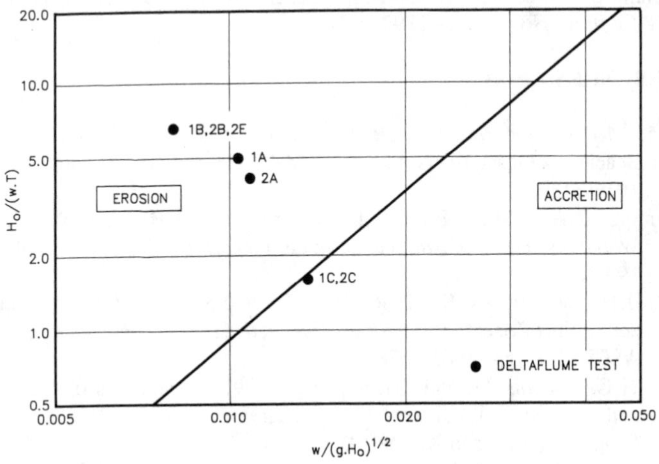

Figure 1. Accretion/erosion predictor (Kraus et al., 1991)

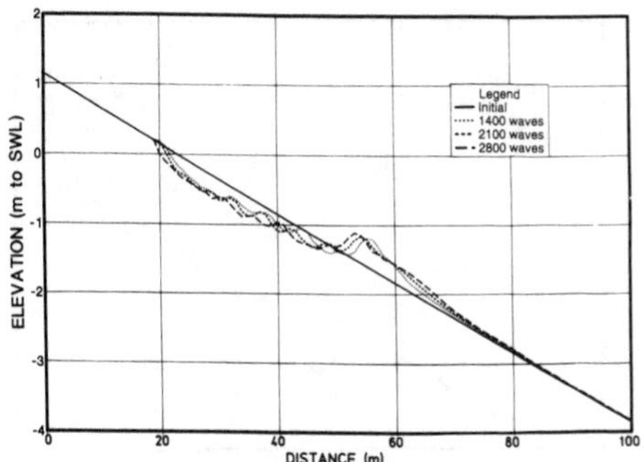

Figure 2a. Typical eroding profile

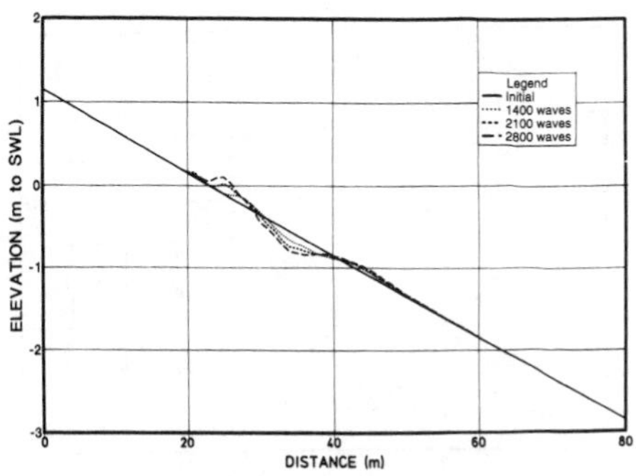

Figure 2b. Typical 'neutral' profile

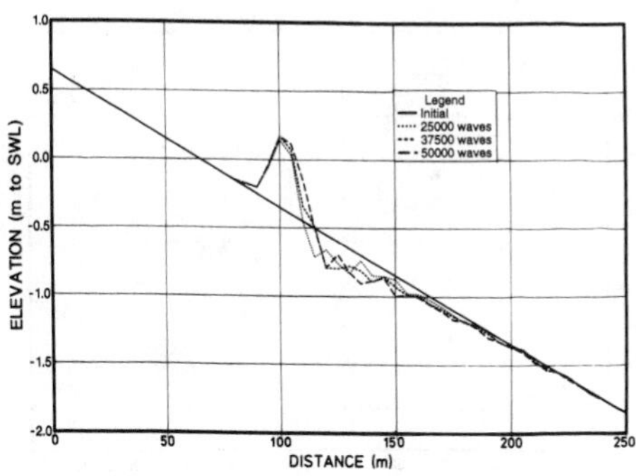

Figure 2c. Typical accreting profile

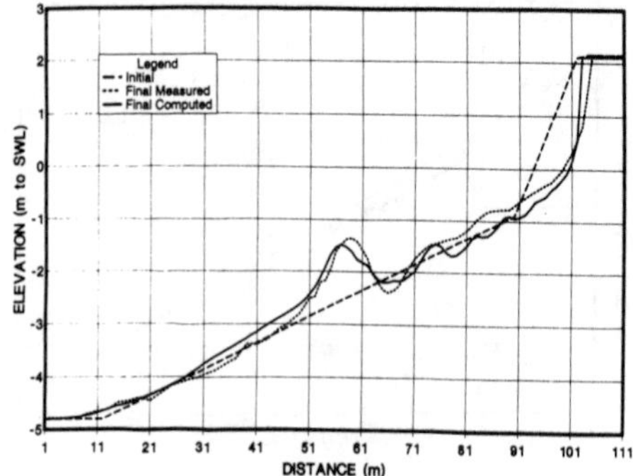

Figure 3. Model calibration : profiles

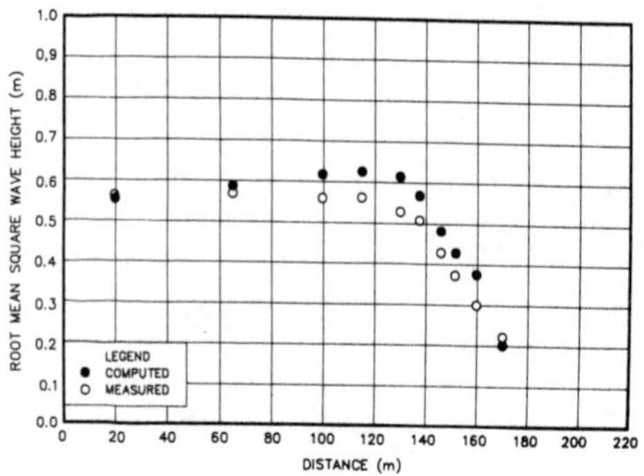

Figure 4a. Model validation : wave heights : run 1

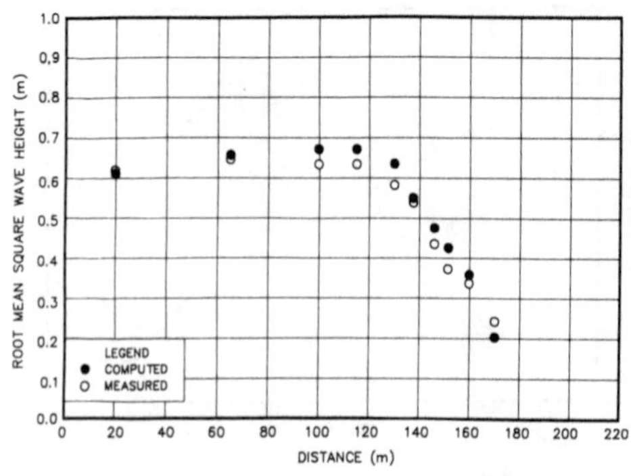

Figure 4b. Model validation: wave heights : run 2

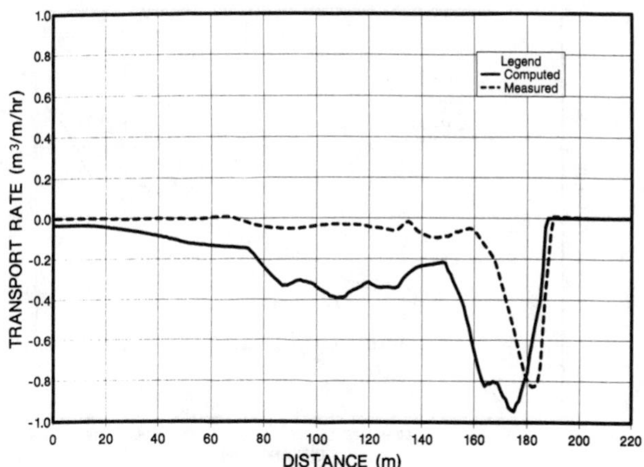

Figure 5. Model validation : transport rates: test 2E

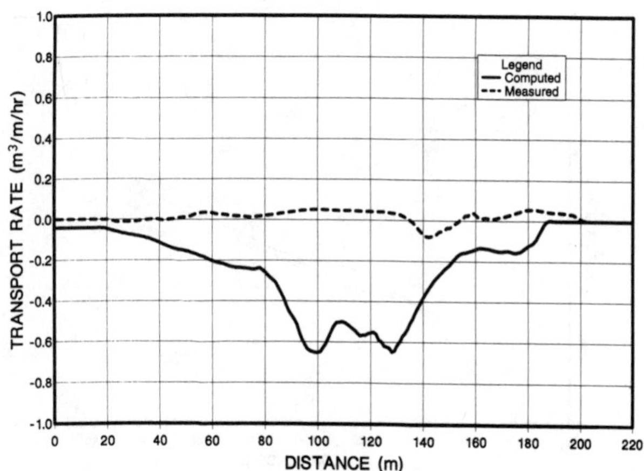

Figure 6. Model validation : transport rates: test 1B

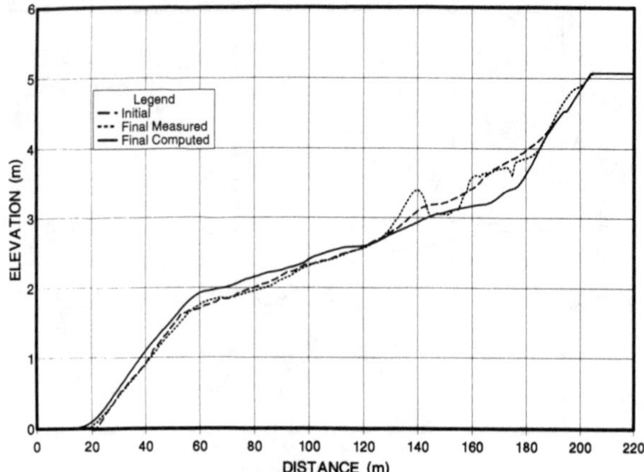

Figure 7a. Model validation : profiles : run 1

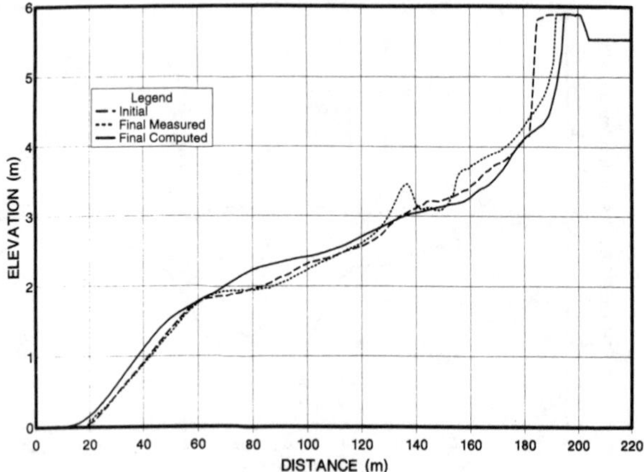

Figure 7b. Model validation : profiles : run 2

IRREGULAR WAVE RUN-UP on COMPOSITE ROUGH SLOPES

Marek SZMYTKIEWICZ[1], Ryszard B. ZEIDLER[2]
and Krystian W. PILARCZYK[3]

ABSTRACT: *IBW PAN* experiments encompassed numerous series of run-up tests on smooth slope, rough slope with cuboidal elements, rough slope with strip elements, two shallow water configurations with smooth and rough composite slope, and bermed slopes, in addition to introductory methodological and comparative runs. Three methods of run-up computation, basing on concepts of water residence on slope and SWL crossing, the latter with or without dummy incident waves not reaching the slope, have been employed. Smooth-slope tests provided reference for rough, bermed and composite slopes. Rough slope tests with cuboidal and strip elements point to the importance of block positioning. The positioning about SWL (rather above than below) provides the best reduction of run-up. The active length on slope should not be smaller than 0.5 R_u (run-up height on smooth slope) and the blocks placed just above SWL play the most important role. The effect of roughness height on run-up reduction (rough vs smooth slope) increases with relative roughness height k/H_{s2} but the most substantial reduction due to blocks occurs at $k/H_{s2} \approx 0.10$ (a certain threshold). Further increase in k/H_{s2} brings about little improvement. Blocks reduce the run-up height by 30% on the average while the respective reduction by strips is only 10%.

1 INTRODUCTION

1.1 IBW PAN Measurements

The phenomenon of wave run-up has been of great concern to coastal engineers for a long time, because of its extraordinary importance to the behaviour and design of coastal structures. Nowadays, when the greenhouse effect poses

[1]Snr Res. Assoc. & [2]Professor, Polish Academy of Sciences' Institute of Hydro-Engineering *IBW PAN*, 80-953 Gdansk, Koscierska 7, Poland

[3]Ir., Rijkswaterstaat, van der Burghweg 1, CS 2628 Delft, the Netherlands

Wave parameters in wave-generator control sets.

No	H_{os} (m)	T_{op} (s)	L_{op} (m)	H_{os}/T_{op}	ξop
2	0.150	2.193	7.50	0.020	1.768
3	0.150	1.790	5.00	0.030	1.443
4	0.150	1.550	3.75	0.040	1.250
5	0.200	1.601	4.00	0.050	1.118
6	0.100	2.532	10.00	0.010	2.500
7	0.100	1.462	3.33	0.030	1.443
8	0.100	1.132	2.00	0.050	1.118

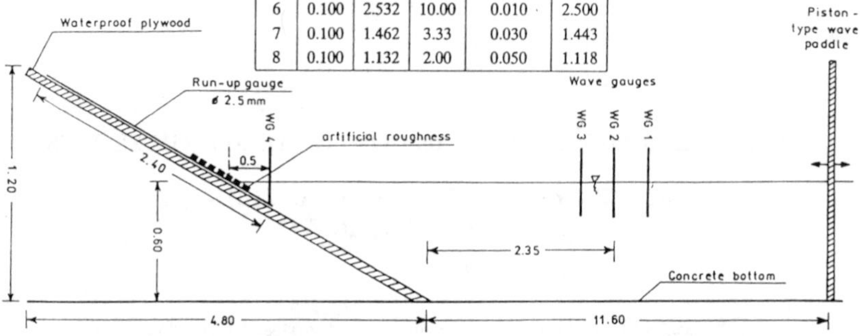

Figure 1. General view of laboratory setup and the range of wave parameters.

hazards to all activities in the coastal zones throughout the world, the problem becomes even more serious.

In 1992 *IBW PAN* initiated an extensive research programme, in close co-operation with *Delft Hydraulics* and *Rijkswaterstaat Delft*. Our laboratory tests on various composite, bermed and artificially roughened slopes (cf. Figures 1 and 2 with our setup and configurations) are almost completed, and the ultimate objective of structure optimization with respect to run-up and other factors will hopefully be reached in 1994. It thus seems worthwhile to formulate some conclusions and share our present experience.

The *IBW PAN* laboratory run-up experiments were conducted in a wave flume 20 m long, 0.5 m wide and 1.5 m high (Fig. 1). The depth of water h during the experiments was 0.50, 0.60 and 0.70 m. The spacing of the wave board and the slope measured along SWL was 14.00 m at h = 60 cm. Visual observations and video taping are facilitated by provision of a glassed wall section of the flume. The wave generator was produced by the *Danish Hydraulic Institute.*

Wave parameters were measured with capacitance-type wave gauges WG. Three gauges WG1, WG2 and WG3 were installed in the flume section with flat bed; they were primarily intended for measurement of the wave reflection factor. After analysis of wave transformation in the wave flume it was decided

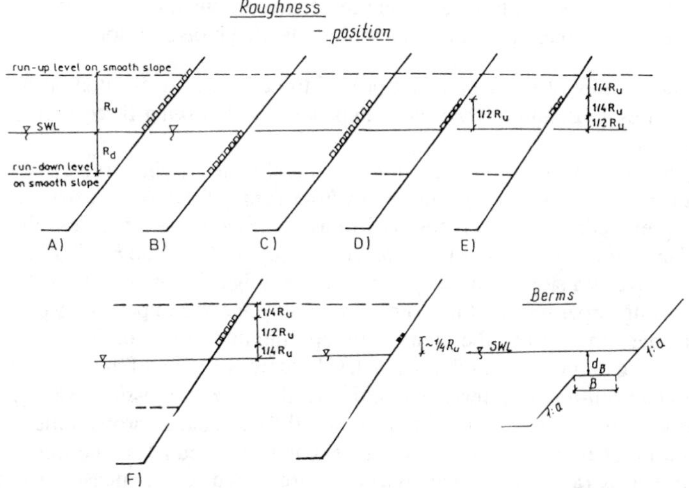

Figure 2. Configurations of *IBW PAN* composite slopes in run-up tests.

that the wave parameters at WG2 provide the most adequate reference for run-up in all tests. Gauges WG1 i WG3 were spaced from WG2 by $\frac{1}{4}L_{op}$, where L_{op} is the peak length for the spectrum produced by the generator. The fourth gauge, WG4, was situated 0.5 m from water line to measure the breaker height. The experiments covered the range of wave parameters specified in the table pasted in Figure 1. — The run-up height was measured directly with a capacitance probe consisting of two parallel strings 2.5 mm in diameter, spaced 12.5 mm and stretching 4–5 mm above slope. Each string was 2.4 m long. Both wave gauges and the run-up probe were practically linear over their entire ranges of measurement.

1.2 Data Processing and Testing

The following conditions have been adopted for the experiments — type of spectrum: Jonswap; duration: recording of 1000 waves on wave gauges; sampling frequency: 20 Hz; depth of water at slope toe: 60 cm. For every and each data series measured we have computed the run-up exceedance curves.

The probability distribution of run-up levels is widely believed to be Rayleigh type, which can be generalized as the Weibull type (van der Meer & Stam 1992):

$$p = Pr(R_u > R_{up}) = \exp\left[-\left(\frac{R_{up}}{R_*}\right)^f\right] \quad (1)$$

in which p = probability of exceedance of the run-up level R_{up}, R_* = scale parameter, f = shape parameter (f = 2 for Rayleigh distribution).

Since the validity of this type is not proved to occur universally and because of uncertainties in definitions we have employed the following three methods of run-up computation:
1. A standard sorting program is used to identify the heighest run-up and the lowest run-down, and arrange all signals from zero (SWL) to the extrema. The resulting exceedance curve is referred to as *residence curve* for it describes the exceedance of given levels by the *water* running up or down and *residing* on slope and *not* waves on slope. The respective run-up height is denoted by $R_u^{(1)}$.
2. A sorting program identifies run-up waves by the zero upcrossing principle and arranges them from the highest run-up amplitude to zero. The reference zero-crossing datum was tried at a few levels about SWL, and finally established as the mean run-up/down level, where the number of zero crossings was highest. Two procedures have been used to compute the exceedance probabilities:
2a. The number of run-up waves was assumed to be equal to the number of incident waves (at WG2). Dummy zeros were added to the measured run-up series, as if some waves were 'invisible' in the data recorded on the slope. This is the procedure adopted by *Delft Hydraulics* (personal communication from de Waal); that run-up height was denoted by $R_u^{(2)}$.
2b. The real number of waves on slope was taken i.e. no dummy waves were added to the measured set. That run-up height was denoted by $R_u^{(3)}$.

2 INTRODUCTORY AND SMOOTH SLOPE EXPERIMENTS

2.1 Check on Definitions

A number of tests were carried out on smooth 1:4 slope prior to our principal experiments. Their primary objective was to determine the accuracy of wave reflection computations; establish repeatability of tests; find out how the type of wave spectrum affects the run-up quantities; look at wave transformation patterns; assess the effect of water depth on the measured data; select best methods for computation of run-up exceedance and compare the data measured elsewhere (at *Delft Hydraulics*) and those studied at *IBW PAN*.

First of all, we have compared the irregular wave run-up exceedance curves obtained under four different assumptions. Some typical examples are ilustrated in Figure 3. Not surprisingly, the exceedance curves defined differently can be quite different, depending on definition and reference (levels, number of waves). The most adequate description seems to be warranted by the residence curve approach and the run-up upcrossing in approach (b) (the crossing reference datum taken as the set-up level on slope, not SWL). However, one must admit that an engineer is interested rather in waves running up and down than in water staying on the slope. For the sake of subsequent completeness of comparison and intercalibration of various datasets, the entire data processing procedure

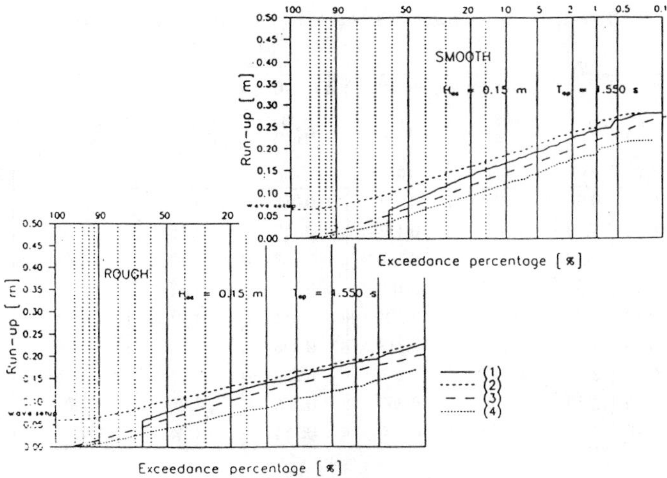

Figure 3. *IBW PAN* irregular run-up data interpreted in four versions: (1) run-up upcrossing, approach (a); (2) run-up upcrossing, approach (b), run-up heights above SWL; (3) residence curve, approach (c), absolute levels above SWL; (4) run-up upcrossing, approach (b), run-up heights above mean run-up/down level.

applied to our tests has incorporated both approaches — (a) and (b).

2.2 Effect of Water Depth

The relative water depth h/H, in a variety of versions, is commonly accepted as a simple though important controlling quantity. A series of experiments with different depths of water were conducted for H_{os}=0.10 m, nine peak periods T_{op}=0.9–2.5 s and three water depths h = 0.50, 0.60 i 0.70 m. In addition, at h = 0.80 m four other experiments were executed for the same H_s and T_{op} = 0.9, 1.1, 1.3 and 1.5 s. The results are summarized in an internal report by *IBW PAN*. Given in Table 1 is $R^{(2)}_{u2\%}$ for different depths of water, computed by method [2a].

From the collected data one can draw the following *tentative* conclusions:

- run-up grows as h increases, in the tested range;
- $R^{(2)}_{u2\%}$ for h = 0.60 m is by some 19–25% greater than $R^{(2)}_{u2\%}$ at h = 0.50 m; and $R^{(2)}_{u2\%}$ at h = 0.70 m is greater by some 4–7% than at h = 0.60 m;
- run-up height differences are the greater the longer the generated waves (slightest differences for T_{op} = 0.9 s, and greatest at T_{op} = 2.5 s).

Table 1. $R^{(2)}_{u2\%}$ computed by method [2a] for three depths of water.

		h=0.50 m	h=0.60 m	h=0.70 m		
H_{os} (m)	T_{op} (s)	$R^{(2)}_{u2\%}$ (m)	$R^{(2)}_{u2\%}$ (m)	$R^{(2)}_{u2\%}$ (m)	$\frac{R^{(2)}_{u2\%(h=0.6)}}{R^{(2)}_{u2\%(h=0.5)}}$	$\frac{R^{(2)}_{u2\%(h=0.7)}}{R^{(2)}_{u2\%(h=0.6)}}$
0.10	0.90	0.072	0.081	0.088	1.13	1.09
0.10	1.10	0.096	0.113	0.119	1.18	1.05
0.10	1.30	0.118	0.140	0.146	1.19	1.04
0.10	1.50	0.139	0.188	0.181	1.35	0.96
0.10	1.70	0.155	0.221	0.211	1.43	0.95
0.10	1.90	0.177	0.213	0.226	1.20	1.06
0.10	2.10	0.193	0.234	0.252	1.21	1.08
0.10	2.30	0.206	0.263	0.266	1.28	1.01
0.10	2.50	0.209	0.260	0.281	1.24	1.08

In view of the above distinctions one is tempted to reconsider the effect of water depth on run-up height in terms of the Ursell number rather than relative depth h/H, whereas the former encompasses wavelength.

2.3 Effect of Spectral Shape and Other Effects

The constant $C_{2\%}$ in the run-up formula $R_{u2\%} = C_{2\%} H_s \xi_p$ is known to assume values from about 1.4–1.5 for narrow-band wave spectra to about 1.8–2.0 for broad-band wave spectra. Our findings are illustrated in Figure 4. The two parameters for Pierson-Moskowitz (PM) were taken as significant wave height and peak period, while one parameter was either of them. Curve (1) deviates strongly in all cases tested. From our data it can be inferred that the one-parameter PM spectrum yields a much lower run-up height. For instance, R_{umax} is smaller by 38%, while $R_{u2\%}$ falls by 50% versus the other types of spectra, all latter having practically the same effect. It can be concluded that close run-up patterns are obtained if spectral shapes are not 'too narrow'. Similar observations have been provided by van der Meer & Stam (1992).

Other effects touched upon in our introductory analysis encompass the *influence of wind, bottom mobility* and the resonance phenomena associated with the uprush-downrush time.

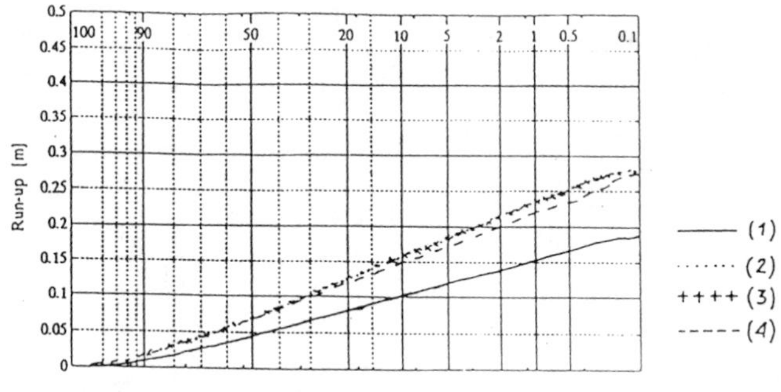

Figure 4. *IBW PAN* run-up data for four types of wave spectrum: (1) one-parameter PM; (2) two-parameter PM; (3) Bretschneider; (4) JONSWAP.

2.4 Basic Findings for Run-up on Smooth Slope

Results of *IBW PAN* run-up measurements on smooth 1:4 slope are illustrated in Figure 5, which depicts results of tests as per programme, i.e. for waves corresponding to control parameters given in Figure 1 and contains additionally the values of $R^{(2)}_{u2\%}$ measured at Delft Hydraulics (van der Meer & Stam, 1991). The drawing summarizes our experimental results for h = 0.50, 0.60, 0.70 and 0.80 m. By and large, the agreement of *DH* and *IBW PAN* data can be deemed satisfactory, so no particular findings deviating from the earlier knowledge have been established. The major purpose of our smooth-slope tests was then understood as provision of reference for rough, bermed and composite slopes.

3 *IBW PAN* IRREGULAR WAVE RUN-UP TESTS on ROUGH SLOPE

3.1 Tests with Cuboidal Roughness Elements

The run–up tests on slope with cuboidal roughness elements (of quadratic base arranged in stagerred manner) were executed for six configurations of the elements (A—F in Figure 2), two densities and three roughness heights. The density ($D_1 = 1/9$; $D_2 = 1/25$) has been defined as the surface area of all roughness elements to the overall area of slope occupied by the elements The densities are identical with those applied in the earlier experiments by Delft Hydraulics. For D_1 we measured run–up on blocks of $k = 22$, 12 and 6 mm in height, while only for $k = 22$ and 12 mm for D_2, which resulted in 30 combinations and 187 test runs (five to seven test series for each combination).

The ratio of roughness height to wave height k/H_s (at WG2) varied from 0.04 to 0.26 and the wave index ξ_p was in the range of 1.0–2.2. The results of these

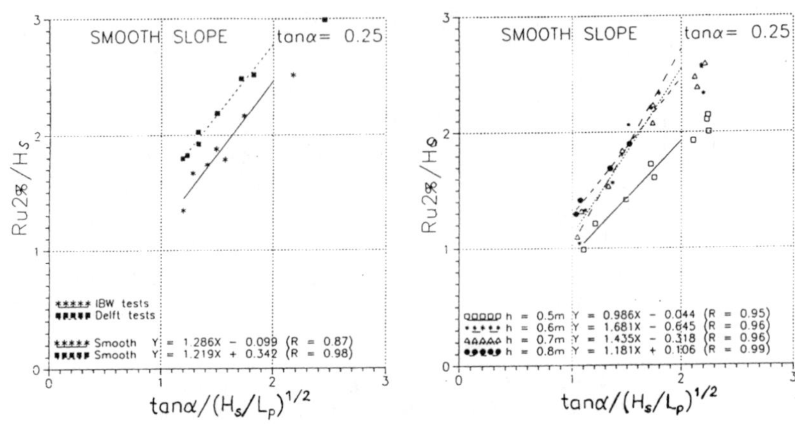

Figure 5. Results of run-up tests on smooth slope.

a and *b* for various block versions.

Configuration	a	b
RG–9–A	0.798	-0.032
RG–9–B	1.115	0.014
RG–9–C	0.761	-0.038
RG–9–D	0.798	-0.040
RG–9–E	1.076	0.006
RG–9–F	0.918	-0.019
RG–25–A	0.879	-0.014
RG–25–B	0.997	-0.001
RG–25–C	0.807	-0.026
RG–25–D	0.870	-0.015
RG–25–E	1.081	0.007
RG–25–F	0.948	-0.005

R2u2% = 2−% wave run−up height as defined by DH
H2s = significant wave height at WG2
Lp2 = peak wavelength at WG2
R2 = regression coefficient

smooth (Y=1.286*X−0.099; R2=0.870)
version A (Y=0.857*X+0.080; R2=0.968)
version B (Y=1.319*X−0.050; R2=0.906)
version C (Y=0.692*X+0.245; R2=0.927)
version D (Y=0.825*X+0.069; R2=0.994)
version E (Y=1.039*X+0.114; R2=0.931)
version F (Y=0.748*X+0.365; R2=0.941)

R2u2%/H2s
ksi=tan(alpha)/sqrt(H2s/Lp2)

Figure 6. *IBW PAN* run-up data for slope with artificial roughness blocks; (H_{2s}) significant wave height at *IBW PAN* wave gauge WG2; ($R_{u2}^{(1)}$) 2-% run-up height as in approach (a).

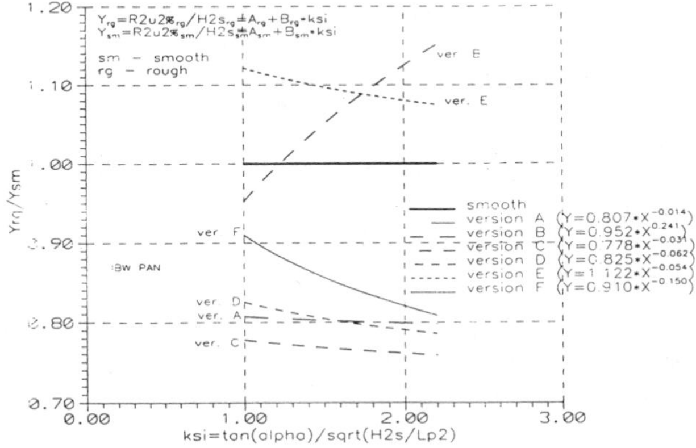

Figure 7. *IBW PAN* run-up data for rough slope with artificial roughness blocks vs. smooth slope.

run-up tests for rough slopes have been compared with our earlier findings for smooth slope. Some results are depicted in Figure 6.

In summary, one can conclude that versions A and C are always advantageous, for all their run–up curves fall below the counterparts for smooth slope. Almost the same can be said about version D, except for 9–6–D, which is slightly above the smooth slope curve. Similarly, one can ascertain that the lowest pull-down of the exceedance curve for rough slope has been achieved for versions 9-22-D (density 1/9 and height k=22 mm) and 25-22-C (density 1/25 and height k=22 mm). The configurations B, E and F have not brought about any substantial trend of run–up due to the presence of blocks. For the configurations 25–12–E, 9–22–B $\approx$ 9–6–F, 25–12–E and 9–12–E, the exceedance curves have been found above their smooth slope counterparts. Among those configurations, causing higher run–up than for smooth slope, one should point to the highest 'amplification' effect for 9–12–E.

Effect of block height on wave run–up

For each configuration A ... F, the measured data has been approximated with power curves of the following form, for the ratio $R_{u2\%}$ of 2-% run-up height, rough over smooth:

$$\frac{[R^{(2)}_{u2\%}]_{rough}}{[R^{(2)}_{u2\%}]_{smooth}} = a \cdot (\frac{k}{H_{s2}})^b \tag{2}$$

with the coefficients a and b estimated individually for each version.

Table 2 summarizes the coefficients a and b for two densities and six versions of roughness elements on slope. It is seen that $R_{u2\%}^{(2)}$ decreases generally with increasing relative roughness k/H_{s2} in versions A (the strongest reduction of run-up), C and D (the weakest yet clear-cut reduction). For some versions, however, the trend was opposite. Inter alia, for version 9–B the run-up height was always higher than the smooth slope counterpart. Here, for the ratio k/H_{s2} increasing from 0.05 to 0.10, the relative run-up increases but further growth of relative k is accompanied by an almost unaltered run-up, as might be judged from the scattered data. Some misleading (bidirectional) data scatter is also noted for versions 9–D, 9–E and 9–F. It can even be claimed that for the versions D, E and F the relative run–up heights are higher about the relative block height of 0.05, and it is only when the block height increases above this threshold of k/H_{s2}=0.05 that the run–up height decreases noticeably.

For the other block density of 1/25, the measurements were executed in the range of relative block height k/H_{s2} = 0.08–0.25. Again, increasing block height brings about general reduction of run–up height for versions A, C and D (in this order of increasing reduction). However, for versions B, E and F the run–up rather increases than decreases.

Conclusions are easier to draw on the basis of the summary drawing (Fig. 7), following the version–by–version arrangement. It appears that the most important factor of run–up reduction should be sought in the block configuration (primarily location above or below SWL and the extension along slope) first and only in the relative block height thereafter. The versions A, C and D are most beneficial. The highest reduction of run–up (always relative to smooth-slope run–up!) has been achieved for the versions 9–C, 9–A, 25–C and 9–D (in this order of decreasing reduction). For the version 9–C one encounters the following reduction of run–up (relative to smooth slope), for the following relative block height: $\frac{k}{H_{s2}} = 0.05 \rightarrow 15\%$; $\frac{k}{H_{s2}} = 0.10 \rightarrow 17\%$; $\frac{k}{H_{s2}} = 0.15 \rightarrow 18\%$; $\frac{k}{H_{s2}} = 0.20 \rightarrow 19\%$; $\frac{k}{H_{s2}} = 0.25 \rightarrow 20\%$.

On the basis of the smoothed data (some scatter smoothed out!) one can claim that the average reduction for versions 9–A and 25–C was weaker by 3–4% than for 9–C and by 5% for 9–D. In general, if the block height is increased by a factor of five (k/H_{s2} from 0.05 to 0.25), the relative run–up will fall by only 5%. The run–up on rough slopes is clearly higher than on smooth slope if blocks are arranged as in versions 9–B, 9–E and 25–E.

Comparison of run–up heights measured as function of ξ has been facilitated by approximating the measured data with straight lines in the form

$$\frac{R_{u2\%}^{(2)}}{H_{s2}} = a \cdot \xi + b \tag{3}$$

The approximation lines do not encompass those obtained for the cases with

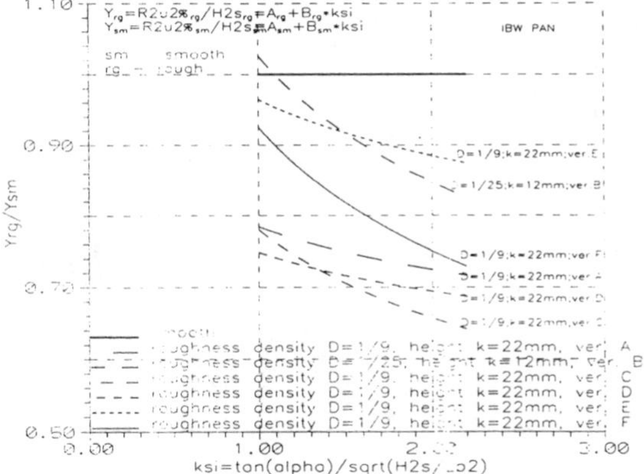

Figure 8. *IBW PAN* best rough slope configurations.

L_{op} = 10 m, for which the wave reflection factor r was about 0.6. Hence the processed data holds for ξ between 1.1 and 2.2. As usual, the smooth slope cases have also been added for comparison.

In order to make the results more conspicuous, the following ratio has been chosen as a measure of the run-up reducing (amplifying) effect, rough slope vs. smooth one:

$$\frac{Y_{rg}}{Y_{sm}} = \frac{(A_{rg} + B_{rg} \cdot \xi)}{(A_{sm} + B_{sm} \cdot \xi)} \tag{4}$$

in which the indices 'rg' and 'sm' denote rough and smooth slope, respectively. The coefficients A and B for smooth slope have been taken as -0.099 and 1.286, respectively.

The computations have been carried out for ξ in the range 1.0 to 2.2, in steps of 0.1. The sets of Y_{rg}/Y_{sm} have been approximated as the following power functions of the dimensionless parameter ξ:

$$\frac{Y_{rg}}{Y_{sm}} = a \cdot \xi^b \tag{5}$$

The results of this transformation are depicted in Figure 7. Analysis of the drawings shows *inter alia* that the highest reduction of run-up height (relative to smooth slope) has been reached in the following configurations, in the descending order of reduction: 9-22-C, 9-22-D, 9-22-A, 9-22-F, and roughly identical 25-12-B and 9-22-E. Figure 8 shows the best roughness configurations permitting the highest reduction of run–up height in the various versions. The two best

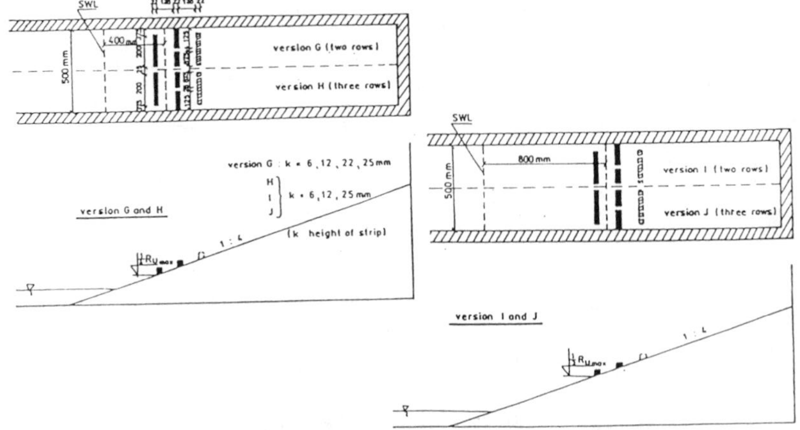

Figure 9. Configuration of strips in *IBW PAN* tests of run-up on rough slope.

options 9-22-C and 9-22-D are described by $Y_{rg}/Y_{sm} = 0.780\xi^{-0.248}$ (particularly for $\xi > 1.3$) and $Y_{rg}/Y_{sm} = 0.749\xi^{-0.109}$ (for $\xi < 1.3$), respectively.

The following remarks summarizing our rough slope tests can be endeavoured:

- The wave reflection factor r has turned out independent of the density, height and position of roughness elements. Its values has ranged from 0.15 to 0.38 in our tests, with the exception of the long–wave cases (10-m wave), where they amounted to 0.55–0.68;

- The effect of roughness height on run–up reduction (rough vs smooth slope) increases with relative roughness height k/H_{s2} but the most substantial part of this reduction occurs for k/H_{s2} about 0.10 (which seems to be a certain threshold value where the run–up reduction effect becomes clear and conspicuous). Further increase in k/H_{s2} brings about little progress. This is illustrated by the finding that a fivefold growth of the roughness height (k/H_{s2} increasing from 0.05 to 0.25) adds more only 5% of run–up reduction. Hence it is always advisable to think twice before embarking on a higher roughness scheme (for the run–up reduction effect can be disproportionally low in terms of the material and construction cost);

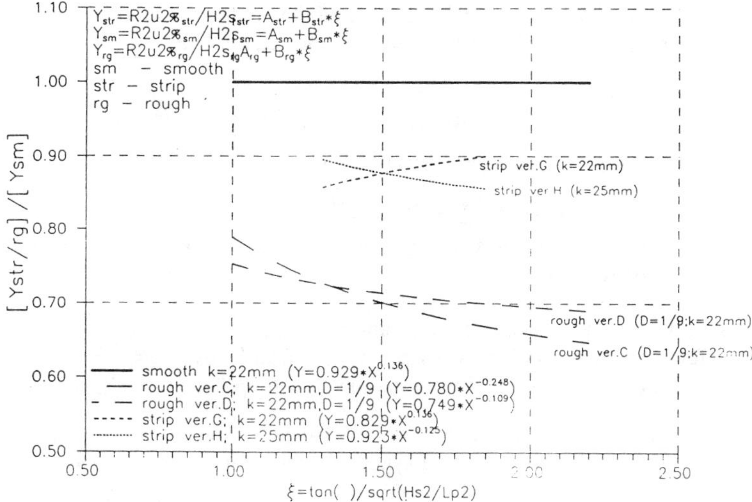

Figure 10. Comparison of run-up findings for strips and cuboidal roughness elements in *IBW PAN* tests.

- The position of blocks on rough slope proves to be the important factor of block effectiveness with regard to run–up damping, the active length of rough slope being of lesser importance. Hence version C is the best option, covering the entire range of parameters tested. The positioning of roughness elements about SWL provides the best reduction of run–up. One should rather place blocks above SLW than below it. From the tests it appears that the active length on slope should not be smaller than 0.5 R_u and that the blocks placed just above SWL play the most important role. The extension of the active length from 0.5 to 1.0 R_u brings about only minor improvement if above SWL but the extension of the active length about SWL, say one half of the active length above SWL and just the same below SWL is quite effective.

3.2 Tests with Strip Roughness Elements

This layout of artificial strips arranged across the slope in parallel rows is illustrated in Figure 9. From the 88 run-up tests for various configurations of strip elements, and upon comparison with their counterparts for smooth slope one can draw the following conclusions:

- The position of strips on slope is a decisive run-up controlling factor;

- the minimum run-up height on slope has been found as follows: version G-22 for $\xi < 1.5$ (k/H_{s2} = 0.148–0.263) and version H-25 for $\xi > 1.5$ (k/H_{s2} = 0.142–0.294);
- the reduction has been formulated mathematically as follows:
  - version G-22 $\rightarrow \frac{Y_{str}}{Y_{sm}} = 0.829 \cdot \xi^{0.136}$;
  - version H-25 $\rightarrow \frac{Y_{rg}}{Y_{sm}} = 0.923 \cdot \xi^{-0.125}$;
  - version G denotes two rows, while version H embodies three rows of strips positioned at $\frac{1}{4}R_{umax(smooth)}$ above still water line SWL;
- upon analysis of the relative run-up height $R^{(2)}_{u2\%(strip)}/R^{(2)}_{u2\%(smooth)}$ as a function of the height of strip k to significant wave height measured at WG2 (k/H_{s2}) one obtains the minimum run-up height for the version H;
- for the range tested, i.e. k/H_{s2} = 0.03–0.29, the results have been approximated by a second-order polynomial, which for version H reads

$$\frac{[R^{(2)}_{u2\%}]_{strip}}{[R^{(2)}_{u2\%}]_{smooth}} = 7.973 \cdot (\frac{k}{H_{s2}})^2 - 3.094 \cdot (\frac{k}{H_{s2}}) + 1.202 \qquad (6)$$

- for version H and the range k/H_{s2} = 0.083–0.29, the run-up heights in the presence of strips have been lower than for the smooth slope. The reduction reads respectively: $\frac{k}{H_{s2}} = 0.10 \rightarrow 3\%$; $\frac{k}{H_{s2}} = 0.15 \rightarrow 8\%$; $\frac{k}{H_{s2}} = 0.20 \rightarrow 10\%$; $\frac{k}{H_{s2}} = 0.25 \rightarrow 7\%$; $\frac{k}{H_{s2}} = 0.29 \rightarrow 2\%$.
- the minimum run-up height for version H has been attained for

$$\frac{k}{H_{s2}} = 0.194 \rightarrow [R^{(2)}_{u2\%}]_{strip}/[R^{(2)}_{u2\%}]_{smooth} = 0.902 \qquad (7)$$

- the effect of strips, stretching across the entire breadth of the slope, on the run-up height is much weaker than that of cuboidal blocks

For the sake of comparison, Figure 10 illustrates the effect of strips and cuboidal blocks on the run-up reduction. To this end, the ratios Y_{str}/Y_{sm} and Y_{rg}/Y_{sm} are given as functions of ξ for the versions providing the highest reduction of run-up height, compared with that on smooth slope. The above reduction (strip or block vs. smooth) reads as follows:

- strip versions
  - G-22 (k/H_{s2} = 0.148–0.263) by 14% at $\xi = 1.25$ and by 10% at $\xi = 1.85$;
  - H-25 mm (k/H_{s2} = 0.142–0.294) by 10% at $\xi = 1.25$ and by 14% at $\xi = 1.85$.

- block versions
  - C-9-22 (k/H_{s2} = 0.154–0.262) → by 20% at $\xi = 1.0$ and by 35% at $\xi = 2.2$
  - D-9-22 (k/H_{s2} = 0.149–0.253) → by 30% at $\xi = 1.0$ and by 30% at $\xi = 2.2$.

Hence one can infer that blocks reduce the run-up height by 30% on the average while the respective reduction by strips is only 10%, thus proving that strips are less effective than blocks in confining run-up.

4 CONCLUDING REMARKS

It remains to be shown whether and to which extent, the small-scale run-up tests carried out at *IBW PAN* and described herein are equivalent to such experiments on large-scale facilities. The comparison of our findings for smooth slope with *DH* counterparts proves promising. Some physical phenomena on slope, such as inertia, gravity and perhaps wave reflection, can be scaled by Froudian law, while the friction inherent in run-up seems to be slightly scale-dependent, for it was likely to has occurred at high Reynolds numbers (and supercritical regimes happen to be similar in small and large scales). Hence, at least in part, our experiments exemplify some cases when small scale research setups can be employed to obtain results comparable with those stemming from large facilities, with an obvious advantage of labour and cost savings.

Acknowledgements The investigations described herein have been carried out under the programme DWW-542, which is gratefully acknowledged.

REFERENCES

van der Meer, J.W. & C-J.M. Stam (1992). Wave run-up on smooth and rock slopes of coastal structures. *Journal of the Waterways, ... Ocean Engineering, ASCE* WW5, Paper 1626.

van der Meer, J.W. & J.P. de Waal (1991). 'Waterbeweging op taluds'. Waterloopkundig Laboratorium WL, *H1256.*

Shore Protection Manual (1984). U.S. Army Coastal Engrg. Res. Center; U.S. Government Printing Office, Washington, D.C.

Zeidler, R.B., M. Szmytkiewicz, J. Kolodko, M. Skaja & R. Jednachowski (1994). Wave run-up on smooth and rock slopes of coastal structures. Discussion. *ASCE Journal of the Waterway, Port, Coastal, and Ocean Engineering* (in press).

REVIEW OF WORKS USING CRIEPI FLUME AND PRESENT WORK

Ryoichi Kajima [1] and Tsutomu Sakakiyama [2]

ABSTRACT: CRIEPI constructed a large wave flume 15 years ago to get large scale experimental data of cross−shore transport due to waves in which scale effect was considered inevitably significant in small scale experiments. After the data were obtained, the flume has been used to investigate the scale effect on the interaction between waves and coastal structures such as stability of armor units, wave reflection and transmission, and also the wave pressure on breakwaters. Recently the interaction of a seawall and waves is a main concern to develop a new design method of seawalls for siting power plants on a man−made island. For this purpose, deformation of an armor layer of seawall under attack of waves higher than those to which the armor blocks are critically stable has been studied including investigation of the change of overtopping rate. Through these experiences, some of the facts about scale effects are shown.

INTRODUCTION

A large wave flume(CRIEPI FLUME) was constructed in 1979 by Central Research Institute of Electric Power Industry,Japan. It is 205m long,6.0m deep,3.4m wide with a 1/15 slope at the end. The purpose at that time was to obtain large scale experimental data of cross−shore sediment transport due to waves in which the scale effect of hydraulic model experiment was considered inevitably significant in small scale experiments. This experiment made a chain of a research project on shore processes,and the results of the project were published (K. Horikawa, ed. 1988).

Afterwards CRIEPI FLUME has been used to investigate the scale effect on the interaction between waves and coastal structures such as the stability of armor units,

1) Associate Vice President, Abiko Research Institute, Central Research Institute of Electric Power Industry (CRIEPI), 1646 Abiko, Abiko−shi, Chiba−ken 270−11, Japan.
2) Senior Research Engineer, ditto.

the wave reflection and transmission, and also the wave pressure on breakwaters.

Recently the interaction of a seawall and waves is a main concern to develop a new design method of seawalls for siting power plants on a man–made island. For this purpose, deformation of armor layers of seawalls under attack of waves higher than those to which the units are critically stable has been investigated including the effect on overtopping rate. Table 1. shows the brief history.

Finally these experiences on scale effects are summarized.

Table 1. Brief History of CRIEPI Flume

	Wave generating system	Main subjects
1979	built	
1980	Regular	Cross–shore transport rate. Profile change. Stability limit of armor units.
1985	Irregular	Energy dissipating monolithic breakwaters.
	Absorbing	Mild sloped seawall blocks. Waves through permeable structure.
1990		Wwave force to seaweed. Reduction of impact force Deformation & overtopping charcteristics
1994		of front mounted armor blocks.

EXPERIMENTS ON CROSS–SHORE SEDIMENT TRANSPORT

Twenty four runs of experiments on cross–shore sediment transport were carried out for four years using two kinds of sand. The wave conditions were as follows:the regular wave period T=3.0 to 12.0s and the deep water wave height H_o =0.3 to 1.78m. The median diameters of sand were 0.47 and 0.27mm. Beach profiles were measured at several stages from the initial to the near equilibrium profile corresponding to each wave and sediment condition. From these profile data, sediment transport rates were calculated with Shields parameter Ψ_m up to 3. The power model of sediment transport rate formulas both onshore and offshore were confirmed on the basis of these large scale (Kajima et al. 1982) and the small scale ones (Watanabe et al. 1980). This is at present included in a more general formula of sediment transport rate due to wave and current (Watanabe et al. 1984).

We are now planning to conduct similar experiments under irregular wave conditions with absorbing wave ganarating system,which will give more useful data of the profile change and cross–shore transport rate.

EXPERIMENTS ON COASTAL STRUCTURES

Stability of Armor Units

From 1985 to 1988, regular wave experiments on the scale effect in stability of armor units were carried out using several sizes of armor unit models for three kinds of blocks. Fig.1 shows the scale effect presenting the correcting factor of critical stability number N_s ,which is proportional to the root cube of K_D value for Hudson's formula, as a function of Reynolds number R_N ,in which the length scale is the size of the model of armor units and the velocity scale is the square root of *gH*,where *g*acceleration due to gravity and *H*wave height (Shimada *et al.* 1986).

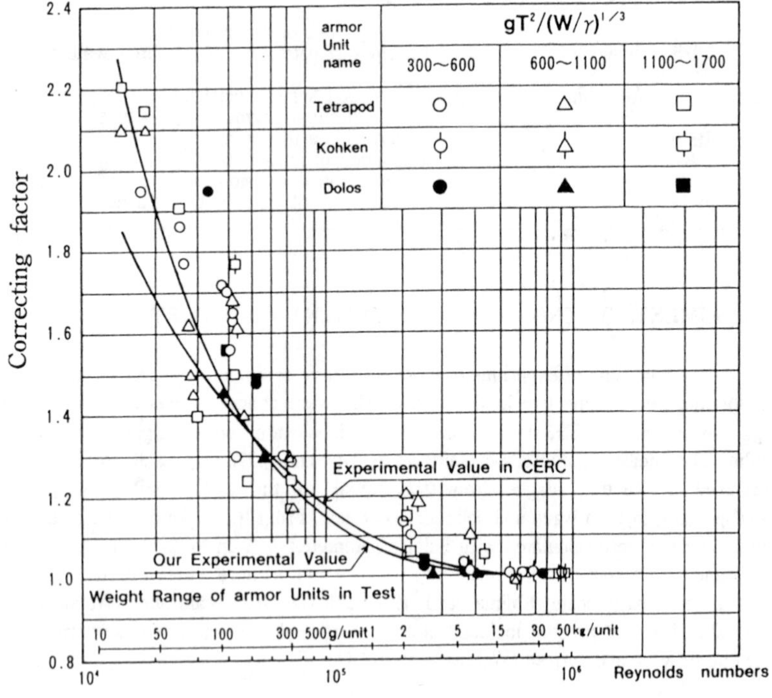

Fig.1 Correcting factor of experimental stability number

$$N_s = \frac{H}{(W/\gamma)^{1/3}(Sr-1)} \quad , \tag{1}$$

$$K_D = \frac{N_s^{\,3}}{\cot\theta} \quad , \tag{2}$$

$$R_N = \frac{(W/\gamma)^{1/3} \cdot (gH)^{1/2}}{\nu} \quad , \tag{3}$$

where H:progressive wave height, W:the weight of an armor unit, γ :unit weight of armor unit, Sr :specific wave of armor unit, θ :the angle of slope of armor layer. The result agrees fairly well with that obtained at CERC(Thomsen et al. 1972).

According to the result, the larger the size of armor unit model is, the more stable the model of armor layer is. Because we often determine the weight of armor units according to some reduced scale model tests, there is a certain margin of stability in reality due to the scale effect. The figure shows that the critical Reynolds number R_N beyond which the scale effect becomes negligible is nearly $3\text{x}10^5$. Sakakiyama & Kajima(1990) explained the mechanism of the scale effect on stability of armor units from the point of view of the scale effect in wave force acting on armor units. It is derived that the parameter K_D is inversely proportional to the third power of the drag coefficient C_D which is a function of Reynolds number.

$$K_D \sim C_D^{-3} . \tag{4}$$

In addition,because $K_D \sim N_S^{-3}$, they considered that the dependence of N_S on R_N could be interpreted almost as that of C_D on the same parameter R_N.

Waves near Permeable Structures

From 1987 to 1990, CRIEPI FLUME was used mainly to investigate the interaction of waves and permeable structures, such as wave reflection by and wave transmission and forces through permeable structures to a solid wall. At the same time, a numerical simulation method to predict two−dimensional wave motion around submerged and/or partly−submerged structures with arbitrary permeability was beeing developed by extending the Navier−Stokes equation (Sakakiyama *et al.* 1992). We had needs to estimate considerably low transmission rate as correctly as possible. For we were engaged in a project of feasibility study of floating plants. In general wave transmission rate through ordinary breakwaters is considered several to 15 % in wave height ratio. On the other hand the required tranquility of a mooring basin for the plant was several percents.

From experimental data of wave reflection and transmion coefficients, the drag and inertia coefficients included in the numerical method were estimated to give best fit to the experiments. The scale effects on the wave reflection,transmission and wave pressure were also investigated. Fig.2 shows a simulated result of wave transformation near and inside a caisson breakwater covered with armor units and stones.

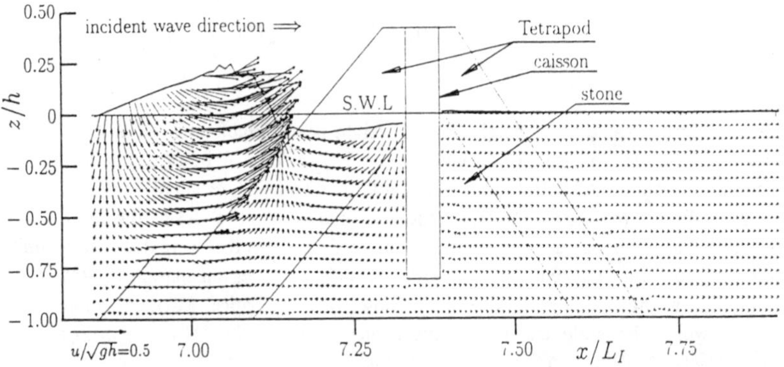

Fig.2 Numerical result of wave transformation around a permeable breakwater

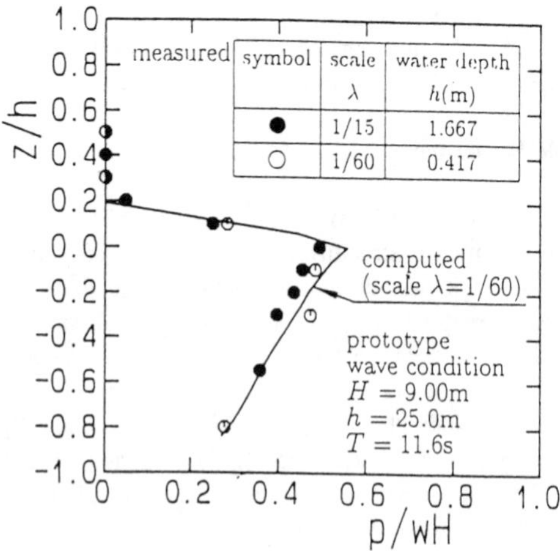

Fig.3 Wave pressure profile along seaward vertical face of inner structures of Fig.2

Fig.3 is the comparison of wave pressures acting on a front wall of a caisson covered with permeable materials, the same structure as shown in Fig.2; the values are obtained by different experiments and a numerical simulation(Sakakiyama *et al.* 1992). The simulated result agrees very well with the experimental results. This figure, in addition, shows a scale effect in wave pressure. There seems to be a tendency that the wave pressure obtained by the 1/60−scaled experiment is slightly larger than that obtained by the 1/15−scaled experiment. The difference is, however, negligible in a practical sense.

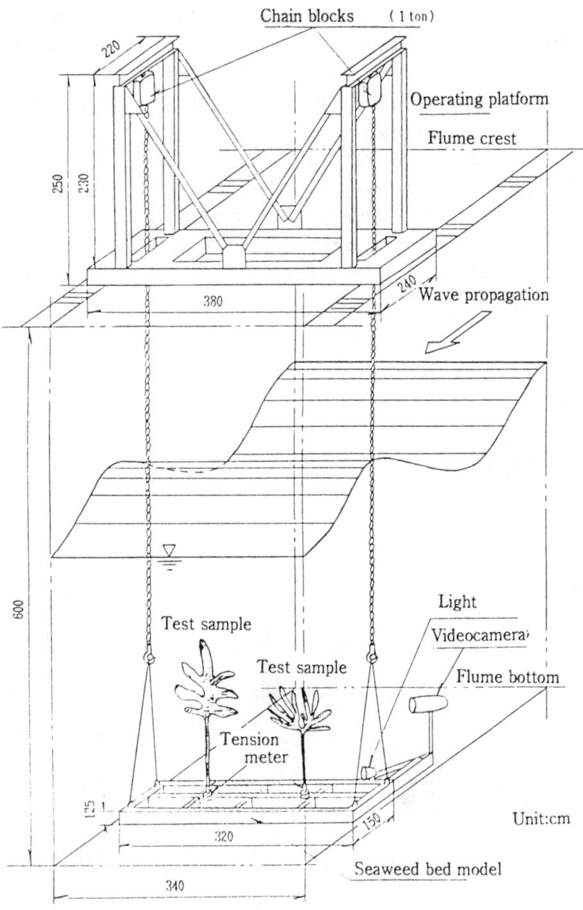

Fig.4 Schematic view of measurement of wave force on real seaweeds

Wave Force on Seaweeds

In a chain of study to develop a method of kelp forest creation at local sea areas, the needs arose to estimate wave forces through seaweeds to surface materials of artifitial kelp bed foundations. The total wave force acting on a single seaweed which grows on a horizontal sea bottom was measured in CRIEPI FLUME.
The test pieces were sampled at a coastal area and transported in sea water tanks to the test site, and cared for near the flume. The measurement was to be finished in 1 hour after a sample was fixed to a measuring platform floor. Because fresh water was used in stead of sea water, the seaweeds were anticipated to be damaged by osmotic pressure,and thus leaves would deform and stems soften. Fig.4 gives the conceptual scheme of measurement, and Fig.5 shows the result for *Eisenia bicyclis*, where F:total wave force, w_0 :unit weight of sea water, and H:wave height. The total wave force to the kind of seaweeds growing in the depth of 4.5 meters increases in proportion with the scale of seaweeds with up to 2000 cm^2 of leaf area, and it saturates for larger seaweeds. This is considered the effect of flexibility of the seaweeds. The wave force on a seaweed is estimated nearly equal to or less than 1/150 of that on a vertical wall with same area. The vertical component was found about 20% of the horizontal component. Thus the upwardforce on each surface block were estimated by the sum of direct fluid force and upward wave force through seaweeds in a supposed full growth stage(Hasegawa *et al.* 1992).

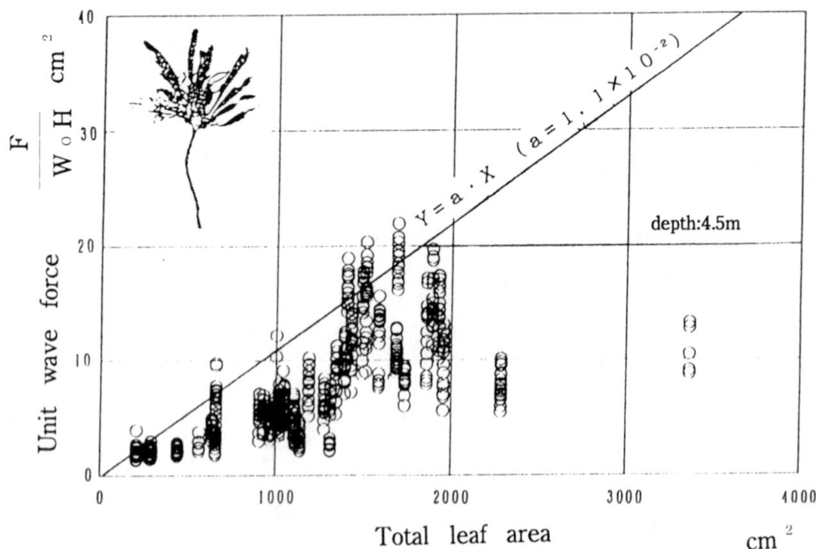

Fig.5 Wave force on a single seaweed measured for *Eisenia bicyclis*

New Design Method of Specifically Important Seawalls

Two–step design. The authors are at present make efforts to develop a rational design method of seawalls of man–made fill islands for nuclear power plants(NPP's). Examples of return periods of design natural conditions adopted in the structures in Japan are shown in Fig.6. The design waves of protective facilities of ports are the waves with return period of 30 to 100 years, among which only the design wave for floating oil storage bases is prescribed in the governmental technical standards as a concrete value of return period. It can be seen that the more important the facilities are, the longer return period waves are adopted for design. In Fig.6 examples of the periods for design seismic motion for nuclear plants in Japan are described as well, which correspond to much longer periods than those for protective facilities of ordinary ports. In this regard, we should say that there are different opinions about the probabilistical description of design earthquakes. But the basic idea of design natural conditions for NPP's in Japan is that they should be the severest which occurred or are supposed to occur in each local history.

This leads to the following situation:
When speaking to the design of an important seawalls of man–made islands for NPP's, we should have a prospect of how to cope with the much higher waves than ordinary design waves, even if there is a difference in characteristics of earthquake load and wave load; the former strikes the heart of plants directly, and the latter does not.

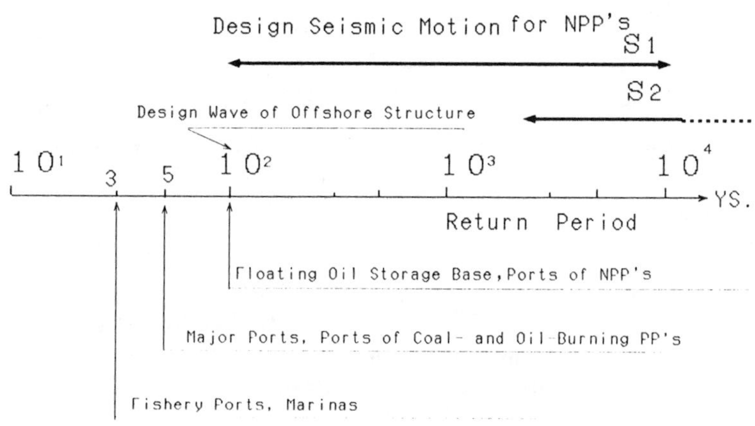

Fig.6 Return periods of design conditions for structures

Because it cannot be denied that there would be any progressive effects that might reach plants at last, we should have a plant safety scenario against much severer waves than ordinary design waves. But on the other hand, those structures need not be stable enough to the severer and rarer waves in the same sense as to ordinary design waves. Fig.7 shows the situation. For a severer design waves, the seawall structure can be said sufficiently sound and strong only if the deformation is limited and repairable in a term not so long, and the overtopping water has no effect on plant safety.

Thus the target is to develop a 2−step design method as follows:

In the first step, the fundamental section of a seawall is designed due to the conventional method to obtain a stable section against waves with 100 years or so of recurrence period;
In the second step, the section should be checked up for severer wave and other natural conditions to maintain minimun necessary function of the seawall.

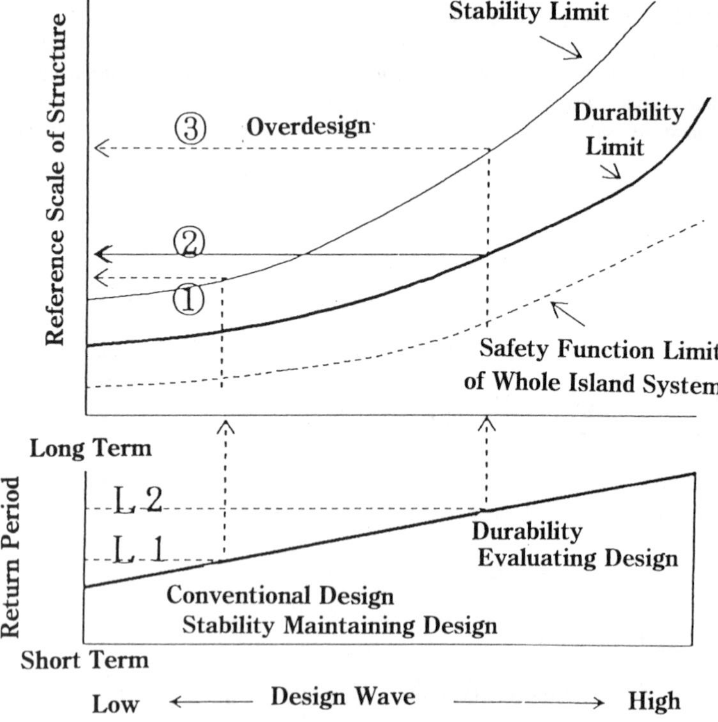

Fig.7 Stability limit design and durability limit design

Characteristics of wave dissipating works beyond their stbility limit. From 1991, the deformation of an armor layer of a seawall and the overtopping rates due to irregular waves higher than critical waves for stability have been investigated to establish the method. We are using both large and small scale model tests: the model scales are λ =1/15.4 and 1/87.5, respectively. The damage level $S=A/D_{n50}$ (A : damaged or eroded sectional area, and D_{n50} : the nominal diameter of the armor unit) defined by van der Meer(1987) is adopted as the level of deformation of an armor layer, which we call the deformation level.

Fig.8 shows the section of the object of study. Figs.9 and 10 are examples of the results which show the scale effects in deformation and wave overtopping rate, respectively(Kajima & Sakakiyama,1993). The scale effect in deformation seems to be more significant for larger deformation corresponding to higher wave height ratio H/H_D, where H_D is the wave height corresponding to stability limit, and larger deformation arises for smaller models. This coincides with the previous results. But no significant difference is seen between small and large scale models around the stability limit. This is contrary to the previous results. Some results for regular wave experiments are plotted in the same figure. Though the reason is not clear, there might be some tendency that the scale effect on the wave force is more remarkable for regular wave condition.

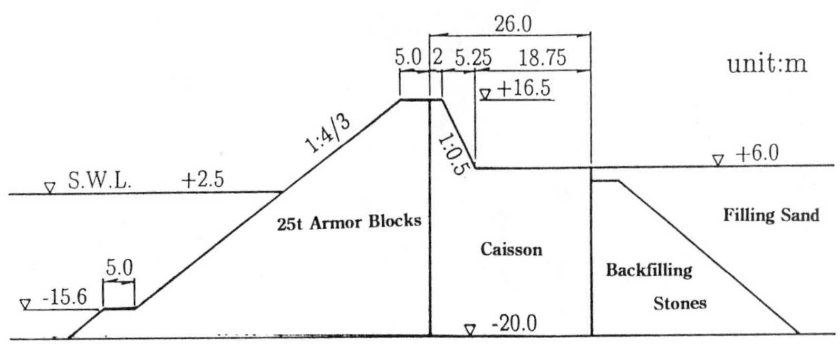

Fig.8 Fundamental components of seawall
Target structure of experiment

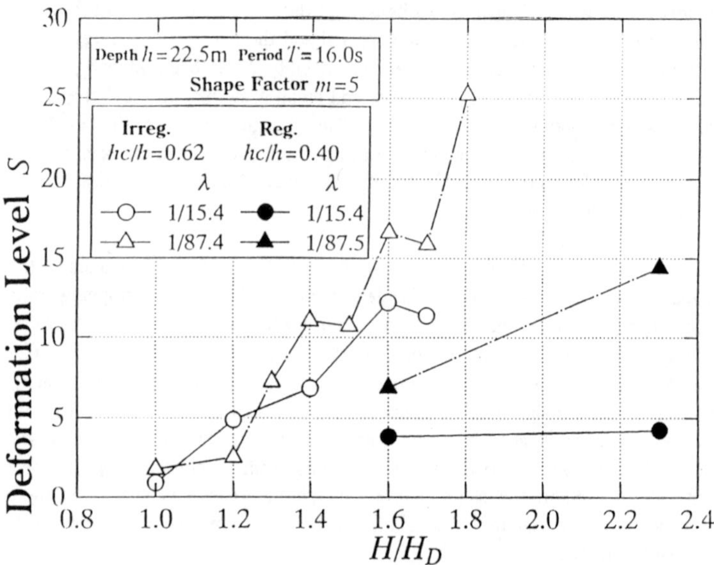

Fig.9 Deformation level S after 3000 waves vz. excess wave height ratio

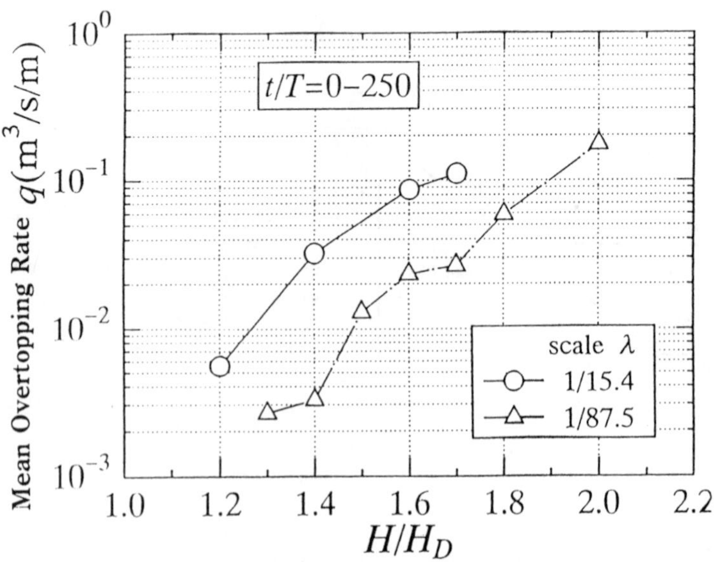

Fig.10 Converted wave overtopping rate vz. excess wave height ratio

The scale effect on overtopping is clearer. There is a tendency that the larger overtopping rates are estimated from larger scale models. Though no data are sited here, the tendency is the same for regular waves. The difference is considered due mainly to the difference of the fluid motion and energy dissipation within the armor layer of different scale models. This shows the possibility to explain the effect with the same drag force effect. The critical Reynolds number can be expected similar to that for armor stability. Thus we are considering that the rate obtained from the larger scale model is approximately the prototype value.

From these experiments, we could know that there is no critical wave height within up to 2.3 times the critically stable wave height beyond which armor mounds would abruptly collapse, and that the deformation level around 20 corresponds to two−layer thickness of the maximum erosion depth of tetrapods. In addition changes in overtopping rate due to deformation of wave dissipating works are within twice the initial value. Thus by preparing sufficient drain capacity with large drainage channel behind parapets, some stability margin can be expected in the target structure, and a good prospect of the two−step design method was obtained.

4. CONCLUDING REMARKS

As for stability limit and deformation of armor layer, the scale effect for irregular waves seems to be less siginificant than for regular waves, especially near stability limit. As the excess ratio of irregular wave height to critical wave height increases,the scale effect becomes clearer and the larger model shows better stability.

As to overtopping, the rate is estimated larger by a larger model. However the authors are expecting that the critical Reynolds number of no significant scale effect is near the critical for the stability of armor layers.

A good prospect of two−step design of specifically important seawalls was obtained. A flow chart of the design method is shown in Fig.11.

In conclusion, although there are data which can obtained only by large scale experiments, most of the large scale experiments have enlightened the significant and important role of medium to small scale experiments by making clear the scale effect quantitatively.

ACKNOWLEDGEMENTS

We are indebted to our former and present colleagues in Hydralics Department of CRIEPI for conducting many projects including large scale experiments.

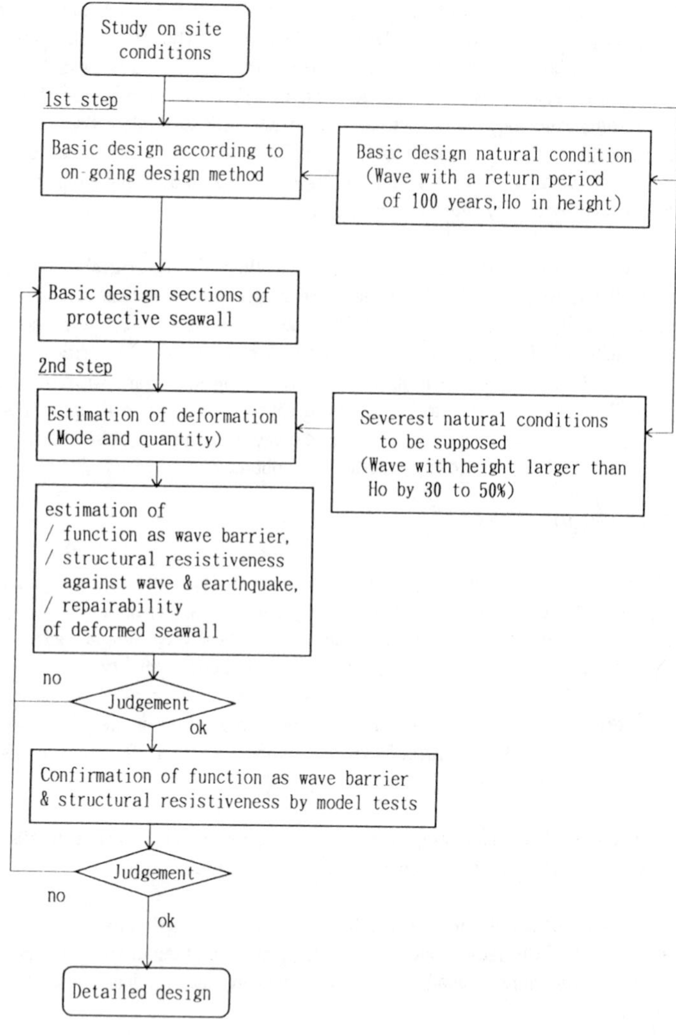

Fig.11 Flow chart of structurally resistve design of specifically important seawall against waves

REFERENCES

Hasegawa,H.,H.Hirakuti,T.Terawaki and Y.Kawasaki(1992) : Structural design of artifitial foundation for kelp bed,Proc. Civil Engineering in the Ocean,JSCE,vol.8,pp.379–384(in Japanese).

Horikawa,K.(ed.)(1988) : Nearshore dynamics and coastal processes –Theory,measurement,and predictive models, University of Tokyo Press,pp.522,Japanese edition was published in 1985.

Kajima,R.,T.Shimizu,K.Maruyama and S.Saito(1982) : Experimental study on cross–shore sediment trans– port by large wave flume (Report No.2) – Sand transport model for coarse sand beach(D50=0.47mm), Proc. 29th JCCE,JSCE,pp.228–232(in Japanese).

Kajima,R. and T. Sakakiyama(1993) : Stability and waveovertopping characteristics of seawall for man–made island against irregular waves higher than stability limit of armor blocks, Proc. 40th JCCE, JSCE, pp. 686–690(in Japanese).

Sakakiyama,T. and R.Kajima(1990) : Scale effect of wave forces on armor units, Proc. 22nd ICCE,ASCE, pp.1716–1729.

Sakakiyama,T. and R.Kajima(1992) : Numerical simulation of nonlinear wave interacting with permeable breakwaters,Proc. 23rd ICCE,ASCE,pp.1517–1530.

Shimada,A.,T.Fujimoto,S.Saito,T.Sakakiyama andH.Hirakuchi(1986) : Scale effects on stability and wave reflection regarding armor units,Proc. 20th ICCE,ASCE,pp.2238–2252.

Thomsen,A.L.,P.E.Wohnt and A.S.Harrison : Rip–rap stability on earth embankment in large and small scale wave tanks,CERC Technical Memorandum No.37.

Van der Meer,J.W.(1987).Stability of breakwater armor layers design formulae.Coastal Engineering,Vol.11,pp.219–239.

Watanabe,A.,Y.Riho and K. Horikawa(1980) : Beach profile and on–offshore sediment transport,Proc.17th ICCE,ASCE,pp.1106–1121.

Watanabe,A.,K.Maruyama,T.Shmizu and T.Sakakiyama(1984) : Numerical model of three–dimensional beach change by coastal structures,Proc. 31st JCCE,JSCE,pp.406–410(in Japanese).

SCALE EFFECT ON IMPULSIVE WAVE FORCES AND REDUCTION BY SUBMERGED BREAKWATER

Tsutomu Sakakiyama[1] **and Masaharu Ogasawara**[2]

Abstract: This paper presents the results of large- and small-scale model tests on impulsive wave forces acting on a caisson breakwater. The effectiveness is evaluated of countermeasure against the impulsive load, constructing a submerged breakwater in front of the caisson. With scaling the testing conditions the scale effects on the impulsive wave forces, the transmitted wave heights over the permeable submerged breakwater and the resultant wave forces are investigated. The wave heights and forces of the longer wave period obtained at the large-scale tests are larger than those at the small-scale ones when the permeable submerged breakwater is constructed. The large-scale model test results are obtained greater than the critical Reynolds number and can be applied for the design according to the Froude law.

1. INTRODUCTION

Impulsive wave pressures acting on a vertical wall of a caisson have been actively investigated by field measurements(Blakmore and Hewson,1984), using a small-scale model test(Hattori and Arami, 1992) and large-scale model ones(Schmit *et al.*, 1992; Oumeraci *et al.*, 1992) and also numerical models(Topliss *et al.*, 1992; Grill *et al.*, 1992). Among the above works, European researchers involving in MAST project aim to develop design guidelines. In Japan, it is recommended that armor units should be provided at the front of the upright section of the caisson to dissipate the impulsive wave force. However, the wave force and the overtopping rate can increase owing to construction of the wave absorbing work because waves

1) Senior research engineer, Central Research Institute of Electric Power Industry(CRIEPI), 1646, Abiko, Abiko-shi, Chiba, 270-11, Japan.
2) Kansai Electric Power Industry Co., 3-3-22, Nakanoshima, Kita-ku, Osaka, 530, Japan

run up on the slope of the armor layer. An alternative countermeasure certainly to reduce the impulsive wave forces is to make the waves break in advance.

This paper describes the efficiency of countermeasure, constructing a submerged breakwater against the impulsive wave forces. The transmitted wave heights and the resultant wave forces are investigated at the small-scale experiments concerning both impermeable and permeable submerged breakwaters.

Also presented is the scale effects on the impulsive wave forces on the vertical wall of the caisson without the submerged breakwater, the transmitted wave height and the resultant wave forces with the permeable submerged breakwater in front of it. Applicability of the results obtained at the large-scale model test for a design is evaluated taking account of the critical Reynolds number for the scale effect.

2. EXPERIMENTS

Two physical scale-model tests were performed by using a large wave flume (CRIEPI FLUME:205m long, 6.0m deep and 3.4m wide) and a small one(55m long, 1.5m deep and 2.0m wide). The model scales λ are $\lambda=1/8$ and $1/50$, respectively(the scale factor of length is 6.25). Fig. 1 shows models of a submerged breakwater on an impermeable mound and a caisson used in the experiment indicating with prototype values.

The distance l between the submerged breakwater and the caisson was determined to be $l/L_{max} \approx 0.3$, where L_{max} is the maximum wave length concerned. When the submerged breakwater is installed farther than this position, the nondimensional wave pressure decreases to $p/wH \approx 1$(Nakaizuni *et al.*). If it constructed closer than $l/L_{max} \approx 0.3$, the wave pressure increases due to the direct attack of the breaker. As will be mentioned later on, the maximum wave period concerned is 15.5s and the water depth is h=10m. We obtained L_{max}=149.1m and chose $l=0.3L_{max} \approx$45m.

The bottom slope was set as 1 on 15 to induce plunging breakers to act on the caisson. A crown height of the caisson is set as h_c=20m above the still water level. It was twice as high as the water depth h=10m in order to obtain data of

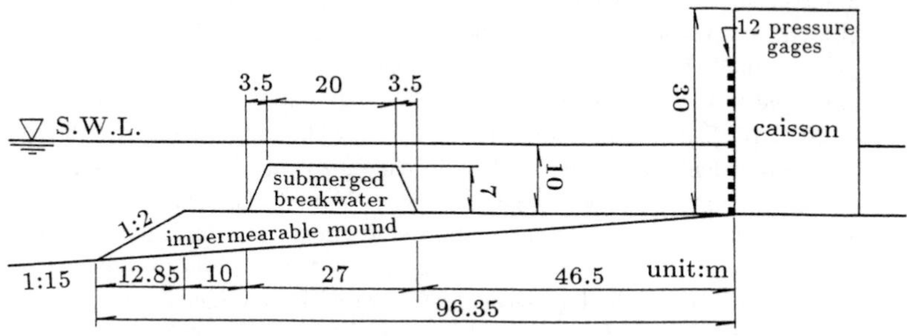

Fig. 1 Models of submerged breakwater and caisson

Table 1 Experimental conditions(λ=1/50)

wave period T(s)	incident wave height H_i(m)	water depth		
		h_i(m)	h(m)	h_r(m)
1.20	0.044–0.215	0.59	0.20	0.06
1.70	0.049–0.182			
2.19	0.062–0.191			

Table 2 Experimental conditions(λ=1/8)

wave period T(s)	incident wave height H_i(m)	water depth		
		h_i(m)	h(m)	h_r(m)
3.01	0.40–1.46	4.25	1.25	0.375
4.24	0.34–1.46			
5.48	0.32–1.44			

the wave pressure under no overtopping condition.

At the first tests, the wave pressures induced by the plunging breakers on the bottom of 1/15-uniform slope were measured without the countermeasure at both the small- and large-scale tests. The wave pressures were measured at twelve levels on the front surface of the caisson as shown in Fig. 1.

At the second tests, the transmitted wave heights over the submerged breakwater and the resultant wave pressure were measured. Both impermeable and permeable submerged breakwaters were used at the small-scale model tests. At the large-scale test only the permeable one was used to investigate the scale effect on the transmitted wave heights and forces.

Twelve 5mm-diameter pressure gages were used at the small-scale tests and 12mm-diameter ones at the large-scale tests. The sampling frequency was 400 Hz at both tests. The data of the large-scale test were recorded at relatively higher frequency than that of the small-scale test(the time scale factor is 2.5). The horizontal wave forces acting on the vertical wall were obtained by integrating the wave pressures vertically.

The permeable submerged breakwater consisted of 7.5t-Tetorapod covered with wire netting not to be deformed by wave action. The impermeable submerged breakwater was made of cement mortar.

Tables 1 and 2 show the experimental conditions. The monochromatic wave periods used in the experiments were T=8.5, 12.0 and 15.5s at prototype values (T=1.20, 1.70, 2.19s at 1/50-model test and T=3.01, 4.24, 5.48s at 1/8-model test). The incident wave heights H_i were chosen so that both nonbreaking and breaking waves attacked the caisson. In Tables 1 and 2, h_i is the water depth of the flat bottom in the wave flume, h the water depth at the submerged breakwater and the caisson, and h_r the depth on the crest level of the submerged breakwater. The water depth h_i at the large-scale test was relatively larger than that at the small-one (T =8.5s: h_i/L_0=0.26 at 1/50-model test, h_i/L_0=0.30 at 1/8-model test and T =15.5s: h_i/L_0=0.079 at 1/50-model test, h_i/L_0=0.091 at 1/8-model test).

Mean values on the wave height, peak pressure and wave force are obtained from eleven waves and are used in the following discussion.

3. WAVE FORCE REDUCTION

Figures 2 through 4 show the comparisons between the progressive wave height on the 1/15-uniform bottom slope(without the countermeasure) and the transmitted wave heights passing over the permeable or impermeable breakwaters obtained at the small-scale experiments (model scale $\lambda=1/50$), where H_0 is the deep-water wave height, and H is the wave height at the water depth h=10m. In these figures, the progressive wave heights on the impermeable mound are also plotted.

The progressive wave height on the 1/15-slope increases proportionally to the deep-water wave height H_0 due to the wave shoaling until the wave breaks. The maximum wave heights for the individual wave period at the water depth h=10m are about H=9.5m at the wave period T=8.5s, H=10.1m at T=12.0s and H=10.5m at T=15.5s without the submerged breakwater.

With constructing the submerged breakwater, waves break on it at the smaller incident wave of which height is greater than about 4m. The transmitted wave over the submerged breakwater disintegrates into free waves with the subharmonic frequencies. The wave height changes spatially as the free waves propagate. The wave height denoted with H is the wave height at the distance of 46m onshore from the submerged breakwater and at the water depth h=10m.

The transmitted wave height decreases to less than 4m at the wave period T=8.5s, to about 5m at the wave period T=12.0s and at T=15.5s to about 5m by the permeable submerged breakwater and to about 6m by the impermeable submerged breakwater.

Figures 5 through 7 show the corresponding comparisons of the wave forces acting on the caisson to Figures 2 and 4, respectively. When any structures are not constructed in front of the caisson, the wave force F_X is proportional to the wave height H when the wave height is less than H=8m. Comparing the wave force at the same incident wave height, clearly seen in Fig. 5, the wave force with the submerged breakwater is greater than that without the submerged breakwater.

The impulsive wave force on the caisson was observed when the wave height becomes greater than 8m. The maximum wave forces per unit meter measured are about 1160 kN/m(120tf/m) at T=8.5s about 1720 kN/m(175tf/m) at T=12.0s, and 7400kN/m(760tf/m) at T=15.5s, respectively. The dependency of the wave force on the wave period extremely high.

The maximum wave forces at T=8.5s reduced by the permeable or impermeable submerged breakwaters were some 590kN/m(60tf/m) which is a half of the maximum wave force without the submerged breakwater. At the wave period T=12.0s, the maximum wave force decreases from 1720kN/m(176tf/m) to around

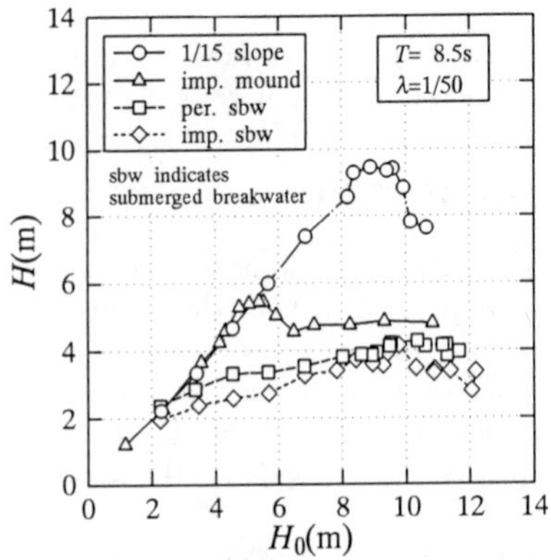

Fig. 2 Transmitted wave heights(T=8.5s)

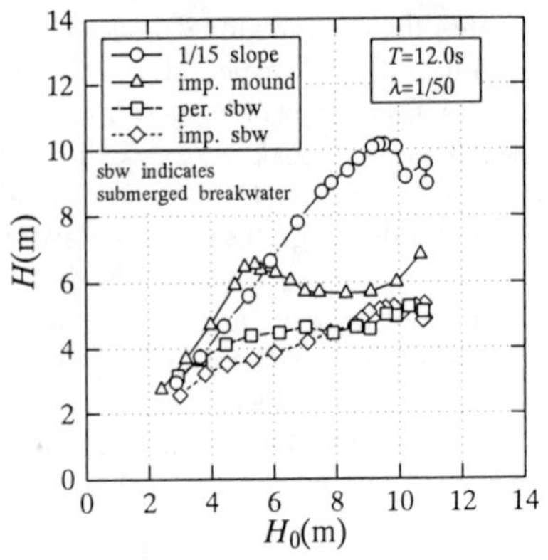

Fig. 3 Transmitted wave heights(T=12.0s)

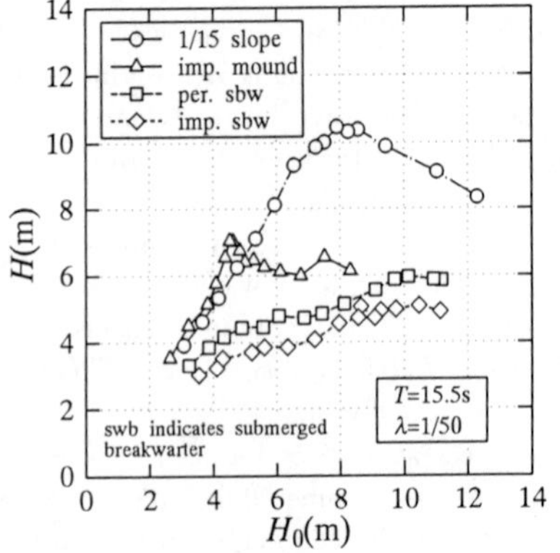

Fig. 4 Transmitted wave heights(T=15.5s)

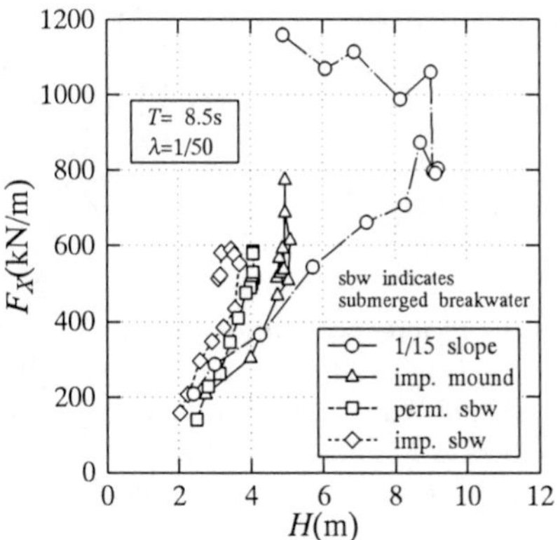

Fig. 5 Wave forces(T=8.5s)

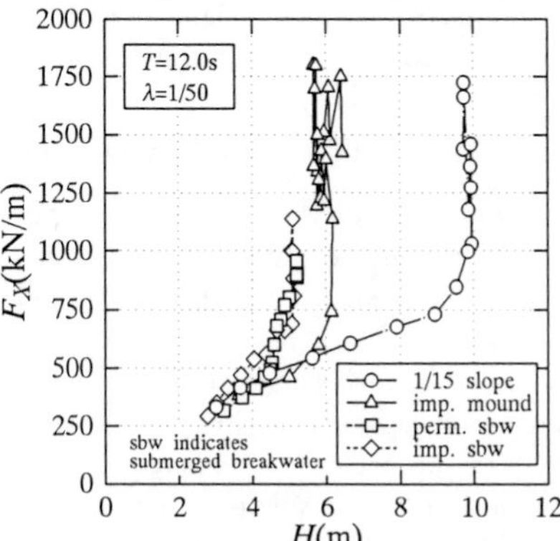

Fig. 6 Wave forces(T=12.0s)

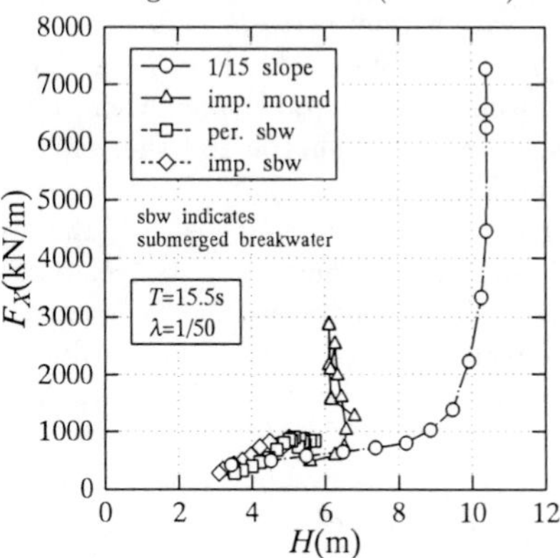

Fig. 7 Wave forces(T=15.5s)

Table 3 Reduction of wave heights and forces

wave period T(s)	wave height H(m)			wave force F_X(kN/m)		
	1/15-slope	perm.	imperm.	1/15-slope	perm.	imperm.
8.5	9.5	4.3	4.0	1160	500	600
12.0	10.1	5.3	5.3	1720	1110	960
15.5	10.5	5.0	6.0	7400	890	900

1100kN/m(112tf/m) of which reduced ration is 0.63. At the wave period T=15.5s, the maximum wave force is reduced to around 890kN/m(100tf/m) which is about one eighth(factor is 0.12) of that without the submerged breakwater. The reduction rate at the wave period T=15.5s is greater than that at T=8.5s. This means that the impulsive wave force at T=15.5s is much greater than those at T=8.5s and T=12.0s.

The effectiveness of the submerged breakwater is summarized in Table 3 from the point of view of the maximum wave height and wave force. The efficiency of the submerged breakwaters to reduce the impulsive wave forces is confirmed.

4. SCALE EFFECT ON IMPULSIVE WAVE FORCES

Fig. 8 shows the comparison of the progressive wave height H without the submerged breakwater between the small- and large-scale experimental results, where the wave height H is normalized with the water depth h at the position of the caisson. The scale effect on the progressive wave height can be due to the bottom friction and the wave energy dissipation during wave breaking in which air bubbles are involved and can affect the wave energy dissipation. The significant difference of the progressive wave height between the small- and large-scale experiments is not seen as a whole except that the maximum wave height of the wave period T = 8.5s of the small-scale experiment. With taking a closer look at Fig. 8, the difference of the wave height between the small- and large-scale ones increases as the wave period increases.

Fig. 9 shows the comparison of the wave force F_X acting on the caisson between the small- and large-scale experimental results, where F_X is normalized with the product of the specific weight of fluid w and the water depth h squared. The results of the wave period T=8.5s are in good agreement each other. These are almost the standing wave forces. The difference of the wave forces at the wave period T = 15.5s is clearly seen. The trend of the relationship between the deep-water wave steepness and the nondimensional wave force slightly differs. At the impulsive wave force region($H_0/L_0 > 0.018$), the results of the small-scale test is greater than those of large-scale one while the relationship of the wave heights is opposite as shown in Fig. 8.

Although aeration was not measured in the present work, the above result may support the statement that an increase of the compressibility of the air and water mixture causes the relatively less shock pressure in the field data by comparison with model data(Crawford *et al.*, 1994).

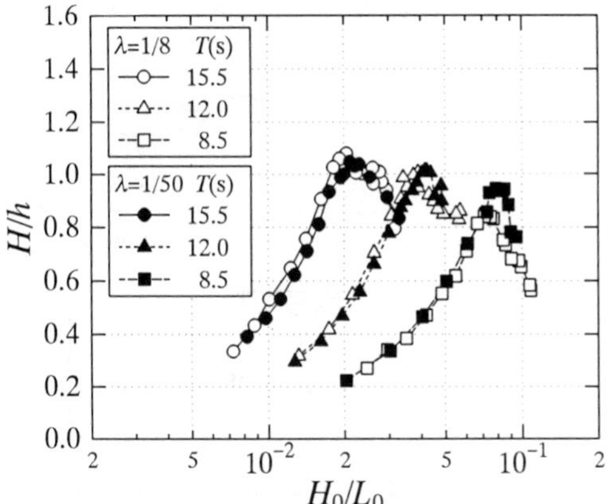

Fig. 8 Scale effect on progressive wave height

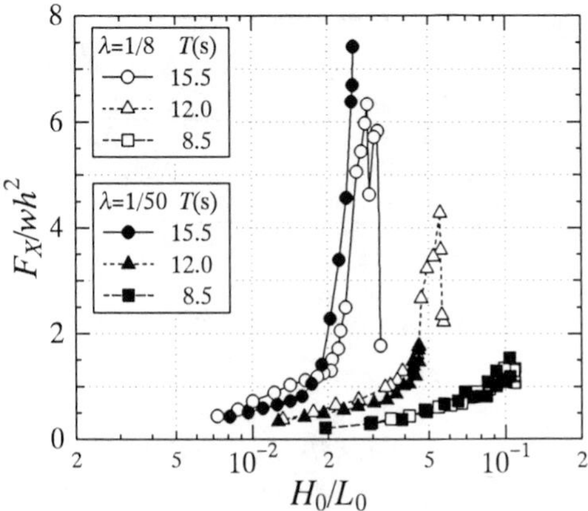

Fig. 9 Scale effect on wave force on caisson

Because a number of variables affect the wave impulsive pressures, especially when including the air bubbles, understanding the scale effect on the impulsive wave pressure is more complicated. It should be back to a fundamental research considering the effects of the compressibility of the fluid which includes the air bubbles.

Next presented is the results of the transmitted wave height over the permeable breakwater and the resultant wave forces. Fig. 10 shows the comparisons of time histories of the surface displacements of the transmitted waves between the small- and large-scale experiments. Those denoted with η are normalized with the water depth h. Fig. 10(a) is the surface displacement of nonbreaking wave and (b) and (c) are those of breaking wave. The waves disintegrated into the free

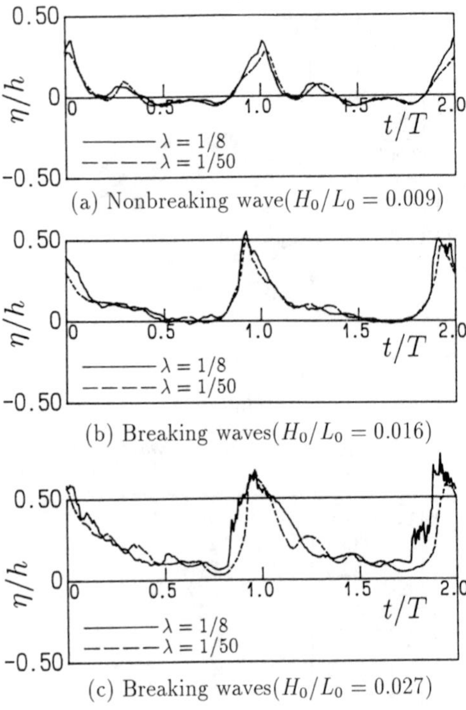

(a) Nonbreaking wave($H_0/L_0 = 0.009$)

(b) Breaking waves($H_0/L_0 = 0.016$)

(c) Breaking waves($H_0/L_0 = 0.027$)

Fig. 10 Comparison of surface displacements

waves after passing over the permeable submerged breakwater. The mean water level increases as the wave steepness H_0/L_0 increases. The surface displacements are very similar to each other but these obtained at the large-scale experiments have larger wave height and higher energy than those at the small-scale ones. Although the sampling frequency of the large-scale test is relatively two and half times higher than that of the small-one, the surface displacement of the small-scale test is smoother than that of the large-one. It means that less air bubbles are contained near the free surface of the transmitted wave at the small-scale test.

Fig. 11 shows the comparison of the transmitted wave height over the permeable submerged breakwater between the small- and large-scale experimental results. The difference between them is small at T=8.5s. As the wave period and wave height increase, the difference become significant. The result of T=15.5s indicated with hollow circles shows that the wave height increases again when the wave steepness H_0/L_0 exceeds 0.03. Such tendency is not seen in other cases within the present experimental conditions.

Fig. 12 shows the corresponding comparison of the resultant wave forces to

Fig. 11. The tendency of the difference between the small- and large-scale tests is similar to that of the transmitted wave heights. As the wave period increases, the difference of the wave force increases: Especially the results of T=15.5s is significant. It is also found that the wave force increases again when the wave steepness H_0/L_0 exceeds 0.03. Due to the scale effect on the transmitted wave height, the wave forces of the large-scale experiment are larger than those of the small-scale ones. Concentrating on the longest wave period, T=15.5s for the design condition, the maximum wave force obtained is F_X/wh^2=1.35 with the 1/8-model test at H_0/L_0=0.022 and $F_X wh^2$=0.9 with the 1/50-model test at the same wave steepness. The wave height divided by the water depth h is H/h=0.62

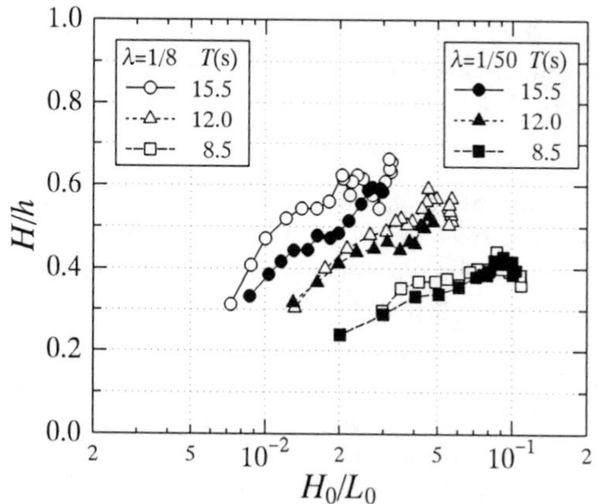

Fig. 11 Scale effect on transmitted wave height

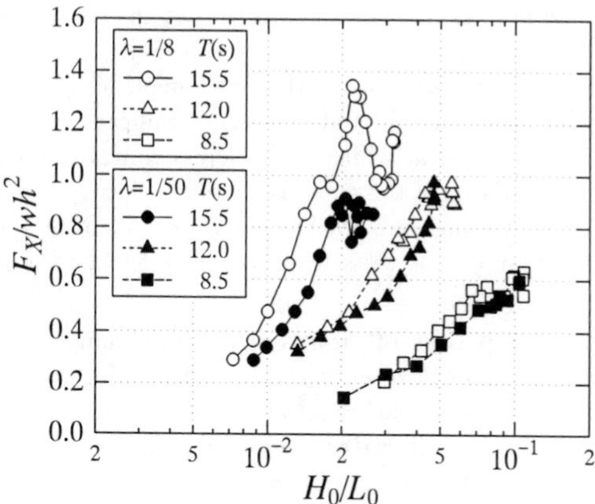

Fig. 12 Scale effect on reduced wave force

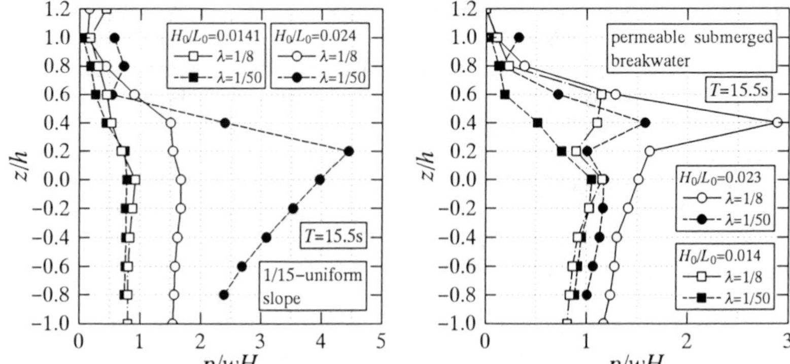

Fig. 13 Profiles of wave pressure without submerged breakwater

Fig. 14 Profiles of wave pressure with submerged breakwater

with the 1/8-model test at H_0/L_0=0.022, while H/h=0.51 with the 1/50-model test. At the same wave steepness of H_0/L_0=0.022, the wave height increases by about 20% and the wave force by some 50%. This is due to the nonlinearity of the wave force on the wave height.

As shown in Fig. 9 the relationship of the wave force and the wave steepness at the wave period T =15.5s changes at H_0/L_0=0.018: When the wave steepness is less than H_0/L_0=0.018, the wave force obtained at the large-scale model test is larger than those at the small-one. Here the wave pressure profiles are compared. Fig. 13 shows comparison of the profiles of the maximum wave pressures without the countermeasure between the large- and small-scale tests results. It is not necessarily meant that the maximum wave pressures at different levels occurred simultaneously. The profiles of the wave pressure at the deep-water wave steepness H_0/L_0 = 0.0141 are in good agreement each other. These are the standing wave pressure without wave breaking at the water depth considered. The largest wave pressure is found at the still water level. Significant difference is seen in the profiles at H_0/L_0 = 0.024 at which the progressive wave breaks at the water depth considered. The result of the small-scale test is clearly the impulsive wave pressure because the wave pressure reaches $p/wH \approx 4.5$ and measured above the still water level. The large-scale tests result seems not to be the impulsive wave pressure.

Fig. 14 shows the comparison of profiles of the maximum wave pressure with the permeable submerged breakwater at H_0/L_0 = 0.014 and at 0.023. Both waves break on the submerged permeable breakwater. Difference is seen in the pressures of H_0/L_0 = 0.014 above the still water level while not seen below the still water level. On the wave steepness of H_0/L_0 = 0.023 the maximum wave pressure of the large-scale test is greater than the small-scale one over the whole levels. Although the deep-water wave steepness is constant, the transmitted wave hight

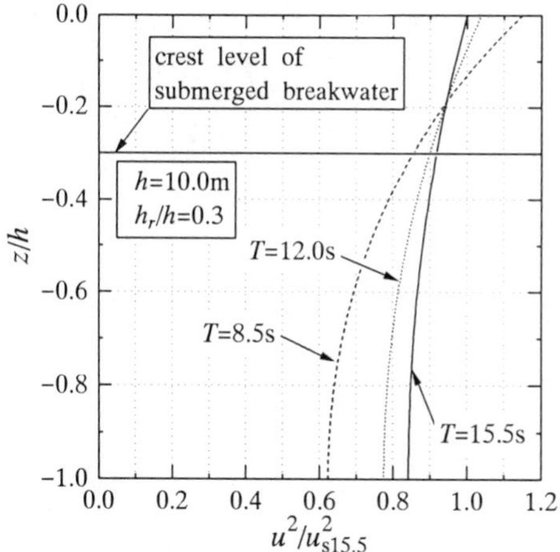

Fig. 15 Profile of horizontal velocity squared(linear wave theory)

obtained at the large-scale model test is greater than that at the small-scale test and the wave pressure increases owing to the nonlinearity on the wave force.

The wave-period dependency on the scale effect on the transmitted wave height can be understood by considering the velocity profile. Fig. 15 shows the profiles of the horizontal velocity squared of three different wave periods calculated with the linear wave theory. In this figure, the horizontal velocity squared u^2 is normalized with $u^2_{s15.5}$, where $u^2_{s15.5}$ is the horizontal velocity squared of the wave period T=15.5s at the still water level. As the wave period increases, the velocity squared to which the wave energy dissipation is proportional increases below the crest level of the submerged breakwater. The longer wave interacts much intensely with the permeable submerged breakwater than the shorter one. Consequently, the scale effect on the wave energy dissipation become predominant as the wave period increases.

Applicability of the data obtained at the large-scale model tests for design according to the Froude law is evaluated considering the critical Reynolds number. Because the critical Reynolds number on the transmitted wave height over permeable submerged breakwaters is not given so far, here that on the stability of armor units is referred to examine the applicability. The mechanism of the scale effect on the stability was explained by Sakakiyama and Kajima(1990). Because both the stability of the armor units and transmitted wave height depend on the drag force, it is reasonable to refer the critical Reynolds number on the stability of armor units. The stability Reynolds number is defined by the following equation:

$$R_N = \frac{\sqrt{gH}(W/\gamma_r)^{\frac{1}{3}}}{\nu} \qquad (1)$$

where H is the wave height, g the acceleration due to gravity, W the weight of armor unit, γ_r the specific weight of an armor unit and ν the kinematic viscosity of the fluid. The critical Reynolds number is proposed by Thomsen *et al.*(1972) and Shimada *et al.*(1986) for regular wave tests of which results were comprehended by Sakakiyama and Kajima(1990) and also proposed by van der Meer(1988) for irregular wave test as follows:

$$R_{NC} = (3 \sim 5) \times 10^5 \quad \text{(regular wave tests)} \tag{2}$$

$$R_{NC} = 3 \times 10^4 \quad \text{(irregular wave tests)} \tag{3}$$

Under the present experimental condition of the large scale model, the maximum Reynolds number is obtained from the weight of Tetrapod W=20kg and the maximum wave height H=1.25m as follows:

$$R_N = 9.5 \times 10^5 \quad \text{(regular wave test)} \tag{4}$$

The value above is slightly greater than the critical Reynolds number $R_{NC} = (3 \sim 5) \times 10^5$. The viscous effect on the transmitted wave height is negligible at the large-scale model. Due to the effect of the air bubbles on the wave pressure, physical model tests overestimate the wave pressure. Less air bubbles are contained in breaking waves at the small-scale test and the density of the fluid is larger than full-scale phenomena then the wave pressure is greater than that of the full-scale. It concluded that the large-scale test results can be applied for a design according to the Froude law.

5. CONCLUSIONS

The small- and large-scale model tests were carried out to investigate the impulsive wave forces on the vertical wall. The submerged breakwater was used as the countermeasure against them and transmitted wave heights and wave forces are examined. The following is conclusion obtained in this work:

1. The wave heights are reduced by 60% with the impermeable submerged breakwater and by 50% with the permeable submerged breakwater compared with the wave height without the submerged breakwaters.

2. The maximum value of the impulsive wave forces on the vertical wall are effectively reduced with the submerged breakwaters by inducing to break waves on them.

3. The wave force due to the transmitted waves broken on the submerged breakwater increases compared with the standing wave force at the same incident wave height.

4. The scale effect on the impulsive wave pressure is found under the long wave conditions. The larger wave forces are obtained at the small-scale test compared with the wave forces at the large-scale test.

5. The transmitted wave height over the permeable submerged breakwater and the resulting wave force obtained at the large-scale experiments are greater than those at the small-scale ones.

6. The large-scale experimental results using the permeable submerged breakwater are obtained at the Reynold number 10^6 which is greater than the critical Reynolds number on the stability of armor units. The viscous effect is negligible and can be applied for the design according to the Froude law for scaling.

REFERENCES

Blakmore, P.A. and P.J.Hewson(1984): Experiments on full-scale wave impact pressures, *Coastal Engineering*, 8, pp.331-346.

Crawford, A.R., M. J. Walkden, J. Griffiths, P.J. Hewson, P.A. Bird and G. N. Bullock(1994): Wave impacts on sea walls and breakwaters, *Book of extended abstract of Coastal Dynamics'94, international conference on the role of large scale experiments in coastal research*, pp.199-202

Hattori,M. and A.Arami(1992): Impact breaking wave pressures on vertical walls, *Proc. 23rd Coastal Engineering Conf.*, Vol.2, ASCE, pp.1785-1798.

Grill, S.T. and M.A.Losada and F.Martin(1992): Wave impact force on mixed breakwaters, *Proc. 23rd Coastal Engineering Conf.*, Vol.1, ASCE, pp.1161-1174.

Oumeraci,H., H.W.Partenscky, S.Kohlhase and P.Klammer(1992): Impact loading and dynamic response of caisson breakwaters *Proc. 23rd Coastal Engineering Conf.*, Vol.2, ASCE, pp.1475-11488.

Nakaizumi,M., M.Yamamot and M.Manabe(1988): Experimental study on wave pressure exerted on breakwater with remote submerged rubble mound, *Proc. 35th Japanese Conf. Coastal Engineering*, JSCE, pp.632-636(in Japanese).

Sakakiyama, T. and R.Kajima(1990): Scale effect of wave force on armor units, *Proc. 22nd Coastal Engineering Conf.*, Vol.2, ASCE, pp1716-1729.

Schmidt,R., H.Oumeraci and H.W.Partenscky(1992): Generation mechanisms of pressure by plunging breakers on vertical structures, *Proc. 23rd Coastal Engineering Conf.*, Vol.2, ASCE, pp.1545-1558.

Shimada, A., T. Fujimoto, S. Saito, T.Sakakiyama and H.Hirakuchi(1986): Scale effects on stability and wave reflection regarding armor units, *Proc. 20th Coastal Engineering Conf.*, Vol.3, ASCE, pp.2238–2252.

Thomsen,A.L.,P.E.Wohlt and A.S.Harrison(1972): Riprap stability on earth embankment tested in large- and small-scale wave tanks, CERC Technical Memorandum, No.37.

Topliss,M.E., M.J.Cooker and D.H.Peregrine(1992): Pressure oscillations during wave impact on vertical wall, *Proc. 23rd Coastal Engineering Conf.*, Vol.2, ASCE, pp.1639-1650.

Van der Meer, J.W.(1988): Rock slopes and gravel beaches under wave attack, Doctoral thesis, Delft University of Technology, The Netherlands.

WAVE REFLECTION: FIELD MEASUREMENTS, ANALYSIS AND THEORETICAL DEVELOPMENTS

Mark A. Davidson[1], Paul A.D. Bird[2], Geoff N. Bullock[3] & David A. Huntley[4]

ABSTRACT: This contribution details analysis of field data collected using an array of 6 pressure transducers in reflective wave fields seawards of both rock island (Elmer, West Sussex) and berm (Plymouth, Devon) breakwaters. Field measurements show that in strongly reflective wave fields the cross-shore distribution of sea-surface elevation variance can be accurately modelled using the shallow water linear wave equations even where the bathymetry is complex. When waves break on or close to the structure the gravity band reflection is inversely related to incident wave steepness. However, when the surfzone is fully developed reflected wave energy is low (<4% of the incident wave energy) and uncorrelated with wave steepness. Due to the influence of the shallow sloping beach and berm seawards of these structures their reflection performance is strongly affected by the local water depth (tidal state), wave reflection increasing towards high water.

INTRODUCTION

This contribution is in two parts. Firstly, an analytical solution for standing waves over complex bathymetry is developed and compared to field data. Secondly, full scale data are used to assess the reflection performance of a rock island and berm breakwater under varied incident wave and tidal conditions.

It is well known that when waves are strongly reflected from coastal structures phase locking between incident and reflected wave components leads to a series of partial nodes and antinodes. It is of great importance to coastal engineers and oceanographers alike to be able to predict the spatial distribution of wave energy due to its relevance to topics such as sediment transport, undermining of coastal structures, numerical modelling and harbour seiching. Solutions for standing waves over simple bathymetry have been available for some time.

[1]Research Associate, School of Civil & Structural Eng., Plymouth Univ., Plymouth, Devon, UK.
[2]Research Associate, School of Civil & Structural Eng., Plymouth Univ., Plymouth, Devon, UK.
[3]Head of School, School of Civil & Structural Eng., Plymouth Univ., Plymouth, Devon, UK.
[4]Reader, Institute of Marine Studies, Plymouth Univ., Plymouth, Devon, Devon, UK.

However, solutions for complicated seabed profiles are not trivial. It is common practice to develop numerical solutions for standing wave fields over complex bathymetry using methods such as the elliptical mild slope equations (Berkhoff, 1972). However, these models are only strictly valid for profiles with low slope angles (up to 1:3, Booij, 1983) and are therefore often inappropriate to reflection from coastal structures. Analytical solutions for standing waves on a sloping shelf were first derived by Lamb (1932) for tides using the shallow water linear wave equations. Suhayda (1974) extended Lamb's solution to a two-component model including a sloping foreshore and flat offshore section. Hotta *et al.* (1981) similarly produced a dual slope model also using the shallow water wave equations. All these solutions were expressed in terms of Bessel functions of the first kind. Both Suhayda and Hotta *et al.* found excellent agreement between their model and field data. This contribution presents an extension of Suhayda's and Hotta *et al.'s* models to a multiple component solution including an arbitrary number of flat or inclined segments. Theoretical predictions are compared with field data.

The reflection performance of coastal structures forms an essential element in the design criterion of modern coastal structures. This fact has prompted many model scale experiments aimed at evaluating the reflection performance of full scale coastal structures. These model scale experiments have lead to a number of predictive equations for wave reflection from full-scale structures. The majority of these equations are formulated around the Iribarren number and are calibrated using empirical calibration coefficients which account for effects such as porosity, surface roughness, breaking offshore of the structure and multiple layers of armour. The potential scaling problems involved in these non-linear processes casts some doubt as to whether these semi - empirical equations will be valid at full scale. Here the reflection performance of two full scale coastal structures are compared with predictive equations resulting from model scale tests.

FIELD MEASUREMENTS

Field measurements were obtained using an array of 6 pressure transducers (Bird *et al.*, 1994) in reflective wave fields fronting a rock island breakwater (Elmer, West Sussex) and berm breakwater (Plymouth, Devon) (Figure 1). Five of the Six pressure transducers were deployed along a shore-normal transect with sensor spacings increasing exponentially in a seawards direction. Data were sampled at 2Hz every 3 hours for a total duration of 73 and 51 days at the Elmer and Plymouth sites respectively. This sampling strategy allowed an assessment of both the cross-shore distribution of wave energy and the reflection performance of the structure in a range of incident wave conditions.

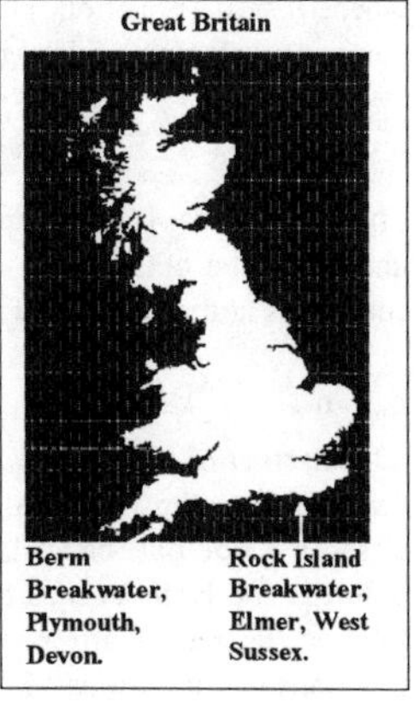

Figure 1. Field Sites.

The rock island breakwater at Elmer is one of a series of eight 110m long shore-parallel breakwaters each separated by a gap of 60m. The breakwaters are comprised of 4-8 tonne carboniferous limestone blocks. The seawards face of the structure has a gradient of 1:1.1 and is fronted by a shallow sloping (1:50) beach with regular bathymetry. The structure stands 5-6m in height and is only overtopped at high water during the largest storm waves. Incident wave spectra at Elmer are dominated by locally wind generated sea with an occasional weak Atlantic swell wave component.

The berm breakwater at Plymouth is straight in plan view and is nearly 2km in length. It is built of nearly 4 million tonnes of limestone. Although complicated in cross-section the effective dissipating area of the breakwater consists of a smooth impermeable sloping foreshore (Figure 3) comprised of dovetailed granite ($\tan\beta = 1:6$) stretching seawards to an approximately horizontal berm. Large 25-100 tonne concrete blocks have been placed on the berm to enhance wave dissipation. In this macro tidal area the berm and blocks are completely submerged to a depth $\approx$3m at high water and completely exposed on the lowest tides. The structure is frequently overtopped by moderate waves at high water.

STANDING WAVES OVER COMPLEX BATHYMETRY: FIELD DATA AND THEORETICAL DEVELOPMENTS

Lamb (1932) derived solutions for the free surface elevation η for standing waves with radian frequency σ and shoreline amplitude a_0 as a function of offshore distance x over a sloping bed with gradient $\tan\beta$:

$$\eta_0 = a_0 J_o\left(2\sigma\sqrt{\frac{x}{g\tan\beta_0}}\right) \qquad \text{Shore Face Solution. (1)}$$

Here g is the acceleration due to gravity, and J_o is a Bessel function of order zero. Suhayda (1974) applied Lambs solution for tides to shallow water incident waves on a two component beach consisting of a sloping shoreface (Equation 1) and a flat offshore section where η is given by;

$$\eta_n = A_n \sin\left(\frac{2\pi x}{L}\right) + B_n \cos\left(\frac{2\pi x}{L}\right) \qquad \text{Flat Bed Solution (2)}$$

L is the shallow water wavelength ($2\pi[gh]^{1/2}/\sigma$) and n represents the profile segment number where $n = 0$ (Equation 1) represents the shoreface. Hotta *et al.* (1981) developed a dual component sloping model where the water surface elevation for $n = 1$ is given by;

$$\eta_n = A_n J_o\left(2\sigma\sqrt{\frac{x}{g\tan\beta_n}}\right) + B_n Y_o\left(2\sigma\sqrt{\frac{x}{g\tan\beta_n}}\right) \qquad \text{Offshore Slope (3)}$$

The coefficients A and B in Equations 2 & 3 have the dimensions of length. Solutions for A and B can be obtained by solving the simultaneous equations resulting from the boundary conditions at the junction of the profile segments;

$$\eta_n = \eta_{n+1}, \quad \frac{\partial\eta_n}{\partial x} = \frac{\partial\eta_{n+1}}{\partial x} \qquad \text{Boundary Conditions (4a, b)}$$

(Davidson *et al.*, 1992). Equations 1 to 4 potentially result in 5 different solutions for A and B depending on the nature of the boundary (e.g. shoreline-offshore slope, shoreline-flat bed, offshore slope-flat bed, flat bed-offshore slope or offshore slope-offshore slope). Here Suhayda and Hotta *et al.'s* two component solutions have been extended to a multiple sloping solution having an arbitrary number of profile segments by solving each of the 5 sets of simultaneous equations. In this way even complex bathymetry can be modelled using numerous linear profile segments. It is assumed here that the reflection coefficient of the structure is unity and that the phase lag between the incident and reflected wave components at the shoreline is zero.

Comparisons Of The Model With Field Data.

An example of the spatial variability of energy with offshore distance can be seen in the 5 spectral estimates recorded at different offshore locations on a shore-normal transect during very low frequency (f_{peak}=0.055Hz) swell conditions at Plymouth Breakwater (Figure 2). Distances in Figure 2 are measured with respect to the mean shoreline position.

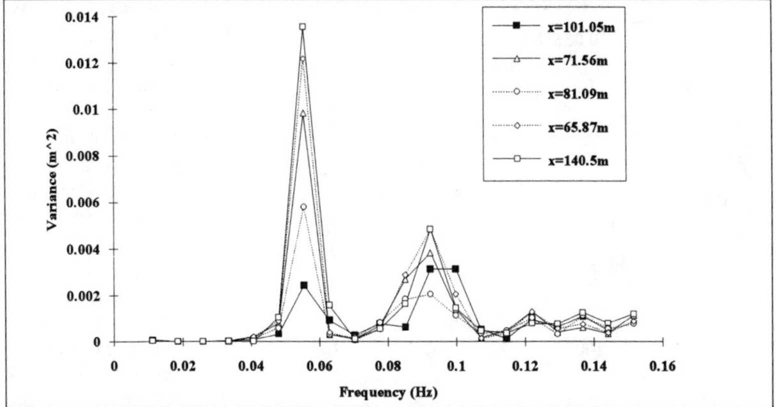

Figure 2. Spatial distribution in surface elevation variance with offshore distance, (Plymouth Breakwater 12th February, 1989).

Assuming a perfectly linear system the observed spatial variability in the surface elevation variance can be examined separately at each frequency estimate and compared to model results. Figure 3 shows a 5-component slope model of the Plymouth breakwater bathymetry along with measured survey data. The erratic hump in the survey data represents one of the large concrete blocks placed on the berm. Figure 4 shows that there is excellent agreement between the model predictions based on the bathymetry shown in Figure 3 and the observations extracted from spectra in Figure 2. Good agreement between the model and data occurs not only at the principal frequency but also at less energetic frequencies. Unfortunately, no runup data were collected in this experiment to facilitate measurement of the

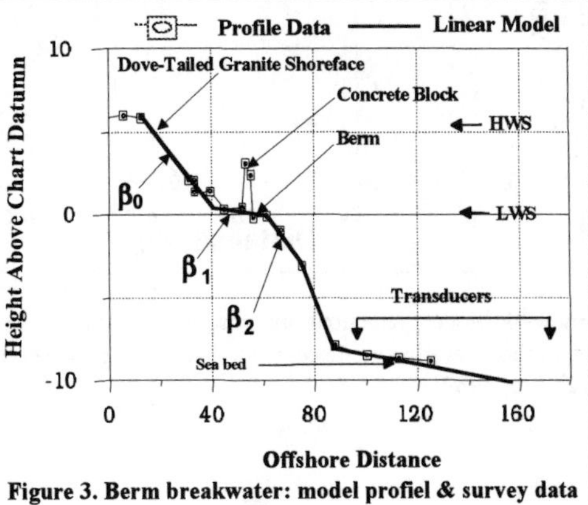

Figure 3. Berm breakwater: model profiel & survey data

shoreline amplitude of the standing waves. Therefore, in order to compare the predicted (plotted here for a_o=1) and observed cross-shore variances it was necessary to translate the measured data with respect to the y-axis by a factor computed by comparison of the most energetic data point and the model at the equivalent offshore distance. In all cases the location of high and low energy in the data representing partial nodes and antinodes is well predicted by the model. Note that although these low frequency waves are strongly reflected the non-unity reflection coefficient leads to non-zero values at locations where nodes are predicted by the model.

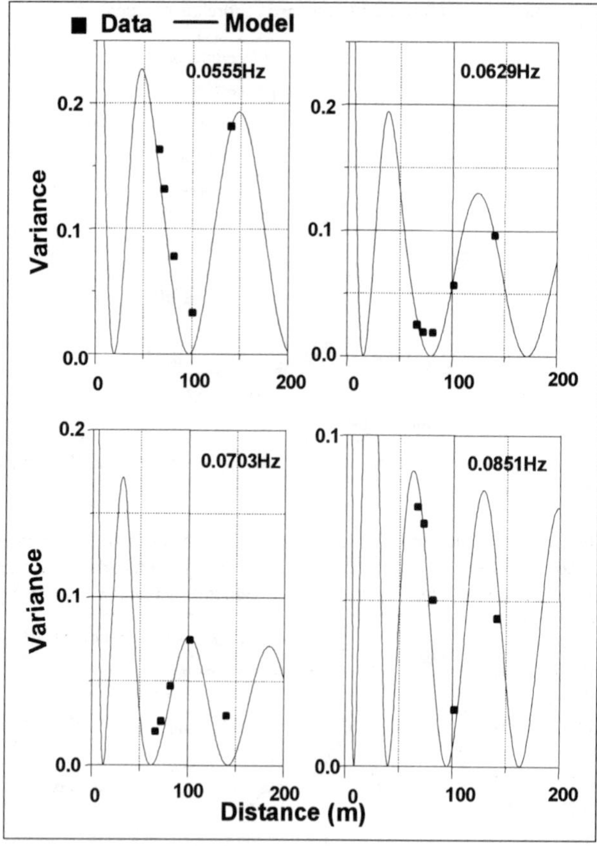

Figure 4. Model predictions and observations of the cross-shore distribution of surface elevation variance with offshore distance, (Plymouth Breakwater, 12th Feb. 1989).

Similarly good matches were obtained for a comparison between model predictions and data for a two component model of the rock island breakwater at Elmer, West Sussex, (Figure 5).

An important result of this work is its ability to provide a value for the "effective reflection point" for the Modified Maximum Likelihood Method for directional spectral analysis, discussed in the companion paper by Bird *et al.*, (1994).

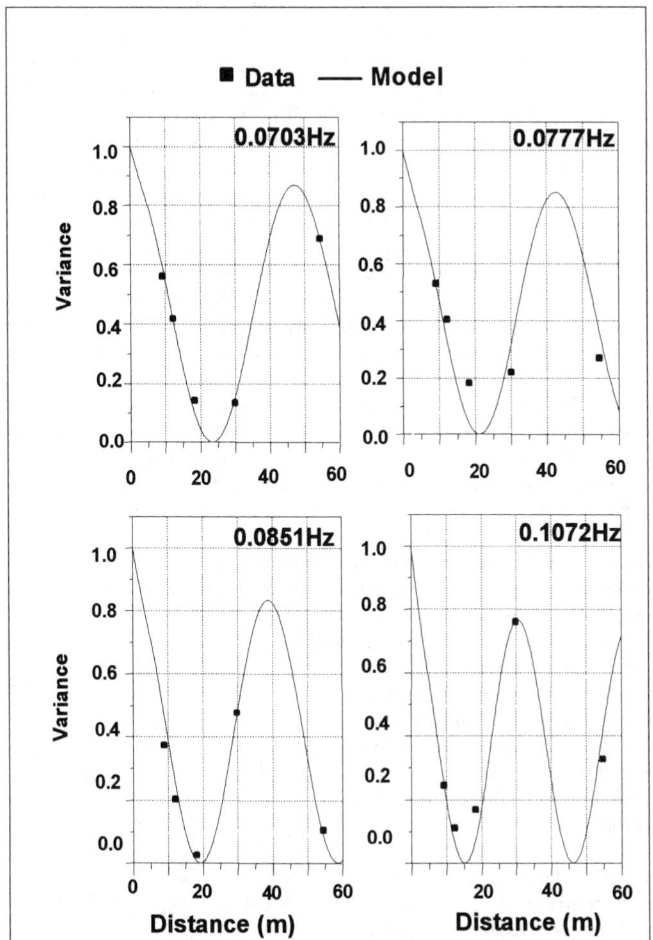

Figure 5. Model predictions and observations of the cross-shore distribution of surface elevation variance with offshore distance, (Elmer, West Sussex, 5th July 1992).

REFLECTION PERFORMANCE OF ROCK ISLAND AND BERM BREAKWATERS

Three of the transducers in the shoreline array were used to compute frequency dependant reflection functions (FDRFs = $K_r(f) = \nu_{rr}(f)/\nu_{ii}(f)$, ν =surface elevation variance, rr=reflected, ii=incident) and frequency averaged reflection coefficients ($K_r = \int \nu_{rr} df / \int \nu_{ii} df$)

using the algorithm derived by Gaillard *et al.,* (1980). This routine although based on Kajima's (1969) method is superior in the sense that it eliminates singularities corresponding to critical probe spacings between sensor pairs. Numerical tests have shown that Gaillard *et al's* 2-dimensional method is fairly insensitive to oblique wave approach. With the array geometry used in these tests r.m.s. errors in the FDRF were <0.05 for wave angles of <30°. Estimates of the characteristic reflection coefficient were very reproducible irrespective of which sensor triplet was selected from the shore-normal array of 5 transducers. Tests on field data have shown that the standard deviation in the frequency averaged reflection coefficient estimate calculated using the 10 possible permutations of transducer triplets is typically of the order 0.01. Reflection coefficients were only evaluated over the energetic portion of the spectrum, at other frequencies low signal to noise ratios leads to overestimates of the reflection coefficient. In order to maximise the accuracy of FDRFs estimates, only data corresponding to frequencies where the incident wave energy was greater than 20% of the spectral peak variance and with high signal coherence (>0.8 between all sensors) were considered.

Typical FDRFs for rock island and berm breakwaters are shown in Figure 6 along with an example from a natural beach at Felpham, West Sussex. Each of the lines in Figure 6 represents a single data set with the arrow marking the spectral peak frequency. Although there is considerable variability in the FDRFs at each site depending on the incident wave conditions Figure 6 illustrates clearly some typical trends observed in the data.

Firstly, both the rock island and rubble mound breakwaters show a general trend of a decrease in wave reflection with an increase in frequency. The natural beach however, where waves break well away from the shoreline shows no clear trend with frequency and almost negligible reflected wave energy (typically < 4% of the incident value) within the gravity band. This result is typical for conditions where a surfzone is fully developed. Equivalent low values were observed by Tatavarti (1989) for gravity band estimates of $K_r(f)$ on 3 different natural beaches.

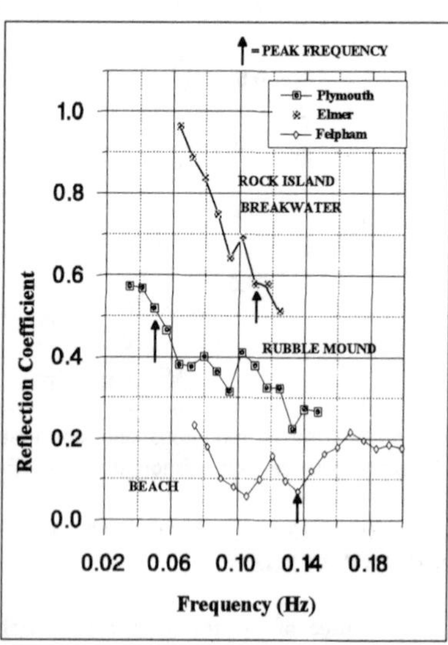

Figure 6. FDRFs Computed for a rock island breakwater, berm breakwater and natural beach.

Secondly, there is a trend for wave reflection to increase with the gradient of the shoreface. The respective shoreface gradients in these examples are 1:1.1, 1:6 and 1:50 for the rock island breakwater, berm breakwater and natural beach respectively.

The reflection performance under varying incident wave conditions may be examined more thoroughly by looking at the variation in the frequency averaged reflection coefficient K_r with the Iribarren number $\left(\xi = \frac{\tan\beta_o}{(H_{ij}/L_o)^{1/2}}\right)$. Here L_o is the deep water wavelength corresponding to the spectral peak frequency, H_{ii} is the incident wave height $\left(H_{ii} = 4\sqrt{\int_{f=0.05Hz}^{f=0.40Hz} \nu_{ii} \ df} \right)$ and $\tan\beta_0$ is the gradient of the shoreface.

Field data from both Plymouth and Elmer were entered into a data base so that the data set could be used to assess the performance of the various predictive equations for K_r which are available to the coastal engineer, and to investigate the dependence of K_r on different parameters. A plot of data from the rock island breakwater collected between the 2/7/92 and 16/7/92 against some commonly used predictive equations appropriate to the structure is given in Figure 7. Each of the points in Figure 7 represents one data run. The data base has been used to isolate data where there is some wave activity ($H_{ii} > .05m$) and include only those data runs where the instruments were fully submerged, (the instruments dry out at low water). Data with multi-peaked energy spectra were also eliminated from analysis in order to avoid ambiguities in selecting the appropriate length scale (L_o) when computing the Iribarren number. Details of the equations used to generate the curves in Figure 7 are given in Table 1.

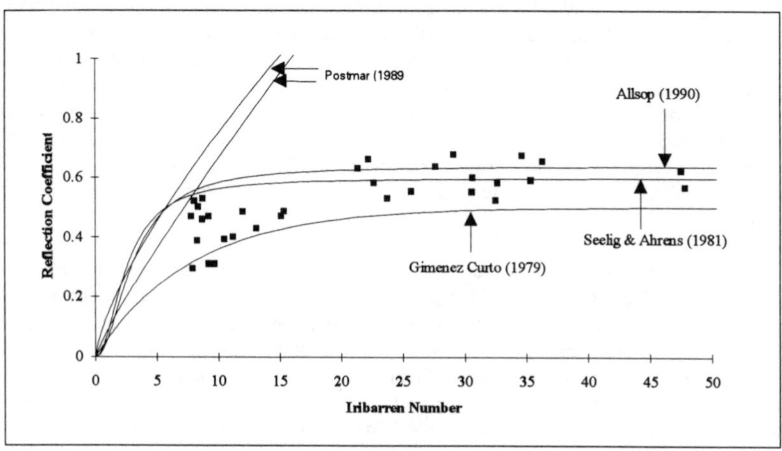

Figure 7. Model predictions versus data collected at the rock island breakwater.

Inspection of Figure 7 shows that some of the observations correspond to Iribarren numbers which are considerably higher than those reported in previous laboratory reflection experiments. These high values result from data runs with very low amplitude waves ($H_{ii} \approx 0.1$m) and consequently low values of H_{ii}/L_o, and the steep gradient of the structure (1:1.1). These high Iribarren numbers although of limited interest to the coastal engineer who is primarily concerned with the impact of larger waves, do illustrate well the trend of the data at the reflective extremes, and the upper saturation value (≈0.6 to 0.65) for the reflection coefficient.

Equation	Coefficients / Comments	Author(s)
$K_r = a(1 - \exp(b\xi))$	a=0.503, b=0.125 (rubble mound)	Gimenez-Curto (1979)
$K_r = a\xi^2/(b + \xi^2)$	a=0.6, b=6.6 Conservative (rubble mound)	Seelig & Ahrens (1981)
$K_r = 0.14\xi^{0.73}$	Conservative estimate. (rubble mound)	Postmar (1989)
$K_r = 0.081P^{-0.14}\cot\beta^{-0.78}(H_{ii}/L_o)^{-0.44}$	P=Permeability factor=0.6 Accurate estimate (rubble mound)	Postmar (1989)
$K_r = a\xi^2/(b + \xi^2)$	a=0.64, b=9.64 (2 layers of large rock).	Allsop (1990)

Table 1. Predictive equations for the reflection performance of rock breakwaters.

The general trend of the data is predicted most accurately by the equations of Gimenez-Curto (G-C), Seelig and Ahrens (S&A) and the modified S&As' equation by Allsop (A). Although, G-C's equation, which was based on the laboratory data of Sollit and Cross (1972) for rubble mound structures, tends to under-predict the value of K_r. S&A's and A's equations predict accurately the upper saturation level of the reflection coefficient and provide a conservative estimate for lower Iribarren numbers. Postmar's equations provide similar (very conservative) estimates of the reflection coefficient to S&A's and A's equations for low Iribarren numbers ($\xi < 8$) but fail to predict the saturation value for the reflective extremes of the data.

There is considerable scatter in the data in Figure 7 corresponding to Iribarren Numbers < 20. The observed spread in the data is due to the effect of local water depth (tidal variations). This can be illustrated more clearly by using the data base to isolate and re-plot data from limited depth bands. Three non-overlapping depth bands have been selected and data from Figure 7 has been re-plotted in Figures 8a, b and c. In these examples the depth is measured relative to the toe of the structure (d_t). Also shown for reference is a dotted line corresponding to S&As' equation (see Table 1, a=0.6, b=6.6), and since this equation best suits the trend of the data a second solid curve has been plotted corresponding to the best fit values of a and b for S&As' equation. S&A comment that the coefficient a is the reflection coefficient reduction factor which empirically accounts for dissipation due to waves breaking offshore of the structure, surface roughness and multiple layers of armour. Effectively, a determines the saturation value of K_r for large ξ. Laboratory tests have indicated that the coefficient b is inversely related to the beach slope and is larger for random waves than monochromatic waves (Seelig & Ahrens, 1981).

Note that the best fit values for a in Figures 8a to c are approximately constant for the full range of water depths. The value of b however, systematically increases with decreasing water depth.

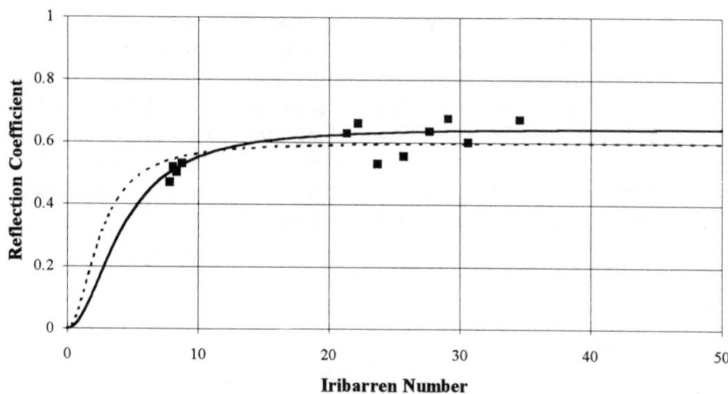

Figure 8a. d_t>3.25*m*, *a*=0.65, *b*=25, (Rock island breakwater).

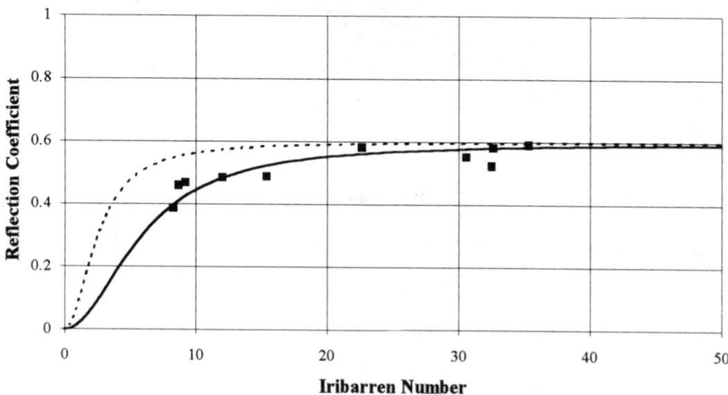

Figure 8b. $2.5 \leq d_t \leq 3.25$, *a*=0.60, *b*=35, (Rock island breakwater).

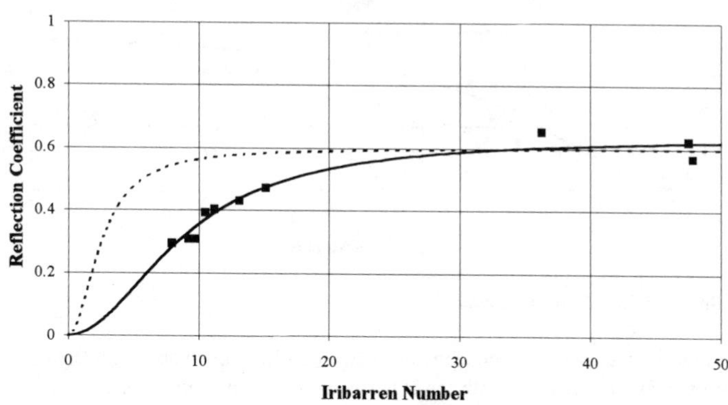

Figure 8c. d_t<2.5*m*, a = 0.64, b = 80 , (Rock island breakwater).

Tidal state is also has a very significant effect on the reflection performance of the berm breakwater. This is illustrated in Figures 9a and b where K_r has again been plotted against the Iribarren number for different depth regimes. In these examples the depth is measured relative to the sea bed at the break in slope between the sloping dovetailed granite shoreface and the berm (see Figure 2). Figure 9 includes data collected between 12/2/89 and 24/2/89. Iribarren numbers computed for the berm breakwater are much lower than those presented previously for the rock island breakwater due to both the more energetic incident wave conditions at this locality and the shallower sloping shoreface ($\tan\beta = 1:6$).

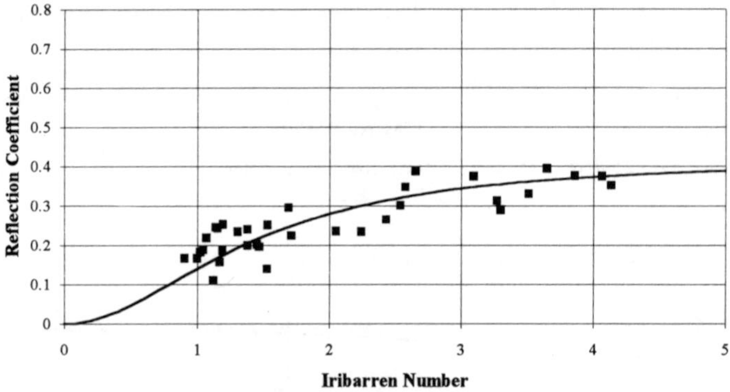

Figure 9a. d_t>3m, a=0.42, b=2.0, (Berm Breakwater).

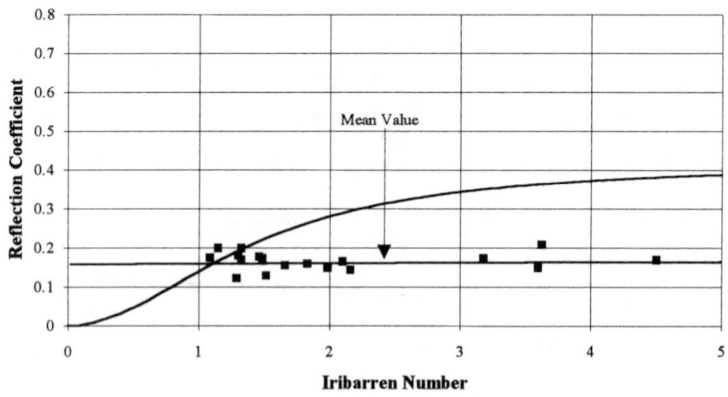

Figure 9b. d_t<2.5m, (Berm Breakwater).

Figure 9a shows that when the tide is high (d_t>3m) the reflection coefficient increases in the usual way in proportion with the Iribarren number. The best fit values for a and b for S&As' equation shown by the solid curve in Figure 9a are 0.42 and 2.0 respectively. However, when the water depth over the berm is less than 2m (Figure 9b) the reflection coefficient is low (K_r<0.2) and uncorrelated with the Iribarren number. The best fit line for d_t>3m has been included again in Figure 9b for reference. The low reflection coefficients (Figure 9b) indicate that less than 4% of the incident wave energy is being reflected by the structure. FDRFs associated with these shallow water (d_t<2m) readings show consistently low reflection ($K_r(f)$<<0.2) across all frequencies similar to those calculated for the natural beach.

DISCUSSION

Figures 4 and 5 illustrate that linear wave theory accurately predicts the cross-shore variation in wave energy over complex bathymetry for highly reflective conditions. However, the fit of the model to data is not so impressive for less reflective (K_r<0.4) conditions (not included here). Lower reflection coefficients indicate higher wave energy dissipation due to breaking and consequently steeper, more non-linear wave conditions. Similar observations were made in laboratory tests carried out by Guza and Bowen (1976) who found a good agreement between the linear shallow water wave equations and laboratory data providing K_r>0.4. They found that as the amplitude and steeepness of the waves increased so enhancing dissipation through breaking the deviation of the observed data from linear wave theory predictions increased. This they attributed partly to increasing non-linear effects resulting from wave steepening and indirectly due to second order effects such as wave setup. Guza and Bowen's (1976) laboratory data also indicated that the phase angle between incident and reflected wave components at the shoreline which is equal to zero for high reflection coefficients (K_r>0.4) increases as the reflection coefficient decreases causing nodes and antinodes to move offshore.

It may be concluded that the excellent fit of the model to the data presented in Figures 4 an 5 is perhaps not surprising considering that the corresponding values of $K_r(f)$ were > 0.4 and that low Ursell Numbers for this data (Elmer Ur=0.090, Plymouth Ur=0.213) indicate that waves were reasonably linear, but care should be exercised when applying the model to less reflective (K_r<0.4) more non-linear (Ur>1) wave conditions.

Comparisons between predictive equations and data recorded at the rock island breakwater indicate that Seelig and Ahrens' (1981) equation for wave reflection provides a reasonable, conservative estimates for wave reflection from the rock island breakwater. If more accurate estimates of the reflection coefficient are required the local water depth must be considered. Figures 8a, b & c showing the variation in the reflection performance of the rock island breakwater with water depth (tide) indicate that saturation values of K_r corresponding to highly reflective condition ($\xi > 20$) do not vary significantly with water depth. Thus, the best fit value for S&As' reflection coefficient reduction factor a which in part empirically accounts for wave breaking offshore of the structure is approximately constant ($0.6 < a < 0.65$).

The constancy of a and the high Iribarren numbers computed for the data set indicate that wave breaking occurred exclusively on the structure itself for these data sets. Conversely, the best fit value of b in S&As' equation exhibits a strong inverse relationship with the water depth (Figures 8a, b & c). This is entirely consistent with S&As' observations that b increases with decreasing shoreface slope. This indicates that as the water level of the toe of the structure is reduced, the "effective slope" of the structure is decreased due to the increasing effect of the shallow sloping beach seawards of the structure.

Reflection estimates for the berm breakwater are consistently less than those obtained for the steeper rock island breakwater. This is due partly to the more dissipative nature of the structure including the shallower shoreface slope, the dissipating effect of the berm and concrete blocks positioned on the berm, and partially due to wave transmission over the structure near high tide. Note also that the scatter in the berm breakwater data (Figure 9a) is slightly higher that for the rock island data set particularly for low Iribarren numbers. This is due to some of the larger waves inconsistently over-topping the breakwater during the high tide data runs.

When the water depth at the toe of the shoreface-slope is less than 2.5m all waves break over the berm irrespective of incident wave steepness leading to low reflection coefficients (mean K_r=0.16) which are uncorrelated with the Iribarren number (Figure 9b). This dissipation is enhanced by the emplacement of the concrete blocks on the berm. Under these conditions the surfzone is fully developed and the reflection characteristics of the structure are analogous to a natural beach (see Figure 6).

CONCLUSIONS

1) Theoretical developments based on linear wave theory for the cross-shore distribution of surface elevation variance for standing waves over complex bathymetry show excellent agreement with field data recorded at two sites in highly reflective conditions (K_r>0.4) (Figures 4 & 5).

2) The reflection coefficient is directly related to the steepness of the incident waves unless a surfzone is fully developed. When waves break offshore gravity band reflections are minimal (K_r<0.2) and uncorrelated with wave steepness (see Figures 6 for the beach FDRF and Figure 9b) .

3) Field observations indicate that a reasonable, conservative estimate of the reflectivity of a rock island breakwater may be obtained from Seelig and Ahrens' (1981) equation, or Allsop's (1990) variation of Seelig and Ahrens' equation (Table 1, Figure 7).

4) The reflection performance of both the rock island and berm breakwaters was found to be very sensitive to tidal state in these macro-tidal environments, with reflectivity of the structures reducing towards low tide (Figures 8 & 9). This observation relates to the decreasing "effective slope" of the structure with reducing water depth resulting from the increasing influence of the shallow sloping beach and berm at the respective sites.

ACKNOWLEDGEMENTS

The authors would like to thank the following people and authorities for their assistance in this work; The Science Environmental Research Council (SERC) for their financial support of this project, Roger Spencer, Ray Trainer *et al.* of Arun District Council for their assistance in the field and William Allsop, Robert Jones *et al.* at Hydraulics Research, Wallingford for the use of their facilities and collaboration in this project.

REFERENCES

Allsop, N.W.H., 1990. "Reflection performance of rock armoured slopes in random waves," Proceedings 22nd Coastal Engineering Conference, ASCE, NY.

Berkhoff, J.C.W., 1972. "Computations of combined refraction-diffraction," Coastal Engineering, 13th, Chapter 24.

Bird, P.A.D., Davidson, M.A., Bullock, G.N. and Huntley, D.A., 1994. "Wave measurements near reflective structures", Proceedings Coastal Dynamics '94, (this edition).

Booij,N., 1983. "A note on the accuracy of the mild slope equation," Coastal Engineering 7, pp. 191-203.

Davidson, M.A., Bird, P.A.D., Bullock, G.N. and Huntley, D.A., 1993. "An analytical. solution for standing shallow water waves over a multiple component sloping sea bed," Internal report 93-002, School of Civil & Structural Engineering, University of Plymouth, UK., 40 pp.

Gaillard, P., Gauthier, M. and Holly, F., 1980. "Method of analysis of random wave experiments with reflecting coastal structures," Proceedings 17th Conference Coastal Engineering, ASCE, NY, pp. 204-220.

Giminez-Curto, L.A., 1979. "Behaviour of rubble mound breakwaters under wave action," Ph.D. Thesis, University of Santander, Spain, (in Spanish).

Guza, R.T. and Bowen, A.J., 1976. "Resonant interactions for waves breaking on a beach," Proceedings 15th Conference Coastal Engineering, ASCE, NY, pp. 560-579.

Hotta, S., Miziguchi, M. and Isobe, M., 1981. "Observations of long period waves in the nearshore zone," Coastal Engineering in Japan, Vol. 24, pp. 41-76.

Kajima, R., 1969. "Estimation of an incident wave spectrum under the influence of reflection," Proceedings 13th Congress of IAHR, Kyoto Japan, Vol. 5.1 pp. 285-288.

Lamb, H., 1932. "Hydrodynamics," Sixth edition, Dover Publications, NY, 738 pp.

Postmar, G.M., 1989. "Wave reflection from rock slopes under random wave attack," Unpub. Ph.D. Thesis, Delft University of Technology, 106 pp.

Seelig, W.N. and Ahrens, J.P., 1981. "Estimation of wave reflection and energy dissipation for beaches, revetments and breakwaters, " CERC Technical paper 81-1, Fort Belvoir, US Army Engineer Waterways Coastal Experiment Station, Vicksburg, Miss., 40 pp.

Sollit, Ch.K. and Cross, R.H., 1972. "Wave reflection and transmission at permiable breakwaters," MIT, R,M. Persons Laboratory Technical Rep., No. 147, 235 pp.

Suhayda, J.N., 1974. "Standing wave on beaches," J. Geophysical Research, Vol. 79, No. 21, pp. 3065-3071.

Tatavarti, R.V.S.N., 1989. "The reflection of waves on natural beaches," Unpub. Ph.D. Thesis, Dalhousie University, Halifax, Nova Scotia, Canada, 175 pp.

WAVE IMPACTS ON SEA WALLS AND BREAKWATERS

A.R.Crawford[1], M.J.Walkden[1], P.A.D.Bird[1],
G.N.Bullock[1], P.J.Hewson[1], J.Griffiths[2]

ABSTRACT: Details are given of both field and model tests to investigate the characteristics of wave impact pressures on steep fronted coastal structures. The pressure records are supplemented by measurements which define both the incident waves and the aeration level in each wave as it breaks on the structure. A new instrument which has been developed to measure aeration levels in full scale waves is described. The behaviour of bubbles in salt and fresh water is considered with reference to the likely effect on the compressibility of the air water mixture. Pressure data obtained from the first field test site is contrasted with data obtained from small and large scale physical models. Evidence is formed to support the hypothesis that ocean wave impact pressures predicted by Froude law scaling of measurements from conventional models are likely to be excessively high. Future work intended to investigate further this scale effect is outlined.

INTRODUCTION

The design of many breakwaters and sea walls could be improved if it were possible to make a more accurate assessment of the hydrodynamic loading due to breaking waves. However field measurements of wave impact pressures on steep fronted coastal structures are rare and difficult to obtain, especially during storm conditions. Consequently the methods used to predict the extreme loading on such structures for design purposes are still in need of full scale verification. Where data does exist, it suggests that the impact pressures are often lower than might be predicted by scaling the results of small scale model tests on the basis of the Froude law (Blackmore and Hewson, 1984).

[1] School of Civil & Structural Engineering, University of Plymouth, Palace Court, Palace Street, Plymouth, PL1 2DE U.K.

[2] Institute of Marine Studies, University of Plymouth, Drake Circus, Plymouth, PL4 8AA U.K.

The apparent discrepancy between prototype and model data is thought to be due, at least in part, to the tendency of ocean breakers to entrain appreciable quantities of air. It is hypothesised that this air causes a significant reduction of impact pressures by increasing the compressibility of the air water mixture and by reducing the celerity of shock waves which changes the pressure time history after impact.

This paper provides an overview of ongoing research at the University of Plymouth in which both field and model tests are being carried out in order to gain a better understanding of wave impact phenomena. Particular attention is being paid to the significance of entrained air and instruments have been developed which enable the void fractions (proportion of bubbles) to be estimated at both full and model scale. Although the work is far from complete, comparison of the results obtained to date at different scales in both fresh and salt water is already shedding new light on the scaling problem.

FULL SCALE FIELD TESTS

The instruments used both to measure impact pressures and to assess void fraction in full scale ocean waves are shown in Figure 1.

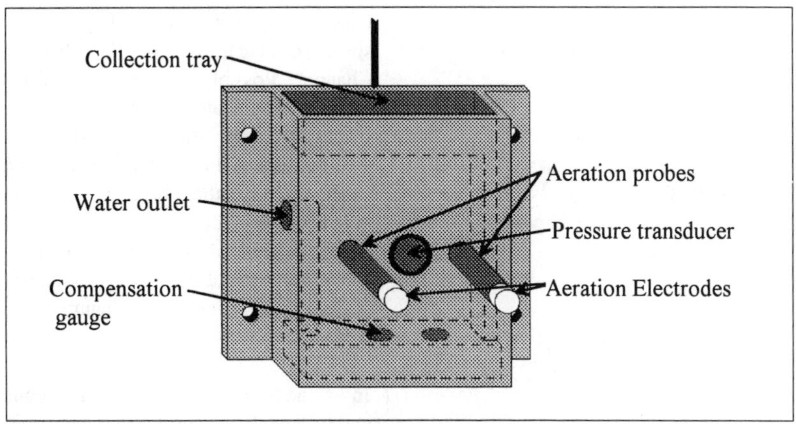

Figure 1, Field instrumentation arrangement used at the Plymouth site

The void fraction in an approximately spherical volume in the vicinity of the pressure transducer is estimated from the measured electrical impedance between the two electrodes (Griffiths and Hewson, 1992). The aeration instrument is designed to have a fast response with circuitry capable of 0.5 ms rise time. Alternating voltages are used so that polarisation does not occur at the electrodes. An internal gauge measures the conductivity of unaerated samples of water to compensate for the changes in salinity and temperature that are likely to have a significant effect on water conductivity during the course of a long field deployment. On the reasonable assumption that most if not all voids within the water are due to the presence of air, the close proximity of the electrodes to the pressure transducer enables the temporal variations in both aeration and pressure to be investigated at almost the same point. In particular the measurements taken immediately before, during and after wave impacts enable the influence of entrained air upon impact pressures to be examined. Information on the spatial distributions of air and pressure can be obtained by the use of an array of gauges. This helps to distinguish between a continuum of bubbles and the air pockets trapped when plunging breakers strike a structure, as illustrated in figure 2. It also provides information on the shape and velocity of the water surface at breaking.

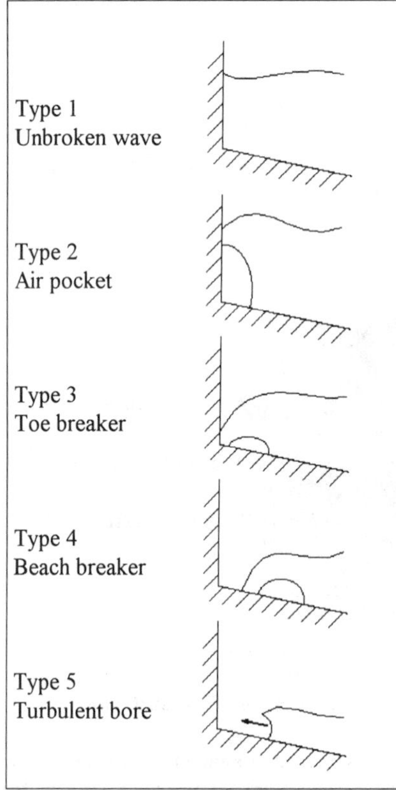

Figure 2, Wave impact classification

An instrument of the type described above has been used to investigate the relationship between impact pressures and aeration levels on the seaward face of a breakwater in Plymouth Sound. The breakwater has a steep face constructed from granite blockwork, and a rubble core. The high forces which act upon it are caused by Atlantic swell propagating up the English Channel. Possible failure mechanisms include displacement of the whole structure, and the expansion of localised damage originating from some weakness in the blockwork. The magnitude and duration of the impact pressures on this and similarly constructed structures is of significance to both designers and maintenance engineers.

In addition to simultaneous measurement of pressure and void fraction both the energy and shape of the waves were recorded. The reflective wave field opposite the structure was measured by means of an array of six sea-bed mounted pressure transducers linked to a self-contained recording system (Bird and Bullock,

Figure 3a, Pressure and Aeration Time Histories of a Field Type Two Wave Impact, Recorded at the Plymouth Test Site, Hi = 1.5 m, F = 0.125 Hz

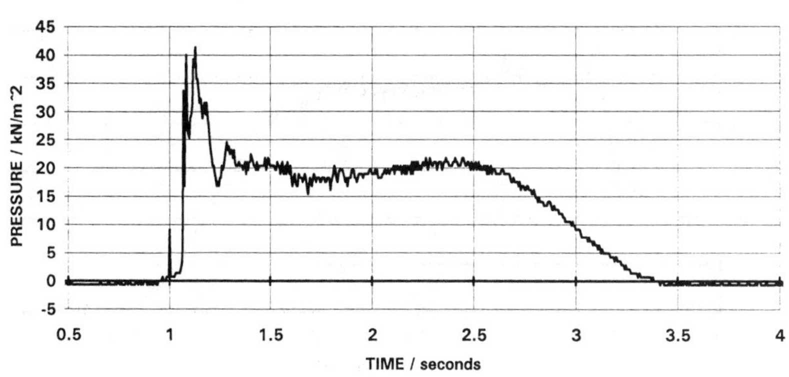

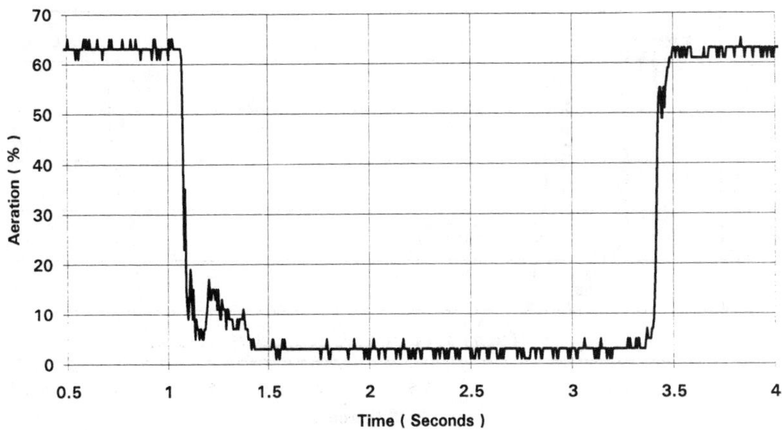

1991). Analysis of the resulting six sets of data enables the compound wave field to be resolved into incident and reflected wave components. It also yields the full directional spectrum so that the influence of the direction of wave approach can be investigated. Information on the shape of the breaking waves was captured by means of a video recorder which was also used to extract measurements of wave height and wave velocity at impact.

Figure 3b, Pressure and Aeration Time Histories of a Field Type Three Wave Impact, Recorded at the Plymouth Test Site

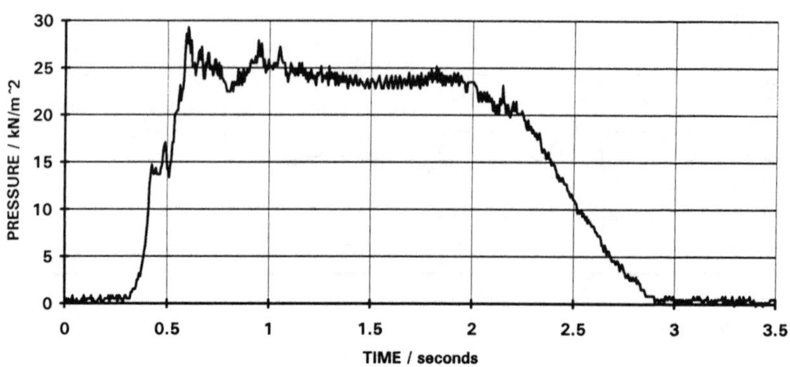

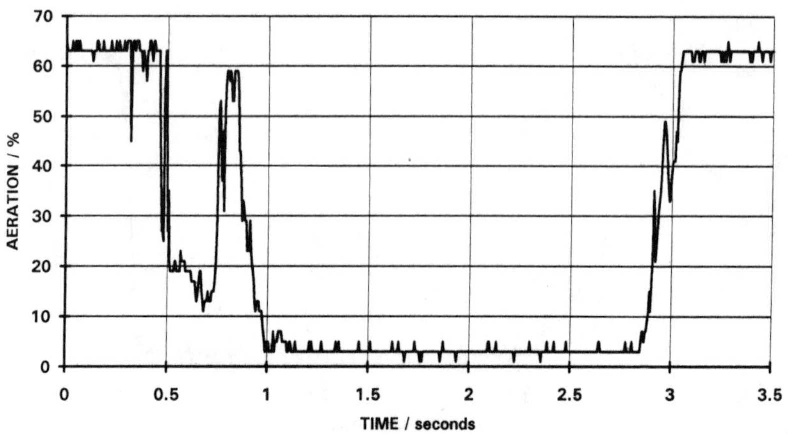

Examples of the temporal variations in pressure and aeration, measured at the field site when there was a fairly regular swell of approximately 8 seconds period approaching the breakwater, are shown in Figure 3. Void fractions tend to be high at this site, due to the waves having to pass over an irregular rocky foreshore. Further tests will indicate whether there is a negative correlation between the maximum impact pressure and the degree of aeration as suggested by the small scale model experiments described in the next section.

Video recordings of both field and laboratory waves indicate that the peak pressures are extremely sensitive to the precise form of the breaking waves. In an attempt to simplify the problem, different types of breaker have been classified as shown in Figure 2 (Graham et al, 1993). The time histories of both impact pressure and aeration for each type of breaker exhibit different characteristics. The field data shown in Figure 3a and 3b were obtained for breaker types 2 and 3 respectively.

PHYSICAL MODEL TESTS

There are many factors which affect the magnitude and form of the impact pressure and aeration time histories at any particular field location. In addition to the height, period and angle of approach of the incident waves, these include breaker type and position relative to the structure, the nature of the foreshore, the geometry of the structure, the ambient level of aeration and various other history effects. Although mathematical analysis can provide some insight into the likely effect of some of these variables, physical model experiments under controlled conditions are necessary both to develop the analytical techniques and to promote recognition of important phenomena.

The influence of aeration on the impact pressures produced by various types of breaker is being investigated as part of the European Union (EU) funded Marine Science and Technology Programme. Most of the tests are being conducted at small scale (Graham *et al*, 1993, Walkden *et al*, 1993) although some data has recently been obtained at large scale in collaboration with the Franzius Institute, University of Hannover.

The small scale tests are being conducted in a 20 m long by 1.2 m wide by 0.9 m deep wave channel (Walkden *et al*, 1993) equipped with an absorbing wavemaker (Bullock and Murton, 1989). In the tests carried out to date, regular waves of up to 100 mm high have been induced to break against a vertical wall by a 1 : 4.5 sloping foreshore. Wave heights are measured by means of a resistance-type wave gauge mounted on a motorised trolley. By traversing the gauge along the channel and recording the envelope of water surface displacements, the waves can be resolved into incident and reflected components (on the basis of linear theory). A multi-probe system has been used for more recent tests, this allows analysis of incident and reflected wave energy for both regular and random waves (Kajima 1969).

Impact pressures and aeration fractions are measured by means of an array of high response miniature pressure transducers and aeration gauges, see Figure 4.

Video records of breaker shape and underwater air/water interaction provide additional information on air entrainment in terms of both content and distribution. The simultaneous records of wave motion, pressure and void ratio enable the influence of aeration on impact pressures to be studied in greater detail than has been previously possible. An example of the temporal variations in pressure and aeration measured in the laboratory for a type 2 wave impact is shown in Figure 5.

Some control over the level of aeration was achieved in the small scale laboratory tests by attaching an artificial aerator to the foreshore in front of the wall. This enabled a cloud of small bubbles to be injected into the passing waves. The effect of the increased aeration is considered in the next section.

Recently a vertical array of pressure transducers and aeration gauges was mounted in the form of an instrumented spar on the face of a model caisson under test in the 324 m long, 5 m wide 7 m deep wave channel of the University of Hannover and Technical University of Braunschweig. This enabled fresh water tests to be carried out, in collaboration with the Franzius Institute, with waves up to 1.1 m high.

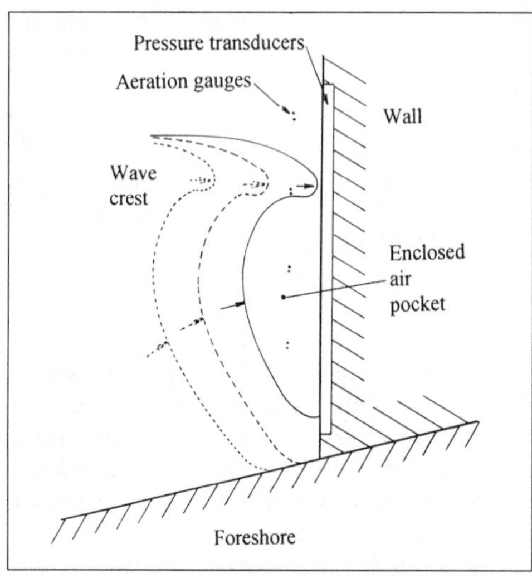

Figure 4, Model scale wave wall

Measurements were taken during 18 sets of regular waves and 6 sets of irregular waves as well as a number of single wave impacts. The wave heights and periods were selected to produce a variety of conditions ranging from reflective waves to well developed plunging breakers. The effect of aeration on impact pressures is most noticeable in waves which tend to produce high impact pressures so these were the most preferred type. Laboratory waves such as these tend to be cleaner (smoother in shape) than field waves and strike the wall normally. This relative perfection of form might itself give rise to higher impact pressures than typical of ocean waves of similar overall size. Be that as it may, when measurements were taken from a pressure transducer mounted rigidly in the caisson a huge peak pressure of 480kN/m^2 with a very fast rise time was recorded. This is higher than almost all the field measurements reported in the literature. Indeed it is only known to be exceeded

Figure 5, Pressure and Aeration Time Histories of Model Scale Laboratory Type Two Wave Impact, Hi = 90 mm, F = 0.75 Hz

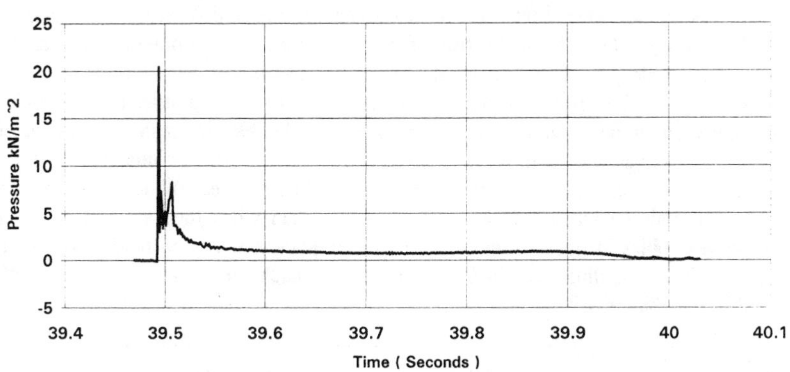

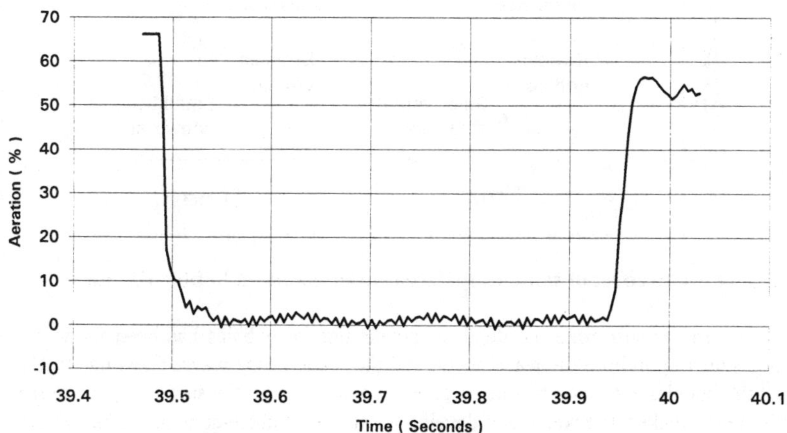

by the 690 kN/m^2 claimed to have been measured by Rouville *et al* at Dieppe in 1937 (Blackmore and Hewson, 1984). It is certainly very much higher than both the impact pressures expected from similar sized ocean waves and the pressures used in the design calculations for most breakwaters.

DISCUSSION OF RESULTS

Although wave shape and velocity can be expected to scale in accordance with the Froude law, this is not true of variables associated with aeration. Thus, for example bubble, size, bubble rise velocity and the celerity of shock waves will be at least in part Weber, Reynolds and Mach number dependent. Another factor which is currently under investigation is the persistence of bubbles in fresh and sea water. Scott (1975) has shown how the presence of salt and surface contaminants in sea water tends to stabilise bubbles. Consequently, when bubbles collide, they are more likely to coalesce when they are in fresh water than when they are in salt water. Because bubbles in fresh water can grow more quickly, they also tend to rise more quickly. Thus, as illustrated in Figure 6, air can reach the surface more quickly in fresh water than in salt water.

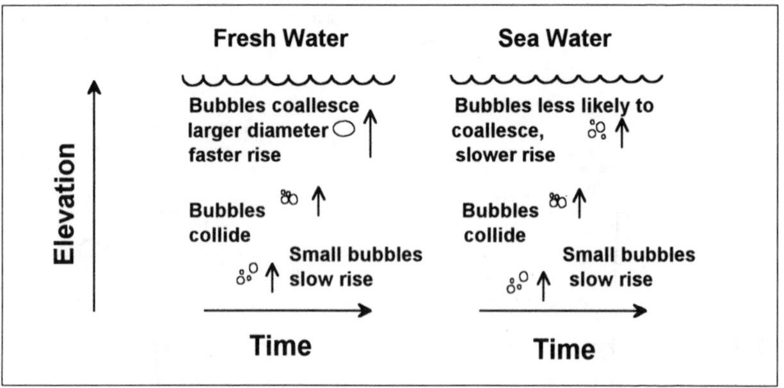

Figure 6 The effect of salt and surface contaminants on bubble rise times

In the surf zone the depth of penetration of bubbles has been found to be dependent upon both the wave height and the shape of the waves (Hwung, 1992). If all the bubbles entrained by one breaker have not risen to the surface by the onset of the next breaker, the water will develop an ambient bubble content. A further effect of the presence of salt and surface contaminants is that bubbles can remain unbroken at the surface for an appreciable period of time and this also adds to the ambient bubble level.

Both the difference in fluid properties and the greater distances that must be travelled will cause bubbles to persist for relatively longer in the field than they do in the typical laboratory wave model which uses fresh water and penetrates to Froude law scales. The resultant difference in the compressibility of the air water mixture is likely to introduce significant scale effects, particularly when high waves are being considered. Thus, both the maximum impact pressures and the rates of pressure increase predicted by physical models will tend to be upper bounds which in many circumstances may be unreasonably high. One of the objectives of the present investigation is to gain a better understanding of these scale effects.

Although no attempt was made to reproduce the field environment in laboratory tests, the pressure records from the type two wave impact reproduced in Figures 3a, 5a, and 7 are broadly similar. Thus, both exhibit a short duration shock pressure followed by the near horizontal 'S' shape associated with the deceleration of the incident wave, the production of the uprush, and the acceleration of the reflected wave. During the laboratory wave impacts the shock pressures were generated by a wave crest breaking against the wall and both trapping and compressing a lens shaped pocket of air against the wall. Air was also trapped against the wall by the full scale wave but the entrainment process was disrupted both by irregularities in the shape of the natural wave and by the high levels of aeration. The relatively less pronounced shock pressure in the field data by comparison with the model data are compatible with the scale effects described above.

The ability to relate pressure and aeration time histories to the separate phenomena which make up the wave breaking process will be useful during the interpretation of future field results.

Another possible indication of scale effects can be gained by noting that the data shown in Figure 5a was obtained with a 0.09 m high wave whereas for Figure 3a the wave was approximately 1.5 m high. The average of the maximum impact pressures recorded for a sequence of 75 waves in the model was 10.5 kN/m^2. If the Froude law is used to scale this value up to predict the level which might be expected from a 1.5 m wave, the result is 175 kN/m^2 which is significantly higher than the 42 kN/m^2 measured in the field. Although a direct comparison of these results is not justified, the possibility of discrepancies of this magnitude can not be excluded from conventional model predictions based on a simple application of the Froude law.

It is also instructive to compare the results obtained from the 0.09 m high waves with those obtained from a sequence of twenty five, 0.9 m high waves in the large scale laboratory tests. The respective wave periods were 1.5 s and 4.5 s. This suggests that on the basis of Froude law scaling, the ratio of the average maximum pressures should be approximately in the ratio of 1:10 and indeed the measured values were 10.5 kN/m^2, and 107.5 kN/m^2 respectively. Thus in this particular case there is good agreement between the results of the two fresh water models which differ by an order of magnitude in scale.

Further insight into the effects of aeration can be gained by comparing the results of small scale tests with and without artificial aeration.

Figure 7, Pressure Time History of a Large Scale Laboratory Type Two Wave Impact, Hi = 0.9 m, F = 0.222 Hz

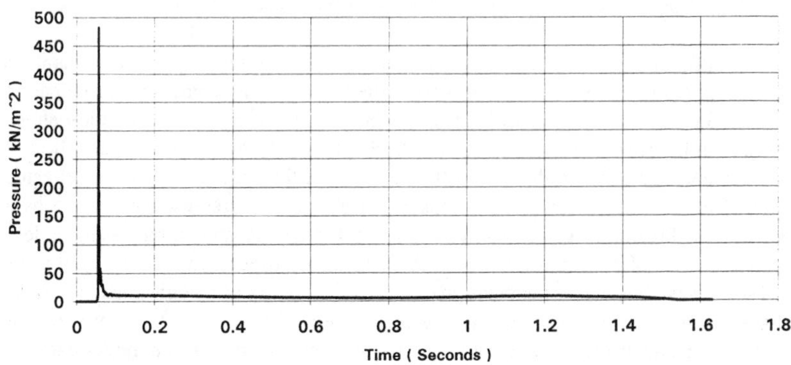

Figures 8a and 8b show results obtained from two sequences of 75 type 2 breakers. This type was selected because it normally generates the highest impact pressures which conceivably make it more sensitive to changes in the compressibility of the air/water mixture. In both cases the waves were 0.09 m high, their frequency was 0.75 Hz and the depth of water at the toe of the wall was 70 mm. In an attempt to reproduce the natural sea condition of an ambient level of aeration prior to breaking, the waves in the second sequence were artificially aerated. The data presented were taken from the pressure transducer which recorded the highest impact pressures.

Figures 8a and 8b show the maximum impact pressures recorded for each wave in both sequences. The mean for the first sequence was 10.5 kN/m^2 (with a standard deviation of 3.0 kN/m^2) which reduced to a mean of 8.0 kN/m^2 (standard deviation = 2.0 kN/m^2) for the aerated sequence. The frequency distributions of these pressures are presented Figure 10, which despite the scatter of results clearly shows the reduction occurring during artificial aeration.

The cushioning effect of aeration is illustrated by Figures 9a and 9b which show the gradients of the pressure rise that occurred during the impact of each wave. This gradient is defined as dP/dt where dP is the pressure difference between 10 % and 90 % of the maximum pressure and dt is the time difference between the occurrence of these pressures. For the naturally aerated waves the mean pressure gradient was 2.47 kN/m^{2}s (standard deviation = 2.3 kN/m^{2}s), which reduced to 1.08 kN/m^{2}s (standard deviation = 1.46 kN/m^{2}s) with the addition of artificial aeration.

The above results clearly show that the injection of additional air into the breaking waves can reduce both the maximum impact pressures and the impact pressure gradients. However, it does not prove that the effect was caused by an

Figure 8a, Maximum impact pressure for a sequence of 75 laboratory wave impacts Hi = 90 mm f = 0.75 Hz

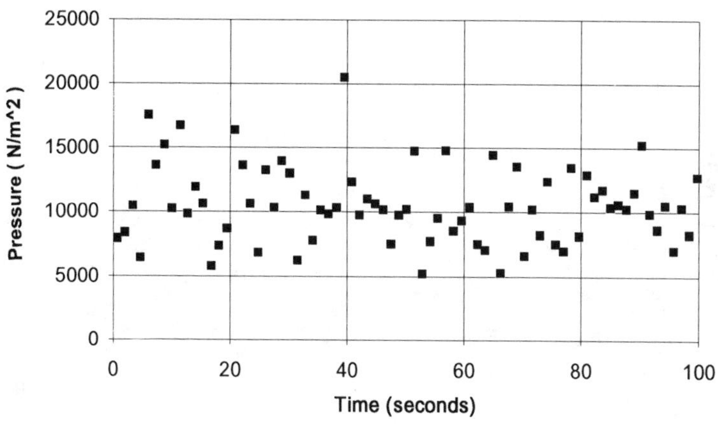

Figure 8b, Maximum impact pressure for a sequence of 75 artificially aerated laboratory wave impacts Hi = 90 mm f = 0.75 Hz

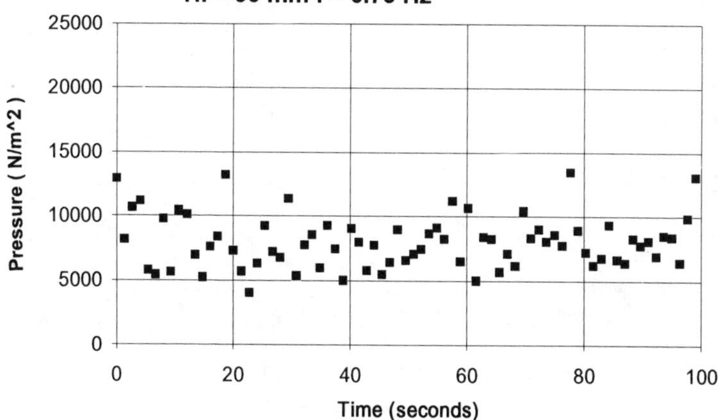

Figure 9a, Pressure gradients for a sequence of 75 laboratory wave impacts Hi = 90 mm f = 0.75 Hz

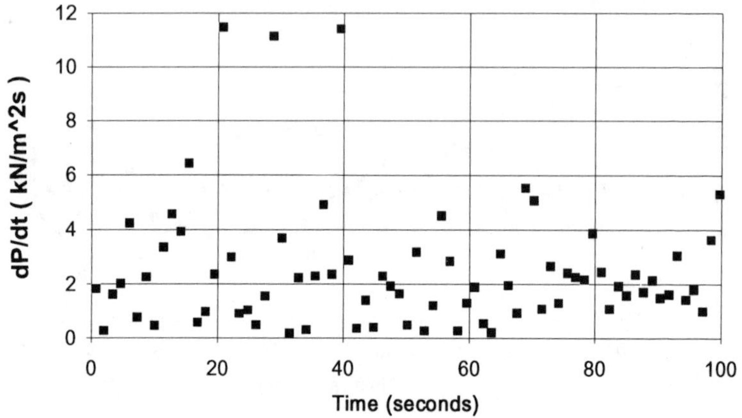

Figure 9b, Pressure gradients for a sequence of 75 artificially aerated laboratory wave impacts Hi = 90 mm f = 0.75 Hz

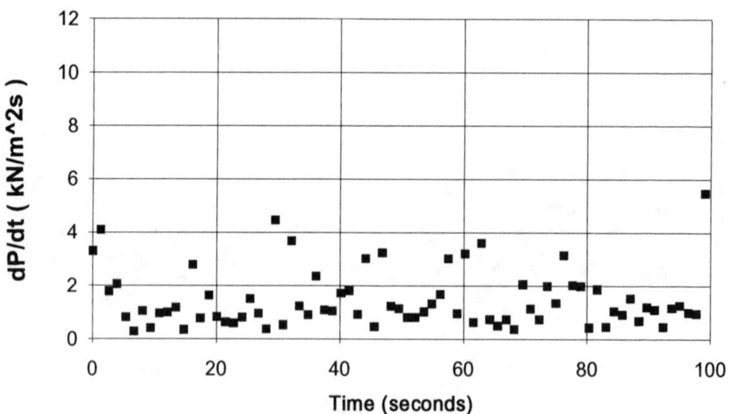

increase in the compressibility of the air/water mixture. It could, for example, have been caused by a change in wave profile although steps were taken to minimise this. As far as can be judged, it seems likely that both the increase in compressibility and the change in shape contributed to the reduction.

Figure 10, Frequency Distribution of Maximum Impact Pressures, F=0.75, Hi=0.09 m

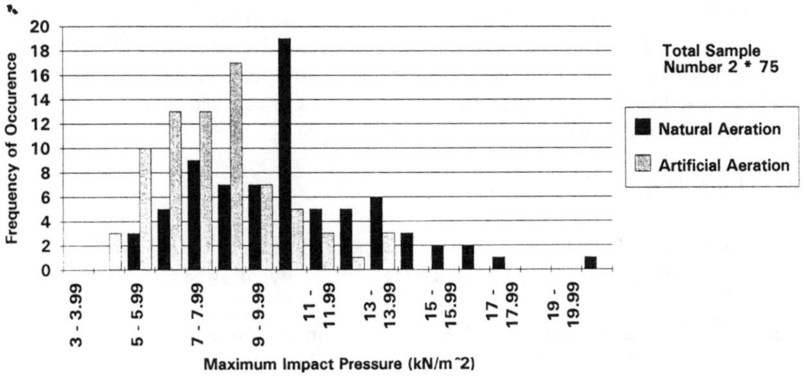

FUTURE WORK

Future field work will make use of vertical arrays of pressure transducers and aeration gauges which are being developed with the benefit of the experience gained from the earlier tests. The resulting improvement in spatial coverage will significantly increase the probability of capturing the highest pressures. It will also provide information on the vertical distributions of pressure and aeration. An arrangement of bi-directional accelerometers is to be mounted on the top of the breakwater to measure the structure's response which will be related to the peak pressures and other parameters. Large quantities of data will be logged without the necessity of either mains power or continuous attention by an operator. This will enable the equipment to be deployed at remote sites where high waves and high impact pressures are expected.

In addition to continuing the laboratory tests described above, small scale laboratory tests will also be carried out using sea water in order to assess the effects of salinity and surface tension on aeration and impact pressure. Drop impact tests will also be continued in which pressure transducers are mounted on a flat plate and dropped on a stationary column of water through which controlled volumes of air bubbles will be passed. The results from these tests will show the effects of bubble size and concentration on sea and fresh water compressibility without the complications of a varying wave shape.

It is hoped on the basis of these model and full scale tests to acquire a better understanding of the physical processes involved and thus improve wave impact pressure predictions based on small scale physical models.

ACKNOWLEDGEMENTS

The work described in this paper was in part carried out under contract MAS2-CT92-0047 of the European Union's Marine Science and Technology Programme, and in part with the support of grants from the UK Science and Engineering Research Council. This, together with co-operation and facilities made available by the Franzius Institut, Universitat Hannover and the Technischen Universitat Braunschweig are gratefully acknowledged.

REFERENCES

Bird, P.A.D., and Bullock, G.N. 1991. "Field Measurements of the Wave Climate," Developments in Coastal Engineering, University of Bristol, March, pp. 25-34.

Blackmore, P.A., and Hewson, P.J. 1984. "Experiments on full-scale wave impact pressures," Coastal Engineering, Vol. 8, pp. 331-346.

Bullock, G.N., and Murton G.J., 1989 "Performance of a Wedge Type Absorbing Wave Maker," Journal of Waterway, Port, Coastal and Ocean Engineering Vol. 115, No'1, pp. 1-20.

Graham, D.I., Hewson, P.J. and Bullock, G.N. 1993. "The Influence of Entrained Air upon Wave Impact Pressures," Paper presented at MAST EUROMAR meeting, Brussels, March.

Griffiths, J., and Hewson, P.J. 1992. "Aeration in waves," Internal Report, School of Civil & Structural Engineering, University of Plymouth, 19p.

Hwung, H.H. 1992. "Energy dissipation and air bubbles mixing inside surf zone," Proceedings of Coastal Engineering, Vol. 1, Ch. 22, pp. 308-321.

Kajima, R. 1969 "Estimation of an incident wave spectrum under the influence of reflection," Proceedings 13th Congress - International Association for Hydraulics Research, pp. 285-289.

Scott, J.C. 1975. "The Role of Salt in Whitecap Persistence," Deep Sea Research, Vol. 22, pp. 653-657.

Walkden, M.J., Hewson, P.J., Bullock, G.N., and Graham, D.I. 1993 "The Effect of Artificially Entrained air upon Wave Impact Pressures," University of Plymouth internal report number SCSE-93-004. Paper presented at MAST workshop Monolithic (Vertical) Coastal Structures MAST II MCS Project, Madrid, October 93, 27p.

PERMEABILITY MEASUREMENTS FOR THE MODELLING OF WAVE ACTION ON AND IN POROUS STRUCTURES

Marcel R.A. van Gent[1]

ABSTRACT: Permeability measurements were carried out in a U-tube tunnel to study flow through coarse granular material. The results are of importance for studying the flow in permeable structures such as rubble-mound structures and gravel beaches. The contributions of laminar and turbulence friction terms have been determined as well as the importance of inertial resistance. Differences between stationary flow and oscillatory flow conditions have been studied. The influence of oscillatory flow conditions on the porous flow friction coefficients have been implemented in new expressions for porous flow friction coefficients. These can be used both to study scale effects in small-scale physical models and for the implementation in numerical models simulating porous flow.

INTRODUCTION

Many coastal structures contain permeable parts. The porous flow for instance in gravel beaches and rubble-mound structures affects the external wave motion considerably. Therefore, porous flow needs to be modelled correctly both in physical models and in numerical models to obtain a representative wave motion. To achieve this, insight in the porous flow friction terms must be acquired.

Permeability tests have been performed to determine the porous flow friction coefficients and to study the different effects caused by stationary and oscillatory flow conditions. Such permeability tests have already been performed with stationary flow but only very few tests were carried out with oscillatory porous flow. To obtain pressure gradients and velocities that are of interest for many

---

1) Delft University of Technology, Department of Civil Engineering, P.O.Box 5048, 2600 GA Delft, The Netherlands.

practical purposes, measurements have been performed in the oscillating water tunnel of DELFT HYDRAULICS. These tests were carried out within the framework of the European MAST-G6 Coastal Structures research project.

Among other formulae, non-stationary porous flow can be described by the extended Forchheimer equation:

$$I = au + bu|u| + c\frac{du}{dt} \tag{1}$$

where I is the hydraulic gradient, u the filter/bulk velocity and a, b and c dimensional coefficients. The first term can be seen as the laminar friction term, the second term as the turbulence friction term and the third term as the time-dependent term (inertial resistance). For the non-dimensional coefficients a, b and c several expressions have been derived. These will be discussed later.

Gu and Wang (1991) discussed the relative importance of inertia for many practical purposes; for material whose size ranges between small gravel and large rock, inertial resistance cannot be neglected. Unlike large material, the laminar contribution cannot be neglected in small-scale physical model tests.

Smith (1991) provided a set of friction coefficients for oscillatory flow with flow conditions that can be compared with small gravel in prototype conditions. However, these tests were mainly performed with spheres. The present tests were mainly performed with rock material and are comparable with porous flow in prototype conditions in large gravel and in small and large rock.

EXPERIMENTAL SET-UP

In Van Gent (1993) the measurements, the test results and the analysis to derive new expressions for porous flow coefficients have been described in detail. In the following, a summary of these activities is given.

The tests were performed in an oscillating water tunnel having the shape of a U-tube. In the horizontal section of the U-tube, 15 m long and 0.30 m wide, a sample (l=0.75 m; h=0.50 m; b=0.30 m) with porous material was placed, see Figure 1. To obtain a sufficient large flow rate through the samples, the cross-section of the horizontal section of the U-tube was reduced. To obtain this, an additional bottom was placed 0.30 m above the bottom of the tunnel and a slope was created on both sides of the test section.

A piston positioned at one of the vertical parts produced oscillating water movements. Various combinations of amplitudes and oscillation periods could be generated. The piston movement was recorded (both the control signal and the actual signal). This enabled assessment of the actual filter velocities through the sample. These velocities were checked with a Laser-Doppler velocity meter positioned above the slope and outside the range of the water particles moving through the sample. Although the control signal that was used for the oscillatory flow tests was sinusoidal, it appeared that the actual displacement did not copy. For each oscillation period the stroke of the piston was increased by steps of about 1% (≈ 0.015 m) of the maximum stroke of the piston (≈ 1.50 m). The step-wise increase in stroke was continued till the moment the maximum pressure of the tunnel was reached. The average maximum piston movement was about 10% of the maximum stroke of the piston which corresponded to a maximum velocity of about 0.5 m/s near the sample. This indicates that the tests were carried out in the lower range of the possible piston displacement.

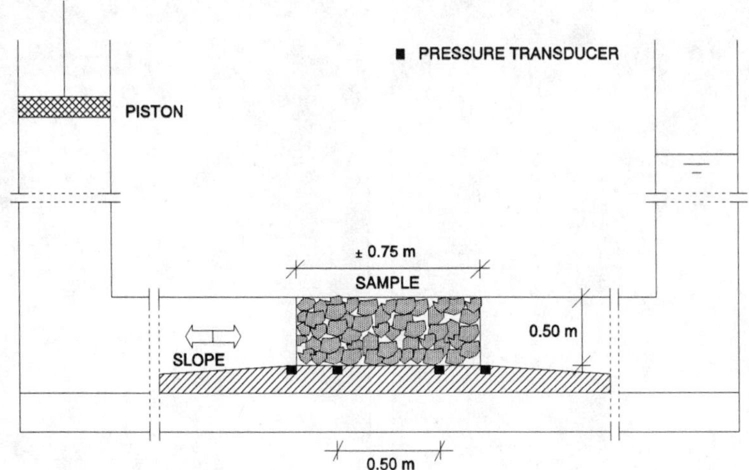

Fig.1 Experimental set-up.

A stationary flow can be produced in the tunnel with a flow rate up to 0.10 m^3/s. After positioning of the sample in the tunnel this flow was imposed on the sample to remove air bubbles. The actual test-runs started with a constant flow of 0.01 m^3/s. The flow rate was increased by steps of 0.01 m^3/s up to a maximum 0.10 m^3/s.

At the bottom of the box pressure transducers and differential pressure transducers were installed, both inside the box and just outside. The distance between the transducers inside the box was 0.50 m. The distance between the transducers installed outside the box was 0.8 m. For the stationary flow tests the flow rate generated by the pump was measured (both the control and the actual flow rate) and checked with the LDV-equipment. All signals where recorded during one minute with a sampling frequency of 100 Hz.

To reduce wall effects halves of spheres having roughly the same size as the tested rock material were glued to the vertical sides of the box containing the samples with rock material, see Figure 2. Although the wall effects were reduced, the filter velocities were multiplied by a factor of 0.95 to account for the remaining wall effects. The velocities as measured from the piston and the measured velocities near the sample differed slightly due to leakage that occurred during the tests. This has also been taken into account, see for details Van Gent (1993).

Fig.2 Reduction of the wall effects. *Fig.3 Sample with spheres.*

TESTED MATERIAL

Five samples with various types of stones have been tested. Their relevant properties are given in Table 1. Test materials denoted with R1, R3 and R4 were provided by Hydraulic Research, Wallingford. Test material R3 was obtained from material R1 which was rounded to get a 5 to 10% weight loss. R4 was obtained from material R1 and rounded to get a 20 to 25% weight loss. A full description is given by Bradbury et al. (1988) and Williams (1992). Material R8 was used as core material in tests at Hannover, see Ouméraci (1991).

The porosity n was assessed by weighing the stone sample in a box with a volume equal to the box placed in the oscillating water tunnel. The volume of the stones was found by division by the stone density. The porosity of the sample with spheres was theoretically derived. Because the wooden spheres expand in water, the actual porosity could have been slightly smaller. The aspect ratio l/t is defined as the average length of the longest axis of the stones (l) divided by the minimum length perpendicular to this axis (t). The rock samples were compacted before testing so that no compaction could take place during execution of the tests.

Table 1 Description of the tested material.

Code	Material	D_{EQ}	D_{n50}	D_{n15}	D_{n85}/D_{n15}	l/t	n
R1	Irregular rock	0.0760	0.0610	0.0525	1.27	1.9	0.442
R3	Semi round rock	0.0607	0.0487	0.0419	1.27	2.0	0.454
R4	Very round rock	0.0606	0.0488	0.0425	1.26	2.2	0.393
R5	Irregular rock	0.0251	0.0202	0.0170	1.03	2.3	0.449
R8	Irregular rock	0.0385	0.0310	0.0230	1.74	2.0	0.388
S1	Spheres	0.0460	0.0460	0.0460	1.0	1.0	0.476

Apart from the five samples with stones, a sample with wooden spheres in a cubic packing arrangement has been tested, see Figure 3. Also three samples with cylinders in "squared packing" arrangements have been tested. An analysis of the results with cylinders can be found in Andersen et al. (1993).

The number of samples is not sufficient for a full parameter study. Parameters as porosity, diameter, grading, aspect ratio and shape (gross shape, roughness and surface texture) have been varied. Since there were more parameters varied than the number of tested samples, existing formulae cannot be extended with more parameters. However, the results can be compared with existing formulae.

STATIONARY FLOW TESTS

The coefficients a and b from the extended Forchheimer equation (eq.1) could be derived using the measured pressure gradients and measured filter velocities and applying linear regression analysis. Assuming that those coefficients are constant for the tested range, a plot I/u versus u would give a straight line. I is the measured hydraulic gradient and u is the calculated filter velocity derived from the piston displacement. Extrapolations of the lines in Figure 4 give the a-values at the vertical axis. The b-values can be derived from the slopes of the lines.

It appears that the assumption that the a and b values for a particular sample are constant within the tested range is valid since the measured data correspond fairly well with the fitted line except for the two lowest measuring points with sample R8 (excluded from further analysis). One data point from sample R4 has also been excluded; this point is supposed to be caused by an error in the data-acquisition or by failure of the equipment. All other data points, resulting in rather straight lines, do not indicate that the validity of the Forchheimer equation (prescribing straight lines in Figure 4) should be questioned. Therefore, these divergent points have not been included in the further analysis.

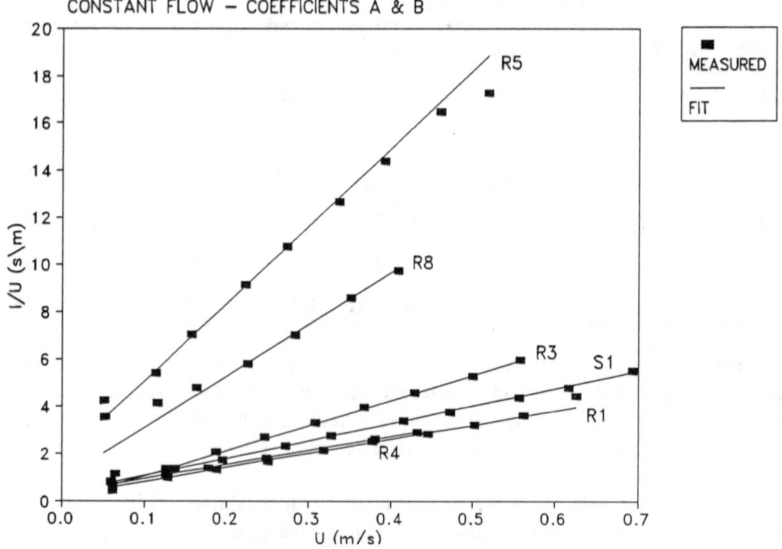

Fig.4 Data points from stationary flow tests and fits to the Forchheimer equation.

For the non-dimensional coefficients a and b, several expressions were proposed. Those by Ergun (1952) could be derived theoretically, see for instance Van Gent (1991). Therefore, these will be used here (α and β are non-dimensional coefficients and ν is the kinematic viscosity).

$$a = \alpha \frac{(1-n)^2}{n^3} \frac{\nu}{g D^2}$$
$$b = \beta \frac{1-n}{n^3} \frac{1}{g D} \tag{2}$$

Table 2 shows the results from the stationary flow tests. The coefficients a and b from the Forchheimer equation (eq.1) are given as well as their standard deviation (std). The values of α and β from equation 2 have been calculated using three characteristic length scales for D: D_{n15}, D_{n50} and D_{EQ}.

Table 2 Forchheimer friction coefficients from stationary flow tests.

Sample	a	std a	α-D_{n15}	α-D_{n50}	α-D_{EQ}	b	std b	β-D_{n15}	β-D_{n50}	β-D_{EQ}
R1	0.23	0.037	1327	1791	2780	6.0	0.076	0.48	0.55	0.69
R3	0	0.016	0	0	0	10.7	0.05	0.75	0.88	1.09
R4	0.34	0.015	808	1066	1644	6.0	0.06	0.25	0.29	0.36
R5	1.81	0.093	1204	1662	2566	32.8	0.75	0.91	1.07	1.33
R8	0.89	0.055	554	1007	1552	21.7	0.4	0.47	0.63	0.78
S1	0.33	0.023	2070	2070	2070	7.4	0.16	0.69	0.69	0.69

The coefficients α and β are not constant, see also tests described by Burcharth and Christensen (1991). This indicates that the expressions for a and b (eq.2) are over-simplified. Probably parameters such as grading, aspect ratio and shape (gross shape, roughness and surface texture) still have to be implemented in the expressions. In Van Gent (1993) an implementation of the influence of the orientation of stones with respect to the mean flow-direction is proposed.

The measured friction coefficients for stationary flow correspond fairly well with the data from Smith (1991), see Van Gent (1993). However, some of the numerous existing empirical expressions for stationary flow conditions, see Hannoura and Barends (1981), overestimate the quadratic friction term considerably.

OSCILLATORY FLOW TESTS

The oscillatory flow tests were performed for relative large Reynolds numbers and for small Keulegan-Carpenter numbers. The ranges of some relevant parameters for the oscillatory flow tests are listed in Table 3. Note that both in the Re-number and the KC-number the maximum pore velocity ($\hat{U}/n$) is taken as the representative velocity. The Ac-number which is a measure for the accelerations in the porous medium, is introduced here. Three wave periods were used: 2, 3 and 4 s. The maximum filter velocities $\hat{U}$ up to 0.50 m/s were used.

The c-term amplifies deviations in a smooth velocity signal causing that the contribution of the a and b-term could not be separated for each tests run individually. Therefore, the assumption has been made that the coefficient a is equal to the a coefficient measured in the stationary flow tests.

As will be shown later, the results of the oscillatory tests show that the term $a{\cdot}u+b{\cdot}u{\cdot}|u|$ for oscillatory flow conditions is larger than for stationary flow. The difference is assumed to be caused by a larger b-value (turbulence contribution) rather than by a larger a-value (laminar contribution). It can be expected that, compared with a stationary flow, an oscillatory movement of the fluid causes extra turbulence rather than extra laminar flow. This indicates that it is more likely that for oscillatory flow the *b*-values are larger than that the *a*-values are larger.

*Table 3 Ranges in oscillatory flow tests**.

Sample	Û	T	Re/1000	KC	Ac*1000
R1	0.13-0.50	2-4	15-66	8-60	7-49
R3	0.12-0.45	2-4	12-46	8-65	7-33
R4	0.12-0.49	2-4	16-58	11-82	7-58
R5	0.05-0.25	2-4	2-10	9-88	3-22
R8	0.09-0.34	2-4	6-25	13-91	7-36
S1	0.07-0.51	2-4	5-38	6-93	7-52

* $Re=\hat{U}D_{EQ}/n\nu$; $KC=\hat{U}T/nD_{EQ}$; $Ac=\hat{U}/nTg$.

The *b* and *c* coefficients were determined with a graphical approach; comparisons were made between the signal from the measured pressure gradient I and the calculated signal using the extended Forchheimer equation. The term $c\cdot\partial u/\partial t$ is supposed to be zero at the peak of the velocity signal ($u=\hat{U}$). Therefore, the term $a\cdot u+b\cdot u\cdot|u|$ could be determined from these maximum velocities. The *c*-term is relatively important around the zero-crossings of the velocity signals ($u\ll\hat{U}$); the *c*-coefficients could be determined by fitting the extended Forchheimer equation to the measured signal near the zero-crossings. Despite the complicate way to derive the *c*-coefficients it is estimated that this approach does not give errors larger than roughly 10%.

*Table 4 Results oscillatory flow tests**.

Sample	*b*(stat)	*b*(avg)	*b*(std)	*b*(range)	*c*(avg)	*c*(std)	*c*(range)
R1	6.0	8.5	1.23	7.2-11.5	0.21	0.14	0-0.40
R3	10.7	13.6	2.02	12-17	0.27	0.14	0-0.45
R4	6.0	9.2	0.84	8.1-12	0.30	0.14	0-0.45
R5	32.8	35	5.23	31-50	0.12	0.16	0-0.40
R8	21.7	23	1.80	21-28	0.31	0.14	0-0.45
S1	7.4	9.3	3.34	6-21	0.15	0.13	0-0.3

* *b*(stat) is the *b*-coefficient from the stationary flow tests; (avg) denotes the average value in the oscillatory flow tests; (std) denotes the standard deviation.

Figure 5 shows an example of measured pressure gradients from a test with rock sample R5 using an oscillation period of 2 s and a maximum filter velocity of $\hat{U}$=0.20 m/s. The fit to the extended Forchheimer equation is shown as well.

In Table 4, the results from the oscillatory flow tests have been summarized. For a comparison, the *b*-values from the stationary flow tests have also been listed. The *b*-values from the oscillatory flow tests are larger. The *b* and *c*-values show large standard deviations. However, as will be shown in the following section, these deviations are not random but rather systematic.

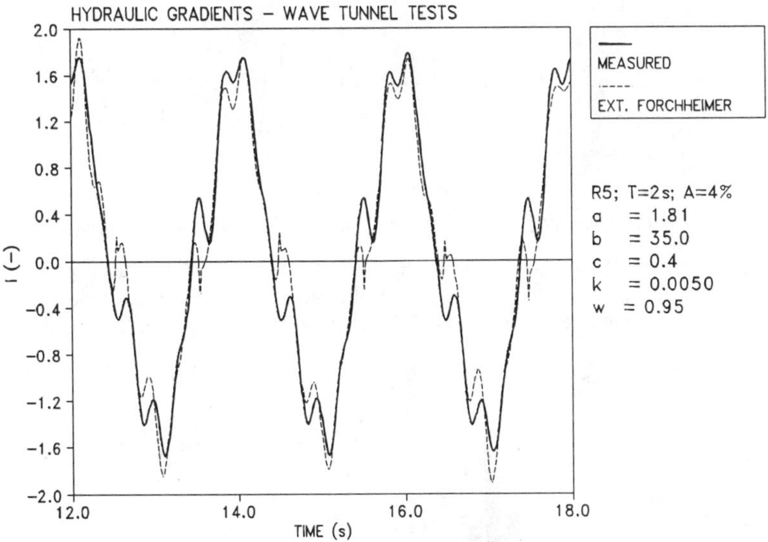

Fig.5 Measured hydraulic gradient (I) and fit to the extended Forchheimer equation.

In Figure 6, the contributions of the *a*, *b* and *c*-terms to the complete signal from the test shown in Figure 5 are made visible. The figure shows that the contribution of the *c*-term is rather limited. Even for a test with a relatively large *c*-value (c=0.4), the contribution of the *c*-term is only of relative importance in a small part of the oscillation period, to be more specific near the zero-crossings.

For each sample the contributions of the three terms from the extended Forchheimer equation are calculated with respect to the maximum hydraulic gradient (I_{MAX}), see Figure 7. The ratio $a \cdot \hat{U}/I_{MAX}$ varied between 0 and 0.41. The term with coefficient *b* is the largest for all samples; $b \cdot \hat{U} \cdot |\hat{U}|/I_{MAX}$ varied between 0.59 and 1.00. The contribution of the term with coefficient *c* reached its maximum just after the zero-crossings. At that point, the ratio $(c \cdot \partial u/\partial t)_{MAX}/I_{MAX}$ reached its maximum

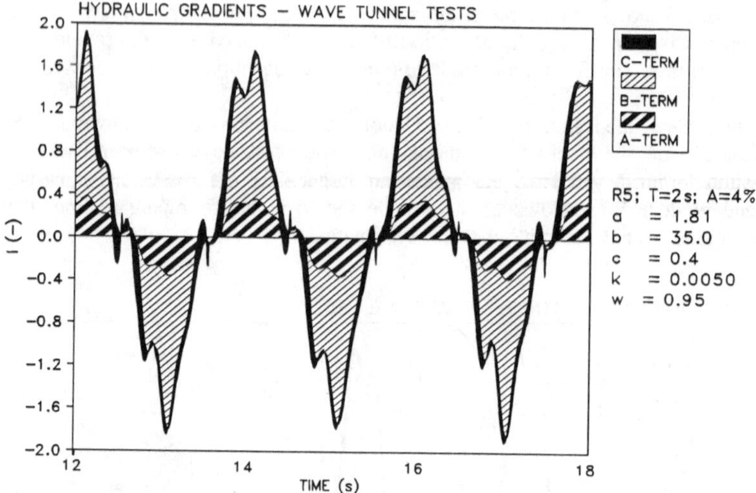

Fig.6 Contribution of each term from the extended Forchheimer equation.

contribution of 40% of I_{MAX} for some tests with sample R1. Note that it may look as if the influence of the *a*-term is of the same order of magnitude as the contribution of the *c*-term. However, the contribution of the *c*-term lasts relatively short.

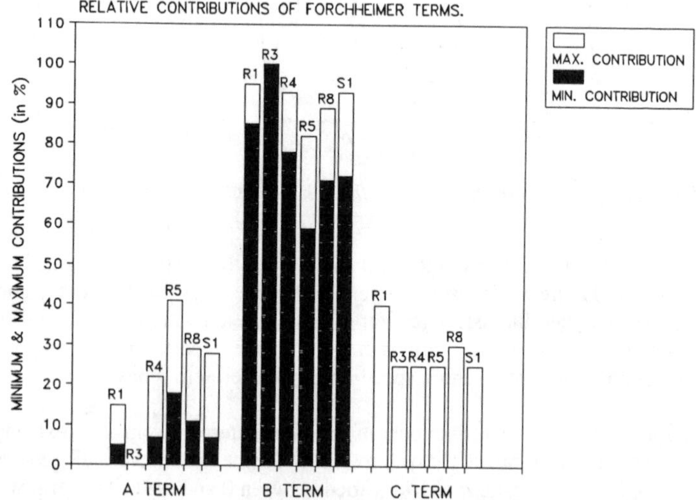

Fig.7 Contributions of each term from the extended Forchheimer equation.

FORMULAE FOR NON-STATIONARY POROUS FLOW

The measurements showed that oscillatory flow conditions lead to larger friction coefficients *b* than under stationary flow conditions, especially for the relatively low KC-numbers. Figure 8 shows this dependency for the tests with the rock samples. The expression for the turbulence friction *b* contains the non-dimensional coefficient β, see equation 2. Here, this coefficient is divided into the stationary flow contribution β_c and an extra resistance β' present in the case of an oscillatory wave motion. In Figure 9 this extra resistance is shown as function of the Keulegan-Carpenter number defined as $KC=\hat{U}T/(nD_{n50})$.

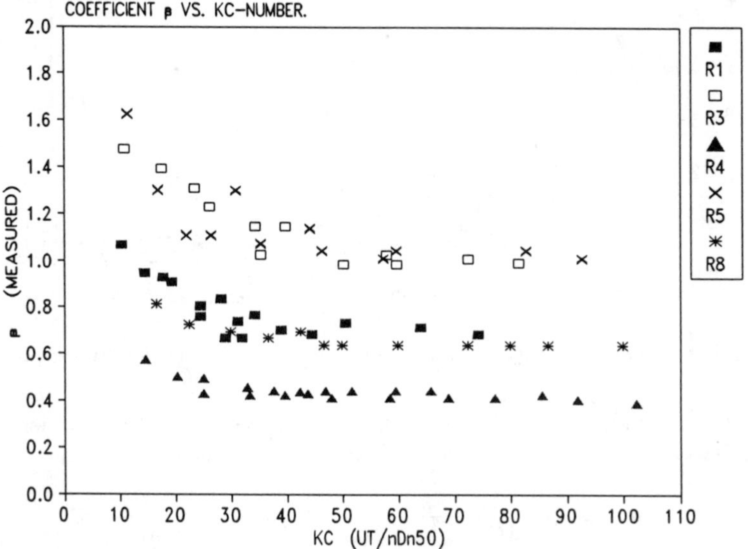

Fig.8 Friction coefficient β as a function of the KC-number.

A dependency on the KC-number was also found for the resistance of a single cylinder in an oscillatory flow, see Chakrabarti (1987). Boundary layers and possibly small eddies, will be destroyed if the flow direction changes. This destruction of the boundary layers requires an extra amount of momentum. The destruction of these boundary layers will be larger if the inertia term, relative to the turbulence term, is larger. This is inversely proportional to the KC-number since the KC-number can be seen as the ratio of the influence of the turbulent term and the influence of inertia. Boundary layers are not developed instantaneously. This causes a kind of history effect in the friction term; the friction at a certain point of time is not directly dependent on the (average) velocity at that point of time. To account for this phenomenon, a characteristic velocity of the flow field is more useful than the momentary velocity. Therefore, this phenomenon can be implemented by taking the maximum bulk/filter velocity Û for the characteristic velocity in the KC-number.

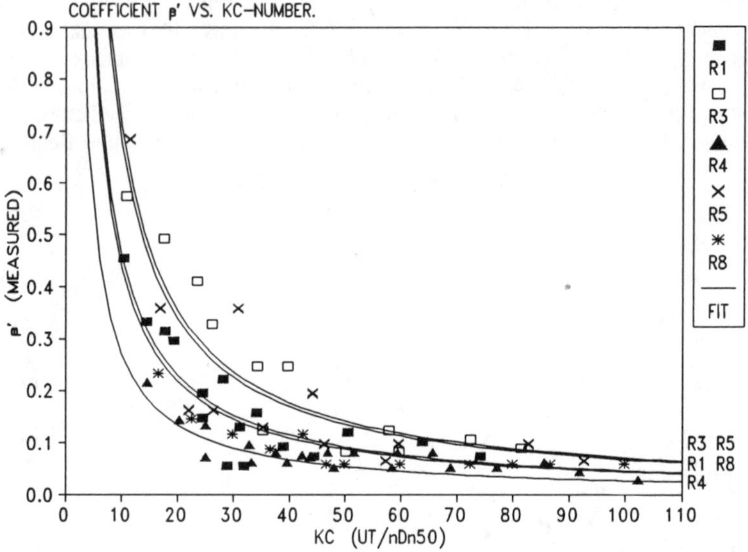

Fig.9 Friction coefficient β' as a function of the KC-number (with fits to eq.3).

The dependency on the KC-number has been incorporated in the expression for *b*. The expression for the Forchheimer coefficient *b* becomes:

$$b = \beta_c \, (1 + \frac{7.5}{KC}) \, \frac{1-n}{n^3} \, \frac{1}{g \, D_{n50}} \quad \text{where} \quad KC = \frac{\hat{U}T}{n D_{n50}} \tag{3}$$

The inertia term from the Forchheimer equation has also been analysed. It was found that this coefficient also depends on the flow field, namely, on the acceleration parameter Ac=Û/nTg, see Figure 10. This indicates that the coefficient *c* also depends on the flow-field.

Gu and Wang (1991) and Van Gent (1991) found an expression for *c* after a theoretical derivation:

$$c = \frac{1 + \gamma \, \frac{(1-n)}{n}}{n \, g} \tag{4}$$

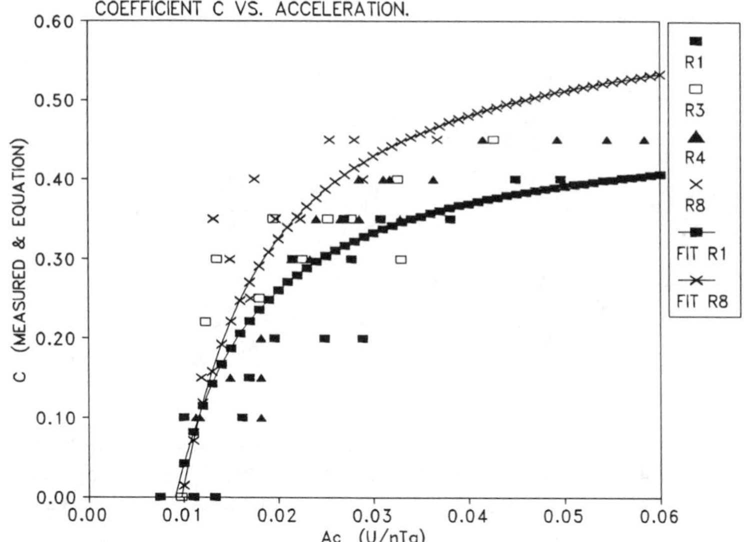

Fig.10 Measured c-coefficients and the expression for c versus the Ac-number.

where γ is a non-dimensional coefficient that accounts for the phenomenon "added mass". To accelerate a certain volume of water a certain amount of momentum is needed. The amount of momentum that is needed to accelerate the same volume of water in a porous medium is larger. This is called "added mass" because the extra amount of momentum suggests that a larger volume of water has to be accelerated.

Equation 4 has been extended with an empirical relation for γ taking the dependency on the flow-field into account. For two samples, the corresponding expressions for c (see equation 5) are shown in Figure 10. The contribution of this term to the total resistance is relatively small, see Figures 6 and 7.

Equation 5 shows the extended Forchheimer equation with the expressions for a, b and c. Although the coefficients α and β_c may still depend on parameters like grading, shape, aspect ratio or the orientation of the stones, for α and β_c 1000 and 1.1 can be used respectively. These expressions for a, b and c can also be applied in equations describing porous flow with a phreatic surface. This was done in Van Gent (1994) where the expressions have been implemented in a numerical wave model including a porous flow model based on long-wave equations. In Van Gent et al. (1994) the expressions have been applied in a numerical 2-DV Navier-Stokes model where a porous flow model is described with adapted Navier-Stokes equations.

$$I = a\,u + b\,u\,|u| + c\frac{\partial u}{\partial t} \quad \text{with}$$

$$a = \alpha\,\frac{(1-n)^2}{n^3}\,\frac{\nu}{g D_{n50}^2}$$

$$b = \beta_c\,(1+\frac{7.5}{KC})\,\frac{1-n}{n^3}\,\frac{1}{g\,D_{n50}} \quad \text{where } KC = \frac{\hat{U}T}{n D_{n50}} \tag{5}$$

$$c = \frac{1 + \frac{1-n}{n}\left(0.85 - \frac{0.015}{Ac}\right)}{ng} \quad \text{with } Ac = \frac{\hat{U}}{ngT} > \frac{0.015}{\frac{n}{1-n} + 0.85}$$

The expressions in equation 5 do not include a possible resistance due to the presence of a convective term. Such a resistance, if any, could be incorporated in the *b*-term because it would probably be quadratic with the velocity, see Van Gent (1991).

CONCLUSIONS

The measurements, performed in a U-tube tunnel to study porous flow, showed differences for stationary and oscillatory flow conditions. For flow regimes where inertia is important (low KC-numbers), the quadratic/turbulence porous-flow friction is larger than under stationary flow conditions. The differences between stationary and oscillatory flow depend on the Keulegan-Carpenter number which means that the resistance is depending on the flow-field. Inertia coefficients have been determined as well. The contribution of the inertial resistance is much smaller than the quadratic/turbulence friction. The test results have been incorporated in the expressions for porous flow friction coefficients. These expressions have already been implemented in several numerical models that are able to simulate wave motion on and in permeable structures such as rubble-mound breakwaters and gravel beaches.

ACKNOWLEDGEMENTS

The financial support by the Commission of the European Communities by way of the MAST G6-Coastal Structures project (contract 0032-C) and by Rijkswaterstaat (Dutch Governmental Water Control and Public Works Department) is gratefully acknowledged. I would like to express my appreciation to the participants in the MAST-G6S project, in particular to Dr. J.W. van der Meer from Delft Hydraulics and Dr. H. den Adel from Delft Geotechnics for their critical reading and valuable comments on the analysis.

REFERENCES

Adel, H. den (1991), *Inventory of existing knowledge on permeability formulae*, Delft Geotechnics, Report MAST-G6 Coastal Structures, project 1.

Andersen, O.H., M.R.A. van Gent, J.W. van der Meer, H.F. Burcharth and H. den Adel (1993), *Non-steady oscillatory flow in coarse granular materials*, Proc. MAST-G6 Coastal Structures workshop-Lisbon, project 1.

Bradbury, A.P., N.W.H. Allsop, J-P. Latham, M. Mannion and A.B. Poole (1988), *Rock armour for rubble mound breakwaters, sea walls and revetments: Recent progress*, H.R.-Wallingford, Report SR 150.

Burcharth, H.F. and C. Christensen (1991), *On stationary and non-stationary porous flow in coarse granular materials*, Aalborg University, Department of Civil Engineering, Report MAST-G6 Coastal Structures, project 1.

Chakrabarti, S.K. (1987), *Hydrodynamics of offshore structures*, Springer verslag, Berlin, Computational Mechanics Publ., Southampton, Boston.

Ergun, S. (1952), *Fluid flow through packed columns*, Chem. Engrg. Progress, Vol.48, No. 2, pp.89-94.

Gent, M.R.A. van (1991), *Formulae to describe porous flow*, Report MAST-G6 Coastal Structures, project 1 and Communications on Hydraulic and Geotechnical Engineering, ISSN 0169-6548 No.92-2, Delft University of Technology.

Gent, M.R.A. van (1993), *Stationary and oscillatory flow through porous media*, Communications on Hydraulic and Geotechnical Engineering, ISSN 0169-6548 No.93-9, Delft University of Technology.

Gent, M.R.A. van (1994), *The modelling of wave action on and in coastal structures*, Coastal Engineering, Vol. 22 (3-4), pp.311-339, Elsevier, Amsterdam.

Gent, M.R.A. van, P. Tönjes, H.A.H. Petit and P. van den Bosch (1994), *Wave action on and in permeable structures*, To be presented at ICCE'94, Kobe.

Gu, Z. and H. Wang (1991), *Gravity waves over porous bottoms*, Coastal Engineering, Vol.15, pp.695-524, Elsevier, Amsterdam.

Hannoura, A.A. and F.B.J. Barends (1981), *Non-Darcy flow: A state of the art*, Proc. EuroMech 143, pp.37-51, Delft.

Ouméraci, H. (1991), *Wave-induced pore pressure in rubble mound breakwaters*, Franzius Institute, Report MAST-G6 Coastal Structures, project 1.

Smith, G. (1991), *Comparison of stationary and oscillatory flow through porous media*, M.Sc.-thesis, Queen's University, Canada.

Williams, A.F. (1992), *Permeability of rubble mound material*, Report MAST-G6 Coastal Structures, project 1.

SCALE AND MODELING EFFECTS IN CONCRETE ARMOR EXPERIMENTS

Jeffrey A. Melby, A.M. ASCE and George F. Turk[1]

ABSTRACT: Small scale concrete armor unit experiments commonly contain a variety of structural and hydraulic scale and modeling effects. The hydraulic scale effects generally result in a conservative design. But no structural measurements have been made that adequately quantify the maximum surface strains occurring on concrete armor for static and dynamic response. Therefore, structural scale and modeling effects are unknown. This paper discusses a recent test series using 26kg dolosse to quantify the structural response for a variety of loading conditions while reducing a majority of the hydraulic scale effects. Over 1000 dry-land and hydraulic flume tests were carried out with the units. Impact response scaling criteria are examined and new insight is gained on the physics of the impact response. Finally, it is found that the dolosse are hydraulically stable for somewhat larger wave heights than indicated by many previous studies, with differences attributed to hydraulic scale effects. Also, impact stresses are shown to be higher than previously thought, with differences due to structural modeling effects.

INTRODUCTION

Concrete armor unit (CAU) hydraulic stability experiments are commonly done at scales from 1:30 to 1:75. Predominant scale effects relating to the hydraulic stability at these scales include Reynolds flow force, inter-unit surface friction, and entrained air. Although many experiments have been done to identify armor Reynolds numbers, above which, scale effects appear negligible, the studies have not separately identified and quantified surface friction and air entrainment effects and therefore, have actually quantified the entire hydraulic scale effect rather than just the Reynolds effects. Also, scale and model effects associated with the entire response, including both hydraulic stability and structural response, have not been adequately quantified. This knowledge

1) Research Hydraulic Engineers, Coastal Engineering Research Center, USAE Waterways Experiment Station, Vicksburg, MS, 39180

is crucial to efficient armor designs because the hydraulic stability and structural response are coupled.

There have been a number of CAU experiments which included strength scaling or structural instrumentation during the last decade. For dolosse, Timco (1983) scaled the material properties for non-impact structural response. As discussed by Burcharth et al. (1991), small-scale strength scaling is nearly impossible because impact response scales differently than the static and non-impact dynamic responses. The material required to accomplish simultaneous scaling of the different response types has not been developed and, if developed, would be extremely difficult to work with. Therefore, the majority of structural measurements of model concrete armor units are done with strain gage instrumentation. For dolosse, Scott et al. (1986), Anglin et al. (1989), Markle (1989), Melby (1992), and Burcharth et al. (1991) all addressed modeling of structural response using small scale structural instrumentation. For tetrapods, Terao et al. (1982), Burger et al. (1989), Nishigori et al. (1986), and Nishigori et al. (1989) discussed measurement of structural response.

The types of CAU structural measurements made to-date can be grouped into three main categories: small scale units with internally-placed load cells utilizing very small strain gages, mid to large scale units with either an internal load cell or surface mounted strain gages, and prototype units with internal strain gages (Figure 1). The following discussion addresses the attributes of several key studies done with structurally instrumented dolosse.

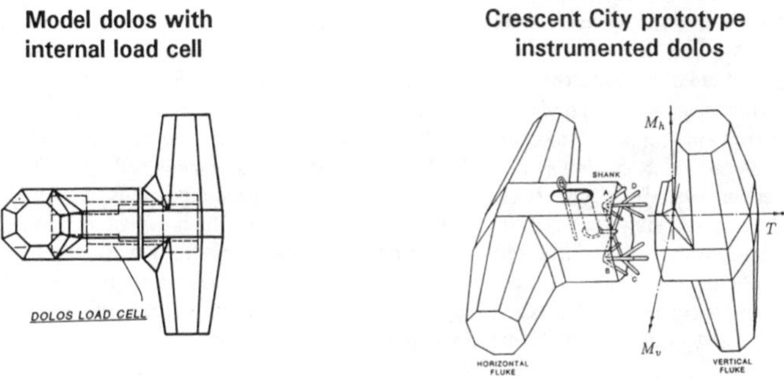

Figure 1. Load cell and prototype dolos structural instrumentation schemes

LOAD CELL INSTRUMENTATION

Small scale CAU structural instrumentation used is predominantly based on load cell technology. The load cell is typically a hollow aluminum cylinder with strain gages affixed to the outer surface of the aluminum and configured as a Wheatstone bridge. The cell is placed in a hollowed section of the armor unit. Because the cell has less stiffness than the original uncut armor section, it flexes more and therefore amplifies the strain signal in the section. This amplification is required because, in small scale armor units, the surface strains are typically too small to be measured with even the most sensitive silicon strain gages. The load cell-instrumented unit is typically calibrated by hanging weights from the end of the unit and determining the relationship between bridge output voltage and theoretical slender beam moments. Several comprehensive studies using load cell-instrumented dolosse are discussed in Anglin et al. (1989), Markle (1989), and Burcharth et al. (1991). Anglin used small scale load cells developed by Scott et al. (1986) and the others used small scale load cells developed by Markle. In addition to using units developed by Markle, Burcharth also used large load cells for 200 kg units.

The load cell instrumentation scheme has limitations and requires several assumptions to arrive at a stress state for the model dolosse (Melby et al. 1989). First, for a particular cross-section, extreme fiber stresses are typically resolved based on simple elastic slender beam theory where the flexure and torsional strains are assumed to be linearly distributed across the section. The maximum principal slender beam surface stress is derived from moments which are determined using internally differenced strains from opposing load cell surfaces. The load cell measurements assume that shear and axial strains are either small and/or compressive and negligible. Because the armor units are not composed of slender members and have complicated geometry in the regions of highest stresses, stress states are not linear and can be quite complicated (Melby et al. 1989, Lillevang and Nickola 1976). The load cells therefore introduce error because actual surface strains are not measured. Using 200 kg load cell-instrumented dolosse, Burcharth et al. (1991) showed that there is little error in neglecting the axial and flexural shear stresses, for design-level stresses. But no measurements of actual surface strains have been done to verify these load cell measurements.

Markle found reasonable agreement between the wave loading-induced response of the prototype Crescent City internally-instrumented dolosse and 200 g load cell-instrumented model armor units. All of these measurements were limited to a single cross section in the dolos, at the shank-fluke interface at one shank end. The non-impact slender beam principal stress determined from the prototype units was reasonably reproduced using small scale load cell-instrumented units. But note that the same slender beam theory was used to determine maximum principal stress from moments for both the prototype and model. Because surface strains were not measured, there is still some question concerning the actual magnitude of surface strains and the effects of flexural shear and axial stresses, as discussed in Melby et al. (1989).

Melby (1992) compared small scale static load cell measurements with Crescent City prototype static measurements and found poor agreement. The disagreement was primarily due to the fact that several of the prototype dolosse had extremely high stresses, and the prototype data set was extremely small. Burcharth et al. (1991) found reasonable agreement between mean stresses from static ramp tests of two load cell model sizes, 200 g and 200 kg, but the standard deviations were very different. So stresses derived from static load cell measurements have not been verified.

The primary problem with load cell measurements is that the load cell distorts the dynamic response under impact loading. This occurs because the measured response will be dominated by the dynamic response of the load cell, which is unlike that of a homogeneous uncut armor unit. Burcharth accounts for this using a fictitious modulus

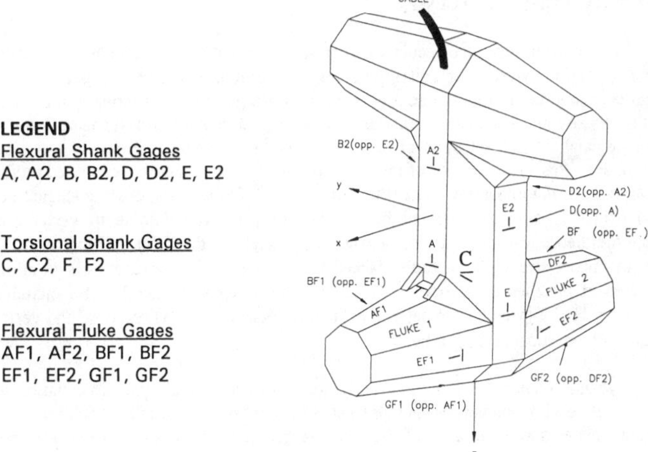

Figure 2. LSDFS dolos structural instrumentation -- strain gage layout

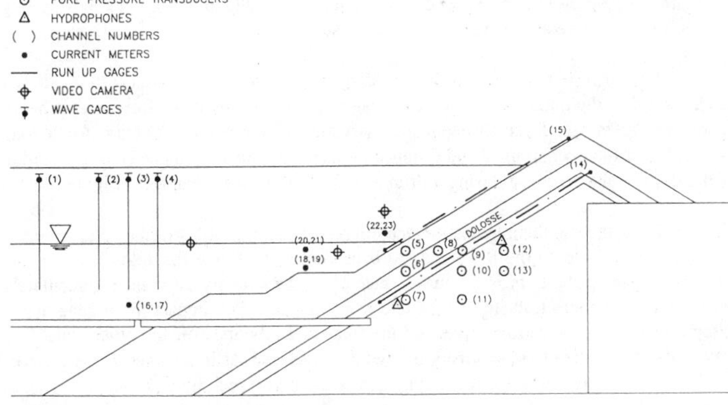

Figure 3. LSDFS flume instrumentation

IMPACT RESPONSE SCALING

Impact strain response occurs when armor units that are unstable and in motion on the slope collide with other units. The unit-to-unit impact stresses can occur in the unit that is in motion or in an underlying unit that is being struck by a moving unit. The impact response can occur in any combination of four predominant modes: shear, flexure, torsion, or axial. Different parts of the same armor unit can have different dominant impact modes. Previous analyses of impact response of dolosse have concentrated on the axial response of the impacting fluke (Burcharth 1981). These analyses were elegant but utilized primarily non-instrumented failure tests of prototype dolosse to verify and calibrate the theoretical relationships between drop height of the dolos and the resulting maximum impact stress. The studies lacked the surface strain measurements to allow verification of the theoretical basis. Therefore, the first part of the LSDFS included idealized dry-land impact tests to determine the physics of the impact response and verify Burcharth's theoretical scaling relations.

Dynamic response scaling is usually done using a general dynamic equation of motion, where each response type is represented by a different term in the equation. But for armor unit response, because of the complex geometry, such an equation does not exist. So for this analysis we will look at each general response type independently. Note that this is not rigorous and will only yield independent empirical relations for each response type. Under idealized conditions we can excite each response type independently; but if the instrumented units are on the breakwater, because the entire response is measured, it is not possible to separate the different impact response types from a typical measured strain signal. And if these predominant impact response types scaled differently, it would be impossible to develop a reasonable and accurate scaling criterion for impact response. But, as will be shown, the different impact responses scale similarly. The intent of this section is to provide a greater understanding of the physics of the impact response of dolosse and verify that it is possible to develop a single scaling criterion for concrete armor unit impact response.

Burcharth (1981) has proposed a scaling criterion for **axial** impact response based on the speed of the impact stress wave propagation through the dolos fluke. The theory quantifies the impact force through a conservation of momentum relation for a long slender beam fully constrained and dropped on its end. This is a simple characterization of the dolos fluke response during a drop test, ie. falling like a hammer (Figure 4).

In this theory, the impulse-momentum relation at impact is given by $F\Delta t = m\Delta V$, where F is the force on the fluke end, Δt is amount of time that the dolos is in contact with the impact surface, m is the mass, and ΔV is the velocity difference immediately before and after impact during the infinitesimal time Δt. The impact force generates a compressive shock wave that travels up the fluke and rebounds off the other end. The wave induces tensile stresses during the return travel and Δt is assumed to be proportional to the time required for this impact stress wave to travel the fluke length twice as follows.

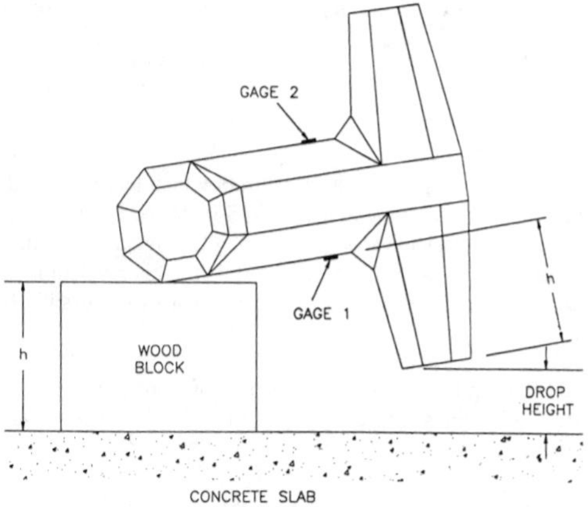

Figure 4. Dolos impact drop test configuration

$$\Delta t = \frac{2C}{c} = \frac{2C}{\sqrt{\frac{E}{\rho}}} = 2C\sqrt{\frac{\rho}{E}} \tag{1}$$

where C is the fluke length, c is the speed of stress wave in the armor unit, E is the dynamic modulus of elasticity, and ρ is the mass density. The time scale for this axial impact duration follows as

$$N_{\Delta t} = N_C\sqrt{\frac{N_\rho}{N_E}} = \sqrt{\frac{N_M}{N_L N_E}} \tag{2}$$

where N is the ratio of prototype to model for the subscripted parameter, M is the characteristic mass and L is the characteristic length. In this case, is the length of the fluke.

In addition to the strain gages shown in Figure 2, drop tests of the LSDFS dolosse were done with a single strain gage mounted within a centimeter of the impacting fluke tip. The fluke tip gage showed a single spike with duration 0.001 sec. This is considered the impact duration. The drop test records from the shank gages showed a damped oscillation with a period of 0.001 sec. The drop tests time series was similar to the flexure vibration time series shown in Figure 7. The impact duration was identical to the flexural vibrational period.

The LSDFS drop tests showed that the speed of the stress wave propagation was approximately 3050 m/sec, which is the same as that given by the theoretical relation $c = (E/\rho)^{0.5}$. Using this stress wave speed, and assuming that the impact duration is governed by the axial vibration of the impacting fluke, the impact duration would be

$$\Delta t = \frac{2C}{c} = \frac{2C}{\sqrt{\frac{E}{\rho}}} = 0.0003\text{sec} \tag{3}$$

which is nearly an order of magnitude shorter duration than that measured. Therefore, drop tests of the instrumented dolosse in the LSDFS indicated that the dolos impact time scale for the drop test configuration was governed by the **flexural** response of the shank rather than the **axial** compressive rebound of the fluke. This revelation complicates the original scaling development because we now have at least two possible time scales for impact duration.

Initial LSDFS drop tests showed that the torsional mode had a natural frequency of 800 Hz (Figure 5), as opposed to the flexural of 1000 Hz. This is yet another time scale for dolos impact response.

But the question remains: Does basing the impact force duration on flexural rather than axial vibration mode change the basic impact stress scaling? Consider the conventional drop test, with the dolos falling like a hammer. The impulse force follows the formulation given above, with the duration of impact determined by the time it takes for the shank to flex through one cycle. The flexural vibrational frequency for a slender beam (which is not the characteristic shape of the dolos but is a reasonable formulation for scaling) is given by

$$f_f = \left(\frac{\lambda}{2\pi L_s^2}\right)\sqrt{\frac{EIL_s}{M}} \tag{4}$$

where λ is a dimensionless constant, L_s is the shank length, E is the modulus of elasticity, I is the shank moment of inertia, and M is the mass of the shank. The time scale follows with $N_I = N_L^4$ as

$$N_{\Delta t} = \frac{1}{N_f} = \sqrt{\frac{N_M}{N_L N_E}} \tag{5}$$

which is the same as the axial scaling as long as the length scales are identical, which they are. If we assume that the natural vibration is **shear** dominated rather than flexural, then the modal frequency is given by

$$f_s = \left(\frac{\lambda}{2\pi L_s}\right)\sqrt{\frac{KG}{\rho}} \tag{6}$$

where $K = f(\vartheta)$ is the shear coefficient and $G = f(E,\ \vartheta\ (poisson's\ ratio))$ is the shear modulus. If we assume that $N_\vartheta = 1$ then $N_K = 1$ and $N_G = N_E$, then the time scale associated with the shear vibrational frequency scaling is identical to Equation 5. Therefore, although the time required for the dolos to vibrate through a single cycle is different for each different impact loading category, it appears that the time scales are identical. It is anticipated then that the different impact loading types will scale similarly.

Now that we have the time scale we can determine the scaling for the impact stress as per Burcharth (1981). The velocity for our falling hammer can be determined from $V = (gh)^{0.5}$, where h is the drop height and g the gravitational constant, and scales according to $N_V = (N_g N_L)^{0.5}$. Substituting the time scale into the impulse-momentum relation, for the impact stress scale yields,

$$N_{\sigma_i} = \frac{N_F}{N_A} = \frac{N_M N_V}{N_{\Delta t} N_L^2} = \sqrt{N_\gamma N_L N_E} \tag{7}$$

where $\gamma = \rho g$ is the specific weight. Therefore, according to this formulation, the impact stress scales according to the square root of the length scale if prototype strength concrete is used as the model material (i.e. $N_\gamma = N_E = 1$).

The previous discussion shows that the impact stress scaling is identical for each of the loading modes: axial, flexure, and shear. This is primarily because the time scale for each of the modes is a function of the single length scale for the three modes. The previous discussion sheds some light on the physics of the armor impact problem. This discussion may have been obvious to the early developers of impact scaling laws, but to the authors' knowledge, has never been discussed in any formal publication.

Accurate description of the scaling laws of the static, pulsating, and impact stresses is essential because, as stated earlier, the static and pulsating response scales are proportional to the length scale while the impact stress scale is proportional to the square root of the length scale. Thus, independent scaling of the impact stress is required to determine the prototype design stress level from model tests measurements.

PRELIMINARY LSDFS OBSERVATIONS

Static Ramp Tests

In the static ramp tests, seventy-eight 26 kg dolosse were placed randomly in a 2.4 m square by 0.6 m deep wooden box at a packing density of 0.83. These dry land tests were done to measure the static stresses during the simulated quasi-static nesting of a dolos armor layer. Eighty-four different tests or buildings were conducted using one instrumented dolos in one of four different positions on the slope (Figure 5). Prior to each test, the instrumented dolos was placed in a no-load support configuration and a static reading was taken. It was then placed, along with the uninstrumented dolosse, on the flat slope, the slope raised to 1V:1.5H, vibrated through 5000 cycles, lowered, and the dolosse removed. At each step during this process, the dolos surface strains were recorded. The total of 84 different tests represented 84 different boundary condition situations.

In Figure 5, the non-dimensional static stress is shown to be a function of position on slope. Here the stress is nondimensionalized by the product of the specific weight and the fluke length. "Mean" is the mean stress for the four dolos positions. As expected, the static stresses are significantly higher in the lowest position on slope, position 2. The effects of vibrating or nesting the slope can also be seen in Figure 5. "Placed" is the initial stress state with the ramp lying flat on the ground, "Unvibrated" is the stress state with the ramp raised to the 1V:1.5H slope, and "Vibrated" is the stress after vibration. The nested stress is consistently larger than the unnested stress and the stress in position 2 shows much higher sensitivity to nesting.

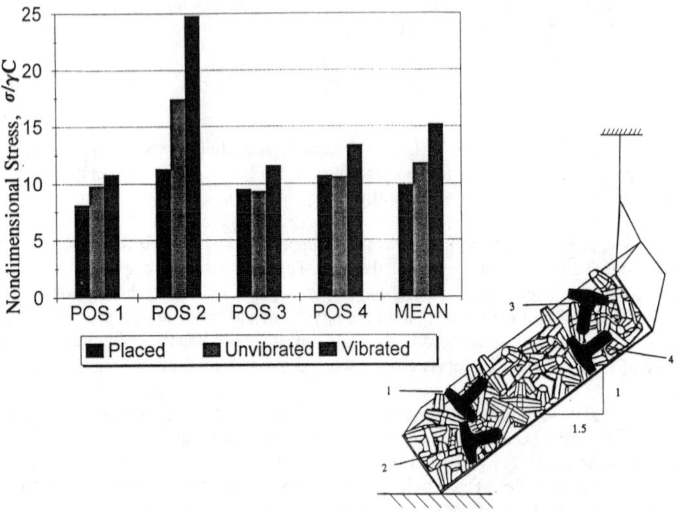

Figure 5. LSDFS static ramp nondimensional stress vs position on slope

Table 1 shows the nondimensional stresses compared to other tests mentioned previously in this article. The LSDFS static ramp tests are slightly lower than Crescent City prototype results and somewhat larger than model load cell results. Note that the Crescent City statistics are based on an extremely small sample size of 12 boundary conditions.

Table 1. Comparison of LSDFS static ramp nondimensional stress to other studies

STUDY	NONDIMENSIONAL TENSILE STRESS	
	MEAN	STANDARD DEVIATION
LSDFS 26kg Concrete	15.2	9.0
Crescent City Prototype	18.6	12.0
Melby (1992) 210g Load Cell	11.1	10.6
Markle (1989) 210g Load Cell	12.0	11.6
Burcharth (1991) 200g Load Cell	12.0-15.0	not available
Burcharth (1991) 200kg Load Cell	8.3	1.8

Drop Tests

Impact drop tests, as discussed previously, were also done with the 26 kg LSDFS instrumented dolosse. For these tests, the dolosse were placed so that the horizontal fluke was lying on a wooden block, the shank was horizontal, and the vertical fluke was on a 0.6 meter thick concrete floor. Metal shims of various thicknesses were pulled quickly from under the vertical fluke to drop the fluke to the floor. The surface strains were recorded at a rate of 12.5 kHz during the drops. The resulting impact stresses are shown in Figure 6, non-dimensionalized by the square root of the product of the dynamic modulus of elasticity, the specific weight, and the fluke length, plotted against the drop height, which is non-dimensionalized by the fluke length. For Figure 6, the drop height is relative to the centroid of the dolos, rather than the fluke tip, to allow direct comparison with previously published results. The results of Burcharth et al. (1991) are shown for comparison. The LSDFS are significantly higher than those of Burcharth. This difference could be due to the fact that Burcharth's measurements are based on uninstrumented prototype destructive dolos field tests. The prototype dolosse were dropped several times before cracks could be seen. Initial cracking and cylinder strength were used to define impact stress. As discussed earlier, there is a great deal of uncertainty in this type of stress determination.

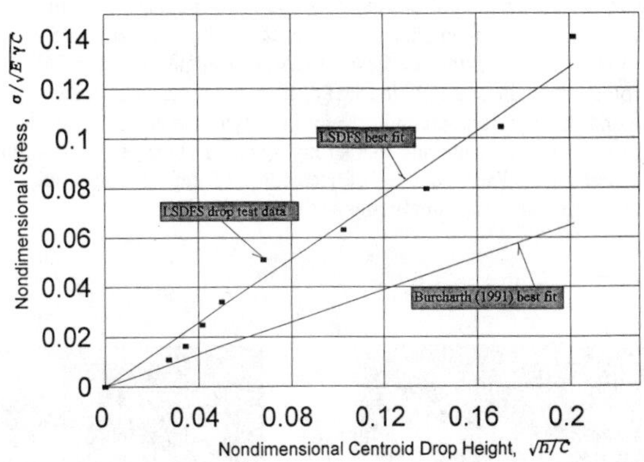

Figure 6. Nondimensional centroid drop height vs nondimensional tensile stress

Flume Tests

Several preliminary observations can be made from the flume tests. First, although the slope was reconstructed completely several times and partially reconstructed many times, the stability changed little from run-to-run for the same wave conditions. For the maximum wave energy ran, the corresponding Hudson equation stability coefficient was K_D= 64. Seven tests of 200 regular waves each were run at this wave energy and the dolosse simply got more stable as the tests wore on. The corresponding small scale tests showed the dolos slope unraveling at stability coefficients approximately half this value, corresponding to scaled waves of 80% the height of those in the large scale. This additional instability in the small scale tests can be attributed to hydraulic scale effects, and the results agree well with previous stability scale effects studies (Sollitt and Debok 1976). Note though that the previous studies discussed these scale effects as flow force or Reynolds effects but it is suspected that surface friction and entrained air contribute substantially to the instabilities.

Some preliminary observations concerning the impact stresses in the dolosse can be made. Figure 7 shows typical flexural and torsional impact signals. It can be seen that the maximum tensile strain for this particular impact is $80\mu\epsilon$, corresponding to a stress of approximately 2.2MPa. This was not an unusual magnitude for an impact stress for the LSDFS tests. If we scale this strength up to that of a 20-tonne unit (length scale of 1:8.5) using Burcharth (1981) impact scaling law, where the impact stress scales like the square root of the length scale, the prototype stress would be 6.4 MPa. This stress would easily exceed most prototype strengths.

During the three-week testing period, 12 dolosse or approximately 4% of all dolosse, broke. One of these broken dolosse was instrumented and data were collected during the failure. Nearly all of these failed armor units were observed to break due to wave-induced impact loading as the dolosse rocked on the slope. But during any given run only 1% to 2% of the dolosse on the slope were rocking. Few impact strains were above the strength of the model dolosse because the concrete's 28-day compressive strength was 55MPa, making the flexural tensile strength around 5.5MPa. This would give a prototype tensile strength of 15MPa, if scaled up to that of a 20-tonne unit. This is far beyond the concrete strength of any prototype structure. The model units were designed to be overly strong and this strength was predicted to be enough to eliminate nearly all breakage. The frequency of occurrence of impact strains large enough to fail the model armor units was simply higher than expected.

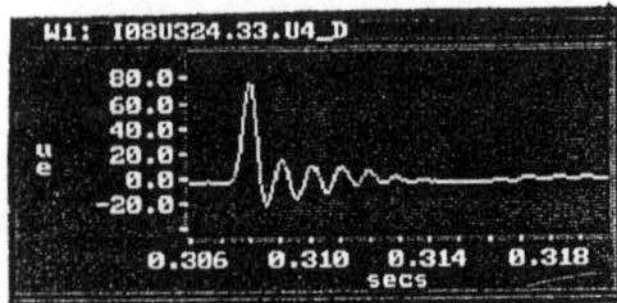

Flexure, Gage D

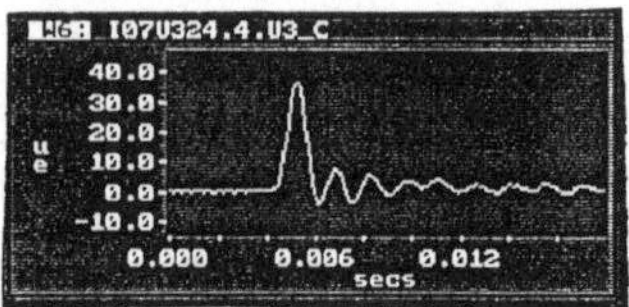

Torsion, Gage C

See Figure 2 for strain gage locations

Figure 7. Typical dolos flexural and torsional strain gage signals in strain*10^{-6}

CONCLUSIONS

The Large Scale Dolos Flume Study appears to have been very successful in measuring surface strains on concrete dolosse. Although most of the data have not yet been analyzed, several preliminary conclusions can be drawn from the discussion in this paper.

1. Impact stress from drop tests is governed by the flexural vibration frequency
2. Previously determined impact scaling laws were verified
3. Stresses from static ramp tests were more accurate but generally agreed with previous measurements
4. Drop tests more accurately determined stress levels than previous studies
5. Impact stresses are somewhat higher than expected in rocking units

ACKNOWLEDGMENTS

Permission to publish this paper was granted by the Chief of Engineers.

REFERENCES

Anglin, C.D., et al., 1989. "The Development of Structural Design Criteria for Breakwater Armor Units," ASCE/WPCOE Sem. on Stres. in Conc. Ar. Un., ASCE, NY, NY.

Burcharth, H.F., et al., 1991. "On the Determination of Concrete Armor Unit Stresses Including Specific Results Related to Dolosse," Coastal Engr. 15 (1991), pp. 107-165.

Burcharth, H.F., 1981. "Full Scale Dynamic Testing of Dolosse to Destruction," Coastal Engr. 4 (1981), pp. 229-251.

Burger, W.W., et al., 1989. "Impact Strain Investigations on Tetrapods: Results of Dry and Hydraulic Tests," ASCE/WPCOE Sem. on Stres. in Conc. Ar. Un., ASCE, NY, NY.

Lillevang, O.J., and Nickola, W.E., 1976 "Experimental Studies of Stresses Within the Breakwater Armor Piece 'Dolos'," Proc. 17th Int. Conf. on Coast. Engrg. ASCE, NY, NY.

Markle, D.G., 1989. "Crescent City Instrumented Dolos Model Study," ASCE/WPCOE Sem. on Stres. in Conc. Ar. Un., ASCE, NY, NY.

Melby, J.A., 1993. "Large Scale Dolos Flume Study Plan," Internal report, USAE WES CERC, Vicksburg, MS.

Melby, J.A. 1992. "Application of dolos design methods ," Proc. Coast Engrg. Practice '92, ASCE, NY, NY.

Melby, J.A., et al., 1989. "An Analytical Investigation of Static Stresses in Dolosse,"

ASCE/WPCOE Sem. on Stres. in Conc. Ar. Un., ASCE, NY, NY.

Nishigori, W., et al., 1986. "On Stress in Tetrapods Under Wave Action," Proc. 18th Int. Conf. on Coastal Engrg., ASCE, NY, NY.

Nishigori, W., et al., 1989. "Similarity Law of Impact Between Model and Prototype Tetrapods," ASCE/WPCOE Sem. on Stres. in Conc. Ar. Un., ASCE, NY, NY.

Scott, R.D., Turke, D.J., Baird, W.F., 1986. "A Unique Instrumentation Scheme for Measuring Loads in Model Dolos Units," Proc. 20th Int. Conf. on Coast. Engr., ASCE, NY, NY.

Sollitt, C.K., and Debok, D.H., 1976. "Large Scale Model Tests of Placed Stone Breakwaters," Proc. 15th Int. Conf. on Coast. Engr., ASCE, NY, NY.

Terao, T., et al., 1982. "Prototype Testing of Dolosse to Destruction," Proc. 18th Int. Conf. on Coastal Engrg., ASCE, NY, NY.

Timco, G.W., 1983. "On the Structural Integrity of Dolos Units under Dynamic Loading Conditions," Coastal Engr, 7 (1983) 91-101.

WAVE MEASUREMENT NEAR REFLECTIVE STRUCTURES

P.A.D.Bird[1], M.A. Davidson[1], G.N. Bullock [1] and D.A. Huntley[2]

ABSTRACT: A wave recording system has been developed for the measurement and analysis of the complex wave fields that occur in close proximity to coastal structures. It detects wave activity at six locations simultaneously with an array of sea bed mounted pressure transducers. It is fully self-contained and may be deployed unattended for long periods. The pressure records are converted using linear theory into corresponding surface elevation records. A cross-comparison exercise with a surface piercing wave staff validated this approach. An estimate of the wave directional spectrum is obtained using the "modified maximum likelihood method". The system has been deployed extensively at several types of coastal structure.

INTRODUCTION

Most coastal structures reflect waves to some degree; some quite markedly. Wave measurements taken in the coastal region must, therefore, take reflection into account if a true estimate of wave activity is to be obtained. In certain cases the incident wave component alone is of interest (for example in relating structural performance to wave attack) and this must be separated from the measured incident-reflected combination. In other cases (for example in determining the reflection performance of a structure) the ratio of reflected to incident wave components must be derived from the measured data. This latter aspect is of increasing interest to the coastal engineer because of the crucial effect of wave reflection on undermining the base of a structure by scour. Also, near harbours and waterways reflection must be minimized to avoid local sea states dangerous to shipping. A clearer understanding of the degree to which various structural forms

[1] School of Civil & Structural Engineering, University of Plymouth, Palace Court, Palace Street, Plymouth, PL1 2DE U.K.

[2] Institute of Marine Studies, University of Plymouth, Drake Circus, Plymouth, PL4 8AA U.K.

reflect waves in a range of sea conditions (which will follow from the measurements described in this paper) is likely to lead to increased effectiveness and cost saving in the design of coastal defences.

Many approaches have been taken to the measurement of sea waves (Bird and Bullock, 1991). They may be classified broadly by the location of the sensors: under the surface, at the surface, and above the surface. The first includes sea-bed mounted pressure and current measurement, and "upward looking" echo sounders. Surface piercing gauges and accelerometer buoys comprise the second category, and "remote sensing" methods the third, either close-in (downward looking infra-red and microwave sensors) or at some distance (radar and stereo-photogrammetry).

At Plymouth our requirement was for a measurement system that would provide a description of a sea state in the form of its directional spectrum: a full distribution over both frequency and angle. Incident and reflected components are then readily identified by knowledge of the layout of the site. We wanted a system that would operate unattended for long periods independently of any shore-based services, and that would survive and accurately record any storm events. The system could then be deployed at a wide range of sites, and would be able to collect data on a representative set of sea conditions. A review of available wave measuring apparatus showed that none fulfilled the requirements.

WAVE RECORDING SYSTEM

The University of Plymouth Wave Recording System (Plate 1 and Figure 1) was developed during 1987 and 1988. The system consists of six pressure transducers and a central signal conditioning and data storage unit. Each is fitted inside a pressure housing designed to be non-corrodable and fouling resistant, and to provide reliable long-term sealing. These features were achieved using materials such as polyacetal, titanium and silicon bronze. The pressure transducers are set out in an array on the sea bed and connected with armoured cable to the central unit. The user retrieves data by visiting the site by boat, recovering a buoyed data cable and downloading to a portable personal computer. The typical location of the equipment - away from the surf zone - gives maximum chance of surviving storm damage, and is also out of view of the public. (At two sites to date, however, the wave recording system had to be in the surf zone, but with the assistance of protective steelwork it remained undamaged.) All the equipment and ground tackle may be deployed by a team of divers in one working day. The four megabyte capacity data store holds up to six weeks' data on a typical measurement schedule (17 minutes operation every three hours), and may be re-used after downloading. Batteries last for approximately four months: only then is it necessary to recover the central unit in which the batteries are housed. The system features high precision measurements (down to 2mm resolution for small waves), automatic ranging circuits (to ensure optimum

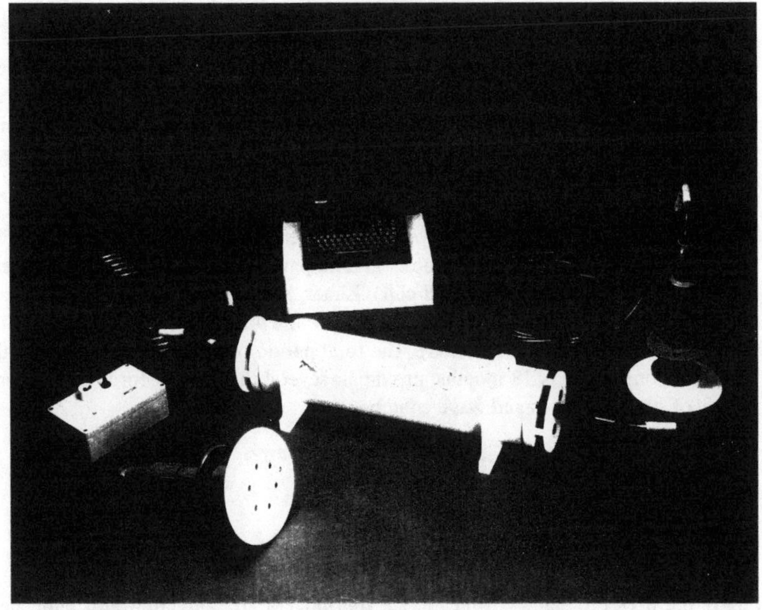

Plate 1. University of Plymouth Wave Recording System

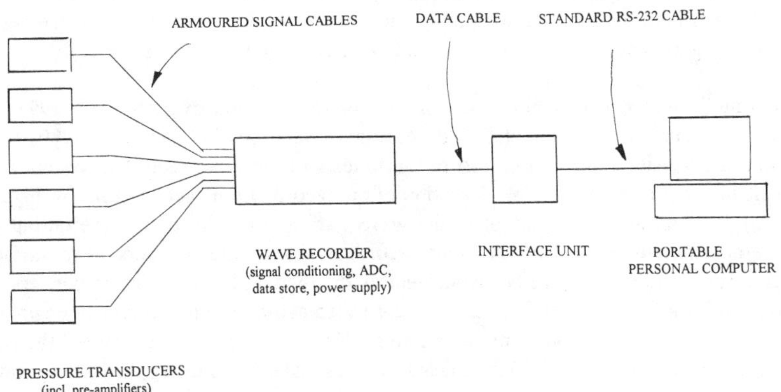

Figure 1. General arrangement of the Wave Recording System

resolution of the wave component within the pressure signal), semi-automatic calibration (to enable compensation during data analysis if any circuit should drift over time - although this has not so far occurred), and comprehensive battery supervision (to give the user warning of declining battery charge, and to prevent spurious operation when the batteries become exhausted). Relatively minor modification of the signal conditioning circuits would permit the use of transducers for quantities other than pressure, such as current meters.

Three units have been built to date and deployments made at five sites in the UK: Plymouth Breakwater (1 in 6 slope, granite-paved rubble mound breakwater); Bovisand Bay, Plymouth (beach and cliff); Elmer, West Sussex (inshore close to a 1 in 1.1 slope rock island breakwater, and at 1 km offshore) and Felpham, West Sussex (beach topped by a sea wall). The total period of operation in service at the time of writing exceeds 18 months, providing several thousand records that cover a variety of structure types and wave conditions.

DATA ANALYSIS PROCEDURES

Surface Elevation from Sub-surface Pressure

Firstly the pressure signals from the six transducers are converted into equivalent surface elevation records. This is done using linear wave theory implemented by a digital filter in the frequency domain (Davidson, 1992a). The validity of linear theory for this conversion has been examined in many studies in the past 25 years; a review is presented by Bird (1993). Many of these, especially the earlier ones, found considerable differences (up to 35%) between surface elevations measured directly and those constructed from pressure records using linear theory. Having considered the mechanisms that give rise to the discrepancies, it is the authors' opinion that they will contribute together an acceptably small error (<< 5%) in the case of the wave recording system operating in depths and wave conditions of interest here.

Further confidence was obtained from a cross-comparison exercise (Ilić, 1993) in which a surface piercing electrical resistance wave staff (Chadwick, 1989) was placed alongside one of the pressure transducers of the wave recording system at a suitable beach site (Felpham). A section of the records from each is shown in Figure 2(a). The full line is the output of the wave staff, and the dotted line the output of the appropriate channel of the wave recording system after conversion to surface elevation. The records are not coincident as the digital filter used in the conversion cuts off at a frequency in the region of 0.4 Hz to avoid applying excessive emphasis to very small pressure fluctuations and noise. Good correspondence between the two signals is seen in Figure 2(b) in which a low pass filter has been applied to the wave staff output. Corresponding spectra are shown in Figure 3. The bed mounted pressure transducer faithfully indicates wave activity of most relevance to the study

a)

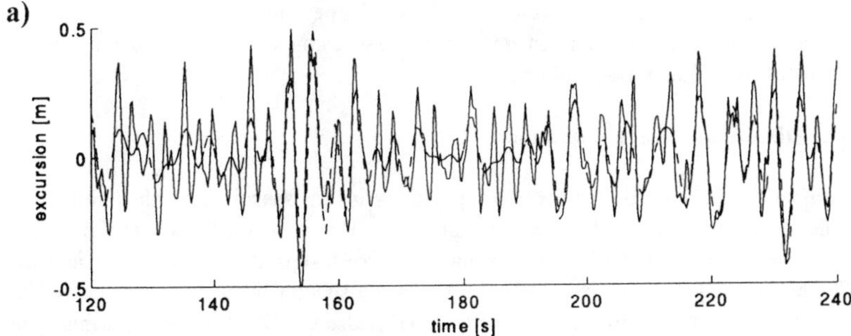

b)

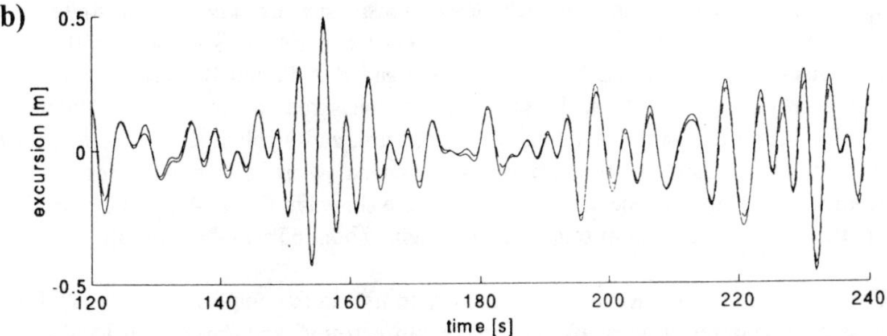

Figure 2. Comparison of the surface elevation measurements from the wave recording system and a wave staff (a) Surface elevations (b) The comparison after low-pass filtering the wave staff signal.

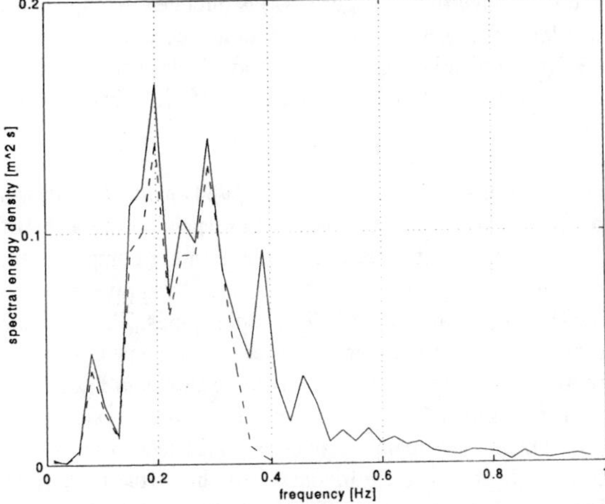

Figure 3. Comparison of the spectra of surface elevation measurements from the wave recording system and a wave staff.

of coastal structures, but as may be expected does not detect higher frequency surface chop or ripples. Adjustment of the pressure conversion filter's characteristics at cut-off would be possible if desired.

The Directional Wave Spectrum

Combining the records from the six positions in the sensor array then follows in order to estimate the full directional spectrum. This is relatively straightforward in the case of shore-normal waves. Kajima's (1969) algorithm, much used in wave flumes, decomposes incident and reflected wave spectra from the surface elevation records taken at two points. Our software (Davidson, 1992b) implements this using the output of three of the sensors to overcome the algorithm's discontinuities at wavelengths which are integer multiples of sensor spacing. Figure 4 illustrates the method with data from the rock island breakwater at Elmer. Spectra from three of the sensors, in a line normal to the breakwater and at 9, 12 and 18 metres from it, are shown in Figure 4(a). It can be seen that the measured spectra are very different. This is due to the partial standing wave structure, and highlights the inadequacy of using a single point surface elevation measurement where reflection is present. The resolved incident and reflected wave spectra are shown in Figure 4(b). The reflection coefficient (as a function of frequency, averaging about 65%) follows in 4(c).

For non shore-normal waves it is necessary to use more complicated methods. Best angular resolution is given by the "maximum entropy" and "maximum likelihood" methods. However these both fail if "phase locked" components interfere within the wave system - as they do with reflection. The latter method is, however, capable of modification to account for these (Isobe and Kondo, 1984), and the resulting "modified maximum likelihood method" (MMLM) is the one normally used in this work. The directional spectrum of Figure 5(a) is obtained by applying the MMLM to another of the Elmer data sets. A section through the contour plot at 0.102 Hz is given in Figure 5(b), indicating the resolution available with a six-sensor array, and producing an estimate of reflection coefficient of 60%. This agrees well with the values in Figure 4(c).

Unlike Kajima's method, the Isobe and Kondo (1984) method requires knowledge of the location of the effective reflection line with respect to the array of sensors. For non-vertical surfaces that location is a strong, and rather complex, function of mean depth and wave length. The location can be obtained by applying methods derived by Hotta *et al.* (1981) and Suhayda (1984) for simple slopes. These have been extended by Davidson *et al.*(1994) to cover more complex slopes, such as those of a breakwater fronted by a berm. The wave envelope shown in Figure 6 is predicted by this analysis for the Elmer breakwater and foreshore, evaluated for the single frequency of 0.1 Hz and assuming a reflection coefficient of unity. The vertical axis scale has been normalized, but the horizontal axis shows the disposition of the nodal structure over cross-shore distance. The five points plotted are measured wave

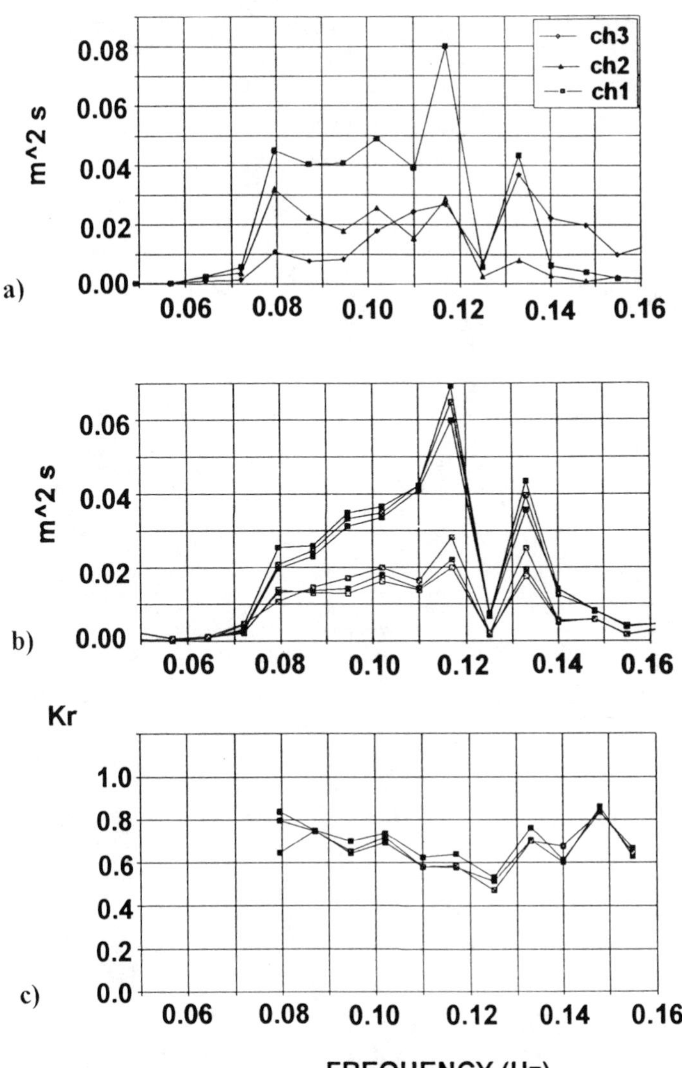

Figure 4. Decomposition of incident and reflected wave components from three sensors in a predominantly shore-normal sea. (a) Autospectra of surface elevation of the three sensors (b) De-composed incident and reflected spectra (c) Frequency dependent reflection function plotted for the energetic parts of the spectrum.

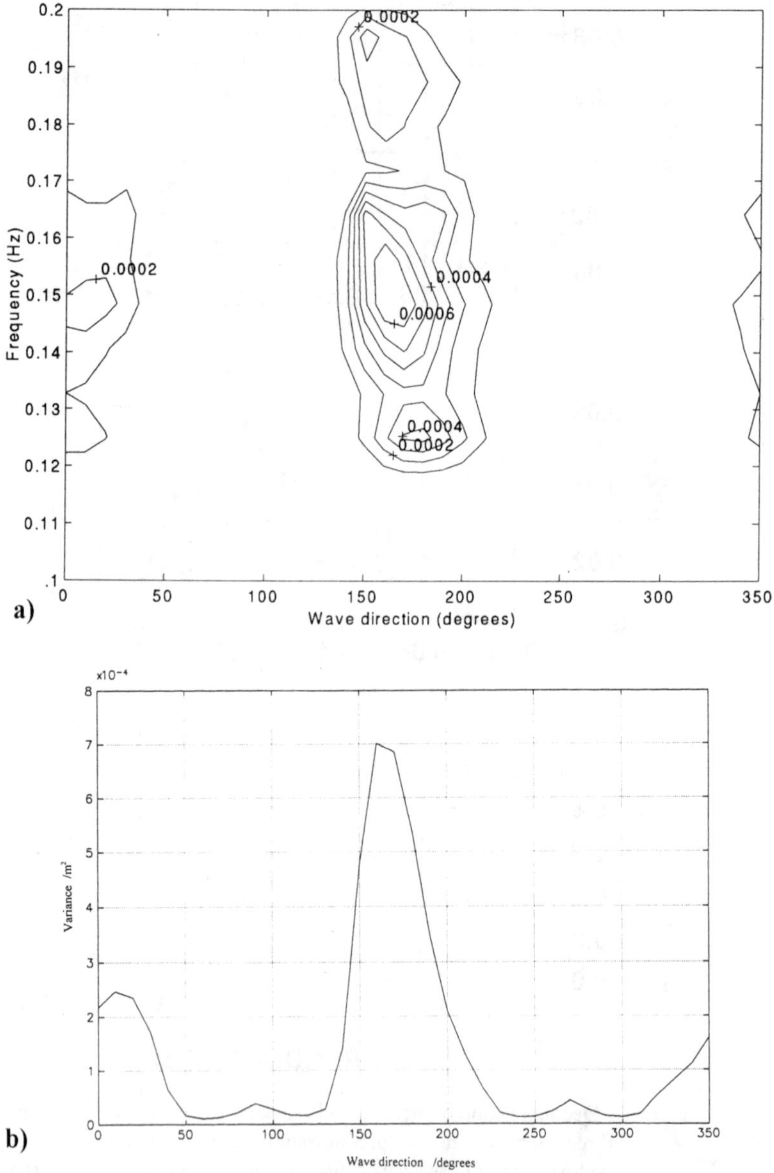

a)

b)

Figure 5. Directional spectrum from the rock island breakwater at Elmer a) As a contour plot b) Directional distribution for the 0.15Hz component.

components at this frequency from the five transducers of the array with differing cross-shore distances. Taking into consideration that in the measured case the reflection coefficient is not unity, there is good agreement between the theory and the data.

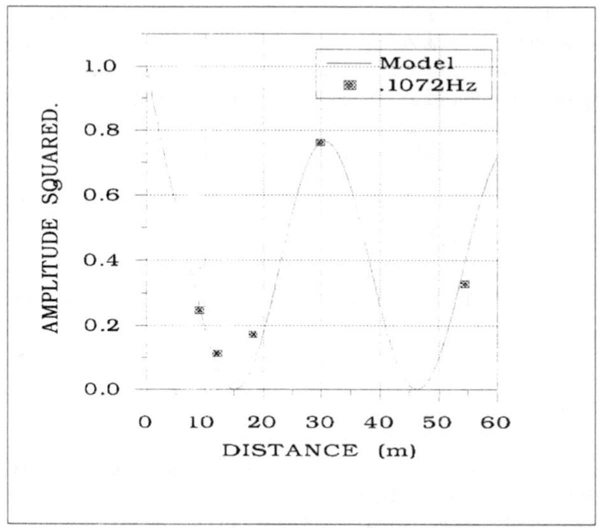

Figure 6 : Distribution of energy over cross-shore distance for a 0.107 Hz wave component: theoretical prediction for $K_r = 1$, and measured values (Elmer).

CONCLUSIONS

A measurement system and associated analysis techniques have been developed to estimate the wave directional spectrum, both with and without reflected components. Near structures that reflect to a significant degree the nodal pattern produced by the interaction of incident and reflected components is clearly seen. A single point measurement of surface elevation in these conditions would lead to a gross error in the estimated wave climate.

The MMLM is identified as the most appropriate method (in the non-shore normal case) for estimating the directional spectrum from simultaneous point measurements of surface elevation. A pre-requisite for its application to non-vertical reflecting structures is knowledge of the location of the effective reflection line relative to the transducer array. New developments in theoretical prediction of the distribution of nodes and anti-nodes given the structure's geometry and adjacent bathymetry (described in more detail in Davidson *et al.*, 1994), together with the analysis of selected data sets, are providing information on the reflection line and its dependency on frequency and depth.

Work is continuing on analyzing the large volumes of data collected with the objective of formulating a predictive scheme for the wave reflection properties of various types of coastal structure. In addition, the measurement equipment will be enhanced by the provision of a telemetry link from a buoy at the wave recorder to a shore station, and a satellite link from a shore station anywhere in Europe to the campus of the University of Plymouth.

ACKNOWLEDGEMENTS

The work described in this paper was supported by the UK Science and Engineering Research Council.

REFERENCES

Bird, P.A.D., and Bullock, G.N., 1991 "Field Measurements of the Wave Climate," *Developments in Coastal Engineering,* University of Bristol, March, 25-34.

Bird, P.A.D., 1993, "Measurement and analysis of sea waves near reflective structures", PhD Thesis, School of Civil and Structural Engineering, University of Plymouth, Plymouth, UK.

Chadwick, A.J.,"The measurement and analysis of inshore wave climate", *Proc Inst Civ Engrs* Part 2 **87**, March, 23-28.

Davidson, M.A., 1992a. "Implementation of linear wave theory in the frequency domain for the conversion of sea bed pressure to surface elevation", Research Report No. SCSE 92-008, School of Civil and Structural Engineering, University of Plymouth, Plymouth, UK.

Davidson, M.A., 1992b. "Development and implementation of a software routine for the analysis of the reflection processes associated with random and monochromatic waves", Research Report No. SCSE 92-003, School of Civil and Structural Engineering, University of Plymouth, Plymouth, UK.

Davidson, M.A., Bird, P.A.D., Bullock, G.N. and Huntley, D.A., 1994, "Wave reflection: field measurements, analysis and theoretical developments", *Proc ASCE Conf: Coastal Dynamics '94* Barcelona, Spain (this edition)

Hotta, S., Mizuguchi, M., and Isobe, M., 1981, "Observations of long period waves in the nearshore zone", *Coastal Engineering in Japan*, **24** 41-76.

Ilić, S., 1994, "The role of offshore breakwaters in coastal defence: comparison of the two measurement systems", Research Report No. SCSE 94-002, School of Civil and Structural Engineering, University of Plymouth, Plymouth, UK.

Isobe, M. and Kondo, K., 1984, "Method for estimating directional wave spectrum in incident and reflected wave field", *Proc ASCE Conf on Coastal Engineering*, Houston, Texas, USA, 467-483.

Kajima, R., "Estimation of incident wave spectrum under the influence of reflection", *Coastal Engineering in Japan*, **12**, 9-16.

Suhayda, J.N., 1974, "Standing waves on beaches", *Jnl Geophys Res* **79** (21) 3065-3071.

SEDIMENT SUSPENSION EVENTS AND SHEAR INSTABILITIES IN THE BOTTOM BOUNDARY LAYER

D.L. Foster[1], R.A. Holman[1] , and R.A Beach[1]

ABSTRACT: The intermittent, rapid suspension of sediment in the surf zone is not well understood. Because the boundary layer is the region of fluid in direct contact with the sea bed, we believe that it plays an important role in the sediment suspension process. In this paper we examine the flow characteristics during the initiation of suspension events using field data from a recent Oregon coast experiment, and propose a hypothesis for the generation of events through a shear instability of a oscillatory bottom boundary layer.

Theoretical predictions suggest that flow in the bottom boundary layer leads that of the free stream, resulting in an inflection point (a necessary condition for a shear instability) in the vertical profile of cross-shore velocity during flow deceleration and reversal. It is during this phase of the flow that small perturbations may grow exponentially to breaking, leading to increased levels of turbulence. Bottom boundary layer shear instabilities, leading to large near bed velocity fluctuations, may be responsible for the rapidly suspended sediment. A simple linear instability analysis predicts if and when small perturbations will become unstable.

The 1993 San Marine field experiment on the Oregon coast investigated suspended sediment concentrations and high frequency near-bed fluid motions. Up to four hot film anemometers, sampled at 2000 Hz for 34 min., were located between 1 and 4 above the bed. Variance was roughly partitioned between high frequency motions (which includes turbulence) and wave motions by examining the time variation of variance within 1/8 second blocks of data. Sixty-five percent of the suspension events are correlated to variance events with an average lag of 0.75 sec. Both the suspension events and variance events are shown to occur during the offshore decelerating and onshore accelerating flow phases.

---

[1]College of Oceanic and Atmospheric Sciences,, Oregon State University, Ocean Admin Bldg #104, Corvallis, OR 97331, USA internet: dfoster@oce.orst.edu

INTRODUCTION

In the surf zone, the process of sediment suspension is punctuated with rapid appearances and disappearances of high sediment concentration of sediment at elevations above typical wave boundary layers O(5~10cm). These rapid and intermittent events cannot easily be explained or modeled as a turbulent diffusive process. Understanding fluid-sediment interactions leading to the sediment suspension process is essential to the development of large scale sediment transport models.

While these sediment suspension events have been readily observed by numerous investigators (Jaffe et.al., 1984; Huntley and Hanes, 1987; Beach and Sternberg, 1988), the fluid forcing responsible for them has not. The rapid appearance and disappearance of sediment at high levels above the bed may be due to simple horizontal advection of a sediment cloud back and forth past a sensor array. However, this does not explain how the material reaches such an elevated position. In the absence of turbulence generated by bedforms or injected from the surficial wave breaking, the explanation for this phenomenon has remained relatively elusive.

The rapid appearance of suspension events may be related to a rapid generation or introduction of near-bed turbulence. Previous research has focused on the introduction of turbulence to the bottom boundary layer through wave breaking at the surface or the generation of turbulence at the bed due to bedforms. Alternatively, we explore the generation of turbulence from within the boundary layer through the breaking of a shear instability wave.

The purpose of this investigation is to characterize the fluid motions associated with individual suspension events using high frequency response, hot film anemometers paired with optical back scatter sensors. These measurements help to evaluate boundary layer shear instability as a mechanism for the rapid generation of turbulence within the bottom boundary layer which, in turn, leads to sediment suspension. The first part of the paper introduces the shear instability theory and requisite boundary conditions and background profile structure. In the second part of the paper, field data from an Oregon coast beach is presented and the viability of the shear instability mechanism for generating suspension events is evaluated.

THEORY

The bottom boundary layer plays an important role in the sediment suspension process, as it is the region of fluid in contact with the bed. In traditional boundary layer theory, the excess pressure above hydrostatic inside the boundary layer is assumed constant in depth. The no slip condition at the bed, causes the fluid close to the bed to have smaller inertia and consequently, responds to free stream cross-shore pressure gradients prior to both the fluid in the upper part of the boundary layer and in the free stream layer. Thus, the boundary layer leads the free stream flow, resulting in an inflection point in the bottom boundary layer profile during flow deceleration, reversal, and subsequent acceleration (Smith,1977). A sketch of a typical velocity profile within a monochromatic wave boundary layer at several wave phases is shown in Fig. 1.

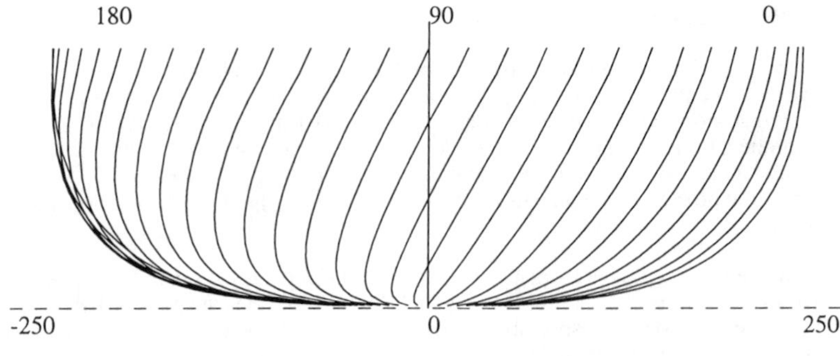

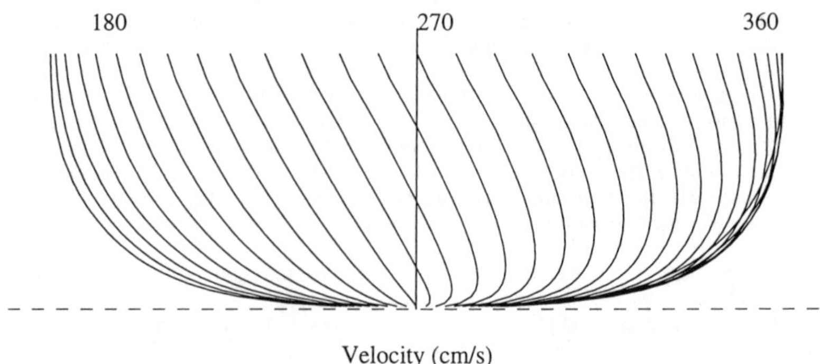

Figure 1. Oscillatory velocities as a function of height above the bed for phase angles from 0° to 360°. Fluid response in the boundary layer leads that of the free stream velocity.

Raleigh (Kundu, 1990) determined that the necessary condition for an instability of inviscid parallel shear flow is that the profile contain maximum vorticity (an inflection point) within the flow region . If a velocity profile contains a large vertical shear and an inflection point, small perturbations may become unstable and grow exponentially in time. If the growth rate of the unstable wave is of large magnitude, the wave will break and generate turbulence. A linear instability is investigated as a possible mechanism for generating large near bed velocity fluctuations leading to a rapid vertical redistribution of sediment.

This derivation will closely follow that of Bowen and Holman, 1989, for instabilities of a mean alongshore current with a cross-shore variation. This analysis will examine instabilities of the bottom boundary layer cross-shore velocity with a vertical variation.

The cross-shore, u', and vertical, w', velocities are decomposed into mean, U and W, and perturbation components, u and w,

$$u'(x,z,t) = U(z) + u(x,z,t)$$
$$w'(x,z,t) = w(x,z,t) \tag{1}$$

where U is the background cross-shore velocity and assumed fixed in x and t. The mean vertical velocity ,W, is defined to be zero; and u and w are the cross-shore and vertical perturbation velocities, respectively. x and z are the cross-shore and vertical coordinates(x is positive offshore and z is positive upwards).

The cross-shore and vertical linear inviscid 2-d equations of motion are

$$u_t + Uu_x + wU_z = -\frac{1}{\rho}P_x$$
$$w_t + Uw_x = -\frac{1}{\rho}P_z - g \tag{2}$$

where P is the pressure; g is gravitational constant; ρ is fluid density. Alongshore homogeneity is assumed. The 2-d conservation of mass equation is

$$u_x + w_z = 0. \tag{3}$$

By cross differentiating and combining with Eq. 2 P is eliminated, resulting in

$$\left(\frac{\partial}{\partial t} + U\frac{\partial}{\partial x}\right)\left(\nabla^2\Psi\right) = U_{zz}\Psi_x \tag{4}$$

where the stream function, Ψ, is defined in terms of **u** as,

$$u \equiv -\Psi_Z$$
$$w \equiv \Psi_X. \tag{5}$$

The perturbation solution is assumed to be

$$\Psi = \Re e\left\{\psi(z)e^{i(kx-\sigma t)}\right\}, \tag{6}$$

where k and σ are defined as the cross-shore perturbation wave number and cross-shore perturbation frequency. k is assumed real, and σ and ψ are assumed complex, $\sigma=\sigma_r + i\sigma_i$ and $\psi=\psi_r + i\psi_i$. After expanding Eq. 4 becomes

$$\left(-\frac{\sigma}{k} + U\right)\left(\psi_{zz} - k^2\psi\right) = U_{zz}\psi. \tag{7}$$

Ψ will grow exponentially in time when σ_i is greater than zero.

MODEL FORMULATION

A simple one dimensional, time-dependent, turbulent diffusion model is employed to approximate the velocity structure within the bottom boundary layer (BBL) (Smith, 1977; Beach and Sternberg, 1992). In calculating the boundary layer velocity profiles, the cross-shore velocity time series (at z=14 cm) is decomposed into a series of onshore/offshore half wave segments. Each half wave segment is identified with the half-period defined as the time between zero crossings, and the amplitude defined as the maximum velocity of the free stream profile filtered to 0.5 Hz. Use the amplitude and period to iteratively approximate the bed shear velocity ($u_o = u*_{max} \ln(\delta/z_o)$), and boundary layer thickness ($\delta = u*_{max}/(2\omega)$). The bottom roughness, z_o is held constant at 0.0766. The eddy viscosity, for each half wave segment, increases linearly away from the bed with an exponential decay above the wave boundary layer (Beach and Sternberg, 1992). The model yields the vertical velocity structure of the cross-shore velocity through the BBL to the free stream for each time within the segment interval. Subsequent profiles from the next half wave period are merged to form a continuous time series over the 34 minute run.

Because the boundary layer model approximates each half wave as a sinusoid, important asymmetries in natural wave accelerations (always larger during the transition to onshore flow) are neglected.

Based on predictions by the boundary layer model, the velocity profile contains an inflection point for up to approximately T/4, where T is the wave period, Fig. 1. As a simple illustration, assume the cross-shore velocity profile is fixed for the entire T/4 secs (which it is not). For a perturbation amplitude to grow to be 100 times its original amplitude, the growth rate, σ_i, would have to be 18.4/T or greater. For a 10 sec wave, the growth rate would be 1.84 Hz.

Using a steady-background instability model on oscillatory background flow problem requires that the perturbation time and cross-shore length scales be much smaller than the mean (in this case oscillatory) flow scales. Neglecting time-dependent background flow terms in the instability model , assumes that the oscillatory cross-shore vertical profile is fixed at each time instant, requiring that the instability grow much faster than the boundary layer changes. The linear instability analysis is performed on each modeled velocity profile at each 1/8 sec time step. The primary goal of this analysis is to determine whether the model will predict instabilities with rapid growth rates over the wavelength range of 2 cm to 50 cm. This wavelength range was selected because we expect instability length scales to be of order boundary layer thickness.

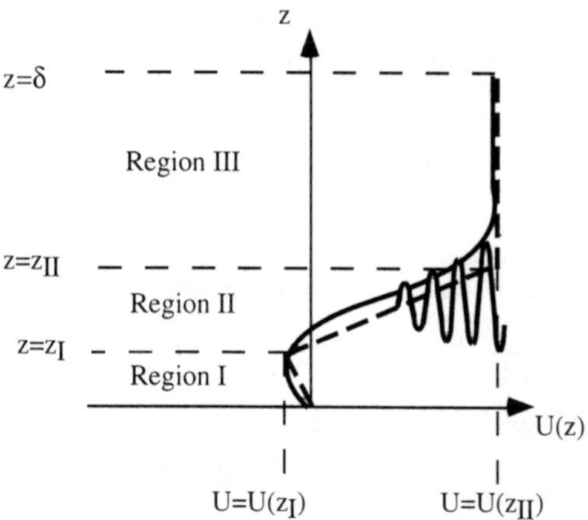

Fig. 2. Boundary layer geometry showing the three regions of the linear analysis.

To model the shear instability, the modeled boundary layer velocity profile is approximated with three linear regions, Fig. 2. Region I is defined at the lower limit by U(0)=0 and at the upper limit by searching upwards for a local extremum, $U(z_I)$ at z_I. The slope of the line in region II is defined as the maximum vorticity (maximum vertical shear), by searching upwards from z_I and the upper limit determined by projecting the line up to the free stream velocity, $U(z_{II})$. The velocity in region III is defined as constant, $U(z_{II})$ and bounded at the boundary layer thickness, δ. In each linear region, Eq 7 reduces to

$$\psi_{zz} - k^2 \psi = 0. \qquad (8)$$

The solution for each region is

Region I: $\psi_I = A_1 \sinh(kz)$

Region II: $\psi_{II} = A_2 \sinh(kz) + B_2 \cosh(kz)$

Region III: $\psi_{III} = A_3 e^{-kz}$

where $A_1, A_2, A_3,$ and B_2 are integration constants and determined with the boundary conditions. The perturbation frequency, σ, may be solved in terms of the perturbation wavenumber, k.

FIELD STUDY

Location

A field study was conducted in the inner surf zone of a dissipative beach on the central Oregon coast during September 21-28, 1993. The San Marine beach runs north-south and is exposed to North Pacific storm and swell conditions. The

September 26,1993 offshore significant wave height and dominant period were 1.5 m and 7 sec, as recorded by wave buoy at Newport (38 km north of San Marine). The average beach slope was 1/50. Previous studies have found the sediment to be very well sorted with a mean grain size of 0.23 mm (Beach and Sternberg, 1988). The sea bed was smooth and without ripples or scour holes.

Instrumentation

Instrumentation consisted of 1 Marsh-McBirney electromagnetic current meter, 1 strain gauge-type pressure transducer, 3 ducted impeller current meters(SM), 3 optical back scatter sensors (OBS), up to 4 hot film anemometers, 1 underwater laser, and 1 underwater video camera. The hot films were sampled at 2000 Hz and all other instruments were sampled at 8 Hz.

The instruments were mounted to a cross bar supported by two pipes jetted in the alongshore with a 6 m separation (Fig. 3). Instrument orientation and maintenance was performed at low tide and data were recorded for approximately 2 hours during high tide. Data from 26 Sept 1993, when two hot films were deployed at 1 and 2 cm from the bed, will be discussed below. Three pairs of OBS(C1, C2, & C3) and current meter (SM1, SM2, & SM3) were located at z=4, 9, and 14 cm, respectively. The electromagnetic current meter (EM1) was 14 cm off the bed.

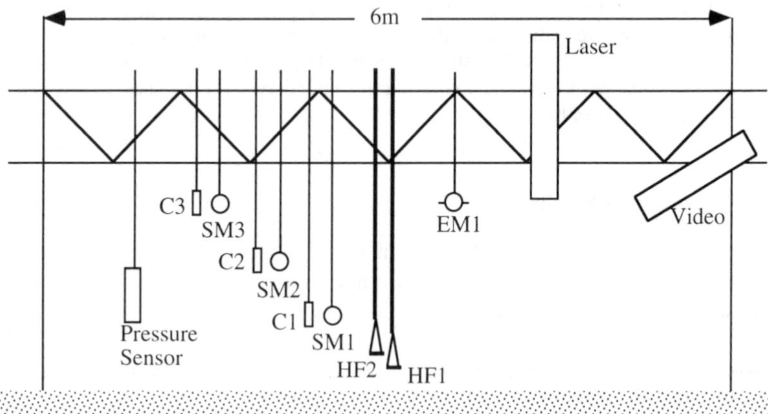

Figure 3. Sketch of instrumentation deployment on 26 Sept 93, shows the configuration of 3 OBS concentration sensors(C1, C2, & C3), 3 ducted impeller current meters (SM1, SM2, & SM3), 2 Hot Film Anemometers (HF1 & HF 2), 1 electromagnetic current meter(EM1),.

Hot Film Calibration

The hot film anemometers were calibrated to the electromagnetic current meter in situ (George, 1992). Each hot film probe was calibrated for 10 minutes at z=14 cm, then lowered to the desired elevation. The output of each hot film sensor, HF1 and HF2, was block averaged to 8 Hz and plotted versus the rectified current meter data, EM1, (z=14 cm). A best fit exponential curve was fit to the paired data and used as the transfer function from output (volts) to speed (cm/s). This technique only calibrates the magnitude of the velocity(speed), as the hot films are rectified and direction is unknown. Maximum observed velocities occasionally exceeded the range of hot films (±187 cm/s, HF1, and ±287 cm/s, HF2).

RESULTS AND DISCUSSION

A five minute time series of the SM1 and EM1 velocities and the C1 and C2 suspended sediment concentration is shown in Fig 4a, 4f, and 4g, respectively. HF1 and HF2 are block-averaged to 8 Hz and are shown in Fig. 4c. For visual comparison to Fig 4c, SM1 is rectified and shown in Fig 4b. Offshore directed flow is positive.

Each 2000 Hz hot film time series is variance-partitioned into an 8 Hz variance time series by merging the variance of subsequent 1/8 sec intervals. The variance time series characterizes the levels of high frequency velocity fluctuations within each 1/8 sec interval, over time (Fig 4d and 4e). A sensitivity analysis determined that partitioning the variance over larger time intervals affected only the magnitude of variance and not the occurrence of a variance event. This investigation is only concerned with the initiation of turbulence events.

An event based analysis defined suspension events to have concentration greater than 10 gm/l at z=4 cm and variance events to have velocity variance exceeding 300 cm^2/sec^2 at z = 2 cm. The 8 Hz time series of velocity variance is similar to the suspended sediment time series as both contain rapid intermittent events. Sixty-five percent of the suspension events are correlated with variance events and lag the variance events by 0.75 secs.

An expanded view of the time series from Fig. 4 is shown in Fig. 5. The concentration event at 8.5 minutes follows a backwash of long duration and occurs near the period of rapid flow reversal. During this period of flow deceleration, reversal, and subsequent acceleration, the velocity sensors at z=1, 2, and 4 (HF1, HF2, and SM1) reverse direction prior to the velocity at sensor z=14 (EM1), and are of large magnitude. This large magnitude signal, at the sensors 4 cm and lower, precludes the variance and suspension events. Bottom boundary layer theory predicts the phase lead, however does not predict the large onshore accelerations which follow. The long backwash and rapid flow reversal are the background velocity conditions which could lead to a shear instability. The large velocity magnitude of the lower sensors could have been a result of exponential growth of a small perturbation, a shear instability.

To characterize the background cross-shore flow associated with events, the velocity versus acceleration at the initiation of suspension events are plotted in Fig. 6a. The initiation of the sediment and variance events is determined by manually searching back in time from the event, as specified earlier, until the sediment concentration is 2 gm/l, at z=4 cm, and the variance is 20 cm^2/sec^2, at z=2 cm. The majority of the

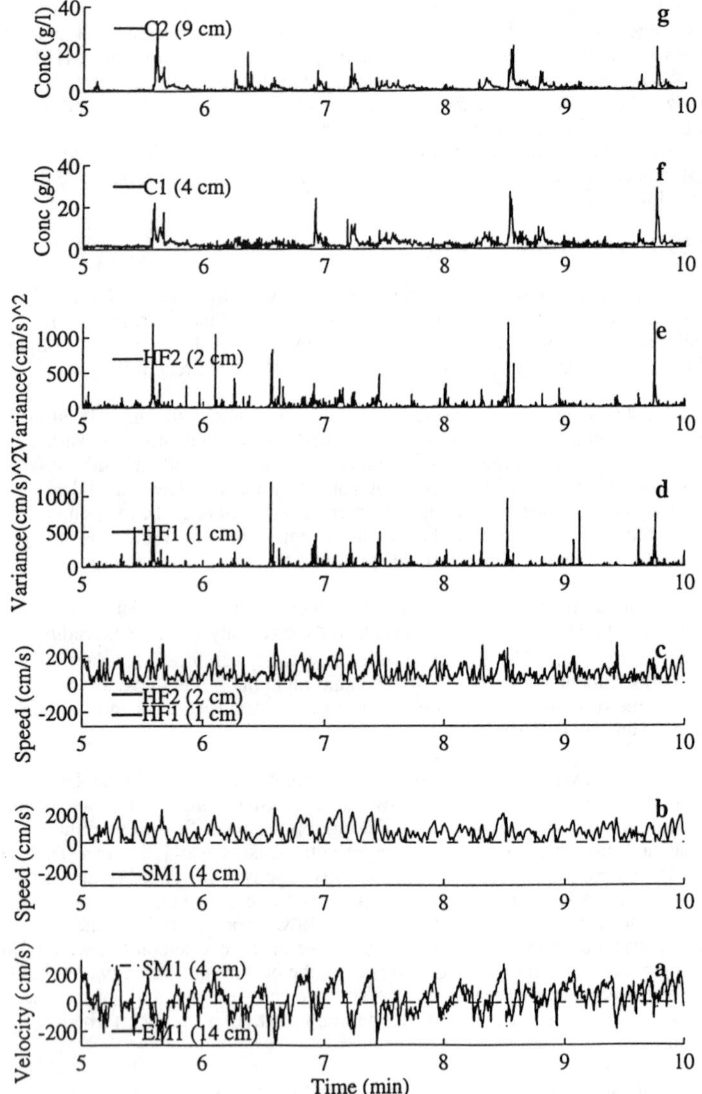

Figure 4. Five minute time series, as recorded on 26 Sept 93, of: a)velocity at SM1 and EM1 b)speed at SM1 (rectified) c)speed at HF1 and HF2 d)variance partitioned at HF1 e)variance partitioned at HF2 f)concentration at C1 g)concentration at C2.

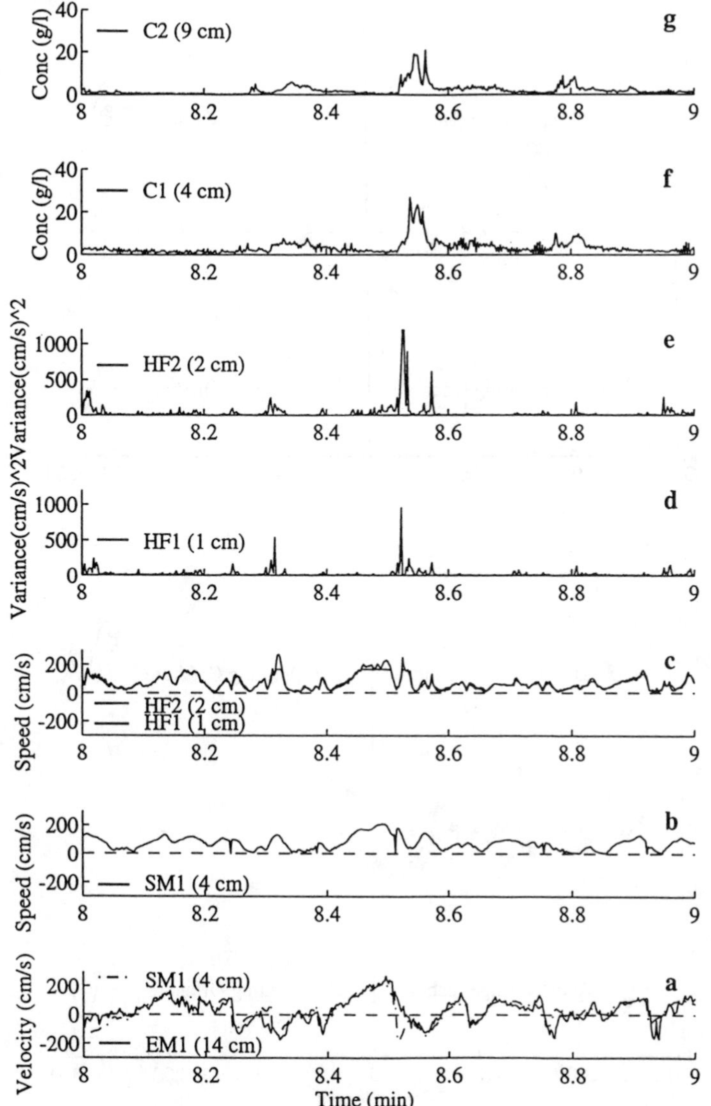

Figure 5. One minute time series, as recorded on 26 Sept 93, of: a)velocity at SM1 and EM1 b)speed at SM1 (rectified) c)speed at HF1 and HF2 d)variance partitioned at HF1 e)variance partitioned at HF2 f)concentration at C1 g)concentration at C2.

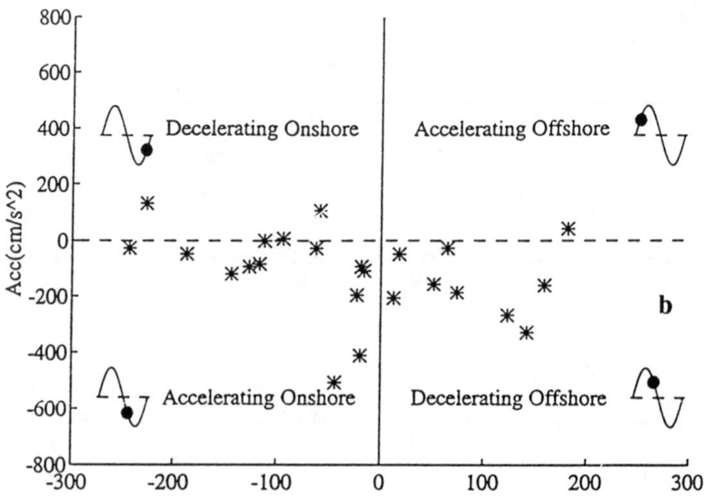

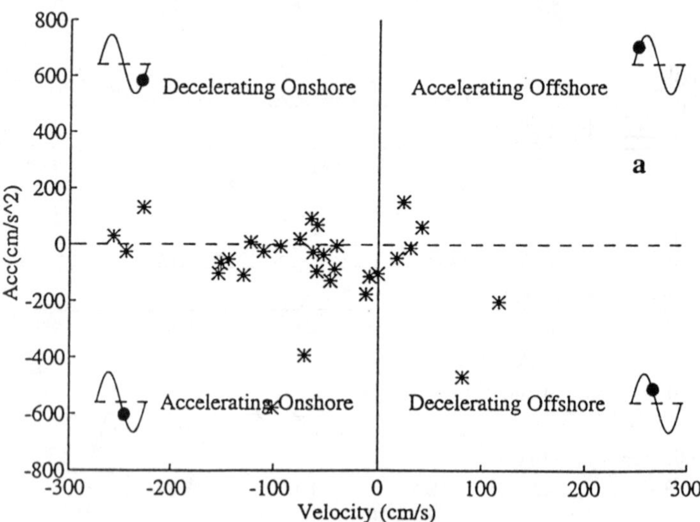

Figure 6. Velocity versus acceleration of 0.5 Hz filtered EM1 sensor at the initiation of: a)suspended sediment events & b)variance events.

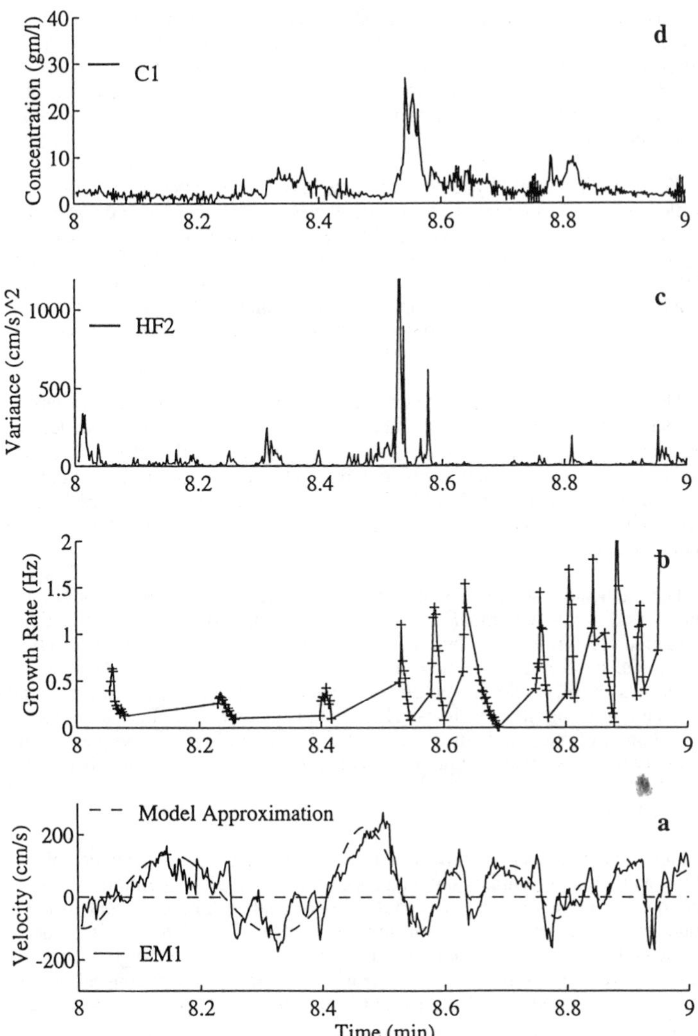

Figure 7. Time series as recorded on 26 Sept 93 of: a)velocity of EM1 & boundary layer model approximation of EM1 b)predicted growth rates c)variance partitioned at HF2 d)concentration at C1.

suspension events occur during the offshore decelerating and during the onshore accelerating phases of the flow. Fig. 6(b) is similar to Fig 6(a) but shows the velocity versus acceleration of the free stream flow at the initiation of variance events, at z=2 cm, which are correlated to the suspension events. These events occur during the offshore decelerating and onshore accelerating phases of the flow, and slightly lead the suspension events.

The shear instability hypothesis predicts instabilities to occur when the velocity profile has an inflection point. Surf zone waves are asymmetric and consequently, the cross-shore deceleration is largest during the transition between offshore to onshore flow (from positive velocity to negative velocity). It is during this flow phase that we expect the shear in the bottom boundary layer to be the largest. If the shear instability hypothesis is valid, we expect the initiation of variance and suspension events to occur during the transition from offshore to onshore flow when the pressure gradients are the largest, Figs. 6(a) & 6(b).

The imaginary frequencies, or growth rates, predicted by the linear instability analysis, when z_I is less than 6 cm, are presented in Fig. 7b. The suspended sediment concentration at z=4 cm is shown in Fig. 7d, the velocity variance at z=2 is shown in Fig. 7c, and the free stream velocity and the modeled velocity (for the boundary layer model) at z=14 are presented in Fig. 7a. Instabilities are predicted more frequently than variance and suspension events occur. The prediction does not guarantee the instability development.

The predicted instabilities from 8.6 to 9 min. are associated with low amplitude small period waves with rapid decelerations. Instability predictions during this time segment are results of the simplifying assumptions required for the bottom boundary layer model and for the linear time-independent instability model. Also, the predicted instabilities are of smaller magnitude or of shorter duration, than required to cause a perturbation to magnify 100 times. However, instability amplitudes are large enough to yield order of magnitude increases in initial perturbations. Consequently, the shear instability remains a viable hypothesis for the generation of large near bed velocities.

SUMMARY

Field observations of sediment concentration in the surf zone can be episodic in nature, with rapid appearances and disappearances of high concentrations above the wave bottom boundary layer. One possible explanation for these events is a bottom boundary layer shear instability during flow reversal. The viability of this mechanism is evaluated with data from the inner surf zone of a dissipative Oregon beach. Data includes paired current meter velocities and suspended sediment concentrations at z=4,9, and 14 cm above the bed; and hot film anemometer speeds at z=1 and 2 cm.

Each 2000 Hz hot film time series is variance partitioned into an 8 Hz variance time series by combining the variance of subsequent 1/8 sec intervals. This variance time series characterizes levels of high frequency fluctuations contained within each segment. The episodic time series is correlated to the suspension events and leads by 0.75 secs.

The boundary layer leads the free stream and results in an inflection point in the vertical cross-shore velocity profile during flow deceleration, reversal, and subsequent acceleration. If the shear contained within the profile during this phase of the flow is large enough, small perturbations may become unstable and grow exponentially in time. Plots of the free stream velocity versus acceleration at the initiation of variance and sediment suspension events show the majority of the events occur during the transition to onshore flow. Wave asymmetries cause the cross-shore decelerations to be larger during the transition to onshore flow.

The existence of instabilities during the period of flow reversal was modeled through a simple linear instability analysis of a time varying bottom boundary layer. At each time, small perturbations were superimposed on a modeled boundary layer cross-shore velocity profile and instability growth rates were predicted over wavelengths of 2 to 50 cm.

Predicted growth rates are large enough to yield an order of magnitude increase in initial perturbation amplitudes. Instabilities are predicted more frequently than variance and suspension events occur. Some of the instabilities which occur under small waves may be attributed to the simplifying assumptions required for the bottom boundary layer model and for the linear time-independent instability model. Nevertheless, the hypothesis is encouraging as the instabilities have reasonable wavelengths and growth rates and occur during the time of flow reversal.

The ultimate goals of this ongoing project are to examine the link between turbulence generation and sediment suspension. Future investigations will utilize more sophisticated bottom boundary layer models, explore time-dependent numerical instability models and compare predicted instability wave numbers and frequencies to an array of the hot films.

ACKNOWLEDGMENTS

This research is being performed in conjunction with the University of Washington and sponsored by the Office of Naval Research Contract # N00014-9310145. The first author, a civil engineering student, is supported with a DOD/ONR fellowship and tolerated (barely) in the Coastal Imaging Lab. As ever, the technical expertise of Rex 'Hot Wire' Johnson and John Stanley was invaluable. Editorial assistance from Tony Bowen, Todd Holland, and Nathaniel Plant was especially appreciated. The first author would like to thank Andrea Ogston, University of Washington, for her field and instrumentation-art assistance, and Ed Thornton, Naval Postgraduate School, for his generosity of temporary office and computing facilities.

REFERENCES

Beach, R.A. and Sternberg, R.W., 1992. Suspended sediment transport in the surf zone: Response to incident wave and long shore current interaction. Mar. Geol., 108: 275-294.

Beach, R.A. and Sternberg, R.W., 1988. Suspended sediment transport in the surf zone: Response to cross-shore-infragravity motion. Mar. Geol., 80: 671-679.

Bowen, A.J. and Holman, R.A., 1989. Shear instabilities of the mean long shore current. J. Geophys. Res., 94: 18023-18030.

George, R. Observations of turbulence in the natural surf zone. Ph.D. Dissert., Scripps Inst. of Ocean., 26 pp.

Huntley, D.A. and Hanes, D.M., 1987. Direct Measurement of Suspended Sediment Transport. Coastal Sediments Conf. Proc., Vol. I

Jaffe, B.E., Sternberg, R.W., and Sallenger A.H., 1984. The role of Suspended Sediment in Shore-Normal Beach Profile Changes. Coastal Engin. Conf. Proc.

Kundu, P. K., 1990. Fluid Mechanics. Academic Press, San Diego.

Smith,J.D., 1977. Modeling of sediment transport on continental shelves. In:e.D. Goldeberg (Editor), The Sea. Wiley, New York, 6: 539-577.

SEDIMENT TRANSPORT UNDER WAVE-CURRENT INTERACTION ON MACRO-TIDAL BEACHES

Franck Levoy[1], Olivier Monfort[1] and Hélène Rousset[1]

ABSTRACT : Coastal engineering on beaches requires taking into account the natural environment, its evolution and its hydrosedimentary behaviour to reduce the inevitably induced negative impacts. In this goal, many *in situ* measurements of hydrodynamic parameters and sediment transport have been carried out on the beaches of the western coast of the Department of La Manche. Several wave-current interaction transport formulas have been implemented. The comparison between measured and computed transport rates allows to choose the best formulation according to the hydrodynamic context.

INTRODUCTION

On the macro-tidal beaches of the western coast of the Cotentin Peninsula (Fig.1), tidal water fluctuations, tidal currents, unbroken waves and currents induced by waves provoke complex sedimentary dynamics.

To model the sediment transport on a long-term basis and using a large scale approach, it is necessary to validate classical formulations and sometimes, to define new empirical coefficients using local environment data.

In the ROMIS program (Levoy and Avoine, 1991), about 50 fields experiments have been carried out to quantify sediment transport under wave and current interactions on macro-tidal sandy beaches, during complete tidal cycles.
Some methods use the measured near-bed sediment concentration to evaluate the global transport rates. In this experimental program, we measured directly the sediment transport rates for periods of several hours, typically, of a tidal cycle, with sand traps and fluorescent tracers.

---

(1) University of CAEN, Centre Régional d'Etudes Côtières, Laboratoire de Géologie Marine, 54 Rue Charcot, 14530 LUC-SUR-MER

During these experiments, the water level fluctuations, the wave characteristics and the cross-shore variations of grain-size have also been recorded to quantify the sediment transport rates using some classical wave-current interaction formulae. The results were then compared with the *in situ* sediment transport measures.

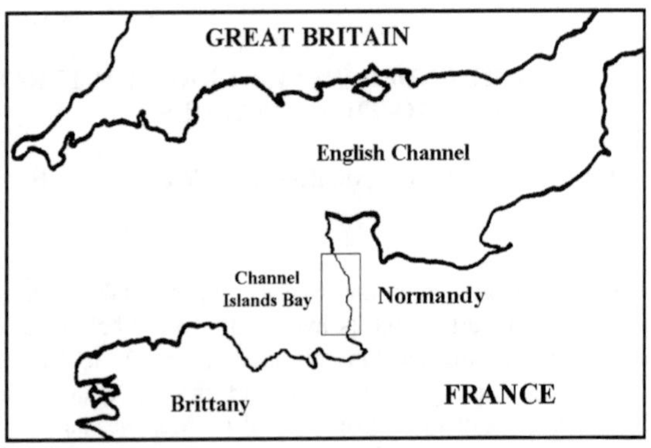

figure 1 Location of the project site

REGIONAL ENVIRONMENT

Hydrodynamic conditions

The Channel Islands Bay is subjected to very complex and specific hydrodynamic conditions. The tidal range, one of the highest in the world, decreases from 15 m in the South of the bay to 12 m in the North at high spring tide (Pingree and Maddock, 1977). These conditions induce very strong tidal currents. The mean velocities can reach 1 m/s along the west coast of Cotentin and 3 m/s in the mouth of small estuaries in this area. Maximum velocities are observed at high tide and the main direction of the water movement is oriented towards the North, parallel to the coastline. Often, on mean spring tide, the longshore tidal current can induce, by itself, sediment transport on the lower foreshore.

Open to the west offshore area, the Channel Islands Bay is subjected to long waves coming from the North Atlantic Ocean. However, the propagation of swells is complicated by the complex continental shelf bathymetry, the Channels Islands and numerous shoals and archipelagos. Near the coast, rocky platforms or tidal deltas modify locally the wave propagation patterns.

The Channel Islands Bay can be viewed as a dissipative embayment with a decreasing gradient of wave height from North to South and from West to East. The yearly significant wave height varies from 4.2 m to 2.8 m along the study area.

In winter, during storm events, wind-wave conditions dominate the wave climate. Peak period is lower than 5 s and the waves come mainly from the South-West to the North-West directions.

The conjunction between spring tides and strong wind-wave conditions induces high energy and coastal sediment motions. This transport, very intense at high tide, can result in important beach morphological changes and in coastline recession.

The coastal morphology

The western coast of Cotentin is a sandy rectilinear coast interrupted by small inlets. Between each of these small estuaries, the intertidal beach can be divided into three zones: a high tide zone, a mid-tide zone and a low tide zone, with a global concave shape (Short, 1991). In the new conceptual model of Masselink and Short (1993), the beaches of the western coast of Cotentin can be considered as ultra-dissipative (RTR>7 and Ω>2).

The high and mid-tide zones are dominated by swash and surf zone processes due to the cross-shore migration of the breaker zone during the tidal cycle. The beach slope (tan β) varied from 0.02 to 0.08. Beach cusps can be sometimes observed but there are no parallel bars. Grain-size is coarse with medium and coarse sand and gravel.

The bottom topography variations are very important: a storm event can induce a two-meter fall. The natural replenishment that leads to the re-establishment of the original beach profile may take several months.

The lower part of the beach is very planar (tan β=0.005) and wet due to the outcropping of the water-table. The sedimentary structures are lacking, except mini-ripples. Grain-size is fine and the sands well-sorted (D50=0.180 mm). The seasonal bottom level variations are weak, less than 10 centimetres.

Shore parallel tidal currents during spring tides induce a longshore sediment transport to the North. But during the storm events, the interactions between the tidal currents and wave-induced orbital currents play a prominent part in the nearshore sediment circulation. The present paper focusses on this phenomenon.

MAIN FORMULAS AVAILABLE TO COMPUTE SEDIMENT TRANSPORT RATES UNDER NON BREAKING WAVES AND TIDAL CURRENT

Introduction

Many studies have been performed to define sediment transport laws when currents and waves act simultaneously. Some of the resulting formulations have been implemented and the calculated transport rates have been compared with field measurements.

The Van-Rijn formula

The total sediment transport rate is computed as the sum of bedload and suspended load transport (Van Rijn, 1990, Van Rijn 1993) :

$$Q_t = Q_b + Q_s$$

Where Q_t: total sediment transport rate(m³/m/s);
Q_b: time-averaged bedload transport (m³/m/s) and Q_s: time-averaged suspended load transport (m³/m/s).

The time averaged bedload transport can be computed by averaging over a wave period the instantaneous bedload transport $q_b(t)$ given in m³/m/s by :

$$q_b(t) = 0.25 \times \alpha \times D_*^{-0.3} \sqrt{\frac{\tau_{cw}}{\rho_w}} \left(\frac{\tau_{cw} - \tau_{cr}}{\tau_{cr}} \right)^{1.5}$$

Where α= calibration factor (non-dimensional);

$D_* = D_{50}\left(\frac{(s-1)g}{\upsilon^2}\right)^{1/3}$ = dimensionless particle parameter (non-dimensional);

D_{50}= median particle diameter of bed material (m);
g= gravitational acceleration (m/s²);
$s-1 = \frac{\rho_s - \rho_w}{\rho_w}$ = relative apparent density of bed material (non-dimensional) ;
ρ_s and ρ_w are the sediment and fluid densities (kg/m³);
υ= cinematic viscosity(m²/s).
τ_{cw}= grain-related instantaneous bed-shear stress due to combined current and waves (N/m²);
$\tau_{cr} = (\rho_s - \rho_w)g \times D_{50} \times \Theta_{cr}$ = critical bed-shear stress according to Shields (N/m²);
Θ_{cr} = Shields parameter (non-dimensional).

The time-averaged suspended load transport is computed by numerical integration over the depth:

in the current direction : $Q_s = \int_{z=a}^{z=h} u(z).C(z).dz$

in the wave direction : $Q_s = \int_{z=a}^{z=h} v(z).C(z).dz$

Where $u(z)$= resultant current velocity at height Z above the bed in the current direction (m/s);
$v(z)$= wave-induced velocity at height Z above the bed in the wave direction (m/s);
$C(z)$= sediment concentration at height z above the bed (non-dimensional);
a= reference level(m);
h= water depth(m).

The sediment concentration profile $C(z)$ is deduced from the integration over the depth of the time-averaged convection-diffusion equation.

Some results of the Van-Rijn formula were compared with flume measurements and field data (Van-Rijn, 1989). For the bedload transport, the ratios between computed and measured rates vary from 1 to 2 at high rates ($\geq 10 g/m/s$) but can reach values of 5 to 10 for lower rates.

We carried out some sensitivity tests of the Van-Rijn formula to different parameters:
* hydrodynamic parameters: current velocity, water depth, wave characteristics, angle between wave and current directions,
* bed characteristics,
* sediment characteristics.

This study shows a strong dependency of the computed rates on the current velocity: a variation of about 10 % of the velocity can result in a multiplication of the rate by a factor of 2. A modification of the water depth produces a variation of the transport in the same proportion. Other parameters are less important for the computed sediment transport rate values.

The Frijlink-Bijker formula

The total load can be computed by (Delft Hydraulics Laboratory, 1976; Migniot 1987):

$$\boldsymbol{Q_t = Q_b \times (1 + 1.83 \times \left(I_1 \times \ln\left(\frac{33h}{r}\right) + I_2 \right)}$$

$\boldsymbol{r}$ = equivalent Nikuradze roughness of the bed (m);
$\boldsymbol{I_1}$ and $\boldsymbol{I_2}$ are integrals given by Einstein (non-dimensional).
The bed-load transport rate $\boldsymbol{Q_b}$ (m^3/m/s) is given by:

$$Q_b = 5 \times D_{50} \times V \times \frac{\sqrt{g}}{C_h} \times \exp\left(-\frac{0.27 \times \rho_w \times D_{50} \times (s-1)}{\mu \times \tau_{wc}}\right)$$

V= mean current velocity (m/s);
C_h= Chézy roughness coefficient ($m^{1/2}/s$) ;
ρ_w= density of water (kg/m^3);
μ= ripple factor: part of the bed-shear that is not used to overcome bed resistance (non-dimensional) ;
τ_{wc}= bed-shear stress in combined waves and currents given by the Bijker approximation ($kg/m/s^2$).

The Frijlink-Bijker formula is an improved version of the original Frijlink formula. It retains the basis concept of a bed shear stress caused by the combined wave orbital velocity and the current velocity field. The choice of an expression of the combined wave-current bed shear stress allows the computation of the bedload transport. The suspended load transport is then computed from integration of the sediment concentration profile given by the Einstein-Rouse equation. The resulting suspended load transport depends very closely on the assumed bedload layer thickness. This is very important in cases where the suspended load transport constitutes the dominant part in the total load.

The results were compared with measurements obtained by Bijker in a basin. The Frijlink-Bijker method yields transport capacities that are clearly higher than measured rates. This overestimation increases with decreasing transport capacities.

The Engelund, Hansen and Swart formula

The initial Engelund-Hansen formula was modified by Swart to take into account combined wave and currents conditions (Delft Hydraulics Laboratory, 1976; Migniot 1987). The initial shear stress expression is modified in the presence of waves, and the total sediment transport rate is given by:

$$Q_t = \frac{0.1}{2\sqrt{g}}\sqrt{s-1}C_h^2 A^{5/2} D^{3/2}$$

$A = \dfrac{\tau_{wc}}{gD(\rho_s - \rho_w)}$ is a non-dimensional shear stress;
D= sediment characteristic diameter (m).

This formula is in good agreement with measured rates when the suspended-load transport is not negligible. However the rates are overestimated when the bed shear stress is low.

The Ackers-White formula

This formula, which gave sediment transport rates by currents, was modified by Swart to take into account combined wave and currents conditions (Ackers-White, 1973, Delft Hydraulics Laboratory 1976; Migniot, 1987).

It is based on a dimensional analysis and physic considerations. The transport is supposed to be mainly due to bedload processes for coarse sediments. The fine sediments are supposed to be transported within the body of the flow where they are put in suspension by turbulence. The empirical parameters which appear in the equation were determined from flume experiments. The total sediment load is expressed as:

$$Q_t = \frac{1}{1-p} \times V \times D_{35} \times \left(\frac{C_h}{\sqrt{g}}\right)^n \times \left(\frac{\tau_{wc}}{\tau_c}\right)^{-n/2} \times \frac{C}{A^m} \times (F_{wc} - A)^m$$

Where D_{35} = bed particle diameter of that fraction of the bed material which is exceeded in size by 65% in weight of the total sample (m);

F_{wc} = sediment mobility in combined wave and current conditions (non-dimensional);

A, C, m, n are coefficients which are functions of the non-dimensional particle parameter $D_* = D_{50}\left(\frac{g \times (s-1)}{\upsilon^2}\right)^{1/3}$;

The Grant-Madsen formula

The authors considered that the wave and current motions in the vicinity of the bed cannot be treated separately (Van-Rijn, 1989). The instantaneous total sediment transport rate is given in m³/m/s by :

$$\vec{Q}_t = 40 W_s D_{50} \Theta^3 \frac{\vec{V}}{\|\vec{V}\|}$$

Where W_s = particle fall velocity (m/s);

$\Theta = \frac{0.5 \times F_w \times U^2}{(s-1) g D_{50}}$ = particle velocity parameter(non-dimensional);

F_w = friction factor(non dimensional);

$\vec{U} = \vec{U}_c + \vec{U}_w$ = instantaneous velocity vector near the bed (m/s);

$\vec{U}_c$ = near-bed current velocity(m/s);

$\vec{U}_w$ = near-bed wave orbital velocity (m/s).

The mean transport rate and direction are computed by time integration of the instantaneous rate over a wave period.

FIELD TECHNIQUES USED TO QUANTIFY SEDIMENT TRANSPORTS

Sediment transport quantification techniques were used on the macrotidal beaches of the western coast of Cotentin to define quantitatively both the direction and the intensity of the observed movements. Two different types of approaches were used:

* Eulerian type measurements: multidirectional sediment traps.

These traps were especially conceived for the macrotidal environment of the western coast of Cotentin where strong tidal currents and stormy wave conditions act simultaneously on the lower foreshore.

Photograph 1: View of a multidirectional sediment trap

To reduce the flow disturbance induced by the traps, only three levels of streamer were put on the trap. The first one, near the bed, traps the bed-load transport. The two others, placed 70 and 140 centimetres above the bed, allow evaluation of the suspended-load transport.

The experiments were conducted over a tidal cycle to estimate the correlation between the transport and the hydrodynamic conditions recorded in the immediate vicinity. Nevertheless, during calm periods, when only tidal currents are active, the transport rate weakness allows us to have the immersion duration of some tidal cycles.

For the transport rate integration over depth, the sea water-level is held constant and equal to the mean level recorded during the trap immersion.
The residual sediment load is computed as the vectorial sum of the rates trapped in the eight directions.

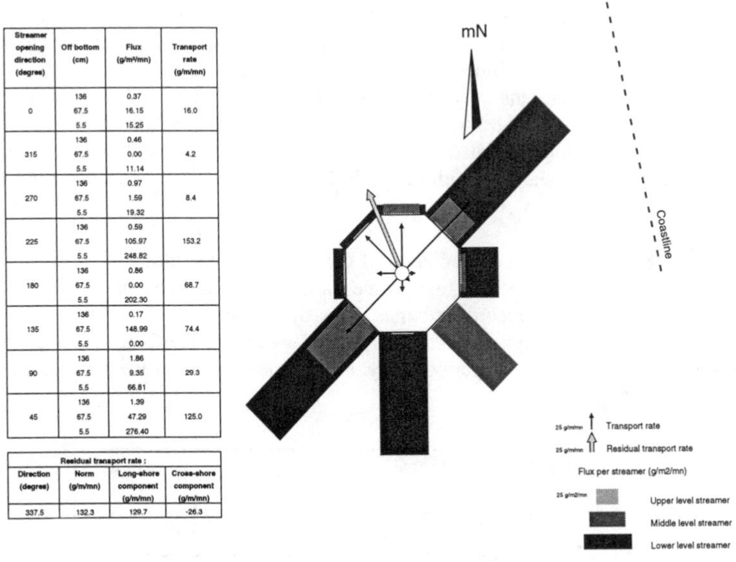

Streamer opening direction (degrees)	Off bottom (cm)	Flux ($g/m^2/mn$)	Transport rate (g/m/mn)
0	136	0.37	16.0
	67.5	16.15	
	5.5	15.25	
315	136	0.46	4.2
	67.5	0.00	
	5.5	11.14	
270	136	0.97	8.4
	67.5	1.59	
	5.5	19.32	
225	136	0.59	153.2
	67.5	105.97	
	5.5	248.82	
180	136	0.86	68.7
	67.5	0.00	
	5.5	202.30	
135	136	0.17	74.4
	67.5	148.99	
	5.5	0.00	
90	136	1.86	29.3
	67.5	9.35	
	5.5	66.81	
45	136	1.39	125.0
	67.5	47.29	
	5.5	276.40	

Residual transport rate :			
Direction (degrees)	Norm (g/m/mn)	Long-shore component (g/m/mn)	Cross-shore component (g/m/mn)
337.5	132.3	129.7	-26.3

Figure 2: Measurement of sediment transport with multidirectional streamer trap on the low tidal zone of Saint-Germain-sur-Ay (Normandy, France) November 1991.

* Lagrangian type measurements: fluorescent tracers.

This method consists in immersing coloured particles that are representative of the study zone bed particle stock, and in monitoring their dispersion from the immersion location.
Different fluorescent particle sampling techniques may be used to quantify sediment load in the main transport direction (Madsen, 1989). For these experiments, the Spatial Integration Method is used, because it is more suitable for emerging zones.
This technique enables a study of sediment mobility for quite short periods (typically, a tidal cycle).

Generally, used during calm conditions, this technique allows a definition of sediment transport laws as stated by Kraus et al (1982).

COMPARISONS BETWEEN MEASURED AND COMPUTED SEDIMENT RATES

Some preliminary numerical tests showed a large scattering of the computed sediment rates depending on the used formula. These results agree with the works of Pattiaratchi and Collins (1985) who concluded that the ratios between computed and measured rates could reach values of 500 for stormy conditions.

In order to select the physical environment most adapted formulation, we decided to compute the sediment rates using the above formulas and the hydrodynamic conditions recorded in field. These computed sediment rates were then compared with those measured with the multidirectional sediment traps and with the fluorescent tracer technique.

With this object, a time integration of the instantaneous rates was necessary. It is based on the assumption of a linear time variation of the transport between two hydrodynamic conditions recording bursts. This time step is about 30 minutes. The recording duration is 18 minutes. The current and wave characteristics are supposed to be constant during the latter period. Only one sediment transport value is computed for every time step.

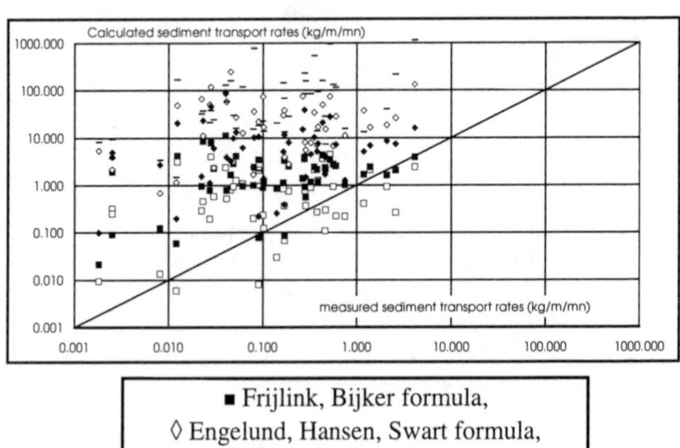

■ Frijlink, Bijker formula,
◊ Engelund, Hansen, Swart formula,
♦ Ackers, White formula,
- Grant, Madsen formula.

Figure 2: Comparison between computed and measured sediment transport rates

Figure 2 shows the measured sediment transport rates and those computed by different formulas for the 48 experiments carried out on the western coast of Cotentin from January 1991 to May 1993. The measured sediment transport rates ranged from 2.5 g/m/mn to 4.11 kg/m/mn.

The Engelund-Hansen-Swart, Ackers-White and Grant-Madsen formulas seem to overestimate systematically the sediment loads. The ratios between computed and measured rates are very important, and these formulas are thus unsuitable for the conditions prevailing on the western Cotentin coast.

The ratios between the results of the Frijlink-Bijker and the measured rates ranged from 1.1 to 800 with a mean value of about 55.
The Van-Rijn formula gives the best agreement with measured rates (Figure 3).

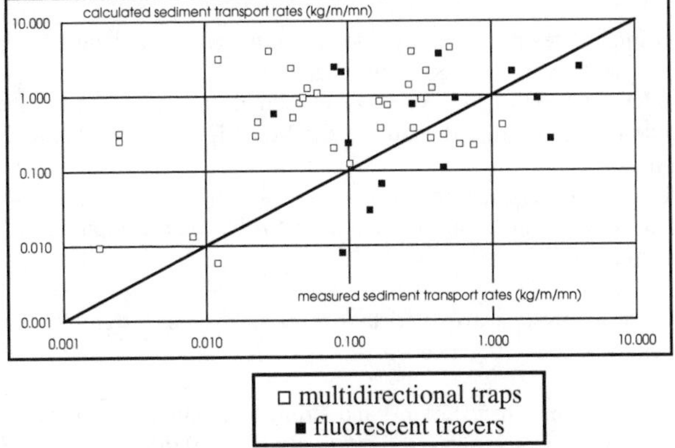

□ multidirectional traps
■ fluorescent tracers

Figure 3: Comparison between the computed Van-Rijn formula and measured sediment transport rates

Nevertheless, the transport rates are overestimated in about 70 % of the cases. The ratios between computed and measured sediment load spread from 0.1 to about 250 with a mean value of about 20.

Figure 3 shows that for the big majority of the sediment traps experiments, the computed sediment loads are higher than the measured values: trap efficiency is probably less than 1. Traps induce a modification of the fluid flows and, probably, a sediment transport rate less important than the values which would have been recorded for the same hydrodynamic conditions if the trap were absent. Moreover, the traps do not catch all the movable sediments. Tests will be carried out in flumes in order to define the hydraulic and sedimentary efficiency of traps.

Table 1 shows the distribution of the ratio between the measured rates and those computed by the Van-Rijn formula for the 48 *in situ* experiments.

Ratio between Qc and Qm	Percentage
1 to 2	17
2 to 5	31
5 to 10	17
10 to 20	19
20 to 30	6
30 to 50	0
50 and more	10

Table 1: Distribution of the ratios between the measured transport rates Qm and the rates computed by the Van-Rijn formula Qc

The sediment rates measured by fluorescent tracers are in better agreement with the values computed by the Van-Rijn formula.

A detailed analysis of the hydrodynamic conditions for each field experiment was done to define the validity domain for the Van-Rijn formula. It reveals that for a conjunction of:

- waves of significant height above 1.5 m and tidal currents, measured 30 cm above the bed, of less than about 0.3 m/s the ratio between computed and measured sediment transport rates varies from 1 to 4;

- moderate to strong waves (0.5 to 1.5 m) and velocities greater than 0.3 m/s, the ratio varies from 10 to 20, and even 30.

- weak waves (less then 0.25 m) and strong tidal currents (above 0.5 m/s), the ratios between computed and measured rates vary from 10 to 20.

Some experiments have been realised with multidirectional sediment traps implanted on beach-rock platforms. In those cases, the ratio is above 50 and the computed sediment load is a potential value. The ratio between the computed and the measured sediment transport rates grows with increasing wave conditions. The lack of sediment disposal, due to the bed nature, produces a high value of Qc/Qm.

The Van-Rijn formula gives better results for combinations between strong waves and weak tidal current velocities than for all the situations where the currents are strong.

To tackle the problem of transport direction, we decided to assume that the instantaneous sediment transport occurs in the direction of the vectorial resultant (Van Rijn 1989):

$$\vec{V}(t) = \vec{U}_\delta(t) + \vec{V}_{r,\delta}$$

Where $\vec{V}(t)$ is a vector collinear to the instantaneous transport direction;
$\vec{U}_\delta(t)$= near-bed wave orbital velocity (m/s);
$\vec{V}_{r,\delta}$= near-bed current velocity (m/s).

The assumption consists in stating that the above relation, written with instantaneous values, is also applicable to time averaged quantities:

$$\vec{V} = \alpha.\vec{U}_o + \vec{V}_{r,\delta}$$

Where $\vec{V}$ is a vector collinear to the transport direction;
$\vec{U}_o$= near bed wave orbital velocity amplitude (m/s).

The factor α that may vary from -1 to +1 transformation of the instantaneous equation into a time-averaged one.

The comparisons between measured transport directions and those computed by the Van-Rijn formula show, once again, a large variability of the results with the value of α. Nevertheless, it clearly appears that, for the hydro-sedimentary context of the lower foreshore of the west Cotentin coast, in 56% of the cases the best agreement between computed and measured transport direction is reached for $\alpha = \mathbf{1}$. In this case the transport is supposed to occur in the direction of the vectorial sum of the near bed tidal current velocity and of the near-bed orbital velocity.

CONCLUSIONS

(1) Several formulas available to compute sediment transport rates under wave-current interaction have been tested. The Van Rijn formula gives the best agreement between the rates computed and those measured during complete tidal cycles.

(2) The ratios between transport rates computed by the Van Rijn formula and those measured were put on a parallel with measured hydrodynamic conditions. This study reveals that a good agreement (factor 1 to 4) between computed and measured sediment transport rates is obtained for conjunctions between strong non-breaking waves (above 1.5 m on the lower part of beaches) and weak tidal current velocities (under 0.3 m/s near the bed). When strong tidal currents are combined with moderate or low wave heights, the Qc/Qm ratio increases.

(3) As for the sediment movement direction, the field observations show that the best agreement is obtained when the computed directional vector is set equal to the vectorial sum between near-bed tidal current velocity and near-bed wave orbital velocity.

(4) The sediment transport rates measured by fluorescent tracers are in better agreement with the calculated values. The multidirectional traps induce scour and turbulence in the fluid motion for high energy conditions.

ACKNOWLEDGEMENTS

This work was undertaken as a part of the ROMIS program. It was funded by the Conseil Général de la Manche, the Conseil Régional de Basse-Normandie and the Secrétariat d'Etat à la Mer to define a global coastal policy of protection of the coastline of Normandy against storm effects.

REFERENCES

ACKERS P., WHITE W.R. (1973) - Sediment Transport: New Approach and Analysis, Journal of the Hydraulics Division. pp. 2041-2060.

DELFT HYDRAULICS LABORATORY (1976) - Coastal Sediment Transport - Computation of Longshore Transport, Report of Investigation, Toegepast Onderzoek Waterstaat. 61 p.

KRAUS N.C., ISOBE M., IGARASHI H., SASAKI T. and HORIKAWA K. (1982) - Field experiments on longshore sand transport rate in the surf zone, Proc. 18th Coastal Eng. Conf. ASCE. pp. 969-988.

LEVOY F., AVOINE J. (1991) - Quantitative approaches to coastal sediment processes, Field measurements to estimate sediment transport in a high-energy coastal area. Proc. Coastal Sediment' 91, Seattle, June 1991, (published separate from proceedings), ASCE.

LEVOY F. et LARSONNEUR C. (1993) - Etude globale de défense contre la mer de la côte du département de la Manche, synthèse des mesures in situ. Conseil Général de la Manche, Université de Caen, rapport 9. 163 p.(in french)

MADSEN O.S. (1989) - Transport determination by tracers, tracer theory. Nearshore Sediment Transport edited by R.J.SEYMOUR. pp 103-114.

MIGNIOT C. - Manuel sur l'hydrodynamique sédimentaire et l'érosion et sédimentation du littoral, Premiére partie : hydrodynamique sédimentaire, Cours enseigné à l'Ecole Nationale des Travaux Publics de l'Etat et à la Faculté des Sciences d'Orsay - Paris-Sud. 159 p.(in french).

MASSELINK G., SHORT A.D. (1993) - The effect of tide range on beach morphodynamics and morphology: a conceptual beach model. Journal of coastal research, 9(3), pp 785-800.

MIGNIOT C. (1987) - Abaques de Transports solides des sédiments non Cohésifs-Transport des Sédiments sous l'action des courants et des houles, Laboratoire Central d'Hydraulique de France. 7p.(in french)

PATTIARATCHI C.B. and COLLINS M.B. (1985) - Sand transport under the combined influence of waves and tidal curents: an assessment of available formulae. Mar. Geol., 67. pp 83-100.

PINGREE, R.D., MADDOCK, L. (1977) - Tidal residuals in the English Channel. Journal of marine biological association of UK, 57: pp 339-354.

SHORT, A.D. (1991) - Macro-meso tidal beach morphodynamics - an overview. Journal of coastal research, 7(2), pp 417-436.

VAN RIJN L.C. (1989) - Handbook Sediment Transport by Currents and Waves, Delft Hydraulics. 307 p.

VAN RIJN L.C.(1990) - Principles of fluid flow and surface waves in rivers, estuaries, seas and oceans, Delft Hydraulics. Aqua Pubications, Amsterdam. 335 p.

VAN RIJN L.C.(1993) - Principles of sediment transport in rivers, estuaries and coastal seas, Delft Hydraulics. Aqua Pubications, Amsterdam. 673 p.

ENERGY DISSIPATION AND SEDIMENT TRANSPORT IN LARGE SCALE TESTS

H. Rahlf [1] and Y. Wu [2]

ABSTRACT: Data from prototype experiments in the LARGE WAVE FLUME and methods on data evaluation and analysis for use in numerical modelling of coastal sediment transport processes are presented.

1. INTRODUCTION

Data from prototype experiments and methods on data evaluation and analysis for use in numerical modelling of coastal sediment transport processes are presented. The data were obtained from two test series with cross shore beach profiles. The investigations were mainly focussed on wave energy dissipation due to surging and spilling breakers and the mobilization of sediments (suspension) across the surf zone at equilibrium beach profile conditions. By these means cross shore sediment transport can be studied based upon data.

2. EXPERIMENTAL ARRANGEMENT

The data were obtained from prototype experiments in the LARGE WAVE FLUME (LWF) in Hannover, a joint central research facility for coastal engineering studies by both the University of Hannover and by Technical University of Braunschweig in Germany. The LARGE WAVE FLUME has a length of about 300 m, a depth of 7 m and a width of 5 m, that means that prototype experiments in coastal engineering are possible with wave heights up to 2.0 m.

[1)] Research engineer, Dipl.-Ing, Federal Waterways Engineering and Research Institute, Wedeler Landstraße 157, D-22559 Hamburg, Germany.

[2)] Visiting assistant scientist, Dr.-Ing.,Techn. Univ. of Braunschweig, Leichtweiss-Institute for Hydraulics, Dept. for Hydromech. and Coast. Eng., Beethovenstraße 51a, D-38106 Braunschweig, Germany.

In both test series (1990 and 1991) theoretical equilibrium beach profiles according to DEAN with a fairly flat slope of 1 to m = 1 to 80 were built in LWF as initial profiles.
In test series I a sand with a mean grain size of D_{50} = 220 μm was used and in test series II a mean grain size of D_{50} = 330 μm. The profiles and the arrangement of measuring positions are illustrated in Fig.1. The waves (water level elevations) were recorded by means of up to 12 resistance type wave gauges across the surf zone. Horizontal and vertical components of orbital velocity were recorded by means of 2D electromagnetic current meters simultaneously at different vertical positions. Suspended sediment concentrations were obtained from up to 6 vertical positions in the water column above the bed at 5m spacings along the equilibrium profile by means of suction methods used by BOSMAN (1987).

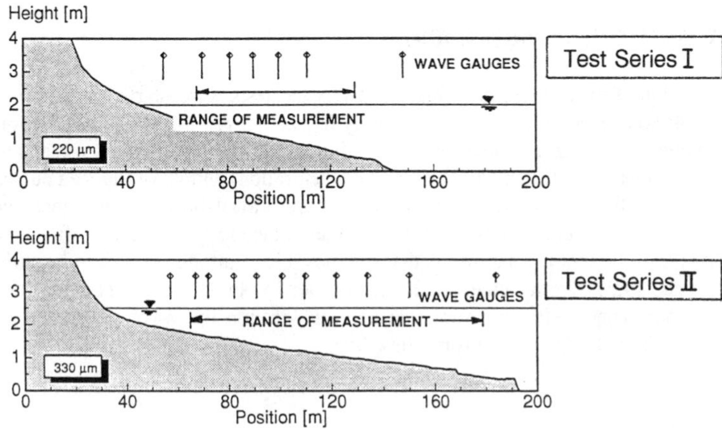

Fig. 1: Initial profiles and arrangement of measuring positions in the LARGE WAVE FLUME.

3. DESIGN WAVE CONDITIONS

Both monochromatic waves as well as wave spectra were used. As wave spectra PM (Pierson Moskowitz) and TMA spectra (Hughes, 1987) were chosen. The significant wave heights varied between H = 0.8 m and H = 1.0 m and the wave periods between T = 4 secs and T = 8 secs. Further details are summarized in Table 1.

Test Series I				Test Series II			
Monochromatic Waves		PM-Spectrum		Monochromatic Waves		TMA-Spectrum	
H [m]	T [sec]	H_s [m]	T_p [sec]	H [m]	T [s]	H_s [m]	T_p [sec]
0.7	4	0.8	6	0.9	4	0.9	6
0.8	6	0.8	8	0.9	6	1.0	8
0.8	8			1.0	8		

Table 1: Wave parameters chosen for Test Series I and II in the LWF.

4. EXPERIMENTAL PROCEDURE

After the initial profile had been built in LWF the tests were started until the real equilibrium occurred. This was controlled by repeated levellings of the beach profile at intervals of 300 wave sequences. When the equilibrium profile was established, the measurements of suspension (sediment concentration) and wave induced currents were started. Since only two pumping systems were available, the suspended sediment measurements could only be carried out simultaneously over the profile at two positions (5m distance). That means that the movable pump device had to be shifted step by step (5m) over a range of 120 m in order to obtain a complete data set all over the surf zone. After measurement at each location the wave generation was stopped in order to carry out profile levellings.

5. RESULTS

5.1. EQUILIBRIUM PROFILE

The initial profiles were shaped according to the formula from KRIEBEL and DEAN (1985), were the depth from water surface y is

$$y = A \cdot X^m$$

'y' is expressed by the product of a parameter A with the shoreline distance X to power m. From earlier tests the value of parameter A was assessed between 0.13 and 0.17, for m = 2/3.

Under wave spectra no significant change of the initial (theoretical) profile was observed.

As to be expected, the profile changes due to monochromatic waves was considerable. Under monochromatic waves with the longer wave periods a bar started to form fairly quickly. Fig. 2 shows the development of equilibrium profiles due to the following wave parameters:

First run with a wave height of H = 0.9 m and a short period of T = 4 secs. After the equilibrium profile was reached the wave height was increased to H = 1.0 m and the wave period at first increased to T = 6 secs and afterwards to T = 8 secs.

The DEAN formula does not describe a bar formation in the beach. Nevertheless, other models, e.g. the cross shore model from M. LARSON can be used for such cases.

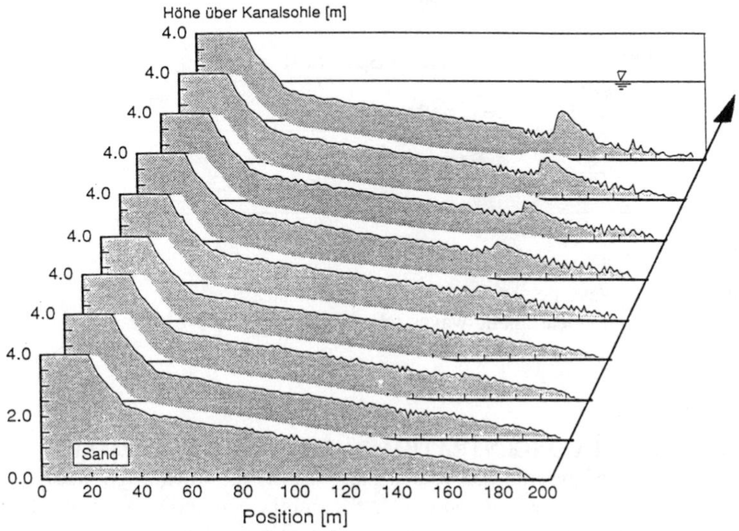

Fig. 2: Profile changes due to tests with monochromatic waves.

5.2. WAVE HEIGHT DECAY AFTER BREAKING

Fig. 3 shows the wave height decay across the surf zone obtained from measurements. According to the closely spaced wave gauges the breaking point becomes visible for tests with monochromatic waves. There is a good agreement between the measured break point and that observed during the tests. The breaker zone varied between position 100 m and 120 m within test series I and between 120 m to 150 m in test series II. Plunging breaker were predominant.

Fig. 3 (below) shows the wave height decay due to wave spectra. No exact break point occurred and a very flat wave height decay across the surf zone is typical for the wave spectra. Most breakers started as plunging breakers, spilling breakers were not observed.

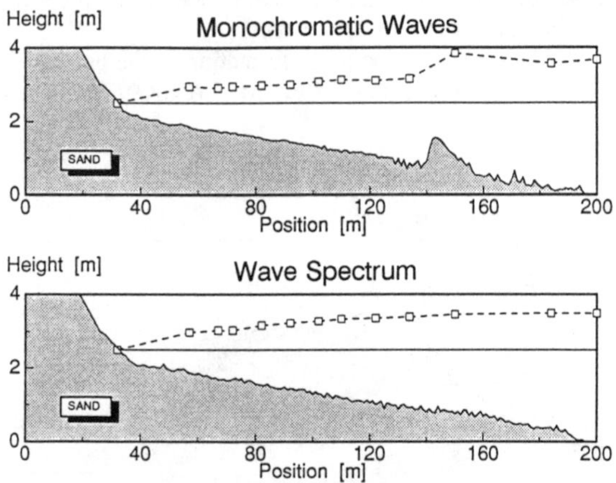

Fig. 3: Wave height decay analyzed from measurements: monochromatic waves (above) and TMA wave spectrum (below).

5.3. SEDIMENT CONCENTRATIONS ACROSS THE SURF ZONE

As an example of experimental results Fig. 4 shows the wave height decay in the surf zone ($H_{rms}(x)$, measured and calculated) and the equilibrium sand profile with the range of measurements. In Table 2 as example one data set of concentration measurements (at up to 6 positions above the bed) throughout the surf zone (17 locations) is listed (TMA-spectrum, H_{rms} = 0.7 m, T_p = 8 sec). The complete data set on wave measurements and on profile changes was prepared for MAST G6 - Coastal Morphodynamics -(DETTE et al. 1992 d).

6. ANALYSIS AND COMPARISON OF RESULTS FROM MEASUREMENTS AND CALCULATIONS

6.1. WAVE HEIGHT DECAY AFTER BREAKING

The breaking of waves across the surf zone is a significant process in coastal dynamics, the formation of breaking and broken waves is the dominant factor in the hydrodynamics of coastal processes, especially for sediment transport in surf zones.

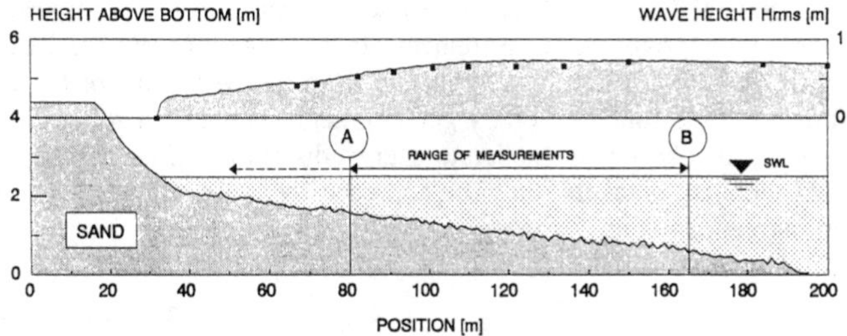

Fig. 4: Wave height decay H_{rms} (measured (■) and calculated) and the equilibrium sand profile.

Pos	Ds = Position below SWL [m] C = Sediment concentration [g/l]						
83	Ds	0.92	0.89	0.84	0.76	0.63	
	C	1.40	1.15	0.67	0.15	0.24	
90	Ds	1.05	1.02	0.97	0.89	0.76	0.55
	C	3.14	1.80	0.99	0.66	0.49	0.35
95	Ds	1.07	1.04	0.99	0.91	0.78	0.57
	C	0.67	0.39	0.61	0.42	0.41	0.22
100	Ds	1.16	1.13	1.08	1.00	0.87	0.66
	C	1.31	0.81	0.52	0.50	0.42	0.36
105	Ds	1.21	1.18	1.13	1.05	0.92	0.71
	C	1.52	1.05	0.94	0.64	0.55	0.27
110	Ds	1.27	1.24	1.19	1.11	0.98	0.77
	C	1.61	1.34	0.91	0.55	0.42	0.28
115	Ds	1.41	1.38	1.33	1.25	1.12	0.91
	C	4.04	4.00	1.27	0.63	0.21	0.27
120	Ds	1.44	1.41	1.36	1.28	1.15	0.94
	C	2.15	1.83	0.88	0.53	0.51	0.41
125	Ds		1.43	1.38	1.30	1.17	0.96
	C		1.07	1.21	0.43	0.45	0.30

Pos	Ds = Position below SWL [m] C = Sediment concentration [g/l]						
130	Ds	1.55	1.52	1.47	1.39	1.26	1.05
	C	2.18	1.20	1.10	0.74	0.47	0.45
135	Ds		1.55		1.42	1.29	1.08
	C		1.72		1.15	0.30	0.56
140	Ds		1.49		1.36	1.23	1.02
	C		0.75		0.95	0.79	0.57
145	Ds	1.64	1.61		1.48	1.35	1.14
	C	3.84	1.75		1.20	0.81	0.50
150	Ds	1.72		1.64	1.56	1.43	1.22
	C	12.8		1.87	1.11	0.92	0.64
155	Ds	1.70	1.67	1.62	1.54	1.41	1.20
	C	3.74	0.97	1.61	0.84	0.74	0.57
160	Ds	1.76	1.73		1.60	1.47	1.26
	C	5.98	4.29		2.94	2.03	1.34
165	Ds	1.75	1.72	1.67	1.59	1.46	1.25
	C	4.57	3.43	3.37	2.60	1.60	0.54
	Ds						
	C						

Table 2: Data set of sediment concentration measurements (TMA-Spectrum; H_{rms} = 0.7 m, T_p = 8 secs).

OELERICH and DETTE (1988) compared several analytical and empirical wave decay models with prototype measurements in the LARGE WAVE FLUME in Hannover as well as with field measurements from the west coast of the Island of SYLT/North Sea, and found generally good agreements between solutions and test data for spilling breakers with moderate energy dissipation (Fig. 5, left). But for predominant plunging breakers, only the approach by Dally et al. (1984) showed a satisfactory agreement (Fig. 5, right).

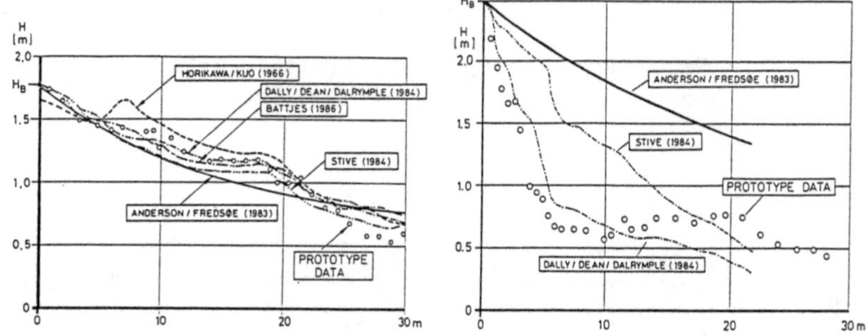

Fig. 5: Wave height decay in the breaker zone after wave breaking (OELERICH and DETTE, 1988).

Both numerical models from Dally (1984), for monochromatic waves and wave spectra, are used by Leichtweiss-Institute to calculate the wave decay in the surf zone. From Fig. 6 the following can be concluded:

(1) For monochromatic waves, the wave decay after breaking is well reproduced with Dally's regular model for all ranges of periods. Especially the breaking-reestablishment-breaking process of large plunging breakers with T = 8 seconds period waves over a bar is well simulated.

(2) For random waves, Dally's numerical model with a matrix of 60x60 discretized bins shows good results for H_{rms} and $H_{1/3}$ across the surf zone, in comparison to measured H_{rms} and $H_{1/3}$.

The results of the spectra analysis show, that the random wave which was produced in the LARGE WAVE FLUME is narrow banded in frequency, so that the randomness in the waves enters mainly through variability in wave height. Therefore an approach to simulate the random wave decay with Monte-Carlo method, as proposed by KRAUS and LARSON (1991), was also used and compared with the test data. Under the assumption that the wave heights in deep water are Rayleigh-distributed, each representative wave within the distribution is transformed across the surf zone

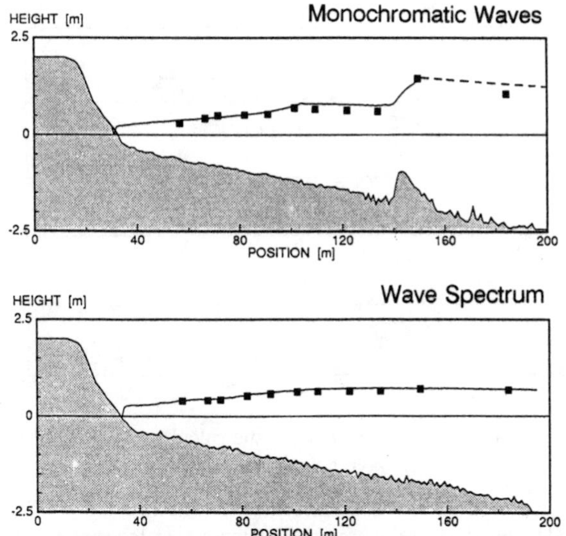

Fig. 6: Comparison of measured (■) and calculated (DALLY-model, 1984) wave height decay after breaking for monochromatic waves (above) and TMA - wave spectrum (below).

as a regular wave according to Dally's regular wave model. By using a large number of single waves, for example, 500 waves, the spectrum in deep water will be described satisfactorily. In the own numerical experiments the random number generator was avoided, because in Dally's wave decay model, wave-wave interactions are not included. So the order of different wave heights is not important. At each point in surf zone, H_{rms} can be easily calculated. $H_{1/3}$ can also be obtained by means of a sorting method at each point.

The own numerical results showed that both methods, Dally's spectrum model (60x60 bins) and also the Monte-Carlo approach (500 waves) give good results with respect to the prediction of H_{rms} wave height across surf zone (see Fig. 7).

The H_{rms} values from Dally's spectrum model lie little higher than the measured H_{rms} and those from the Monte-Carlo method (500 waves) a little below the measured H_{rms}. It should be pointed out, that the computing time by Monte-Carlo approach has the advantage of very short computation times.

The verification of different wave decay models with prototype and field data is needed for further studies, especially with regard to the development of a sediment

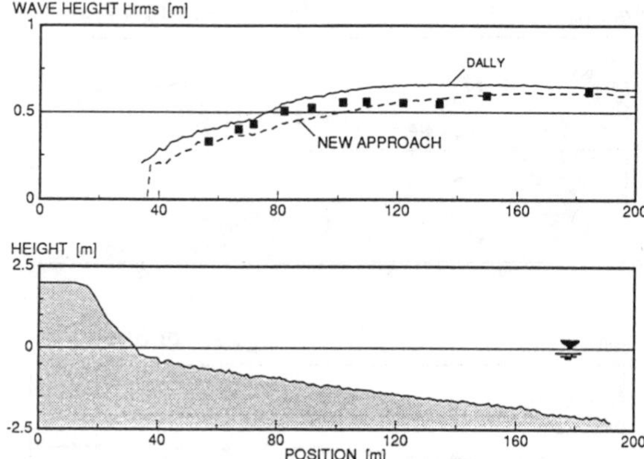

Fig. 7: Comparison of measured (■) and calculated (with DALLY-model and new approach) wave height decay after breaking.

transport model, in which the energy dissipation across the surf zone is considered as an important component of the model.

6.2. SEDIMENTTRANSPORT

A computer simulation program is treated in the Leichtweiss-Institute (LWI) with the aim to develop an as simple as possible method for engineering applications (DETTE and RAUDKIVI, 1991 and RAUDKIVI and DETTE, 1993).

In many models the CERC-formula is used for the calculation of longshore sediment transport. In other sophisticated models often the used parameters are not well verified by data yet. Some 90 % of the longshore sediment transport occurs in the form of suspended sediment. It would be a progress when the calculation of suspended sediment could be based on good data.
Outside the surf zone the calculation of orbital velocities and superimposed currents is an easier task. That means that there is a relative "exact" input for longshore sediment transport calculations of velocities and bed shear stresses.

The hydrodynamics in the surf zone are in comparison to the seaward area fairly complex. Large amounts of wave energy are dissipated over a relative small area through a multitude of breaker forms. The wave energy is converted into turbulence, but very little is known about the characteristics of this turbulence thus making an accurate calculation of specific values difficult. Under the assumption that the turbulence is proportional to wave energy dissipation, the suspension will increase with the incoming wave energy flux. The transport of grain-water-mixture is, however, proportional to the longshore component of this energy flux, i.e one may have high suspension and weak current or vice versa. Correspondingly the littoral transport has to be described as a function of both the total energy flux and the longshore component of energy flux and not as a function of only the longshore component of energy flux as by the CERC formula.

That shows that it is desirable to link the total wave energy dissipation at a point in the surf zone with sediment concentration parameters. This is the approach in the LWI models.

In this sense DETTE and RAUDKIVI (1991) developed a conceptual model for the estimation of suspended sediment transport due to wave and tide induced currents outside and inside the surf zone. The development of this semi-empirical model is in the first stage based on experimental data from investigations in the LARGE WAVE FLUME.

Briefly the calculation of sediment concentrations is based on the assumptions that the sediment concentration distribution can be described by

$$C = C_0 \cdot e^{(-a \cdot z)}$$

were C is temporal mean concentration at elevation z, C_0 is a reference concentration at the bed and 'a' is a decay parameter for the concentration with elevation above the bed.

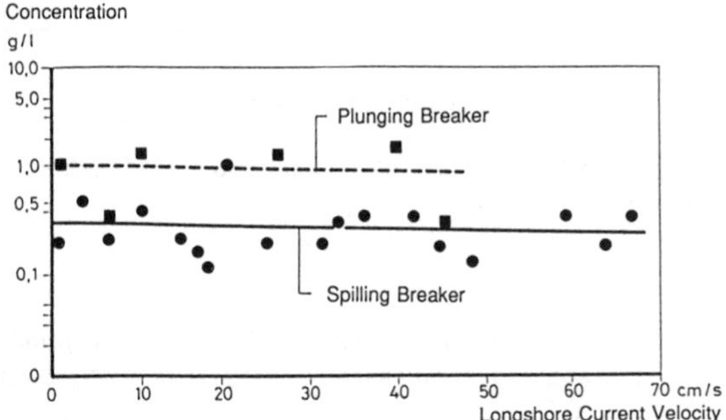

Fig. 8: Field data from KANA (1979) showing the independence of sediment concentration at a given elevation from the longshore velocity.

Another assumption is, that the suspended sediment inside the surf zone is mobilized due to turbulence by wave energy dissipation and transported by the littoral current. This assumption is supported by field data from KANA (1979), which show that the concentration at a given elevation is not a function of the coastal current velocity in the surf zone, but is mainly dependent on the breaker form (Fig. 8).

Inside the surf zone the empirical parameter 'C_0' and 'a' were fitted by data from the LARGE WAVE FLUME as

$$C_0 = 0.08 \cdot (D \cdot T)^{0.55}$$

and

$$a/T = 7 \cdot (D \cdot T)^{-0.7}$$

were D [W/m^2] is the energy dissipation and T [sec] the wave period (RAUDKIVI and DETTE, 1993). First comparisons between calculated and measured sediment concentrations are illustrated below. The measured data from earlier prototype experiments are well fitted by the calculated sediment concentration distribution (Fig. 9).

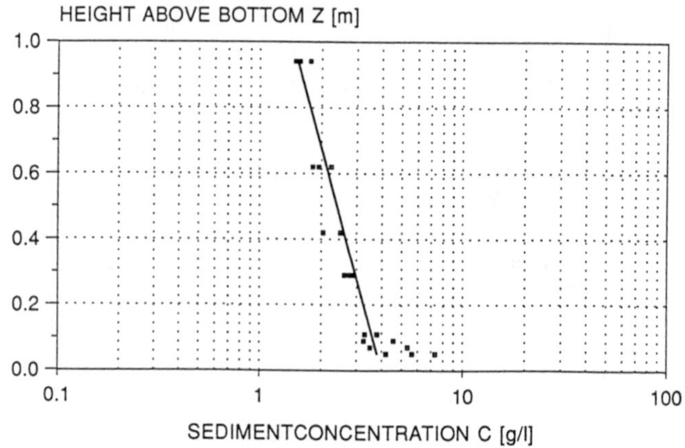

Fig. 9: Comparison of measured (■) and calculated sediment concentration distribution.

Outside the surf zone the empirical parameters 'C_0' and 'a' are fitted in dependence of the orbital velocity above the bed (RAUDKIVI and DETTE, 1991).

More field measurements are necessary in order to enable an extension of the conceptual model, e.g., the influence of grain-size to be included more reliably than at present.

7. OTHER STUDIES

The studies mentioned here are only a small part of work by LWI with the data from large scale experiments inside the LARGE WAVE FLUME.

During the tests outside the surf zone sheet flow conditions above a horizontal bed were studied by means of a high speed camera with 50 frames per seconds. Because of the high resolution of the 3 X 4 cm object window it could be observed that with increasing velocity near the bed, ripples became gradually flatter until only a relatively thin layer of grains was moved to and fro on the bed on orbital velocities of approx. 1.0 m/s outside the surf zone. The layer was equivalent to the thickness of few grain diameters, approx. 1 mm. From the film the upper and lower layer in which the grains were moved within the passage of one wave could be analyzed as well as the distribution of nearbed velocity distribution, which was determined from the movement of single grains. From the high speed film a video tape has been prepared and is available for study (DETTE et al.,1991).

8. CONCLUSION

Large scale experiments have been carried out in the LARGE WAVE FLUME in Hannover. This paper conveys the experimental arrangement, test procedures and some selected data sets.
It was shown how the experimental results are used to develop models for coastal sediment transport and surf zone dynamics.

In the present study with the experiments using DEAN-type beach profiles, some experimental data sets are available for verification of computer models.

9. ACKNOWLEDGEMENT

This work has been undertaken as part of MAST G6 Research Programme -Coastal Morphodynamics-. It was funded partly by the Commission of European Communities, Directorate General for Science, Research and Development under MAST Contract No. 0035-C and partly by national fundings through Deutsche Forschungsgemeinschaft (DFG) within the Research Programme "Coastal Engineering" of Sonderforschungsbereich (SFB) 205 at the University of Hannover and by the Ministry of Research and Technology (BMFT), Germany. The support is greatfully acknowledged.

10. REFERENCES

Bosman, J.J et al. (1987). Sediment Concentration Measurements by Transverse Suction. *Coastal Engineering, Vol. 11, 353 - 370.*

Dally, W.R.; Dean, R.G.; Dalrymple, R.A. (1984). A Model for Breaker Decay on Beaches. *Proc. 19th. Conf. on Coast. Eng., pp. 82-98, Houston, USA*

Dette, H.H.; Raudkivi, A.J. (1991). Model for littoral transport. *Proc. of the Int. IAHR - Symposium on the transport of suspended sediments and it's mathematical modelling, Florence, Italy.*

Dette, H.H.; Uliczka, K.; Rahlf, H. (1991). Beach Erosion in the Big Wave Flume. Research film B1791, prepared for the Research Programme "Coastal Engineering" of Sonderforschungsbereich (SFB) 205 Project A6 at the University of Hannover, *Institut für den Wissenschaftlichen Film, Göttingen, 1989-1991*

Dette, H.H.; Rahlf, H. (1992 a). Time - dependent dune and beach transformations. - Prototype experiments with monochromatic waves -. *Leichtweiß-Institut, Report No. 734* (unpublished, prepared for MAST G6)

Dette, H.H.; Oelerich, J.; Peters, K. (1992 b). Time - dependent dune and beach transformations. - Prototype experiments with JONSWAP-Spectrum -. *Leichtweiß-Institut, Report No. 735* (unpublished, prepared for MAST G6)

Dette, H.H.; Rahlf, H.; Peters, K. (1992 c). Suspension Measurements Outside the Surf Zone. - Prototype experiments in the Large Wave Flume -. *Leichtweiß-Institut, Report No. 739* (unpublished, prepared for MAST G6)

Dette, H.H.; Rahlf, H.; Wu, J.; Peters, K. (1992 d). Wave Measurements across the Surf Zone at Equilibrium Beach Profile -Prototype experiments (1990/91) with monochromatic waves and wave spectra-. *Leichtweiß-Institut, Report No. 762* (unpublished, prepared for MAST G6)

Dette, H.H.; Rahlf, H.; Peters, K. (1992 e). Suspension Measurements Across the Surf Zone. - Prototype experiments (1990/91) with monochromatic waves and wave spectra -. *Leichtweiß-Institut, Report No. 763* (unpublished, prepared for MAST G6)

Hughes, S.A.; Miller, H.C. (1987). Transformation of Significant Wave Heights. *Journal of Waterway, Port, Coastal and Ocean Eng., Vol. 113, No.6, pp 588-605*

Kana, T.W. (1979). Suspended Sediment in Breaking Waves. *Univ. of South Carolina, Dept. for Geology, Tech. Rep. No. 18-CRD*

Kraus, N.C.; Larson, M.; Kriebel, D.L. (1991). Evaluation of Beach Erosion and Accretion Processes. *Proc. Coast. Sediments '91, ASCE, pp. 572-587*

Kriebel, D.L.; Dean R.G. (1985). Numerical Simulation of Time Dependent Beach and Dune Erosion. *Coastal Engineering, Vol. 9, pp 221-245*

Oelerich, J.; Dette, H.H. (1988). About the Energy Dissipation over Barred Beaches. *Proc. 21th Conf. on Coast. Eng., pp 292-306, Malaga, SPAIN.*

Raudkivi, A.J.; Dette, H.H. (1993). Ein vereinfachtes Verfahren zur Ermittlung der Suspensionsfracht in der Brandungszone. *Die Hansa, 1993.*

ABOUT THE RANGE OF VALIDITY OF DIFFERENT SPECTRAL MODELS FOR WIND–GENERATED GRAVITY WAVES

Germán Rodríguez R.* & José Jiménez, Q.†

ABSTRACT: This paper is concerned with investigating the range of applicability of two spectral models, frequently used to modelize the wind-generated gravity waves in practical applications, by using synthetic and field data representing various sea state conditions. Furthermore, we explore the effects of the computational methods, employed to estimate the spectral density function and to develop the curve-fitting procedure, on the relative goodness of fit between measured and theoretical spectra.

INTRODUCTION

Sea waves are random in nature and hence can only be described by their statistical properties in the time domain and by means of the spectral density function $S(f)$, usually named wave spectrum, which reveals the wave energy distribution in the frequency domain.

The shapes and scales of wave spectra presents a large variety depending on the geographical location, wind speed, wind duration and fetch, sea state development and background swell presence, among other factors. However, the shape of wave spectra is not completly arbitrary and certain general features can be distinguished. One of such general properties is the existance of an upper bound, or equilibrium range, on the tail of the spectrum having frequencies well above the peak frequency, f_p, proportional to a given inverse power relation of the frequency.

As an attempt to obtain a useful description of the wave field in the frequency domain, taking into account the known general features of the wind-generated gravity waves, a large number of one-dimensional spectral models have been proposed in the literature.

The development of these spectral models is based on empirical fits to wave spectra, measured under reasonably well defined conditions. The main

*Departamento de Física, Facultad de Ciencias del Mar, Universidad Las Palmas de G. C., P.O.Box 550, 35017 Las Palmas, Spain. Tlf:+34-28-451289, Tlfax:+34-28-452922

†Dept. Enginyeria Hidràulica, Marítima i Ambiental, Univ. Politècnica de Catalunya, 08034 Barcelona, Spain. Tlf:+34-3-4016469, Tlfax:+34-3-4017357

theoretical ideas used to determine its analytical expressions rest on the notion of similarity in the spectral shape, which central concept is the above mentioned existance of a saturation range in the high frequency region.

Based on dimensional arguments, Phillips (1958) argued that the high frequency range of the spectral density, $S(f)$, were governed by wave breaking due to gravitational instabilities and were independent of wind speed. Thus, assuming that saturation conditions were determined exclusively by the frequency and the gravitational acceleration, g, he proposed the following formulation to modelize the saturation range:

$$S(f) = \frac{\alpha g^2}{(2\pi)^4} f^{-5} \qquad \text{for} \qquad f \gg f_p \tag{1}$$

where α is a dimensionless constant and f is the wave frequency in hertz.

According to similarity hypothesis (Kitaigorodskii, 1962), in general, the wave spectrum over all frequency range takes the form:

$$S(f) = \frac{\alpha g^2}{(2\pi)^4} f^{-5} \psi \left(\frac{f}{f_p} \right) \tag{2}$$

where ψ is a function which explicit forms are usually developed on the basis of experimental studies.

Two commonly used spectral models of this type are the Pierson-Moskowitz (1964) fully developed spectrum, which analytic form is:

$$S(f) = \frac{\alpha g^2}{(2\pi)^4} f^{-5} \exp \left[-\frac{5}{4} \left(\frac{f}{f_p} \right)^{-4} \right] \tag{3}$$

and the *JONSWAP* fetch-limited form (Hasselmann, et al., 1973), represented by:

$$S(f) = \frac{\alpha g^2}{(2\pi)^4} f^{-5} \exp \left[-\frac{5}{4} \left(\frac{f}{f_p} \right)^{-4} \right] \gamma^{\exp\left[\frac{-(f-f_p)^2}{2\sigma^2 f_p^2} \right]} \tag{4}$$

where γ is a factor that enhances the spectral peak and σ is a peak width factor, with $\sigma = \sigma_a$ for $f \leq f_p$ and $\sigma = \sigma_b$ for $f > f_p$.

The *JONSWAP* spectral density function has been considered as a good model to describe the energy content of the wave fields, not only for developing seas but even for decaying seas (Ewans and Kibblewhite 1990). However, like the rest of the proposed unimodal spectral functions, it fails trying to describe wave fields resultant from the coexistence of swell and wind waves or due to suddenly changing winds, which estimated spectra are often multimodal. To solve this problem, a few spectral models have been proposed so that they are capable to represent bimodal spectra (Ochi and Hubble 1976, Soares 1984). All of these models are based on the original idea due to Strekalov and Massel (1971), to separate the double peak wave spectra into two parts; one for the low frequency band of the spectra, corresponding to the background swell components, and the other for the high frequency band, related to wind wave components.

The most commonly used bimodal spectral formulation was developed by Ochi and Hubble (1976). These authors proposed to describe the bimodal spectra by superposition of two modified *Pierson-Moskowitz* spectra, each one with three parameters, giving place to a six-parameter spectral model. The six parameters are, the significant wave height, H_{m_o}, peak frequency and a spectral shape parameter, λ, for each of the two frequency bands. That is:

$$S_i(f) = \frac{\left(\frac{4\lambda_i+1}{4} f_{p_i}^4\right)^{\lambda_i}}{4\Gamma(\lambda_i)} \frac{H_{m_{o_i}}^2}{f^{4\lambda_i+1}} \exp\left[-\frac{4\lambda_i+1}{4}\left(\frac{f_{p_i}}{f}\right)^4\right] \tag{5}$$

where the significant wave height is given by:

$$H_{m_0} = 4\sqrt{m_0} \qquad \text{and} \qquad m_n = \int_0^\infty f^n S(f) df$$

Γ is the gamma function and $i = 1$ and $i = 2$ refer to the swell and wind sea frequency bands, respectively. Then, combining these two expressions, they established the final six-parameter spectral formulation.

The JONSWAP range

In this paper, we use the *JONSWAP* and the *Ochi-Hubble* formulations to modelize unimodal and bimodal spectra, respectively. These models are fitted to the computed spectra to contrast the relative validity of its ranges of applicability, as function of the significant wave height H_{m_0} and the peak period T_p, which are two of the most important sea state parameters from the practical point of view. To achieve this goal, we follow the procedure suggested by Torsethaugen et al. (1984). According to these authors, the spectral variability over a wide range of sea states, characterized as function of H_{m_0} and T_p, may be subdivided into three zones (see fig.1). The zone 1 corresponds to sea states dominated by wind sea but significantly influenced by some swell components, the zone 2 to more or less pure wind seas, or possible eventual swell components located well inside the wind sea frequency band, and zone 3 to sea states dominated by swell but significantly influenced by a local wind sea.

The *JONSWAP* spectrum is considered to be a reasonably good model for wave spectra only in a given zone of the whole (H_{m_o}, T_p) space. This zone, called the *JONSWAP range*, is specifyed by the inequalities:

$$11.3 < T_p\sqrt{\frac{g}{H_{m_o}}} < 15.7 \tag{6}$$

which lead immediately to the limits for the subspace (H_{m_o}, T_p) where the *JONSWAP* model is assumed to be acceptably valid (see fig. 2).

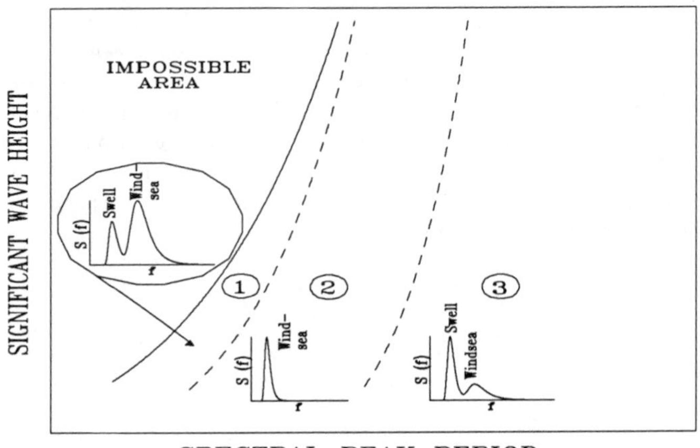

Figure 1: *Qualitative indication of spectral variability (from Torsethaugen et al. 1984)*

WAVE DATA DESCRIPTION

The data sets analyzed in the present study were obtained from field measurements conducted off Canary Islands and at the North Sea, and by means of a numerical simulation procedure which will be described hereafter.

Wave data measurements

Both Waverider buoys, instaled at Las Canteras Bay (Canary Islands) and Stajford Oil Field (Norway), were placed at deep waters and the recorded signals were digitized with a sampling frequency of 2 Hz. Each time serie was subjected to quality check to eliminate possible outliers, non-physical accelerations and polynomial trends. To confirm the validity of the stationarity basic hypothesis, needed for the correct application of the spectral analysis techniques, each record was submitted to the *run* and *trend* tests for the mean, variance and autocorrelation function. A total amount of 800 measured wave records with a duration of 17 minutes were analyzed. From these, less than a 2% was rejected for presenting an evolutionary behaviour. Besides, 25 wave records were simulated. Thus, the study was carried out for a total of 810 wave records.

Used field data sets corresponds to the periods: December 1986–December 1988 (Las Canteras Bay) and January 1989 (Statfjord oil field).

Wave data simulation

The basic idea underlying in the proposed method of simulation is the possibility of generate random time series following a Normal distribution, with

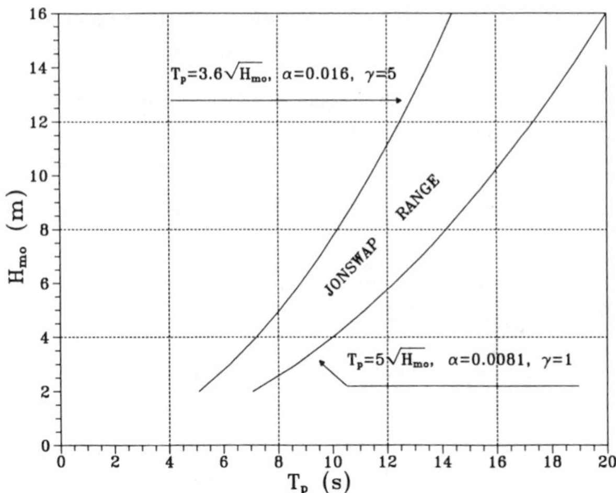

Figure 2: *JONSWAP range as a function of* H_{m_o} *and* T_p *(from Torsethaugen et al. 1984)*

zero mean and selected variance, so that we can control the H_{m0} values. Besides, the random signal may be forced to have a prefixed autocorrelation function $R(\tau)$, in such a way that the values of T_p and the spectral bandwidth are implicit in the proposed model for $R(\tau)$ (Næss 1985, Rodríguez 1992, Rodríguez et al. 1992a). The main steps involved in the numerical procedure are the following:

[1]–Generate N random numbers with an U(0,1) distribution. This array is operated on by Box-Müller method changing its probability distribution to an N(0,1).

[2]–Especify the values of the requiered autocorrelation $R(n\Delta t)$. To compute the desired values of $R(\tau)$ we have used three different procedures:

a) Calculating the autocorrelation function corresponding to the *JONSWAP* and *Ochi-Hubble* spectral models, by invoking the Wiener-Kintchine theorem, so that:

$$R(\tau) = \int_0^\infty S(f) \cos(2\pi f \tau)\, df \tag{7}$$

b) Modeling the autocorrelation function of real sea wave records by means of a simple model:

$$R(\tau) = R(0) \cdot e^{-\alpha\tau} \cos(\beta\tau) \tag{8}$$

where $R(0)$ represents the variance of the wave record and hence the value of H_{m_0}, α is a bandwidth measure and β determines the peak frequency.

c) Approximating $R(\tau)$ to the autocorrelation function of a real wave record by sums of complex exponentials:

$$R(\tau) = R(n\Delta t) = \sum_{i=1}^{N} a_i \cdot e^{\lambda_i n \Delta t} \tag{9}$$

where a_i and λ_i are complex unknowns which must be estimated numerically (Daubisse, 1981).

[3]–Transform the set of random numbers $x(n)$ into a new discrete series $\eta(n)$ according to:

$$\eta(n) = \sum_{k=1}^{N} c_k \cdot x(n-k) \tag{10}$$

The autocorrelation $R(n\Delta t)$ of the random values $\eta(n)$ is taken into account through the coefficients c_k. These coefficients must be determined by solving the following system of non-linear algebraic equations, writted in matrix form:

$$\begin{bmatrix} R(0) \\ R(1) \\ \vdots \\ R(N-2) \\ R(N-1) \end{bmatrix} = \begin{bmatrix} c_1 & c_2 & \cdots & c_{N-1} & c_N \\ c_2 & c_3 & \cdots & c_N & 0 \\ \vdots & \vdots & \ddots & \vdots & \vdots \\ c_{N-1} & c_N & \cdots & 0 & 0 \\ c_N & 0 & \cdots & 0 & 0 \end{bmatrix} \begin{bmatrix} c_1 \\ c_2 \\ \vdots \\ c_{N-1} \\ c_N \end{bmatrix} \tag{11}$$

[4]–Finally, upon sustitution of the coefficients c_k into equation (10), we obtain the discrete ordinates of the random normal stationary process $\eta(n)$, with the requiered autocorrelation function.

SPECTRAL DATA ANALYSIS

Spectral density function has been computed using the conventional techniques, Blackman–Tukey and Fast Fourier Transform, and the Maximum Entropy Method, henceforth called BT, FFT and MEM.

In the BT method, we used the well known "*rule of thumb*" ($N/10$) to select the number of lags for computing the autocorrelation function (ℓ), where N is the number of points in the wave record. Smooth values of the spectral density were obtained by applying Hanning's spectral window.

For the FFT algorithm, the raw spectrum has been smoothed by means of a moving average over 8 raw spectral estimates, ($\tilde{N}_f$).

To compute the spectra with the MEM method the only one parameter which must be carefully choosen is the order of the autoregressive model to be fitted,

(P). The selected value in this study is 26, which seems to be an appropiate value to represent one and two-peaked spectra (Rodríguez et al. 1992b).

The parameters ℓ, $\tilde{N}_f$ and P, control the budget between the resolution and the statistical stability in BT, FFT and MEM, respectively. Therefore, these have been selected trying to get the closest agreement between the various spectral parameters obtained from the wave spectrum, computed by means of different techniques, mainly H_{m_o} and f_p, which are two very important inputs to the fitting procedures.

Usually, differences among the values of the wave parameters obtained by different spectral methods are not too large. However, small variations may get an important role on the effectiveness of the fitting algorithms.

Figure 3 shows two examples of wave spectra computed by means of the three techniques above introduced. It is of interest to note the commented variability for H_{m_o} and f_p.

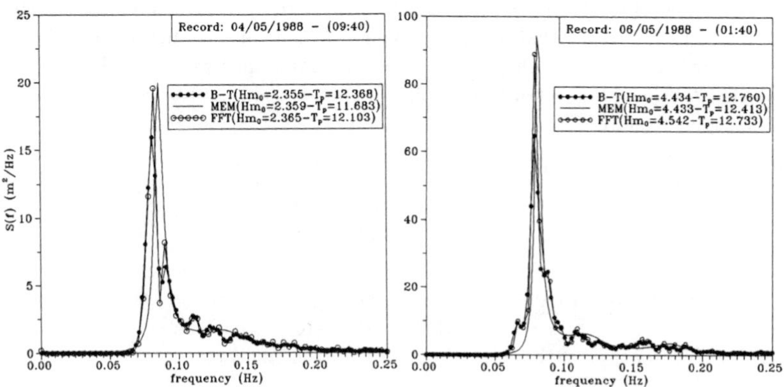

Figure 3: *Comparison of wave spectra computed with BT, FFT and MEM.*

METHODS TO FIT SPECTRA

The estimated spectra are fitted to the *JONSWAP* model by means of five algorithms. Among these it is included an improved non-linear least squares method, which is also used to fit the *Ochi-Hubble* spectrum.

Algorithms to fit the JONSWAP model

LeBlond method: LeBlond et al.(1982), expressed the *JONSWAP* model in the following general form:

$$S(f) = \frac{A}{f^5} \exp^{-B/f^4} \gamma^{\exp\left[\frac{-(f-f_p)^2}{2\sigma^2 f_p^2}\right]} \tag{12}$$

and adopted the mean values, found by Hasselman et al.(1973), for σ. That is, $\sigma_a = 0.07$ and $\sigma_b = 0.09$.

Then, there are only three parameters left for fitting, A, B and γ. The parameters A and B are defined, for the *Pierson-Moskowitz* model, as:

$$A = \frac{5}{16} H_{m_0}^2 f_p^4 \qquad \text{and} \qquad B = \frac{5}{4} f_p^2$$

The relationship between B and f_p is also valid for the *JONSWAP* spectrum. However, the relation between A and H_{m_0} is more complicated for this model, because of the presence of the peak enhacement factor γ. In the range $1 < \gamma < 4$, which includes most values encontered, Mitsuyasu et al.(1980) have obtained the relation:

$$A \approx \frac{5 H_{m_0}^2 f_p^2}{16 \gamma^{(1/3)}} \tag{13}$$

which allow us to identify H_{m_0}, f_p and γ as the three free parameters to be fitted. With f_p estimated as that value of f where $S(f)$ attains its maximum value.

Therefore, matching the height of the observed spectral peak $S(f_p)$ to that of the *JONSWAP* spectrum, the following expression is obtained for the peak enhacement factor:

$$\gamma = \left[\frac{S(f_p)}{\frac{5 H_{m_0}^2}{16 f_p} \exp(-5/4)} \right]^{3/2} \tag{14}$$

Torsethaugen method: Torsethaugen, et al. (1984), assume that σ is given by its mean values or by $\sigma_a = \sigma_b = 0.08$ and relate the parameters α, γ and f_p to each other, for a given value of H_{m_0}, through the equation:

$$\gamma = \exp\left\{ 3.484 \left(1 - 0.1975 \alpha \frac{T_p^4}{H_{m_0}^2} \right) \right\} \tag{15}$$

specifying that this equation shoul only be applied within the *JONSWAP range*. Taking into account the relationship between α and f_p suggested by Mitsuyasu et al.(1980), and considering that α varies as a linear function of T_p over this region, it is possible to express the Phillips' parameter as:

$$\alpha = 0.036 - 0.0056 \frac{T_p}{\sqrt{H_{m_0}}} \tag{16}$$

The corresponding value of γ is then obtained by introducing equation (16) into equation (15).

S. Gran method: According to S.Gran(1986), the *JONSWAP* parameters may be estimated in the *JONSWAP range* approximately, from H_{m_0} and the average period between zero up-crossings T_z, with σ taking its mean values. Thus, for a given set of H_{m0} and T_z, α is determined from:

$$\alpha = \frac{4\pi^3 H_{m0}^2}{g^2 T_z^4} \qquad \text{with} \quad \alpha > 0.00315 \quad \text{and} \quad T_z = \sqrt{\frac{m_0}{m_2}} \tag{17}$$

Then, the spectral peakedness parameter γ, can be estimated by:

$$\gamma = 7\left(1 - \frac{0.0027}{\alpha}\right) \tag{18}$$

Müller method: This method, developed by Müller (1976) seems to be the most commonly used. It consists of the following steps:

[1]–Determination of the frequency peak, using a least square parabolic fit to the three highest values of the spectral density (the highest spectral estimate, $S(f_p)$, and the two estimates to either side).

[2]–The Phillips' equilibrium "coefficient", α, is determined by an average of $f^5 S(f)$ over the frequency range $1.35 f_p \leq f \leq 2.0 f_p$, well outside the vecinity of the spectral peak . It is assumed that over this range the spectrum can be approximated by a *Pierson-Moskowitz* spectral model. Thus:

$$\alpha = \frac{(2\pi)^4 g^{-2}}{n_2 - n_1 + 1} \sum_{n=n_1}^{n_2} S(f_n) f_n^5 \exp\left[\left\{\frac{5}{4}\left(\frac{f_p}{f_n}\right)^4\right\}\right] \tag{19}$$

where

$$n_1 = 1.35\frac{f_p}{\Delta f} \quad \text{and} \quad n_2 = 2.0\frac{f_p}{\Delta f}$$

[3]–The peak enhancement factor γ is determined by taking the ratio of the estimated spectral density corresponding to the frequency peak, computed in [1], to the maximun of the *Pierson-Moskowitz* spectrum, with the same values of f_p and α calculated in steps [1] and [2]. That is:

$$\gamma = \frac{S(f_p)}{S_{PM}(f_p, \alpha)} \tag{20}$$

[4]–The left hand and right hand peak widths σ_a and σ_b are defined only if $\gamma > 1$; if $\gamma < 1$ the peaks widths remains unspecified. For $\gamma > 1$ these are determined in the frequency range $0.8 f_p \leq f \leq 1.35 f_p$ by using the equations:

$$\left.\begin{aligned} &\sum_{n=n_0}^{n_1} \left[S_J(f_n) - S(f_n)\right] = 0 \\ &\sum_{n=n_0}^{n_1} (f_n - f_p)\left[S_J(f_n) - S(f_n)\right] = 0 \end{aligned}\right\} \tag{21}$$

with

$$n_0 = \left(0.8\frac{f_p}{\Delta f}\right) + 1 \quad \text{and} \quad n_1 = 1.35\frac{f_p}{\Delta f}$$

Equations (21) are solved by Newton's method of succesive approximations. That is, the difference between the observed and fitted spectra is minimized by iteration, in terms of the zeroth and first spectral moment, in such a way that σ_a and σ_b are selected to give the same area under the spectrum and the same first moment about f_p as the measured spectrum in the frequency range used.

The proposed Method is a Marquardt non-linear type method improved by obtaining the best possible initial values to begin the iterative process as first step. To reach this end, we compute f_p by the Delf Method:

$$f_{pD8} = \frac{\int_{f_1}^{f_2} fS(f)df}{\int_{f_1}^{f_2} S(f)df}$$

where f_1 and f_2 are the upper and lower frequencies corresponding to the intersections of $S(f_1)$ and $S(f_2)$ with the threshold value $0.8S(f_p)$.

α and γ are calculated applying two expressions proposed by Yamaguchi (1984) and Goda (1985). The first one is a formula relationing m_0, f_p, α and γ, in such a way that, having determined m_0 and f_p, it is possible to obtain α from γ, and viceversa, conserving the wave energy:

$$\alpha = \frac{(2\pi)^4 m_0 f_p^4}{(0.065\gamma^{0.8} + 0.135)g^2} \tag{22}$$

Goda (1985) shows that the *JONSWAP* model can be rewritten, in an approximate form, as function of H_{m_0} and T_p, with the following relationship between α and γ:

$$\alpha \approx \frac{0.0624}{0.230 + 0.0336\gamma - 0.185(1.9+\gamma)^{-1}} \tag{23}$$

Therefore, once the total energy m_0 is determined, it is possible to estimate all the initial free parameters by solving equations (22) and (23), and adopting the mean values of σ_a and σ_b. The unknown parameters are found by inserting these values in an iterative precedure, which minimizes a function Ψ, following a non-linear least squares criterion:

$$\Psi = \sum_{i=1}^{N_f} \left(\tilde{S}(f_i) - S^*(f_i)\right)^2$$

where $\tilde{S}(f_i)$ are the estimated values, $S^*(f_i)$ are the values computed from the model to be fitted and N_f is the number of used frequencies.

Algorithm to fit the Ochi-Hubble model

To fit the *Ochi-Hubble* spectrum we apply the proposed non-linear least square algorithm. To start the iterative procedure, we accept that both spectral parts are narrow banded and, as a consequence:

$$H_{m0} = \sqrt{H_{m0_1}^2 + H_{m0_2}^2} \tag{24}$$

Besides, differenciating equation (5) with respect to f and setting the result equal to zero to determine the value of f at which $S(f)$ is maximum f_p, it is possible to define the following function (Borgman 1991):

$$G(\lambda_i) = \frac{S(f_{p_i})f_{p_i}}{H_{m0_i}^2} = \frac{\left[\frac{4\lambda_i+1}{4}\right]^{\lambda_i} \exp\left[-\frac{4\lambda_i+1}{4}\right]}{4\Gamma(\lambda_i)} \tag{25}$$

expression that is easy to solve by Newton-Raphson iteration, for the value of λ_i requiered to achieve a given value of $G(\lambda_i)$.

Thus, estimating f_p and $S(f_p)$ to each part of the computed spectrum, we can use equations (24) and (25) to compute the approximate values of λ_i and H_{m0_i}. Therefore, the six parameters estimated by the above procedure are the initial input to the proposed non-linear method, which permits to reach a better set of parameters.

DISCUSSION

To obtain an assessment of the accuracy of fitness for both models, we use the deviation index (DI), (Liu 1983), which is expressed as:

$$DI = \sum_{i=1}^{N_s} \left[\frac{\left|\tilde{S}(f_i) - S(f_i)^*\right|}{\tilde{S}(f_i)} \times 100\right] \left[\frac{\tilde{S}(f_i)\Delta f}{m_0}\right] \tag{26}$$

Thus, if the fitness is perfect the DI value will be equal to zero and it will be increased for worst agreements.

The results obtained fitting the *JONSWAP* model, lead us to conclude that if the pairs (H_{m0} , T_p) falls inside the *JONSWAP range*, the Müller, LeBlond and Torsethaugen methods give good enough results. While if these couples are placed outside of this region, the Torsethaugen, et al. method presents high values for DI, compared with those given by the other two commented procedures. This confirms the restriction indicated by its authors. This fact is shown in figure 3, which ilustrates the measured spectra for two wave records, one computed by the MEM method, for a pair (H_{m0} , T_p) falling outside the *JONSWAP zone* (plate A), and another calculated through the FFT, inside of such zone (plate B), together with the spectra obtained by means of the five fitting methods previously described.

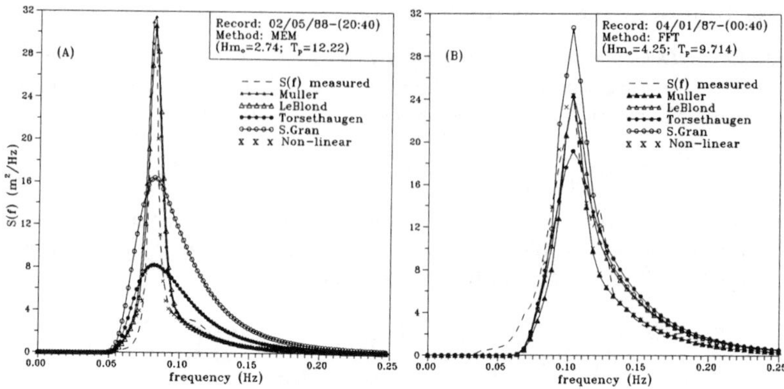

Figure 4: *Comparison of fitness for spectra outside and inside the JONSWAP range to JONSWAP model.*

Inspecting the results for the total of records analized during the study (see table 1), it is clear that the goodness of fit for these three methods is so good and that the DI values becomes to be lower, when the measured spectrum is computed by MEM.

Moreover, we must make reference to the relatively low severity of the wave data recorded in Las Canteras Bay, in which the highest recorded value for the significant wave height was 6.2 meters. These data represent more than 65% of the total data set analized. Consequently, many of the estimated spectra were associated to values of H_{m_0} and f_p falling outside the *JONSWAP range*. Perhaps, this may be the reason to explain the high values of DI given, frequently by the S.Gran method, such as can be seen in table 1.

According to values displayed in table 1, is easy to comprobe that the proposed non-linear method gives the better results. Besides, due to the insertion of a good enough initial estimate of the solution as starting point, the method exhibits a correct and quick convergence and seldom presents problems.

With respect to the degree of fitness as function of the technique employed to calculate the spectrum it is of interest to comment that some times the value of DI is relatively high for the BT anf FFT methods, although a visual inspection of graphics reflects a good enough agreement between the spectra computed with these techniques and the spectral models. This fact is obviously due to the statistical inestabilities present in the spectra computed by BT and FFT methods, which are a direct cause of the truncation procedure realized on the autocorrelation function or the time series, respectively, due to the use of a finite record. This problem is not present in the MEM method because it is a data adaptative method, capable of generating higher resolution spectral estimates from shorter wave records, minimizing the disturbations on the neighbour frequencies. These random fluctuations in the BT and FFT methods, with respect to the fitted model, are not canceled mutually in the DI expression, due to the absolute value in the numerator, giving unexpected high values of DI.

Table 1: *Average DI results for the JONSWAP model fitness*

	LeBlond	**Torseth.**	**S.Gran**	**Müller**	**N-linear**
BT	35 ± 14	38 ± 21	98 ± 21	33 ± 08	27 ± 18
FFT	34 ± 09	38 ± 13	96 ± 28	31 ± 15	29 ± 07
MEM	32 ± 17	36 ± 16	85 ± 19	29 ± 12	23 ± 11

Figure (5) shows a bimodal spectrum fitted to the *JONSWAP* and *Ochi-Hubble* models, using the proposed non-linear methodology. It can be clearly seen that the *JONSWAP* model is not useful to modelize this kind of spectra, with more than one peak or with a considerable amount of energy forming a plateau in the range of high frequencies far away from the spectral peak. However, we have found that the *Ochi-Hubble* formulation is useful to modelize both unimodal and bimodal spectra with a considerable accuracy

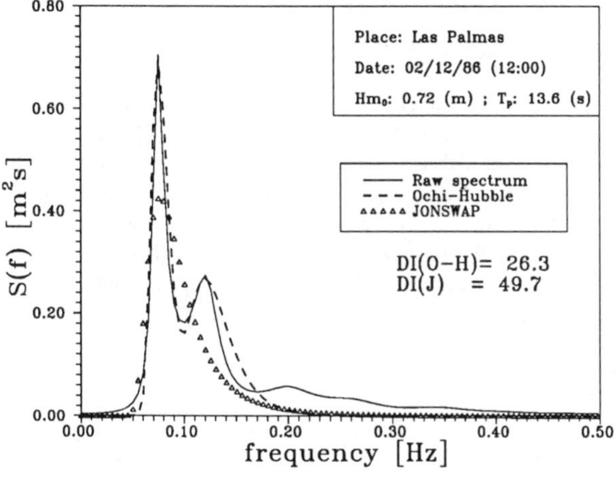

Figure 5: *JONSWAP and Ochi-Hubble models fitted to a bimodal spectrum.*

CONCLUSION

As summary of the above commented results we can conclude that the goodness of fit between the spectra computed from real and simulated wave records and the *JONSWAP* and *Ochi-Hubble* spectral models, for each couple

(H_{m_0}, T_p), shows a rather well agreement with the three zones defined by Torsethaugen et al. However, there are some spectra placed at the zones 1 and 3, often near the bounds, which can be modelized by means of the *JONSWAP* spectrum with enough precision and some spectra with (H_{m_0}, T_p) inside this zone which can not be correctly represented by this model. These results suggests that the limits proposed by these authors, which confines the named *JONSWAP range*, should be more flexible, or be acompanied by some confidence limits. This fact is been studied actually by the authors. On the other side, the *Ochi-Hubble* model permits the representation of a more general sea estate than the *JONSWAP* formulation.

The proposed improved Marquardt method gives the better results for estimating the free parameters required to fit the *JONSWAP* spectral function to the measured spectra, mainly when these are computed by using the MEM technique. Similar results are found when this curve-fitting algorithm is employed to fit the *Ochi-Hubble* spectrum following the procedure above described.

With respect to the other procedures for fitting the *JONSWAP* spectrum is remarkable the good behabiour displayed by the LeBlond et al. and the Torsethaugen et al. methods, if we consider sea states inside the *JONSWAP range* for the latter. The widely used Müller method gives similar results to these methods. But it is important to take into account the simplicity of the LeBlond and Torsethaugen methods, in particular for practical applications.

ACKNOWLEDGMENTS

This paper was prepared partly during a visit by the first author to the Department of Civil Engineering, Division of Structural Engineering (N.I.T.) at Trondheim. He wishes to thank Professor Arvid Næss and to other members of that Department for their hospitality and assistance. He would also like to express thanks to Professors Knut Torsethaugen, of the Norwegian Hydrotechnical Laboratory, and Dag Myrhaugh, of the Division of Marine Hydrodynamics (N.I.T.) for providing information and advice. We are indebted to Programa de Clima Marítimo (Spain) and STATOIL Company (Norway) for providing data used in this study.

References

[1] Borgman, L., 1991. "Irregular ocean waves: Kinematics and Forces," in *Ocean Engineering Science*, The Sea, Vol.9, part 4, pp. 121-168.

[2] Daubisse, J., 1981. "Some results on approximation by exponential series applied to hydrodynamic problems," *Proc. 3th Int. Conf. on Numerical Ship Hydrodynamics*, pp. 551-558.

[3] Goda, Y., 1985. "Random Seas and Design of Maritime Structures," *University of Tokio Press.*

[4] Gran, S., 1986. "Discussion of Morisson's equation," *Veritas Research, report*, No. 86-2011.

[5] Guedes Soares, C., 1984. "Representation of double-peaked sea wave spectra," *Ocean Engineering*, Vol. 11, pp. 185-207.

[6] Hasselmann, K. et al., 1973. "Measurements of wind-wave growth and swell decay during the Joint North Sea Wave Project (JONSWAP)," *Deutsche Hydrograph. Z.*, (A8 No.12).

[7] Kitaigorodskii, S.A., 1962. "Applications of the theory of similarity to the analysis of wind-generated wave motion as a stochastic process," *Bull. Acad. Sci. USSR. Geophys. Ser.*, Vol. 1, pp. 105-117.

[8] LeBlond, P.H., Calisal, P., and Isaacson, M., 1982. "Wave spectra in Canadian waters," *report Can. Hydrog. and Ocean Sciences*, 6.

[9] Liu, P., 1983. "A representation for the frequecy spectrum of wind generated waves," *Ocean Engineering*, Vol. 10, pp. 429-441.

[10] Mitsuyasu, H., Tasai, F., Suhara, T., Mizuno, S., Ohkusu, M., Honda, T., and Rikiishi, K., 1980. "Observations of the power spectrum of waves using a cloverleaf buoy," *J. Phys. Oceanogr.*, Vol. 10, pp. 286-296.

[11] Müller,P., 1976. "Parametrization of one dimensional wind wave spectra and their dependence on the state of development," *Hamburger Geophys. Einz. Hamburg*, 31.

[12] Næss, A., 1985. "On the distribution of crest-to-trough wave heights," *Ocean Engineering*, Vol. 12, pp. 221-234.

[13] Ochi, M.K., and Hubble, E.N., 1976. "On six-parameter wave spectra," *Proc. 15th Coastal Eng. Conf.*, ASCE, NY, pp. 301-328.

[14] Phillips, O.M., 1958. "The equilibrium range in the spectrum of wind-gererated waves," *J. Fluid Mech.*, Vol. 4, pp. 426-434.

[15] Pierson, W.J. and Moskowitz, L. (1964): A proposed spectral form for fully developed wind seas based on the similarity theory of S.A. Kitaigorodskii. *J.Geophys.Res*, **69**, No. 24, pp. 5181–5190.

[16] Rodríguez, G.R., 1992. "A study about the dependence on bandwidth and nonlinearity of the sea wave heights, with simulated records from the Wallops spectrum," *Hydrosoft IV, Fluid Flow Modelling*, pp. 549-560.

[17] Rodríguez, G.R., Alejo, M., and Tejedor, L., 1992. "Numerical study on the range of applicability of the JONSWAP spectrum and its relations with the autocorrelation function features," *Proc. I J. Españolas de Ing. Ocean. y Costas*, pp. 307-319. (in Spanish)

[18] Rodríguez, G.R., Grisolía, D.S., and Díaz, M.M., 1992. "On the statistical variability of some spectral bandwidth and nonlinearity parameters of wind generated gravity waves," *Proc. Sixth IAHR Int. Symp. on Stochastic Hydraulics*, pp. 353–360.

[19] Strekalov, S., and Massel, S., 1971. "Niektore zagadnienia widmowej analizy falowania wiatrowego," *Arch. Hydrot*, Vol. 28, pp. 457-485. (in Polish)

[20] Torsethaugen, K., Faanes, T., and Haver, S., 1984. "Characteristica for extreme sea states on the Norwegian Continental Shelf," *NHL report*, No. 2-84123.

[21] Yamaguchi, M., 1984. "Approximate expressions for integral properties of the JONSWAP spectrum", (in Japanesse).

LARGE SCALE MEASUREMENTS OF NEAR SURFACE ORBITAL VELOCITIES IN LABORATORY WAVES BY USING A SURFACE FOLLOWING DEVICE

Stefan Woltering[1] and Karl F. Daemrich[2]

ABSTRACT: Large scale measurements of horizontal orbital velocity components were performed using a surface following velocity probe. Results of several calculation methods are compared with the measurements, and a conclusive method of wave kinematics interpretation (LAGRANGEian approach) is introduced. With LAGRANGEian approach it is possible to analyse the real mass transport in regular waves from EULERian velocity measurements.

INTRODUCTION

Measurements of orbital velocities are important for the understanding of wave kinematics and prediction of forces on coastal structures using wave theories. Especially near the surface uncertainties exist with respect to adequate theoretical approaches. Comparisons of measurements of orbital velocities in irregular waves with various wave theories (e.g. Stokes wave theories, empirical calculation methods, transfer function methods) often contain tendencies of over- or underprediction of the extreme values at crests and troughs. Uncertainties exist further with respect to channel effects (e.g. backflow) and the wave induced mass transport velocity. Concerning sediment transport, the wave induced mass transport velocity is needed for the estimation of transport direction and magnitude.

1, 2) both Sonderforschungsbereich 205, Franzius-Institute, University Hannover, Nienburger Straße 4, 30167 Hannover, Germany.

To clear up this inconsistencies in the application of theoretical approaches to irregular waves, a series of tests in regular waves was carried out. It was the primary aim of the test program to check the experimental validity of the Stokes wave theories for the calculation of wave kinematics. The investigations were concentrated on steep waves from deep water to nearly shallow water.

EXPERIMENTAL SET-UP

Test Facilities

The tests were carried out in the wave channel "Großer Wellenkanal" (GWK) in Hannover. The GWK is a joint central institution of both, the University of Hannover and the Technical University Braunschweig. The channel is 324m long, has a width of 5m and a water depth up to 5m. It is equipped with a mechanical wave maker, used in piston type mode for the described investigations.

Wave Generation

Regular waves generated by a sinusoidal motion of a piston type wave generator do not always result in a stable wave profile along the wave channel. An unstable wave profile results from superposed free waves with the frequencies of the bound higher harmonic components of the basic wave.

As the wave theories require that no free waves are present, an appropriate regular wave generation should be a pre-condition for an evidential check of the experimental validity of a wave theory. To avoid free wave contamination, modified higher order control signals were used as input for the wave generator. The modification was done empirically for each wave parameter, but there is a strong correlation to theoretical approaches given for 2nd order theory (e.g. Buhr Hansen et al. 1975).

Movable Instrument Carriage (MIC)

The near surface measurements were carried out using a surface following velocity probe. Fig. 1 shows a principle sketch of the Movable Instrument Carriage (MIC).

The vertical movement of the MIC, and therewith the position of the velocity probe, is controlled by the signal of a near by wave gauge (1), located at the wave channel wall. A potentiometer transducer, fixed at the gear box of the motor, gives the actual position of the MIC (3), and therewith the position of the velocity probe, during the tests. Time series of the surface elevation (1), the orbital velocities (4) and the position of the velocity probe (3) were sampled simultaneously in a cross section about 110m in front of the wave paddle.

For the velocity measurements an inductive type Colnbrook sensor (Electro-Magnetic-Current-Meter) was used. The velocity probe was fixed to the movable part of the MIC. The submerged depth of the probe below the wave surface was y/d=-0.05 during all tests, where y *depth below wave surface*, in contrast to z *depth below still water level.*

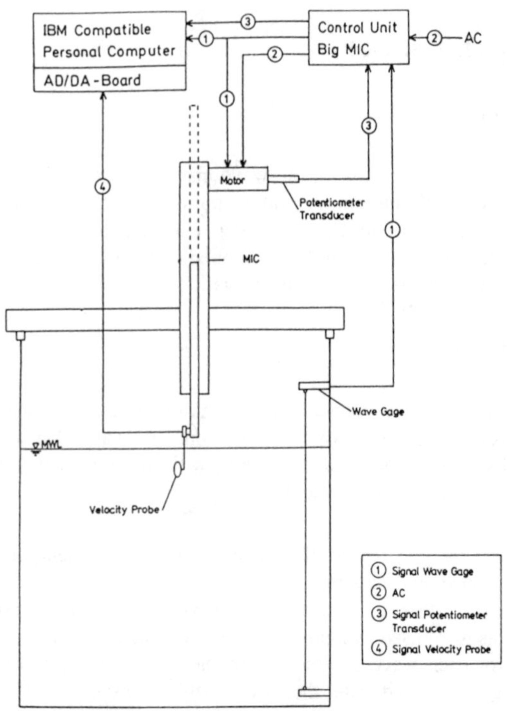

Figure 1. Principle sketch of the surface following device (MIC)

Fig. 2 shows time series of the surface elevation, the probe position and the difference of both for one example of measurements in a shallow water wave. Looking at the difference curve of surface elevation and probe position it can be seen, that due to the not perfect control characteristics of the electronic and mechanic equipment, the position of the velocity probe was not exact constant below the moving surface.

However, the comparison of measurement and theory is not influenced by this, because the actual measured time series of the probe location was used as input for the respective formulae of velocity calculation.

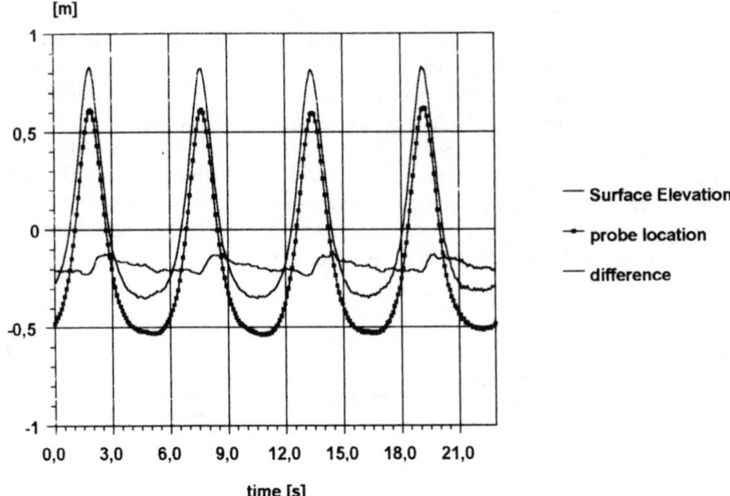

Figure 2. Time series of surface elevation, probe position and the difference of both (T = 5,76s, H = 1,20m, d = 3,00m, d/L = 0.10, y/d = -0.05)

THEORY

All theoretical results compared with the measurements are based on Stokes 1st, 2nd or 3rd order theory in different ways of interpretation:

- an EULERian approach,
- a co-ordinate stretching method, similar to Wheeler-Stretching (Wheeler 1969),
- a LAGRANGEian interpretation of Stokes wave theories.

The considered approaches will be explained in the following for linear wave theory exemplary.

EULERian Approach

First the Stokes theories were applied in the traditional EULERian way. In an EULERian calculation the horizontal orbital velocity results from the derivation of the potential function Φ at a given level z below the still water level (SWL).

$$u_{EULER}(z,\theta) = \frac{\partial \phi}{\partial x} = \frac{gHT}{2L} \cdot \frac{\cosh 2\pi(z+d)/L}{\sinh 2\pi d/L} \cdot \cos\theta \qquad (1)$$

where u_{EULER} = horizontal orbital velocity; L = wave length; T = wave period; H = wave height; d = water depth; θ = phase angel; g = gravity acceleration constant; z = immersed depth of velocity probe with respect to SWL; x = co-ordinate of length;

Solving this equation for a constant level of z below SWL, the magnitude of the velocity below the wave crest is equal to the magnitude of the velocity below the wave trough. Assuming waves of finite height, linear wave theory used in EULERian way already includes a wave induced mass transport velocity. The EULERian mass transport profile over depth, analysed as the time mean over a wave period or as the dc-value of a Fourier analysis, is shown in Fig. 3a. The integral of the EULERian mass transport profile is equal to the integral of the LAGRANGEian mass transport profile (Fig. 3b), which gives the real movement of the water particles. The LAGRANGEian mass transport profile, correct to Stokes 2nd order theory, can be calculated using formulae of orbital paths as given by linear wave theory.

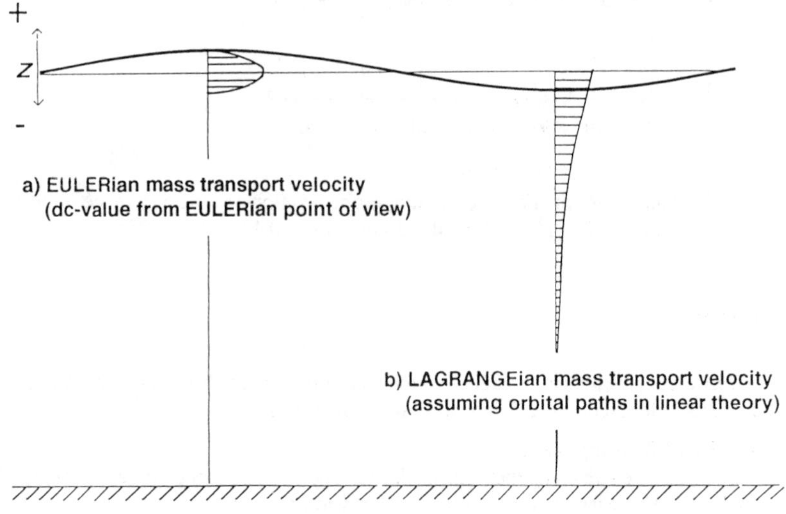

Figure 3. EULERian (a) and LAGRANGEian (b) mass transport profile (Linear Wave Theory)

Wheeler Stretching

Secondly the results of a stretching method, similar to the one that Wheeler introduced for linear wave theory (Wheeler 1969) were compared to the measurements. Wheeler introduced a stretching factor with respect to the probe position relative to the surface.

$$Stretching\ Factor = S' = \frac{d}{d + \eta(x,t)} \qquad (2)$$

where $\eta(x,t)$ = surface elevation; t = co-ordinate of time;

The value of the stretched horizontal orbital velocity is given by equation (3) for linear wave theory.

$$u_{WHEELER}(z,\theta) = \frac{\partial\phi}{\partial x} = \frac{gHT}{2L} \cdot \frac{\cosh 2\pi(S'(z+d)/L)}{\sinh 2\pi d/L} \cdot \cos\theta \qquad (3)$$

Calculating the horizontal orbital velocity at the free surface, the magnitude of the maximum velocity at wave crest is equal to the magnitude of the minimum velocity at wave trough. For the velocity at the wave crest, the velocity is calculated at SWL with EULERian approach, and then stretched to the free surface, vice versa for the velocity at wave trough. It can be stated, concerning the extreme values, that the co-ordinate stretching factor compensates for the mass transport velocity. This should be one contribution to the often mentioned underprediction of crest velocities by co-ordinate stretching methods. For the here shown comparison of results from the stretching method with measurements, the mass transport velocity according Stokes 2nd order theory was added to the stretched time series.

LAGRANGEian Approach

As the results from both methods were finally not conclusive, a LAGRANGEian approach of Stokes wave theories was introduced. Fig. 4 shows a principal sketch of the LAGRANGEian approach.

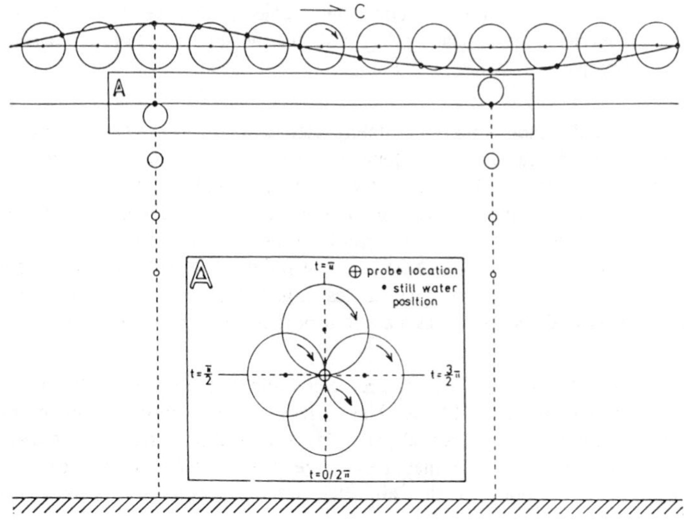

Figure 4. Principal sketch of LAGRANGEian approach

The orbital velocity in a LAGRANGEian sense is the velocity of a water particle along its orbital path. The time dependent location of a water particle is to be calculated from the respective formulae of the Stokes theories, and therewith a function of the still water position (x_0, z_0) of respective particles passing the velocity probe during a wave period. Equations (4) and (5) are the formulae for the half axis of the orbital paths given by linear wave theory.

$$\zeta(x,t) = -\frac{H}{2}\cdot\frac{\cosh 2\pi(z_0+d)/L}{\sinh 2\pi d/L}\cdot\sin 2\pi\left(\frac{x_0}{L}-\frac{t_i}{T}\right) \tag{4}$$

$$\eta(x,t) = \frac{H}{2}\cdot\frac{\sinh 2\pi(z_0+d)/L}{\sinh 2\pi d/L}\cdot\cos 2\pi\left(\frac{x_0}{L}-\frac{t_i}{T}\right) \tag{5}$$

where $\zeta(x,t)$ = horizontal half axis; $\eta(x,t)$ = vertical half axis.

The orbital velocity is the time derivation of the closed orbital path as described by Equations (4) and (5) for linear wave theory and is equivalent to the EULERian velocities in x_0, z_0. Using this method for wave kinematics calculation, the velocity below wave crest corresponds to the smaller orbit below the velocity probe, the velocity below wave trough relates to the wider orbit above the velocity probe, vice versa for the time steps beyond crest and trough (Fig. 4)

By Fourier analysis of this theoretical time series, a negative dc-value and higher harmonic components are appearing. Fig. 5 shows the results of Fourier analysis for a theoretical example.

The negative dc-value is not to be interpreted as a real mass transport velocity or drift velocity, as the calculation is done in the first step, by definition, with closed orbits. The magnitude of the LAGRANGEian dc-value, calculated as described above using closed orbits of linear wave theory, is equal to the Stokes drift velocity according to 2nd order theory. For mass conservation the wave induced mass transport has to be added to the LAGRANGEian time-series. Only by this the equation of continuity is satisfied. This again gives a clear argument, that linear wave theory used in EULERian way, already includes mass transport correct to 2nd order.

The LAGRANGEian approach for deep water and with linear wave theory is equal to the wave theory of GERSTNER. The surface elevation results from the particle movement on the most upper orbital paths. For the mentioned deep water wave case with linear movement on circular paths, the LAGRANGEian surface elevation contains all higher harmonics as they are calculated e.g. according to Stokes 5th order theory.

A

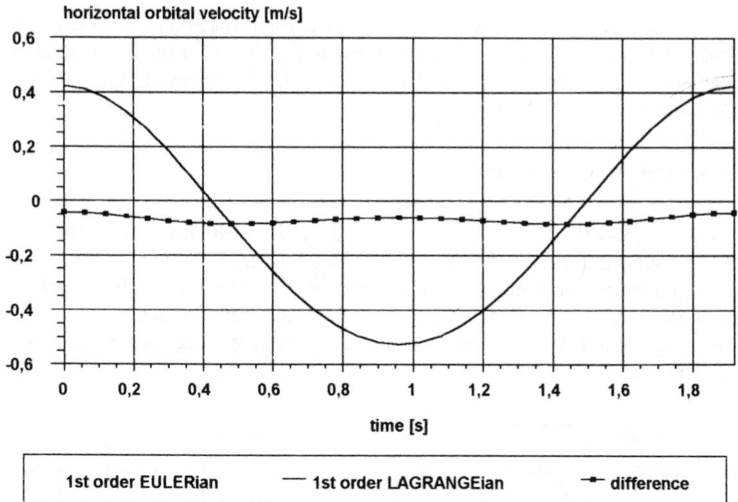

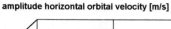

B

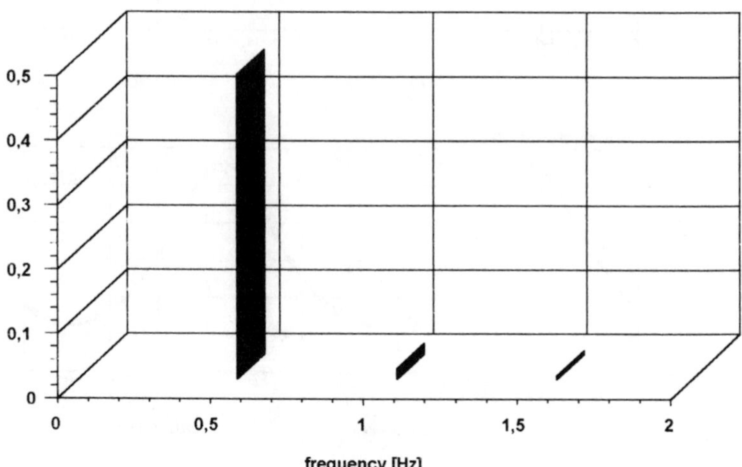

Figure 5. Theoretical example for LAGRANGEian approach with linear wave theory; A: time domain, B: frequency domain (T = 1,92s, H = 0,30m, d=1,00m, z= -0,20m)

PRESENTATION AND DISCUSSION OF RESULTS

The analysis of all data was done in a time window without reflected waves. In principle the measured wave parameter were used as input for the formulae of velocity calculation. The backflow, generally attributed to the closed channel system, was taken into account in all cases.

Horizontal Orbital Velocity

Fig. 6 shows the results of 1st and 2nd order EULERian approach for a wave in intermediate water depth with d/L = 0.21. As mentioned before the measurement was taken in a relative submerged depth of the velocity probe y/d = -0.05, which is 15cm below the moving surface in this case. Deviations are mainly appearing in the crest and trough region, although in general the agreement between measurement and theories is not too bad.

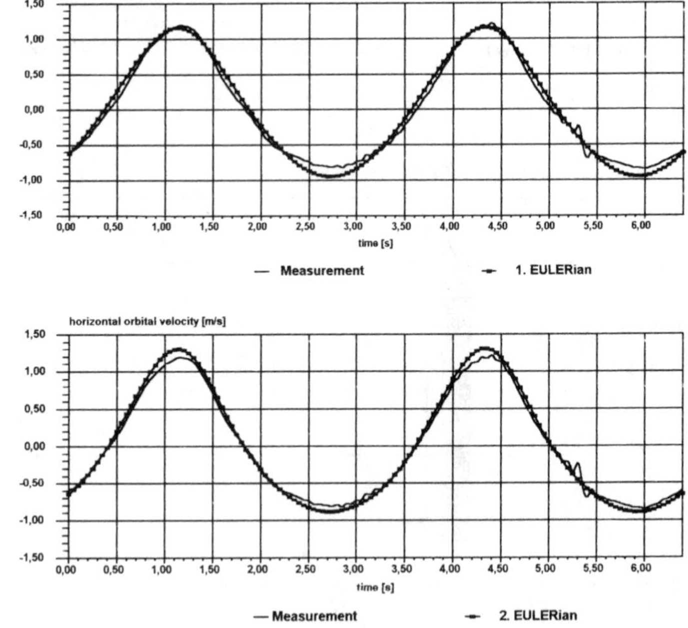

Figure 6. Measurement and EULERian approach
(T = 3,20s; H = 1,00m; d = 3,00m; d/L = 0,21; y/d = -0,05)

Co-ordinate stretching (Fig. 7) shows a good correspondence of crest and trough velocities with the measured velocity extremes. Strong deviations can be seen in the zero-crossing area. Concerning the co-ordinate stretching, the mass transport velocity according to Stokes 2nd order theory was added to the theoretical time series. Only by this good results can be achieved. Fig. 8 gives an estimation of the error in velocity calculation with the stretching technique, if mass transport velocity and backflow are not taken into account.

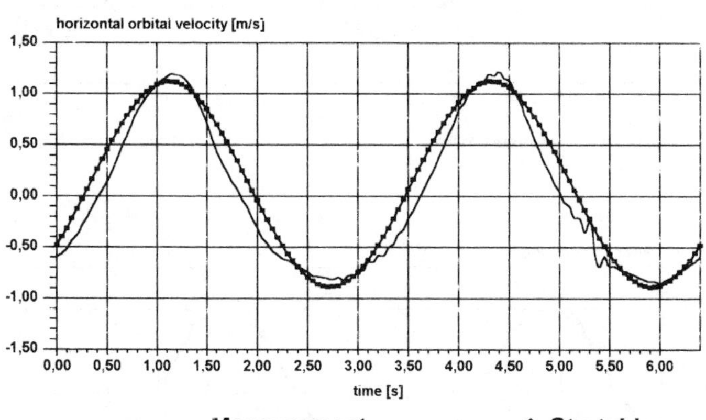

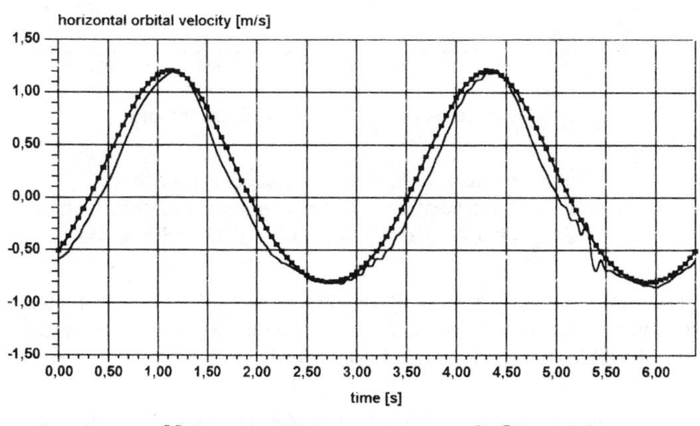

Figure 7. Measurement and Stretching approach
(T = 3,20s; H = 1,00m; d = 3,00m; d/L = 0,21; y/d = -0,05)

A

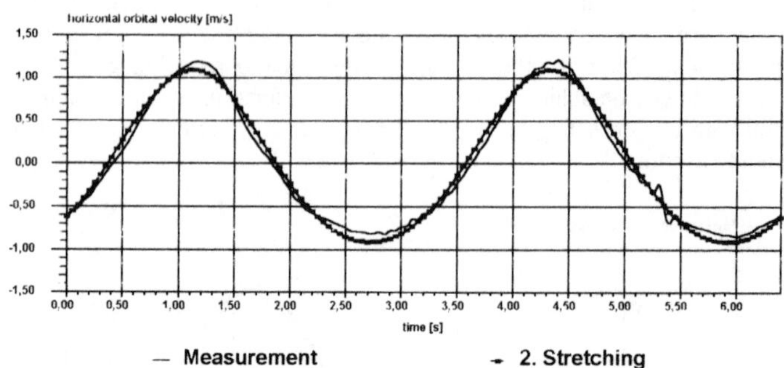

B

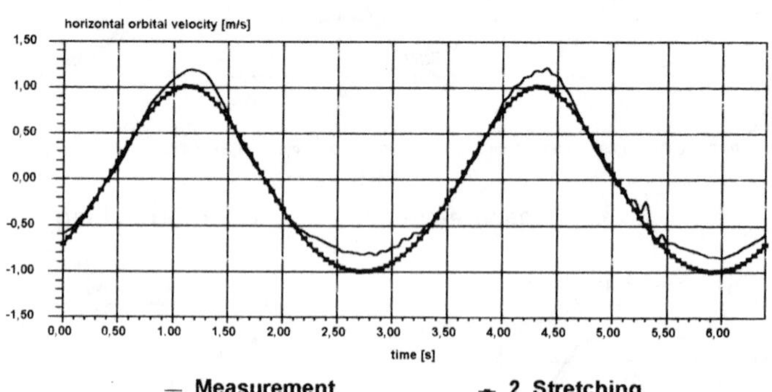

Figure 8. Measurement and Stretching approach
A: without mass transport and backflow | B: without mass transport, with backflow
(T = 3,20s; H = 1,00m; d = 3,00m; d/L = 0,21; y/d = -0,05)

Fig. 9 shows results of the LAGRANGEian approach. A much better overall agreement can be seen, especially for 2nd order theory. However, the results of 1st order LAGRANGEian approach are also not too bad.

The tendency of good results with LAGRANGEian approach holds over a wide range of wave parameter. Fig. 10 shows the results with various 3rd order approaches for a wave in more shallow water with d/L=0.10. Looking at the trough region of the calculated time series, however, the need for a higher order theory is obvious. A consequence of a higher order theory would be higher velocities under the crest, too. That would improve the results of LAGRANGEian approach and worsen the results of EULERian approach. Stretching shows similar deviations as in the example before.

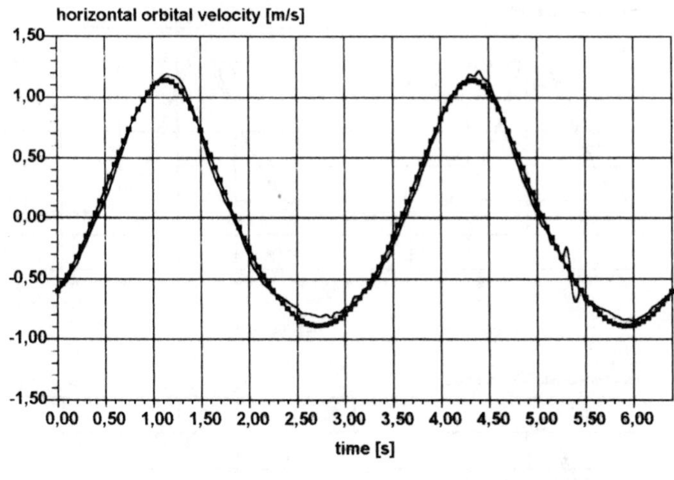

horizontal orbital velocity [m/s]
1,50
1,00
0,50
0,00
-0,50
-1,00
-1,50
0,00 0,50 1,00 1,50 2,00 2,50 3,00 3,50 4,00 4,50 5,00 5,50 6,00
time [s]
— Measurement
2. LAGRANGEian

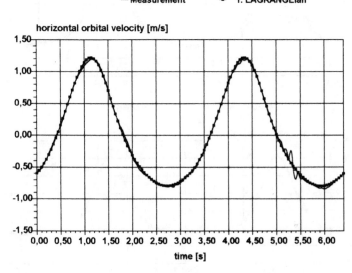

Figure 9. Measurement and LAGRANGEian approach
(T = 3,20s; H = 1,00m; d = 3,00m; d/L = 0,21; y/d = -0,05)

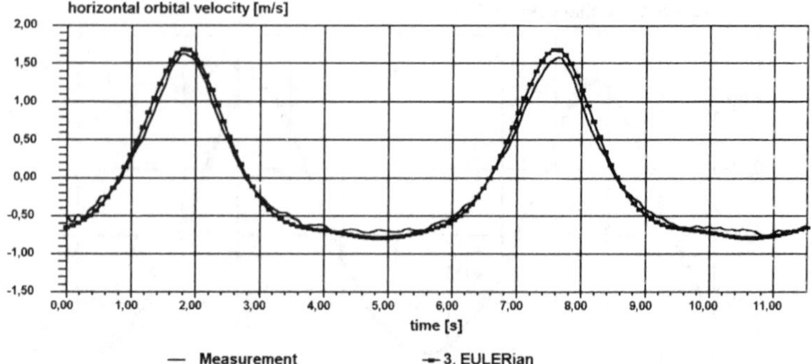

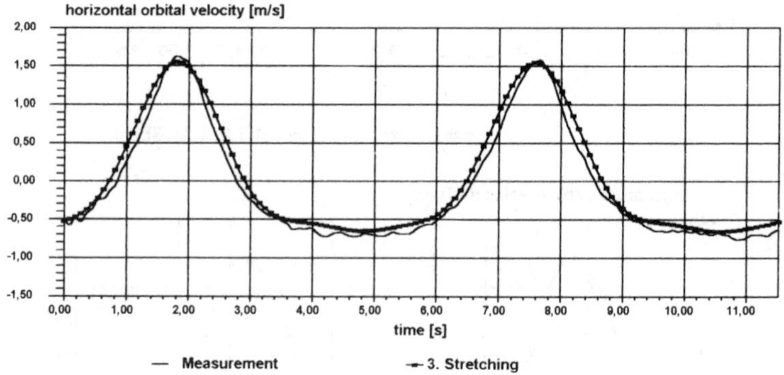

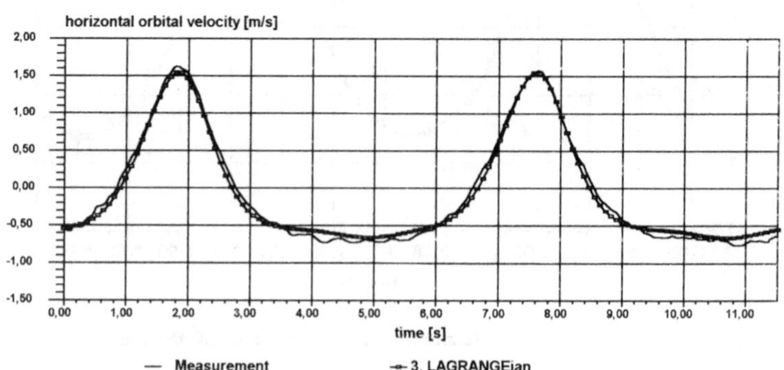

Figure 10. Measurement and comparison with
EULERian approach, Stretching approach, LAGRANGEian approach
(T = 5,76s; H = 1,20m; d = 3,00m; d/L = 0,10; y/d = -0,05)

Mass Transport Velocity

With the LAGRANGEian approach the orbital velocities are calculated with respect to the still water position of the water particles. Due to this a transformation of the measured dc-values to the real mass transport velocity is possible. Fig. 11 shows measured dc-values and theoretical mass transport profiles plotted as depth distribution. The depth of the probe is normalised by the water depth *d*.

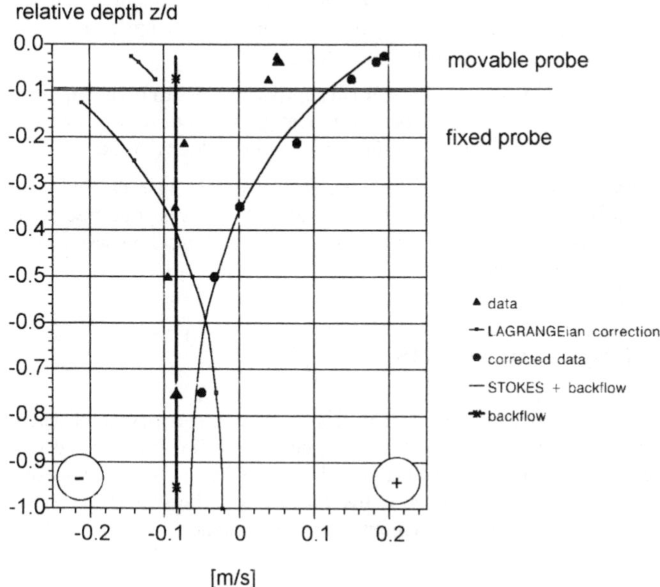

Figure 11. Mass transport velocity with LAGRANGEian approach (T = 3,20s; H = 1,20m; d = 4,00m; d/L = 0,27)

The triangles above the horizontal line are the dc-values as seen by a surface following velocity probe. Below the horizontal line, measurement and calculation is done in the sense of a fixed velocity probe. For the measurements with fixed probes a fair agreement with the constant over depth considered backflow profile can be seen. The measurements taken with a surface following velocity probe can be interpreted neither as backflow nor as drift velocity. Only by the LAGRANGEian approach a real mass transport velocity and backflow can be analysed from the data. The analysed data in LAGRANGEian sense give a conclusive profile of the drift velocity according to Stokes 2nd order theory, including a backflow for a zero-net mass transport in the closed wave channel.

CONCLUSIONS

Measurements of horizontal orbital velocities of regular waves are analysed based on different interpretations of Stokes wave theories. A LAGRANGEian approach is introduced by the authors. The analysis of the data result in the following conclusions:

- Stokes theories are a good basis for the calculation of wave kinematics from deep water to nearly shallow water.
- The introduced LAGRANGEian approach on the basis of Stokes 1st, 2nd or 3rd order theory gives good results concerning velocity components and also concerning real mass transport velocity.
- Calculation with LAGRANGEian approach allows to transform the measured dc-values (time mean velocities) to real mass transport velocities, especially for measurements done with a surface following velocity probe.
- Stretching techniques of the Wheeler type are not the right physical tool for the calculation of wave kinematics. The Stretching gives good results for the velocity extremes, however, mass transport velocity and backflow has to be taken into account necessarily.

ACKNOWLEDGEMENTS

The investigation described in this paper was carried out as part of the research programme of the Sonderforschungsbereich 205 (SFB 205) at the FRANZIUS-Institute for Hydraulic Research and Coastal Engineering, University of Hannover. The authors gratefully acknowledge the Deutsche Forschungsgemeinschaft (DFG) for their financial support throughout the investigations.

REFERENCES

Buhr Hansen J., and Schiolten, P., 1975. "Laboratory Generation of Waves of Constant Form", *Institute of Hydrodynamics and Hydraulic Engineering, Technical University of Denmark*, Series Paper 9

Daemrich, K.-F., Eggert, W.D., and Cordes H., 1982. "Investigations on Orbital Velocities and Pressures in Irregular Waves", *Proceedings 18th ICCE*, Cape Town, pp. 297-311.

Daemrich, K.-F., and Götschenberg, A., 1986. "Near-Surface Orbital Velocities in Irregular Waves," *Proceedings 20th ICCE*, Taipei, pp. 97-108.

Wheeler, J.D., 1969. "Method for Calculating Forces produced by Irregular Waves", *Offshore Technology Conference (OTC)*, Pre-prints 1970, Vol. 1, pp. I-71 / I-82.

Woltering, S., and Daemrich, K.-F., 1994. "Investigations on Regular Wave Kinematics in Wave Channels", to be published in : *Schriften des Vereins der Freunde und Förderer des GKSS-Forschungszentrums Geesthacht e.V.*.

ON SOME HYDRODYNAMIC EFFECTS FOUND IN LARGE SCALE EXPERIMENTS

Nikolai S. Speranski [1)]

ABSTRACT: Several hydrodynamical phenomena which have been found in field experiments are under the consideration. They are as follows: secondary waves-1 (decomposed waves), secondary waves-2, velocity equalization and anomaly dispersion of phase velocity.

INTRODUCTION

One of the field experiment possible purposes is discovery of new phenomena, which existence was not predicted by theory. Such investigations stimulate understanding of natural processes more detail as well as theory development. The aim of the report is to consider some nonlinear, dispersive and dissipative effects taking place in shoaling and breaking waves and stydied by auther in field experiments.

MEASUREMENTS

Measurements were conducted at the research pier "Shcorpilovcsi" of Institute of Oceanology of Bulgarian Academy of Science in 1979-1988. The facility is located 4 km far from the edge of 12-km straight sandy beach at

---

1) Ph.D., Senior Researcher of P.P.Shirshov Institute of Oceanology RAS, Krasicova str.23, Moscow 117 851, Russia.

the Black sea coast near Varna-city. The 250 meter length research pier reached 5 m of depth. The average value of the bottom slope was equaled to 0.02. Surface oscillations were measured by resistance type wave gauges. Inductive dynamical pressure gauges were used as velocimeters (Kuznetsov, Speranski, 1986).

RESULTS

The secondary waves-1 (decomposed waves)

Waves of this type appear when the main wave system passes over top of underwater bar and runs over relatively deep water in a trough between the bar top and beach (Fig.1). Height of the decomposed waves achieves

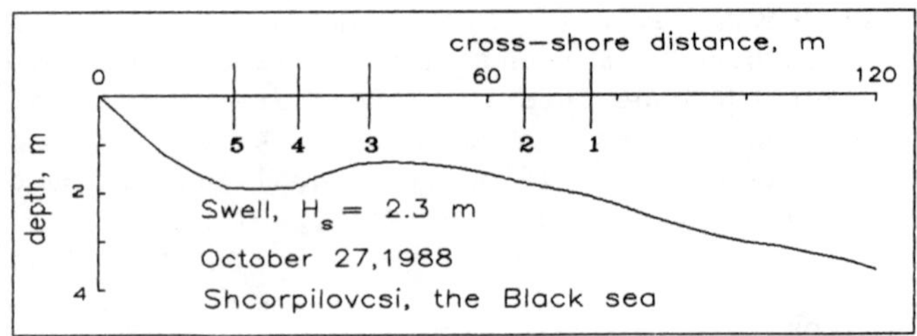

Fig.1. Cross-shore profile and wave gauges location.

40% and more of the main waves height (Longinov, 1963; Kuznetsov, Speranski, 1986; Beji et al., 1992). The necessary condition for the decomposed waves appearance is strong deformation of the main waves, i.e. existence of powerful high harmonics which couple with the primary and each other. In that case the phenomenon is observed when waves either break over bar or pass it without breaking.

The "secondary" and the "tertiary" waves were observed in the field conditions (Fig.2). It was found that the 2nd and the 3rd harmonics celerity are approximately equal to correspondence values of visible secondary and "tertiary" waves. Therefore the mechanism of decomposed waves formation may be as follows. Passed bar top area and running over trough, waves occur in

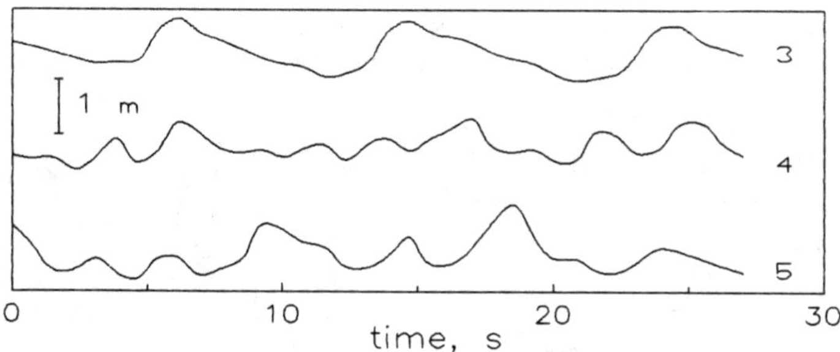

Fig.2. Decomposed waves at points 3,4,5 in Fig.1.

relatively strong dispersive media, where high harmonics move slower than the primary. As a result of their velocity lag the main waves decompose on the several visible waves-harmonics. Generation of the secondary waves-1 is an example of **nonlinear-dispersion phenomenon** in coastal zone.

The secondary waves-2

They are observed on some wave records were made in surf zone at plane slopping bottom. These waves are noticed well in laboratory conditions (Fig.3), where new

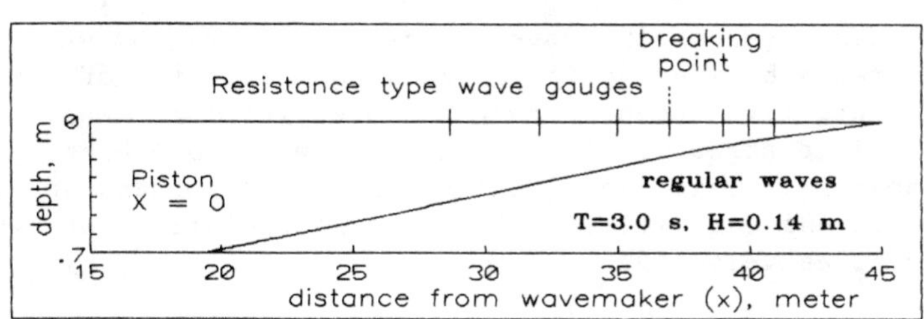

Fig.3. Scheme of measurements in wave flume of Qingdao Ocean University, China.

waves with period 0.6 s appear at troughs of main waves (Fig.4). The spectrum of these oscillations shows high frequency peaks to be grow intensively (Fig.5). These data suggest that the secondary waves-2 are "visible"

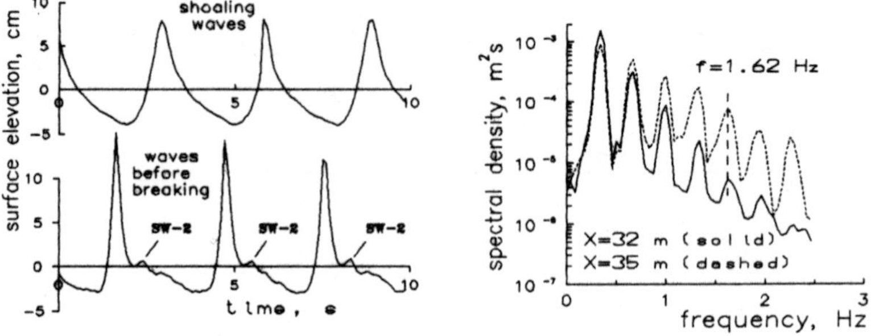

Fig.4. Surface oscillations in points X=32 m and X=35 m.
Fig.5. Spectra of surface oscillations.

high harmonics (in our case, the 5-th harmonic). Energy of this harmonic are getting more as a result of nonlinear interactions , and smoothness of wave surface is disturbed. Thus, new visible waves seems to be a result of **nonlinear** transfer of energy into high-frequency domain.

Velocity equalization phenomenon

Measurements in wave flume (ref. Fig.3) show that deference between positive amplitudes (crests of waves) and negative ones (troughs) are getting smaller in bore-like structure in comparison with breaking waves (Table 1). This "equalization effect" was found in field conditions as well, but only in relation of water particles velocity.

Table 1. Relation of positive (H^+) and negative (H^-) amplitudes in different structures

Coordinate of measuring point, m	32	35	37	39	40	41
H^+/H^-	2.2	5.2	3.4	2.8	3.2	3.2
Structure	Sh	Pl	B	B	B	B

Measurements of eulerian velocity horizontal component were made by inductive dynamical pressure gauges. Sensors were positioned in shoaling zone, in relatively narrow breaking zone and in bore-structure (Fig.6). The significant height of waves (swell) and the

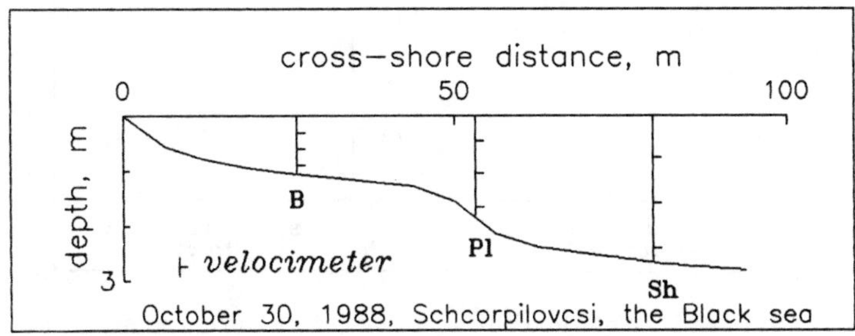

Fig.6. Location of velocimeters in shoaling (Sh) and plunging (Pl) waves and in bore-like structure (B).

primary harmonic frequency were equal to 1.46 m and 0.12 Hz correspondingly. Waves propagated in cross-shore direction. Each record was digitized to get series of 8500 points with Δt=0.1s.

Empirical distribution functions of absolute value of horizontal velocity $|u|$ were analysed. Calculated integral probability function $F(|u|/\langle|u|\rangle)$, where $\langle\rangle$ denotes a mean value, was approximated by Weibull distribution formulae

$$F=\exp(-a|u|/\langle|u|\rangle)^{b}.$$

Parameters **a** and **b** were found by the least squares method. Relations **a**(**n**) and **b**(**n**) were studied (**n** is a number of subdivisions), and it was showed that optimal **n** to be equal 50 (Speranski, 1989). Fig.7 demonstrates how the parameters vary, when waves dynamics changes. Value of **a**-parameter decreases from 0.95 in shoaling waves to 0.80 in bore-like structure. At the same time **b**-parameter increases from 1.5 to 1.9. Statistical errors of **a** and **b** calculations were found as 0.02 and 0.05 correspondingly. The variation of the parameters means that empirical function tends to uniform-distributed one. For example, the repetition of events when $|u|/\langle|u|\rangle = 2.5$ is 2.2 times less in bore-like structure than in shoaling waves (Fig.8).

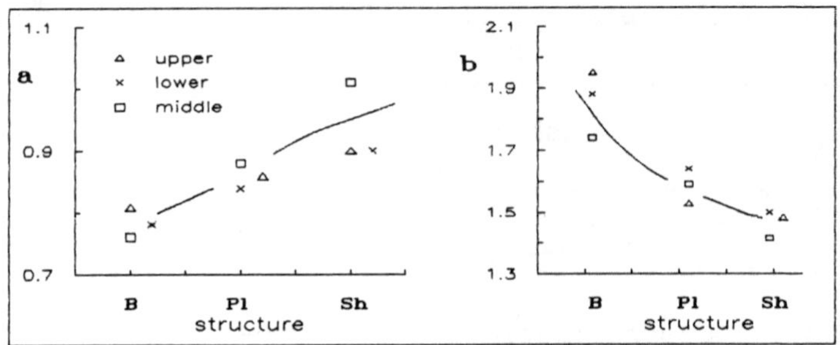

Fig.7. Variation of Weibull distribution parameters.

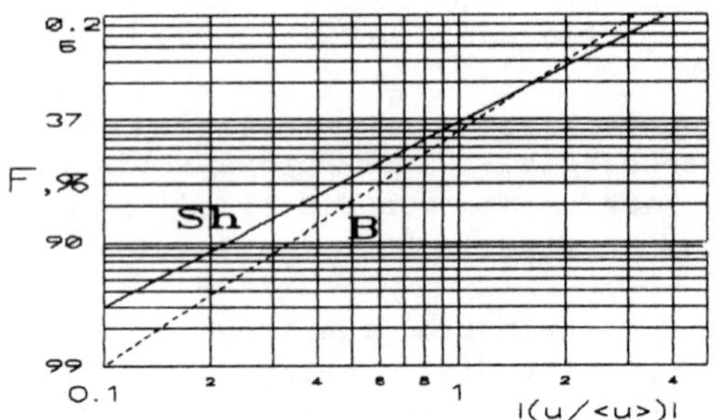

Fig.8. Measured integral distribution functions.

This set of measurements in conditions when all structures were expressed well, demonstrates velocity equalization. The similar analysis for the case, when structures were expressed not so clear (swell propagated over monotonic profile, demonstrates very week effect. Thus, the equalization effect is associated with continuous breaking of majority of waves, when they propagate in form of bore-like structure. This causes decrease of waves height and consequently decrease of the maximal velocity values under the crests. As a result the difference becomes smaller between the maximal velocities in crests and those in troughs. This phenomenon may be classified as a **dissipative effect**.

It is of interest to note as a collateral result of

It is of interest to note as a collateral result of the study that Weibull distribution function is a good approximation for empirical distributions of horizontal velocity absolute values. Its parameters depend on their own initial values and bottom profile.

Anomalous dispersion of phase velocity

This phenomenon was found in field measurements at surf zone (Buesching, 1978). This second order effect appears as high harmonic celerity increase when frequency grows. The only anomalous celerity spectrum was registered at surf zone in field experiment of Elgar and Guza (1985). However in the detail field measurements of Thornton and Guza (1982), conducted in depth range 7.2-1.1 m this effect was not observed.

Several times anomalous dispersion effect was revealed in measurements at the research pier Scorpilovcsi" at the outer part of surf zone (Kuznetcov, Speranski, 1990) The celerity spectra measured showed intensity of the effect (value of $[c(f_2)-c(f_1)]/c(f_1)$, where c is a celerity, f is a frequency), to be up to 20% but more often about 10% (Fig.9). The effect was observed

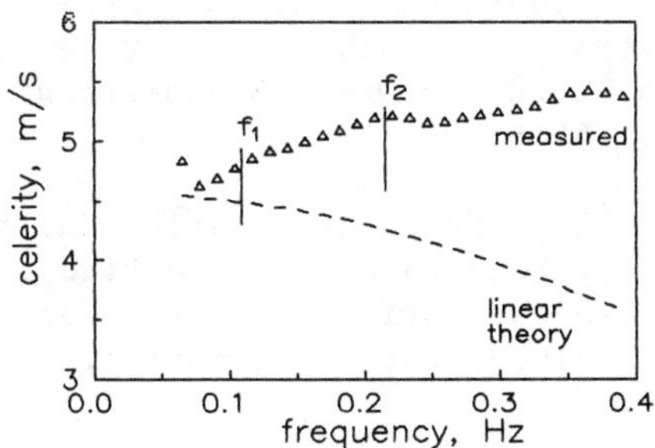

Fig.9. Celerity spectra between points 1 and 2 in Fig.1.

inside of outer part of surf zone. Measurements of author made in wave flume of Qingdao Ocean University, China confirmed the effect existence. The maximal intensity was found as 23%.

Thus anomalous dispersion effect is fixed reliably by

property of alternation. Moreover it has some paradoxical features. Really, accepted the principal possibility of celerity growing, we enter in contradiction with the general theory of waves. At the other hand, if celerity can not increase then observed effect may be connected only with phase change due to nonlinear interactions. But in that case propagation of harmonics is not adequate to their phases changes, and the concept of phase velocity has no meaning. Although the paradox is not solved, the second version seems more probable.

CONCLUSION

Mechanisms of secondary waves generation and velocity equalization were found. The mechanism of high-frequency anomaly dispersion is unknown, but the problem connected with this effect is formulated.

All considered effects have been found in meso-scale experiments as well. By this reason meso-scale measurements may be used for preliminary detecting of unknown phenomena in shoaling and breaking waves.

REFERENCES

Beji S.,Ohyama T.,Battjes J.A.,and Nadaoka K., 1992. "Transformation of nonbreaking waves over a bar", **Proceedings 23rd Coastal Engineering Conference**, ASCE, NY, pp.51-61.

Buesching, F.,1978. "Wave deformation due to decreasing water depth", Sonderd ruck aus Heft der Mitteilungen des Leichtweiss-Instituts fur Wesser bau der Technischen Universitat Braunschweig, 63: pp.168-217 (in German).

Elgar, S., Guza, R.T., 1985. " Shoaling gravity waves: comparisons between field observations, linear theory and a nonlinear model", **J. Fluid Mech.**., 161(2): pp.425-448.

Kuznetsov, S.Yu. and Speranski, N.S., 1986. "Dispersion properties of secondary waves in the surf zone", **Oceanology**, 26(3): pp.315-316.

Kuznetsov, S.Yu. and Speranski, N.S., 1986. " Dynamic characteristics of the VDK sensor as a velocity meter", **Oceanology**, 26(2): pp.248-253.

Longinov V.V., 1963. " Coastal zone dynamics at nontidal seas", Nauka, Moscow: pp 379.(in Russian).

Speranski, N.S.,1989. "Effect of orbital velocities equalization in surf zone",. In N.A.Aibulatov (Editor), Problems of coasts development. Nauka, Moscow: pp.22-26 (in Russian).

Thornton, E., Guza, R.T., 1982. "Energy saturation and phase speed measured on a natural beach", **J. Geophys. Res.**, C86(5): pp. 4149-4160.

VELOCITY MOMENTS ON A MACRO-TIDAL INTERMEDIATE BEACH

Yolanda L. M. Foote[1] and David A. Huntley[2]

ABSTRACT: Energetics-based models have been widely used for shoreline evolution prediction modelling, although verification of the suitability of such models is lacking. This paper examines cross-shore current and suspended sediment data, collected on a macro-tidal intermediate beach over a range of incident wave conditions, with the aim of estimating the usefulness of certain sand transport predictors. Overall, prediction for mean and long period contributions was good, whilst prediction of the incident wave contribution was generally poor, probably due to bedform effects.

INTRODUCTION

Sand suspension under shoaling waves is an integral feature of any nearshore sediment transport study, and the interaction between suspended sand concentrations and the nearshore velocity field has been extensively investigated (*e.g*. Downing *et al*., 1981; Jaffe *et al*., 1984; Huntley and Hanes, 1987; Russell, 1993). The development of optical backscatter sensors and electromagnetic current meters, for deployment on natural beaches, has enabled the reliable monitoring of suspended sediment concentrations and current velocities, respectively, and consequently the ability to test sediment transport models.

Energetics-based sediment transport models (*e.g*. Bowen, 1980; Bailard, 1981) developed from uni-directional, steady flow theory, generate expressions for the local cross- and long-shore transport rates which incorporate various moments of the velocity field, including the mean, variance and skewness. Accurate estimation of these transports requires a knowledge of the spatial distributions of the relative magnitudes of the relevant moments, which we call spatial 'shape functions' (Foote *et al*., 1994). In this paper, moment shape functions are compared with suspended sediment measurements in order to assess the use of moment-based predictions of sand transport rates, in the natural beach environment.

---

1) Researcher, Institute of Marine Studies, University of Plymouth, Drake Circus, PL4 8AA, UK.
2) Reader, Institute of Marine Studies, address as above.

FIELD SITE AND DATA COLLECTION

The data described in this paper were collected as part of the B-BAND (British Beach and Nearshore Dynamics group) field experiment at Spurn Head, UK (Russell *et al.*, 1991; Davidson *et al.*, 1992). Spurn Head is a sand spit some 5km in length, located on the Northeast English coast, between the North Sea and the Humber Estuary (Figure 1). Beach slope varied across the deployment site, from 0.023 at the inter-tidal terrace, comprising medium well sorted sands (D_{50} = 0.35mm), to a value of 0.0975 at the steeper high tide beach face, comprising mostly fine to medium gravel (Hoad, 1991; Davidson *et al.*, 1993). Figure 2 illustrates the beach profile at the Spurn Head field site.

The deployment comprised four sensor stations in a square array, positioned approximately 200m offshore of the high water level. Each sensor rig typically comprised, three optical backscatter sensors, three electromagnetic current meters and one pressure sensor. Data from one sensor station are examined here. Ten tidal cycles of data, in total, were collected during 16-25 April, 1991, but this paper focuses on two tidal cycles with wave conditions ranging from a 2m swell (18 April, Tide 184PM), just after a violent storm event, to a calmer regular groupy swell (23 April, Tide 234PM).

CROSS-SHORE SUSPENDED SAND TRANSPORT RATES

Bowen (1980) and Bailard (1981, 1987) have constructed a simple beach sediment transport model, based on Bagnold's (1966) **energetics** approach, which provides predictions based on time-averaged **moments** of the velocity field. Guza and Thornton (1985) used field measurements of velocities to show that the dominant moments within Bailard's theory (assuming that cross-shore currents dominate) are:

Bedload Transport

$$\psi_1 = \frac{\overline{u^3}}{\left(\overline{u^2}\right)^{3/2}} \tag{1}$$

Suspended Load Transport

$$\psi_2 = \frac{\left(\overline{|u|^3 u}\right)}{\left(\overline{u^2}\right)^2} \tag{2}$$

where u = cross-shore current velocity (ms^{-1}); | | = modulus and time-averaging is denoted by an overbar ($^{-}$).

The primary objective of the present study is to compare moment *predictions* with field *observations* and since purely suspended sediment concentration data were available only the suspended load transport term can be considered here. Examining the suspended load term (ψ_2) in more detail, it is first necessary to sub-divide the term in order to understand the relative contributions of the incident waves, long period surf beat and mean flows. However, there is no simple way to split ψ_2 into these different components and a simplifying assumption is needed. In this case, the we assume $u_s \gg \overline{u}, u_L$; this generates three further terms in ψ_2:

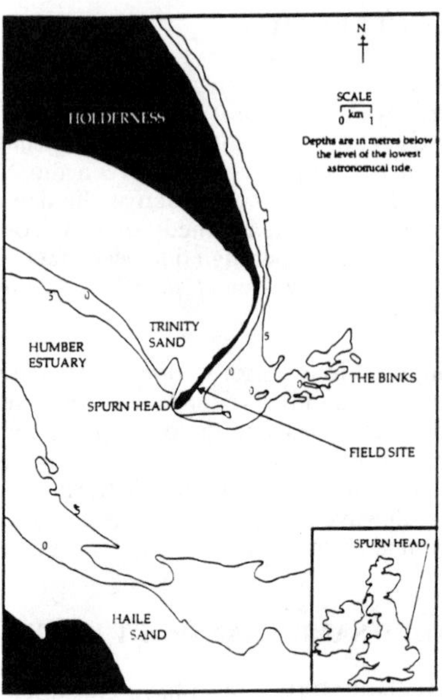

Figure 1. Location of the field site at Spurn Head

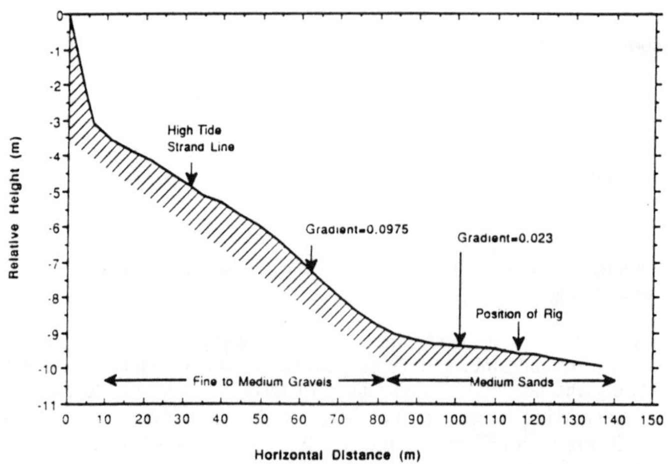

Figure 2. Beach profile at Spurn Head (23 April, 1991)

Short Waves $\overline{\left(u_s^{\,2}\right)^{3/2} u_s}$ (3)

Long Waves $\overline{\left(u_s^{\,2}\right)^{3/2} u_L}$ (4)

Mean Flow $\overline{\left(u_s^{\,2}\right)^{3/2}}\,\bar{u}$ (5)

where u_S = incident wave flows; u_L = long period flows; $\bar{u}$ = mean flows and u = cross-shore current velocity (ms^{-1}). Eqns. (3) to (5) show the un-normalised velocity terms. The cross-shore velocity time series were filtered with a 0.05Hz cut-off, to separate incident wave and long period wave flows. The mean flows are taken as the averages over the 17 minute runs.

Comparison of these three velocity sub-terms with corresponding suspended sediment flux term measurements is the next step. This is achieved using concurrent velocity (EMCM) and suspended sediment concentration (OBS) measurements, 10cm above the seabed, in order to calculate the local sediment flux:

$$\overline{uc} = \overline{u_S c_S} + \overline{u_L c_L} + \bar{u}\,\bar{c} \qquad (6)$$

Total = Short + Long + Mean

where c = suspended sediment concentration (gl^{-1}). Note that the predictors (Eqns. 3 to 5) relate to total, depth-integrated sand transport whereas the observed fluxes (Eqn. 6) are measured only at a single height. However, we make the simplifying assumption that the measured fluxes are representative of the total transport, based upon previous observations of suspended sediment profiles (*e.g.* Russell, 1993).

Mean Flow Contribution

Each of the three velocity terms and suspended sand flux terms were calculated for all data runs of the three tidal cycles. Figures 3 and 4 show mean flow contributions of the velocity <u>moment</u> vs. those of the <u>measured</u> sand flux. Figures 3 and 4 provide a comparison between predicted and observed values for Tide 184PM (18 April, 1991) and Tide 234PM (23 April, 1991), respectively. An arbitrary relative scale has been applied to the data in Figures 3 and 4 so that it is the <u>shapes</u> of the terms which are being compared. Generally there is very good agreement between moment and measurement, with offshore transport (negative values) inside the surf zone, and onshore transport (positive values) outside the surf zone in deeper water.

When plotted against mean water depth, the mean flow component of the suspended load term, and the mean flow component of the sand flux measurements, also agree very well. Figure 5, for example, shows data from Tide 234PM. Here the predictions can be seen to 'track' the behaviour of the sand flux measurements. The moment values and measured flux values have been plotted against each other in order to provide an estimate of the degree of correlation between the two parameters. In Figure 6 the regression fit is shown for the mean flow contribution. It is interesting to note, in Figure 6, the high correlation coefficient (R^2 = 0.87)

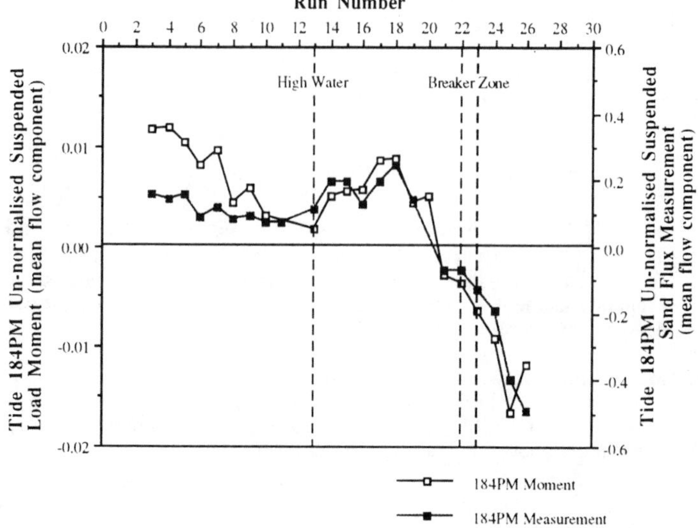

Figure 3. Moment & measurement transport vs run no. for mean flow (184PM)

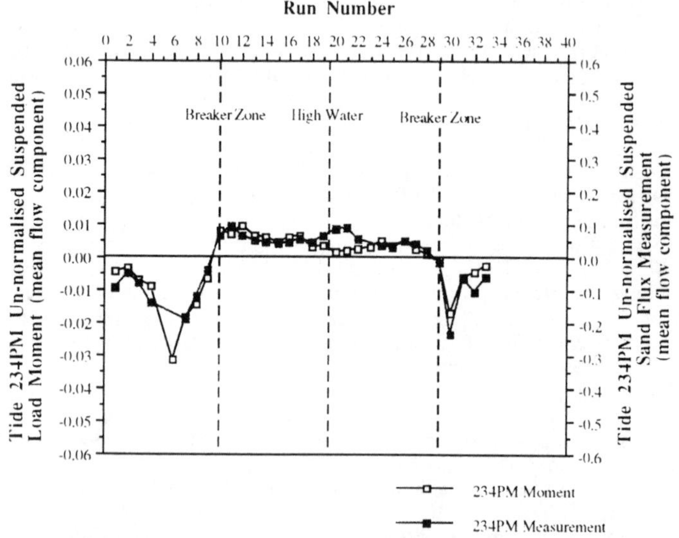

Figure 4. Moment & measurement transport vs run no. for mean flow (234PM)

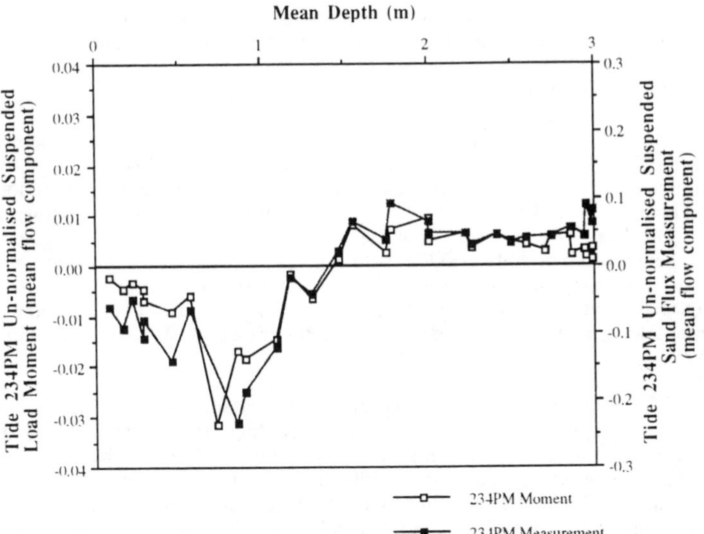

Figure 5. Moment & measurement transport vs mean depth for mean flow (234PM)

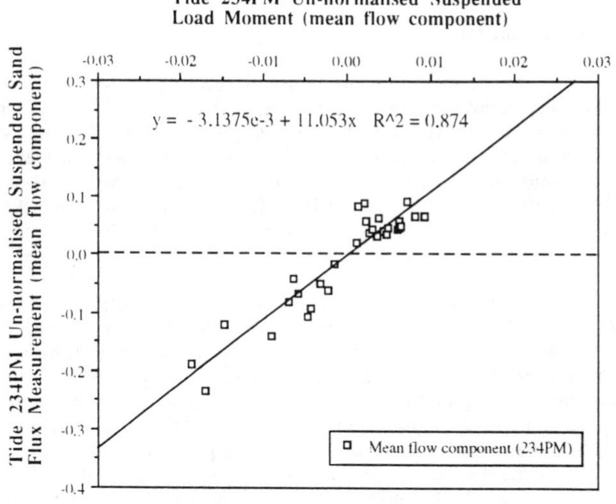

Figure 6. Moment vs measurement for mean flow contribution (234PM)

which suggests that the suspended load velocity moment term does have some predictive capability, at least for describing the mean flow component.

Long Wave Contribution

The long wave contribution (wave periods greater than twenty seconds) are compared in Figures 7 and 8. Figure 7 shows the variation of the long wave component for both velocity moment and sand flux measurement, with run number, for Tide 184PM (18 April, 1991). Figure 8 represents the same variables, but for Tide 234PM (23 April, 1991). For the two days of data chosen there appears to be, once again, a remarkably good fit between the velocity term predicted values and the observed suspended sand flux values. In contrast to the mean flow contribution, discussed previously, the longer period wave-induced flows are of a much smaller magnitude, tending to play a minor role in the transport of suspended sand on this beach.

Comparing the long wave contribution for both moment and measurement with mean depth creates more scatter in the data (Figure 9), but this is to be expected with a very small long wave contribution. Overall, there is reasonable agreement between the data, although the measurements do perhaps suggest rather more onshore flux in the shallow water than the predicted moments. This long wave driven transport shows slightly positive values (onshore direction), at the shoreline, progressing to negative values (offshore direction) in the deeper water offshore of the break line. The regression fit between the moment and measurement long period flow components, Figure 10, shows more scatter than the mean flow contribution and a lower regression value ($R^2 = 0.48$) but there is still a clear trend in the data.

Short Wave Contribution

Figures 11 and 12 show the variation with run number for the incident wave contribution to suspended load transport for Tides 184PM and 234PM, respectively. One of the most noticeable features in these two diagrams is a distinct asymmetry between flood and ebb tides with significantly higher suspended sediment transport rates during the ebb half of the tide. (The flood tide occurred around Runs 1-13 for Tide 184PM and Runs 1-19 for Tide 234PM; the ebb tide passed over the sensors approximately between Runs 14-26 for Tide 184PM and Runs 20-33 for Tide 234PM).

However, unlike the mean flow and long wave contributions, the predictions and observations of the short wave contributions do <u>not</u> agree well. The suspended load term is generally positive (*onshore* direction) for both Tide 184PM and 234PM, increasing linearly from the shoreline to a maximum at, or just beyond, the breaker line, then decreases slowly offshore. Conversely, the measured suspended sand flux values reveal large rates of *offshore* transport. These predominant offshore transport rates, driven by the incident wave motions, reach a peak during the ebb part of the tidal cycle with a considerable degree of variability over the earlier half of the tidal cycle (*i.e.* the flood tide), with sand fluxes both offshore and onshore in direction.

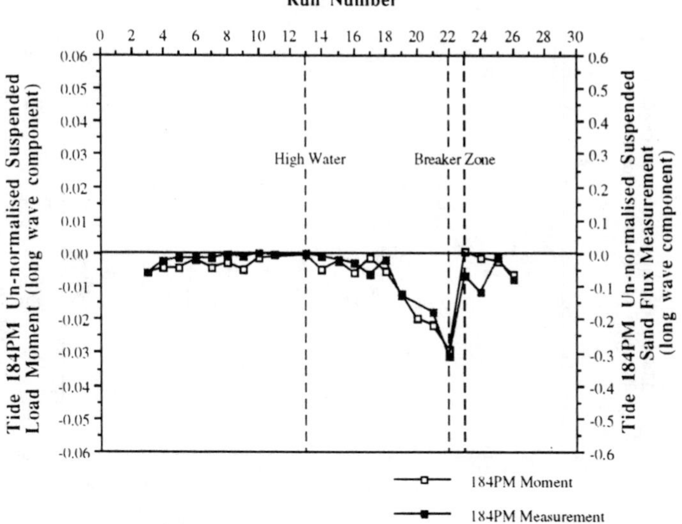

Figure 7. Moment & measurement transport vs run no., long wave flow (184PM)

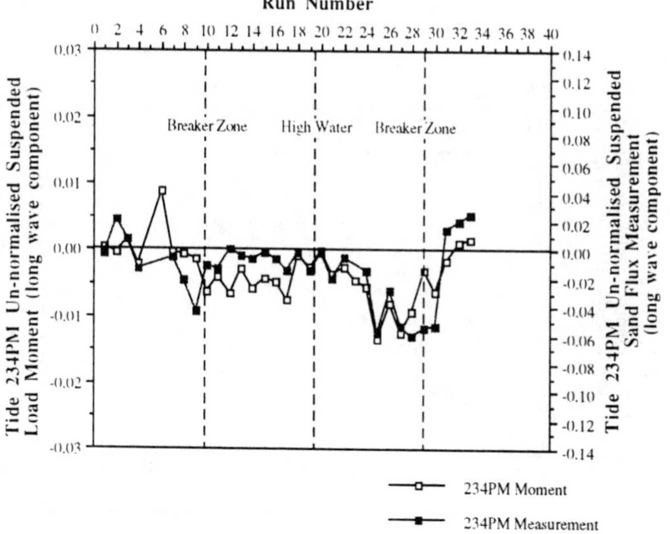

Figure 8. Moment & measurement transports vs run no., long wave flow (234PM)

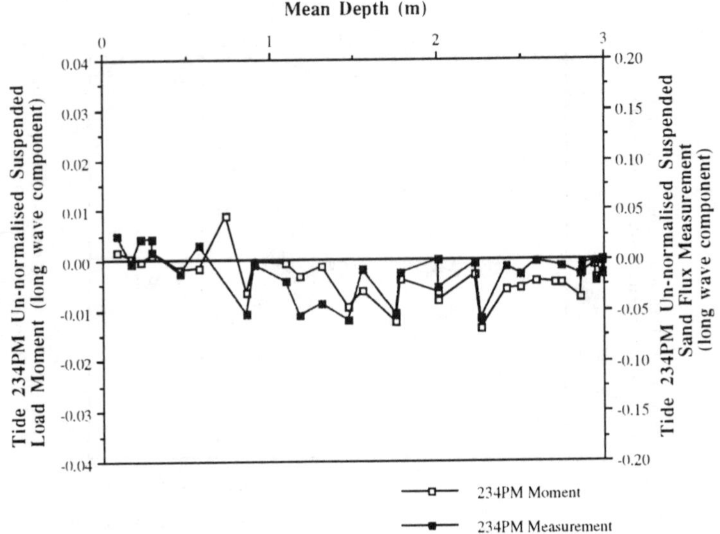

Figure 9. Moment & measurement transport vs mean depth for long wave (234PM)

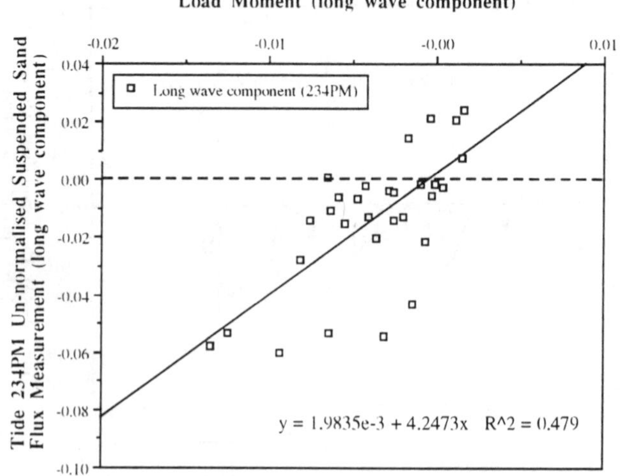

Figure 10. Moment vs measurement for long wave contribution (234PM)

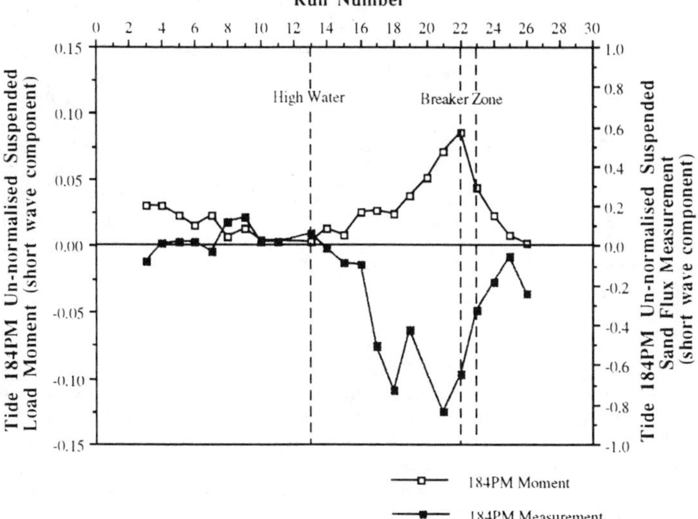

Figure 11. Moment & measurement transport vs run no., short wave flow (184PM)

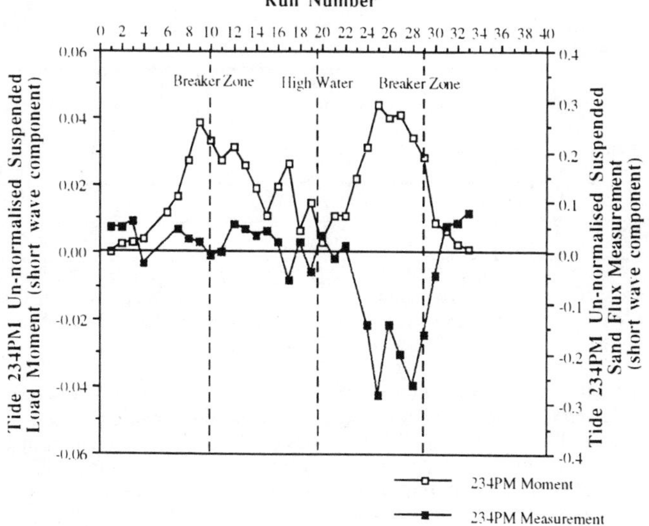

Figure 12. Moment & measurement transport vs run no., short wave flow (234PM)

Figures 13 and 14 show the separate behaviour of the moment predictions and the sand flux observations for the flood and ebb of Tide 234PM, respectively. The main difference seen here is the irregular distribution of data points and relatively small values of measured flux on the flood tide. On the ebb tide, however, the incident wave contribution is much greater and the distribution of data points is much more consistent. However, the behaviour of the moments is opposite to that of the sand flux measurements. Correlation between the moment predictions and the sand flux measurements for the separated flood, and ebb, is shown in Figures 15 and 16, respectively, for Tide 234PM. Examining this regression analysis emphasises the negligible degree of correlation between moment and measurement on the flood tide ($R^2 = 0.05$) and the very high degree of correlation between moment and measurement on the ebb tide ($R^2 = 0.89$).

DISCUSSION

A direct comparison between suspended sand flux terms and velocity moments provides a test of the capabilities of energetics-based velocity moments or spatial 'shape functions' as predictors of suspended sand transport. Linear regression fits between suspended sand flux and moment predictors, shown previously, for mean flow, long period and incident wave contributions include both correlation (R^2) values and regression fit slope values for the two contrasting tidal cycles. Table 1 contains the linear regression fit parameters for the compared flux measurements and moment predictions:

Table 1. Linear Regression Fits: Flux vs Moments

FREQUENCY BAND	DATE				
	18/4/91			23/4/91	
	R^2	S (slope)	Corrected S*	R^2	S
Mean	0.85	22.4	9.9	0.87	11.1
Low	0.76	9.2	4.1	0.48	4.3
Incident	0.53	-9.7	-4.3	0.45	-4.8
Incident - Flood	0.41	-4.2	-1.9	0.053	-0.64
Incident - Ebb	0.53	-7.9	-3.5	0.89	-7.3

*Corrected S = values adjusted to 10cm height above seabed, using Nielsen's (1988) diffusion model for concentration profile. For 18 April, 1991 (Tide 184PM) $h_b = 1.3$m; OBS sensor height = 6.5cm. For 23 April, 1991 (Tide 234PM) $h_b = 0.8$-1.0m; OBS sensor height = 10.0cm.

For the mean flow contribution, in Table 1, the correlation between suspended sand flux measurements and velocity moment predictions is very high for both Tides 184PM (0.85) and 234PM (0.87). The slope values from the linear regression fits, however, differ by approximately a factor of 2, from 22.4 (184PM) to 11.1 (234PM). It is likely that this difference in slope is the result of the different sensor heights on the two days, with the smaller height for 184PM resulting in a larger measured local flux. In fact it is possible to make an approximate adjustment to the slope by assuming a theoretical concentration profile. Using Nielsen's (1988) diffusion model for the concentration profile, the regression fit slope values, for Tide 184PM, have been adjusted to 10cm height and thus yields the corrected S

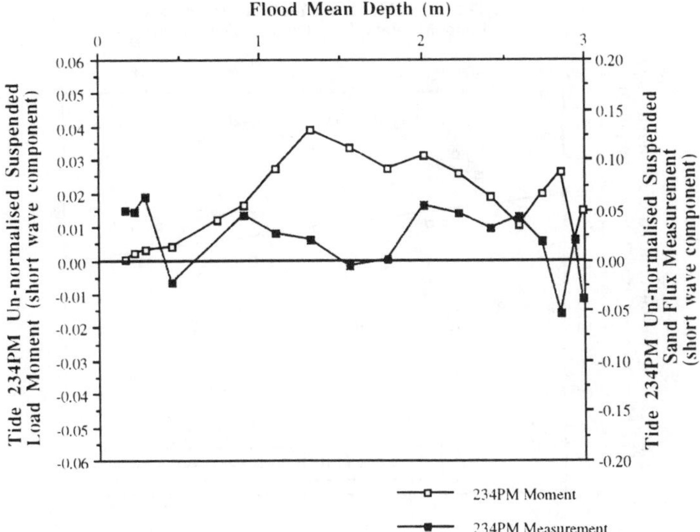

Figure 13. Moment, measurement, mean depth for short wave flow (Flood Tide)

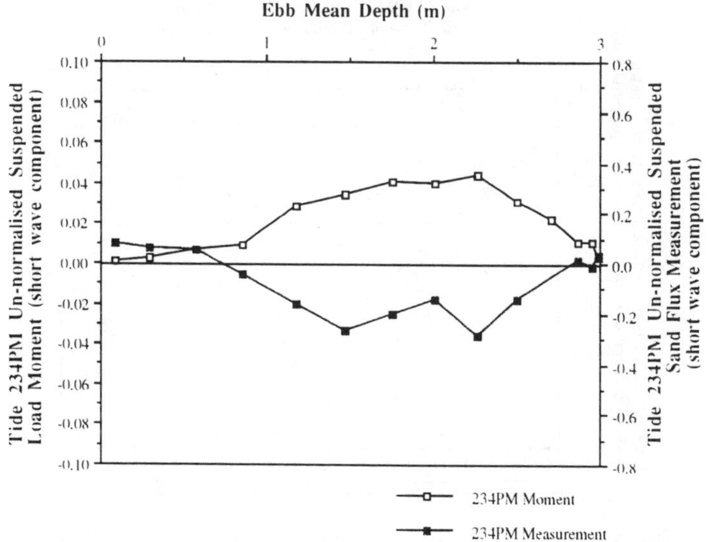

Figure 14. Moment, measurement, mean depth for short wave flow (Ebb Tide)

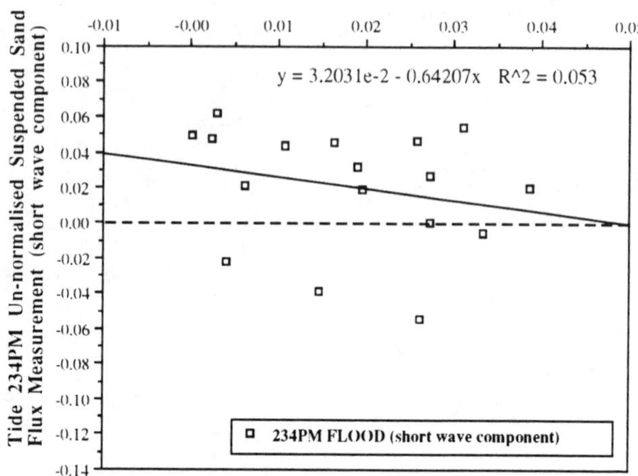

Figure 15. Moment vs measurement for short wave contribution (Flood Tide)

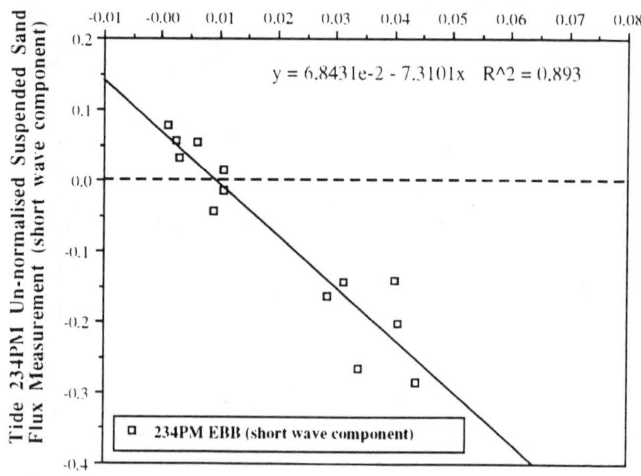

Figure 16. Moment vs measurement for short wave contribution (Ebb Tide)

values in Table 1. The "correction" to the standard height is necessarily approximate but suggests a high degree of consistency for the correlation slopes on these two days.

Flux/moment correlation values for the long period contribution are generally lower (R^2 = 0.76, Tide 184PM; R^2 = 0.48, Tide 234PM), but still significant and, once again, the corrected slopes agree very well from day to day. However, there is a factor of two difference between the mean flow and long period contributions which would suggest that the energetics model is not accurately predicting the relative weighting of these two modes of suspended sand transport.

Overall, the correlation between predicted and observed values, in Table 1, for the short wave component is fairly low (R^2 = 0.53, Tide 184PM and R^2 = 0.45, Tide 234PM) and the S values, as expected, have a negative *sign* (predicted and observed in opposite directions). The division of the incident wave terms into flood and ebb tides, in Table 1, shows a poor degree of correlation and erratic slope values for the flood, and significantly higher correlation values but dissimilar slope values (a factor of two difference) for the ebb. Nevertheless, slope magnitudes on the ebb are of similar values to those for the low frequency flows, prompting the wild suggestion that the energetics model might provide reasonable predictions under these conditions if the sign of the incident wave effect was reversed! This asymmetric behaviour between flood and ebb has been discussed by Davidson *et al.* (1993) and Foote *et al.* (1994). It is probable that the anomalous behaviour on the ebb is due to bedforms. The energetics model for the influence of incident waves clearly breaks down in the presence of bedforms, and this creates a significant limitation for its application to coastal prediction.

CONCLUSIONS

1. Correlation coefficients between fluxes and moments are generally very high.

2. Slopes of the correlation fits for separate days, with significantly different wave conditions, are remarkably similar.

3. Slopes for the mean flow driven transport are about twice those for long period related transport, implying that the energetics model does not wholly predict the relative importance of these modes.

4. For the incident wave transport, the sign of the slope is negative and, therefore, the predicted and observed transports are in opposite directions! This effect is compounded by the asymmetry of the sediment response between flood and ebb tides, which almost certainly were caused by bedforms.

5. Overall, prediction within a factor of two appears plausible for the mean and long period contributions (and for the short wave contributions if the sign of the predictions is reversed!).

6. Prediction of the dominant incident wave transport term is generally very poor, probably due to bedform effects.

REFERENCES

Bagnold, R. A., 1966. "An Approach to the Sediment Transport Problem from General Physics". *U. S. Geological Survey Professional Paper*, 422-I, 37pp.

Bailard, J. A., 1981. "An Energetics Total Load Sediment Transport Model for a Plane Sloping Beach". *Journal of Geophysical Research*, 86, 10, pp. 10938-10954.

Bailard, J. A., 1987. "Surfzone Wave Velocity Moments". *Proceedings of the Coastal Hydrodynamics Conference*, University of Delaware, pp. 328-342.

Bowen, A. J., 1980. "A Simple Model of Nearshore Sedimentation: Beach Profiles and Longshore Bars". In: *The Coastline of Canada*. S. B. McCann (editor), Geological Survey of Canada, Ottawa, pp. 1-11.

Davidson, M. A., Russell, P. E., Huntley, D. A., Hardisty, J. and Cramp, A., 1992. "An Overview of the British Beach And Nearshore Dynamics (B-BAND) Programme". *Proceedings of the 23rd International Conference on Coastal Engineering*, Venice, ASCE, NY, pp. 1987-2000.

Davidson, M. A., Russell, P. E., Huntley, D. A. and Hardisty, J., 1993. "Tidal Asymmetry in Suspended Sand Transport on a Macrotidal Intermediate Beach". *Marine Geology*, 110, pp. 333-353.

Downing, J. P., Sternberg, R. W. and Lister, C. R. B., 1981. "New Instrumentation for the Investigation of Sediment Suspension Processes in the Shallow Marine Environment". *Marine Geology*, 42, pp. 19-34.

Foote, Y., Huntley, D. A. and O'Hare, T., 1994. "Sand Transport on Macro-tidal Beaches". *Proceedings of the Euromech Conference*, Le Havre, (in press).

Guza, R. T. and Thornton, E. B., 1985. "Velocity Moments in Nearshore". *Journal of Waterway, Port, Coastal and Ocean Engineering*, 111 (2), pp. 235-256.

Hoad, J., 1991. "Monitoring of the Response of the Inter-Tidal Beach Profile to Tidal and Wave Forcing." *Proceedings of the Coastal Sediments '91 Conference*, Seattle, pp. 385-395.

Huntley, D. A. and Hanes, D. M., 1987. Direct Measurements of Suspended Sediment Transport. *Proceedings of the Coastal Sediments '87 Conference*, New Orleans, pp. 723-737.

Jaffe, B. E., Sternberg, R. W. and Sallenger, A. H., 1984. "The Role of Suspended Sediment in Shore-Normal Beach Profile Changes." *Proceedings of the 19th International Conference on Coastal Engineering*, Houston, ASCE, NY, pp. 1983-1996.

Nielsen, P., 1988. "Three Simple Models of Wave Sediment Transport". *Coastal Engineering*, 12, pp. 43-62.

Russell, P. E., 1993. "Mechanisms for Beach Erosion During Storms." *Continental Shelf Research*, 13 (11), pp. 1243-1265.

Russell, P., Davidson, M., Huntley, D., Cramp, A., Hardisty, J. and Lloyd, G., 1991. "The British Beach And Nearshore Dynamics (B-BAND) Programme". *Proceedings of the Coastal Sediments '91 Conference*, Seattle, pp. 371-383.

OBJECTIVES OF PROPOSED LARGE SCALE EXPERIMENTAL FACILITY (ESTEX)

T. M. Parchure [(1)], Member, ASCE, W. H. McAnally, Jr.[(2)], Member, ASCE, G. Nail [(3)]

Abstract: A large experimental facility will be built at the U. S. Army Engineer Waterways Experiment Station, (WES), Vicksburg, MS. It will consist of two main components. One will be a 152 m long, 1.8 m deep flume basin with a total width of 21.3 m. This wide basin will include one flume with a fixed 3 m width. The remaining 18.3 m width could be used either as one flume, 18.3 m wide, or could be divided into two flumes of varying widths. The flume will have the capability of producing both steady and unsteady flow. The other facility will be a towing tank, 152 m long, 4.6 m wide, and 1.8 m deep. It will have an electrically driven carriage over its full width. Some of the laboratory studies in the past are suspected to be vitiated due to small size, vertical distortion, side wall effects or low Reynolds Number resulting from the limitations of the available experimental facilities. These include arrested saline wedge studies, stability of density-induced interface under current and waves, deposition of sediment in stratified flows, precise calibration of current meters and sediment gages, diffusion of buoyant and non-buoyant jets, and vessel effects. The paper provides details of the large research facility to be built at WES. The facility will provide an opportunity for research workers to quantify the side wall and scale effects which have been an important issue in past hydraulic investigations, and will provide new experimental results that will advance understanding of important hydraulic processes.

---

(1) Research Hydraulic Engineer, U. S. Army Engineer Waterways Experiment Station, Vicksburg, MS, 39180, USA. (2) Chief, Estuaries Division, U. S. Army Engineer Waterways Experiment Station, Vicksburg, MS, 39180, USA. (3) Research Hydraulic Engineer, U. S. Army Engineer Waterways Experiment Station, Vicksburg, MS, 39180, USA.

WHAT IS ESTEX

ESTEX is experimental research and development facility that will be built at the U. S. Army Engineer Waterways Experiment Station, (WES), Vicksburg, MS. The facility will be used for generating unsteady, non-uniform flow. The large size of ESTEX will permit physical model investigations and laboratory experiments at scales large enough to minimize or even eliminate the scale effects and sidewall effects. ESTEX will also provide an opportunity for research workers to quantify the sidewall and scale effects which have been an important issue in past hydraulic investigations. New experimental results will advance the understanding of important hydraulic processes.

ESTEX will have two main components, a large shallow subdivided basin and a separate tow tank. The large basin will be a 152 m long, 1.8 m deep flume basin with a total width of 21.3 m. This basin will accommodate one flume with a fixed 3 m width. The remaining 18.3 m width could be used either as one flume, 18.3 m wide, or could be divided into two flumes of varying widths. Each flume will have the capability of producing both steady and unsteady flow. The towing tank, 152 m long, 4.6 m wide, and 1.8 m deep, will have an electrically driven carriage over its full width of 4.6 m. Details of the proposed facility are reported by Nail and McAnally (1993).

NEEDS OF EXPERIMENTAL HYDRAULICS

In addition to the facilities for conducting basic hydraulics research, large size physical models were traditionally utilized in the past for solving coastal engineering problems. For instance, the Forth Estuary model at University of Strathclyde, UK reproduced a reach of 35 miles at scales of 1:400 horizontal and 1:60 vertical. This model reproduced salinity distribution, fresh water flows and thermal releases from power stations. Over the recent years, one of the objectives of scientific research has been to develop generalized predictive capabilities which could be applied to specific problems, with some modifications and relevant input. Such capabilities are being developed for a variety of purposes from personal accounting to study of global atmospheric circulation. For example, numerical models have been developed for weather forecasting, industrial growth, population changes and so on for societal requirements. In the field of coastal engineering, pollutant transport, sediment movement in estuaries and equilibrium beach profiles are just a few illustrations of problems which utilize numerical models. Considerable advances have been made over the past years in hydrodynamic numerical modeling. Three dimensional visualization of flow field and pressure distribution in turbine impellers, around submerged objects and along airline wings is now done through numerical models. Spectacular advances in numerical modeling in every field of science and technology might create an impression that the days of physical modeling are over. It is probably true that new physical models of estuaries or bays covering reaches of several miles will seldom be constructed. The aspects such as flow pattern and salinity distribution, as well as their seasonal and spatial variations can now be visualized most conveniently and quickly on a computer screen using numerical models. Similarly, effects of parameters such as wind, air temperature and cloud cover can be simulated more easily on numerical thermal models than controlling these parameters under laboratory conditions for a physical model, which would be highly impractical.

An effective predictive model consists of two components; the first is understanding the physical processes involved and the second is mathematical formulation of these processes. All numerical models invariably include terms related to physical properties and processes. Assigning numerical values to these terms is not an easy task and requires deep insight into the physical, and sometimes chemical and biological processes which may be coupled. Assuming that a complete mathematical representation is possible, the upper limit of success of any numerical model is limited to the level of understanding of the physical processes.

Understanding of processes could be achieved through field studies or through a faithful simulation in a laboratory environment. Field measurements can be expensive, laborious and time consuming. Although an ideal microcosm, simulating every parameter involved, may not be economical and practical, the role of each predominant parameter can certainly be studied on physical models. For instance, the process of flocculation of fine sediment particles is influenced by several parameters including the scale of turbulence, particle size distribution, mineral composition, and chemical constituents in water (salinity, dissolved organic and inorganic substances). At the present time, laboratory experimentation is the only option available for understanding the nature and extent of influence of each one of the above parameters in the total and complex process of flocculation. Thus, hydrodynamic modeling, in some form, remains a viable technique for gaining insight into basic physical processes. The traditional role of physical models played in coastal engineering in the past has changed, however, the importance and necessity of experimental hydraulics continues to prevail. The new role consists of obtaining numerical values for mathematical terms which characterize physical properties and processes and this task is essential for providing representative input to the numerical models.

NEED FOR ESTEX

Some of the laboratory studies in the past had to be conducted on very small size or vertically distorted scale models, simply because the available experimental facilities did not have the required capability or their size was not adequate for conducting studies on geometrically similar models of preferred scale. Several laboratory investigations conducted in the past are suspected to have been vitiated by side wall effects or low Reynolds Number. It is necessary to determine the extent of any errors caused by scale effects in such studies. The limitations of existing facilities are restricted in terms of their length, breadth, depth, or flow velocity. These restrictions in turn limit the reliability of model results in making quantitative predictions and thus adversely affect the very purpose of conducting model studies.

An example of scale limitations may be found in the U.S.B.R. study using 1:72 scale, undistorted, density-stratified physical model of a thermal barrier for limiting the supply of warm water from a reservoir. It has been stated by Johnson and Vermeyen (1993) that "Unfortunately, an appropriate model scale, that allows inclusion of critical topography, yields flow depths and velocity that generate Reynolds numbers of 1500 to 3000. As a consequence, viscous effects reduce mixing intensity and make model evaluation of curtain performance qualitative." Such frustration can be avoided by making available a large size facility where proper scale models could be constructed. ESTEX will offer this facility where the Reynolds number could be as high as 1,000,000.

The results obtained in different laboratories are often device-specific which is a serious limitation on their universal applicability. For instance, the flow fields in a small rotating cylinder, a large rotating channel and in a long rectangular flume are not only different from each other, but they may also be far from the real conditions in nature. The maximum size of an eddy in these devices is limited by the available water depth. Since transfer of momentum depends on the size of eddies, the results of soil erosion obtained from these devices are likely to be different from the reality in nature. A larger size facility such as ESTEX can reproduce conditions closer to reality. It has been estimated that compared to the existing facilities, an order of magnitude increase is needed in width to depth ratio for eliminating sidewall effects. Hence ESTEX has been designed to give a width to depth ratios ranging upwards to 15:1.

POSSIBLE USES

Minimizing resuspension during dredging operations is essential not only for environmental considerations, but it could be the only way if dredging has to continue during varying physical conditions. This would permit dredging over a longer season and result in substantial cost savings. Large scale controlled tests will be carried out in ESTEX on the performance of different dredge heads and for the design improvements aimed at minimizing resuspension.

Studies on an arrested saline wedge and the stability of density-induced interface were conducted in the past with a relatively small (26,000) Densimetric Froude Number (F_D). This is significantly lower than that found in nature. Further investigations on the mechanics and control of saline wedges could be made in ESTEX with a high F_D on the order of 400,000. The capabilities of simulating stratified flows can also be used for i) study of deposition of suspended sediments under stratified flow conditions, ii) selective withdrawal through cooling water intakes of thermal and nuclear power stations and iii) dispersion of buoyant jets in semi-infinite depth-limited environment.

Optical back scatter (OBS) devices are reliable for field use only when calibrated using the same sediment which prevails at site. These will be calibrated in ESTEX.

In addition, depositional characteristics of fine sediments will be determined by using a series of OBS sensors over a vertical traverse. Different types of bed sediment gages and suspended sediment devices can also be calibrated by simulating relevant site conditions.

No facility in the USA is adequate for calibrating large field velocity instruments to the required accuracy. This work would be accommodated in ESTEX.

Success of wetland modeling depends on reliable input of site-specific friction coefficients. These could be evaluated experimentally by reproducing vegetation and benthic characteristics of site on a large scale.

Physical model requirements for navigational studies have been described in detail by Martin, (1993) and the need for large-scale, high Reynolds number experiments in stratified flow have been described by Jirka, (1993). Both these studies will be conducted to the desired scale in ESTEX.

ACKNOWLEDGMENT

Permission was granted by the Chief of Engineers to publish this information..

REFERENCES

Johnson P. L., and T. B. Vermeyen, A flexible curtain structure for control of vertical reservoir mixing generated by plunging inflows, Hydraulic Engineering, ASCE, Proc. of 1993 National Conference on Hydraulic Engineering and International Symposium on Engineering Hydrology, vol. 2, p.2371-76.

Jirka G. H., The need for large-scale, high Reynolds number experiments in stratified flow, Proc. First International Conference on Hydro-Science and Engineering, Ed. Sam S. Y. Wang, Center for Computational Hydro-Science and Engineering, Univ. of Mississippi, p. 1917-23.

Martin S. K. , 1993, Physical model requirements for navigation effects studies, Proc. First International Conference on Hydro-Science and Engineering, Ed. Sam S. Y. Wang, Center for Computational Hydro-Science and Engineering, University of Mississippi, p. 1924-30.

Nail G. H. and W. H. McAnally, 1993, ESTEX: A hyperflume for unsteady flow experimentation, Proc. First International Conference on Hydro-Science and Engineering, Ed. Sam S. Y. Wang, Center for Computational Hydro-Science and Engineering, University of Mississippi, p.1909-16.

ESTEX FLUME PLAN VIEW

(ALL DIMENSIONS IN FEET)

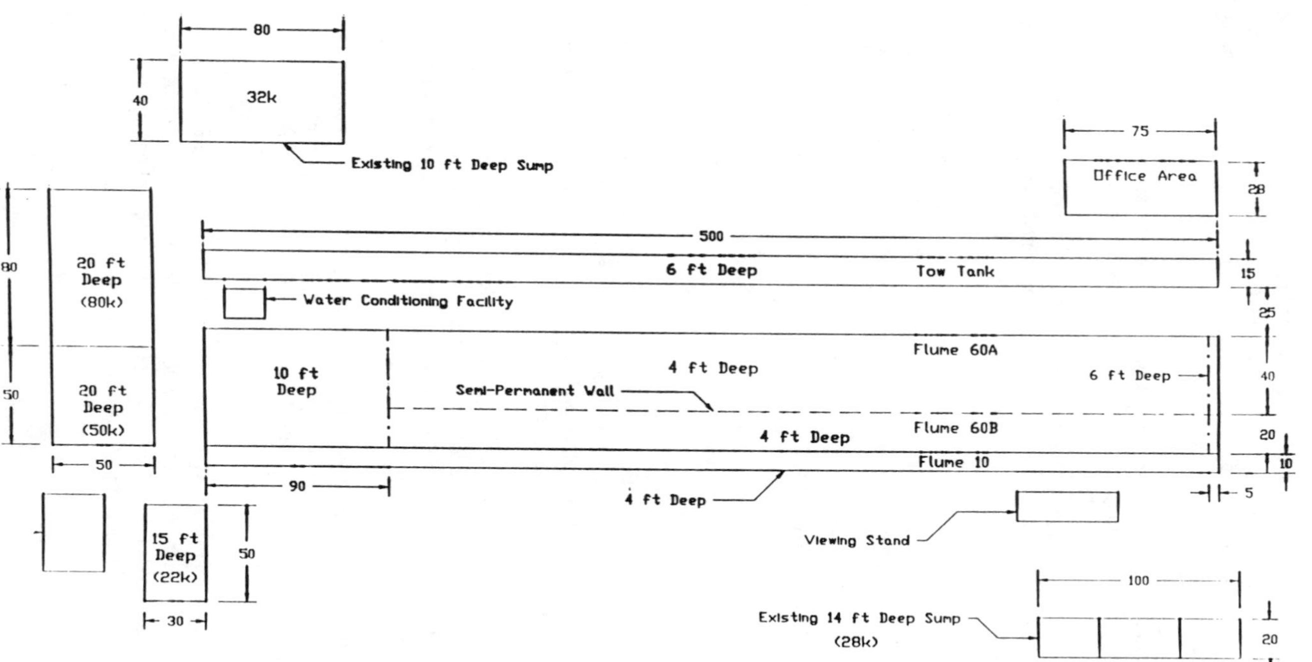

THE DUCK94 NEARSHORE FIELD EXPERIMENT

William. A. Birkemeier[1], M.ASCE and Edward B. Thornton[2], M.ASCE

ABSTRACT: Two ambitious nearshore field experiments are being planned in the United States to further fundamental understanding of nearshore processes with an emphasis on sediment transport. DUCK94, scheduled for August and October 1994, is a pilot experiment for the more ambitious and comprehensive SANDYDUCK in 1997. Both experiments will be hosted at the US Army Engineer, Waterway's Experiment Station's Field Research Facility located in Duck, North Carolina.

INTRODUCTION

A pair of nearshore experiments known as DUCK94 and SANDYDUCK is being hosted by the US Army Engineer Waterways Experiment Station's, Coastal Engineering Research Center at the Field Research Facility (FRF) located in Duck, NC. These experiments are jointly sponsored by the US Army Corps of Engineers, the Office of Naval Research, the Naval Research Laboratory, and the United States Geological Survey. The purpose of this paper is to summarize the scientific objectives and implementation of these experiments.

The overall objective of SANDYDUCK is to advance fundamental understanding of sediment transport in the nearshore zone through an integrated program of field measurements and numerical modeling. SANDYDUCK is scheduled for October 1997. DUCK94, scheduled for August and October 1994, is a preliminary effort to SANDYDUCK designed to give the investigators an opportunity to field test their theories and equipment.

---

[1] U.S. Army Engineer Waterways Experiment Station, Coastal Engineering Research Center, 3909 Halls Ferry Road, Vicksburg, MS, 39180-6199, USA.
[2]Department of Oceanography, Naval Postgraduate School, Monterey, CA, 93943-5100, USA

A long term goal of nearshore processes research is to predict the evolution of the bathymetry of a natural beach given the initial bathymetry, sediment characteristics, and the temporal variation of the wind, tide, and incident wave field. Such predictions are not presently possible because we do not understand the complex and interacting fluid and sediment processes, particularly small scale boundary layer processes and the three-dimensional circulation on complex bathymetry. Recent U.S. field experiments (NSTS, SUPERDUCK, DELILAH and others) focused on measurements of wave-induced flows in mid-water column over simple topography (either no sand bar or a predominantly linear bar). Models of sediment response remain primitive and empirical. SANDYDUCK will expand our knowledge of the nearshore by properly sampling processes on more complex, three-dimensional bathymetries. The three focus areas for SANDYDUCK are:

a. Small and medium scale sediment transport and morphology.
b. Wave shoaling, breaking, and nearshore circulation, to put small and meso-scale in context and to test modelling capabilities.
c. Swash processes and sediment motion, to test capabilities of hydrodynamic swash models.

The experimental approach is to measure sediment flux, bathymetry, and fluid flow at spatial scales ranging from a few cm to 100 m, and at temporal scales ranging from seconds to weeks. Models will relate changes in large-scale bathymetry to spatial gradients in time-averaged sediment flux. Topographic processes which may influence waves and currents span a range of scales including the formation and migration of ripples, mega-ripples, beach cusps, bars, channels and other morphologic features. Observations are designed to test existing large and small-scale sediment transport models and to provide information about processes (e.g., effects of mega-ripples on bottom stress and sand transport) for which there are as yet no models.

FIELD EXPERIMENTS

The field experiments will be conducted at the FRF, the site of several previous experiments. The site offers a variable wave climate and a beach that is known to have substantial bathymetric changes in response to fluid forcing. The FRF also provides unparalleled logistical support. Critical to the research capabilities of the FRF is the Coastal Research Amphibious Buggy or CRAB, a unique three-wheeled, 10-m-high vehicle used for accurate surveying of the nearshore zone and for instrument deployments. Position of the CRAB is determined with an auto-tracking survey system which tracks a light-reflecting prism mounted on the CRAB. Other FRF vehicles include a 10-m long LARC V amphibious vehicle, four-wheel drive forklift, inflatable boats, and other special purpose equipment useful for conducting nearshore research. The Duck site is an extremely well studied site with over 10 years of wave and large-scale morphologic observations, which greatly aids in the design of comprehensive experiments.

DESIGN CRITERIA AND LAYOUT FOR DUCK94

A necessary objective of the DUCK94 pilot experiment is to determine the relevant measurement space-time scales required to address the science objectives of SANDYDUCK. Using a continuous 12-day velocity data set from DELILAH at nine locations in the cross-shore, and using a Bowen/Bailard model of Bagnold's formulation of sediment transport as a framework to plan a sediment transport experiment, the following was suggested:

a. Approximately 10-15 m spacing between current meter locations in the cross-shore is required to properly describe the divergence of the sediment flux.
b. The influence of the tide on the hydrodynamic forcing (owing to changing water level) is O(1), suggesting that measurements need to be made continuously.
c. Mean, shortwave, and longwave velocity forcing were found to be of the same order within the surf zone.
d. Predicted transport was primarily due to suspended transport owing to the slow fall velocity of the dominantly fine sand within the surf zone.
e. Measurement of alongshore changes in sediment flux is important, particularly during times of three-dimensional morphology.

Based on the above considerations for spatial coverage in the cross-shore and the alongshore, and the available measurement locations proposed, a layout was designed (Figure 1). The attempt is to obtain measurements of the spatial inhomogeneity with four instrumented cross-shore arrays. The experiment plan is based on historical wave, weather, and morphologic data. In summary, the waves are generally from the south in August (longshore currents to the north) and variable in October with usually the largest waves from the north owing to Northeasters. The morphology tends to be three-dimensional with crescentic bars except during times of large waves when the inner bar is linear. Based on five years of statistics relative to FRF coordinates, the mean shorelines in August and October are located at 104 ± 6m and 110 ± 8m with the mean inner bar locations from the shoreline at 61 ± 22m and 82 ± 26m. The longshore influence of the FRF pier (located at 516m on Figure 1) on the shoreline in the experiment area varies from year to year, but is greatest during August, a result of the consistently southerly approaching waves. The influence is much less during October following the corrective action of waves from the northeast. The mean grain size during these months is a bimodal mixture of coarse and medium sand on the foreshore, fine in the surf zone (0.2 mm) and finer offshore.

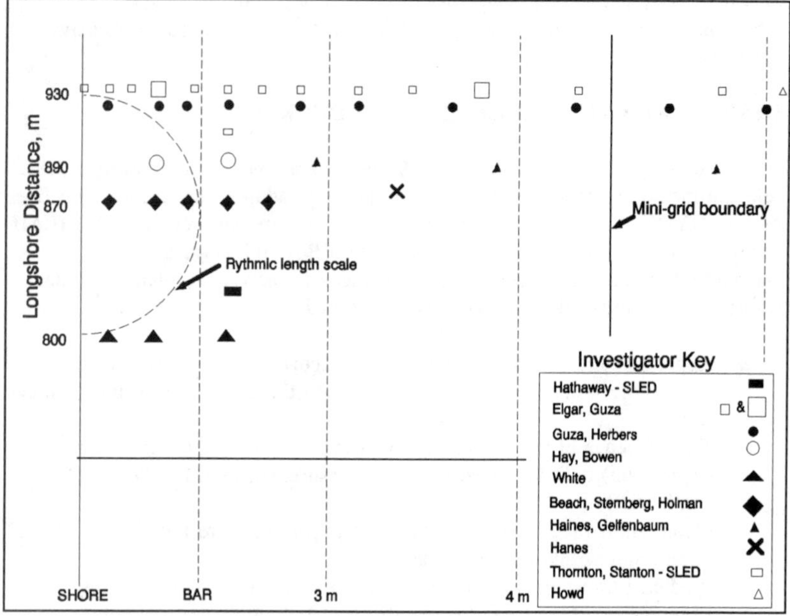

Figure 1. DUCK94 nearshore instrument layout. Open squares are combination sonic altimeter/pressure gauge/current meters. Solid circles are pressure sensors. Other symbols indicate vertical sediment flux (SUV), and pressure sensors.

INVESTIGATIONS

DUCK94 has evolved as a series of independently developed investigations which have been integrated into the larger experiment. The scientific themes for DUCK94 are described in the following.

Shoaling and transformation of the directional wave spectrum will be measured with two intensive arrays within the surf zone (see Figure 1) with a pressure sensor array extending all the way to the shelf edge in 200m of water. Three direction wave buoys will be used to measure the waves in deep water. A 16 element directional wave array is permanently operated in 8m depth. The combined array will be used to investigate the role of surface gravity waves in generating a broad spectrum of fluid motions on the continental shelf including infragravity waves and the importance of swell and low-frequency waves reflected from the beach.

The nearshore circulation will be measured with the combined arrays in the nearshore. Most of the measurement locations will measure velocities over the vertical to obtain initial information on three dimension flow over a large array synoptically. Considerable effort has been made to make all measurements synchronous and accurately timed. Two moveable sleds instrumented with vertical arrays of current meters and pressure sensor arrays will be located where most appropriate.

New instrumentation will be implemented in the field to measure near bed flows to include the boundary layer. Stacked arrays of close mounted electromagnetic current meters have been designed. New doppler acoustic velocity meters will be implemented. Stacks of hot film anemometers will be mounted very close to the bed. The FRF has the Sensor Insertion System (SIS), a large moveable crane which will be used to deploy instruments from the pier, across the surf zone and close to the bed.

Swash processes will be measured remotely using stereo video. Using this technique, timeseries of swash at many locations alongshore can be measured simultaneously along with the elevation of the bed (accurate to approximately 2 cm).

A primary effort is to measure sediment transport. Sediment flux will be measured in three cross-shore arrays using vertical stacks of optical backscatter sensors (OBS) and electromagnetic current meters. In addition, many of the arrays contain acoustic backscatter devises to measure sediment concentration over the vertical. A new doppler acoustic sediment flux meter will be used off a moveable sled.

The geology of the area will be examined using vibra-coring and surface samples collected several times during the experiment. Examination of the cores in combination with the sediment flux measurements will enhance the understanding of the overall processes occurring during the experiment.

The morphology will be measured both temporally and spatially over a range of scales from cm's to 100 m's. The CRAB will provide daily bathymetry over a 1 by 0.35 Km grid. Side-scan sonar and a sonic altimeter attached to the CRAB will measure the small scale morphology within this grid. An array of sonic altimeters will continuously measure changes in the bottom at a 2 Hz sample rate. Rotating side scan and sonic altimeters will be used at four locations to measure changes in morphology approximately every minute in an 8 m diameter circle.

Ten universities, four government agencies, and two private companies are involved in DUCK94 (Table 1). Very significant resources are required for this experiment. There will be an average of 70-100 participants at Duck during the October phase. There are plans to bring a total of 60 PC's and workstations

Table 1
DUCK94 Principal Investigators

Government Agencies

USACE Waterways Experiment Station: William Birkemeier, Kent Hathaway, Robert Jensen, Charles Long, Herman Miller, J. Bailey Smith, Donald Stauble, Thomas White
United States Geological Survey: John Dingler, Guy Gelfenbaum, John Haines
Naval Research Laboratory: Richard Bennett, Casey Church, Ming-Yang Su, Dennis Trizna, Charles Walker
Naval Postgraduate School: Thomas Herbers, Tom Lippmann, Timothy Stanton, Edward Thornton

Universities

Duke University: Peter Howd
Oregon State University: Reginald Beach, Todd Holland, Rob Holman
Scripps Institution of Oceanography: Thomas Drake, Robert Guza, Bradley Werner
University of Delaware: Ib Svendsen
University of Florida: Daniel Hanes
University of Miami: Hans Graber
University of Washington: Richard Sternberg
Washington State University: Steven Elgar
Dalhousie University: Anthony Bowen
Memorial University of Newfoundland: Alexander Hay

Private Companies

Arete Assocaiates: John Dugan
Neptune Sciences, Inc.: Marshall Earle

which will be networked together. Space will be provided for electronic installation and setup, data acquisition, and equipment staging, along with work space for the scientists. More than 100 pipes (to mount 95 electromagnetic current meters, 55 pressure gages, 68 OBS, and 33 sonar altimeters) will be jetted into the bottom. A large network of cables will connect the sensors to shore based data acquisition systems. Data collection will run August 8-19, and October 3 to 21, 1994.

SUMMARY

This paper has provided an overview of DUCK94 and SANDYDUCK, two nearshore experiments to be conducted in the United States in 1994 and 1997 respectively. These experiments seek to significantly improve our fundamental knowledge of nearshore sediment transport, hydrodynamics and bathymetric response through a combination of field measurements and numerical modeling.

ACKNOWLEDGEMENTS

DUCK94 and SANDYDUCK will take place because of the interest and encourgament of the program managers of the sponsoring agencies including Dr. Thomas Kinder from the Office of Naval Research, Dr. Asbury Sallenger from the United States Geological Survey, and Dr. Linwood Vincent and Ms. Carolyn Holmes of the US Army Engineer Waterways Experiment Station.

Permission to publish this paper was granted by the Chief of Engineers.

TIDE- AND STORMDRIVEN SEDIMENT TRANSPORT ON THE INNER-SHELF ALONG THE DUTCH COAST

Jan W.H. van de Meene[1] and Leo C. van Rijn[1,2]

ABSTRACT

Based on field observations, this paper discusses the hydrodynamic processes dominating the sediment mobility and sediment transport directions in the inner-shelf along the Dutch coast. Sediment transport observations during fair-weather conditions were obtained with a Total Load Sampler, that was operated from a ship. The sediment transport processes under storm conditions were analyzed using instantaneous velocity and suspended sediment concentration data obtained with a stand-alone platform. These data allow for the determination of the contributions of the mean, low- and high-frequency oscillatory sediment fluxes to the total flux (e.g. Wright et al., 1991).

INTRODUCTION

The near-bed water and sediment motion on the inner-shelf along the Dutch coast during fair-weather and storm conditions were studied using a Total Load Sampler and a stand-alone instrumented tripod. The objective of the study was to obtain insight in the processes that dominate the sediment mobility in the study area under different conditions. The field experiments discussed here form part of a larger research program that aims at establishing the formation and maintenance of a set of linear sand banks found on the inner-shelf along the central part of the Dutch coast. This program is carried out within the framework of the large Dutch multi-disciplinary research project 'Coastal Genesis'.

---

[1] Institute for Marine and Atmospheric Research Utrecht (IMAU), Netherlands Centre for Coastal Research (NCK), Utrecht University, P.O. Box 80.115, 3508 TC Utrecht, The Netherlands.

[2] Delft Hydraulics, P.O. Box 152, 8300 AD Emmeloord, The Netherlands.

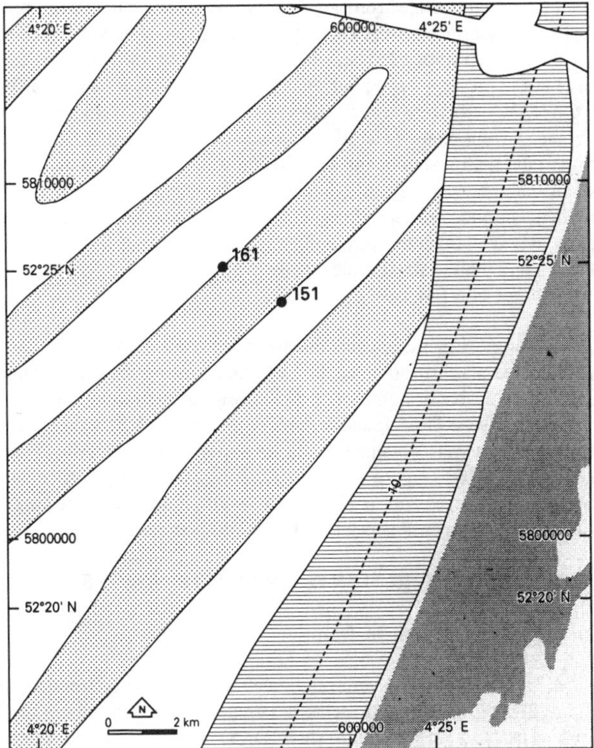

Figure 1 Location of the instrument deployments.

STUDY AREA

The inner-shelf along the central part of the Dutch coast is covered by a system of large linear sand banks (length 30 km, width 2-4 km, height 2-6 m), attached to the shoreface. The measurement sites were located near one of the ridges, in waterdepths between 14 and 16 m (Figure 1). The mean grain size of the sediments in the area is between 250 and 300 μm. The tidal range is between 1.5 and 2 m, giving near-bed tidal currents between 0.2 and 0.5 m/s. The mean annual wave height is 1.1 m, the wave height during typical winter storms (Bf. 8-10) is between 3 and 5 m. These conditions are typical for a mixed-energy shelf in a semi-enclosed sea. They indicate that sediment transport processes in the study area may be affected by both tidal currents and waves.

FIELD EXPERIMENTS

Sediment transport measurements during fair-weather conditions were done with a Total Load Sampler operated from a ship (Van Rijn and Gaweesh, 1992). This Total Load Sampler consists of a bedload trap and a set of intake nozzles, connected to pumps, to measure time-averaged suspended sediment concentrations. In addition, three current meters (two Ott's and one electromagnetic flow meter) were mounted on the Total Load Sampler. The measurements were carried out on the top of one of the shoreface-connected ridges, between locations 151 and 161 (Figure 1), on 14 and 15 August 1990. The water depth at the location varied between 13 and 15 m. Maximum near bottom currents (0.45 m above the bed) were around 0.45 m/s, while depth-averaged currents obtained with an Elmar current meter reached values up to 0.8 m/s. During the measurements, the weather was fair, with a significant wave height of 0.3 and 0.8 m and a wave period of 4 s. The small scale morphology superimposed on the large ridges was determined from echosoundings and video-observations. The small scale bedforms consisted of megaripples with an average length of 10 m and an average height of 0.2-0.3 m and, superimposed on these small dunes, small scale ripples with an estimated length of 0.2 m and a height in the order of a few cm.

Sediment transport measurements during storm conditions were done with a stand-alone instrumented platform, built at the Department of Physical Geography, Utrecht University. An extensive description of this measurement frame (the 'EMF-frame') is given by Van de Meene (1994), and is summarized in the next section. The measurements presented here were obtained during an experiment in December 1991, at deployment site MP 161 (Figure 1). During the measurement period a couple of storms passed by with wind speeds up to 20 m/s (Bf. 8-9), significant wave heights up to 4 m, peak wave periods between 8 and 9 s and near bed wave orbital velocities up to 1.2 m/s.

THE EMF-FRAME: INSTRUMENTS AND CALIBRATIONS

Description of the measurement frame

The EMF-frame is a tripod with relatively thin and massive iron legs spaced wide apart (Figure 2). It is deployed in a classic U-mooring fashion, with a marking buoy used for retrieval. The Data Acquisition System (DAS) consists of a data logger (Campbell, model CR10, with a SM716 storage module) and a final data storage facility (a single board computer with 40MB harddisk). The data logger has been programmed to sample data in a burst mode. The power supply for the instrumentation consists of two sets of dry batteries, one set for the sensors and a separate set for the data logger and computer.

Electromagnetic flow meters (EMF's) were used to measure the instantaneous current velocity. These EMF's are two-axis flow meters, with a spherical head with a diameter of 4.0 cm. They have been produced by Delft Hydraulics. Suspended sediment concentrations were measured with optical backscatter sensors (OBS's) produced by D&A instruments (Downing et al., 1981). A pressure sensor (Keller) was used to register pressure fluctuations due to surface gravity waves and tidal water level variations (sampling frequency at 2 Hz.). An echosounder (HE-720 TS Seaview ultrasonic ranger) was mounted on the frames to monitor bed level changes. Tiltmeters were used to assess the position of the measurement frame relative to the sea bed and to verify whether the frames are positioned (approximately) level at the seabed. A compass (Plessy) was used to get the orientation of the electromagnetic flow meters relative to the magnetic north and also to record movements in the tripod.

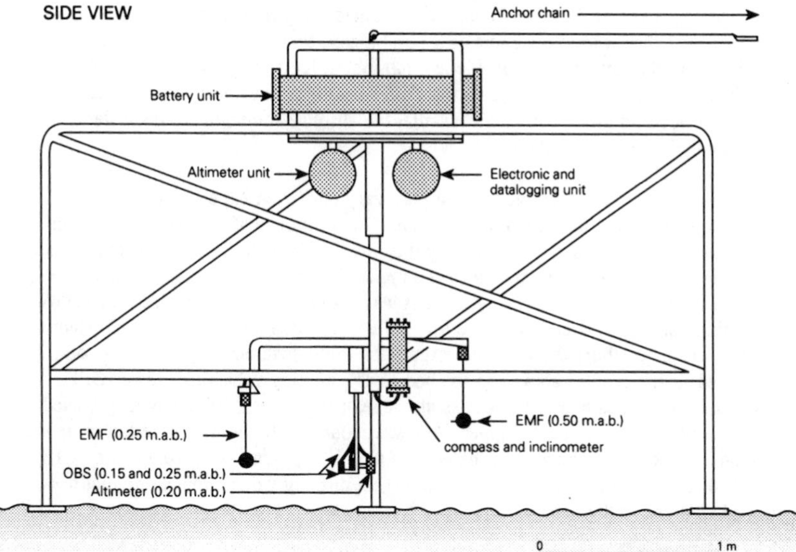

Figure 2 EMF-frame.

Calibration of validation of the EMF and OBS data

The electromagnetic flow meters were calibrated before and after the experiment in a towing tank at Delft Hydraulics. The calibrations were conducted for each axis (x and y) independently. The calibration curves are linear within the range of ± 2 m/s, with a high correlation coefficient ($R^2 > 0.999$). The variation in calibration curves for the sensors during different calibrations was negligible.

The OBS's were tested and calibrated in a circulation tank built at the Physical Geography Laboratory (Van de Meene, 1994). The sediments for these calibrations were obtained from the seabed at the deployment site and consisted of medium grained (d_{50} = 270 μm), well sorted sand. The calibrations were repeated three times. They show a linear trend over the entire range of concentrations (0-10 kg/m^3, linear fits with $R^2 >$ 0.999). The gain factors show little variation (coefficient of variation 3 to 6%), indicating that the calibration tank gives reproducible results.

In contrast with the gain factor of the OBS's, which can be determined accurately in the laboratory, the offset of the sensors during the measurements is difficult to establish. This offset consists of an electronic offset, inherent to the sensor, and a physical offset, caused by small background concentrations of mud, silt, or organic matter. It is not possible to distinguish between these two types of offset. Due to variations in background concentrations, the offset will vary in time, both in the calibration tank and in nature. The offset is ideally determined in the field, by in-situ calibrations. However, this was not possible in the present study, as the OBS's were mounted on a stand-alone frame. Alternatively, the total offset was estimated by the minimum reliable value encountered in each individual time-series. In each time-series, this minimum value was subtracted from all observations. Doing this, it was implicitly assumed that the suspension process is intermittent, and that at least once during a burst a zero sand concentration was observed. Figure 7 reveals that this is a reasonable assumption, at least in a sandy inner-shelf environment. The absolute minimum in the OBS-series is always very close to the average of all local minima. This indicates that the absolute minimum forms a stable estimate for the base-level of the record.

The height of the OBS's above the bed may vary during an experiment, due to settling of the frame and migration of bedforms. Due to failure of the echosounder during the experiment, the exact measurement height of the OBS's is not known. To overcome this problem, an estimate of the height of the upper OBS was obtained using the degree of burial in the sea bed of the lower sensor. It was assumed that the amount of overflow data (missing values) registered with the lower OBS provided a measure for the extent to which this sensor was buried. With this assumption three different height classes could be distinguished: i. lower OBS above the sea bed (0-1% missing values); ii. lower OBS approximately at the sea bed (1-99% missing values); iii. lower OBS completely buried (100% missing values). Since the upper OBS was mounted 10 cm above the lower, these three height classes imply heights of the upper OBS: i. >10 cm; ii. ≈ 10 cm; iii. <10 cm. In all three cases the upper OBS had almost no missing values, indicating that it was positioned well above the bed.

Only the data from the upper OBS in height class 2 were selected for further analysis. Data falling in this height class have the best defined measurement height (approximately 0.1 m above the bed). Data from height class 1 appeared less accurate and less interesting, since they were obtained during much quieter conditions. As a result of these relatively quiet conditions, small background concentrations of organic matter, mud or algae may obscure the observed sand concentrations. The data falling in the third class were considered inaccurate due to the unknown but small height above the bed. Of a total of 74 usable bursts, 20 fell in height class 2. These data are discussed in the present paper.

FAIR-WEATHER SEDIMENT TRANSPORT

Data analysis

The bedload transport rates were determined using:

$$q_b = \frac{e(G_s - G_o)}{bT} \tag{1}$$

where q_b = bedload transport rate (kg/ms); G_s = dry weight of the total sand catch obtained with the bedload trap (kg); G_o = dry weight of the sand catch related to the initial and scooping effect (kg); b = width of the sampler mouth (=0.096 m) (m); T = sampling period (s); e = efficiency factor (-). The efficiency factor is assumed to be unity (Van Rijn et al., 1991). The dry weight (G_o) of the sand catch related to the initial and scooping effects was estimated by series of measurements with a zero-sampling period.

The suspended load transport (q_s) is defined as (cf. Van Rijn et al., 1991):

$$q_s = \int_a^h u \cdot c \, dz \tag{2}$$

where q_s = suspended load (kg/ms); a = thickness of the bedload layer (a = 0.04 m); h = water depth (m); u = flow velocity at height z above the bed (m/s); c = sand concentration at height z above the bed (kg/m^3).

A distinction between bedload and suspended load was made on practical grounds: the sediment caught in the bedload sampler is defined as bedload, the rest as suspended load. Tests were done in a laboratory flume to determine thickness of the layer of sediment that was trapped in the bedload bag. This height (a) above the bed appeared to be 0.04 m for low transport rates (filling percentage of the bag on average between 0 and 25 percent). The bedload transport is then the transport of the particles in a layer with thickness a = 0.04 m, the suspended load transport the transport of particles above that level. The thickness of the bedload layer corresponds reasonably well with the height of the observed small scale ripples. This suggests that the observed bedload transport occurs in the form of migration of these small scale ripples.

The velocity and concentration profile data measured by the impellor meters and by the suspended load sampler were extrapolated down to the height of the bedload layer (a = 0.04 m), using the three different extrapolation methods given by Van Rijn (1993). These methods gave comparable results, indicating that the obtained estimates are sufficiently reliable. The observed bedload and suspended load transport rates were compared with the prediction method of Van Rijn (1993).

Results and discussion

A selection of the current velocity, bedload and suspended load transport data is given in Figure 3 and 4 (15 August 1990, morning). The sampling period for the measurements presented here was 10 minutes. The median particle size of the bedload catches agreed well with the median grain size of the bed material (around 280 μm in both cases). The bedload transport rate is given in Figure 3. The measured transport rates were small and occurred only during a period of about 2 hours around maximum flow of the flood phase. Near-bed current velocities (0.45 m above the bed) were around 0.4 m/s during this period. The suspended sediment concentrations at z = 0.07, 0.11 and 0.15 m are given in Figure 4. The measured suspended sediment concentrations were very low, with a maximum value of 40 mg/l at 0.07 m above the bed. The depth-integrated suspended load transport rates were calculated using the measured sediment concentrations and current velocities (eqn. 2). These calculated suspended load transport rates are given in Figure 3 as well. The suspended load transport rates are slightly smaller than the bedload transport rates. The period during which sediment is transported in suspension is also shorter than the bedload transport period (1 hour). Based on this, it is concluded that bedload is dominant at low tidal velocities, whereas the total fair-weather transport is highly episodic.

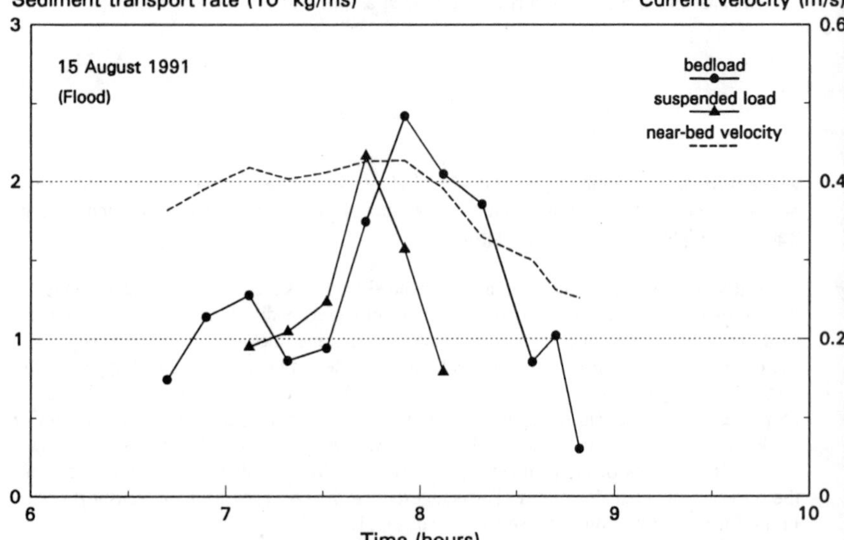

Figure 3 Near-bed current velocity, bed load and suspended load transport during springtide flood (15 August 1990).

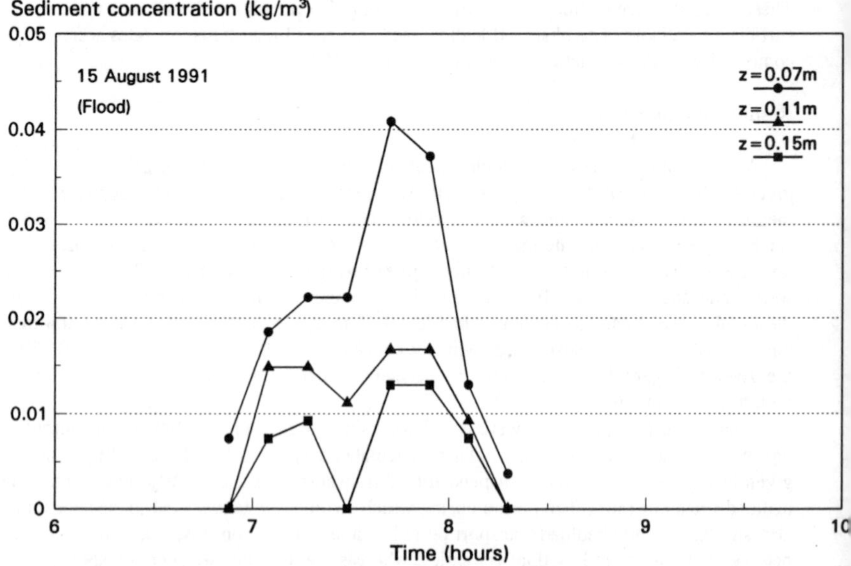

Figure 4 Suspended sediment concentrations during springtide flood (15 August 1990).

Table 1 Comparison measured and computed bed load and suspended load transport rates

$\bar{U}$ [1]	$\bar{U}_{corr}$ [2]	$\bar{U}_{0.5}$ [3]	Sediment transport rate			
			measured			computed
				n[4]	vc[5]	
			(10^{-3}kg/ms)	(-)	(-)	(10^{-3}kg/ms)
Bed load						
0.41	0.35	0.25	0.3 ± 0.6	3	1.0	0.001
0.56	0.48	0.35	0.8 ± 1.7	19	0.5	0.5
0.71	0.60	0.43	2.8 ± 2.1	9	0.3	2.8
Suspended load						
0.78	0.62	0.46	1.0 ± 0.5			3.6

Input parameters Transpor

h	= 14.4	(waterdepth)
d_{50}	= 276 μm	(median grainsize)
d_{90}	= 332 μm	(90% grainsize)
d_{ss}	= 276 μm	(suspended sediment size)
k_s	= 0.05 m	(current related roughness)
T	= 19° C	(water temperature)
Sa	= 30‰	(salinity)

[1] Measured depth-averaged velocity

[2] Depth-averaged velocity used as input for Transpor

[3] Measured near-bed current velocity (0.5 m above the bed). The depth-averaged velocity used as input for Transpor has been chosen so that measured and calculated near-bed current velocities match.

[4] Number of samples.

[5] Variation coefficient (= $\sigma/\mu\sqrt{n}$, where σ, μ = standard deviation and average value of the measured sediment transport rate respectively; cf. Van de Meene, 1994)

For a comparison of the observed bedload and suspended load transport rates with the calculated transport rates using Van Rijn's (1993) transport model Transpor, the bedload measurements were lumped into three different near-bed velocity classes (0.2-0.3 m/s, 0.3-0.4 m/s, 0.4-0.46 m/s), while all suspended load measurements were lumped into one velocity class (0.42-0.5 m/s). The measured and computed bedload and suspended load transport rates for each velocity class are given in Table 1. The input conditions used for the Transpor calculations are given in Table 1 as well.

Sediment transport will occur when the effective bed shear stress τ_b' exceeds the critical bed shear stress $\tau'_{b,cr}$. Using the Shields diagram, initiation of motion for the given circumstances (mean water depth = 14.4 m; d_{50} = 276 μm; d_{90} = 332 μm) is predicted at a depth-averaged current velocity of 0.40 m/s (Van Rijn, 1993). The measurements (Table 1) show that this threshold velocity does not correspond with zero transport. This has been observed in laboratory experiments also (Van Rijn, 1993) and agrees with bedload observations in the river Waal (Van Rijn et al., 1991). It indicates that the method fails to predict the sediment transport rates at low current velocities correctly. At higher current velocities, the bedload sediment transport rates are predicted reasonably well. The suspended load transport rate is overpredicted by a factor 4.

SEDIMENT TRANSPORT DURING STORMS

Data analysis

The suspended sediment transport data were used to achieve a qualitative insight in the mechanisms that dominate the sediment transport processes during storms. The analysis followed the instantaneous approach, looking at the transport processes at an intra-wave time-scale. This concept has been applied in the nearshore zone by e.g. Jaffe et al. (1985), Osborne et al. (1990), and Davidson et al. (1993), and in the shoreface environment by Wright et al. (1991).

The instantaneous measurements of velocity and suspended sediment concentration can be decomposed into a mean and an oscillating component; the oscillating part, in addition, can be split into short and long periodic components:

$$u = \bar{u} + \tilde{u}_s + \tilde{u}_L \quad [m/s] \tag{3}$$

$$v = \bar{v} + \tilde{v}_s + \tilde{v}_L \quad [m/s] \tag{4}$$

$$c = \bar{c} + \tilde{c}_s + \tilde{c}_L \quad [kg/m^3] \tag{5}$$

where: u, v = cross- and along-bank wave orbital velocities (m/s); c = sediment concentration (kg/m^3); $\bar{u}$, $\bar{v}$, $\bar{C}$ = time-averaged values; $\tilde{u}_S$, $\tilde{v}_S$, $\tilde{C}_S$ = high-frequency oscillations; $\tilde{u}_L$, $\tilde{v}_L$, $\tilde{C}_L$ = low-frequency oscillations.

As a result, the total time-averaged cross- and along-bank sediment fluxes $<cu>$ and $<cv>$ can be split into a mean flux component and several oscillating flux components:

$$<u \cdot c> = \bar{u} \cdot \bar{c} + <\tilde{u}_s \cdot \tilde{c}_s> + <\tilde{u}_L \cdot \tilde{c}_L> + <\tilde{u}_s \cdot \tilde{c}_L> + <\tilde{u}_L \cdot \tilde{c}_s> \quad [kg/m^2 s] \tag{6}$$

$$<v \cdot c> = \bar{v} \cdot \bar{c} + <\tilde{v}_s \cdot \tilde{c}_s> + <\tilde{v}_L \cdot \tilde{c}_L> + <\tilde{v}_s \cdot \tilde{c}_L> + <\tilde{v}_L \cdot \tilde{c}_s> \quad [kg/m^2 s] \tag{7}$$

In the present study, the brackets $< >$ indicate time-averaging over 10 minutes. The long and short periodic components were obtained by low- and high-pass filtering the instantaneous time-series, with a cutoff frequency of 0.05 Hz. This is twice the maximum observed peak incident wave period. In total a set of 6 fluxes is obtained: a total flux, consisting of a mean and four oscillatory fluxes. The mean flux is, by definition, directed in the direction of the mean current. In the inner-shelf environment, this mean current is a combination of tide, wind- and density-driven flows, with the Coriolis effect acting on all these currents. The short wave flux represents the net effect of the sediment fluxes induced by the incident wave orbital motion. This short wave flux may be directed with the propagation direction of the waves, when the waves are asymmetric, while it may be directed against the wave propagation direction, when strong vortex motions over steep ripples dominate, giving a net backward transport (e.g. Van Rijn, 1993). The long wave oscillatory flux is associated with the bound long waves, associated with grouped wind waves (e.g. Roelvink, 1993). The short-long wave interaction terms were found to be zero, indicating that $\tilde{u}_S$ and $\tilde{u}_L$ are uncorrelated with $\tilde{c}_L$ and $\tilde{c}_S$ respectively. The sediment fluxes, calculated according to equations 7 and 8 are represented by depicting the fluxes in x and y direction in vector diagrams.

The asymmetry ratio for the near-bed orbital velocity was defined by:

$$A_V = \frac{V_{s,crest}}{V_{s,crest} + V_{s,trough}} \tag{8}$$

where: A_v = asymmetry ratio (-); $V_{s,crest}$ = significant current velocity under the wave crest (onshore, m/s); $V_{s,trough}$ = significant current velocity under the wave troughs (m/s). This ratio equals 0.5 for symmetric waves and takes a value between 0.5 and 1 for (shoreward) asymmetric waves. Similarly, an asymmetry factor was defined for the oscillatory fluxes ($<u_S \cdot c_S>$ etc.).

Results

The asymmetry ratio's for the wave orbital velocities and for the oscillatory fluxes, as a function of $V_{s,crest}$, are given in Figure 5. The orbital velocity asymmetry ratio is between 0.50 and 0.55 and independent of $V_{s,crest}$ (Figure 5a). This indicates that waves were slightly asymmetrical during the observed storm conditions. The measured wave orbital velocities could be reproduced well with linear wave theory.

The oscillatory flux asymmetry increases with increasing peak velocity, albeit with considerable scatter (Figure 5b). The range of flux asymmetries varies between 0.25 and 0.75. Here, asymmetries < 0.5 indicate sediment fluxes directed against the propagation direction of the incident waves and asymmetries > 0.5 indicate fluxes directed with the incident waves. This corresponds with a seaward sediment flux component for an asymmetry ratio < 0.5 and a landward sediment flux component for an asymmetry ratio > 0.5. The transition between seaward and landward directed oscillatory fluxes is around $V_{s,crest}$ = 0.7 m/s (Figure 5b).

A selection of vectorplots showing the measured sediment fluxes at 0.1 m above the bed during two different bursts, is given in Figure 6. In general, the mean fluxes $\bar{u} \cdot \bar{c}$ and $\bar{v} \cdot \bar{c}$ are dominant. This indicates that wave stirring and advection by the mean current is the dominant sediment transporting process. The short wave oscillatory fluxes are variable in magnitude and direction and may cause significant differences between the

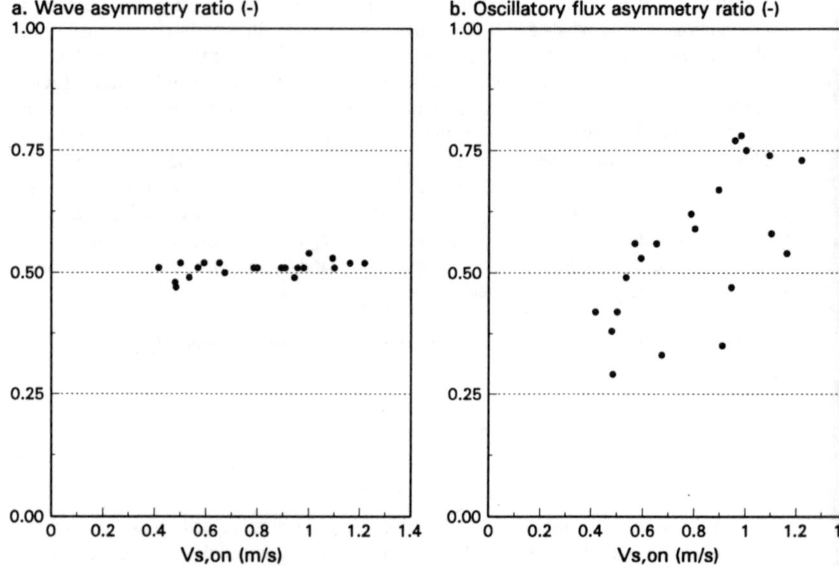

Figure 5 Asymmetry ratios as a function of $V_{s,on}$. a). wave orbital velocities; b). wave orbital fluxes.

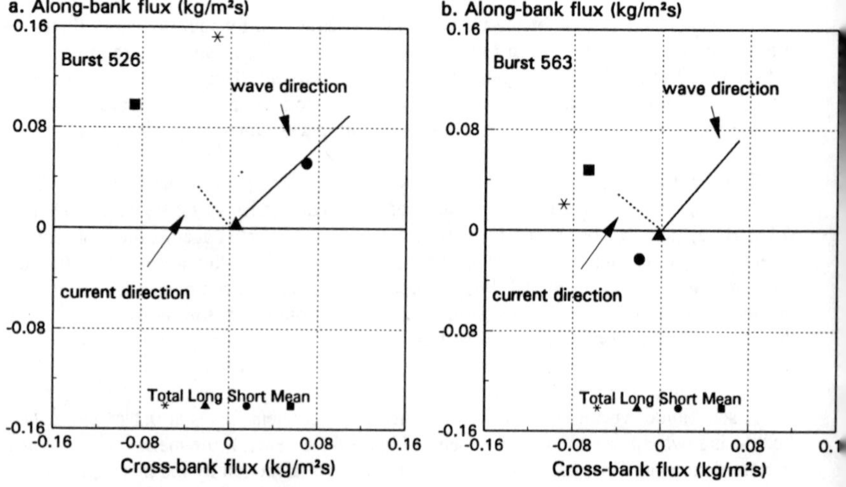

Figure 6 Vector plots of sediment fluxes. a). burst 526; b). burst 563. Dotted line represents current direction, solid line represents wave direction.

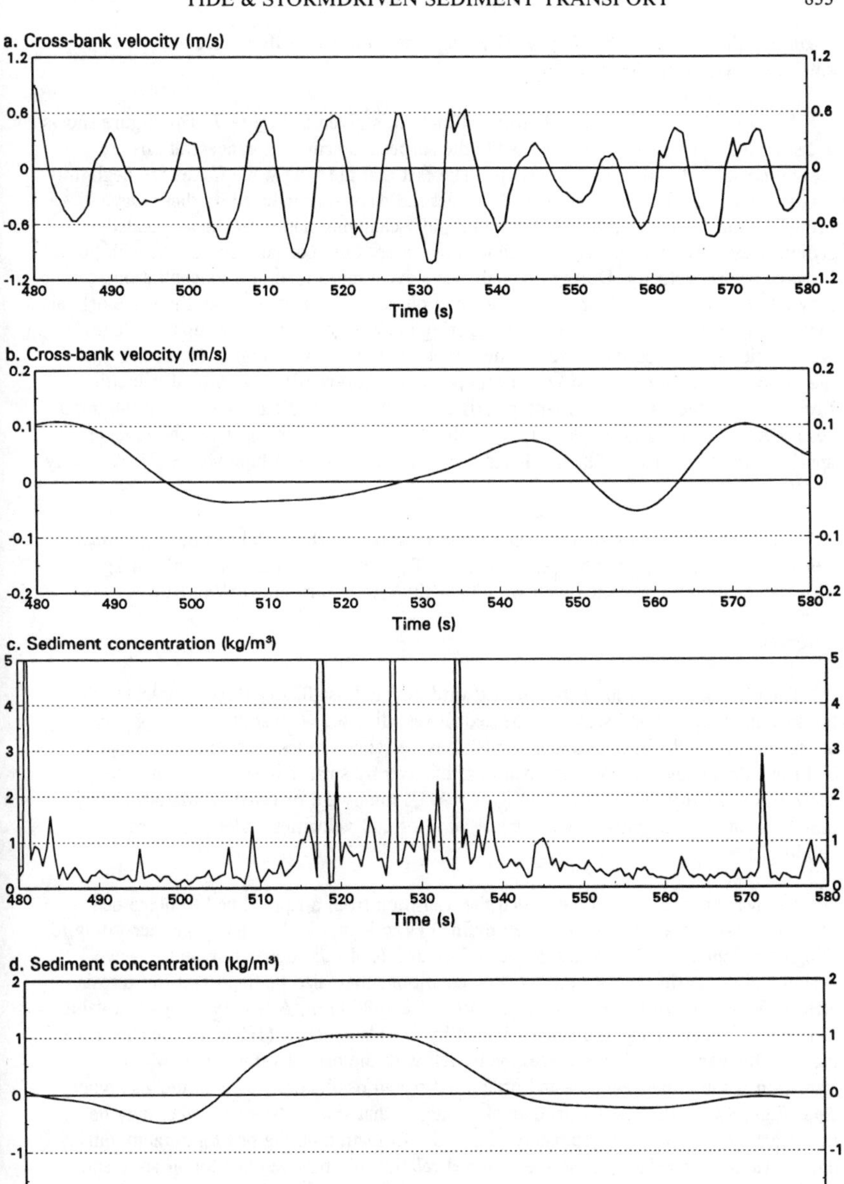

Figure 7 Detail of water and sediment motion during burst 526 (18 December 1991; 13.00 CET, Jday 352; H_s = 3.2 m; T_p = 8 s). a) cross-bank velocity; b) cross-bank velocity, low-pass filtered signal (f_c = 0.05 Hz); c) sediment concentration; d) sediment concentration, low-pass filtered signal (f_c = 0.05 Hz).

mean and the total sediment fluxes. The long wave oscillatory fluxes are always very small and may be neglected.

An example of the water and sediment motion is given in Figure 7. This figure shows a distinct groupiness of the waves, while the suspended sediment concentrations are clearly correlated with the wave groups. The fact that the long wave fluxes are negligibly small (Figure 6), indicates that there is no net sediment flux induced by these wave groups. Figure 7 shows the irregular and rapidly changing nature of the suspended sediment events. Peaks in sediment concentration are only partially correlated with peaks in wave orbital velocity. During one half-wave cycle more than one concentration peak may be observed. In addition, there are small phase-differences between the wave-orbital motion and the sediment concentration. During burst 526, the wave oscillatory fluxes were positive, directed landward, in the direction of the waves (Figure 6). In this case the largest suspension peaks occur during positive (shoreward) wave orbital velocities. Inspection of other time-series with positive wave oscillatory fluxes shows that the trend described above is valid in general; when the oscillatory flux is negative, the reverse applies. Small or zero oscillatory fluxes occur when the wave orbital velocity is relatively small.

The observed mean sediment concentrations at 0.1 m above the bed during typical storm conditions (U_{tide} = 0.6 m/s, H_s = 3 m, T_p = 7 s) was in the order of 0.4 kg/m^3. This value could be predicted well with Van Rijn's transport model (Van Rijn, 1993).

Discussion

The observed transition between landward directed oscillatory fluxes (directed with the incoming waves) and seaward directed fluxes (directed against the incoming waves), at wave orbital velocities around 0.7 m/s, may be related to the transition of a rippled bed towards a plane bed. Vortex shedding, induced by small ripples, may cause net seaward oscillatory fluxes (e.g. Van Rijn, 1993). Landward directed oscillatory fluxes may be related to (slightly) asymmetrical wave orbital velocities under plane-bed conditions.

The wave orbital velocity at which the transition from a rippled bed to plane bed conditions takes place is not very well defined (Van Rijn, 1993). The range, according to Dingler & Inman (1977), varies between 0.46 and 1.14 m/s, depending on the wave period and the grain size. Preceding flow conditions may also be important (hysteresis effects, Van Rijn, 1993). For grainsizes around 200-300 μm, Allen (1984) gives a value of 0.7 m/s, which is also the value adopted by Davidson et al., (1993) for their measurements. This value of 0.7 m/s corresponds well with the orbital velocities at which a transition is found between on- and offshore directed oscillatory fluxes in the Zandvoort data (Figure 5b). The Zandvoort data also suggest that some hysteresis effect may be important (Figure 5b). Unfortunately, detailed information on the bed topography during the measurements is lacking, so the assumed relationship between bed topography and oscillatory flux direction remains speculative. The results stress the importance of registering the bedforms during the measurements, although this is virtually impossible in practice, especially during storm conditions (Vincent et al., 1991).

Wright et al. (1991) have done an extensive study on sediment transport processes in the sandy shoreface environment along the east coast of U.S.A., in water depths generally around 10 m and with a maximum of 17 m. The measurements described here can be compared with their study; the measurement techniques in both studies are roughly similar.

Comparable with the data presented in this paper, Wright et al. (1991) find the mean sediment fluxes to be the most important contributor to the total sediment flux. They found this to be the case for a variety of different conditions, ranging from fair-weather to storms. Incident waves were found to be important for stirring up the sediment. As in the present study, oscillatory fluxes were directed onshore as well as offshore. According to Wright et al. (1991), this variety in direction may be explained by phase lags between u(t) and c(t), caused by the presence of small scale ripples. Low-frequency effects were measurable, but not dominant. Low-frequency fluxes were directed landward as well as seaward. Part of their data set covers a storm (dataset obtained near Duck, N.C. in 1985), allowing a more specific comparison with the data presented here. Their storm data were obtained at the shoreface at a water depth of 8 m. The significant wave height was between 1 and 1.4 m, the significant wave period around 8 s, while the observed wave orbital velocities were in the order of 1 m/s. These conditions were slightly quieter, but further reasonably comparable with the conditions under which the Zandvoort data were obtained. The median grainsize (d_{50}) of the bed material near Duck, N.C. was 125 μm, which is finer than the Zandvoort sediments (d_{50} around 270 μm).

There is good agreement between the results from both experiments. Although in both studies the mean sediment fluxes were dominant, the type of mean flow was different: during the Duck experiment the mean flow was dominated by seaward directed downwelling, while during the Zandvoort measurements the mean flow consisted of tidal and wind-driven currents. The sediment fluxes are of comparable magnitude, with mean sediment fluxes in the order of 0.02-0.50 kg/m^{2}s (Zandvoort) and 0.05-0.30 kg/m^{2}s (Duck), and oscillating fluxes in the order of 0-0.01 kg/m^{2}s (Zandvoort) and 0-0.04 kg/m^{2}s (Duck).

CONCLUSIONS

Sediment transport observations during spring tidal, fair-weather conditions have shown that the sediment transport rates under these conditions are very low and occur only during a period of about 2 hours around maximum tidal flow. Bedload transport is slightly dominant.

The sediment transport measurements during storm conditions have revealed a large variability in transport directions. Although the mean fluxes dominate the sediment mobility during storms, the contribution of short wave oscillatory fluxes to the total flux cannot be neglected. Long wave oscillatory fluxes are negligible.

Calculations with the Transpor model (Van Rijn, 1993) agree reasonably well with the observed sediment transport rates and suspended sediment concentrations, under fair-weather and under storm conditions.

ACKNOWLEDGEMENTS

This study was done within the framework of the Coastal Genesis Programme and was partly funded by the National Institute for Coastal and Marine Management (RIKZ) of the Ministry of Transport, Public Works and Water Management. Technical assistance for these measurements was provided by the Department of Physical Geography, Utrecht University. The crew of RV *Professor Lorentz* skilfully handled the instruments at sea.

REFERENCES

Allen, J.R.L. (1984). Sedimentary structures. Their characters and Physical basis. *Developments in Sedimentology*, 30. I: 593 pp, II: 663 pp. Elsevier Publisher B.V., Amsterdam.

Davidson, M.A., P.E. Russell, D.A. Huntley & J. Hardisty (1993). Tidal asymmetry in suspended sand transport on a macrotidal intermediate beach. *Marine Geology*, 110: 333-353.

Dingler, J.R. & D.L. Inman (1977). Wave-formed ripples in nearshore sands. *Proceedings 15th Conference on Coastal Engineering*, ASCE, New York: 2109-2126.

Downing, J.P., R.W. Sternberg & C.R.B. Lister (1981). New instrumentation for the investigation of sediment suspension processes in the shallow marine environment. *Marine Geology*, 42: 19-34.

Jaffe, B.E., R.W. Sternberg and A.H. Sallenger (1985). The role of suspended sediment in shore-normal beach profile changes. *Proceedings 19th Conference on Coastal Engineering*, ASCE, New York, 1983-1996.

Osborne, P.D., B. Greenwood & A.J. Bowen (1990). Cross-shore suspended sediment transport on a non-barred beach: the role of wind waves, infragravity waves and mean flows. *Proceedings of the Canadian Coastal Conference 1990*: 349-361.

Roelvink, J.A. (1993). Surf beat and its effect on cross-shore profiles. Thesis, Technical University Delft, 116 pp..

Van de Meene, J.W.H. (1994). The shoreface-connected ridges along the central Dutch coast. Ph.D-thesis. *KNAG/Netherlands Geographical Studies 174*, Utrecht, The Netherlands, 222 pp..

Van Rijn, L.C. (1993). Principles of sediment transport in rivers, estuaries and coastal seas. Aqua Publications, Amsterdam, The Netherlands.

Van Rijn, L.C., B. Kornman & B. Gehrels (1991). The Delft Nile Sampler. Bedload transport measurements in the river Waal. Report Q1300.02/h840.21. Delft Hydraulics, Delft, The Netherlands.

Van Rijn, L.C. and M. Gaweesh (1992). A new total load sampler. *Journal of Hydraulic Engineering*, Vol. 118, No. 112.

Vincent, C.E. & D.M. Hanes & A.J. Bowen (1991). Acoustic measurements of suspended sand on the shoreface and the control of concentration by bed roughness. *Marine Geology*, 96: 1-18.

Wright, L.D., J.D. Boon, S.C. Kim and J.H. List (1991). Modes of cross-shore transport on the shoreface of the Middle Atlantic Bight. *Marine Geology*, 96: 19-51.

SEDIMENT TRANSPORT MEASUREMENTS IN COMBINED WAVE-CURRENT FLOWS

I. Katopodi[1], J. S. Ribberink[2], P. Ruol[3] and C. Lodahl[4]

ABSTRACT: A series of experiments were carried out in the Large Oscillating Water Tunnel of Delft Hydraulics in the end of 1993. The aim of the investigation was to study the behaviour of the sediment and water flow in the wave-current boundary layer in the sheet flow regime. Four conditions with sinusoidal waves and net currents were realized. A selected part of the new experimental data is presented. The attention is focused on vertical profiles of time-averaged and time-dependent suspended sediment concentrations and velocities as well as net sediment transport rates.

INTRODUCTION

The understanding of the physical processes of the near-bed dynamics under the influence of waves and currents, crucial for an adequate prediction of the sea bed changes and coast line evolution, is very limited. The validity of the various mathematical models can hardly be assessed due to the low number of measurements available, especially at prototype scale (see Horikawa, 1988). Especially scarce are the measurements in the wave boundary layer near the sea bed.

The Large Oscillating Water Tunnel of Delft Hydraulics is a large scale experimental facility in which wave- and current-related near-bed flow and sediment transport phenomena can be simulated in the scale of nature (1:1). During the preceding years the tunnel became an important source of

---

1) Scientific Officer, Democritus University of Thrace, Department of Civil Engineering, 67100 Xanthi, Greece.
2) Senior Researcher, Delft Hydraulics, P.O. Box 152, 8300 AD Emmeloord, the Netherlands.
3) Researcher, University of Padova, Faculty of Engineering, Institute of Maritime Constructions, Via Ognissanti 39, 35129, Padova, Italy.
4) Ph.D student, Technical University of Denmark, ISVA, Building 115, Lyngby, DK-2800 Denmark.

data for the development of sediment transport models under linear and non-linear waves (Ribberink and Al Salem, 1991, 1992; Al Salem, 1993). In 1992 the tunnel was extended with a recirculation flow system and during a first experimental programme time-averaged sediment transport phenomena in various wave-current flows were examined (Ramadan, 1993).

The present experiments, carried out in October and November 1993, focus on the time-dependent behaviour of the sediment and water flow in the wave current boundary layer under a number of sinusoidal wave and net current combinations in the sheet flow regime. The obtained data can be used for the verification of existing and the development of new intra-wave sediment transport formulations (see Al Salem, 1993 and Fredsøe, 1993 for reviews).

EXPERIMENTAL SET-UP

The Large Oscillating Water Tunnel (LOWT) has the shape of a vertical U-tube with a long rectangular horizontal section and two cylindrical risers on either end. The desired oscillatory water motion inside the test section is imposed by a steel piston in one of the risers. The other riser is open to the atmosphere. The test section is 14 m long, 1.1 m high and 0.3 m wide. A 30 cm thick sand bed can be brought into the test section, leaving 0.8 m free for the oscillatory flow above the bed. The range of the oscillatory velocity amplitudes is 0.2-1.8 m/s and the range of periods is 4-15 s. The LOWT, extended with a recirculation flow system for the generation of a steady current, is shown in Fig. 1. The maximum capacity of the two pumps is 100 l/s and 20 l/s. The maximum superimposed mean current velocity in the test section of the tunnel is 0.5 m/s. A sand trap consisting of a 12 m long pipe is part of the recirculation system and can be used for sediment transport measurements.

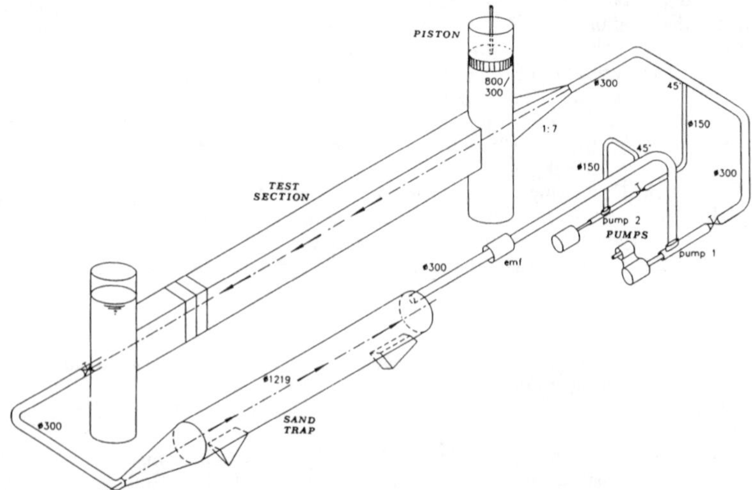

Fig. 1 The Large Oscillating Water Tunnel with its recirculation system

Four experimental conditions with sinusoidal waves/net current (coded E1...E4) were realized in the tunnel by imposing the piston movement and the pump discharge. Table 1 shows for each condition the pump discharge, the average net current velocity and the velocity amplitude and period of the oscillatory flow. With increasing condition number (E1...E4) the imposed net current increases and the amplitude of the (sinusoidal) oscillatory flow decreases. Due to the net current all conditions concern asymmetric flow. The asymmetry increases with condition number. All four conditions were in the sheet flow regime (plane bed). Sand with characteristics $D_{10}=0.15$ mm, $D_{50}=0.21$ mm and $D_{90}=0.32$ mm was used.

Ribberink and Al Salem, (1991) showed that there is a very consistent relation between the net sediment transport rate and the third order velocity moment of the horizontal oscillatory flow above the wave boundary layer (for asymmetric waves). Based on this, the four conditions were chosen with almost the same third order velocity moments so that similar sediment transport rates would occur.

Condition	Net current		Oscillatory flow	
	pump discharge (m3/s)	average velocity in tunnel (m/s)	velocity amplitude in tunnel (m/s)	period (s)
E1	0.036	0.15	1.65	7.2
E2	0.048	0.20	1.50	7.2
E3	0.070	0.29	1.15	7.2
E4	0.103	0.43	0.96	7.2

Table 1 Experimental conditions

During 5 series of experiments, 115 tests (tunnel runs) were carried out with different measuring techniques. The flow velocities were measured with a laser-doppler system (LDFM) and an electromagnetic flow meter (EMF). The grain velocities near the bed were extracted from detailed high-speed video recordings. For the measurement of the suspended sediment concentration a transverse suction and an optical method (OPCON) were used. A concentration conductivity meter (CCM) was used for the measurement of the very large sediment concentrations in the sheet flow layer. The net sediment transport rates were computed with a mass conservation technique using sand trap volumes and measured bed level changes.

In the present paper, concentrations and velocities in the suspension layer and total net sediment transport rates are presented and discussed. Information about all the measurements can be found in Katopodi at al (1994).

EXPERIMENTAL RESULTS

Net sediment transport rates

For the computation of the net sediment transport rate four tests were realized per condition, simultaneously with the transverse suction measurements. The bed level along the test section was measured before and after the experiment (also the weight of the sand collected in the sand traps) and the distribution of the net sediment transport rate along the test section was computed solving the mass balance equation twice for each test, starting from either the left or the right sand trap (mass conservation technique). The mean value of the two computations in the center of the test section was used.

In Table 2 the net sediment transport rates averaged over all the tests of each condition are presented. Moreover, the standard deviation, the relative error and the relative error of the averaged transport rate are also shown. Given the rather small number of repetitions the errors remained at an acceptable level. The net sediment transport rates are, as designed, of comparable magnitude for all four conditions. Some grouping though can be observed. The net transport rates are almost the same for conditions E1 and E2 as well as for E3 and E4.

test	$<q_{(s)avg}>$ (10^{-6} m^2/s)	σ (10^{-6} m^2/s)	r (%)	$\frac{r}{\sqrt{N}}$ (%)
E1	92.4	6.63	6.5	3.8
E2	96.4	7.61	7.53	3.8
E3	69.1	8.45	11.7	5.8
E4	71.0	7.01	9.3	4.6

Table 2 Net sediment transport rates

Time-averaged suspended sediment concentration profiles

For each experimental condition, five separate (wave-averaged) concentration profiles were measured with a transverse suction technique (Bosman et al, 1987). Two of the profiles were measured in the middle of the test section, x=0.0 m, and three at x=2.35 m downstream. The concentration was measured simultaneously at ten elevations (from ~25 cm down to ~1 cm above the average bed). Sand/water mixture was extracted from the flow for about 8 min and the concentration was measured with the help of calibrated tubes.

In Fig. 2 the concentration profiles are plotted for all conditions using a log-log x-y scale. The straight best fit lines imply that a power law for the concentration is valid (Ribberink and Al Salem 1992, Al Salem, 1993). The lines are more or less parallel for the four conditions with a small deviation in E3 where the tests at x=0.0 m had to be discarded (the profiles at x=0.0 m were somewhat steeper than those at x=2.35 m).

We can also observe that as the wave amplitude decreases (from E1 to E4), a general reduction of the suspended sediment concentration occurs. Apparently, the suspension is dominated by the waves in the chosen conditions. At the same time the net current decreases and, despite the reduction of the suspended load, the net sediment transport rates are approximately the same (see previous section).

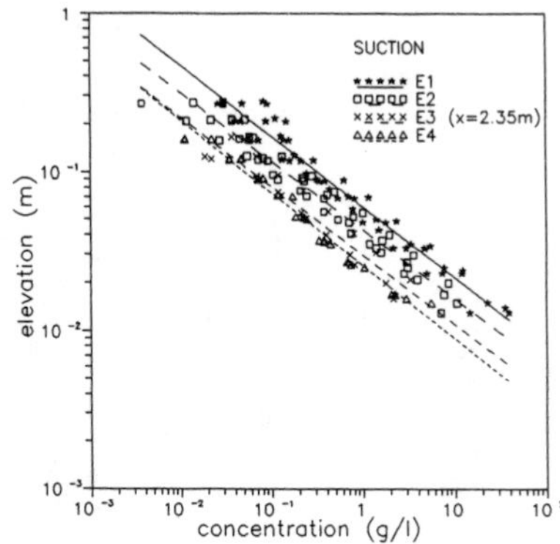

Fig. 2 Time-averaged suspended sediment concentration profiles

For every second transverse suction test, the collected sand samples were stored and their median diameter D_{50} was determined later with the help of a settling tube. In the layer $0 < z < 3$ cm a strong reduction in grain size occurred (from D_{50}=0.21 mm to ~ 0.18 mm); above this layer the distribution is more uniform. The vertical sorting behaviour was more or less the same for the four conditions.

Time-dependent suspended sediment concentrations

An optical/electronic instrument (OPCON) was used for the measurement of the intra-wave suspended sediment concentration. The height of its sensing volume is 2.6 mm and it was operated with a sampling frequency of 40 Hz. The instrument was positioned at x=2.00m. The range of elevations of the measurements was from 10 cm from the bed down to 0.5 cm. Ten tests per condition were realized. During each test (~ 9 min) the concentration was measured at three elevations (three minutes per elevation, ~ 25 waves). The OPCON calibration was adjusted for grain size variations along the vertical profile using the transverse suction results given above.

The concentrations at various elevations above the average bed level are plotted in Fig. 3 (E1 and E2) and Fig. 4 (E3 and E4). The plots show the concentration during one wave cycle (obtained by ensemble-averaging of 21 to 24 waves). In the upper part of the figures the velocity measured at 20 cm above the bed as well as vertical position of the piston (driving the oscillatory flow) are shown.

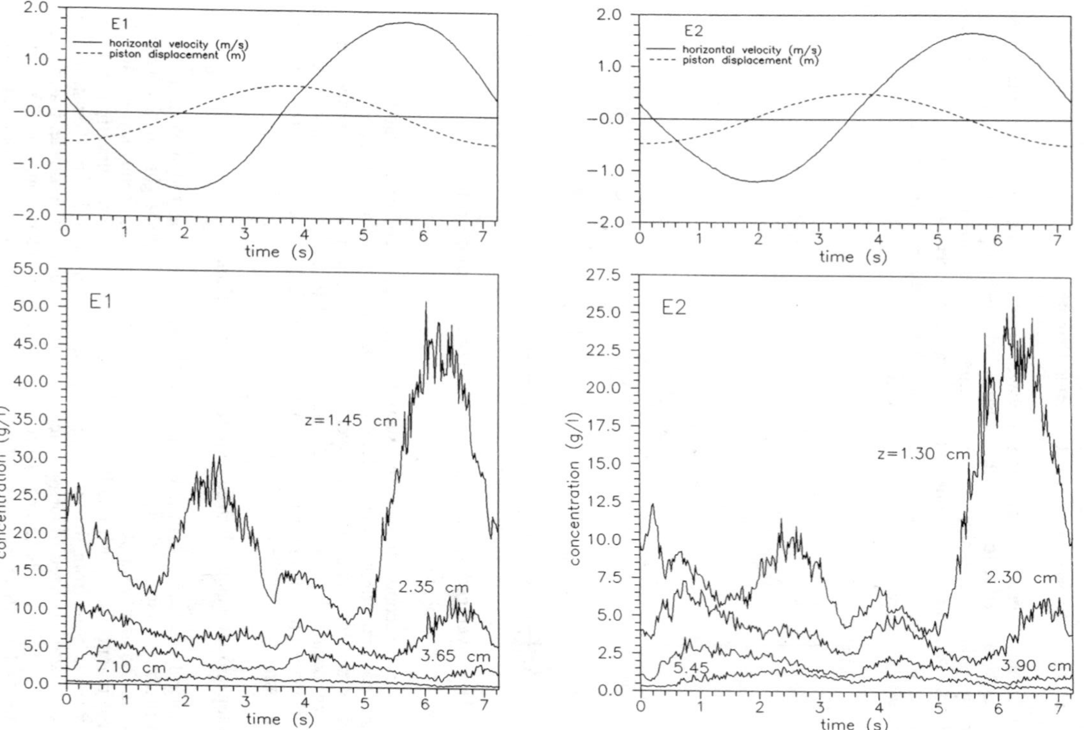

Fig. 3 Time-dependent concentrations measured with OPCON at various elevations (E1, E2)

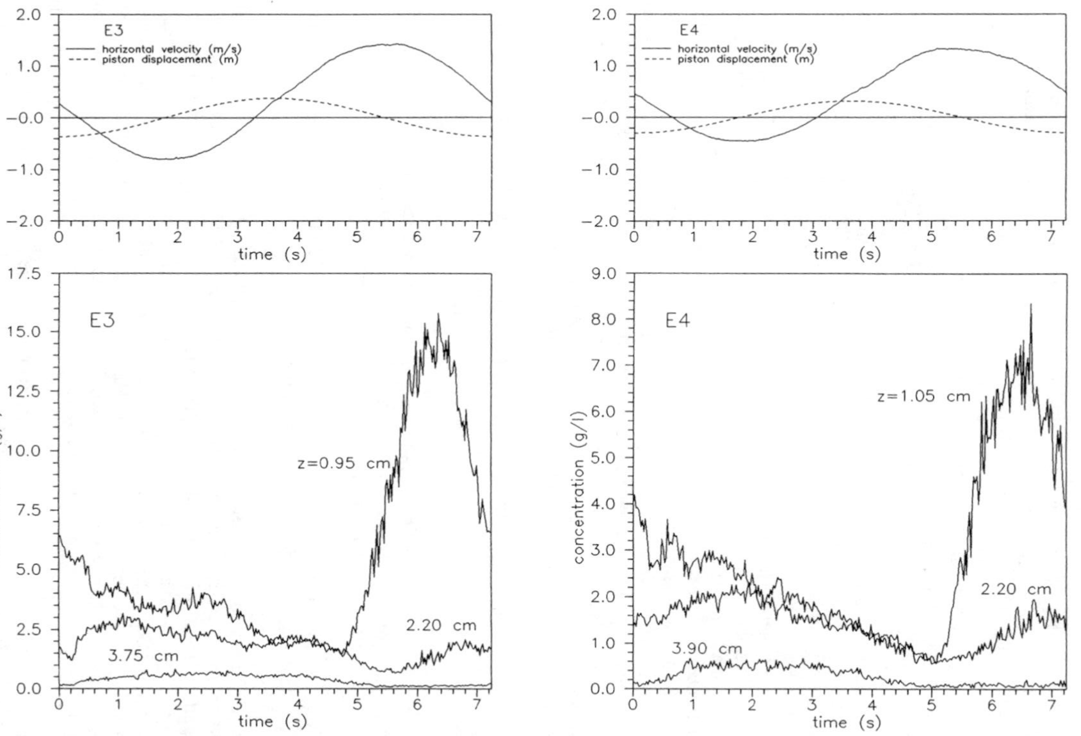

Fig. 4 Time-dependent concentrations measured with OPCON at various elevations (E3, E4)

It can be observed in Fig.3 (E1 and E2) that near the bottom the concentration exhibits two peaks that roughly coincide with the maximum and minimum velocities. The two peaks are not of equal magnitude. The concentration is higher at the maximum (downstream or "crest") velocity, while it is smaller at the minimum (upstream or "trough") velocity. This asymmetry increases with the asymmetry of the flow (and increasing condition number). For the conditions E3 and E4 (Fig. 4) the peak at minimum velocity can very roughly be recognized and it looks that the concentration is dominated by the previous half cycle. This behaviour was observed before by Ribberink and Al Salem (1992), see also Al Salem (1993), for asymmetric waves.

Two other generally smaller concentration peaks occur near the moments of flow reversal. Their intensity is decreasing with condition number. For E3 the small peaks at flow reversal can still be recognized and have approximately the same magnitude as the peak at minimum velocity. In E4 the peaks at flow reversal disappear. It looks that they are suppressed by the current. The concentration peaks at flow reversal were also observed by Al Salem (1993) for waves only (both sinusoidal and asymmetric). A possible explanation for their existence may be found in Foster et al (1994) where it is argued that the suspension events at flow reversal may be the result of a shear instability of the bottom boundary layer.

As we go away from the bottom, the concentration decays rapidly with elevation and the maxima occur at a later moment (time lag effect). At higher elevations the maximum concentrations occur even after the next flow reversal (e.g. from z = 3.65 cm on for condition E1). For a depth range up to ~ 4 cm from the bottom the concentration at the velocity maximum/minimum decays faster than the concentration peaks at flow reversal (although the latter disappear at higher elevations).

In Fig. 5 the energy density spectrum of the OPCON signal at elevations near the bed is shown for the four conditions.

For E1, near the bottom (z=1.45 cm), the energy is concentrated in the second harmonic (f = 0.28 Hz), as it was expected from the ensemble averaged concentration signal that presents two peaks (at the moments of max and min velocity). Due to the asymmetry of the concentration signal (the two peaks are not of equal magnitude), also the first harmonic appears in the spectrum (f = 0.14 Hz). The latter is clearly a result of the presence of the current. The two smaller concentration peaks at flow reversal give rise to the fourth harmonic (f=0.55 Hz).

For condition E2, the power spectrum of the concentration near the bottom (z=1.30cm) shows a similar behaviour. As the current (and velocity asymmetry) is getting stronger (relative to E1), the first harmonic gains energy over the second. The two concentration peaks at flow reversal are less pronounced now, but the fourth harmonic is still recognizable.

For E3, near the bed (z=0.95), the first harmonic dominates the second. The fourth harmonic has disappeared and the third harmonic (f=0.41) is rising. For E4, near the bed (z=1.05) the first harmonic is even more dominant over the second harmonic. The third harmonic is higher than in E3.

In Fig. 6 the power spectra of the concentration at two more elevations for condition E1 are shown. As we go higher from the bed (z=3.65cm) the first harmonic gains energy over the second (compare with z=1.45 cm, Fig. 5). Even higher (z=7.10cm) the second harmonic almost disappears. This is natural once a diffusion process of both the bottom generated turbulence and the suspended sediment entrained from the sand bed takes place over the vertical. In such processes it is expected that the higher harmonics are attenuating faster as we go away from the boundary. For the conditions E2, E3 and E4 the change of the relative weight of the harmonics over the vertical was the same as for E1.

In general, we can observe that the sinusoidal part of the flow is expressed by the even harmonics (second and fourth) of the concentration power spectrum while the asymmetry caused by the superposition of the current is expressed by the odd harmonics (first and third for the chosen conditions). The latter become more dominant as the current (and flow asymmetry) increases from E1 to E4.

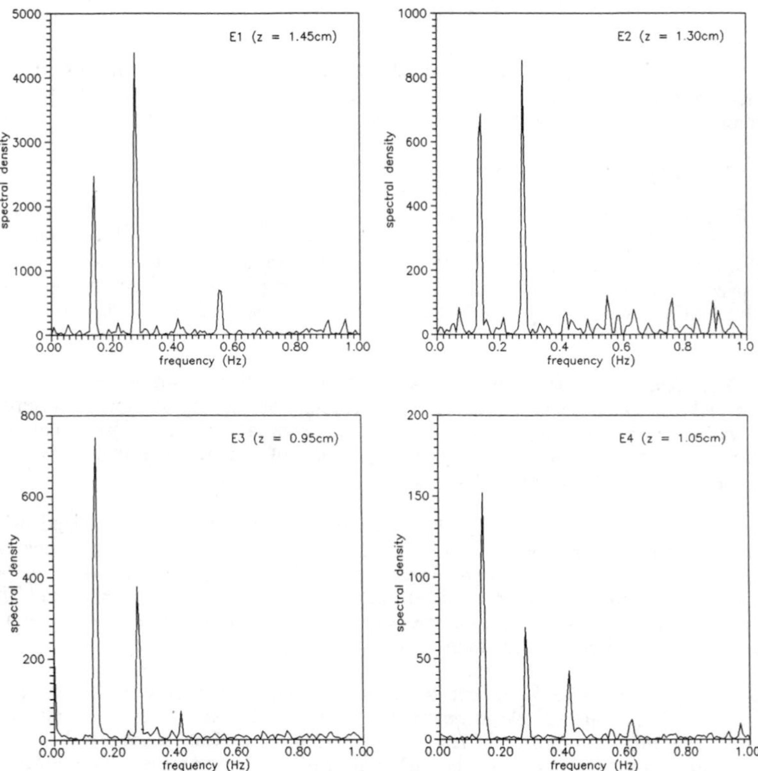

Fig. 5 Energy density spectra of the near-bed concentration (E1...E4) (energy density in $(g/l)^2/Hz$)

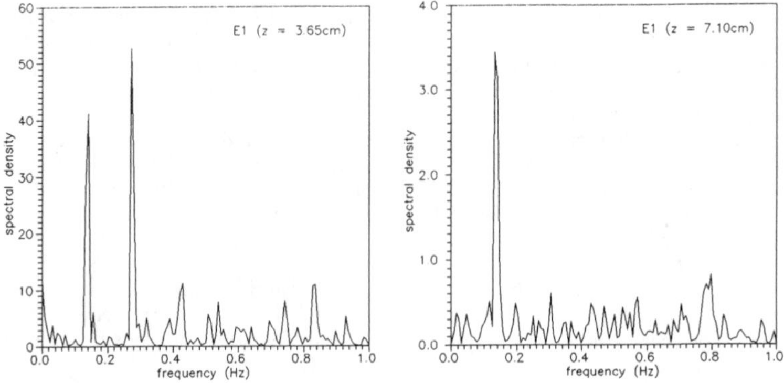

Fig. 6 Energy density spectra of the concentration at two elevations (E1) (energy density in $(g/l)^2/Hz$)

The time-dependent OPCON signal was averaged over the duration of the measurement for each elevation. The time averaged concentration profiles compared very well with the ones measured by transverse suction.

Time-dependent velocities

Time-dependent velocities were measured by a forward scatter laser system (LDFM) with a sensing volume height of 0.3 mm. Moreover, an electromagnetic flow meter (EMF) was used with a probe of disk type with horizontal dimensions 11 x 33 mm and a sensing volume height of 3-5 mm. The EMF was used near the bottom where heavy suspension blocked the laser beams.

Four tests per condition with a duration of 10 min were realized with the EMF. During each test the horizontal velocities at three different elevations were measured with sampling frequency 10 Hz. The instrument was positioned at x=2.0 m and the elevation range of the measurements was from 9 cm down to 1.5 cm above the bed.

Eight tests per condition with a duration of 12 min were realized with the LDFM. During each test, the horizontal and vertical velocities at three different elevations were measured, with a sampling frequency of 40 Hz. The instrument was positioned either at x=0.0 m or at x=2.0 m. The elevation range of the measurements was from 20 cm down to 3 to 4 cm above the bed (depending on the condition).

In Fig. 7 the time-dependent horizontal velocities measured with the EMF are shown at various elevations for conditions E1 and E3. The velocities were ensemble-averaged over 20 - 24 wave cycles. The phase lead in the boundary layer can clearly be distinguished by comparing the velocities near the bed (z=1.45 cm for E1 and z=1.65 cm for E3) with the velocities at higher elevations. For both conditions the velocity profile is more or less uniform for the part of the cycle from minimum (negative) velocity to maximum (positive) velocity.

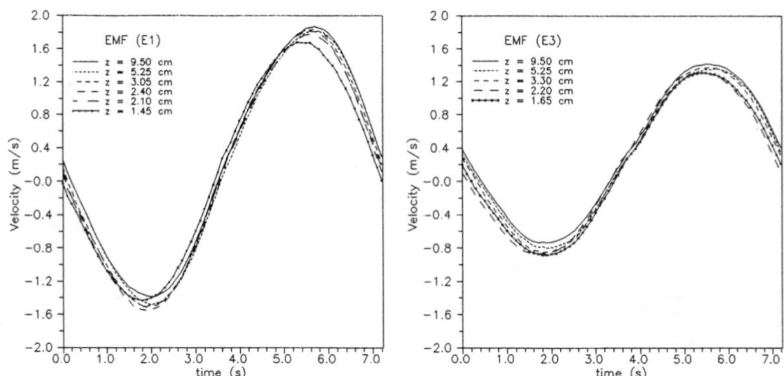

Fig. 7 Time-dependent horizontal velocities measured with EMF (E1, E3)

The wave-averaged velocity profiles (computed from the EMF signal) had a logarithmic shape for all conditions. Two slopes could be distinguished, one for the layer from 10 cm to ~ 3 cm from the bed and another for elevations lower than 3 cm. As expected, due to wave-current interaction, the slope near the bottom (wave-current boundary layer) was steeper than in the upper part (outer flow). See e.g. van Kesteren and Bakker (1984).

In Fig. 8 the time-dependent flow velocity profiles during the wave cycle are presented with steps of 0.3 s (15 degrees) for condition E1 (EMF). The overshooting in the wave-current boundary layer is clear. Inflection points in the vertical can be distinguished at t =0, 3.62 and 7.23 s (0, 180 and 360 degrees).

The time-averaged velocity profiles measured by the LDFM were also logarithmic and showed a uniform behaviour along the tunnel (measurements at x=0.0 m and at x=2.0 m).

In Fig. 9 the time-dependent horizontal velocities measured with the LDFM are shown at various elevations for conditions E1 and E3. The velocities were ensemble-averaged over 20 - 24 waves. The measurements were realized outside the wave boundary layer and as a result the phase lead of the velocity can not be distinguished. We can observe that the velocities are the same at all elevations for the part of the cycle from the maximum (positive) to the minimum (negative) velocity. Exactly the opposite occurred during the EMF measurements (see Fig. 7).

Comparisons of measurements conducted by the two instruments at the same elevation revealed the presence of a difference between the measured

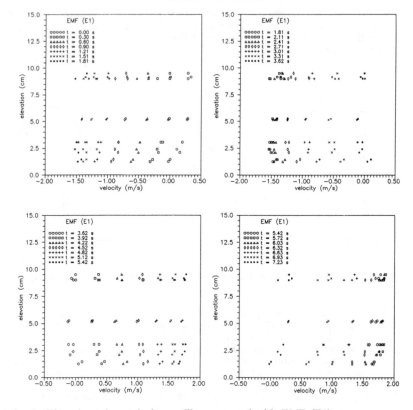

Fig. 8 Time-dependent velocity profiles measured with EMF (E1).

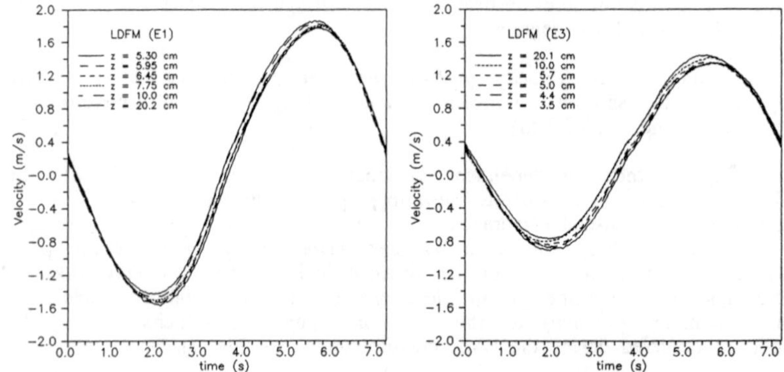

Fig.9 Time-dependent horizontal velocities measured with LDFM (E1, E3)

velocities that is variable during the wave cycle. Considerable is the difference in the part of the cycle when the water is moving upstream (negative velocities) and minor when it is moving downstream (positive velocities). The bigger difference occurs around trough velocity (maximum ~ 9 cm/s). This asymmetric behaviour of the difference during the flow cycle (i.e. the fact that EMF measures systematically smaller negative velocities) adds up to considerably higher wave-mean velocities over the whole depth. For more extensive considerations about this EMF/LDFM difference, reference is made to the original data report (Katopodi et al, 1994).

CONCLUSIONS

A data set of time dependent concentrations and velocities as well as net sediment transport rates was obtained during a series of experiments (series E) in the Large Oscillating Water Tunnel of Delft Hydraulics, under four combined sinusoidal wave/net current conditions. During all the conditions the bed was plane (sheet flow). For an extensive data report with detailed results see Katopodi et al (1994).

The net sediment transport rates, obtained with a mass conservation technique, were comparable in magnitude for all four conditions indicating that the relation between the net sediment transport rate and the third order velocity moment is not only valid for asymmetric waves (Ribberink and Al Salem, 1991) but also holds for combined wave/current conditions.

The distribution over the depth of the time-averaged concentrations measured by transverse suction appeared to follow a power law (Ribberink and Al Salem, 1992). The concentration best fit lines were more or less parallel for the four flow conditions. The concentration reduced strongly from E1 to E4, indicating that the suspension is dominated by the waves.

The time-averaged concentrations computed from the time dependent OPCON signal generally showed a good agreement with the ones measured by transverse suction.

The ensemble-averaged concentrations measured by OPCON in the suspension layer showed the following behaviour:

1. During the wave cycle two maxima occurred more or less in phase with the maximum and minimum free stream velocities. The two maxima were not symmetric. The concentrations were larger at (the larger) crest velocity and smaller at (the smaller) trough velocity. This asymmetry increased with the asymmetry of the flow (and increasing condition number).
2. Approximately at the moments of flow reversal two concentration peaks appeared (probably due to an instability in the wave-current boundary layer). The peaks were sharper and more intense near the bottom. The peaks got smaller with increasing condition number and appeared to be suppressed by the increasing current.
3. The main concentration as well as the peaks at flow reversal decayed rapidly with elevation and lagged behind the velocity. The peaks at flow reversal seemed to disappear at elevations outside the boundary layer.

Power spectral analysis of the OPCON signal showed that only the first four harmonics were dominant for the given four flow conditions. It appeared that the sinusoidal part of the concentration is expressed by the even harmonics (second and fourth) of the power spectrum while the asymmetry caused by the superposition of a net current is expressed by the odd harmonics (first and third). Odd harmonics became more and more dominant as the current (and flow asymmetry) increased from E1 to E4. As the distance from the bottom increased, the lower harmonics became more and more important, with the first harmonic being the dominant one at the highest measured elevations. This behaviour is typical for the vertical diffusion process of turbulence and suspended sediment.

The net current profiles, computed from the time dependent LDFM and EMF signals, showed a logarithmic shape. Due to the blocking of the laser beams by suspended sediment only velocities outside the wave boundary layer could be measured by the LDFM. The EMF measurements showed the effects of the oscillatory flow (wave boundary layer) on the net velocity profile (wave/current interaction). The velocity profiles were uniform along the tunnel.

In the time-dependent velocity profiles (measured with the EMF), the overshooting in the boundary layer as well as the phase lead of the near bed flow could be clearly distinguished. Also the ensemble-averaged velocities at various elevations revealed this phase lead in the boundary layer.

A discrepancy occurred between the time-dependent velocities measured by LDFM and EMF that adds up to considerable differences in the net current. Due to the wide use of electromagnetic flow meters in the lab and field, this matter needs further investigation.

Further data analysis will include the derivation of the sediment fluxes combining the measured velocities and concentrations (suspension and sheet flow). Moreover, the cross-spectra of concentrations and velocities are of importance. Turbulence quantities can be extracted from the LDFM measurements.

Finally, the data will support further developments in mathematical modelling of intra-wave sediment transport in wave-current flows.

ACKNOWLEDGEMENTS

The experimental investigation was part of the EC programme "Access to Large-scale Facilities and Installations" and was funded by the Commission of the European Communities, Directorate General for Science, Research and Development, contract no. GE1*-CT91-0032 (HSMU). It was executed by an international team with members, apart from the authors, also R.Koelewijn, S. Longo, A. Crosato and H.Wallace. The data analysis and reporting was partly done in the framework of the Mast G8, Coastal Morphodynamics research programme of the Commission of the European Communities, contract no. Mas2-CT92-0027.

REFERENCES

Al-Salem, A.A., 1993. "Sediment transport in oscillatory boundary layers under sheet flow conditions". Ph.D. thesis, Delft Univ. of Technology.

Bosman, J.J., E.T.J.M. van der Velden and C.H. Hulsbergen, 1987. "Sediment concentration measurement by transverse suction. *Coastal Engineering*, vol.11, pp. 353-370.

Foster, D., R. Holman and R. Beach, 1994. "Correlation between sediment suspension events and shear instabilities in the bottom boundary layer of the surf zone". *Proceedings Coastal Dynamics '94*, ASCE, Barcelona, Spain.

Fredsøe, J., 1993. "Modelling of non-cohesive sediment transport processes in the marine environment", *Coastal Engineering*, Vol.21, pp. 71-103.

Horikawa, K. (ed), 1988. "Nearshore dynamics and Coastal Processes", University of Tokyo Press, Tokyo, 522 pp.

Katopodi, I., J.S. Ribberink, P. Ruol, R. Koelewijn, C. Lodahl, S. Longo, A. Crosato and H. Wallace, 1994. "Intra-wave sediment transport in an oscillatory flow superimposed on a mean current", Delft Hydraulics Report H1570.10, Part III.

Ramadan, K.A.H., 1993. "Time-averaged sediment transport phenomena in combined wave-current flows". Delft Hydraulics Report H 1889.11, Part I

Ribberink, J.S. and A. Al-Salem, 1991. "Near-bed sediment transport and suspended sediment concentrations under waves". *Proc. Int. Symp. on "The transport of suspended sediments and its mathematical modelling"*, IAHR, Florence, Italy, pp.375-388.

Ribberink, J.S. and A. Al-Salem, 1992. "Time dependent sediment transport phenomena in oscillatory boundary layer flow under sheet flow conditions". Delft Hydraulics Report H840.20 Part VI.

Van Kesteren, W.G.M. and W.T. Bakker, 1984 "Near bottom velocities in waves with a current; Analytical and numerical computations". *Proc. 19th Coastal Engineering Conference*, ASCE, Houston, USA, pp. 1161-1177.

HYDRODYNAMIC PROCESSES ON THE LOWER SHOREFACE OF THE DUTCH COAST

Piet Hoekstra[1] and Klaas T. Houwman[1]

ABSTRACT: The lower shoreface forms the link between the surfzone (or upper shoreface) on the landward side and the inner shelf on the seaward margin. Oscillatory flow processes and (mean) currents on the lower shoreface are responsible for an exchange of sediment with the coast. The lower shoreface near Egmond aan Zee is subject to wind-driven circulations. Near-bed orbital velocities are generally symmetric on the lower shoreface and even during storm conditions the near-bed wave asymmetry is limited. In the transition zone from shoreface to surfzone, asymmetry ratios near the bed increase to values of about 0.62 (significant onshore orbital velocities are 1.6 times the offshore velocities) during a heavy storm. In these conditions though the effect of wave asymmetry is totally overruled by offshore-directed near bottom flows due to wind set-up, breaking-induced undertow and possibly dynamic rip current systems.

INTRODUCTION

Littoral systems have commonly been divided in a series of sub-environments. Each subenvironment (e.g. the surfzone) is characterized by a dominant set of hydrodynamic processes and a certain morphology. In a cross-shore direction boundaries were assumed to exist between e.g. the inner shelf, the lower shoreface and the surfzone. At present though one realizes that there is a strong coupling and interaction between these subsystems (Niedoroda et al. 1985; Wright 1987 and Wright et al. 1991). In reality there is a continuum of processes and a transfer of water and sediment across the artificial boundaries of the consecutive systems. The frequency and intensity of these processes

---

[1]Netherlands Centre for Coastal Research (NCK)
Institute for Marine and Atmospheric Research Utrecht, Utrecht University, P.O. Box 80.115, 3508 TC UTRECHT, the Netherlands

changes in time and space and the influence of one process relative to another may vary substantially in a cross-shore direction. The shoreface is a typical transition zone in which both oscillatory and more steady flows act upon the substratum.

In the last decade shoreface systems of the Mid-Atlantic Bight (USA) have been subject of relatively extensive research. Hydrodynamic and sediment transport studies are reported by Swift et al. (1986) and Wright et al. (1991,1994). The role of shoreface dynamics in barrier island evolution, i.e. from a mixed hydrodynamical/geological point of view is discussed by e.g. Swift et al. (1985). Liu and Zarillo (1990,1993) have studied the substratum of the shoreface in relation to hydrodynamic sorting mechanisms and existing sources and sinks for sediment. From a summary, given by Wright (1987) it is clear that shoreface dynamics is determinded by (quasi-steady) tide-, wind- and density-driven currents, oscillatory flow processes (e.g. short wave asymmetry and wave-wave interaction), the interaction of waves and currents and the force of gravity. The inner shelf and shoreface of the Mid-Atlantic Bight is a typical storm-dominated environment, exposed to the Atlantic ocean with significant swell and with a small tidal range. Conditions on the Dutch coast are markedly different. The North Sea is a shallow shelf sea in a relatively enclosed tidal basin in which both waves and tides are probably of equal importance for coastal behaviour. Therefore concepts of shoreface dynamics developed for the US East coast are expected to be not generally applicable along the Dutch North Sea coast. For the Dutch coast there is only limited knowledge with respect to the hydrodynamic processes on the (lower) shoreface and the direction and relative magnitude of longshore and, especially cross-shore sediment fluxes. A first attempt to estimate sediment fluxes on the shoreface by modelling was undertaken by Roelvink and Stive (1990). A strongly schematized model was applied to compute the net-annual longshore and cross-shore sediment transport rates (by using a Bailard approach) due to combined currents and waves. Van Rijn (1993), evaluating cross-shore sediment transport mechanisms, presented new model computations for transport across the -20 and -8 m isobath. Both model studies suggest a dominantly longshore and onshore directed sediment transport on the shoreface.

SHOREFACE RESEARCH EXPERIMENT

To extend our knowledge of shoreface processes acting along the Dutch coast, a field research programme was initiated in 1991. The study is focussing on an analysis of the sediment exchange processes between surfzone and lower shoreface. The research project, commissioned by the Tidal Waters Division of the Ministry of Transport, Public Works and Water Management (at present RIKZ, National Institute for Coastal and Marine Management) is carried out in the framework of the Dutch Coastal Genesis Programme. The general aims of the shoreface project are:

* to determine the dominant hydrodynamic processes that generate suspended sediment fluxes in a longshore, and especially a cross-shore direction on the shoreface;
* to determine the dominant meteorological and hydrodynamical conditions for which one can observe a significant net sediment transport on the shoreface.

This paper discusses the current and wave-driven water motion on the lower shoreface. Emphasis is given to hydrodynamic processes acting in a cross-shore direction and their potential for inducing suspended sediment fluxes for average fair weather and storm conditions.

STUDY AREA

The experiment has been carried out near the village of Egmond aan Zee, province of North Holland, the Netherlands.

The coast of Egmond is oriented approximately NNE-SSW and is fully exposed to the North Sea. The morphology of the surfzone is characterized by an inner and outer breaker bar (Fig. 1). The top of the outer bar is located at an average depth of about -4 m below NAP (Dutch Ordnance Datum). Median grain sizes in the surfzone vary from 200 to 450 μm. The position of the outer bar marks the transition to the upper part of the lower shoreface, starting at a depth of -4 m and extending towards the -12 m isobath. This part of the shoreface has a straight slope of about 1:120 (= 0.0085 m/m). The lower part of the lower shoreface covers the area between the -12 m and -16 m isobath and the seafloor gradually merges into the inner shelf. The concave profile has an average slope of about 1:250 (= 0.004 m/m). Median grain sizes on the lower shoreface are finer than in the surfzone: D_{50} = 145 to 165 μm. Tides near Egmond aan Zee are semidiurnal with a neap tidal range of about 1.3 - 1.6 m and a spring tidal range of 1.8 - 2.1 m. There is a clear tidal asymmetry: rising tides are observed for about 4 hours, falling tides for about 8 hours. The associated flood and ebb tidal currents generally also are of different duration. The flood tidal current heading in a northward direction, parallel to the coast, is often stronger than the southward directed ebb tidal current. The local wave climate is characterized by waves incident from directions varying between SSW and NNW. The average annual significant wave height is about 1.2 m. Both wind waves with periods of 5-6 s and swell (8-9 s) are present.

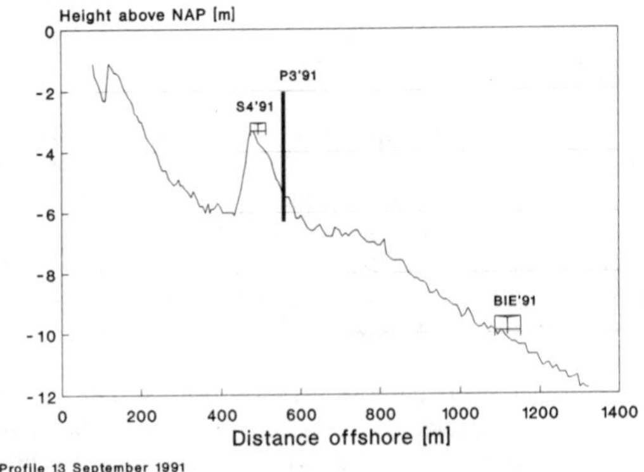

Figure 1. Nearshore morphology and measuring locations in 1991.

INSTRUMENTS

Measurements have been carried out in the autumn and winter of 1991 and 1992-1993 in the so-called Egmond transect (see e.g. Kroon 1994). This Egmond transect consists of a cross-shore array of instruments, partly covering the surfzone and shoreface. In 1991 the instrument array consisted of a measuring frame (type: Interocean S4) located on top of the outer bar (water depth ca. 3.5 m), a measuring pole on the seaward side of the outer bar (depth ca. 5.5 m) and an instrumented tripod deployed at a water depth of 10 m (Fig. 1). During the 1992 campaign the programme was somewhat reduced. The more landward located part was now only covered by the measuring pole, still located at the seaward side of the outer bar, although at a lesser depth (ca. 4m), due to the seaward migration of the outer bar. At -7 m depth contour an instrumented tripod was placed on the seafloor. Both in 1991 and 1992 a wave directional buoy (WAVEC), located at 15 m water depth, recorded the offshore wave conditions (wave height, period and direction). The buoy has a sampling frequency of 1.28 Hz and a burst length of 20 minutes. The S4 frame measured long- and cross-shore velocities at a height of about 0.5 m above the bed. The data were collected in a burst-mode sampling scheme with a burst of 40 minutes during every 3 hours, the sampling frequency was 1 Hz. The measuring pole was equipped with a capacitance wire and a pressure sensor to measure water level elevations (waves, tides and wind set-up). The burst length, interval and sampling frequency were 40 min., 1 hour and 4 Hz respectively. The instrumented tripods at greater depths (-7 and -10 m) measured long- and cross-shore flows at about 0.25 and 0.5 m above the bed. A pressure sensor in the frame recorded wave height and wave period. The instruments in these tripods measured with a sampling frequency of 2 Hz, a burst length of 34 minutes and a burst interval of 1 hour. A compass and tiltmeters recorded the orientation of each tripod. In addition to the near-bed velocities measured with tripods, ship-borne flow measurements (13 hours tidal measurements) have been carried out on December 22 (1992), in order to determine the vertical distribution of longshore and cross-shore flows.

RESULTS FIELD MEASUREMENTS

Wave regime

Wave conditions in 1991 and 1992 are characterized by both fair weather and storm events. In 1991 (Fig. 2) a heavy storm (Bft 9-10) between hour 40 and 140 resulted in offshore significant wave heights with a maximum of about 5 m and a period in the order of 9-10 s. During the peak of the storm, the tripod at -10 m isobath failed to operate and maximum recorded storm conditions correspond with an offshore wave height of about 3.5 - 3.8 m. The S4 data show an almost continuous record in 1991 (Fig. 2). In 1992 the maximum recorded wave height was about 2.3 m during a small storm.

Mean flows

Mean (and quasi-steady) cross-shore currents can be tide-, wind- and density-induced. Density-driven flows are presumably small near Egmond (compare Van Alphen et al. 1988 and Van der Giessen et al. 1990) and our data records do not contain information about the salinity distribution. Therefore, in this paper tide-, wind- and wave-driven (cross-shore) flows are considered.

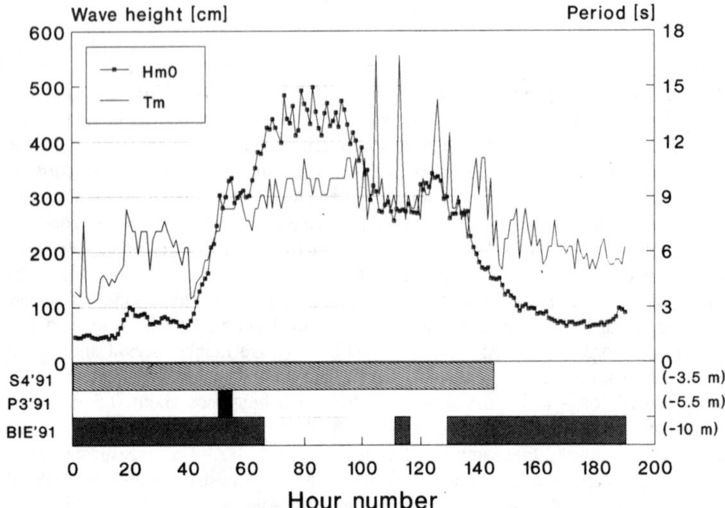

Figure 2. Offshore wave climate during the campaign of 1991. The bar-diagram shows operational hours of the tripods and the measuring pole

Tidal current measurements have been executed in the water column during a complete ebb-flood cycle, 2 days after neap tide. Measurements were performed during fair weather conditions (wind < 4 m/s; $H_s < 0.5$ m) at an average water depth of about 8 m. The tidal current ellipses (surface and bottom) are very flat and oriented shore parallel (Houwman and Hoekstra 1994). At the surface the current vector rotates in a clockwise direction in time; at the bottom an anticlockwise rotation is measured. A similar pattern is observed by Van der Giessen et al. (1990) further offshore (5 km). Peak flood flows at both surface and bottom are stronger than peak ebb flows, the ebb current has a longer duration. The horizontal tide though is far less asymmetric than the vertical tide. Most relevant observation is, of course, the absence of a significant tide-induced cross-shore current. Tidal flow in the nearshore zone is simply shore-parallel, characteristic for flow in the friction-dominated zone.

In the friction-dominated zone, the scale of the friction forces is much larger than the scale of the Coriolis effect. As a result currents tend to flow parallel to the direction of the forcing mechanism (compare Swift et al. 1985). For conditions with moderate to strong winds, the surface currents are aligned with the surface wind stress. A direct coupling, however, between wind shear and current velocities cannot always be determined. Regional variations in atmospheric pressure or persistent winds during a couple of days, will result in large-scale horizontal pressure gradients which will drive a flow as well.

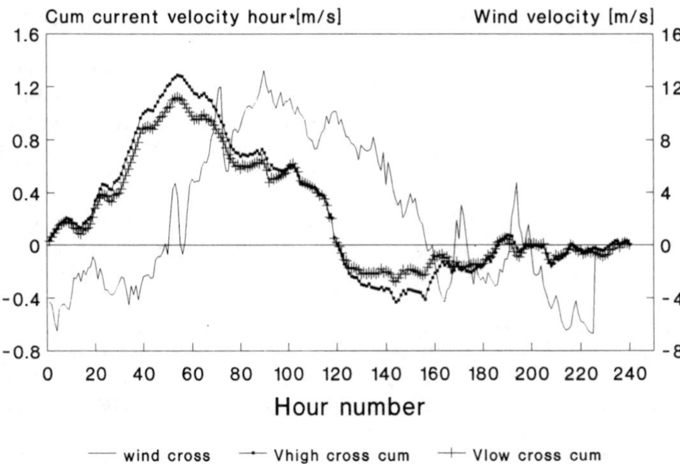

Figure 3. Cross-shore wind vector and cross-shore cumulative current measured near the bottom at the -7 m isobath (data 1992)

The effect of cross-shore wind patterns on cross-shore circulations is commonly determined by computing the residual flow over a great number of tidal cycles with a specific wind regime. Electro-magnetic flow meters though are generally not accurate enough to estimate the strength of small residual currents. One of the problems is the existence of a small offset. Consequently, a different approach is developed to determine the impact of the cross-shore components of the wind in the friction-dominated zone (data 1992; tripod at -7 m). Fig. 3 shows the cumulative cross-shore currents measured by 2 EMF-probes at different heights above the bed and in response to onshore and offshore winds over a period of 240 hours. The burst-averaged cross-shore current was added to the preceding burst values, resulting in a cumulative flow. To avoid any effects of systematic offsets, the average residual current over the entire time interval (240 h.) is assumed to be zero. In Fig. 3 a positive value for the wind corresponds with an onshore wind, likewise onshore flows also have a positive sign. A rising curve (or positive tendency) corresponds to an onshore flow component which is added to the ever present (small) tidal current. Time series of the cumulative current of both EMF-probes show a similar development in time, demonstrating the consistency of the data set. Onshore directed wind components are expected to yield onshore directed surface currents, building up a slope in the mean surface elevation. This offshore directed pressure gradient throughout the water column generates an offshore directed near bottom flow (downwelling; hour 5 to 125, Fig. 3). Offshore wind components (hour 0 to 55; 160 to 240) create an upwelling circulation. The pattern is not always consistent and may have been affected by the presence of waves. The period between hour 55 to 140 is characterized by wave heights (Hm0) varying from 1.0 m (55 h.) to 2.3 m (around h. 90) and decreasing again to 1.5 m (after 140 h.). In a water depth of 7 m some wave breaking may have taken place, producing an undertow and enhancing the offshore directed flow.

Remarkably though is that around the moment of maximum wave height (90 h.) the average offshore directed near bottom current is reduced in magnitude (hour 80 to 112; V=0.01 m/s) instead of showing an increase. Another complicating factor in this respect is the effect of the vertical tide, having influence on the downwelling mechanism and the magnitude of the undertow. For offshore wind (set-down), onshore near-bottom flows can have a magnitude of 1-2 cm/s. For moderate onshore winds (≤ 12 m/s; set-up), offshore near bottom flows reach already a strength of about 2-5 cm/s (Fig. 3).

Measurements on top of the outer bar, taken by the S4 frame (1991), demonstrate the existence of strong offshore-directed near bottom flows (Fig. 4). For the first 40 h. there is a more or less weak tidal oscillation. After hour 40 the influence of a heavy storm becomes visible. Wind velocities (initially from the SW) reach a maximum of 28-30 m/s (Bft 9-10) and the wind gradually veers to the north. The maximum observed offshore significant wave height (WAVEC) is 5.3 m. When the storm is at its height, the maximum measured longshore current (0.5 m above the bed) is 1.3 m/s and most probably consists of a tidal component and a strong wave- and wind-driven component. Especially in shallow water (here: 3.5 m) the local wind stress will have a great influence. The maximum cross-shore current is measured at hour 53 (0.72 m/s). There is not yet a strong onshore wind: the cross-shore wind does not become important before hour 60 and the downwelling component must still be small (Fig. 4). Meanwhile, the increase in offshore directed currents is accompanied by an increase in offshore wave heights but the maximum in offshore wave height doesn't correspond with the peak offshore flows. This partly rules out a predominant role for the breaking-induced undertow. The magnitude of the undertow is expected to vary with the energy (= wave height) of the incoming wave fields. Wave saturation - wave height everywhere limited by depth - may offer an explanation in this case. Another striking observation is the appearance of large fluctuations in the cross-shore current from one hour to the next. There seems to be no relationship with the fluctuations in the longshore current or the incoming wave height. And neither a return flow as result of wind set-up nor a breaking-induced undertow seems to be able to produce such a strongly pulsating and fluctuating cross-shore flow. Even values within a burst do show large fluctuations (Houwman and Hoekstra 1994).

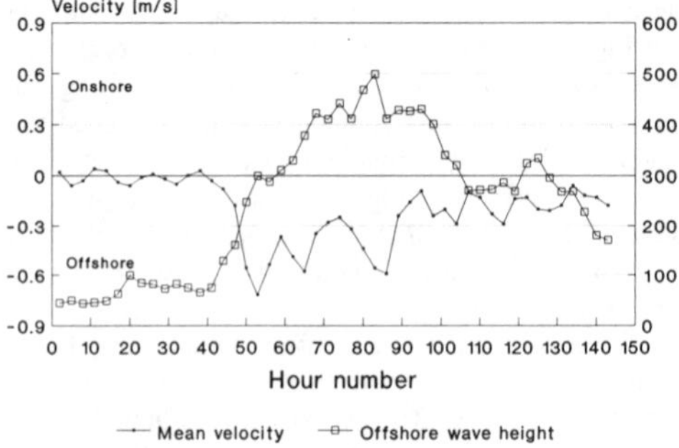

Figure 4. Cross-shore near-bottom velocities measured on top of the outer bar (S4 frame), in relation to offshore wave height and asymmetry of the near-bed orbital velocities

This has lead to the hypothesis that part of the cross-shore flow is generated by a dynamically shifting and pulsating rip-current system above the outer bar. Such a rip-current system has no stable character and moves across the outer bar without significantly eroding the bed. Any morphological evidence for its existence is therefore lacking. In conclusion, the offshore near-bottom flow on the outer bar is considered to be the result of 3 major components: a breaking-induced undertow, a highly dynamic rip-current system and a wind-driven circulation due to wind set-up.

Oscillatory flows: wave asymmetry

Oscillatory flow processes such as wave asymmetry and the streaming mechanism (Longuet-Higgins) can result in a net wave-driven transport. The streaming mechanism is hard to verify in the field and no reliable information in this respect is available. The role of wave asymmetry in onshore transport is not as obvious as one may expect, as discussed by Wright (1987). Even if the near-bed orbital velocities show an asymmetry, suspended sediment transport is not necessarily in an onshore direction. This is due to wave-wave interaction (bound long waves relative to short waves) and the effect of bed forms and corresponding phase lags. In model studies the estimated role of wave asymmetry very much depends on the applied wave theory.

Waves propagating into shallow water are subject to wave transformation. The mean shape of the wave changes: the crest becomes steeper and higher and the trough will be longer with a smaller amplitude. The onshore directed peak orbital velocity is larger than the offshore directed peak orbital velocity. The asymmetry (A_η) of the wave surface profile is described by the following ratio:

$$A_\eta = \frac{\eta_{s,cr}}{\eta_{s,cr} + \eta_{s,tr}} \quad (1)$$

where:

$\eta_{s,cr}$ = significant (average of highest third part of the) crest amplitude
$\eta_{s,tr}$ = significant (average of highest third part of the) trough amplitude

The asymmetry of the oscillating wave velocity is defined in the same way by replacing crest amplitude by the significant onshore velocity ($V_{1/3,on}$) and trough amplitude by the significant offshore velocity ($V_{1/3,off}$). The asymmetry of the oscillating pressure under a wave is defined in a similar way. In theory there is a difference between the wave asymmetry of a total (demeaned) time-series and the wave asymmetry of the high-frequency ($f > 0.05$ Hz) time-series due to the presence of bound long waves. In both series the asymmetry is generally dominated by the short waves. Due to the presence of bound long waves, the peak value of the crests - relative to the zero level - is reduced and the peak value of the troughs has increased. The (demeaned) time-series with bound long waves is characterized by a lower wave asymmetry than the high-frequency series. During high energy conditions bound long waves are responsible for a significant part of the infragravity energy (Houwman and Hoekstra 1994). On the shoreface though the contribution of this infragravity energy to the total amount of wave energy is only small. This is confirmed by the calculation of wave asymmetries of both the (demeaned) time series and the asymmetry of the corresponding high-frequency time series (pole data 1993, shoreface - outer surfzone; period 22 days). From these computations it is obvious that the reduction of wave asymmetry by the bound long wave is only of minor importance; there hardly is a difference between both asymmetry-ratios (Houwman and Hoekstra 1994). In the following analysis the wave asymmetry of the (demeaned) time series has been used. It is essentially the difference between the onshore peak velocity and the offshore peak velocity, the wave asymmetry of the demeaned time series, which determines the integrated sediment transport. To compute the actual asymmetry from an individual burst the mean current (or average water level or pressure) has to be removed from this time series; one obtains a demeaned time series. If a standard regression polynoom is used to remove the mean value, the sum of squares will be minimized. In case of asymmetrical waves the sum of the squared values is dominated by the crest values. When this regression line is used to remove the mean current (or water level, pressure) the burst mean value will not be zero (but higher) and the asymmetry of the orbital motion will be underestimated. To avoid this problem a burst is divided in 20 subseries and mean values are calculated from these subseries. A second order polynoom is calculated using these mean values, and the polynoom is taken to remove the mean current from the time series. A zero-crossing method is applied to define the individual crests and troughs.

Computations of the mean wave shape also give an idea of the degree of asymmetry and provide information about the validity and applicability of several wave theories in describing the wave field. The wave shape is calculated from the time series by following a stepwise approach. This approach includes e.g. the removal of the low frequency motions ($T \geq 20$ s.), the selection of the highest third part of the waves, the definition of a spline-function to describe these waves and a computation of the normalized wave amplitudes. Details of the computational procedure are given by Houwman and Hoekstra (1994). In order to make a fair comparison between the individual measurements, the wave asymmetry is expressed as a function of the near bed significant onshore velocities ($V_{1/3,on}$; see Fig. 5 and 6). To have an additional scaling factor for maximum local wave

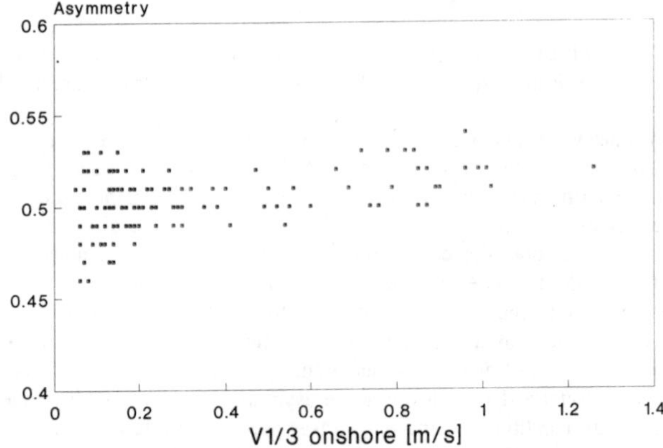

Figure 5. Wave asymmetry of the near-bed cross-shore velocities as a function of the significant onshore velocity in a water depth of 10 m (1991)

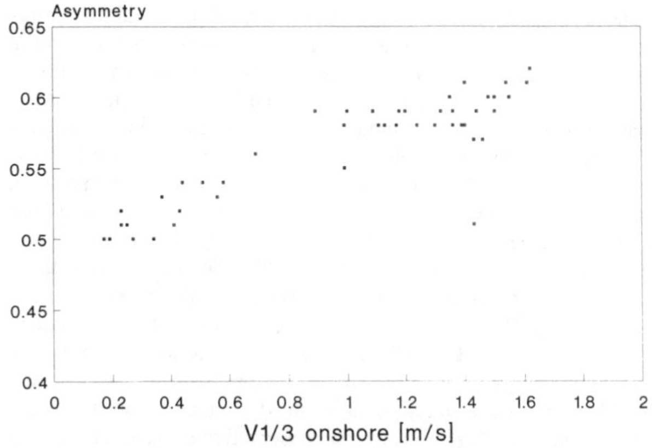

Figure 6. Wave asymmetry of the near-bed cross-shore velocities as a function of the significant onshore velocity in a water depth of 3.5 m (1991)

conditions, the relative wave height - the wave height to water depth ratio - is used. The distribution of relative wave height over the shoreface, in relation to offshore wave conditions, is given in Table 1. The estimated local relative wave heights are based on model computations of the wave height transformation across the shoreface (UNIBEST-model; Delft Hydraulics).

Table 1. Offshore wave conditions and relative wave height (H/h) distribution across the shoreface

$H_{s,off}$ (m)	Period (s)	Relative wave height -10 m	-7 m	-4 m
2.3	6.7	0.23	0.33	0.48
2.5	7.6	0.25	0.35	0.47/0.44*
3.5	7.6	0.34	0.44	0.50

*) Measurements

Wave asymmetry ratios of the cross-shore velocity signals, measured for variable wave conditions and at different depths are presented in Fig. 5 (-10 m) and 6 (-4 m). As illustrated in Fig. 5 at low energy conditions ($V_{1/3,on}$ < 0.25 m/s) the asymmetry ratio fluctuates between 0.46 and 0.53. For high energy conditions ($V_{1/3,on}$ > 1 m/s, offshore wave height 3.5-3.8 m and H/h about 0.34) the ratio never exceeds a value of 0.53. The oscillating motions near the bottom are still nearly symmetrical. Computations of the mean wave shape during storm conditions showed a similar tendency, the oscillating velocity has an almost sinusoidal shape. The strong fluctuations for low energy conditions seem to be related to tidal conditions, i.e. the vertical tide and the longshore tidal current. For obliquely incident, low-energy waves the wave shape will be affected by both a partly opposing current and a current with a flow component in the direction of wave propagation. For more energetic wave conditions, this mechanism is largely suppressed. Wave asymmetry ratios of the cross-shore velocity signals, recorded at the -7 m depth contour, confirm the trends observed at -10 m. For maximum offshore significant wave heights of 2.3 m, the wave asymmetry again reaches a maximum value of about 0.53. Based on relative wave heights (H/h=0.33; Table 1) the local conditions are comparable with those measured at the -10 m isobath. Fig. 6 illustrates the asymmetry ratios, based on measurements on top of the outer bar (S4 frame). Maximum measured significant onshore velocities of about 1.6 m/s correspond with maximum offshore wave heights of about 5 m (Fig. 2). The cross-shore velocity signals of the S4 frame have been sampled with a frequency of 1 Hz. Unfortunately, for waves with sharp crested peaks as expected during high energy conditions, this sampling frequency is not sufficient for accurately reproducing the wave shape. As a result, the actual asymmetry is underestimated and the data presented in Fig. 6 do give a conservative estimate of the existing wave asymmetry. The wave asymmetry of the near bed orbital velocities already becomes important for relatively small onshore velocities (0.3-0.4 m/s). Maximum asymmetry (0.62) is reached during maximum storm conditions for H/h is about 0.50 (Table 1). Wave-breaking is expected for waves in the range H/h = 0.4 - 0.8. Therefore, many asymmetric waves probably will no longer exist at this point, because they have broken further offshore. This is supported by observations of the mean offshore directed near bottom flows, recorded by the S4 frame (Fig. 4) during the storm, and the assumed relation with the breaking-induced undertow component.

Wave shape and asymmetry, this time of the wave surface profile instead of the near bed orbital velocities, is also determined from the measuring pole data (outer bar). The capacitance wire measured the actual water surface elevation whereas the pressure gauge recorded the pressure at about 2.4 m below the mean water surface. The local water depth near the pole varied from 4.8 m in 1991 to 4.0 m in 1993. An impression of the computed mean wave shape, showing the asymmetry of the wave field, is given in Fig. 7 (hour 50, 1991; H/h=0.44, beginning of the storm). In 1991 the maximum measured asymmetry of the surface waves is 0.67, meaning a crest height which is twice the trough height. For the 1993 data set for a somewhat smaller depth and for a local bottom slope $\tan \beta = 0.02$, the wave asymmetry of the surface elevation for a total period of 520 hours was calculated. In this case the maximum asymmetry is 0.72, corresponding with a crest height which is 2.6 times the trough height. In the latter case there also appears to be a strong coupling between the wave asymmetry of the surface waves and the relative wave height. For H_s/h equal or larger than 0.1 there is a strong linear relationship between both parameters:

$$A = 0.5 + 0.4 * \left(\frac{H_s}{h} - 0.1 \right) \tag{2}$$

For $H_s/h < 0.1$ this relation is less evident since for low values of H_s there generally is a large scatter in the asymmetry.

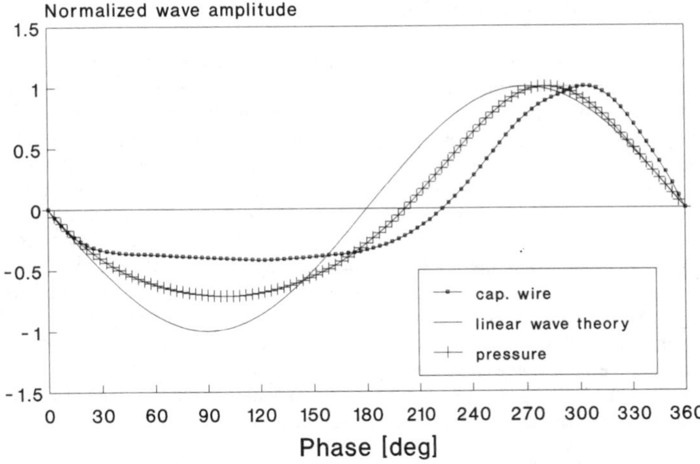

Figure 7. Computed mean wave shape during hour 50 (1991), at the beginning of the storm, and based on measuring pole data

A comparison of the data collected with the capacitance wire and the pressure gauge (Fig. 7) shows a significant difference in the degree of asymmetry: the shape of the waves computed from the pressure gauge is far less asymmetric than from the capacitance wire. It clearly demonstrates the principle of decay of especially the high-frequency oscillations from the surface to the bottom. In theory, an asymmetrical wave can be decomposed into a harmonic sinusoidal wave and several waves with a higher frequency. The individual frequencies show a differential decay of orbital velocity and pressure with depth. According to simple linear wave theory, the higher the frequency, the greater the decay with depth. This implies that the contribution of high-frequency waves to the resulting wave shape diminishes with depth in the water column. The asymmetry of the orbital motion near the bottom is less than at the surface. In reality there will be an interaction between the different wave frequencies and non-linear wave theories (e.g Stokes, 2nd order Cnoidal or Rienecker and Fenton) have to be applied to estimate the actual decay of wave asymmetry with depth. The decay with depth is of great importance in water depths ranging from 3 to 15 m. At greater depths the waves are commonly not subject to shoaling and remain sinusoidal. For shallow water (< 3 m) the difference in decay of the distinct frequencies becomes negligible. Summarizing the facts, the wave asymmetry of the near-bottom orbital velocities (or pressure) not only depends on the relative wave height (and offshore wave climate), as illustrated by e.g. Fig. 5 and 6 and eq. 2, but is also a function of the total water depth in relation to the differential decay of especially higher wave frequencies.

DISCUSSION AND CONCLUSIONS

Wave asymmetry has always been considered as an important process in generating onshore sediment fluxes on the shoreface. The observations in the Egmond transect indicate that on the upper part of the lower shoreface (measurements at -10 and -7 m) the effect of wave asymmetry is limited. For fair weather conditions (H_s = 0.5-1.0 m) the asymmetry-ratio is varying around 0.5, corresponding with symmetric and sinusoidal waves. During storm conditions (H_s > 2.0 m; locally H/h=0.33-034) asymmetries do not exceed the value of 0.53 ($V_{1/3,cr} = 1.10 * V_{1/3,tr}$) and waves are still nearly symmetrical. In most cases linear wave theory can still be applied to calculate the near bed orbital velocities. This means that the mechanism of onshore sediment transport by asymmetrical wave orbital fluid stresses during non-storm conditions, as described by e.g. Niedoroda et al. (1985), cannot play an important role at this location on the Dutch shoreface. In the transition zone from shoreface to surfzone, on top of the outer bar, again for fair weather conditions there hardly is any asymmetry. During heavy storm, however, the asymmetry increases to 0.62 ($V_{1/3,cr} = 1.6 * V_{1/3,tr}$; a conservative estimate). In these conditions wave asymmetry may be responsible for a significant onshore transport.

The asymmetry of surface waves clearly is no criterion for the asymmetry of the near bed orbital velocities, especially for intermediate water depths ($3<h<15$ m). The degree of asymmetry decreases with depth in the water column due to the differential decay of higher wave frequencies with depth. This effect can only be predicted by non-linear wave theories such as the 2nd and 3rd order Stokes, the Cnoidal (1 and 2) and Rienecker and Fenton. An appropriate wave theory for describing cross-shore sediment fluxes in the nearshore zone should be able to reproduce the wave shape, asymmetry and its decay with depth with a sufficient degree of accuracy.

In previous research (e.g. Swift et al. 1985 and Van Rijn 1993) the role of wave asymmetry has probably been emphasized due to the application of wave models that overestimate the wave asymmetry. Swift et al. (1985) calculated asymmetrical orbital stresses using second order Stokes theory at a water depth of 10 m. He found a strong onshore sediment flux as result of shoaling waves. Van Rijn, again applying second order Stokes theory for calculating the sediment transport across the -8 m isobath, found asymmetry ratios varying from 0.5 to 0.56 for wind waves and around 0.6 for swell. A comparison of wave asymmetry, computed with the 2nd order Stokes theory, and measurements from the Egmond data set clearly shows that the Stokes theory overpredicts the asymmetry of the near-bed orbital velocities at these depths (Houwman and Hoekstra 1994).

With respect to mean cross-shore currents a large part of the shoreface near Egmond aan Zee, up to at least a level of about -7 m NAP, is affected by upwelling and downwelling mechanisms. Moderate to strong onshore winds, occurring most frequently along the Dutch coast and generating low to high-energy waves, result in offshore directed near-bottom flows. During offshore winds generally coincident with the absence of waves, there is a weak onshore flow. In the area where wave asymmetry becomes important, in the transition zone from lower shoreface to surfzone, strong offshore directed near-bottom flows do exist. These flows are the combined result of breaking induced undertow and a wind set-up and may be temporarily (and locally) enhanced by dynamic and pulsating rip current systems.

In conclusion, on the lower shoreface near Egmond aan Zee, mean cross-shore flow processes are probably dominant. As soon as wave asymmetry becomes important its effect on suspended sediment transport can be totally overruled by the existence of strong offshore-directed near bottom flows.

ACKNOWLEDGMENTS

This research project is part of the Coastal Genesis Programme (contracts DG-319 and DG-475) and is funded by the Tidal Waters Division (at present RIKZ, National Institute for Coastal and Marine Management) of the Ministry of Transport, Public Works and Water Management. Important logistic support was offered by the Directorate North-Holland of the same Ministry. Technical and financial assistance was also obtained from the Department of Physical Geography, Utrecht University.

REFERENCES

Houwman, K.T., and Hoekstra, P., 1994. "Shoreface Hydrodynamics," Institute for Marine and Atmospheric Research Utrecht, Utrecht University, Report IMAU R-94.2, 29 pp.

Kroon, A., 1994. "Sediment transport and morphodynamics of the beach and nearshore zone near Egmond, the Netherlands," thesis, Utrecht University.

Liu, J.T., and Zarillo, G.A., 1990. "Shoreface dynamics, evidence from bathymetry and surficial sediments," *Marine Geology*, Vol. 94, pp. 37-53.

Liu, J.T., and Zarillo, G.A., 1993. "Simulation of grain-size abundances on a barred upper shoreface," *Marine Geology*, Vol. 109, pp. 237-251.

Niedoroda, A.W., Swift, D.J.P., and Hopkins, T.S., 1985. "The shoreface," in: Coastal Sedimentary Environments (ed. R.A. Davis), Springer Verlag, New York, pp. 533-624.

Roelvink, J.A., and Stive, M.J.F., 1990. "Sand transport on the shoreface of the Holland Coast," *Proceedings 22nd Coastal Engineering Conference*, ASCE, New York, pp. 1909-1921.

Swift, D.J.P., Niedoroda, A.W., Vincent, C.E., and Hopkins, T.S, 1985. "Barrier island evolution, Middle Atlantic Shelf, U.S.A. Part 1: Shoreface Dynamics," *Marine Geology*, Vol. 63, pp. 331-361.

Swift, D.J.P., Han, G. and Vincent, C.E., 1986. "Fluid processes and sea-floor response on a modern storm-dominated shelf: Middle Atlantic Shelf of North America. Part 1: The storm-current regime," in: Shelf Sands and Sandstones (eds. R.J. Knight and J.R. Maclean), Canadian Society of Petroleum Geologists, Memoir II, pp. 99-119.

Van Alphen, J.S.L.J., Ruijter, W.P.M. de, and Borst, J.C., 1988. "Outflow and three-dimensional spreading of River Rhine water in the Netherlands coastal zone," in: Physical Processes in Estuaries (eds. J. Dronkers and W. van Leussen), Springer Verlag, pp. 70-92.

Van der Giessen, A., Ruijter, W.P.M. de, and Borst, J.C., 1990. "Three-dimensional current structure in the Dutch Coastal zone," *Netherlands Journal of Sea Research*, Vol. 25, pp. 45-55.

Van Rijn, L.C., 1993, "Considerations on cross-shore sediment transport," Netherlands

Centre for Coastal Research - Delft Hydraulics, Report H840, 25 pp.

Wright, L.D., 1987. "Shelf-surfzone coupling: diabathic shoreface transport," *Proceedings Coastal Sediments 87*, ASCE, New York, pp. 25-40.

Wright, L.D., Boon, J.D., Kim, S.C., and List, J.H., 1991. "Modes of cross-shore sediment transport on the shoreface of the Middle Atlantic Bight," *Marine Geology*, Vol. 96, pp. 19-51.

Wright, L.D., Madsen, O.S., and Chisholm, T.A., 1994. "Inter continental shelf bottom boundary layer processes and across-shelf sediment transport: Field experiments in the Middle Atlantic Bight," *Proceedings Coastal Dynamics '94*, ASCE, New York, this volume.

INNER CONTINENTAL SHELF TRANSPORT PROCESSES: THE MIDDLE ATLANTIC BIGHT

L.D. Wright[1], O.S. Madsen[2], T.A. Chisholm[1], and J.P. Xu[1]

ABSTRACT: A field study focused on across-shelf sediment transport has been conducted on the inner continental shelf off the U.S. Army Corps of Engineers Field Research Facility at Duck, North Carolina. Data were obtained from depths of 7 to 17 m using in situ bottom boundary layer instrumentation systems over the period 1985 - 1992 and under the full range of conditions from fairweather to severe storms. Wind-driven seaward flows prevail near the bed during storms and appear to be the dominant mode of offshore sediment flux. Incident waves under moderate energy conditions and during the waning phase of storms dominate onshore transport. Bed roughness variations affect sediment resuspension and transport. Observations during the extreme "Halloween storm" of 1991 showed large seaward transports to have been associated with high waves, downwelling flows and sheet flow roughness conditions. Modeling results suggest that the fluxes of suspended sediment past the 8 m and 13 m isobaths differed only slightly and, hence, that this depth region serves more as a pathway than a sink for sediments eroded from the nearshore zone. At other times, however, inner shelf bed level fluctuations of up to 18 cm testify to significant flux divergence.

---

1) School of Marine Science, Virginia Institute of Marine Science, College of William and Mary, Gloucester Point, VA 23062, USA.

2) Ralph M. Parsons Laboratory, Massachusetts Institute of Technology, Cambridge, MA 02139, USA.

INTRODUCTION

Predicting large scale coastal stability or change requires that we understand the morphodynamic processes that operate on the entire inner continental shelf (depth < 30 m). Surf zone and beach behavior is closely coupled with the inner shelf particularly on decadal time scales. The inner shelf is a friction-dominated realm in which surface (air - sea) and bottom boundary layers overlap and commonly occupy the entire water column (Nittrouer and Wright 1994). Understanding and modeling inner shelf morphodynamic behavior requires knowledge of: (1) inner shelf physical oceanography; (2) bottom boundary layer processes and bed micromorphodynamics; (3) sediment suspension and transport; and (4) the sediment flux divergences that cause erosion or accretion. In an ongoing field observation program that began in 1985 we have been examining all four aspects of this problem as they operate in the southern part of the Middle Atlantic Bight. In this paper we offer a short overview of some of the more significant results to date.

FIELD OBSERVATIONS

Field data were obtained on the inner shelf seaward of the U.S. Army Corps of Engineers Field Research Facility at Duck, North Carolina in the southern part of the Middle Atlantic Bight (Lat. 36° 07' N; Long. 75° 39' W). The upper part of the inner shelf profile, from the surf zone to the 20 m isobath, is concave upward; a field of sand ridges prevails at depths of 20 to 30 m (Figure 1). Instrumented tripods or tetrapods were deployed at depths of 7 m to 17 m, often in pairs, in seven field campaigns conducted in autumn 1985 (fairweather and storm), summer 1987 (fairweather), winter 1988 (swell-dominated), summer 1991 (post-hurricane fairweather), autumn 1991 (severe storm), late winter 1992 (strong offshore winds), and autumn 1992 (fairweather and mild storms). The instrumentation systems utilized 4 or 5 electromagnetic current meters in near-bed (z < 125 cm) arrays to measure velocity profiles, 5 optical backscatterance (OBS) turbidity sensors to measure suspended sediment concentration profiles, pressure transducers, and digital sonar altimeters. Descriptions of the instrumentation can be found in Wright *et al.* 1991 and Wright 1993.

PHYSICAL OCEANOGRAPHIC PROCESSES

The physical oceanographic regime of the Middle Atlantic Bight inner shelf is characterized as storm dominated. It is extratropical northeasterly storms that generate the largest waves and the strongest currents. The largest waves and associated bed stresses are accompanied by strong, southerly setting currents with downwelling across-shelf flows. Figure 2 illustrates the relationship among wave height and along-shelf and across shelf flows as

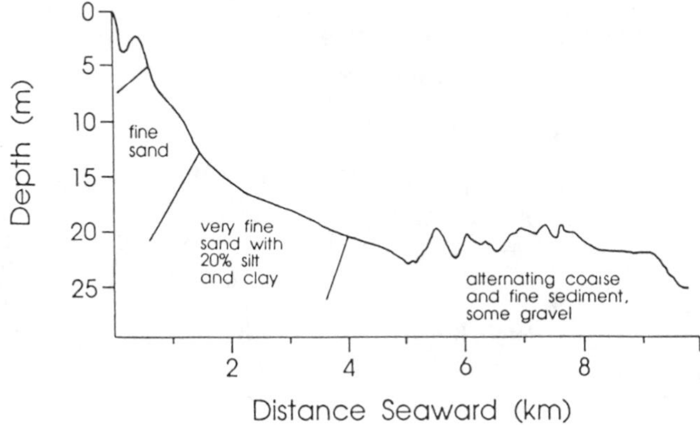

Figure 1. Cross sectional profile of the inner shelf off Duck, North Carolina.

observed in autumn 1992. Notably, the highest waves (H ~ 1.5 m in this example) coincide with southerly setting wind-driven near bottom current velocities of over 20 cm s^{-1} and seaward flows of 5 cm s^{-1}. Data obtained from the 13 m isobath during the severe "Halloween storm" showed a qualitatively similar relationship. However, during that storm wave height reached 6 m and wind-driven near bottom (z = 125 cm) current speeds reached 50 cm s^{-1} (Madsen *et al.* 1993; Wright *et al.* 1994). The across-shelf component of mean flow attained a maximum seaward velocity of 10 cm s^{-1} during the rising phase of the storm but, near the end of our record, when wave-induced bed stresses approached a maximum, flows turned onshore (Wright *et al.* 1994). Probably because of frequent breaking, the wind-drag coefficient over the inner shelf during the Halloween storm had a high value of $C_a \sim 4.7 \times 10^{-3}$ (Madsen *et al.* 1993).

Strong infragravity pulsing of the mean current, particularly of the across-shelf component, caused significant intensification of seaward flows with maxima reaching 20 cm s^{-1} (Wright *et al.* 1994). In many of the bursts from the Halloween storm the seaward pulses coincided with packets of high waves, consistent with the model of group-bound long waves. However, this relationship did not hold consistently.

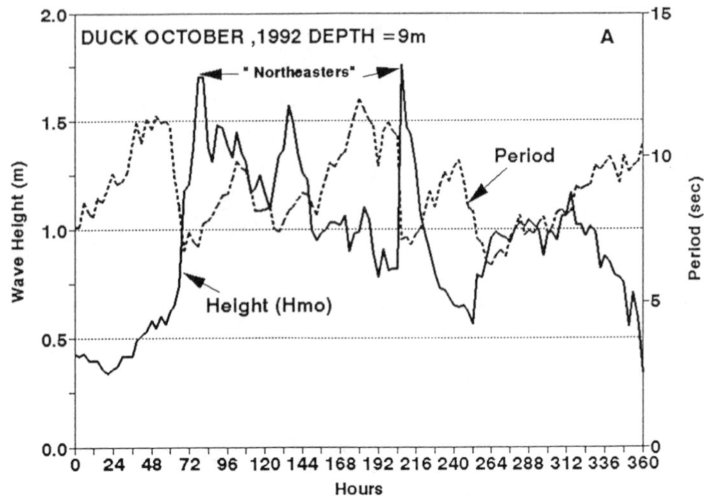

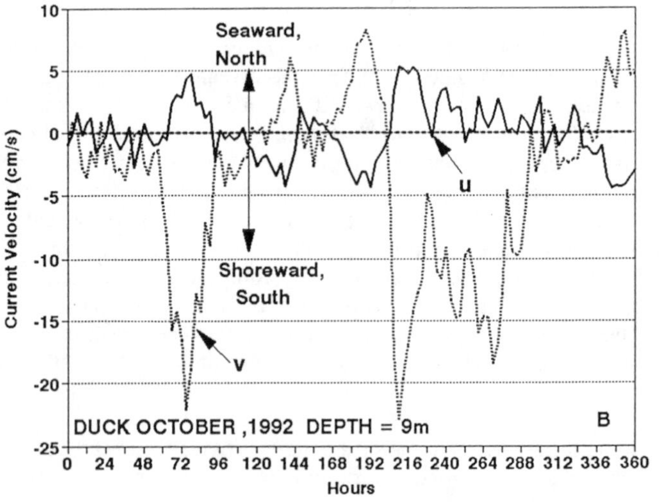

Figure 2. Waves (A) and currents (B) measured over the period 27 October - 11 November 1992 on the 9 m isobath off Duck, NC.

A data set was obtained during a period of strong westerly and southwesterly winds (offshore) in March 1992. The aim was to evaluate the extent to which these upwelling-favorable conditions contribute to shoreward transport. As hypothesized northerly setting alongshelf flows were accompanied for part of the time by shoreward across-shelf flows near the bed. Although the across-shelf flows attained speeds of up to 10 cm s^{-1}, they were not exclusively shoreward. Furthermore, since the offshore winds were accompanied by low waves, total bed stress was low. The conclusion from these observations was that offshore winds probably make a small contribution to across-shelf transport.

BOTTOM BOUNDARY LAYER CHARACTERISTICS

We utilized log-layer velocity profiles obtained from the tripods to estimate the apparent hydraulic roughness heights, z_o', under different conditions. We relied on sediment profiling camera imagery and diver observations for field checking modelled ripple geometries. The model of Grant and Madsen (1986) was used to estimate the friction velocities related to total skin friction (u_* skin friction), total combined wave and current friction (u_{*cw}), wave friction (u_{*wm}) and the friction related to the current (u_{*c}). Temporal variability of these quantities is large. Table 1 summarizes the roughness characteristics and boundary layer quantities observed during different experiments. Included in the table are some results reported by Wright (1993).

Bed roughness varied considerably ranging from large ripples during fairweather to a plane moving bed (sheet flow) during the Halloween storm. Velocity profiles indicate that the lowest effective hydraulic roughness prevailed during a post-hurricane fairweather period in August 1991 when a hummocky irregular bed was mantled by a layer of freshly-deposited mud. Although effective hydraulic roughness estimated from mean current profiles was enhanced by wave-current interactions during the severe storm, removal of the wave-current boundary layer effects revealed a relatively low "sheet flow" roughness length, k_n, equal to about 15 times the median sediment diameter (Madsen *et al.* 1993). The largest roughness in the data set occurred in winter 1988 when rounded but significant wave ripples coexisted with a relatively thick wave boundary created by long-period swell.

SEDIMENT SUSPENSION

Application of conventional Shields criteria to estimate the "threshold" or critical skin friction shear stress required to initiate motion of the silty fine sand on the Duck inner shelf yields a value of 0.15 to 0.16 Pa. However, deployment of an annular seabed flume to obtain direct in situ measurement of the critical stress at which particles are suspended into the water column gave a

TABLE 1 SUMMARY OF BOTTOM BOUNDARY LAYER CHARACTERISTICS: DUCK INNER SHELF

Parameters	Summer fairweather July,1987	Post-hurricane fairweather August,1991	Winter swell-dominated January,1988	Moderate energy October,1992	Extra-tropical storm October,1991
Bed roughness characteristics	Large ripples & biogenic	Ripples on mounds & holes	Small ripples, Irregular	Small ripples, Irregular	Highly mobile plane bed
u*skin fric.	1.2-1.6 cm/s	1.0-1.4 cm/s	2.9-4.5 cm/s	1.5-2.5 cm/s	3.6-7.7 cm/s
u*cw	2.8-6.5 cm/s	4.0-5.0 cm/s	7.9-8.2 cm/s	4.2-8.2 cm/s	7.8-12.2 cm/s
u*wm	2.8-6.4 cm/s	4.0-5.0 cm/s	7.8-8.0 cm/s	3.5-8.2 cm/s	7.7-11.9 cm/s
u*c	0.7-2.0 cm/s	0.1-1.4 cm/s	0.5-1.5 cm/s	0.2-3.2 cm/s	1.0-4.3 cm/s
z'o	1.09+/-0.28 cm	0.83+/-0.15 cm	2.50+/-0.52 cm	1.01+/-0.65 cm	1.45+/-0.68 cm
Cd(1m) x 10^-3	8.17+/-1.05	7.04+/-0.54	12.30+/-1.38	7.59+/-1.35	8.10+/-2.10

somewhat higher value of 0.22 Pa (Maa *et al.* 1993). Under fairweather conditions the higher level rarely is exceeded at depths greater than 12 m. However, under moderate and high energy conditions the entire inner shelf is activated. Figure 3 shows skin friction shear stresses (estimated by the model of Grant and Madsen 1986) at the 9 m and 14 m instrument sites during the autumn 1992 experiment period. The two threshold levels, also shown on Figure 3, are seen to have been exceeded for most of this period. Typical suspended sediment concentrations within the wave boundary under these moderate conditions were over 5 g l^{-1}; concentrations at an elevation of 20 cm reached 1 g l^{-1}.

During the 1991 Halloween storm, concentrations of over 1 g l^{-1} extended to elevations as high as 147 cm above the bed. Within and just above the wave boundary layer concentrations exceeded 5 g l^{-1} (Madsen *et al.* 1993; Wright *et al.* 1994). These high concentrations were responses to skin friction shear stresses that reached over 3 Pa near the end of our records and remained above 1 Pa throughout the event. From the observed skin friction shear velocities and suspended sediment concentration profiles, Madsen *et al.* (1993) estimated that the resuspension coefficient, γ_o, for the storm-driven sheet flow conditions was ~ 4 x 10^{-4} with the reference concentration assumed to be at 7 grain diameters above the bed.

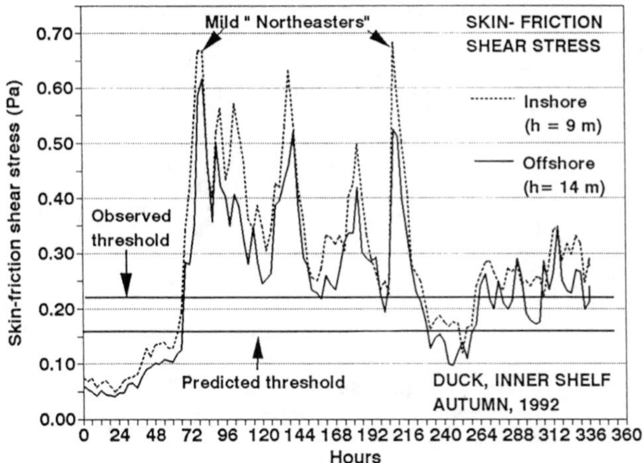

Figure 3. Skin friction shear stresses at h = 9 m and h = 14 m over the period 27 October - 11 November 1992.

ACROSS-SHELF SEDIMENT FLUX

Across-shelf sediment fluxes were determined in two ways: (1) by time averaging the instantaneous product of elevation-dependent concentration (C) and across-shelf flow (u); and (2) by application of a numerical model to observed of wind and wave conditions. Both approaches indicate that the largest flux rates occur during storms and are largely associated with wind-driven, downwelling, near bottom currents (Wright *et al.* 1991; 1994). However, cospectral analyses show that important, but secondary roles are also played by infragravity oscillations and by waves. During the Halloween storm, observed seaward fluxes ($<u(t)\ C(t)>$) at z = 27 cm above the bed reached 350 g m^{-2} s^{-1}; higher rates are assumed to have prevailed closer to the bed (Wright *et al.* 1994). Fluxes at infragravity frequency were seaward for most of the time and those at incident wave frequency were most commonly shoreward. Under moderate energy wave conditions, particularly when the onshore component of wind is weak, waves appear to dominate the shoreward transport of sediment. The shoreward transport by waves was especially evident in data from the first of the two "northeasters" (Figures 2 and 3) in autumn 1992. This transport is well illustrated by the u - C cospectral analysis shown in Figure 4.

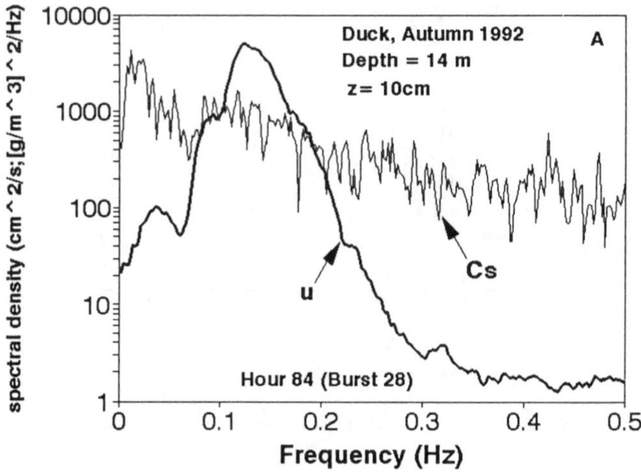

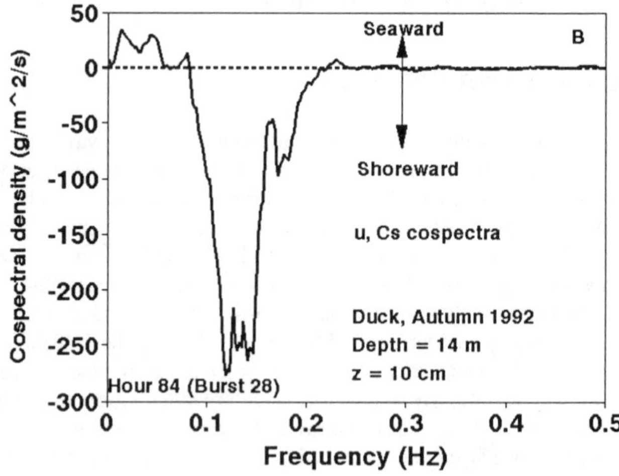

Figure 4. Power spectra (A) and cospectra of suspended sediment concentraction, C, and across-shelf flow, u, at a depth of 14 m during an onshore transport event, 31 October 1992.

The observed wind and bed drag coefficients and γ_o value (Madsen *et al.* 1993) were used as input to a numerical model that was applied to predict the depth integrated alongshelf and across-shelf transports at the 8 m and 13 m isobaths over the duration of the Halloween storm (Chisholm 1993). The model is based on the coastal ocean flow model of Jenter and Madsen (1989) and the Grant and Madsen (1986) wave-current boundary layer model. It utilizes a two-layer Rouse-type suspended sediment model for estimating suspended sediment concentration profiles and a two-layer eddy viscosity model for across-shelf flow (Chisholm 1993). Across-shelf transport results for the 8 m and 13 m isobaths are shown in Figure 5.

Over the 144 hour period to which the model was applied, total seaward transports past the 8 m and 13 m isobaths are predicted to have been 29.9 m^3 m^{-1} and 27.6 m^3 m^{-1} respectively. Nearshore surveys by personnel at the Duck Field Research Facility indicated losses of 30 to 50 m^3 m^{-1} so model results seem reasonable. An important implication of the model results is that very little of the transported sediment is predicted to have been deposited between the 8 m and 13 m isobaths; the sediment was presumably bypassed to deeper water. However, most of the modelled sediment transport turns out to have been fine sediment advected by mean flows. Coarser material transported within the lower part of the wave-current boundary layer may well have been deposited at depths shallower than 13 m.

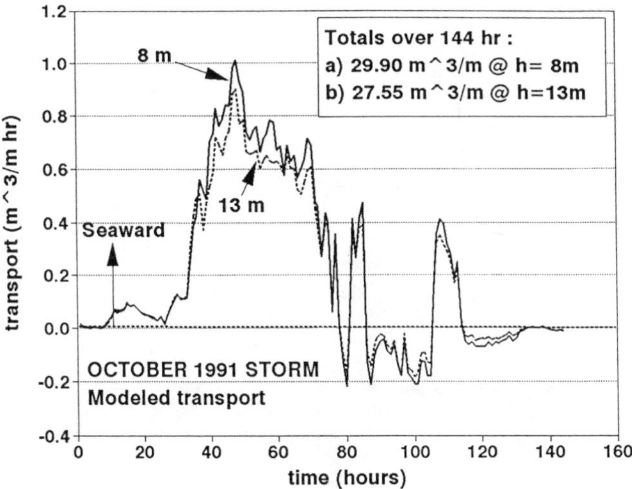

Figure 5. Across-shelf sediment transport at h = 8 m and h = 13 m as predicted by the model of Chisholm 1993.

SEDIMENT FLUX DIVERGENCE AND BED LEVEL CHANGE

The model results just described suggest that the shallower regions of the inner shelf may serve more as a pathway than a sink for sediments eroded from the nearshore zone during storms. On the other hand, significant gradients in skin friction shear stress exist across the region separating the 15 m isobath and the surf zone under more moderate storms (Figure 3) and flux divergences and convergences should be significant. Furthermore, sediment returning shoreward from deeper water after a storm has past must transit the entire shoreface and presumably reside there for a time.

Whatever the cause, our digital sonar altimeter data show bed level fluctuations of up to 18 cm on time scales of hours to days. Measurements of rapid bed accretion and slower erosion made in 1985 at a depth of 8 m were reported by Wright *et al.* (1986). Similar results obtained from the 14 m isobath during the autumn 1992 deployment are shown in Figure 6. Notably, Figure 6 shows bed accretion to have taken place after the passage of a mild event and at a time when net sediment flux was observed to be directed shoreward.

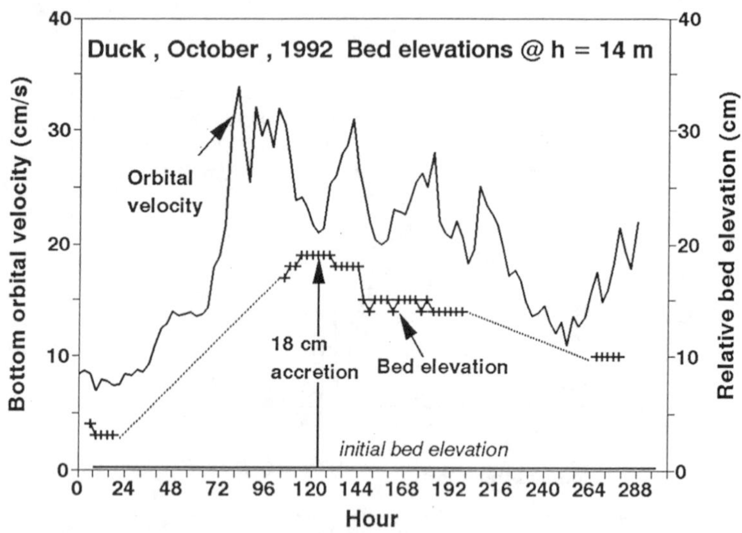

Figure 6. Bed level fluctuations and associated root-mean-squared (RMS) bottom orbital velocities observed at a depth of 14 m, 27 October - 8 November 1992.

CONCLUSIONS

The results just described are specific to the storm-dominated inner shelf regime of the Middle Atlantic Bight. The general conclusions from our field study may be summarized as follows:

1. Skin friction shear stresses during storms exceed those of fairweather by 100 times.

2. Wind-driven downwelling flows cause large seaward fluxes of sediment during storms.

3. Seaward flux is significantly enhanced by flow intensifications at infragravity frequencies.

4. Shoreward transport is effected by moderate waves and high energy swell, but not by fairweather waves.

5. The shallower portion ($h < 15$ m) of the inner shelf appears to serve more as a pathway to deeper water than a sink for nearshore sediments eroded during severe storms.

6. Sediment flux divergences cause bed level fluctuations of 10 to 20 cm on time scales of hours.

ACKNOWLEDGEMENT

The study summarized in this paper has been supported by the National Science Foundation, Marine Geology and Geophysics, Grant OCE-9017828.

REFERENCES

Chisholm, T.A., 1993. *Analysis of Field Data on Near Bottom Turbulent Flows and Suspended Sediment Concentrations*, Civil Engineer's Degree Thesis, Massachusetts Institute of Technology.

Grant, W.D. and Madsen, O.S., 1986. "The Continental Shelf Bottom Boundary Layer," *Annual Review of Fluid Mechanics*, Vol. 18, pp. 265-305.

Jenter, H.L. and Madsen, O.S., 1989. "Bottom Stress in Wind-Driven Depth Averaged Coastal Flows," *Journal of Physical Oceanography*, Vol. 19(7), pp. 962-974.

Maa, J.P.-Y., Wright, L.D., Lee, C.-H., and Shannon, T.W., 1993. "VIMS Sea Carousel: A Field Instrument for Studying Sediment Transport," *Marine Geology*, Vol. 115, pp. 271-287.

Madsen, O.S., Wright, L.D., Boon, J.D., and Chisholm, T.A., 1993. "Wind Stress, Bed Roughness, and Sediment Suspension on the Inner Shelf During an Extreme Storm Event," *Continental Shelf Research*, Vol. 13, pp. 1303-1324.

Nittrouer, C.A. and Wright, L.D., 1994. "Transport of Particles Across Continental Shelves," *Reviews in Geophysics*, Vol. 32, pp. 85-113.

Wright, L.D., 1993. "Micromorphodynamics of the Inner Continental Shelf: A Middle Atlantic Bight Case Study," *Journal of Coastal Research*, Special Issue Number 15, pp. 93-124.

Wright, L.D., Boon, J.D., Green, M.O., and List, J.H., 1986. "Response of the Mid Shoreface of the Southern Mid-Atlantic Bight to a "Northeaster"," *Geo-Marine Letters*, Vol. 6, pp. 153-160.

Wright, L.D., Boon, J.D., Kim, S.C., and List, J.H., 1991. "Modes of Cross-Shore Sediment Transport on the Shoreface of the Middle Atlantic Bight," *Marine Geology*, Vol. 96, pp. 19-51.

Wright, L.D., Xu, J.P., and Madsen, O.S., 1994. "Across-Shelf Benthic Transports on the Inner Shelf of the Middle Atlantic Bight During the Halloween Storm of 1991," *Marine Geology*, Vol. 188 (in press).

HYDRODYNAMICS OF RANDOM WAVE BOUNDARY LAYERS

B.A O'Connor[1], H. Kim[2] and J.J. Williams[3]

ABSTRACT: The paper presents hydrodynamic results obtained from the deployment of a bed boundary layer rig (STABLE) at the Middelkerke Bank off the Belgian coastline during a moderate storm in February 1993 as part of a multi-disciplinary EC-funded (MAST2) study of the role of nearshore sand banks in modifying coastal processes (the CSTAB Project). Results from STABLE were compared with a range of mathematical models, including a one-dimensional (1DV) random wave boundary layer model, and shown to give realistic values. In addition, the results from both STABLE and the models suggest that near-bed processes are dominated by wave action during storms leading to increased bed roughness and shear stresses along with reduced near-bed tidal current profiles and current-induced bed forms.

INTRODUCTION

Despite much past activity, relatively little is known about both the local and global effects of nearshore coastal banks (Off 1963). Consequently in 1992 a detailed study of the effect of the Middelkerke Bank, located off the Belgian coastline, was started as part of the EC's MAST2 Research Programme. The new project, CSTAB, involves eleven European research institutes from the UK, Denmark, Holland, Germany and Portugal and brings together engineers, geographers, oceanographers and geologists in a multi-disciplinary study, see also (O'Connor 1993). In addition, the CSTAB Group works collaboratively with a similar multi-disciplinary group, STARFISH, co-ordinated by researchers from the University of Ghent, see (de Moor et al. 1993a).

[1] Department of Civil Engineering, University of Liverpool, L69 3BX, U.K.

[2] Marine Environmental Engineering Group, KORDI, Ansan, P.O. Box 29, Seoul, 425-600 Korea.

[3] Proudman Oceanographic Laboratory, Bidston Observatory, Merseyside, L34 7RA, U.K.

The CSTAB Project involves the deployment of a wide range of instruments and tracers both around the Middelkerke Bank and on adjacent beaches. In addition, a range of computer models are being developed to study local scale processes due to tides and sediment movements in the vicinity of the Bank and on adjacent beaches, see also (O'Connor 1993).

The present paper presents initial results obtained from the CSTAB Project, and in particular from a large autonomous boundary layer rig (STABLE) deployed at the northern end of the Bank in February 1993. STABLE data is also compared with a range of models to assess its accuracy and help interpret local processes.

Field Data

As part of the CSTAB Project, the Proudman Oceanographic Laboratories' (POL) bed boundary layer rig STABLE was deployed at the northern end of Middelkerke Bank in 20 m of water during the period 25th February - 3rd March 1993. During this time a moderate storm passed through the area generating significant wave heights (H_s) of 3 m. A range of other equipment was also deployed at the same time as STABLE, including two wave-rider buoys (WRB) for the measurement of wave heights; precision water level recorders for the measurement of water depth; and ADCP's and vertical arrays of Aanderra and S4 current meters (CMS) to provide additional information on the depth-distribution of tidal currents, see (Williams 1993). Figure 1 shows the location of both STABLE and Middelkerke Bank in relation to the coastline.

The STABLE rig is a tripod-type frame, which is able to carry a variety of instruments (Humphrey 1987). For the Middelkerke deployment, the rig carried 4 electro-magnetic (ECM) torroidal open-head current meters (Valeport Series 2000) mounted in two pairs at 80 cm and 40 cm above the seabed, and arranged so that each meter measured two components of the instantaneous near-bed flow in orthogonal vertical planes (x-z, y-z). Additional instrumentation included 4 rotary current meters, which were located at 91 cm, 73 cm, 55 cm and 37 cm above the seabed on a vertical support at the centre of the rig; two acoustic back-scatter (ABS) suspended sediment probes (1 MHz and 2.5 MHz), which covered the lower 1.2 m of the water column; and two pressure transducers at 1.69 m above the seabed (Digiquartze) to measure water levels and wave heights, see Figure 2.

An on-board data-logger recorded data in two ways. Firstly, the ECM's, ABS and wave sensors were sampled for 20 mins. every hour at 8Hz, producing 58 data sets (called Bursts) over the deployment period. Given a typical wave period of 6 s. at the site, each Burst data set provided almost 50 data points per individual wave period of which some 200 were recorded over each Burst period. Secondly, readings from the rotor current meters, tide sensor and rig orientation sensors were averaged over a period of 1 min. Such "mean" data eliminates high frequency

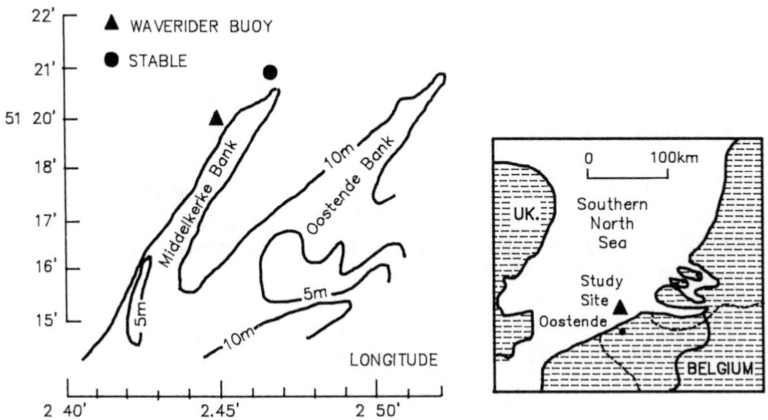

Figure 1. Location of STABLE measurements.

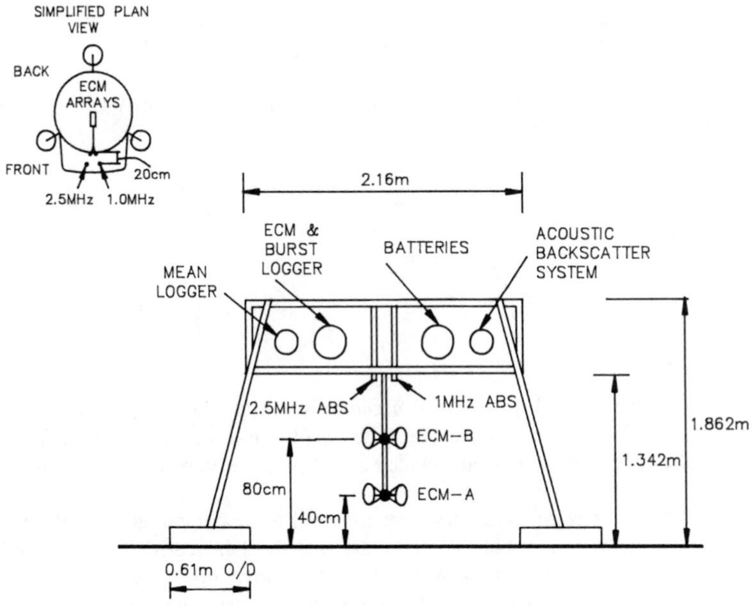

Figure 2. STABLE instrumentation.

velocity components to permit study of tidal flows.

Models

(a) 1DV Boundary Layer Model

A 1DV random wave boundary layer model was proposed for use in the CSTAB Project to assess the accuracy of the STABLE data. The model provides details of horizontal (x) and lateral (y) velocity and shear stress distributions over the water column at the STABLE site.

The model solves simplified Reynolds momentum equations in the x and y co-ordinate directions, see (O'Connor et al. 1993). The relevant equations are given as:-

$$\frac{\partial u}{\partial t} = \frac{1}{\rho}\frac{\partial p}{\partial x} + \frac{1}{\rho}\frac{\partial \tau_{xz}}{\partial z}; \quad \frac{\partial v}{\partial t} = \frac{1}{\rho}\frac{\partial p}{\partial y} + \frac{1}{\rho}\frac{\partial \tau_{yz}}{\partial z} \tag{1}$$

were u, v are the cartesian velocity components in the x, y co-ordinate directions, respectively (m/s); t is time (s); ρ is the fluid density (kg/m^3); p is pressure (N/m^2); τ_{xz}, τ_{yz} are the horizontal and lateral components of the Reynolds stresses in the x and y directions respectively (N/m^2); and z is a vertical cartesian co-ordinate measured positive upwards from the seabed (m).

The shear stress terms are replaced using an eddy viscosity concept with a mixing length hypothesis. Thus:-

$$\tau_{xz} = \rho\epsilon_z \frac{\partial u}{\partial z}; \quad \tau_{yz} = \rho\epsilon_z \frac{\partial v}{\partial z} \tag{2a}$$

$$\epsilon_z = \ell^2 \sqrt{\left(\frac{\partial u}{\partial z}\right)^2 + \left(\frac{\partial v}{\partial z}\right)^2} \quad ; \quad \ell = \kappa z\sqrt{1 - \frac{z}{d}}, \quad for\ z \geq z_0 \tag{2b}$$

where ϵ_z is the vertical distribution of eddy viscosity (m^2/s); ℓ is a mixing length (m); d is the wave-period-average water depth (m); and z_o is the equivalent Nikuradse roughness of seabed, divided by 30 (m); κ is von Karman's constant.

The pressure gradient terms are assumed to be composed of contributions from both the waves (orbital acceleration) and the tidal currents (water surface slope, assumed steady for each Burst). The wave orbital velocities can be synthesised from an assumed directional spectrum or measured values can be used at the 80 cm level; the mean water surface slope being adjusted to give the correct tidal depth-mean velocity, see (O'Connor et al. 1993).

Using the measured 80 cm velocities from STABLE, together with an assumed z_o value, the model is operated until sufficient time has elapsed to generate a stable log-profile in the upper part of the water column. The model then produces the vertical distribution of u and v velocity components and shear stresses (τ_{xy}, τ_{xz}) every time step (typically 8Hz) over each Burst period. By subjecting the velocities or shear stresses to a spectral analysis, it is possible to produce frequency spectra for both velocities and shear stresses as well as a range of statistical parameters, including mean and significant values. By taking average values of the velocities and shear stresses over the Burst period, it is also possible to produce the depth distribution of wave-enhanced tidal current shear stress (τ_{wc}); the vertical profile of the wave-enhanced tidal current and its deviation over the flow depth due to wave-current interaction; and the wave-enhanced roughness of the seabed (z_a) from a semi-log plot of the tidal current profile.

(b) Bed Roughness Models

In order to operate the 1DV model, it is necessary to determine an appropriate value of seabed roughness (z_o) at the STABLE site. If the model is operated with an assumed directional spectrum it is necessary to select an appropriate z_o value. However, if real data is used, for example, the velocity values at 80cm, it is possible to adjust the z_o value until a reasonable match is obtained between the calculated and observed velocities at the 40 cm level. Such a method is very time consuming and consequently was only done for conditions near the peak of the storm. The bed roughness was then kept at this lattermost value for other model runs during the storm.

In order to check on the z_o value obtained from the 1DV model, use was made of two approaches. Firstly, a bed form model was used (O'Connor 1992). This lattermost model equalizes the drag from the tidal flow with that from bed forms and uses information on the seabed grain-size (d_{35}, d_{50}, where the subscripts indicate % finer values); grain relative submerged density (Δ_s); depth-mean maximum tidal velocity (U_{max}) and local depth-mean tidal velocity (U); tidal mean water depth; the local water depth (d); and the orbital wave velocity (u_o) to determine the size of wave-induced and current-induced ripples as well as the size of any associated mega ripples.

The seabed roughness is then calculated from the equations, (Van Rijn 1989)

$$z_o = n\Delta \; (\Delta/\lambda) \quad or \quad z_o = m\Delta \tag{3}$$

where n, m typically have values of 0.53-0.83 and 0.1-0.13 and Δ, λ are the height and wavelength of the bed forms, respectively (m).

The second approach (MEA) used empirical equations (Madsen et al. 1991), which relate the bed roughness under spectral waves to the orbital amplitude of a representative wave of the wave spectrum, and the relative grain friction Shields parameter (Grain Froude Number) in relation to critical erosion conditions (S_r). Thus:-

$$z_o = A_o\, S_r^{-2.5} \quad for \quad S_r > 1.2 \tag{4}$$

where $A_o = u_o/(20\omega)$, (m); and u_o, ω are the representative orbital wave velocity and frequency for the wave spectrum, respectively (m/s, 1/s).

In order to see that the values of wave-enhanced seabed roughness (z_a) given by the direct analysis of STABLE results and from the 1DV model were realistic values, use was also made of an empirical equation, (Van Rijn 1989). The equation (VR) relates the enhanced roughness to the ratio of orbital velocity (u_o) and depth-mean tidal velocity (U) as well as the intersection angle (ϕ) between the wave and current directions. Thus:-

$$z_a = z_o \exp\,(\gamma\, u_o/U) \tag{5}$$

where $\gamma = 0.75$ for $\phi \leq 90°$ and 1.1 for $\phi = 180°$.

(c) **Enhanced Shear Stress Models**

In order to check the values of wave-enhanced tidal current shear stresses found by STABLE and the 1DV model (τ_{wc}), use was made of two approaches. Firstly, a relatively simple bed friction model was used, (O'Connor and Yoo 1988), which incorporates improved coefficients (Yoo 1989, 1991). The model (OCY) uses information on z_o, d, U, ϕ and u_o and has been shown to give realistic predictions of τ_{wc} for a range of laboratory and field conditions.

The second approach (LP1) assumes that the tidal current simply sees the interaction between the waves and currents as an enhanced seabed roughness, so that logarithmic flow profiles exist over the majority of the water column. The enhanced current shear stress τ_{wc} is thus given by the simple equation:-

$$\tau_{wc} = \rho U^2/(5.75 \log_{10}(12d/k_a))^2 \tag{6}$$

where $k_a = 30\,z_a$. In the absence of waves, $k_a = 30\,z_o$, (m), and equation (6) gives values of the seabed shear stress (τ_c) due to the tidal currents (N/m^2).

Stable Data Analysis

The recorded data from STABLE were calibrated, screened and corrected for zero drift, tides and sensor mis-alignment to the principal flow direction. 'Instantaneous' u, v, w velocity components were then obtained taking into account the orientation of the rig to the cartesian axes x, y, z. Burst-mean values were then obtained from the ECM's to determine tidal velocities and flow directions. Data from the STABLE pressure sensor and ECM's and from the wave-rider buoy were next analyzed to produce wave orbital velocities (u_o) and the variation of significant wave height over the full deployment period, see Figure 3.

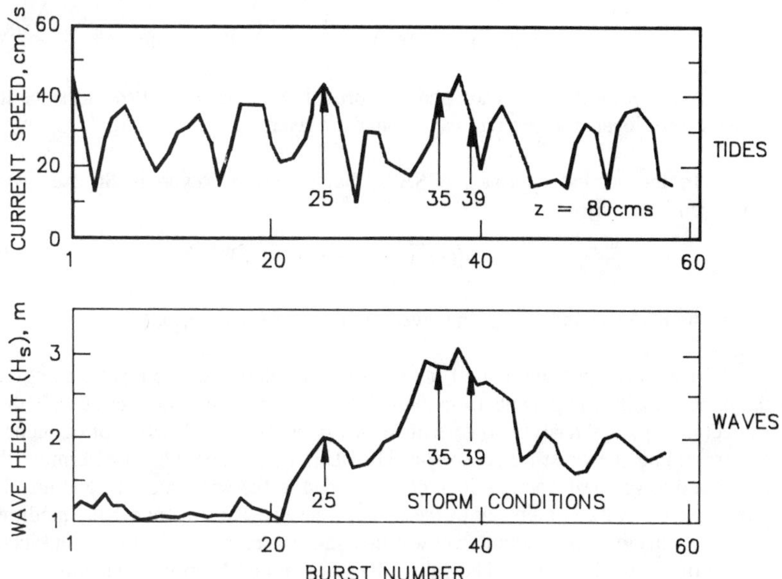

Figure 3. Current speed, wave heights for each burst.

Data from the rotors was analyzed to determine estimates of enhanced bed shear stress (τ_{wc}) and apparent bed roughness (z_a) by fitting log profiles to the four velocity values. Similar estimates were also made from the ECM velocities at 40 cm and 80 cm levels, assuming a log profile (LP2), that is:-

$$\tau_{wc} = \rho \kappa^2 B^2 / \ell n \ (z_{80}/z_{40}) \tag{7}$$

where z_a = exp (A/B), (m); A = ($u_{80}\ell n$ (z_{40}) - $u_{40}\ell n$ (z_{80})), (m/s); B = u_{80} - u_{40}, (m/s); u_{40}, u_{80} represents the ECM velocity values at 40 cm and 80 cm, respectively (m/s); and z_{40}, z_{80} are the distances above the seabed to the nominal 40 cm and 80 cm levels (m/s).

Three further estimates of τ_{wc} were next determined from the ECM's, referred to as Reynold Stress Approaches (RSA1, RSA2) and the Turbulent Kinetic Energy Approach (TKE), respectively. For the first approach (RSA1), Burst-mean velocities were removed from the u, v, w 'instantaneous' values so that only wave-induced and turbulent velocities remained. The resulting time series were then multiplied together to provide time series estimates of τ_{xz} (= $-\rho$ u´ w´) and τ_{yx} (= $-\rho$ u´ v´). The value of τ_{wc} was then determined using all instantaneous Reynolds stress values from the equation:-

$$\tau_{wc} = \rho[\overline{(-u'w')^2 + (-u'v')^2}]^{0.5} \quad \textbf{(8)}$$

on the assumption that any wave-induced orbital velocities cancelled out since the ECM's were located above the wave boundary layer.

In the second approach (RSA2), burst-average Reynolds Stresses were combined in the form:-

$$\tau_{wc} = \rho[(-\overline{u'w'})]^2 + (-\overline{v'w'})^2]^{0.5} \quad \textbf{(9)}$$

where the overbars indicate a time average over the Burst period.

The third approach (TKE) involved the calculation of the kinetic energy of the three velocity components from the ECM's. The resulting time series were then subject to a spectral analysis to determine energy density as a function of frequency. By examining the pressure sensor on STABLE, it was possible to determine the upper and lower frequencies which corresponded to the wave motion, and thereby to remove the wave component of the kinetic energy. The area under the modified energy spectrum of each component was then taken to represent the turbulent kinetic energy of the tidal current. The value of τ_{wc} was found from the equation:-

$$\tau_{wc} = 0.19\rho T \quad \textbf{(10)}$$

where T is half the area under the sum of the energy spectrum for each component (m^2/s^2); and the constant (0.19) is given by (Soulsby 1983).

Finally, the pressure records from STABLE were subjected to spectral analysis using Hashimoto's MEM approach, (Kobune and Hashimoto 1986). This approach enables determination of H_s, T_p, (peak period), T_z (zero-crossing period), mean wave direction and degree of angular spreading about the mean wave direction, respectively. Figure 4 shows typical results for Burst 25. Peak energy is clearly associated with a particular direction (approximately from the North) and directional spreading is confined to $\pm$ 90° of the mean wave direction. Plots of the instantaneous u and v burst velocities (Fig. 5) show a similar degree of wave spreading as the theoretical Goda spectrum results of Fig 6b.

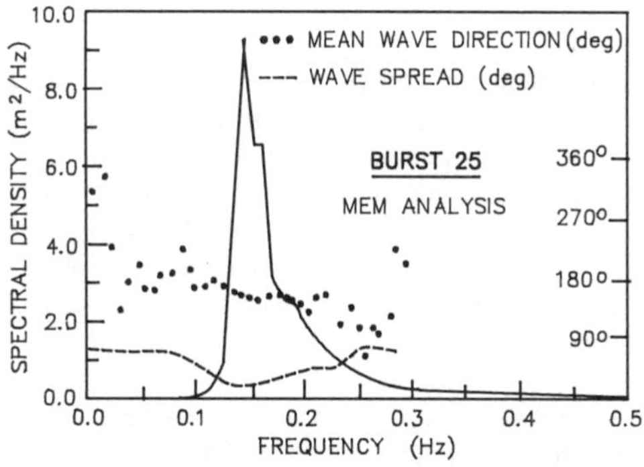

Figure 4. Frequency spectrum.

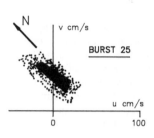

Figure 5. U_{80} Velocity vectors.

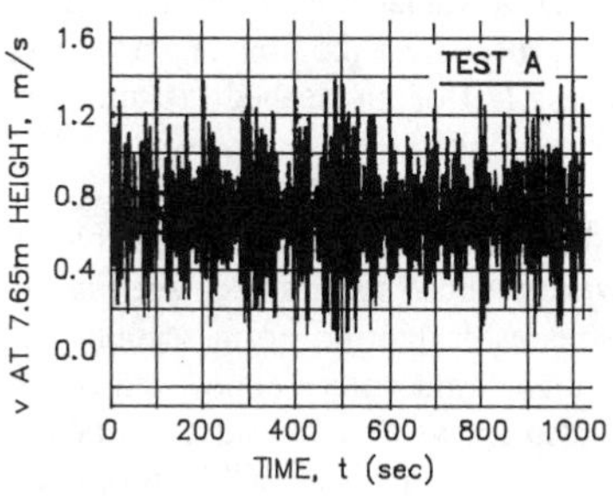

Figure 6a. Depth–mean velocities.

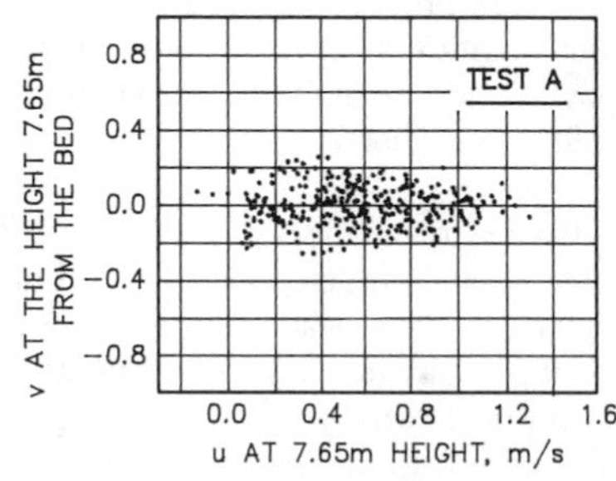

FIGURE 6b. Velocity vectors.

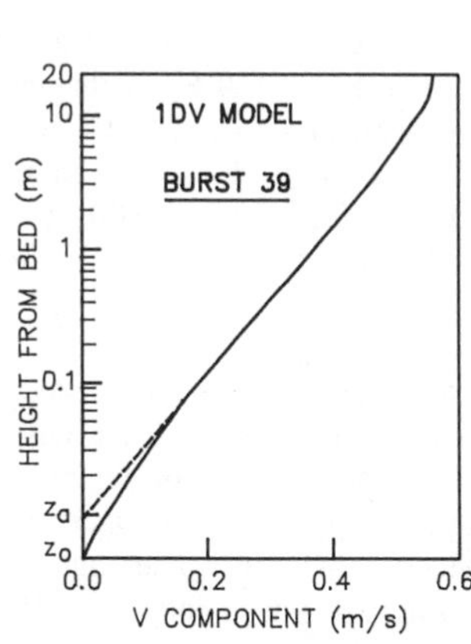

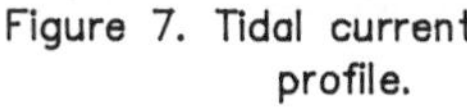

Figure 7. Tidal current profile.

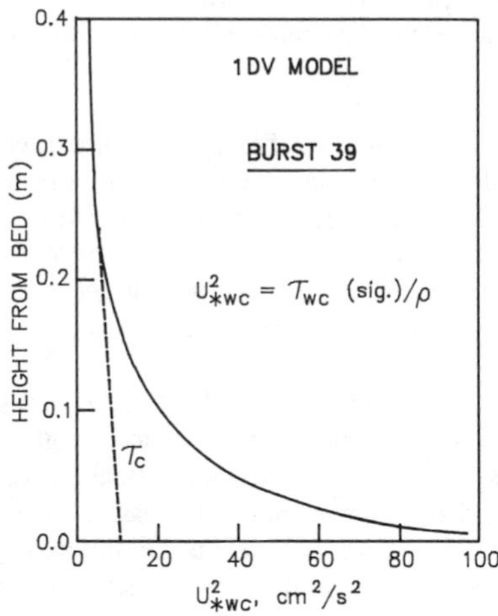

Figure 8. Shear velocity profile.

Model Analysis

The 1DV model was first operated with rough estimates for conditions appropriate to the peak of the storm, see Fig 3. Typical values of model parameters were z_o = 5 mm; d = 20 m; U = 0.67 m/s; significant wave height (H_s) = 3.07 m; wave period (T_z) = 6.44 s; co-linear waves and currents; and a Goda Directional Wave Spectrum with a spreading function (s) = 20. Model results (Test A) show the typical groupy nature of the mid-depth velocities and the spreading effect of the directional wave spectrum, see Figures 6a, b. The raw velocity data from the STABLE ECM's showed similar results.

In order to compare directly with Burst data, the 1DV model was next driven using the u, v velocity components at the 80 cm level for Burst 39. By adjusting model z_o values until the Burst-mean model velocity at the 40 cm level matched that from the ECM probe, it was possible to predict the z_o value of the seabed (5 mm). By fitting equation (6) to the Burst-mean velocity profile over the flow depth, it was then possible to determine the apparent roughness (z_a) of the seabed (10 mm for Burst 39), see Figure 7.

By spectral analysis of the 1DV model shear stresses, it was also possible to determine τ_{wc} and the significant value of the wave-induced shear stress, see Figure 8 for Burst 39. It is clear that wave-induced stresses during the storm account for most of the disturbance of bed sediments with significant values almost ten times larger than tidal current values (τ_c). Figure 8 also shows that the wave boundary layer is confined to the lowest 30 cm of the water column. The ECM probes are located, therefore, well clear of the wave boundary layer itself. However, the interaction between the wave and tidal current boundary layers still produces lower near-bed tidal currents compared to the situation without waves, see Figure 7.

In order to operate the bed form model, it is necessary to have information on U, U_{max} and u_o. Since U was not directly measured by STABLE, use has been made of an adjacent Aanderra current meter (CMS), which recorded at 10 m below the water surface. The value of u_o was obtained from the u_{80}, u_{40} ECM's. Sediment characteristics were taken from published data, see (de Moor et al. 1993b), giving typical d_{35}, d_{50} values for the surface material of 270μm and 310μm, respectively. A value of Δ_s = 1.65 was assumed.

The bed form model showed that small-scale roughness was dominated by wave-induced ripples with typical height (Δ) and wavelength (λ) of 7 cm and 40 cms, respectively, but that the tidal currents produced mega-ripples of height 0.32 m and 10 m wavelength. Increasing the grain size to d_{50} = 410 μm produced only marginal increases (3%) in the dimensions of the wave-induced ripples but more major increases in the size of the mega-ripples (0.3 m to 1.3 m in height and 10 m to 26 m in wavelength, respectively). In the absence of waves, the tidal currents would produce slightly larger height mega-ripples (0.6-1.7 m), with superimposed current ripples of some 1-3 cm height and 30-40 cm wavelength.

Since the size of wave-induced ripples will be affected by the groupy nature of the random sea, the bed form model was operated for a range of wave heights to simulate a typical group (BFM(WG)), assuming that the ripples had sufficient time to react within each local wave period.

The results were then averaged to produce mean ripple dimensions for the wave group, see Table 1 for Burst 39. Such an approach allows the larger waves in the group to influence bed conditions.

Table 1. Wave-Induced Ripple Dimensions (BFM)

Wave Ht (m):	0.82	:	1.63	:	3.26	:	4.89
Period (s):	4	:	5	:	6	:	7
Δ (cm):	1.3*	:	1.4	:	7.4	:	1.9
λ (cm):	31	:	8	:	44	:	34
Mean Δ (cm):	3.5	:		:	Mean λ (cm)	:	28

* current ripples dominate

A comparison of all the various methods for z_o, z_a and τ_{wc} from both model and STABLE data is given in Table 2 for Bursts 39, 35 and 25 along with various estimates of environmental conditions. Details of the veering angle (ψ) between the surface tidal flow direction and the flow direction near the seabed from the 1DV model is also included, since STABLE cannot measure this quantity accurately.

DISCUSSION

Examination of the results in Table 2 shows a number of points:-

(i) the 1DV model produces depth-average current velocities that agree well (within 4% on average) with the CMS values;

(ii) the MEM and WRB results agree well for wave heights (within 3%) but are less good for peak periods, perhaps due to a larger attenuation of the higher frequency wave components than allowed in the theory;

(iii) the 1DV model seems to produce realistic estimates of the seabed roughness

Table 2.Comparison of Various Models and STABLE Values (Burst 39, 25, 35)

Param.	Method	Burst 39	Burst 25	Burst 35	Units
d	STABLE	21.2	20.6	20.7	m
U	Current Meters (CMS)	53	66	36	cm/s
U	1DV	53	72	37	cm/s
u_o(Sig)	STABLE (ECM)	31	15.8	28	cm/s
H_s	Wave rider (WRB)	3.0	2.3	2.9	m
T_s	WRB	5.4	4.8	5.5	s
H_s	Analysis (MEM)	3.2	2.3	2.9	m
T_p	Analysis (MEM)	8.5	6.8	7.5	s
T_p	Linear Theory (u_o)	6.2	5.4	6.0	s
ϕ	STABLE (ECM)	45	52	29	deg.
ψ	1DV @ z_{40}	1.38	0.85	0.77	deg.
ψ	1DV @ z = 8 mm	4.57	3.67	3.46	deg.
z_o	1DV	5	5	5	mm
z_o	BFM(EM,u_o)	5.9-9.2	3.1-4.7	5.9-9.2	mm
z_o	BFM(WG)	2.2-4.6	1.2-4.2	2.6-8.2	mm
z_o	MEA.Eqn.4	0.5-0.6	0.9-1.1	0.5-0.7	mm
z_a	1DV	10	7.8	13.6	mm
z_a	VR Eqn.5	7.6	5.9	9.0	mm
z_a	STABLE (ECM)	9.4	9.6	7.5	mm
τ_c/ρ	LP1(z_o)	8.2	12.7-15.2	3.92	cm^2/s^2
τ_{wc}/ρ	1DV	11	17.8	5.7	cm^2/s^2
τ_{wc}/ρ	STABLE (LP2)	13.7	21.2	16.8	cm^2/s^2
τ_{wc}/ρ	STABLE (RSA1@z_{40})	52.0	33.4	28	cm^2/s^2
τ_{wc}/ρ	STABLE (RSA1@z_{80})	52.8	32.3	35	cm^2/s^2
τ_{wc}/ρ	STABLE (RSA2@z_{40})	7.4	5.1	3.1	cm^2/s^2
τ_{wc}/ρ	STABLE (RSA2@z_{80})	2.6	5.9	9.3	cm^2/s^2
τ_{wc}/ρ	STABLE (TKE @ z_{40})	6.2	7.3	3.8	cm^2/s^2
τ_{wc}/ρ	STABLE (TKE @ z_{80})	15	15.3	8.0	cm^2/s^2
τ_{wc}/ρ	OCY	14	15.6-18.2	7.2	cm^2/s^2

during the storm. In general the calculation of z_o values based on bedform roughness and the use of H_s, T_p overestimates the bed roughness (within 27%, on average) while the wave-group approach underestimates it (within 43%). The MEA approach seems to totally underestimate z_o values. The bed form model confirms that, in general, wave-induced bed ripples dominate the local bed roughness during the storm and that the effect of the waves it to produce lower height mega-ripples (20-50% reduction in height);

(iv) both the 1DV model and the STABLE ECM results demonstrate the enhanced bed roughness (z_a) that occurs due to interaction between the wave and current fields. Very good agreement is shown between the model and field values (within 6%) for Bursts 25 and 39. However, results for Burst 35 are less good. Van Rijn's approach, based on laboratory data, produces sensible results, which are consistently lower than the 1DV model results, perhaps indicating that a larger γ value should be used;

(v) both model (1DV and OCY) and STABLE results confirm the enhanced bed shear stress (τ_{wc}) found due to interaction between the wave and current fields. The OCY bed friction model consistently produced marginally greater values (19%, on average) than the 1DV model, as did the TKE approach for the z_{80} ECM results (within 24%, on average). The average of the TKE approach for the z_{40} and z_{80} levels produces very good agreement with the 1DV model for Bursts 39 and 35 (within 2%) but a less good comparison for Burst 25 (within 37%). Clearly, the OCY approach and the TKE approach at z_{80} also produces a very good comparison (within 5%, on average).

Of the other STABLE methods, the LP2 approach consistently overestimates the 1DV results, as does the RSA1 approach. The RSA2 approach generally underestimates the 1DV model results but gives much better results than the RSA1 approach owing to the exclusion of wave effects. However, they are less good on average than the TKE method;

(vi) the 1DV model shows that interaction between the wave and current fields extends throughout the wave boundary layer with tidal current flow directions being deviated nearly 4° from those at the water surface. Coriolis effects will, of course, produce additional veering action. However, the 1DV model does not contain Coriolis terms. Unfortunately, the veering aspects of the 1DV model cannot be checked by STABLE since it records data above the wave boundary layer. Typical variations in flow direction between u_{40} and u_{80} amounted to between 1.9-4.7°, which suggests that the effect of interactions in the boundary layer has a similar magnitude to the Coriolis effect.

CONCLUSIONS

Comparison of a 1DV random wave boundary layer model and other bed friction and bedform models with data from STABLE shows that sediment movements and flow hydrodynamics at the site are controlled by the interaction between waves and currents. Tidal current profiles are modified by the presence of waves such that directional veering and reduced flow occurs in the wave boundary layer together with enhanced seabed shear stresses (τ_{wc}) and bed roughness (z_a). Local seabed shear stresses are dominated by wave-induced values (Fig. 8) which produce wave-induced ripples and also a reduction in size of existing tidally-induced mega-ripples.

Values of actual seabed roughness (z_o) determined by the 1DV model from velocities measured by STABLE, were found to be in reasonable agreement with values based on wave-induced bed forms. The effect of wave groups was shown to produce lower z_o values than methods based on H_s and T_p. However, the MEA approach appears to give too drastic a reduction and may suggest that the laboratory-derived coefficients need adjustment for field conditions. Van Rijn's approach for the prediction of apparent roughness values yields sensible answers but again the laboratory-derived coefficients may need adjustment for field values.

Comparison of the various models with analysis of STABLE's ECM current meter data suggests that the TKE method provides the best approach to the determination of wave-enhanced tidal shear stresses whereas methods based on Reynolds stresses need careful evaluation to avoid errors.

Finally, the various model and STABLE comparisons suggest that STABLE has produced a useful set of reliable field data with which to further study seabed processes at the Middelkerke site.

ACKNOWLEDGEMENTS

This work was undertaken as part of the CSTAB research programme. It was funded by the Commission of the European Communities, Directorate-General for Science, Research and Development under MAST Contract No. MAS2-CT92-0024C

REFERENCES

de Moor, G., Lanckneus, J., Berne, S., Chamley, H., de Batist, M., de Putter, B., Marsett, T., Van Sielegham, J., Stolk, A., Terwindt, J., Vincent, C.1993a. "Sediment transport and bed form mobility in a sandy shelf environment STARFISH)", In MAST days and EUROMAR market, eds. (Barthel, K.G., Bohle-Carbonell, M., Fragakis, C. and Weydert, M.), 15-17th March 1993, pp 209-313.

de Moor, G., Lanckneus, J., Berne, S., Chamley, H., de Batist, M., Houthuys, R., Stolk, A., Terwindt, J., Trentesaux, A., Vincent, C. 1993b. "Relationship between sea floor currents and sediment mobility in the southern northern sea". In MAST days and EUROMAR market, eds. (Barthel, K.G., Bohle-Carbonell, M., Fragakis, C. and Weydert, M.), 15-17th March 1993, pp 193-208.

Humphrey, J.D. 1987. "STABLE - an instrument for studying current structure and sediment transport in the benthic boundary layer". 5th Int. Conf. on Electronics for Ocean Technology, 24-26th March, Heriot-Watt University, Edinburgh, Institute of Electronic and Radio Engineering, Pub. No. 72, pp 57-62.

Kobune, K. and Hashimoto, N. 1986. "Estimation of directional spectra from the maximum entropy principle, Proceedings 5th International Offshore Mechanical and Arctic Engineering Synopsium, ASCE, 1, pp 80-85.

Madsen, O.S., Mathisen, P.P. and Rosengaus, M.M. 1991. "Moveable bed friction factors for spectral waves". In Coastal Engineering 1990, ed. (B.L. Edge), ASCE, New York, 1, pp 420-429.

O'Connor, B.A., 1992. "Prediction of seabed sand waves". In Computer Modelling of Seas and Coastal Regions, ed. (P.W. Partridge), Computational Mechanics/Elsevier Applied Science, pp 321-338.

O'Connor, B.A. 1993. "Circulation and sediment transport around banks: the CSTAB Project". In MAST days and EUROMAR market, eds. (Barthel, K.G., Bohle-Carbonell, M., Fragakis, C. and Weydert, M.), 15-17th March 1993, II, pp 214-221.

O'Connor, B.A. and Yoo, D.H. 1988. "The mean bed friction of combined wave-current flow". Coastal Engineering, 12, pp 1-21.

O'Connor B.A., Harris, J.M., Kim, H., Wong, Y.K., Oebuis, H. and Williams, J.J. 1993. "Bed boundary layers". In Coastal Engineering 1992, ed. (B.L. Edge), ASCE, New York, 2, pp 2307-2320

Off, T., 1963. "Rhythmic linear sand bodies caused by tidal currents". Bull Am. Soc. Petrol. Geol., 47, pp 324-341.

Soulsby, R.L. 1983. "The bottom boundary layer in shelf seas". In Physical Oceanography of Coastal and Shelf Seas, ed. (B. Johns), Elsevier, Amsterdam, pp 189-266.

Van Rijn, L.C. 1989. "Handbook of sediment transport by currents and waves". Delft Hydraulics, Delft, The Netherlands.

Williams, J.J. 1993. "CSTAB Cruise Report". Proudman Oceanographic Laboratory, Cruise Report, No. 16, pp 53.

Yoo, D. 1989. "Explicit modelling of bottom friction in combined wave-current flow", Coastal Engineering, 13, pp 325-340.

Yoo, D. 1991. "Bottom friction of wave-current flow on a natural beach", Proceedings 3rd Conference on Coastal Engineering, Korean Society of Coastal and Ocean Engineers, Seoul, Korea, pp 1-4.

SHINGLE MOVEMENT UNDER WAVES & CURRENTS: AN INSTRUMENTED PLATFORM FOR FIELD DATA COLLECTION

George Voulgaris, Michael P. Wilkin and Michael Collins

ABSTRACT: An autonomous benthic layer instrumented platform has been developed for the measurement of flow structure over shingle beds. The natural noise created by the inter-collision of shingle particles is concurrently measured by the system. Laboratory calibration enables the conversion of the recorded noise into shingle transport rates. Autonomous operation of the system combined with telemetric transmission of the data, ensures continuous monitoring of performance and retrieval of the data collected. Preliminary results are presented from a deployment off the southern coastline of England.

INTRODUCTION

Free standing structures used as instrument platforms in shelf boundary layer and sediment transport research have been utilised since the early seventies (Sternberg, 1971). Most of these systems have been used for measurements over sand/silt beds (e.g. Cacchione and Drake, 1979; Lyne at al.,1990; Green, 1992). An understanding of shingle (gravel) transport is based mainly upon river or laboratory experiments, under steady flow conditions. No information is available on the mobility of shingle beds under the combined action of waves and currents.

The present investigation describes an instrument-mounting platform (TOSCA, Transport Of Sediment Combined Apparatus), designed and developed at the University of Southampton, for the investigation of:
(i) the hydrodynamic characteristics of the benthic boundary layer;
(ii) determination of the threshold of shingle movement, under various combinations of waves and currents; and
(iii) derivation of empirical formulae for the prediction of shingle transport rates.

The concept of Self Generated Noise (SGN) is used for the remote sensing of shingle

---

Department of Oceanography, The University, Southampton, SO17 1BJ, U.K.

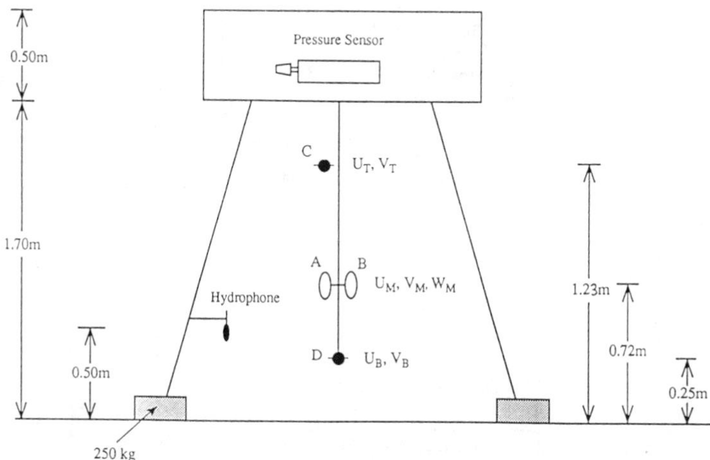

Figure 1. TOSCA tripod and instrument arrangement (A to D em heads).

movement. The system records the rms level of noise generated by the inter-collision of shingle particles, together with other hydrodynamic parameters (i.e instantaneous wave and current conditions). Laboratory calibration enables the conversion of the recorded rms noise into sediment transport rates.

The developed system utilises radio telemetry for the almost "real time" monitoring of operational status and data collection. Hence, it combines advantages of both unattended systems with those tethered to the shore or a vessel.

Typical examples of averaged and instantaneous data are presented below; these were collected during a deployment of the system off the south coast of England.

INSTRUMENTATION AND METHODS

The TOSCA rig consists of an underwater tripod, with most of the hardware located within the upper (i.e cylindrical part) of the body. Only the sensors and support legs are located in the free flow area (Fig. 1).

An onboard PC-based micro-controller (32 bit 80386 processor) and an A/D card (with 4 input/output control lines) are utilised for sensor control, data collection and transmission. Sampling schemes can be created easily through a control file, enabling burst mode sampling, at any frequency, and at regular or irregular (predefined) time intervals between bursts. The sequence of operations executed by the data-logger, during deployment is shown in the form of a schematic flow diagram on Figure 2.

TOSCA is connected to a surface buoy, by an armoured cable. The buoy functions as: (i) a marker buoy; (ii) the power supply; and (iii) a data transmission system. Data collected by the sensors are saved on board the tripod on a hard disc and subsequently sent to the surface

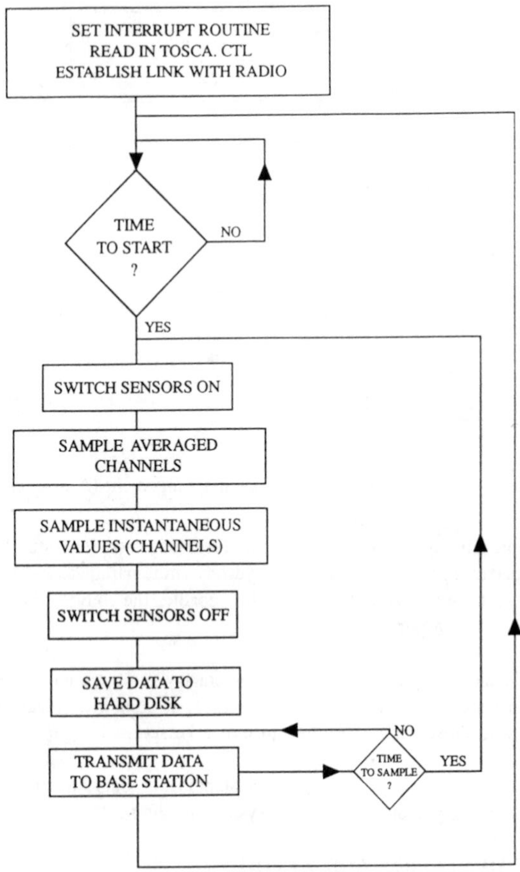

Figure 2. Schematic flow diagram of the sequence of operations executed by the data-logger.

buoy, from where they are transmitted to the base station at the adjacent coastline (Fig. 3).

Instrumentation

The instrument suite of the system consists of 4 x 2-axis electromagnetic (em) current meter sensors. Horizontal components of currents are measured at two elevations above the sea bed (0.25 and 1.23m) with two spherical em sensors (5.5cm in diameter), whilst all three components are measured at 0.72m above the sea bed, using two annular em sensors (11.5cm in diameter) arranged in an orthogonal configuration (Fig. 1). The arrangement of the current sensors was established on the basis of existing methods for the estimation of bed shear stress

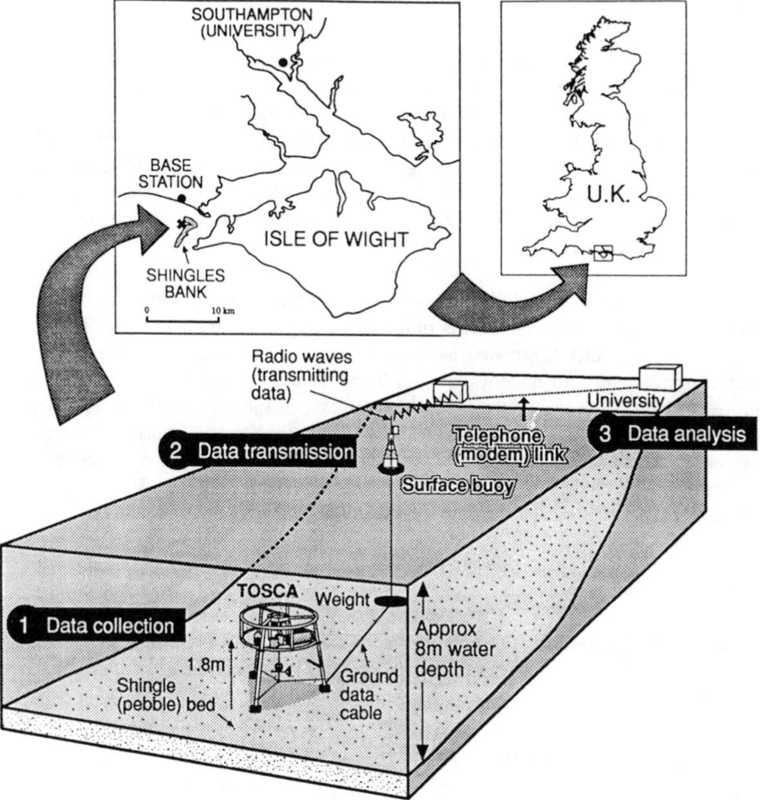

Figure 3. Deployment site and artist's impression of deployment configuration and mode of data transmission.

(i.e. law of the wall (Drake and Cacchione, 1991); inertia dissipation method (Huntley, 1988); Reynolds stress (Gross et al., 1986); Turbulent Kinetic Energy (Soulsby and Humphery, 1990)).

Instantaneous sea surface elevation and mean water depth are derived from the measurements of a pressure transducer. The noise levels created by the inter-collision of the moving shingles is recorded using a hydrophone (Fig. 1), with a specially developed underwater processor. This processor is used for: (i) filtering the raw signal, allowing only the frequency band corresponding to SGN processes to be passed (Voulgaris et al., 1993); and (ii) converting the raw signal into a true rms level, which is then recorded together with the hydrodynamic measurements.

A fluxgate compass and two tilt-meters complete the instrument suite of the system.

SGN & Sediment Transport

Background. Acoustic noise generated by moving non-cohesive marine sediments (i.e. sand and gravel) was observed initially by sonar users; it has been studied since by Harden-Jones and Mitson (1982), Thorne and Foden (1988), and Thorne (1986a; 1986b; 1990). The use of this noise to quantify sediment transport rates has been examined in both the laboratory (Johnson and Muir, 1969) and the field (Tywonick and Warnak, 1973; Jonys, 1976). Recently, the method has been used in a tidally-dominated area (West Solent, southern England), with the results published in a series of papers (Williams et al., 1989; Thorne et al., 1983; Thorne et al., 1989).

In principle, when shingle is transported as bedload, particles within the moving layer collide with each other, and particles of the underlying stationary layer. Such inter-collision generates noise, which is proportional to the mass of moving particles and the velocity of the impacting particles. An underwater rigid body radiation theory has been applied to such processes by Thorne and Foden (1988), for the case of spherical particles. The Hertz 'law of contact' was used to describe the acceleration time-history, during impact. It was demonstrated that the theory could be applied to describe radiation arising from the collision of gravel particles. In the theory, the resonant pressure wave of SGN is given by a frequency (F_t), arising from the duration time of the contact of the particles:

$$F_t = 0.0855 \cdot \left(\frac{E}{\rho_s(1-\sigma^2)}\right)^{0.4} \cdot \left(\frac{U_s^{0.2}}{d}\right) \tag{1}$$

where E = Young's modulus (N/m^2), ρ_s = solid particle density (kg/m^3), σ = Poisson's ratio (nondimenional); U_s = impact velocity (m/s), and d = sphere diameter (m). The acoustic pressure peak is given by:

$$P_p = 1.29 \times 10^{-6} \cdot E \cdot \left(\frac{d}{2 \cdot R}\right)^{1.07} \cdot \left(\frac{U_s}{c}\right)^{1.25} \cdot \sqrt{\frac{Z_W}{Z_A}} \tag{2}$$

where R = distance from the source (m); c = speed of sound in water (m/s); and Z_W, Z_A = acoustic impedance of water and air, respectively (N·s/m^3). Equations (1) and (2) reveal that SGN spectra are dependent upon particle size (d, which is proportional to mass) and impact velocity (U_s). These quantities both define sediment transport rate.

Laboratory Calibration. In the present study, a laboratory calibration was undertaken for the conversion of the SGN into shingle sediment transport rates. An oscillating bed flume was used, with the bed of the flume oscillating backwards and forwards at set periods. The bed was covered with particles of gravel from the study area. Representative particles were attached then to the end of strings fixed to the frame of the flume. As the flume oscillates, the particles are forced to move in relation to the bed simulating saltation processes. The immersed weight transport for each of the experimental runs is defined as:

$$I_b = \left(1 - \frac{\rho_o}{\rho_s}\right) \cdot g \cdot \frac{M \cdot U}{A} \tag{3}$$

where M = total mass of the material attached to strings (moving shingle) (kg); U the

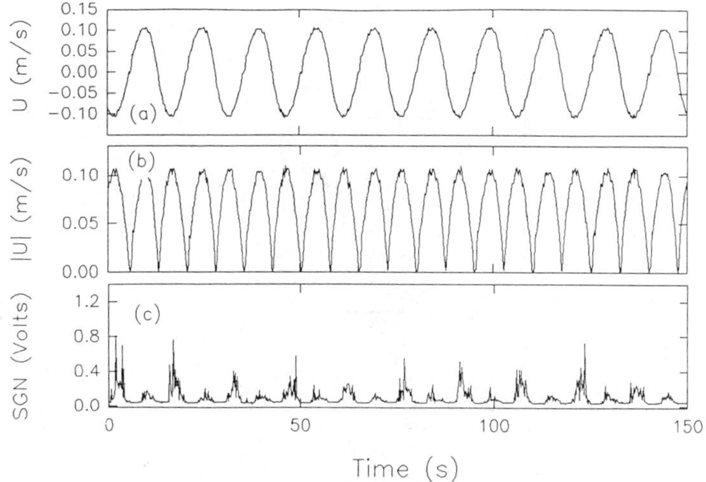

Figure 4. Example of time-series collected during the SGN calibration.

velocity of movement (m/s); A = area of the flume bed overlain by moving particles (m^2); g = acceleration of gravity (m/s^2); and ρ_o = density of the water (kg/m^3). A typical example of time-series collected during the SGN laboratory calibration of the hydrophone is shown on Figure 4. In this particular run, the transported mass of gravel was 55.7g, the period of oscillation 15s and the orbital semi-excursion 0.258m. The current speed is shown on Figure 4(b), whilst the SGN is shown on Figure 4(c). The flat areas between the peaks in the SGN time-series correspond to times of directional change, where there was a time lag before the attached gravel particle started moving. Asymmetry in the SGN peaks is due to increased noise from mechanical vibrations, occurring during half the wave cycle (this was corrected in subsequent analyses).

Transport rates derived using the above calibration technique have been related to the recorded SGN levels (Fig. 5) using the equation:

$$I_b = 1.154 \cdot P_o^{0.891} \qquad (where\ P_o\ in\ Pa) \tag{4}$$

Equation (4) was derived under laboratory conditions, where the agitated sediment is limited into a small area (A) of equivalent seabed. In the field, noise sources extend over a wider area of seabed. To compensate for the differences, the theoretical expression regarding the amplitude of the pressure created by the moving particles inter-collision (Thorne and Foden, 1988) is combined with equation (4) and, after integration over the whole area of the seabed, yields, (Voulgaris et al., 1994):

$$I_b = 1.958 \cdot \left[\frac{P_o \cdot R_f}{\pi}\right]^{0.891} \tag{5}$$

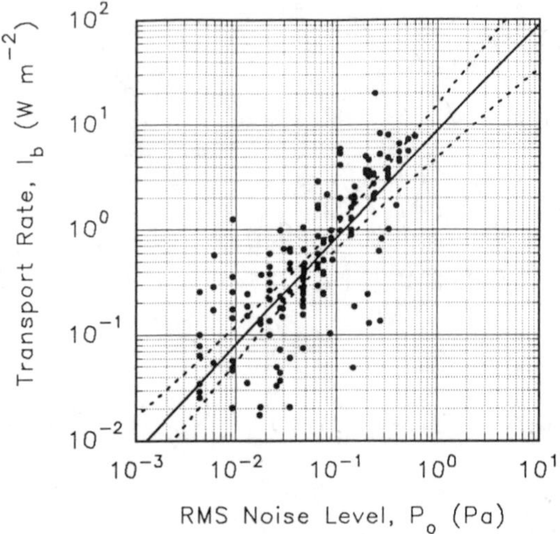

Figure 5. Calibration curve for conversion of SGN into shingle transport rate.

where P_o = pressure level obtained in the field (Pa (=N/m^2)); and R_f = height of the hydrophone above the sea bed (m).

Additional experiments were carried out to establish the relationships between acoustic levels and hydrodynamic noise around the hydrophone sensor. The results are used to correct the recorded SGN values prior to their conversion into transport rates using equation (5). Also, an approximate correction for ambient noise can be performed using the diagram presented by Thorne (1986b), assuming that wind conditions are known throughout the experiment.

DEPLOYMENT / DATA COLLECTION

The system was deployed from 20th April to 8th May, 1993 (Julian Days 110 - 128), in Christchurch Bay (southern England), in a mean water depth of 8m (Fig. 3). The sea bed was covered by loose shingle with a mean particle size of 1.70cm. Mean spring and neap tidal ranges for the area are 2.0 and 0.9m, respectively. Dominant wave approach is from the southwest. Some 130 bursts of data were collected, with a sampling frequency of 5 Hz and approximately 10 min duration (3072 data points). The time interval between subsequent bursts was 2 hours.

Data collection was interrupted twice during the deployment due to power supply problems. The radio telemetry enabled early identification of the problem, however, so that in both cases rectification and continuation of the data collection program was achieved.

The exact morphology of the sea bed, in the area adjacent to the location of the tripod deployment is shown on Figure 6. The information displayed on this Figure is a compilation of divers' observations and side-scan sonar survey results. TOSCA was located about 10m

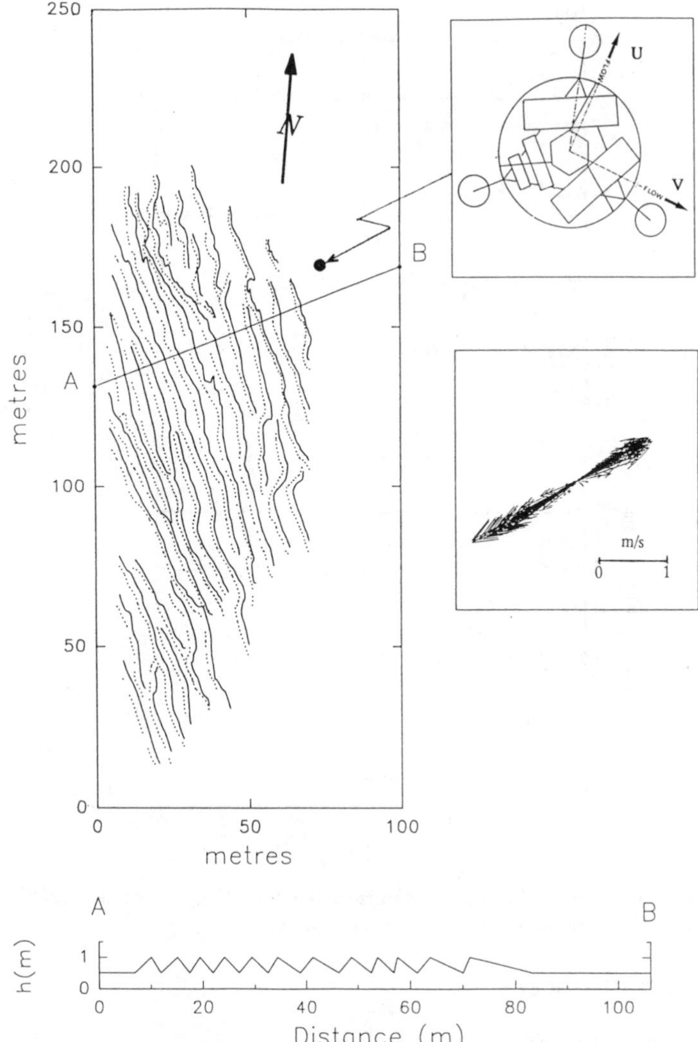

Figure 6. Micro-topography of the deployment site, together with the tripod's orientation and prevailing tidal vectors.

to the east of a gravel wave field, with wave height (η) and length (λ) of 0.5 and 10m, respectively. The outlines of the crests (solid lines) and the troughs (dotted lines) are shown on the Figure, together with the orientation of the tripod and the tidal current vectors (see insets on Figure 6).

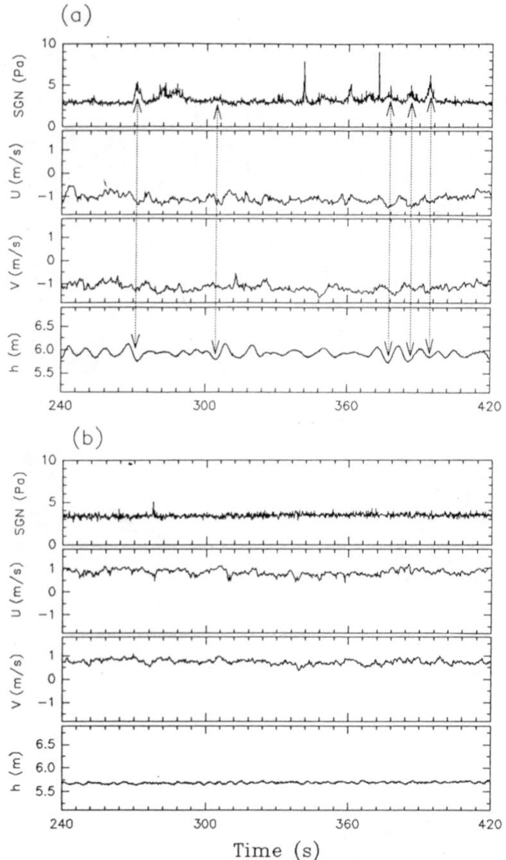

Figure 7. Time-series of instantaneous horizontal currents (U,V), sea surface elevation (h) and SGN, collected during two different burst cycles.

PRELIMINARY RESULTS

Instantaneous Time-Series

Two examples of time-series, obtained from burst data sets, are shown on Figure 7. In both cases, strong quasi-steady tidal flows (1.5 m/s) are present in the records. The signal of one of the bursts (Fig 7(a), ebb flow) is characterised by the superimposition of fair wave activity (H_{sig}=0.37m), which appears mainly as an amplitude-modulated wave (wave group). Such wave oscillations, although clear on the elevation time-series, appear to be distorted in the current records due to the high turbulence signal present there. Nevertheless, the SGN signal exhibits a background level of 3 Pa, with certain peaks present. Most of the peaks are associated with wave troughs (see the marked events on Fig. 7(a)). This correlation appears to be due to an enhancement of the instantaneous current speed, as the main tidal currents and the wave-induced orbital current act in the same direction during this particular phase of the wave cycle.

In contrast, the records shown on Figure 7(b) (flood flow) are free of any significant wave oscillations. Current variance here is due to turbulence only, with the SGN signal remaining steady at a level equivalent to about 3.5 Pa.

From the analysis of these and other records, it has been revealed that the shingle exhibits a response to variations of the mean flow at time scales much smaller than those of the wave periods, but greater than those of turbulence variation. This observation is in agreement with Thorne et al. (1989), who concluded that horizontal velocity is primary responsible for shingle transport in a tidally-dominated environment.

Burst-Averaged Time-Series

Additional analyses involve the computation of mean values of hydrodynamic parameters and sediment transport rates derived using equation (5). Hydrodynamic parameters such as mean water depth (h), wave height (H_{sig}), period (T_z), mean current speed ($|U_z|$) for the three elevations (z) of current measurement) and mean shingle transport rates (immersed weight, I_b), for the period of deployment are shown on Figure 8. The zero axis for the current speed on Figure 8(b) is shown by a dashed line, shifted by 1m/s, for current measured at different elevations above the sea bed.

As can be seen on Figure 8, data have been collected for both neap and spring tidal periods with measured tidal ranges of between 1.30 and 2.30 m, respectively. The measured currents, at 1.23m above the sea bed, reach speeds of 1.0 and 1.8 m/s for neap and spring tides, respectively. Moderate wave activity was recorded during the 5 days of deployment (Julian Days 111 to 116) whilst for the remaining period the waves were from almost non-existent (Julian Days 117-120) to fair (Julian Days 123-128). A characteristic feature of the record is the modulation of wave height with mean water depth. This is due to wave energy dissipation over a shallow bank to the southeast of the deployment location (Fig. 3), during low water. Shingle transport rates (Fig. 8(a)), appear to be varying according to the tidal currents declaring the latter as being the driving mechanism for the period of the experiment.

Flow Structure

Shear stresses have been derived from both the upper and lower em sensors using the inertia dissipation method (Huntley, 1988). An example of wavenumber energy spectrum of the downstream turbulent velocity, measured at 1.23m above the sea bed, is shown on Figure 9. Some 132 shear stresses have been obtained from the upper sensor, whilst from the lower sensor only 53 estimates have been calculated because of the proximity of the sensor to sea bed (i.e. inertia subrange out of the measured spectra range). These estimates have been related to the mean currents using the quadratic stress law analysis (Sternberg, 1972), on Figure 10. Drag coefficients have been derived from the ebb and flood phase of the tide, using best-fit regression analysis (Table 1).

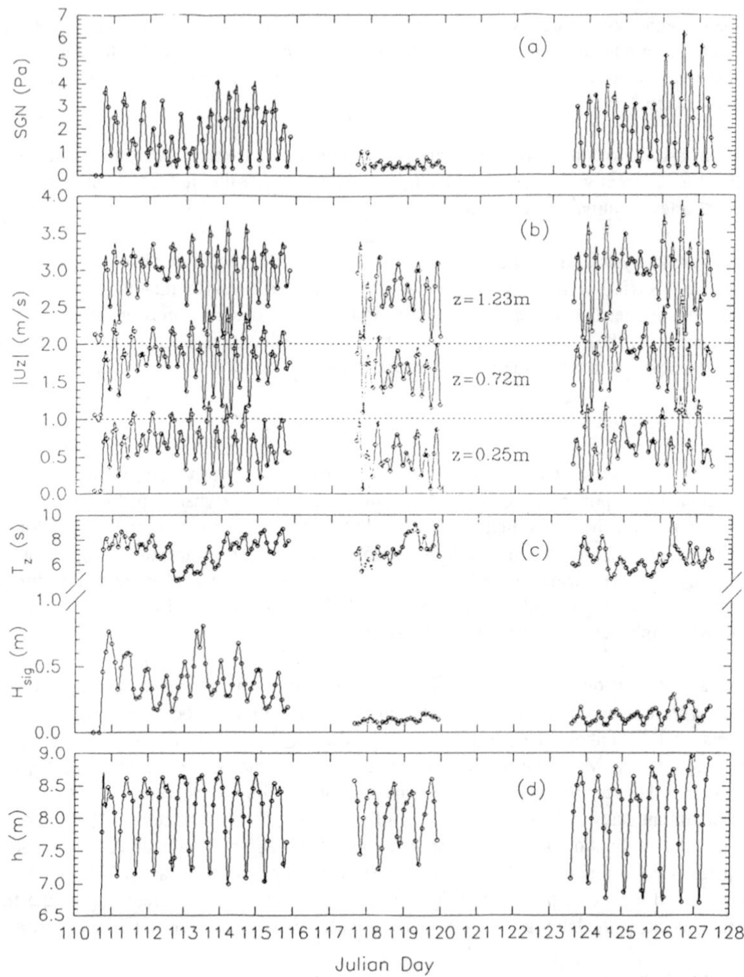

Figure 8. Time series of mean values of 'burst data', collected during the deployment period.

Thus, it is observed a mean value of drag coefficient (at z=1m) $3.5x10^{-3}$ and $5.6x10^{-3}$ can be used for relating bed shear stresses to mean current, for ebb and flood flow, respectively. The observed differences in the values of the drag coefficient with tidal phase is due to the presence of the bedforms upstream of the tripod's location, during the flood flow (Fig. 6). Using the above values, the roughness parameter (z_o) was found to be 1.24mm during the ebb (flat bed, skin friction predominant) and 4.90mm during flood flow (gravel waves, with skin friction and form-drag).

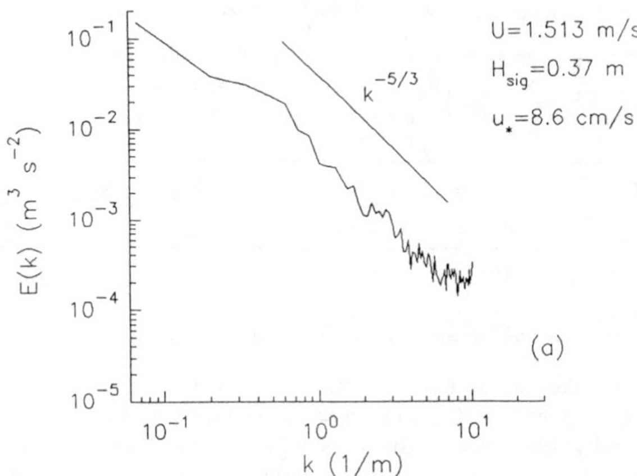

Figure 9. Example of downstream velocity turbulence spectra used for derivation of bed shear stress with the inertia dissipation method.

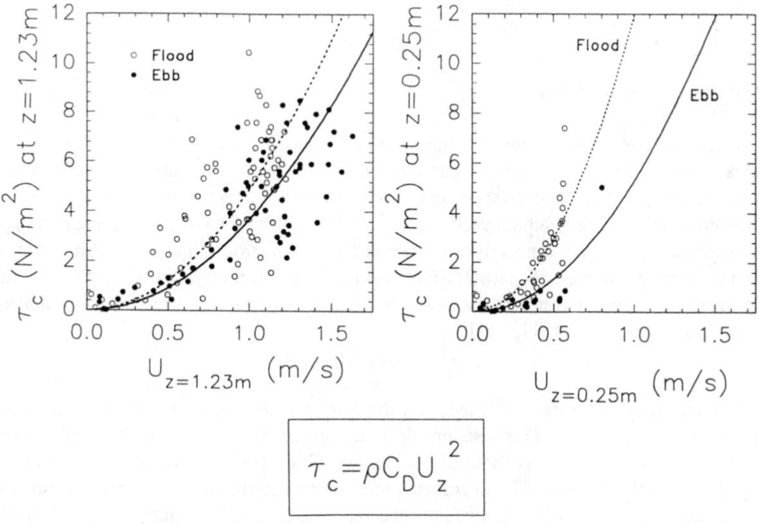

Figure 10. Quadratic stress law analysis applied to the mean current and shear stress data.

Table 1. Results of the quadratic stress law 'analysis' shown on Figure 7.

EM Head Elevation z (m)	Tidal Phase	No of Points	Drag Coefficient, C_D at z ($x10^{-3}$)	Variability of C_D (%)	Drag Coefficient, C_D at z=1m ($x10^{-3}$)
1.23	Ebb	63	3.597	4.8	3.813
	Flood	69	4.769	4.9	5.129
0.25	Ebb	17	5.141	13.2	3.298
	Flood	36	11.601	6.5	6.155

The drag coefficient derived above (for skin friction) is slightly lower than that suggested by Sternberg (1972) for gravel ($4x10^{-3}$). Comparing the physical roughness (z_o) to that of 3mm suggested by Soulsby (1983), those derived from the present study are both lower and higher (for skin friction and total roughness, respectively) suggesting that Soulsby's value expresses an 'average' situation.

Combining the above values with shingle particle size (d_{50}) and bedform dimensions (η and λ), the following expressions are suggested for the physical roughness due to skin friction (k_d) and form-drag (k_B):

$$k_d = 2 \cdot d_{50} \quad and \quad k_B = 4.4 \cdot \eta \cdot \frac{\eta}{\lambda} \tag{6}$$

Sediment Transport Rates

Shear stresses due to waves (τ_w) have been calculated following the method described by Sleath (1991) where the roughness parameter used was $2 \cdot d_{50}$. The quadratic stress equation, with the derived drag coefficients, is used for deriving the mean current shear stress (τ_c). The magnitude (τ_{wc}) of the resultant vector shear stress is compared then to the sediment transport rates obtained. The latter have been corrected for both hydrodynamic noise and offset due, to the noise from rocking particles (see Voulgaris and Collins, 1994). A relationship is obtained between the resultant maximum wave and current shear stress and the immersed weight shingle transport (Fig. 11):

$$I_b = 4.444 \cdot 10^{-3} \cdot \tau_{wc}^3 \tag{7}$$

The equation above states that sediment (bedload) transport is a function of the 3rd power of the bed shear stress. This relationship is in agreement with the stochastically derived models for bedload, such as those of Madsen and Grant (1976; based on Einstein's (1950) early work) and Zanke (1987). In contrast, most of the deterministically derived expressions (i.e Meyer-Peter and Muller, 1948; Engelund and Hansen, 1967; Williams et al., 1989) relate bedload to bed shear stress through the power of 3/2.

No threshold value can be derived directly from the results presented on Figure 11. It is suggested that shingle transport is a continuous mechanism, whilst the definition of a threshold criterion is a function of defining a minimum sediment transport rate (Paintal, 1971).

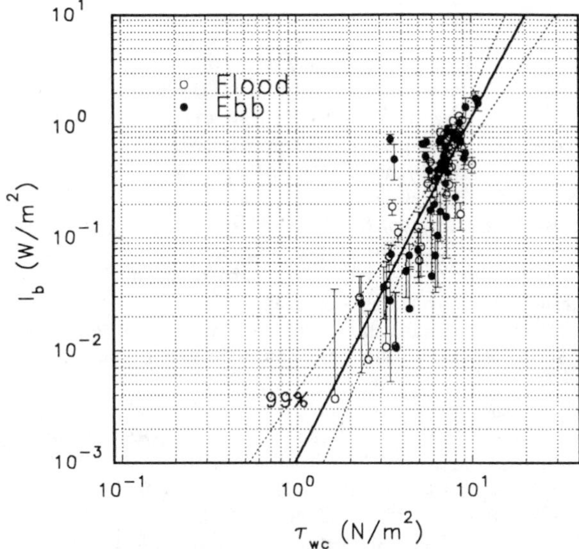

Figure 11. Relationship between total shear stress and shingle transport rates.

CONCLUSIONS

A new autonomous system has been presented for reliable hydrodynamic and sediment transport measurements, in the benthic boundary layer over shingle beds. The concept of the Self-Generated Noise, a passive acoustic and non-intrusive technique for measuring shingle transport at times comparable to turbulence variations, has been proved to work under wave conditions in the open marine environment. However, more investigations are required to examine the conversion of laboratory SGN calibrations to field applications.

Preliminary results from a deployment off the south coast of England have shown the there to be a a difference in flow structure over a flat gravel (d_{50}=1.70cm) bed, than over a gravel wave field (η=0.5m, λ=10m). Roughness parameters (z_o) of 1.24 and 4.90mm, respectively, have been identified.

Shingle sediment transport has been shown to be proportional to bed shear stress to a power of 3; this is in agreement to stochasticlly-derived formulae.

ACKNOWLEDGEMENTS

The authors would like to thank J. Davis for his assistance with the development and deployment of the platform. Also, the diving team of the Department of Oceanography, University of Southampton, are acknowledged for their assistance during the deployment. The research was funded by the Ministry of Agriculture Fisheries and Food (Coastal and Flood Defence Division), U.K. (R & D Contract CSA 1839).

REFERENCES

Cacchione, D.A. and Drake, D.E., 1979. "An Instrument System to Investigate Sediment Dynamics on Continental Shelves", *Marine Geology*, Vol. 30, pp. 299-312.

Drake, D.E. and Cacchione, D.A., 1991. " Wave-Current Interaction in the Bottom Boundary Layer During Storm and non-Storm Conditions: Observations and Model Predictions", *Continental Shelf Research*, Vol. 12, pp. 1331-1352.

Einstein, H.A., 1950. "The Bedload Function for Sediment Transportation in Open Channel Flows. Soil Conservation Service, U.S. Department of Agriculture, Tech. Bull. 1026, 78pp.

Engelund, F. and Hansen, E., 1967. "A Monograph on Sediment Transport in Alluvian Streams", Technisk Vorlag. Copenhagen, 62 pp.

Green, M.O., 1992. " Spectral Estimates of Bed Shear Stress at Subcritical Reynolds Numbers in a Tidal Boundary Layer", *Journal of Physical Oceanography*, Vol. 22, pp. 903-917.

Gross, T.F., Williams III, A.J. and Grant, W.D., 1986. Long-Term In-Situ Calculations of Kinetic Energy and Reynolds Stress in a Deep Sea Boundary Layer", *Journal of Geophysical Research*, Vol. 91, pp. 8461-8469.

Harden-Jones, F.R. and Mitson, R.B. 1982. "The Movement of Noisy Sand Waves in the Strait of Dover", *J. Cons. Int. Mer.*, Vol. 40, pp. 53-61.

Huntley, D.A., 1988. "A Modified Inertial Dissipation Method for Estimating Seabed Stresses at Low Reynolds Numbers with Application to Wave/Current Boundary Layer Measurements", *Journal of Physical Oceanography*, Vol. 18, pp. 339-346.

Johnson, P. and Muir, T.C., 1969. Acoustic Detection of Sediment Movement", *Journal of Hudraulic Research*, Vol. 7., pp. 519-540.

Lyne, V.D, Butman, B. and Grant, W.D., 1990. "Sediment Movement Along the U.S. East Coast Continental Shelf - 1. Estimates of Bottom Stress using the Grant-Madsen Model and Near-Bottom Wave and Current Measurements", *Continental Shelf Research*, Vol. 10, pp. 397-428.

Madsen, O.S. and Grant, W.D., 1976. "Sediment Transport in the Coastal Environment", Ralph M. Parsons Lab. MIT Report No. 209, 105pp.

Meyer-Peter, E. and Muller, R., 1948. "Formulas for Bed Load Transport", *Proceedings 2nd Congress IAHR, Stockholm, Sweden*, Vol. 2, pp. 39-64.

Paintal, A.S., 1971. "A Stochastic Model of Bed Load Transport", *Journal of Hydraulic Research*, Vol. 9, No 4, pp. 527-550.

Sleath, J.F.A., 1991. Velocities and Shear Stresses in Wave-Current Flows", *Journal of Geophysical Research*, 96, pp. 15237-15244.

Soulsby, R.L., 1983. "The Bottom Boundary Layer of Shelf Seas", In: B. Johns (ed), *Physical Oceanography of Coastal and Shelf Seas*, Elsevier, pp. 1989-266.

Soulsby, R.L., and Humphery, J.D., 1990. "Field Observations of Wave-Current Interaction at the Sea Bed", In: A. Torum and O.T. Gudmestad (eds), *Wave Water Kinematics*, Kluwer Academic, pp. 413-428.

Sternberg, R.W., 1972. "Predicting Initial Motion and Bedload Transport of Sediment Particles in the Shallow Marine Environment", In: D.J.P. Swift, D.B. Duane and O.H. Pilkey (eds), *Shelf Sediment Transport, Process and Pattern*, pp. 61-82.

Sternberg, R.W., 1971. "Measurements of Incipient Motion of Sediment Particles in the Marine Environment", *Marine Geology*, Vol. 10, pp. 113-119.

Thorne, P.D., 1990. "Seabed Saltation Noise", In: B.Kermen (ed), *Natural*

Physical Sources of Underwater Sound, Kuwer Academic Publishers (in press).

Thorne, P.D., 1986a. " Acoustic Techniques for Underwater Sediment Transport Studies", *Proceedings of The Institute of Acoustics*, Vol. 8., pp. 51-65.

Thorne, P.D., 1986b. " An Intercomparison Between Visual and Acoustic Detection of Seabed Gravel Movement", *Marine Geology*, Vol. 72, pp. 11-31.

Thorne, P.D., Williams, J.J. and Heathershaw, A.D., 1989. "In Situ Acoustic Measurement of Marine Gravel Threshold and Transport", *Sedimentology*, Vol. 36, pp. 61-74.

Thorne, P.D. and Foden, D.J., 1988. "Generation of Underwater Sound by Colliding Spheres", *Journal of Acoustical Society of America*, Vol. 84, pp. 2144-2152.

Tywoniuk, N. and Warnock, R.G., 1973. "Acoustic Detection of Bedload, Fluvial Processes and Sedimentation", *Proceedings 9th Canadian Hydrology Symposium, Ottawa*, pp. 728-749.

Zanke, U., 1987. "Sedimenttransportformlen fur Bed-load im Vergleich", *Mitt. Franzius Inst. Wasserbau und Kusteningenieurswesen der Universitat Hannover*, Vol. 64, pp. 327-411.

Voulgaris, G. and Collins., M.B., 1994. "Storm Damage - South Coast Shingle Study", Technical Report to MAFF, Southampton University, Oceanography Deapartment (in preparation).

Voulgaris, G., Wilkins, M.P. and Collins, M.B., 1994. " The In-Situ Passive Acoustic Measurement of Shingle Movement under Waves and Currents: Instrument (TOSCA) Development and Preliminary Results", *Continental Shelf Research* (in press).

Voulgaris, G., Collins, M.B. and Wilkin, M.P., 1993. "Storm Damage - South Coast Shingle Study", Technical Report SUDO/TEC/93/8/C, Southampton University, Oceanography Department, 44 pp plus Appendices.

Williams, J.J., Thorne, P.D. and Heathershaw, 1989. "Comparisons Between Acoustic Measurements and Predictions of the Bedload Transport of Marine Gravels", *Sedimentology*, Vol. 36, pp. 973-979.

SOME FEATURES OF THE INITIAL STAGE OF SEDIMENT MOTION IN WATER FLOWS

Svetlana G. Beloshapkova[1], Alexander V. Beloshapkov[2], Sergey M. Antsyferov[3]

ABSTRACT: The detachment mechanics of individual particle placed on the permeable bed is considered. Some features of particles detachment and transport which are in contradiction with traditional model of the initial stage of sediment motion are discussed. The new model of particles detachment which connects the detachment with the action of coherent structures and bed permeability and considers the noted features is offered.

According to this model the mechanics of particles detachment appears to be different for fine sand and for coarse sand, gravel, pebble. For fine particles it is caused by the percolation suspending of the upper layer of sediment and occurs at the areas of coherent structures - bottom interaction. For coarse particles the mechanism of percolation suspending is not effective. In this case the detachment is realized by turn over mechanism. Quantitative estimations of the conditions of particles detachment calculated by new model and their comparison with experimental data are shown. The influence of bed permeability, percolation effects and particles shape are evaluated.

INTRODUCTION

The understanding of the mechanics of the detachment and of the initial stage of particles motion is important for most of scientific and applied problems connected with the investigation of the sediment transport. It becomes particularly vital in cases when hydrodynamic regime can not maintain intensive sediment transport.The great attention was given to the problem of initial stage of sediment motion. It gave possibility to establish fairly reliable formula for the critical conditions of sediment movement. The traditional model of particles detachment is following.

1) Researcher, State Oceanographic Institute, 6, Kropotkinsky per., 119838 Moscow, Russia.
2) Senior Researcher, State Oceanographic Institute, 6, Kropotkinsky per., 119838 Moscow, Russia.
3) Senior Researcher, Institute of Oceanology, 23, Krasikova st., 117851 Moscow, Russia.

The particle, placed on the bed is acted by horizontal directed drag force F_d and upward directed lift force F_l. They are resisted by the particle weight in water F_o and the friction force F_f (Fig.1). In case of local accelerations in the flow it is necessary to consider the inertia force F_v. The detachment of particles is a result of violation of forces or moments equilibrium balance. The bed is considered to be impermeable and the magnitude of the vertical component of the flow velocity V_z on the bottom is set equal to zero.

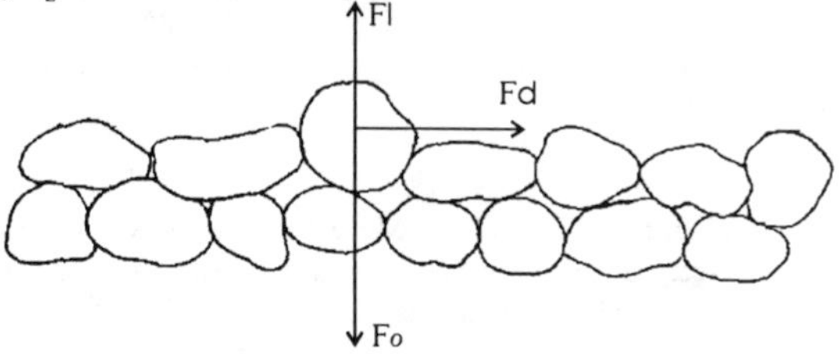

Figure 1. Traditional model of particles detachment

The real conditions on the bed usually differ from pattern. It was found that the bed surface compounded by non-cohesive ground is permeable for steady currents and wave flows. Wave flow penetrate in the ground at depth sufficiently greater then the mean sediment diameter **d** (Debolsky et al. 1990). The magnitude of the vertical velocity component V_z at the surface of gravel bed can reach 8 - 12% of the horizontal component V_x (Longinov et al. 1991).

Turbulent pulsations of the flow velocity V'_x, V'_z causing pulsations of drag force and lift force are usually considered to be the immediate reason the particles detachment. The present physical model is traditional for qualitative description of the detachment process of particles of non-cohesive ground by water flow. The most of publications on the problem of sediment transport contains this model. However this model has several essential defects in view of the conversion to quantity evaluations. For example, it is a well known fact that particles detachment begins at mean flow velocity which corresponds to shear stress 7-10 times less than it would be enough to move particles of such diameter (White, 1940). In this situation the momentary values of flow velocity have to be 2.5 - 3 times greater then average magnitude. Some estimations of detachment conditions with described physical model show that the probability of particles detachment for normal distribution of flow pulsations is negligible small. Consequently, this model can not be the main reasons of particles detachment (Beloshapkova, et al. 1994).

The models which explain the detachment of the particles by means of interaction between rest and mobile particles are often suggested as an alternative. This mechanism really makes great contribution to the bed-load sediment transport, particularly in case of their organized motion (Drake et al, 1989). But it can not exist by itself. The presence of moving particles in the flow for its existence is necessary. But in that case question arises - what is the reason for these moving particles detachment if probability of grain ground detachment for the describing above model is negligible small?

So, traditional concepts can not give physically based quantity evaluations of particles detachment conditions so that for practice some empiric formulas are used. Further, traditional model can not explain some characteristic features of the interaction flow with eroded bottom some of which were detected during experimental investigations. There are:

- the specific character of fine sand particles detachment conform to the shape of thin jets of suspending particles (Grass 1982);
- the different influence of percolation flow for conditions of coarse and fine particles detachment (Mikhaylova et al. 1978, Murakami et.al. 1990);
- the organized movement of gravel and shingle sediment in which intensity of sediment transport considerably exceed to its mean value (Drake et al. 1988);
- differences in detachment conditions of particles on permeable and impermeable bed (Mikhaylova et al. 1980);

and some others.

Obviously, to explain these features of sediment mechanics and to receive of physically substantiated quantity estimations it is necessary to apply some new hypotheses. The results of investigations of bottom boundary layer hydrodynamic was used for this.

Really, concepts of turbulent structure of bottom boundary layer had been changed during last 10 - 20 years. There was found the existence of liquid movement regular elements coherent structures (CS). CS have statistically stable size, shape and internal structure and they are much more regular than ordinary turbulent eddies (Cantwell 1981, Gyr 1983) (fig.2). Coherent structures are immediately connected with bursting phenomena in bottom boundary layer. This phenomena cause intensive flow-bottom interaction concentrated in small parts of bed. Common area of CS - bottom interaction usually is not more then 1 percent of bottom area. But despite this fact these zones are responsible for the 80-90% of the Reynolds stress.

The high intensity of hydrodynamic action in CS - bottom interaction zones stimulated the scientific research and in some works the particles detachment had been related with the action of CS. Employment of this hypothesis enabled one to give qualitative explanation of some features of sediment motion, but the attempts to make some quantitative evaluations were not so successful. Receiving of quantitative estimations was hindered because of the absence of suitable CS - permeable bed interaction model.

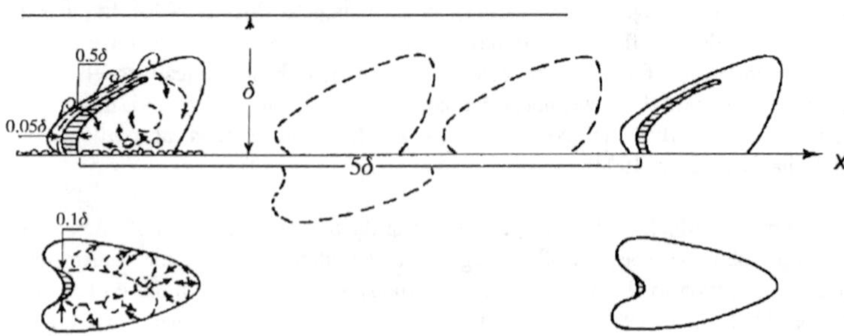

Figure 2. Coherent structures of bottom boundary layer by (Gyr 1983)

Creating of such model is sufficiently difficult problem. Really, the scale of CS - bed interaction areas often are comparable with sediment characteristic diameter. On these areas the significant values of flow velocity and its gradient take place, the vertical flow component is not equal to zero. In case of permeable bed the CS action can induce in it percolation flows.

CS-bed interaction processes are nonlinear and they can be described in terms of probability methods only. However for the probability models development it is necessary to use the results of special experimental investigations. In this paper the model which uses the linear approach is suggested. Used experimental results are based on authors experimental data or had been taken from published works.

PHYSICAL MODEL OF PARTICLES DETACHMENT

The model of particles detachment by the action of CS consists of two parts (Beloshapkova 1992). There are "percolation suspending" model for fine particles (fine and medium sand) and "turn over" model for coarse sand, gravel and pebble. The mechanics of these models is shown on fig. 3. The reasons caused the separation of models for fine and coarse particles detachment will be described below.

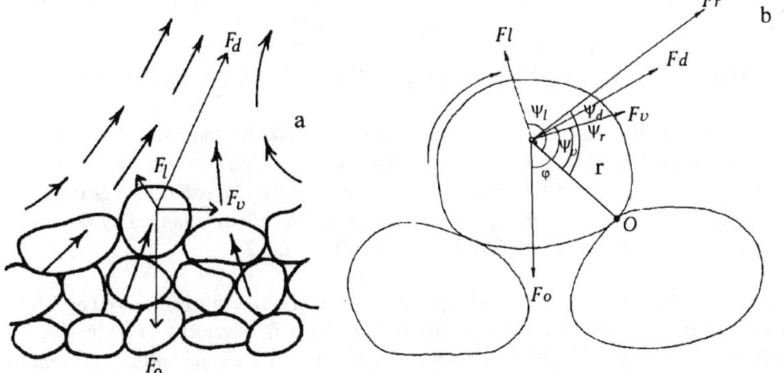

Figure 3. Models of particles detachment: a - percolation suspending model, b -turn over model

According to percolation suspending model the detachment of particles is caused by the action of percolation flows connected with the ejection areas, placed in the rear side of CS. These flows act in upper layer of permeable bed. By the turn over model the particles detachment is caused by moments equilibrium disturbance. These disturbances are connected with velocity and pressure pulsations in CS-bottom interaction areas.The local character of areas on which the particles detachment has place and the movement of this areas above the bed with the velocity near mean flow velocity is common for both models. This generality is connected with some features of CS-bottom interaction. So the same hydrodynamic model of this interaction is used in percolation suspending and turn over models.

The scheme of flow circulation in CS - permeable bottom is shown on fig. 4. Horizontal and vertical components of flow velocity on the bottom surface and the characteristic points of CS - bottom interaction are shown there too. This scheme had been developed on the flume experimental data (Gyr, 1983) and on the investigations of boundary layer turbulence in wind tunnel (Antonia et al. 1989). According to this scheme the most intensive CS - bottom interaction can take place in a region between **E** and **F** points, which are points of vertical and horizontal velocity peak values.

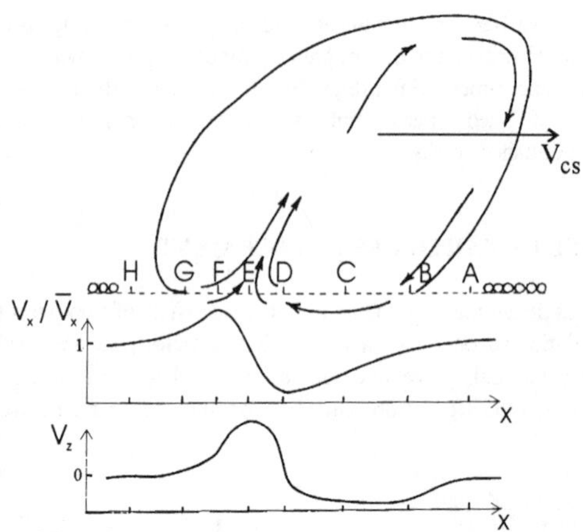

Figure 4. The scheme of flow circulation in CS -permeable bed interaction zone

Let us consider the permeable flat bed consists of non-cohesive particles with porosity ε and permeability coefficient **k**. The flow, having mean velocity V and friction velocity V_* acts on the particles of surface layer. The thickness of bottom boundary layer δ can be estimated by (Cantwell 1981). The value of δ for some experimental investigations is limited by flow depth H. Shape and size of CS can be estimated by (Gyr 1983), see fig. 2.

Designing the hydrodynamic model of liquid circulation in permeable bed caused by CS we can use the model worked out for estimations of percolation flow velocity in permeable bed caused by surface waves (Saks,1987). According to this model the **z** component of flow on the surface of permeable bed can be calculated in terms of velocity **x** component value, wave period **T**, permeability coefficient **k** and porosity ε.

$$V_z / V_x = \frac{\varepsilon \tanh(\kappa \xi)}{(1 + \beta^2)^{0.5}} \qquad (1)$$

where $\beta = \varepsilon \; g / \omega \; k$ is the nondimensional parameter, ξ - permeable layer thickness (m), **g** - gravity constant (m/sec^2), $\omega = 2\pi/T$, κ -wave number (1/m).

Let us consider, that the action of pulsation component of flow is similar to wave one and evaluate $V'_{zs} = V'_z / \varepsilon$ - maximum of seeping flow velocity vertical component caused by CS. So we can consider CS as a wave moving above the bed with average velocity V_{cs} and use (1).

The characteristic size **l** of CS ($l = 0.5\;\delta$) can be chosen as a length of such wave **L**, the wave period can be estimated as $T = \delta / 2\; V_{cs}$ and $\omega = 4\;\pi\; V_{cs}/\delta$. According to (1) we can get evaluation for V'_{zs}:

$$V'_{zs} = \frac{\Delta V_x \tanh(4\;\pi\;\xi/\;\delta)}{[1 + (\varepsilon\; g\; \delta \;/\; 4\;\pi V_{cs}\; k)^2]^{0.5}} \qquad (2)$$

where ΔV_x is the variations of horizontal velocity on CS -bed interaction areas. Using the results of experimental data (Antonia et al. 1989, Habahpasheva et al. 1976) it is possible to show that near the bed the horizontal velocity pulsations distribution differs from normal law. Maximal values of ΔV_x having probability 0,001 -0,01 can be reach from 4 to 6 of root-mean square of its mean value. The magnitude of $<V_x'>/V_x$ is about 0.25 - 0,40 in dependence of particles size and flow depth.

The average porosity of sand surface layer can be estimated as $\varepsilon = 0.5$. Permeability coefficient **k** depends on shape and size of particles, their sorting and packing, the content of silt. We use **k** estimates for non-cohesive medium sorted sands and gravel with corresponding average diameter, which are typical for beach and alluvial sediments.

Percolation suspending model

The estimations of percolation velocity vertical component in sediment surface layer shows, that V'_{zs} is small in compare with particles settling velocity **W**. So be some questions may arise how can slow percolation flows detach the particles. Really, the percolation flow with $V'_{zs} << \mathbf{W}$ can not detach a single particle, but it can cause the suspending of sediments of upper layer in CS - bed interaction zone. It is well known fact, that the percolation velocity value **W'**, which can suspend the sediments upper layer can be 10 - 20 times less then **W** (Happel et al. 1965). The value of **W'** depends on particles average diameter **d**, porosity ε and percolation regime.

Let us consider the conditions of particles detachment according to the percolation suspending model. For suspending of a particle the vertical component of forces trying to disturb it had to exceed particles weight in water:

$$F_r \sin \psi - F_o > 0 \tag{3}$$

where $F_r = F_d + F_l + F_v$ and ψ is the angle between F_r and horizontal direction.

At considering conditions the drag force is not directed horizontally, it is at some angle ψ_d to horizontal direction, so F_d can be written as:

$$F_d = F_{dx} + F_{dz} \tag{4}$$

F_{dx} is the same as in the expression for F_d in traditional form:

$$F_{dx} = 0.5\, \rho\, \mathbf{C_d}\, \mathbf{A_2}\, \mathbf{d}^2\, V_x^2 \tag{5}$$

here ρ- liquid density (kg/m³), $\mathbf{C_d}$ - drag coefficient, $\mathbf{A_2}$ - second coefficient of particles shape ($\mathbf{A_2} = \pi/4$ for sphere). F_{dz} is generated by percolation flow and is vertically directed. It can be calculated as:

$$F_{dz} = 0.5\, \rho\, \mathbf{C_d}'\, \mathbf{A_2}\, \mathbf{d}^2\, V'^2_{zs} \tag{6}$$

The value of $\mathbf{C_d}'$ - the drag coefficient for percolation flow depends on Reynolds number $\mathbf{Re}_{dw}$ calculated by particles diameter **d** and settling velocity **W** and from ground porosity. For fine particles $\mathbf{C_d}'$ can be significantly more than $\mathbf{C_d}$. The high magnitudes of $\mathbf{C_d}'$ cause the percolation suspending phenomena at low values of V'_{zs}.

Horizontal component of water flow velocity near the bed is much more than vertical, so it is possible to use the traditional equation for F_l calculation:

$$F_l = 0.5\ \rho\ C_l\ A_2\ d^2\ V_x^2 \tag{7}$$

The magnitude of inertia force F_v depends on flow acceleration. Assuming the CS transport velocity V_{cs} to be constant it becomes possible to write the value of flow acceleration as $dV_x/dt = V_{cs}\ dV_x/dx$ and the equation for F_v as:

$$F_v = \rho\ C_v\ (\pi/6)\ d^3\ V_{cs}\ dV_x/dx \tag{8}$$

The estimations being done shown, that the influence of lift and inertia forces on the fine particles detachment is negligible small, the action of F_{dx} can not cause it detachment too (Sleath 1984, Nielsen 1992). Therefore we can assume that in this case the condition $F_{dz} > F_o$ is necessary for particles detachment. And the last condition demands that vertical component of percolation flow V'_{zs} has to be more higher than the velocity which cause the suspending of sediment **W'**.

For the verification of percolation suspending model the experimental data of vertical velocity component pulsations measurements are necessary. Authors had no possibility to carry out the experiment. So the model testing was done on the base of the works provided in physical laboratory of Moscow State University in which the influence of percolation effects on the particles detachment magnitude had been investigated (Mikhaylova et al. 1978).

This experimental study had been conducted in hydraulic flume 7m long, 0.4 high and 0.2m width. The averaged injection and ejection percolation flow through permeable bed which had been made from natural material of 20 sm thickness it had been possible to produce. The mean flow velocity had been measured by Pito pipe, the intensity of particles detachment - by high speed filming, sediment transport - by sand traps. The flat bed regime had been maintained in experiments. There had been carried out some runs with fine sand, coarse sand and gravel in various percolation regimes (fig.5).

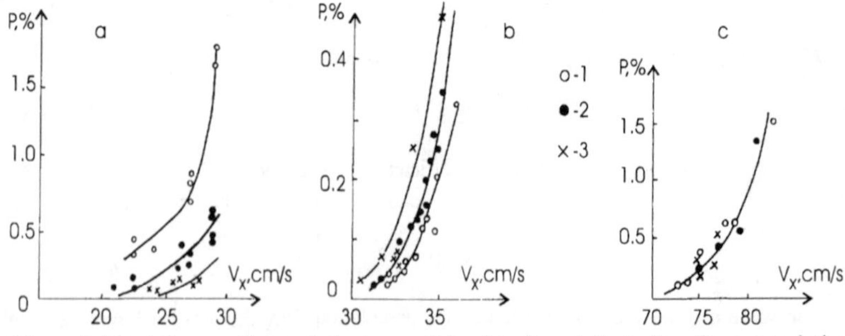

Figure 5. The influence of percolation on particles detachment. 1-ejection, 2-no percolation, 3-injection. a- fine sand, b- coarse sand, c- gravel, P - probability of particles detachment.

Estimations of V'_{zs} values calculated by equation (2) for initial stage of sediment motion and magnitudes of **W'** are presented in the table 1.

Table 1. Particles detachment by percolation suspending model.

Material	d, mm	W,cm/s	$\mathbf{Re}_{dw}$	W'/W	δ, cm	V_{cs}, cm/s	k, cm/s	V'_{zs},cm/s
fine sand	0.25	2.7	6.75	0.04	2	20	0.02	0.13
coarse sand	0.7	9.0	63	0.08	3	30	0.11	0.97
gravel	7.0	27.0	1890	1.0	6	70	0.63	13

The analysis of data shows that **W'** is about 5% of settling velocity **W** for fine sand and about 8% of settling velocity for coarse sand. So the suspending of particles can occur than velocity of seepage is sufficiently less than their settling velocity. In conditions of experiments with fine and coarse sand V'_{zs} values were 1,1-1,3 times greater than **W'**. The conditions of suspending were accomplished for both fine and coarse sand on CS passing by.

The further examination of considering experimental data gives the possibility to investigate some features of particles detachment in the case if the averaged percolation flow acts on bed particles besides the CS. Apparently the interaction between CS caused and averaged percolation flows is nonlinear, but in a first approximation we can consider it to be linear and the particles of the upper layer are acted by the percolation velocity $V_{zs} = V_s + V'_{zs}$

For fine sand the action of percolation from the bed to flow (ejection) with the percolation velocity $V_s = 0.05$ cm/s is equal to increasing of average velocity in flume on 20 - 25%. Or, if one use the (2) formula, to increasing of the V'_{zs} on 0. 05 - 0.065 cm/c. The percolation into the bed (injection) with $V_s = -0.07$cm/s has the opposite influence. It is equal to decreasing of mean flow velocity on 4-5 cm/s and to decreasing of V'_{zs} on 0.04 - 0.05 cm/s (see fig.5).

Therefore, injection as well as ejection type percolation in conditions of constant probability of particles detachment can be compensated by mean flow velocity V_x changes and corresponding changes of V_{zs}. The value of $V_{zs} = V_s + V'_{zs}$ remains almost the constant (V_{zs} = **W'**) in spite of the action of injection percolation is rather less because of percolation flow non-uniformity.The same estimations were executed for coarse sand. The influence of ejection (V_s = 0.13 cm/s) and injection (V_s = - 0.26 cm/s) percolation took place, but it was not so significant. This result can be explained by the reason that the percolation suspending is not the one mechanism of coarse sand detachment. No influence of percolation flow on the detachment of gravel particles had been ascertained even for V_s = 9.5cm/s.

The obtained results shows that the percolation suspending model can explain the reasons and conditions of fine particles detachment but it is not effective for coarse ones. This conclusion is rather natural because the percolation conditions changes in dependence to $\mathbf{Re}_{dw}$ For $\mathbf{Re}_{dw} < 2 - 5$ the percolation is laminar, if $2 - 5 < \mathbf{Re}_{dw} < 140 - 210$ there is transition regime and if $\mathbf{Re}_{dw} > 140 - 210$ percolation becomes turbulent. In experiments with fine sand the value of $\mathbf{Re}_{dw}$ was about 0.3 - 0.4, in experiments with coarse sand - 5 - 7 and in experiments with gravel it was about 900. So we can note that in experiments with fine sand it was the laminar regime of percolation, coarse sand experiments fall within a beginning of transition regime and experiments with gravel fall into turbulent regime.

Using characteristic values of particles density, ground porosity and percolation coefficient **k** it is possible to show, that in conditions of initial stage of sediment transport $\mathbf{Re}_{dw}$ = 2 - 5 equals to $\mathbf{Re}_*$ = 5 - 15 and $\mathbf{Re}_{dw}$ = 140 - 210 equals to $\mathbf{Re}_*$ = 70 - 100. By other words, the beginning of transitional regime of percolation induced by CS action corresponds to the minimum of Shields curve. The beginning of turbulent percolation regime falls to beginning of curves part, which is characterized by constant values of the Shields parameter $\Psi = \rho V_* / (\rho_s - \rho)\, g\, d$, where ρ_s is particles density.

The Darsy law is not applicable for the turbulent percolation regime and the use of (2) is debated. In these conditions **W** and **W'** values becomes nearly equal and the percolation suspending model becomes not effective. The detachment by turn over model is more effective for coarse particles. This mechanism becomes significant for $\mathbf{Re}_*$ > 10.

Turn over model

Let us consider the equilibrium of moments acting on a particle, placed on the bottom surface. The moment, which tends to turn over the particle is determined by acting of the force $F_r = F_d + F_l + F_v$. The resisting moment is formed by the weight in fluid F_o. The conditions of particle detachment can be written as:

$$|F_r|\,\mathbf{r}\cos(\varphi-\psi_r) - |F_o|\,\mathbf{r}\sin(\varphi) > 0 \qquad (9)$$

where **r** is characteristic particles radius. In this conditions the particle starts to turn over decreasing the angle φ and the magnitude of resisting moment. Transforming the expression (9) in non-dimensional form we can write it as:

$$\frac{|F_d|\,\mathbf{r}\cos(\varphi-\psi_d)}{|F_o|\,\mathbf{r}\sin(\varphi)} + \frac{|F_l|\,\mathbf{r}\cos(\varphi-\psi_l)}{|F_o|\,\mathbf{r}\sin(\varphi)} + \frac{|F_v|\,\mathbf{r}\cos(\varphi-\psi_v)}{|F_o|\,\mathbf{r}\sin(\varphi)} > 0 \qquad (10)$$

Let us apply equation (10) for particles detachment estimations in different conditions for permeable and impermeable bed. The Shields curve can be used as a criteria of motion initiation. For coarse particles ($\mathbf{Re}_*$ > 70) it corresponds to the constant magnitude of Shields parameter Ψ = 0.05. Let us assume bed to be homogeneous, permeability of the bed to be small, so F_d and F_v are directed horizontally and F_l –vertically (Beloshapkova et al. 1994). In this condition $\psi_d = \psi_v = \pi/4 - \varphi$, $\psi_l = -\varphi$.

The values of forces acting on a particle placed on the surface of the impermeable bed composed by the same particles can be written in terms of Shields parameter:

$$\frac{|F_d|}{|F_o|} = \frac{3}{4}\,\frac{\rho}{\rho_s - \rho}\,\frac{C_d}{\mathbf{gd}}\,\bar{V}_x^2 = 0.75\, C_d\, \bar{V}_{x+}^2 \Psi \qquad (11)$$

$$\frac{|F_l|}{|F_o|} = \frac{3}{4}\frac{\rho}{\rho_s - \rho}\frac{C_l}{gd}\bar{V}_x^2 = 0.75\, C_l\, \bar{V}_{x+}^2\Psi \qquad (12)$$

$$\frac{|F_v|}{|F_o|} = \frac{\rho}{\rho_s - \rho}\frac{C_v}{g}\frac{|d\, V_x|}{dt} \qquad (13)$$

where $V_{x+} = V_x/V_*$ is nondimensional velocity. For **z** equal to bed roughness V_{x+} is equal to 6.65. For $\mathbf{Re_*} > 70$ $\mathbf{C_d} = 0.4$ and taking into account the influence of the bed it can be assumed $\mathbf{C_d} = 0.48$. Lift coefficient $\mathbf{C_l}$ in this conditions can be evaluated as $\mathbf{C_l} = 0.4$ and $\mathbf{C_v} = 1.5$ (Grishin 1982, Sleath 1984, Nielsen 1992).

The value dV/dt we can estimate as $d\, V_x / dt = V_x\, d\, V_x / dx$ and write down the equation (13) in form:

$$\frac{|F_v|}{|F_o|} = C_v\Psi\, \frac{d}{V_x}V_{x+}^2\, \frac{d\, V_x}{dx} \qquad (14)$$

Estimations of particles stability showed that if particle is on the surface on the bed, when $\eta = 0.8$ it's state is not stable for $\Psi = 0.05$. This conclusion is not unexpected - in laboratory experiments it was found that in this condition particles start to move when Shields parameter value is about 5 times less. The detachment of spherical particles in conditions of $\eta = 0.1$ - 0.15 corresponds to Shields curve. Here η - particles relative intrusion. It shows how high the particles top is above the bed mean level in part of particles diameter. The value of η as well as shape of particles is one of most significant factors which influence on detachment conditions, because they defines the value of angle jo and forces F_d, F_l and F_v. In the case when spherical particle turns over another one $\varphi = \cos^{-1}(\eta)$ and when it turns over between two other ones:

$$\varphi = tg^{-1}\,[(1 - \eta^2)^{0.5} / 2\eta] \qquad (15)$$

Now let us consider the situation when η is small and the particle is acted by CS which cause intensive pulsations of horizontal flow velocity - **N** times higher than it's root - mean - square value $<V'_x>$. This particle had less exposure to flow then one placed on the surface of the bed. The equations 11-14 in this case can be written in form:

$$\frac{|F_d(\eta)|}{|F_o|} = 0.75\, C_d\, \bar{V}_{x+}^2\Psi\, \kappa_d(\eta).(1 + N<V'_x>/\bar{V}_x)^2 \qquad (16)$$

$$\frac{|F_l(\eta)|}{|F_o|} = 0.75\, C_l\, \bar{V}_{x+}^2\Psi\, \kappa_l(\eta).(1 + N<V'_x>/\bar{V}_x)^2 \qquad (17)$$

$$\frac{|F_v(\eta)|}{|F_o|} = C_v\Psi\, \frac{d}{V_x}\bar{V}_{x+}^2\, \frac{d\, V_x}{dx}\, \kappa_v(\eta) \qquad (18)$$

Nondimensional parameters $K_d(\eta)$, $K_l(\eta)$ and $K_v(\eta)$ shows the ratio of forces acting on the particle with a particle placed on the surface of the bed.

The generalization of experimental results in which the dependence of F_d and F_l on η had been studied (Chen et al. 1973, Shen 1980) showed that in impermeable bed condition the value of $K_l(\eta) = F_d(\eta)/F_d(\eta = 0.8)$ had the linear dependence on η. The dependence for $F_l(\eta)$ is more complex. When η increases from 0 to 0. 75 the $K_l(\eta)$ magnitude raises linearly. After, that η reach 0.8, $K_l(\eta)$ decreases about two times and remains its magnitude for higher η values, see fig.6. We have no data about $K_v(\eta)$ and have to assume, that it is possible to take it similar to $F_d(\eta)$.

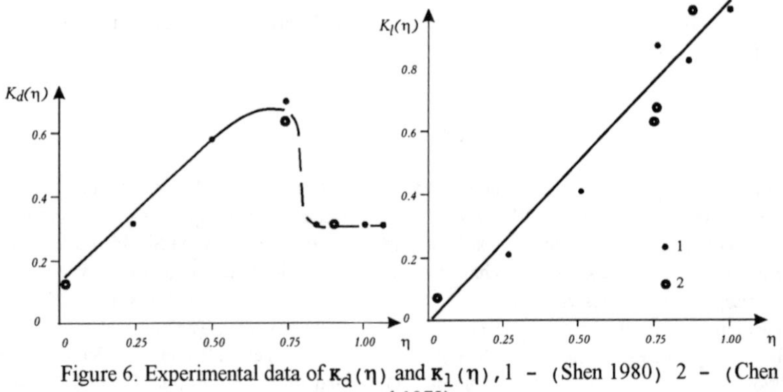

Figure 6. Experimental data of $K_d(\eta)$ and $K_l(\eta)$, 1 - (Shen 1980) 2 - (Chen et al.1973)

There are no estimations of $K_d(\eta)$ and $K_l(\eta)$ for permeable bed. Taking into account some features of velocity distribution above and below the permeable bed average level (Debolsky et al. 1990) it is possible to assume that $K_d(\eta)$ is about two times higher but $K_l(\eta)$ changes slightly. We have also note that the changes of $K_d(\eta)$ relates to F_{dx}component only. Using this evaluations of $K_d(\eta)$, $K_l(\eta)$ and $K_v(\eta)$ some estimations of particles detachment conditions have been done by equations (10) and (16-18). The results obtained in the case of spherical particles and impermeable bed shows, that for characteristic values of Shields parameter $\Psi = 0.05$, turbulent intensity $<V'_x>/V_x = 0.25 - 0.3$ and typical magnitudes of $\eta = 0.1 - 0.15$ the detachment of particles occurs for the pulsations of horizontal component of velocity 3 - 4 times greater it's root-mean-square value. This result is similar to the evaluations of the pulsation velocities at the CS zones and the data of Fenton and Abbott (Fenton et al. 1977) experiments with spherical particles and impermeable bed, see fig.7.

According to provided estimations the lift force made the main contribution to the detachment of particles. The action of drag force F_d was less sufficient and its contribution to the equation was not exceed 10 -20 %. The contribution of inertial force F_v was also slight. But let us keep in mind that these conclusions are true only for spherical particles. For natural particles contribution of F_l can be 2 - 3 times less. It depends on their shape. Near the point **E** (see fig.4) maximal value of volume force F_v takes place. The magnitude of drag and lift forces there are 4-5 times less then near **F** point. The influence of drag force vertical component above the impermeable bed is not significant. So the action of drag and lift forces can exceed 25-30% of value which can turn over the particle.

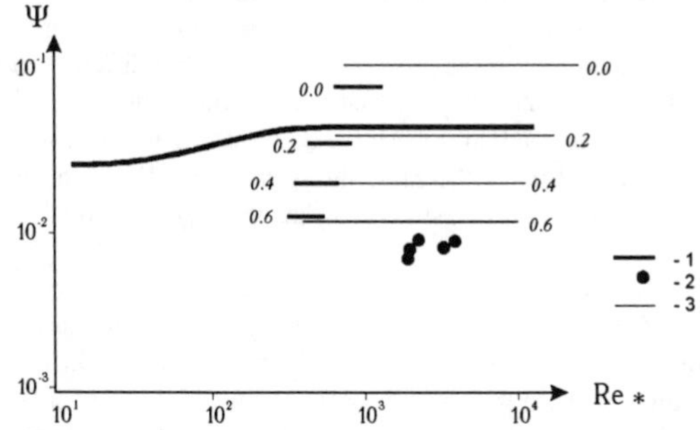

Figure 7. Estimations of particles detachment conditions plotted on Shields curve.
1 - Fenton and Abbott experimental data $\eta = 0, 0.2, 0.4, 0.6$ (Fenton et al. 1977),
2 - Fenton and Abbott $\eta = 0, 82$, 3 - calculations with turn over model

Let us evaluate the value of F_v/F_o. We can write $d\,V_x/dx$ as $\Delta V_x/\Delta \mathbf{x}$ and to assume that $\Delta V_x = 4 \langle V'_x \rangle = 1.2\, V_x$ and $\Delta \mathbf{x}$ as $\alpha\, \mathbf{d}$. In this assumption the detachment of particles become possible if

$$\frac{|F_v(\eta)|}{|F_o|}\, \kappa_v(\eta)\ \mathrm{ctg}\varphi = 1.2\, \frac{C_v \Psi V_{x+}^2}{\alpha}\, \kappa_v(\eta)\ \mathrm{ctg}\ \varphi > 0.75 \quad (19)$$

The estimations done by (16) shows that even in condition of the utmost possible magnitude of velocity gradient the inertial force can not detach the particle placed on impermeable bed and the detachment of particles can take place near the point **F** only.

Let us remove the limitations on the bed permeability. It will not make some strong changes to the mechanism of detachment at the point **F**, because the vertical component of velocity is close to zero at this point. The only one difference is connected with some increasing of drag force due to higher values of $\kappa_d(\eta)$. Thus in case of permeable bed near point **F** the conditions of particles detachment are a little easier than at impermeable bed, but at low η magnitudes the difference is not significant.

But at point **E** in case of permeable bed the vertical component of F_d appears, force F_v increases nearly twice and magnitudes of angles changes. As a result of this fact in cases when the area of the zone of CS - bed interaction becomes close to the particle diameter, and exactly reaches 2 - 3 particles diameters the detachment becomes possible. So, the conditions of detachment may occurs in case of permeable bed as distinct from impermeable one. The detachment is also possible in any point between **E** and **F**. Therefore in case of permeable bed the detachment of the particles is possible not only in the vicinity of point **F** but can cover large areas of bed. This phenomena cause the decreasing of critical velocity and stresses comparing with impermeable bed.

The influence of particles shape can act on the detachment conditions by two ways - by the drag and lift coefficients changes and by the changes in surface layer packing which changes the average value of η. The influence of this factors causes the significant changes in detachment conditions of particles having the same mean diameter but different shape. It was found that particles which are characteristic to beach and alluvial sediments and have form coefficient $\theta = 0.8$ start to move in conditions when the flow velocity is 1.4 -1.6 times higher, then for spherical particles ($\theta = 1.0$) of the same mean diameter. The detachment of flat particles begins when flow velocity is 2 - 2.5 times more than in case of spherical particles.

Assuming $\eta = 0.1$-0.5 as a characteristic magnitude for natural particles it is possible to give the following explain of such phenomena. The value of $\mathbf{C_d}$ has no significant changes due to shape variations (Grishin 1982), so the changes of $\mathbf{C_l}$ can take place. In particular, the estimates showed above can be reached if $\mathbf{C_l} = 0.15$ for $\theta = 0.8$ and $\mathbf{C_l} = 0.08$ for $\theta = 0.6$ takes place. This feature caused by particles shape takes place as for permeable as impermeable bed.

CONCLUTION

Concluding we can resume, that considered model enables to make qualitative description and to obtain quantitative estimations of particles detachment. It also makes possible to detect some particular features of these conditions induced by bed permeability and particles shape. This work is based on the generalization of experimental data, which were obtained by the authors before and the data of published papers. That is why some thesis of suggested model need experimental confirmation. It is necessary to carry out some experimental investigations of the forces acting on particles in case of permeable bed and create more exact models, describing the hydrodynamic of turbulent formations - permeable bed interaction and use stochastic methods of detachment processes description to develop this model further.

ACKNOWLEDGMENTS

Participation of one the authors of this work in the conference "Coastal dynamics-94" become possible thanks to the financial support of the John D. and Catherine T. MacArthur Foundation. The paper design was done with support of ECO-System company.

REFERENCES

Antonia, R., and Bisset, D.K. 1989 "Spanwise structure in the near-wall region of a turbulent boundary layer", J. Fluid Mech., Vol. 210, pp.437-458.

Beloshapkova, S.G. 1992. "Causes of sediment movement initiation", Oceanology, Vol.32, No.2, pp.233-237.

Beloshapkova, S.G., and Beloshapkov, A.V. 1994. "Mechanics of particles detachment under the the action of coherent structures in water flows (percolation suspending model)", Oceanology, Vol.34, No.1, pp.127-132. (in Russian)

Cantwell, B.J. 1981 "Organized motion in turbulent flow", Ann. Rev. Fluid Mech., Vol.13, pp.457-515.

Chen, C.N., and Carstens, M.R., 1973. "Mechanic of removal of a spherical particle from a flat bed", in: Proc.of XY Congr. IAHR, Istambul, Vol.1, pp.247-252.

Debolsky, V.K., and Gubeladze, D.O. 1990. "Hydraulic resistance in a river-bed flow in the presence of infiltration",Meteology and Hydrology, No.9, pp.118-126. (in Russian)

Drake, T.G., Shreve, R.L., Dietrich, W.E., Whiting, P.J., and Leopold, L.B. 1988. "Bed transport of fine gravel observed by motion - picture photography", J. Fluid Mech., Vol.192, pp. 193-217.

Fenton, A.D., and Abbott, I.E. 1977. "Initial movement of grains on a stream bed: the effekt of relative intrusion", Proc. Roy. Soc., London, A, Vol.352, pp.523-537.

Grass, A.J. 1982. "The influence of boundary layer turbulence of the mechanics of sediment transport", Euromech. 156. Mechanics of sediment transport. Istanbul. pp.3-17.

Grishin, N.N. 1982. "Mechanics of bottom sediments", Nauka, Moscow, 160 p (in Russian)

Gyr, A. 1983. "Towards a better definition of the types of sediment transport", J. Hydraulic Res., Vol.21, No.1, pp.1-14.

Habahpasheva, E.M., and Mikhaylova, E.C. 1976. "Turbulence investigation in water flow with parallel walls", in: Experimental investigations of near-wall turbulence and viscous layer, Novosibirsk, pp.33-56. (in Russian)

Happel, J., and Brenner H. 1965 "Low Reynolds number hydrodynamics", Prentice-hall, 630p.

Longinov, V.V., and Saks, S.E. 1991. "Experimental investigation of wave movements of flow on the surface and inside of permeable bed", in: Engeneering and geologycal conditions of shelf and methods of these investigations, Riga, pp.35-42. (in Russian)

Mikhaylova, N.A., and Beloshapkova, S.G. 1980. "The influence of mofing rougness on the particles movement characteristics in turbulent flow", Vestnik of Moskow State University, Physics and Astronomy, Vol.20, No.6, pp.69-73. (in Russian)

Mikhaylova, N.A., and Mulykova, N.B. 1978. "The influence of percolation flow on intensity of particles detachment in conditions instability of flow", Hydraulics Civil Engineering, No.11, pp.28-31. (in Russian)

Murakami, S., Tsujiumoto T. and Hakagawa H. 1990. "Bed-load transport influenced by transportation thtough a bottom of fluvial open-channel", Proc.JSCE, No.423, pp.53-62. (in Japanese)

Nielsen, P. 1992. "Coastal bottom boundary layers and sediment transport", World Scientific, Singapore, New Jersey, London, Hong Kong, 324p.

Saks, S.E. 1987. "Gravity waves in heavy liquid flow with permaeble bed", Izvestiya of Academy of Sciences of USSR, Mechanics of flow and and gas, No.2, pp.115-118. (in Russian).

Shen, H.W. 1980. " Considerations for the construction of stochastic sediment bed load models for flat bed", Proc. 3rd Symp. Stochastic Hydraul., Tokyo, pp.29-45.

Sleath, J.F. A. 1984. "Sea bed mechanics", New York, by J. Wiley & Sons, 335p.

White, C.M. 1940. " The equilibrium of grains on the bed of a steam", Proc.Roy.Soc., Ser.A, Vol174, No.958, pp.322-338.

PHYSICAL MODELLING OF THE RESPONSE OF SHINGLE BEACHES IN THE PRESENCE OF CONTROL STRUCTURES

Tom T. Coates[1]

ABSTRACT: Large wave basin and flume facilities have been utilized by HR Wallingford to undertake an ongoing research programme on the response of shingle (gravel) beaches to wave action. This paper reviews the programme and discusses the recent work on the influence of groynes on the cross-shore distribution of transport and the plan shape development of the beaches. The experimental work investigated groyne spacings, lengths, crest profiles and construction materials under a range of random wave conditions in a 1:50 scale basin.

The physical model results will be used to enhance existing parametric models of beach response. In particular, the physical model study indicates the importance of cross-shore transport adjacent to groynes and the potential for suspension of shingle within the breaker zone. Modifications to the numerical models have not yet been developed as further model tests and field verification are required.

BACKGROUND

Much of the research on beach response concentrates on sand beaches. However, in the U.K. shingle beaches (10 mm < D_{50} < 50 mm) form an important element of the coastline. The shingle is derived from either relict glacial deposits or coastal cliff erosion. Attempts to stabilize cliff erosion, combined with other forms of human intervention, such as the construction of harbour breakwaters, have reduced the availability of shingle. The result has been the progressive erosion of the existing beaches.

---

[1] **Project Manager, Coastal Group, HR Wallingford, Wallingford, Oxfordshire, OX10 8BA, UK**

Traditionally, closely spaced timber groynes have been constructed to control erosion by intercepting longshore transport to retain a reservoir of beach material. In many cases these groynes have been unsuccessful due to poor design and they have often led to rapid erosion of downdrift beaches. More recent beach management practice has shifted to construction of rock mound groynes or detached breakwaters, combined with beach nourishment. However, coastal engineers have continued to rely on past experience, site specific physical models and personal judgement to predict the potential response of shingle beaches to the local wave and water level regime in the presence of control structures. In order to improve this situation HR Wallingford undertook an ongoing research programme utilizing physical models with the aim of developing design tools and guidelines.

This paper presents a review of the programme, including the development of numerical models and then discusses some aspects of the recent work on multiple groyne systems, with particular reference to potential improvements to existing design models. It should be noted that all dimensions are quoted in prototype terms unless otherwise indicated.

RESEARCH REVIEW

The research programme has investigated the response of shingles beaches in a series of physical model studies, as summarized in Table 1 below:

Table 1: Shingle Beach Physical Model Programme

Model Type	Purpose	Variables	Reference
Wave flume (1:17 scale)	Cross-shore Response	- H_s, T_p - Duration - D_{50} and D_{85} / D_{15} - Initial beach slope - Depth of mobile layer	Powell 1990
	Influence of vertical sea walls	- H_s, T_p, SWL - Initial beach profile	Sayers 1993
Wave basin (1:17 scale)	Open beach response	- H_s, T_p - Wave direction	Coates & Lowe 1993

Table 1: Shingle Beach Physical Model Programme - Continued

Model Type	Purpose	Variables	Reference
	Influence of vertial sea walls	- H_s, T_p, SWL - Initial beach profile - Wave direction	-
	Beach response with groynes	- H_s, T_p - Wave direction - Groyne type	Coates & Lowe 1993
Random wave multi-flume (1:40 scale)	Cross-shore response with dissimilar sediments	- H_s, T_p, SWL - Sediment grading	Powell 1992
Wave basin (1:50)	Beach response with groynes	- H_s, T_p, SWL - Wave direction - Construction materials - Groyne spacing, length, elevation & crest profile	Coates 1994
	Beach response with detached breakwaters	- T_p, SWL - Structure length, free board, offshore distance and spacing	Coates 1994

Based on these physical model studies, a number of numerical models have been developed. The original flume study resulted in a parametric profile response model known as SHINGLE (Powell, 1990). Results from the open beach, sea wall interaction and multiple flume studies have been developed as potential add-ons to this model (Coates and Lowe 1993, Sayers 1993, Powell 1992). Some of the results of the flume and larger scale basin studies have been incorporated into a longshore beach development model known as BEACHPLAN.

BEACHPLAN has evolved over many years. It is described most recently by Brampton and Goldberg (1991). Briefly, the model combines the one-line planshape model of Brampton and Motyka (1984) with elements of the profile response model SHINGLE, and assumes the cross-shore distribution of longshore transport which is described in Coates and Lowe (1993). Some of the requirements for the design of a groyne controlled beach renourishment are dealt with by the model, but it is limited by its overly simplistic approach to transport around and over groynes. This limitation has been partially addressed by Huang (1993) using results from the 1:17 scale wave basin model. However, the present research indicates that further work is required before the numerical model can produce reliable results.

Figure 1 illustrates the relevant part of Huang's model, which assumes that shingle is transported as bed load and that the cross-shore distribution of transport typical of an open beach is maintained in the presence of a groyne.

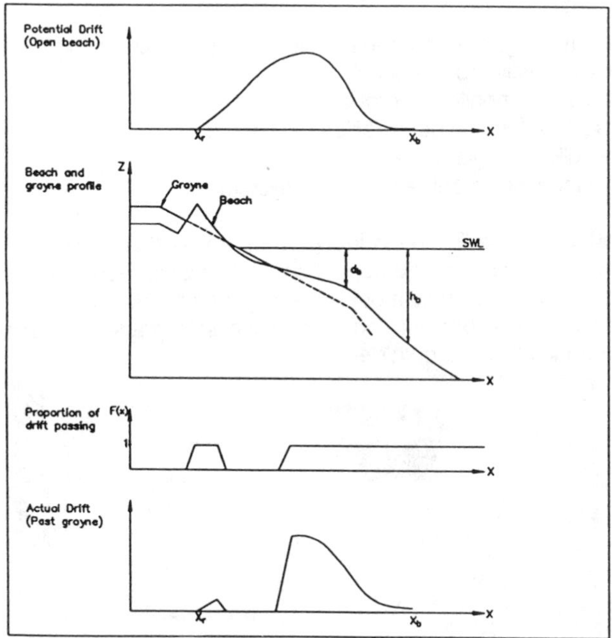

Figure 1. Cross-groyne transport assumptions used by Huang (1993)

The upper figures present the cross-shore distribution of transport for an open beach between the run-up limit and the maximum depth of significant transport (Coates and Lowe 1993) and a possible beach profile at a groyne. The lower figures illustrate Huang's predicted redistribution of transport over the groyne. The open beach distribution is assumed where the beach is above the groyne, while zero drift is assumed where the beach is below the groyne; according to this model the groyne would have to be completely covered to allow the open beach drift rate to resume.

This simple model is a useful step in predicting beach response. However, field observations and the recent wave basin model results do not appear to support the assumptions about shingle transport.

Model Studies

The test programme was conducted at HR Wallingford in a 23 m by 24m wave basin with a maximum working depth of 0.4 m. The model was designed at an undistorted scale of 1:50, according to Froude's relationships. The basin was equipped with:

- a 15 m long, mobile wave paddle capable of generating waves at up to 45° relative to the beach;
- nine wave monitoring probes;
- a semi-automatic bed profiler;
- an oblique angle camera;
- a manual sediment recirculation system.

The model layout is illustrated in Figure 2. It was designed to simulate a typical UK sand lower/shingle upper beach. The lower portion of the beach was constructed from a rigid moulding at a slope of 1:50. The upper portion comprised a mobile bed formed of crushed and graded anthracite coal formed at an initial slope of 1:7½.

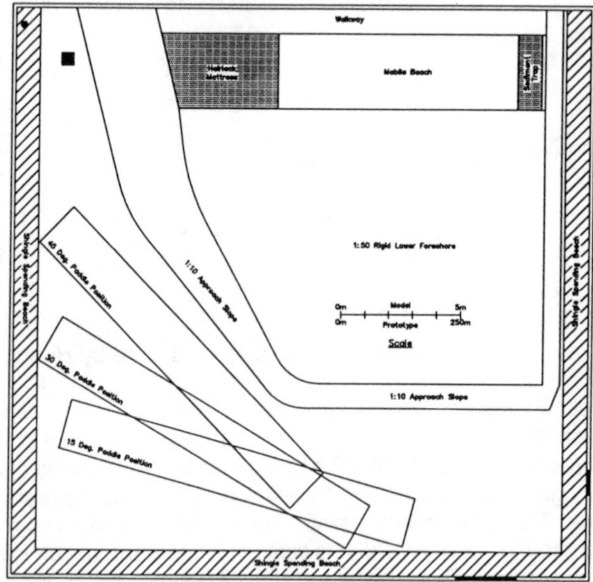

Figure 2. The wave basin facility

The mobile bed was designed to be similar to those used in previous model studies within the programme and to simulate typical UK beaches. The scaling relationships used in selecting the model beach material are based on a well established method which attempts to satisfy three criteria:

- the permeability, which governs slope;
- the relative magnitude of onshore or offshore movement, which determines whether erosion or accretion will occur;
- the threshold velocity of particle motion.

The methods published by Yalin (1963) which relate slope to a non-linear function of the voids Reynolds Number are used to satisfy the permeability requirement. The fall velocity parameter H_b/wT proposed by Dean (1973) is used to satisfy the onshore/offshore criteria while the relationship of Komar and Miller is used for the threshold of motion. At the model scale of 1:50, an assumed prototype D_{50} of 15 mm and a density of 2.65 T/m^3 then the required model sediment should have a D_{50} of 2.5 mm. and a density of 1.41 T/m^3. Anthracite coal has a density of 1.39 T/m^3 and can be obtained in a range of grades. It is therefore an appropriate material for use as the mobile bed.

The sea state variables used during the investigation included:

- offshore significant wave height
- offshore wave steepness (H_s/L_m)
- water depth at the toe of the initial mobile beach
- wave direction relative to the beach

The model was calibrated for prototype offshore wave heights of 1m, 2m and 3m at wave steepnesses of 0.02, 0.04 and 0.06, with water depths of 2m, 3m and 4m, and directions of 15°, 30° and 45°. Due to time constraints, testing concentrated on the 2m waves at steepnesses of 0.02 and 0.06. Water depths were restricted to 3m and 4m and most tests were run with a 30° wave direction, apart from a final series of five tests at 15°. JONSWAP wave spectra were used for all tests.

The structural variables investigated included:-

- the construction materials, either rock or timber
- groyne spacing
- groyne length, from a pre-determined base line
- crest profile
- groyne head elevation

The groyne types are illustrated by Figure 3. A total of 48 tests were run.

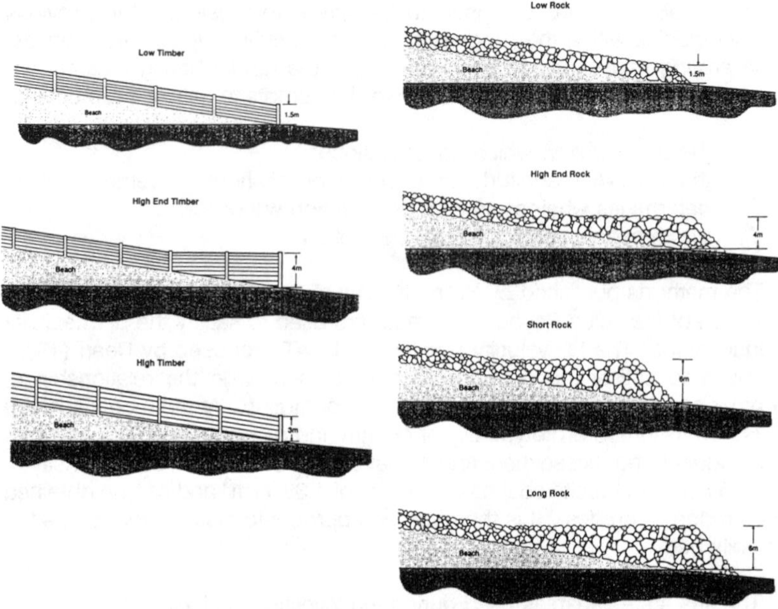

Figure 3. Groyne configurations investigated

During calibration, the model was run with an open beach, fed by an unlimited supply of sediment. The transport rates and cross-shore distribution of transport were measured for each wave and water level combination. The transport rates were used to determine standard sediment input rates for all subsequent tests; these are referred to as the calibrated rates. Some of the test layouts were run initially with the calibrated input rates, followed by a test with no input. This allowed comparisons to be made between fully developed and fully eroded beaches.

Following the calibration stage, the groyne system was installed. The first series of tests concentrated on the low sloping timber groynes, as these were considered to be the most simple and are also the most common form around the UK. Further tests were then run on the six other configurations.

During the course of each test, the downdrift output rate was monitored continuously. Beach plan shape measurements were taken at regular intervals to monitor the position of the crest, SWL and toe, and therefore the beach volume. As the model beach reached stability, measurements were taken of the cross-shore distribution of transport along the line of the central groyne using an array of wire mesh traps. The tests were run until the output drift rate and the beach volume stabilized. Beach profiles were measured within three groyne bays along lines set 7.5 m either side of the groynes and along the centre lines of the bays.

The full results of the study are published in Coates (1994). They include:

- the change in groyne effectiveness over time for different groyne types and spacings;
- the impact of zero sediment input relative to the calibrated inputs;
- the potential variation in cross-shore transport distribution over and around groynes of different types;
- the variation in beach profiles along a groyne bay.

The results were complex and, in some respects, inconclusive. Despite the length of the test programme, several important aspects of beach response were not fully explored. However, several interesting results were obtained which will be illustrated by comparing measurements obtained from tests of two of the seven groyne configurations. The first of these is a rock barrier with a horizontal end 2m above the maximum SWL. The second is a timber groyne with the seaward end set on the maximum SWL (Figures 3b and f). Both of these groyne layouts were spaced at 95m, or an approximate length to spacing ratio of 1:1½.

Several terms used in the discussion of the results need to be defined. *Groyne efficiency* is a measure of the percentage of the calibrated open beach drift that that is intercepted by the groynes. 100% efficiency indicates that the groynes allow no bypassing, while 0% efficiency indicates that the groynes are not influencing the longshore drift. The *pinch point* of a groyne bay is the narrowest point measured from the beach head to the crest. If erosion causes the pinch point to reach the beach head, then the beach is no longer providing complete protection. Figure 4 illustrates these terms.

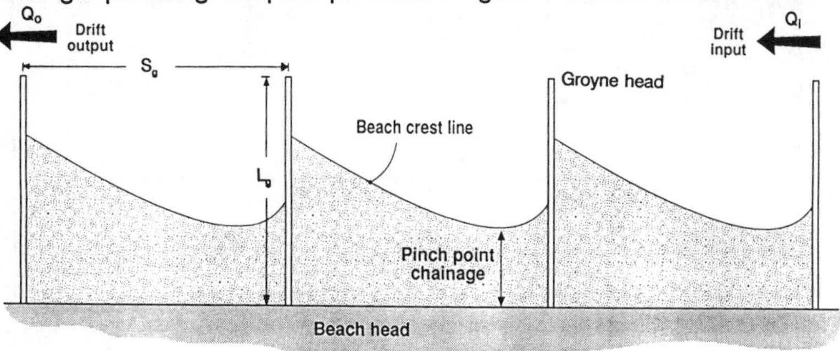

$$\% \text{ Efficiency} = \left(1 - \frac{Q_o}{Q_I}\right) \times 100$$

Figure 4. Definition of efficiency and pinch point

DISCUSSION OF THE RESULTS

The high end timber and short rock barrier groyne configurations were spaced at 95m and had effective lengths of 65 m. They were tested under 2m H_s waves at an offshore steepness of 0.02 and a 30° direction, with a water depth of 4m at the toe of the initial beach. For each configuration the calibrated drift input was added until the beach development stabilized and the efficiency dropped to zero; profiles were then taken. The test was then continued with zero input until the drift rates had risen to about 80% and the rate of beach erosion had stabilized.

Figure 5 indicates that the beaches stabilized with very similar pinch points under the calibrated input condition, while under the zero input conditions the pinch point for the higher rock groyne was significantly greater than for the lower timber groyne. This difference in the zero input response for the two groyne types is also illustrated by the updrift profiles presented in Figure 6. The calibrated (100%) input profiles are similar for both groynes, while the zero input profiles are very different.

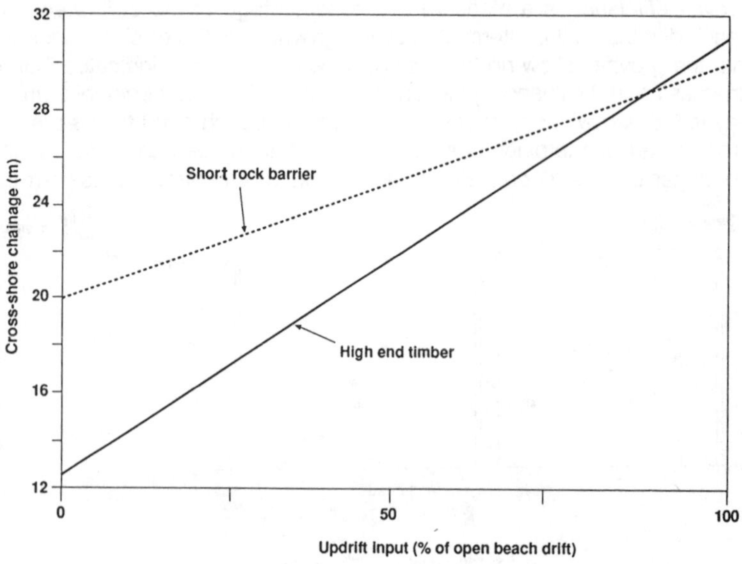

Figure 5. Effect of sediment input on final pinch point chainages

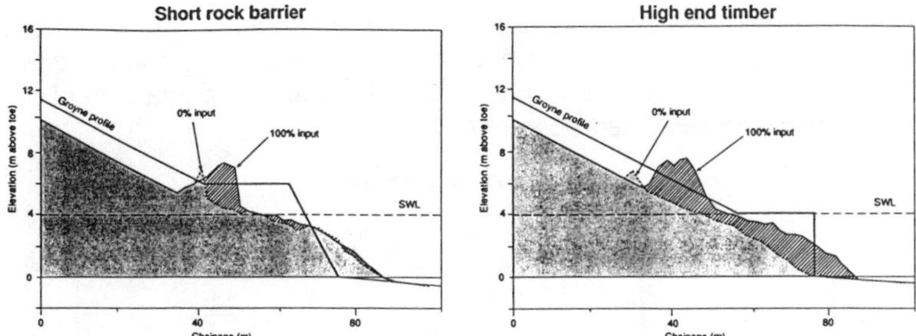

Figure 6. Effect of sediment input on beach profiles updrift of groynes

The groyne efficiency and transport distribution curves provide an explanation for these differences. Figure 7 contrasts the efficiency curves over time for the calibrated input tests. The high rock barriers retained nearly 100% of the available material until the beach volume reached a threshold point, after which efficiency dropped rapidly to below zero and then recovered. The timber groynes initially retained only about 75% of the drift, after which the efficiency dropped gradually to zero. Although the pinch point and profile figures indicate that the final volumes of material within the groyne systems were similar, it is apparent that the way in which the beaches developed over time was significantly different.

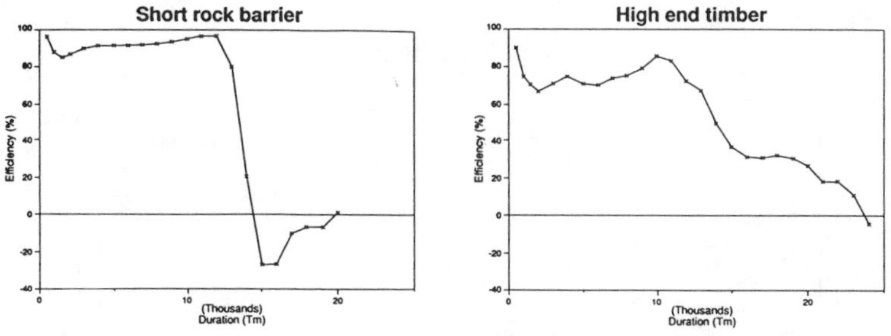

Figure 7. Groyne efficiency curves for calibrated input tests

Figure 8 compares the cross-shore distribution of transport over or around the groynes, relative to the open beach transport distribution. The plotted distributions are the mean of four sets of measurements taken after the central groyne bays had stabilized during the calibrated input tests. In view of the potential errors in the method, the drift distribution for the timber groyne test is regarded as being similar to that of an open beach, suggesting that the groyne is no longer having a significant influence on either the rate or the distribution of transport. In contrast, the higher rock groyne provides a complete barrier to sediment overtopping and the drift is forced to bypass the seaward end. In both instances, the total rate of drift was equal to the calibrated open beach rate.

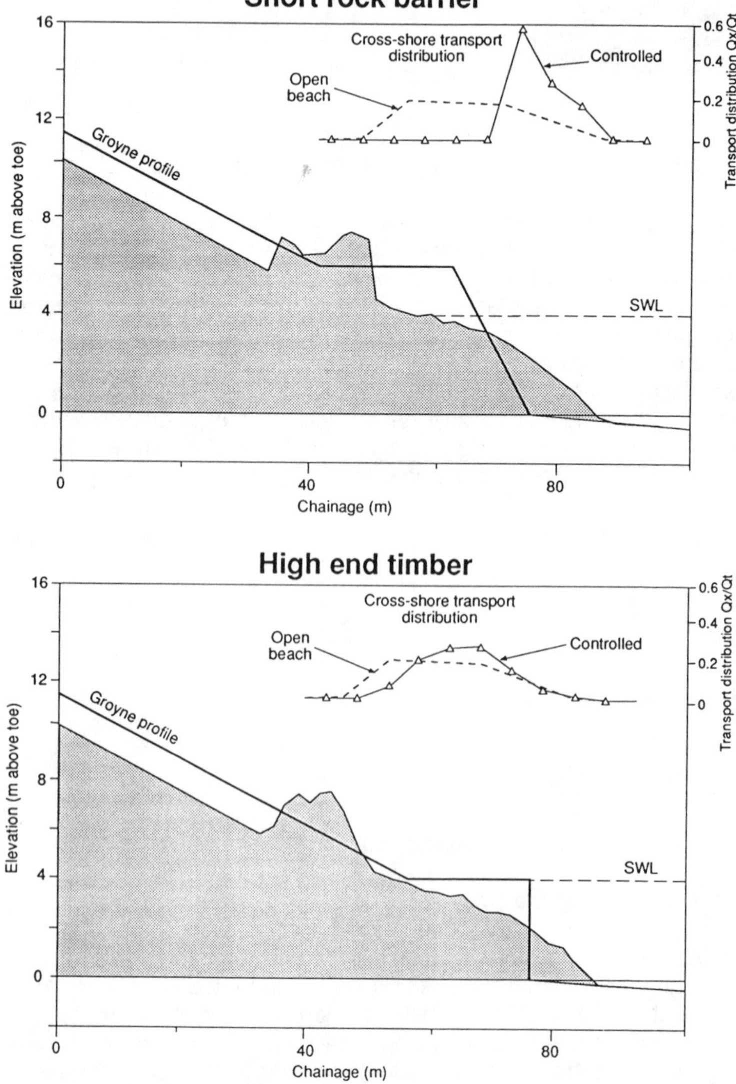

Figure 8. Cross-shore distribution of transport at the groynes

These results do not support the assumptions made by Brampton and Goldberg (1991) or Huang (1993). In the case of the lower groyne, the beach profile is below the level of the groyne profile, except at the beach crest. This suggests that shingle can be put into suspension to a much greater extent than assumed in the numerical models, which treat shingle transport as bed load. This concept is further supported by the eroded profile of Figure 6, which suggests that significant transport can occur, under the model test conditions, until the groyne profile is about 1m above the beach profile.

The transport distribution around the high rock groyne refutes the numerical model assumption that the cross-shore distribution of transport for an open beach is maintained for a groyned beach. Clearly, cross-shore transport modifies the transport processes at the groyne, causing longshore transport to be concentrated lower down the beach profile around the groyne head.

Both of these observations suggest that the existing numerical models over-predict the volume of material that will be retained by a groyne system. Unfortunately, the results of this research study are not sufficiently comprehensive or conclusive to allow further model developments.

The distribution of transport over and around the groynes also has implications for the effectiveness of the two different groyne structures. The observation that shingle is transported over the lower groynes until the beach profile is well below the structure crest indicates that these groynes will allow the beach plan shape and, therefore, volume, to vary over a wide range. This potential variation would make it difficult to estimate renourishment maintenance volumes and would require that a wider berm is left in front of the beach head to ensure that an adequate reservoir of material is retained in each groyne bay.

The rock barrier profiles under the calibrated and zero input conditions (Figure 6) show a more complex response than observed for the timber groynes. Under the zero input conditions the toe of the beach remained unchanged relative to the fully developed profile, while the beach crest moved landward. This subtle change in profile is associated with the shift from near 100% efficiency to zero efficiency as shown in Figure 7. The beach crest position, and more importantly, the beach volume will vary over a much narrower range on a beach controlled by high barriers rather than lower groynes. Comparison of the two hatched areas of Figure 6 shows the difference in volume change associated with the two groyne types.

CONCLUSIONS

The physical model study of shingle beach response described is part of an ongoing research programme at HR Wallingford. The aim of the programme has been to develop parametric numerical models and design guidelines for coastal engineers working with shingle beaches.

The study was inconclusive in its intended purpose of developing guidelines for designing groynes on renourished beaches. However, it did demonstrate the complexity of shingle beach response in the presence of groynes and the limitations of the existing numerical models of beach development. In particular, the numerical model assumptions that shingle is transported primarily as bed load and that the cross-shore distribution of transport for an open beach remains unchanged at groynes are both shown to be significant over simplifications.

The research programme has not yet produced sufficient data to allow prediction of the range of responses that may result from different groyne types and dimensions, nor have the results been verified against good quality field data. Until these two deficiencies can be rectified groyne design for beach renourishment schemes will continue to rely on site specific physical model testing combined with the careful use of the existing numerical models.

ACKNOWLEDGEMENTS

This work was carried out with the support of the UK Ministry of Agriculture Fisheries and Food under Research Commission FD0701

REFERENCES

Brampton, A H and Motyka, J M (1984). Modelling the planshape of shingle beaches, In: *Lecture notes on coastal and estuarine studies*, Vol 12, Springer Verlay, Berlin.

Brampton, A H and Goldberg, D G (1991). Mathematical model of groynes on shingle beaches. HR Wallingford Report SR 276.

Coates, T T and Lowe, J P (1993). Three dimensional response of open and groyned shingle beaches. HR Wallingford Report SR 276.

Coates T T (1994). The response of shingle beaches in the presence of control structures. HR Wallingford Report SR 387 (in preparation).

Huang, Z (1993). Mathematical model of shingle beaches with a groyne system. HR Wallingford Report IT 399.

Komar, P D and Miller, M C (1973). The threshold of sediment movement under oscilatory water waves. *J. Sediment Petrol.*, 43 : 1101 - 10.

Powell, K A (1990). Predicting short term profile response for shingle beaches. HR Wallingford Report SR 219.

Powell, K A (1992). Study of dissimilar sediments. *MAFF Conf. of River and Coastal Eng.*, Loughborough, UK.

Dean, R G (1973). Heuristic models of sand transport in the surf zone. *Conference on Engineering Dynamics in the Coastal Zone*.

Sayers, P B (1993). Beach profile response in front of vertical sea walls. HR Wallingford Report SR 327.

Yalin, S (1963). A model shingle beach with permeability and drag forces reproduced. Proc. 10th IAHR Congress 1963. London p.169.

STUDY ON THE SAND DEPOSITED BEHIND THE DETACHED BREAKWATERS

Wen-Juinn Chen[1] and Ching-Ton Kuo[2]

ABSTRACT: Two series of experiments were carried out in this paper. One is to check the efficiency of sand entrapment with the detached breakwater designed by Taiwan Water Resource Bureau. The second is to study the effects of offshore distance on the sand entrapment behind the breakwaters. There shows the cross section designed by TWRB is stability under the storm wave actions, but there just a little sand entrapped into the sheltered region of the breakwater. The permanent tombolo will be developing at the ratio of X/B=0.7~1.0. If the offshore distance is smaller than half the breakwater length then the most sands will deposit in the gap, therefore shoreline growing was limited. The maximum volume of sands deposits behind the sheltered region of breakwater and gap(Q_{b+g}) occurred at the ratio of X/B=0.9.

INTRODUCTION

Mi-Tou is a small town of Kaoshong county where locates on the west-southern coast in Taiwan, there exists a berm and wide beaches behind the original shoreline for reduce the incident wave energy and to preserve the safety of seawalls previously.

---

1) Associate professor, Department of Civil Engineering, National Chia-Yi Institute of Agriculture, Chia-Yi, Taiwan, 60083, R.O.C.
2) Professor, Department of Hydraulic & Oceanic Engineering, National Cheng Kung University, Tainan, Taiwan, 70101, R.O.C.

After a liquid nature gas receiving terminal was built at 500m near it in the northern direction then the wave pattern in the nearshore was changed. This change causes the beach to erode and seawalls were damaged when the typhoon wave is coming. Through, there have many seawalls constructed for shore protection, beach erosion still occur continually.

There has been proved that the detached breakwater has the efficiency in wave energy attenuation and trapping the sand deposited behind the breakwater to form a salient or tombolo. Therefore Taiwan Water Resource Bureau(TWRB) has decided to build some detached breakwater in Mi-Tou coast for reduce the incident wave intensity and preserve the stability of seawall and beach. Due to the complexity of coastal processes, still now there haven't any design criteria for reference on the design of the detached breakwater. It was relied heavily on the experience came from practical experiences and the judgment of designer. However, the cost to construct the detached breakwater is more expensive than seawall or other landward structures, so the effective design must consider.

For understand the efficiency of shore protection with the detached breakwater that was designed by TWRB a series of model tests was performed in the wave basin. Also, a series of basic experiments was doing in this study to discuss the effect of sand entrapment behind the breakwater with different offshore distance.

EXPERIMENTAL PROCEDURE

The experiments were acted in a 40m long, 40m wide and 1m high wave basin that a 10HP wave generator is installed inside the basin for generate the regular wave. A sandy beach with 13m long, 11m wide was placed in the wave basin for study the topography changes affected by the detached breakwaters. It consisted of well-sorted sand which median diameter d_{50}= 0.23mm and specific gravity γ_s=2.65. The contour and detached breakwaters were modeled according to the sea map and cross section that offers from TWRB with the scale of 1/40. Tab. 1 is the test condition of case 1.

Tab. 1 Experimental Condition of Case 1

CASE	h	X/B	H(m)	T(sec)	t(hrs)	block	stone	Tide (m)
A	-3m	0.5	5.6	9.5	4,8	yes	yes	2.2
B	-2m	0.3	5.6	9.5	4,8	yes	yes	2.2
C	-3m	0.5	2.24	7.6	4~20	no	yes	0.55
D	-2m	0.3	2.24	7.6	4~20	no	yes	0.55
E	-3m	0.5	5.6	9.5	4~16	no	yes	2.2
F	-2m	0.3	5.6	9.5	4~16	no	yes	2.2
G	-3m	0.5	5.6	9.5	4~16	no	no	2.2
H	-3m	0.7	5.6	9.5	4~16	no	no	2.2

Fig. 1 shows the sketch of wave basin using in this paper. The model detached breakwaters were made of broken stone about 1.5 cm and closed by the wire-entanglements. Fig. 2 shows the model detached breakwater used in this experiment. The arrangement of the detached breakwater in case 1 shows in Fig. 3.

For discuses the effects of sand trapped capacity under the different offshore distance, we do the second series tests i.e. case 2. A gentle sandy beach with 1/30 slope was placed in the same wave basin for experiment. Fig. 4 shows the arrangement of the detached breakwater here. Tab. 2 is the test condition of case 2.

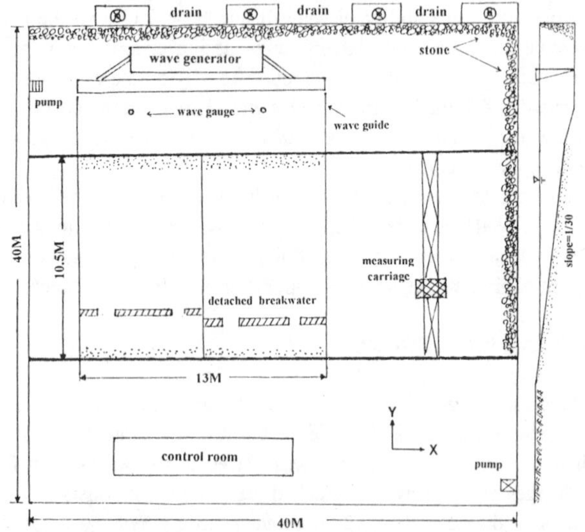

Fig. 1 Sketch of the experimental wave basin

Fig. 2 Model breakwater used in case 1.

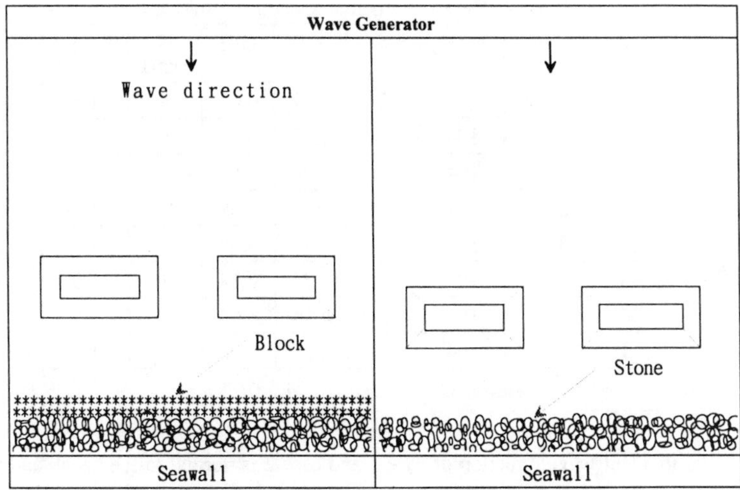

Fig. 3 The arrangement of breakwaters using in case 1.

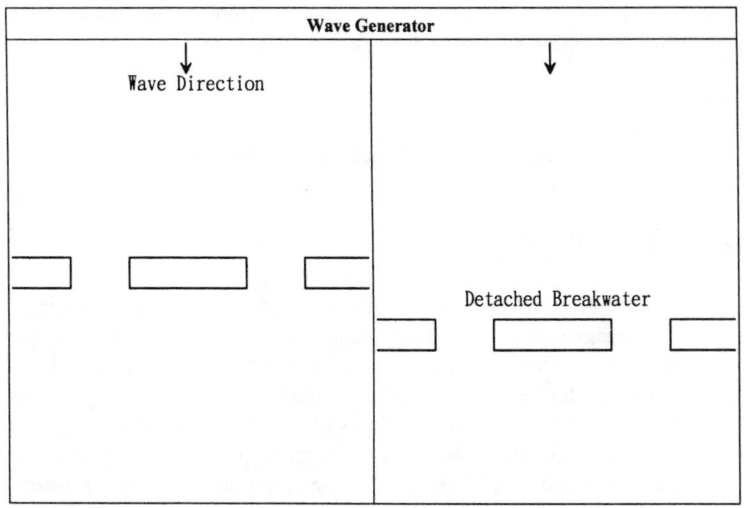

Fig. 4 The arrangement of breakwater using in case 2.

Table 2. The Experimental Conditions of Case 2

CASE	X (cm)	B (cm)	G (cm)	X/B	G/B	H (cm)	T (sec)
A	150	194	131	0.77	0.68	8	1.5
B	225	194	131	1.16	0.68	8	1.5
C	97	194	131	0.5	0.68	8	1.5
D	271.4	194	131	1.4	0.68	8	1.5
E	310.4	194	131	1.6	0.68	8	1.5
F	48.5	194	131	0.25	0.68	8	1.5
G	73.72	194	131	0.38	0.68	8	1.5
H	174.6	194	131	0.9	0.68	8	1.5
Beach slope =1/30 d_{50}=0.23 mm							

The incident wave direction of case 1 and case 2 is normal to the beach and wave heights were measured by the capacitance wave gauge at the distance shoreward approximately 3m from the wave generator. Here, we take four hours as one wave cycle, when the duration of wave action was reach to a cycle; then the morphological changes were measured by the profile meter continuously in the cross-shore directions and 25cm space in longshore direction. All the data were collected by the data acquisition system and a video camera was used to picture the flow pattern around the breakwater for discus the reason of topography changes.

RESULTS AND DISCUSSIONS

All the experimental data of case 1 and case 2 in this paper were discussed detail in this section, and the results are mentioned below.

Results of Case 1

From the experiment of case 1, we found the cross-section of detached breakwater and seawall designed by TWRB is safety under the storm wave actions. Most energy of the incident wave was diminished by the detached breakwater and decayed again by the blocks in the forward of seawall, so there hasn't any overtopping occurred. Therefore, we attempt to remove the blocks from the seawall and test again with the same wave condition. The result shows overtopping has been occurred at some place where just behind the gap, if we heighten the elevation of seawall or decrease the slope, the overtopping will be avoiding.

Fig. 5 and Fig. 6 show the shoreline change behind the detached breakwater of some test. We found there just a little sand entrapped into the region behind the breakwater and there hasn't any tombolo developed in the experiment of case 1. These results show that the detached breakwater designed by TWRB may haven't the effective capacity to trap the sand into the sheltered area of the breakwater to form the salient or tombolo, and the beach may be eroded continuously when the typhoon wave is coming. The reason of those defects is the cross section of detached breakwater is too big, then the rate of sand percolates into the sheltered area of the breakwater from seaward may be decreased. The gap width near the bottom between two breakwaters is too narrow and prevents the capacity of bed load sediments to entrap into the shadowed region behind the detached breakwater. Another important reason is the distance between original shoreline and the breakwater is too short; therefore there haven't enough space to let the incident wave energies that diffract into the breakwater to dissipate sufficiently for fall down the sediments; Then the amounts of sand were transport seaward by the rip currents that generated by the reflection induced from seawall. From the results of case 1, we suggest to decrease the cross section of the breakwater that designed by TWRB and extend the offshore distance i.e., move the breakwaters seaward to an appreciate depth for accumulate the sand behind the breakwaters.

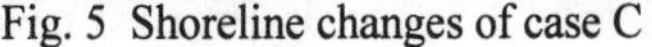

Fig. 5 Shoreline changes of case C

Fig. 6 Shoreline changes of case G

Results of Case 2

For study the effects of offshore distance on sand entrapment behind the breakwater, we carried out the second experiment (case 2) in the same wave basin of case 1. Fig. 7 shows the representation of the geometric characteristics used in this paper. Where G is the width of the gap between two breakwaters, X is the distance from original shoreline to breakwater, B is the length of the detached breakwater. The symbols of A_b, A_g represents the area above the still water level where accumulated in the shadowed region of breakwater and gap respectively; X_b is the distance from the original shoreline to the breaking point and Y_b is the growing distance of shoreline from the original shoreline.

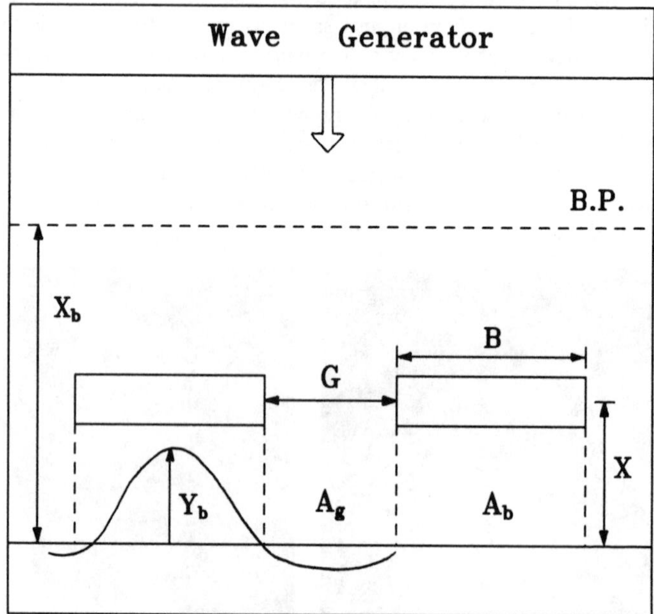

Fig. 7 Sketch of the geometric symbols.

Fig. 8 is the relationship of Q/H^2G versus X/B, where H is the incident wave height. The symbols of Q_b, Q_g in the figure represents the cumulative volume of sand behind the sheltered region of the breakwater and gap respectively. Q_{b+g} is the sum of Q_b plus Q_g. This figure shows that the sand deposited behind the breakwater(Q_b) increases with the enlargement of X/B and has a maximum value at X/B=1.4. When X/B was larger than 1.4 then Q_b begins to decrease. The sand deposit behind the sheltered region of gap(Q_g) has a maximum at X/B=0.9. After this value it decreased rapidly and the erosion is appearing when the offshore distance is larger than length of the breakwater. The reason is that when X/B>1 there has enough space to let the wave erode the sand behind the gap and transport those erosive sands to the sheltered area of the breakwater between both sides. There shows when X/B is smaller than 0.5 then

Q_g didn't change with the increasing of X/B. This was due to the sand trapping from seaward are accumulated mostly at the gap and just the small sand can be transported into the sheltered region. The total volume of sand deposit behind the sheltered region of breakwater and gap(Q_{b+g}) is direct proportional to the ratio of X/B and reaches a maximum value at X/B=0.9. When X/B is larger than 0.9 then Q_{b+g} was decreased, this phenomenon is influenced by the drastic erosion occurred in the sheltered region behind the gap.

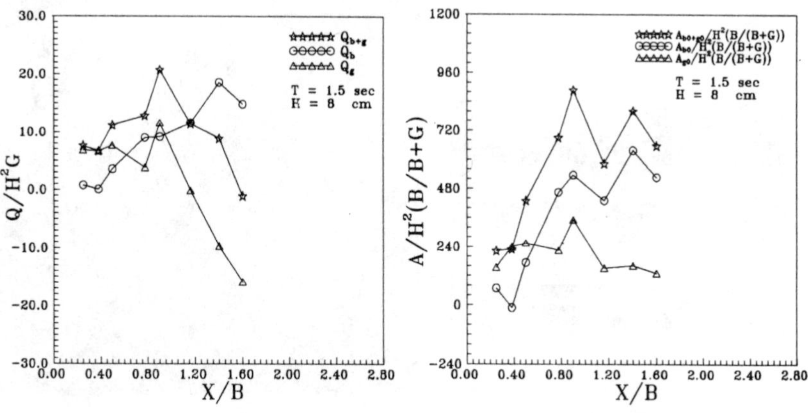

Fig. 8 Q/H^2G versus X/B

Fig. 9 $A/H^2(B/B+G)$ versus X/B

The relationships between nondimensional cumulate area above the still water level $A/H^2(B/B+G)$ and X/B is showing in Fig. 9. There shows the cumulative area above the still water level behind the shadowed region of the breakwater (A_b) are increasing with the enlargement of the ratio of X/B and got a maximum area at the X/B=1.4. If X/B is larger than this value, the cumulative area begins to decreasing. Reason of these phenomena is the offshore distance was enlarged, the intensity of the circulation and wave energy will be weakened and the capacity to transport the sediments toward the sheltered region is decreased. The area above the still water level where accumulated behind the sheltered region of gap (A_g) is almost unaffected with the increment of the ratio of X/B. However the sands accumulated in this region are decreasing rapidly when X/B increased (see Fig. 8); This mean when the offshore distance is enlarged, the subaqueous sands accumulated in this region have eroded evidently. The total cumulated area above the still water level ($Ab._{+g}$) also increasing with the increment of X/B, and got a maximum area at X/B=0.9.

Fig. 10 and Fig. 11 shows the salient and tombolo formed behind the breakwater in this experiment respectively. Fig. 13 shown as the nondimensional shoreline changes Y_b/X versus X/B. Where Y_b is the distance of shoreline moving seaward from the original shoreline behind the detached breakwater, if $Y_b/X=1$ mean there has a tombolo developing behind the breakwater. This figure shows the best conditions to form a tombolo behind the breakwater is at X/B=0.7~1.0, if the ratios of X/B outside this range there haven't any tombolo formed permanently under the wave actions. Due to the lack of sand which tapping from seaward to accumulate behind the breakwater, the shoreline just has a small growing distance in the ratio of X/B <0.5 i.e.; there haven't any tombolo or salient development behind the breakwater of those tests.

Fig.10 Tombolo behind the breakwater

Fig. 11 Salient behind the breakwater

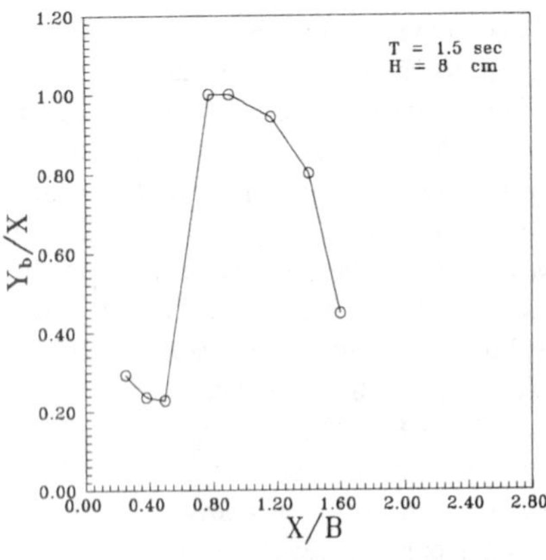

Fig. 12 Y_b/X versus X/B

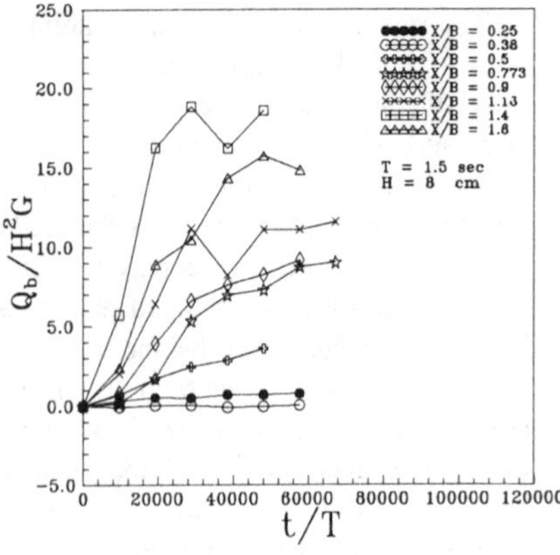

Fig. 13 Q_b/H^2G versus t/T

Fig. 13 and Fig. 14 shown as the Q_b/H^2G and Q_g/H^2G versus t/T respectively, where T is the wave period, t is the duration of wave action. In Fig. 13 it could be seen sand deposit behind the breakwater increase rapidly before twenty thousand waves' acts.

When the duration is larger than twenty thousand wave periods the cumulative capacity increased gradually and an equilibrium state may be reach. The phenomenon mentioned above was different with the conditions of X/B<0.5. In these cases almost haven't any sands entrapped into the sheltered region of the breakwater, therefore it didn't change with the time of wave action. Fig. 14 shows the sand entrapped into the sheltered region of the gap is proportional to the increase of t/T and reaches an equilibrium state at t/T=60,000. However, the cases of X/B=1.4, 1.6 show the erosive volume were increasingly as the duration of wave actions enlarged.

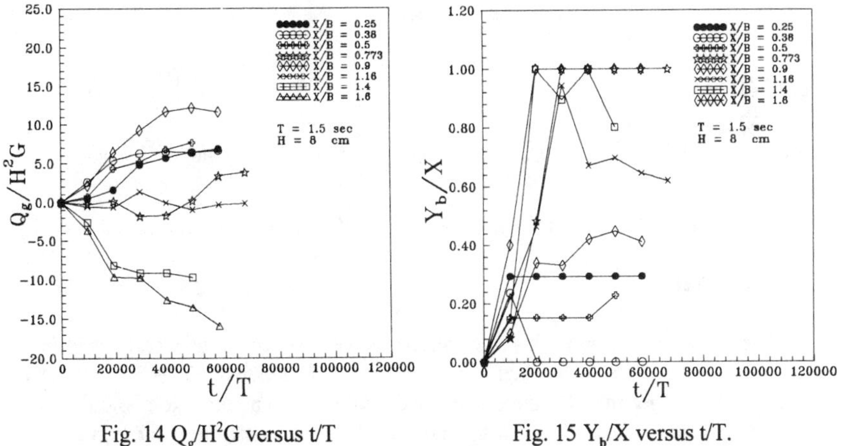

Fig. 14 Q_g/H^2G versus t/T

Fig. 15 Y_b/X versus t/T.

Fig. 15 showed as the shoreline development behind the detached breakwater Y_b/X versus t/T, it shows when X/B<0.5 the shoreline movements behind the breakwater are very small. The permanent tombolo was developed when the duration of wave action is twenty thousand times of wave period at the case of X/B=0.7~1.0. A temporary tombolo grew at the case of X/B=1.16,1.4 and finally a salient will form in these cases.

Comparison of Case 1 and Case 2

Fig. 16 shows the comparison of sand deposit behind the breakwater with case 1 and case 2. From this figure we could find clearly that the capacity of catches the sands into the sheltered region behind the breakwater of case 1 were smaller than case 2 at the same ratio of X/B. This may be influenced by the improper cross section of the breakwater and the offshore distance was too small. For get the more efficiency in sand entrapments, modify the section of the breakwater and extend the offshore distance to a suitable place is needed.

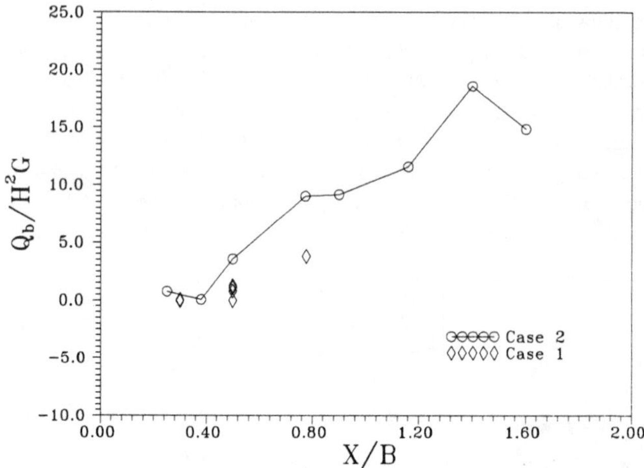

Fig. 16 Comparison of Q_b/H^2G versusX/B with case 1 and case 2

CONCLUSION

The results obtain from this study shown as the best geometric conditions to form a tombolo behind the breakwater permanently is at the range of X/B= 0.7~1.0. If the ratio X/B is larger than 1.0 there may have a salient or temporary tombolo develop. When X/B was smaller than 0.5 the most sands are deposited near the gap and the growth of shoreline was limited, therefore there wasn't any salient or tombolo formed in this experiment. The nondimensional volume of sand deposited behind the sheltered region of the breakwater, i.e., Q_b/H^2G increased rapidly when the duration of wave action is smaller than twenty thousand of wave period; then it increases gradually and reaches an equilibrium state at t/T is larger than sixty thousand. The result of case 2 also shows that Q_b/H^2G is proportional to the ratio of X/B and has a maximum value at X/B=1.4. However, the maximum total volume of sand deposited behind the sheltered region of gap and breakwater, i.e., Q_{b+g}/H^2G is at X/B=0.9. Due to the complexity of coastal processes on sediment transport and the phenomena's concern to the capacity of shore protection with the detached breakwater are uncertainly, the results giving above are our initial study. More detail consideration of other parameters such as width of gap G, length of breakwater B and the beach slope etc. will be doing continually in the further for obtains a reliable criterion on the design of the detached breakwater.

ACKNOWLEDGEMENTS

We give our acknowledgments to the national science council for the faience support of this study. The contract number is NSC 82-0414-P-006-015-B.

REFERENCES

Harris, M.M. and Herbich, I.B., 1986. "Effects of breakwaters spacing on sand entrapment," *Journal of Hydraulic Research*, Vol. 24, No. 5, pp. 346-357.

Kuo, C.T., Chen, W.J., Shei, W.K. and Lin, P.Y., 1993. "Study on the Mee-Toa Shore Protection Methods(II)," Report CKHOPR-93-007, Department of Hydraulics and Ocean Engineering, National Cheng-Kong University, Taiwan, R.O.C.(in Chinese)

Mimura, M. and Shimizu, T., 1983. "Laboratory study on the influence of detached breakwater on coastal change," *Proceeding of Coastal Structrue' 83*, ASCE, pp. 740-752.

Pope, J. and Dean, R.G., 1986. "Development of design criteria for segmented breakwaters," *Proceedings 20th Coastal Engineering Conference*, ASCE, pp. 2144-2158.

Suh, K. and Dalrymple, R.A., 1986. "Offshore breakwaters in laboratory and field," *Journal of Waterway, Port, Coastal and Ocean Engineering*, Vol. 113, No. 3, pp. 105-121.

Walker, J.R., Dlark, D. and Pope, J.,1982. "A detached breakwater system for beach protection," *Proceedings 18th Coastal Engineering Conference*, ASCE, pp.1968-1987.

PERFORMANCE EVALUATION OF OFFSHORE BREAKWATERS A FIELD AND COMPUTATIONAL STUDY

Andrew J Chadwick[1], Christoper Fleming[2], Jonathan Simm[3], Geoffrey N Bullock[4]

ABSTRACT: An ongoing field measurement and computational model study of wave transformations and shingle beach morphodynamics near a new offshore breakwater scheme is described. Details of the breakwater design and physical model studies are summarised. Two computational models of wave transformations are described and some particular aspects of their capabilities identified for evaluation from field measurements. The field measurement programme is summarised with particular attention focused on the spectral and directional analysis techniques employed and the new instrumentation developed for measuring nearhsore and shoreline directional wave spectra. Finally some of the initial field measurements of storm waves within the surf zone and in a region of wave diffraction are presented together with the specification for the proposed database.

INTRODUCTION

Offshore breakwaters are used as a means of sea defence or coast protection and can also provide an environmentally sensitive solution for the creation or preservation of beaches. The first such scheme in the UK which is fully documented is that at Leasowe Bay (Barber & Davies 1985). More recently, the

1 Dr, Reader in Coastal Engineering, School of Civil & Structural Engineering, University of Plymouth, Palace Court, Palace Street, Plymouth PL12DE, England

2 Dr, Director of Coastal, Sir William Halcrow and Partners Ltd, Burderop Park, Swindon, Wiltshire, SN4 0QD, England

3 Manager, Engineering Support Section, Coastal, HR Wallingford, Hydraulics Research Ltd, Wallingford, Oxfordshire, OX10 8BA, England

4 Professor, Head of School, School of Civil & Structural Engineering, University of Plymouth, Palace Court, Palace Street, Plymouth PL1 2DE, England

emphasis in the design of coastal protection works has shifted from hard to soft defences as described, for example, by Fleming (1991). Offshore breakwaters are categorised as forming part of a soft defence strategy. However, design data detailing their effects on coastlines and wave climate are currently inadequate. This is an unsatisfactory state of affairs since in the UK alone the capital value of coastal protection and sea defence works has been estimated to be more than £4 billion (Institution of Civil Engineers 1985).

Comprehensive prototype measurements of the effects of such breakwaters on the inshore *directional* wave climate have never before been made. Directional wave measuring techniques have recently been developed at the University of Plymouth (Bird & Bullock 1991) for waves near breakwaters and sea walls (which take account of reflections) and at the University of Brighton (Chadwick et al (in prep) & Chadwick (1989)) for waves incident on beaches (see plate 1). A new offshore Breakwater scheme at Elmer, Sussex, UK has provided a timely opportunity for the detailed measurement of offshore and concurrent inshore directional wave measurements together with regular surveys of the beach response.

One of the main aims of the research is to address the basic question "How effective are physical and computational models in assessing the performance of offshore breakwater schemes?" In this context breakwater performance is restricted to their influence on nearshore wave transformations and beach evolution. The research programme has been designed accordingly.

The research programme involves researchers at the Universities of Plymouth and Brighton and also draws upon other expertise in computational modelling, physical modelling and coastal engineering design from Sir William Halcrow & Partners and HR Wallingford.

DESIGN AND PHYSICAL MODEL STUDIES

The Elmer coastal defence scheme (see Fig. 1) was commissioned jointly by Arun District Council and the National Rivers Authority to protect a 2km headland area of the Sussex coast, UK at risk from coastal erosion and flooding. The natural shingle beach in this area had eroded and there was a need to replace the beach to prevent seawall undermining and to reduce wave overtopping and consequent flooding of hinterland housing. Historically, timber groynes had been used on the frontage but had failed to retain the beach and wave control structures were found to be necessary, shore parallel rock breakwaters proving to be the most cost effective.

Three central large breakwaters were designed by consultants Robert West & Partners with crest elevations of +4.5m OD, about one metre above highest 100 year surge tide following early flume tests at HR Wallingford. Breakwater gap widths were also set quite small to ensure a minimum shingle beach crest width of 5m at all locations, thereby eliminating all but minor wave overtopping. 6-10 tonne armourstone at side slopes of 1 in 2 were found to be necessary for stability reasons in the light of a full range of design conditions and adjustments were therefore necessary to two temporary breakwaters which had been placed by Arun District Council with front face slopes of between 1 in 1 and 1 in 1.5.

The initial design was confirmed in a 1 in 80 scale physical model including a mobile anthracite based beach to represent the shingle. The model enabled the "tapering-out" of the scheme either side of the 3 large breakwaters in the headland area to be correctly optimised with a series of 5 shorter breakwaters. (Two of these breakwaters at the Eastern end of the scheme had a reduced crest elevation of +3.0m and were only required to ensure that a natural shingle bank here was not breached under extreme conditions.) The scheme was tested from wave directions south-west, south and south-east using conditions established from "wave rider" calibrated offshore predictions and a wave refraction study. Morphologically averaged beach response was also simulated by use of a 5 times per year south-westerly wave condition run for a duration representing the net effect of all waves from all directions.

On completion of the main design model tests, a further series of tests were carried out in which wave measurements were made around the model breakwaters for comparison with the proposed field measurement and computational modelling.

Commencement of construction work started in the Summer of 1992 and was completed in August 1993.

COMPUTATIONAL MODELLING

Computational modelling of offshore breakwaters and other nearshore structures poses severe problems. It is necessary to represent the effects of refraction, diffraction, reflection and breaking of directional wave spectra in shallow water. No single model can approximate all these, though some combinations of these processes can be modelled satisfactorily and economically over small areas.

Two wave models will be tested during the course of the research project. The first uses the multigrid method to solve the elliptic form of the mild slope equation for irregular propagation (Li and Anastasiou, 1992). The effects of shoaling, refraction, diffraction and wave breaking are included together with the ability to represent wave-current interaction (Li and Fleming, 1993a) and directional wave spectra (Li, Reeve and Fleming, 1993b). The multigrid method involves transforming the mild slope equation into a new form in terms of amplitude and phase, after which the requirement of 8 to 10 computational points per wavelength commonly required for parabolic solutions involving wave diffraction, is no longer necessary. The model can be used to transform either formulated or digitised deep water wave spectra by discretising the spectrum into components and has been found to be extremely stable even with wave directions in excess of 80 deg to the primary axis of the grid.

Whilst the multigrid model described above should provide a detailed description of wave heights landward of the breakwaters it does not have the capacity to represent wave reflection. Therefore an alterative model will be tested. This will be a model based on the hyperbolic solution to the mild slope equation as derived by Copeland (1985). This model includes internal boundaries of arbitrary reflective properties and a driving boundary of transmissive and reflective waves. For this model there remains the requirement to maintain a very high grid resolution of at least 10 points per wave length. This provides a particularly strong visualisation of the wave patterns due to the propagation solution.

Both the models described have different strengths and will be tested against the field data collected at various points around the site.

FIELD AND COMPUTATIONAL STUDIES PROGRAMME

A multiphase study commenced in January 1993 comprising:

(a) a field measurement programme which includes:

(i) offshore directional wave measurements located approximately 500 m seaward of the breakwaters in a depth of about 2 m below lowest astronomical tide and a further set of directional measurements immediately seaward of one of the breakwaters;

(ii) inshore directional wave measurements located landward of the breakwaters at the lower end of the tidal range supplemented by two further point measurements in the lee of the same breakwater;

(iii) regular and frequent beach surveys and some hydrographic surveys; video recordings and photographs of wave action and beach features. Some aerial video and photographic work is also planned.

(b) the analysis of these field measurements to relate the inshore wave energy and directional spectra to offshore conditions, including the delineation of wave reflection effects.

(c) a computational model study in which selected subsets of the field data will be used to evaluate the capability of the current computational models, including and excluding wave reflections, and thus provide fresh insights and knowledge of the predictive performance of these models for future design purposes. It is anticipated that the current methods in the computational models to predict inshore directional spectra from offshore conditions will be thoroughly tested.

(d) All data, results and analyses will be comprehensively archived, using a proprietary database system, for future use.

SPECTRAL AND DIRECTIONAL ANALYSIS TECHNIQUES

Although spectral and directional analysis techniques are now well established, their practical application requires careful specification. In this study, in which wave measurement devices from two institutions are being deployed, it was also important to use a common set of wave analysis techniques. The research teams at Brighton and Plymouth have therefore agreed such a common set of analysis techniques, based on their combined knowledge and experience. In outline these techniques consist of the following elements.

For spectral analysis , the time series of instantaneous depth is first detrended using a second order polynomial applied between the first and last upward zero crossing points to remove the tidal component. The Fast Fourier Transform algorithm is then used , in conjunction with a full width window (Welch window) to overlapping subsets of the data. Ensemble averaging of the resulting Fourier coefficients is then applied, together with spectral smoothing. Finally the spectral energies are normalised to the time domain variance. Further details may be found in Chadwick et al (in prep).

For directional analysis, the Maximum Likelihood Method (MLM) is being used for the offshore and inshore locations and the Modified Maximum Likelihood Method (MMLM) is being used for the wave recorder immediately seaward of the breakwaters, where significant wave reflection occurs. The MLM has been shown to provide accurate estimates of the true incident directional spectrum for the two wave recorder systems in use (see Chadwick et al (in prep) for further details). The MMLM has also been shown to provide similarly accurate estimates for combined incident and reflected wave fields (see Bird (1993) for further details). However it should be noted that both the MLM and MMLM are quite sensitive to the spectral analysis techniques and are subject to instability for some data sets, for reasons that have yet to be identified.

COMPARISON OF WAVE RECORDER SYSTEMS

As two wave recorder systems are in use (one using pressure sensors and the other using resistive wave poles) it was felt desirable to determine whether these two devices where compatible. To this end, field trials where carried out in June 1993 in which the two systems where deployed alongside one another. Analysis of this data demonstrated very high compatibility of results, even in shallow water. Complete details of this exercise may be found in Ilic (1994).

SOME RESULTS RECORDED AT THE INSHORE LOCATION

By way of example of the spectral results obtained to date, figure 2 shows the spectral energy densities recorded at the inshore locations on the 15th December 1994. At this time a major storm event was occurring. At the approximate time of high water, the waves propagating shoreward between the breakwaters was depth limited with an H_{mo} value of 2m in a depth of 3.9m (denoted as channel 5 in Figure 2). Further shoreward of this position, in the region of wave diffraction, the wave height had decreased to about 1.5m at channel 4 and to about 0.9m at the inshore wave climate monitor (IWCM channels 0 to 3), situated in the lee of one of the breakwaters. The dramatic attenuation of wave energy in the diffracted region (particularly at low frequencies) is clearly shown in Figure 2.

The complimentary directional energy densities recorded by the IWCM for this particular event is shown in Figure 3. Here it is evident that the directional spectrum has a very peaky distribution with peak energies over all frequencies occurring within the same direction increment. A small amount of wave reflection at low frequencies is also apparent in Figure 3. Although this may be expected in reality, the MLM is not capable of correctly analysing phase locked reflected waves from beaches and thus this aspect of Figure 3 should be treated with caution.

DRAFT SPECIFICATION FOR PROPOSED DATABASE

One of the project objectives is to form a database of all the field measurements. This will serve several purposes, including selecting particular events on which to evaluate the capability of the computational models. The final database will be available to other research groups and other interested parties (at a date yet to be specified) The current specification for this database includes three components as follows:

1 Wave data

For each of the three wave measurement locations:

Date and time of measurement
Mean water depth
Signal variance (σ^2)
Mean zero crossing wave height and period (H_z, T_z) and maximum wave height (H_{max})
Significant wave height and period (H_s,T_s)
Spectral moments (m_0 to m_4), from which most spectral wave parameters may be derived
Spectral wave parameters (f_p , H_{mo})
Principal wave direction (θ_p) (defined as the frequency integrated peak value)
Reflection coefficient (R(f)) : Breakwater position only

2 Morphological Response

Bathymetry : at specified dates
Beach profiles : at specified dates

3 Other data storage

All raw wave data will be stored on magnetic tape and documented.

TYPICAL GRAPHICAL OUTPUT FROM THE DATABASE

Currently the database is incomplete, but some of the field results have been entered. To illustrate the usefulness of such a database in selection of events for further analysis and in using the database to find relationships between various wave parameters Figures 4 to 6 are presented.

Figure 4 shows a time series plot of H_{mo} for December 1993, at the three inshore locations (Series 1 is for channel 4, Series 2 is for channel 5 and Series 3 is for channel 0). This was a month in which a succession of storm events prevailed. The largest recorded event in this month was selected for the spectral and directional analyses presented in Figures 2 and 3.

Figure 5 shows H_{mo} plotted against mean water depth for channel 5 for the period October 1993 to mid January 1994, about 400 data records in all. This figure clearly shows that depth limited waves were experienced on a large number of occasions and therefore indicates that those waves were recorded within the surf zone. The limiting condition for the data presented in this plot is H_{mo}=0.54h_b, which is in reasonable agreement with theoretical predictions of depth limited waves with a Rayleigh distribution of wave heights.

Finally Figure 6 shows a scatter plot of T_p versus H_{mo} for channel 5. This figure demonstrates that a wide range of wave periods (about 3 to 15 seconds) and wave heights (about 0.1 to 2 metres) where recorded. It is also evident that the large storm events were producing peak periods of about 10 seconds and that the larger wave periods were associated with swell waves (probably propagating along the English Channel from the Atlantic Ocean).

CONCLUSIONS

A unique high quality field data set is being assembled which includes:

a Three sets of concurrent directional wave spectra, recorded over a one year period, near a new offshore breakwater scheme (one of which is in a region of diffracted surf zone waves and another in a reflective wave field immediately seaward of the breakwaters).

b The wave reflection properties of rock offshore breakwaters.

c Bathymetry and beach evolution data which will be used in a morphodynamic study.

The performance of currently used computational, hydrodynamic models of wave transformations will be evaluated directly from field data. Particular attention will be focused on their ability to represent random directional sea states and reflective sea states near coastal structures.

A database of all the field measurements will be produced and made available to other researchers and practitioners for their own studies.

ACKNOWLEDGEMENTS

The financial support of the Science and Engineering Research Council through research grants GR/H74360 GR/J58947 and GR/H82969 is gratefully acknowledged. The project also has the active support and assistance of Arun District Council, The Standing Conference on Problems Associated with the Coastline and the National Rivers Authority - Southern Region. The contributions of D J Pope, D M McDowell, P A D Bird, D A Huntley, S Ilic and P Axe to this research project are also gratefully acknowledged.

REFERENCES

Barber P C, Davies C D, 1985. "Offshore breakwaters - Leasowe Bay", *Proc Instn Civ Engrs* Part 1, 77, Feb, 85-109.

Bird, P A D, 1993. *"Measurement and analysis of sea waves near a reflective structure"*, PhD Thesis, University of Plymouth.

Bird P A D, Bullock G N, 1991. "Field measurements of the wave climate" pp 13-24 in Peregrine D H, Loveless J H, (eds) *Developments in Coastal Engineering* March 1991 Bristol University.

Chadwick A J, 1989. "The measurement and analysis of inshore wave climate", *Proc Instn Civ Engrs* Part 2, 1989, 87, March, 23-28.

Chadwick A J, Pope D J, Borges J, Ilic S. "Shoreline directional wave spectra, part 1: an investigation of spectral and analysis techniques", submitted to *Proc Instn Civ Engrs, Water Maritime and Energy.*

Chadwick A J, Pope D J, Borges J, Ilic S. "Shoreline directional wave spectra, part 2: instrumentation and field measurements", submitted to *Proc Instn Civ Engrs, Water Maritime and Energy.*

Copeland G A, 1985. "A practical alternative to the mild slope equation". *Coastal Engineering*, 9, 125-149.

Fleming C A, 1991. "The development of wider perspectives in coastal erosion control", pp 79-92 in Peregrine D H Loveless J H (1991) *Developments in Coastal Engineering.* Symp University of Bristol March 1991.

Ilic S, 1994. *"The role of offshore breakwaters in coastal defence - comparison of the two measurement systems"*. University of Plymouth, School of Civil & Structural Engineering, Internal Report SCSE-94-002

Institution of Civil Engineers, 1985. *"Research requirements in coastal engineering"*, Thomas Telford, London.

Li B and Anastasiou K, 1992. "Efficient elliptic solvers for the mild-slope equation using the multigrid technique", *Coastal Engineering,* 16, 245-266

Li B and Fleming C A, 1993a. "Multigrid Model of Wave-Current Interaction, *SUT Advances in Underwater Technology", Ocean Science and Offshore Engineering*, Vol 29, Wave Kinematics and Environmental Forces, Klawer Academic Press.

Li B, Reeve D E and Fleming C A 1993b. "Numerical Solution of the Elliptic Mild-Slope Equation for Irregular Wave Propagation", *Coastal Engineering,* 20, 85-100.

Plate 1. The Inshore Wave Climate Monitor in position between the two original breakwaters at the Elmer site

Figure 1.

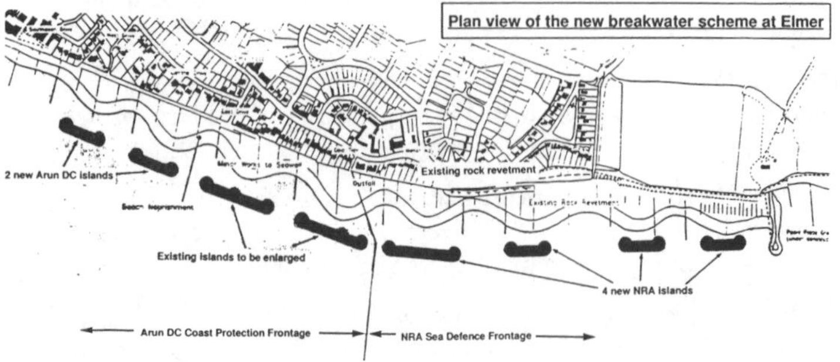

Figure 2 Spectral energy density curves recorded at the inshore locations on 15th December 1993

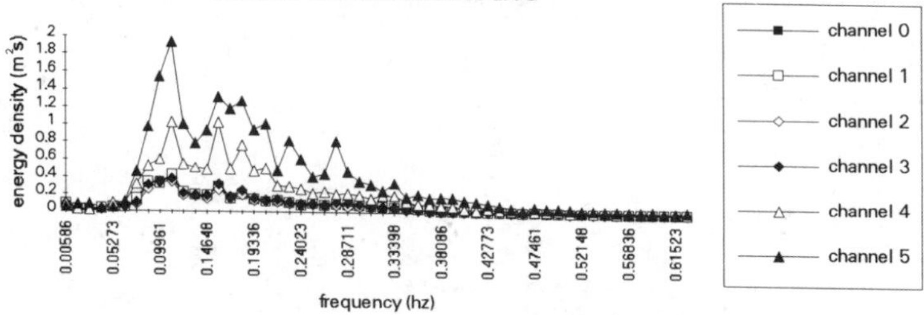

Figure 3 Directional energy density plot recorded at the inshore location on 15th December 1993

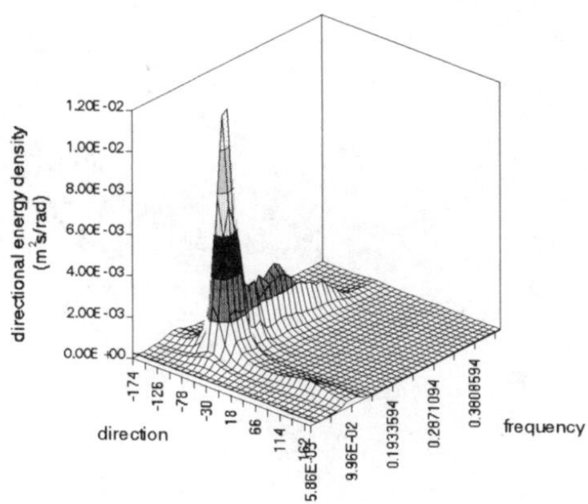

Figure 4 Time series plot of Hmo for the inshore locations for December 1993

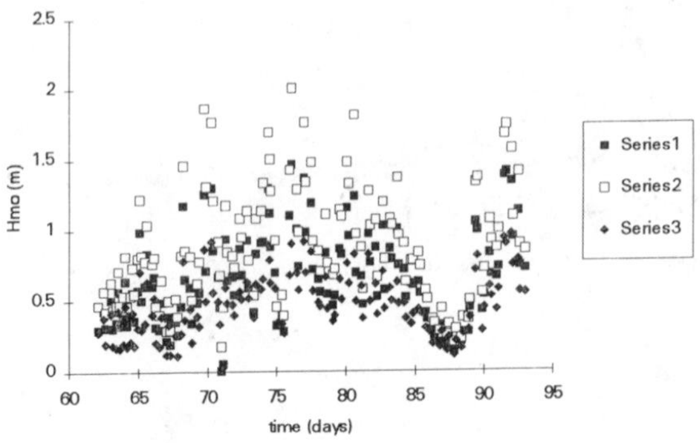

Figure 5 All recorded values of Hmo versus depth for channel 5

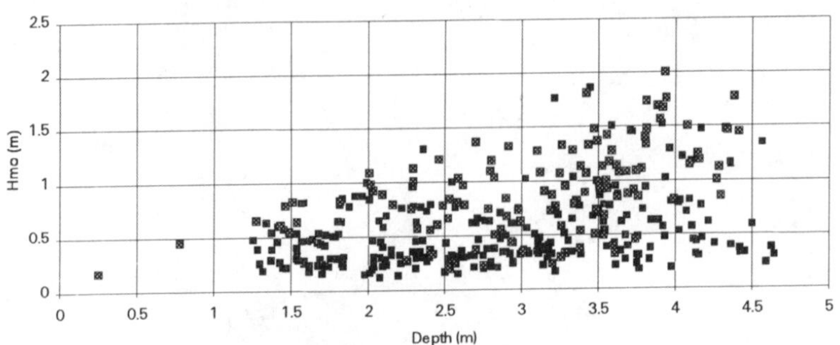

Figure 6 Scatter plot of Tp versus Hmo for channel 5

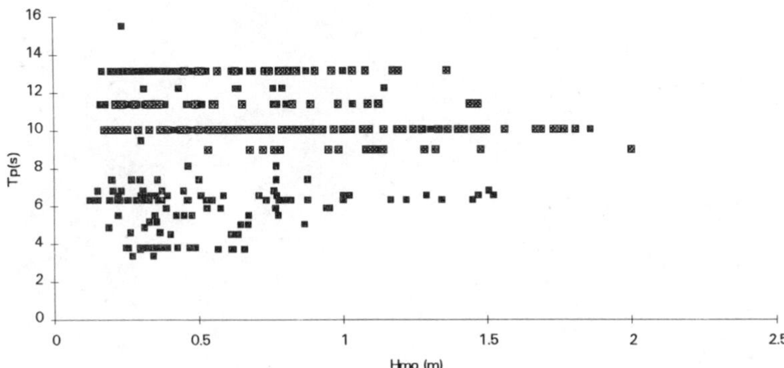

SPACE-TIME VARIABILITY OF SEDIMENT SUSPENSION IN THE NEARSHORE ZONE

A. E. Hay[1] and A. J. Bowen[2]

ABSTRACT: Suspension patterns obtained with modern acoustic instruments are intriguingly complex. Only in the most general terms do they accord with existing theoretical models based on a one-dimensional balance in the vertical. Negative suspension gradients may persist for several seconds; and sediment disappears at rates far faster than can be accounted for by the particle settling velocity. The obvious rationalisation is that there is lateral advection of plumes of suspended sand through the acoustic beams. This implies that the sediment is suspended in compact clouds with spatial scales which are short relative to a wave orbital diameter. In addition, the observations of suspension during wave groups suggest a strong, but complex, dependence on the previous time history of the waves.

INTRODUCTION

Large amplitude fluctuations, with time scales much shorter than a wave period, are seen in all recent measurements of the suspended sediment concentration in the nearshore zone. These were first reported by Downing et al. (1981), in measurements made with optical backscatter sensors (OBS's), but similar fluctuations are also seen in data from acoustic backscatter systems (Hanes et al. 1988, Hay and Sheng 1992).

---

1) Department of Physics and Ocean Sciences Centre, Memorial University of Newfoundland, St. John's, Newfoundland, Canada, A1B 3X7

2) Department of Oceanography, Dalhousie University, Halifax, N.S., Canada B3H 4J1

The fluctuations are of general interest as indicators of the basic physical processes which suspend sediment. In addition, the fluctuations being discussed here may well contribute significantly to the net sediment flux: they represent sediment concentrations an order of magnitude or more greater than the mean, extend well beyond the wave boundary layer, and can remain phase-locked to the waves for several wave periods (Hay and Bowen 1994).

These observations indicate that there must be length scales, other than the ripple wavelength, that lead to strong horizontal inhomogeneity in the concentration field. 'Small scale' variability, on time scales less than a wave period, is seen at spatial scales less than one orbital diameter and can be 'understood' in terms of advection by the waves of 30-cm scale structures in the suspension past closely spaced instruments (Hay and Bowen 1994). These structures are too large to be the separate vortices shed by individual ripples observed in laboratory measurements (Nakato et al. 1977, Sleath 1984). It is important recognize, however, that sediment suspension processes in the field are complicated by the possible existence of bedforms of different types and horizontal scales (Clifton 1976, Hay and Wilson 1994). Larger scale variations are seen with more widely separated sensors. The simplest view would be that the hydrodynamic forcing due to the waves propagates across the nearshore region at the wave group velocity. This effect is well documented due to the very obvious association of major suspension events with large wave groups, but there are clearly complicating factors such as wave breaking, large-scale bedforms (megaripples), and the presence of other wave motions and currents in the nearshore region.

METHODS

Field measurements were carried out at Stanhope Lane Beach, Prince Edward Island in 1989. The acoustic backscatter measurements were made with the Remote Acoustic Sediment TRANsport (RASTRAN) system (Hay et al. 1988). RASTRAN was deployed on the seaward flank of a linear bar, about 200 m offshore in a mean water depth of 2.2 m. The median size of the bottom sediments at this location was 170 μm. Acoustic sounders, flowmeters, and optical backscatter sensors (OBS's) were mounted on an aluminum frame. A complete description of the instrumentation is given elsewhere (Hay and Sheng 1992).

The data presented here are mainly from two acoustic sounders operating at 2.25 MHz, separated in the cross-shore direction by a distance of 1.45 m (Figure 1). Data are also presented from an electromagnetic flowmeter at a nominal height above bottom of 35 cm, and an OBS at 5-cm nominal height. (The actual heights of point sensors changed during the experiment as the bed elevation changed, due to bedform and bar adjustment.)

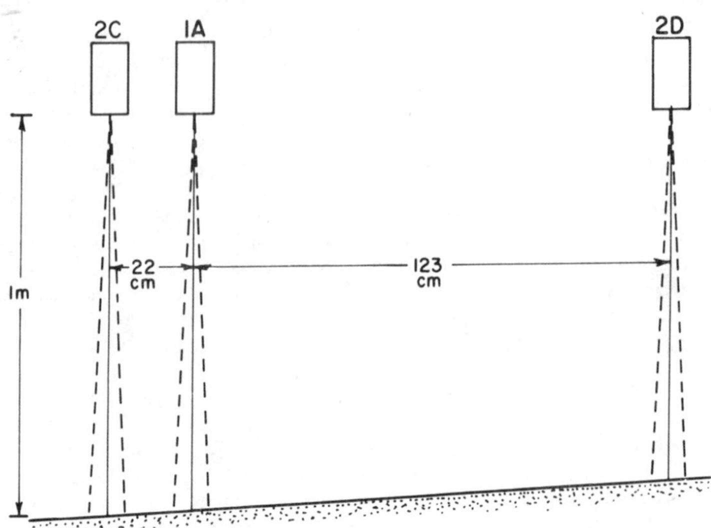

Figure 1. Cross-shore configuration of acoustic sounders. Shoreward direction is to the right. Sounder designations and frequencies are: 1A, 1 MHz; 2C and 2D, 2.25 MHz.

RESULTS

In presenting the results on sediment concentration variability, a distinction is made between short and long time scales, and between small and large space scales. By 'short' time scales, we mean variability occurring on time scales less than a wave period, typically 4-5 s for these data. 'Long' time scales are those associated with variations over intervals longer than a wave period, and are associated mainly with wave groups in these data. 'Small' space scales represent variations on scales much less than a wave orbital diameter. 'Large' space scales are those which just exceed, or are comparable to, the wave orbital diameter. Orbital diameters were O(1m) for the data presented here.

Small Separations, Short Timescales

Figure 2 shows a series of large amplitude acoustic backscatter fluctuations, associated with the first few wave cycles in a wave group, at 10-cm height above bottom from sounders 1A (1 MHz) and 2C (2.25 MHz). These two sounders were separated by 22 cm in the shorenormal direction (Figure 1), which separation is much less than the wave orbital excursions. The association of peak fluctuations with wave crests and troughs is apparent, but the concentration peaks do not occur simultaneously in the two records. Under wave troughs (fluid motion offshore), the peak fluctuations from sounder 1A occur earlier than those from sounder 2C, located farther offshore. The observed time lag is in the opposite sense under wave crests (fluid motion onshore), consistent with the advection of the plumes by the wave orbital velocities. At 243s (arrow), a pronounced broad peak occurs in the 2C record, more or less centred at the zero crossing in horizontal velocity at the end of a shoreward excursion. This has no obvious counterpart in the record from sounder 1A. This plume has been advected in from deeper water and just reached 2C on this particular orbit.

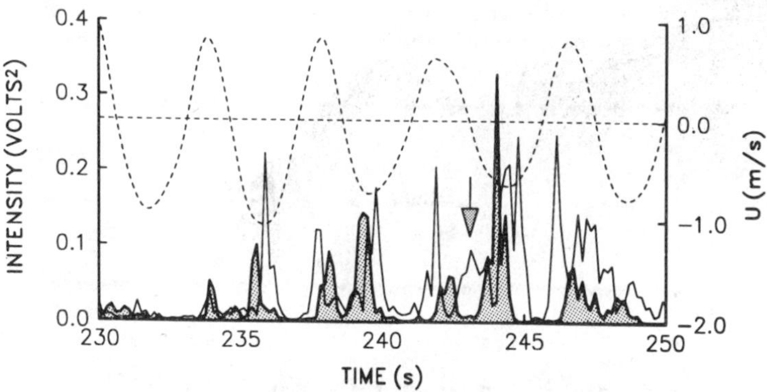

Figure 2. Short-separation backscatter intensities at 10 cm height from sounders 1A (shaded) and 2C (unshaded). Dashed line shows cross-shore velocity component, positive onshore. Run 299.027. (from Hay and Bowen 1994).

High concentration clouds of suspended sediment passing through the beam of one of these sounders, should usually be seen by the other sounder during the same wave cycle, with a time delay determined by the horizontal component of the wave velocity. While this behaviour is a fairly obvious consequence of the advection of anything by the wave orbital velocities, it seriously complicates the analysis and interpretation of the data. For example a sediment plume may: (a) just reach sensor 1 (seen once) and reverse so it is not seen by sensor 2; (b) pass 1, not quite reach 2, return past 1 (seen twice by 1); (c) pass 1 (seen twice) and just reach 2 (seen once by 2 halfway); (d) pass both sensors, first going by 1 then by 2, but on the way back it will be seen first by sensor 2 and then by 1.

It is clear that this behaviour is not well described by a simple cross spectral analysis, both the frequencies and the phases may be seriously aliased by the wave motion. As an extreme example, a single, very long-lived cloud could produce a long and complicated signal all by itself. An important general conclusion is that all reasonably long-lived plumes are usually seen twice during a wave cycle, if they are seen at all. (The complication is the plume that just reaches an instrument at the extreme end of its orbital motion.)

Large Separations, Long Time Scales

Figure 3 shows backscatter amplitude as a function of range (r) for two 2.25 MHz sounders: 2A (CH2) and 2D (CH4), separated by 1.45 m in the cross-shore direction (Figure 1). Sounder 2D (CH4) was closest to shore. The data are from the same run as Figure 2. Note that in Figure 3, and in subsequent plots like it, the acoustic data are shown for all range bins, including the bin nearest the transducer. The signal in this bin is saturated, and it appears as a solid black line. The bottom echo, at 90-95 cm range, is also saturated much of the time. There is little indication in this and later records of bubbles due to breaking waves which, when present, produce high backscatter near the transducer.

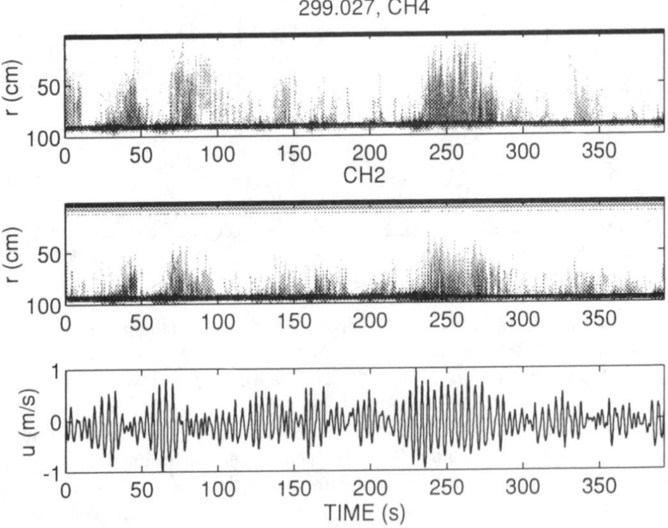

Figure 3. 'Large'-separation backscatter intensities, as a function of range, from sounders 2D (CH4, top) and 2C (CH2, middle). Run 299.027.

The Figure shows the association between suspension events and wave groups, and that group-generated suspension events are coherent at this separation. The development sequence during a suspension event can be characterized by near-bottom 'stirring' during the first 2-3 waves (e.g. near 30, 60 and 230 s), followed by more rapid growth of the suspension cloud, or 'pumping up'. This 'pumping up' continues after passage of the largest waves in the group, and in some cases well after entire group has passed (40-50 s). Rapid dissipation of the sediment cloud may then occur, much faster than the settling time scale. For example, in the CH2 (2C) record at 40-50 s, the cloud reaches 30 cm height, then appears to dissipate very rapidly, within a few seconds at most. In contrast, the settling time scale is 15 s (30 cm height, 2 cm/s settling velocity). Dissipation of the clouds does sometimes occur on a time scale consistent with settling (e.g. CH4, 90-120 s).

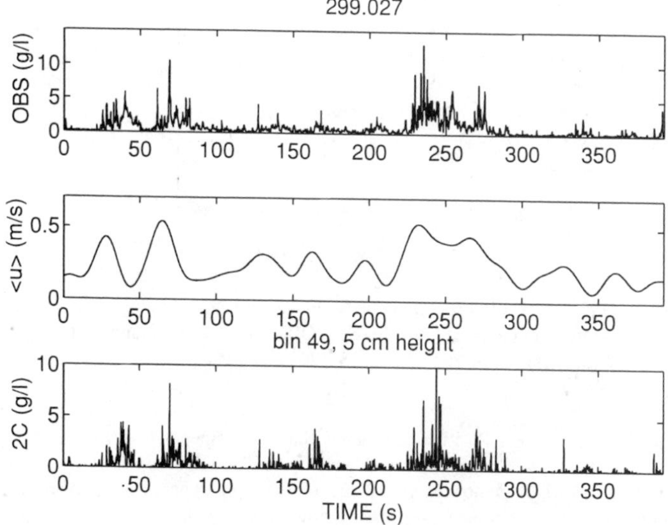

Figure 4. Top: OBS suspended sediment concentrations at 5 cm height. Middle: wave orbital velocity envelope. Bottom: 'nominal' concentrations from sounder 2C (CH2), at 5 cm height. Run 299.027.

Figure 4 shows OBS suspended sediment concentrations at 5-cm height, 'nominal' concentrations from sounder 2C at 5 cm height, and the wave orbital velocity envelope. This envelope was obtained by low pass filtering (20-s cutoff period) the absolute value of the velocity signal after removing the mean, and represents the envelope of the wave groups. By 'nominal' concentrations from sounder 2C, it is meant that the acoustic signals have not been corrected for the attenuation due to suspended sediments along the path between the transducer and 5-cm height range bin. These data show that acoustic and OBS concentration estimates are correlated at group time scales. (Note that the acoustic concentrations are lower on average than those from the OBS, which is expected as no correction has been made for scattering attenuation.) There is a noticeable lag between suspension events and wave groups.

Figure 5 shows the lagged cross-correlation function between the group-envelope and the low-passed (20-s cutoff) OBS or acoustic backscatter signals at 5-cm height. Correlation functions are presented for both 2.25 MHz sounders. The Figure shows high peak correlations, with R^2 values above 0.8. This means that the OBS and acoustic signals are all highly correlated with the group-envelope velocity signal, and therefore with each other. The lags at peak

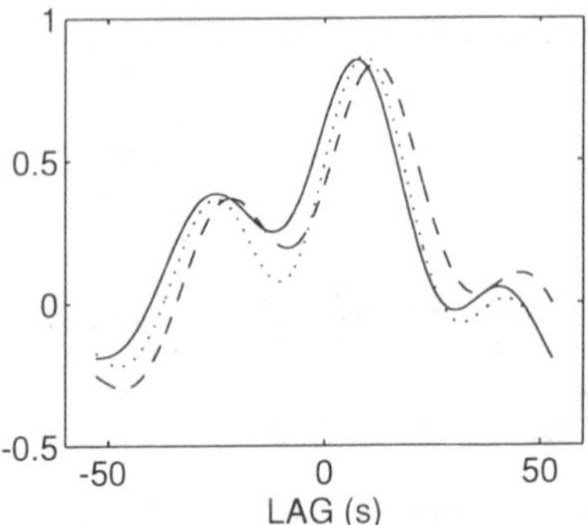

Figure 5. Cross-correlation functions between wave orbital velocity envelope and suspended sediment signals at 5-cm height: OBS (solid line); sounder 2C (dots); sounder 2D (dashes). Run 299.027.

correlation are 5-15 s. Positive lag means that the suspended sediment sensor (OBS or sounder) lags the wave group envelope. The OBS appears to lead the acoustic signals. This is most probably due to the strong dependence of peak lags on height in the near bottom region (see below).

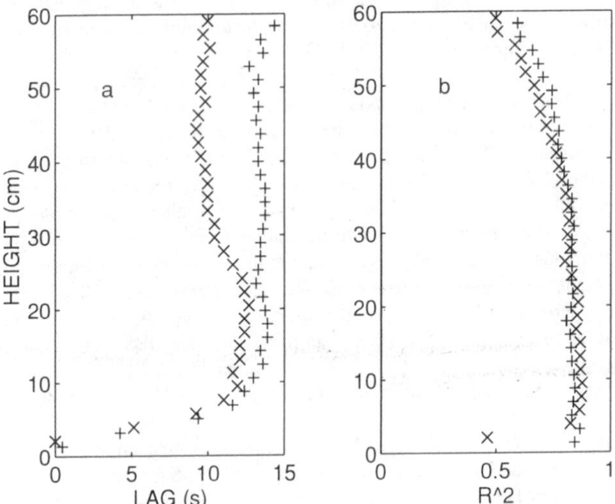

Figure 6. (a) Lags and (b) R^2 at peak cross-correlation as a function of height for sounders 2C (CH2, x) and 2D (CH4, +). Run 299.027.

Figure 6 shows the lags and R^2 values at peak cross-correlation as a function of height above bottom for sounders 2C (CH2) and 2D (CH4). Note the increasing lag with height in the near-bottom 10 cm, which at least partially explains the apparent lead between the OBS the acoustic backscatter signals in Figure 5. Note also that 10 cm is roughly the wave boundary layer thickness. Above 10 cm, the lags are roughly constant at 10-15s, particularly for sounder 2D. These results suggest that it takes 2-3 wave cycles (10-15 s) for sediment to diffuse far enough away from the boundary to allow more rapid growth of cloud. This is consistent with the picture, suggested by Figure 4, of an initial near-bottom stirring phase, followed by 'pumping up'.

Figure 7 shows data from another run (299.065), later on the same day. Again, the main suspension events are associated primarily with wave groups. Also, on group time scales, suspension is coherent at the 1.45-m separation between sounders, and 'pumping up' continues for several wave cycles after the passage of the largest waves in the group.

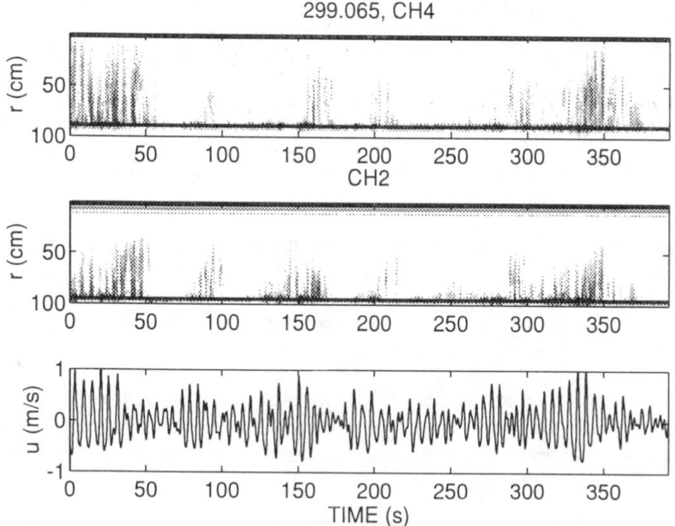

Figure 7. 'Large'-separation backscatter intensities from sounders 2D (CH4, top) and 2C (CH2, middle). Run 299.065.

Large Separations, Short Time Scales

In Figure 7, the primary sediment clouds generated during wave groups are more distinctly separate from one wave to the next than in Figure 3. In Figure 8, the first wave group from Figure 7 is shown on an expanded time scale. Ignoring the substructure of the suspension events, which in most cases consists of two or

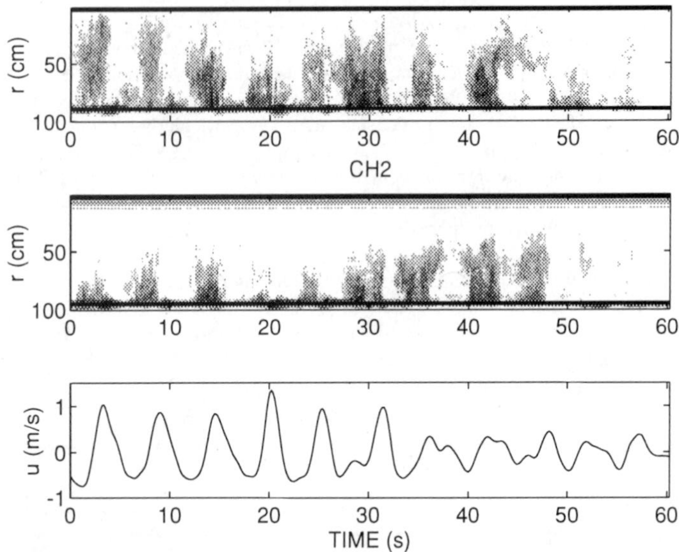

Figure 8. 'Large'-separation backscatter intensities, for the first 60 s of run 299.065.

more plumes as described previously, there is one main event per wave cycle. Differences between the two runs Figures in 3 and 7, which might explain the different response to group forcing on the time scale of individual waves, include the longer wave period and greater wave asymmetry, or skewness, for run 299.065 compared to run 299.027 (see Table 1).

Table 1: Flow field parameters: T_p = period at the wave spectral peak; U, V are the mean cross- and long-shore currents; u_{rms} is the square-root of the variance of the cross-shore velocity; S = skewness, or the ratio of the third moment of the cross-shore velocity to u_{rms}^3.

Run, s	T_p, m/s	U, m/s	u_{rms}, m/s	S	V, m/s
299.027	4.3	-0.09	0.312	0.22	0.10
299.065	5.1	-0.03	0.332	0.41	0.08
302.038	5.1	-0.06	0.307	0.24	0.06
302.042	5.1	-0.09	0.302	0.24	0.01

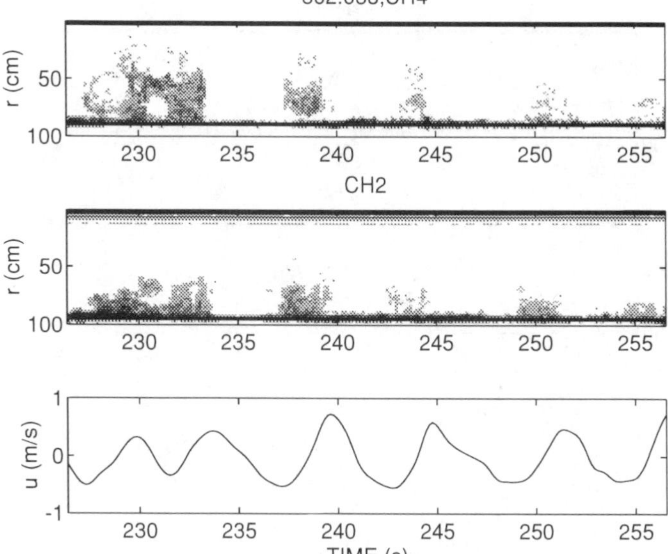

Figure 9. 'Large'-separation backscatter intensities, for a segment from run 302.038.

The clouds in Figure 8 are mostly in phase at either end of the frame except at the end of the run, from 45-55 s, when they are in antiphase. Other examples of similar behaviour are shown in Figures 9 and 10. In Figure 9, from run 302.038 three days later, the primary events once more occurred once per wave cycle (again ignoring substructure), and were in phase at either end of the frame. In Figure 10 (from a run about 1 h after that in Figure 9), however, the main suspension events at the two sounders are in distinct antiphase.

Successive events sometimes join at intermediate heights above bottom, forming 'holes' in the suspension pattern: for the CH4 record in Figure 9 at 231 s, and that in Figure 10 at 146 s, for example. For each of the 'holes' in Figures 9 and 10, the bridge between clouds at intermediate heights spans a flow reversal. If the holes are caused by advection of the cloud pattern past the sensors, as seems likely, then the cross-shore dimension of the clouds must increase with height. Or the clouds must lean over in the cross-shore direction, seaward in Figure 9, landward in Figure 10. Or, depending upon the details of the vertical structure of displacement amplitude and phase, both.

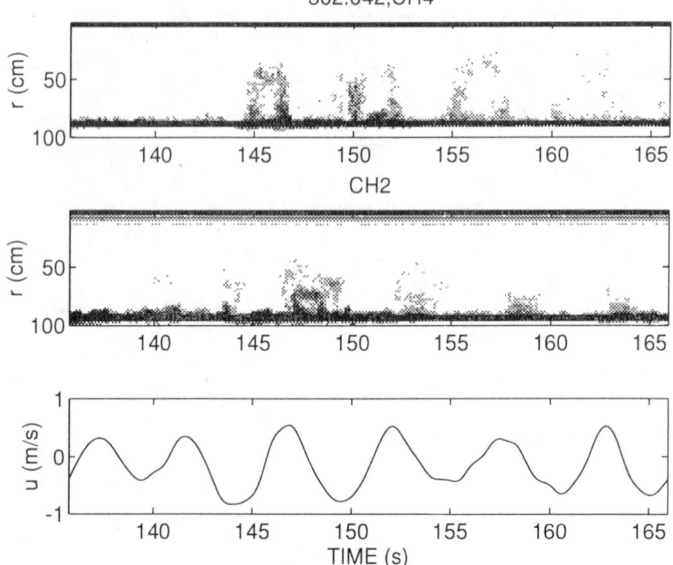

Figure 10. 'Large'-separation backscatter intensities. Data segment from run 302.042.

DISCUSSION

Table 1 shows the flow parameters for the four runs, as determined from the electromagnetic flowmeter records. The rms wave orbital velocities were comparable (0.30 to 0.33 m/s) for all runs. Wave periods were longer during the later runs (5.1 s), than for 299.027 (4.3 s). The skewness for run 299.065 was twice that of the other runs. The mean longshore currents were small, 10 cm/s or less, roughly similar to the mean cross-shore currents, which were directed offshore.

Figure 11 shows the cross-shore displacements, obtained by integrating the flowmeter record, for run 299.065. The upper panel shows the displacements for the entire run, with that due to the mean cross-shore flow removed. A 100-s period low frequency component is apparent. The lower panel shows the displacement during the first 60s of the run, with the low frequencies removed (20 s cutoff period). The orbital excursions during this time were about 1.3 m, similar to the 1.45 m distance between the two sounders. Comparing the lower panel in Figure 11 with the acoustic records in Figure 8, it can be seen that the suspension clouds are centred on the times of maximum offshore orbital displacement. This means that the spatial centre of the clouds must have been located shoreward of each sounder, but at a distance substantially less than the 0.6-m orbital semiexcursion and determined by the (two-plume) substructure of the clouds, because, while the duration of the clouds was roughly half a wave period, they appeared at the two sounder locations roughly simultaneously. This interpretation provides an explanation for the previously mentioned shift to antiphase behaviour at 45-50s: the weaker orbital motions late in the wave group were not sufficient to displace the cloud landward of the frame back as far as the position of sounder 2D.

The advective velocities at frequencies below the incident gravity wave spectrum were O(10 cm/s) or less for the runs presented here. This includes the mean motions (Table 1), and the infragravity band (not shown here, except in Figure 10). Because the low frequency motions were weak, the rapid disappearance of clouds produced by wave groups (at 50 s in Figure 3 for example) leads to the conclusion that the suspended sediment clouds reaching 30 cm height or more are compact in the horizontal, of order 50 cm or so wide. This is consistent with the picture put forward by Hay and Bowen (1994), on the basis of the short-separation measurements.

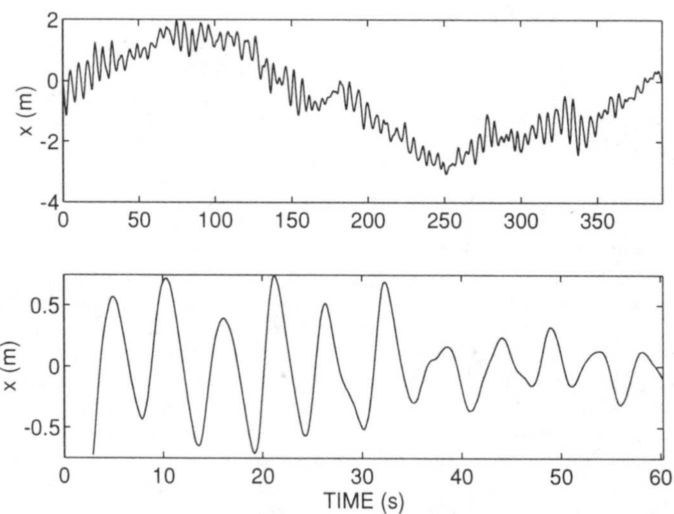

Figure 11. Cross-shore displacements, run 299.065. Top panel, incident wave and low-frequency passband, less displacement due to mean current. Bottom panel, incident gravity wave band only, for first 60s.

CONCLUSIONS

At time scales less than a wave period, and for separations less than a wave orbital diameter, the spatial coherence pattern of suspension events reaching heights of 15-30 cm or more is dominated by sediment clouds advected by the wave orbital velocities. While the suspension events can remain phase-locked to the waves for several wave cycles, there is no general relation between the phase of the wave and the events. Successive events appear to be produced by the same cloud being repeatedly advected by the wave orbital motion past the sensor. For comparable orbital velocities, the detailed suspension pattern at short time scales appears to be a strong function of wave period, and perhaps also of wave asymmetry.

At the 1.45 m separation, which normally exceeded the orbital diameter, and at the longer time scales associated with wave groups, high peak cross-correlations are observed between the group orbital velocity envelope and the low-frequency variations in sediment suspension. The lag between peak cross-correlations and the group envelope increases from zero near the bottom to several wave periods at 10-15 cm height, and then remains essentially constant farther from the bed.

The development of suspension during wave groups begins with near-bed stirring during the first few wave cycles, followed by more rapid growth of the cloud to higher elevations, a process we call 'pumping up'. This upward growth persists well after the largest waves in the group have passed. Disappearance of the cloud occurs after the passage of the wave group, and can be quite rapid, much faster than the settling time scale. This rapid disappearance is consistent with spatially-compact clouds.

We have been concerned here mainly with suspension events reaching heights of 15-30 cm or more: that is, heights well above the wave boundary layer. The observations identify a number of things which need explanation. Principal among these are the mechanisms responsible for 'pumping up' during wave groups, and for producing spatially compact clouds of order 30-50 cm vertical and horizontal extent, and with 1-2 m cross-shore spacing. Some of the possibilities which could produce sediment plumes at these scales, and which need to be examined further, are: (1) the bedform field; (2) surface-injected vortices; and (3) the sensor support structure. With regard to (1), we need to know the range of bedforms present, and to consider the possible effects of multiple-neighbour interactions among vortices shed by 10-cm scale anorbital ripples. Point-vortex shedding models (Longuet-Higgins 1981, Hansen et al. 1991) could be a useful approach to this problem. Surface-injected vorticity and turbulence is associated mainly with breaking waves, and produces both 2- and 3-dimensional vortices at these scales (Nadaoka et al. 1989). Direct measures of wave breaking are needed in field experiments. The potential effects of instrument frames on the suspension process at these scales need to be investigated, and minimized. These effects may be direct (e.g. vortex shedding from the frame legs: Hay and Bowen 1994), or indirect (e.g. frame-generated bedforms: Hay and Wilson 1994), and could be usefully investigated in suitable laboratory experiments.

ACKNOWLEDGEMENTS

These measurements were made as part of the Canadian Coastal Oceanography and Sediment Transport project (C*COAST) with financial support from the Natural Sciences and Engineering Research Council of Canada.

REFERENCES

Clifton, H. E., 1976. "Wave-Formed Sedimentary Structures-A Conceptual Model", In *Beach and Nearshore Sedimentation*, R. A. Davis, Jr. and R. L. Ethington, eds. SEPM, Special Pub. No. 24: 126-148.

Downing, J. P., Sternberg, R. W., and Lister, C. R. B., 1981. "New Instrumentation for the Investigation of Sediment Suspension Processes in the Shallow Marine Environment", *Mar. Geol.* 42, 19-34.

Hanes, D. M., Vincent, C. E., Huntley, D. A., and Clarke, T. L., 1988. "Acoustic Measurements of Suspended Sand Concentration in the C^2S^2 Experiment at Stanhope Lane, Prince Edward Island", *Mar. Geol.* 81, 185-196.

Hansen, E. A., Fredsoe, J., and Deigaard, R., 1991. "Distribution of Suspended Sediment Over Wave Generated Ripples", *Int. Symposium on the Transport of Suspended Sediment and its Mathematical Modelling*, Florence, Italy, September 2-5, pp. 111-128.

Hay, A. E., Huang, L., Colbourne, E. B., Sheng, J., Bowen, A. J., 1988. "A High-Speed Multichannel Data Acquisition System for Remote Acoustic Sediment Transport Studies", *Proc. Oceans '88*, 413-418.

Hay, A. E. and Sheng, J., 1992. "Vertical Profiles of Suspended Sand Concentration and Size in the Nearshore Zone from Multifrequency Acoustic Backscatter", *J. Geophys. Res.* 97, 15,661-15,677.

Hay, A. E. and Bowen, A. J., 1994. "On the Spatial Coherence Scales of Wave-Induced Suspended Sand Concentration Fluctuations", *J. Geophys. Res.* (in press, June issue).

Hay, A. E. and Wilson, D. J., 1994. "Rotary Sidescan Images of Nearshore Bedform Evolution During a Storm". *Mar. Geol.* (in press).

Longuet-Higgins, M., 1981. "Oscillating Flow Over Steep Sand Ripples", *J. Fluid Mech.* 107, 1-35.

Nadaoka, K., Hino, M., and Koyano, Y., 1989. "Structure of the Turbulent Flow Field Under Breaking Waves in the Surf Zone". *J. Fluid Mech.* 204, 359-387.

Nakato, T., Locher, F. A., Glover, J. R., and Kennedy, J.F., 1977. "Wave Entrainment of Sediment from Rippled Beds", *J. Waterw. Port Coastal Ocean Div.* 103 (WW1), 83-99.

Sleath, J. F. A., 1984. *Sea Bed Mechanics*, John Wiley & Sons, New York, 335 pp.

Subject Index

Page number refers to first page of paper

Author Index

Page number refers to the first page of paper